Guidebook to the

Extracellular Matrix, Anchor, and Adhesion Proteins

Other books by Sambrook & Tooze Publications

Guidebook to the Cytoskeletal and Motor Proteins
Edited by Thomas Kreis and Ronald Vale

Guidebook to the Homeobox Genes
Edited by Denis Duboule

Guidebook to the Secretory Pathway
Edited by Jonathan Rothblatt, Peter Novick, and Tom Stevens

Guidebook to Cytokines and Their Receptors
Edited by Nicos A. Nicola

Guidebook to the Small GTPases
Edited by Marino Zerial and Lukas A. Huber

Guidebook to the Calcium-Binding Proteins
Edited by Marco Celio; co-edited with Thomas Pauls and Beat Schwaller

Guidebook to Protein Toxins and Their Use in Cell Biology
Edited by Rino Rappuoli and Cesare Montecucco

Guidebook to Molecular Chaperones and Protein-Folding Catalysts
Edited by Mary Jane Gething

Guidebook to the
Extracellular Matrix, Anchor, and Adhesion Proteins

Second Edition

Edited by

Thomas Kreis
University of Geneva,
Geneva,
Switzerland

and

Ronald Vale
University of California,
San Francisco,
USA

A SAMBROOK & TOOZE PUBLICATION
AT OXFORD UNIVERSITY PRESS
1999

OXFORD
UNIVERSITY PRESS
Great Clarendon Street, Oxford OX2 6DP

Oxford University Press is a department of the University of Oxford and furthers the University's aim of excellence in research, scholarship, and education by publishing worldwide in

Oxford New York
Athens Auckland Bangkok Bogotá Buenos Aires Calcutta
Cape Town Chennai Dar es Salaam Delhi Florence Hong Kong Istanbul
Karachi Kuala Lumpur Madrid Melbourne Mexico City Mumbai
Nairobi Paris São Paulo Singapore Taipei Tokyo Toronto Warsaw

and associated companies in Berlin Ibadan

Published in the United States
by Oxford University Press Inc., New York

A catalogue record for this book is available from the British Library

Library of Congress Cataloging in Publication Data
Guidebook to the extracellular matrix and adhesion proteins / edited by Thomas Kreis and Ronald Vale. — 2nd ed.
"A Sambrook & Tooze publication at Oxford University Press."
1. Extracellular matrix proteins. 2. Cell adhesion molecules.
I. Kreis, Thomas. II. Vale, Ronald.
QP552.E95G85 1999 572′.6—dc21 98–51826

ISBN 0 19 859959 5 (Hbk)
ISBN 0 19 859958 7 (Pbk)

Typeset by
EXPO Holdings, Malaysia
Printed in Great Britain by
The Bath Press, Avon.

This book is dedicated to Eric St George-Kreis

Thomas and Eric, 1998

Thomas Erhard Kreis
(1952–1998)

In the final stages of editing this Guidebook, Thomas Kreis lost his life in a tragic airplane crash off the coast of Nova Scotia, in which all passengers perished. Thomas received his Ph.D. from Zurich in 1981 and performed postdoctoral studies with Dr B. Geiger at the Weizmann Insitute and Dr H. Lodish at MIT. He was a research group leader at the European Molecular Biology Laboratory in Heidelberg from 1983 to 1992 and was a professor at the University of Geneva from 1992 until his death.

Throughout his career, Thomas was a world-class scientist with numerous innovative contributions in the areas of membrane trafficking and the cytoskeleton. Thomas was one of the first individuals to follow the dynamics of fluorescently-labelled cytoskeletal proteins within living cells; he discovered important proteins involved in vesicle trafficking (β-COP) and in linking vesicles to microtubules (CLIP-170), and he advanced important ideas concerning the role of the cytoskeleton in the transport and positioning of membrane organelles. Thomas also devoted enormous time to the scientific community and was on several editorial boards.

Above all, Thomas was a wonderful human being, who is much beloved and deeply missed by all that knew him. This Guidebook is a testimony to Thomas' enormous scientific and personal energies, which ended so abruptly and far too soon.

Preface

The biology of the 1990s will likely come to be remembered as the era of protein discovery. Recent advances in molecular biology, genetics, protein purification, and NMR/X-ray crystallography have conspired to accelerate the rate at which the amino acid sequences, three-dimensional structures, and biological functions of cellular proteins are being elucidated. Moreover, genome sequencing efforts are producing complete inventories of the proteins that govern the workings of several organisms.

Now that the floodgate of protein discovery has been opened wide, the amount of new information on cellular proteins is exceeding the capacity of assimilation of even the best-read scientist. At the same time, it has become imperative for researchers to expand their knowledge base, since interactions between previously unconnected sets of proteins are being uncovered at a rapid pace and are leading to important biological discoveries.

These considerations motivated us to compile the *Guidebook to the Cytoskeletal and Motor Proteins* and the *Guidebook to the Extracellular Matrix, Anchor, and Adhesion Proteins*, which are designed to serve both seasoned scientists and students alike. Each class of proteins is prefaced by a general introduction that describes their overall functions and some of the interesting questions that challenge workers in the field. The biological functions, localization, structural attributes, purification methods, reagents (e.g. cDNA clones and antibodies), and relevant medical information are described concisely for individual proteins (or groups of related proteins) by leading individuals who participated in their discovery or characterization.

The progress over the past five years, which is chronicled in this revised Guidebook, is startling. Many proteins that were first described only at the beginning of this decade are now known to belong to large superfamilies, and in many cases, a detailed understanding of protein mechanism and biological function has emerged. In addition to complete revisions of prior entries, 57 new proteins and introductions have been added, which were not contained in the earlier editions of these Guidebooks. In a change from the first editions, the section on 'Cytoskeleton-associated anchor and signal transduction proteins' now appears here rather than in the *Guidebook to the Cytoskeletal and Motor Proteins*. This reorganization emphasizes the communication that occurs between the outside and the inside of cells.

The primary emphasis of this volume is on well-characterized structural proteins, although several critical regulatory proteins (e.g. kinases and G proteins) are also included. Given the rapid pace of science, we acknowledge that the coverage of all relevant proteins is, regrettably, incomplete, and we apologize to the investigators whose protein of interest is not found in this volume.

This project was motivated by a spirit of assisting the general scientific community. We are indebted to the many authors who contributed considerable time to this effort. Special thanks are extended to Drs A. Bershadsky, W. Birchmeier, B. Geiger, L. Reichardt, and M. Sternlicht for their special assistance and advice.

August 1998
Geneva
San Francisco

T.K.
R.V.

Contents

Part 1. Cytoskeleton-associated anchor and signal transduction proteins

Part 2. Cell adhesion and cell–cell contact proteins

Part 3. Extracellular matrix molecules and their receptors

ECM molecules

ECM proteinases

List of Plates

The plates appear between pp. 12 and 13

List of Contributors

Steven M. Albelda, Pulmonary and Critical Care Division, Department of Medicine, University of Pennsylvania Medical Center, Philadelphia, PA 19104-4283, USA.

W. Scott Argraves, Department of Cell Biology and Anatomy, Medical University of South Carolina, Charleston, SC 29425-2204, USA.

Spyros Artavanis-Tsakonias, Howard Hughes Medical Institute and Department of Cell Biology and Biology, Boyer Center for Molecular Medicine, Yale University, New Haven, CT 06536-0812, USA.

Kenneth A. Beck, Department of Cell Biology and Human Anatomy, University of California, Davis, CA 95616, USA.

Mary C. Beckerle, Department of Biology, University of Utah, Salt Lake City, UT 84112, USA.

J. Behrens, Max Delbrück Center for Molecular Medicine, Robert-Rossle Strasse 10, 13122 Berlin, Germany.

Alexey Belkin, Department of Cell Biology and Anatomy, University of North Carolina at Chapel Hill, Chapel Hill, NC 27599, USA.

Vann Bennett, Howard Hughes Medical Institute and Department of Cell Biology, Duke University Medical Center, Durham, NC 27710, USA.

Merton Bernfield, Department of Pediatrics, The Children's Hospital, Harvard Medical School, Boston, MA, USA.

Mark Berryman, Section of Biochemistry, Molecular and Cell Biology, Cornell University, Ithaca, NY 14853, USA.

Alexander Bershadsky, Department of Molecular Cell Biology, The Weizmann Institute of Science, Rehovot 76100, Israel.

Peter Besmer, Molecular Biology Program, Memorial Sloan-Kettering Cancer Center, 1275 York Avenue, New York, NY 10021, USA.

Walter Birchmeier, Max-Delbrück Center for Molecular Medicine, 13125 Berlin, Germany

Elayne A. Bornslaeger, Departments of Pathology and Dermatology, 303 East Chicago Avenue, Northwestern University Medical School, Chicago, IL 60611, USA.

Claude Boucheix, INSERM U268, Hôpital Paul Brousse, 94807 Villejuif Cedex, France.

William J. Boyle, Department of Cell Biology, Amgen Inc., 1840 DeHavilland Drive, Thousand Oaks, CA 91320, USA.

Damir Bozic, Department of Biophysical Chemistry, Biozentrum University Basel, Klingelbergstrasse 70, CH-4056 Basel, Switzerland.

Anthony Bretscher, Section of Biochemistry, Molecular and Cell Biology, Cornell University, Ithaca, NY 14853, USA.

Eric J. Brown, Division of Infectious Diseases, Washington University School of Medicine, St Louis, MO 63110, USA.

Thomas Brümmendorf, Max-Delbrück-Center for Molecular Medicine, Robert-Rössle Strasse 10, 13122 Berlin, Germany.

Joy Burchell, Imperial Cancer Research Fund, London, UK.

Kevin P. Campbell, Howard Hughes Medical Institute, Department of Physiology and Biophysics and Department of Neurology, University of Iowa College of Medicine, Iowa City, IA 52242, USA.

Laurent Caron, Howard Hughes Medical Institute and Department of Cell Biology and Biology, Boyer Center for Molecular Medicine, Yale University, New Haven, CT 06536-0812, USA.

Coralie A. Carothers Carraway, Department of Biochemistry and Molecular Biology, University of Miami School of Medicine, PO Box 016960, Miami, FL 33101, USA.

Kermit L. Carraway, Depatment of Cell Biology and Anatomy, University of Miami School of Medicine, PO Box 01690, Miami, FL 33101, USA.

S. Chakravarti, Departments of Medicine and Genetics, Case Western Reserve University, 10900 Euclid Avenue, Cleveland, OH 44106-4952, USA.

Ruth Chiquet-Ehrismann, Friedrich Miescher Institute, PO Box 2543, CH-4002 Basel, Switzerland.

Sandra Citi, Dipartimento di Biologia, Università di Padova, Italy; and Departement de Biologie Moleculaire, Universite de Geneve, Switzerland.

David R. Colman, Brookdale Center for Developmental and Molecular Biology, The Mount Sinai School of Medicine, One Gustave Levy Place, New York, NY 10029, USA.

Pam Cowin, Department of Cell Biology, New York University Medical Center, 550 First Avenue, New York, NY 10016, USA.

Patricia Crisanti, CNRS, Development and Immunity of the Central Nervous System, Université Paris VI, Faculte de Médecine Broussais, 15 Rue de L'Ecole de Médecine, 75270 Paris Cedex 06, France.

David R. Critchley, Department of Biochemistry, University of Leicester, University Road, Leicester LE1 7RH, UK.

Dorothy E. Croall, Department of Biochemistry, Microbiology and Molecular Biology, University of Maine, Orono, ME 04469-5375, USA.

Joseph G. Culotti, Division of Molecular Immunology and Neurobiology, Samuel Lunenfeld Research Institute, Mount Sinai Hospital, Toronto M5G 1X5, Canada.

Tom Curran, Department of Developmental Neurobiology, St Jude Children's Research Hospital, Memphis, TN, USA.

Michael J. Dans, Cellular Biochemistry and Biophysics Program, Memorial Sloan-Kettering Cancer Center, New York, NY 10021, USA.

Gabriella D'Arcangelo, Department of Developmental Neurobiology, St Jude Children's Research Hospital, Memphis, TN, USA

Guido David, Laboratory for Glycobiology and Developmental Genetics, Center for Human Genetics, University of Leuven and Flanders Interuniversity Institute for Biotechnology, Campus Gasthuisberg O & N, Herestraat 49, 3000 Leuven, Belgium.

Horace M. DeLisser, Pulmonary and Critical Care Division, Department of Medicine, University of Pennsylvania Medical Center, Philadelphia, PA 19104-4283, USA.

Kurt J. Doege, Shriners Hospital Research, Department of Cell and Developmental Biology, Oregon Health Sciences University, Portland, OR, USA.

Beth E. Drees, Department of Biology, University of Utah, Salt Lake City, UT 84112, USA.

Robert J. Dunn, Center for Research in Neuroscience, Montreal General Hospital Institute, Montreal, Canada H3G 1A4.

Donatella D'Urso, Neurologische Klinik, Neurochemisches Labor, University of Dusseldorf, Moorenstrasse 5, 40255 Dusselfdorf, Germany.

Michael L. Dustin, Department of Pathology, Washington University School of Medicine, St Louis, MO 63110, USA.

Martin Eigenthaler, Medizinische Universitätsklinik, Klinische Biochemie, 97080 Würzburg, Germany.

Jürgen Engel, Department of Biophysical Chemistry, Biozentrum University Basel, Klingelbergstrasse 70, CH-4056 Basel, Switzerland.

R. Alan B. Ezekowitz, Laboratory of Developmental Immunology, Department of Pediatrics, Massachusetts General Hospital, Harvard Medical School, Boston, MA, USA.

J. Faix, Abteilung Zellbiologie, MPI für Biochemie, 82152 Martinsried, Germany.

Marie T. Filbin, Department of Biology, Hunter College, 695 Park Avenue, New York, NY 10021, USA.

Larry W. Fisher, Craniofacial and Skeletal Diseases Branch, NIDR, NIH, Room 228, Building 30, Bethesda, MD 20892, USA.

Marilyn L. Fitzgerald, Department of Pediatrics, The Children's Hospital, Harvard Medical School, Boston, MA, USA.

Geoffrey Flood, Department of Biochemistry, University of Leicester, University Road, Leicester LE1 7RH, UK.

Andrew J. W. Furley, Developmental Genetics Programme, Department of Biomedical Science, University of Sheffield, Sheffield S10 2TN, UK.

Benjamin Geiger, Department of Molecular Cell Biology, The Weizmann Institute of Science, Rehovot 76100, Israel.

Cecilia M. Giachelli, Pathology Department, University of Washington, Seattle, WA, USA.

Filippo G. Giancotti, Cellular Biochemistry and Biophysics Program, Memorial Sloan-Kettering Cancer Center, New York, NY 10021, USA.

Paul F. Goetinck, Cutaneous Biology Research Center, Massachusetts General Hospital, Harvard Medical School, Building 149, 13th Street, Charleston, MA 02129, USA.

Daniel A. Goodenough, Department of Cell Biology, Harvard Medical School, 240 Longwood Avenue, Boston, MA 02115, USA.

Alexander Gow, Brookdale Center for Developmental and Molecular Biology, Box 1126, The Mount Sinai School of Medicine, One Gustave Levy Place, New York, NY 10029, USA.

Kathleen J. Green, Departments of Pathology and Dermatology, 303 East Chicago Avenue, Northwestern University Medical School, Chicago, IL 60611, USA.

Alan Hall, MRC Laboratory for Cell Biology, University College London, Gower Street, London WC1E 6BT, UK.

Deborah E. Hall, Department of Neurology, Rm C215, Box 0114, 505 Parnassus Avenue, San Francisco, CA 94143-0114, USA.

J. R. Hassell, Shriners Hospital for Children, 12502 North Pine Drive, Tampa, FL 33612-9466, USA.

Martin E. Hemler, Dana-Farber Cancer Institue, 44 Binney Street, Boston, MA 02115, USA.

Michael D. Henry, Howard Hughes Medical Institute, Department of Physiology and Biophysics and Department of Neurology, University of Iowa College of Medicine, Iowa City, IA 52242, USA.

Lindsay Hinck, Department of Biology, University of California, Santa Cruz, CA, USA.

Susan Hockfield, Section of Neurobiology, School of Medicine, Yale University, New Haven, CT 06520-8001, USA.

Jennifer Hodge-Dufour, Wistar Institute, Philadelphia, PA, USA.

Michael Hortsch, Department of Anatomy and Cell Biology, University of Michigan, Ann Arbor, MI 48109-0616, USA.

Otmar Huber, Max-Planck-Institut für Immunbiologie, Stübeweg 51, 79108 Freiburg, Germany.

Tony Hunter, The Salk Institute, 10010 North Torrey Pines Road, La Jolla, CA, USA.

Richard Hynes, Howard Hughes Medical Institute and Center for Cancer Research, Department of Biology, Massachusetts Institute of Technology, MA, USA.

Heinz Jacobs, Basel Institute for Immunology, Grenzacherstrasse 487, CH-4005 Basel, Switzerland.

Thomas Jarchau, Institute for Clinical Biochemistry and Pathobiochemistry, University of Würzburg, Versbacher Strasse 5, D-97078 Würzburg, Germany.

Fernando Jiménez, Centro di Biología Molecular, CSIC-UAM, Madrid, Spain.

Judith P. Johnson, Institute for Immunology, Goethestrasse 31, 80336 Munich, Germany.

Keith R. Johnson, Department of Biology, University of Toledo, Toledo, OH 43606, USA.

Rolf Kemler, Max-Planck-Institut für Immunobiologie, Stübeweg 51, 79108 Freiburg, Germany.

John Kendrick-Jones, MRC Laboratory of Molecular Biology, Hills Road, Cambridge CB2 2QH, UK.

Brigitte H. Keon, Department of Neurobiology, Harvard Medical School, 240 Longwood Avenue, Boston, MA 02115, USA.

Adam M. Koppel, Department of Neuroscience, University of Pennsylvania School of Medicine, Philadelphia, PA 19104, USA.

Helmut Krämer, Department of Cell Biology and Neuroscience, University of Texas Southwestern Medical Center at Dallas, 5323 Harry Hines Boulevard, Dallas, TX 75235-9111, USA.

Geoffrey W. Krissansen, Department of Molecular Medicine, School of Medicine and Health Sciences, University of Auckland, Auckland, New Zealand.

Janet Kurjan, Department of Microbiology and Molecular Genetics and the Vermont Cancer Center, University of Vermont, Burlington, VT 05405, USA.

Arthur D. Lander, Department of Developmental and Cell Biology, University of California at Irvine, Irvine, CA 92697-2275, USA.

Jack Lawler, Department of Pathology, Beth Israel Deaconess Medical Center, Research North, Room 270C, 99 Brookline Avenue, Boston, MA 02215, USA.

Klaus Lehnert, Department of Molecular Medicine, School of Medicine and Health Sciences, University of Auckland, Auckland, New Zealand.

E. David Leonardo, Howard Hughes Medical Institute and Departments of Anatomy and of Biochemistry and Biophysics, University of California, San Francisco, CA 94143-0452, USA.

Euphemia Leung, Department of Molecular Medicine, School of Medicine and Health Sciences, University of Auckland, Auckland, New Zealand.

Pat Levitt, Department of Neurobiology, University of Pittsburgh School of Medicine, Pittsburgh, PH 15261, USA.

Diane C. Lin, Department of Developmental and Cell Biology, University of California at Irvine, Irvine, CA 697-1450, USA.

Shin Lin, Department of Developmental and Cell Biology, University of California at Irvine, Irvine, CA 697-1450, USA.

Frederick P. Lindberg, Division of Infectious Diseases, Washington University School of Medicine, St Louis, MO 63110, USA.

Peter N. Lipke, Department of Biology, Hunter College of City University of New York, New York, NY 10021, USA.

Sergey V. Litvinov, Department of Pathology, Leiden University, Leiden, The Netherlands.

Roy R. Lobb, Biogen Inc., Cambridge, MA, USA.

Elizabeth J. Luna, University of Massachusetts Medical Center, Worcester Foundation Campus, Shrewsbury, MA, USA.

Yoichiro Matsuoka, Howard Hughes Medical Institute and Department of Cell Biology, Duke University Medical Center, Durham, NC 27710, USA.

L. McKerracher, Department of Pathology, University of Montreal, Montreal, Canada.

U. J. McMahan, Department of Neurobiology, Stanford University School of Medicine, Stanford, CA, USA.

Robert P. Mecham, Washington Unversity School of Medicine, Department of Cell Biology and Physiology, Box 8228, 660 South Euclid Avenue, St Louis, MO 63110, USA.

Carolyn Moores, MRC Laboratory of Molecular Biology, Hills Road, Cambridge CB2 2QH, UK.

Jon S. Morrow, Department of Pathology, Yale Medical School, New Haven, CT, USA.

Bernhard Moser, Theodor-Kocher Institute, University of Bern, Freiestrasse 1, CH-3012 Bern, Switzerland.

Diana G. Myles, Section of Molecular and Cellular Biology, University of California at Davis, Davis, CA 95616, USA.

W. James Nelson, Department of Molecular and Cellular Physiology, Beckman Center, B121, Stanford University School of Medicine, Stanford, CA 94305-5426, USA.

Hoàng-Oanh Nghiêm, Institut Pasteur, Neurobiologie Moléculaire, 25 Rue du Dr Roux, 75724 Paris Cedex 15, France.

Yoshifumi Ninomiya, Department of Molecular Biology and Biochemistry, Okayama University Medical School, 2-5-1 Shikata-cho, Okayama 700, Japan.

B. Öbrink, Department of Cell and Molecular Biology, Medical Nobel Institute, Karolinska Institute, PO Box 285, S-17177, Stockholm, Sweden.

Åke Oldberg, Department of Cell/Molecular Biology, PO Box 94, University of Lund, S221 00 Lund, Sweden.

Bjorn Reino Olsen, Department of Cell Biology, Harvard Medical School and Harvard-Forsyth Department of Oral Biology, Harvard School of Dental Medicine, Boston, MA, USA.

Laurelee Osborn, Biogen Inc., Cambridge, MA, USA.

Katsushi Owaribe, Unit of Biosystems, Graduate School of Human Informatics, Nagoya University, Nagoya 464-1, Japan.

Liliana Pedraza, Brookdale Center for Developmental and Molecular Biology, The Mount Sinai School of Medicine, One Gustave Levy Place, New York, NY 10029, USA.

David L. Paul, Department of Neurobiology, Harvard Medical School, 240 Longwood Avenue, Boston, MA 02115, USA.

Leslie Petch, Department of Cell Biology and Anatomy, University of North Carolina at Chapel Hill, Chapel Hill, NC 27599, USA.

Aurea Pimenta, Department of Neurobiology, University of Pittsburgh School of Medicine, Pittsburgh, PA 15261, USA.

Michel Piovant, Laboratoire de Génétique et Biologie Cellulaire, CNRS, Marseille, France.

Hidde L. Ploegh, Department of Pathology, Harvard Medical School, 200 Longwood Avenue, Boston, MA 02115, USA.

Pascal Pomiès, Department of Biology, University of Utah, Salt Lake City, UT 84112, USA.

Paul Primakoff, Department of Cell Biology and Human Anatomy, University of California at Davis, Davis, CA 95616, USA.

Ellen Puré, Wistar Institute, 3601 Spruce Street, Philadelphia, PA 19104, USA.

Jonathan A. Raper, Department of Neuroscience, University of Pennsylvania School of Medicine, Philadelphia, PA 19104, USA.

Alexander Redlitz, Department of Biological Structure, University of Washington, Box 357420, Seattle, WA 98195-7420, USA.

Louis F. Reichardt, Department of Physiology and Howard Hughes Medical Institute, 513 Parnassus Avenue, University of California, San Francisco, CA 94143, USA.

Kathrin Reinhard, Institute for Clinical Biochemistry and Pathobiochemistry, University of Würzburg, Versbacher Strasse 5, D-97078 Würzburg, Germany.

Matthias Reinhard, Institute for Clinical Biochemistry and Pathobiochemistry, University of Würzburg, Versbacher Strasse 5, D-97078 Würzburg, Germany.

Steven Rosen, Department of Anatomy and Program in Immunology, University of California, San Francisco, CA 94143-0452, USA.

Eric Rubenstein, INSERM U268, Hôpital Paul Brousse, 94807 Villejuif Cedex, France.

Zaverio M. Ruggeri, Roon Research Center for Arteriosclerosis and Thrombosis, Departments of Molecular and Experimental Medicine and of Vascular Biology, The Scripps Research Institute, 10550 North Torrey Pines Road, La Jolla, CA 92037, USA.

E. Helen Sage, Department of Biological Structure, University of Washington, Box 357420, Seattle, WA 98195-7420, USA.

Takako Sasaki, Max-Planck-Institut für Biochemie, D-82152 Martinsried, Germany.

Kedarnath N. Sastry, Department of Pathology, Boston University School of Medicine, Boston, MA, USA.

Michael D. Schaller, Department of Cell Biology and Anatomy, University of North Carolina at Chapel Hill, Chapel Hill, NC 27599, USA.

Tito Serafini, Department of Molecular and Cell Biology, University of California, Berkeley, CA 94270-3200, USA.

Lawrence Shapiro, Program in Structural Biology, Department of Physiology, The Mount Sinai School of Medicine, One Gustave Levy Place, New York, NY 10029, USA.

Sanford J. Shattil, Department of Vascular Biology, The Scripps Research Institute, 10550 North Torrey Pines Road, La Jolla, CA 92037, USA.

Yu Shen, Department of Cell Biology and Anatomy, University of North Carolina at Chapel Hill, Chapel Hill, NC 27599, USA.

Barry D. Shur, Department of Cell Biology, Emory University School of Medicine, Atlanta, GA, USA.

Suzanne Simon, The Salk Institute, 10010 North Torrey Pines Road, La Jolla, CA, USA.

Peter Sonderegger, Department of Biochemistry, University of Zurich, Winterthurerstrasse 190, CH-8057 Zurich, Switzerland.

Timothy A. Springer, Center for Blood Research and Department of Pathology, Harvard Medical School, Boston, MA 02115, USA.

Mark D. Sternlicht, University of California at San Francisco, Department of Anatomy, LR-208, 3rd & Parnassus Avenues, San Francisco, CA 94143-0452, USA.

Craig M. Story, Department of Pathology, Harvard Medical School, 200 Longwood Avenue, Boston, MA 02115, USA.

Xin Sun, Howard Hughes Medical Institute and Department of Cell Biology and Biology, Boyer Center for Molecular Medicine, Yale University, New Haven, CT 06536-0812, USA.

Masatoshi Takeichi, Department of Biophysics, Faculty of Science, Kyoto University, Kyoto 606-8502, Japan.

Palmer Taylor, Department of Pharmacology 0636, University of California, San Diego, La Jolla, CA 92093, USA.

Joyce Taylor-Papadimitriou, Imperial Cancer Research Fund, London, UK.

Marc Tessier-Lavigne, Howard Hughes Medical Institute and Department of Anatomy, University of California, San Francisco, CA 94143-0452, USA.

Jean Paul Thiery, CNRS UMR 144 and Institut Curie, 26 rue d'Ulm, 75248 Paris Cedex 05, France.

Rupert Timpl, Max-Planck-Institut für Biochemie, D-82152 Martinsried, Germany.

Bryan P. Toole, Department of Anatomy and Cellular Biology, Tufts University Health Science Schools, 136 Harrison Avenue, Boston, MA 02111, USA.

Shachiko Tsukita, College of Medical Technology, Kyoto University Faculty of Medicine, Shogoin-Kawahara, Sakyo-ku, Kyoto 606, Japan.

Shoichiro Tsukita, Department of Cell Biology, Kyoto University Faculty of Medicine, Yoshida-Konoe, Sakyo-ku, Kyoto 606, Japan.

Christopher E. Turner, Department of Anatomy and Cell Biology, SUNY Health Science Center at Syracuse, 750 East Adams Street, Syracuse, NY 13210, USA.

Judith A. Varner, University of California at San Diego, Department of Medicine/Cancer Center, 9500 Gilman Drive, La Jolla, CA 92093-0684, USA.

Dietmar Vestweber, Institute of Cell Biology, ZMBE University of Münster, Technologiehof, Mendelstrasse 11, D-48149 Münster, Germany.

Mark Veugelers, Laboratory for Glycobiology and Developmental Genetics, Center for Human Genetics, University of Leuven and Flanders Interuniversity Institute for Biotechnology, Campus Gasthuisberg O & N, Herestraat 49, 3000 Leuven, Belgium.

Hansjürgen Volkmer, Max-Delbrück-Center for Molecular Medicine, Robert-Rössle Strasse 10, 13122 Berlin, Germany.

Bruce G Wallace, Department of Physiology and Biophysics, University of Colorado Health Sciences Center, USA.

Ulrich Walter, Institute for Clinical Biochemistry and Pathobiochemistry, University of Würzburg, Versbacher Strasse 5, D-97078 Würzburg, Germany.

Jerry Ware, Roon Research Center for Arteriosclerosis and Thrombosis, Departments of Molecular and Experimental Medicine and of Vascular Biology, The

Scripps Research Institute, 10550 Torrey Pines Road, La Jolla, CA 92037, USA.
Paul M. Wassarman, Department of Cell Biology and Anatomy, Mount Sinai School of Medicine, New York, NY 10029-6574, USA.
Zena Werb, University of California at San Francisco, Department of Anatomy, LR-208, 3rd & Parnassus Avenues, San Francisco, CA 94143-0452, USA.
Gary Wessel, Department of Molecular and Cell Biology and Biochemistry, Box G, Brown University, Providence, RI 02912, USA
Margaret J. Wheelock, Department of Biology, University of Toledo, Toledo, OH 43606, USA.
Gerhard Wiche, Vienna Biocenter, Institute for Biochemistry and Molecular Cell Biology, Vienna, Austria.
Dieter R. Zimmermann, Research Laboratory, Institute of Clinical Pathology, Department of Pathology, University of Zürich, Schmeizbergstrasse 12, 8091 Zürich, Switzerland.

1

Cytoskeleton-associated anchor and signal transduction proteins

Vinculin in cell–cell adherens and cell–extracellular matrix adhesion sites of chicken lens cells. (Courtesy of B. Geiger.)

Introduction

Membrane–cytoskeleton interaction

In this chapter heterogeneous families of molecules will be discussed, including cytoplasmic proteins that are involved in the physical linkage of cytoskeletal filaments to the membranes and different cytoskeleton-associated signalling molecules. This reflects the growing body of evidence that many cytoskeleton-bound proteins are also involved in signalling processes, in particular, in adhesion-dependent signal transduction.[1–5,112–115] The precise definition of the 'anchor proteins' is not as straightforward as it may seem. Ideally, one might expect them to exhibit specific binding affinity to both the exposed cytoplasmic moieties of integral membrane constituents and to specific cytoskeletal components. However, most known membrane–cytoskeleton associations consist of many potential links, forming complex multi-level networks,[2,4] so that some typical anchor proteins (for example, vinculin) may have no direct connections with transmembrane receptors. In addition, the currently available information on the properties and binding specificity of many putative anchor molecules is still rather limited. Therefore, an operational definition will be adopted here according to which cytoskeletal anchor proteins are identified by their apparent association with membrane–cytoskeleton interfaces, irrespective of whether they interact primarily with the membrane or the cytoskeleton, directly or indirectly. This broad definition immediately includes in this group many proteins associated with specific domains of the membrane–cytoskeleton interface but not necessarily having direct linking functions. It is significant that the majority of these proteins are protein kinases, phosphatases, and adapter proteins involved in a variety of signal-transduction pathways related to growth regulation, development, and oncogenesis. Both 'true' anchor proteins and associated signal transduction proteins do not have to be connected permanently to the plasma membrane and may partly be diffusely distributed in the cytoplasm or associated with other regions along the cytoskeletal network. Moreover, recent studies have revealed a new class of molecular triggers (rho family GTPases) whose activation is indispensable for the assembly of specific cytoskeleton–membrane anchorage domains, but which are not necessarily localized to these domains. Therefore, in addition to the *bona fide* anchors of this section, other anchors may be found in the different sections of the *Guidebook to the Cytoskeletal and Motor Proteins* (e.g. brush border myosin may also play an important role in linking the core bundle of microfilaments to the microvillar membrane in intestinal epithelial cells).

Notably, the cellular functions of cytoskeletal anchor and associated signal transduction proteins are highly diversified, ranging from the control of assembly of cytoskeletal networks and the transmembrane or transcellular transduction of mechanical forces, to the local modulation of membrane dynamics and signal transduction. Such activities are involved in major cellular activities including motility, adhesion, and morphogenesis as well as regulation of growth and differentiation.

Involvement of cytoskeletal anchor proteins in cell adhesion

One of the most prominent functions of cytoskeletal anchor proteins is their participation in the adhesive interactions of cells; many of the proteins reviewed in this chapter are associated with specialized cell–matrix or cell–cell junctions (Fig. 1). Such cellular interactions are characteristic of metazoan organisms and are responsible for the assembly of individual cells into functional tissues and organs. Attempts to elucidate the molecular mechanisms underlying cell adhesion usually address two distinct aspects, namely the specificity of interaction and the short- and long-range morphogenetic response triggered by it. The former depends primarily on the nature of the transmembrane receptors which directly mediate the adhesion (see the chapters in Part 2, Cell adhesion and cell–cell contact proteins) and the latter involves the assembly of the relevant force-generating cytoskeletal structures and their attachment to the cytoplasmic faces of the newly formed adhesions. The proper assembly of the entire transmembrane 'adhesion complex' is of critical physiological importance for a wide variety of cellular processes including cell spreading and locomotion, embryonic cell sorting, mesenchymal-to-epithelial transformation or vice versa, tissue assembly, formation of transcellular barriers, establishment of transcellular forces, and so on.

Molecular and functional diversity of cell adhesions and their activity as 'signal transducing units'

Studies on cell adhesion, carried out over the past several decades, have indicated that this process is not only molecularly complex, involving many extrinsic and intrinsic components, but also highly diversified at the cellular and functional levels. Thus, cell adhesions cannot be regarded as one uniform entity but rather as a heterogeneous family of related structures. There are many distinct 'extracellular ligands' providing the adhesive surface to which cells may attach (Fig. 1). These include the various components of the extracellular matrix (ECM) and basement membranes as well as specialized integral membrane-bound cell adhesion molecules. Interactions

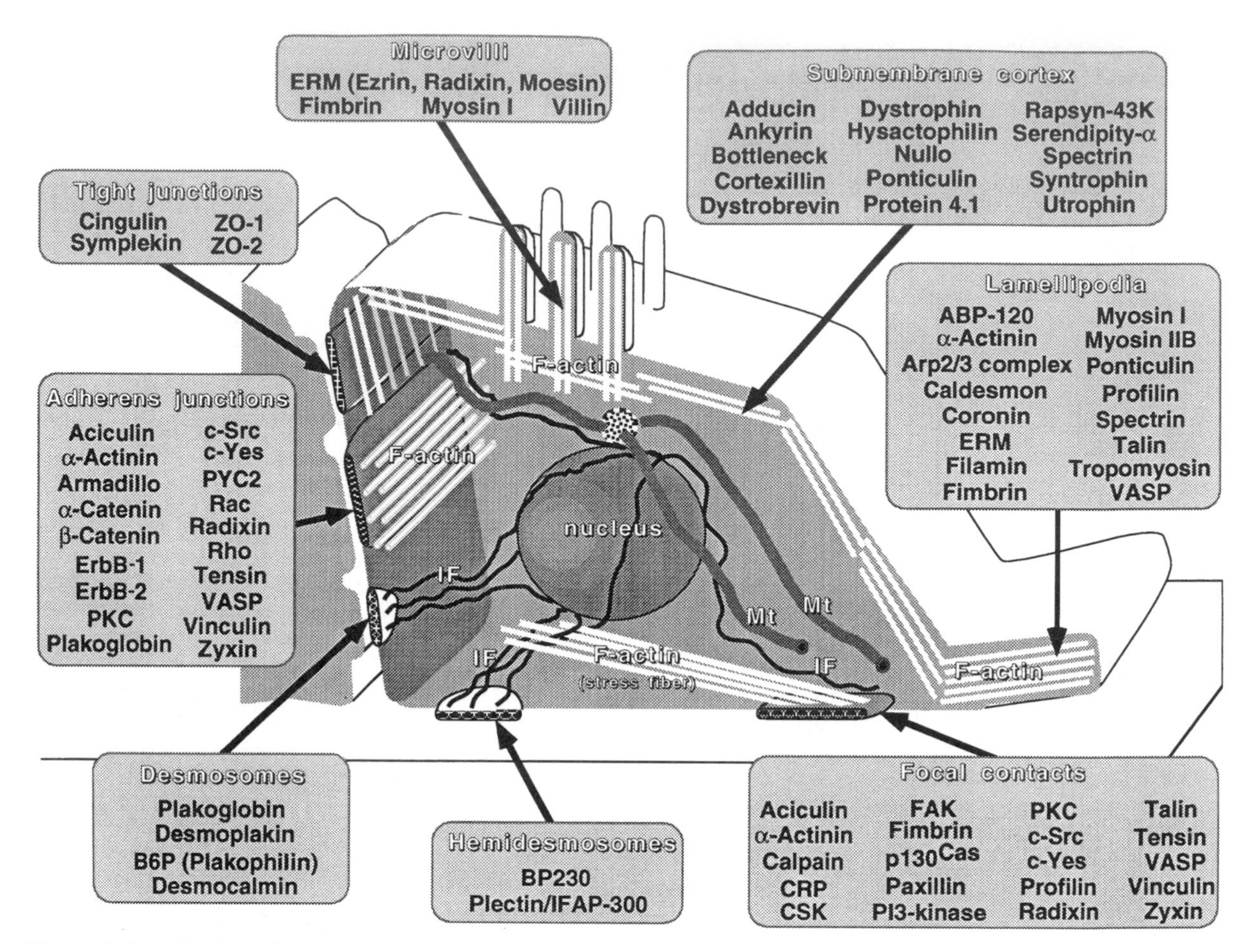

Figure 1. Localization of cytoskeleton-associated anchor proteins in different submembrane domains. The scheme shows an idealized 'universal' cell having all types of junctions with the neighbouring cell (on the left) and with different ECM components (bottom); microvilli, lamellipodia, and specialized submembrane cortexes of several types are also presented. Anchor and signal transduction proteins localized in each of these domains are listed in alphabetical order in the corresponding boxes. It should be emphasized that we do not imply that all the proteins listed in each box are indeed present in the same cellular compartment, since the list consists of proteins of different species, from amoebae to higher vertebrates and of different cell types. In particular, the box entitled 'Submembrane cortex' contains components of submembrane cortexes of different types: the spectrin-based network of red blood cells, the dystrophin/utrophin-based cortex of muscle cells, a protein involved in the anchorage of acetylcholine receptor (rapsyn-43k), *Drosophila* anchor proteins involved in the cellularization process, and cortical proteins from Dictyostelium. Mt, microtubules; IF, intermediate filaments.

with these ligands trigger the formation of the two major classes of cell adhesions, namely, cell–matrix and cell–cell junctions, respectively. Furthermore, within each family there are several structurally and functionally distinct types of cell adhesions, identified, mainly, by electron microscopy and immunocytochemistry.

Specialized matrix adhesions include focal adhesions[6] and hemidesmosomes.[7] Focal adhesions are mediated by integrin transmembrane receptors associated with the actin cytoskeleton inside the cell. Two major linkers performing this association are talin and α-actinin, which were shown to interact with the cytoplasmic domain of the $\beta1$ chain of integrin and with actin. Vinculin binds to talin, α-actinin, and actin and apparently stabilizes the entire structure.[2,6] Hemidesmosomes that mediate adhesion of epithelial cells to the underlying basement membrane are associated with the network of intermediate filaments. Transmembrane receptors of the hemidesmosomes include at least one special type of integrin, $\alpha6\beta4$, and protein BP180 (bullous pemphigoid antigen 2). The main anchor proteins in the hemidesmosomal plaque are BP230 (bullous pemphigoid antigen 1) and plectin, which can connect the $\beta4$ cytoplasmic domain of integrin with intermediate filaments.[7,116]

Intercellular adhesion in vertebrate organisms is mediated through four major types of junctions, namely *tight junctions, adherens junctions, desmosomes,* and *gap junctions* (Fig. 2). Recent molecular studies have indicated that each of these contact sites contains a unique set of proteins which interact with each other, conferring on the particular adhesion region its unique structure and properties (Fig. 2). Tight junctions,[8] which usually occupy the most apical position in the junctional complex of epithelia, contain the transmembrane proteins claudin-1 and -2 and occludin,[117] and the plaque proteins ZO-1, ZO-2, cingulin, symplekin, and 7H6 antigen.[9] The ZO-1 was suggested to connect the C-terminal cytoplasmic domain of occludin with the tetrameric form of spectrin and

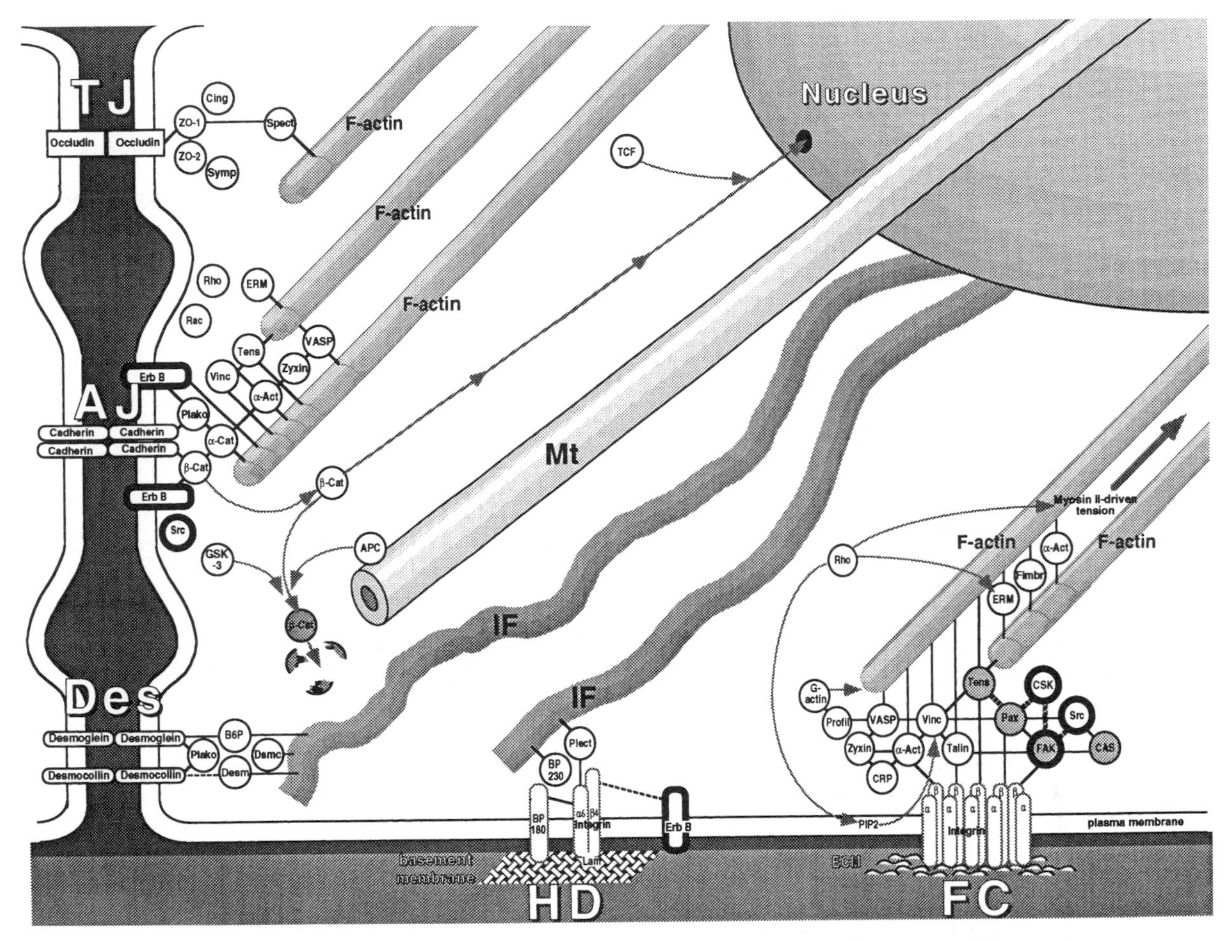

Figure 2. Molecular interactions in different types of cell junctions. The scheme depicts possible protein–protein interactions in cell–cell and cell–ECM junctions. An idealized cell having three types of cell–cell junctions: tight junctions (TJ), adherens junctions (AJ), and desmosomes (Des) and two types of cell–ECM junctions: hemidesmosomes (HD) and focal contacts (FC) is shown. Cytoplasmic proteins are represented by circles, transmembrane proteins as boxes. A possibility of interactions between two proteins (based primarily on *in vitro* binding experiments) is represented by a straight solid line connecting the corresponding proteins. If the circle corresponding to the particular protein is connected to several other circles it means that this protein can interact with several other proteins, but does not imply that it can interact with all of them simultaneously. Therefore this scheme does not represent the real topography of protein–protein interactions in the junctions. The indirect distant interactions and dynamic processes are represented by grey arrows. Tyrosine kinases are represented by circles or boxes with thick outlines. Proteins that undergo tyrosine phosphorylation in response to cell adhesion to the ECM are represented by grey circles. SH2 domain–phosphotyrosine interactions are represented by thick broken lines. Putative protein–protein interactions are shown as thin broken lines. Abbreviations (in alphabetic order): α-Act, α-actinin; α-Cat, α-catenin; β-cat, β-catenin; B6P, band 6 protein (plakophilin); BP180 and BP230, bullous pemphigoid 180 kDa and 230 kDa polypeptides; Cing, cingulin; Desm, desmoplakin; Dsmc, desmocalmin; ERM, ezrin/radixin/moesin protein family; Fimbr, fimbrin; Lam, laminin; Pax, paxillin; Plako, plakoglobin; Plect, plectin; Profil, profilin; Spect, spectrin; Symp, symplekin; Tens, tensin; Vinc, vinculin. Mt, microtubules; IF, intermediate filaments.

consequently with the actin cytoskeleton.[8] Tight junctions are usually believed to be responsible for transepithelial resistance as well as for the formation and maintenance of membrane polarity (apical vs, basolateral) in simple epithelial cells.

Adherens junctions, characterized by their specific association with the contractile microfilament system[2,10] contain 'classical' cadherins (i.e. E, N, or P-type) as transmembrane receptors. β-Catenin and γ-catenin (plakoglobin), as well as their *Drosophila* homologue, armadillo protein, can bind to the conserved cytoplasmic domain of the cadherins,[11] while α-catenin can link either β-catenin or plakoglobin[12] to the actin filament system, through its binding to α-actinin,[13] vinculin[118] or to actin itself.[14] Other actin-associated components of the adherens junctions include ERM family proteins[15] as well as tensin and zyxin which are also present in focal adhesions.[2,4] In cadherin-based cell–cell contacts of non-epithelial cells ZO-1 was found to participate in cross-linking between a cadherin/catenin complex and the actin cytoskeleton.[16]

The third class of cell–cell contacts to be discussed here are desmosomes,[3,17] which are characterized by cadherins of the desmoglein and desmocollin type. These desmosomal cadherins interact with the intermediate filament system via plakophilin (band 6 protein) and desmoplakin. Plakoglobin binds desmosomal cadherins too[8,19] and is an indispensable component of the desmosomal plaque.[20,21] Both plakoglobin and plakophilin may bind to desmoplakin.[119]

The three junctions mentioned above (as well as gap junctions, which will not be discussed here) are distinct, both structurally, topologically, and molecularly, yet they retain some characteristic spatial relationships and functional interdependence. In epithelial junctional complexes, tight junctions occupy the most apical position, followed by adherens junctions and, then, desmosomes. Their interdependence is demonstrated by the fact that modulation of adherens junctions by antibodies to uvomorulin leads also to breakdown of the neighbouring tight junctions.[22] There is also a cross-talk between the formation of adherens junctions and the formation of desmosomes.[23] It has also been found that the formation of focal adhesions and of cell–cell adherens-type junctions are often reciprocally related, probably due to competition on the shared anchor proteins.[24,25]

The inter-relationships between adhesion and the regulation of cell motifity, growth, and differentiation are among the most exciting and challenging questions in this field of research. It is now becoming apparent that the modulation of adherens junctions has dramatic effects on cell behaviour and that changes in the signal transduction machinery (for example, modulation of protein phosphorylation) may dramatically affect cellular adhesion interactions. This new role of cell–cell and cell–matrix junctions came into view when it was realized that seemingly all types of junctional specializations contain, in their cytoplasmic part, certain proteins known to participate in signal transduction pathways (Fig. 2).

Thus, both focal adhesions and cell–cell adherens junctions contain protein tyrosine kinases of the src family. Focal adhesions contain a specific tyrosine kinase, FAK (focal adhesion kinase), which is not found in the cell–cell junctions; recently, however, a homologue of FAK, cell adhesion kinase β,[26] also known as PYK2[27] and RAFTK,[28] was localized to cell–cell junctions after expression in COS-7 cells.[26] The adapter protein shc, involved in the initiation of the ras/MAP kinase pathway in response to a variety of external signals, was recently shown to associate with some integrins[29] as well as with cadherins.[30]

Receptor tyrosine kinases of the ErbB family were found to associate specifically with cell–cell adherens junctions in epithelial cells. In particular, erbB1 (EGF receptor) and erbB2 proteins bind β-catenin and can phosphorylate it.[31–33] EGF receptor can apparently interact also with the hemidesmosomal $\alpha6\beta4$ integrin, and by phosphorylating the $\beta4$ subunit may induce deterioration of hemidesmosomes.[34] An analogue of the EGF receptor in *C. elegans*, LET-23 protein, was shown to localize to the junctions between vulval precursor cells during vulval induction.[35,36] One of the proteins responsible for this localization, LIN-2A, belongs to the MAGUK family of cell junction proteins that include the vertebrate tight junction proteins ZO-1 and ZO-2.[37] Protein tyrosine phosphatases were found both in focal adhesions[38] and in cell–cell adherens junctions in association with the β-catenin/cadherin complex.[39,40]

Serine/threonine kinase PKCα, involved in many aspects of cell signalling, is localized to the focal adhesions and adherens junctions (reviewed in ref. 41). In addition, a new serine/threonine kinase specifically associated with $\beta1$ integrin (integrin-linked kinase, ILK) was found recently.[42] There is some evidence for the junctional localization of phospholipase Cγ and phosphatidylinositol 3-kinase (reviewed in refs 2, 4).

Some structural components of adhesion plaques can have dual function and participate in the signal transduction processes in addition to their structural role. For example, the 'typical' anchor protein β-catenin and its *Drosophila* homologue armadillo were shown to play an essential role in the transduction of signals initiated by wnt growth factors ('wingless' in *Drosophila).* The key feature of armadillo/β-catenin which provides functioning of this signalling pathway is their ability to interact with a transcription factors termed TCF (or LEF-1) in vertebrates[120] and dTCF (or pangolin) in *Drosophila (*for a review see ref. 43). A major parameter involved in the regulation of this signalling pathway seems to be the amount of β-catenin/armadillo available for the interactions with the corresponding transcription factor. This amount can be altered due to ubiquitin-dependent proteolysis of catenin.[44] This proteolysis is controlled by β-catenin interactions with the tumour suppressor protein APC, with the serine/threonine kinase GSK-3 (reviewed in ref. 45) and with the protein axin (conductin).[121–124] On the other hand, β-catenin can be sequestered and protected from proteolysis by binding to cadherin at the junctional plaque.[46,125] The complex interplay between the interactions of β-catenin with cadherin, APC/GSK-3, and the transcription factors has, most probably a central effect on wnt signalling. It is noteworthy that another homologue of armadillo, namely the desmosomal and adherens junctions component plakoglobin, may also participate in wnt signalling modulation.[47,126] Interestingly, genetic defects that result in upregulation of β-catenin may play a role in neoplastic transformation,[48,49] most probably due to constitutive transcriptional activation of target genes by the β-catenin–TCF complex.[50] Thus, β-catenin may act as an oncogene,[51] while several other junctional plaque proteins, including vinculin,[52] α-actinin,[53] α-catenin,[54] and plakoglobin[55,56] were shown to operate as tumour suppressors.

In conclusion, cell adhesions appear to be major sites for cytoskeletal anchor protein organization in cells. It appears likely that in these membrane domains anchor proteins play major roles in the mechanical stabilization of the junction structure and in the assembly of different adhesions into coherent junctional complexes. These complexes appear also be major sites of localization of proteins involved in transduction of external signals inside the cell.

Assembly of cell adhesions

Formation of the adhesion junctions is a multi-stage process that by itself depends on adhesion-dependent signalling. It is not easy to distinguish the signal pathways leading to the assembly of junction plaques from those involved in downstream adhesion-dependent cell regulation, mainly, since the assembly of junctions is usually necessary for downstream signalling. In this and the following paragraphs, we will discuss the processes of junction formation, while other aspects of integrin and cadherin-dependent signalling are discussed in Part 2, Cell adhesion and cell–cell contact proteins.

Formation of cellular adhesions depends on transmembrane interactions with the cytoskeleton mediated transmembrane adhesion receptors and cytoplasmic 'anchor proteins' (Fig. 2). It has been shown that integrin or cadherin chains missing parts of their cytoplasmic moieties (following transfection with mutant cDNAs) fail to mediate adhesion to the external surface.[57–61] Clustering and occupancy of the transmembrane adhesion molecules trigger the formation of a submembrane plaque and induce local cytoskeletal organization which, in turn, mechanically stabilizes the contact area, increases the overall avidity by supporting multivalent interactions, and recruits new receptors and anchor proteins to the nascent contact site.

There are several experimental systems in which the formation of adhesion molecular complexes has been studied. One can study the attachment of cells to a flat substrate covered with various ligands to the cell adhesion receptors. In some cases it is more convenient to study the adhesion of beads covered with these ligands to the cell surface. In this case, recruitment of cytoskeletal components and proteins involved in signalling to the cytoplasmic surface of the bead–cell interface can be studied, as well as surface motility of the beads. Formation of cell–cell adherens junctions can be followed after addition of Ca^{2+} to the culture in which these junctions were disrupted by incubation in low-calcium medium. Similarly, cells of some types lose their focal adhesions and associated actin filament bundles after incubation in serum-free medium, despite being adherent to the proper ECM. Formation of focal adhesions in such cells can be stimulated by addition of serum or purified growth factors.

Experiments with the beads covered with anti-integrin antibodies has shown that clustering of β1 integrins is sufficient for the recruitment of FAK and tensin to the nascent adhesion site while with an intact actin cytoskeleton, many additional proteins involved in signal transduction become associated with the clustered integrins.[62,63] The major structural proteins of adhesion plaques are recruited to the beads only when, in addition to integrin clustering, there is also binding of the appropriate ligand (i.e. fibronectin) to the integrin receptor. When this occurs, talin, α-actinin and vinculin associate with the clustered integrin molecules in a process that is not sensitive to actin-cytoskeleton disruption by cytochalasin D. If the integrity of the actin cytoskeleton is not disturbed, paxillin, the actin-binding protein filamin, and actin itself complete the formation of the entire focal adhesion complex.[62,63] Under these conditions, a directed centripetal movement of the surface-attached beads occurs.[64] The hierarchy of molecular events in the formation of cell–cell adherens junctions has not yet been elucidated.

The dynamics of the assembly of junctional complexes is still poorly understood. Numerous immunofluorescence observations suggest that elongated oval-shaped focal adhesions are the products of 'maturation' of smaller dot-like adhesions localized usually to the cell edge. The rationale for this suggestion is that the dot-like contacts appear first in the course of cell spreading on ECM-covered substrate; the dot-like contacts are also abundant under conditions in which the formation of 'mature' contacts is hindered by some means (disruption of the actin cytoskeleton by cytochalasins, incubation of cells in serum-free medium, decrease of tyrosine phosphorylation by some inhibitors; see below). Some components of focal contacts, namely the actin-binding proteins VASP and profilin, are thought to be associated mainly with nascent focal adhesions.[6] In late (24–36 h) cultures, when a sufficient amount of ECM components has been presented, in addition to the focal adhesions formed at the interface between the cell and the solid surface of a culture dish, special ECM adhesions are formed with ECM fibrils, located either at the dorsal or ventral cell surface.[65] The dynamics of formation of these 'fibrillary adhesions' and their relation to regular focal adhesions are worth studying.

Future studies of the dynamics of focal adhesion components in living cells using either fluorescently tagged anchor proteins or GFP-fusion constructs may elucidate the time course of assembly events.

Tyrosine phosphorylation of the components of adhesion plaques

Focal adhesions and cell–cell adherens junctions appear to contain a high level of tyrosine phosphorylated proteins, a hallmark of signalling molecules[24,66] (Fig. 2). Formation of focal adhesions is normally accompanied by the tyrosine phosphorylation of FAK and several other components of the focal adhesions, including paxillin, tensin, p130^{CAS} and cortactin (reviewed in refs 2, 4, 5, 67). FAK seems to play a major role in this tyrosine phosphorylation cascade. In contrast to members of the Src family and other kinases, FAK can be recruited to nascent adhesion sites even when tyrosine phosphorylation is blocked by genistein.[63] Biochemical studies suggest that autophosphorylation of FAK may induce the SH2 domain-driven recruitment and phosphorylation of other components of focal adhesions.[68–70] In fact, loading of cells with the dominant negative FAK construct[71] or overexpression of naturally occurring FAK homologue lacking the kinase domain (FRNK)[72] leads to a significant reduction of the level of tyrosine phosphorylation at focal adhesions, but apparently does not inhibit their formation.

Genistein and herbimycin A, broad specificity inhibitors of protein tyrosine phosphorylation, efficiently prevent the formation of focal adhesions induced by a variety of stimuli in different experimental systems and, also, disintegrate existing focal adhesions in cultured cells.[63,73–75] Accordingly, an inhibitor of a class of protein tyrosine phosphatases, phenylarsine oxide (PAO), induces assembly of focal adhesions in serum-starved cells.[76] Thus, tyrosine phosphorylation of some target(s) is a critical event in focal adhesion formation. It is still not clear whether tyrosine phosphorylation of some proteins outside focal adhesions is essential for focal adhesion assembly or whether local phosphorylation of the focal adhesion components is the more critical process.

It is noteworthy that tyrosine phosphorylation may also be necessary for cell detachment and release of focal adhesions.[77] This is supported by the observation that suppression of FAK function increases the size of focal adhesions and reduces cell motility.[71,78] It was proposed long ago that excessive tyrosine phosphorylation of focal qdhesion components in RSV-transformed fibroblasts is responsible, at least in part, for the destabilization of cell–matrix adhesions typical of the transformed phenotype,[79] and that inhibition of tyrosine phosphorylation by specific tyrphostins restores focal adhesion formation in these cells.[25]

The effect of tyrosine phosphorylation on cell–cell adherens junctions was also first noted for RSV-transformed cells in which phosphorylation by pp60$^{v\text{-}src}$ was accompanied by deterioration of cell–cell adherens junctions.[24,80–82] Here, as well as in several other types of transformed cells, tyrosine phosphorylation of β-catenin seems to be an important parameter leading to adherens junction breakdown (reviewed in ref. 127). Tyrosine phosphorylation of β-catenin decreases its interactions with E-cadherin and disrupts the integrity of adherens junctions.[82,83] Besides src family members, some receptor tyrosine kinases (in particular, the RPTK of the erbB family) may also phosphorylate β-catenin, destabilizing cell–cell junctions.[127]

Reduction of tyrosine phosphorylation by various inhibitors fails to disrupt adherens junctions.[84] On the contrary, it has been demonstrated that treatment by specific tyrphostins leads to the restoration of cell–cell junctions in transformed cells.[25] More recently, however, it was shown that while excessive tyrosine phosphorylation has a detrimental effect on junction structure, a certain basal level of phosphorylation promotes adhesion and junction assembly.[85] Moreover, it was demonstrated that 'waves' of junctional tyrosine phosphorylation induced in serum-starved cells by EGF stimulation or by pervanadate treatment were accompanied by waves in the junctional levels of vinculin, actin, and plakoglobin, followed by later changes in the recruitment of cadherins to these sites.[86]

Thus, tyrosine phosphorylation-dependent regulation of junction assembly via interaction of SH2 and PTB domains of some proteins with the corresponding phosphotyrosine residues appears to be a central mechanism, although the precise roles of specific tyrosine kinases and their target proteins are still unclear.

Regulation of anchor protein interactions by rho-family small G proteins

One of the most important regulators of the assembly of focal adhesions and actin microfilament bundles associated with them is the small G protein, rho (Fig. 2). This protein is not associated constitutively with focal adhesions,[87] but its function is necessary for their formation and maintenance in ECM-attached cells[88,89] (reviewed in refs 90,91). Other members of the rho family also affect the membrane-associated actin filament system. Rac is involved in the formation of lamellar actin-rich cell extensions, ruffles, and lamellipodia,[92] while cdc42 is required for the formation of finger-like filopodia or microspikes.[93,94] The activation of the rho family G proteins requires extracellular ligands such as serum factors and other hormones including lysophosphatidic acid (LPA), a major serum component which activates rho,[88] PDGF and EGF, which activate rac,[92] and bradykinin which activates cdc42.[93]

Since rho and rho family proteins interact with a variety of target molecules, several hypotheses have been proposed for the possible role of downstream components in rho-dependent signalling that may result in the assembly of focal adhesions and associated actin filament bundles. Schwartz and co-workers[95] provided evidence suggesting the involvement of rho in the regulation of 4,5-PIP2 production by the activation of PI4-phosphate 5-kinase. Rac also was shown to induce rapid 4,5-PIP2 synthesis.[96] Increased 4,5-PIP2 production could be relevant to formation of focal adhesions, as this signalling phospholipid is enriched in focal adhesions[97] and α-actinin and vinculin are 4,5-PIP2-binding proteins.[97] It was further suggested that 4,5-PIP2 induces a conformational change in vinculin which exposes cryptic sites and allows interaction with actin, α-actinin, and talin.[98] In addition, it was reported[96] that D3 and D4 polyphosphoinositides, including 4,5-PIP2, can uncap actin filaments and promote actin polymerization in cells.

Actin polymerization is crucial for the rac-dependent formation of membrane protrusions, while rho-induced formation of actin bundles does not require massive assembly of actin filaments.[99] Therefore, the target of rho-dependent regulation which is responsible for actin filament movement in the cell might be most important in rho-induced actin reorganization. In this connection the effects of rho-activated serine/threonine kinase, ROK (or rho kinase) are most interesting. It was shown that rho-kinase microinjection can mimic the effect of rho activation on actin bundles and focal adhesion formation.[100] Rho kinase phosphorylates (and activates) the regulatory light chain of myosin II[101] and, together with rho, inactivates the myosin light chain phosphatase by phosphorylating it.[102] This, in turn increases the level of phosphorylation of the myosin II regulatory light chain and thereby stimulates cell contractility. It has been observed in several cellular systems that rho activation, in fact, induces cell contraction.

Could this development of contraction stimulate the formation of focal adhesions and actin filament bundles? Several lines of evidence favour such a hypothesis. It was demonstrated that a burst of cell contractility induced by disruption of cytoplasmic microtubules in serum-starved cells induces the assembly of focal adhesions and tyrosine phosphorylation of FAK and paxillin.[75] These effects required cell adhesion to the ECM and did not occur when myosin II-driven contractility was inhibited by several myosin light chain kinase inhibitors.[75] The impeding of myosin II activity by inhibitors blocks the rho-induced formation of focal adhesions and stress fibres.[103] Moreover, it was shown that the strength of the links between fibronectin-coated beads and the cytoskeleton can be controlled by application of external forces.[104] In agreement with the view that increased contractility in firmly ECM-attached cells invariably increases the tension at focal adhesions,[105] it is conceivable that focal adhesions are 'tension-sensing devices' converting cell contraction into protein modification (tyrosine phosphorylation?) and assembly events. This, in turn, may locally induce the assembly of stress fibres and focal contacts and induce more distal responses leading to the transition from G_1 into S phase.

More recently it was shown that both rac and rho activities are necessary also for the assembly and maintenance of cell–cell adherens junctions.[106,107] In contrast to the situation with focal contacts, the presence of both GTPases at cell–cell adhesion sites is documented.[87,107,108] Blocking endogenous rho or rac selectively removed cadherin complexes from the junctions, while desmosomes were not perturbed. Rac activity is sufficient for the recruitment of actin to clustered cadherin receptor while both rac and rho are necessary for the formation of complete junctions.[107] The mechanism of small GTPase involvement in the formation of junctional structures, as well as possible mechanisms of adhesion-dependent activation of the rho-family GTPases, are the subjects for future studies.

Adhesion-independent cytoskelal anchor proteins

The high abundance of junction-associated proteins listed here as 'cytoskeletal anchor proteins' may create the biased impression that cell adhesions are the only prominent sites of such proteins. The main reason for this bias is the fact that junctions are easily recognizable membrane domains and thus more amenable to immunolocalization studies. However, long- or short-term anchorage of cytoskeletal structures occurs widely also in extrajunctional sites and is mediated, most likely, by specific anchor proteins. Among these are members of the spectrin family, ankyrin, ABP120 of *Dictyostelium*, brush border myosin I, and many additional proteins classified in the *Guidebook to cytoskeletal and motor proteins* under different categories. The function of these proteins is, in fact, very similar to that of junctional proteins: creation and maintenance of lateral heterogeneity in membrane structure.[109,110] A variation on the same theme is the support of mechanical stability of planar membranes as well as three-dimensional membrane specializations (microvilli, stereocilia, endocytotic vesicles, etc.). It is not surprising therefore that some anchorage proteins, for example, members of the ERM-family, participate both in junctional structures and in non-junctional membrane specializations.

Adhesion-independent submembranous networks are often not less complex than the adhesion plaques, and their formation may also be a signal-dependent multistaged process. Thus, tyrosine phosphorylation of ezrin, for example, is probably the signalling event that triggers the formation of microvilli in some cell types;[111] activation of rac and cdc42, as mentioned above,[92–94] is a signal for formation of the actin network associated with membrane protrusions of different types.

Concluding remarks

The limited, yet significant, information available on the structures of different cytoskeletal anchor proteins points to considerable molecular diversity. Sequence homologies are limited and the size of proteins varies significantly. Beside this, one generalization may be added – that these molecules are expected to contain multiple distinct functional domains. This notion is suggested by the mere fact that, by definition, anchor proteins are effectively 'multidomain adapter proteins' with binding sites to one or more components of the cytoskeleton, to the membrane, and to additional anchor proteins. These complex interactions are apparently important for the formation of three-dimensional networks through which membrane–cytoskeleton attachments occur. With the harnessing of molecular genetic approaches and the development of numerous immunochemical reagents it appears likely that within the next few years much more complete molecular information will become available on structure–function relationships in these proteins. This includes a broader knowledge of the diverse types of anchor proteins and a deeper understanding of their mode of action at the molecular level.

References

1. Hitt, A. L. and Luna, E. J. (1994). *Curr. Opin Cell Biol.*, **6**, 120–30.
2. Geiger, B., Yehuda-Levenberg, S., and Bershadsky, A. D. (1995). *Acta Anat. (Basel)*, **154**, 46–62.
3. Cowin, P. and Burke, B. (1996). *Curr. Opin. Cell Biol.*, **8**, 56–65.
4. Yamada, K. M. and Geiger, B. (1997). *Curr. Opin. Cell Biol.*, **9**, 76–85.
5. Ben-Ze'ev, A. and Bershadsky, A. D. (1997). *Adv. Mol. Cell Biol.*, **24**, 125–63.
6. Jockusch, B. M., Bubeck, P., Giehl, K., Kroemker, M., Moschner, J., Rothkegel, M., *et al.* (1995). *Ann. Rev. Cell Dev. Biol.*, **11**, 379–416.
7. Borradori, L. and Sonnenberg, A. (1996). *Curr. Opin. Cell Biol.*, **8**, 647–56.
8. Tsukita, S., Furuse, M., and Itoh, M. (1996). *Cell Struct. Funct.*, **21**, 381–5.
9. Mitic, L. L. and Anderson, J. M. (1998). *Annu. Rev. Physiol.*, **60**, 121–42.
10. Kemler, R. (1993). *Trends Genet.*, **9**, 317–21.
11. Aberle, H., Schwartz, H., and Kemler, R. (1996). *J. Cell. Biochem.*, **61**, 514–23.
12. Obama, H. and Ozawa, M. (1997). *J. Biol. Chem.*, **272**, 11017–20.
13. Knudsen, K. A., Soler, A. P., Johnson, K. R., and Wheelock, M. J. (1995). *J. Cell Biol.*, **130**, 67–77.
14. Rimmn, D. L., Koslov, E. R., Kebriaei, P., Cianci, C. D., and Morrow, J. S. (1995). *Proc. Natl Acad. Sci. USA*, **92**, 8813–17.
15. Tsukita, S., Yonemura, S., and Tsukita, S. (1997) *TIBS*, **22**, 53–8.
16. Itoh, M., Nagafuchi, A., Moroi, S., and Tsukita, S. (1977). *J. Cell Biol.*, **138**, 181–92.
17. Garrod, D., Chidgey, M., and North, A. (1996), *Curr. Opin. Cell Biol.*, **8**, 670–8.
18. Witcher, L. L., Collins, R., Puttagunta, S., Mechanic, S. E., Munson, M., Gumbiner, B., and Cowin, P. (1996). *J. Biol. Chem.*, **271**, 10904–9.
19. Wahl, J. K., Sacco, P. A., McGranahan-Sadler, T. M., Sauppe, L. M., Wheelock, M. J., and Johnson, K. R. (1996). *J. Cell Sci.*, **109**, 1143–54.
20. Ruiz, P., Brinkmann, V., Ledermann, B., Behrend, M., Grund, C., Thalhammer, C., *et al.* (1996) *J. Cell Biol.*, **135**, 215–25.
21. Bierkamp, C. Mclaughlin, K. J., Schwarz, H., Huber, O., and Kemler, R. (1996). *J. Dev. Biol.*, **180**, 780–5.
22. Gumbiner, B. and Simons, K. (1986). *J. Cell Biol.*, **102**, 457–68.
23. Lewis, J. E., Wahl J. K., Sass, K. M., Jensen, P. J., Johnson, K. R., and Wheelock, M. J (1997). *J. Cell Biol.*, **136**, 919–34.
24. Volberg, T., Geiger, B., Dror, R., and Zick, Y. (1991). *Cell Regul.*, **2**, 105–20.
25. Volberg T., Zick Y., Dror, R., Sabanay, I., Gilon, C., Levitzki, A., and Geiger, B. (1992). *EMBO J.*, **11**, 1733–42.
26. Sasaki, H., Nagura, K., Ishino, M., Tobioka, H., Kotani, K., and Sasaki, T. (1995). *J. Biol. Chem.*, **270**, 21206–19.
27. Lev, S., Moreno, H., Martinez, R., Canoll, P., Peles, E., Musacchio, J. M., *et al.* (1995). *Nature*, **376**, 737–45.
28. Avraham, S., London, R., Fu, Y., Ota, S., Hiregowdara, D., Li, J., *et al.* (1995). *J. Biol. Chem.*, **270**, 27742–51.
29. Wary, K. K., Mainiero, F., Isakoff, S. J., Marcantonio, E. E., and Giancotti, F. G. (1996). *Cell*, **87**, 733–43.
30. Xu, Y., Guo, D.-F., Davidson, M., Inagami, T., and Carpenter, G. (1997). *J. Biol. Chem.*, **272**, 13463–6.
31. Hoschuetzky, H., Aberle, H., and Kemler, R. (1994). *J. Cell. Biol.*, **127**, 1375–80.
32. Kanai, Y., Ochiai, A., Shibata, T., Ushijima, S., Akimoto, S., and Hirohashi, S. (1995). *Biochem. Biophys. Res. Commun.*, **208**, 1067–72.
33. Shibata, T., Ochiai, A., Kanai, Y., Akimoto, S., Gotoh, M., Yasui, N., *et al.* (1996). *Oncogene*, **13**, 883–99.
34. Mainiero, F., Pepe, A., Yeon, M., Ren, Y., and Giancotti, F. G. (1996). *J. Cell Biol.*, **134**, 241–53.
35. Simske, J. S., Kaech, S. M., Harp, S. A., and Kim, S. K. (1996). *Cell*, **85**, 195–204.
36. Lambie, E. J. (1996) *Curr. Biol.*, **6**, 1089–91.
37. Hoskins, R., Hajnal, A. F., Harp, S. A., and Kim, S. K. (1996). *Development*, **122**, 97–111.
38. Serra-Pages, C., Kedersha, N. L., Fazikas, L., Medley, Q., Debant, A. and Streuli, M. (1995). *EMBO J.*, **14**, 2827–38.
39. Brady-Kalnay, S. M., Rimm, D. L., and Tonks, N. K. (1995). *J. Cell Biol.*, **130**, 977–86.
40. Balsamo, J., Leung, T. C., Ernst, H., Zanin, M. K. B., and Hoffman, S. (1996). *J. Cell Biol.*, **134**, 801–13.
41. Jaken, S. (1996). *Curr. Opin. Cell Biol.*, **8**, 168–73.
42. Hannigan, G. E., Leung-Hagesteijn, C., Fitz-Gibbon, L., Coppolino, M. G., Radeva, G., Filmus, J., *et al.* (1996). *Nature*, **379**, 91–6.
43. Cavallo, R., Rubenstein, D. and Peifer, M. (1997). *Curr. Opin. Genet. Dev.*, **7**, 459–66.
44. Aberle, H., Bauer, A., Stappert, J., Kispert, A., and Kemler, R. (1997). *EMBO J.*, **16**, 3797–804.
45. Gumbiner, B. (1997). *Curr. Biol.*, **7**, R443–6.
46. Fagotto, F., Funayama, N., Glück, U., and Gumbiner, B. (1996). *J. Cell Biol.*, **132**, 1105–14.
47. Merriam, J., Rubenstein, A., and Klymkowsky, M. W. (1997). *Dev Biol.*, **185**, 67–81.
48. Morin, P. J., Sparks, A. B., Korinek, V., Barker, N., Clevers, H., Vogelstein, B., and Klinzler, K. (1997). *Science*, **275**, 1787–90.
49. Rubinfeld, B., Robbins, P., El-Gamil, M., Albert, I., Porfiri, E., and Polakis, P. (1997). *Science*, **275**, 1790–2.
50. Korinek, V., Barker, N., Morin, P. J., van Wichen, D., de Weger, R., Kinzler, K. W., *et al.* (1997). *Science*, **275**, 1784–7.
51. Peifer, M. (1997). *Science*, **275**, 1752–3.
52. Rodríguez Fernández, J. L., Geiger, B., Salomon, D., Sabanay, I., Zöller, M., and Ben-Ze'ev, A. (1992). *J. Cell Biol.*, **119**, 427–38.
53. Glück, U., Kwiatkowski, D. J., and Ben-Ze'ev, A. (1993). *Proc. Natl Acad. Sci. USA*, **90**, 383–7.
54. Bullions, L. C., Notterman, D. A., Chung, L. S., and Levine, A. J. (1997). *Mol. Cell. Biol.*, **17**, 4501–8.
55. Aberle, H., Bierkamp, C., Torchard, D., Serova, O., Wagner, T., Natt, E., *et al.* (1995). *Proc. Natl Acad. Sci. USA*, **92**, 6384–8.
56. Simcha, I., Geiger, B., Yehuda-Levenberg, S., Salomon, D., and Ben-Ze'ev, A. (1996). *J. Cell Biol.*, **133**, 199–209.
57. Nagafuchi, A. and Takeichi, M. (1988). *EMBO J.*, **7**, 3679–84.
58. Hayashi, Y., Haimovich, B., Reszka, A., Boettiger, D., and Horwitz, A. (1990). *J. Cell Biol.*, **110**, 175–84.
59. Geiger, B., Ginsberg, D. Salomon, D., and Volberg, T. (1990). *Cell Diff. Dev.*, **32**, 343–53.
60. Marcantonio, E. E., Guan, J.-L., Trevithick, J. E., and Hynes, R. O. (1990). *Cell Regul.*, **1**, 597–604.
61. Ozawa, M., Ringwald, M., and Kemler, R. (1990). *Proc. Natl Acad. Sci. USA*, **87**, 4246–50.
62. Miyamoto, S., Akiyama, K., and Yamada, K. M. (1995). *Science*, **267**, 883–85.

63. Miyamoto, S., Teramoto, H., Coso, O. A., Gutkind, J. S., Burbelo, P. D., Akiyama, S. K., and Yamada, K. M. (1995). *J. Cell Biol.*, **131**, 791–805.
64. Felsenfeld, D. P., Choquet, D., and Sheetz, M. P. (1996), *Nature*, **383**, 438–40.
65. Chen, W.-T., and Singer, S. J. (1982). *J. Cell Biol.*, **95**, 205–22.
66. Maher, P. A., Pasquale, E. B., Wang, J. Y., and Singer, S. J. (1985). *Proc. Natl Acad. Sci. USA*, **82**, 6576–80.
67. Craig, S. W., and Johnson, R. P. (1996). *Curr. Opin. Cell Biol.*, **8**,74–85.
68. Richardson, A., and Parsons, J. T. (1995). *BioEssays*, **17**, 229–36.
69. Ilic, D., Damsky, C. H., and Yamamoto, T. (1997). *J. Cell Sci.*, **110**, 401–7.
70. Hanks, S. K., and Polte, T. R. (1997). *BioEssays*, **19**, 137–45.
71. Gilmore, A. P., and Romer, L. H. (1996). *Mol. Biol. Cell*, **7**, 1209–24.
72. Richardson, A., and Parsons, J. T. (1996). *Nature*, **380**, 538–40.
73. Burridge, K., Turner, C. E., and Romer, L. H. (1992). *J. Cell Biol.*, **119**, 893–903.
74. Ridley, A. J., and Hali, A. (1994). *EMBO J.*, **13**, 2600–10.
75. Bershadsky, A. D., Chausovsky, A., Becker, E., Lyubimova, A., and Geiger, B. (1996). *Curr. Biol.*, **6**, 1279–89.
76. Retta, S. F., Barry, S. T., Critchley, D. R., Defilippi, P., Silengo, L., and Tarone, G. (1996). *Exp. Cell Res.*, **229**, 307–17.
77. Crowley, E., and Horwitz, A. F. (1995). *J. Cell Biol.*, **131**, 525–37.
78. Ilic, D., Furuta, Y., Kanazawa, S., Takeda, N., Sobue, K., Nakatsuji, N., *et al.* (1995). *Nature*, **377**, 539–43.
79. Rohrschnieder, L.R. (1980). *Proc. Natl Acad. Sci. USA*, **77**, 3514–18.
80. Matsuyoshi, N., Hamaguchi, M., Taniguchi, S., Nagafuchi, A., Tsukita, S., and Takeichi, M. (1992). *J. Cell. Biol.*, **118**, 703–14.
81. Hamaguchi, M., Matsuyoshi, N., Ohnishi, Y., Gotoh, B., Takeichi, M., and Nagai, Y. (1993). *EMBO J.*, **12**, 307–14.
82. Behrens, J., Vakaet, L., Friis, R., Winterhager, E., Van Roy, F., Marcel, M. M., and Birchmeier, W. (1993). *J. Cell. Biol.*, **120**, 757–66.
83. Kinch, M. S., Clark, G. J., Der, C. J., and Burridge, K. (1 995) *J. Cell Biol.*, **130**, 461–71.
84. Kinch, M. S., Petch, L., Zhong, C. and Burridge, K. (1997). *Cell Adhesion Commun.*, **4**, 425–37.
85. Michalides, R., Volberg, T., and Geiger, B. (1994). *Cell Adhesion Commun.*, **2**, 481–90.
86. Ayalon, O. and Geiger, B. (1997). *J. Cell Sci.* **110**, 547–56.
87. Adamson, P., Paterson, H., and Hall, A. (1992). *J. Cell Biol.*, **119**, 617–27.
88. Ridley, A.J. and Hall, A. (1992). *Cell*, **70**, 389–99.
89. Hotchin, N. A. and Hall, A. (1995). *J. Cell Biol.*, **131**, 1857–65.
90. Hall, A. (1998). *Science*, **279**, 509–14.
91. Tapon, N. and Hall, A. (1997). *Curr. Opin. Cell Biol.*, **9**, 86–92.
92. Ridley, A. J. (1994). *BioEssays*, **16**, 321–7.
93. Kozma, R., Ahmed, S., Best, A., and Lim, L. (1995). *Mol. Cell. Biol.*, **15**, 1942–52.
94. Nobes, C. D. and Hall, A. (1995) *Cell*, **81**, 53–62.
95. Chong, L. D., Traynor-Kaplan, A., Bokoch, G. M., and Schwartz, M. A. (1994). *Cell*, **79**, 507–13.
96. Hartwig, J. H., Bokoch, G. M., Carpenter, C. L., Janmey, P. A., Taylor, L. A., Toker, A., and Stossel, T. P. (1995). *Cell*, **82**, 643–53.
97. Fukami, K., Endo, T., Imamura, M., and Takenawa, T. (1994). *J. Biol. Chem.*, **269**, 1518–22.
98. Gilmore, A. P., and Burridge, K. (1996). *Nature*, **381**, 531–5.
99. Machesky, L. M. and Hall, A. (1997). *J. Cell Biol.*, **138**, 913–26.
100. Amano, M., Chihara, K., Kimura, K., Fukata, Y., Nakamura, N., Matsuura, Y., and Kaibuchi, K. (1997). *Science*, **275**, 1308–11.
101. Amano, M., Ito, M., Kimura, K., Fukata, Y., Chihara, K., Nakano, T., *et al.* (1996). *J. Biol. Chem.*, **271**, 20246–9.
102. Kimura, K., Ito, M., Amano, M., Chihara, K., Fukata, Y., Nakafuku, M., *et al.* (1996). *Science*, **273**, 245–8.
103. Chrzanowska-Wodnicka, M. and Burridge, K. (1996). *J. Cell Biol.*, **133**, 1403–15.
104. Choquet, D., Felsenfeld, D. P., and Sheetz, M. P. (1997). *Cell*, **88**, 39–48.
105. Ingber, D. E., Dike, L., Hansen, L., Karp, S., Liley, H., Maniotis, A., *et al.* (1994). *Int. Rev. Cytol.*, **150**, 173–224.
106. Kotani, H., Takaishi, K., Sasaki, T., and Takai, Y. (1997). *Oncogene*, **14**, 1705–13.
107. Braga, V. M., Machesky, L. M., Hall, A., and Hotchin, N. A. (1997). *Cell Biol.*, **137**, 1421–31.
108. Takaishi, K., Sasaki, T., Kameyama, T., Tsukita, S., Tsukita, S., and Takai, Y. (1995). *Oncogene*, **11**, 39–48.
109. Jacobson, K., Sheets, E. D., and Simson, R. (1995). *Science*, **268**, 1441–2.
110. Kusumi, A. and Sake, Y. (1996). *Curr. Opin. Cell Biol.*, **8**, 566–74.
111. Berryman, M., Gary, R., and Bretscher, A. (1995). *J. Cell Biol.*, **131**, 1231–42.
112. Schwartz, M. A., Schaller, M. D., and Ginsberg, M. H. (1995). *Annu. Rev. Cell Dev. Biol.*, **11**, 549–99.
113. Clark, E. A. and Brugge, J. S. (1995). *Science*, **268**, 233–9.
114. Barth, A. I., Nathke, I. S., and Nelson, W. J. (1997). *Curr. Opin. Cell Biol*, **9**, 683–90.
115. Howe, A., Aplin, A. E., Alahari, S. K., and Juliano, R. L. (1998). *Curr. Opin. Cell Biol.*, **10**, 220–31.
116. Rezniczek, G. A., de Pereda, J. M., Reipert, S., and Wiche, G. (1998). *J. Cell Biol.*, **141**, 209–25.
117. Furuse, M., Fujita, K., Hiiragi, T., Fujimoto, K., and Tsukita, S. (1998). *J. Cell Biol.*, **141**, 1539–50.
118. Weiss, E. E., Kroemker, M., Rudiger, A. H., Jockusch, B. M., and Rudiger, M. (1998). *J. Cell Biol.*, **141**, 755–64.
119. Smith, E. A. and Fuchs, E. (1998). *J. Cell Biol.*, **141**, 1229–41.
120. Behrens, J., von Kries, J. P., Kuhl, M., Bruhn, L., Wedlich, D., Grosschedl, R., and Birchmeier, W. (1996). *Nature*, **382**, 638–42.
121. Ikeda, S., Kishida, S., Yamamoto, H., Murai, H., Koyama, S., and Kikuchi, A. (1998). *EMBO J.*, **17**, 1371–84.
122. Sakanaka, C., Weiss, J. B., and Williams, L. T. (1998). *Proc. Natl Acad. Sci. USA*, **95**, 3020–3.
123. Hart, M. J., de los Santos, R., Albert, I. N., Rubinfeld, B., and Polakis, P. (1998). *Curr. Biol.*, **8**, 573–81.
124. Behrens, J., Jerchow, B. A., Wurtele, M., Grimm, J., Asbrand, C., Wirtz, R., Kuhl, M., Wedlich, D., and Birchmeier, W. (1998). *Science*, **280**, 596–9.
125. Simcha, I., Shtutman, M., Salomon, D., Zhurinsky, J., Sadot, E., Geiger, B., and Ben-Ze'ev, A. (1998). *J. Cell Biol.*, **141**, 1433–48.
126. Salomon, D., Sacco, P. A., Roy, S. G., Simcha, I., Johnson, K. R., Wheelock, M. J., and Ben-Ze'ev, A. (1997). *J. Cell Biol.*, **139**, 1325–35.
127. Daniel, J. M. and Reynolds, A. B. (1997). *BioEssays*, **19**, 883–91.

■ *Alexander Bershadsky and Benjamin Geiger*
Department of Molecular Cell Biology,
The Weizmann Institute of Science,
Rehovot 76100,
Israel

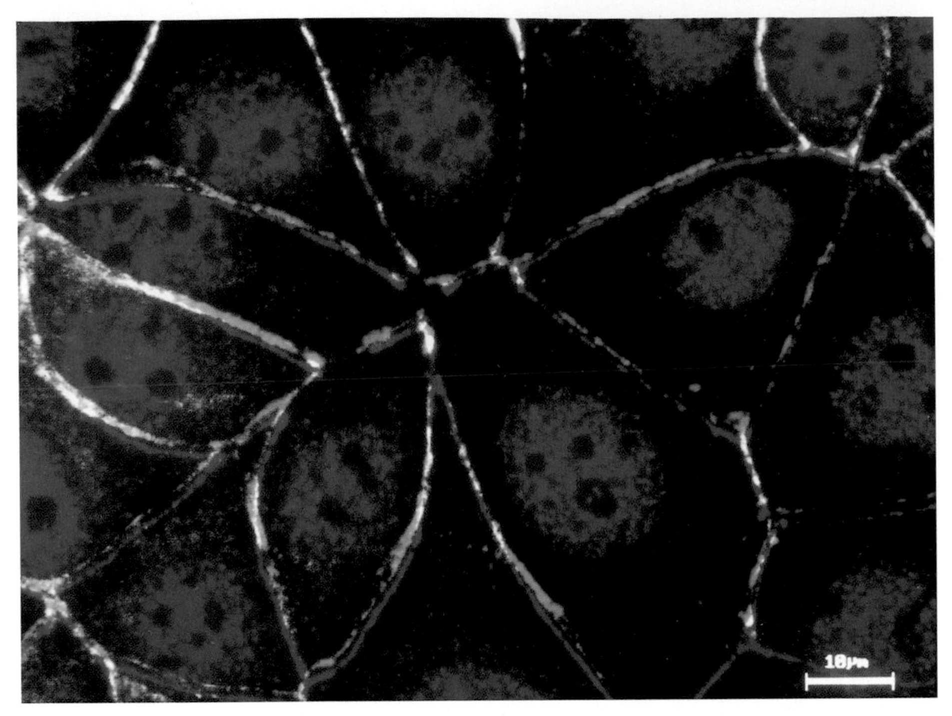

Plate 1. Confocal microscopy image of double-label immunofluorescence comparison of symplekin (red fluorescence) with E-cadherin (green fluorescence) in a monolayer culture of human mammary carcinoma cell line, MCF-7. Symplekin and E-cadherin do no co-localize, but shadow one another, resulting in the distinction of the zonula occludens (red contour line) from the zonula adherens (green line). Using a protocol designed to minimize losses and redistributions of soluble proteins, an intense and specific nuclear reaction of symplekin is observed. Bar represents 10 μm. Courtesy of Dr B. Keon (Harvard Medical School).

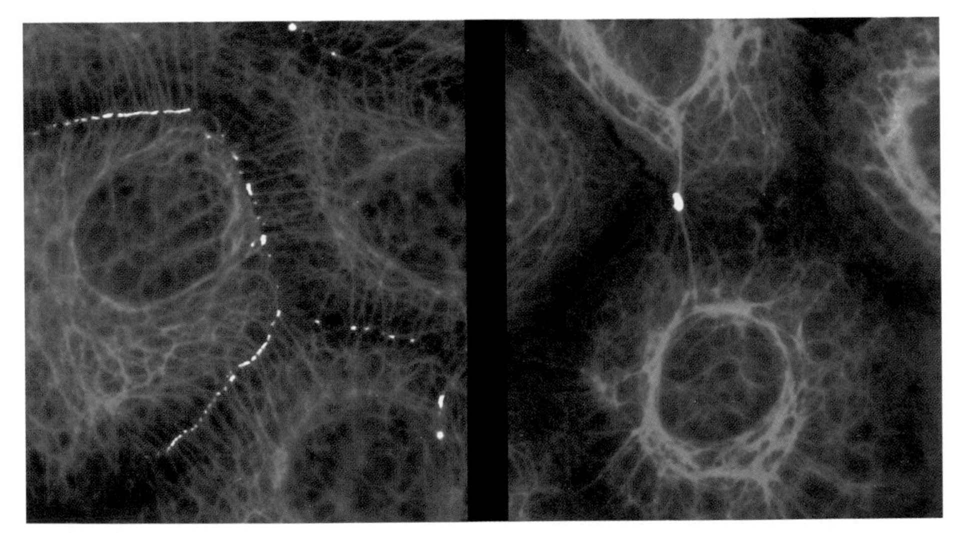

Plate 2. Function of desmoplakins. A dominant-negative desmoplakin mutant was stably expressed in desmosome-bearing A-431 epithelial cells, leading to competition with endogenous desmoplakin for binding sites in the desmosomal plaque and loss of keratin intermediate filament anchorage. Control cells (left) or mutant-expressing cells (right) were prepared for immunofluorescence and reacted with antibodies against keratin (rhodamine) and endogenous desmoplakin (fluorescein). Note that only areas still expressing endogenous desmoplakin act as sites for intermediate filament attachment, suggesting that desmoplakin is required for cytoskeletal anchorage. Courtesy of Dr Kathleen Green (Northwestern University Medical School).

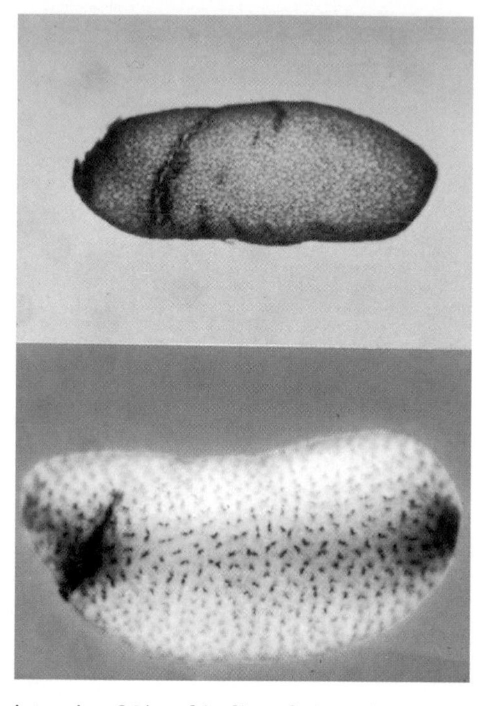

Plate 3. Localization patterns of the integrin $\alpha 8\beta 1$ and its ligands in embryonic day 13.5 mouse kidney. An antibody specific for the integrin $\alpha 8$ subunit has been utilized to detect the expression pattern of the integrin $\alpha 8\beta 1$, visualized as semilunar-like crescents in the mesenchymal tissues surrounding the actively branching tubules derived from the ureteric bud. A chimeric integrin consisting of the extracellular domains of each subunit of the integrin $\alpha 8\beta 1$ fused to alkaline phosphatase has been utilized to identify the ligand for this integrin localized in the basal lamina separating the tubular epithelium and surrounding mesenchyme. (Photomicrograph contributed by Sumiko Denda, Ulrich Müller, and Louis F. Reichardt, University of California, San Francisco. For more details, see Müller, U., Wang, D., Denda, S., Meneses, J. J., Pedersen, and Reichardt, L. F. (1997). Integrin $\alpha 8\beta 1$ is critically important for epithelial-mesenchymal interactions during kidney morphogenesis. *Cell*, 88, 603–13.)

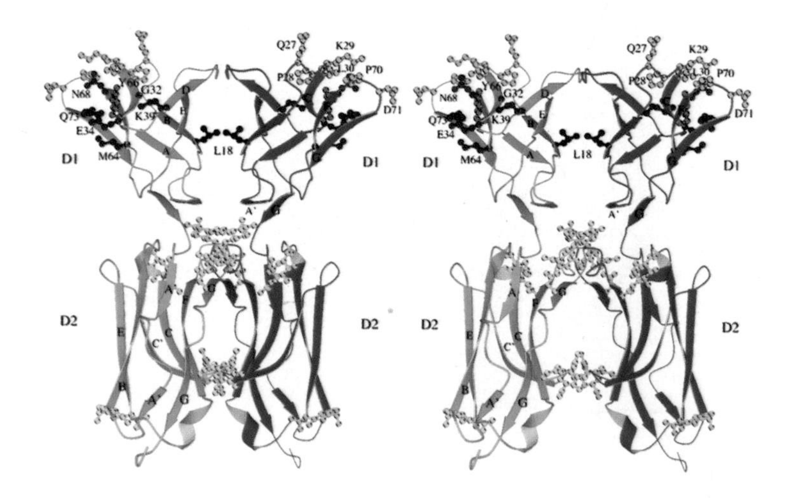

Plate 4. Human ICAM-1 (domains 1 and 2) dimer with LFA-1 (red and orange), rhinovirus (yellow and orange) and *Plasmodium falciparum* (blue) binding residues highlighted. Courtesy of Michael Dustin (Washington University, St Louis) and Tim Springer (Harvard Medical School).

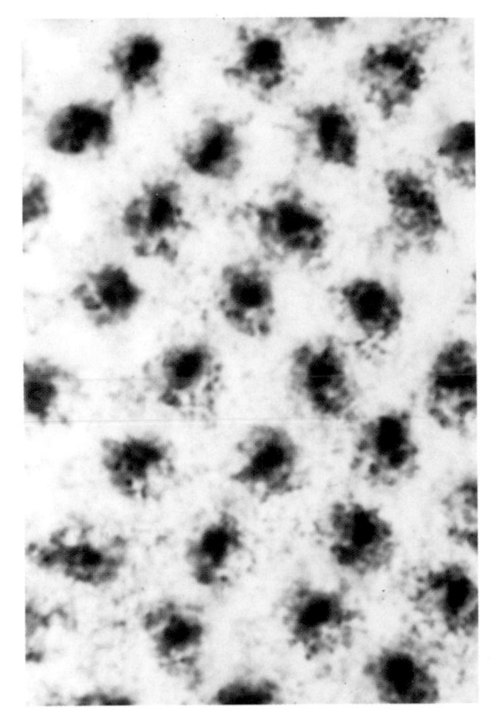

Plate 5. A *Drosophila* eye imaginal disc, double-labelled to visualize the expression pattern of Sevenless (brown) and Boss (purple) proteins. Courtesy of Dr H. Kramer (University of Texas Southwestern Medical Center at Dallas).

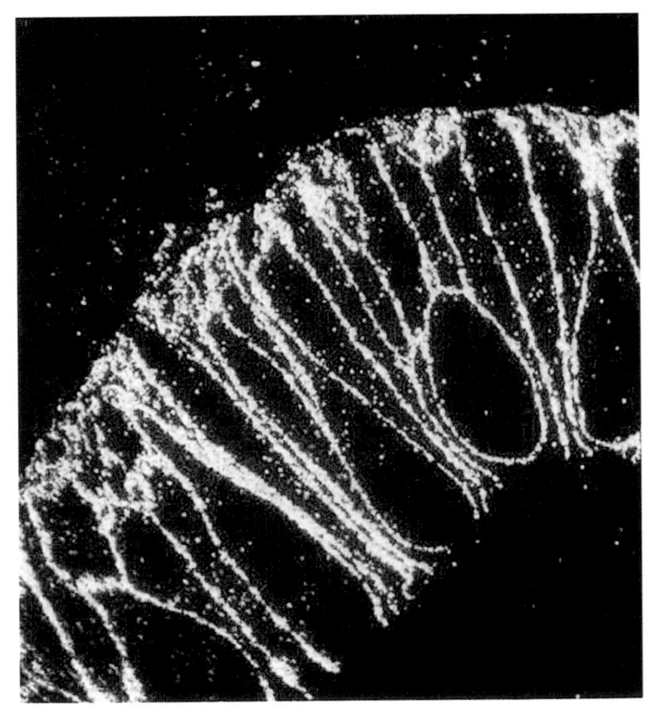

Plate 6. The presence of EpCAM in the lateral domains of human colonic epithelium, as detected on ultrathin tissue sections using immunocytochemistry. Courtesy of Dr Litinov (Leiden University).

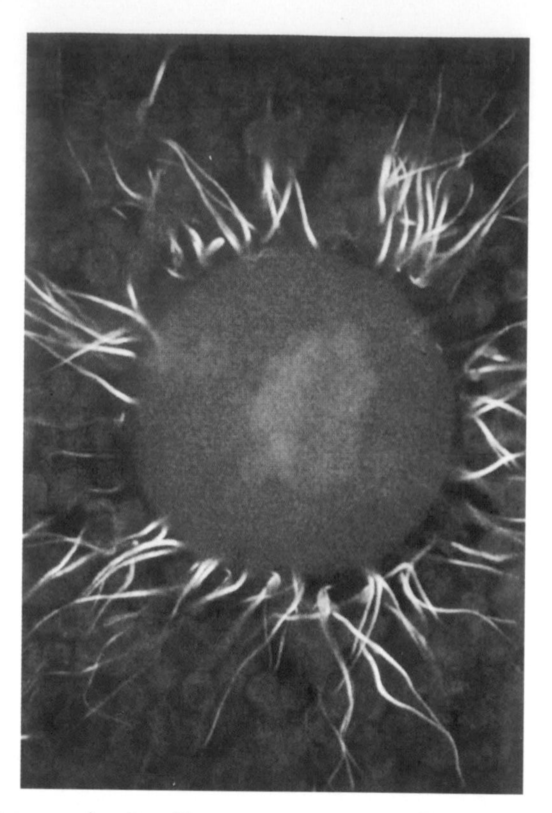

Plate 7. Mouse sperm binding to an egg *in vitro*. Many sperm are seen adhering to the egg coat (zona pellucida) after traversing a path through the extracellular matrix of the surrounding cumulus cells, visible as faint, ghostly-looking cells around the egg. The sperm mitochondria, which are clustered in the anterior sperm tail, have been labelled with the fluorescent dye Rhodamine 123. (Micrograph taken by Ying Lin in the laboratory of Paul Primakoff and Diana Myles.)

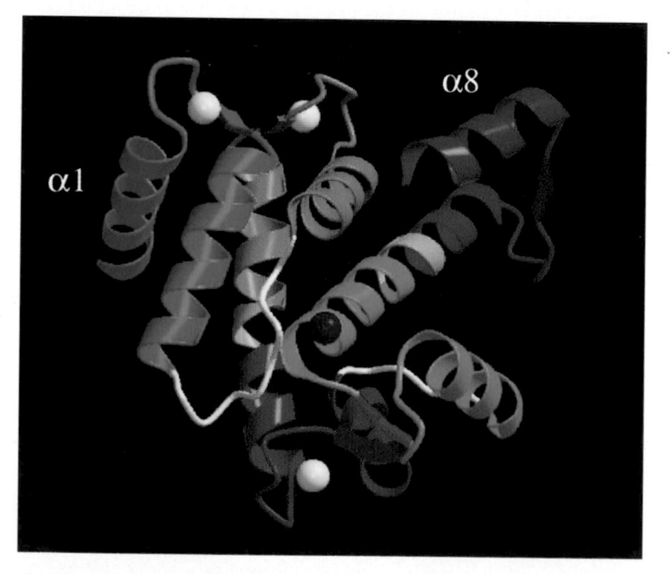

Plate 8. Structure of the small subunit domain VI of rat calpain. A ribbon representation of the helix–loop–helix composition of domain VI shows three calcium ions (yellow), and a fourth (purple) bound only at high concentrations and in an unusual site. The ribbon model is based on the crystal structure described by Blanchard *et al.* (1997), *Nature Struct. Biol.*, **4**, 532–8. (Figure kindly provided by Dr M. Cygler, Biotechnology Research Institute, National Research Council, Montreal, Canada.)

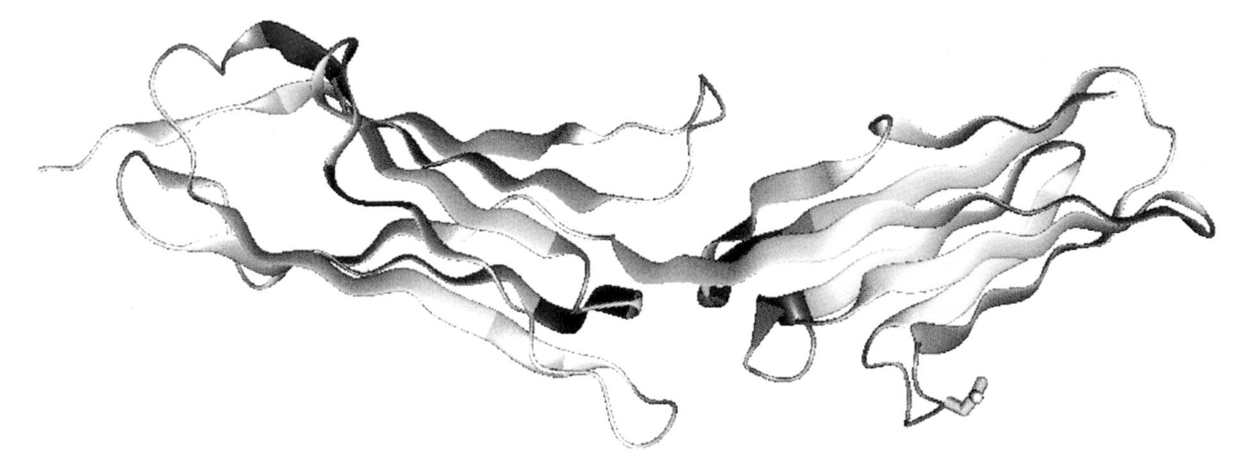

Plate 9. Secondary structure representation of the crystal structure of the N-terminal two domain fragment of human VCAM-1. Note (bottom right) the loop structure in domain 1 with its extended aspartic acid, presumed to interact with the integrin counter-ligand. Courtesy of Drs Lobb and Osborn (Biogen Inc., Cambridge, MA).

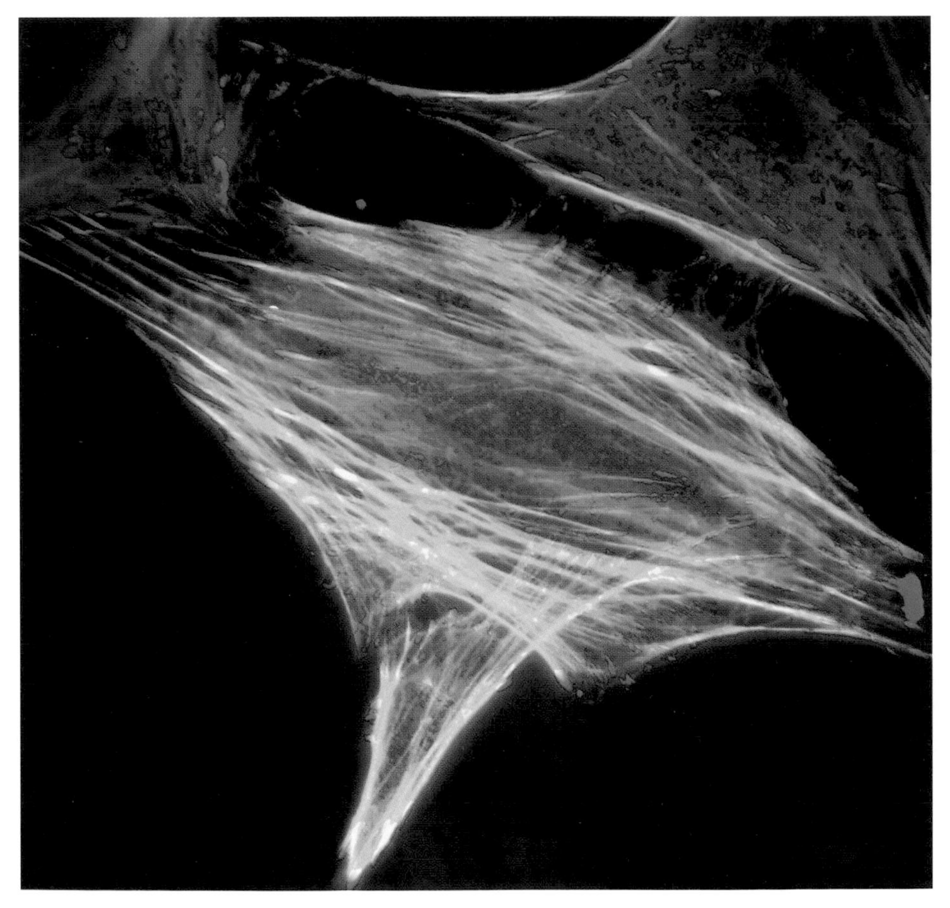

Plate 10. Immunofluorescence labelling of vinculin (red) and actin (green) reveals focal contact sites. Contributed by Benjamin Geiger, Department of Chemical Immunology, The Weizmann Institute of Science, Rehovot, Israel.)

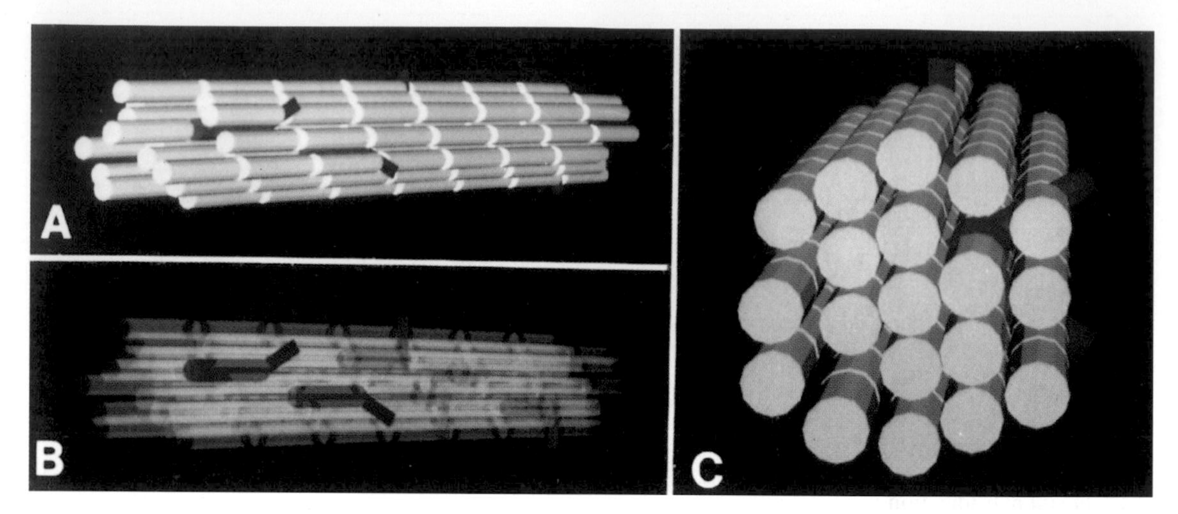

Plate 11. Different views of a heterotypic type I/V collagen fibril, showing type V collagen molecules with amino-terminal propeptide domains among completely processed type I collagen molecules. It is believed that the presence of the type V amino propeptide on the surface of the fibrils prevents additional fibril growth due to steric hindrance. (From Linsenmayer *et al.* (1993) *J. Cell Biol.*, **121**, 1181–9.)

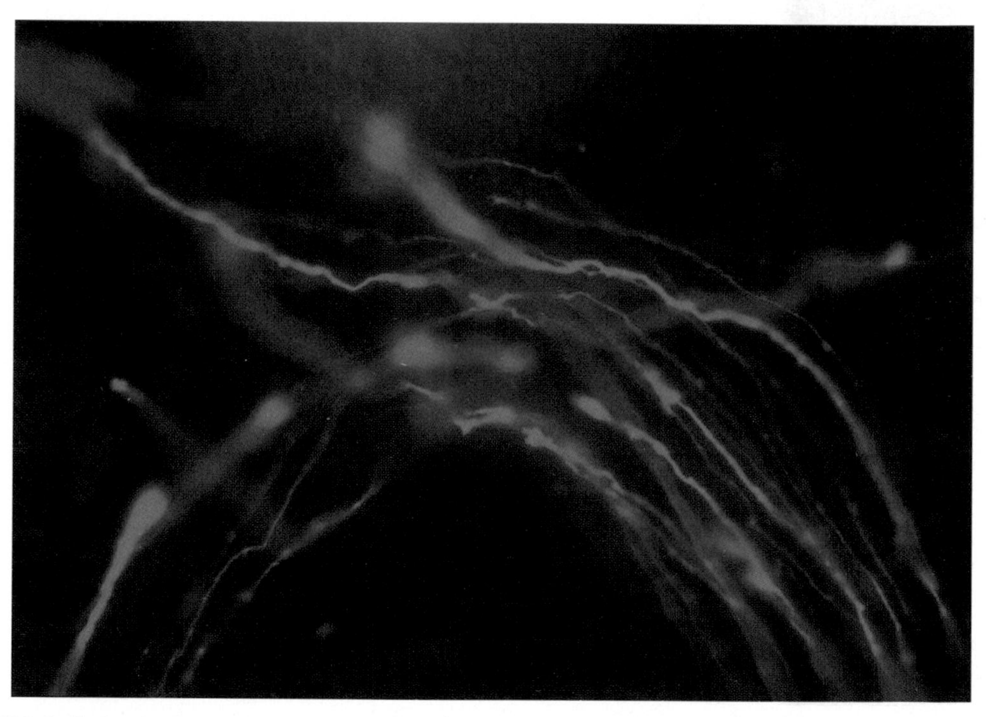

Plate 12. DiI-labelled retinal ganglion cell axon projections in a day 12 mouse embryo. Crystals of DiI were injected into both eyes of a mouse embryo to label retinal ganglion cells and their projections. Visible in this photomicrograph are the axons and growth cones of the initial group of these axons at the midline of the ventral hypothalamus where the majority of these fibres cross to reach the contralateral side of the brain. Individual growth cones utilize information present in the embryonic neuroepithelium to either cross the midline to enter the contralateral optic tract or turn to enter the ipsilateral optic tract. Cell surface-associated and extracellular matrix-associated cues are believed to direct these growth cones. Photomicrograph contributed by David W. Sretavan and Louis F. Reichardt, University of California, San Francisco. For more information, see Sretavan, D. R. and Reichardt, L. F. (1993). Time-lapse video analysis of retinal ganglion cell axon pathfinding at the mammalian optic chiasm: growth cone guidance using intrinsic chiasm cues. *Neuron*, **10**, 761–77.

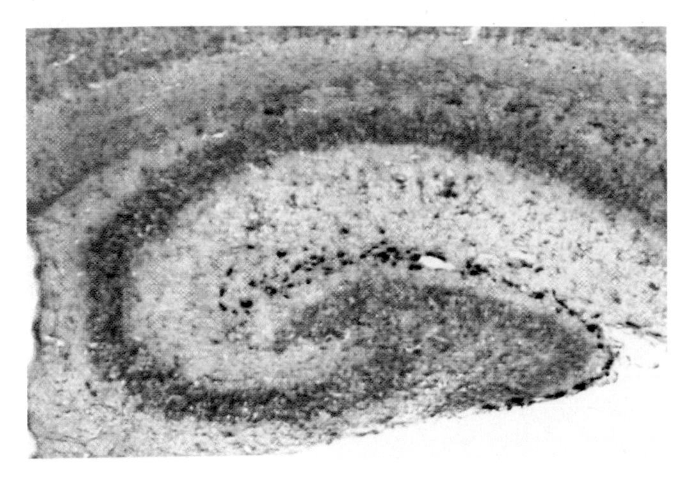

Plate 13. *In situ* hybridization of a mouse hippocampal section at postnatal day 6 with a digoxigenin-labelled *reelin* riboprobe. *reelin* mRNA is expressed in superficial layers (outer molecular layer of the dentate gyrus and stratum lacunosum-moleculare of the hippocampus proper) by Cajal-Retzius-like cells. Photography by Dennis S. Rice.

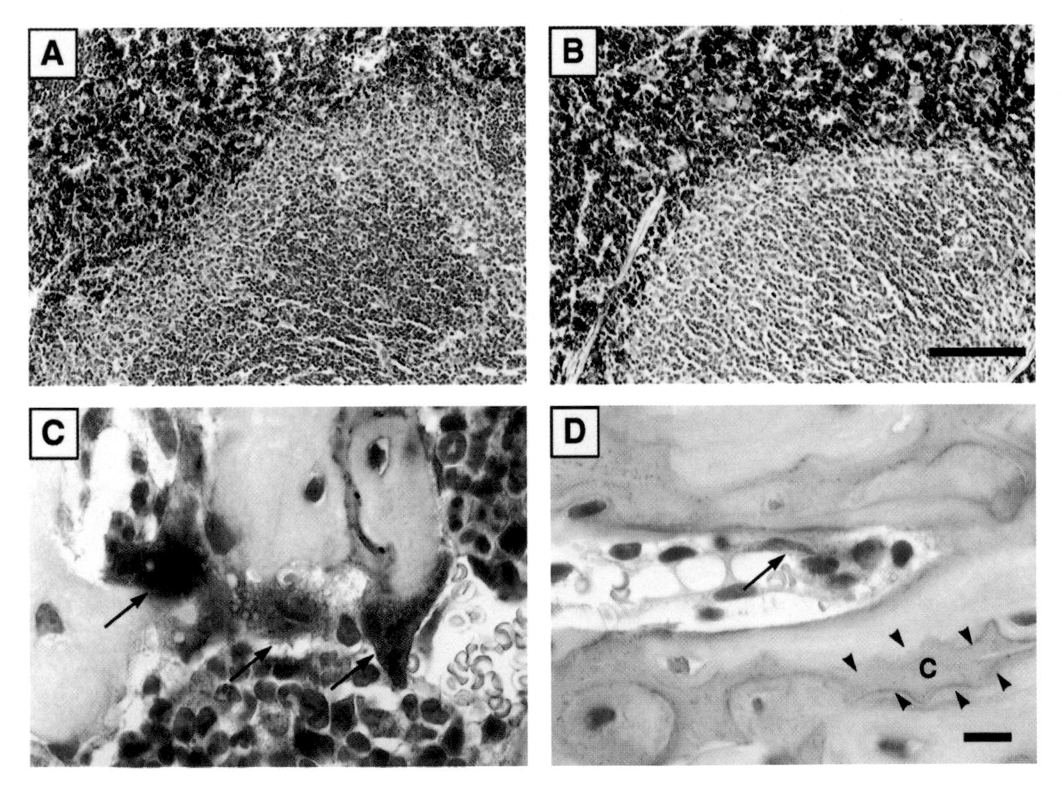

Plate 14. Osteoprotogerin (OPG) transgenics appear to have a defect in the terminal stages of osteoclast formation. Osteoclasts (arrows) in the distal femoral metaphysis were identified by TRAP cytochemistry. Many large, multinucleate osteoclasts were present on the surface of metaphyseal trabecular bone in normal mice (C), while in the OPG transgenic mice (D) only a few small TRAP-positive cells were seen in the same region. A cartilage remnant is marked 'c' and its edge marked by arrowheads (bar represents 10 μm). To confirm the presence of tissue macrophages, which share a common precursor with osteoclasts, immunohistochemistry was performed using anti-F480 antibodies, which recognize a cell surface antigen on murine macrophages. The bottom panels are photomicrographs of spleens from either normal (A) or OPG transgenic (B). Numerous F480-positive cells (brown staining) are detected in both spleens (bar represents 100 μm). Courtesy of Dr William Boyle, (Department of Cell Biology, Amgen, CA).

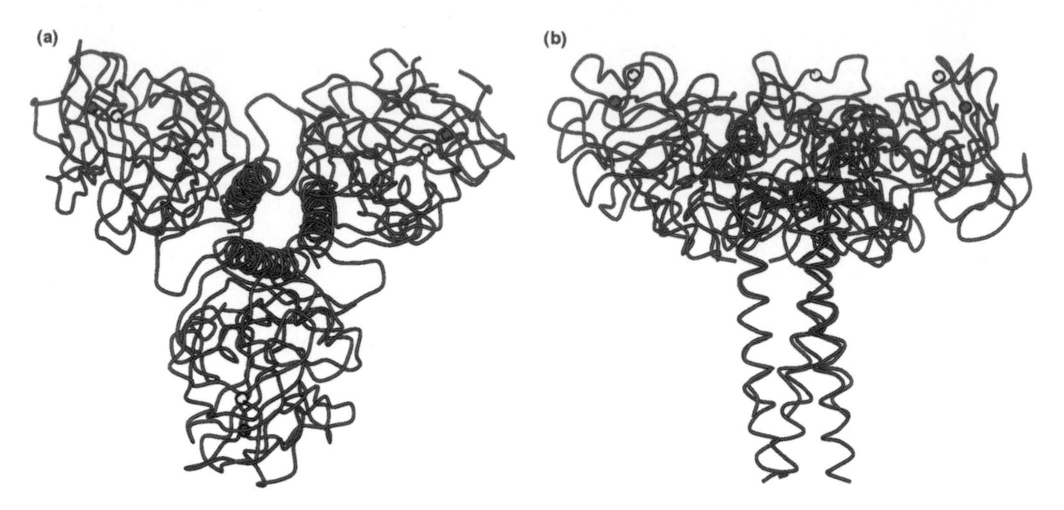

Plate 15. Superposition of human (red), and rat (blue) trimeric MBPs. (a) View down the three-fold axis of superposition of α helical coiled coils of rat MBP, and human MBP trimers. The oligosaccharide-binding (Ca 2) is shown as red spheres for human, and yellow spheres for rat. (b) View 90° to that in (a) showing that human MBP CRDs are pointed slightly more 'up' than those in the rat structure. (a) and (b) show that when the α helical coiled coils are superimposed, the CRDs do not superimpose. Coordinates from (22°°, 23°°). Superposition was performed with the program ALIGN by G. H. Cohen (Satow *et al.* (1986) *J. Mol. Biol.*, **190**, 593–604). Figure produced by MOLSCRIPT (Kaulis (1991) *J. Appl. Crystallogr.*, **24**, 946–50).

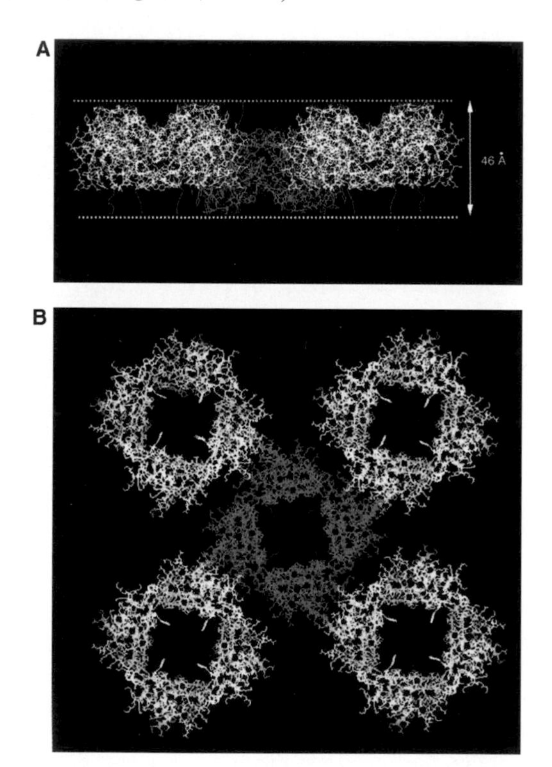

Plate 16. Combined P_0 interactions observed in the crystal lattice. (A) P_0 extracellular domain tetramer structures joined by twofold symmetric putative adhesive interfaces. These tetramers are depicted here as they might emanate from their respective membrane surfaces (schematically indicated with dotted lines) at the C termini, with the blue-coloured tetramers protruding from the blue surface, and the yellow-coloured tetramers from the yellow surface. The Trp28 side-chains, which may intercalate into the opposite membrane bilayer, are shown in red, and a model for main-chain atoms for the five residues of the disordered linker to the membrane is shown in purple. The c axis is vertical, and the a axis horizontal. (B) Perpendicular view of this layer of the crystal lattice. Courtesy of Hayasaka, K. *et al.* (1993) *Genomics*, **17**, 755–8.

Linkers to actin filaments

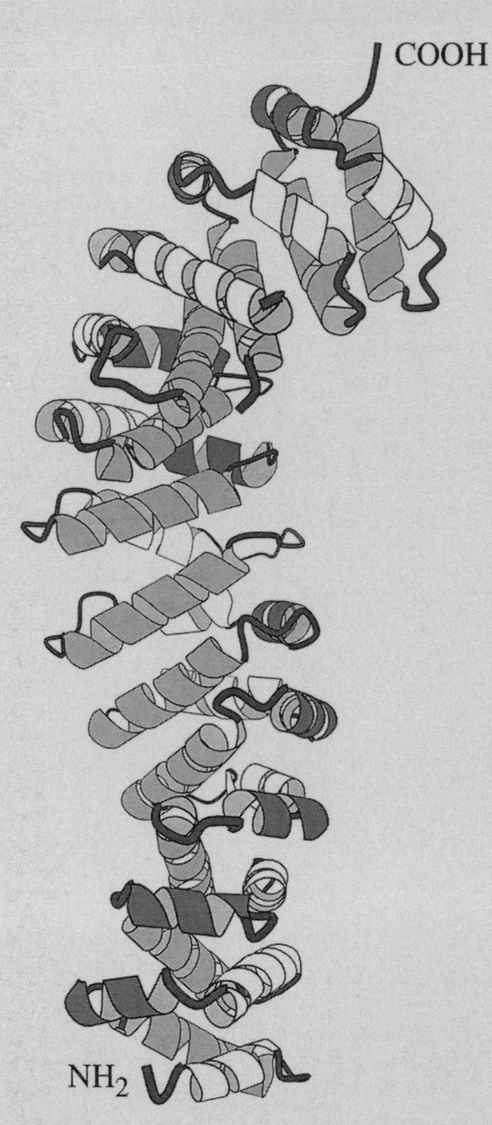

Ribbon diagram of the armadillo repeat region of β-catenin (residues 150–665). Tandem repeats, each consisting of three α-helices (dark grey, light grey, white) pack together to form a superhelix. A shallow, positively charged groove runs along the superhelix and may be the site of interaction with several binding partners of β-catenin. (Courtesy of A. Huber and W. Weis, Stanford University School of Medicine.) Reproduced from Huber *et al.* (1997) *Cell*, **90**, 871–82.)

Aciculin

Aciculin[1–6] is a phosphoglucomutase-related cytoskeletal protein localized in cell–matrix and cell–cell adherens junctions. Aciculin is associated with dystrophin and/or utrophin in various tissues and cell types.

Protein properties

Aciculin (phosphoglucomutase-related protein,[5] PGM-RP) is a member of the phosphoglucomutase superfamily,[5,7–15] which contains several forms of the glycolytic enzyme phosphoglucomutase, characterized in various species, and parafusin from *Paramecium* (Table 1).The amino acid identity levels between aciculin and the other members of the superfamily vary from 38 per cent to 70 per cent.[5] The aciculin (PGM-RP) gene is less conserved than other mammalian PGM1 genes. The divergence in amino acid sequence is greatest in the so-called N-terminal domain 1, which is 30 per cent shorter in human aciculin, and in the C-terminal domain 4, which contains twice as many amino acid differences as the rest of the sequence.[5]

Aciculin (PGM-RP) isolated from mammalian smooth muscle tissue[1] is a 60/63 kDa monomeric protein. Depending on tissue type, aciculin occurs either as a 63 kDa polypeptide or as 60 and 63 kDa variants that probably represent alternatively spliced products or protein isoforms produced by post-translational modification.[1,2]

Aciculin is particularly abundant in visceral and vascular smooth muscle tissues.[1] Substantial amounts of aciculin are synthesized in cardiomyocytes and skeletal muscle where its expression markedly increases during differentiation.[1,2] Most non-muscle tissues express very low levels or no aciculin. Among cultured cells, aciculin is abundantly expressed in smooth muscle cells and differentiating skeletal muscle myotubes. Aciculin is synthesized in skin fibroblasts, astrocytes, and keratinocytes. However, aciculin cannot be detected by immunochemical methods in platelets, endothelial cells, hepatocytes, and neurones.[1–4]

In cultured fibroblasts and smooth muscle cells aciculin is localized at focal adhesions and periodically along stress fibres.[1] Its intracellular localization is most similar to those of α-actinin and utrophin.[1,4] Notably, aciculin is absent from some initial focal adhesions adjacent to cell margins, but its distribution is often extended beyond the length of mature focal adhesions covering a considerable or the entire length of an actin bundle[1] (Fig. 1). Aciculin is found in cell–cell adherens-type junctions of certain epithelial cells.[1,4] In muscle tissues, aciculin is concentrated in dense plaques of smooth muscle, intercalated discs of cardiomyocytes, skeletal muscle myotendinous junctions, and striated muscle costameres.[1–4] However, aciculin is not present at neuromuscular junctions *in vivo*.[2]

Aciculin is not a phosphoprotein *in vivo* like its homologue PGM1 and does not display any detectable phosphoglucomutase enzymatic activity *in vitro*.[1,5] Aciculin does not interact directly with F-actin or G-actin. Although its exact biological function remains to be determined, aciculin probably represents a structural component of the cytoskeletal-matrix transmembrane link at sites where actin filaments terminate at the plasma membrane.

Purification

Aciculin is purified from low ionic strength extracts of smooth muscle using a modification of the original procedure employed for the isolation of vinculin, metavinculin,

Table 1 Nucleotide and amino acid identities between human aciculin (PGM-RP)[5] and other members of the phosphoglucomutase gene family.

	Amino acid identity (%)									
Nucleotide identity (%)	H. aciculin (PGM- RP)	HPGM1	HPGM1 a	OPGM1	OPGM1 a	RPGM1	SPGM1	SPGM2	PFUS	APGM
H. aciculin (PGM-RP)		68	69	69	70	69	38	39	40	46
HPGM1	70		93	97	90	96	46	47	45	51
HPGM1a				91	96	91	45	46	44	50
OPGM1	70	87			93	97	46	47	45	51
OPGM1a	70	83		95		92	45	46	45	51
RPGM	70	88		88	84		46	47	45	51
SPGM1	55	54		56	56	53		78	36	44
APGM	59				58					

HPGM1, human PGM1;[7] HPGM1a, human fast muscle PGM1 isoform;[7] OPGM1, rabbit PGM1;[8,13] OPGM1a, rabbit muscle-specific PGM1 isoform;[8,14] RPGM1, rat PGM1;[9] SPGM1 and SPGM2, *Saccharomyces cerevisiae* PGM1 and PGM2;[10,15] PFUS, *Paramecium* parafusin;[12] APGM, PGM from *Agrobacterium tumefaciens*.[11]

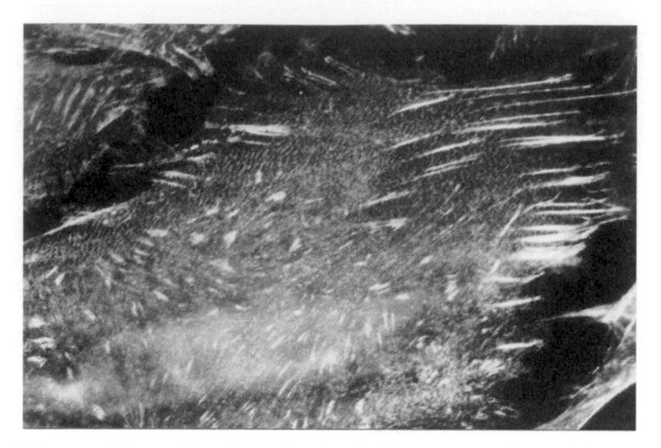

Figure 1. Localization of aciculin in rat embryo fibroblasts. Aciculin is present at focal adhesions, periodically along stress fibers and is enriched at terminal parts of actin bundles.

filamin, and α-actinin.[1,16] A combination of conventional chromatographies enables 1–2 mg of aciculin to be purified from 100 g of smooth muscle tissue.[1]

Activities

The only reported activity of aciculin is its interaction with dystrophin and/or utrophin (dystrophin-related protein, DRP) *in vitro*.[3]

Antibodies

Rabbit polyclonal antibody and several mouse monoclonal antibodies against aciculin are available.[1] They cross-react with many species from *Xenopus* to humans. These antibodies can be used for immunoblotting, immunoprecipitation, and immunostaining.[1–6] Some of them also recognize phosphoglucomutase (PGM1), while two monoclonal antobodies, XIVF8 and XIVD11, are specific for aciculin.[1] Monoclonal antibody XIVF8 is sold by Chemicon Int. (cat. no. MAB1640).

Genes

Several cDNA clones, including full length A111 clone, are available for human smooth muscle aciculin (PGM-RP).[5] Human aciculin genomic clones are also available. There is a single gene for aciculin (PGM-RP) in the human genome. It is located on chromosome 9qcen–q13.[5] The aciculin (PGM-RP) sequence data are available from GSDB/EMBL/NCBI sequence data banks under accession number L40933 and the sequences of exons 2, 3, 5, 8, and 9 of the mouse gene are available under accession numbers L42902–L42905 and L42907, respectively.[5] The positions of the boundaries of exons 2, 3, 5, 8, and 9 in the mouse aciculin gene are identical to those of the equivalent exons in the PGM1 gene. These data suggest that the aciculin gene evolved by gene duplication from PGM1 and has subsequently diverged to encode a cytoskeletal protein with a structural rather than enzymatic role.

Mutant phenotype/disease states

No information is available.

Structure

No information is available.

References

1. Belkin, A. M., Klimanskaya, I. V., Lukashev, M. E., Lilley, K., Critchley, D. R., and Koteliansky, V. E. (1994). *J. Cell Sci.*, **107**, 159–73.
2. Belkin, A. M. and Burridge, K. (1994). *J. Cell Sci.*, **107**, 1993–2003.
3. Belkin, A. M. and Burridge, K. (1995). *J. Biol. Chem.*, **270**, 6328–37.
4. Belkin, A. M. and Burridge, K. (1995). *Exp. Cell Res.*, **221**, 132–40.
5. Moiseeva, E. P., Belkin, A. M., Spurr, N. K., Koteliansky, V. E., and Critchley, D. R. (1996). *Eur. J. Biochem.*, **235**, 103–13.
6. Belkin, A. M. and Smalheiser, N. (1996). *Cell Adhesion Commun.*, **4**, 142–56.
7. Whitehouse, D. B., Putt, W., Lovegrove, J. U., Morrison, K., Hollyoake, M., Fox, M. F., Hopkinson, D. A. and Edwards, Y. H. (1992). *Proc. Natl Acad. Sci. USA*, **89**, 411–15.
8. Lee, Y. S., Marks, A. R., Gureckas, N., Lacro, R., Nadal-Ginard, B., and Kim, D. H. (1992). *J. Biol. Chem.*, **267**, 21080–88.
9. Rivera, A. A., Elton, T. S., Dey, N. B., Bounelis, P. and Marchase, R. B. (1993). *Gene (Amst.)*, **133**, 261–6.
10. Boles, E., Liebetrau, W., Hofmann, M., and Zimmermann, F. K. (1994). *Eur. J. Biochem.*, **220**, 83–96.
11. Uttaro, A. D. and Udalge, R. A. (1994). *Gene (Amst.)*, **150**, 117–22.
12. Subramanian, S. V., Wyroba, E., Andersen, A. P., and Satir, B. H. (1994). *Proc. Natl Acad. Sci. USA*, **91**, 9832–6.
13. Ray, W. J. Jr, Hermodson, M. A., Puvathingal, J. M., and Mahoney, W. C. (1983). *J. Biol. Chem.*, **258**, 9166–74.
14. Putt, W., Ives, J. H., Hollyoake, M., Hopkinson, D. A., Whitehouse, D. B., and Edwards, Y. H. (1993). *Biochem. J.*, **296**, 417–22.
15. Hofmann, M., Boles, E., and Zimmermann, F. K. (1994). *Eur. J. Biochem.*, **221**, 741–7.
16. Feramisco, J. R. and Burridge, K. (1980). *J. Biol. Chem.*, **255**, 1194–9.

Alexey Belkin
Department of Cell Biology and Anatomy,
University of North Carolina at Chapel Hill,
Chapel Hill, NC 27599,
USA

α-Actinins

α-Actinin is a rod-shaped antiparallel homodimer (subunit M_r 94–103 kDa) which cross-links and bundles actin filaments. In addition, it binds to the cytoplasmic domain of a number of integral membrane proteins and to signalling molecules such as PIP2, PI3-kinase, and PKN. Skeletal, smooth, and non-muscle isoforms have been identified, and the actin gelation activity of non-muscle isoforms is inhibited by calcium. α-Actinin is localized in structures to which actin filaments are anchored including the Z disc in skeletal muscle, dense bodies in smooth muscle, and adherens-type membrane junctions in smooth, cardiac, and non-muscle cells.

Homologous proteins

α-Actinin displays homology with erythroid and non-erythroid spectrins, dystrophin, and utrophin. These proteins all share a similar domain structure, with an N-terminal actin-binding domain, a variable number of triple α-helical spectrin-like repeats, and a pair of C-terminal EF hands. The actin-binding domain of α-actinin is homologous to that in filamin, fimbrin, and ABP 120, and there is a more limited match with the N-terminal domain of calponin.

Protein properties

α-Actinin was originally discovered as a component of the Z disc in skeletal muscle, and was subsequently found in the membrane-associated dense plaques and cytoplasmic dense bodies of smooth muscle, the intercalated discs in cardiac muscle, and adherens-type junctions in non-muscle cells.[1] It is a rod-shaped antiparallel homodimer with a subunit molecular mass of approximately 100 kDa. It can be divided into three domains, an N-terminal actin-binding domain, four spectrin-like repeats, and a C-terminal region containing two EF-hand motifs (Fig. 1).[1,2] Proteolytic cleavage of chicken smooth muscle α-actinin liberates the actin-binding domain as a monomer, and the repeats as a rod-shaped dimer which dissociates only under denaturing conditions. Therefore, formation of the α-actinin homodimer would appear to be mediated by the spectrin-like repeats. Interpretation of EM data obtained on two-dimensional crystals of α-actinin has given rise to a model in which the subunits are staggered with respect to each other.[3] Such a model raises the possibility that repeat 1 (or repeat 4) is unpaired and may not contribute to dimer formation. However, deletion of either of the terminal repeats (1 or 4) markedly reduces dimer formation, a finding more consistent with a model in which interactions between repeats 1 and 4 and repeats 2 and 3 of adjacent subunits are responsible for dimeriztion.[4] The dimensions of the α-actinin rod vary between species from 3–4 nm × 30–40 nm (chicken smooth muscle isoform) to 44 nm for *Acanthamboeba* α-actinin. The length of skeletal muscle α-actinin also varies markedly with salt concentration (40 nm in high salt, 74 nm in low salt) suggesting that the protein may exist in both a compact and an extended conformation.

Actin binding

The interaction between α-actinin and actin is complex. At low α-actinin concentrations, isotropic networks are formed, whereas at higher concentrations, α-actinin bundles actin filaments. The k_d values for binding of *Acanthamoeba* and chicken smooth muscle α-actinin to F-actin have been estimated to be 4.7 μM and 0.6 μM respectively.[5] However, analysis of binding kinetics is complicated by the formation of actin bundles. The recombinant monomeric actin-binding domain from chick

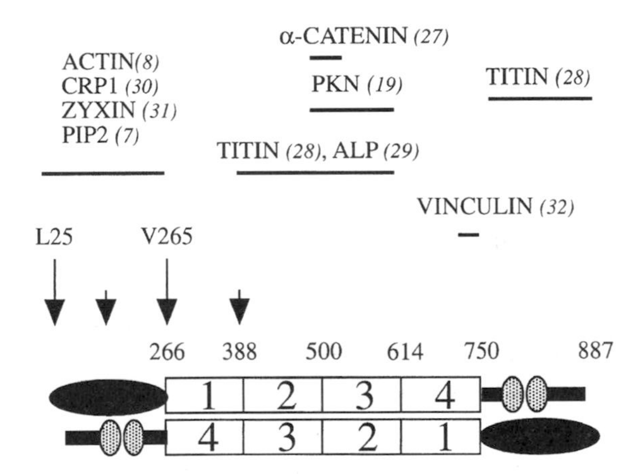

Figure 1. Structure of α-actinin. The diagram shows the relative positions of the actin-binding domain (amino acid 1–245), the spectrin-like repeats (266–749), the C-terminal region (750–887) containing two EF-hand motifs of chicken smooth muscle α-actinin, plus the possible orientation of the α-actinin subunits in the anti-parallel homodimer. The positions of the two major thermolysin cleavage sites (large arrows) plus to two less susceptible sites (small arrows) are shown. Thermolysin digestion of α-actinin gives rise to a monomeric 27 kDa fragment capable of binding actin, and a dimeric rod-shaped 55 kDa fragment containing the spectin-like repeats. Regions of α-actinin which interact with various binding partners are indicated. References are in italics.

smooth muscle α-actinin binds actin with a k_d of 2.4 μM at 25°C.[6] Binding affinity is reduced at higher temperatures, and is inhibited by tropomyosin.[1] Phosphatidyl inositol 4,5-bisphosphate (PIP2) dramatically increases the gelation activity of chicken smooth muscle α-actinin, and a PIP2 binding site has been identified within the actin-binding domain of the chicken skeletal muscle isoform.[7] Three discrete regions of the actin-binding domain of α-actinin have been implicated in binding to actin.[8] Peptide inhibition studies suggest that α-actinin binds to the opposite face of actin sub-domain 1 to that recognized by myosin.[9]

The gene encoding chicken smooth muscle α-actinin gives rise to a non-muscle isoform by alternative splicing; the two proteins differ only in the C-terminal region of the first EF hand. This may account for the fact that actin binding by the non-muscle isoform is calcium regulated whereas the muscle isoform is calcium insensitive. Chicken skeletal muscle α-actinin is approximately 80 per cent identical to the smooth muscle isoform and is encoded by a separate gene which again gives rise to an alternatively spliced transcript varying only in the first EF hand.[10] In humans, two skeletal muscle α-actinin genes have been identified.[11] Genetic analysis of *Dictyostelium discoideum* α-actinin has shown that EF hand 1 is required for actin cross-linking, and has a much lower affinity for calcium than EF hand II. Calcium regulation is thought to be mediated by EF hand I, whereas EF hand II has a high affinity for calcium, and is probably non-regulatory.[12] The idea that EF hand I mediates calcium regulation of non-muscle α-actinins is consistent with the observation that this is a region subject to alternative splicing.[10] However, not all non-muscle α-actinins would appear to be calcium sensitive.

α-Actinin binding proteins

As well as binding actin, α-actinin has been shown to bind to the cytoskeletal proteins zyxin,[31] vinculin,[32] and α-catenin[27] and to the cytoplasmic domains of various cell surface receptors including L-selectin,[14] β1, β2, and β3 (CD18) integrins,[13] I-CAM-1 (CD54), I-CAM-2,[15] the N- and E-cadherins probably via α-catenin,[16] and the NMDA receptor.[17] These findings support a model in which α-actinin links integral membrane proteins either directly or indirectly to the actin cytoskeleton. α-Actinin also binds PIP2 and phosphatidyl inositol 3-kinase,[18] both of which are involved in signal transduction pathways. It has recently been reported to bind to the rho effector kinase PKN in a PIP2-dependent manner.[19] α-Actinin also binds to several proteins important in muscle differentation. Titin binds to two distinct sites in α-actinin, one in repeats 2 and 3, and one in the C-terminal region of the protein.[28] The Z-disc protein ALP (α-actinin-associated LIM protein) also binds to α-actinin repeats 2 and 3 via an N-terminal PDZ domain in ALP.[29] A second LIM-domain protein, CRP-1, binds to the actin binding domain of α-actinin.[30]

Purification

α-Actinin can be purified from smooth and skeletal muscle with yields of 10–20 mg from 100 g of tissue.[20]

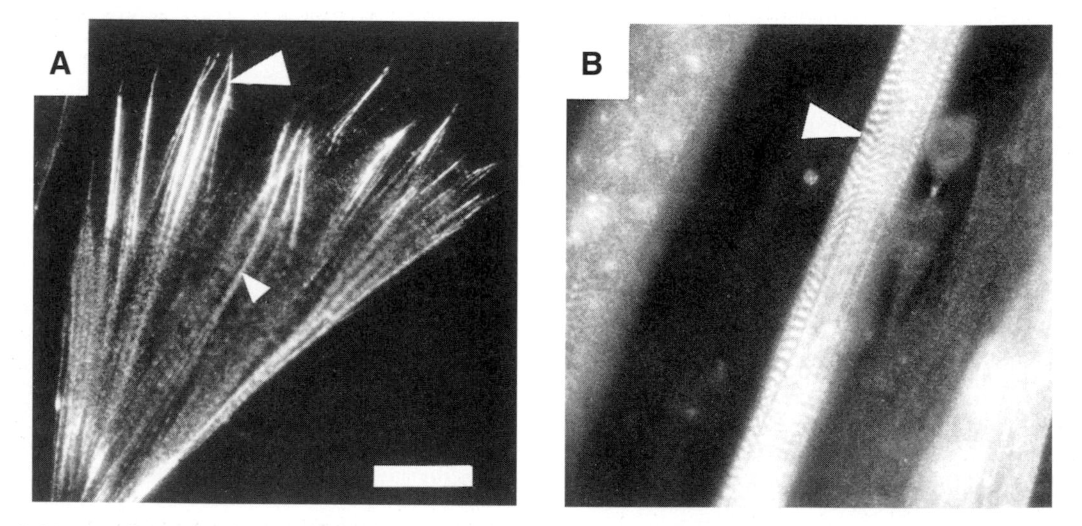

Figure 2. Immunofluorescence localization of α-actinin in chicken myoblasts (A), and mouse myotubes (B). The α-actinin in (A) was detected with an antiserum to the chicken smooth muscle protein; note that α-actinin is localised in both focal adhesions, and along actin stress fibres where the staining appears periodic in some cells. The α-actinin in (B) was detected with an antiserum to chicken skeletal muscle α-actinin; note the periodic staining associated with the Z-disc. Magnification bar 20 μM. (T. Parr, and D. R. Critchley, unpublished).

Much lower yields are obtained from non-muscle tissue (~1 mg/100 g).

Activities

The actin-binding properties of α-actinin have been investigated by numerous techniques including sedimentation, electron microscopy, viscometry, and fluorescence.[16]

Antibodies

Rabbit antibodies to chick skeletal and smooth muscle α-actinin have been produced, and antibodies to the latter are commercially available (Sigma). The antibodies are cross-reactive but can also be rendered isoform specific by adsorption. They also cross-react with mammalian α-actinins as detected by indirect immunofluorescence. Monoclonal antibodies to *Dictyostelium* α-actinin[21] and chicken smooth muscle α-actinin[22] have been reported, and Serotec (UK) market a monoclonal anti-α-actinin antibody (BM 75.2).

Genes

cDNAs encoding chicken skeletal (X13874), smooth muscle (J03486), and non-muscle (M74143), as well as *Drosophila* (X51753) and *Dictyostelium* (Y00689) α-actinins have been sequenced. Additionally, a chicken lung α-actinin (D26597) with reduced calcium sensitivity, and alternatively spliced chicken skeletal muscle isoforms (X68797–68802) have been characterized. In humans the non-muscle isoform (AACT1) is on chromosome 14q22–24 (M31300), whilst the two characterized skeletal muscle isoforms (AACT 2, 3) are located on chromosomes 1q42–q43 (M86406) and 11q13–q14 (M86407) respectively. AACT2 is expressed in both skeletal and cardiac muscle, but AACT3 is only expressed in skeletal muscle. An additional human non-muscle α-actinin gene (D89980) has recently been identified (AACT4).[31]

Mutant phenotypes/disease states

Knock-out of the α-actinin gene in *Drosophila* results in abnormal flight muscles,[23] although no non-muscle phenotype was observed. In addition, the loss of α-actinin in *Dictyostelium* does not result in any distinguishable phenotype, although when both α-actinin and ABP-120 are deleted, a number of abnormal characteristics were found.[24] An inherited human muscle disease, nemalin rod myopathy, is characterized by abnormal α-actinin expression, and results in skeletal muscle containing disordered Z lines and rod-shaped bodies containing α-actinin. However, there is no evidence that the defect resides in the known α-actinin genes.[25] A deficiency in the skeletal muscle α-actinin *AACT3* gene product has been reported in merosin-positive congenital muscular dystrophy.[26]

Structure

Although no detailed structural information is available for α-actinin, X-ray crystallography and NMR spectroscopy have been performed on single repeat fragments of spectrin. These have confirmed predictive models for the structure of the α-helical repeat motifs of the spectrin family.

References

1. Blanchard, A., Ohanian, V., and Critchley, D. R. (1989). *J. Muscle Res. Cell Motil.*, **10**, 280–9.
2. Baron, M. D., Davison, M. D., Jones, P., and Critchley, D. R. (1987). *J. Biol. Chem.*, **262**, 17623–9.
3. Taylor, K. A. and Taylor, D. W. (1993). *J. Mol. Biol.*, **230**, 196–205.
4. Flood, G., Kahana, E., Gilmore, A. P., Rowe, A. J., Gratzer, W. B., and Critchley, D. R. (1995). *J. Mol. Biol.*, **252**, 227–34.
5. Wachsstock, D. H., Scwartz, W. H., and Pollard, T. D. (1993). *Biophys. J.*, **65**, 205–14.
6. Kuhlman, P. A., Ellis, J., Critchley, D. R., and Bagshaw, C. R. (1994). *FEBS Lett.*, **339**, 297–301.
7. Fukami, K., Sawada, N., Endo, T., and Takenawa, T. (1996). *J. Biol. Chem.*, **271**, 2646–50.
8. Winder, S. J., Hemmings, L., Maciver, S. K., Bolton, S. J., Tinsley, J. M., Davies, K. E., *et al.* (1995). *J. Cell Sci.*, **108**, 63–71.
9. Lebart, M-C., Mejean, C., Roustan, C., and Benyamin, Y. (1993). *J. Biol. Chem.*, **268**, 5642–8.
10. Parr, T., Waites, G. T., Patel, B., Millake, D. B., and Critchley, D. R. (1992). *Eur. J. Biochem.*, **210**, 801–9.
11. Beggs, A. H., Byers, T. J., Knoll, J. H. M., Boyce, F. M., Bruns, G. A. P., and Kunkel, L. M. (1992). *J. Biol. Chem.*, **267**, 9281–8.
12. Witke, W., Hofmann, A., Koppel, B., Schliecher, M., and Noegel, A. A. (1993). *J. Cell Biol.*, **121**, 599–606.
13. Jockusch, B. M., Bubeck, P., Giehl, K., Kroemaker, M., Moschner, J., Rothkegel, M., *et al.* (1995). *Ann. Rev. Cell. Dev. Biol.*, **11**, 379–416.
14. Pavalko, F. M., Walker, D. M., Graham, L., Goheen, M., Doerschuk, C. M., and Kansas, G. S. (1995). *J. Cell Biol.*, **29**, 1155–64.
15. Heiska, L., Kantor, C., Parr, T., Critchley, D. R., Vilja, P., Gahmberg, C. G., and Carpen, O. (1996). *J. Biol. Chem.*, **271**, 26214–19.
16. Knudsen, K. A., Soler, A. P., Johnson, K. R., and Wheelock, M. J. (1995). *J. Cell Biol.*, **130**, 67–77.
17. Wyszynski, M., Lin, J., Ra, A., Nigh, E., Beggs, A. H., Craig, A. M., and Sheng, M. (1997). *Nature*, **385**, 439–42.
18. Sibaski, F., Fukami, K., Fukui, Y., and Takenawa, T. (1994). *Biochem. J.*, **302**, 551–7.
19. Mukai, H., Toshimori, M., Shibata, H., Takanaga, H., Kitagawa, M., Miyahara, M., *et al.* (1997). *J. Biol. Chem.*, **272**, 4740–6.
20. O'Halloran, T., Molony, L., and Burridge, K. (1986). *Meth. Enzymol.*, **134**, 69–77.
21. Schleicher, M., Noegel, A., Schwarz, T., Wallraff, E., Brink, M., Faix, J., *et al.* (1988). *J. Cell Sci.*, **90**, 59–71.
22. Jackson, P., Smith, G., and Critchley, D. R. (1989). *Eur. J. Cell Biol.*, **50**, 162–9.
23. Roulier, E. M., Fyrberg, C., and Fyrberg, E. (1992). *J. Cell Biol.*, **116**, 911–22.
24. Witke, W., Schleicher, M., and Noegel, A. A. (1992). *Cell*, **68**, 53–62.
25. Hashimoto, K., Shimuza, T., Nonaka, I., and Mannen, T. (1989). *J. Neurosci.*, **93**, 199–209.

26. North, K. N. and Beggs, A. H. (1996). *Neuromusc. Disorders*, **6**, 229–35.
27. Nieset, J. E., Redfield, A. R., Jin, F., Knudsen, K. A., Johnson, K. R., and Wheelock, M. J. (1997). *J. Cell Sci.*, **110**, 1013–22.
28. Young, P., Ferguson, C., Banuelos, S., and Gautel, M. (1998). *EMBO J.*, **17**, 1614–24.
29. Xia, H. H., Winokur, S. T., Kuo, W. L., Altherr, M. R., and Bredt, D. S. (1997). *J. Cell Biol.*, **139**, 507-15.
30. Pomies, P., Louis, H. A., and Beckerle, M. C. (1997). *J. Cell Biol.*, **139**, 157–68.
31. Honda, K., Yamada, T., Endo, R., Ino, Y., Gotoh, M., Tsuda, H., Yamada, Y., Chiba, H., and Hirohashi, S. (1998). *J. Cell Biol.*, **140**, 1383–93.
32. Crawford, A. W., Michelsen, J. W., and Beckerle, M. C., (1992). *J. Cell Biol.*, **116**, 1381-93.
33. McGregor, A., Blanchard, A. D., Rowe, A. J., and Critchley, D. R. (1994). *Biochem. J.*, **301**, 225–33.

David R. Critchley and Geoffrey Flood
Department of Biochemistry,
University of Leicester,
University Road,
Leicester LE1 7RH,
UK

Adducin

Adducin is a membrane-skeletal protein localized at spectrin–actin junctions[1] that was first purified from human erythrocytes based on calmodulin-binding activity.[2] Adducin also is a substrate for protein kinases A and C.[3-5] Adducin caps the fast-growing end of actin filaments[6] and promotes association of spectrin with actin.[7,8] Adducin capping activity may target spectrin to the ends of actin filaments and help stabilize the short actin filaments that comprise the spectrin–actin network of erythrocytes.

Protein properties

Erythrocyte adducin is comprised of α and β subunits closely related in amino acid sequence and domain organization.[9,10] α adducin is expressed in most tissues, while β adducin has a more restricted pattern of expression.[10] The human β adducin gene is composed of at least 13 exons varying in size from 81 to 1729 bp. A characteristic 3'-end alternative splicing in combination with a very complex differential splicing in the internal exons generates eight splicing variants of the β adducin gene.[11] γ adducin, which is similar in sequence to α and β adducin, is a likely companion for α adducin in cells lacking the β subunit.[12] Each adducin subunit has three distinct domains: a 39 kDa N-terminal globular protease-resistant head domain including a proline-rich N terminus, connected by a 9 kDa 'neck' domain to a C-terminal protease-sensitive tail domain. C termini of all three subunits contain a highly basic stretch of 22 amino acids with sequence similarity to the myristoylated alanine-rich C kinase substrate (MARCKS)[13] (Fig. 1(a)). Erythrocyte adducin in solution is a mixture of heterodimers and tetramers with α and β subunit head domains in contact to form a globular core, and interacting α and β adducin tails extended away from the core.[14] Tail domains of both the α and β adducin subunits are responsible for binding to actin filaments and spectrin–actin complexes,[14] and have been proposed to form lateral contacts involving several actin subunits and the β subunit of spectrin[15] (Fig.1(b)).

Adducin is localized at lateral cell borders of epithelial cells including MDCK cells, keratinocytes, and intestinal

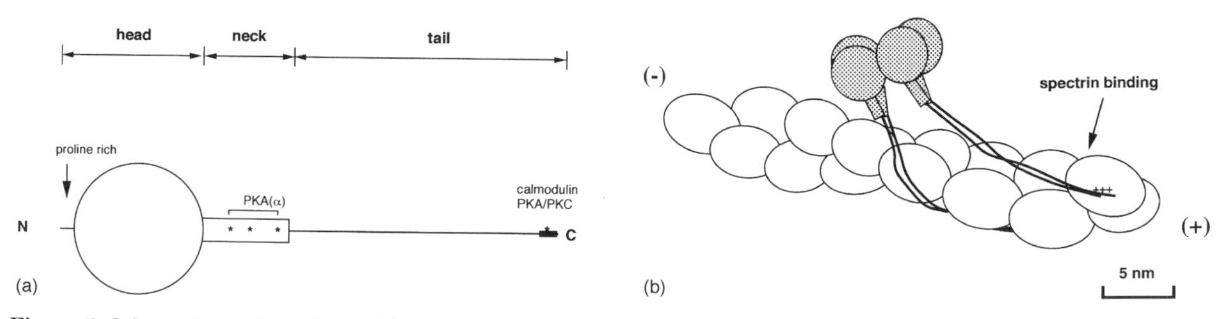

Figure 1. Schematic models of an adducin monomer and an adducin–actin filament complex. (a) The C-terminal basic domain (i.e. MARCKS-related domain) corresponding to the major calmodulin-binding site is indicated as a closed box and the major sites of phosphorylation by PKA and PKC are indicated by asterisks. (b) The C-terminal basic domain of adducin caps the fast-growing end of the actin filament. This domain recruits a second spectrin molecule to the filament end.

epithelium[5] as well as pre- and postsynaptic regions of neurones[16] (Fig. 2). Adducin is also abundant in all blood cells, which do not form cell–cell junctions but do contact each other on some occasions. In mouse oocytes, adducin is localized to chromosomes in metaphase I and II.[17] An adducin-like protein is associated with the *hts* phenotype of defective oocyte development in *Drosophila*. The *hts* gene product is localized to ring canals, which have an abundance of actin filaments, and which represent a specialized region of cell–cell contact.[18] A *C. elegans* homologue of the adducin gene has also been cloned and its product found to be highly concentrated in the nerve ring, a structure rich in synapses.[19]

The MARCKS-related domain of adducin contains the major calmodulin-binding site as well as a major phosphorylation site common to protein kinases A and C[20] (Fig. 1(a)). This domain is the dominant functional domain for actin binding, actin capping, and spectrin recruitment activities of adducin. These activities are regulated by binding of calmodulin and by phosphorylation of the MARCKS-related domain.[6,20]

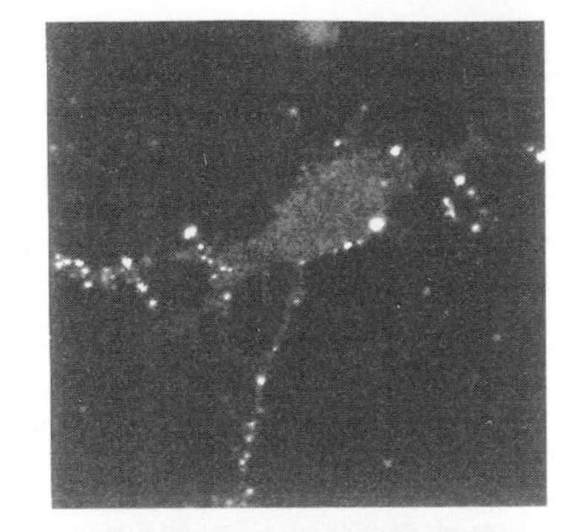

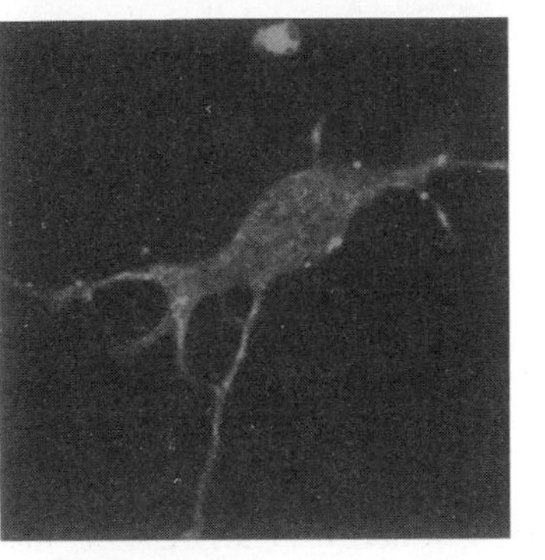

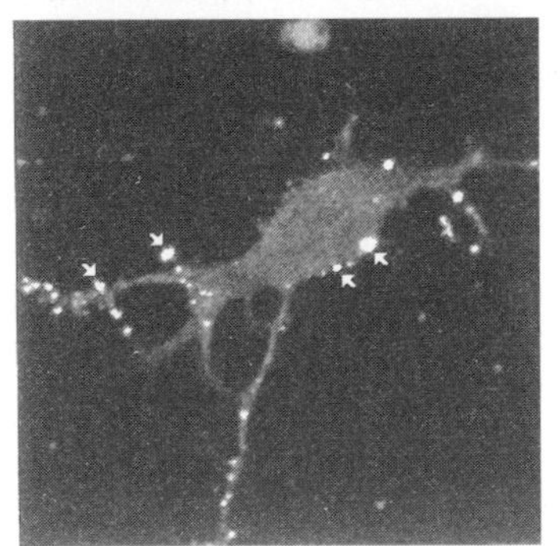

Figure 2. Colocalization of adducin with glutamate receptor in dendritic spines of hippocampal neurones. Cultured hippocampal neurones were double stained with monoclonal anti-glutamate receptor subunit R2/4 antibody (GluR2/4) and anti-MARCKS-domain serum (Adducin). Adducin is highly concentrated in dendritic spines (composite, arrows) where glutamate receptors also are abundant.

Purification

Functional adducin is extracted from erythrocyte membranes under low ionic-strength conditions[21] and purified by conventional column chromatography. Typically, adducin is separated from spectrin (the other major structural component of the erythrocyte membrane skeleton) by cation exchange chromatography through S-Sepharose. Pure erythrocyte adducin can be obtained by further anion exchange chromatography.[14]

Activities

Adducin caps the fast-growing ends of actin filaments,[6] bundles them,[22,23] and promotes association of spectrin with actin filaments preferentially at the fast-growing ends.[24] Adducin also binds calmodulin with K_d = < 200 nM.[2]

Dimer(s) of adducin are required for maximal activities of actin capping and spectrin recruitment. The C-terminal basic domain makes contact with the fast-growing end of the actin filament, caps it, and promotes binding of spectrin, whereas the neck domains interact with each other to form a dimer.[24]

Antibodies

Polyclonal antibodies against human erythrocyte adducin (α and β adducin) and against rat α and γ adducin have been characterized. Polyclonal sera raised against human erythrocyte adducin cross-react with adducin from a number of other mammalian species. A polyclonal antibody specific to the phosphorylation state of the serine residue in the C-terminal basic domain (MARCKS-related domain) has been produced in our laboratory.[25]

Genes

α adducin has been cloned and sequenced from human[10] (EMBL/GenBank/DDB X58141) and rat[26] (EMBL/GenBank Z49082). β adducin has been cloned and sequenced from human[10] (EMBL/GenBank/DDB X58199) and a spliced form from rat[27] (EMBL/GenBank M63894). γ adducin has also been cloned and sequenced from human (EMBL/GenBank U37122) and rat[12] (EMBL/GenBank U35775). A *Drosophila* homologue of mammalian adducin has been cloned and sequenced[18] (EMBL/GenBank L05016).

References

1. Derick, L. H., Liu, S.-H., Chishti, A. H., and Palek, J. (1992). *Eur. J. Cell Biol.*, **57**, 317–20.
2. Gardner, K. and Bennett, V. (1986). *J. Biol. Chem.*, **261**, 1339–48.
3. Palfrey, H. and Waseem, A. (1985). *J. Biol. Chem.*, **260**, 16021–9.
4. Ling, E., Gardner, K., and Bennett, V. (1986). *J. Biol. Chem.*, **261**, 13875–8.
5. Kaiser, H. W., O'Keefe, E., and Bennett, V. (1989). *J. Cell Biol.*, **109**, 557–69.
6. Kuhlman, P. A., Hughes, C. A., Bennett, V., and Fowler, V. (1996). *J. Biol. Chem.*, **271**, 7986–91.
7. Gardner, K. and Bennett, V. (1987). *Nature*, **328**, 359–62.
8. Bennett, V., Gardner, K., and Steiner, J. P. (1988). *J. Biol. Chem.*, **263**, 5860–9.
9. Joshi, R. and Bennett, V. (1990). *J. Biol. Chem.*, **265**, 13130–6.
10. Joshi, R., Gilligan, D. M., Otto, E., McLaughlin, T., and Bennett, V. (1991). *J. Cell Biol.*, **115**, 665–75.
11. Tisminetzky, S., Devescovi, G., Tripodi, G., Muro, A., Bianchi, G., Colombi, M., *et al.* (1995). *Gene*, **167**, 313–16.
12. Dong, L., Chapline, C., Mousseau, B., Fowler, L., Ramsay, K., Stevens, J. L., and Jaken, S. (1995). *J. Biol. Chem.*, **270**, 25534–40.
13. Blackshear, P. J. (1993). *J. Biol. Chem.*, **268**, 1501–4.
14. Hughes, C. A. and Bennett, V. (1995). *J. Biol. Chem.*, **270**, 18990–6.
15. Li, X. and Bennett, V. (1996). *J. Biol. Chem.*, **271**, 15695–702.
16. Seidel, B., Zuschratter, W., Wex, H., Garner, C. C., and Gundelfinger, E. D. (1995). *Brain Res.*, **700**, 13–24.
17. Pinto-Correia, C., Goldstein, E. G., Bennett, V., and Sobel, J. S. (1991). *Dev. Biol.*, **146**, 301–11.
18. Yue, L. and Spradling, A. C. (1992). *Genes Dev.*, **6**, 2443–54.
19. Moorthy, S. and Bennett, V. (1998).
20. Matsuoka, Y., Hughes, C. A., and Bennett, V. (1996). *J. Biol. Chem.*, **271**, 25157–66.
21. Bennett, V. (1983). *Meth. Enzymol.*, **96**, 313–24.
22. Mische, S. M., Mooseker, M. S., and Morrow, J. S. (1987). *J. Cell Biol.*, **105**, 2837–45.
23. Taylor, K. A. and Taylor, D. W. (1993). *J. Mol. Biol.*, **230**, 196–205.
24. Li, X., Matsuoka, Y., and Bennett, V. (1998). *J. Biol. Chem.*, (In press).
25. Matsuoka, Y., Li, X., and Bennett, V. (1998). *J. Cell Biol.*, **142**, 485–97.
26. Tripodi, G., Casari, G., Tisminetzky, S., Bianchi, G., Devescovi, G., Muro, A., *et al.* (1995). *Gene*, **166**, 307–11.
27. Tripodi, G., Piscone, A., Borsani, G., Tisminetzky, S., Salardi, S., Sidoli, A., *et al.* (1991). *Biochem. Biophys. Res. Commun,.* **177**, 939–47.

Yoichiro Matsuoka and Vann Bennett
Howard Hughes Medical Institute and Department of Cell Biology, Duke University Medical Center, Durham, NC 27710, USA

Ankyrins

Ankyrins are a large family of cytosolic adapter proteins that link membrane proteins to the spectrin-based submembrane skeleton. They are conserved in evolution from mammals to insects and are expressed in a wide variety of cell types including erythrocytes, epithelia, muscle, and neurones. The globular amino terminal domain is conserved and generally comprises up to 24 characteristic 33 amino acid 'ankyrin' repeats (membrane binding domain) and a spectrin binding region. Ankyrins bind to a variety of membrane proteins, including ion channels and transporters and cell adhesion molecules. The carboxyl terminus is variable and, depending on the isoform, may comprise a 'death' domain, a regulatory domain, or a predicted extended random coil. Ankyrins are localized to the plasma membrane, Golgi complex, and lysosomes and function in membrane stability and the spatial organization of membrane proteins into structural and functional membrane domains.

Protein properties

Ankyrin was originally identified as a component of the human erythrocyte membrane skeleton, but has subsequently been found in all cells in the body in mammals[1] as well as in *Caenorhabditis elegans*[2] in *Drosophila.*[3] The ankyrin gene family in mammals comprises the three principal members described below, although recent studies indicate further divergence.

1. ANK1 (also called ankyrin$_R$; M_r ~206 000 Da) is expressed in erythrocytes[4,5] and a subset of neurones in the brain[6] (e.g. Purkinje cells and granule cells of the cerebellum, motor neurones of the spinal cord, and a small subset of neurones in the hippocampus).
2. ANK2 (also called ankyrin$_B$) is the major isoform of ankyrin in the brain[7]: it exists as a M_r ~220 000 Da isoform in the adult, which is broadly distributed in all neuronal cell bodies and dendrites and glia; and a

M_r ~440 000 Da isoform[8] which is expressed specifically in unmyelinated axons and dendrites during neonatal development.

3. ANK3 isoforms[9] are widely expressed in epithelia, macrophages, cardiac, smooth and striated muscle, initial segments of neurones, and bodes of Ranvier (also called ankyrin$_G$[10]); isoforms of M_r ~480 000 Da (ankyrin$_G$), 215 000, 200 000, 170 000, 120 000, and 105 000 Da have been identified.

In general, ANK1, 2, and 3 have two, highly conserved structural domains in the amino-terminus globular region, and a variable carboxyl-terminus domain that contains a 'death' domain: [1]

1. The amino-terminal globular domain comprises a 89–95 kDa membrane binding domain and a ~57–62 kDa spectrin binding domain (see Fig. 1). In general, different ankyrin isoforms exhibit a relatively high degree of sequence homology in the membrane binding domain (>70 per cent). This domain has a characteristic structure of up to 24 copies of a 33 amino acid 'ankyrin' repeat (see Fig. 1). 'Ankyrin' repeats have been found in many other proteins, including transmembrane signalling proteins, cell cycle control proteins, and transcription factors.[11] The 33 amino acid repeats fold to form four subdomains, each of which is composed of six repeats; this arrangement of repeats probably facilitates coordinate binding of ankyrin to more than one membrane protein.[12] Many membrane proteins have been shown to interact with ankyrin (see Activities), although detailed analyses of the binding sites on ankyrin only for the anion exchanger, Na/K-ATPase, and neurofascin have been reported.[12,13] These proteins bind with high affinity to more than one set of tandem repeats in the membrane binding domain (k_D ranging from 10 nM to 25 nM). There is not a simple linear

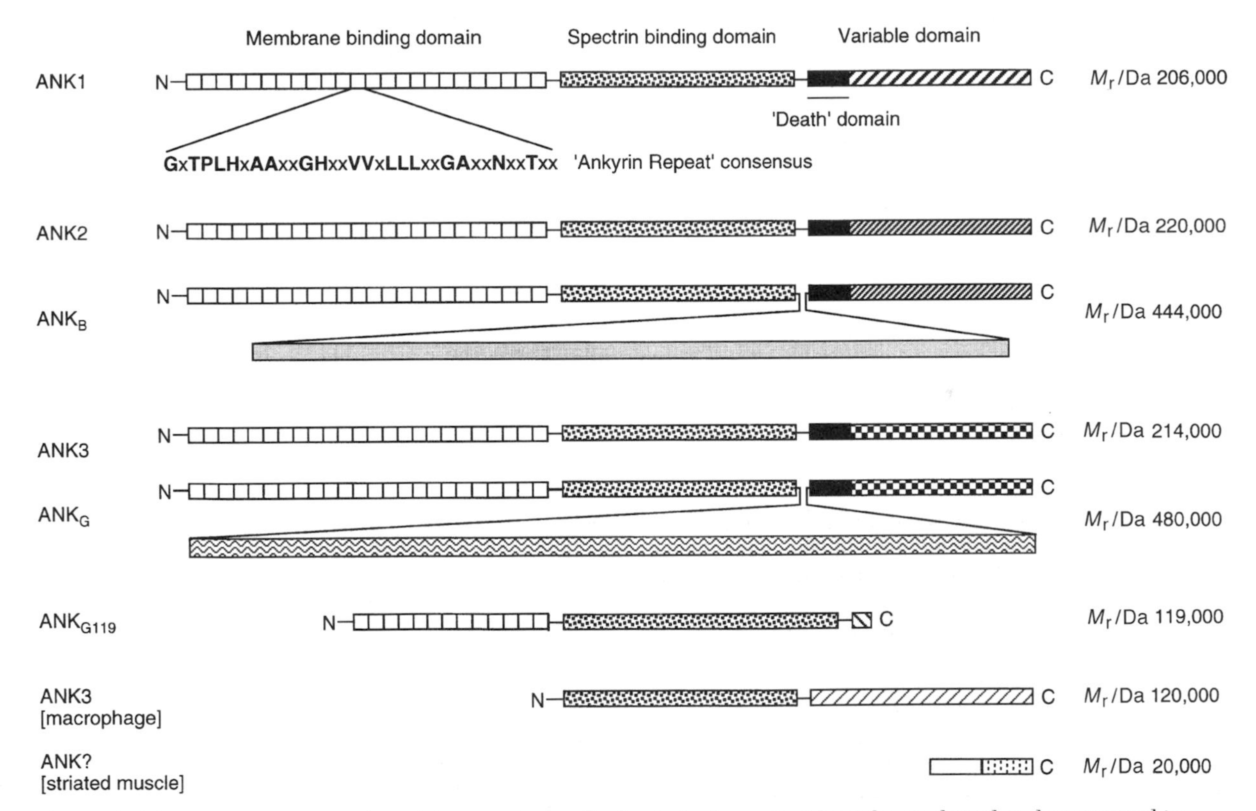

Figure 1. Schematic representation of the structure of ankyrin isoforms that have been cloned and sequenced to date (not to scale). ANK1, 2, and 3 represent the three major ankyrins, ANK$_B$ and ANK$_G$ are isoforms of ANK2 and 3, respectively, and contain large inserts between the globular amino-terminus domains and the 'death' domain. The membrane binding domain comprises up to 24 repeats that have a characteristic consensus sequence ('ankyrin repeat') and is highly conserved between different isoforms (>75 per cent). The spectrin binding domain is also conserved between different isoforms (>60 per cent). With the exception of the 'death' domain, the carboxyl-terminus domain is highly variable between different isoforms. Three novel ankyrins have been described recently (ANK$_{G119}$, ANK3 [macrophage], ANK? [muscle]) and have domain structures different from other ankyrins.

consensus sequence for the binding site on ankyrin for these membrane proteins. The *spectrin binding domain* contains binding sites for the *b*-subunit of the $(\alpha b)_2$ spectrin (fodrin) heterotetramer. This domain also contains a high affinity binding site for Na/K-ATPase.[13] In general, ankyrin isoforms exhibit a relatively high degree of sequence homology in the spectrin binding domain (>65 per cent).

2. 'The death' domain (~12 kDa) has been identified based on sequence homology to proteins (e.g. FAS, tumour necrosis factor receptor) involved in apoptosis.[14] The function of the 'death' domain in ankyrin is unknown, although it has been suggested that it may play a role in protein–protein interactions.
3. The variable carboxyl-terminal domain exhibits the highest sequence divergence between different ankyrin isoforms (<30 per cent homology) due to tissue-dependent and developmentally regulated alternative splicing. This domain comprises the regulatory domain in ANK1, which modulates the binding affinity of ankyrin to the anion exchanger and spectrin.[15] ANK2 comprises a spliced variant that contains a 6 kb (~220 kDa) insert between the spectrin binding domain and the 'death' domain/carboxyl terminus.[16] This insert is characterized by 15 serine-rich 12 amino acid repeats and many potential phosphorylation sites. Analysis of the structure of this insert indicates an extended random coil. Because of its estimated length of 0.5 μm, this insert could link proteins bound to the amino terminus membrane/spectrin binding domain at one end to proteins bound to the 'death' domain at the other end over considerable distances in the cell. ANK3 also comprises an isoform (ankyrin$_G$) that contains a large insert between the membrane/spectrin binding domain and the 'death' domain.[11] This insert is serine rich (35 per cent of residues), contains *O*-glycosylated residues, and is also predicted to be a random coil.

Recently, three new members of the ankyrin family have been identified from novel cDNAs and several novel spliced forms of ANK3 have been identified by Western and Northern blotting, but the sequences of the corresponding cDNAs are not available. These are described below.

1. ANK$_{G119}$[17] is a small ankyrin localized to the Golgi complex. It contains a truncated membrane binding domain comprising 13 'ankyrin' repeats, a spectrin binding domain, and a very short carboxyl terminus regulatory domain (see Fig. 1).
2. ANK3 (macrophages)[18] is an isoform localized to lysosomes in macrophages. It does not contain an amino-terminus membrane binding domain or 'ankyrin' repeats, but contains a spectrin binding domain and a regulatory domain (see Fig. 1).
3. ANK? (muscle)[19] is a very small ankyrin localized to the sarcoplasmic reticulum and comprising a short segment of the carboxyl terminus of the regulatory domain and a novel amino terminus sequence (see Fig. 1).
4. Novel ANK3 isoforms[9] have been identified by Western and Northern blotting. These isoforms vary in size: M_r ~215 000, 200 000, 170 000, 120 000, and 105 000 Da. Finally, a 195 kDa ankyrin has been identified with an antibody to ANK1 and is localized exclusively to the Golgi complex,[20] but the cDNA and sequence of this isoform are also unknown at present.

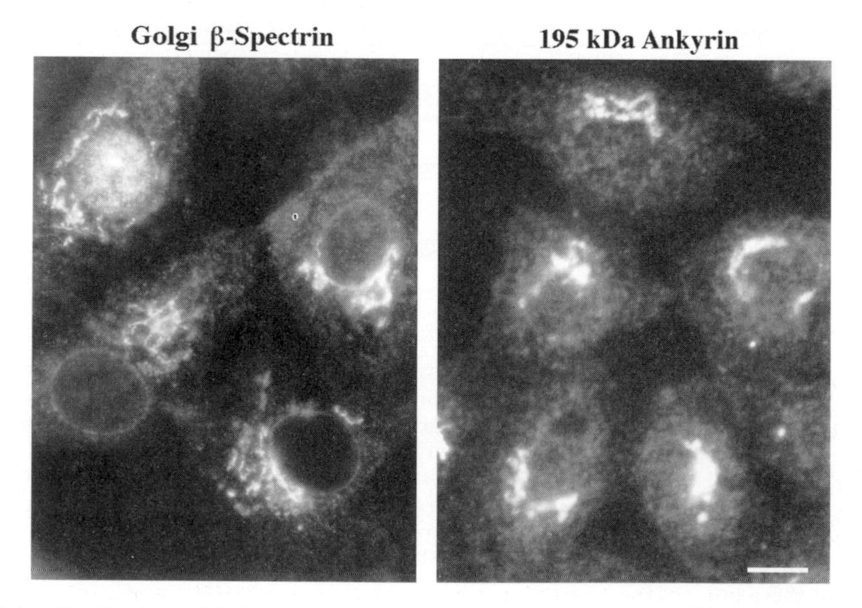

Figure 2. Subcellular distributions of Golgi-localized froms of spectrin and ankyrin.[20,24] Bar represents 10 μm.

Purification

Functional ankyrin can be purified in large quantities from out-dated human erythrocytes (ANK1)[21] or from pig brain (ANK2)[22] using differential protein extraction from cells/tissue and conventional column chromatography. With the availability of cDNAs to most of the major isoforms (see below), functional domains of ankyrin can now be expressed in bacteria and purified by affinity and ion exchange chromatography.

A large number of proteins contain 'ankyrin' repeats,[1,11], each of which comprise a characteristic 33 amino acid motif (see Fig. 1). Proteins related through their content of 'ankyrin' repeats have diverse functions, including transmembrane ligand/receptor signal transduction, cell cycle control and transcription factors.[11] 'Ankyrin' repeats are thought to be involved in protein–protein interactions.

Activities

The erythrocyte isoform of ankyrin (ANK1 or ankyrin$_R$) is the best studied in terms of protein biochemistry and function; components of the erythrocyte membrane skeleton can be purified to homogeneity, and purified erythrocyte membranes (inside-out ghosts), depleted of specific proteins, can be used as a template to examine protein–protein binding sites *in situ*. In the erythrocyte,[1] ankyrin binds directly to the carboxyl terminus of the transmembrane anion exchanger (also called band 3 or Cl^-/HCO_3^--exchanger) through its membrane binding domain (see above), and to the β subunit of the spectrin $(\alpha\beta)_2$ heterotetramer through its spectrin-binding domain (see above). Spectrin heterotetramers (~200 nm long) are linked together at their ends by a ternary complex of adducin, a short actin oligomer, protein band 4.1, and several other proteins, which form a cross-linked protein lattice underneath the plasma membrane. The ankyrin/spectrin lattice provides strength to the plasma membrane, which is important for the reversible deformability of erythrocytes during blood circulation. The lattice also immobilizes the anion exchanger in the plane of the lipid bilayer; disruption of the ankyrin/spectrin lattice (e.g., by mutation or deletion of specific proteins) results in a 50-fold increase in the diffusion coefficient of the anion exchanger.

In non-erythroid cells, ankyrins associate with the plasma membrane through binding to specific membrane proteins and bind non-erythroid forms of spectrin (also called fodrin). Ankyrin has been shown to bind[1] ion channels and exchangers (voltage-dependent Na^+ channel, Na/K-ATPase, Na/Ca-exchanger, H/K-ATPase), cell adhesion proteins (the hyaluronic receptor, CD44; members of the Ig/FnIII superfamily neurofascin, L1, NrCAM, and NgCAM), and the inositol 1,4,5-triphosphate (IP_3) receptor.

Despite detailed knowledge of ankyrin binding partners in non-erythroid cells, less is known about the functional significance of these interactions. Interestingly, in some cell types (e.g., neurones and epithelia) isoforms of ankyrin and specific membrane proteins are not randomly distributed in the plasma membrane. Ankyrin$_G$ (M_r ~480 000 Da isoform of ANK3, see above) is localized in myelinated neurones to the initial segments and is clustered in specialized sites along the axon, termed nodes of Ranvier, together with voltage-dependent Na^+ channels and the cell adhesion molecule neurofascin.[10] In polarized kidney epithelial cells, the distributions of an ANK3 isoform (M_r ~215 000 Da) and Na/K-ATPase are restricted to the basal-lateral membrane domain, and excluded from the apical membrane domain.[23] Thus, ankyrin isoforms that bind specific membrane proteins *in vitro* colocalize with the same proteins in specialized plasma membrane domains in polarized cells. Furthermore, in both neurones and epithelial cells, restriction of ankyrin isoforms and specific membrane proteins to these membrane domains during development of cell polarity is coincident in time and space. Thus, ankyrin and the associated spectrin (fodrin) lattice may play roles in the establishment and maintenance of structurally and functionally distinct membrane domains by corralling membrane proteins within specific regions of the plasma membrane.

Recent studies also show that isoforms of ankyrin associate with the Golgi complex[17,20] and lysosomes in macrophages.[18] Although the function of these ankyrins is not known, it is interesting that the Golgi ankyrins associate with a Golgi-specific form of spectrin.[17,20,24] Furthermore, Golgi-spectrin is found in a complex with centractin,[25] a protein complex implicated in microtubule-based motility of organelles; it is noteworthy that early studies demonstrated that ankyrin binds and bundles microtubules.[26] Together, these observations raise interesting possibilities that organelle-localized forms of ankyrin and spectrin may provide a template for the attachment of motor proteins and may play roles in the subcellular distribution of organelles and the organization in and, perhaps, sorting of proteins between different membrane compartments.[27]

Antibodies

A number of polyclonal and monoclonal antibodies against specific determinants on different ankyrin isoforms have been characterized. In general, antibodies raised against either human, mouse, chicken, or dog ankyrins cross-react with the appropriate ankyrin in other species.

Genes

ANK1 has been cloned and sequenced from human (EMBL/GenBank/DDBJ M28880). Ankyrin$_B$/ANK2 cDNA encoding the M_r ~220 000 Da isoform has been cloned and sequenced from human (EMBL/GenBank/DDBJ X5658) and the large alternative insert (to generate the M_r ~440 000 Da isoform) has been cloned and sequenced from human (EMBL/GenBank/DDBJ X56957); the complete M_r ~440 000 Da form from human has also been cloned and sequenced (EMBL/GenBank/DDBJ Z26634). ANK3 has been cloned and sequenced from mouse (EMBL/GenBank/DDBJ L40631, 5 kb isoform; EMBL/GenBank/DDBJ L40632, 7 kb isoform). The M_r ~480 000 Da isoform of ANK3 (ankyrin$_G$) has been cloned and sequenced from human (EMBL/GenBank/DDBJ U13616). The isoform of ANK3 associated with lysosomes in macrophages has been cloned and sequenced from mouse (EMBL/GenBank/DDBJ U89275). ANK$_{G119}$, which associates with the Golgi complex, has been cloned and sequenced from human (EMBL/Genbank/DDBJ U43965). The very small ankyrin isoform found in muscle has been cloned and sequenced from mouse (EMBL/GenBank/DDBJ U73972).

Mutant phenotype/disease states

Mutant phenotypes are associated with abnormalities in expression of ankyrin isoforms in many species. In humans, hereditary spherocytosis is the result of decreased expression or expression of mutated forms of ANK1 (ankyrin$_R$).[28] In mice, the *nb/nb* mutation is due to nearly complete loss of ANK1 (ankyrin$_R$) expression and results in severe anaemia with degeneration of a subset of Purkinje cell neurones and subsequent cerebellar dysfunction.[29] The *unc44* mutation in *Caenorhabditis elegans* is due to mutations within the ankyrin gene that lead to abnormal axonal guidance in development.[2]

Structure

Neither NMR nor X-ray crystal structures of ankyrins are available.

References

1. Bennett, V. (1993). *J. Biol. Chem.*, **267**, 8703–6.
2. Otsuka, A. J., Franco, R., Yang, B., Shim, K. H., Tang, L. Z., Zhang, Y. Y., *et al.* (1995). *J. Cell Biol.*, **129**, 1081–92.
3. Dubreuil, R. R. and Yu, J. (1994). *Proc. Natl Acad. Sci., USA*, **91**, 10285–9.
4. Lux, S. E., John, K. M., and Bennett, V. (1990). *Nature*, **344**, 36–42.
5 Lambert, S., Yu, H., Prchal, J., Lawler, J., Ruff, P., Speicher, D., *et al.* (1990). *Proc. Natl Acad. Sci., USA*,. **87**, 1730–4.
6. Lambert, S. and Bennett, V. (1993). *J. Neurosci.*, **13**, 3725–35.
7. Otto, E., Kunimoto, M., Mclaughlin, T., and Bennett, V. (1991). *J. Cell Biol.*, **114**, 241–53.
8. Chan, W, Kordeli, E., and Bennett, V. (1993). *J. Cell Biol.*,**123**, 1463–73.
9. Peters, L. L., John, K. M., Lu, F. M., Eicher, E. M., Higgins, A., Yialamas, M., *et al.* (1995). *J. Cell Biol.*, **130**, 313–30.
10. Kordeli, E., Lambert, S., and Bennett, V. (1995). *J. Biol. Chem.*, **270**, 2352–9.
11. Bork, P. (1993). *Proteins*, **17**, 363–74.
12. Michaely, P. and Bennett, V. (1995). *J. Biol. Chem.*, **270**, 31298–302.
13. Davis, J. Q. and Bennett, V. (1990). *J. Biol. Chem.*, **265**, 17252–6.
14. Cleveland, J. L. and Ihle, J. N. (1995). *Cell*, **81**, 47–482.
15. Davis, L. H., Davis, J. Q., and Bennett, V. (1992). *J. Biol. Chem.*, **267**, 18966–72.
16. Chan, W., Koerdeli, E., and Bennett, V. (1993). *J. Cell Biol.*, **123**, 1463–73.
17. Devarajan, P., Stabach, P. R., Mann, A. S., Ardito, T., Kashgarian, M., and Morrow, J. S. (1996). *J. Cell Biol.*, **133**, 819–30.
18. Hoock, T. C., Peters, L. L., and Lux, S. E. (1997). *J. Cell Biol.*, **136**, 1059–70.
19. Zhou, D., Birkenmeier, C. S., Williams, M. W, Sharp, J. J., Barker, J. E., and Bloch, R. J. (1997). *J. Cell Biol.*, **136**, 621–31.
20. Beck, K. A., Buchanan, J., and Nelson, W. J. (1997). *J. Cell Sci.*, (In press.)
21. Bennett, V. (1983). *Meth. Enzymol.*, **96**, 313–24.
22. Bennett, V., Baines, A. J., and Davis, J. Q. (1986). *Meth. Enzymol.*, **134**, 55–69.
23. Morrow, J. S., Cianci, C. D., Ardito, T., Mann, A. S., and Kashgarian, M. (1989). *J. Cell Biol.*, **108**, 455–65.
24. Beck, K. A., Buchanan, J., Malhotra, V., and Nelson, W. J. (1994). *J. Cell Biol.*, **127**, 707–23.
25. Holleran, E. A., Tokito, M. K., Karki, S., and Holtzbaur, L. F. (1996). *J. Cell Biol.*, **135**, 1815–29.
26. Bennett, V. and Davis, J. Q. (1981). *Proc. Natl Acad. Sci., USA*, **78**, 7550–4.
27. Beck, K. A. and Nelson, W. J. (1996). *Am. J. Physiol. (Cell)*, **270**, C1263–70.
28. Peters, L. L. and Lux, S. E. (1993). *Seminars Hematol.*, **30**, 85–118.
29. Peters, L. L., *et al.* (1991). *J. Cell Biol.*, **114**, 1233–41.

W. James Nelson
Department of Molecular and Cellular Physiology, Beckman Center, B121, Stanford University School of Medicine, Stanford, CA 94305–5426, USA

Kenneth A. Beck
Department of Cell Biology and Human Anatomy, University of California, Davis, CA 95616, USA

Calpains (EC 3.4.22.17)

The calpains are non-lysosomal calcium-dependent proteolytic enzymes that constitute a family within the papain clan (superfamily) of cysteine proteinases. At least two isoforms are ubiquitous in vertebrates, calpain-1 (micro-, mu– or μ-) and calpain-2 (milli- or m-). Tissue specific calpains have been described for skeletal muscle (calpain-3) and stomach. Cleavage site specificity, *in vivo* regulation and physiological substrates remain poorly defined.

Synonomous names

'Kinase activating factor'; calcium activated neutral proteinases (CANPs); calcium dependent proteases (CDPs).

Homologous proteins

cDNA sequences identified in *Drosophila,*[1] *Schistosoma,* and gene sequence within *C. elegans* have several patches of identity with well characterized calpains. Some sequences have been called 'homologues' solely on the basis of the catalytic residues where resemblance is closer to calpain than to other subfamilies within the papain clan. Those lacking demonstrated proteinase activity or mechanisms for calcium binding are not considered here (see ref. 30 for recent review).

Protein properties

Calpains 1 and 2 are heterodimeric enzymes with catalytic subunits of approximately 80 kDa and a small subunit of approximately 28 kDa. The enzymes have a native molecular weight of approximately 110 kDa. Each subunit has a C-terminal domain with some relationship to the calmodulin family of calcium binding proteins (Fig. 1; see recent reviews in refs 2–4). Two of the originally predicted EF hands bind calcium and an additional calcium binding site precedes the two EF hands in each subunit. How calcium binding regulates enzyme conformation and activity is not yet known. Evidence of a significant conformational change induced by calcium comes from the Ca^{2+} dependent interaction of calpain with its substrates and its physiological inhibitor, calpastatin. Autoproteolytic processing of each subunit occurs through both intra- and intermolecular processes.[2–6] Cleavage of the N terminus of the catalytic subunit results in a reduced calcium ion requirement;[6] processing of the small subunit may be a prerequisite for cleavage of exogenous substrates.[2] Acidic phospholipids may reduce the calcium requirement for autoproteolysis.[3] Homologues of the small subunit have not yet been described in *Drosophila*, *C. elegans*, or *Schistosoma*. The novel or tissue specific forms may function as monomers.[4]

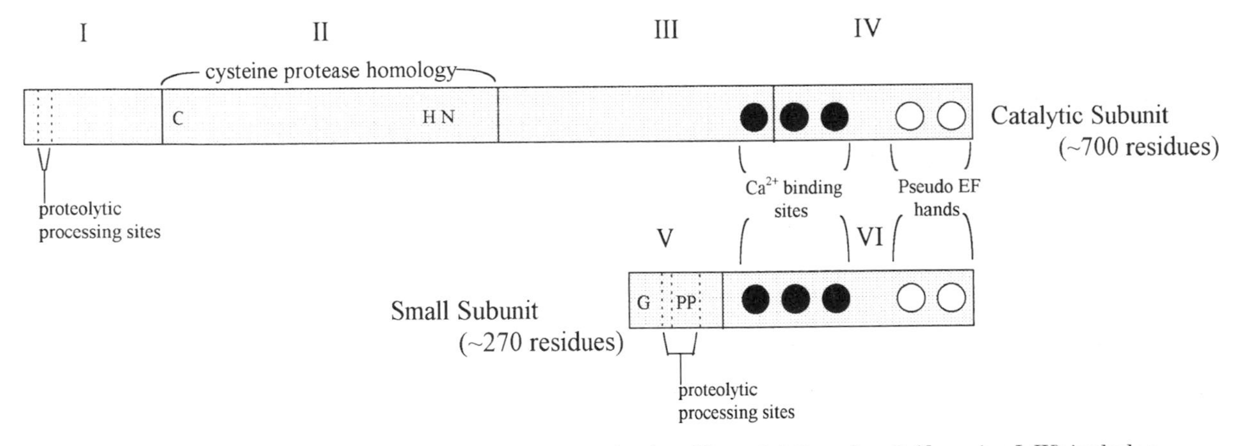

Figure 1. Schematic summary of calpain sequence organization. The catalytic subunit (domains I–IV) includes autoproteolytic processing sites (domain I), the catalytic residues (domain II) and five EF hands three bind calcium, two are 'pseudo' sites). The last four have traditionally been recognized as a 'calmodulin like' domain. The small subunit includes domain V, an N-terminal glycine rich region, with several autoproteolytic processing sites, a proline rich 'hinge', and domain VI, which is highly homologous to domain IV of the catalytic subunit.

Calpains 1 and 2 are ubiquitous in cells of vertebrates although their absolute and relative concentrations are cell type specific. A specific endogenous inhibitor protein, calpastatin, coexists in all cells known to contain calpain. Both cytosolic and nuclear distributions have been described for the enzymes and calpastatin. In some cells calpain appears to be localized to focal adhesions. Changes in immunolocalization of calpain in developing embryos of *Drosophila melanogaster* suggest that calpain function relates to reorganization of the actin cytoskeleton.[7] Immunolocalization with most antisera reveals total cellular calpain, only a small fraction of which may be active at any given time. It is widely thought, but not proven, that transient association with membranes may juxtapose calpain with both target substrates and the calcium fluxes required for activity.[2–4]

Cellular regulation of calpain mediated proteolysis and its consequences in physiological processes are not clearly defined. Studies of aggregating platelets first suggested that limited cleavage of cytoskeletal or membrane skeletal proteins (talin, filamin, spectrin) by calpain-1 may be regulated by integrin dependent signalling pathways.[2,8,31,32] Studies designed to inhibit calpains, either through overexpression of calpastatin[32] or treatment with calpain inhibitors,[9,31] further support that this may be their general role. Numerous other cytoskeletal proteins are reported to be calpain susceptible including α-actinin, adducin, ankyrin, ezrin, microtubule associated protein 2 (MAP-2), some of the unconventional myosins (e.g. I and V), and paxillin.[2,3] Cells selected on the basis of resistance to an irreversible calpain inhibitor were characterized by a prolonged G^1 phase of the cell cycle; an apparent result of reduced calpain-1 activity.[9] Surprisingly, calpain-2 appeared to be unaffected in these cells, although the inhibitor does not discriminate between isoforms *in vitro*. Many recent reports link cell cycle regulation with integrin regulated signalling, adhesion dependent cytoskeletal organization, and mitogen activation.

Calpain may participate in cellular damage cascades contributing to necrosis resulting from anoxic/ischaemic damage and in some pathways of apoptosis. Improved survival of hepatocarcinoma cells after anoxia is attributed to a defect in a phospholipase dependent activation pathway for calpain.[10] Experimental calpain inhibitors have been shown to minimize post ischaemic damage in animal models of brain trauma, stroke, and spinal cord injury. Spectrin, MAP-2, tau, tubulin, myelin basic protein, neurofilament proteins, adducin, and ankyrin are reported to be calpain susceptible.[9,11,12] In neuronal cells immunodetection of specific cleavage products of α-spectrin can be diagnostic for calpain activity. The use of calpain inhibitors, including calpastatin, supports a role for calpain in apoptotic pathways of T lymphocytes when the pathway for apoptosis requires transcriptional activation; examples include induction by glucocorticoids or irradiation and T-cell receptor dependent induction of FAS-ligand.[13] Cleavage of c-jun, c-fos, I-κB, and p53 by calpain have been reported.[3]

Calpain may play a direct role in regulating calcium signalling through feedback control of calcium channels and pumps including the plasma membrane Ca^{2+}-ATPase, L-type calcium channels, and receptors for IP3 and ryanodine.[2–4,14] Modification of the erythrocyte plasma membrane Ca^{2+}-ATPase by calpain renders the activity calmodulin independent.[15] The IP3 receptor is downregulated in neuroblastoma cells chronically stimulated with carbachol[14] however, the proteasome also may participate in its degradation. Several other enzymes, such as IP3 kinase, protein kinase C, protein tyrosine phosphatase 1B, pp60$^{c\text{-}src}$ are also substrates for calpain, at least *in vitro*.[2,3]

Purification

Naturally occurring enzymes

The two ubiquitous isoforms have been purified from a variety of tissues and cell types. Both proteins are

1. acidic at neutral pH and bind to anion exchangers;
2. sufficiently hydrophobic to bind phenyl sepharose at low to moderate ionic strength (e.g. 0.05–0.5 M KCl);
3. are reasonably stable at 4°C in the absence of calcium as long as reducing agents are present; and
4. enzyme protein concentration is 0.2 mg/mL.

Calpain-1 (μ-) is the dominant isoform in erythrocytes and human platelets, sources commonly used for its purification.[2,15,16] Isolation procedures from brain or cardiac muscle are also published. The co-elution of calpain-1 with calpastatin and the bulk of cellular proteins, at least in many tissues, makes its purification more tedious than that of m-calpain. Hydrophobic interaction chromatography (e.g. phenyl–sepharose) or size exclusion chromatography in the presence of chelators will separate calpain-1 from calpastatin. Kidney and muscle (skeletal, smooth, or cardiac) are frequently used sources of calpain-2 (μ-). Anion exchange chromatography, dye-affinity chromatography on reactive red agarose and gel filtration are, in general, sufficient to purify calpain-2 from muscle.[2] Commercial sources for one or both of the ubiquitous forms include Calbiochem, Nacalai Tesque, and Sigma. Purification of the tissue specific forms has been problematic. Calpain-3 is reported to undergo rapid autoproteolysis. Its low protein concentration and susceptibility to autoproteolysis complicate purification efforts.[17]

Recombinant enzymes

Recovery of active recombinant calpain has required either coexpression of the two subunits or denaturation and renaturation of the individually expressed subunits *in vitro*. Calpains 1, 2, and 3 have each been produced through recombinant cDNA and expression systems.[16–19] Mutation of the active site residues allows production of quantities of stable enzyme suitable for crystallization.[18,20] There may be solubility problems for calpain-1 in *E. coli* expression systems, but some activity can be reconstituted after solubilization under denaturing conditions and refolding. Calpain-2 (rat) can be expressed in

E coli. Several mutants have been characterized to confirm identity of catalytic residues;[17,20] alter calcium sensitivity;[4] and prevent autoproteolysis of the N terminus of the catalytic subunit.[6] Purification has been aided by C-terminal hexa-His tagging of the catalytic subunit. N-terminal extensions and truncations resulted in variations in levels of expression and activities for calpain-2.[6] Four calpain–3 recombinant mutants designed to inactivate the active site (Cys) have allowed purification of the inactive enzyme.[17]

Activities

In vitro

Hydrolysis of peptide bonds is dependent upon Ca^{2+}, requires a reducing environment, and is optimal at pH 7–7.5. Activity is measured with a variety of protein substrates or fluorogenic peptides.[2] Casein is most commonly used due to calpain's ability to cleave casein to acid soluble peptides; using radiolabelled or fluorescent tagged casein the assay is sensitive (nanogram amounts of enzyme). In contrast, most protein substrates (e.g. α-spectrin, talin, caldesmon, calponin, ezrin, myosin V, MAP-2 and tau, Ca^{2+}-ATPase, protein kinase C, protein tyrosine phosphatase 1B, EGF receptor, integrin B-3) are not extensively hydrolysed but cleaved at only one or a limited number of sites. Hydrolysis of these substrates is usually monitored by SDS–PAGE or Western/immunoblotting. Fluorescent tagging of MAP-2 has provided a continuous assay of a protein substrate.[5] One recent study has evaluated the significance of the P2 residue in determining the calpain susceptibility of spectrin α2.[11] Although cleavage patterns produced by calpains 1 and 2 are very similar, if not identical, the specific activities of the two calpains against individual substrates can vary widely. For example, calpain-2 has approximately twice the specific activity of calpain 1 when using casein as a substrate; the reverse is true if using α-spectrin as substrate. By comparison with papain, peptide substrates are poorly hydrolysed by calpain, but do provide convenient assays. *In vitro* calpain's preferred peptide substrates include succinyl–Leu–Leu–Val–Tyr–7 aminomethylcoumarin (AMC) and succinyl–Leu–Tyr–AMC.

In vivo, in situ

Some cell-permeant peptide substrates (e.g boc–leu–met–7–amino chloromethylcoumarin) can be used to monitor calpain activity in intact cells.[22] Antibodies specific for calpain produced cleavage products are also used to monitor specific breakdown products (e.g. of spectrin or filamin). This approach should be adaptable to other putative *in vivo* substrates.

Inhibitors

The only inhibitor truly specific for the calpains is the biological inhibitor protein calpastatin.[2] Recombinant expressed domains of calpastatin and synthetic peptide mimics are effective, specific, and commercially available. No inhibitor is known to discriminate between calpain isoforms. Numerous reversible and irreversible inhibitors are available, but their specificities vary. General inhibitors of cysteine proteinases include N-ethylmaleimide, iodoacetate, and mercurials (e.g. mersalyl). Peptide based inhibitors include epoxysuccinyl peptides (e.g. E-64 and derivatives), leupeptin, calpeptin, and MDL28179.[23] The irreversible inhibitor benzoylcarboxy-Leu–Leu–Tyr–CHN2, although also effective against cathepsins B and L, is reasonably cell permeable and has been used *in vitro* and *in situ*.[9,22] A mercaptoacrylate derivative (e.g. PD150606) provides a novel mechanism for inhibition of calpain.[12] It binds to the C-terminal domains and the crystal structure of a dimer domain VI with bound inhibitor has been solved.[24] Inhibitors with potential therapeutic use are being developed by industrial sources (e.g. Alkermes, Cephalon, Cortex, Merrell Dow, Parke-Davis, and SmithKline-Beecham) and may be useful for both biochemical and physiological investigations.

Antibodies

Numerous polyclonals have been raised against native calpains 1 and 2. Without efforts to remove cross reactive species, most sera recognize both subunits of both isoforms.[2] Antisera raised against holo-calpains can be used for Western blotting and immunolocalization. Synthetic peptides mimicking the N termini of the catalytic subunits are reasonably isoform specific,[16,25] and may be designed to be specific for the autoproteolysed calpain.[16,26] Additional epitope-specific polyclonals have been raised against a conserved segment of the catalytic domain (22) this serum is cross-reactive with most calpains tested. Antipeptide sera are excellent for Western blotting but not all are useful for immunolocalization. Several reports include monoclonal antibodies but most remain uncharacterized as to the specific epitope recognized. Commercial sources do exist for anti-calpain and/or anti-calpastatin (Chemicon, Pierce).

Genes

Genomic clones sequenced include those for the catalytic subunit from chicken calpain; catalytic subunit for calpain 1 (human chromosome 11, mouse 7); catalytic subunit calpain 2 (human chromosome 1); calpain 3 (human chromosome 15) and the shared small subunit (human chromosome 19, rat, mouse). Intron/exon boundaries are conserved between species where data are available; there is conservation of many boundaries within the family of catalytic subunits. Three *C. elegans* genes have homology related to (at least) the catalytic domain of calpain.[3,4]

cDNAs have been cloned from numerous sources. These include catalytic subunits from chicken, rabbit calpain-1 and -2, rat calpain-1, -2, and calpain-3; human calpain-1 and -2; *Drosophila* and *Schistosoma* calpains; and the small subunit from bovine, chicken, human, mouse, porcine, rabbit, and rat calpains.

Mutant phenotypes/disease states

Mutations in calpain-3 are linked to the autosomal recessive limb-girdle muscular dystrophy type 2A.[28] At least 58 unique mutations throughout the length of the gene are described at this time. No reports have yet described transgenic mice lacking calpain although this approach is currently being pursued in several laboratories.

Structure

Structures have been resolved by X-ray crystallography for the C-terminal domain of the small subunit of both rat[29] and porcine[24] calpain see Plate 8. Dimeric structures suggest that the heterodimers associate via their homologous C-terminal domains. Structures determined in the presence of calcium indicate that only the first two of the four EF hands bind calcium and identify a novel calcium-binding site immediately preceding domains IV and VI. Thus 6 moles of calcium are expected to bind to the heterodimeric enzyme. Surprisingly little change in structure occurs in the presence versus the absence of calcium for the isolated domains. Crystallization of the heterodimers of each of the ubiquitous isoforms is in progress.

References

1. Theopold, U., Piner, M., Daffre, S., Tryselius, Y., Freidrich, P., Nassel, D. R., and Hultmark, D. (1995). *Mol. Cell Biol.*, **15**, 824–34.
2. Croall, D. E. and DeMartino, G. N (1991). *Physiol. Rev.*, **71**, 813–47.
3. Kawashima, H. and Kawashima, S. (1996). *Mol. Membrane Biol.*, **13**, 217–24.
4. Suzuki, K., Sorimachi, H., Yoshizawa, T., Kinbara, K., and Ishiura, S. (1995). *Biol. Chem. Hoppe-Seyler*, **376**, 523–29.
5. Baki, A., Tompa, P. Alexa, A., Molnar, O., and Friedrich, P. (1996). *Biochem. J.*, **318**, 897–901.
6. Elce, J. S., Heagdorn, C., and Arthur, J. S. C. (1997). *J. Biol. Chem.*, **272**, 11268–75.
7. Emori, Y. and Saigo, K. (1994). *J. Biol. Chem.*, **269**, 25137–42.
8. Fox, J. E. B., Taylor, R. G., Taffarel, M., Boyles, J. K., and Goll, D. E. (1993). *J. Cell Biol.*, **120**, 1501–7.
9. Mellgren, R. L., Lu, Q.,Zhang, W., Lakkis, M., Shaw, E., and Mericle, M. (1996). *J. Biol. Chem.*, **271**, 15568–74.
10. Arora, A. S., deGroen, P., Croall, D. E., Emori, Y., and Gores, G. (1996). *J. Cell. Physiol.* **167**, 434–42.
11. Stabach, P. R., Cianci, C. D., Glantz, S. B., Zhang, Z., and Morrow, J. S. (1997). *Biochemistry*, **36**, 57–65.
12. Wang, K. K. W., Nath, R., Posner, A., Raser, K. J., Buroker-Kilgore, M., Hajimoham-madreza, I., *et al.* (1996). *Proc. Natl Acad. Sci., USA*, **93**, 6687–92.
13. Squier, M. K. T. and Cohen, J. J. (1997). *J. Immunol.*, **158**, 3690–7.
14. Wojcikiewicz, R. J. H. and Oberdorf, J. A. (1996). *J. Biol. Chem.*, **271**, 16652–5.
15. Molinari, M., Anagli, J., and Carafoli, E. (1995). *J. Biol. Chem.*, **270**, 2032–5.
16. Meyer, S. L., Bozyczko-Coyne, D., Mallya, S. K., Spais, C. M., Bihofsky, R., Kawooya, J. K., *et al.* (1996). *Biochem J.*, **314**, 511–9.
17. Sorimachi, H., Kinbara, K., Kimura, S., Takahashi, M., Ishiura, S., Sasgawa, N., Sorimachi, N., *et al.* (1995). *J. Biol. Chem.*, **270**, 31158–62.
18. Elce, J. S., Hegadorn, C., Gauthier, S., Vince, J. W., and Davies, P. L. (1995). *Prot. Engineering,* **8**, 843–8.
19. Graham Siegenthaler, K., Gauthier, S., Davies, P. L., and Elce, J. S (1994). *J. Biol. Chem.*, **269**, 30457–60.
20. Arthur, J. S. C., Gauthier, S., and Elce, J. S. (1995). *FEBS Lett.* **368**, 397–400.
21. Arthur, J. S. C. and Elce, J. S. (1996). *Biochem. J.*, **319**, 535–41.
22. Anagli, J., Hagman, J., and Shaw, E. (1992). *Biochem. J.*, **274**, 497–502.
23. Wang, K. K. W. and Yuen, P.-W. (1994). *Trends Pharm. Sci.*, **15**, 412–19.
24. Lin, G-D., Chattopadhyay, D., Maki, M., Wang, K. K. W., Carson, M., Jin, L., Yuen, P., Takano, E., Hatanaka, M., DeLuca, L. J., and Narayana, S. V. L. (1997). *Nature Struct. Biol.*, **4**, 539–47.
25. Croall, D. E., Slaughter, C. A., Wortham, H. S., Skelly, C. M., DeOgny, L., and Moomaw, C. (1992). *Biochim. Biophys. Acta*, **1121**, 47–53.
26. Saido, T. C., Suzuki, H., Yamazaki, H., Tanoue, K., and Suzuki, K. (1993). *J. Biol. Chem.*, **268**, 7422–6.
27. Spencer, M. J., Croall, D. E., and Tidball, J. G. (1995). *J. Biol. Chem.*, **270**, 10909–14.
28. Richard, I., Broux, O., Allamand, V., Fougerousse, F., Chiannilkulchai, N., Bourg, N., *et al.* (1995). *Cell*, **81**, 27–40.
29. Blanchard, H., Grochulski, P., Li, Y., Arthur, J. S. C., Davies, P. L., Elce, J. S., and Cygler, M. (1997). *Nature Struct. Biol.*, **4**, 532–8.
30. Sorimachi, H., Ishiura, S., and Suzuki, K. (1997). *Biochem. J.*, **328**, 721–32.
31. Huttenlocher, A., Palecek, S. P., Lu, Q., Zhang, W., Mellgren, R. L., Lauffenburger, D. A., Ginsburg, M. H., and Horwitz, A. F. (1997). *J. Biol. Chem.*, **272**, 32719–22.
32. Potter, D. A., Tirnauer, J. S., Jansenn, R., Croall, D. E., Hughes, C. N., Fiacco, K. A., Mier, J. W., Maki, M., and Herman, I. M. (1998). *J. Cell Biol.*, **141**, 647–62.

Dorothy E. Croall
Department of Biochemistry,
Microbiology and Molecular Biology,
University of Maine,
Orono, ME 04469–5375,
USA

α-Catenin

α-Catenin is a subunit of the cadherin–catenin adhesion complex that mediates the anchorage of classical cadherins to the actin microfilament network in epithelial cells. Based on the homology of the primary structures of α-catenin and vinculin, their interactions with the actin cytoskeleton may be regulated in a similar way. Impaired expression of α-catenin results in dysfunction of the cadherin cell–cell adhesion system, leading to invasive cell behaviour.

Protein properties

Originally α-catenin was identified in immunoprecipitation experiments as a 102 kDa protein coprecipitating with antibodies against E-cadherin.[1,2] Primary structure analysis of α-catenin revealed homology to vinculin,[3,4] a protein first described as a component of focal adhesions, and suggested the existence of a vinculin-related family of proteins that mediates cytoplasmic anchorage of cell–cell and cell-substrate adhesion molecules to the cytoskeleton. There are three regions of homology, an N-terminal, a central and a C-terminal domain; in vinculin these domains have been shown to contain binding sites for talin, α-actinin, paxillin, and actin. Initial biochemical analysis has already indicated that α-catenin represents the key component for linking the cadherin adhesion complex to the actin filament network.[5] A direct interaction of α-catenin with F-actin[6] and α-actinin[7,8] has now been reported. Thereby α-actinin may play a cooperative and/or regulatory role. Interestingly, the N-terminal half domain of the tight junction protein ZO-1 expressed ectopically in EL-cells was found specifically associated with α-catenin suggesting that ZO-1 is a cross-linker between α-catenin and actin filaments in non-epithelial cells and also presumably during initial steps of junction formation in epithelial cells.[9]

In vitro reconstitution experiments demonstrated that α-catenin is unable to bind directly to the cytoplasmic domain of cadherins, but does bind to a specific domain in either β-catenin or plakoglobin (γ-catenin),[10] thereby generating in a mutually exclusive way, two distinct cadherin–catenin complexes within a cell.[11,12] The corresponding binding site for α-catenin is located at a homologous N-terminal region of β-catenin, plakoglobin,[10,13] and armadillo.[14] Conversely the binding site for β-catenin and plakoglobin in α-catenin was shown to be identical and was mapped within the N terminus of α-catenin.[8,15,16] Results of mutational analyses suggest that α-catenin and β-catenin/plakoglobin/armadillo associate by hydrophobic interactions between two α helices.[13,16]

Several reports describe impaired expression of α-catenin in human cancer cell lines and tissues[17–19] because of gene deletion or aberrant mRNA or protein. PC9 human lung carcinoma cells deleted in the α-catenin gene form only weak cell–cell contacts, despite expressing high levels of E-cadherin and β-catenin. Transfection with α-catenin cDNA resulted in re-establishment of intercellular adhesion, accompanied by a redistribution of other junctional proteins,[20] indicating that cell–cell adhesion may not only be impaired by mutations of E-cadherin itself. Ov2008, a human ovarian carcinoma

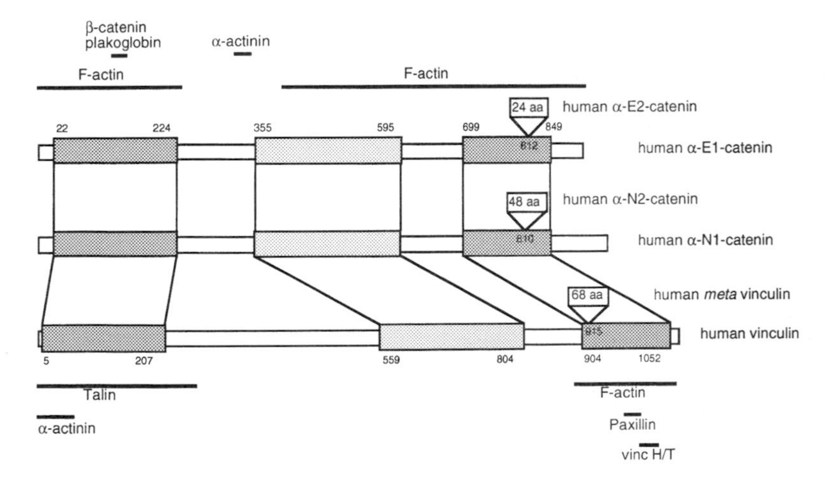

Figure 1. Schematic comparison of human α-E- and α-N-catenin with human vinculin (modified from ref. 24). Regions of homology between α-catenin and vinculin are shaded. Insertions generated by alternative splicing events are indicated. Binding sites for interaction partners are presented at the top for α-catenin and at the bottom for vinculin (vinc H/T: vinculin intramolecular head to tail association).

derived cell line, expresses a mutant α-catenin with the β-catenin binding site deleted.[21] Similarly, re-expression of full-length α-catenin restored epithelial cell morphology and was shown to affect growth and tumorigenicity, identifying α-catenin as a growth regulatory and candidate tumor suppressor gene[21]. Expression of an E-cadherin–α-catenin fusion protein in L-cells was sufficient to induce cell adhesion activity, although cell motility and the downregulation of adhesion during cytokinesis were significantly suppressed.[22]

A subtype of α-catenin, termed α-N-catenin was found associated with N-cadherin and is mainly expressed in the nervous system but also is able to associate with E-cadherin, as shown by transfection experiments in cell lines.[23]

The classical human α-catenin (α-E-catenin) has two isoforms, generated by alternative splicing, as does mouse α-N-catenin. These exhibited an insertion in the C-terminal domain of 24 amino acids in α-E-catenin[24] and 48 amino acids in α-N-catenin.[25] The biological function of these insertions is at present unknown.

During development, α-E-catenin is already provided maternally as mRNA and protein, both in *Xenopus*[26] and mouse,[27] and later exhibits a rather ubiquitous distribution, appearing in many cell types.[28] α-N-catenin is first detected in nerve fibres of cranial and dorsal root ganglia and also in early neurones in the neural tube.[25] An α-E-catenin gene trap mutation has clearly defined its function in mouse preimplantation embryos;[29] homozygous mutant embryos show a disruption of the trophectoderm epithelium at the blastocyst stage (Fig. 2), a phenotype which parallels the defect observed in E-cadherin mutant embryos. Similarly, depletion of maternal α-catenin mRNA in *Xenopus* embryos results in disaggregated blastulae.[30]

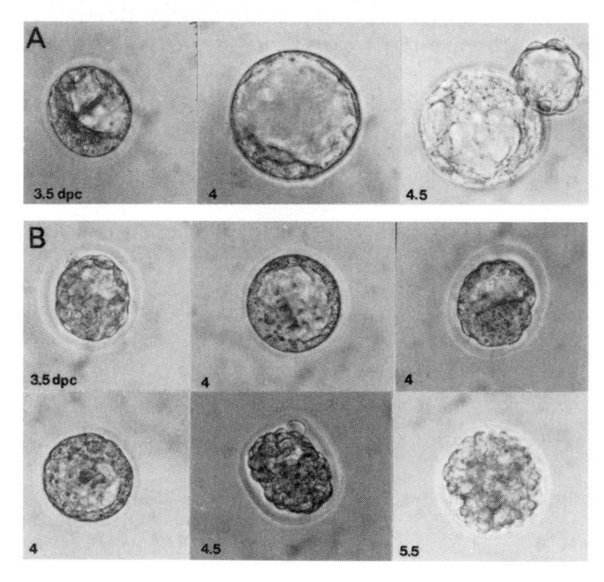

Figure 2. Mutant phenotype of α-E-catenin blastocysts. (a) Three time points of *in vitro* development of a wild-type blastocyst. 3.5 d.p.c., early blastocysts; 4 d.p.c., expanded blastocyst; 4.5 d.p.c., hatching blastocyst. (b) Six time points of *in vitro* development of an abnormally developing embryo derived from a cross between α-E-catenin heterozygous animals (from ref. 29).

Purification

α-Catenins can be purified by immunoprecipitation and affinity column chromatography using poly- or monoclonal antibodies. Soluble recombinant α-catenin has also been obtained using bacterial expression systems.[10]

Activities

The major biological function of α-catenin consists in linking the cadherin adhesion complex to the actin-based cytoskeleton. Changes in α-catenin mRNA levels or gene deletions have been observed in carcinoma cells, suggesting that α-catenin is subjected to alterations during tumorigenesis.

Genes

The complete coding sequences for α-E-catenin from mouse (GenBank/EMBL X59990), human (GenBank/EMBL L23805 and U03100 for α-E1 and α-E2-catenin respectively) and *Drosophila* (GenBank/EMBL D13866) are available. In addition, a human α-E-catenin pseudogene[31] has been identified (GenBank Z37994). α-N-catenin has been isolated from chicken (GenBank/EMBL D11090), mouse (GenBank/EMBL D25281 and D25282 for α-N1-catenin and α-N2-catenin respectively) and human[32] (GenBank/EMBL M94151). The gene coding for α-E-catenin (*Catna 1*) localizes to mouse chromosome 18[33] and human chromosome 5q31[31,34] and the α-E-catenin pseudogene was mapped to human chromosome 5q22.[31] The α-N-catenin gene has been assigned to mouse chromosome 6[25,32] and human chromosome 2.[32]

References

1. Ozawa, M., Baribault, H., and Kemler, R. (1989). *EMBO J.*, **8**, 1711–17.
2. Nagafuchi, A. and Takeichi, M. (1989). *Cell Regul.*, **1**, 37–44.
3. Herrenknecht, K., Ozawa, M., Eckerskorn, C, Lottspeich, F., Lenter, M. and Kemler, R. (1991). *Proc. Natl. Acad. Sci. USA*, **88**, 9156–60.
4. Nagafuchi, A., Takeichi, M., and Tsukita, S. (1991). *Cell*, **65**, 849–57.
5. Ozawa, M. and Kemler, R. (1989). *J. Cell Biol.*, **116**, 989–96.
6. Rimm, D. L., Koslov, E. R., Kebriaei, P., Cianci, C. D., and Morrow, J. S. (1995). *Proc. Natl. Acad. Sci. USA*, **92**, 8813–17.
7. Knudsen, K. A., Soler, A. P., Johnson, K. R., and Wheelock, M. J. (1995). *J. Cell Biol.*, **130**, 67–77.
8. Nieset, J. E., Redfield, A. R., Jin, F., Knudsen, K. A., Johnson, K. R., and Wheelock, M. J. (1997). *J. Cell Sci.*, **110**, 1013–22.
9. Itoh, M., Nagafuchi, A., Moroi, S., and Tsukita, S. (1997). *J. Cell Biol.*, **138**, 181–92.

10. Aberle, H., Butz, S., Stappert, J., Weissig, H., Kemler, R., and Hoschützky, H. (1994). *J. Cell Sci.*, **107**, 3655–63.
11. Butz, S. and Kemler, R. (1994). *FEBS Lett.*, **355**, 195–200.
12. Näthke, S. I., Hinck, L., Swedlow, J. R. Papkoff, J., and Nelson, W. J. (1994). *J. Cell Biol.*, **125**, 1341–52.
13. Aberle, H., Schwarz, H., Hoschützky, H., and Kemler, R. (1996). *J. Biol. Chem.*, **271**, 1520–6.
14. Pai, L. M., Kirkpatrick, C., Blanton, J., Oda, H., Takeichi, M., and Peifer, M. (1996). *J. Biol. Chem.*, **271**, 32411–20.
15. Obama, H. and Ozawa, M. (1997). *J. Biol. Chem.*, **272**, 11017–20.
16. Huber, O., Krohn, M., and Kemler, R. (1997). *J. Cell Sci.*, **110**, 1759–65.
17. Shimoyama, Y., Nagafuchi, A., Fujita, S., Gotoh, M., Takeichi, M., Tsukita, S., and Hirohashi, S. (1992). *Cancer Res.*, **52**, 5770–4.
18. Morton, R. A., Ewing, C. M., Nagafuchi, A., Tsukita, S., and Isaacs, W. B. (1993). *Cancer Res.*, **53**, 3585–90.
19. Kadowaki, T., Shiozaki, H., Inoue, M., Tamura, S., Oka, H., Doki, Y., *et al.* (1994). *Cancer Res.*, **54**, 291–6.
20. Watabe, M., Nagafuchi, A., Tsukita, S., and Takeichi, M. (1994). *J. Cell Biol.*, **127**, 247–56.
21. Bullions, L. C., Notterman, D. A., Chung, L. S., and Levine, A. J. (1997). *Mol. Cell. Biol.*, **17**, 4501–8.
22. Nagafuchi, A., Ishihara, S., and Tsukita, S. (1994). *J. Cell Biol.*, **127**, 235–45.
23. Hirano, S., Kimoto, N., Shimoyama, Y., Hirohashi, S., and Takeichi, M. (1992). *Cell*, **70**, 293–301.
24. Rimm, D. L., Kebriaei, P., and Morrow, J. S. (1994). *Biochem. Biophys. Res. Commun.*, **203**, 1691–9.
25. Uchida, N., Shimamura, K., Miyatani, S., Copeland, N. G., Gilbert, D. J., Jenkins, N. A., and Takeichi, M. (1994). *Dev. Biol.*, **163**, 75–85.
26. Schneider, S., Herrenknecht, K., Butz, S., Kemler, R., and Hausen, P. (1993). *Development*, **118**, 629–40.
27. Ohsugi, M., Hwang, S.-Y., Butz, S., Knowles, B. B., Solter, D., and Kemler, R. (1996). *Dev. Dynamics.*, **206**, 391–402.
28. Butz, S. and Larue, L. (1995). *Cell Adhesion Commun.* **3**, 337–52.
29. Torres, M., Stoykova, A., Huber, O., Chowdhury, K., Bonaldo, P., Mansouri, A., *et al.* (1997). *Proc. Natl. Acad. Sci. USA*, **94**, 901–6.
30. Kofron, M., Spagnuolo, A., Klymkowsky, M., Wylie, C.,and Heasman, J. (1997). *Development*, **124**, 1553–60.
31. Nollet, F., Hengel, Y., Berx, G., Molemans, F.,and Roy, F. (1995). *Genomics*, **26**, 410–13.
32. Claverie, J. M., Hardelin, J.P., Legouis, R., Levilliers, J., Bougueleret, L., Mattei, M. G. and Petit, C. (1993). *Genomics*, **15**, 13–20.
33. Guénet, J.-L., Simon-Chazottes, D., Ringwald, M.,and Kemler, R. (1995). *Mamm Genome*, **6**, 363–6.
34. Furukawa, Y., Nakatsuru, S., Nagafuchi, A., Tsukita, S., Muto, T., Nakamura, Y., and Horii, A. (1994). *Cytogenet. Cell Genet.*, **65**, 74–8.

■ *Otmar Huber and Rolf Kemler*
Max-Planck-Institut für Immunbiologie,
Stübeweg 51, 79108 Freiburg,
Germany

Cysteine-rich protein family

Members of the cysteine-rich protein family are highly related proteins that exhibit two tandemly arrayed LIM domains and are associated with the actin cytoskeleton. Members of this family have been implicated in muscle differentiation and in cell proliferation. Because the LIM domain is a protein-binding interface, CRP family members are thought to function as adapter molecules that couple proteins in functional complexes.

■ Protein properties

CRP1, CRP2, and the Muscle LIM Protein MLP, also referred to as CRP3, are the three members of the cysteine-rich protein family described in vertebrates[1] (Fig. 1). These three closely related proteins exhibit two copies of a structural motif called the LIM domain. The LIM domain has been shown to coordinate two zinc atoms[2] and to

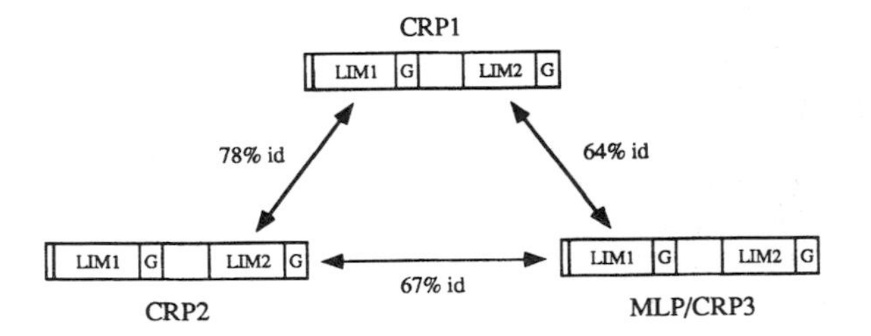

Figure 1. Structure and amino acid sequence comparison of the three chicken CRP family members. The percentage of identity (id) is indicated as well as the localization of the LIM domains (LIM1 and LIM2) and the glycine-rich repeats (G).

mediate specific protein–protein interactions.[3,4] Each LIM domain is followed by a conserved glycine-rich repeat. Each CRP family member in chicken is between 192 and 194 amino acids in length. Avian CRP1 and CRP2 have been isolated and they both exhibit an apparent molecular mass of 23–24 kDa on SDS–polyacrylamide gels.

CRP1, previously referred to as the chicken CRP (cCRP), is a developmentally regulated LIM domain protein, expressed prominently in smooth muscle cells.[5] In fibroblasts, CRP1 is localized along the actin stress fibres and in some adhesion plaques[5,6] (Fig. 2). It has been shown that overexpression of CRP1 in C2 cells in culture promotes their myogenic differentiation[7] and that CRP1 expression is strongly decreased in transformed fibroblasts.[1] CRP1 is encoded by an immediate early response gene that exhibits an expression profile similar to that of c-*myc* after serum induction.[8]

CRP2 is nearly 80 per cent identical in amino acid sequence to CRP1.[1] CRP2 is a developmentally regulated protein, preferentially expressed in arterial smooth muscle cells.[9] The level of transcripts encoding CRP2 has been shown to be reduced dramatically in oncogenically and chemically transformed fibroblasts and to be correlated directly with the acquisition of the transformed phenotype.[1,10]

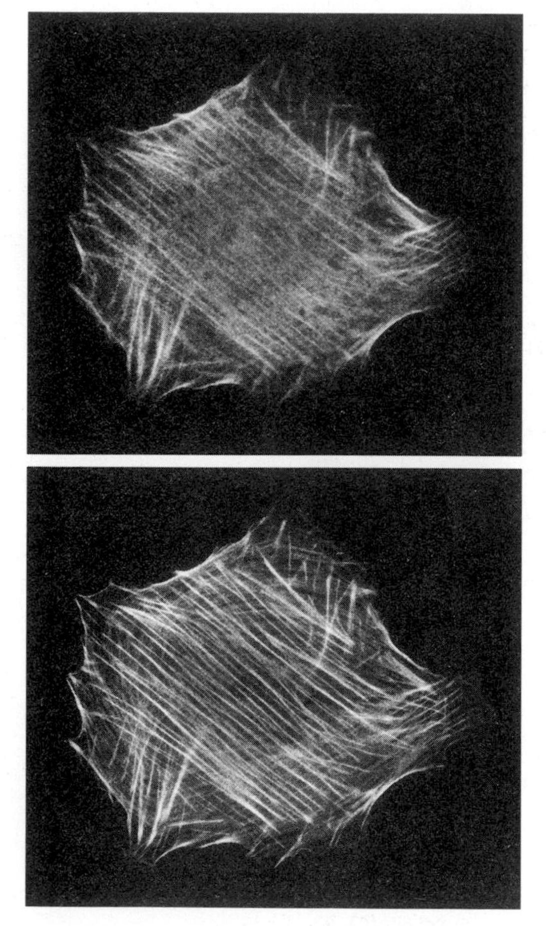

Figure 2. Immunolocalization of CRP1 along the actin stress fibres in a cultured chicken embryo fibroblast (upper panel). Pattern of filamentous actin in the same cell (lower panel).

MLP/CRP3, expression of which is enriched in striated and cardiac muscle, is an essential positive regulator of myogenic differentiation.[7] Overexpression of MLP/CRP3 potentiates the myogenic differentiation of cultured C2 cells and suppression of MLP/CRP3 mRNA expression in C2 cells, by antisense RNA technology, prevents muscle differentiation. It has been shown that MLP/CRP3 accumulates along actin-based filaments in the cytosol,[11,12] one report also suggests that it concentrates in cell nuclei.[11] Furthermore, MLP/CRP3-deficient mice exhibit a disruption of myofibrils leading to alteration of cardiac cytoarchitectural organization and heart failure.[13]

Two members of the CRP superfamily, referred to as muscle LIM proteins or Mlps, have been identified in *Drosophila*.[7,14] Mlp60A and Mlp84B exhibit LIM–glycine repeats very similar to those in vertebrate CRPs. A single LIM–glycine repeat is found in Mlp60A; five tandemly arrayed LIM–glycine repeats are found in Mlp84B. Both Mlp60A and Mlp84B gene expression is muscle-specific, and is developmentally regulated throughout the *Drosophila* life cycle.[14] Moreover, *Drosophila* Mlps localize to the actin cytoskeleton when they are expressed in vertebrate cells.[14]

Purification

CRP1 can be extracted in low ionic strength buffer from either fresh or frozen chicken gizzard smooth muscle and can be purified after ammonium sulphate fractionation, followed by chromatography on DEAE–cellulose, phenyl-sepharose CL-4B, and sepharose CL-6B columns.[5] The purified protein is a monomer with an apparent molecular mass of 23 kDa and can be detected on high percentage SDS–polyacrylamide gels, as shown in Fig. 3. Approximately 20 mg of CRP1 can be purified from 300 g of fresh chicken gizzard. CRP1 can also be purified in a soluble form after bacterial expression.[2]

Activities

An *in vitro* interaction has been demonstrated between CRP1 and the cytoskeletal LIM domain protein, zyxin.[6] More precisely, CRP1 interacts with the N-terminal LIM domain of zyxin.[4] CRP1 and zyxin exhibit overlapping subcellular distributions in fibroblasts along the actin stress fibres near where they terminate at adhesion plaques.[5,6] CRP1 has also been shown to interact *in vitro* and *in vivo* with the actin cross-linking protein, α-actinin.[15] CRP1 binds specifically to α-actinin *in vitro* with a moderate affinity (kd ~10^{-6}M). The binding site for CRP1 on α-actinin has been located to the 27 kDa globular head of α-actinin, whereas α-actinin interacts with the N-terminal part of CRP1 containing the N-terminal LIM

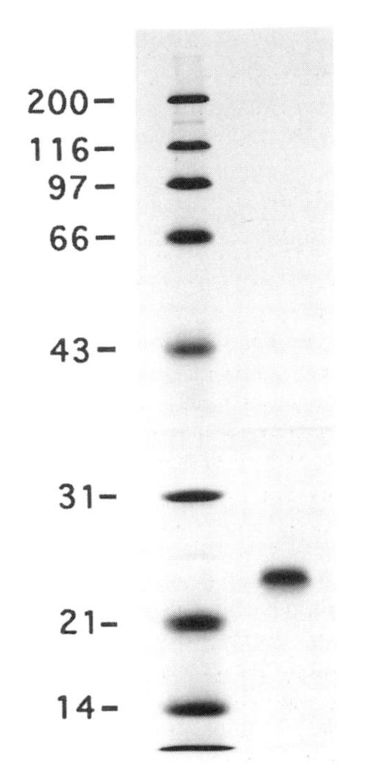

Figure 3. Coomassie-blue stained gel showing CRP1 isolated from chicken gizzard smooth muscle.

domain of the protein followed by the first glycine-rich repeat.

Recently, CRP2 and MLP/CRP3 have been shown to interact *in vitro* with zyxin and α-actinin.[12]

An interaction between MLP/CRP3 and three muscle-specific basic helix–loop–helix transcription factors. MyoD, MRF4, and myogenin, has been reported.[16] The N-terminal LIM domain of MLP/CRP3 and the basic domain of MyoD are required for the association of MLP/CRP3 with MyoD.

Antibodies

Two polyclonal antibodies against chicken gizzard CRP1 (B31 and B32) were used by Western immunoblot and indirect immunofluorescence to characterize CRP1 in chicken embryo fibroblasts.[6] Isoform specific antibodies directed against avian CRP family members have been generated.[12] No commercial antibodies against CRP family members are available.

Genes

Nucleotide sequences available: quail CRP1[1] (accession number Z28333), chicken CRP1[5] (X73831), human CRP1[8] (M76375–M76378), quail CRP2[10] (Z21643), chicken CRP2[1] (X84264), rat CRP2[9] (U44948), human CRP2[9] (U46006), *Drosophila* Mlp60A[7,14] (X91244), *Drosophila* Mlp84B[14] (X91245), rat MLP/CRP3[7] (X81193).

The genes encoding CRP family members are given the designation, *CSRP*. The human *CSRP1*, *CSRP2* and *CSRP3* genes are localized to chromosome subbands 1q24–1q32, 12q21.1, and 11p15.1, respectively.[17–19]

Mutant phenotype/disease states

MLP/CRP3-deficient mice have been generated.[13] These mice develop major alterations in the cytoarchitecture of cardiac muscle cells and thus may provide an animal model for studies of dilated cardiomyopathy and heart failure.

Structure

The three-dimensional solution structure of the C-terminal LIM domain of the chicken CRP1 and the quail CRP2 have been defined by nuclear magnetic resonance spectroscopy.[20,21] The global fold of the C-terminal LIM domain of CRP1 is very similar to that of the C-terminal LIM domain of CRP2 and it shows a novel three-dimensional structure that contains two zinc atoms bound independently in two tightly packed CCHC and CCCC modules. Interestingly, the C-terminal metal binding module folds in a manner very similar to that observed for the DNA-binding modules of the steroid hormone receptor and GATA-1 transcription factors. The structural analysis raises the possibility that the LIM domain may function in nucleic acid binding in addition to its previously described function in protein–protein interaction.

References

1. Weiskirchen, R., Pino, J. D., Macalma, T., Bister, K., and Beckerle, M. C. (1995). *J. Biol. Chem.*, **270**, 28946–54.
2. Michelsen, J. W., Schmeichel, K. L., Beckerle, M. C., and Winge, D. R. (1993). *Proc. Natl Acad. Sci., USA*,. **90**, 4404–8.
3. Feuerstein, R., Wang, X., Song, D., Cooke, N. E., and Liebhaber, S. A. (1994). *Proc. Natl Acad. Sci., USA*,. **91**, 10655–9.
4. Schmeichel, K. L. and Beckerle, M. C. (1994). *Cell*, **79**, 211–19.
5. Crawford, A. W., Pino, J. D., and Beckerle, M. C. (1994). *J. Cell Biol.*, **124**, 117–27.
6. Sadler, I., Crawford, A. W., Michelsen, J. W., and Beckerle, M. C. (1992). *J. Cell Biol.*, **119**, 1573–87.
7. Arber, S., Halder, G., and Caroni, P. (1994). *Cell*, **79**, 221–31.
8. Wang, X., Lee, G., Liebhaber, S. A., and Cooke, N. E. (1992). *J. Biol. Chem.*, **267**, 9176–84.
9. Jain, M. K., Fujita, K. P., Hsieh, C.-M., Endege, W. O., Sibinga, N. E. S., Yet, S.-F., *et al.* (1996). *J. Biol. Chem.*, **271**, 10194–99.
10. Weiskirchen, R. and Bister, K. (1993). *Oncogene*, **8**, 2317–24.
11. Arber, S. and Caroni, P. (1996). *Genes Dev.*, **10**, 289–300.
12. Louis, H. A., Pino, J. D., Schmeichel, K. L., Pomiès, P., and Beckerle, M. C. (1997). *J. Biol. Chem.*, **272**, 27484–91.
13. Arber, S., Hunter, J .J., Ross, J., Hongo, M., Sansig, G., Borg, J., *et al.* (1997). *Cell*, **88**, 393–403.
14. Stronach, B. E., Siegrist, S. E., and Beckerle, M. C. (1996). *J. Cell Biol.*, **134**, 1179–95.
15. Pomiès, P., Louis, H. A. and Beckerle, M. C. (1997) *J. Cell Biol.*, **139**, 157–68.

16. Kong, Y., Flick, M. J., Kudla, A. J., and Konieczny, S. F. (1997). *Mol. Cell. Biol.*, **17**, 4750–60.
17. Wang, X., Ray, K., Szpirer, J., Levan, G., Liebhaber, S. A., and Cooke, N. E. (1992). *Genomics*, **14**, 391–7.
18. Weiskirchen, R., Erdel, M., Utermann, G., and Bister, K. (1997) *Genomics*, **44**, 83–93.
19. Fung, Y. W., Wang, R. X., Heng, H. H. Q., and Liew, C. C. (1995). *Genomics*, **28**, 602–3.
20. Perez-Alvarado, G. C., Miles, C., Michelsen, J. W., Louis, H. A., Winge, D. R., Beckerle, M. C. and Summers, M. F. (1994). *Nature Struct. Biol.*, **1**, 388–98.
21. Konrat, R., Weiskirchen, R., Krautler, B., and Bister, K. (1997). *J. Biol. Chem.*, **272**, 12001–7.

Pascal Pomiès and Mary C. Beckerle
Department of Biology,
University of Utah,
Salt Lake City, UT 84112,
USA

Dystrophin/utrophin

Dystrophin and utrophin are homologous members of the spectrin superfamily of cytoskeletal proteins. They are thought to form a structural linkage between the actin-based cytoskeleton and the extracellular matrix, via a multimeric membrane-associated glycoprotein complex, which acts to maintain the integrity of the cell membrane. The absence of dystrophin in humans causes a lethal progressive myopathy, Duchenne muscular dystrophy, but the absence of utrophin has not been associated with any disease.

Homologous proteins

Utrophin: dystrophin-related protein (DRP). The dystrophin and utrophin genes are subject to complex transcriptional and splicing controls, both from 5′ promoters and from promoters within the genes themselves, so that a variety of tissue specific isotypes and shorter isoforms may be expressed[1,2] (see Table 1). All except one of the isoforms so far identified share the characteristically unique C-terminal domains of dystrophin and utrophin. The tissue distribution associated with each isoform differs and their functions remain unclear. Other more distantly related proteins are dystrobrevin,[3] found predominantly at the neuromuscular junction, and the so-called DRP2 protein.[4]

Protein properties

Dystrophin and utrophin are homologous, cytoskeletal proteins with 46 per cent sequence identity, consisting of four major domains[5] (Fig. 1).

Table 1 Comparison of dystrophin and utrophin

	Dystrophin	Utrophin
Human chromosome location	Xp21	6q24 (mouse chromosome 10)
Locus	*DMD*	*UTRN*
Gene size	2.5 Mb	0.9 Mb
Protein size	427 kDa	395 kDa
Isoforms	Full length isotypes C-dystrophin M-dystrophin P-dystrophin Others Dp260 Dp140 Dp116 Dp71 Dp45	G-utrophin (113 kDa) 62 kDa 'apo-utrophin' (proposed to include N terminus)
Tissue distribution of full length protein	Muscle and cells of neuronal origin	Ubiquitous expression but especially in vascular tissue
Distribution in normal muscle	Throughout the sarcolemma and in the troughs of the neuromuscular junctional folds; myotendinous junction	At the crests of the neuromuscular junctional folds; myotendinous junction
Pathological absence	Duchenne muscular dystrophy	None known

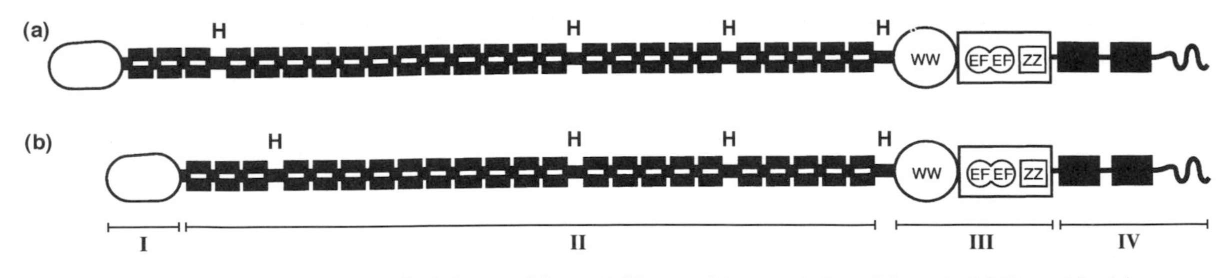

Figure 1. The domain organization of (a) dystrophin, and (b) utrophin, consisting of domain I followed by 24 (dystrophin), and 22 (utrophin) triple helical repeats, interrupted by hinge regions (H); domain III contains the WW, EF hand, and ZZ motifs, and domain IV contains 2 regions of predicted α-helical coiled coil.

Domain I, the 30 kDa N-terminal domain, shows 67 per cent sequence identity between dystrophin and utrophin and a high degree of sequence similarity to other members of the spectrin superfamily of F-actin binding proteins. This domain consists of two calponin-homology (CH) domains[32] — these are structural motifs, found in a number of functionally unrelated proteins. Both CH domains of dystrophin and utrophin are required for full protein function. Recombinant domain I of both proteins co-sediments with F-actin[6,7] and has a higher affinity for non-muscle actin than for skeletal actin, suggesting that both dystrophin and utrophin bind to the submembranous actin cytoskeleton rather than directly to the contractile apparatus in muscle. NMR studies,[8] deletion analysis,[9] and site-directed mutagenesis of this domain have identified three highly conserved sequences, known as actin binding sites 1, 2, and 3, which are strongly implicated in the interaction with actin, although the exact nature of the interaction remains unclear.

Domain II, which accounts for 75 per cent of the molecules, consists of a series of spectrin-like[10] nested triple helical repeats which are proposed to form an extended, flexible region, interrupted by several proline-rich hinges. Twenty-four of these triple helical repeats are found in dystrophin and 22 in utrophin, each about 120 amino acids long. Comparison of all the repeats in dystrophin and utrophin has shown that sequence conservation between repeats, in contrast to those in spectrin, is fairly minimal and insertions and irregularities are a feature of these regions.[11] These observations strongly suggest that, unlike the repeats in α-actinin and β-spectrin, those in dystrophin and utrophin will not support dimerization of the molecules. Physical and biochemical analyses of expressed regions of the domain II from dystrophin indicate that they are monomeric,[12,33] which further supports this conclusion. Furthermore, repeats 12–16 from dystrophin have actin-binding capabilities,[13] suggesting a new model for the interaction of dystrophin (and possibly utrophin) with F-actin (Fig. 2).

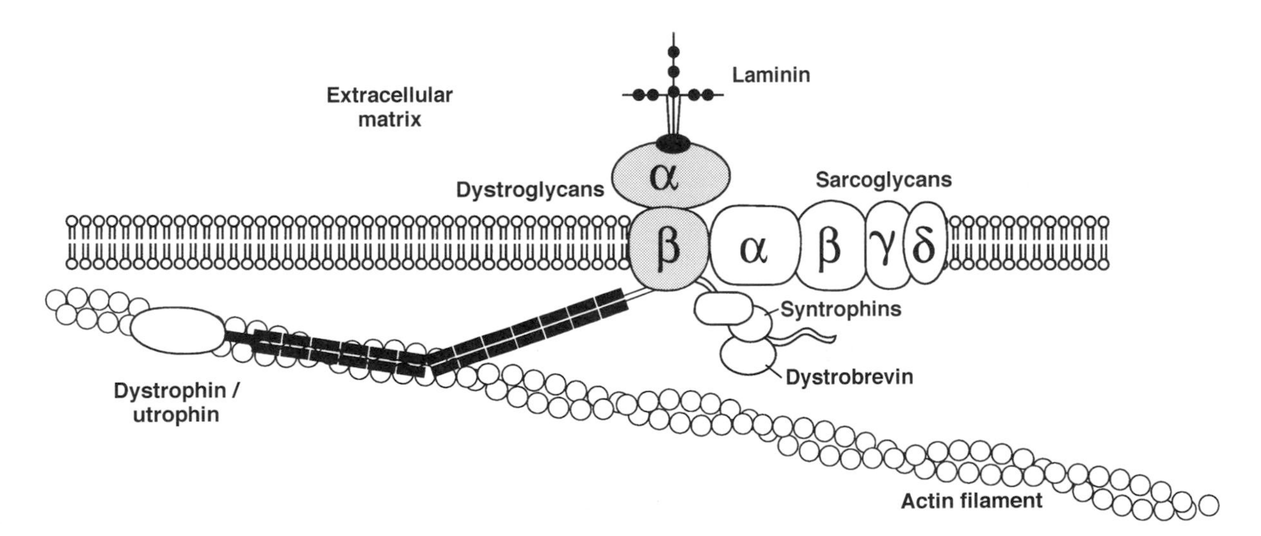

Figure 2. The current model for the association of dystrophin/utrophin with F-actin. Domain I, and a portion of domain II are thought to bind to the submembranous network of F-actin. Domain III interacts with β-dystroglycan, which in turn links through the membrane to laminins in the extracellular matrix via α-dystroglycan. Also present in the glycoprotein complex are the sarcoglycans. Domain IV is potentially involved in a number of interactions with members of the syntrophin family, and with dystrobrevin.

Domains III and IV make up the last 600 amino acids and are extremely conserved between utrophin and dystrophin. Domain III, a cysteine-rich domain, has strong similarities to the carboxyl-terminal domain of α-actinin, and domain IV is unique to dystrophin and its related proteins. Almost all the information about these domains is derived from sequence comparisons and little functional information is currently available; the sequence motifs which have been identified are assumed to be involved in protein–protein interactions. In domain III, the first recognizable sequence motif is the WW motif, characterized by two highly conserved tryptophans and found in a variety of apparently unrelated proteins.[14] The structure of the WW domain from human Yes-associated protein (YAP) has been solved and is known to bind proline-rich peptides *in vitro*.[15] Next, two putative EF hands have been identified, which in other proteins have been shown to bind calcium. However, those in dystrophin seem to lack sufficient coordinating ligands to be able to chelate the metal ion, although recombinant fragments of the domain appear to do so *in vitro*.[16] Finally, a putative zinc binding motif (ZZ) has been identified with some sequence similarities to the DNA-binding zinc finger family of proteins.[17] The metal ion may play some fundamental role in the protein structure but the nature of this role in the cell is unknown. Domain IV contains two regions of predicted coiled coil separated by a proline-rich linker, the first of which may bind the syntrophins. The involvement of these coiled coil regions in protein–protein interactions is thought to be favoured by hydrophobic forces in a manner analogous to that seen in leucine zipper proteins.[18]

Dystrophin is localized to the cytoplasmic face of the sarcolemma in muscle, at costameres and myotendinous junctions in skeletal muscle, and at Z discs in cardiac muscle. Utrophin, its autosomal homologue, is expressed ubiquitously, also at the plasma membrane, but is localized particularly to the neuromuscular and myotendinous junctions in adult skeletal muscle and at the intercalated discs in cardiac muscle. These locations suggest that both proteins may act to buffer the muscle cell membrane against the stresses that result from contraction. Indeed, studies on whole muscle provide evidence that dystrophin (and by extension, utrophin) does act in this way.[19,20] However, the fact that these molecules are expressed in tissues other than muscle implies that the situation is rather more complex than we are currently aware.

At the membrane, dystrophin and utrophin bind β-dystroglycan, a component of the glycoprotein complex (consisting of dystroglycans and sarcoglycans) via sequences in domain III.[21,22] β-dystroglycan binds α-dystroglycan which, in turn, interacts with the extracellular matrix family of laminins, thus forming the buffering link between the F-actin cytoskeleton and the surrounding tissue. This linkage is known to be important for maintaining the integrity of the sarcolemma, since the absence of dystrophin causes both muscle cell pathology[23] (with symptoms which suggest a defect in the sarcolemma) and reduced levels of the other components of the glycoprotein complex. Dystrophin and utrophin are also known to bind to a family of cytoplasmic proteins called the syntrophins, and also to dystrobrevin, via domain IV.

Expression of both proteins appears to be developmentally regulated and reciprocal, suggesting that utrophin may be the fetal isoform of dystrophin.[2] If this is the case, it might explain the apparent upregulation of utrophin throughout the sarcolemma of regenerating muscle in several inflammatory myopathies, although the mechanism of this upregulation is uncertain.

Purification

Both dystrophin and utrophin purification depends primarily on their attachment to the glycoprotein complex, permitting isolation of the whole complex by lectin affinity chromatography. Dystrophin has been purified from skeletal muscle[24] and utrophin from lung,[25] where there are high levels due to its expression in endothelial cells. The quantity obtained of each is limited because they make up such a small proportion of the total tissue protein – dystrophin constitutes 0.002 per cent of the total muscle protein. Most workers have therefore chosen to express individual domains in bacteria to obtain sufficient quantities for detailed biochemical and structural analysis.

Antibodies

A number of monoclonal antibodies are available commercially against all regions of both dystrophin and utrophin, from Sigma (anti-dystrophin D8043 and D8168) and Novacastra (anti-dystrophin NCL-DYS1, NCL-DYS2, NCL-DYS3 and anti-utrophin NCL-DRP1, NCL-DRP2). Individual groups have also generated their own panels of antibodies (monoclonal and polyclonal) to various portions of these proteins.[26–28]

Genes

Full cDNA sequences are available for human dystrophin (EMBL M18533), chicken dystrophin (X13369), *Torpedo* dystrophin (M37645), and mouse dystrophin (SwissProt P11531). The cDNA sequences for human (X69086) and mouse utrophin (U43520) are also available.

Mutant phenotypes/disease states

Duchenne muscular dystrophy (DMD) is a lethal, X-linked muscle wasting disorder resulting from the absence of dystrophin.[23] With a frequency of 1 in 3500 live born males, it is the second most common human inherited disorder; the size of the gene is thought to account for the high sporadic mutation rate. Becker muscular dystrophy (BMD) is a rarer allelic disorder with similar, but milder symptoms, in which patients express truncated and thus semi-functional dystrophin. This explanation of the difference in phenotypes in DMD and BMD, known as the frameshift hypothesis, provides a correlation between patient genotype and phenotype in more than 90 per cent of cases.

Mouse (*mdx*), cat (hypertrophic feline muscular dystrophy, HFMD), and dog (canine X-linked muscular dystrophy, CXMD) animal models for DMD exist; all are dystrophin-deficient but have distinct clinical features. The cat and dog are severely affected, the dog in particular showing clear parallels with the human disease, but the *mdx* mouse shows a less severe phenotype and has a normal life expectancy. The mouse model has, however, been the focus of most study (in the disease context) as a trial organism for potential therapies and as the subject of transgenic experiments.

Therapeutic strategies, developed over the ten years since the primary defect in DMD and *mdx* was identified,[1] have included dystrophin-positive myoblast injection, direct injection of DNA coding for dystrophin, and the use of viral vectors for delivery of a minigene (encoding the shortest functional molecule possible). However, the observed upregulation of utrophin and its localisation at the sarcolemma of dystrophin-deficient muscle fibres, which may prevent loss of the glycoprotein complex by binding in the place of dystrophin, suggests that artificial upregulation of utrophin may be a potential therapy. In fact, the generation of an *mdx* transgenic mouse overexpressing utrophin has led to the amelioration of its symptoms.[29] The search is now on for the means by which this up-regulation could be achieved in humans.

Absence of utrophin has not been associated with any human or animal diseases and utrophin null mutant mice are reported to be virtually normal,[30,31] suggesting an intriguing redundancy in the molecules involved in the architecture of cell membranes. However, mice with both dystrophin and utrophin genes disrupted display a pattern of symptoms very similar to those seen in DMD patients.[34,35]

Structure

Not available.

References

1. Ahn, A. H. and Kunkel, L. M. (1993). *Nature Genet.*, **3**, 283–91.
2. Blake, D. J., Tinsley, J. M., and Davies, K. E. (1996). *Brain Pathology*, **6**, 37–47.
3. Blake, D. J., Nawrotzki, R., Peters, M. F., Froehner, S. C., and Davies, K. E. (1996). *J. Biol. Chem.*, **271**, 7802–10.
4. Roberts, R. G., Freeman, T. C., Kendall, E., Vetrie, D. L. P., Dixon, A. K., Shaw-Smith, C., *et al.* (1996). *Nature Genet.*, **13**, 223–6.
5. Winder, S. J., Knight, A. E., and Kendrick-Jones, J. (1997). In *Dystrophin: gene, protein and cell biology*, (ed. J. A. Lucy, and S. C. Brown), pp. 27–53. Cambridge University Press, Cambridge.
6. Fabbrizio, E., Bonet-Kerrache, A., Leger, J. L., and Mornet, D. (1993). *Biochemistry*, **32**, 10457–63.
7. Winder, S. J., Hemmings, L., Maciver, S. K., Bolton, S. J., Tinsley, J. M., Davies, K. E., *et al.* (1995). *J. Cell Sci.*, **108**, 63–71.
8. Levine, B. A., Moir, A. J. G., Patchell, V. B., and Perry, S. V. (1992). *FEBS*, **298**, 44–8.
9. Corrado, K., Mills, P. L., and Chamberlain, J. S. (1994). *FEBS*, **344**, 255–60.
10. Yan, Y., Winograd, E., Viel, A., Cronin, T., Harrison, S. C., and Branton, D. (1993). *Science*, **262**, 2027–30.
11. Winder, S. J., Gibson, T. J., and Kendrick-Jones, J. (1995). *FEBS Lett.*, **369**, 27–33.
12. Kahana, E., Flood, G., and Gratzer, W. B. (1997). *Cell Motil. Cytoskel.*, **36**, 246–52.
13. Rybakova, I. N., Amann, K. J., and Ervasti, J. M. (1996). *J. Cell Biol.*, **135**, 661–72.
14. Einbond, A. and Sudol, M. (1996). *FEBS Lett.*, **384**, 1–8.
15. Macias, M. J., Hyvönen, M., Baraldi, E., Schultz, J., Sudol, M., Saraste, M., and Oschikinat, H. (1996). *Nature*, **382**, 646–9.
16. Milner, R. E., Busaan, J., and Michalak, M. (1992). *Biochem. J.*, **288**, 1037–44.
17. Ponting, C. P., Blake, D. J., Davies, K. E., Kendrick-Jones, J., and Winder, S. J. (1996). *TIBS*, **21**, 11–13.
18. Blake, D. J., Tinsley, J. M., Davies, K. E., Knight, A. E., Winder, S. J., and Kendrick-Jones, J. (1995). *TIBS*, **20**, 133–5.
19. Pasternak, C. Wong, S., and Elson, E. L. (1995). *J. Cell Biol.*, **128**, 255–361.
20. Petrof, B. J., Shrager, J. B., Stedman, H. H., Kelly, A. M., and Sweeney, H. L. (1993). *Proc. Natl Acad. Sci. USA*, **90**, 3710–14.
21. Suzuki, A., Yoshida, M., Hayashi, K., Mizuno, Y., Hagiwara, Y., and Ozawa, E. (1994). *Eur. J. Biochem.*, **220**, 283–92.
22. Rafael, J. A., Cox, G. A., Corrado, K., Jung, D., Campbell, K. P., and Chamberlain, J. S. (1996). *J. Cell Biol.*, **134**, 93–102.
23. Emery, A. E. H. (1993). *Duchenne muscular dystrophy.*, Oxford University Press, Oxford.
24. Ervasti, J. M., Kahl, S. D., and Campbell, K. P. (1991). *J. Biol. Chem.*, **266**, 9161–5.
25. Matsumara, K., Shasby, D. M., and Campbell, K. P. (1993). *FEBS*, **326**, 289–93.
26. Nicholson, L. V. B., Davison, K., Falkous, G., Harwood, C., O'Donnell, E., Slater, C. R., and Harris, J. B. (1989). *J. Neurol. Sci.*, **94**, 125–36.
27. thi Man, N., Cartwright, A. J., Morris, G. E., Love, D. R., Bloomfield, J. F., and Davies, K. E. (1990). *FEBS*, **262**, 237–40.
28. thi Man, N., Ellis, J. M., Love, D. R., Davies, K. E., Gatter, K. C., Dickson, G., and Morris, G. E. (1991). *J. Cell Biol.*, **115**, 1695–700.
29. Tinsley, J. M., Potter, A. C., Phelps, S. R., Fisher, R., Trickett, J. I., and Davies, K. E. (1996). *Nature*, **384**, 349–53.
30. Grady, R. M., Merlie, J. P., and Sanes, J. R. (1997). *J. Cell Biol.*, **136**, 871–82.
31. Deconinck, A. E., Potter, A. C., Tinsley, J. M., Wood, S. J., Vater, R., Young, C., Metzinger, L., Vincent, A. Slater, C. R., and Davies, K. E. (1997). *J. Cell Biol.*, **136**, 883–94.
32. Stradel, T., Kranewitter, W., Winder, S. J., and Gimona, M. (1998). *FEBS Lett.*, **431**, 134–7.
33. Chan, Y. and Kunkel, L. M. (1997). *FEBS Lett.*, **410**, 153–9.
34. Deconinck, A. E., Rafael, J. A., Skinner, J. A., Brown, S. C., Potter, A. C., Metzinger, L., Watt, D. J., Dickson, J. G., Tinsley, J. M., and Davies, K. E. (1997). *Cell*, **90**, 717–27.
35. Grady, R. M., Teng, H. B., Nichol, M. C., Cunningham, J. C., Wilkinson, R. S., and Sanes, J. R. (1997). *Cell*, **90**, 729–38.

Carolyn Moores and John Kendrick-Jones
MRC Laboratory of Molecular Biology,
Hills Road, Cambridge CB2 2QH,
UK

ERM proteins: ezrin, radixin, moesin

The ERM proteins[1–4] are closely related (~75 per cent identity) molecules encoded by different genes, and are involved in the attachment of microfilaments to the plasma membrane in cell surface structures such as microvilli. Radixin is also present in adherens junctions and the cleavage furrow.[3,5] The N-terminal domains are believed to associate with membrane proteins, and the C-terminal domains contain an F-actin-binding site.[6] The differential expression of ERM proteins in various cell types throughout the body[10,11] and differences in their abilities to serve as substrates for certain protein kinases[12–16] suggest that these proteins have related but distinct functions. It is believed that the function of ERM proteins is regulated through phosphorylation,[12,15,16,45–47,55] the formation of higher order ERM oligomers,[17] and phosphoinositide metabolism.[18] ERM proteins are also believed to be involved in the Rac and Rho signalling pathways.[48–50]

Synonymous names

Ezrin was also known as cytovillin[19] until the identity of the two proteins was appreciated.

Homologous proteins

The N-terminal ~300 residues show ~35 per cent identity to the corresponding domain of erythrocyte band 4.1, which defines members of the band 4.1 superfamily.[20] Other members include talin, merlin, and several protein tyrosine phosphatases.

Protein properties

The ERM family members have subunit molecular masses of about 68–69 kDa; but they migrate on SDS gels with apparent molecular weights of 81, 80, and 77 kDa, respectively. Although existing largely as monomeric proteins *in vivo*, the ERM proteins are capable of regulated homo- and heterotypic associations through their N- and C-terminal domains, known as N- and C-ERMADs (ERM-association-domain);[21–25] at least in the case of ezrin in placenta, these oligomeric molecules are enriched in the microvilli.[17] The monomers are relatively globular in shape, whereas the oligomers are highly extended.[17,21]

The proteins have a clear domain organization (Fig. 1). The N-terminal ~300 residues constitute a protease-resistant domain[13] in which resides the homologies to band 4.1,[4,26–28] the N-ERMAD activity,[24] and binding sites for PIP2,[18,29] CD44,[18] CD43,[51] ICAM-2,[51] ICAM-3,[52] EBP50,[53] and Rho-GDI.[48] This is followed by a domain predicted to be largely α helical and in which resides a binding site for the regulatory subunit of the type II A-kinase.[30] In ezrin and radixin, but not moesin, this is followed by a short stretch very rich in prolines. The C-terminal ~110 residues contain the C-ERMAD[24] and the F-actin-binding activity.[6,8,9] The nature of the F-actin binding activity remains controversial: for ezrin, it has been suggested to be direct[8,9,31,54] and indirect,[32] for radixin to be a barbed end

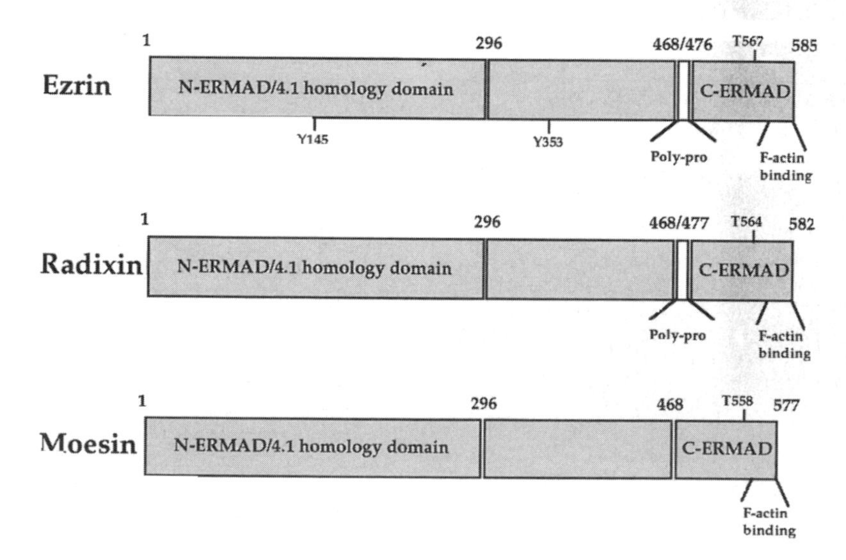

Figure 1. Domain organization of ERM proteins.

capping activity,[3] and for all members to be side binding.[9] The two major sites of EGF receptor tyrosine phosphorylation in ezrin[33] and of threonine phosphorylation in all members are shown.[16,46,55]

In cultured cells, the ERM proteins are specifically enriched in actin-based structures, including microvilli, ruffling membranes, and filopodia.[1,2,6,13,15,19,23,34,35] In addition, radixin has been reported to be concentrated in cell–cell and cell–substrate adhesions, and in the cleavage furrow during cytokinesis.[3,5,20,36] The suppression of ERM protein expression with antisense oligonucleotides has been shown to cause disruption of cell–cell and cell–substrate adhesion as well as the loss of microvilli.[37] Although all three ERM proteins are co-expressed in most cells grown in culture,[20,34] their expression pattern in different cell types in tissues varies considerably; for example, ezrin is enriched in epithelial cells whereas moesin is enriched in endothelia[10] (Fig. 2). Ezrin undergoes a spectacular reorganization during polarization of the early mouse embryo,[38] and moesin is translocated to the cell cortex immediately following fertilization in the sea urchin.[39] Immunoelectron microscopy has shown that ezrin is concentrated specifically on the region of the apical plasma membrane that covers the microvilli[10,17] (Fig. 3). Although originally isolated from hepatic adherens junctions,[3] radixin appears to be enriched in hepatocyte microvilli.[40]

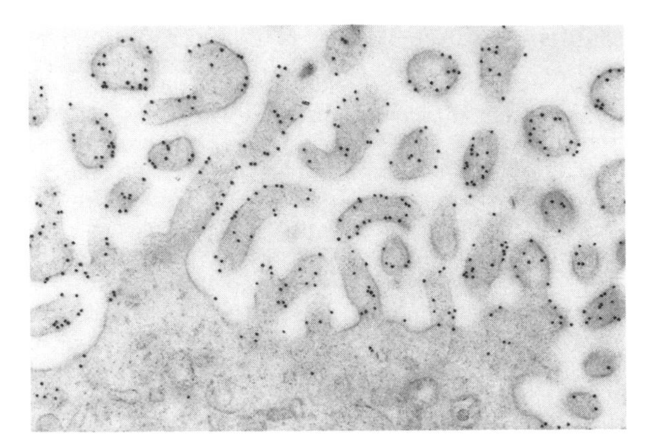

Figure 3. Immunoelectron microscopy of ezrin in human placenta demonstrating specific staining of the syncytiotrophoblast microvilli and the close proximity of ezrin to the plasma membrane.[10,17]

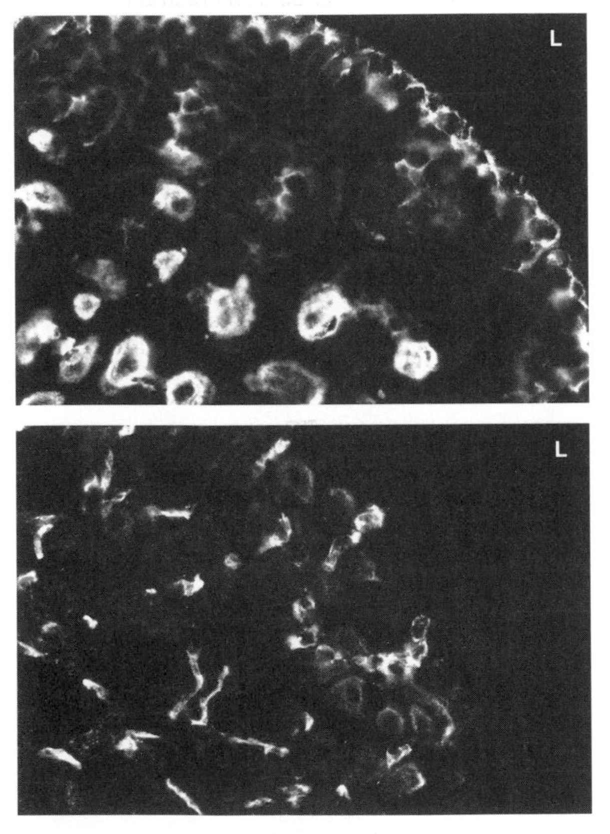

Figure 2. Immunofluorescence localization of ezrin (top), and moesin (bottom) in mouse stomach. Ezrin is concentrated in the parietal cells and in the mucous epithelial cells whereas moesin is enriched in endothelial cells.[10] The lumen (L) is indicated.

Based on their sequence identities, subcellular distributions, biochemical studies, and transfection experiments,[6] ERM proteins are believed to be membrane–microfilament linking proteins. This is supported by the finding that their C-terminal domains contain an F-actin binding site and their N-terminal domains bind to CD44. However, the subcellular localizations of CD44 and ERM proteins often do not correspond in tissues,[10,17,41] indicating the existence of additional membrane binding partners. Among these are CD43,[51] ICAM-2,[51] ICAM-3,[52] and EBP50.[53,56] EBP50 is a PDZ domain protein that links ERM members to the cytoplasmic tails of specific transmembrane proteins, such as the cystic fibrosis transmembrane conductance regulator.[57] The functions of ERM proteins are probably controlled by several different regulatory mechanisms. Ezrin is a substrate of the activated EGF receptor, PDGF receptor, and pp60$^{c\text{-}src}$ tyrosine kinases[2,12,45] as well as the cAMP-dependent protein kinase;[14] radixin is a substrate of the activated PDGF receptor;[12] moesin becomes phosphorylated in thrombin-activated platelets[16] and by protein kinase C-θ from leukocytes.[55] The interaction between ERM family members and CD44 occurs through a Rho-dependent signalling pathway and requires PIP_2.[18] The self-association of ERM proteins may involve regulation of intra- or intermolecular interactions between the N- and C-ERMADs, as suggested by transfection experiments[35,36] and biochemical analyses.[17,21,24,25] Although the oligomerization state of ezrin is enhanced by activation of the EGF receptor,[17] it is not known if this is related to its phosphorylation. The two major tyrosine phosphorylation sites in ezrin are known to be required for the morphogenetic response to treatment of LLC-PK1 cells with HGF.[45]

Purification

Ezrin,[1,15,31,32] radixin,[3] and moesin[4,15] have all been purified from appropriate tissues and cell lines. All have the useful property of binding tightly to hydroxyapatite resin through their N-terminal domain. Ezrin and moesin are readily isolated from placenta using a procedure that involves chromatography on hydroxyapatite, gel filtration (to separate monomers from oligomers), and anion and cation exchange chromatography.[21] The proteins have been expressed and isolated from bacterial and insect cells.[18,23,27]

Activities

There are no standardized assays to monitor biochemical activities associated with the ERM proteins. However, several protein–protein interactions have been discovered using blot overlays in which ERM proteins are subjected to SDS–PAGE and transferred to a membrane. Purified labelled ERM proteins will associate with great specificity to each of the immobilized ERM proteins.[22,24] Labelled skeletal muscle F-actin also binds to ERM proteins immobilized on a blot.[9] In addition, purified labelled recombinant A-kinase regulatory subunit associates with ezrin immobilized on blots.[30] Ezrin purified from parietal cells binds to β- but not α-F-actin in cosedimentation assays.[31] The N-terminal domains of ERM proteins also bind to Rho-GD1[48] and EBP50[53,56] and the cytoplasmic domain of CD44.[18,51]

Antibodies

Multiple antibodies, both polyclonal and monoclonal, have been made to all ERM members, which work well for blotting, immunoprecipitations, and localizations. Before the close relationship between ERM members was appreciated, some antibodies were used which unknowingly recognized more than one family member. However, a number of antibodies specific for each protein have now been described. A commercial monoclonal antibody raised against the last ten residues of human ezrin recognizes all three proteins on immunoblots, but in our hands does not work well for localization or immunoprecipitation of the native proteins: this is probably because the extreme C terminus is sterically hidden.[24]

Genes

cDNAs are available for human[4,26,27] and murine[20] ERM proteins; sea urchin[39] and *Drosophila*[42] moesin are also available. No alternative splicing has been reported for any of the cDNAs. The human genes for ezrin,[27] radixin,[43] and moesin[44] are located on chromosomes 6,11, and 10, respectively.

Mutant phenotype/disease states

No mutants have been described having defective or missing ERM proteins.

References

1. Bretscher, A. (1983). *J. Cell Biol.*, **97**, 425–32.
2. Gould, K. L., Cooper, J. A., Bretscher, A., and Hunter, T. (1986). *J. Cell Biol.*, **102**, 660–9.
3. Tsukita, S., Hieda, Y., and Tsukita, S. (1989). *J. Cell Biol.*, **108**, 2369–82.
4. Lankes, W. T. and Furthmayr, H. (1991). *Proc. Natl Acad. Sci. USA*, **88**, 8297–301.
5. Sato, N., Yonemura, S. Obinata, T., Tsukita, S., and Tsukita, S. (1991). *J. Cell Biol.*, **113**, 321–30.
6. Algrain, M., Turunen, O. A., Vaheri, A., Louvard, D., and Arpin, M. (1993). *J. Cell Biol.*, **120**, 129–39.
7. Tsukita, S., Oishi, K., Sato, N., Sagara, J., Kawai, A., and Tsukita, S. (1994). *J. Cell Biol.*, **126**, 391–401.
8. Turunen, O., Wahlstrom, T., and Vaheri, A. (1994). *J. Cell Biol.*, **126**, 1445–53.
9. Pestonjamasp, K., Amieva, M. R., Strassel, C. P., Nauseef, W. M., Furthmayr, H., and Luna, E. J. (1995). *Mol. Biol. Cell*, **6**, 247–59.
10. Berryman, M., Franck, Z., and Bretscher, A. (1993). *J. Cell Sci.*, **105**, 1025–43.
11. Schwartz-Albiez, R., Merling, A., Spring, H., Moller, P., and Koretz, K. (1995). *Eur. J. Cell Biol.*, **67**, 189–98.
12. Fazioli, F., Wong, W. T., Ullrich, S. J., Sakaguchi, K., Appella, E., and Di Fiore, P. P. (1993). *Oncogene*, **8**, 1335–45.
13. Franck, Z., Gary, R., and Bretscher, A. (1993). *J. Cell Sci.*, **105**, 219–31.
14. Hanzel, D., Reggio, H., Bretscher, A., Forte, J. G., and Mangeat, P. (1991). *EMBO J.*, **10**, 2363–73.
15. Bretscher, A. (1989). *J. Cell Biol.*, **108**, 921–30.
16. Nakamura, F., Amieva, M. R., and Furthmayr, H. (1995). *J. Biol. Chem.*, **270**, 25324–7.
17. Berryman, M., Gary, R., and Bretscher, A. (1995). *J. Cell Biol.*, **131**, 1231–42.
18. Hirao, M., Sato, N., Kondo, T., Yonemura,. S, Monden, M., Sasaki, T., *et al.* (1996). *J. Cell Biol.*, **135**, 37–51.
19. Pakkanen, R. and Vaheri, A. (1989). *J. Cell. Biochem.*, **41**, 1–12.
20. Sato, N., Funayama, N. Nagafuchi, A. Yonemura, S. Tsukita, S. and Tsukita, S. (1992). *J. Cell Sci.* **103**, 131–43.
21. Bretscher, A., Gary, R., and Berryman, M. (1995). *Biochemistry*, **34**, 16830–7.
22. Gary, R. and Bretscher, A. (1993). *Proc. Natl Acad. Sci. USA*, **90**, 10846–50.
23. Andreoli, C., Martin, M., Le Borgne, R., Reggio, H., and Mangeat, P. (1994). *J. Cell Sci.*, **107**, 2509–21.
24. Gary, R. and Bretscher, A. (1995). *Mol. Biol. Cell*,. **6**, 1061–75.
25. Magendantz, M., Henry, M. D., Lander, A., and Solomon, F. (1995). *J. Biol. Chem.*, **270**, 25324–7.
26. Gould, K. L., Bretscher, A. Esch, F. S., and Hunter, T. (1989). *EMBO J.*, **8**, 4133–42.
27. Turunen, O., Winqvist, R. Pakkanen, R. Grzeschik, K. Wahlstrom, T. and Vaheri. A. (1989). *J. Biol. Chem.*, **264**, 16727–32.
28. Funayama, N., Nagafuchi, A., Sato, N., Tsukita, S., and Tsukita, S. (1991). *J. Cell Biol.*, **115**, 1039–48.
29. Niggli, V., Andreoli, C., Roy, C., and Mangeat, P. (1995). *FEBS Lett.*, **376**, 172–6.

30. Dransfield, D. T., Bradford, A. J., Smith, J., Martin, M., Roy, C., Mangeat, P. H., and Goldenring, J. R. (1997). *EMBO J.*, **16**, 35–43.
31. Yao, X. and Forte, J. G. (1996). *J. Biol. Chem.*, **271**, 7224–9.
32. Shuster, C. B. and Herman, I. M. (1995). *J. Cell Biol.*, **128**, 837–48.
33. Krieg, J. and Hunter, T. (1992). *J. Biol. Chem.*, **267**, 19258–65.
34. Amieva, M. R. and Furthmayr, H. (1995). *Exp. Cell Res.*, **219**, 180–96.
35. Martin, M., Andreoli, C., Sahuquet, A., Montcourrier, P., Algrain, M., and Mangeat, P. (1995). *J. Cell Biol.*, **128**, 1081–93.
36. Henry, M. D., Gonzalez Agosti, C., and Solomon, F. (1995). *J. Cell Biol.*, **129**, 1007–22.
37. Takeuchi, K., Sato, N., Kasahara, H., Funayama, N., Nagafuchi, A., Yonemura, S., *et al.* (1994). *J. Cell Biol.*, **125**, 1371–84.
38. Louvet, S., Aghion, J., Santa-Maria, A., Mangeat, P., and Maro, B. (1996). *Dev. Biol.*, **177**, 568–79.
39. Bachman, E. S. and McClay, D. R. (1995). *J. Cell Sci.*, **108**, 161–71.
40. Amieva, M. R., Wilgenbus, K. K., and Furthmayr, H. (1994). *Exp. Cell Res.*, **210**, 140–4.
41. Nakamura, H. and Ozawa, H. (1996). *J. Bone Miner. Res.* **11**, 1715–22.
42. McCartney, B. M., and Fehon, R. G. (1996). *J. Cell Biol.*, **133**, 843–52.
43. Wilgenbus, K. K., Milatovich, A., Francke, U., and Furthmayr, H. (1993). *Genomics*, **16**, 199–206.
44. Wilgenbus, K. K., Hsieh, C. L., Lankes, W. T., Milatovich, A., Francke, U., and Furthmayr, H. (1994). *Genomics*, **19**, 326–33.
45. Crepaldi, T., Gautreau, A., Comoglio, P. M., Louvard, D., and Arpin, M. (1997). *J. Cell Biol.*, **138**, 423–34.
46. Matsui, T., Maida, M., Doi, Y., Yonemura, S., Amano, M., Kaibuchi, K., Tsukita, S., and Tsukita, S. (1998). J. Cell Biol., **140**, 647–57.
47. Shaw, R. J., Henry, M., Solomon, F., and Jacks, T. (1998). *Mol. Cell. Biol.*, **9**, 403–19.
48. Takahashi, K., Sasaki, T., Mammoto, A., Takaishi, K., Kameyama, T., Tsukita, S., Tsukita, S., and Takai, Y. (1997). *J. Biol. Chem.*, **272**, 23371–5.
49. Mackay, D. J. G., Esch, F., Furthmayr, H., and Hall, A. (1997). *J. Cell Biol.*, **138**, 927–38.
50. Fukata, Y., Kimura, K., Oshiro, N., Saya, H., Matsuura, Y., and Kaibuchi, K. (1998). *J. Cell Biol.*, **141**, 409–18.
51. Yonemura, S., Hirao, M., Doi, Y., Takahashi, N., Kondo, T., Tsukita, S., and Tsukita, S. (1998). *J. Cell Biol.*, **140**, 885–95.
52. Serrador, J. M., Alonso-Lebrero, J. L., del Pozo, M. A., Furthmayr, H., Schwartz-Albiez, R., Calvo, J., Lozano, F., and Sanchez-Madrid, F. (1997). *J. Cell Biol.*, **138**, 1409–23.
53. Reczek, D., Berryman, M., and Bretscher, A. (1997). *J. Cell Biol.*, **139**, 169–79.
54. Roy, C., Martin, M., and Mangeat, P. (1997). *J. Biol Chem.*, **272**, 20088–95.
55. Pietromonaco, S. F., Simons, P. C., Altman, A., and Elias, L. (1998). *J. Biol. Chem.*, **273**, 7594–603.
56. Reczek, D. and Bretscher, A. (1998). *J. Biol. Chem.*, **273**, 18378–84.
57. Short, B., Trotter, K. W., Reczek, D., Kreda, S. M., Bretscher, A., Boucher, R. C., Stutts, M. J., and Milgram, S. L. (1998). *J. Biol. Chem.*, **273**, 19797–801.
58. Martin, M., Roy, C., Montcourrier, P., Sahuquet, A., and Mangeat, P. (1997). *Mol. Biol. Cell.*, **8**, 1543–57.

Anthony Bretscher and Mark Berryman
Section of Biochemistry,
Molecular and Cell Biology.
Cornell University, Ithaca NY 14853,
USA

Focal adhesion kinase (FAK)

The focal adhesion kinase,[1–4] FAK or pp125^{FAK} (also called FadK or Emsk), is a non-receptor protein tyrosine kinase (PTK), which localizes principally in focal adhesions. FAK is functionally linked to the integrins, becoming tyrosine phosphorylated and enzymatically activated upon integrin engagement with their ligands. It can also become tyrosine phosphorylated in response to transformation by the *src* oncogene and a variety of other stimuli such as platelet-derived growth factor, bombesin, and lysophosphatidic acid. FAK may be involved in regulating cell spreading, cell migration, and in the transmission of an adhesion dependent anti-apoptotic signal.

Homologous proteins

A FAK-related PTK has recently been isolated and is known as CAKβ, PYK2, RAFTK, CADTK, and FAK2. This PTK exhibits the same domain structure as FAK and its amino acid sequence is 45 per cent identical to that of FAK. The highest degree of homology is in the catalytic domain (60 per cent) and C-terminal 150 amino acids (61 per cent).[5,6]

Protein properties

FAK[1–4] is a non-receptor PTK containing 1052 residues and exhibiting a M_r of 125 kDa. It is expressed in most tissues and cell types except in certain B- and T-cell populations and is highly conserved evolutionarily (91 per cent identity between *Xenopus*, avian, and human FAK). It contains large N- and C-terminal non-catalytic domains (~400 residues in length) flanking the central catalytic domain. Immunofluorescent staining indicates that FAK localizes primarily in focal adhesions. The subcellular localization of FAK is mediated by amino acids 853–1052, the focal adhesion targeting (FAT) sequence. The FAT

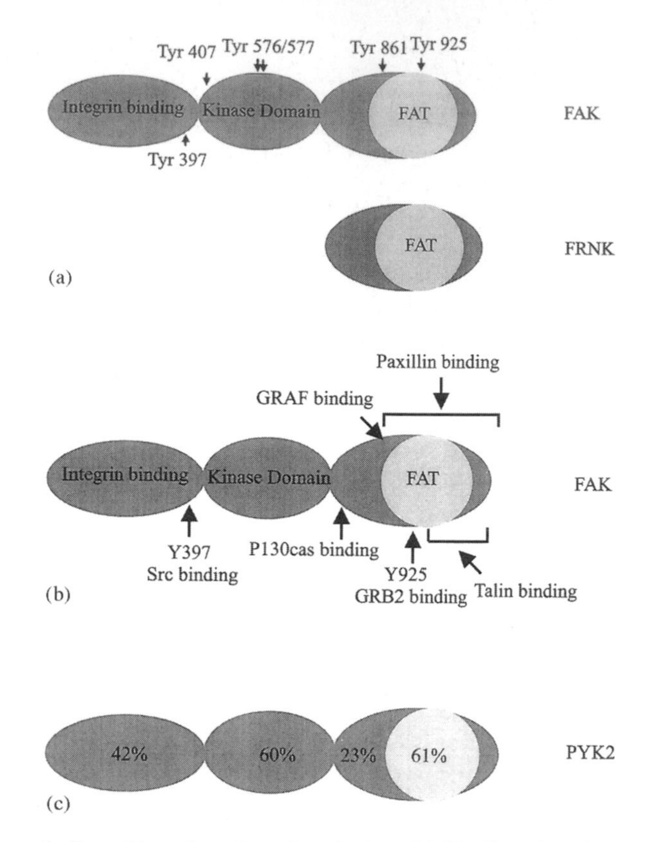

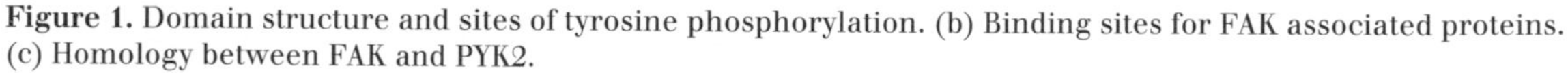

Figure 1. Domain structure and sites of tyrosine phosphorylation. (b) Binding sites for FAK associated proteins. (c) Homology between FAK and PYK2.

sequence also contains binding sites for talin and paxillin. Binding to paxillin has been proposed as the mechanism of localization of FAK to focal adhesions,[7] but this conclusion is controversial.[8] The C-terminal domain of FAK has several proline rich motifs that serve as docking sites for the SH3 domains of p130^{cas}, a tyrosine phosphorylated focal adhesion protein, and GRAF, a GTPase activating protein for the rho family of GTP-binding proteins. Phosphatidyl inositol 3' kinase (PI3K) has also been shown to bind to FAK. The N-terminal domain of FAK contains a putative integrin binding site, but the physiological role of this interaction remains speculative.

Post-translational modification, especially tyrosine phosphorylation, plays an important role in regulation of FAK function.[1–4] FAK contains six sites that can be tyrosine phosphorylated. The major site of autophosphorylation is tyrosine 397. The remaining sites, tyrosines 407, 576, 577, 861, and 925 can be phosphorylated by the src protein tyrosine kinase. Phosphorylation of tyrosines 397 and 925 create binding sites for the SH2 domains of src and grb-2 respectively. Phosphorylation of tyrosines 576 and 577 enhances the enzymatic activity of FAK two-fold. The function of phosphorylation at residues 407 and 861 is unknown.

Several alternative forms of FAK have been described. In some cells, alternative mRNA splicing leads to the autonomous expression of the C-terminal non-catalytic domain, called FRNK. FAK^{+}, an isoform of FAK with an insertion of three amino acids in the focal adhesion targeting region, is expressed preferentially in brain.[9] A FAK-related kinase, referred to as fakB, is expressed in T and B cells and may be involved in antigen-induced lymphocyte signalling.[10]

Tyrosine phosphorylation of FAK can be induced by cell–extracellular matrix interactions, transformation by the *src* oncogene and a variety of other stimuli such as PDGF, bombesin, anandamide, Alzheimer's β- peptide, activated rho, vasopressin and endothelin.[1,2,4,11] It is critical for the cytoskeleton to be intact for tyrosine phosphorylation of FAK. Disruption of the cytoskeleton by cytochalasin D abolishes the tyrosine phosphorylation of FAK induced by all these stimuli. Presumably, a primary biochemical function of pp125^{FAK} is to control tyrosine phosphorylation of downstream substrates. p130^{cas}, paxillin, and tensin are potentially downstream components of FAK signalling. They colocalize with FAK in focal adhesions and become tyrosine phosphorylated in coordination with tyrosine phosphorylation of FAK. Tyrosine phosphorylation of these substrates may be directly mediated by FAK or alternatively, FAK could function to recruit a second PTK, such as src, which is then responsible for phosphorylating these substrates.[1,2,4,11]

The biological functions of FAK are beginning to be elucidated. FAK is an essential gene and *fak*$^{-/-}$ mice die as embryos.[12] The function of FAK in cells has been addressed using fibroblasts derived from the *fak*$^{-/-}$ embryos and strategies designed to perturb normal FAK signalling (i.e. dominant negative approaches) or enhance FAK signalling (i.e. using constitutively active FAK variants). These results suggest that FAK controls cell spreading, the rate of formation of focal adhesions, and cell migration.[12–15] Recent evidence also suggests that FAK may function in the transmission of a cell adhesion dependent signal that is required for cell survival.[16,17] Abnormal FAK signalling may also be important biologically. As an accomplice of the *src* oncogene, FAK may play a role in the development of cancer. The integrins, regulators of FAK, have also been implicated in cancer.[18,19] Finally, FAK itself has been found to be amplified in several types of human tumours.[20–22]

Purification

No protocols available. Most analysis/assays are conducted upon immunoprecipitated protein.

Activities

FAK is a PTK, which is an enzyme that specifically transfers a phosphate from ATP to a tyrosine residue on a target protein. The enzymatic activity of FAK can be measured *in vitro* in an immune-complex PTK assay.[23,24] Since tyrosine phosphorylation of FAK is important for its activity and for binding to other signalling molecules, the phosphotyrosine content of FAK is frequently used as an indirect indicator of stimulation. This is monitored by Western blotting with phosphotyrosine antibodies. Biologically, FAK may be involved in regulating cell migration and in rescuing cells from apoptosis.

Antibodies

The C-terminal and the catalytic domains of FAK have been used successfully to generate antibodies.[25,26] Both monoclonal and polyclonal antibodies have been raised and some are commercially available (e.g. from Transduction Laboratories, Lexington, KY and Upstate Biotechnology Inc., Lake Placid NY). Antipeptide antibodies that specifically recognize fakB or FAK^{+} have also been described.[10,27] Generally, both polyclonal and monoclonal antibodies recognize FAK from different species and can be used for immunoprecipitation, Western blotting, and immunofluorescence.

Genes

The complete cDNAs encoding the murine (GenBank M95408),[26] avian (GenBank M86656),[25] human (GenBank L13616),[28] and *Xenopus* (GenBank L33920)[29] FAK have been cloned. A partial cDNA encoding rat FAK has also been isolated (GenBank U43940, 43941, 43942). FAK is highly conserved between species (~91 per cent amino acid identity between species). The mouse gene of FAK has been mapped to chromosome 15, distal to the myelocytomatosis proto-oncogene (*myc*). The human homologue lies on chromosome 8. On the basis of synteny of the mouse and human chromosome maps, the human FAK gene has been assigned to human chromosome 8q24–qter.[30] The human CAKβ gene was mapped to chromosome 8p11.2–p22.[31]

References

1. Schaller, M. D. (1996). *J. Endocrinol.*, **150**, 1–7.
2. Richardson, A. and Parsons, T. (1995). *Bioessays*, **17**, 229–36.
3. Schwartz, M. A., Schaller, M. D., and Ginsberg, M. H. (1995). *Ann. Rev. Cell Dev. Biol.*, **11**, 549–99.
4. Clark, E. A. and Brugge, J. S. (1995). *Science*, **268**, 233–9.
5. Lev, S., Moreno, H., Martinez, R., Canoll, P., Peles, E., Musacchio, J. M., *et al.* (1995). *Nature*, **376**, 737–45.
6. Sasaki, H., Nagura, K., Ishino, M., Tobioka, H., Kotani, K., and Sasaki, T. (1995). *J. Biol. Chem.*, **270**, 21206–19.
7. Tachibana, K., Sato, T., D'Avirro, N., and Morimoto, C. (1995). *J. Exp. Med.*, **182**, 1089–100.
8. Hildebrand, J. D., Schaller, M. D., and Parsons, J. T. (1995). *Mol. Biol. Cell*, **6**, 637–47.
9. Burgaya, F. and Girault, A. (1996). *Mol. Brain Res.*, **37**, 63.
10. Kanner, S. B., Aruffo, A., and Chan, P.-Y. (1994). *Proc. Natl Acad. Sci., USA*, **91**, 10484–7.
11. Matsumoto, K., Ziober, B. L., Yao, C. C., and Kramer, R. H. (1995). *Cancer Metastasis Rev.* **14**, 205–17.
12. Ilic, D., Furuta, Y., Kanazawa, S., Takeda, N., Sobue, K., Nakatsuji, N., *et al.* (1995). *Nature*, **377**, 539–44.
13. Gilmore, A. and Romer, L. H. (1996). *Mol. Biol. Cell*, **7**, 1209–24.
14. Richardson, A. and Parsons, J. T. (1996). *Nature*, **380**, 538–40.
15. Cary, L. A., Chang, J. F., and Guan, J. L. (1996). *J. Cell Sci.*, **109**, 1787–94.
16. Hungerford, J. E., Compton, M. T., Matter, M. L., Hoffstrom, B. G., and Otey, C. A. (1996). *J. Cell Biol.*, **135**, 1383–90.
17. Frisch, S. M., Vuori, K., Ruoslahti, E., and Chan-Hui, P.-Y. (1996). *J. Cell Biol.*, **134**, 793–9.
18. Juliano, R. L. and Varner, J. A. (1993). *Curr. Opin. Cell Biol.*, **5**, 812–18.
19. Albelda, S. M. (1993). *Lab. Invest.*, **68**, 4–17.
20. Weiner, T. M., Liu, E. T., Craven, R. J., and Cance, W. G. (1993). *Lancet*, **342**, 1024–5.
21. Xu, L. H., Owens, L. V., Sturge, G. C., Yang, X. H., Liu, E. T., Craven, R. J., and Cance, W. G. (1996). *Cell Growth Diff.*, **7**, 413–18.
22. Owens, L. V., Xu, L. H., Dent, G. A., Yang, X. H., Sturge, G. C., Craven, R. J., and Cance, W. G. (1996). *Ann. Surg. Oncol.*, **3**, 100–5.
23. Guan, J.-L. and Shalloway, D. (1992). *Nature*, **358**, 690–2.
24. Lipfert, L., Haimovich, B., Schaller, M. D., Cobb, B. S., Parsons, J. T., and Brugge, J. S. (1992). *J. Cell Biol.*, **119**, 905–12.
25. Schaller, M. D., Borgman, C. A., Cobb, B. S., Vines, R. R., Reynolds, A. B., and Parsons, J. T. (1992). *Proc. Natl Acad. Sci., USA*, **89**, 5192–6.
26. Hanks, S. K., Calalb, M. B., Harper, M. C., and Patel, S. K. (1992). *Proc. Natl Acad. Sci., USA*, **89**, 8487–9.
27. Derkinderen, P., Toutant, M., Burgaya, F., Bert, L. M., Siciliano, L. C., Franciscis, V., *et al.*, (1996). *Science*, **273**, 1719–23.

28. Andre, E. and Becker-Andre, M. (1993). *Biochem. Biophys. Res. Commun.*, **190**, 140–6.
29. Hens, M. D. and DeSimone, D. W. (1995). *Dev. Biol.*, **170**, 274–88.
30. Fiedorek, F. T. and Kay, E. S. (1995). *Mamm. Genome*, **6**, 123–6.
31. Herzog, H., Nicholl, J., Hort, Y. J., Sutherland, G. R., and Shine, J. (1995). *Genomics*, **32**, 484–6.

Yu Shen and Michael D. Schaller
Department of Cell Biology and Anatomy,
University of North Carolina at Chapel Hill,
Chapel Hill, NC 27599,
USA

Paxillin

Paxillin localizes to actin–membrane attachment sites at cell–extracellular matrix junctions. Tyrosine phosphorylation of paxillin is stimulated by integrin-mediated cell adhesion and in response to growth factors. In addition, paxillin binds several structural and regulatory proteins. Together, these observations suggest a role for paxillin as a molecular adapter important in recruiting signalling components to the cytoskeleton–membrane interface.

Protein properties

Paxillin localizes to the cytoplasmic face of focal adhesions in many cultured cell types (Fig. 1).[1] These discrete sites of cell attachment provide a structural link between the extracellular matrix and the actin cytoskeleton and also serve as signal transduction centres, involved in regulating cellular processes ranging from cell migration to gene expression.[2] Paxillin is enriched in similar cell–extracellular matrix junctional complexes *in vivo* including the dense plaques of smooth muscle and the myotendinous junctions of skeletal muscle[3]. It is also found at the post-synaptic face of the neuromuscular junction.[3] Paxillin is absent from cell–cell junctions.[1] While paxillin is not readily detected in brain tissue homogenates[4] it is quite abundant in several cultured neuronal cell lines.[5] Paxillin is relatively abundant in leukocytes[6] but, unlike many other focal adhesion proteins, has not been detected in platelets (Turner; unpublished observations).

Paxillin migrates on SDS–PAGE as a broad band with an apparent molecular mass of between 68 and 75 kDa and exhibits several isoelectric variants (p*I* range 6.31–6.85).[1] The cDNA (chicken) encodes for a protein of 559 amino acids with an actual molecular mass of 61.2 kDa.[7] While the diffuse banding pattern seen on SDS–PAGE is due to phosphorylation at multiple sites of primarily tyrosine and serine residues, the retarded migration of the unphosphorylated protein is likely to be due to a relatively high (9 per cent) proline content.

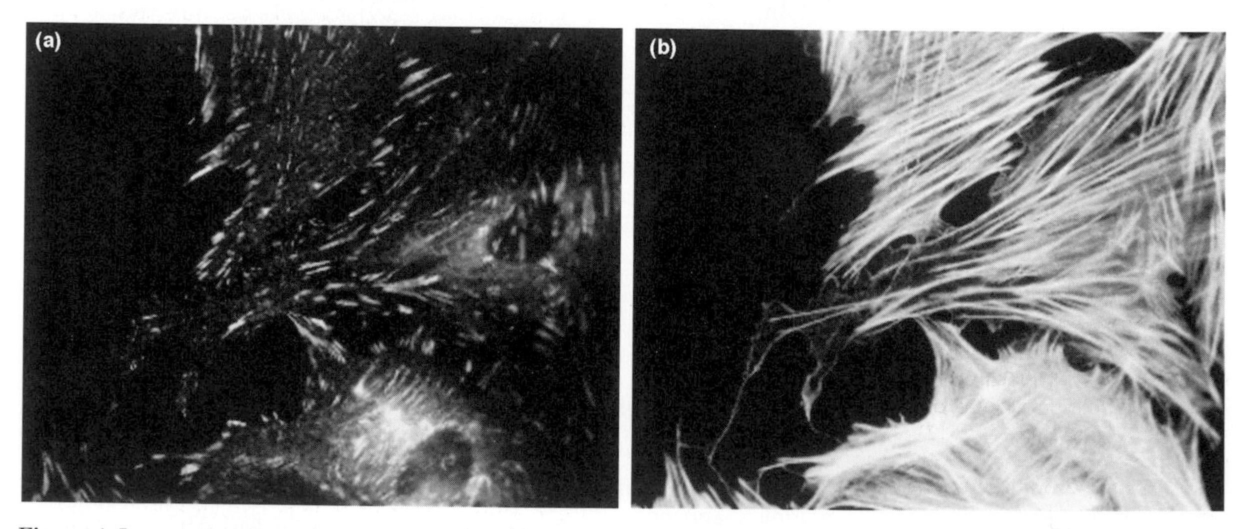

Figure 1. Immunolocalization of paxillin (a) to focal adhesions in rat embryo fibroblasts. Cells were double labelled with rhodamine–phalloidin to visualize the actin filament-containing stress fibres (b) that terminate at the paxillin-rich focal adhesions.

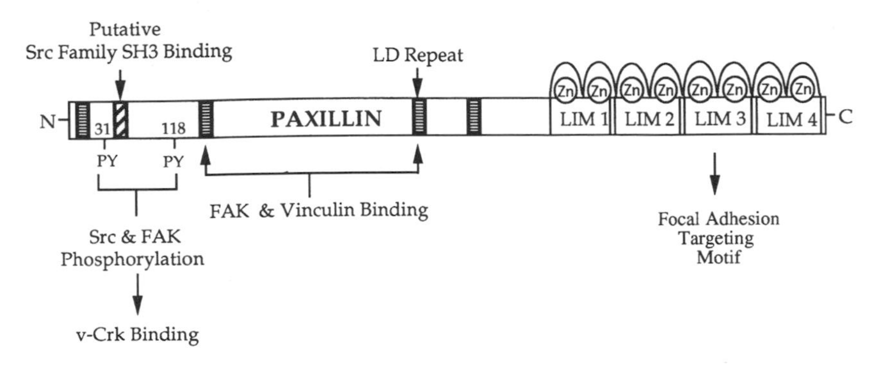

Figure 2. Schematic representation of the structural and functional domains of paxillin.

The cDNA for paxillin predicts a protein comprising several distinct structural domains.[7] This information, together with *in vitro* binding data, suggests that paxillin functions as a molecular adaptor (Fig. 2), serving to recruit structural and regulatory proteins into a large signalling complex (i.e. focal adhesions). The first 325 amino acids of paxillin contains several protein-binding sites including a proline-rich region (a.a. 46–55) believed to be involved in binding the SH3 domain of pp60^{src} (src).[8] This portion of the protein also contains binding sites for vinculin[1,7,9] and the focal adhesion kinase (FAK).[7] Interaction between these proteins is mediated through conserved paxillin binding sequences (PBS) on vinculin and FAK[9,10,11] and leucine-rich sequences on paxillin (termed LD motifs).[12] Of the four LD motifs in paxillin, vinculin binding can be assigned to LD2 while both LD2 and LD3 contribute to the binding of FAK.[12] The LD domain may represent a protein-binding motif of general importance since the papilloma virus oncoprotein E6 binds paxillin LD1 (and possibly LD2) in addition to binding the E6-associated protein (E6-AP) via an LD-like domain on E6-AP.[13]

The C-terminus of paxillin (a.a. 326–559) contains four LIM domains. These are histidine- and cysteine-rich stretches of approximately 50 amino acids that chelate two zinc ions, folding each LIM domain into two zinc fingers.[14] LIM domains have been implicated in protein–protein interactions and, in the case of paxillin, LIM3 is essential for targeting the protein to focal adhesions.[12] Phosphorylation of the LIM domains regulates paxillin localization to focal adhesions.[31] In contrast, the vinculin- and FAK-binding sites are neither necessary nor sufficient for localization of paxillin to focal adhesions.[12] Binding to vinculin and FAK may prove more important in stabilizing the association of paxillin with focal adhesions. Paxillin shares a high level of structural similarity with the protein Hic 5, implicated in cellular senescence.[15] A *Dictyostelium* homologue of paxillin, Pax B, has been identified (GenBank 015817) as well as a leukocyte-specific form, leupaxin.[32] Both of these family members share extensive homology with the LD motifs and LIM domains of paxillin.

A striking feature of paxillin biology is its propensity for phosphorylation by tyrosine kinases. In this regard, paxillin is tyrosine phosphorylated in an integrin-dependent manner during cell adhesion to extracellular matrix,[16] following growth factor stimulation of growth-arrested cells,[17] in cells transformed by tumor viruses,[5,18] and during embryonic development.[19] The coordinate phosphorylation and activation of FAK during these processes,[20] in conjunction with more direct *in vitro* kinase assays, implicates FAK in the phosphorylation of paxillin.[21,22] However, phosphorylation of paxillin by src kinase, which is known to associate with FAK, and csk, a negative regulator of src activity, has also been reported,[23] suggesting that phosphorylation of paxillin results from a complex interplay between several tyrosine kinases.

Although tyrosine phosphorylation of paxillin has been associated with cytoskeletal remodelling[16] and results in the generation of SH2 binding sites (Y31 and Y118)[21,22] that support v-crk binding in cultured cells[5] and perhaps other SH2-containing proteins such as tensin and csk,[23] the precise function of paxillin phosphorylation remains unclear. Recent reports indicate that paxillin is also phosphorylated heavily on serine residues in response to adhesion.[24,25] Paxillin contains numerous consensus sequences for phosphorylation by several Ser/Thr kinases including protein kinase C, MAP kinase, protein kinase A, but the evidence for the involvement of any of these kinases in phosphorylating paxillin *in vivo* remains indirect.[24]

Purification

Paxillin is readily extracted in low ionic strength buffer from chicken gizzard smooth muscle and can be purified by a combination of conventional ion exchange and antibody affinity chromatography.[1] Due to the relatively low abundance of paxillin, yields are generally 50–100-fold lower than for another focal adhesion protein, talin. More recent analyses have utilized bacterial expression of paxillin either as glutathione *S*-transferase (GST) or maltose binding protein (MBP) fusions.[7] While the full length protein is poorly soluble when expressed in *E. coli* as a GST-fusion protein (the full length MBP fusion is more soluble), smaller regions of the protein are readily soluble when expressed in this form and exhibit several functions associated with the native protein, such as vinculin and FAK binding.[7,12,21]

Activities

Paxillin, either purified from smooth muscle or expressed as bacterial fusion proteins, has been shown to interact directly *in vitro* with vinculin (k_d 6 × 10^{-8} M),[1] FAK,[10,11] and the SH3 domain of Src.[8] Coprecipitation studies using cellular lysates also suggest associations between paxillin and integrin cytoplasmic domain peptides,[26,27] the nerve growth factor receptor trkA,[28] and the FAK homologue Pyk2, although these have not been confirmed as direct interactions. Paxillin interacts in a tyrosine phosphorylation dependent manner with the SH2 domains of v-crk, crkl, and csk.[5,23] Tyrosine phosphorylation of paxillin can be induced in cultured cells by plating on extracellular matrix components[16] or by stimulating serum-starved cells with a wide variety of growth factors.[2,17] Pretreatment of cells with the actin cytoskeleton disrupting drug, cytochalasin D, blocks paxillin phosphorylation.[17] Paxillin also associates with the tyrosine phosphatase PTP-PEST.[33]

Antibodies

Several mouse monoclonal antibodies raised against paxillin have been reported.[1,18] In general, these antibodies, raised against chicken paxillin, cross-react with paxillin from other mammalian species. Monoclonal antibody 165 also cross-reacts with *Xenopus laevis* paxillin.[3] Monoclonal antibodies are available from several commercial sources including Sigma and Transduction Laboratories. Polyclonal antibodies have been described for both chicken[7,12,21] and human paxillin.[29] The chicken antibody does not cross-react with paxillin from other species, as assayed by immunofluorescence microscopy.[12,21] The human paxillin polyclonal antibody has been reported to cross-react widely.

Genes

The paxillin cDNA was originally cloned from chicken[7] (EMBL/GenBank L30099) and subsequently from human[29] (GenBank 14588). The gene has been mapped to chromosome 12q24 in humans. Limited expression of two additional isoforms of human paxillin have been reported.[30] Paxillin family members include Hic5 (GenBank L22482), Pax B (GenBank 015817) and leupaxin (GenBank AF0620775).

References

1. Turner, C. E., Glenney, J. R., and Burridge, K. (1990). *J. Cell Biol.*, **111**, 1059–68.
2. Schwartz, M. A., Schaller, M. D., and Ginsberg, M. H. (1995). *Ann. Rev. Cell Biol.*, **11**, 549–99.
3. Turner, C. E., Kramarcy, N., Sealock, R., and Burridge, K. (1991). *Exp. Cell Res.*, **192**, 651–655.
4. Turner, C. E., Schaller, M. S., and Parsons, J. T. (1993). *J. Cell Sci.*, **105**, 637–45.
5. Birge, R. B., Fajardo, J. E., Reichman, C., Shoelson, S. E., Songyang, Z., Cantley, L. C., and Hanafusa, H. (1993). *Mol. Cell Biol.*, **13**, 4648–56.
6. Graham, I. L., Anderson, D. C., Holers, V. M., and Brown, E. J. (1994). *J. Cell Biol.*, **127**, 1139–47.
7. Turner, C. E. and Miller, J. T. (1994). *J. Cell Sci.*, **107**, 1583–91.
8. Weng, Z., Taylor, J. A., Turner, C. E., Brugge, J. S., and Seidel-Dugan, C. (1993). *J. Biol. Chem.*, **268**, 14956–63.
9. Wood, C. K., Turner, C. E., Jackson, P., and Critchley, D. R. (1994). *J. Cell Sci.*, **107**, 709–17.
10. Hildebrand, J. D., Schaller, M. D., and Parsons, J. T. (1995). *Mol. Biol. Cell.*, **6**, 637–47.
11. Tachibana., K., Sato, T., Avirro, N. D., and Morimoto, C. (1995). *J. Exp. Med.*, **182**, 1089–100.
12. Brown, M. C., Perrotta, J. A., and Turner, C. E. (1996). *J. Cell Biol.*, **135**, 1109–24.
13. Vande Pol, S., Brown, M. C., and Turner, C. E. (1998) *Oncogene*, **16**, 43–52.
14. Sanchez-Garcia, I. and Rabbits, T. H. (1994). *Trends Genet.*, **10**, 315–20.
15. Shibanuma, M., Mashimo, J.-I., Kuroki, T., and Nose, K. (1994). *J. Biol. Chem.*, **269**, 26767–74.
16. Burridge, K., Turner, C. E., and Romer, L. (1992). *J. Cell Biol.*, **119**, 893–903.
17. Rozengurt, E. (1995). *Cancer Surv.*, **24**, 81–96.
18. Glenney, J. R. and Zokas, L. (1989). *J. Cell Biol.*, **108**, 2401–8.
19. Turner, C. E. (1991). *J. Cell Biol.*, **115**, 201–7.
20. Guan, J.-L. and Chen, H.-C. (1996). *Int. Rev. Cytol.*, **168**, 81–121.
21. Bellis, S. L, Miller, J. T., and Turner, C. E. (1995). *J. Biol Chem.*, **270**, 17437–41.
22. Schaller, M. D. and Parsons, J. T. (1995). *Mol. Cell Biol.*, **15**, 2635–45.
23. Sabe, H., Shoelson, S. E., and Hanafusa, H. (1995). *J. Biol. Chem.*, **270**, 31219–24.
24. De Nichol, M. O. and Yamada, K. M. (1996). *J. Biol. Chem.*, **271**, 11016–22.
25. Bellis, S. L., Perrotta, J. A., Curtis, M. S., and Turner, C. E. (1997). *Biochem. J.*, 325, 375–81.
26. Schaller, M. D., Otey, C. A., Hildebrand, J. D., and Parsons, J. T. (1995). *J. Cell Biol.*, **130**, 1181–7.
27. Tanaka, T., Yamaguchi, R., Sabe, H., Sekiguchi, K., and Healy, J. M. (1996). *FEBS Lett.*,. **399**, 53–8.
28. Melamed, I., Turner, C. E., Aktories, K., Kaplan, D. R., and Gelfand, E. W. (1995). *J. Exp. Med.* **181**, 1071–9.
29. Salgia, R., Li., J.-L., Lo, S. H., *et al.* (1995). *J. Biol Chem.*, **270**, 5039–47.
30. Mazaki, Y., Hashimoto, S., and Sabe, H. (1997). *J. Biol Chem.*, **272**, 7437–44.
31. Brown, M. C., Perrotta, J. A., and Turner, C. E. (1998). *Mol. Biol. Cell*, (In press).
32. Lipsky, B. P., Beals, C. R., and Staunton, D. E. (1998). *J. Biol., Chem.*, **273**, 11709–13.
33. Shen, Y., Schneider, G., Clouier, J., Viellette, A., and Schaller, M. (1998). *J. Biol. Chem.*, **273**, 6474–81.

Christopher E. Turner
Department of Anatomy and Cell Biology,
SUNY Health Science Center at Syracuse,
750 East Adams Street, Syracuse,
NY 13210, USA

Plakoglobin, β-catenin, and the ARM family

Plakoglobin, β-catenin and armadillo associate with the cytoplasmic domain of cadherins and are concentrated at intercellular adhesive junctions. They are also found in distinct protein complexes in the cytoplasm and nucleus which participate in *Wnt/wingless* signal transduction pathways that play important roles in early embryonic patterning and cancer.

Protein properties

Role in adhesive cell–cell junctions

Plakoglobin (M_r ~86 000 Da),[1,2] β-catenin (M_r ~92 000 Da),[3,4] and armadillo (M_r ~99 000–103 000 Da),[5,6] the protein product of a *Drosophila* segment polarity gene, are enriched in the electron-dense submembranous mats of *adherens* junctions.[7] They share ~65 per cent sequence identity and a common tripartite structure consisting of a central domain containing 13 imperfect 42 amino acid 'ARM' repeats flanked by less conserved, non-repetitive, terminal domains. Plakoglobin, β-catenin, and armadillo associate through a central block of their ARM repeats with the highly conserved cytoplasmic domains of classical cadherins.[8–10] Concurrently, they bind through their N-terminal domain and first repeat to α-catenin and thereby provide an essential link in a chain of proteins that tether cadherins to the cytoskeleton (Fig. 1).[8–12] This linkage is essential for cell adhesion and can be antagonized by association of the central domains of plakoglobin or β-catenin to ligand occupied EGFR and erbB2 and consequent phosphorylation of their termini.[13,14] Plakoglobin and β-catenin are also tyrosine phosphorylated in v-*src* transformed cells[15,16] and their stability can be negatively regulated by serine phosphorylation (see below). There are several reports of association of the cadherin–catenin complex with a variety of receptor tyrosine phosphatases (PTPκ,[17] PTPμ,[18] PTP-1b,[19] and LAR[20]), although this is currently a controversial topic.[21]

Plakoglobin is unique in its ability to form additional distinct protein complexes at desmosomes.[1] It binds to regions of the cytoplasmic domains of desmogleins and desmocollins that share sequence similarity with the catenin-binding domain of classical cadherins.[22–24] The N-terminal domain and first three repeats of plakoglobin bind desmoglein whereas both ends, and possibly all, of the central repeat region of plakoglobin are required for desmocollin association.[11,25] These binding sites overlap those for classical cadherins and α-catenin, suggesting that steric hindrance may account for the mutually exclusive plakoglobin complexes found at desmosomes and adherens junctions. The C-terminal repeats 11–13 of plakoglobin are reported to bind repeats 5–10 and to contain additional cryptic cadherin-binding domains.[25] Weak but direct interactions, detected by yeast two-hybrid and gel overlay assays, have been described between plakoglobin and the N-terminus of desmoplakin, an intermediate filament linker protein as well as to the keratin 5 head domain.[91,92] These interactions suggest that plakoglobin functions to link desmosomal cadherins to the cytoskeleton. β-catenin is reported to bind to the actin bundling protein fascin and to the tight junction associated protein ZO-1.[26,27] The role of these interactions is currently obscure.

Mice null for β-catenin and plakoglobin show different phenotypes. β-catenin null mice are early embryonic lethals (E7) show defects in embryonic ectoderm forma-

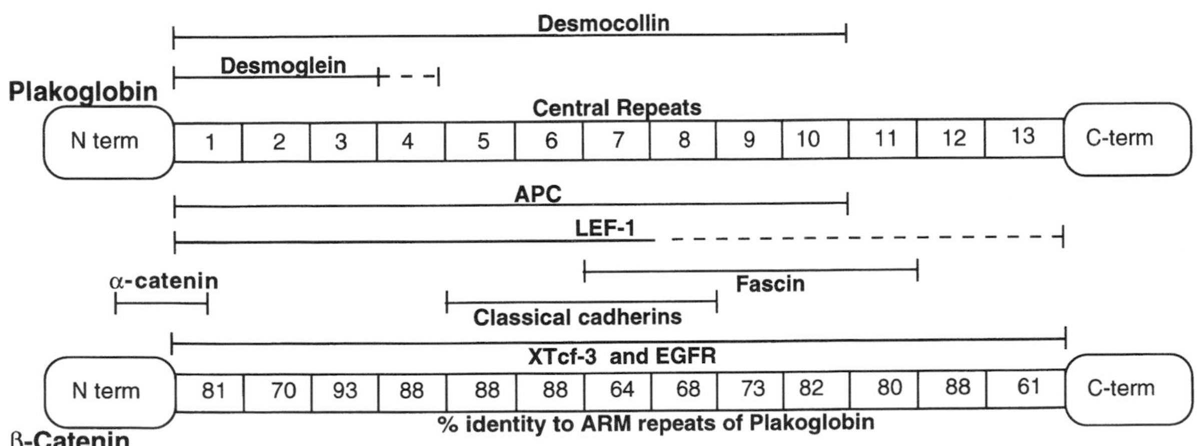

Figure 1. Binding sites for partners of plakoglobin and β-catenin.

tion and fail to form mesoderm, suggesting that the primary defect results from lack of β-catenin signalling (see below).[28] Plakoglobin-null mutants die later (E15) from loss of heart integrity and, those that survive longer, show defects in skin integrity.[29–32] In these mice, desmosomal and adherens proteins become intermingled in abnormal junctions and the desmosomes are reduced and fail to attach intermediate filaments.

Role in *Wnt/wingless* signalling pathways

Armadillo mutants show defects in segment polarity similar to that of another segment polarity gene *wingless* that encodes a secreted morphogen. Wingless, acting through *Drosophila* Frizzled 2, a putative receptor on neighbouring cells, antagonizes *zeste-white* 3 (the *Drosophila* homologue of the serine-threonine kinase, GSK3-β) and causes post-transcriptional upregulation of armadillo protein.[33] *Wnt-1*, the vertebrate homologue of *wingless*, exerts a similar effect on plakoglobin and β-catenin levels in cultured mammalian cell lines.[34,35] Furthermore, injection of *Xenopus* embryos with mRNA encoding *wnt-1*, a dominant negative mutant of GSK3-β, plakoglobin, or β-catenin, produces the same phenotype of axis duplication, suggesting that the *wnt/wingless* pathway is conserved in vertebrates.[36–39] Many different *wnts* and a correspondingly large family of *frizzled* genes exist in mammals;[33,40,41] however, it is likely that they all act through plakoglobin/β-catenin. There is one report that a membrane-anchored form of plakoglobin can signal, although its mechanism is not presently clear.[42,43] Cytosolic plakoglobin, β-catenin and armadillo have been shown to interact with the APC tumour suppressor gene product, GSK, axin and conductin.[93,94] These interactions act to oppose *wnt* signalling by promoting the phosphorylation and ubiquitination of plakoglobin and β-catenin and their consequent degradation.[95,96] In the presence of wnt, plakoglobin and β-catenin escape this process and interact with the HMG-box transcription factors L*ef*-1, T*cf*3, T*cf*4, and pangolin and to translocate to the nucleus as a complex and modulate transcription.[44–48] L*ef*-1 is known to act by bending DNA, thereby altering the relative configuration and interactions of other transcription factors bound to the target gene promoters.[49,50] It has been proposed that the C-terminus of armadillo may act as a transactivator in the complex.[51] However, in *Xenopus* studies, mutants lacking the C terminus of β-catenin retain axis-duplicating ability.[39] Downstream targets of plakoglobin/β-catenin/armadillo nuclear protein complexes include the homeobox containing *siamois* gene in *Xenopus*;[52] *engrailed*, *eyelid* and *ultrabithorax* in *Drosophila*,[33,53] and E-cadherin and keratin genes in mouse.[45,54] It is likely that many other downstream genes will be discovered.

Role of plakoglobin and β-catenin in cancer

Three sets of observations suggest that upregulation of cytosolic plakoglobin/β-catenin correlates with hyperproliferation.

1. *Wnt-1* was discovered as a gene that when activated by proviral insertion caused mammary tumours in mice.[55] This suggested that downstream effects of wnt-1, including upregulation of plakoglobin and β-catenin, might be oncogenic. Changes in genes further back in the putative *wnt/wingless* pathway that are also predicted to lead to upregulation of plakoglobin and β-catenin, such as inactivation of *patched* and amplification of *gli*, are linked to basal cell cancer of the skin and glioma, respectively.[56–58]
2. Mutations in the tumour suppressor gene *APC* which inactivate the ability of the APC/GSK3-β-complex to target plakoglobin and β-catenin for degradation are linked to familial adenopolyposis coli, an inherited hyperplasia of the colonic epithelium.[59]
3. N-terminally truncated forms of β-catenin transform NIH-3T3 cells and have been isolated from cancer cells.[60,61] Mutations in Ser-37 residue of β-catenin have been found in colonic tumour cells and melanoma cells.[62,63] These deletions and mutations result in highly stable protein[62] by removing critical sites of serine phosphorylation that target β-catenin and plakoglobin for ubiquitination and degradation by the proteosome.[95,96]

In *Xenopus*, the axis-duplicating signals of cytoplasmic plakoglobin and β-catenin can be suppressed by co-expression of desmosomal or classical cadherins.[38,64] These proteins are thought to act either by sequestering plakoglobin/β-catenin at the membrane or by sterically hindering association of these proteins with their downstream signalling partners. Cadherins and catenins therefore mutually regulate each other: catenins are necessary for the stability of cadherin-mediated adhesion and cadherins in turn suppress the cytosolic signals of catenins that are potentially oncogenic. Therefore, mutations that disrupt interactions between cadherins and catenins are also likely to be oncogenic. Null mutations in cadherins have been linked to lobular breast cancer;[65] deletions of the α-catenin binding region of β-catenin[61] and a deletion in the fourth arm repeat of plakoglobin that inhibits its association with desmoglein, classical cadherins and APC have been reported in cancer cell lines.[66]

Recently, overexpression of plakoglobin has been reported to displace β-catenin from cadherin association and promote its degradation.[97] It has also been shown to retard the growth and tumorigenicity of highly transformed cell lines, suggesting that it may act in certain cellular contexts as a tumour suppressor.[98]

The 'ARM' family

Many more distantly related proteins have been described which contain varying numbers and arrangements of ARM repeats (Table 1). Some are clearly functionally related to plakoglobin/β–catenin and Armadillo. For example, the p120^{ctn} group of proteins, first described as substrates of src and growth factor receptors, bind to classical cadherins.[16,67] p120 lacks an α-catenin binding

Table 1 'ARM' proteins

ARM protein	Percentage identity to ARM consensus	Ref.	Chromosome	Gen/EMBL accession no
Plakoglobin		1, 2	17q21	Z68228, M90365, M95593
β-catenin	60–70	3, 4	3p22–p21.3	X87838, M90364,
M77013				
Armadillo		5, 6		X54468
p120		16, 67	11q11	Z17804
ARV		72	22q11	U51269
p0071		69	2	X81889
p6542	33–60	69		
Plakophilin 1/B6P		68, 70		S75710, Z73678
Plakophilin 2		71		X97675
Importin-α		81	6, 3	U93240, L36339
SRP1		80		U28386
RCH1	48–62	76		U09559
Pendulin/OHO 31		83		U12269, U34229
PF16 microtubule-binding flagellar protein		79		U40057
APC tumour suppressor gene		59, 82	5q21	M74088, M88127
smgGDS nucleotide exchange factor		85		X63465, M63325
SMAP smgGDS-binding protein		75	1	U59919
Protein phosphatase 2A	< 25	77	11	J02902, M9889
Elongation factor 3		78		J05197
β-adaptin		83	17q11–q12	L13939, M34176, X75910
SpKAP115, subunit of kinesin II		86		U38655

These proteins contain repeat sequences that conform to a consensus derived from comparisons of the 13 consecutive 42 amino acid repeats first noticed in the armadillo protein.[5,6] The presence of ARM repeats in many of these proteins was first described in refs[83,84]

domain and therefore impairs the ability of the cadherins to promote stable cell adhesion. Five proteins (band 6/plakophilin 1, plakophilin 2, ARV, p0071, and p6542) closely related to p120 have recently been cloned.[68–71] Band 6/plakophilin 1 bind to desmosomal cadherins and intermediate filaments.[22,68] Plakophilin 2 and p0071 localize to desmosomes, the cytoplasm and the nucleus. p0071 may also associate with adherens junctions.[69] ARV has been linked to velo-cardio-facial syndrome.[72]

APC, the colonic tumour suppressor gene product that functions to target plakoglobin and β-catenin for degradation, also contains ARM repeats, although it does not use them in its interactions with plakoglobin and β-catenin.[73,74] In contrast, smgGDS, a guanine nucleotide-exchange factor, uses its ARM repeats to bind to SMAP, another ARM protein of unknown function.[75] Other ARM family members are engaged in a wide variety of cellular functions including nuclear import, cytoskeletal stability, cell cycle control, and serine–threonine dephosphorylation (see Table 1).[76–86]

Structure and function of ARM repeats

Structural determination of a synthetic APC arm repeat showed the presence of a pair of antiparallel α-helices.[87] Helical wheel and extensive mutational analysis of PP2A-A ARM repeats has led to a model in which each repeat folds to form two amphipathic interacting helices. The inter-repeat and intra-repeat loops projecting from either surface form the binding sites for its many partner proteins.[88] X-ray crystallography of the central repetitive region of β-catenin essentially confirms the PP2A-A model and shows that the repeats stack to form an elongated superhelix with the inter-repeat sequences forming a central positively charged groove that probably forms the interactive surface.[99] Thus arm proteins may be viewed as docking platforms for protein–protein interactions (Fig. 2). Plakoglobin, β-catenin and armadillo, like PP2A, bind to many different partners. In the latter protein it has been demonstrated that partners compete and cooperate to bind in specific combinations of targeting and regulatory subunits that confer specific cellular localization and thus different targets on this phosphatase. As the common denominator for many competing partners, ARM proteins act as integrators of a number

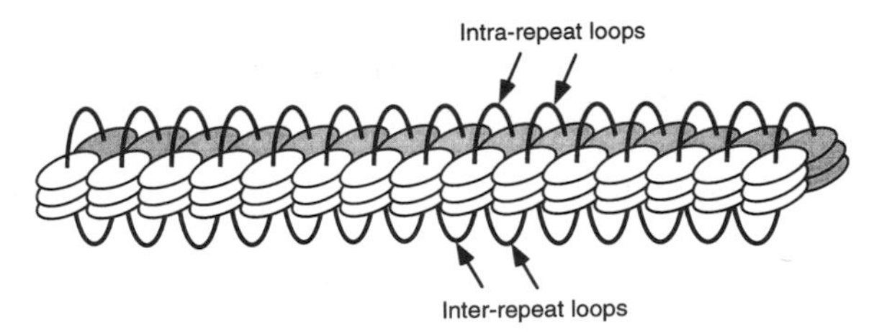

Figure 2. Current model of an ARM protein based on a model proposed for PP2A-A by Ruedigger *et al.*[88]

of cellular processes that must be orchestrated. For example, plakoglobin and β-catenin appear coordinately, to regulate cellular adhesion and cell proliferation and are therefore likely to be the regulators of contact-inhibition of cell proliferation.

Purification

Partial purification protocols for soluble plakoglobin and β-catenin have been published.[89,90]

Antibodies

Polyclonal and monoclonal antibodies that are specific for plakoglobin and β-catenin and armadillo have been described.[1,3,6]

References

1. Cowin, P., Kapprell, H.-P., Franke, W. W., Tamkun, J., and Hynes, R. O. (1986). *Cell*, **46**, 1063–73.
2. Franke, W. W., Goldschmidt, M. D., Zimblemann, R., Mueller, H. M., Schiller, D. L., and Cowin, P. (1989). *Proc. Natl Acad. Sci., USA*, **86**, 4027–31.
3. McCrea, P. D., Turck, C. W., and Gumbiner, B. (1991). *Science*, **254**, 1359–61.
4. Butz, S., Stappert, J., Weissig, H. and Kemler, R. (1992). *Science*, **257**, 1142–3.
5. Riggleman, B., Wieschaus, E., and Schedl, P. (1989). *Genes Dev.*, **3**, 96–113.
6. Peifer, M. and Weischaus, E. (1990). *Cell*, **63**, 1167–78.
7. Cowin, P. and Burke, B. (1996). *Curr. Opin. Cell Biol.*, **8**, 56–65.
8. Hulsken, J., Birchmeier, W., and Behrens, J. (1994). *J. Cell Biol.*, **127**, 2061–71.
9. Sacco, P. A., McGranahan, M., Wheelock, M. J., and Johnson, K. R. (1995). *J. Biol. Chem.*, **270**, 20201–6.
10. Orsulic, S. and Peifer, M. (1996). *J. Cell Biol.*, **134**, 1283–300.
11. Witcher, L., Collins, R., Puttagunta, S., Mechanic, S., Munson, M., Gumbiner, B., and Cowin, P. (1996). *J. Biol. Chem.*, **271**, 10904–9.
12. Aberle, H., Schwartz, H., Hoschuetzky, H., and Kemler, R. (1996). *J. Biol. Chem.*, **271**, 1520–6.
13. Hoschuetzky, H., Aberle, H., and Kemler, R. (1994). *J. Cell Biol.*, **127**, 1375–81.
14. Kanai, Y., Ochai, A., Shibata, T., Oyama, T., Ushijima, S., Akimoto, S., and Hirohashi, S. (1995). *Biochem. Biophys. Res. Commun.*, **208**, 1067–72.
15. Matsuyoshi, N., Hamaguchi, M., Taniguchi, S., Nagafuchi, A., Tsukita, S., and Takeichi, M. (1992). *J. Cell Biol.*, **118**, 703–14.
16. Reynolds, A. and Daniel, J. (1997). In *Cytoskeletal–membrane interactions and signal transduction* (ed. P. Cowin and M. Klymkowsky), pp. 31–48. R.G. Landes, Austin.
17. Fuchs, M., Muller, T., Lerch, M. M., and Ulrich, A. (1996). *J. Biol. Chem.*, **271**, 16712–19.
18. Brady-Kalnay, S. M., Rimm, D. L., and Tonks, N. K. (1995). *J. Cell Biol.*, **130**, 977–86.
19. Balsamo, J., Leung, T., and Ernst, H. (1996). *J. Cell Biol.*, **134**, 801–13.
20. Kypta, R. M., Su, H., and Reichardt, L. F. (1996). *J. Cell Biol.*, **134**, 1519–29.
21. Zondag, G. C. M., Moolenaar, W. H., and Gebbink, M. (1996). *J. Cell Biol.*, **134**, 1513–17.
22. Mathur, M., Goodwin, L., and Cowin, P. (1994). *J. Biol. Chem.*, **269**, 14075–80.
23. Troyanovsky, S. M., Troyanovsky, R. B., Eshkind, L. G., Krutovskikh, V. A., Leube, R. E., and Franke, W. W. (1994). *J. Cell Biol.*, **127**, 151–60.
24. Troyanovsky, S., Troyanovsky, R. B., Eshkind, L. G., Leube, R. E., and Franke, W. W. (1994). *Proc. Natl Acad. Sci., USA*, **91**, 10790–4.
25. Chitaev, N. A., Leube, R. E., Troyanovsky, R. B., Franke, W. W., and Troyanovsky, S. (1996). *J. Cell Biol.*, **133**, 359–69.
26. Tao, Y. S., Edwards, R. A., Tubb, B., and McCrea, P. (1996). *J. Cell Biol.*, **134**, 1271–81.
27. Ravjasarakaran, A. K., Hojo, M., Huima, T., and Rodriguez-Boulan, E. (1996). *J. Cell Biol.*, **132**, 451–63.
28. Haegel, H., Larue, L., Oshugi, M., Fedorov, L., Herrenknecht, K., and Kemler, R. (1995). *Development*, **121**, 3529–37.
29. Bierkamp, C., McLaughlin, K. L., Schwarz, H., Huber, O., and Kemler, R. (1996). *Dev. Biol.*, **180**, 780–5.
30. Ruiz, P., Brinkmann, V., Ledermann, B., Behrend, M., Grund, C., *et al.* (1996). *J. Cell Biol.*, **135**, 215–25.
31. Plantier, J.-L. and Kemler, R. (1997). In *Cytoskeletal–membrane interactions and signal transduction* (ed. P. Cowin and M. Klymkowsky), pp. 1–11. R.G. Landes, Austin.
32. Gelderloos, J., Witcher, L., Cowin, P., and Klymkowsky, M. (1997). In *Cytoskeletal–membrane interactions and signal transduction* (ed. P. Cowin and M. Klymkowsky), pp. 12–30. R.G. Landes, Austin.
33. Hacker, U. and Perrimon, N. (1997). In *Cytoskeletal–membrane interactions and signal transduction* (ed. P. Cowin and M. Klymkowsky), pp. 61–72. R.G. Landes, Austin.
34. Bradley, R. S., Cowin, P., and Brown, A. M. C. (1993). *J. Cell Biol.*, **123**, 1857–65.
35. Hinck, L., Nelson, W. J., and Papkoff, J. (1994). *J. Cell Biol.*, **124**, 729–41.
36. Dominguez, I., Itoh, K., and Sokol, S. Y. (1995). *Proc. Natl Acad. Sci., USA*, **92**, 8498–502.
37. McMahon, A. P. and Moon, R. T. (1989). *Cell*, **58**, 1075–84.
38. Karnovsky, A. and Klymkowsky, M. (1995). *Proc. Natl. Acad. Sci.*, **92**, 4522–26.
39. Funayama, N., Fagotto, F., McCrea, P., and Gumbiner, B. M. (1995). *J. Cell Biol.*, **128**, 959–68.
40. Wang, Y., Macke, J. P., Abella, B. S., Anderson, K., Worley, P., Gilbert, D. J., *et al.* (1996). *J. Biol. Chem.*,. **271**, 4468–76.
41. Bhanot, P., Brink, M., Samos, C. H., Hsieh, J.-C., Wang, Y., Macke, J. P., *et al.* (1996). *Nature*, **382**, 225–30.
42. Klymkowsky, M. (1998). Bioessays (In press.)
43. Merriam, J. M., Rubenstein, A. B. and Klymkowsky, M. W. (1997). *Dev. Biol.*, **185**, 67–81.
44. Behrens, J., von Kries, J. P., Kuhl, M., Bruhn, L., Wedlich, D., Grosschedl, R., and Birchmeier, W. (1996). *Nature*, **382**, 638–42.
45. Huber, O., Korn, R., McLaughlin, J., Ohsugi, M., Herrmann, B. G., and Kemler, R. (1996). *Mech. Dev.*, **59**, 3–10.

46. Molenaar, M., van de Wetering, M., Oosterwegel, M., Peterson-Maduro, J., Godsave, S., Korinek, V., *et al.* (1996). *Cell*, **86**, 391–9.
47. Korinek, V., Barker, N., Morin, P. J., van Wichen, D., de Weger, R., Kinzler, K., *et al.* (1997). *Science*, **275**, 1784–7.
48. Brunner, E., Oliver, P., Schweizer, L., and Basler, K. (1997). *Nature*, **385**, 829–33.
49. Love, J. L., Li, X., Case, D., Giese, K., Grosschedl, R., and Wright, P. E. (1995). *Nature*, **376**, 791–5.
50. Giese, K., Kingsley, C., Kirshner, J. R. and Grosschedl, R. (1995). *Genes Dev.*, **9**, 995–1008.
51. van de Wetering, M., Cavallo, R., Dooiles, D., von Beest, M., van Es, J., *et al.* (1997). *Cell*, **88**, 789–99.
52. Carnac, I., Kodjabachian, L., Gurdon, J. B., and Lemaire, P. (1996). *Development*, **122**, 3055–65.
53. Riese, J., Yu, X., Munnerlyn, A., Eresh, S., Hsu, S., Grosschedl, R., and Bienz, M. (1997). *Cell*, **88**, 777–87.
54. Zhou, P., Byrn, C., Jacobs, J., and Fuchs, E. (1995). *Genes Dev.*, **9**, 570–83.
55. Tsukamoto, A., Grosschedl, R., Guzman, R., Parslow, T., and Varmus, H. (1988). *Cell*, **55**, 619–25.
56. Hahn, H., *et al.* (1996). *Cell*, **85**, 841–51.
57. Johnson, R. L., Rothman, A. L., Xie, J., Goodrich, L. V., Bare, J. W., Bonifas, J. M., *et al.* (1996). *Science*, **272**, 1668–71.
58. Kinzler, K., Ruppert, J. M., Bigner, S. H., and Vogelstein, B. (1988). *Nature*, **332**, 371–374.
59. Polakis, P. (1995). *Curr. Opin. Genet. Dev.*, **5**, 66–71.
60. Whitehead, I., Kirk, H., and Kay, R. (1995). *Mol. Cell. Biol.*, **15**, 704–10.
61. Kawanishi, J., Kato, J., Sasaki, K., Fujii, S., Watanabe, N., and Niitsu, Y. (1995). *Mol. Cell. Biol.*, **15**, 1175–81.
62. Rubinfeld, B., Robbins, P., El-Gamil, M., Albert, I., Porfiri, E., and Polakis, P. (1997). *Science*, **275**, 1790–2.
63. Morin, P. J., Sparks, A. B., Korinek, V., Barker, N., Clevers, H., Vogelstein, B., and Kinzler, K. (1997). *Science*, **275**, 1787–90.
64. Heasman, J., Crawford, A., Goldstone, K., Garner-Hamrick, P., Gumbiner, B., McCrea, P., *et al.* (1994). *Cell*, **79**, 791–803.
65. Berx, G., Cleton-Jansen, A. M., Nollet, F., de Leeuw, W. J., van de Vijver, M., Cornelisse, C., and van Roy, F. (1995). *EMBO J.*, **14**, 6107–15.
66. Ozawa, M., Terada, H., and Pedraza, C. (1995). *J. Biochem.*, **118**, 1077–82.
67. Reynolds, A. B., Daniel, J., McCrea, P., Wheelock, M., Wu, J., and Zhang, Z. (1994). *Mol. Cell. Biol.*, **14**, 8333–42.
68. Hatzfeld, M., Kristjansson, G. I., Plessmann, U., and Weber, K. (1994). *J. Cell Sci.*, **107**, 2259–70.
69. Hatzfeld, M. (1997). In *Cytoskeletal–membrane interactions and signal transduction* (ed. P. Cowin and M. Klymkowsky), pp. 49–60. R.G. Landes, Austin.
70. Heid, H. W., Schmidt, A., Zimbelmann, R., and Franke, W. W. (1994). *Differentiation*, **58**, 113–31.
71. Mertens, C., Kuhn, C., and Franke, W. W. (1996). *J. Cell Biol.*
72. Sirotkin, H., O'Donnell, H., Dasgupta, R., Halford, S., St. Jore, B., Puech, A., *et al.* (1997). *Genomics*, **41**, 75–83.
73. Su, L. K., Vogelstein, B. and Kinzler, K. W. (1993). *Science*, **262**, 1734–7.
74. Rubinfeld, B., Souza, B., Albert, I., Muller, O., Chamberlain, S., Masiarz, F. R., *et al.* (1993). *Science*, **262**, 1731–4.
75. Shimizu, K., Kawabe, H., Minami, S., Honda, T., Takaishi, K., Shirataki, H., and Takai, Y. (1996). *J. Biol. Chem.*, **271**, 27013–17.
76. Cuomo, C. A., Kirch, S. A., Gyuris, J., Brent, R., and Oettinger, M. A. (1994). *Proc. Natl Acad. Sci., USA*, **91**, 6156–60.
77. Hemmings, B., Adams-Pearson, C., Maurer, F., Muller, P., Goris, J., Merlevede, W., *et al.* (1990). *Biochemistry*, **29**, 3166–73.
78. Qin, S., Xie, A., Bonato, C. M., and McLaughlin, C. S. (1990). *J. Biol. Chem.*, **265**, 1903–12.
79. Smith, E. F. and Lefebvre, P. A. (1996). *J. Cell Biol.*, **132**, 359–70.
80. Yano, R., Oakes, L., Tabb, M., and Nomura, M. (1994). *Proc. Natl Acad. Sci., USA*, **91**, 6880–4.
81. Belanger, K. D., Kenna, M. A., Wei, S., and Davis, L. (1994). *J. Cell Biol.*, **126**, 619–30.
82. Kinsler, K. W., *et al.* (1991). *Science*, **253**, 661–9.
83. Kussell, P. and Frasch, M. (1995). *J. Cell Biol.*, **129**, 1491–507.
84. Peifer, M., Berg, S., and Reynolds, A. (1994). *Cell*, **76**, 789–91.
85. Kikuchi, A., Kaibuchi, K., Hori, Y., Nonaka, H., Sakoda, T., Kawamura, M., *et al.* (1992). *Oncogene*, **7**, 289–93.
86. Gindhart, J. G. and Goldstein, L. S. (1996). *Trends Biochem. Sci.*, **21**, 52–3.
87. Hirschl, D., Bayer, P., and Muller, O. (1996). *FEBS Lett.*, **383**, 31–6.
88. Ruediger, R., Hentz, M., Fait, J., Mumby, M., and Walter, G. (1994). *J. Virol.*, **68**, 123–9.
89. Kapprell, H.-P., Cowin, P., and Franke, W. W. (1987). *Eur. J. Biochem.*, **166**, 505–17.
90. McCrea, P. D. and Gumbiner, B. M. (1991). *J. Biol. Chem.*, **266**, 4514–20.
91. Kowalczyk, A. P., Bornslager, E., Borgwardt, J. E., Palka, H., Avinder, S. D., Corcoran, C. M., Denning, M., and Green, K. J. (1997). *J. Cell Biol.*, **139**, 773–84.
92. Smith, E. A. and Fuchs, E. (1998). *J. Cell Biol.*, **141**, 1229–41.
93. Zeng, L., Fagotto, F., Zhang, T., Hsu, W., Vasicek, T. J., Perry, W. L., Lee, J. J., Tilghman, S. M., Gumbiner, B. M., and Costantini, F. (1997). *Cell*, **90**, 181–92.
94. Behrens, J., Jerchow, B.-A., Wrtele, M., Grimm, J., Asbrand, C., Wirtz, R., Kuhl, M., Wedlich, D., and Birchmeier, W. (1998). *Science*, **280**, 596–9.
95. Orford, K., Crockett, C., Jensen, J. P., Weissman, A. M., and Byers, S. (1997). *J. Biol. Chem.*, **272**, 24735–8.
96. Aberle, H., Bauer, A., Stappert, J., Kispert, A., and Kemler, R. (1997). *EMBO J.*, **16**, 3797–804.
97. Simcha, I., Geiger, B., Yehuda-Levenberg, S., Salomon, D., and Ben-Ze'ev, A. (1996). *J. Cell Biol.*, **133**, 199–209.
98. Salomon, D., Sacco, P., Roy, S., Simcha, I., Johnson, K. R., Wheelock, M. J., and Ben-Ze'ev, A. (1997). *J. Cell Biol.*, **139**, 1325–35.
99. Huber, A. H., Nelson, W. J., and Weis, W. I. (1997). *Cell*, **90**, 871–82.

Pam Cowin
Department of Cell Biology,
New York University Medical Center,
550 First Avenue, New York, NY 10016,
USA

Ponticulin

Ponticulin is an integral membrane protein that binds directly to the sides of actin filaments and nucleates actin assembly at the cytoplasmic surface of the plasma membrane. Cells lacking ponticulin exhibit defects in pseudopod dynamics, chemotaxis, and multicellular morphogenesis.

Synonyms and related proteins

None known.

Protein properties

Although an immunologically cross-reactive protein of similar size has been observed in human polymorphonuclear leukocytes,[1] only ponticulin from the soil amoeba, *Dictyostelium discoideum*, has been characterized. Purified *Dictyostelium* ponticulin consists of six major 17-kDa isoforms with pIs ranging from 4.2 to 5.2 on two-dimensional SDS–PAGE.[2] In addition, minor polypeptides at 19 kDa (five isoforms, pI range 4.4 to 5.2) and 15 kDa (two isoforms, pIs 4.3 and 4.7) co-isolate with ponticulin on F-actin affinity columns.[2,3] Because all these isoforms are absent from cells in which the single-copy gene encoding ponticulin has been disrupted, the ponticulin isoforms appear to arise from differential post-translational processing.[4] Purified ponticulin in octylglucoside micelles has a sedimentation coefficient of 2.7 S and a Stokes radius of 3.6 nm, suggesting that this protein is an elongated monomer in detergent.[5]

Ponticulin is a transmembrane protein. Extracellular sites are detected by side-specific labelling with biotin and by binding of the lectin concanavalin A.[3] Also, the deduced amino acid sequence predicts a protein with a cleaved amino-terminal signal sequence and a carboxyl-terminal glycosyl anchor,[4] modifications that require processing in the lumen of the endoplasmic reticulum, that is the topological equivalent of the extracellular membrane surface. These predictions are supported by amino acid sequencing of mature ponticulin[6] and by metabolic labelling with glycosyl anchor components,[4] respectively. The presence of an intracellular domain in ponticulin is indicated by the existence of a discontinuous epitope that is accessible to antibody in plasma membranes and permeabilized cells, but not in intact cells.[3,4] A cytoplasmically oriented actin-binding domain is also required for the demonstrated role of ponticulin in binding actin to the plasma membrane *in vivo*.[7] Both antibody- and actin-binding activities are abolished by thiol agents, suggesting that disulphide bond(s) are required for proper folding. Despite the evidence for sequences on both sides of the membrane, no α-helical membrane-spanning domains are apparent in the deduced amino acid sequence for ponticulin.[4] Instead, several hydrophobic and/or sided β-strands, each long enough to traverse the membrane, are predicted. Thus, the mature ponticulin molecule apparently contains both a membrane-spanning β-barrel, possibly stabilized by disulphide bonding, and a glycosyl anchor (Fig. 1). Such dual anchoring in the membrane bilayer makes ponticulin one of the founding members of a new class of membrane proteins, called type VI.[4,8]

Ponticulin binds directly to the sides of actin filaments in F-actin blot overlays.[2] This protein also nucleates and stabilizes F-actin at the surfaces of isolated *Dictyostelium* plasma membranes,[9] an activity that can be stored by incorporating purified ponticulin into vesicles containing *Dictyostelium* membrane lipids.[10] Ponticulin is responsible for 90–96 per cent of the high-affinity actin–membrane binding observed *in vitro* and *in vivo*.[3,7] Relatively abundant, ponticulin constitutes 0.7 ± 0.4 per cent of the total

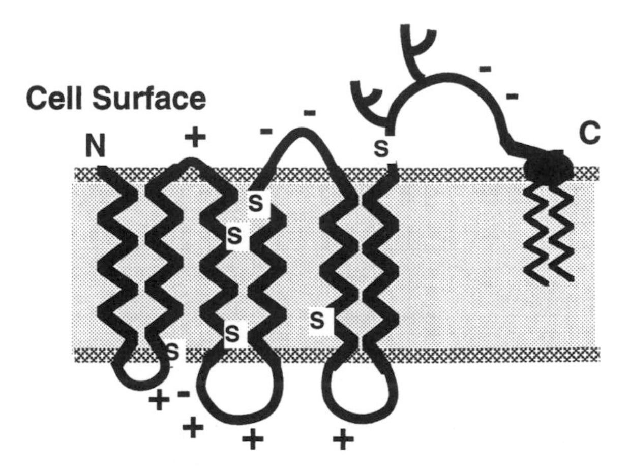

Figure 1. Hypothetical model of ponticulin structure. Because the amino- (N), and carboxy- (C) termini of ponticulin are both processed in the lumen of the endoplasmic reticulum, transmembrane beta-strands must exist in pairs. The six putative membrane-spanning beta-strands shown presumably form a beta-barrel within the membrane. Approximate locations of positively-charged residues (+), negatively-charged residues (–), and cysteines (S) are shown. This model places most of ponticulin's positively-charged residues on the cytoplasmic surface of the membrane, consistent with the observation that the interaction with negatively-charged actin is predominantly electrostatic.[3]

Dictyostelium plasma membrane protein, or ~1.2×10^6 copies per cell, and has an estimated surface density of around 300 monomers per μm^2 of membrane, or approximately that of actin oligomers in the erythrocyte.[5]

By immunofluorescence microscopy shows that *Dictyostelium* ponticulin is concentrated at the plasma membrane and in cytoplasmic punctae suggestive of internal membrane pools undergoing biosynthesis and/or recycling[1] (Fig. 2). The amount of ponticulin in the plasma membrane increases two- to three-fold when amoebae form aggregation streams,[11] a time at which message levels are plummeting,[4] suggesting a dramatic rearrangement within the cell. A role for ponticulin function during multicellular development is further supported by analyses of *Dictyostelium* amoebae lacking ponticulin (see below). The human 17 kDa ponticulin-analogue is localized in plasma membranes of polymorphonuclear leukocytes and is especially evident in actin-rich cell extensions.[1]

Regulation of ponticulin activity *in vivo* is unknown. Possibilities include targeting to specialized regions of the cell membrane, perhaps mediated by the glycosyl anchor, and inhibition of actin-binding activity by disulphide reduction in the lumens of endocytotic vesicles.

Purification

Ponticulin can be purified by F-actin affinity chromatography of detergent-solubilized *Dictyostelium* plasma membranes.[6] However, plasma membrane purification is time consuming and results in the loss of considerable material.[12] For instance, the intracellular stores of ponticulin (Fig. 2) are not recovered. In an improved purification procedure,[3] 8-fold more ponticulin is recovered in one-third the time by extracting whole cells with 1 per cent Triton X-114 in a high salt buffer, followed by detergent phase partitioning, F-actin affinity chromatography, and hydrophobic interaction chromatography with elution by a gradient of octylglucoside.[5]

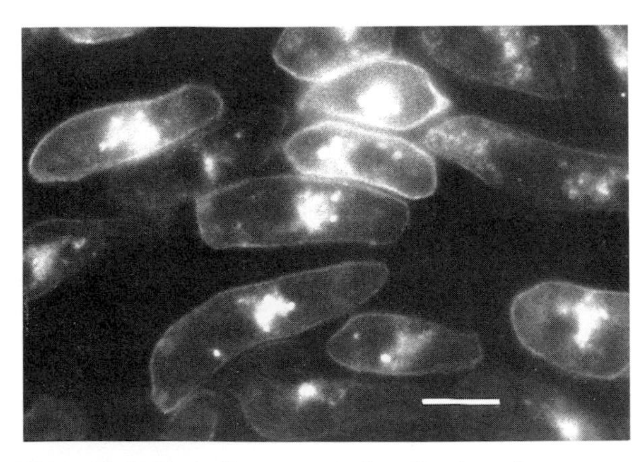

Figure 2. Immunofluorescence localization of ponticulin in aggregating *Dictyostelium* amoebae. Bar, 5 μm.

Activities

Ponticulin binding to actin can be measured by co-sedimentation binding assays[9] and by F-actin blot overlays of SDS–PAGE purified protein after electrotransfer and fixation on to nitrocellulose blots.[2] A 'slot blot' assay employing non-denatured protein yields similar results.[2] Pyrene–actin nucleation assays with ponticulin-containing membrane vesicles[5,10] are performed as described for soluble proteins.[13]

Antibodies

A number of antibodies against ponticulin have been generated in the author's laboratory. The first antibody, and the only one useful for immunofluorescence or immunoprecipitation of nondenatured protein, is a rabbit polyclonal antibody called adsorbed R67 IgG or adsorbed anti-$NaIO_4$-PM IgG.[3,4] As the latter name suggests, this antiserum was generated against plasma membranes that had been sequentially extracted with NaOH (to remove peripheral proteins) and $NaIO_4$ (to oxidize potentially immunogenic carbohydrates). After adsorption against intact, fixed *Dictyostelium* amoebae, remaining antibodies were directed almost solely against the native conformation of ponticulin. Affinity purification on columns of endogenous ponticulin and competition experiments with octylglucoside-solubilized, functional ponticulin verified that the immunofluorescence signal generated with adsorbed R67 IgG was specific for ponticulin. R67 IgG recognizes ponticulin on immunoblots if SDS solubilization is performed for short periods at 70°C, but not if the protein is boiled or if thiol-reducing agents are included in the solubilization buffer.

Other antibodies have been generated against peptides encoded by the ponticulin gene or against fusion proteins containing much or all of the most hydrophilic sequences in ponticulin (unpublished data). Most of these antibodies specifically recognize denatured ponticulin on immunoblots, but none gives a signal in immunofluorescence microscopy. Immunoblot staining with these antibodies is generally improved by boiling and/or reduction of disulphide bonds with thiol reagents. Thus, it appears that antibodies against *Dictyostelium* ponticulin are generally specific either for the endogenous, membrane-bound conformation or for linear epitopes. None of the antibodies against *Dictyostelium* ponticulin reacts strongly with the apparent vertebrate analogues, suggesting that the most immunogenic epitopes are only weakly conserved.

Genes

A full length cDNA for *Dictyostelium* ponticulin (GenBank/EMBL Data Bank accession number Z36535) has been published.[4] Upstream genomic sequences are also known (accession number Z36534). In addition, an EST (Shimizu and Urushihara, GenBank accession number C22774) that potentially encodes a protein with structural similarities to ponticulin has been cloned from a *Dictyostelium* sexual stage library.

Mutant phenotypes

Dictyostelium amoebae lacking ponticulin have been generated by disrupting the single-copy ponticulin gene using homologous recombination.[7] These cells, which lack detectable amounts of ponticulin message and protein, also are deficient in high affinity actin–membrane binding. Furthermore, plasma membranes from ponticulin-null cells exhibit little or no binding or nucleation of exogenous actin *in vitro*. The absence of ponticulin correlates the loss of pseudopod positional stabilization relative to the substratum; pseudopods in ponticulin-null cells 'slip' towards the rear of the cell at rates up to seven-fold greater than the velocity of forward cell movement.[14] As an apparent consequence, the ponticulin-null cells are relatively inefficient at chemotaxis in spatial gradients.[14]

Ponticulin also plays a role during multicellular development. Under conditions that reduce cell–cell adhesion in wild-type cells (highly dispersed cells in low concentrations of magnesium ions), ponticulin-null cells aggregate into mounds faster than do wild-type cells during starvation-induced development.[7] Subsequent morphogenesis of the mutants proceeds asynchronously and is much slower, on average, although viable spores do form. Because the fast-aggregating aspect of this phenotype is the opposite of that predicted for inefficiently chemotaxing cells, another developmental event appears to be abnormal in these mutants. One possibility is that ponticulin normally serves as a negative modulator of initial cell–cell adhesion interactions, a modulation that is required for proper synchronous development.

Structure

Not available.

Web sites

http://www.ummed.edu/dept/cellbio/luna.html
http://www.otus.oakland.edu/biology/staff/hitt.htm

References

1. Wuestehube, L. J., Chia, C. P., and Luna, E. J. (1989). *Cell Motil. Cytoskel.*, **13**, 245–63.
2. Chia, C. P., Hitt, A. L., and Luna, E. J. (1991). *Cell Motil. Cytoskel.*, **18**, 164–79.
3. Wuestehube, L. J. and Luna, E. J. (1987). *J. Cell Biol.*, **105**, 1741–51.
4. Hitt, A. L., Lu, T. H., and Luna, E. J. (1994). *J. Cell Biol.*, **126**, 1421–31.
5. Chia, C. P., Shariff, A., Savage, S. A., and Luna, E. J. (1993). *J. Cell Biol.*, **120**, 909–22.
6. Wuestehube, L. J., Speicher, D. W., Shariff, A., and Luna, E. J. (1991). *Meth. Enzymol.*, **196**, 47–65.
7. Hitt, A. L., Hartwig, J. H., and Luna, E. J. (1994). *J. Cell Biol.*, **126**, 1433–44.
8. Howell, S. and Crine, P. (1996). *TIBS*, **21**, 171–2.
9. Schwartz, M. A. and Luna, E. J. (1988). *J. Cell Biol.*, **107**, 201–9.
10. Shariff, A. and Luna, E. J. (1990). *J. Cell Biol.*, **110**, 681–92.
11. Ingalls, H. M., Barcelo, G., Wuestehube, L. J., and Luna, E. J. (1989). *Differentiation*, **41**, 87–98.
12. Goodloe-Holland, C. M., and Luna, E. J. (1987). *Meth. Cell Biol.*, **28**, 103–28.
13. Cooper, J. A., Walker, S. B., and Pollard, T. D. (1983). *J. Musc. Res. Cell Motil.*, **4**, 253–62.
14. Shutt, D. C., Wessels, D., Wagenknecht, K., Chandrasekhar, A., Hitt, A. L., Luna, E. J., and Soll, D. R. (1995). *J. Cell Biol.*, **131**, 1495–506.

Elizabeth J. Luna
University of Massachusetts Medical Center,
Worcester Foundation Campus,
Shrewsbury, MA,
USA

p58^{gag}

p58^{gag} is a membrane- and microfilament-associated protein isolated from branched microvilli of a highly metastatic rat mammary adenocarcinoma. Its most striking feature is a high degree of homology to retroviral Gag proteins, although it is truncated at the C terminus and, consequently, lacks the nucleic acid-binding sequence. p58^{gag} contains an N-terminal myristoylation consensus sequence, predicted from its ability to bind phospholipid in a headgroup-independent manner. The Gag protein, which binds microfilaments and acts as a noncleaving, capping protein, is proposed to stabilize membrane–microfilament interactions and restrict membrane glycoprotein mobility, a property correlated with xenotransplantability of p58-expressing rat ascites tumour cell sublines into mice. It also binds TMC-gp65, a component of a large glycoprotein complex which forms the core of a signal transduction particle that contains p185^{neu}, p60^{src}, and c-Abl.

Synonymous names

58 K.[1]

Homologous proteins

p60^{gag} of MAIDS virus,[2,3] also a truncated retroviral Gag protein; p65^{gag} precursor, and related precursors of all retroviral Gag proteins. These proteins are all related to the retroviral Gag protein precursor for the proteins of the viral capsid structure.[4]

Protein properties

p58^{gag} is a monomeric protein with an asymmetric structure (frictional ratio 1.9).[1] By combined gel filtration and velocity sedimentation analyses, its native molecular weight was estimated to be 57 kDa. Its molecular weight by SDS–PAGE under reducing conditions was estimated to be 58 kDa. These values are considerably higher than the 48.3 kDa value calculated from sequence analysis.[5] This variance probably results from two factors: a high proline content (12 per cent), which would increase its apparent molecular weight on SDS–PAGE, and posttranslational phosphorylation. The latter is further suggested by the fact that the product of *in vitro* transcription and translation of a full length cDNA is only 55 kDa by SDS–PAGE. *In vitro* and *in vivo* phosphorylation of p58^{gag} were demonstrated by two dimensional electrophoresis.[6,7] Phosphoamino acid and antiphosphotyrosine immunoblot analyses in the absence of tyrosine phosphatase inhibitor showed serine phosphate >> tyrosine phosphate and no threonine phosphate.[7,8] In the presence of inhibitor, tyrosine and serine phosphate were approximately equivalent and phosphothreonine was present.[7]

The sequence of p58^{gag} is highly homologous to that of the Gag precursor from rat leukaemia virus. Two variant stretches are probably due to sequencing errors in the p58^{gag}, based on sequence comparisons with retroviral Gag sequences from other species. The major difference is the truncation at the C terminus of p58^{gag}.[5] This truncation is reminiscent of the structure of p60^{gag} of the mouse acquired immunodeficiency syndrome (MAIDS),[2,3] which is the causative agent of the syndrome. These observations raise questions about the origin of p58^{gag}. One possibility is that it is a defective viral product carried by a productive infectious virus with the tumour cells from which it was isolated. A second possibility is that it is derived from an endogenous retroviral sequence which is specifically expressed in the tumour cells.

Regardless of the origin, p58^{gag} appears to have profound effects on the 13762 ascites tumour cell subline from which it was isolated.[9] Sublines expressing p58 are xenotransplantable into mice, a property correlated with unusual branched microvilli morphology[10] and with a loss of membrane mobility of cell surface glycoproteins.[11,12] These phenomena are associated with a stabilization of the microvilli to disruption by agents such as cytochalasin D and hypotonic swelling.[10] Microvillus stability has been attributed to stabilization of membrane–microfilament interaction sites by p58^{gag} association both with membranes and microfilaments, as demonstrated by fractionation studies on the microvilli.[13] Additional fractionation studies under microfilament depolymerizing conditions have shown that p58^{gag} is associated with a stable, multimeric complex of glycoproteins containing the growth factor receptor p185^{neu}/ErbB2[14] and actin (transmembrane complex),[15,16] postulated to be both a microfilament binding site at the microvillus membrane and the scaffolding for a p185^{neu}-containing signal transduction particle.[17] These properties are consistent with a role for p58^{gag} in stabilizing the microfilament–membrane interaction and in tumour progression in these ascites cells.

Binding studies on purified or *in vitro* translated p58^{gag} are also consistent with this role. It will bind F-actin, but not G-actin.[1] Under actin polymerizing conditions p58^{gag} acts as a capping protein without severing activity. It will bind phospholipid vesicles with little head group specificity,[1] probably due at least in part to fatty acid linked to the N-terminal glycine myristoylation site.[5] *In vitro* translated p58^{gag} binds to gp65[16] of the glycoprotein complex isolated from the ascites cell microvilli.[15–17] A model for the role of p58^{gag} in stabilizing the interaction of microfilaments with the microvillus transmembrane glycoprotein complex is shown in Fig. 1.

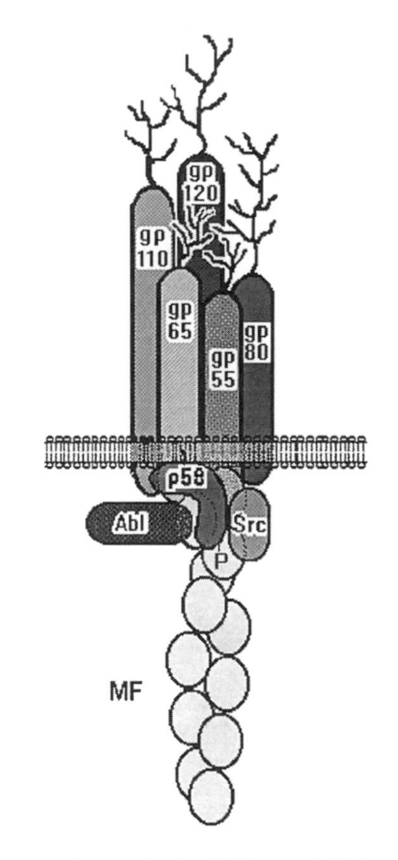

Figure 1. Model for role of p58^{gag} in stabilizing microfilament–membrane interactions in 13762 ascites tumour cell microvilli.

Purification

$p58^{gag}$ has been purified from microvilli isolated from the MAT-C1 ascites subline of the 13762 rat mammary adenocarcinoma.[1] Microvillar microfilament cores are prepared by Triton X-100 extraction and centrifugation. The cores are stripped of peripheral proteins with EGTA/EDTA and solubilized in 1 M NaCl. Dialysis under actin polymerization conditions followed by high speed centrifugation yields a pellet containing a complex of $p58^{gag}$, actin, and associated glycoproteins. The pellet is solubilized again in 1 M NaCl and fractionated on a combined hydroxyapatite–Sephadex G-150 column to yield purified $p58^{gag}$.

Activities

$p58^{gag}$ will bind and cap microfilaments during actin polymerization.[1] Binding can be measured by adding G-actin to $p58^{gag}$ in solution and initiating polymerization. Binding of $p58^{gag}$ is demonstrated by high speed centrifugation of the resulting microfilaments and SDS–PAGE of the pellet. Capping activity is measured using pyrenyl-actin polymerization in the presence of varying concentrations of $p58^{gag}$. Binding of $p58^{gag}$ to phospholipid vesicles can be measured by incubating vesicles with $p58^{gag}$, followed by high speed centrifugation and SDS–PAGE of the pellets.[1] Transfection of Cos-7 and MDCK cells with $p58^{gag}$ cDNA resulted in rearrangement of microfilament structure, as predicted for a microfilament-associated protein.[5] The $p58^{gag}$ sequence has proline-rich regions which are similar to motifs reported to bind Src and Abl, and $p58^{gag}$ binds both full length and recombinant SH3 domain proteins of $p60^{src}$ and c-Abl and is phosphorylated by each. The Gag protein is hypothesized to be a scaffolding protein which recruits these cytoplasmic kinases to the signal transduction particle in the tumour cells.[18] Consistent with this hypothesis is the observation by confocal microscopy that much of the c-Abl of the ascites cells is delocalized from the nucleus to the plasma membrane in p58-expressing, but not in non-expressing ascites cells.[19]

Antibodies

Polyclonal antibodies have been made that recognize $p58^{gag}$ using two methods. In the first method a crude complex of microvillus microfilament-associated glycoproteins, actin and $p58^{gag}$ was injected into rabbits.[1] $p58^{gag}$-specific antibodies were blot-purified from the resulting antiserum and used for immunofluorescence and immunoblot analyses. For the second method, purified $p58^{gag}$ was injected into mice.[5] The resulting antiserum was used for immunoprecipitation, immunofluorescence, and immunoblot analyses.

Genes

A full length cDNA has been isolated.[5] No genomic clones are available.

Mutant phenotypes/disease states

No information available.

References

1. Liu, Y., Carraway, K. L., and Carraway, C. A. C. (1989). *J. Biol. Chem.*, **264**, 1208–14.
2. Jolicoeur, P. (1991). *FASEB J.*, **5**, 2398–405.
3. Morse, H. C. III, Chattopadhyay, S. K., Makino, M., Fredrickson, T. N., Hugin, A. W., and Hartley, J. W. (1992). *AIDS*, **6**, 607–21.
4. Wills, J. W. and Craven, R. C. (1991). *AIDS*, **5**, 639–54.
5. Juang, S. -H., Huang, J., Li, Y., Salas, P. J. I., Fregien, N., Carraway, C. A. C., and Carraway, K. L. (1994). *J. Biol. Chem.*, **269**, 15067–75.
6. Liu, Y. (1988). Ph. D. thesis, University of Miami.
7. Juang, S. -H., Carvajal, M. E., Whitney, M., Liu, Y., and Carraway, C. A. C. (1996). *Oncogene*, **12**, 1033–42.
8. Juang, S. -H. (1993). Ph. D. thesis, University of Miami.
9. Sherblom, A. P., Huggins, J. W., Chesnut, R. W., Buck, R. L., Ownby, C. L., Dermer, G. B., and Carraway, K. L. (1980). *Exp. Cell Res.*, **126**, 417–26.
10. Carraway, K. L., Huggins, J. W., Cerra, R. F., Yeltman, D. R., and Carraway, C. A. C. (1980). *Nature*, **285**, 508–10.
11. Huggins, J. W., Trenbeath, T. P., Yeltman, D. R., and Carraway, K. L. (1980). *Exp. Cell Res.*, **127**, 31–46.
12. Howard, S. C., Hull, S. R., Huggins, J. W., Carraway, C. A. C., and Carraway, K. L. (1982). *J. Natl Cancer Inst.*, **69**, 33–40.
13. Carraway, C. A. C., Cerra, R. F., Bell, P. B., and Carraway, K. L. (1982). *Biochim. Biophys. Acta*, **719**, 126–39.
14. Carraway, C. A. C., Carvajal, M. E., Li, Y., and Carraway, K. L. (1993). *J. Biol. Chem.*, **268**, 5582–7.
15. Carraway, C. A. C., Jung, G., and Carraway, K. L. (1983). *Proc. Natl Acad. Sci. USA*, **80**, 430–4.
16. Carraway, C. A. C., Fang, H., Ye, X., Juang, S. -H., Liu, Y., Carvajal, M., and Carraway, K. L. (1991). *J. Biol. Chem.*, **266**, 16238–46.
17. Li, Y., Carraway, K. L., and Carraway, C. A. C. (Manuscript submitted.)
18. Huang, J., Li, Y., Juang, S. -H., Mayer, B. J., Carraway, K. L., and Carraway, C. A. C. (Manuscript submitted.)
19. Huang, J., Price-Schiavi, S., Van Elten, R., Mayer, B. J., and Carraway, C. A. C., (Manuscript in preparation).

Coralie A. Carothers Carraway
Department of Biochemistry and Molecular Biology, University of Miami School of Medicine, PO Box 016960, Miami FL 33101, USA

Kermit L. Carraway
Department of Cell Biology and Anatomy, University of Miami School of Medicine, PO Box 016960, Miami FL 33101, USA

p130^{CAS}

p130^{CAS} is a recently identified focal adhesion protein[1,2] which consists of a single SH3 (src-homology 3) domain and multiple SH2- (src-homology 2) and SH3-binding motifs.[3] Its localization to focal adhesions and its tyrosine phosphorylation in response to diverse stimuli such as cell adhesion, G-protein coupled receptor activation, receptor and non-receptor tyrosine kinase activation, and cell transformation, suggest that p130^{CAS} may play an important role in tyrosine kinase-dependent signalling pathways that regulate cell growth and morphology.

Synonymous names

p130 src-substrate.

Homologous proteins

Two p130^{CAS}-related proteins have recently been identified. These are HEF-1/Cas-L (for human enhancer of filamentation 1, or lymphocyte-type Cas),[4,5] and Efs/Sin (for embryonal fyn-associated substrate, src interacting or signal integrating protein).[6,7] These proteins share a common domain structure with p130^{CAS}: an N-terminal SH3 domain followed by a 'substrate domain' which consists of multiple putative SH2-binding sites with the sequence YXXP, and a C-terminal SH2-binding site for src family kinases. These proteins exhibit significant homology within the SH3 domain and the C terminus.

Protein properties

p130^{CAS} has the structural characteristics of adapter proteins involved in tyrosine kinase-dependent signalling pathways. It consists of an amino terminal SH3 domain followed by a proline rich region and a cluster of putative SH2 binding motifs (Fig. 1). Many of these sites are suitable targets for phosphorylation by cytoplasmic protein tyrosine kinases. This region has been designated the 'substrate domain'. The C terminus of p130^{CAS} contains both an SH3- and an SH2-binding site for src.[3,8]

Two forms of p130^{CAS} have been cloned from both rat and mouse cDNA libraries: a long form encoding a 968 residue protein and an alternately spliced short form which lacks a 282 bp region of the 5'end. This results in an in-frame N-terminal deletion which gives rise to an 874 residue protein.[3,9] The predicted molecular weight of these two forms of p130^{CAS} is approximately 94 kDa (short form) and 104 kDa (long form). On SDS–PAGE, however, p130^{CAS} migrates as multiple species which appear as a broad band of 115–130 kDa. The higher molecular weight forms of p130^{CAS} can be converted to the lower molecular weight forms by treatment with phosphatases, suggesting that the divergent gel mobility is due to multiple phosphorylation states.[3]

By Northern blot analysis, p130^{CAS} is expressed as a single 3.2 kb band, indicating that one of the cDNA species (presumably the long form) is predominantly expressed. Its mRNA is widely expressed in all tissues, with particularly high levels found in the testis, intestine

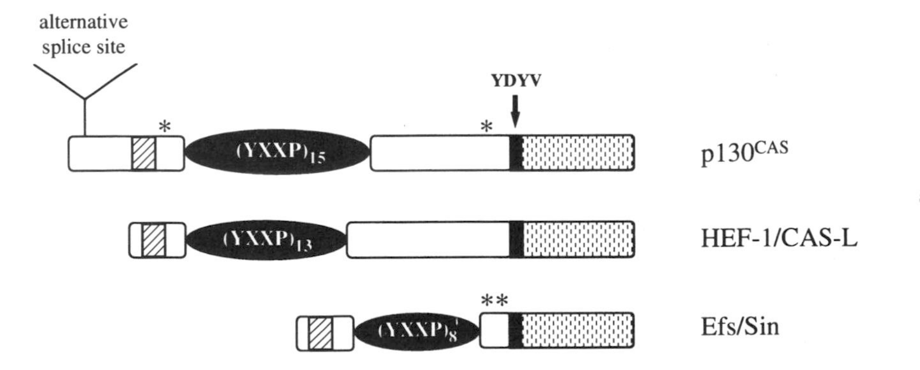

Figure 1. Domain structure of p130^{CAS} and related family members. Regions of homology are designated by shaded or hatched areas. The cross-hatched box at the amino terminus represents the SH3 domain of these proteins. The substrate domain, which contains variable numbers of the SH2-binding motif YXXP, is represented by a black oval. The arrow labelled YDYV denotes the conserved binding site for the src SH2 domain. The asterisks represent the proline-rich regions present in p130^{CAS} and the Efs/Sin protein, but absent in HEF-1/Cas-L.

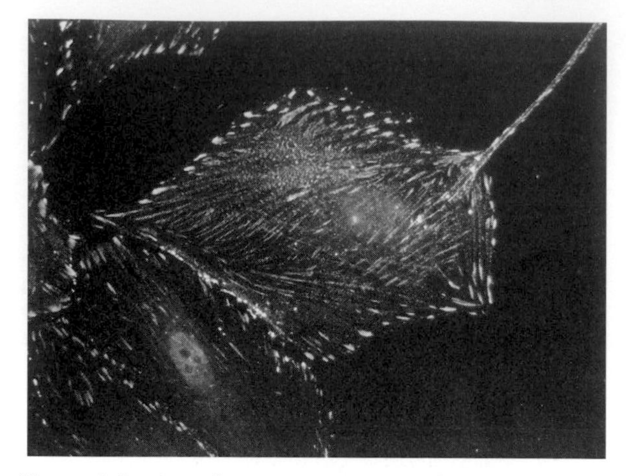

Figure 2. Immunofluorescent staining of rat embryo fibroblast cells (REF52) with the monoclonal antibody 4F4 directed against p130^{CAS}. In these cells p130^{CAS} localizes to both focal adhesions and stress fibres.

and lung.[3] This is in contrast to the other p130^{CAS} family members which exhibit a more restricted pattern of expression.[4,6,7]

Early reports suggested that p130^{CAS} localizes to the cytoplasm and the nucleus.[3,10] More recent studies have shown that p130^{CAS} is not found in the nucleus, but rather localizes to the cytoplasm and to focal adhesions[1,2] (Fig. 2). In focal adhesions, it interacts directly with the focal adhesion kinase pp125^{FAK}.[2,9] Conditions which lead to the tyrosine phosphorylation of p130^{CAS} lead to its translocation to the cytoskeleton and elevated levels of a FAK/p130^{CAS} complex.[11,12] An association with another focal adhesion protein, tensin, has also been reported.[13]

In normal fibroblasts, integrin clustering or adhesion of cells to extracellular matrix proteins leads to an increase in the level of tyrosine phosphorylated p130^{CAS}.[1,2,14,15] Given the localization of p130^{CAS} to focal adhesions and its association with FAK, these data suggest a role for p130^{CAS} in integrin-mediated signalling. Consistent with this, recent studies indicate that p130^{CAS} plays a role in the regulation of cell migration on extracellular matrix proteins. In CHO cells, p130^{CAS} was shown to mediate the FAK-promoted migration of cells on fibronectin,[34] whereas in FG human pancreatic carcinoma cells, expression of p130^{CAS} alone was sufficient to promote cell migration on vitronectin, collagen, and fibronectin.[35] The latter response was dependent on the tyrosine phosphorylation of p130^{CAS} and the assembly of a p130^{CAS}/Crk adaptor protein complex.

Tyrosine phosphorylation of p130^{CAS} has also been observed upon mitogenic stimulation of quiescent cells. Activation of growth factor receptor tyrosine kinases, such as the PDGF receptor,[16] as well as activation of G protein coupled receptors by agents such as lysophosphatidic acid (LPA),[17] bombesin, vasopressin, endothelin,[18] bradykinin,[19] carbachol (through the m1 muscarinic receptor),[20] and melanoma growth stimulatory activity (MGSA), which signals through the class II IL-8 G-protein coupled receptor,[21] all lead to enhanced phosphorylation of p130^{CAS}. Thus, p130^{CAS} may also play a role in mitogenic signalling pathways or, perhaps, serve to link adhesion-induced signalling to growth promoting pathways.

In many instances, tyrosine phosphorylation of this protein correlates with the transformed phenotype of cells. p130^{CAS} is hyperphosphorylated in v-*src*[10,22] and v-*crk*[23,24] transformed cells as well as in haematopoietic cells transformed by the BCR/ABL oncogene.[25] In macrophages, overexpression of the c-*fps/fes* proto-oncogene also leads to enhanced tyrosine phosphorylation of p130^{CAS},[26] as does transformation of fibroblasts by c-Ha-*ras* or by overexpression of ornithine decarboxylase (ODC).[27] In contrast, suppression of v-crk-, v-src-, and ras-induced transformation by overexpression of the protein tyrosine phosphatase PTP1B, correlates with a decrease in p130^{CAS} tyrosine phosphorylation as well as a reduction in the association of p130^{CAS} with v-crk.[36] Furthermore, in the case of v-*src*, c-Ha-*ras*, and ODC transformed cells, expression of antisense mRNA for p130^{CAS} results in reversion of the transformed phenotype suggesting that p130^{CAS} may be required for transformation.[27] Additionally, these data suggest that the oncogenes mentioned above may cause transformation by disrupting or bypassing integrin-mediated signalling events such as the phosphorylation of p130^{CAS}.

The tyrosine kinase(s) responsible for the phosphorylation of p130^{CAS} *in vivo* are the subject of intense study. The fact that p130^{CAS} has been found in a complex with FAK,[2,9] RAFTK (also known as CAKβ, PYK2, and CAD-TK),[37] src,[8,10] abl kinase,[25] and the kinase encoded by the c-*fps/fes* proto-oncogene,[26] and at least three of these (FAK,[31] src,[3] and abl[28]) have been shown to phosphorylate p130^{CAS} *in vitro*, suggests a role for these kinases in the *in vivo* phosphorylation of p130^{CAS}. However, a critical role for src-family kinases in p130^{CAS} phosphorylation is suggested by the fact that, despite normal levels of FAK phosphorylation, the levels of p130^{CAS} tyrosine phosphorylation are somewhat reduced in fyn$^-$ and yes$^-$ cells, and significantly reduced in src$^-$ cells.[29,30]

The domain structure of p130^{CAS} implies a role for this protein in the assembly of signalling complexes. In addition to the kinases mentioned above, a number of SH2 domain-containing proteins have been shown to associate with p130^{CAS} including crk, crkl, nck, grb-2, PLC-γ, and PI 3-kinase.[25,30,31] p130^{CAS} has also been shown to associate directly with, and be a substrate for, the protein tyrosine phosphatase PTP-PEST.[32] Similarly, a direct SH3-mediated association between p130^{CAS} and the protein tyrosine phosphatase PTP1B has been demonstrated.[38] The localization of p130^{CAS} to focal adhesions and its association with cytoskeletal proteins as well as signalling molecules involved in mitogenic signalling suggest that p130^{CAS} probably plays an important role in signalling pathways involved in the regulation of anchorage-dependent growth.

Purification

A complete purification scheme for this protein has not been worked out, but a sequential immunoaffinity purification protocol resulting in the partial purification of p130^{CAS} from v-*crk* transfected rat 3Y1 cells has been reported.[33] In these cells, p130^{CAS} is the major tyrosine phosphorylated protein. An initial purification step, based on the association of p130^{CAS} with the crk oncoprotein, is achieved using a monoclonal antibody against the viral *gag* protein (mAb 1A1) coupled to CNBr-activated Sepharose 4B. This is followed by a second immunoaffinity step using Sepharose 4B coupled to an antiphosphotyrosine antibody (mAb 4G10). Bound protein is eluted using 0.1 M phenylphosphate.

Activities

p130^{CAS} has no measurable activity, but rather appears to function as an adapter protein which presumably plays a role in the assembly of signalling complexes in response to diverse signals such as adhesion of cells to extracellular matrix, G-protein coupled receptor stimulation, cytokine stimulation, activation of receptor and non-receptor tyrosine kinases, and cell transformation by v-*src*, v-*crk*, BCR/abl, and c-Ha-*ras* (see above).

Antibodies

A monoclonal antibody raised against a peptide fragment which contains the src SH2-binding motif and which corresponds to amino acid residues 644–819 near the C terminus of rat p130^{CAS} is commercially available from Transduction Laboratories. Under certain circumstances (presumably those which lead to the tyrosine phosphorylation of the src SH2-binding motif, thereby altering the conformation of the epitope) recognition of p130^{CAS} by this antibody is reduced. This antibody also reportedly crossreacts with the p130^{CAS} related protein HEF-1.[4]

Two polyclonal antibodies, one raised against an N-terminal peptide corresponding to amino acid residues 103–119, and one raised against a C-terminal peptide corresponding to amino acid residues 949–968 of mouse p130^{CAS}, are commercially available from Santa Cruz Biotechnology.

In addition to the commercially available antibodies, a number of polyclonal antibodies have been raised by individual laboratories.[2,3]

Genes

Complete cDNA sequences for p130^{CAS} from rat (GenBank/EMBL/DDBJ/NCBI accession number D297660)[3] and mouse (GenBank accession number U28151)[9] have been reported. In both rat and mouse two forms of p130^{CAS} have been isolated, a short form and a long form, which arise from alternative splicing near the 5'end of the cDNA.

Mutant phenotype/disease states

No information available.

Structure

Not available.

References

1. Petch, L. A., Bockholt, S. M., Bouton, A., Parsons, J. T., and Burridge, K. (1995). *J. Cell Sci.*, **108**, 1371–9.
2. Harte, M. T., Hildebrandt, J. D., Burnham, M. R., Bouton, A., and Parsons, J. T. (1996). *J. Biol. Chem.*, **271**, 13649–55.
3. Sakai, R., Iwamatsu, A., Hirano, N., Ogawa, S., Tanaka, T., Mano, H., *et al.* (1994). *EMBO J.*, **13**, 3748–56.
4. Law, S. F., Estojak, J., Baolin, W., Mysliwiec, T., Kruh, G., and Golemis, E. A. (1996). *Mol. Cell. Biol.*, **16**, 3327–37.
5. Minegishi, M., Tachibana, K., Sato, T., Iwata, S., Nojima, Y., and Morimoto, C. (1996). *J. Exp. Med.*, **184**, 1365–75.
6. Ishino, M., Ohba, T., Sasaki, H., and Sasaki, T. (1995). *Oncogene*, **11**, 2331–8.
7. Alexandropoulos, K. and Baltimore, D. (1996). *Genes Dev.*, **10**, 1341–55.
8. Nakamoto, T., Sakai, R., Ozawa, K., Yazaki, Y., and Hirai, H. (1996). *J. Biol. Chem.*, **271**, 8959–65.
9. Polte, T. R. and Hanks, S. K. (1995). *Proc. Natl Acad. Sci., USA*, **92**, 10678–82.
10. Kanner, S. B. Reynolds, A. B., Wang, H-C. R., Vines, R. R., and Parsons, J. T. (1991). *EMBO J.*, **10**, 1689–98.
11. Polte, T. R. and Hanks, S. K. (1997). *J. Biol. Chem.*, **272**, 5501–9.
12. Nivers, M. G., Birge, R. B., Greulich, H., Verkleij, A., Hanafusa, H., and van Bergen en Henegouwen, P. M. P. (1997). *J. Cell Sci.*, **110**, 389–99.
13. Lo, S. H., Weisberg, E., and Chen, L. B. (1994). *Bioessays*, **16**, 817–23.
14. Nojima, Y., Morino, N., Mimura, T., Hamasaki, K., Furuya, H., Sakai, R., *et al.* (1995). *J. Biol. Chem.*, **270**, 15398–402.
15. Vuori, K. and Ruoslahti, E. (1995). *J. Biol. Chem.*, **270**, 22259–62.
16. Rankin, S. and Rozengurt, E. (1994). *J. Biol. Chem.*, **269**, 704–10.
17. Seufferlein, T. and Rozengurt, E. (1994). *J. Biol. Chem.*, **269**, 9345–51.
18. Zachary, I., Sinnett-Smith, J., and Rozengurt, E. (1992). *J. Biol. Chem.*, **267**, 19031–4.
19. Tippmer, S., Bossenmaier, B. and Häring, H. (1996). *Eur. J. Biochem.*, **263**, 953–9.
20. Gutkind, J. S. and Robbins, K. C. (1992). *Biochem. Biophys. Res. Commun.*, **188**, 155–61.
21. Schraw, W. and Richmond, A. (1995). *Biochemistry*, **34**, 13760–7.
22. Kanner, S. B., Reynolds, A. B., Vines, R., and Parsons, J. T. (1990). *Proc. Natl Acad. Sci. USA*, **87**, 3328–32.
23. Matsuda, M., Mayer, B. J., Fukui, Y., and Hanafusa, H. (1990). *Science*, **248**, 1537–39.
24. Birge, R. B., Fajardo, J. E., Mayer, B. J., and Hanafusa, H. (1992). *J. Biol. Chem.*, **267**, 10588–95.
25. Salgia, R., Pisick, E., Sattler, M., Li, J.-L., Uemura, N., Wong, W.-K., *et al.* (1996). *J. Biol. Chem.*, **271**, 25198–203.
26. Areces, L. B., Dello Sbarba, P., Jücker, M., Stanley, E. R., and Feldman, R. A. (1994). *Mol. Cell. Biol.*, **14**, 4606–15.
27. Auvinen, M., Paasinen-Sohns, A., Hirai, H., Andersson, L. C., and Hölttä, E. (1995). *Mol. Cell. Biol.*, **15**, 6513–25.

28. Mayer, B. J., Hirai, H., and Sakai, R. (1995). *Curr. Biol.*, **5**, 296–305.
29. Bockholt, S. M. and Burridge, K. (1995). *Cell Adhesion Commun.* **3**, 91–100.
30. Vuori, K., Hirai, H., Aizawa, S., and Ruoslahti, E. (1996). *Mol. Cell. Biol.*, **16**, 2606–13.
31. Schlaepfer, D. D., Broome, M. A., and Hunter, T. (1997). *Mol. Cell. Biol.*, **17**, 1702–13.
32. Garton, A. J., Flint, A. J., and Tonks, N. K. (1996). *Mol. Cell. Biol.*, **16**, 6408–18.
33. Sakai., R., Iwamatsu, A., Hirano, N., Ogawa, S., Tanaka, T., Nishida, J., *et al.* (1994). *J. Biol. Chem.*, **269**, 32740–6.
34. Cary, L. A., Han, D. C., Polte, T. R., Hanks, S. K., and Guan, J.-L. (1998). *J. Cell Biol.* **140**, 211–21.
35. Klemke, R. L., Leng, J., Molander, R., Brooks, P. C., Vuori, K., and Cheresh, D. A. (1998). *J. Cell Biol.*, **140**, 961–72.
36. Liu, F., Sells, M. A., and Chernoff, J. (1998). *Mol. Cell. Biol.*, **18**, 250–59.
37. Astier, A., Avraham, H., Manie, S. N., Groopman, J., Canty, T., Avraham, S., and Freedman, A. S. (1997). *J. Biol. Chem.*, **272**, 228–32.
38. Liu, F., Hill, D. E., and Chernoff, J. (1996). *J. Biol. Chem.*, **271**, 31290–95.

Leslie Petch
Department of Cell Biology and Anatomy,
University of North Carolina,
Chapel Hill, NC 27599, USA

pp60$^{c\text{-}src}$

pp60$^{c\text{-}src}$ (c-Src), the 60 kDa product of the c-*src* gene, is a protein-tyrosine kinase (PTK), which is highly conserved throughout the vertebrate kingdom. In mammals c-Src is widely expressed, with the highest levels being found in neurones, osteoclasts, and platelets. c-Src is a member of a family of closely related PTKs, which are all anchored to the inner face of cytoplasmic membranes via an N-terminal lipid modification. Src family PTKs function as signal-transducing subunits of cell surface receptors that lack their own catalytic domain; they are required for signalling by many types of receptors, including receptor PTKs, G-protein-coupled receptors, integrins, cytokine receptors, and multichain antigen receptors. The Src family PTKs are composed of four main functional domains–an N-terminal membrane-association domain linked to a hypervariable (unique) region, a regulatory domain, a protein kinase catalytic domain, and a C-terminal tyrosine phosphorylation site.

Protein properties

Structure and expression

The c-*src* gene is a member of a family of closely related PTK genes, which in mammals currently has eight members (the c-*src*, c-*yes*, c-*fgr*, *lck*, *hck*, *fyn*, *lyn*, and *blk* genes) (for reviews see ref 1, 2). There are related genes in simpler eukaryotic organisms such as *Drosophila*, *C. elegans*, and *Hydra*. All the *src* family genes encode PTKs of about 525 amino acids in length with very similar structural organizations. These proteins are ~80 per cent identical over their C-terminal 450 residues, which includes the catalytic domain, but, apart from the myristoylation signal at the very N terminus, diverge almost completely in their N-terminal 80 residues (exon III in human c-*src*) (Fig. 1). Because the human (536 residues), mouse (535 residues), and chicken (533 residues) c-Src proteins differ slightly in length, and because much of the early work on c-Src defining important residues in the protein were done with chicken c-Src, chicken c-Src numbering is used throughout (Fig. 1). The additional three residues in human c-Src and two residues in mouse c-Src lie in the region between residues 25 and 29 in chicken c-Src. The conserved N-terminal Gly is myristoylated, this modification is essential for membrane association. The N-terminal seven residues of c-Src provide a necessary and sufficient signal for myristoylation, with a Gly at position 1, Ser (or Cys) at position 5 being important. Cysteine residues within the first five residues of src family PTKs are commonly palmitoylated, but c-Src lacks a cysteine in this region. By interacting with phospholipid head groups, alternating lysines within the first 15 residues of src family PTKs cooperate with the myristoyl and palmitoyl groups to stabilize membrane association. Distinct sequences within the N-terminal half have been identified that are responsible for localizing c-Src to different regions of the cytoplasm.

To the C-terminal side of the unique region lie two regulatory domains, SH3 and SH2 (SH = Src homology) (Fig. 1).[3,4] Analysis of the properties of mutant c-Src proteins and the crystal structure of a large fragment of human c-Src including the SH3, SH2, catalytic domain, and the C-terminal domain have helped to define both the enzymatic function and the intramolecular regulation of the c-Src protein.[5] The structures of the individual SH3 and SH2 domains bound to cognate peptides have also been determined.[6,7] The SH3 domain (residues 83–142) plays a negative role in regulating c-Src kinase activity, but has a positive role in downstream signalling through its ability to bind to substrates in a sequence-specific fashion and to mediate association with the actin

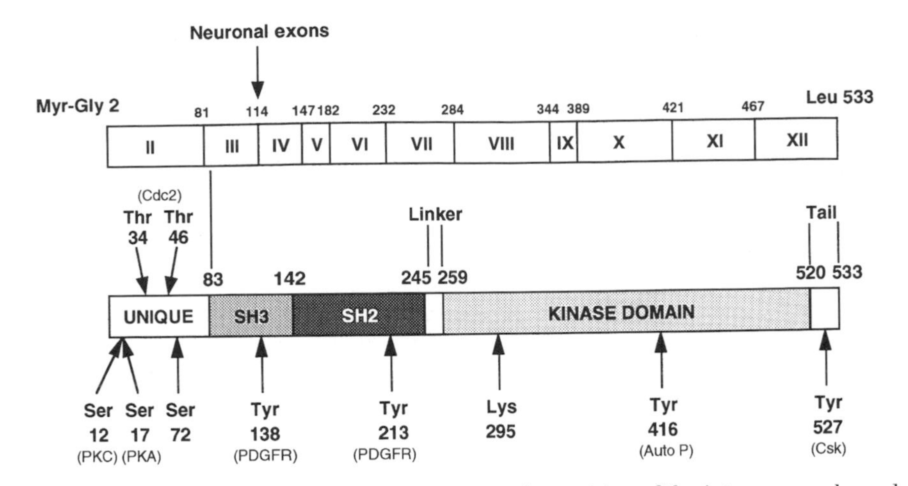

Figure 1. Schematic depiction of c-Src. The upper bar indicates the position of the intron–exon boundaries in the human c-*src* gene, using chicken c-Src residue numbers. The lower bar indicates the boundaries of the unique, SH3 and SH2 domains, the linker, the kinase domains, and the C-terminal tail, using chicken c-Src residue numbers. The known phosphorylation sites, and the protein kinases that phosphorylate them are also indicated.

cytoskeleton. The c-Src SH3 domain is related in sequence to SH3 domains in the Csk, Abl, Btk and Fes PTK families, and in many other proteins, including regulatory enzymes and cytoskeletal proteins. All SH3 domains recognize short proline-rich motifs with a PXXP core in a sequence-specific fashion, being able to bind in either orientation to this motif, which adopts a left-handed polyproline type II helical conformation when bound. As determined by using degenerate peptide libraries and phage display libraries, the c-Src SH3 domain preferentially binds two distinct proline-containing consensus sequences: RXLPPLP (type I) and XPPLPXR (type II), which bind in opposite orientations.[6]

The SH2 domain (residues ~148–245) is also regulatory. By recognizing and binding intramolecularly to P.Tyr-527 at its own C terminus, the SH2 domain exerts a negative influence on c-Src kinase activity; conversely it plays a positive role in substrate recognition and phosphorylation as a result of its ability to bind to specific phosphotyrosine-containing sequences in other proteins. The Src family SH2 domain is related to similar domains in Csk, Abl, Btk and Fes PTK families, as well as in several signalling and adaptor proteins, such as phospholipase C-γ, the regulatory subunit of phosphatidylinositol 3′ kinase, and the adaptor protein Grb-2. The use of degenerate phosphopeptide libraries has established a consensus for the binding of the Src SH2 as pY–E(D/T)–E(N)–I(M/L).[8]

The crystal structure of the inactive form of c-Src phosphorylated at Tyr-527 has revealed the manner in which the SH2 and SH3 domains repress the activity of the catalytic domain (Fig. 2). In the inactive structure the SH2 domain is bound intramolecularly to P.Tyr-527 and the SH3 domain is bound via its ligand-binding surface to Pro-250 and Gln-253 in the short linker region between the SH2 domain and the catalytic domain (residues ~246–259), which adopts a polyproline type II helical conformation. Both the SH2 and SH3 interactions occur on the back side of the catalytic domain and inactivation occurs as a result of interactions between residues in the linker and the RT and N-Src loops of the SH3 domain and specific residues in the N-terminal lobe of the catalytic domain, which cause a distortion of the N-terminal lobe and the outward rotation of the C helix. This allows Glu-310 in the C helix, which in the active state would form a salt bridge with Lys-295 in the ATP-binding site, to interact instead with Arg-385 in the catalytic loop in the C-terminal lobe. Phosphorylation of Tyr-416 in the activation loop combined with the displacement of the SH3 domain causes a conformational switch in which the phosphate on Tyr-416 binds to Arg-385 and the C helix rotates inward, allowing Glu-310 to interact with Lys-295, thus aligning all the catalytic residues in the proper configuration for phosphate transfer from ATP. Peptide binding studies for both the SH2 and SH3 domains indicates that the affinity for both the intramolecular ligands is lower than that for optimal intermolecular ligands. The combination of both of the weak SH2 and SH3 intramolecular interactions is required to maintain the inactive state, which means that c-src can be activated through interaction with high affinity ligands for either the SH2 or SH3 domain.

There are two additional forms of c-Src generated by alternative splicing in neuronal cells in the CNS. These contain six (exon NI) or 17 (exon NI + exon NII) additional residues inserted at position 114. While the function of these forms is unknown, the additional residues are inserted into the N loop of the SH3 domain and alter the ligand-binding specificity of the SH3 domain. Two-hybrid screening using the SH3 domain of neuronal Src as a bait has identified a brain-specific ion channel protein BCNG-1[32] and suggests that these neuronal forms of c-Src may have distinct targets.

Three oncogenically activated forms of the c-*src* gene have been found in acutely transforming avian retroviruses (RSV, S1, and S2), which all cause sarcomas. The

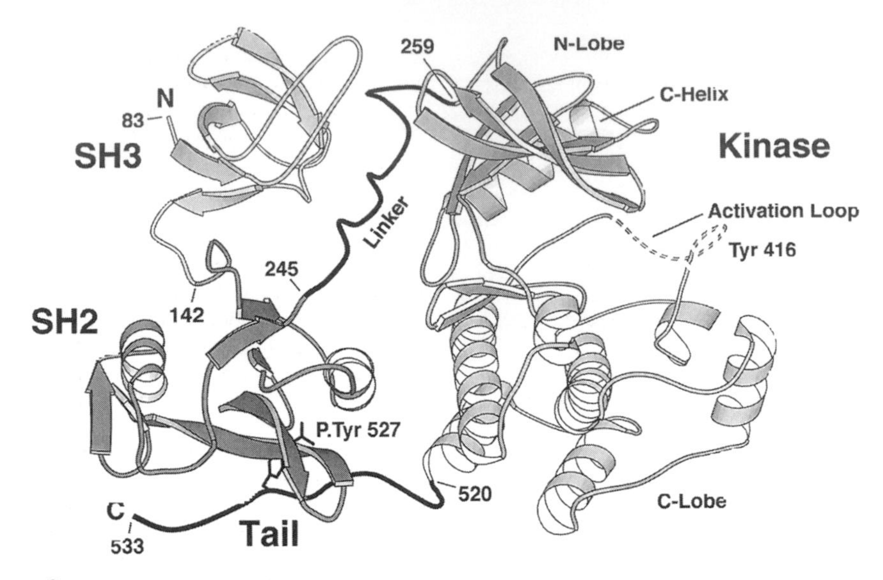

Figure 2. Human c-Src structure. The three-dimensional structure of residues 83–533 of human c-Src phosphorylated at Tyr-527 in the inactive conformation. The structure shows how the SH2 domain binds to P.Tyr-527 intramolecularly, and how the SH3 domain, and the linker interact with the back side of the N-terminal lobe of the kinase domain to hold it in the inactive state. (Figure modified, with permission, from Xu *et al.*[5])

v-Src proteins encoded by these v-*src* genes have been activated by different mutations, but in every case the C-terminal tail, including the negative regulatory Tyr-527 site, has been deleted and replaced by a different sequence. Additional point mutations have been incurred in each case, some of which cause oncogenic activation in their own right (Arg-95 to Trp; Thr-338 to Ile; Glu-378 to Gly; Ile-441 to Phe). The activating effects of many of these mutations can now be explained by their predicted effects in blocking the formation of the inactive conformation. Several different temperature-sensitive mutants of v-Src have been characterized. A number of different point mutations and small deletions can confer temperature dependence on v-src PTK and transforming activities (e.g. mutation of Arg-480 to His in LA24 v-Src and Gly-478 to Asp in LA31 v-Src). Like activated c-Src, v-Src associates with the cytoskeleton. Unlike c-Src, newly synthesized v-Src forms a relatively stable ternary complex with Hsp90 and Cdc37, which together act as a protein kinase chaperone.

In vertebrates, c-Src is expressed in most cell types from a single ~4 kb mRNA. c-Src is localized to cytoplasmic membranes as deduced by cell fractionation and immunofluorescence staining. In most cell types, c-Src is less than 0.005 per cent of total cell protein. c-Src is broadly distributed on the cytosolic face of cytoplasmic membranes, but is enriched at focal contacts and in the perinuclear region, with a specific population being associated with the MTOC and centrosomes.[9] v-Src is distributed in a fashion similar to c-Src as well as at sites of cell–cell contact, although by virtue of its deregulated PTK activity, many subcellular structures, such as focal contacts, are disorganized in cells expressing v-Src. In the animal, c-Src is present at the highest level in platelets and osteoclasts, and in neurones, where c-Src is enriched in axon terminals and growth cones. *In situ* hybridization for c-*src* mRNA and immunolocalization of the neuronal form of c-Src in the brain show that high concentrations are present in the hippocampus, mesencephalon, pons, medulla, olfactory bulb, and cerebellum.

Phosphorylation and regulation

c-Src is phosphorylated at multiple sites.[1,2] Some, but not all, of these sites have been identified, many lie in the unique N-terminal domain. The known sites of phosphorylation and the consequences of phosphorylation are (chicken c-Src numbering): Ser-12 – protein kinase C, effect unknown, but may affect the stability of membrane association, which requires interaction of basic residues at the N terminus with phospholipid head groups; Ser-17 – cAMP-dependent protein kinase, phosphorylation at this site releases c-Src from membranes, possibly by reducing its interaction with phospholipid head groups; Thr-34, Thr-46 (chicken only), and Ser-72 – cyclin B/cdc2 – effect unknown, but c-Src from mitotic cells shows increased PTK activity, possibly due to decreased phosphorylation at Tyr-527, which may result from phosphorylation of the N-terminal sites, and an increase in SH2 accessiblity to bind to substrates; Ser-72 – phosphorylated specifically in neuronal cells by an unidentified neuronal cell protein kinase – effect unknown; Tyr-138, Tyr-213 – PDGF receptor, phosphorylation on Tyr-138 reduces SH3 binding, whereas phosphorylation at Tyr-213 reduces SH2 binding to the regulatory Tyr at 527; Tyr-416 – major autophosphorylation site (but may also be phosphorylated *in trans* by other Src family PTKs and also by other PTKs) – causes a conformational change in the structure of the activation loop resulting in increased activity; Tyr-527-Csk – phosphorylation negatively regulates c-Src PTK activity through binding of P.Tyr-527 to the SH2 domain. Mutations have been made in all these sites; mutation of the N-terminal Ser-12 and Ser-17 does not appear to affect the transforming activity of an activated Phe-527 mutant form of c-Src. However, mutation of Ser-17 to Ala results in an increase in phosphorylation at Ser-12 and a 2–3 fold increase in kinase activity. Mutations of Tyr-416 and Tyr-527 have a significant effect on activity and transformation capacity. Mutation of Tyr-527 to Phe, which prevents negative regulation, increases PTK activity about 10-fold and converts c-Src into a transforming protein. Mutation of Tyr-416 decreases transforming activity of a Phe-527 mutant, by decreasing the ability of the catalytic domain to adopt the active conformation.

In principle, the activity of c-Src can be positively regulated by protein-tyrosine phosphatases (PTP) that dephosphorylate P.Tyr-527. It is not known whether there is a specific P.Tyr-527 PTP, but there is evidence that RPTPα and SHP2 can dephosphorylate P.Tyr-527 *in vivo*.

Purification

c-Src has been partially purified from chick brains[10] and human platelets[11,12] by conventional means, involving several sequential column chromatographies. v-Src has also been purified conventionally from RSV-induced tumours and RSV-transformed cells. c-Src has been expressed in *E. coli* (largely insoluble), in budding and fission yeast[13] (toxic at high levels unless csk is coexpressed), in insect cells using a baculovirus vector (~10 per cent is membrane bound and the rest is soluble),[14] and in chicken and mammalian cells using a variety of vectors. It has been purified from these sources both by mAb 327 affinity chromatography, affinity chromatography on poly(Glu–Tyr) (4:1) and conventional column chromatography. The fraction of c-Src that is soluble and active varies greatly between the different sources. In general the baculovirus-expressed protein is most active, having a low level of phosphorylation at Tyr-527, and has proven most useful for crystallization studies.

Activities

c-Src probably has many different functions, acting in a partly redundant fashion with the closely related Fyn and c-Yes Src family PTKs. Src family PTKs play a critical role downstream of many types of cell surface receptors, including receptor PTKs, G-protein-coupled receptors, integrins, cytokine receptors, multichain antigen receptors, and Fc receptors, as well as neurotransmitter receptors and ion channels.[1] c-Src is rapidly activated in response to ligand binding to these receptors; where tested, expression of dominant-negative, kinase-inactive forms of c-Src or microinjection of anti-Src antibodies abolishes the response to the cognate ligands. c-Src is also activated at mitosis, and injection of anti-Src antibodies or the c-Src SH3 domain blocks entry into mitosis, establishing that Src family kinases play an essential role in the G_2 to M transition.

The c-*src* gene has been disrupted in the mouse and a complete absence of c-*src* results in a relatively mild phenotype, namely osteopetrosis. The osteopetrosis is due to a defect in osteoclast function, which points to an essential role for c-Src in this cell type. Double knockout of c-*src* in combination with either the *fyn* or c-*yes* Src family PTK genes results in perinatal death, suggesting that Fyn and Yes, which are commonly coexpressed with c-Src, may substitute functionally for c-Src.[15,16] While the lack of c-Src expression alone has a limited effect on development in the mouse, mice lacking the gene encoding Csk, which plays a critical role in the negative regulation of c-Src and other Src family PTKs, display a lethal phenotype, suggesting that proper negative regulation of c-Src kinase activity is critical during development.

A fraction of the c-Src population is localized to the inner face of plasma membranes, where it is in a position to act as the catalytic subunit of a cell surface receptor lacking its own catalytic domain, as is the case for Lck, which interacts via its unique domain with the cytoplasmic tails of CD4 and CD8 in T cells. The nature of the interaction between c-Src and the receptors that activate it has not been established in every case, and in some cases may not be direct. The only known instance where the unique region of c-Src interacts with a receptor is the NMDA-type glutamate receptor, which binds to residues 40–58 in c-Src. c-Src interacts with the autophosphorylated PDGF and CSF-1 receptors via its SH2 domain and with the human Kv1.5 potassium channel via its SH3 domain. In other instances, the interaction with c-Src is indirect. For instance, c-Src associates via its SH2 domain with P.Tyr-397 in FAK, once FAK it has been activated through its interaction with the cytoplasmic domains of integrin β subunits. *Src*-related genes have been identified in *Drosophila* and *C. elegans*. Genetic analysis has shown that the *Drosophila* gene *Dsrc64*[33] is required for ovarian ring canal formation.

In vitro, c-src PTK activity can be measured either in immunoprecipitates or in solution using a variety of peptide and protein substrates. c-Src can use ATP (k_M ~5 μM), dATP, or GTP. PTK activity is optimal at 2–3 mM Mn^{2+} (or 10 mM Mg^{2+}), and at pH 7.0–7.5. [Val^5]-angiotensin II, acid-denatured rabbit muscle enolase, and poly(Glu–Tyr) (4:1) are the most commonly used substrates. The catalytic domain of Src shows a sequence specificity for phosphorylation of substrates E/D–E/D–I/V–Y–G/E–E–F(I/L/V)–D, which is similar on the C-terminal side of the target tyrosine to the sequence pY–E(D/T)–E(N)–I(M/L) recognized by the Src SH2 domain. This may allow processive phosphorylation of a substrate that contains several c-Src phosphorylation sites. Physiological substrates of c-Src include focal adhesion kinase (FAK), AFAP110, paxillin, and p130^{CAS} (this is a critical target in osteoclasts). Some of the proteins that bind to the c-Src SH3 domain such as AFAP110, sam68 (only at mitosis), and p130^{CAS}, are in turn phosphorylated on tyrosine by c-Src. In contrast the c-Src SH3-interacting proteins dynamin and synapsin 1a and 1b do not appear to be substrates for phosphorylation. Several potentially relevant v–Src substrates have also been identified, including PI3′ kinase, p120*ras* GAP, and fish, and the cytoskeletal proteins, annexin I and II, tensin, talin, paxillin, vinculin, ezrin, and cortactin, and connexin 43, clathrin heavy chain, caveolin, and integrin β subunits. These proteins may also be substrates for c-Src.

New methods for identifying relevant Src substrates include engineering the Src ATP-binding site in the catalytic domain to specifically accept a unique ATP analogue and transfer labelled phosphate from it to c-Src substrates[34] and the use of anti-phosphotyrrosine antibodies to screen λgt-11 expression libraries phosphorylated by Src.

Inhibitors of c-Src tyrosine kinase activity include antibodies against residues 498–512 and a non-substrate peptide corresponding to residues 137–157 of c-Src. There are also an increasing number of synthetic and natural non-peptide inhibitors of Src family PTK activity, which are membrane permeant and can be used in whole cell studies. These inhibitors are more or less selective for Src family PTKs compared to other PTKs, and generally act by competing with ATP binding. They include herbimycin A[17] and PP1[18] (available from Calbiochem). A number of programmes using combinatorial approaches are under way to develop specific synthetic non-peptide inhibitors of Src, which may yield inhibitors more selective for src than the currently available ones. In addition, allele-specific cell-permeable inhibitors have been synthesized to match Src kinases engineered to accept a unique ATP analogue.

Antibodies

Numerous antibodies have been raised against Src. Anti-Src antibodies were first detected in the serum of RSV tumour-bearing rabbits. Most of these anti-tumour sera recognize both v-Src and chicken c-Src; some of these sera cross-react with mammalian c-Src. A number of polyclonal sera have been raised against intact recombinant c-Src and against separate N- and C-terminal fragments. [19] Likewise, many monoclonal antibodies (mAbs) against c-Src have been obtained.[20,21] The immunodominant epitopes in c-Src lie in the N-terminal 120 residues, most of the mAb-binding sites map to this region. Two main classes of mAb have been identified binding to residues 28–38 (e.g. EB7, EC10, 19A6) and to residues 92–128 (e.g. GD11, EB8, 327, 16E6) (mAb 327 available from Calbiochem). One unusual mAb recognizes both N- and C-terminal sequences in a combined epitope (R2D2).[22] Anti-C-terminal peptide sera specific for either c-Src or v-Src have been generated. Some anti-c-Src C-terminal peptide sera stimulate c-Src kinase activity.[23] Anti-peptide antibodies against other regions have also been made including the N-terminal residues 2–17 (mAb LA022) (Microbiological Associates), the neuronal NI exon (antipeptide and anti-idiotype), to the sequence surrounding the Tyr-416 autophosphorylation site containing either phosphorylated or unphosphorylated Tyr-416 that respectively recognize the phosphorylated or unphosphorylated states, to C-terminal residues 498–512 (inhibits PTK activity),[24] residues 529–532 (mAb clone 28),[25] which selectively recognizes the active, Tyr-527 dephosphorylated form of c-Src, and to a peptide containing P.Tyr-527, which selectively recognize the P.Tyr-527 form of c-Src.

Most of the antibodies can be used for immunoprecipitation and several different anti-tumour sera, anti-recombinant c-Src sera, and mAbs have been used for immunofluorescence staining. Only some of the antibodies are suitable for immunoblotting (e.g. mAbs 327, 19A6, 16E6, LA022).

Genes

Genomic and cDNA clones and sequences for *c-src* have been obtained from humans,[26] mice, chickens,[27] fish, and *Xenopus* (related genes have been cloned from *Drosophila* (*Dsrc64*[28] and *Dsrc41*,[35] *C. elegans*,[29] *Spongilla*,[30] and *Hydra*,[31] but these genes are equally closely related to all of the Src family genes and should formally be called orthologues). The human *c-src* gene has 11 coding exons,[26] with two alternatively spliced exons being utilized in certain neuronal cells in the CNS and at least two 5′ non-coding exons. The promoter region of *c-src* has not been fully characterized, and there may be more than one promoter. The human *c-src* gene maps to chromosome 20q13.3.

References

1. Thomas, S. M. and Brugge, J. (1997). *Ann. Rev. Cell. Biol.*, (In press.)
2. Brown, M. A. and Cooper, J. A. (1996). *Biochem. Biophys. Acta*, **1287**, 121–49.
3. Pawson, T. (1995). *Nature*, **373**, 573–80.
4. Cohen, G. B., Ren, R., and Baltimore, D. (1995). *Cell*, **80**, 237–48.
5. Xu, W., Harrison, S. C., and Eck M. (1997). *Nature*, **385**, 595–602.
6. Feng, S., Chen, J. K., Yu, H., Simon, J. A., and Schreiber, S. L. (1994). *Science*, **266**, 1241–7.
7. Waksman, G., Shoelson, S. E., Pant, N., Cowburn, D., and Kuriyan, J. (1993). *Cell*, **72**, 779–90.
8. Songyang, Z., Shoelson, S. E., Chaudhuri, M., Gish G., Pawson, T., Haser, W. G., *et al.* (1993). *Cell*, **72**, 767–78.
9. David-Pfeuty, T., and Nouvian-Dooghe, Y. (1990). *J. Cell Biol.*, **111**, 3097–116.
10. Purchio, A. F., Erikson, E., Collet, M. S., and Erikson, R. L. (1981). *Cold Spring Harbor Conf. Cell Prolif.*, **8**, 1203–15.
11. Presek, P., Reuter, C., Findik, D., and Bette, P. (1988). *Biochim. Biophys. Acta*, **969**, 271–80.
12. Feder, D. and Bishop, J. M. (1990). *J. Biol. Chem.*, **265**, 8205–11.
13. Weijland, A., Neubauer, G., Courtneidge, S. A., Mann, M. Wierenga R. K., and Superti-Furga, G. (1996). *Eur. J. Biochem.*, **240**, 756–64.
14. Broome, M. and Hunter, T. (1997). *Oncogene*, **14**, 17–34.

15. Soriano, P., Montgomery, C., Geske, R., and Bradley, A. (1991). *Cell*, **62**, 693–702.
16. Stein, P. L., Vogel, H., and Soriano, P. (1994). *Genes Dev.*, **8**, 1999–2007.
17. Uehara, Y., Murakami, Y., Mizuno, S., and Kawai, S. (1988). *Virology*, **164**, 294–8.
18. Hanke, J. H., Gardner, J. P., Dow, R. L., Changelian, P. S. Brissette, W. H., Weringer, S. J., *et al.* (1996). *J. Biol. Chem.*, **271**, 695–701.
19. Resh, M. D. and Erikson, R. L. (1985). *J. Virol.*, **55**, 242–5.
20. Lipsich, L., Lewis A. J., and Brugge, J. S. (1983). *J. Virol.*, **48**, 352–60.
21. Parsons, S. J., McCarley, D. J., Elu, C. M., Benjamin, D. C., and Parsons, J. T. (1984). *J. Virol.*, **51**, 272–82.
22. McCarley, D. J., Parsons, J. T., Benjamin, D. C., and Parsons, S. J. (1987). *J. Virol.*, **61**, 1927–37.
23. Cooper, J. A. and King, C. S. (1986). *Mol. Cell. Biol.*, **6**, 4467–77.
24. Gentry, L. E., Rohrschneider, L. R., Casnellie, J. E., and Krebs, E. G. (1983). *J. Biol. Chem.*, **258**, 11219–28.
25. Kawakatsu, H., Sakai, T., Takagaki, Y., Shinoda, Y., Saito, M., Owada, M. K., and Yano, J. (1996). *J. Biol. Chem.*, **271**, 5680–5.
26. Tanaka, A., Gibbs, C. P., Arthur, R. R., Anderson, S. K., Kung, H., and Fujita, D. (1987). *Mol. Cell. Biol.*, **7**, 1978–83.
27. Takeya, T. and Hanafusa, H. (1983). *Cell*, **32**, 881–90.
28. Simon, M., Drees, B., Kornberg, T.,and Bishop, J. M. (1985). *Cell*, **42**, 831–40.
29. Goddard,J. M., Weiland, J. J., and Capecchi, M. R. (1986). *Proc. Natl Acad. Sci. USA*, **83**, 2172–6.
30. Ottilie, S., Raulf, F., Barnekow, A., Hannig, G., and Schartl, M. (1992). *Oncogene*, **7**, 1625–30.
31. Bosch, T. C. Unger, T. F., Fisher, D. A., and Steele, R. E. (1989). *Mol. Cell. Biol.*, **9**, 4141–51.
32. Santoro, B., Grant, S. G. N., Bartsch, D., and Kandle, E. R. (1997). *Proc. Natl Acad. Sci. USA*, **94**, 14815–20.
33. Guarnieri, D. J., Dodson, S., and Simon, M. A. (1998). *Mol. Cell*, **1**, 831–40.
34. Liu, Y., Shah, K., Yang, F., Witucki, L., and Shokat, K. M. (1998). *Curr. Biol.*, **5**, 91–101.
35. Takahashi, F., Endo, S., Kohima, T., and Salgo, K. (1996). *Genes Dev.*, **10**, 1645–56.

Tony Hunter and Suzanne Simon
The Salk Institute,
10010 North Torrey Pines Road,
La Jolla, CA, USA.

43K-Rapsyn

The 43K-rapsyn protein is a synaptic peripheral protein that co-distributes with the nicotinic acetylcholine receptor (AChR) in electrocyte postsynaptic membrane and at the vertebrate neuromuscular junction. It plays a critical role in the clustering of AChR and in the specialization of the AChR-rich postsynaptic domains of the motor endplates, possibly via the agrin signalling cascade. The 43K-rapsyn protein has also been proposed to constitute a link between the AChR and the subsynaptic cytoskeleton.

Synonymous names

43 K or 43 kDa protein; or ν1; Rapsyn (receptor-associated protein at the synapse).

Homologous proteins

The sequence of *Torpedo* 43K-rapsyn protein does not share homologies with any known protein.

Protein properties

First reported in electric organ as the major companion of AChR during postsynaptic membrane purification,[1] the 43K-rapsyn protein is also present at the vertebrate (rodent[2] and frog[3]) neuromuscular junction. In SDS–PAGE it migrates at 43 kDa,[1] between actin and muscle creatine kinase. It is a peripheral protein[4] tightly associated with the AChR-rich postsynaptic membrane at its inner face,[5] (Fig. 1), and in an approximately equimolar ratio with the AChR.[6] Its sequence is characterized by a high content of cysteine for an intracellular protein.[7] Its N and C termini are highly conserved from one species to the other.[7–10] The protein is co-translationally myristoylated at its N terminus on a glycine residue[11,12] and contains several putative phosphorylation sites.[7] Two cDNA differing by an extension at the 3′ end have been reported in the *Torpedo* electric organ.[8] At its C terminus, just ahead of a potential cAMP-dependent kinase site (KRSS in electrocytes),[7] the long protein isoform has a tandem zinc finger motif[13] with cysteine residues spaced similarly to the regulatory domain of protein kinase C.[7] The putative short isoform contains the first zinc finger motif. The protein is phosphorylated in electrocyte postsynaptic membrane[14,15] and phosphorylatable on serine residues.[15] The three to five isovariants (pI 6.4–7.8[15,16]) are possibly related to different states of phosphorylation.[15]

Cytosolic and membrane associated pools of the 43K-rapsyn protein coexist in the electrocytes throughout its development[17,18] (Fig. 2). In early embryonic electrocyte, AChR-rich postsynaptic membranes contain little 43K-rapsyn protein.[17–19] Upon maturation of the synapse there is an abrupt rise[18] of the membrane-associated form up to the 1:1 (AChR: 43K-rapsyn) molar ratio observed in

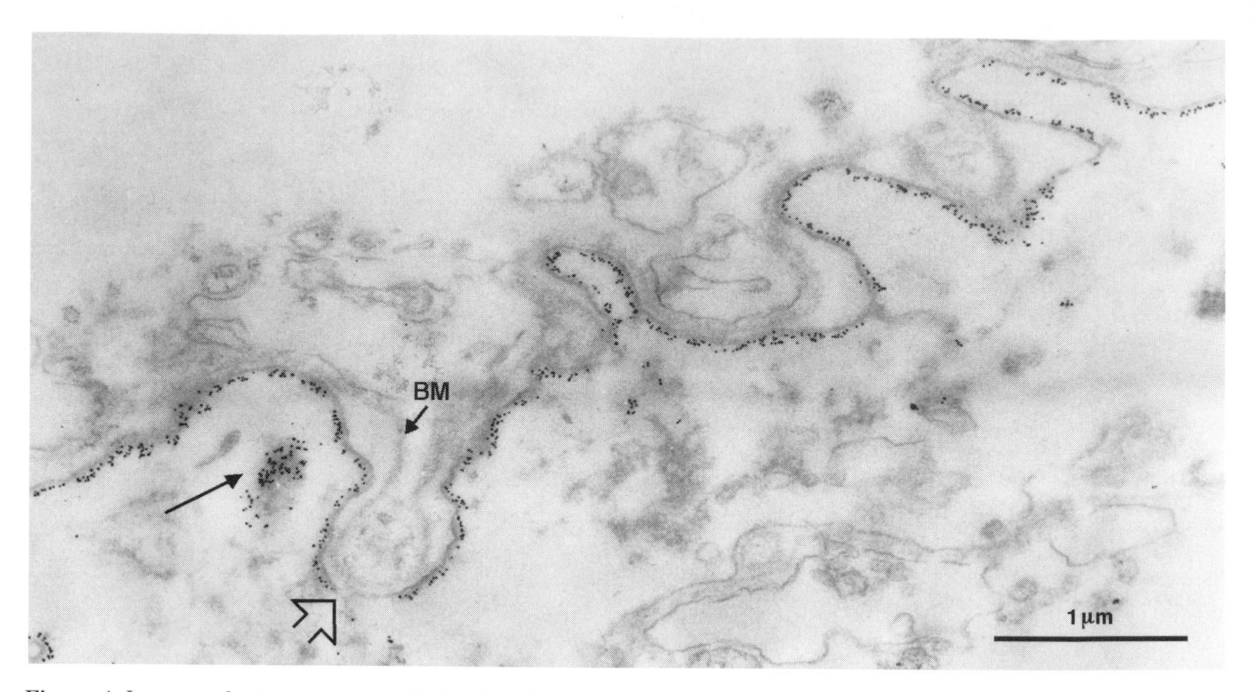

Figure 1. Immunoelectron micrograph showing the cytoplasmic face of the *Torpedo* electrocyte decorated with an anti-43K-rapsyn protein antibody (immunogold staining). The distribution of 43K-rapsyn protein is coextensive with that of AChR. The aggregate of gold particles (arrow) corresponds to a tangential section of an invagination of the postsynaptic membrane. Note the absence of 43K-rapsyn staining in the bottom of a fold which is devoid of AChR (open arrow). BM: basement membrane. (From ref. 5.)

adult tissue.[6,18] This strongly suggests a late role for the 43K-rapsyn in AChR clustering and in the formation and maintenance of postsynaptic structure specializations.[18] Indeed, the early trigger signal most probably comes from motor neurone agrin present in the synaptic cleft.[20]

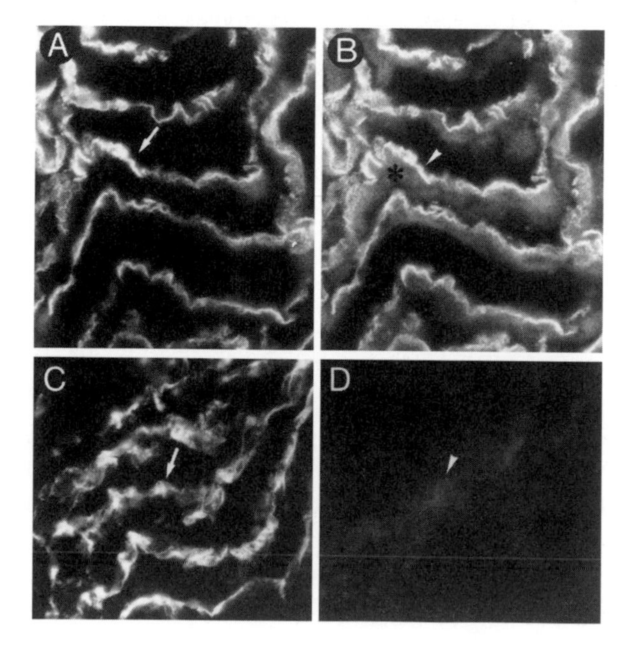

The 43K-rapsyn protein mainly accumulates at the cytoplasmic face[5,21–24] of the postsynaptic domain in adult electrocyte in a co-extensive distribution with the AChR[5,23] (Fig. 1). This distribution is also observed in vertebrate neuromuscular junctions (rodent[2] and batracian[3]) and at AChR clusters on rat[25] and *Xenopus*[26,27] cultured myotubes. Close proximity of 43K-rapsyn protein and AChR has been suggested by images of two-dimensional-crystal arrays of tubular electrocyte AChR-rich postsynaptic membrane vesicles[28] and by the chemically induced cross-linking of 43K-rapsyn with the AChR β subunit.[29] Conflicting results on the coordination of regulation of 43K-rapsyn and AChR subunit transcripts have been reported.[10,30,31]

Figure 2. Double immunofluorescence labelling of 43K-rapsyn protein in adult *Torpedo marmorata* electrocyte. (A, C) fluorescein-labelled α-bungarotoxin staining, showing the AChR-rich postsynaptic face of the electrocyte. (B) rhodamine-indirect staining with an anti-43K-rapsyn showing the presence of 43K-rapsyn mainly associated with the AChR-rich postsynaptic ventral face of the electrocytes but also partly associated with the cytoplasmic domain of the cells. (C, D) The specificity of the antibody staining is demonstrated by quenching of the staining (D) of the postsynaptic membrane (C) after preabsorption of the anti-43K-rapsyn antibodies. (From ref. 18.)

Removal of 43K-rapsyn protein from electrocyte postsynaptic membranes[4] or cultured rat myotubes[25] is accompanied by a significant increase in the susceptibility of AChR to heat denaturation[32] and proteolytic degradation[33] and in their translational[34] and rotational[35] mobility within the membrane plane. The AChR properties (kinetics of binding of acetylcholine, ion-gated channel opening, binding of local anaesthetics) however are not altered.[4]

Clusters of AChR can be induced in quail fibroblasts[36] stably transfected with the AChR upon transient expression of recombinant 43K-rapsyn protein and in *Xenopus* oocytes[37] injected with AChR and 43K-rapsyn protein transcripts. Expression of mutated cDNA in heterologous systems suggested that the full length 43K-rapsyn is important for AChR clustering.[13,38]

Mice defective in the 43K-rapsyn gene fail to form AChR clusters and functional postsynaptic structures.[39]

In transfected cells, the 43K-rapsyn causes clustering of dystroglycan, the agrin-binding component of the dystrophin glycoprotein complex.[40] It also induces clustering and activation of MuSK, a synapse-associated muscle specific kinase[41] and component of the agrin–MuSK–MASC signalling complex responsible for the AChR clustering.[42] This activation results in the phosphorylation of the AChR β subunit.[41]

The 43K-rapsyn protein thus is not required for the function of the AChR as a ligand-gated channel but plays a critical role (probably after agrin triggering) in AChR clustering and in the specialization, maintenance and stabilization of a functional postsynaptic domain. Modulation of phosphorylation has been suggested to be a possible mechanism in signalling mechanism. The 43K-rapsyn protein has also been proposed to constitute a link between the AChR and the subsynaptic cytoskeleton.

Purification

The 43K-rapsyn protein can be extracted from postsynaptic membranes by alkaline pH,[4] detergents[43] or anhydrides.[44] The electrocyte is the only cell rich in the 43K-rapsyn protein. Purification[7,15] has been achieved by elution of the protein from SDS-gels of electrocyte postsynaptic membranes or of the detergent or alkaline extracts.

Activities

No biological assays are available. Estimation of the protein can be achieved by immunochemical methods (Elisa,[6,18] or immunoprecipitation[30]).

Antibodies

Monoclonal,[3,5,24,26] polyclonal,[2,11,18] and anti peptide[17,18] antibodies against electrocyte 43K-rapsyn have been described. Some of them cross-react with mouse,[11] rat,[2] frog[3] and *Xenopus*[26,27] 43K-rapsyn protein. Immunocytolocalization of the 43K-rapsyn at the inner face[5,23] of the postsynaptic membrane has been reported with monoclonal anti-43K-rapsyn antibodies (Figs.1, 3). Immunoblot, immunoprecipitation and immunofluorescence studies have demonstrated the early presence of 43K-rapsyn in the cytosol[18] of embryonic electrocyte.

Genes

cDNA[8] coding for two 43K-rapsyn proteins which differ by an extension of 23 amino acids at the C-terminus (possibly generated by alternative splicing) have been described for electric organ (GenBank J02952, J02953). A single gene is present in the mouse genome and the complete 43K-rapsyn gene sequence for mouse muscle[9] is available (GenBank J03962). Mouse genomic DNA has been cloned and mapped to the central region of chromosome 2.[45] There exists sequence for *Xenopus* 43K-rapsyn.[10] Human cDNA 43K-rapsyn[46] has been sequenced

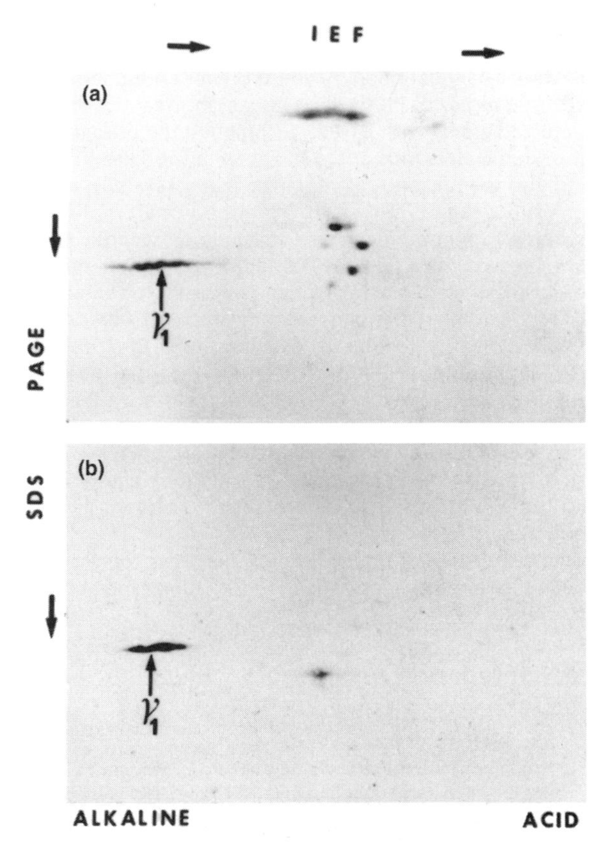

Figure 3. Immunostaining of the 43K-rapsyn isoforms of AChR-rich postsynaptic membranes of *Torpedo marmorata* by monoclonal anti-43K-rapsyn antibodies. Membrane proteins were separated by 2D-gel then immunoblotted with mAb anti-43K-rapsyn antibodies. (a) Ponceau red staining of the separated membrane fraction. (b) the 43K-rapsyn isoforms (p*I* 6.4–7.8) were labelled with the antibodies. (From ref. 5.)

(GenBank/EMBL Z33905) and mapped to chromosome 11P11.2-P11.1.

Mutant phenotype

Targeted disruption[39] of the 43K-rapsyn gene in mutant mice led to an increase of the AChR receptor in muscle but a failure of the formation of AChR clusters and of specialized postsynaptic domains. Such defective mice died perinatally. Some components of the postsynaptic cytoskeleton failed also to cluster at their usual postsynaptic site.

Structure

X-ray and NMR structures of the 43K-rapsyn protein are not available.

Addendum: recent research

In chick ciliary ganglion,[48] rodent brain[49] and superior cervical ganglion (SCG),[50] 43K-rapsyn transcripts are detectable only with RT-PCR techniques: the protein form is undetectable with anti-43K-rapsyn antibodies suggesting also the presence of very little if any 43K-rapsyn.

Although not critical for interneuronal cholinergic synapse formation (both synaptic and non-synaptic AChRs are present in SCGs of 43K-rapsyn –/– mutant mice),[50] 43K-rapsyn can cluster some cotransfected neuronal AChRs.[50,51] It can also cluster cotransfected $GABA_A$ receptors.[49] Contrary to muscle AChR clusters, the neuronal 43K-rapsyn-induced AChR clusters are found mostly at the intracellular level.[50,51] Rapsyn increases the half-life of cotransfected AChRs.[51,52] Distinct domains are responsible for 43K-rapsyn membrane targeting, self-association and 43K-rapsyn-induced AChR clustering.[38,53] Multiple 43K-rapsyn splice variants are present in brain and in muscle.[48]

Possible association of rapsyn and β-dystroglycan has been suggested by cross-linking and binding experiments.[54]

References

1. Sobel, A., Weber, M. and Changeux, J. P. (1977). *Eur. J. Biochem.*, **80**, 215–44.
2. Froehner, S. C., Gulbandsen, V., Hyman, C., Jeng, A. Y., Neubig, R. R., and Cohen, J. B. (1981). *Proc. Natl. Acad. Sci. USA*, **78**, 5230–4.
3. Froehner, S. C. (1984). *J. Cell Biol.*, **99**, 88–96.
4. Neubig, R. R., Krodel, E. K., Boyd, N. D., and Cohen J. B. (1979). *Proc. Natl. Acad. Sci. USA*, **76**, 690–4.
5. Nghiêm, H. O., Cartaud, J., Dubreuil, C., Kordeli, C., Buttin, G., and Changeux, J. P. (1983). *Proc. Natl. Acad. Sci. USA*, **80**, 6403–7.
6. La Rochelle, W. J. and Froehner, S. C. (1986). *J. Biol. Chem.*, **261**, 5270–4.
7. Carr, C., McCourt, D., and Cohen J. B. (1987). *Biochemistry*, **262**, 7090–102.
8. Frail, D. E., Mudd, J., Shah, V., Carr, C., Cohen, J. B., and Merlie, J. P. (1987). *Proc. Natl. Acad. Sci. USA*, **84**, 6302–6.
9. Frail, D. E., McLaughlin, L. L., Mudd, J., and Merlie, J. P. (1988). *J. Biol. Chem.*, **263**, 15602–7.
10. Baldwin, T. J., Theriot, J. A., Yoshihara, C. M., and Burden, S. J. (1988). *Development*, **104**, 557–64.
11. Musil, L. S., Carr, C., Cohen, J. B., and Merlie, J. P. (1988). *J. Cell Biol.*, **107**, 1113–21.
12. Carr, C., Tyler, A. N., and Cohen, J. B. (1989). *FEBS Lett.*, **243**, 65–9.
13. Scotland, P. B., Colledge, M., Melnikova, I., Dai, Z., and Froehner, S. C. (1993). *J. Cell Biol.*, **123**, 719–28.
14. Saitoh, T. and Changeux, J. P. (1980). *Eur. J. Biochem.*, **105**, 51–62.
15. Hill, J. A., Nghiêm, H. O., and Changeux, J. P. (1991). *Biochemistry*, **30**, 5579–85.
16. Gysin, R., Wirt, M., and Flanagan, S. D. (1981). *J. Biol. Chem.*, **256**, 11373–6.
17. Kordeli, E., Cartaud, J., Nghiêm, H. O., Devillers-Thiéry, A., and Changeux, J. P. (1989). *J. Cell Biol.*, **108**, 127–39.
18. Nghiêm, H. O., Hill, J. A. and Changeux, J. P. (1991). *Development*, **113**, 1059–67.
19. La Rochelle, W. J., Witzemann, V., Fiedler, W., and Froehner, S. C. (1990). *J. Neurosci.*, **10**, 3460–7.
20. McMahan U. J. (1990). *Cold Spring Harbor Symp. Quant. Biol.*, **LV**, 407–18.
21. Wennogle, L. P. and Changeux, J. P. (1980). *Eur. J. Biochem.*, **106**, 381–93.
22. Saint-John, P. A., Froehner, S. C., Goodenough, D. A., and Cohen, J. B. (1982). *J. Cell Biol.*, **92**, 333–42.
23. Sealok, R., Wray, B. E., and Froehner, S. C. (1984). *J. Cell Biol.*, **98**, 2239–44.
24. Bridgman, P. C., Carr, C., Pedersen, S. E., and Cohen, J. B. (1987). *J. Cell Biol.* **105**,1829–46.
25. Bloch, R. J. and Froehner, S. C. (1987). *J. Cell Biol.*, **104**, 645–54.
26. Burden, S. J. (1985). *Proc. Natl. Acad. Sci.USA*, **82**, 8270–3.
27. Peng, B. and Froehner, S. C. (1985). *J. Cell Biol.*, **100**, 1698–705.
28. Toyoshima, C. and Unwin, N. (1988). *Nature*, **336**, 247–50.
29. Burden, S. J., Depalma, R. L., and Gottesman, G. C. (1983). *Cell*, **35**, 687–92.
30. Frail, D. E., Musil, S., Buonanno, A., and Merlie, J. P. (1989). *Neuron*, **2**, 1077–86.
31. Froehner, S. C. (1989). *FEBS Lett.*, **249**, 229–33.
32. Saitoh, T., Wennogle, L. P., and Changeux, J. P. (1980). *FEBS Lett.*, **108**, 484–94.
33. Klymkowsky, M. W., Heuser, J. E., and Stroud, R. M. (1980). *J. Cell Biol.*, **85**, 823–38.
34. Lo, M. M. S., Garland, P. B., Lamprecht, J., and Barnard, E. A. (1980). *FEBS Lett.*, **111**, 407–12.
35. Rousselet, A., Cartaud, J., Devaux, P. F., and Changeux, J. P. (1982). *EMBO J.*, **1**, 439–45.
36. Phillips, W. D., Kopta, C., Blount, P., Gardner, P. D., Steinbach, J. H., and Merlie, J. P. (1991). *Science*, **251**, 568–70.
37. Froehner, S. C., Luetje, C. W., Scotland, P. B., and Patrick, J. (1990). *Neuron*, **5**, 403–10.
38. Phillips, W. D., Maimone, M. M., and Merlie, J. P. (1991). *J. Cell Biol.*, **115**, 1713–23.
39. Gautam, M., Noakes, P.G., Mudd, J., Nichol, M., Chu, G. C., Sanes, J. R., and Merlie, J. P. (1995). *Nature*, **377**, 232–6.
40. Apel, E. D., Roberds, S. L., Campbell, K. P., and Merlie, J. P. (1995). *Neuron*, **15**, 115–26.
42. Gillepsie, S. K. H., Balasubramanian, S., Fung, E. T., and Huganir, R. L. (1996). *Neuron*, **16**, 953–62.
43. Glass, D. J. *et al.* (1996). *Cell*, **85**, 513–23.

44. Elliott, J., Blanchard, S. G., Wu, W., Miller, J., Strader, C. D., Hartig, P., Moore, H. P., Racs, J., and Raftery, M. A. (1980). *Biochem. J.*, **185**, 667–77.
45. Eriksson, H., Liljeqvist, G., and Heilbronn, E. (1983). *Biochim. Biophys. Acta*, **728**, 449–54.
46. Gautam, M., Mudd, J., Copeland, N. G., Gilbert, D. G., Jenkins, N. A., and Merlie, J. P. (1994). *Genomics*, **24**, 366–9.
47. Buckel, A., Beeson, D., James, M., and Vincent, A. (1996). *Genomics*, **35**, 613–6.
48. Burns, A. L., Benson, D., Howard, M. J., and Margiotta, J. F. (1997). *J. Neurosci.*, **17**, 5016–26.
49. Yang, S. H., Armson, P. F., Cha, J., and Phillips, W. D. (1997) *Mol. Cell. Neurosci.*, **8**, 430–38.
50. Feng, G. P., Steinbach, J. H., and Sanes, J. R., (1998). *J. Neurosci.*, **18**, 4166–76.
51. Kassner, P. D., Conroy, W. G., and Berg, D. K. (1998). *Mol. Cell. Neurosci.*, **10**, 258–70.
52. Phillips, W. D., Vladeta, D., Han, H. and Noakes, P. G. (1997). *Mol. Cell. Neurosci.*, **10**, 16–26.
53. Ramarao, M. K. and Cohen, J. B. (1998). *Proc. Natl Acad. Sci. USA*, **82**, 8270–73.
54. Cartaud, A., Coutant, S., Petrucci, T. C., and Cartaud, J. (1998). *J. Biol. Chem.*, **273**, 11321–26.

■ *Hoàng-Oanh Nghiêm*
Institut Pasteur, Neurobiologie moléculaire,
CNRS UA D1284, 25 rue du Dr Roux,
75724 Paris, Cedex 15, France

Rho GTPase family

Overview of the family

Members of the rho GTPase family act as molecular switches to regulate the organization of the actin cytoskeleton in response to extracellular stimuli. Rho itself controls the assembly of actin stress fibres and associated integrin adhesion complexes, while rac and cdc42, two other family members, control the formation of lamellipodia and filopodia, respectively. Rho, rac, and cdc42 also regulate other intracellular activities such as the JNK and p38 MAP kinase cascades and the phagocyte-associated NADPH oxidase complex. Rho GTPases coordinately control a programme of intracellular activities associated with the actin cytoskeleton.

Over 60 ras-related GTPases have been described to date in mammalian cells.[1] These small (around 21 kDa), monomeric GTP-binding proteins act as molecular switches to control a wide variety of intracellular processes. The rho subfamily currently has eight distinct members: rho[2,3] (A, B, C isotypes), rac[4] (1, 2 isotypes), cdc42[5,6] (cdc42Hs and G25K isotypes), rhoG,[7] rhoD,[8] rhoE,[9] TC10,[10] and TTF.[11] These eight proteins are between 50 and 60 per cent identical to each other in sequence and around 30 per cent identical to Ras. Isotypes are at least 85 per cent identical to each other.

Like all GTP-binding proteins, the rho-related GTPases exist in an inactive, GDP-bound and an active, GTP-bound conformation and their interconversion is regulated by guanine nucleotide exchange factors (GEFs) and by GTPase activating proteins (GAPs) (see Fig. 1).[12] A large family (over 20) of GEFs has been identified and each GEF contains a DH (dbl homology) domain (that catalyses nucleotide exchange on the GTPase) and a PH (pleckstrin

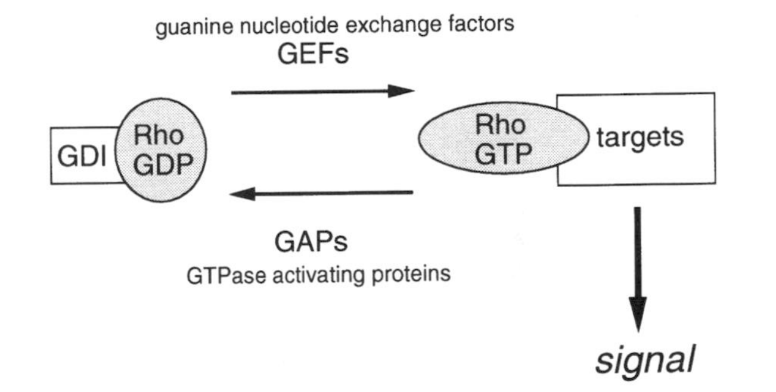

Figure 1. The rho GTPase molecular switch. In the resting state, GDP-bound rho is complexed to GDI. Activation of a GEF catalyses release of GDP and GDI and rho binds GTP. GTP-bound rho adopts an active conformation which can interact with target molecules to produce a signal. The lifetime of the active conformation is determined by GAP proteins.

homology) domain.[13] Similarily, a large family (over 10) of GAPs has been identified.[14] In addition, several guanine nucleotide dissociation inhibitors (GDIs) for the rho family have been identified; their function is not entirely clear though it has been suggested that they control nucleotide-dependent membrane association/ dissociation of the GTPases.[15] The reason for such a complex array of regulatory molecules is not clear – it may in part reflect the complex nature of the biological processes controlled by the rho GTPases, such as cell movement.

Rho GTPases each terminate in a 'CAAX box', i.e. a signal for isoprenylation at the cysteine 'C', proteolysis of 'AAX', and carboxymethylation. All rho GTPases are predicted to be geranylgeranylated (C20 lipid) except for rhoE[9] which should be farnesylated (C15 lipid); however somewhat unexpectedly, rhoB has been shown to contain a mixture of C20 and C15 lipids.[16,17]

Upon sequence alignment with Ras, all the rho GTPases, except rhoE,[9] have a conserved glycine residue at codon 12 and a glutamine at codon 61 (Ras codon numbers). These are sites of oncogenic mutation in Ras and amino acid substitutions at either codon in the rho GTPases confer insensitivity to GAP proteins.[18] Although such changes have not been found in human cancers (as is the case with Ras), rho GTPases with substitutions at codon 12 or 61 are constitutively activated (GTP bound) when introduced into cells and they have been used extensively for analysing GTPase function. RhoE already has atypical amino acids at both codons 12 and 61 and as expected this 'wild type' protein appears constitutively in the GTP-bound state,[9] it is not at all clear how this protein is regulated *in vivo*, although one possibility is that a rhoE-specific GAP exists.

The rho GTPases have a threonine residue at a position corresponding to codon 17 of Ras. An asparagine substitution at this position in Ras results in a protein that has a strong preference for GDP over GTP and N17Ras behaves as a dominant negative protein that has been extensively used experimentally. Similar alterations in rho (N19), rac (N17) and rdc42 (N17) result in specific dominant negative proteins and, although their mechanism of action is not clear, they have been used extensively to examine the role of these GTPases in signalling pathways.[19]

The function of rho, rac, and cdc42 has been examined in greatest detail[19] and will be dealt with separately. Transfection experiments with rhoE produced no clear phenotype,[9] while a recent paper using rhoD suggests that this protein can cause distinctive cell surface protrusions and can affect endocytosis.[8]

References

1. Zerial, M. and Huber, L. A. (1995). *Guidebook to the small GTPases*, Oxford University Press.
2. Yeramian, P., Chardin, P., Madaule, P., and Tavitian, A. (1987). *Nucleic Acids Res.*, **15**, 1869.
3. Chardin, P., Madaule, P., and Tavitian, A. (1988). *Nucleic Acids Res.*, **16**, 2717.
4. Didsbury, J., Weber, R. F., Bokoch, G. M., Evans, T., and Snyderman, R. (1989). *J. Biol. Chem.*, **264**, 16378–82.
5. Shinjo, K., Koland, J. G., Hart, M. T., Narasimham, V., and Johnson, D. L. (1990). *Proc. Natl Acad. Sci. USA*, **87**, 9853–7.
6. Munemitsu, S., Innis, M. A., Clark, R., McCormick, F., Ulrich, A., and Polakis, P. (1990). *Mol. Cell. Biol.*, **10**, 5977–82.
7. Vincent, S., Jeantuer, P., and Fort, P. (1992). *Mol. Cell. Biol.*, **12**, 3138–48.
8. Murphy, C., Saffrich, R., Grummt, M., Gournier, H., Rybin, V., Rubino, M., *et al.* (1996). *Nature*, **384**, 427–32.
9. Foster, R., Hu, K. Q., Nolan, K. M., Thissen, J., and Settleman, J. (1996). *Mol. Cell. Biol.*, **16**, 2689–99.
10. Drivas, G. T., Shih, A., Coutavas, E., Rush, M. G., and D'Eustachio, P. (1990). *Mol. Cell. Biol.*, **10**, 1793–8.
11. Dallery, E., Galiegue, Z. S., Collyn-d'Hooghe, M., Quief, S., Denis, C., Hildebrand, M. P., *et al.* (1995). *Oncogene*, **10**, 2171–8.
12. Boguski, M. S. and McCormick, F. (1993). *Nature*, **366**, 643–54.
13. Cerione, R. A. and Zheng, Y. (1996). *Curr. Opin. Cell Biol.*, **8**, 216–22.
14. Lamarche, N. and Hall, A. (1994). *Trends Genet.*, **10**, 436–40.
15. Isomura, M., Kikuchi, A., Ohga, N., and Takai, Y. (1991). *Oncogene*, **6**, 119–24.
16. Adamson, P., Marshall, C. J., Hall, A. and Tilbrook, P. A. (1992). *J. Biol. Chem.*, **267**, 20033–8.
17. Armstrong, S. A., Hannah, V. C., Goldstein, J. L. and Brown, M. S. (1995). *J. Biol. Chem.*, **270**, 7864–68.
18. Self, A. J. and Hall, A. (1995). *Meth. Enzymol.*, **256**, 67–76.
19. Hall, A. (1994). *Ann. Rev. Cell Biol.*, **10**, 31–54.

Alan Hall
MRC Laboratory for Cell Biology,
University College London,
Gower Street, London WC1E 6BT, UK

Rho

The rho GTPase regulates the assembly of actin stress fibres and focal adhesion complexes in response to a variety of extracellular agonists, but in particular lysophosphatidic acid (LPA). The active, GTP-bound form of rho interacts with target proteins to trigger the cytoskeletal changes as well as other cellular activities such as signalling to the nucleus. Rho is the target of several bacterial exoenzymes and toxins.

Protein properties

Three isotypes of mammalian rho have been identified, rhoA, rhoB and rhoC; A and C are 193 amino acids, B is 196 amino acids and they differ from each other in around 30 residues.[1,2] In *S. cerevisiae*, rho1p is the yeast homologue and cells lacking *RHO1* can be rescued by the mammalian *rho* genes.[3] The nomenclature for the rho GTPases is somewhat confusing; rhoD, rhoE, and rhoG for example, are each around 50 per cent identical to rhoA/B/C and they are not, therefore, isotypes. Similarly, rho2p, rho3p and rho4p are additional members of the rho family in yeast and not isotypes of rho1p.

RhoA appears to be ubiquitously expressed and has been purified from smooth muscle cytosol,[4] whereas rhoB has been purified from brain membrane fractions.[5] RhoB mRNA was identified as an immediate early transcribed sequence in Rat-2 fibroblasts treated with serum.[6] Using tagged expression constructs, each of the three proteins can be found at the plasma membrane, but the bulk of rhoA and rhoC is found in the cytosol and the majority of rhoB is found on early and late endosomes.[7] The significance of these cellular localizations is not known. The proteins are post-translationally modified at their C termini: rhoA, rhoB and rhoC are good substrates for the geranylgeranyl transferase type I while rhoB also appears to be a good substrate for farnesyl transferase.[4,8,9] The lipids attached to rhoB *in vivo*, however, have not yet been identified.

All three rho isotypes are thought to be monomeric, but may be complexed to regulatory molecules *in vivo*. They each bind GDP and GTP and have an intrinsic GTPase activity.[10] A large number of GAPs and GEFs have been identified and, although not all of these have been examined in detail, some appear to act on multiple members of the rho family, while others seem to be relatively specific.[11,12] Lbc, for example, specifically activates rho both *in vitro* and *in vivo*.[13] To date, there is no information on the *in vivo* mechanism by which GEFs or GAPs are themselves regulated.

Rho is the target of a number of bacterial exoenzymes and toxins. C3 transferase from *Clostridium botulinum* ADP-ribosylates rhoA, B, and C at an asparagine residue at codon 41.[14] Although other members of the rho family have the same residue at this position, they are not substrates[15]. ADP ribosylation by C3 transferase inactivates rho function and C3 has been used as a specific rho inhibitor.[15] Rho, rac, and cdc42 are each inactivated by *Clostridium difficile* toxin B through glucosylation on threonine 37 (rho) or 35 (rac and cdc42).[16]

A major role of rho in fibroblasts is to regulate the assembly of actin/myosin stress fibres and integrin-based focal adhesion complexes. Its effects are most easly demonstrated in quiescent Swiss 3T3 fibroblasts, since upon serum starvation these cells lose almost all organized actin structures.[15] Microinjection of constitutively activated recombinant rho protein (V14Rho or L63Rho) leads to assembly of stress fibers and focal adhesions within about 15–30 min (see Fig. 1). RhoA, rhoB, and rhoC behave identically in this assay.[15] These same cytoskeletal changes can be induced by

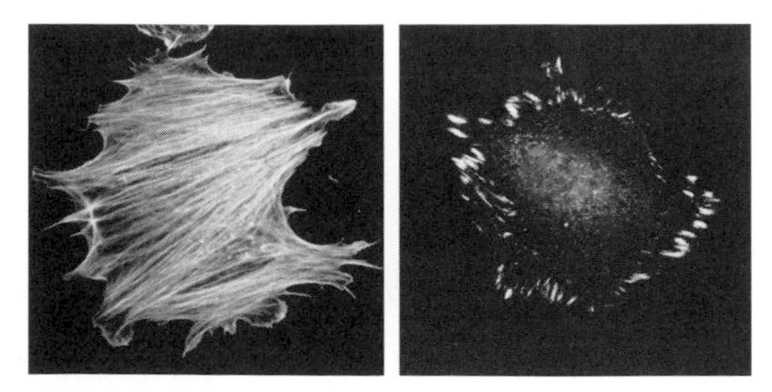

Figure 1. Rho activation in Swiss 3T3 cells triggers assembly of actin stress fibres (left panel; visualized with fluorescently labelled phalloidin) and focal adhesion complexes (right panel; visualized with an anti vinculin antibody).

adding LPA to the quiescent cells and this can be shown to be mediated by endogenous rho proteins, since this effect of LPA is blocked by pre-injection of cells with the rho inhibitor, C3 transferase.[15] Activation of rho by LPA is not pertussis toxin sensitive (unlike LPA activation of Ras) nor is it Ras-dependent; however, it does appear to involve a tyrosine kinase and possibly $G\alpha_{12}$ or $G\alpha_{13}$.[17,18] Rho can be activated by other G protein-coupled receptors, such as those for bombesin or thrombin, though these two agonists also activate the rac GTPase.[15,19] LPA has been shown to induce rho-dependent changes in actin/myosin filaments in neuronal-like cells.[20]

Other activities reported for rho include:

1. activation of the transcription factor, SRF (serum response factor),[21]
2. formation of the contractile ring during cytokinesis,[22]
3. regulated secretion in mast cells,[23]
4. inhibition of endocytosis,[24]
5. G_1 cell cycle progression.[25]

A number of candidate downstream targets that interact specifically with the active, GTP-bound form of rho have been identified by yeast two-hybrid and affinity chromatography techniques; these include two families of Ser/Thr kinases, PKN (also known as PRK), and $p160^{ROCK}$ (also known as $p160^{ROK}$), as well as several structural proteins with no obvious catalytic activity.[26] In addition, both the GDP and the GTP forms of rho have been reported to interact with a type I phosphatidylinositol 4-phosphate 5-kinase (PIP 5-kinase), although it is not clear that binding to the lipid kinase is direct.[27] Recent work suggests that $p160^{ROCK}$ is the crucial rho target for induction of stress fibres and focal adhesions, since these structures can be induced in a rho-independent manner by overexpression of this kinase.[28] Whether this is the only activity required is not clear.

Rho1p in *S. cerevisiae* has been shown to regulate a number of activities, including bud morphogenesis, the activity of glucan synthase, and activation of Pkc1p, a protein kinase related to mammalian PKCs.[29,30]

Purification

Recombinant rho proteins can readily be expressed in *E. coli* and the wild-type proteins are obtained predominantly in a GDP-bound form.[31] Constitutively activated proteins, typically V14 or L63, are purified as GTP-bound proteins, since they have low intrinsic GTPase activity. It has not proved possible to purify a dominant negative (N19) rho from *E. coli*. Proteins purified in this way are not post-translationally modified, but undergo modification after microinjection into cells. Post-translationally modified rho proteins have been obtained using Sf9 insect cells.[32]

Activities

The GTP- or GDP-binding activities of recombinant rho proteins can readily be determined in a nitrocellulose filter binding assay, although this appears to underestimate the amount of protein present.[31] GTP-loaded, wild type rho (produced in *E. coli*) is converted spontaneously to the GDP form with a half life of around 20 min in Mg^{2+} and this can be dramatically decreased by addition of GAP proteins.[10] Rho GTPases are sensitive to proteolysis of C-terminal residues when produced in *E. coli* and, although this does not affect nucleotide binding or GTP hydrolysis, these proteins are biologically impaired.[31]

For expression studies, N-terminally *myc* epitope tagged cDNAs have been used.[7] These can be microinjected into cell nuclei or transfected into cells and the effects on F-actin or focal adhesion assembly monitored by immunofluorescence.

Antibodies

Commercial antibodies that recognize rhoA or rhoB are available and a monoclonal antibody specific for rhoA has been reported.[33] Immunoprecipitation of nucleotide-bound rho has been reported.[34]

Genes

cDNA clones for human rhoA (GenBank accession number X05026), rhoB (GenBank accession number X06820), and rhoC (GenBank accession number X06821) are available. To date, most work has been carried out with rhoA. For expression studies, N-terminally *myc* epitope tagged cDNAs have been used.[7]

Mutant phenotype

Deletion of RHO1 in yeast is lethal.[3] Expression of rho in the *Drosophila* eye causes a dose-dependent disruption of normal eye development.[35] A mutation in the *C. elegans let-502* gene (homologue of the rho target kinase, p160ROCK) leads to disruption of early embryonic morphogenesis.[36]

Structure

Not available.

References

1. Yeramian, P., Chardin, P., Madaule, P., and Tavitian, A. (1987). *Nucleic Acids Res.*, **15**, 1869.
2. Chardin, P., Madaule, P., and Tavitian, A. (1988). *Nucleic Acids Res.*, **16**, 2717.
3. Yamochi, W., Tanaka, K., Nonaka, H., Maeda, A., Musha, T., and Takai, Y. (1994). *J. Cell Biol.*, **125**, 1077–93.
4. Katayama, M., Kawata, M., Yoshida, Y., Horiuchi, H., Yamamoto, T., Matsuura, Y., and Takai, Y. (1991). *J. Biol. Chem.*, **266**, 12639–45.
5. Yamamoto, K., Kondo, J., Hishida, T., Teranishi, Y., and Takai, Y. (1988). *J. Biol. Chem.*, **263**, 9926–32.
6. Jahner, D. and Hunter, T. (1991). *Mol. Cell. Biol.*, **11**, 3682–90.
7. Adamson, P., Paterson, H. F., and Hall, A. (1992). *J. Cell Biol.*, **119**, 617–27.
8. Adamson, P., Marshall, C. J., Hall, A., and Tilbrook, P. A. (1992). *J. Biol. Chem.*, **267**, 20033–8.
9. Armstrong, S. A., Hannah, V. C., Goldstein, J. L., and Brown, M. S. (1995). *J. Biol. Chem.*, **270**, 7864–8.
10. Self, A. J. and Hall, A. (1995). *Meth. Enzymol.* **256**, 67–76.
11. Cerione, R. A. and Zheng, Y. (1996). *Curr. Opin. Cell Biol.*, **8**, 216–22.
12. Lamarche, N. and Hall, A. (1994). *Trends Genet.* **10**, 436–40.
13. Zheng, Y., Olson, M. F., Hall, A., Cerione, R. A., and Toksoz, D. (1995). *J. Biol. Chem.*, **270**, 9031–4.
14. Sekine, A., Fujiwara, M., and Narumiya, S. (1989). *J. Biol. Chem.*, **264**, 8602–5.
15. Ridley, A. J. and Hall, A. (1992). *Cell*, **70**, 389–99.
16. Just, I., Selzer, J., Wilm, M., Eichel-Streiber, C., Mann, M., and Aktories, K. (1995). *Nature*, **375**, 500–3.
17. Nobes, C. D., Hawkins, P., Stephens, L., and Hall, A. (1995). *J. Cell Sci.*, **108**, 225–33.
18. Buhl, A. M., Johnson, N. L., Dhanasekaran, N., and Johnson, G. L. (1995). *J. Biol. Chem.*, **270**, 24631–4.
19. Ridley, A. J., Paterson, H. F., Johnston, C. L. Diekmann, D., and Hall, A. (1992). *Cell*, **70**, 401–10.
20. Jalink, K., van Corven, E. J., Hengeveld, T., Morii, N., Narumiya, S., and Moolenaar, W. H. (1994). *J. Cell Biol.*, **126**, 801–10.
21. Hill, C. S., Wynne, J., and Tresiman, R. (1995). *Cell*, **81**, 1159–70.
22. Mabuchi, I., Hamaguchi, Y., Fujimoto, H., Morii, N., Misshima, M., and Narumiya, S. (1993). *Zygote*, **1**, 325–31.
23. Price, L. S., Norman, J. C., Ridley, A. J., and Koffer, A. (1995). *Curr. Biol.*, **5**, 68–73.
24. Lamaze, C., Chuang, T. H., Terlecky, L. J., Bokcoh, G. M., and Schmid, S. L. (1996). *Nature*, **382**, 177–80.
25. Yamamoto, M., Marui, N., Sakai, T., Morii, N., Kozaki, S., Ikai, K., *et al.* (1993). *Oncogene*, **8**, 1449–55.
26. Tapon, N. and Hall, A. (1997). *Curr. Biol.* **9**, 86–92.
27. Chong, L. D., Kaplan, A. T., Bokoch, G. M., and Schwarz, M. A. (1994). *Cell*, **79**, 507–13.
28. Leung, T., Chen, X. Q., Manser, E., and Lim, L. (1996). *Mol. Cell. Biol.*, **16**, 5313–27.
29. Drgnova, J., Drgon, T., Tanaka, K., Kollar, R., Chen, G. C., Ford, R. A., *et al.* (1996). *Science*, **272**, 277–81.
30. Nonaka, H., Tanaka, K., Hirano, H., Fujiwara, T., Kohno, H., Umikawa, *et al.* (1995). *EMBO J.*, **14**, 5931–8.
31. Self, A. J. and Hall, A. (1995). *Meth. Enzymol.*, **256**, 3–10.
32. Page, M. J., Hall, A., Rhodes, S., Skinner, R. M., Murphy, V., Sydenham, M., and Lowe, P. N. (1989). *J. Biol. Chem.*, **264**, 19147–54.
33. Lang, P., Gesbert, F., Thiberg, J. M., Troalen, F., Dutartre, H., Chavrier, P., and Bertoglio, J. (1993). *Biochem. Biophys. Res. Commun.* **196**, 1522–8.
34. Laudanna, C., Campbell, J. J., and Butcher, E. C. (1996). *Science*, **271**, 981–3.
35. Hariharan, I. K., Hu, K. Q., Asha, H., Quintanilla, A., Ezzell, R. M., and Settleman, J. (1995). *EMBO J.*, **14**, 292–302.
36. Wissmann, A., Ingles, J., McGhee, J. D. and Mains, P. E. (1997). *Genes Dev.* **11**, 409–22.

Alan Hall
MRC Laboratory for Cell Biology,
University College London,
Gower Street, London WC1E 6BT, UK

Rac

The rac GTPase regulates both the polymerization of actin at the cell periphery to form lamellipodia and membrane ruffles and the assembly of associated integrin adhesion complexes. Rac is strongly activated by the tyrosine kinase receptors for PDGF and insulin in fibroblasts and by thrombin in platelets. In addition, rac regulates phagocytic NADPH oxidase activity and has been reported to activate the JNK MAP kinase pathway. It is likely that rac plays an important role in controlling cell movement.

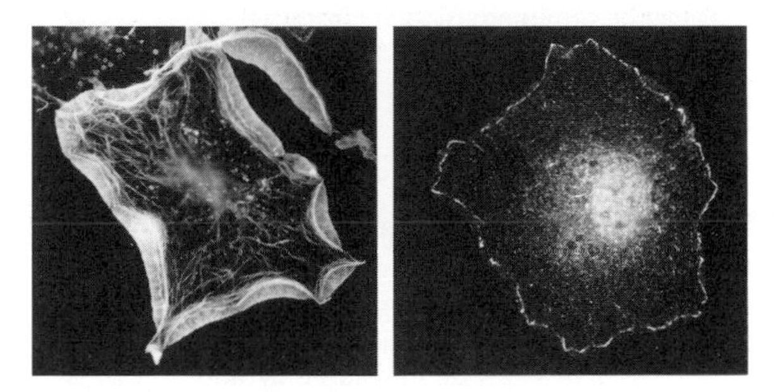

Figure 1. Activation of *rac* in Swiss 3T3 cells triggers the formation of actin-rich lamellipodia (left panel; visualized with fluorescently labelled phalloidin) and focal complexes (right panel; visualized with an anti vinculin antibody).

Protein properties

Two isotypes of rac have been reported, rac1 and rac2.[1] Rac1 appears to be ubiquitously expressed, whereas rac2 is expressed mainly in haematopoietic cells.[1,2] Rac is around 50 per cent identical to rho, but slightly closer in sequence to Cdc42. Although the yeast *S. cerevisiae* has close relatives of rho and Cdc42 (rho1p and Cdc42p), it does not have a rac homologue.

Rac1 and rac2 are geranylgeranylated[3] and in resting neutrophils, Rac2 is found in a cytosolic complex with rhoGDI.[4] Stimulation of neutrophils causes 10 per cent of the rac2 to dissociate from rhoGDI and to move to a membrane fraction and although this is consisitent with the rhoGDI/rac complex being inactive, rhoGDI/rac has been purified as an active complex.[5,6] Rac is not a target for the C3 transferase, but it is a target for *C. difficile* toxin B.[7]

Several members of the GEF and GAP families are active on rac; dbl, the founding member of the GEF family, and vav are active on rac *in vitro* and *in vivo*, though they also appear to be active on rho and Cdc42.[8,9] *Tiam-1*, which was isolated as a gene capable of inducing an invasive phenotype on a non-invasive T lymphoma cell line, is a GEF that is specific for rac.[10] It has been shown that the generation of PIP^3, through activation of PI 3-kinases, is a potent activator of rac,[11] but the target of this lipid (perhaps a GEF) is not yet clear. Several GAPs have activity on rac, in particular the product of the break-point cluster region (*bcr*) gene, which is rearranged in chronic myeloid leukaemia.[12]

A major role of rac is to regulate actin polymerization at the plasma membrane to produce membrane ruffles and lamellipodia.[13] Microinjection of a constitutively activated rac into fibroblasts leads to rapid lamellipodial extensions (see Fig. 1), which are associated with adhesion complexes.[13,14] These integrin complexes are morphologically distinct from classical focal adhesions, but appear to contain many of the same constituents, such as FAK and vinculin.[14] Endogenous rac can be activated by many tyrosine kinase receptors such as PDGF and insulin, but also by G-protein coupled receptors such as bombesin.[13] Interestingly, rac is activated by oncogenic versions of Ras and it has recently been shown that this is an essential downstream signal required for Ras-induced malignant transformation.[13,15] In platelets, rac is rapidly activated by thrombin.[16]

Other activities of rac that have been described are

1. regulation of the NADPH oxidase enzyme complex in phagocytic cells,[5]
2. activation of the JNK and p38 MAP kinase cascades,[17,18]
3. regulation of secretion in mast cells,[6,19]
4. inhibition of endocytosis,[20]
5. induction of malignant transformation and invasive phenotype.[15,21]

A number of candidate target proteins have been identified that interact with the GTP form of rac, including three families of Ser/Thr kinase, $p65^{PAK}$, MLK, and $p160^{ROCK}$, as well as several structural proteins.[22–24] In addition, rac activates a PIP 5-kinase activity *in vitro*[25] and in permeabilized platelets,[16] in platelets this is a key signal for the induction of actin polymerization.[16] The roles of the other target molecules are not clear at this stage.[23,24,26]

Purification

Recombinant rac proteins can readily be expressed in *E. coli* and the wild type proteins are obtained predominantly in a GDP-bound form.[27] Rac2 can be purified from human neutrophils.[28] Constitutively activated recombinant proteins, typically V12 or L61, are purified as

GTP-bound proteins using *E. coli* expression vectors, since they have low intrinsic GTPase activity. A dominant negative protein (N17rac) has been purified, which after microinjection inhibits agonist-stimulated rac activation.[13] This protein, however, has a half life of only around one h after injection. Post-translationally modified rac proteins can be obtained using Sf9 insect cells.[29]

Activities

The GTP- or GDP binding activities of recombinant rac proteins can readily be determined in a nitrocellulose filter binding assay, although this appears to underestimate the amount of protein present. GTP-loaded, wild-type rac (produced in *E. coli*) is converted spontaneously to the GDP form with a half life of around 10 min in Mg^{2+} and this can be decreased by addition of GAP proteins.[30] Rac GTPases are sensitive to proteolysis of C-terminal residues when produced in *E. coli* and, although this does not affect nucleotide binding or GTP hydrolysis, these proteins are biologically impaired. Biological activity can be determined by microinjection of recombinant protein into cells,[13] or using an *in vitro* NADPH oxidase assay.[31]

For expression studies N-terminally *myc*-epitope tagged cDNAs have been used.[23] These can be microinjected into cell nuclei or transfected and the effects on F-actin monitored by immunofluorescence. Rac will also activate JNK reporter plasmids when cotransfected into cells.[17,18]

Antibodies

Commercial antibodies that recognize rac1 or rac2 are available.

Genes

cDNA clones for human rac1 (GenBank accession number J05038), murine rac1 (GenBank accession number X57277), and human rac2 (GenBank accession number M29871) are available. To monitor expression, N-terminally *myc*-tagged cDNAs have been used.[23]

Mutant phenotypes

Although *rac* itself is not mutated, the genes for the four components of the NADPH oxidase (two cytochrome subunits, $p47^{phox}$ and $p67^{phox}$) are loci of mutations in human chronic granulomatous disease (CGD).[32] The *bcr* gene that encodes a GAP for rac is rearranged in Philadelphia chromosome-positive leukaemias.[12] Mutant versions of *rac* expressed in *Drosophila* or mouse neurones cause axon outgrowth defects without affecting dendritic outgrowth.[33,34] Mutant rac expressed in *Drosophila* wing epithelium leads to loss of organized actin at adherens junctions,[35] while expression in the *Drosophila* ovary blocks migration of border cells.[36]

Structure

The three-dimensional structure of human rac1 has been determined by X-ray crystallography.[37]

References

1. Didsbury, J., Weber, R. F., Bokoch, G. M., Evans, T., and Snyderman, R. (1989). *J. Biol. Chem.*, **264**, 16378–82.
2. Moll, J., Sansig, G., Fattori, E., and van der Putten, H. (1991). *Oncogene*, **6**, 863–6.
3. Didsbury, J. R., Uhing, R. J., and Snyderman, R. (1990). *Biochem. Biophys. Res. Commun.* **171**, 804–12.
4. Quinn, M. T., Evans, T., Loetterle, L. R., Jesaitis, A. J., and Bokoch, G. M. (1993). *J. Biol. Chem.*, **268**, 20983–7.
5. Abo, A., Pick, E., Hall, A., Totty, N., Teahan, C., and Segal, A. W. *Nature*, **353**, 668–70.
6. O'Sullivan, A. J., Brown, A. M., Freeman, N. M. and Gomperts, B. D. (1996). *Mol. Biol. Cell*, **7**, 397–408.
7. Just, I., Selzer, J., Wilm, M., Eichel-Streiber, C., Mann, M., and Aktories, K. (1995). *Nature*, **375**, 500–3.
8. Cerione, R. A. and Zheng, Y. (1996). *Curr. Opin. Cell Biol.*, **8**, 216–22.
9. Olson, M. F., Pasteris, G. N., Gorski, J. L., and Hall, A. (1996). *Curr. Biol.*, **6**, 1628–33.
10. Michiels, F., Habets, G. G., Stam, J. C., van der Kammen, R. A., and Collard, J. G. (1995). *Nature*, **375**, 338–40.
11. Nobes, C. D., Hawkins, P., Stephens, L., and Hall, A. (1995). *J. Cell Sci.*, **108**, 225–33.
12. Diekmann, D., Brill, S., Garrett, M. D., Totty, N., Hsuan, J., Monfries, C., *et al.* (1991). *Nature*, **351**, 400–2.
13. Ridley, A. J., Paterson, H. F., Johnston, C. L. Diekmann, D., and Hall, A. (1992). *Cell*, **70**, 401–10.
14. Nobes, C. D. and Hall, A. (1995). *Cell*, **81**, 53–62.
15. Qiu, R. Q., Chen, J., Kirn, D., McCormick, F., and Symons, M. (1995). *Nature*, **374**, 457–9.
16. Hartwig, J. H., Bokoch, G. M., Carpenter, C. L., Jamney, P. A., Taylor, L. A., Toker, A., and Stossel, T. P. (1995). *Cell*, **82**, 643–53.
17. Coso, O. A., Chiariello, M., Yu, J. C., Teramoto, H., Crespo, P., Xu, N., *et al.* (1995). *Cell*, **81**, 1137–46.
18. Minden, A., Lin, A., Claret, F. X., Abo, A., and Karin, M. (1995). *Cell*, **81**, 1147–57.
19. Price, L. S., Norman, J. C., Ridley, A. J. and Koffer, A. (1995). *Curr. Biol.*, **5**, 68–73.
20. Lamaze, C., Chuang, T. H., Terlecky, L. J., Bokoch, G. M., and Schmid, S. L. (1996). *Nature*, **382**, 177–80.
21. van Leeuwen, F. N., van der Kammen, R. A., Habets, G. G. M., and Collard, J. G. (1995). *Oncogene*, **11**, 2215–21.
22. Tapon, N. and Hall, A. (1997). *Curr. Biol.*, **9**, 86–92.
23. Lamarche, N., Tapon, N., Stowers, L., Burbelo, P. D., Aspenstrom, P., Bridges, T., *et al.* (1996). *Cell*, **87**, 519–29.
24. Joneson, T., McDonough, M., Bar-Sagi, D., and van Aelst, L. (1996). *Science*, **274**, 1374–6.
25. Tolias, K. F., Cantley, L. C., and Carpenter, C. L. (1995). *J. Biol. Chem.*, **270**, 17656–9.
26. Westwick, J. K., Laambert, Q. T., Clark, G. J., Symons, M., van Aelst, L., Pestell, R. G., and Der, C. J. (1997). *Mol. Cell. Biol.*, **17**, 1324–35.
27. Self, A. J. and Hall, A. (1995). *Meth. Enzymol.*, **256**, 3–10.
28. Knaus, U. G. and Bokoch, G. M. (1995). *Meth. Enzymol.*, **256**, 25–32.
29. Heyworth, P. G., Knaus, U. G., Xu, X., Uhlinger, D. J., Conroy, L., and Bokoch, G. M. (1993). *Mol. Biol. Cell.*, **4**, 261–9.
30. Self, A. J. and Hall, A. (1995). *Meth. Enzymol.*, **256**, 67–76.

31. Abo, A. and Segal, A. W. (1995). *Meth. Enzymol.*, **256**, 268–78.
32. Segal, A. W. and Abo, A. (1993). *Trends Biochem. Sci.*, Feb. 43–7.
33. Luo, L., Liao, Y. J., Jan, L. Y. and Jan, Y. N. (1994). *Genes Dev.* **8**, 1787–802.
34. Luo, L., Hensch, T. K., Ackerman, L., Barbel, S., Jan, L. Y., and Jan, Y. N. (1996). *Nature*, **379**, 837–40.
35. Eaton, S., Auvinen, P., Luo, L., Jan, Y. N., and Simons, K. (1995). *J. Cell Biol.*, **131**, 151–64.
36. Murphy, A. M. and Montell, D. J. (1996). *J. Cell Biol.*, **133**, 617–30.
37. Hirshberg, M., Stockley, R. W., Dodson, G., and Webb, M. R. (1997). *Nature Struct. Biol.*, **4**, 147–52.

Alan Hall
MRC Laboratory for Cell Biology, University College London, Gower Street, London WC1E 6BT, UK

Cdc42

The cdc42 GTPase regulates both the assembly of actin filaments at the cell periphery to form filopodia and the assembly of associated integrin adhesion complexes. Cdc42 is activated by bradykinin and the cytokines TNF-α and IL-1. In addition, cdc42 has been reported to activate the JNK MAP kinase pathway. It is likely that cdc42 plays an important role in cell movement and in growth cone migration in neurones.

Protein properties

Two isotypes of cdc42 have been reported, cdc42Hs and G25K and both proteins appear to be widely expressed and are particularly high in platelets.[1,2] The yeast *S. cerevisiae* has a close relative cdc42p and mammalian *CDC42* can complement yeast *CDC42* gene defects.[3] *S. pombe* has a single *cdc42* gene.[4] Cdc42 and G25K are both geranylgeranylated.[5] They are not targets for the C3 transferase, but are targets for *C. difficile* toxin B[6]. Cdc42 appears to be localized predominantly on the Golgi/endoplasmic reticulum,[7] although some can be found at the plasma membrane; the significance of this is not clear.

Several members of the GEF and GAP families are active on cdc42, such as dbl and vav which are also active on rac and rho.[8,9] The product of the human genetic disease locus, *FGD1* (facio-genital dysplasia, Aarskog–Scott syndrome) encodes a GEF that is specific for cdc42.[10] Several GAPs have activity on cdc42, including the product of the breakpoint cluster region (*bcr*) gene, which is also active on rac but not rho.[11]

A major role of cdc42 is to regulate the assembly of actin filaments at the plasma membrane to produce finger-like surface protrusions, filopodia.[12,13] Microinjection of a constitutively activated cdc42 into fibroblasts leads to rapid filopodial formation, which is accompanied by subsequent lamellipodial extensions (see Fig. 1). It has been demonstrated that cdc42 is a potent activator of rac in

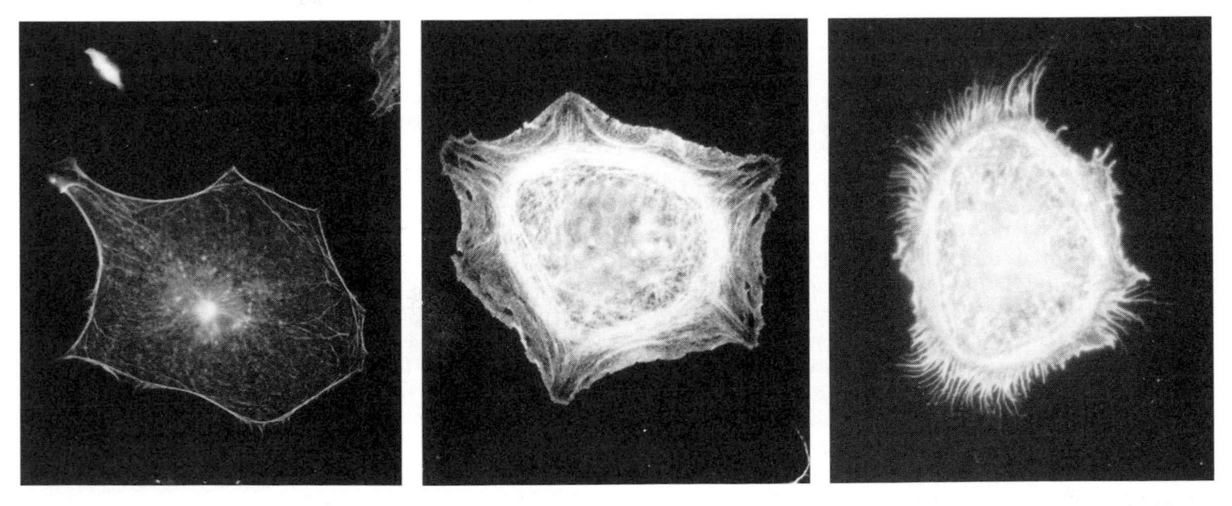

Figure 1. Constitutively activated cdc42 induces filopodia and (via rac) lamellipodia formation (middle panel) when introduced into a quiescent Swiss 3T3 cell (left panel). In the presence of dominant negative rac, cdc42 induces only filopodia (right panel). Polymerized actin visualized with fluorescent phalloidin in each panel.

fibroblasts leading to coordinated filopodia/lamellipodia production.[12,13] This suggests that these two GTPases may be important regulators of cell movement and axonal growth cone guidance. Filopodia induced by cdc42 are associated with integrin-dependent adhesion complexes, which are morphologically distinct from classical focal adhesions, but appear to contain many of the same constitutents, such as FAK and vinculin.[12] Endogenous cdc42 can be activated by bradykinin[13] and by the cytokines TNF-α and IL-1 (Nobes, C.D. and Hall, A.; unpublished data).

Other roles of cdc42 that have been described include:

1. activation of the JNK and p38 MAP kinase cascades,[14,15]
2. formation of a contractile ring during cytokinesis,[16]
3. *Salmonella* invasion,[17]
4. polarization of T cells.[18]

A number of candidate target proteins have been identified that interact with the GTP form of cdc42, including two families of Ser/Thr kinase, $p65^{PAK}$ and MLK, as well as several structural proteins, including the Wiskott-Aldrich syndrome protein (WASP);[19,20] but their cellular functions are not yet clear.[21]

Purification

Recombinant cdc42 proteins can readily be expressed in *E. coli* and the wild type proteins are obtained predominantly in a GDP-bound form.[22] Constitutively activated proteins, typically V12 or L61, are purified as GTP-bound proteins, since they have low intrinsic GTPase activity. A dominant negative protein (N17cdc42) has been purified, which after microinjection into fibroblasts, inhibits agonist-stimulated cdc42 activation.[13] Post-translationally modified cdc42 proteins have been obtained using Sf9 insect cells.[5]

Activities

The GTP- or GDP binding activities of recombinant cdc42 proteins can readily be determined in a nitrocellulose filter binding assay, although this appears to underestimate the amount of protein present.[23] GTP-loaded, wild-type cdc42 is converted spontaneously to the GDP form with a half life of around 10 min in Mg^{2+} and this can be decreased by addition of GAP proteins.[23] Cdc42 GTPases are sensitive to proteolysis of C-terminal residues when produced in *E. coli* and, although this does not affect nucleotide binding or GTP hydrolysis, these proteins are biologically impaired. Biological activity can be determined by microinjecting recombinant protein into cells.[12,13]

For expression studies, N-terminally *myc*-epitope tagged cDNAs have been used. These have been microinjected into cell nuclei or transfected into cells and the effects on F-actin monitored by immunofluorescence.[21] Cdc42 will also activate JNK reporter plasmids when co-transfected into cells.[14,15]

Antibodies

Commercial antibodies that recognize cdc42 are available.

Genes

cDNA clones for huuman cdc42Hs and G25K are available. For expression studies, N-terminally *myc*-epitope tagged cDNAs have been used.[21]

Mutant phenotypes

Although mutations have not been found in the human *cdc42* gene, an exchange factor for cdc42, FGD1, is the locus for facio-genital dysplasia (Aarskog–Scott) syndrome.[24] A target of cdc42, WASP, is the locus for Wiskott–Aldrich syndrome.[25] Expression of cdc42 mutants in *Drosophila* wing epithelium disrupts cell polarization.[26]

Structure

Not available.

References

1. Shinjo, K., Koland, J. G., Hart, M. T., Narasimham, V., and Johnson, D. L. (1990). *Proc. Natl Acad. Sci. USA*, **87**,9853–7.
2. Munemitsu, S., Innis, M. A., Clark, R., McCormick, F., Ulrich, A., and Polakis, P. (1990). *Mol. Cell. Biol.*, **10**, 5977–82.
3. Posada, J., Miller, P. J., McCullough, J. M., Ziman, M., and Johnson, D. L. (1995). *Meth. Enzymol.*, **256**, 281–90.
4. Chang, E. C., Barr, M., Wang, Y., Jung, V., Xu, H. P., and Wigler, M. H. (1994). *Cell*, **79**, 131–41.
5. Cerione, R. A., Leonard, D., and Zheng, Y. (1995). *Meth. Enzymol.*, **256**, 11–14.
6. Just, I., Selzer, J., Wilm, M., Eichel-Streiber, C., Mann, M., and Aktories, K. (1995). *Nature*, **375**, 500–3.
7. Erickson, J. W., Zhang, C., Kahn, R. A., Evans, T., and Cerione, R. A. (1996). *J. Biol. Chem.*, **271**, 26850–4.
8. Hart, M. J., Eva, A., Evans, T., Aaronson, A. A., and Cerione, R.A. (1991). *Nature*, **354**, 311–14.
9. Cerione, R. A. and Zheng, Y. (1996). *Curr. Opin. Cell Biol.*, **8**, 216–22.
10. Olson, M. F., Pasteris, G. N., Gorski, J. L., and Hall, A. (1996). *Curr. Biol.*, **6**,1628–33.
11. Lamarche, N. and Hall, A. (1994). *Trends Genet.*, **10**, 436–40.
12. Nobes, C. D. and Hall, A. (1995). *Cell*, **81**, 53–62.
13. Kozma,R., Ahmed,S., Best,A., and Lim, L. (1995). *Mol. Cell. Biol.*, **15**, 1942–52.
14. Coso, O. A., Chiariello, M., Yu, J. C., Teramoto, H., Crespo, P., Xu, N., Miki, T., and Gutkind, S. (1995). *Cell*, **81**, 1137–46.
15. Minden, A., Lin, A., Claret, F. X., Abo, A., and Karin, M. (1995). *Cell*, **81**, 1147–57.
16. Drechsel, D. N., Hyman, A. A., Hall, A., and Glotzer, M. (1997). *Curr. Biol.* **7**, 12–23.

17. Chen, L. M., Hobbis, S., and Galan, J. E. (1996). *Science*, **274**, 2115–18.
18. Stowers, L., Yelon, D., Berg, L. J., and Chant, J. (1995). *Proc. Natl Acad. Sci. USA*, **92**, 5027–31.
19. Tapon, N. and Hall, A. (1997). *Curr. Biol.*, **9**, 86–92.
20. Aspenstrom, P., Lindberg, U., and Hall, A. (1996). *Curr. Biol.*, **6**, 70–5.
21. Lamarche, N., Tapon, N., Stowers, L., Burbelo, P. D., Aspenstrom, P., Bridges, T., *et al.* (1996). *Cell*, **87**, 519–29.
22. Self, A. J. and Hall, A. (1995). *Meth. Enzymol.*, **256**, 3–10.
23. Self, A. J. and Hall, A. (1995). *Meth. Enzymol.*, **256**, 67–76.
24. Pasteris, N. G., Cadle, A., Logie, L. J., Porteous, M., Schwarz, C. E., Stevenson, R. E., *et al.* (1994). *Cell*, **79**, 669–78.
25. Villa, A., Notarangelo, L., and Vezzoni, P. (1995). *Nature Genet.*, **9**, 414–17.
26. Eaton, S., Auvinen, P., Luo, L., Jan, Y. N. and Simons, K. (1995). *J. Cell Biol.*, **131**, 151–64.

■ *Alan Hall*
RC Laboratory for Cell Biology,
University College London,
Gower Street, London WC1E 6BT, UK

Spectrins

Spectrin denotes a family of acidic, largely α-helical, high molecular weight, multifunctional, actin binding proteins usually found in association with the plasma membrane of mature cells.[1–4] Some forms are also recognized on internal membrane compartments including the Golgi and other intracellular vesicles.[5,6,30] Plasma membrane associated forms exist as heterodimers (α,β) and as heterotetramers $(\alpha,\beta)_2$, the latter being the most common. Larger oligomers also can exist (e.g. $(\alpha,\beta)_3$), and homopolymeric forms of β spectrin have been postulated. All spectrins share a repetitive 106 amino acid motif, with non-homologous sequences at each terminus and within the repeats region marking sites of functional specialization. The spectrins bestow mechanical stability on the membrane by controlling the distribution of integral membrane proteins, and link transmembrane proteins (and perhaps phospholipids) to the cytoskeleton. In non-erythroid cells, spectrins may be confined to distinct membrane domains where they are thought to mediate receptor organization and vesicle trafficking.[1,4,7]

Synonyms/related proteins

First identified in the mammalian erythrocyte, spectrins are now recognized as ubiquitous proteins present in all metazoan organisms including plants. Because of this great diversity, a bewildering array of names, such as fodrin, calspectin, brain spectrin, spectrin G, lung spectrin, γ spectrin, TW260/240, and so on, have been applied to what are essentially the same or very similar molecules. Alternative mRNA splicing also generates additional isoform diversity. To resolve this confusion, mammalian spectrins are now named based on their gene of origin, and whether they are an 'alpha' or 'beta' spectrin.[2,4] Isoforms arising by alternative transcription are designated as 'subtypes' using the symbol Σ. A summary of this nomenclature is listed in Table 1. Non-mammalian spectrins are named by reference to the known mammalian spectrins, or if unique, by designation as a special type of spectrin such as the large 430 kDa spectrin identified in *Drosophila*, and designated β_H.[8]

Distant homology exists with a-actinin, dystrophin, and kalirin.

Protein properties

The first spectrin to be isolated, and the one most divergent, is that from human erythrocyte ghosts. In humans this protein consists of an α subunit (αIΣ1 spectrin) of 280 000 Da (2429 residues)[9] and a β subunit (βIΣ1 spectrin) of 246 000 Da (2137 residues).[10] Both subunits migrate anomalously by SDS–PAGE, with apparent sizes of 240 000 and 220 000 Da respectively. Both subunits display a characteristic 106 residue repeat structure (Fig. 1). The repeat-to-repeat identity is 10–30 per cent. Alternative transcription of the βI gene in skeletal muscle and brain generates βIΣ2 spectrin, in which 22 C-terminal residues of βIΣ1 spectrin are replaced by 213 new residues that include a unique membrane association domain (MAD2) and a pleckstrin homology (PH) domain. *In vitro* reconstitution studies and direct visualization *in situ* indicate that spectrin and actin form an anastamosing and stoichiometric planar array at the cytoplasmic face of the membrane. In erythrocytes, and presumably other tissues as well, this array is linked to the anion transporter (band 3) and to other integral membrane glycoproteins by ankyrin and protein 4.1 (Fig. 2). Less well characterized attachments of spectrin to the membrane also exist, involving at least three membrane association domains (MAD1, MAD2, and MAD3).[11,12] Formation of the spectrin–actin lattice requires competent spectrin self-association and spectrin-F-actin binding. Several haemolytic diseases with enhanced erythrocyte fragility and shape abnormalities arise from point mutations or deletions in spectrin that alter its self-association or protein 4.1-, ankyrin-, or actin-binding ability[13] (Fig. 2).

Table 1 Summary of mammalian spectrins

Gene	Isoforms	Chromosome	GenBank	Comment
αI	Σ1	1q21	M61877	No splice forms identified. Found in red cells and brain
αII	Σ1 to Σ8 (see ref. 14)	9q34.1	U83867 (Σ1) J05243 (Σ2)	Found in most tissues; three regions of alternative splicing, some tissue specific Σ6–Σ8 believed to exist but not proven[14]
βI	Σ1 and Σ2, possibly Σ3	14q23–q24	J05500 (Σ1) M37884 (Σ2)	Predominant in red cells, muscle Σ2 isoform adds MAD2/PH domain and PIP_2 binding ability Putative βIΣ3 may be Golgi form, but this is unproven
βII	Σ1 and Σ2	2p21	M96803 (S1)	Found in most tissues βIIS2 partially characterized[11] Other splice forms may exist, but not well documented
βIII	Σ1	??	AB002300	Only identified by EST match Tissue distribution and validity as a bona fide spectrin not established

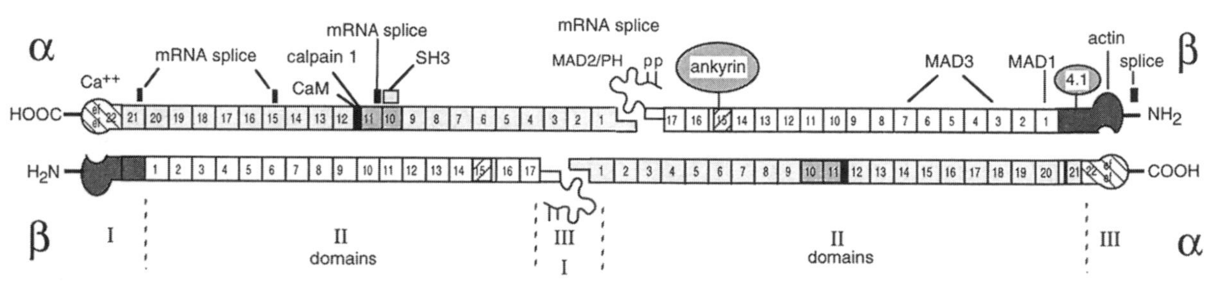

Figure 1. Schematic diagram of spectrin tetramer structure. The two subunits are arranged antiparallel to each other, and each is composed of multiple, approximately 106 residue, repeats. Each displays a tripartite domain structure with non-homologous termini (domains I/III), and a central domain II of repeating units. Non-homologous segments are marked in black or shaded. The sites of major binding proteins, and regions of alternative mRNA transcription are indicated. Direct membrane association domains (MAD1–3) are also shown.

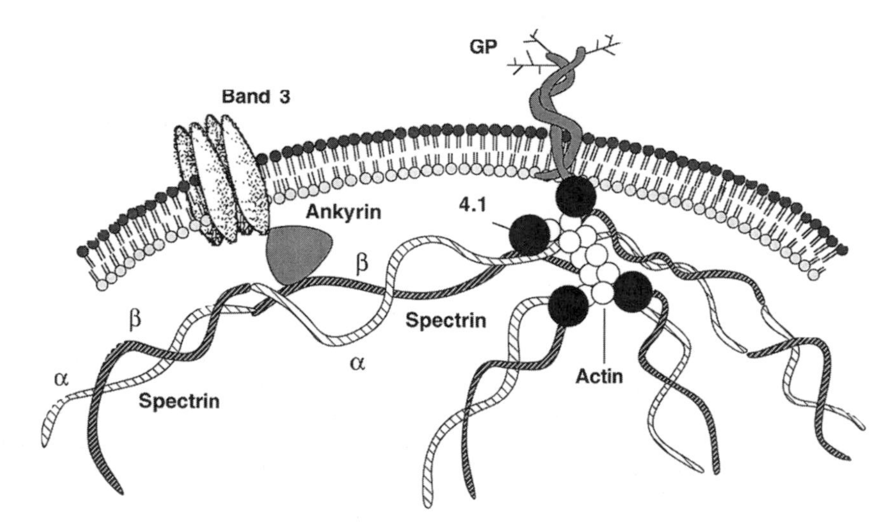

Figure 2. Organization of spectrin in the erythrocyte membrane cytoskeleton. A two dimensional planar lattice is formed by the ability of spectrin to self-associate at the amino terminal end of the α-subunit, and bind actin at the amino terminal end of the β-subunit. This array is linked to integral membrane proteins by means of ankyrin, and protein 4.1, as well as by direct spectrin-membrane interactions mediated by the MAD domains (not shown).

The most general class of spectrins are the so-called *non-erythroid* spectrins (fodrin), now termed αII spectrin and βII spectrin. Human αIIΣ1 spectrin has a molecular weight of 284 000 Da (2472 residues) and is 54 per cent identical to human erythrocyte α spectrin, 63 per cent identical to *Drosophila* αII spectrin, and 96 per cent identical to chicken αII spectrin. In mammals, eight isoforms of αII spectrin (αIIΣ1–8) arise by alternative mRNA splicing.[14] Two splice forms of βII spectrin are recognized (βIIΣ1 and βIIΣ2); these differ in their N-terminal region and display different assembly kinetics for the formation of the αII/βII spectrin heterodimeric complexes,[15] reflecting a contribution from the N-terminal region of βII spectrin to the nucleation of heterodimer formation.[15,16] Most of the specific protein–protein interactions that have been identified involve the β subunit of spectrin $\alpha\beta$ multimers.

Additional spectrins also exist. The terminal web (TW) spectrin of avian intestinal brush borders, called TW260/240, shares a common αII subunit (in avians) with other βII spectrins, but has a unique β subunit (estimated 260 000 Da by SDS–PAGE). This protein cross-links microvillar core actin bundles but does not associate intimately with the plasma membrane. A similar TW-specific isoform of spectrin is not present in mammals. Another variant β spectrin is that of *Drosophila* (β_H), which is unusual since its size (430 kDa) is similar to that of dystrophin and it appears not to exist in association with a paired α-spectrin subunit.[8] And finally, β spectrins (as well as other components of the spectrin-based skeleton) have been identified in association with the Golgi of mammalian cells.[5,6,30] While the precise nature of Golgi spectrin remains to be elucidated, it shares considerable immune and functional cross-reactivity with βI spectrin, suggesting that it may be a third spliciform of this gene (e.g. βIΣ3). Alternatively, it may arise from yet a third β spectrin gene (a putative βIII spectrin); the novel β spectrin gene recently identified as an expressed sequence tag[17] is one likely candidate for Golgi spectrin, or possibly for the unusual βI-like spectrin localized with clustered ACH receptors in myocytes.[4] The spectrins interact cooperatively with many proteins and small molecule ligands, including PIP2, and are the target of several post-translational regulatory events including covalent phosphorylation, Ca^{2+} and calmodulin binding, and proteolysis.

Purification

Erythrocyte spectrin is prepared by extraction of fresh erythrocyte ghosts at low ionic strength with EDTA, followed by gel filtration in isotonic buffers on a large pore sepharose (CL-2B or 4B) column.[18] Non-erythroid spectrins are typically prepared from brain membranes by extraction with either high[19] or low[20] ionic strength buffer, followed by ion exchange chromatography. The high-salt procedure yields αII/βII spectrin of better purity and therefore is the method of choice by most workers. Recombinant spectrin peptides that retain functional activity are also commonly prepared.[11] By most measures, spectrin regains a native conformation after urea denaturation, facilitating its manipulation as small or recombinant peptides.[4]

Activities

Spectrin has no intrinsic enzymatic activity. It associates *in vitro* with itself, ankyrin, F-actin, protein 4.1, adducin, calmodulin, calpain, dynamin, casein kinase 2, α-catenin, and probably other cytoskeletal, membrane, and cytosolic proteins.[1,4] Its ability to self-associate is most easily detected by non-denaturing gel electrophoresis (PAGE)[21] or by velocity sedimentation. Ankyrin, adducin, and F-actin binding are best demonstrated by either co-precipitation or velocity sedimentation. In these assays, the binding of spectrin to actin can be stimulated by protein 4.1 and inhibited under some conditions by Ca^{2+} and calmodulin.

Antibodies

Polyclonal and monoclonal antibodies to a variety of spectrins have been reported.[19,22–24] Antibodies to spectrin are also commercially available from Sigma, Chemicon, and others. Erythrocyte spectrin (αI/βI) antibodies generally react poorly with non-erythroid (αII/βII) spectrins, and *vice versa*. Most αII/βII-spectrin antibodies react well across species lines. Polyclonal preparations typically react more strongly with α spectrin than β spectrin. Most (but not all) antibodies to βI spectrin identify the spectrin associated with the Golgi, and less so the ACH receptor.

Genes

Five human spectrin genes have been identified (Table 1). These include full length cDNA clones for human αI spectrin,[9] human βI spectrin,[10] human αII spectrin,[25] and human βII spectrin.[26] Reports of partial sequences have also appeared. Recently, an unconfirmed EST fragment suggests the existence of βIII spectrin.[17]

Structure

Intact spectrin has not been crystallized. Insights into its structure have derived from studies of individual functional domains, as well as individual repeat units. The individual 106 residue spectrin repeat consists of a triple helical motif.[27] Functional specializations in spectrin are often created by non-homologous sequences inserted either between repeat units, or in some cases, within one of the helices of the repeat unit.[14,28,29] Presumably, these non-homologous and functionally active sequences form independently folding domains that do not disrupt the backbone structure of spectrin.

References

1. Bennett, V. and Gilligan, D. M. (1993). *Ann. Rev. Cell. Biol.*, **9**, 27–66.
2. Winkelmann, J. C. and Forget, B. G. (1993). *Blood*, **81**, 3173–85.
3. Goodman, S. R., *et al.* (1995). *Brain Res. Bull.*, **36**, 593–606.
4. Morrow, J. S., *et al.* (1997). In *Handbook of physiology*, (eds. J. Hoffman and J. Jamieson), pp. 485–540. Oxford, London.
5. Beck, K. A., Buchanan, J. A., Malhotra, V., and Nelson, W. J. (1994). *J. Cell Biol.*, **127**, 707–23.
6. Devarajan, P., *et al.*(1996). *J. Cell Biol.*, **133**, 819–30.
7. Devarajan, P. and Morrow, J. S. (1996). In *Membrane protein–cytoskeleton complexes: protein interactions, distributions and functions*, (ed. W. J. Nelson), pp. 97–128. Academic Press, New York.
8. Dubreuil, R. R., Byers, T. J., Stewart, C. T., and Kiehart, D. P. (1990). *J. Cell Biol.*, **111**, 1849–58.
9. Sahr, K. E., *et al.* (1990). *J. Biol. Chem.*, **265**, 4434–43.
10. Winkelmann, J. C., *et al.* (1990). *J. Biol. Chem.*, **265**, 11827–32.
11. Lombardo, C. R., Weed, S. A., Kennedy, S. P., Forget, B. G., and Morrow, J. S (1994). *J. Biol. Chem.*, **269**, 29212–19.
12. Davis, L. H. and Bennett, V. (1994). *J. Biol. Chem.*, **269**, 4409–16.
13. Lux, S. E. and Palek, J. (1995). In *Blood: principles and practice of hematology*, (eds. R. I. Handin, S. E. Lux, and T. P. Stossel), pp. 1701–1816. JB Lippincott, Philadelphia.
14. Cianci, C. D., Zhang, Z., and Morrow, J. S. (1998). (Submitted.)
15. Pradhan, D., Stabach, P. R., Lombardo, C., and Morrow, J. S. (1997). *Mol. Biol. Cell*, **8**, 61a.
16. Speicher, D. W., Weglarz, L., and DeSilva, T. M. (1992). *J. Biol. Chem.*, **267**, 14775–82.
17. Nagase, T., *et al.* (1997). *DNA Res.*, **4**, 141–50.
18. Bennett, V. (1983). *Meth. Enzymol.*, **96**, 313–24.
19. Glenney, J. R. Jr. and Glenney, P. (1983). *Cell*, **34**, 503–12.
20. Davis, J. and Bennett, V. (1983). *J. Biol. Chem.*, **258**, 7757–66.
21. Morrow, J. S. and Haigh, W. B., Jr (1983). *Meth. Enzymol.*, **96**, 298–304.
22. Zagon, I. S., Higbee, R., Riederer, B. M., and Goodman, S. R. (1986). *J. Neurosci.*, **6**, 2977–86.
23. Yurchenco, P. D., Speicher, D. W., Morrow, J. S., Knowles, W. J., and Marchesi, V. T. (1982). *J. Biol. Chem.*, **257**, 9102–7.
24. Harris, A. S., *et al.*(1986). *J. Cell. Biochem.*, **30**, 51–70.
25. Moon, R. T. and McMahon, A. P. (1990). *J. Biol. Chem.*, **265**, 4427–33.
26. Hu, R. -J., Watanabe, M., and Bennett, V. (1992). *J. Biol. Chem.*, **267**, 18715–22.
27. Yan, Y., *et al.* (1993). *Science*, **262**, 2027–30.
28. Musacchio, A., Noble, M., Pauptit, R., Wierenga, R., and Saraste, M. (1992). *Nature*, **359**, 851–5.
29. Stabach, P. R., Cianci, C. D., Glantz, S. B., Zhang, Z., and Morrow, J. S. (1997). *Biochemistry*, **36**, 57–65.
30. DeMatteis, M. A. and Morrow, J. S. (1998). *Curr. Opin. Cell Biol.*, **10**, (In press).

Jon S. Morrow
Department of Pathology,
Yale Medical School,
New Haven, CT, USA

Talin

Talin is a high molecular weight protein localized in adherens-type junctions with the extracellular matrix (ECM), although it is also localized at the junctions between T-helper and antigen presenting cells. Talin binds *in vitro* to the cytoplasmic domains of the integrin family of extracellular matrix receptors, to vinculin, actin, and the protein tyrosine kinase $pp125^{FAK}$. It has been proposed to be one of several proteins linking actin filament bundles to the cytoplasmic face of integrins.

Protein properties

Talin shows only limited regions of sequence homology to other proteins.[1] Residues 165–363 within the N-terminal region of the mouse protein shows homology with the N-terminal domain in the ezrin/radixin/moesin (ERM) family of cytoskeletal proteins. The extreme C-terminal region of talin shows sequence similarity to the yeast actin-binding protein Sla2p. The overall domain structure of talin and the ERM proteins is broadly similar in that they are all thought to associate with the plasma membrane lipid bilayer via the N-terminal domain and to bind actin via C-terminal sequences.

Talin has been isolated and studied from chicken gizzard smooth muscle[2] and human blood platelets.[3] The apparent molecular mass on SDS–PAGE is 225–235 kDa. Analysis of the complete chicken talin sequence[1] indicates a true M_r of 271 881 Da with 2541 amino acids. Talin is readily cleaved into two domains by many proteases; the Ca^{2+}-activated protease calpain II cleaves gizzard talin between residues 433–434.[4] On SDS–PAGE the resultant fragments have M_r of 47 kDa and 190–200 kDa. Cleavage of talin has been shown to occur in focal adhesions in CHO cells[5] and is observed during platelet aggregation, although the significance of this has not been resolved. The N-terminal 47 kDa fragment is important in targeting the protein specifically to cell–matrix junctions[6] and can bind to charged lipids.[7] The large C-terminal domain of talin is predicted to be α helical and to contain a large number of short (approximately 34 residue) alanine-rich repeats.[1] This region is reported to bind to the $\beta1$ and $\beta3$ integrin cytoplasmic domains[1,8] and contains three vinculin binding sites,[1] which overlap sequences required for

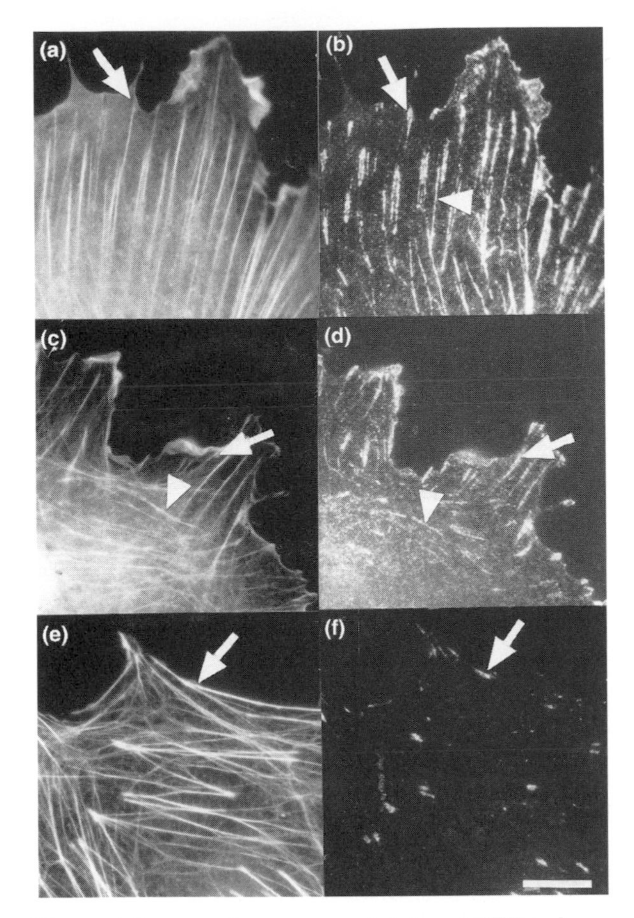

Figure 1. Immunofluorescence staining of CEF and mouse Swiss 3T3 cells with anti-talin monoclonal antibodies.[20] CEF (a, b, c, d) and Swiss 3T3 fibroblasts (e, f) were grown on glass coverslips and immunostained for talin using the anti-talin antibodies TA205 (b) and TD77 (d, f). Cells were co-stained for F-actin with FITC–phalloidin (a, c, e). Bar represents 5 μm.

targeting the protein to focal adhesions. Talin is an actin nucleating protein which binds G-actin with a k_d of 0.25 μM. The nucleation activity is associated with the 190 kDa C-terminal fragment.[7] Studies with recombinant talin polypeptides have identified three non-overlapping regions which bind F-actin with low affinity (5–17 μM), each of which is adjacent to a vinculin-binding site.[1] Talin can cross-link F-actin into both networks and bundles at pH 6.4 in low salt, but bundling, then cross-linking activity is progressively lost as the pH is raised to 7.3.[9] Talin enhances F-actin cross-linking by α-actinin.[1]

Electron microscopy of rotary shadowed talin molecules reveals an elongated, flexible protein, about 60 nm long, consisting of a series of globular masses, like beads on a string.[2,10,11] Talin adopts a more globular configuration in low ionic strength buffers whereas the 190 kDa polypeptide is rod-shaped under both conditions.[2] Interestingly, the 190 kDa talin polypeptide has been reported to bind vinculin more tightly than talin, raising the possibility that the vinculin-binding sites are partly obscured in the intact protein.[1] Below 0.7 mg/ml talin exists as a monomer, but above this concentration it begins to self-associate to form dimers in which the subunits are antiparallel.[30] Analytical ultracentrifugation yields an apparent M_r of 412 kDa and a sedimentation coefficient of $s_{20,w}$ = 11.2 S. Chemical cross-linking indicates that the 190 kDa fragment is solely responsible for dimerization.[11]

In many types of cultured cells, talin is concentrated in focal adhesions[8] (focal contacts, adhesion plaques), regions where bundles of actin filaments attach to the cytoplasmic face of the plasma membrane and where the external face of the membrane adheres most tightly to the underlying substratum. Talin is also found in ruffling membranes and subjacent to bundles of ECM on the cell surface.[12] Talin is concentrated *in vivo* at the cytoplasmic face of the plasma membrane where cells interact with and transmit tension to the ECM; for example, it is enriched at the myotendinous junctions of skeletal muscle[13] and the dense plaques of smooth muscle.[14] Together with other focal adhesion proteins, it is found

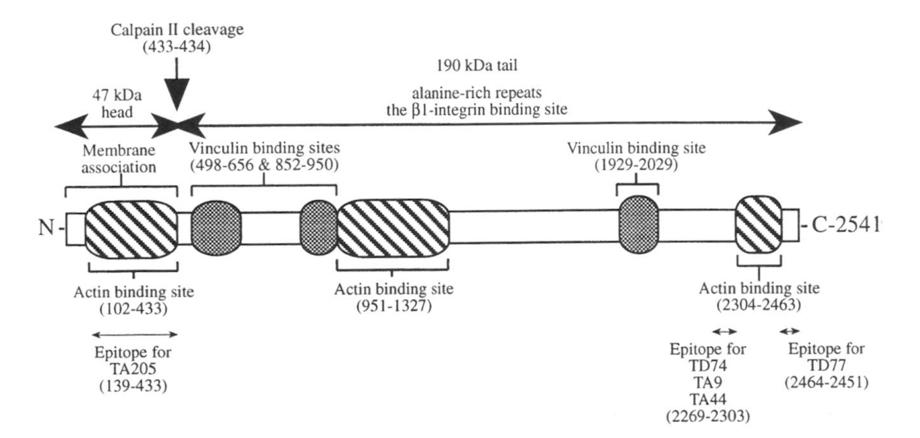

Figure 2. Domain structure of talin.[20] The epitopes recognized by a series of monoclonal antibodies raised against human platelet talin are shown.

at the postsynaptic face of some but not all neuromuscular junctions.[15] Talin is present but less concentrated at the basal surface of epithelial cells where they adhere to the basement membrane.[14] It is very abundant in platelets accounting for more than 3 per cent of total platelet protein.[3] In resting platelets, talin is distributed diffusely throughout the cytoplasm, but in response to activation it is redistributed to the cortex underlying the plasma membrane.[16] Although talin is present in cell-ECM junctions, it is absent from many cell–cell junctions, such as the zonula adherens junctions of epithelial cells.[17] However, microinjection of the 190 kDa C-terminal fragment of talin into epithelial cells in culture results in an accumulation of the talin polypeptide in the zonula adherens junctions of these cells.[6] Talin is concentrated in the cell–cell adhesions made by lymphocytes, for example in cytotoxic lymphocytes where these are adhering to target cells.[18] Notably, these lymphocyte cell–cell adhesions are also sites where integrins are concentrated. The amount of talin appears to be low in neuronal cells, although it has been reported in nerve growth cones.[19]

Evidence from a wide variety of experimental approaches supports the conclusion that talin plays a key role in cell adhesion to the ECM. Microinjection of polyclonal anti-talin antibodies into chicken embryo fibroblasts plated on to fibronectin inhibits cell spreading and causes cell rounding. Similarly, monoclonal antibodies which recognize epitopes at the N- and C-terminal regions of talin disrupt actin stress fibres and focal adhesions, and inhibit cell motility in human fibroblasts.[20] Down regulation of talin in HeLa cells using antisense technology slows cell spreading.[21] Other experiments suggest that talin is involved in the early events associated with cell adhesion, including the formation of filopodia and lamellipodia. In *Dictyostelium discoideum*, talin accumulates at the tips of filopodia formed in response to cAMP[22] and microscale chromophore-assisted laser inactivation of talin in neuronal growth cones results in the temporary cessation of filopodial extension.[19] Talin is localized in membrane ruffles in chick embryo fibroblasts unlike vinculin, and it is recruited to newly forming focal adhesions before vinculin.[1] Studies with *C. elegans* mutants show that the localization of talin to focal adhesion-like structures is dependent on the presence of β-integrin, but not vinculin.[23] In addition, mouse F9 teratocarcinoma cell lines lacking vinculin are able to assemble talin-containing focal adhesions, although the adhesions are smaller, and the cells more motile. The fact that talin can bind integrins[8] and nucleate,[7] cross-link, and bundle actin filaments[9] is consistent with a model in which it serves to link the actin cytoskeleton to the cytoplasmic domain of integrins. Its ability to bind vinculin,[1] and other actin-binding proteins, may be important in stabilising adhesions. Vinculin can bind VASP, which can in turn bind to a profilin/G-actin complex, offering a possible mechanism by which actin monomers might be delivered to sites of filament assembly.

Rather little is known about factors which regulate the activity of talin. It is a substrate for protein kinase C *in vitro*[24] and the level of talin phosphorylation is elevated in cells treated with tumour promoters which simulate protein kinase C. This is associated with the disruption of actin stress fibres and focal adhesions. In L6 myocytes, PDGF has been shown to stimulate tyrosine phosphorylation of talin which correlates with the disruption of actin stress fibres and the loss of vinculin from focal adhesions.[25] However, integrins and talin remain colocalized in focal adhesions. The result raises the possibility that tyrosine phosphorylation of talin inhibits its ability to bind actin and vinculin, but not integrins. Talin phosphorylation is also increased four-fold in thrombin-activated platelets,[26] and this is associated with the relocation of talin from the cytosol to a submembranous location.

Purification

Talin has been purified from low ionic strength extracts of chicken gizzard smooth muscle[2] and from Triton X-100 lysates of human platelets.[3] The purification involves conventional chromatography in ion exchange resins and gel filtration. The extreme sensitivity of talin to cleavage by Ca^{2+}-dependent proteases can be a problem and the removal of calcium by chelation is recommended, as well as the inclusion of protease inhibitors such as leupeptin.

Activities

Talin has been shown to bind to integrin *in vitro*[8] with a low affinity ($k_d \sim 10^{-6}$M),[1] to vinculin with a higher affinity ($k_d \sim 10^{-8}$M)[1] and to pp125^{FAK}.[27] It nucleates actin filament assembly, and binds to G-actin with a $k_d \sim 0.25 \times 10^{-6}$M.[7]

Antibodies

Polyclonal antibodies against chicken gizzard talin[12] cross-react with amphibian talin, but bind only poorly to mammalian talin. Monoclonal antibodies are available against chicken gizzard talin (one of which cross-reacts with mammalian talin[28]). Polyclonal antibodies against human platelet talin have been generated, but the human platelet protein is a very poor immunogen. Monoclonal antibodies against human talin are marketed by Serotec[20] and Sigma.

Genes

Mouse (X56123),[4] *C. elegans* (L46861),[22] and *Dictyostelium discoideum* (U14576)[22] sequences are available in databases and the sequence of chicken talin has been published recently.[1] Partial human cDNAs have been isolated and used to assign the gene to chromosome 9p.[29]

Mutant phenotypes

Dictyostelium talin null mutants show defects in phagocytosis and cell–cell and cell–substrate interactions. Cytokinesis was slightly impaired.[31]

Structure

No three dimensional-structural information available.

References

1. Hemmings, L., Rees, D.J.G., Ohanian, V., Bolton, S.J., Gilmore, A.P., Patel, B., *et al.* (1996). *J. Cell Sci.*, **109**, 2715–26.
2. Molony, L., McCaslin, D., Abernethy, J., Paschal, B., and Burridge, K. (1987). *J. Biol. Chem.*, **262**, 7790–5.
3. Collier, N.C. and Wang, K. (1982). *J. Biol. Chem.*, **257**, 6937–43.
4. Rees, D.J.G., Ades, S.E., Singer, S.J., and Hynes R.O. (1990). *Nature*, **347**, 685–9.
5. Tranqui, L., and Block, M.R. (1995). *Exp. Cell Res.*, **217**, 149–56.
6. Nuckolls, G.H., Turner, C.E., and Burridge, K. (1990). *J. Cell Biol.*, **110**, 1635–44.
7. Niggli, V., Kaufmann, S., Goldmann, W.H., Weber, T., and Isenberg, G. (1994). *Eur. J. Biochem.*, **224**, 951–7.
8. Knezevic, I., Leisner, T.M. and Lam, S. C-T. (1996). *J. Biol. Chem.*, **271**, 16416–21.
9. Zhang, J., Robson, R.M., Schmidt, J.M., and Stromer, M.H. (1996). *Biochem. Biophys. Res. Commun.*, **218**, 530–7.
10. Winkler, J., Lunsdorf, H., and Jocjusch, B.M. (1997). *Eur. J. Biochem.*, **243**, 430–6.
11. Goldmann. W.H., Bremer, A., Haner, M., Aebi, U., amd Isenberg, G. (1994). *J. Struct. Biol.*, **112**, 3–10.
12. Burridge, K. and Connell, L. (1983). *J. Cell Biol.*, **97**, 359–67.
13. Tidball, J.G., O-Halloran, T., and Burridge, K. (1986). *J. Cell Biol.*, **103**, 1465–72.
14. Drenckhahn, D., Beckerle, M., Burridge, K., and Otto, J. (1988). *Eur. J. Cell Biol.*, **46**, 513–22.
15. Rochlin, M.W., Chen, Q., Tobler, M., Turner, C.E., Burridge, K., and Peng, H.B. (1989). *J. Cell Sci.*, **92**, 461–72.
16. Beckerle, M.C., Miller, D.E., Bertagnolli, M.E., and Locke, S.J. (1989). *J. Cell Biol.*, **109**, 3333–46.
17. Geiger, B., Volk, T., and Volberg, T. (1985). *J. Cell Biol.*, **101**, 1523–31.
18. Kupfer, A., Singer, S.J. and Dennert, G. (1986). *J. Exp Med.*, **163**, 489–98.
19. Sydor, A.M., Su, A.L., Wang, F-S., Xu, A., and Jay, D.G. (1996). *J. Cell Biol.*, **134**, 1197–207.
20. Bolton, S.J., Barry, S.T., Mosley, H., Patel, B., Jockusch, B.M., Wilkinson, J.M., and Critchley, D.R. (1997). *Cell Motil. Cytoskel.*, **36**, 363–76.
21. Albiges-Rizo, C., Frachet, P., and Block, M.R. (1995). *J. Cell Sci.*, **108**, 3317–29.
22. Kreitmeier, M., Gerisch, G., Heizer, C., and Muller-Taubenberger, A. (1995). *J. Cell Biol.*, **129**, 179–88.
23. Moulder, G.L., Huang, M.M., Waterston, R.H., and Barstead, R.J. (1996). *Mol. Biol. Cell.*, **7**, 1181–93.
24. Watters, D., Garrone, B., Gobert, G., Williams, S., Gardiner, R., and Lavin, M. (1996). *Exp. Cell Res.*, **229**, 327–35.
25. Tidball, J.G. and Spencer, M.J. (1993). *J. Cell Biol.*, **123**, 627–35.
26. Bertagnolli, M.E., Locke, S.J., Hens;er, M.E., Bray, P.F., and Beckerle, M.C. (1993). *J. Cell Sci.*, **106**, 1189–99.
27. Chen, H.-C., Appeddu, P.A., Parsons, J.T., Hildebrand, J.D., Schaller, M.D., and Guan, J-L. (1995). *J. Biol. Chem.*, **270**, 16995–9.
28. Otey, C., Griffiths, W. and Burridge, K. (1990). *Hybridoma*, **9**, 57–62.
29. Gilmore, A.P., Ohanian, V., Spurr, N.K., and Critchley, D.R. (1995). *Hum. Genet.*, **96**, 221–4.
30. Isenberg, G. and Goldmann, W. H. (1998). *FEBS Lett.*, **426**, 165–70.
31. Niewohner, J., Weber, I., Maniak, M., Muller-Taubenberger, A., and Gerisch, G. (1997). *J. Cell Biol.*, **138**, 349–61.

David R. Critchley
Department of Biochemistry,
University of Leicester, University Road,
Leicester LE1 7RH, UK

Tensin

Tensin (M_r = 150–220 kDa), first purified as an actin capping protein from chicken gizzard, has been immunolocalized to many types of adherens junctions and at or near the Z lines of muscle. The protein has several actin binding and actin capping domains, an SH2 domain, and regions showing sequence homology with other cytoskeletal proteins, protein kinases, protein phosphatases, and a tumour suppressor gene product. Tyrosine phosphorylation of tensin is increased with viral transformation, adhesion to extracellular matrix, and stimulation by platelet derived growth factor. These properties of tensin strongly suggest that the protein plays a role in actin–membrane association and signal transduction.

Synonymous names and homologous proteins

No synonymous names, but tensin does have many regions showing sequence homology with other proteins.

Protein properties

The discovery of tensin stems from studies on the interaction of F-actin with vinculin, a protein found at focal adhesions of fibroblasts. In experiments in the early 1980s, vinculin preparations from chicken gizzard inhibited actin polymerization in a manner suggesting a direct interaction of the protein with the barbed ends of actin filaments (i.e. capping activity).[1] Subsequent studies indicated that this phenomenon is attributable to the activity of a heterogeneous group of peptides (designated HA1; M_r = 20–45 kDa) contaminating the vinculin preparations.[2] Antibodies raised against HA1 were shown to cross-react with high molecular weight bands (M_r of about 150–220 kDa) in immunoblots of proteins from different tissues, suggesting that HA1 peptides are proteolytic fragments of a larger protein.[3] In a 1987 report, the purification of a cross-reactive polypeptide of about 150 kDa from chicken gizzard was described and the name 'tensin' was given on the basis of a putative role in maintaining tension in actin filaments by linking them to other structures.[4] This name is now generally used to include immunologically related polypeptides with molecular weight around 215–220 kDa found in many types of animal cells and tissues. The degree to which the 150 kDa and the higher molecular weight polypeptides are related genetically, structurally, and functionally remains unclear.

Immunofluorescence localization studies with anti-HA1 antibodies and anti-tensin antibodies showed that cross-reactive material is generally found at locations where actin filaments are associated with the cell membrane:[3,5–8] focal adhesions and extracellular matrix contacts of fibroblasts (Fig. 1), cell–cell contacts of epithelial cells, intercalated discs of cardiac muscle, dense plaques of smooth muscle, costameres of skeletal and cardiac muscle, and myotendonous and neuromuscular junctions. In addition, fluorescence labelling is also seen at Z lines[3,5] (Fig. 2), or at the I bands of striated muscle.[7] In studies on differentiating chicken chondrocytes[9] and osteoclasts[10] in culture, tensin expression and localization correlated with substrate adhesion. Tensin expression was also found to be regulated during cartilage cell differentiation *in vivo*.[9] In another study, tensin was shown to be expressed in many different tissues during murine embryogenesis, although tensin null mice developed normally and appeared healthy postnatally for at least several months.[11] Over time, however, these mice developed multiple large cysts in the proximal kidney tubules, with disruption of cell–matrix junctions and lack of polarity in the tubule cells.

A preparation of polyclonal antibodies against chicken gizzard 150 kDa tensin was used to screen a cDNA library prepared from chicken embryo fibroblasts[12] and two libraries from adult chicken heart.[13,14] Overlapping cDNA clones covering enough nucleotides to encode a polypeptide of around 200 kDa were obtained from each of these libraries and the nucleotide sequences derived from these clones are very similar. Recently, a tensin cDNA from a mouse lung library also showed about 90 per cent identity to chicken heart tensin cDNA at the level of deduced amino acid sequence.[11]

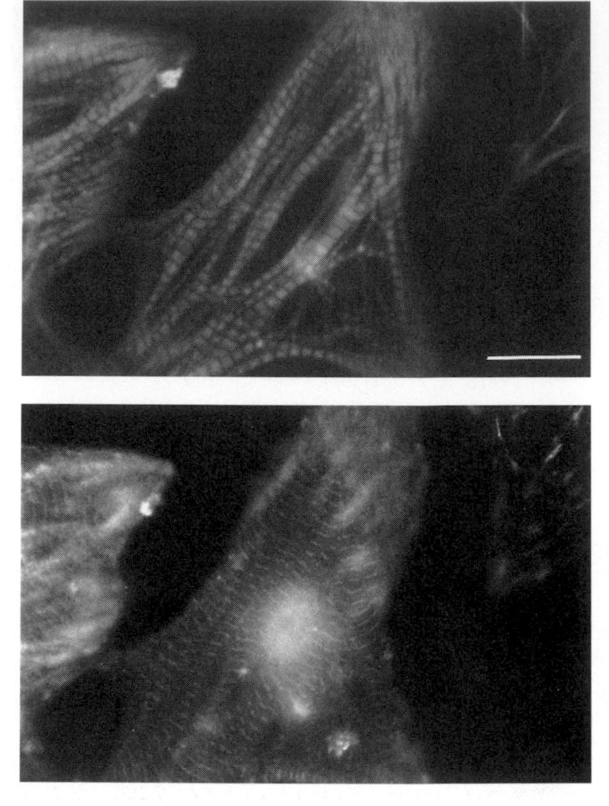

Figure 1. (a) Fluorescence staining of F-actin in chicken embryo fibroblast using fluorescein-labelled phalloidin. Bar, represents 20 μm. (b) Indirect immunofluorescence staining of the same cell using a monoclonal antibody to tensin (TL1)[14] followed by rhodamine-labelled goat anti-mouse antibody.

A study on tensin fragments expressed as fusion proteins revealed that the N-terminal half of tensin contains two low-affinity actin binding regions detectable with a co-sedimentation assay (k_d in micromolar range).[15] A third region near the center of the protein was reported to inhibit barbed end actin polymerization,[15] but another study was unable to confirm this result.[14] Instead, the latter study showed that the adjacent region of S1061–H1145 contains high-affinity actin capping activity similar to that of purified chicken gizzard tensin (k_d in nanomolar range).[14] In a related study, fusion proteins containing tensin regions on the N-terminal and C-terminal sides of the actin capping domain were found to bind with saturability to vinculin and integrin *in vitro*, respectively.[16] These results suggest that that tensin may function as a link between actin filaments and membrane-associated components at adherens junctions. This notion is supported by the report that tensin is the only cytoskeletal protein (together with focal adhesion kinase)

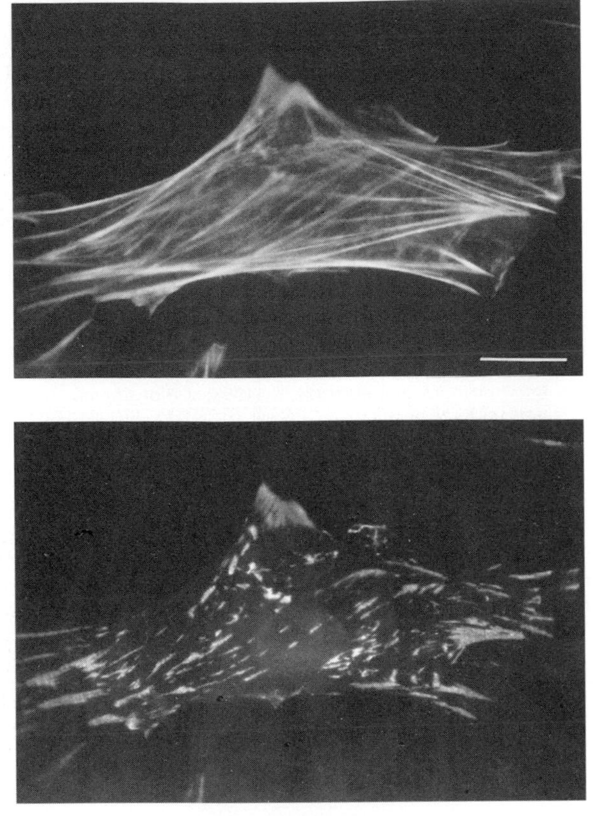

Figure 2. (a) Fluorescence staining of F-actin in cultured chicken cardiac myocytes (left and centre) using fluorescein-labelled phalloidin. Bar represents 20 μm. (b) Indirect immunofluorescence staining of the same cells with a monoclonal antibody against tensin (TL1)[14] followed with rhodamine-labelled goat anti-mouse antibody.

closely associated with clusters of integrin cross-linked with immobilized antibodies in fibroblasts.[17]

Analysis of the deduced amino acid sequence of tensin revealed a number of regions with homology to other cytoskeletal proteins (numbering system according to ref. 14): (a) M49–T78 and L673–H707 with homology to actin, (b) L419–N443 with homology to the actin binding site of actin cross-linking proteins, (c) D1739–E1767 with homology to the actin binding site of myosin, and (d) the first 300 or so residues at the N terminus of tensin with homology to the N-terminal region of auxilin. Interestingly, the last region also has sequences homologous to the middle part of a cyclin G associated kinase,[18] the catalytic site of a protein tyrosine phosphatase (although missing two of the four essential residues for catalytic activity),[19] and a tumour suppressor gene product (designated as P-TEN or MMAC1) associated with different types of cancer.[20,21]

Several lines of evidence indicated that tensin might also be involved in signal transduction. First, the protein contains an SH2 domain (W1520–P1628),[12] which appears to mediate indirectly its association with phosphoinositide 3-kinase.[22] Second, fibroblast tensin contains phosphotyrosine, the level of which increases with transformation by v-*src*,[12] with cell adhesion to extracellular matrix proteins,[23] and with stimulation by platelet-derived growth factor.[24] Third, tensin has been reported to associate with p130^{CAS}, a protein implicated in integrin signalling.[25] Finally, as mentioned above, the N-terminal region of tensin has sequences homologous to a tumour suppressor gene product and to a protein tyrosine phosphatase. All of the known properties of tensin, therefore, strongly suggest that it has an important role in linking actin-based cytoskeletal functions with signal transduction in normal and cancerous cells.

Purification

Tensin was originally purified from a high-salt extract of chicken gizzard with the use of a series of different chromatographic columns.[4] Purification of native tensin is difficult due to the large number of PEST sequences leading to extreme susceptibility to proteolytic degradation.[14] Most studies, therefore, have been performed on recombinant full length or truncated tensin.[14,15]

Activities

Co-sedimentation and pyrene–actin polymerization assays have been used to measure the actin binding and actin capping activities of tensin.[14,15]

Antibodies

Polyclonal and monoclonal antibodies to chicken tensin[3,7,13–15] and polyclonal antibodies to mouse tensin[11] have been described.

Genes

Chicken tensin cDNA clones from several tissues[9,12–14] and mouse tensin cDNA and genomic clones[11] have been described. Many tensin sequences can be found in GenBank.

Mutant phenotype/disease state

Tensin null mice develop normally, but over time, they exhibit signs of renal failure. It was concluded that loss of tensin may lead to a weakening rather than severing of focal adhesions in the kidney.[11]

Structure

Not available.

References

1. Wilkins, J. A. and Lin, S. (1982). *Cell*, **28**, 83–90.
2. Wilkins, J. A. and Lin, S. (1986). *J. Cell Biol.*, **102**, 1085–92.
3. Wilkins, J. A., Risinger, M. A., and Lin, S. (1986). *J. Cell Biol.*, **103**, 1483–94.
4. Wilkins, J. A., Risinger, M. A., Coffey, E., and Lin, S. (1987). *J. Cell Biol.*, **130a**.
5. Risinger, M. A., Wilkins, J.A., and Lin, S. (1987). *J. Cell Biol.*, **130a**.
6. Lin, S., Risinger, M. A., and Butler, J. A. (1989). *Cytoskeletal and extracellular matrix proteins*, (ed. U. Aebi and J. Engel), Vol. **3**, pp. 341–4. Springer Series in Biophysics, Springer, London.
7. Bockholt, S. M., Otey, C. A., Glenney, J. R. Jr, and Burridge, K. (1992). *Exp. Cell Res.*, **203**, 39–46.
8. North, A. J., Galazkiewicz, B., Byers, T. J., Glenney, J. R Jr, and Small, J. V. (1993). *J. Cell Biol.*, **120**, 1159–67.
9. van de Werken, R., Gennari, M., Tavella, S., Bet, P., Molina, F., Lin, S., *et al.* (1993). *Eur. J. Biochem.*, **217**, 781–90.
10. Hiura, K., Lim, S. S., Little, S. P., Lin, S., and Sato, M. (1995). *Cell Motil. Cytoskel.* **30**, 272–84.
11. Lo, S. H., Yu, Q. C., Degenstein, L., Chen, L. B., and Fuchs, E. (1997). *J. Cell Biol.*, **136**, 1349–61.
12. Davis, S., Lu, M.L., Lo, S. H., Lin, S., Butler, J. A., Druker, B. J., *et al.* (1991). *Science*, **252**, 712–5.
13. Lo, S. H., An, Q., Bao, S., Wong, W. K., Liu, Y., Janmey, P. A., *et al.* (1994). *J. Biol. Chem.*, **269**, 22310–19.
14. Chuang, J. Z., Lin, D. C., and Lin, S. (1995). *J. Cell Biol.*, **128**, 1095–109.
15. Lo, S. H., Janmey, P. A., Hartwig, J. H., and Chen, L. B. (1994). *J. Cell Biol.*, **125**, 1067–75.
16. Lin, S. and Lin, D. C. (1996). *Mol. Biol. Cell.*, **7**, 389a.
17. Miyamoto, S., Akiyama, S. K., and Yamada, K. M. (1995). *Science*, **267**, 883–5.
18. Kanaoka, Y., Kimura, S. H., Okazaki, I., Ikeda, M. and Nojima, H. (1997). *FEBS Lett.* **402**, 73–80.
19. Haynie, D. T. and Ponting, C. P. (1996). *Prot. Sci.*, **5**, 2643–6.
20. Li, J., Yen, C., Liaw, D., Podsypanina, K., Bose, S., Wang, S. I., *et al.* (1997). *Science*, **275**, 1943–7.
21. Steck, P. A., Pershouse, M. A., Jasser, S. A., Yung, W. K., Lin, H., Ligon, A. H., *et al.* (1997). *Nature Genetics*, **15**, 356–62.
22. Auger, K. R., Songyang, Z., Lo, S. H., Roberts, T. M., and Chen, L. B. (1996). *J. Biol. Chem.*, **271**, 23452–7.
23. Bockholt, S. M. and Burridge, K. (1993). *J. Biol. Chem.*, **268**,14565–7.
24. Jiang, B., Yamamura, S., Nelson, P. R., Mureebe, L., and Kent, K. C. (1996). *Surgery*, **120**, 427–31.
25. Salgia, R., Pisick, E., Sattler, M., Li, J.L., Uemura, N., Wong, W. K., *et al.* (1996). *J. Biol. Chem.*, **271**, 25198–203.

Shin Lin and Diane C. Lin
Department of Developmental and Cell Biology, University of California, Irvine, CA 697–1450, USA

VASP

VASP, a proline-rich protein substrate of both cAMP- and cGMP-dependent protein kinases, is expressed in most mammalian cell types and tissues. Particularly high VASP levels are detected in human platelets. In cultured cells, VASP is associated with focal adhesions, cell–cell contacts, microfilaments, and highly dynamic membrane regions such as the leading edge. From *in vitro* binding data VASP has been suggested to link profilin to zyxin, vinculin, and the *Listeria spp.* surface protein ActA. Functional evidence indicates that VASP is a crucial factor involved in the enhancement of actin filament formation.

Synonymous name

Vasodilator stimulated phosphoprotein.

Homologous proteins

Drosophila Enabled[1] (Ena), mouse Mena[2] (mammalian Enabled), Mouse Evl[2] (Ena-VASP-like protein).

Protein properties

VASP was initially purified and characterized as a 46 kDa/50 kDa substrate of cAMP- and cGMP-dependent protein kinases in platelets, cultured fibroblasts, and other cells. Since the protein is phosphorylated in response to cylic nucleotide-regulating vasodilators such as cAMP-elevating prostaglandins and cGMP-elevating NO donors it has been named vasodilator-stimulated phosphoprotein,[3–5] with VASP as an acronym. Three distinct phosphorylation sites have been identified:[6] Ser-157, Ser-239, and Thr-278. Whereas cAMP-dependent protein kinase phosphorylates Ser-157 first, followed by Ser-239, cGMP-dependent protein kinase I shows opposite site preferences, which is particularly evident *in vitro*.[6] Phosphorylation of Ser-157 leads to a marked mobility shift, on SDS–PAGE, from 46 to 50 kDa.[3,6] This mobility shift has been widely used as an indicator for cyclic nucleotide-dependent protein kinase activity in intact cell systems (for a review see ref. 7). Moreover, a monoclonal antibody, 16C2, recognizes VASP only when Ser-239 is phosphorylated. This antibody allows the detection of VASP Ser-239 phosporylation and cGMP-dependent protein kinase activation in intact cells.[29] Upon removal of the stimulus, VASP is rapidly dephosphorylated.[5] *In vitro*, VASP is a substrate for protein phosphatases (PP) 2A, 2B, and 2C, with PP2A being the most likely candidate as the phosphatase involved in VASP dephosphorylation in human platelets.[8] A close correlation of VASP phosphorylation at Ser-157 and the inhibition of fibrinogen binding to integrin $\alpha_{\text{IIb}}\beta_3$ of human platelets has been described.[9]

Molecular cloning of human and canine VASP[10] predicted highly homologous proteins of 380 or 384 amino acids, respectively, with prospective N- and C-terminal domains separated by a low complexity proline-rich central segment containing a characteristic GPPPPP motif in single copy and as a three-fold tandem repeat (Fig. 1(a)). The hydrodynamic properties of VASP (sedimentation coefficient 4.5 S; Stokes radius 8 nm; frictional ratio 2.3) argue for a homotetrameric complex with an elongated structure.[10] VASP is the founding member of a new family of proline-rich proteins, which also includes Enabled (Ena), a dose-dependent suppressor of *Drosophila* Abl- and Disabled-dependent phenotypes,[1] its mammalian homologue Mena[2] and the Ena-VASP-like protein Evl.[2] These proteins all share an identical overall domain organization comprising highly homologous N-terminal EVH1 (Ena-VASP homology 1) and C-terminal EVH2 domains of about 110 and 130–190 amino acids respectively, which are separated by low complexity regions including a proline-rich central segment of 60 to 90 amino acids in length (Fig. 1(a)). In agreement with the significant homology between VASP and Ena, VASP has recently been shown to rescue the embryonic lethality associated with loss of Ena function in *Drosophila*.[33] *Drosophila* AE33[11] (a transcriptional target of Rough and Glass), and the putative gene products identified by the human expressed sequence tags T52235 and D80468 appear to have in common a similar but distinct EVH1 domain as compared with the Ena-VASP protein family.

VASP is expressed in a wide variety of cells and tissues with the highest levels being found in platelets.[12] Subcellularly VASP is highly concentrated at focal adhesions and stress fibres, which show a periodic decoration by VASP (Fig. 2(b)). Moreover, VASP is localized at cell–cell contacts of certain cultured cells (Fig. 2(a)) and is associated with highly dynamic membrane structures such as the leading edge. In cells infected with the bacterium *Listeria monocytogenes*, which exploits an actin polymerization-based system to promote its intracellular motility, VASP is found at the interface between the moving bacterium and its actin tail, that is at the site where actin polymerization is thought to take place.[13] Here, VASP colocalizes with profilin, Mena, and the listerial surface protein ActA, the only *Listeria* protein that is both necessary and sufficient for actin-based intracellular bacterial motility.[2,13–16] For two of these proteins, profilin[17,18] and ActA,[13] a direct interaction with VASP has been demonstrated *in vitro*. The VASP interaction with profilin involves the three-fold tandem repeat of the GPPPPP motif in VASP.

Listerial ActA,[13] zyxin,[19] and vinculin[20,21] share another proline-rich motif (E/DFPPPPXD/E)[30] that has been characterized in detail as a functional VASP and Mena binding site.[30] Zyxin and ActA contain three or four of these motifs, while vinculin has a single VASP binding motif within its hinge region. Oligomerization of vinculin has been suggested to compensate for the presence of a

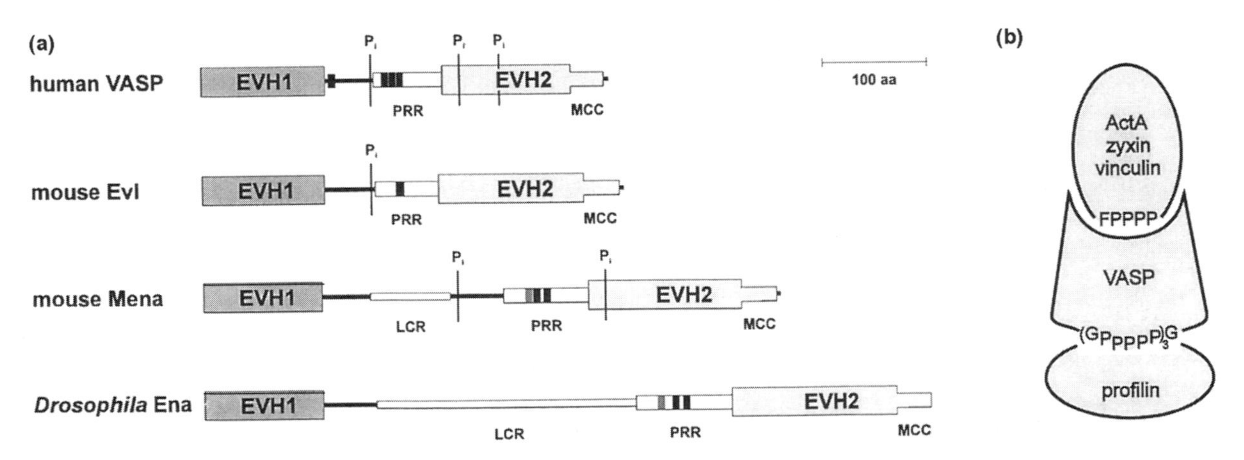

Figure 1. (a) Domain organization of members of the Ena-VASP protein family. EVH1/2: Ena-VASP homology domains 1, and 2; PRR: proline-rich region with GPPPPP motifs; MCC: repetitive mixed-charge cluster at the C-terminal end of the EVH2 domain; LCR: low complexity regions rich in glutamine or glutamic acid, arginine, and leucine in Ena, and Mena respectively; P_i: cyclic nucleotide kinase-dependent serine/threonine phosphorylation sites characterized in VASP. (b) Scheme depicting VASP as an adapter protein linking profilin (bound to the VASP GPPPPP motifs) to the FPPPP-containing proteins vinculin, zyxin or ActA.

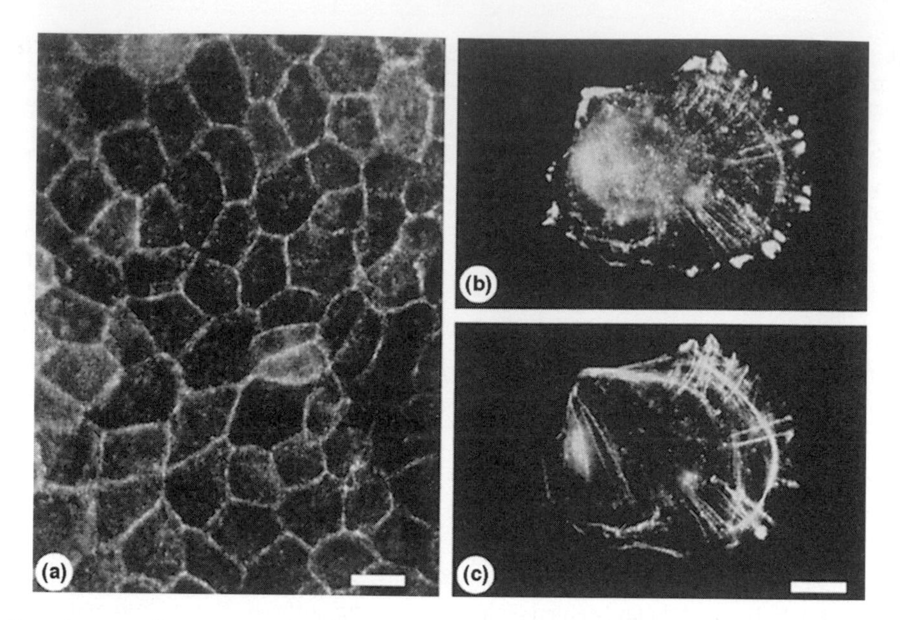

Figure 2. (a) Indirect immunofluorescence localization of VASP in MDCK cells (a), and a human skin fibroblast (b); comparison to F-actin staining (b′). Bar in a: 20 μm, in b′: 10 μm.

single binding site per polypeptide chain only.[20] Indeed, PIP_2 induces vinculin oligomerization and also enhances VASP–vinculin complex formation.[31] In conclusion, VASP has properties of an adaptor protein linking profilin to vinculin, zyxin, or ActA (Fig. 1(b)). In accordance with this concept, profilin localization to the distal pole of *Listeria* is dependent on the presence of the ActA VASP binding sites.[22]

There are two lines of evidence suggesting a function of Ena-VASP family members in enhancing actin filament formation and actin-based motility: microinjection of peptides that are known to interfere with VASP and Mena binding to their FPPPP-containing ligands[2,20,21] displaces VASP[23] and Mena[2] from focal adhesions and leads to cessation of listerial motility and retraction of membrane protrusions.[24] Moreover, deletion constructs of ActA transfected into eukaryotic cells[23] or expressed in bacteria[22,25] revealed that the proline-rich binding sites for Ena-VASP family proteins are required for efficient but not for basal ActA-induced F-actin recruitment and actin-based motility.

Purification

VASP can be purified from human and porcine platelets, with a typical yield from several hundred micrograms (starting from human platelet concentrates equivalent to ~15 l of blood) up to a few milligrams (from 60 l of porcine blood). In a modification[19] of the original protocol,[3] VASP is extracted from human platelet homogenates with 250 mM NaCl, and after dialysis the extract is passed through Q-sepharose at pH 7.5. The flow-through is chromatographed sequentially on Orange A and MonoS HR.

Purification from porcine platelets involves S-sepharose FF chromatography, ammonium sulphate precipitation (0–30 per cent saturation), followed by chromatography on hydroxy/apatite, Orange A, MonoS HR, and an optional second hydroxy/apatite column.[19] Biologically active VASP protein, as assayed by its substrate function for cGMP-dependent protein kinase and its binding activity for ActA or profilin, can also be expressed and purified from *Escherichia coli* and baculovirus-infected insect cells.[26]

Activities

Binding to zyxin, vinculin, and the Listeria surface protein ActA

VASP binds to purified vinculin[20,21], zyxin,[19] and ActA[13] in solid phase binding assays with VASP as a soluble ligand. The interaction with the latter two proteins can also be probed in blot overlays of purified proteins or total cell lysates using ^{32}P-labelled phospho-VASP.[13,19]

Binding to profilin

VASP binding to profilin has been shown in solid phase binding assays with immobilized profilin.[17] Moreover, VASP and an 81 kDa fragment of a ~160 kDa protein are the major platelet proteins, besides actin, retained by a profilin affinity column. Available microsequences of the 81 kDa polypeptide[17] reveal that it is a fragment of human Diaphanous 1.[32] Both proteins can specifically be eluted by a VASP peptide comprising the three-fold tandem repeat of the GPPPPP motif.[17] Mammalian profilin isoforms show different affinities for this proline-rich sequence.[18]

VASP binding to actin

VASP binds[12] to and crosslinks (M.R. and U.W.; unpublished) purified F-actin *in vitro*, as assayed in co-sedimentation experiments. In addition, a putative G-actin binding motif in the EVH2 domain of Ena-VASP family proteins was predicted[2] from sequence alignments.

Antibodies

M4 polyclonal rabbit antiserum

Most of the original work on VASP has been done with the M4 rabbit antiserum[4] produced against VASP purified from human platelets. The polyclonal antibodies recognize VASP from a wide variety of species, including human, rat, mouse, pig, dog, chicken, and marsupials, and have been used in immunoblotting, immunofluorescence, and immunoprecipitation applications. Especially with rodent cells and tissues, there may be a reactivity with an 80 kDa species of yet unknown identity.

Monoclonal VASP antibodies

A whole panel of monoclonal VASP antibodies[27] against human and porcine VASP has been produced, epitope mapped, and tested for interference with VASP protein–protein interactions, phosphorylation, and dephosphorylation (K. Reinhard *et al.*, in preparation). Monoclonal VASP antibodies have also been applied in immunoblotting, immunofluorescence, and immunoprecipitation (K. Reinhard *et al.*; in preparation). In addition, a monoclonal antibody specific for Ser-239 phosphorylated VASP has been developed.[29]

Commercial sources

The polyclonal rabbit-anti-VASP serum M4 and the monoclonal VASP antibody IE273 are commercially available from immunoGlobe GmbH, Grossostheim, Germany (http://www.immunoGlobe.com/).

Genes

VASP cDNAs[10] have been cloned from HL60 (accession no Z46389) and canine MDCK cells (accession no Z46388). Both human (accession no X98476, X98533, X98534) and mouse (accession no X98476) genes encoding VASP have been identified and were shown to be composed of 13 exons.[28] The promoter regions are characterized by a prominent CpG island and several SP1 and AP1 sites, indicating a housekeeping function of VASP.[28] The human gene maps[28] to chromosome 19q13.2–q13.3, about 92 kb distal to ERCC1 and about 300 kb proximal to the gene for myotonic dystrophy protein kinase.

Mutant phenotypes/disease states

No information available.

Structure

Not available.

Web site

http://www.klin-biochem.uni-wuerzburg.de/~VASP/.

References

1. Gertler, F. B., Comer, A. R., Juang, J. -L., Ahern, S. M., Clark, M. J., Liebl, E. C., and Hoffmann, F. M. (1995). *Genes Dev.*, **9**, 521–33.
2. Gertler, F. B., Niebuhr, K., Reinhard, M., Wehland, J., and Soriano, P. (1996). *Cell*, **87**, 227–39.
3. Halbrügge, M. and Walter, U. (1989). *Eur. J. Biochem.*, **185**, 41–50.
4. Halbrügge, M., Friedrich, C., Eigenthaler, M., Schanzenbächer, P., and Walter, U. (1990). *J. Biol. Chem.*, **265**, 3088–93.
5. Nolte, C., Eigenthaler, M., Schanzenbächer, P., and Walter, U. (1991). *J. Biol. Chem.*, **266**, 14808–12.
6. Butt, E., Abel, K., Krieger, M., Palm, D., Hoppe, V., Hoppe, J., and Walter, U. (1994). *J. Biol. Chem.*, **269**, 14509–17.
7. Walter, U., Eigenthaler, M., Geiger, J., and Reinhard, M. (1993). *Adv. Exp. Med. Biol.*, **344**, 237–49.
8. Abel, K., Mieskes, G., and Walter, U. (1995). *FEBS Lett.*, **370**, 184–8.
9. Horstrup, K., Jablonka, B., Hönig-Liedl, P., Just, M., Kochsiek, K., and Walter, U. (1994). *Eur. J. Biochem.*, **225**, 21–7.
10. Haffner, C., Jarchau, T., Reinhard, M., Hoppe, J., Lohmann, S. M., and Walter, U. (1995). *EMBO J.*, **14**, 19–27.
11. DeMille, M. M. C., Kimmel, B. E., and Rubin, G. M. (1996). *Gene*, **183**, 103–8.
12. Reinhard, M., Halbrügge, M., Scheer, U., Wiegand, C., Jockusch, B. M., and Walter, U. (1992). *EMBO J.*, **11**, 2063–70.
13. Chakraborty, T., Ebel, F., Domann, E., Niebuhr, K., Gerstel, B., Pistor, S., *et al.* (1995). *EMBO J.*, **14**, 1314–21.
14. Pollard, T. D. (1995). *Curr. Biol.*, **5**, 837–40.
15. Smith, G. A., Portnoy, D. A., and Theriot, J. A. (1995). *Mol. Microbiol.*, **17**, 945–51.
16. Kocks, C., Marchand, J. B., Gouin, E., d'Hauteville, H., Sansonetti, P. J., Carlier, M. F., and Cossart, P. (1995). *Mol. Microbiol.*, **18**, 413–23.
17. Reinhard, M., Giehl, K., Abel, K., Haffner, C., Jarchau, T., Hoppe, V., Jockusch, B. M., and Walter, U. (1995). *EMBO J.*, **14**, 1583–9.
18. Lambrechts, A., Verschelde, J. -L.,Jonckheere, V., Goethals, M., Vandekerckhove, J., and Ampe, C. (1997). *EMBO J.*, **16**, 484–94.
19. Reinhard, M., Jouvenal, K., Tripier, D., and Walter, U. (1995). *Proc. Natl Acad. Sci. USA*, **92**, 7956–60.
20. Reinhard, M., Rüdiger, M., Jockusch, B. M., and Walter, U. (1996). *FEBS Lett.*, **399**, 103–7.
21. Brindle, N. P., Holt, M. R., Davies, J. E., Price, C. J., and Critchley, D. R. (1996). *Biochem. J.*, **318**, 753–7.
22. Smith, G. A., Theriot, J. A., and Portnoy, D. A. (1996). *J. Cell Biol.*, **135**, 647–60.

23. Pistor, S., Chakraborty, T., Walter, U., and Wehland, J. (1995). *Curr. Biol.*, **5**, 517–25.
24. Southwick, F. S. and Purich, D. L. (1994). *Proc. Natl Acad. Sci. USA*, **91**, 5168–72.
25. Lasa, I., David, V., Gouin, E., Marchand, J. B., and Cossart, P. (1995). *Mol. Microbiol.*, **18**, 425–36.
26. Jarchau, T., Mund, T., Reinhard, M., and Walter, U. (1998). *Methods Enzymol.*, **298**, 103–13.
27. Abel, K., Lingnau, A., Niebuhr, K., Wehland, J., and Walter, U. (1996). *Eur. J. Cell Biol.*, **69**, (Suppl. 42), 39a.
28. Zimmer, M., Fink, T., Fischer, L., Hauser, W., Scherer, K., Lichter, P., and Walter, U. (1996). *Genomics*, **36**, 227–33.
29. Smolenski, A., Bachmann, C., Reinhard, K., Hönig-Liedl, P., Jarchau, T., Hoschuetzky, H., and Walter, U. (1998). *J. Biol. Chem.*, **273**, (In press.)
30. Niebuhr, K., Ebel, F., Frank, R., Reinhard, M., Domann, E., Carl, U. D., Walter, U., Gertler, F. B., Wehland, J. and Chakraborty, T. (1997). *EMBO J.*, **16**, 5433–44.
31. Hüttelmaier, S., Mayboroda, O., Harbeck, B., Jarchau, T., Jockusch, B. M., and Rüdiger, M. (1998). *Curr. Biol.*, **8**, 479–88.
32. Lynch, E. D., Lee, M. K., Morrow, J. E., Welcsh, P. L., Leon, P. E., and King, M. C. (1997). *Science*, **278**, 1315–18.
33. Ahern-Djamali, S. M., Comer, A. R., Bachmann, C., Kastenmeier, A. S., Reddy, S. K., Beckerle, M. C., Walter, U., and Hoffmann, F. M. (1998). *Mol. Biol. Cell*, (In press).

■ *Matthias Reinhard, Thomas Jarchau, Kathrin Reinhard, and Ulrich Walter*
Medizinische Universitätsklinik,
Institut für Klinische Biochemie und Pathobiochemie, Versbacher Strasse 5, D-97078 Würzburg, Germany

Vinculin

Vinculin is an anchor protein which is specifically associated with cell–ECM adhesions such as focal contacts as well as with adherens type cell–cell junctions. In both sites, vinculin participates in the formation of a submembrane 'plaque' structure which is responsible for the attachment of actin filaments to the plasma membrane. Biochemical and molecular genetic analyses of intact vinculin or vinculin fragments suggested that the molecule can interact with a variety of cytoskeletal and anchor proteins. Furthermore, it is shown that vinculin may exist in folded (head-to-tail) and unfolded states and that the unfolding process, induced by acidic phospholipids, may lead to the exposure of its different binding sites.

Synonymous names

130 kDa protein.[1]

Homologous proteins

α-Catenin has been shown to bear homology to vinculin, mainly to its N- and C-terminal regions which contain the major established binding sites of the molecule.[2,3]

Protein properties

Vinculin is a 117 kDa microfilament-associated protein located at the cytoplasmic aspects of focal contacts (FC) and cell–cell adherens type junctions (AJ). These adhesion sites are specialized membrane domains through which cells form tight and stable interactions with either specific extracellular matrix (ECM) molecules or with adhesion molecules on neighbouring cells. The 'receptors' in both these types of adhesions were shown to be members of the integrin and cadherin families, respectively.[4] Based mainly on its ubiquitous localization in FC and AJ, it was proposed that vinculin is involved in linking actin to the junctional membrane and thus promote the establishment and stabilization of cell adhesions.[1,5]

Vinculin was originally isolated from chicken gizzard extract as a by-product during the purification of α-actinin.[1]. Immunolabelling of a variety of cultured cells as well as sections prepared from different tissues indicated that the protein is ubiquitously associated with both cell–cell and cell-ECM adhesions.[5,6] Consequently, vinculin is commonly used as a hallmark for these families of adhesion sites (Fig. 1). Tissue surveys confirmed that vinculin is indeed a ubiquitous protein, present in a wide variety of cell types, including many mesenchymal and epithelial cells, as well as lymphoid cells and platelets which do not form stable, long-term adhesions.

Microinjection of fluorescently tagged protein into cells indicated that vinculin is associated with two cytoplasmic pools; a diffusible fraction and a junctional membrane-associated fraction, which maintain a dynamic equilibrium between them.[7,8]

The general cellular functions of vinculin and its role in the formation of adhesion sites were characterized by modulating the expression levels of the protein in cultured cells. It was shown that reduction in vinculin levels by antisense transfection strongly suppresses cell adhesion[9] while overexpression enhances cell adhesiveness and focal contact assembly.[10]. It was also demonstrated that knock-out of the vinculin gene from F-9 teratocarcinoma cells leads to a reduction in the adhesive properties

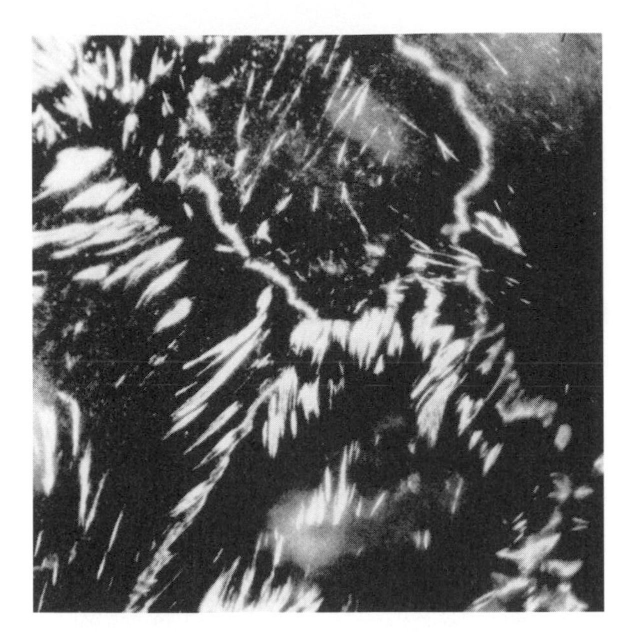

Figure 1. Immunofluorescence labelling for vinculin of cultured chicken lens cells displays both cell–extracellular matrix (mostly focal contacts) and cell–cell adherens-type junctions. Magnification ×890.

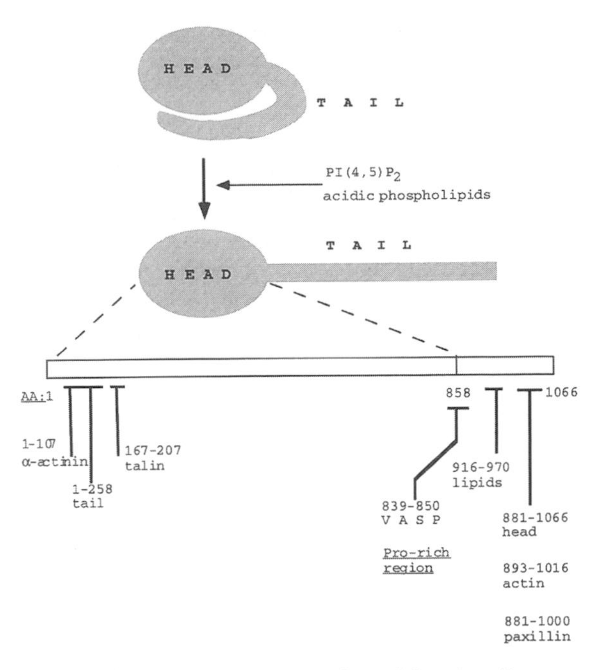

Figure 2. Schematic representation of the vinculin molecule in its folded and extended states. Some of the binding sites in the head and tail regions are marked (for additional information and references, see text).

of the cells and higher migratory activity, compared to the parental cells.[11–13] Interestingly, it was shown that the formation of focal adhesions is not completely blocked in the knocked-out cells.[14] Reduction in vinculin levels was reported for several tumour cell lines, as well as metastatic human tumours.[15,16] Restoration of vinculin levels in the cultured tumour cells by transfection resulted in the formation of normal focal adhesions and markedly suppressed the tumorigenic activity of the cells.[16]

Vinculin binding proteins

Attempts to elucidate the specific role of vinculin in cell adhesion involved binding or co-sedimentation assays of purified vinculin, or recombinant fragments corresponding to its various domains, with different anchor or cytoskeletal proteins (Fig. 2). Among the vinculin binding molecules detected in such experiments are talin,[17–19] α-actinin;[20,21] paxillin,[22,23] actin,[24,25] phosphoinositol 4,5-bisphosphate (PIP2) and acidic phospholipids,[26–29] and vasodilator-stimulated phosphoprotein (VASP).[30,31] In addition vinculin can form either inter-molecular interactions with other vinculin molecules,[18,19,32,33] or intramolecular head-to-tail interactions (Fig. 2 and refs 34–36). Interestingly, some of these sites might be cryptic in the folded state of the molecule and are only exposed following extension of the tail domain (Fig. 2, see ref. 35). This conformational change can be induced by PIP2 or acidic phospholipids.[36,37] It was further proposed that transmembrane signalling may result in local accummulation of such lipids, leading to the unfolding of the vinculin molecule and exposure of its cryptic binding sites to proteins, such as talin, α-actinin, and actin.[21,36,38] The presence of multiple binding sites on vinculin, suggesting that it might interact simultaneously with several junctional molecules in adhesion sites, is also supported by cDNA transfection experiments in which different non-overlapping segments (C- and N-terminal) of the molecule were shown to associate independently *in vivo* with adhesion sites[39] and the microfilament system.

The availability of recombinant vinculin and vinculin fragments and mutants made it possible to assign specific binding sites to specific regions along the molecule (Fig. 2). This determinant mapping suggested that the various interactions of vinculin are mediated via two clusters of sites located along the N-terminal or C-terminal 200 amino acids of the protein. (The exact location of each site, as suggested in Fig. 2, is still rather crude and based either on direct binding to the specified segment of the molecule or the loss of binding following deletion of the marked region.) It is also noteworthy that some interactions may selectively affect binding at adjacent sites (for example, tail–head interaction inhibits vinculin binding to actin or to talin, but, apparently, does not exert a similar effect on paxillin binding).

Molecular structure and heterogeneity of vinculin

Sequencing of vinculin cDNA from various sources indicated that the protein has a molecular weight of approximately 117 000 Da with a large N-terminal domain that contains the talin binding site and three

~110 amino acids repeats. This region is separated from the C-terminal region by several stretches of prolines in which the major proteolytic cleavage sites are located.[40–42] Electron microscopic work confirmed the presence of a globular N-terminal 'head' and a rod-like, C-terminal 'tail'.[32,33,43]

Two-dimensional gel electrophoresis revealed the presence of three major isoelectrophoretic forms of vinculin (denoted α,β, and γ-vinculin), all with the same apparent molecular weight. The two latter forms are specifically expressed in muscle cells and the α form appears to be phosphorylated.[44] In addition, a higher molecular weight variant of vinculin was detected in muscle cells and denoted meta-vinculin.[45–47] Sequence data indicated that this molecule contains an extra segment of 68 amino acids located near the proline-rich region, between the N-terminal globular head and the C-terminal tail.[48] The functional significance of this isoform heterogeneity is still unclear.

Vinculin was shown to be post-translationally modified by phosphorylation on either serines (by protein kinase C[49]) or tyrosines (by pp60^{src}[50] or other focal contact-associated tyrosine-specific kinase).

Purification

The most common source of vinculin is chicken smooth muscle, from which it is usually purified as described by Geiger[1] or by Burridge and Feramisco.[7] In addition, vinculin has been purified, following modifications of these procedures from other sources such as human and pig smooth muscle.

Activities

The presence of multiple binding sites to different anchor and cytoskeletal proteins along the vinculin molecule and the cellular consequences of the modulation of its expression, suggest that vinculin is a junctional 'adaptor protein' which plays a central role in the assembly and stability of cell adhesions.

Antibodies

Antibodies originally used for vinculin localization were raised in rabbits and guinea pigs by repeated injection of the purified chicken gizzard protein.[1,5] In recent years many monoclonal and polyclonal antibodies have been produced which react with vinculins from a large variety of species from *C. elegans* to humans. Many of these antibodies are currently available from commercial sources.

Genes

Vinculin cDNA from different sources has been cloned and sequenced including that of chicken,[40–42] human,[51] mouse,[52] *Caenorhabditis elegans,*[53] and another nematode, Brugia malayi (Dissanayake *et al.* unpublished data). The specific sequences of the extra piece found in chicken, *Xenopus*, human, and pig meta-vinculin have been derived from protein and DNA sequence analysis.[48,54].Genomic sequences are available for the *C. elegans*, human, mouse, and chicken vinculins. Details are available from GenBank/EMBL databases.

References

1. Geiger, B. (1979). *Cell*, **18**, 193–220.
2. Herrenknecht, K., Ozawa, M., Eckerskorn, C., Lottspeich, F., Lenter, M. and Kemler, R. (1991). *Proc. Natl Acad. Sci., USA*, **88**, 9156–60.
3. Nagafuchi, A., Takeichi, M., and Tsukita, S. (1991). *Cell*, **65**, 849–57.
4. Yamada, K. M. and Geiger, B. (1997). *Curr. Opin. Cell Biol.*, **9**, 76–85.
5. Geiger, B., Tokuyasu, K. T., Dutton, A. H., and Singer, S. J. (1980). *Proc. Natl. Acad. Sci. USA*, **77**, 4127–31.
6. Otto, J. J. (1990). *Cell Motil. Cytoskel.* **16**, 1–6.
7. Burridge, K. and Feramisco, J. R. (1980). *Cell*, **19**, 587–95.
8. Kreis, T. E., Avnur, Z., Schlessinger, J., and Geiger, B. (1985). *Molecular biology of the cytoskeleton*, Cold Spring Harbor Symposium. (ed. G. Borisy, D. Cleveland, and D. Murphy), pp. 45–57.
9. Rodriguez Fernandez, J. L., Salomon, D., Ben-Ze'ev, A. and Geiger, B. (1993). *J. Cell Biol.*, **122**, 1285–94.
10. Rodriguez Fernandez, J. L., Geiger, B., Salomon, D. and Ben Ze'ev, A. (1992). *Cell Motil. Cytoskel.*, **22**, 127–34.
11. Coll, J. L., Ben-Ze'ev, A., Ezzell, R. M., Rodriguez-Fernandez, J. L., Baribaut, H., Oshima, R. G., and Adamson, E. D. (1995). *Proc. Natl Acad. Sci. USA*, **92**, 9161–5.
12. Goldmann, W. H. and Ezzell, R. M. (1996). *Exp. Cell Res.*, **226**, 234–7.
13. Goldmann, W. H., Schindl, M., Cardozo, T. J., and Ezzell, R. M. (1995). *Exp. Cell Res.*, **221**, 311–9.
14. Volberg, T., Geiger, N., Kam, Z., Pankov, R., Simcha, I., Sababay, I., *et al.* (1995). *J. Cell Sci.*, **108**, 2253–60.
15. Lifschitz-Mercer, B., Czernobilsky, B., Feldberg, E., and Geiger, B. (1998). (In press.)
16. Rodriguez Fernandez, J. L., Geiger, B., Salomon, D., Sabanay, I., and Ben-Ze'ev, A. (1992). *J. Cell Biol.*, **119**, 427–38.
17. Burridge, K. and Mangeat, P. (1984). *Nature*, **308**, 744–6.
18. Belkin, A. M. and Koteliansky, V.E. (1987). *FEBS Lett.*, **220**, 291–4.
19. Otto, J. J. (1983). *J. Cell Biol.*, **97**, 1283–7.
20. Wachsstock, D. H., Wilkins, J. A., and Lin, S. (1987). *Biochem. Biophys. Res. Commun.*, **146**, 554–60.
21. Kroemker, M., Rudiger, A. H., Jockusch, B. M., and Rudiger, M. (1994). *FEBS Lett.*, **355**, 259–62.
22. Turner, C. E., Glenney Jr., J. R., and Burridge, K. (1990). *J. Cell Biol.*, **111**, 1059–68.
23. Woods, C. K., Turner, C. E., Jackson, P., and Critchley, D. R. J. (1994). *J. Cell Sci.*, **107**, 709–17.
24. Gilmore, A. P. and Burridge, K. (1996). *Nature*, **381**, 531–5.
25. Menkel, A. R., Kroemker, M., Bubeck, P., Ronsiek, M., Nikolai, G., and Jockusch, B. M. (1994). *J. Cell Biol.*, 1231–40.
26. Niggli, V., Dimitrov, D. P., Brunner, J., and Burger, M. M. (1986). *J. Biol. Chem.*, **261**, 6912–8.
27. Burn, P. and Burger, M. M. (1987). *Science*, **235**, 476–9.
28. Fukami, K., Endo, T., Imamura, M., and Takenawa, T. (1994). *J. Biol. Chem.*, **269**, 1518–22.

29. Tempel, M., Goldmann, W. H., Isenberg, G., and Sackmann, E. (1995). *Biophys. J.*, **69**, 228–41.
30. Reinhard, M., Rudiger, M., Jockusch, B. M., and Walter, U. (1996). *FEBS Lett.*, **399**, 103–7.
31. Brindle, N. P., Holt, M. R., Davies, J. E., Price, C. J., and Critchley, D. R. (1996). *Biochem. J.*, **318**, 753–7.
32. Miliam, L. M. (1985). *J. Mol. Biol.*, **184**, 543–5.
33. Molony, L. and Burridge, K. (1985). *J. Cell. Biochem.*, **29**, 31–6.
34. Johnson, R. P. and Craig, S. W. (1994). *J. Biol. Chem.*, **269**, 12611–19.
35. Johnson, R. P. and Craig, S. W. (1995). *Nature*, **373**, 261–4.
36. Gilmore, A. P. and Burridge, K. (1996). *Nature*, **381**, 531–5.
37. Weekes, J., Barry, S. T., and Critchley, D. R. (1996). *Biochem. J.*, **314**, 827–32.
38. Johnson, R. P. and Craig, S. W. (1995). *Biochem. Biophys. Res. Commun.*, **210**, 159–66.
39. Bendori, R., Salomon, D., and Geiger, B. (1989). *J. Cell Biol.*, **108**, 2383–94.
40. Price, G. J., Jones, P., Davison, M. D., Patel, B., Eperon, I. C., and Critchley, D. R. (1987). *Biochem. J.*, **245**, 595–603.
41. Price, G. J., Jones, P., Davison, M. D., Patel. B., Bendori, R., Geiger, B., and Critchley, D. R. (1989). *Biochem. J.*, **259**, 453–61.
42. Coutu, M. D. and Craig, S. W. (1988). *Proc. Natl. Acad. Sci. USA*, **85**, 8535–9.
43. Winkler, J., Lunsdorf, H., and Jockusch, B. M. (1996). *J. Struct. Biol.* **116**, 270–7.
44. Geiger, B. (1982). *J. Mol. Biol.*, **159**, 685–701.
45. Siliciano, J. D. and Craig, S. W. (1987). *J. Cell Biol.*, **104**, 473–82.
46. Belkin, A. M., Ornatsky, O. I., Kabakov, A. E., Glukhova, M. A., and Koteliansky, V. E. (1988). *J. Biol. Chem.*, **263**, 6631–5.
47. Gimona, M., Small, J. V. Moeremans, M., Van Damme, J., Puype, M., and Vandekerckhove, J. (1988) *Protoplasma*, **145**, 133–40.
48. Gimona, M., Small, J. V., Moeremans, M., Van Damme, J., Puype, M., and Vandekerckhove, J. (1988). *EMBO J.*, **7**, 2329–334.
49. Werth, D. K., Niedel, J. E., and Pastan, I. (1983). *J. Biol. Chem.*, **258**, 11423–6.
50. Sefton, B., Hunter, T., Ball, E., and Singer, S. J. (1981). *Cell*, **24**, 165–74.
51. Weller, P. A., Ogryzko, E. P., Corben, E. B., Zhidkova, N. I., Patel, B., Price, G. J., *et al.* (1990). *Proc. Natl. Acad. Sci. USA*, **87**, 5667–71.
52. Ben-Ze'ev, A., Reiss, R., Bendori, R., and Gorodecki, B. (1990). *Cell Regul.*, **1**, 621–36.
53. Barstead, R. J. and Waterston, R. H. (1989). *J. Biol Chem.*, **264**, 10177–85.
54. Koteliansky, V. E., Ogryzko, E. P., Zhidkova, N. I., Weller, P. A., Critchley, D. R., Vancompernolle, K., *et al.* (1992). *Eur.J. Biochem.*, **204**, 676–772.

Benjamin Geiger
Department of Molecular Cell Biology,
The Weizmann Institute of Science,
Rehovot 76100, Israel

Zyxin

Zyxin[1,2] is an adherens junction component that was originally identified by characterization of a nonimmune rabbit serum that stained focal contacts by indirect immunofluorescence. Zyxin is present in low abundance relative to the other adhesion plaque components vinculin, talin, and α-actinin. Zyxin displays a tandem array of domains that mediate specific protein–protein interactions and its subcellular localization suggests that zyxin may act as a molecular scaffold to facilitate assembly of functional complexes at sites of actin–membrane interaction.

Protein properties

Zyxin is a cytoplasmic protein that is found in a number of distinct types of adherens junctions including the adhesion plaques or focal contacts of cultured cells, the dense plaques of smooth muscle cells, and the apical junctional complex of pigmented retinal epithelial cells. The term 'zyxin' is derived from a word root meaning 'a joining' and refers to the fact that this protein is localized extensively at areas where actin filaments are joined to the plasma membrane at sites of cell adhesion. The distribution of zyxin in cultured cells (Fig. 1) is particularly interesting since the protein is found at the adhesion plaques, sites of very close cell–substratum contact defined by interference reflection microscopy, as well as along the actin filament bundles (stress fibres) near where they terminate at the adhesion plaques. Zyxin colocalizes with the majority of the actin filament bundles, shows a periodic distribution along stress fibres, and is concentrated at the leading edge of lamellipodial extensions. The protein is notably absent from actin-containing arcs. The molecular basis for the heterogeneous distribution of zyxin along actin filament arrays remains to be determined.

Avian zyxin has been isolated from smooth muscle and many of its biochemical properties have been determined.[2] Avian zyxin is a 542 amino acid protein which migrates with an apparent molecular mass of 82 kDa on SDS-polyacrylamide gels. Characterization of the physical properties of zyxin shows that it exists as a monomer of 69 kDa and is asymmetric in shape. It has an average pI of 6.9, but displays a number of isoelectric variants in the range of 6.4–7.2. The protein is phosphorylated on multi-

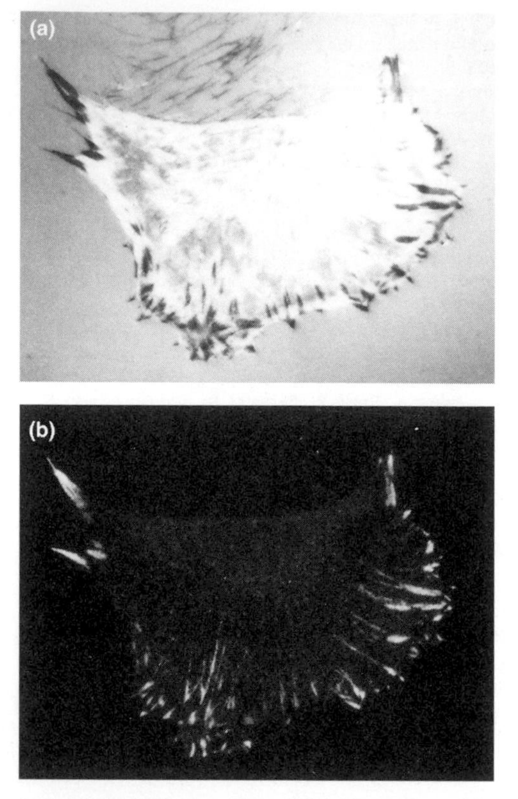

Figure 1. (a) A chicken embryo fibroblast viewed by interference reflection microscopy to visualize the adhesion plaques, which appear black by this approach. (b) Indirect immunofluorescence localization of zyxin.

ple sites *in vivo*; this post-translational modification may contribute to its heterogeneity in isoelectric focusing gels. Zyxin is present in a wide variety of avian tissues and is most prominent in smooth muscle and fibroblasts.

Human zyxin is a 572 amino acid protein; sequence analysis suggests that its biochemical properties are similar to those of avian zyxin.[3,4] An anti-peptide antibody raised against human zyxin recognizes a polypeptide that migrates with an apparent molecular mass of 84 kDa in SDS-polyacrylamide gels. By indirect immunofluorescence, human zyxin also localizes to sites of cell–substrate adhesion. Northern blot analysis of the tissue distribution of human zyxin mRNA shows that zyxin is ubiquitously expressed, with the mRNA levels highest in lung, colon, and blood.[3] The expression of human zyxin is reduced in virally transformed fibroblasts.[4]

Genes

Sequence analysis of avian, human, and mouse cDNA clones has revealed that zyxin contains some extremely interesting structural features.[3,5] Zyxin exhibits an unusually high proline content, with one 200 amino acid region of the protein consisting of > 30 per cent proline. The prolines are concentrated in the N-terminal two thirds of the protein, arranged in numerous arrays of up to seven contiguous proline residues. The interaction of vav with zyxin is mediated by the binding of the $p95^{vav}$ SH3 domain to sequences in this region.[6] In addition, three contiguous LIM repeats[7,8] are found at the C terminus of the protein. The LIM domain is a cysteine- and histidine rich zinc finger motif which is found in certain transcription factors as well as in proteins thought to be important in cell differentiation. As predicted by the presence of such a domain, zyxin is a zinc-binding metalloprotein.[5] The first LIM domain of zyxin has been shown to mediate the interaction of zyxin and the cysteine-rich proteins (CRPs), demonstrating the ability of the LIM domain to act as a protein binding interface.[9]

Comparison of the amino acid sequences of human and avian zyxin is illustrated in Fig. 2. Human zyxin shows 58 per cent identity and 70 per cent similarity overall with the avian zyxin sequence. The highest degree of sequence identity (76 per cent) is present in the LIM domains, with LIM3 being the most highly conserved (92 per cent identity). An interesting difference is the insertion of a 37 amino acid glutamine-rich sequence (amino acids 169–291 in human zyxin) that is not present in avian zyxin.[3]

A leucine-rich sequence that is conserved between mammalian and avian zyxin has been shown to function as a nuclear export signal (NES) (Fig. 2).[10] Deletion of part of this conserved sequence (amino acids 322–331) from avian zyxin results in the accumulation of zyxin in cell nuclei, and recent work has shown that zyxin does indeed shuttle between the nucleus and sites of cell adhesion.

Zyxin contains proline-rich sequences that are remarkably similar to sequences present in the ActA protein of the intracellular pathogen *Listeria monocytogenes*.[11] These sequences are conserved in both human and avian zyxin (Fig. 2); chicken zyxin contains three of these repeat regions (a.a. 101–120), while human zyxin exhibits four (a.a. 69–122). The significance of these sequences is addressed below.

Related proteins

Two other human proteins that bear a striking resemblance to zyxin have been identified. The LPP (lipoma

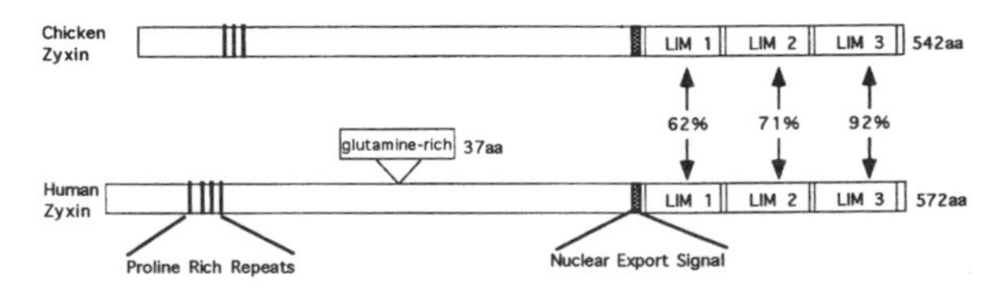

Figure 2. Schematic comparison of avian and human zyxin. The degree of identity within the LIM domains is shown, and conserved features are shown, including the ActA-like proline-rich repeat regions and the nuclear export signal sequence.

preferred partner) protein[12,13] which is frequently found fused to the DNA-binding region of the HMGI-c transcription factor in human lipomas, has a proline-rich N-terminal domain as well as LIM domains which show 62 per cent amino acid identity to those in human zyxin (GenBank accession number U49957). Zyxin is also related to the Trip6 (thyroid receptor interacting protein),[14] a ligand-dependent binding partner for the thyroid hormone receptor (GenBank accession number L40374). The available Trip6 sequence corresponds to the C-terminal region of the protein and specifies two complete and one partial LIM domain, which show 65 per cent amino acid identity to those in human zyxin.

A relationship between zyxin and the *Listeria* ActA protein has been proposed based on the presence of strikingly similar proline-rich repeat regions in both proteins. In addition, antibodies raised against the ActA protein are able to recognize human zyxin, a clear indication that the proteins are structurally related.[15] ActA is responsible for initiation and organization of actin filament assembly into a comet tail at the *Listeria* cell surface, which allows the bacterium to propel itself through the cytoplasm of infected cells.[11] The proline rich repeats in ActA are required for the recruitment of the VASP/profilin complex to sites of actin assembly, and appear to be important in controlling the rate of actin assembly and the speed of bacterial movement.[16,17] The presence of similar proline-rich repeat sequences in zyxin, as well as its association with cytoskeletal regulatory proteins, such as VASP and α-actinin, suggest a role for zyxin in control of actin dynamics. In addition, targeting of zyxin to the plasma membrane in cultured cells leads to the assembly of short actin filaments at these sites, as well as disruption of endogenous actin stress fibres,[15] demonstrating that zyxin is able to recruit the cellular machinery necessary for site-specific actin assembly.

An attractive model for describing zyxin's function is that of a molecular scaffold. Zyxin might direct the assembly of functionally distinct protein complexes through interactions with different groups of binding partners. This model is consistent with the evidence that zyxin is involved in a number of diverse cellular activities. Zyxin's interaction with cytoskeletal regulatory proteins, as well as its structural and functional similarity to *Listeria* ActA, indicate a role for zyxin in specification of sites of actin assembly and recruitment of the necessary machinery to these sites. The interaction of CRP family members with zyxin raises the possibility of its participation in the regulation of muscle cell differentiation. In addition, the ability of zyxin to shuttle between the adhesion plaque and the nucleus suggests that zyxin may be involved in nuclear–cytoplasmic communication. A recent review provides a more extensive analysis of zyxin function.[18]

Purification

Zyxin can be purified from either fresh or frozen chicken gizzard, but yields are significantly increased (up to five-fold) when fresh tissue is used. The protein is extracted from smooth muscle homogenates under low ionic strength, alkaline pH conditions. Subsequent ammonium sulphate fractionation, followed by chromatography on DEAE–cellulose, phenyl sepharose CL-4B and hydroxyapatite results in purified protein, as illustrated in the silver-stained gel shown in Fig. 3. This procedure[2] yields approximately 20 mg of purified zyxin from 300 g of fresh smooth muscle. Human zyxin has been purified from platelets using a similar approach.[19] In addition, zyxin has been purified from porcine platelets.[20] In this case, porcine platelet lysate was fractionated by cation-exchange chromatography on S-sepharose FF, and zyxin-containing fractions were identified based on their ability to interact with the vasodilator-activated phosphoprotein (VASP) in a blot overlay assay. The fraction with VASP-binding activity was precipitated by ammonium sulphate fractionation and further purified by hydroxyapatite and Zn^{2+} chelation chromatography. Using this approach, 10–20 μg of zyxin could be isolated from 8×10^{12} platelets.

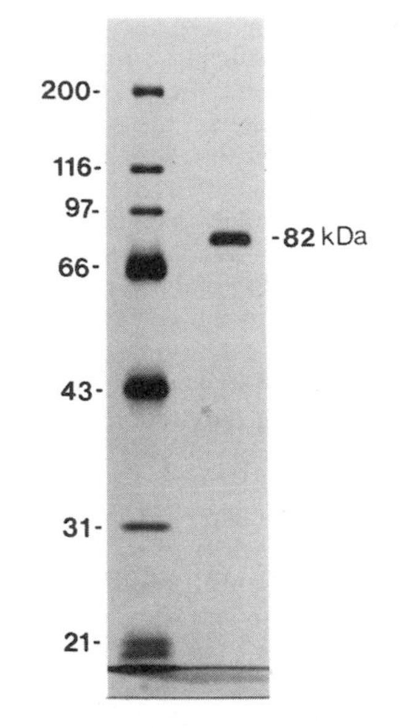

Figure 3. A silver-stained SDS-polyacrylamide gel showing purified zyxin from avian smooth muscle.

Activities

A number of binding partners have been identified for zyxin. Characterization of these interactions provides support for the idea that zyxin may play important roles in signalling events and cytoskeletal regulation at sites of cell adhesion.

Zyxin has been shown to interact with α-actinin.[21] This interaction has been demonstrated by solution and solid phase binding assays. Zyxin and α-actinin are extensively colocalized *in vivo*, raising the possibility that zyxin is targeted to particular regions in the cell at least to some extent by virtue of its association with α-actinin.

Zyxin also interacts with the CRPs, a highly conserved family of proteins that consist of two LIM domains.[22] The interactions between zyxin and the 23 kDa chicken homologue of human CRP[23] have been demonstated using blot overlay and solution binding studies.[5,9] Indirect immunofluorescence revealed colocalization of CRP with zyzin *in vivo*. Three CRP family members have been identified in chicken and shown to interact with zyxin.[24] Expression of the CRPs is developmentally regulated and family members have been shown to be essential in regulation of myogenic differentation.[22,24–26]

The proto-oncogene product, p95^{vav}, has also been shown to interact with zyxin.[6] Vav is required for antigen-dependent signalling in T and B cells and can act as an exchange factor for the rho GTPase.[27] Vav binding to zyxin was demonstrated in solution binding experiments and by experiments using the yeast two-hybrid system.

Additional binding partners which have been identified for zyxin are the vasodilator stimulated phosphoprotein (VASP) and the related proteins mena and evl.[20,27] The interaction of zyxin and VASP was demonstrated in by solution and solid phase binding experiments, and mena binding to zyxin has been characterized in solution binding studies. *In vivo*, VASP and Mena colocalize with zyxin in focal contacts. VASP and mena are profilin binding proteins and have been implicated in the control of actin cytoskeletal dynamics.[27,28]

Antibodies

The original nonimmune rabbit serum[1] is not an efficient reagent for most purposes; it appears to recognize only an avian antigen and has a fairly low titre. Bacterially expressed B-galactosidase–zyxin fusion proteins were used to generate polyclonal antibodies specific for avian zyxin.[5] A polyclonal antiserum raised against a synthetic peptide corresponding to a sequence in human zyxin shows cross-reactivity with other mammalian zyxins by Western blot and indirect immunofluorescence.[3]

The GenBank accession numbers for zyxin cDNA sequences are X69190 (chicken zyxin), Y07711 and X99063 (murine zyxin), and X94991 (human zyxin). Human zyxin maps to chromosome 7q32–q363.[3]

Mutant phenotype/disease states

No information available.

Structure

Crystallographic and NMR structures are not available for zyxin.

References

1. Beckerle, M. C. (1986). *J. Cell Biol.*, **103**, 1679–87.
2. Crawford, A. W. and Beckerle, M. C. (1991). *J. Biol. Chem.*, **266**, 5847–53.
3. Macalma, T., Otte, O., Hensler, M. E., Bockholt, S. M., Louis, H. A., Kalff-Suske, M., *et al.* (1996). *J. Biol. Chem.*, **271**, 31470–8.
4. Zumbrunn, J. and Trueb, B. (1995). *Eur. J. Biochem.*, **241**, 657–63.
5. Sadler, I., Crawford, A. W., Michelsen, J. W. and Beckerle, M. C. (1992). *J. Cell Biol.*, **119**, 1573–87.
6. Hobert, O., Schilling, J. W., Beckerle, M. C., Ullrich, A., and Jallal, B. (1996). *Oncogene*, **12**, 1577–81.
7. Freyd, G., Kim, S. W., and Horvitz, H. R. (1990). *Nature*, **344**, 876–9.
8. Karlsson, O., Thor, S., Norberg, T., Ohlsson, H. and Edlund, T. (1990). *Nature*, **344**, 879–82.
9. Schmeichal, K. L. and Beckerle, M. C. (1994). *Cell*, **79**, 211–9.
10. Nix, D. A. and Beckerle, M. C. (1997). *J. Cell Biol.*, **138**, 1139–47.
11. Kocks, C., Gouin, E., Tablouret, M., Berche, P., Ohayon, H., and Cossart, P. (1992). *Cell*, **68**, 521–31.
12. Schoenmakers, E. F. P. M., Wanshura, S., Mols, R., Bullerdiek, J., von den Berge, H., and Van de Ven, W. J. M. (1995). *Nature Genet.*, **10**, 436–43.
13. Ashar, H. R., Fezjo, M. S., Tkachenko, A., Zhou, X., Fletcher, J. A., Weremowisc, S., *et al.* (1995). *Cell*, **82**, 57–65.
14. Lee, J. W., Choi, H., Gyuris, J., Brent, R., and Moore, D. D. (1995). *Mol. Endocrinol.*, **9**, 243–54.
15. Golsteyn, R. M., Beckerle. M. C., Koay, T., and Friederich, E. (1997). *J. Cell Sci.*, **110**, 1893–906.
16. Smith, G. A., Theriot, J. A., and Portnoy, D. A. (1996). *J. Cell Biol.*, **135**, 647–60.
17. Gerstrel, B., Grobe, L., Pistor, S., Chakraborty, T., and Wehland, J. (1996). *Infect. Immun.*, **64**, 1929–36.
18. Beckerle, M. C. (1997). *BioEssays*, **19**, 949–57.
19. Hensler, M. and Beckerle, M. C. Abstract presented at the 5th International Conference on Cell Biology, 1992.
20. Reinhard, M., Jouvenal, K., Tripier, D., and Walter, U. (1995). *Proc. Natl Acad. Sci., USA*, **92**, 7956–60.
21. Crawford, A. W., Michelsen, J. W., and Beckerle, M. C. (1992). *J. Cell Biol.*, **116**, 1381–93.
22. Weiskirchen, R., Pino. J. D., Macalma, T., Bister, K., and Beckerle, M.C. (1995). *J. Biol. Chem.*, **270**, 28946–54.
23. Liebhaber, S. A., Emery, J. G., Urbanek, M., Wang, X., and Cooke, N. E. (1990). *Nucleic Acids Res.*, **18**, 3871–9.
24. Louis, H. A., Pino, J. D., Schmeichel, K. L., Pomies, P., and Beckerle, M. C. (1997). *J. Biol. Chem.*, **272**, 27484–91.
25. Crawford, A. W., Pino, J. D., and Beckerle, M. C. (1994). *J. Cell Biol.*, **124**, 117–27.
26. Arber, S., Halder, G., and Caroni, P. (1994). *Cell*, **79**, 221–31.
27. Crespo, P., Schuebel, K.-D., Ostrom, A. A., Gutkind, J. S., and Bustelo, X. R. (1997) *Nature*, **385**, 169–72.
28. Gertler, F. B., Niebuhr, K., Reinhard, M., Wehland, J., and Soriano, P. (1996). *Cell*, **87**, 227–39.
29. Reinhard, M., Giehl, K., Abel, K., Haffner, C., Jarchau, T., Hoppe, V., *et al.* (1992). *EMBO J.*, 14, 1583–9.

Beth E. Drees and Mary C. Beckerle
Department of Biology,
University of Utah, Salt Lake City,
UT 84112, USA

Linkers to intermediate filaments

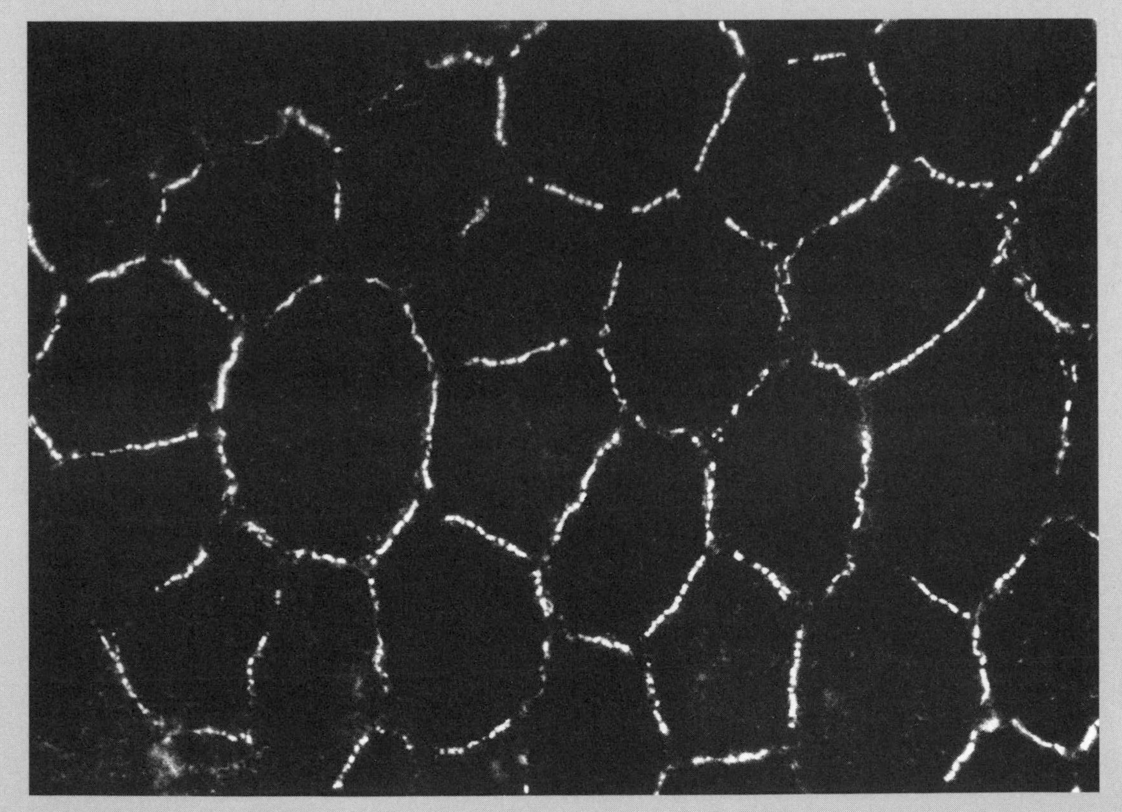

The *zonula adherens* of confluent retina pigment epithelial cells visualized with antibodies against plakoglobin. (Courtesy of Dr Jürgen Kartenbeck, DKFZ, Heidelberg, Germany.)

Desmocalmin

Desmocalmin is a calmodulin-binding high molecular mass protein localizing in desmosomal plaques. It also binds to reconstituted keratin filaments *in vitro*, but not to actin filaments.

Desmosomes are specialized cell–cell junctions[1] where the plasma membranes of neighbouring cells are closely associated via cell-adhesion molecules, such as desmoglein and desmocollin. Intermediate filaments are associated with the cytoplasmic face of desmosomes through well developed plaque structures, extending deep into the cytoplasm.

Desmocalmin is a desmosomal plaque-constitutive protein with a molecular mass of 240 kDa.[2] Desmocalmin was first identified as a calmodulin-binding high molecular mass protein that is concentrated in desmosomes isolated from bovine muzzle. Judging from its isoelectric point and antigenicity, desmocalmin is distinct from another high molecular mass desmosomal protein, desmoplakin. Desmocalmin is concentrated at the desmosomal plaque of stratified epithelial cells (Fig. 1(a)). The purified desmocalmin molecule looks like a flexible rod ~100 nm in length consisting of two polypeptide chains lying side by side (Fig. 1(b)).

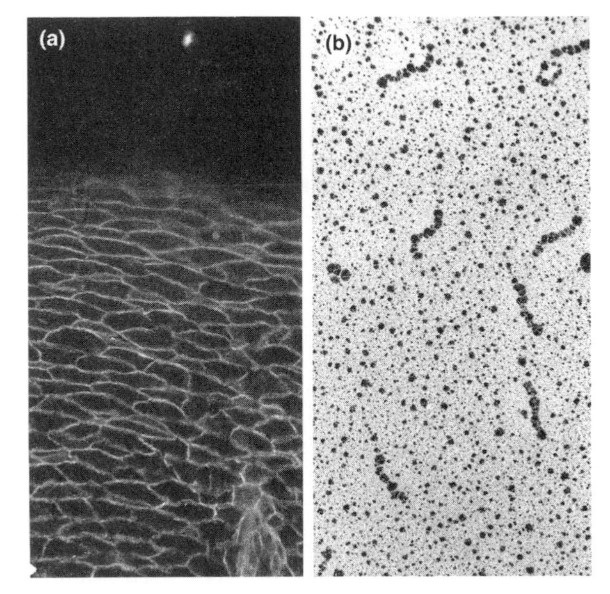

Figure 1. (a) Immunofluorescence localization of desmocalmin in muzzle epidermis. ×230.
(b) Morphology of desmocalmin molecules in rotary-shadowed EM preparations. ×120 000.

Purification

Desmocalmin is enriched in the desmosomal fraction isolated from muzzle epidermis, and effectively solubilized with a low-salt solution at pH 9.5–10.5. Subsequent ammonium sulphate fractionation and gel filtration on Sepharose CL-4B yields a fraction enriched in desmocalmin. The final steps in purification are ammonium sulphate fractionation and Affigel–calmodulin column chromatography.[2]

Activities

Purified desmocalmin binds to calmodulin in a Ca^{2+}-dependent manner. It also binds to reconstituted keratin filaments *in vitro* in the presence of Mg^{2+} but not to actin filaments.

Antibodies

One polyclonal antibody against bovine purified desmocalmin has been characterized.[2]

Genes

Information is not available.

References

1. Staehelin, A. L. (1974). *Int. Rev. Cytol.*, **39**, 191–283.
2. Tsukita, S. and Tsukita, S. (1985). *J. Cell Biol.*, **101**, 2070–80.

Shachiko Tsukita
College of Medical Technology,
Kyoto University Faculty of Medicine,
Shogoin-Kawahara, Sakyo-ku,
Kyoto 606, Japan

Desmoplakins

Desmoplakins are major cytoplasmic components of intercellular adhesive junctions known as desmosomes. Desmoplakins, which belong to a growing family of related cytoskeletal linking molecules, bind directly to intermediate filaments and are thought to provide an essential connection between this cytoskeletal system and the plasma membrane, thus providing structural integrity to the tissues in which they reside.

Protein properties

Desmoplakins (DPs) are the most abundant constituents of the cytoplasmic plaque of desmosomes (macula adherens), intercellular junctions that are particularly prominent in tissues that undergo mechanical stress. DPs have been localized to the interface between intermediate filaments (IF) and the cytoplasmic domains of the transmembrane desmosomal cadherins and are involved in linking IF with the transmembrane adhesive complex.[1-3]

Desmoplakin I (~240–285 kDa by migration on SDS–PAGE; 332 kDa from cDNA sequence) is a constitutive non-glycosylated component of all desmosomes, whereas desmoplakin II (~210–225 kDa by migration on SDS–PAGE; 259 kDa from predicted cDNA sequence) is a smaller less widely expressed form derived from the same gene by alternate RNA splicing.[1,4] Desmoplakin I (DP I) is proposed to form a dumbbell-shaped homodimer with a central α-helical coiled-coil rod domain ~130nm in length, flanked by two globular end domains.[4-6] This structure is consistent with rotary shadowed images of purified DP.[7] (Fig. 1).

The desmoplakin (DP) family, now termed the 'plakin' gene family, includes DP, BP230 (also known as BPAG1 or bullous pemphigoid antigen 1), plectin (possibly the same as IFAP300), envoplakin, and periplakin.[6,8-10] The common structural features and sequence motifs uniting this family were first described for DP,[4,5] and then observed in the other family members.[6,9] Functionally, a common characteristic of this family is their interaction with IF, but certain family members (e.g. plectin and neuronal forms of BPAG1) also have actin-binding domains at their N termini and are therefore thought to interconnect various cytoskeletal systems.[9,11]

DP I is an obligate constituent of all desmosomes described so far and is expressed widely in epithelial tissues, particularly complex epithelia such as epidermis. DP I is also found in the myocardial and Purkinje fibre cells of the heart, arachnoidal cells of meninges, dendritic reticulum cells of germinal centres in lymph nodes and certain glial cells of lower vertebrates.[2] This molecule has also recently been reported in the adherens-type junctions called 'complexus adhaerentes', found in certain endothelial cells containing VE-cadherin (vascular endothelial cadherin, or cadherin 5).[2,12] DP II is expressed in many of the same tissues as DPI, but is present at lower levels in some simple epithelial cells and is altogether absent from heart.[13]

At the light microscope level, DP exhibits a discontinuous punctate distribution at cell–cell borders (Fig. 2(b)). IF extend out from the nuclear surface and appear to attach at these sites (Fig. 2(a)). At the ultrastructural level, DP has been localized to the innermost portion of the desmosomal plaque, in the region where IF are anchored. The type of IF that associate with desmosomes is tissue dependent. Keratin IF associate with desmosomes in epithelial tissues, whereas desmin IF associate with intercalated disc desmosomes in cardiac muscle, and vimentin IF are anchored at meningeal and dendritic cell desmosomes.[2] Although the majority of DP in normal cells is located in a Triton-insoluble junction-associated pool, there is also a soluble cytoplasmic pool of DP that greatly increases in cells maintained in low extracellular calcium concentrations to inhibit junction assembly.[2,14,15]

DP associates with IF networks[16,17] and is required for IF anchorage in cultured epithelial cells.[18] The DP C-terminus binds directly to type II epidermal keratin IF[17,19] as well as simple epithelial keratins, desmin, and vimentin,[19] as assessed by *in vitro* biochemical and/or yeast two-hybrid techniques. The mechanisms of binding between DP and various IF polypeptides are fundamentally different. For instance, the N terminus of type II epidermal keratins is necessary and sufficient for binding to the DP C terminus, whereas simple epithelial keratins and type III IF proteins require the coiled-coiled rod of the IF polypeptide for binding to DP.[17,19] Furthermore, the last 68 residues of DP appear to be required for association with keratin but not type III IF (e.g. vimentin).[16,19] In addition, DP has been reported to be highly phosphorylated[20] and the association of the DP C terminus with IF is inhibited by phosphorylation of a serine residue 23 amino acids from the end of DP.[21] The N terminus of DP contains sequences required for association with plaque components, such as plakoglobin, and its incorporation into the desmosomal plaque.[16,18,22]

Purification

Enriched fractions of the major desmosomal proteins, of which desmoplakins are ~35 per cent by weight, can be prepared from bovine tissues such as the stratified epithelia of snout or tongue as described first by Skerrow and Matoltsy[23] and reviewed in ref. 24. Desmoplakins I and II proteins can be purified to homogeneity from keratomed porcine tongue epithelium or bovine tissues, using extensive extraction in 4 M urea followed by DEAE–cellulose and gel filtration chromatography.[7]

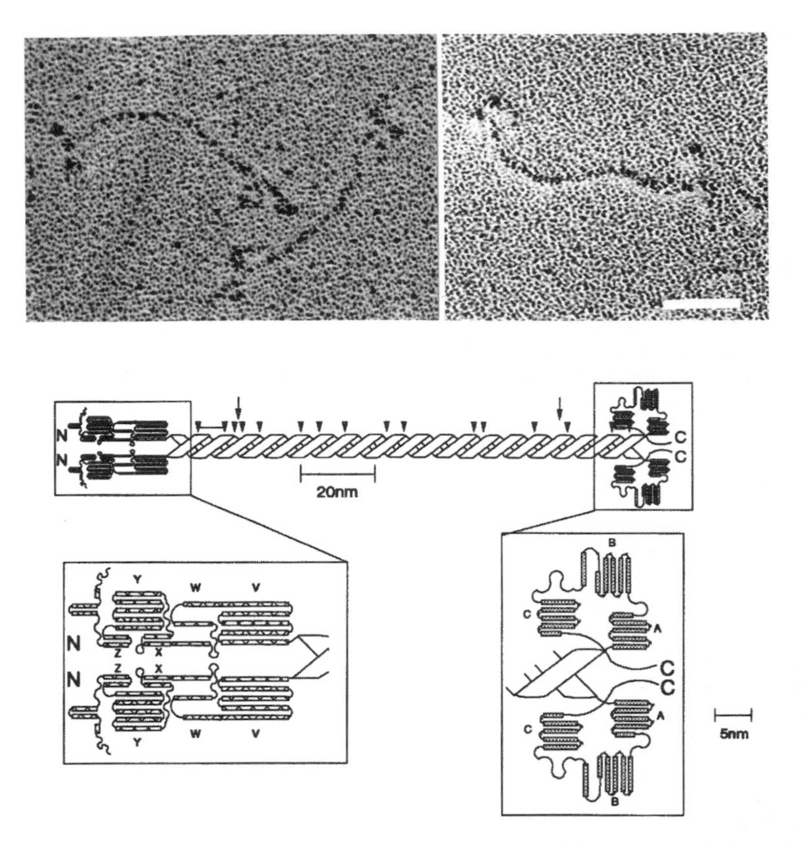

Figure 1. Structure of a desmoplakin homodimer. The predicted model of DP structure (below) is shown in comparison with rotary shadowed images of purified DP I (above). Note that the rod domain consists of a coiled-coil α helix ~130 nm in length. Possible break points and stutters predicted by the amino acid sequence are designated by small arrows above the rod domain, and an extended region of predicted flexibility is designated by a horizontal bar. Large arrows define the boundaries of the DP I specific region. Rectangles in the carboxyl terminus comprise 38 residue repeating motifs predicted to turn upon themselves, forming bundles that are stabilized by ionic interactions.[5] Rectangles in the amino terminus have a heptad substructure and represent predicted α-helical regions that are also predicted to form bundles. Rotary shadowed images of DP I were prepared as described in ref. 7. Note the existence of an extended, flexible central rod flanked by two more globular ends. (Rotary shadowed EM images kindly provided by Dr. Ed O'Keefe; figure adapted with permission from Fig. 5 of Kowalczyk *et al.* (1994) *Biophys. Chem.*, **50**, 97–112.

Activities

Dominant negative studies suggest that DP is required for linking IF to the desmosomal plaque.[18] The ability of the DP C-terminus to bind to IF or IF networks has been assessed by transfection into cultured cells,[17,25] direct binding of purified recombinant DP C-terminal domain or native DP to IF in solid phase binding[17,19] and yeast two-hybrid assays.[19] See also Plate 2. The DP N terminus provides the link to the desmosomal cadherin–plakoglobin complex and is responsible for incorporation into desmosomes.[16,18,22]

Antibodies

Monoclonal antibodies that recognize DP I and II (clones DP2.15 and 11–5F) or DP I only (clone DP2.17) are commercially available. These are reported to cross-react with a wide variety of species.[26] In addition, a number of polyclonal antibodies also with a wide range of cross-reactivity have been reported.[15,18,27,28]

Genes

A partial bovine cDNA clone[1] and complete human cDNA clone (GenBank M77830) have been reported. DP I and II are encoded by a single gene that gives rise to mRNAs of ~9.5 and ~7.5 kb as assessed by Northern analysis, and predicted polypeptides of ~332 and 260 kDa, respectively.[4,5,16] The smaller DP II molecule appears to be generated from an alternatively spliced message via an internal donor, and is missing approximately two thirds of the central coiled-coil rod domain.[4] Two genes of the plakin

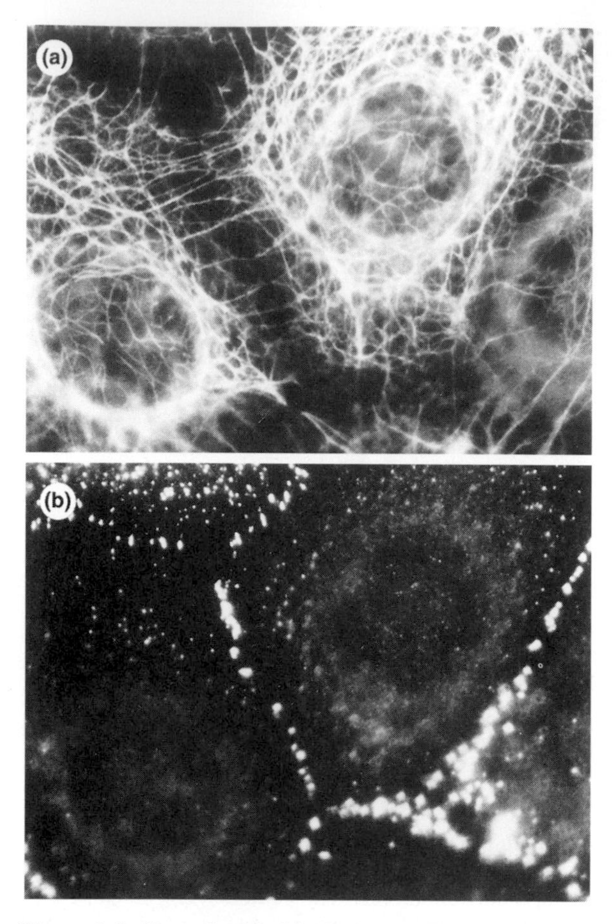

Figure 2. Indirect double label immunofluorescence to detect endogenous DP and keratin IF in A431 cells. Cells were reacted with KSB17.2 (Sigma), directed against keratin 18 (a) and the polyclonal antibody NW6[13] directed against the DP C terminus, to detect DP (b). Note keratin IF bundles impinging upon areas of punctate DP staining.

family have now been mapped to human chromosome 6, including BP230 (6p11–6p12) and desmoplakin (6p21-ter).[29] In addition, plectin has been mapped to 8q24 and envoplakin to 17q25.[31]

Plakin family members exhibit a similar genomic organization characterized by a high concentration of introns in the N-terminal region and a dearth of introns in the C-terminal region[9] (Green *et al.*; unpublished data).

Mutant phenotype/disease states

To date no human diseases or mouse mutants have been attributed to mutations in DP. However, mutations in the family member plectin give rise to a blistering skin disorder (epidermolysis bullosa) with late-onset muscular dystrophy.[8] In addition, a mouse knockout of BPAG1 displays minor blistering of the skin upon mechanical trauma, but more dramatically, severe neurodegeneration and dystonia also seen in dystonia musculorum (Dt/Dt) mice. The explanation for this is that there are both epithelial- and neurone-specific transcripts that arise from the dystonia locus through the use of alternative promoters.[11]

Structure

Not available.

Web sites

This is not meant to be a comprehensive list of sites, but is a sampling of those readily available through the common search engines.
Investigators working on desmosomes and/or desmoplakin:
Green Lab: http://dicty.cmb.nwu.edu/green.html
Troyanovsky Lab:
http://dermatology.wustl.edu/Troyanovsky.html
Sugrue Lab:
http://www.med.ufl.edu/anatomy/sugrue.html
Fuchs Lab: http://www.hhmi.org/science/cellbio/fuchs.htm
O'Keefe Lab:
http://www.med.unc.edu/wrkunits/2depts/derm/okeefe.html
Klymkowsky Lab:
http://spot.colorado.edu/~klym/home.html
Goldman Lab: http://www.nums.nwu.edu/~igp/facidex/GoldmanR.html
Coulombe Lab:
http://www.med.jhu.edu/bcmb/faculty/coulombe.html
Cowin Lab: http://www.med.nyu.edu/Research/P.Cowin-res.html
Magee Lab: http://www.nimr.mrc.ac.uk/resdivn/mbi
Cadherin Web Site: http://156.26.157.6/cadherin/catgories.html
Companies specializing in desmosome antibodies:
http://www.progen.de/

References

1. Schwarz, M. A., Owaribe, K., Kartenbeck, J., and Franke, W. W. (1990). *Ann. Rev. Cell Biol.*, **6**, 461–91.
2. Schmidt, A., Heid, H. W., Schafer, S., Nuber, U. A., Zimbelmann, R., and Franke, W. W. (1994). *Eur. J. Cell Biol.*, **65**, 229–45.
3. Garrod, D., Chidgey, M., and North, A. (1996). *Curr. Opin. Cell Biol.*, **8**, 670–8.
4. Virata, M. L. A., Wagner, R. M., Parry, D. A. D., and Green, K. J. (1992). *Proc. Natl Acad. Sci. USA*, **89**, 544–48.
5. Green, K. J., Parry, D. A. D., Steinert, P. M., Virata, M. L. A., Wagner, R. M., Angst, B. D., and Nilles, L. A. (1990). *J. Biol. Chem.*, **265**, 2603–12.
6. Green, K. J., Virata, M. L. A., Elgart, G. W., Stanley, J. R., and Parry, D. A. D. (1992). *Int. J. Biol. Macromol.*, **14**, 145–153.
7. O'Keefe, E. J., Erickson, H. P., and Bennett, V. (1989). *J. Biol. Chem.*, **264**, 8310–18.

8. Uitto, J., Pulkkinen, L., Smith, F. J. H., and McClean, W. H. I. (1996). *Exp. Dermatol.*, **5**, 237–46.
9. Ruhrberg, C., and Watt, F. M. (1997). *Curr. Opin. Genet. Devel.*, **7**, 392–7.
10. Ruhrberg, C., Hajibagheri, M. A. N., Parry, D. A. D., and Watt, F. M. (1997). *J. Cell Biol.*, **139**, 1835–49.
11. Bousquet, O., and Coulombe, P. A. (1996). *Curr. Biol.*, **6**, 1563–6.
12. Valiron, O., Chevrier, V., Usson, Y., Breviario, F., Job, D., and Dejana, E. (1996). *J. Cell Sci.*, **109**, 2141–9.
13. Angst, B. D., Nilles, L. A., and Green, K. J. (1990). *J. Cell Sci.*, **97**, 247–257.
14. Duden, R., and Franke, W. W. (1988). *J. Cell Biol.*, **107**, 1049–63.
15. Pasdar, M., and Nelson, W. J. (1988). *J. Cell Biol.*, **106**, 677–85.
16. Stappenbeck, T. S., Bornslaeger, E. A., Corcoran, C. M., Luu, H. H., Virata, M. L. A., and Green, K. J. (1993). *J. Cell Biol.*, **123**, 691–705.
17. Kouklis, P. D., Hutton, E., and Fuchs, E. (1994). *J. Cell Biol.*, **127**, 1049–1060.
18. Bornslaeger, E. B., Corcoran, C. M., Stappenbeck, T. S., and Green, K. J. (1996). *J. Cell Biol.*, **134**, 985–1002.
19. Meng, J.-J., Bornslaeger, E. A., Green, K. J., Steinert, P. M., and Ip, W. (1997). *J. Biol. Chem.*, **272**, 21495–505.
20. Mueller, H. and Franke, W. W. (1983). *J. Mol. Biol.*, **163**, 647–71.
21. Stappenbeck, T. S., Lamb, J. A., Corcoran, C. M., and Green, K. J. (1994). *J. Biol. Chem.*, **269**, 29351–4.
22. Kowalczyk, A. P., Bornslaeger, E. A., Borgwardt, J. E., Palka, H. L., Dhaliwal, A. S., Corcoran, C. M., Denning, M. F., and Green, K. J. (1997). *J. Cell Biol.*, **139**, 773–84.
23. Skerrow, C. J. and Matoltsy, A. G. (1974). *J. Cell Biol.*, **63**, 515–23.
24. Hertzberg, E., Tsukita, S., Green, K. J., and Stevenson, B. (1992). In *Cell–cell Interations: A practical approach* (ed. W. Gallin, D. Paul, and B. Stevenson), pp. 111–142, Oxford University Press, Oxford.
25. Stappenbeck, T. S., and Green, K. J. (1992). *J. Cell Biol.*, **116**, 1197–1209.
26. Weimer, R. V. (ed.), (1995). *MSRS Catalog: manufacturers' specifications and reference synopsis* (3rd edn), Primary antibodies. Aerie Corp., Birmingham, Michigan.
27. Cowin, P. and Garrod, D. (1983). *Nature*, **302**, 148–50.
28. Cowin, P., Kapprell, H.-P., and Franke, W. W. (1985). *J. Cell Biol.*, **101**, 1442–54.
29. Buxton, R. S., Magee, A. I., King, I. A., and Arnemann, J. (1994). In *Molecular biology of desmosomes and emidesmosomes* (ed. J. E. Collins and D. R. Garrod), pp. 1–131. R.G. Landes, Austin.
30. Liu, C. G., Maercker, C., Castanon, M. J., Hauptmann, R., and Wiche, G. (1996). *Proc. Natl Acad. Sci. USA*, **93**, 4278–83.
31. Ruhrberg, C., Williamson, J. A., Sheer, D., and Watt, F. M. (1996). *Genomics*, **37**, 381–5.

■ *Kathleen J. Green and Elayne A. Bornslaeger*
Departments of Pathology and Dermatology,
303 East Chicago Avenue,
Northwestern University Medical School,
Chicago, IL 60611, USA

Pemphigoid antigens

The pemphigoid antigens are polypeptides that react with the sera of patients with an autoimmune skin-blistering disease, bullous pemphigoid. It is now known that they are at least two distinct antigens of 230 kDa and 180 kDa and that they are major components of hemidesmosomes in the basal cells of stratified and complex epithelia.

Synonymous names

230 kDa antigen: BP230, BPAG1, BPAG1e (the epithelial type; the neural type is designated as BPAG1n). 180 kDa antigen: BP180, BPAG2, type XVII collagen.

Protein properties

The pemphigoid antigens were originally defined as target molecule(s) that were recognized by autoimmune sera of patients suffering from diseases of the pemphigoid group, the most common of which is *bullous pemphigoid* (BP), characterized by subepidermal blistering. The antigens were found to be present in the epidermal basement membrane zone (BMZ) of these patients. It is now known that these antigens are normal components of certain epithelial BMZ and also are present in cultured keratinocytes.[1] They have been found in the hemidesmosome of stratified epithelial cells (epidermal, oesophageal, corneal, etc.) and complex epithelial cells (tracheal, bladder, glandular myoepithelial, etc.), but not in simple epithelial cells that lack typical hemidesmosomes, including vascular endothelial cells and cardiac cells.[1,2]

The BP autoimmune sera recognize 230 kDa (BP230) and/or 180 kDa polypeptides.[3,4] Antibodies specific to BP230 and BP180 can be affinity purified separately and they do not cross-react by immunoblot analysis.[5] This indicates that the two BP antigens are different proteins. BP230 and BP180 are so far the only proteins recognized as BP or pemphigoid antigens.

BP230 is localized in hemidesmosomal plaques of stratified and complex epithelial basal cells.[5,6] The cDNA of human BP230 (9 kb) encodes a polypeptide of 2649 amino acids.[7] Genomic DNA analysis has revealed that the

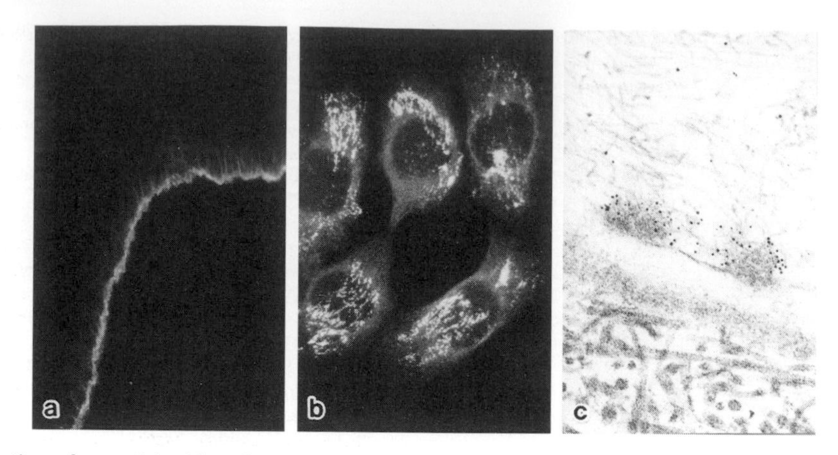

Figure 1. Localization of pemphigoid antigens. (a) Frozen section of bovine skin stained with monoclonal antibody to BP180, showing specific staining of BMZ and basal cell peripheries. (b) Cultured FRSK cells, a rat keratinocyte cell line, stained with BP autoantibodies. (c) Immunoelectron microscopy of bovine cornea stained with BP230 autoantibodies showing specific decoration of hemidesmosomal plaques.

BP230 protein is encoded by a single copy gene, localized to chromosome 6p11–12, and consisting of 22 exons and 20 kb in size.[8] Analysis of the deduced amino acid sequence predicts an α-helical coiled-coil rod in the central portion, flanked by terminal globular domains. The predicted amino acid sequence shows high homology with desmoplakin, a desmosomal plaque protein, and plectin, an intermediate filament binding protein.[9]

BP180 is a major transmembrane glycoprotein,[10] of which epitopes are found in intracellular, extracellular, or both regions of hemidesmosomes, depending on the sera used. The cDNA and its deduced amino acid analyses indicate that BP180 is a 155 kDa type II transmembrane protein with a collagenous carboxyl-terminal extracellular domain and a non-collagenous amino-terminal cytoplasmic domain.[11,12] Genomic DNA analysis has revealed that *BP180* is localized to chromosome 10q24.3,[13] and consists of 56 exons and is 52 kb in size. The extracellular domain of BP180 consists of 15 collagen domains in human and 13 in mouse. These collagenous domains are interrupted by non-collagenous sequences. BP180 exists as a triple-helical molecule.[14] An immunodominant epitope of BP180 is found in the non-collagenous portion just outside the cell membrane.[15] Rabbit antibodies to this region cause epidermal–dermal separation when injected into neonatal mice.[16]

Purification

BP230 has been purified from bovine tongue mucosa by column chromatography in the presence of 9.5 M urea.[17] BP180 has been purified from cultured cells by immunoaffinity chromatography.[14] Hemidesmosome-enriched fractions can be prepared from bovine corneal epithelial cells, in which both BP230 and BP180 are major components on SDS–PAGE.[6]

Activities

In transfected cells, the carboxyl-terminal tail of BP230 is associated with intermediate filament networks[18] and a mutant form of BP180 lacking the collagenous domains interacts with α6 integrin.[19]

Antibodies

Because BP is rather common in the field of dermatology, BP autoimmune sera can be obtained from dermatologists. The sera usually recognize either BP230 or BP230 and BP180, and rarely only BP180. They usually show wide species cross-reactivity. Some polyclonal and monoclonal antibodies to each protein have been reported.[5,6,10,20,21]

Genes

The full length cDNAs for BP230 (human: M69225[7]) and BP180 (human: M91669,[11] mouse: L08407[12]) have been published. The sequences of neural forms of BP230 (BPAG1n) are also available (human: U31850[22]) mouse: U22452, U25158.[23]

Mutant phenotype/disease states

BP230 knockout studies have demonstrated that hemidesmosomes lack inner plaques and attached keratin filaments; the mutant mice produce sensory neurone degeneration identical to that of *dystonia musculorum* (*dt/dt*) mice.[24] Mutations in the *BP180* gene are found in generalized atrophic benign epidermolysis bullosa, which is a clinical variant of junctional epidermolysis bullosa characterized by universal alopecia and atrophy of the skin.[25–27]

Other pemphigoids, *cicatricial pemphigoid* (CP), *vesicular pemphigoid* and *pemphigoid* (or *herpes*) *gestationis,* are clinically classified as diseases different from BP, but some of them may well involve the same antigens, BP230 and/or BP180.[1,28–30] Approximately a third of BP patients have significant mucosal lesions, some of which are similar to those in CP, suggesting some overlap in the two diseases.

Structures

Not available.

References

1. Stanley, J. R. (1993). *Adv. Immunol.*, **53**, 291–325.
2. Owaribe, K., Kartenbeck, J., Stumpp, S., Magin, T. M., Krieg, T., Diaz, L. A. and Franke, W. W. (1990). *Differentiation*, **45**, 207–20.
3. Cook, A. L., Hanahoe, T. H. P., Mallett, R. B., and Pye, R. J. (1990). *Br. J. Dermatol.*, **122**, 435–44.
4. Robledo, M. A., Kim, S. C., Korman, N. J., Stanley, J. R., Labib, R. S., Futamura, S., and Anhalt, G. J. (1990). *J. Invest. Dermatol.*, **94**, 793–7.
5. Klatte, D. H., Kurpakus, M. A., Grelling, K. A., and Jones, J. C. R. (1989). *J. Cell Biol.*, **109**, 3377–90.
6. Owaribe, K., Nishizawa, Y., and Franke, W. W. (1991). *Exp. Cell Res.*, **192**, 622–30.
7. Sawamura, D., Li, K., Chu, M.-L., and Uitto, J. (1991). *J. Biol. Chem.*, **266**, 17784–90.
8. Tamai, K., Sawamura, D., Do, H. C., Tamai, Y., Li, K., and Uitto, J. (1993). *J. Clin. Invest.*, **92**, 814–22.
9. Green, K. J., Virato, M. L. A., Elgart, G. W., Stanley, J. R., and Parry, D. A. D. (1992). *Int. J. Biol. Macromol.*, **14**, 145–53.
10. Nishizawa, Y., Uematsu, J., and Owaribe, K. (1993). *J. Biochem.*, **113**, 493–501.
11. Giudice, G. J., Emery, D. J., and Dias, L. A. (1992). *J. Invest. Dermatol.*, **99**, 243–50.
12. Li, K., Tamai, K., Tan, E. M. L., and Uitto, J. (1993). *J. Biol. Chem.*, **268**, 8825–34.
13. Sawamura, D., Li, K. and Uitto, J. (1992). *J. Invest. Dermatol.*, **98**, 942–3.
14. Hirako, Y., Usukura, J., Nishizawa, Y., and Owaribe, K. (1996). *J. Biol. Chem.*, **271**, 13739–45.
15. Giudice, J., Emery, D.J., Zelickson, B. D., Anhalt, G. J., Liu, Z., and Dias, L. A. (1993). *J. Immunol.*, **151**, 5742–50.
16. Liu, Z., Diaz, L., Troy, J. L., Taylor, A. F., Emery, D. J., Fairley, J. A., and Giudice, G. J. (1993). *J. Clin. Invest.*, **92**, 2480–8.
17. Klatte, D. H. and Jones, J. C. R. (1994). *J. Invest. Dermatol.*, **102**, 39–44.
18. Yang, Y., Dowling, J., Yu, Q.-C., Kouklis, P., Cleaveland, D. W., and Fuchs, E. (1996). *Cell*, **86**, 655–65.
19. Hopkinson, S. B., Baker, S. E., and Jones, J. C. R. (1995). *J. Cell Biol.*, **130**, 117–25.
20. Sugi, T., Hashimoto, T., Hibi, T., and Nishikawa, T. (1989). *J. Clin. Invest.*, **84**, 1050–5.
21. Tanaka, T., Korman, N. J., Shimizu, H., Eady, R. A. J., Klaus-Kovtun, V., Cehrs, K., and Stanley, J. R. (1990). *J. Invest. Dermatol.*, **94**, 617–23.
22. Brown, A., Dalpe, G., Mathieu, M. and Kothary, R. (1995). *Genomics*, **29**, 777–80.
23. Brown, A., Bernier, G., Mathieu, M., Rossant, J. and Kothary, R. (1995). *Nature Genet.*, **10**, 301–6.
24. Guo, L., Degenstein, L., Dowling, J., Yu, Q.-C., Wollmann, J., Perman, B., and Fuchs, E. (1995). *Cell*, **81**, 233–43.
25. Jonkman, M. F., de Jong, M. C. J. M., Heeres, K., Pas, H. H., van der Meer, J. B., *et al.* (1995). *J. Clin. Invest.*, **95**, 1345–52.
26. McGrath, J. A., Gatalica, B., Christiano, A. M., Li, K., Owaribe, K., McMillan, J. R., *et al.* (1995). *Nature Genet.*, **11**, 83–6.
27. McGrath, J. A., Darling, T., Gatalica, B., Pohla-Gubo, G., Hintner, H., Christiano, A. M., *et al.* (1995). *J. Invest. Dermatol.*, **106**, 771–4.
28. Bernard, P., Prost, C., Lecerf, V., Intrator, L., Combemale, P., Bedane, C., *et al.* (1990). *J. Invest. Dermatol.*, **94**, 630–5.
29. Kelly, S. E., Bhogal, B. S., Wojnarowska, F., Whitehead, P., Leigh, I. M., and Black, M. M. (1990). *Br. J. Dermatol.*, **122**, 445–9.
30. Balding, S. D., Prost, C., Diaz, L. A., Bernard, P., Bedane, C., Aberdam, D., and Giudice, G. J. (1996). *J. Invest. Dermatol.*, **106**, 141–6.

Katsushi Owaribe
Unit of Biosystems, Graduate School of Human Informatics,
Nagoya University,
Nagoya 464–1, Japan

Plectin

Plectin[1,2] is a versatile cytoskeletal linker protein of high molecular weight (>500 000 Da) that is abundantly expressed in a wide variety of mammalian tissues and cell types. It interacts with all major cytoskeletal filament networks and is a major constituent of plasma membrane-associated junctional complexes (e.g. hemidesmosomes of epithelia, Z lines, dense plaques and intercalated discs of various types of muscle, and focal contacts). Plectin molecules seem to play an important role in strengthening cells and tissues against mechanical stress, as recently confirmed by findings that link defective plectin expression to human disease (epidermolysis bullosa simplex with muscular dystrophy) and by the observation of similar phenotypes in plectin gene knockout mice. An emerging concept postulates tissue-specific expression of functionally different isoforms of the protein.

Synonymous names

There is speculation that proteins referred to as HD-1[3] or IFAP 300[4] represent isoforms of plectin generated by differential splicing of the PLEC1 gene. Whether this is true awaits cloning and sequencing of these proteins.

Homologous proteins

Plectin has several marked sequence similarities, especially in its carboxyl-terminal domain, to that of desmoplakins (see pp. 102–5), the neuronal (alias dystonins) and epithelial bullous pemphigoid antigen (BPAG) 1 isoforms (pp. 105–7), and envoplakin. Similar to plectin, these proteins possess coiled-coil central rod domains followed by globular carboxyl-terminal domains containing three (desmoplakin), two (BPAG1), or one (envoplakin) tandem repeat(s) homologous to the six repeats found in plectin (Fig. 1). Together with plectin these molecules form a novel family of structurally and most probably functionally related cytoskeletal linker proteins.

Protein properties

Plectin, copurifying with intermediate filaments (IFs) from various sources, migrates in SDS–PAGE as a single or double band of ~300 000 Da.[2] Ultrastructurally, 300 kDa plectin species resemble dumb-bells consisting of a ~190 nm long rod section flanked by two globes of ~9 nm diameter[5] (Fig. 2(b)).

By immunohistochemistry plectin has been identified over a wide range of different tissues and cell types.[2,6,7] Depending on the type of cell, it has been localized either throughout the cytoplasm, where it partially codistributes with IFs (Fig. 2(a)) and/or microfilaments,[8,9] or in peripheral regions at cellular junctions or membrane attachment sites of IFs (Fig. 2(d), (g)) and microfilaments (Fig. 2(e), (f)), or at both locations. Plectin molecules bridging vimentin filaments with actin filaments[9] or microtubules[10] have been visualized using whole mount electron microscopy of cells in culture.

On the molecular level the 300 kDa plectin species has been shown to interact with a number of IF proteins, including the IF subunit proteins vimentin, desmin, GFAP, skin cytokeratins, and the neurofilament proteins (NF-L, NF-M, and N-FH), as well as the nuclear lamina protein lamin B.[2,11] It also binds with high affinity to the plasma membrane-associated proteins α-spectrin and its analogue fodrin,[12] and shows interaction with high molecular weight microtubule-associated proteins (MAP1 and MAP2)[12] and integrin $\beta4$.[31] In addition, plectin molecules interact with themselves forming complex network arrays.[5] A binding site of plectin for IFs residing within its carboxyl-terminal domain[13] has been mapped to a stretch of ~50 amino acid residues within the terminal repeat 5 domain (Fig. 1), and a basic amino acid residue cluster within a nulear targeting sequence motif has been identified as one of its essential elements.[14] A highly conserved and functional actin binding domain (ABD) of the type found in dystrophin, β-spectrin, and other actin binding superfamily members has been identified in the amino-terminal region of the molecule[15,16] (Fig. 1). Furthermore, one of the isoforms of plectin identified contains amino-terminal sequences with high homology to sequences of ribosomal protein S10 that have been implicated in RNA binding.[16]

Plectin serves as a target for several protein kinases *in vivo* and *in vitro*, and its interactions with vimentin and lamin B are differentially regulated by protein kinases A and C.[11,17] Furthermore, during mitosis, plectin is specifically phosphorylated by p34^{cdc2} kinase at a unique site in its carboxyl-terminal tail region.[18] This phosphorylation event correlates with a decreased binding affinity of the molecule to IF structures, as demonstrated both *in vitro* and *in vivo*.[19] Furthermore, 35 out of 95 of the amino acid residues making up the carboxyl-terminus of plectin molecules are hydroxyamino acids, some of which are part of well characterized phosphorylation consensus sequence motifs and serve as prominent phosphoacceptors upon incubation with various kinases *in vitro*.[18]

Based on the diversity and large variety of proteins that have so far been identified as specific interaction partners, the histochemical evidence, and the phenotypes observed in human patients with defects in plectin expression and in plectin gene knock-out mice, the following molecular functions can be proposed for the protein:

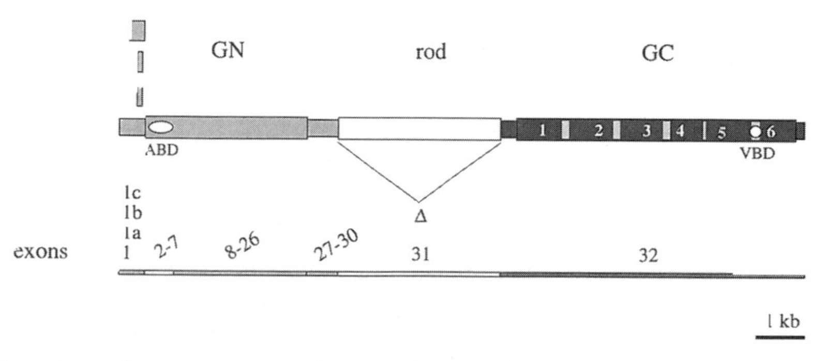

Figure 1. Predicted domain, and gene structure of plectin. Upper bar, schematic representation of the molecule with alternative starts, globular amino-terminal (GN), α-helical coiled-coil rod, and globular carboxy-terminal (GC) domains. Subdomains 1–6 correspond to the core of each of 6 sequence repeats, which consist of tandem repeats of a 19-amino acid residue motif. Actin binding domain (ABD) comprising exons 2–7, and vimentin binding domain (VBD) in exon 32[14] are indicated. Δ, deleted exon 31 of rodless isoforms. Lower bar, positions of exons including 4 alternative first coding exons.

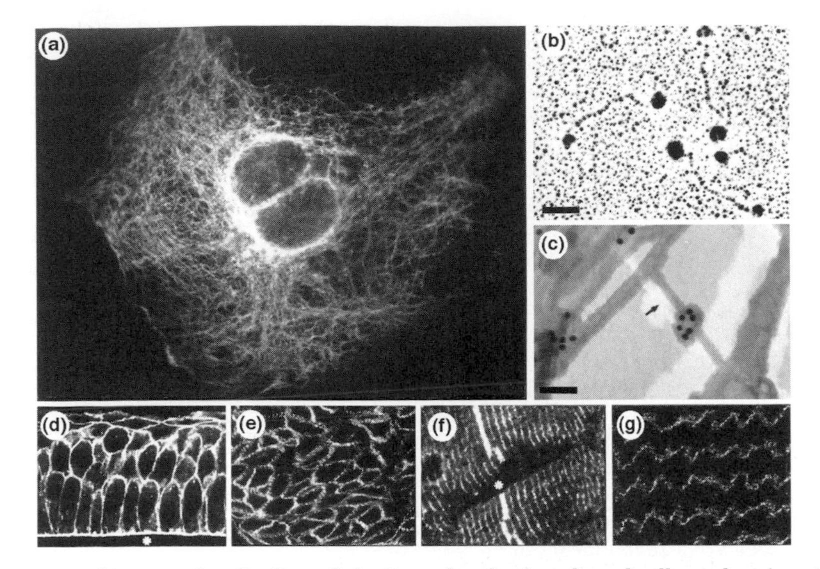

Figure 2. Ultrastructure, and immunolocalization of plectin molecules in cultured cells, and various tissues. (a) Immunofluorescence microscopy of CHO cells after extraction with Triton-X100, and staining with antiserum to plectin. (b) Rotary shadowing electron microscopy of 300 kDa plectin purified from rat glioma C_6 cells.[5] Bar, 60 nm. (c) Immunogold electron microscopy of whole mount cytoskeletons prepared from rat glioma C6-D8 cells after decoration with a monoclonal antibody to plectin (6B8) and secondary antibodies coupled to 10 nm gold particles.[9] The rod domain of a plectin molecule is visualized as thin filament (arrow) connecting a vimentin filament with other cellular structures. *Note*, antibody complexes are bound to the centre portion of the rod, in agreement with epitope mapping.[24,25] Bar, 50 nm. (d–g) Immuofluorescence misrocopy of frozen sections through human cornea (d), rat smooth muscle (e), rat cardiac muscle (f), and rat eye lens (g). Asterisks denote strong staining of basal cell surface membrane in basal cell layer in (d), and intercalated discs in (f). Magnifications (approx.): ×1000 (a and f), ×400 (d and e), and ×500 (g).

(1) interlinking of IFs;
(2) anchorage of IFs at the plasma membrane and the nuclear envelope;
(3) bridging of IFs with microtubules and microfilaments;
(4) linkage of various structural components forming junctional complexes in the interior (e.g. Z lines) and at the periphery of cells (e. g. hemidesmosomes); and
(5) cytomatrix formation through self-association.

Thus, plectin molecules are likely to play a key role in stabilizing cells against mechanical stress, such as pulling or frictional forces occurring inside of cells or along their surfaces. Furthermore, they probably contribute to the inforcement of tissue layers at the interfaces of tissues and fluid-filled cavities, including, among others, liver canaliculi,[6] kidney glomeruli,[20] and the blood brain barrier.[21] Moreover, cells may regulate viscoelastic properties of their cytoplasmic compartments during differentiation and development by selective expression and/or cellular targeting of plectin isoforms of distinct linker qualities. Due to their varied molecular interfaces, different plectin isoforms may also serve as filament-associated or plasma membrane-based docking sites for proteins involved in signal transduction.

Purification

The 300 kDa variant of plectin has been purified from Triton X-100/0.6 M KCl-insoluble residues (crude IF preparations) of various cultured cell lines or from bovine eye lens tissue.[2] Solubilization of such residues in 8 M urea or 1 per cent sodium lauroylsarcosinate[22] followed by gel permeation chromatography yields plectin preparations of > 90 per cent purity.

Activities

Assays used to demonstrate specific binding of the purified 300 kDa variant or recombinant plectin mutant proteins to various other proteins include overlays of nitrocellulose-immobilized[2,11,12] or microtitre plate-immobilized[14,31] proteins with radiolabelled,[2,11,12,31] antibody-detectable,[2,11,12] or Eu^{3+}-labelled[14,31] probes, coimmuno- precipitation,[23] co-sedimentation with vimentin filaments,[2,19] and visualization of plectin molecules decorating and interlinking vimentin filaments assembled *in vitro* by electron microscopy.[2,14] *In vitro* network formation through self-association of dumb bell-shaped plectin molecules,[5] and inhibition of vimentin filament assembly by recombinant plectin fragments (unpublished work), have been monitored by negative staining and rotary shadowing electron microscopy. No enzymatic activities of plectins are known.

Antibodies

Rabbit antisera raised against gel-purified rat 300 kDa plectin are cross-reactive with antigens in a variety of cultured cell lines derived from human, monkey, rat, mouse, hamster, cow, and rat kangaroo,[2] as well as with antigens from various tissues of human, rat, mouse, and cow[2] (and unpublished data). A series of monoclonal antibodies raised against 300 kDa plectin from rat shows staining characteristics similar to those of the antisera, but some of the antibodies show restricted cross-reactivity between species.[24] mAb 7A8 is commercially available from Sigma and has successfully been used in a number of studies (see e.g.ref. 10).

Genes

Plectin has been cloned and sequenced both from rat[16,25] and human.[15,26] The plectin cDNA has a coding capacity for a protein of over 4600 amino acid residues (527 kDa), whose sequence is consistent with a three-domain structural model in which a long central rod domain, having mainly an α-helical coiled-coil conformation, is flanked by globular amino- and carboxyl-terminal domains (Fig. 1). The globular C-terminal domain has a prominent six-fold tandem repeat, with each repeat having a strongly conserved central region based on nine tandem repeats of a 19-residue motif.[25]

Analysis of the human[15] and rat[16] gene loci revealed a complex organization of 35 exons spanning over more than 31 kb of genomic DNA. The positions of splice junctions identified in rat and man exactly matched each other, but slight differences were found in the sizes of introns. The most striking feature of the plectin gene structure is that two of the major domains of the molecule, the rod domain and the carboxyl-terminal globular domain, are encoded by single very large exons of > 3 kb and > 6 kb, respectively (Fig. 1). The plectin gene locus (*PLEC1*) has been mapped to the telomeric region (24q) of human chromosome 8.[15]

There is evidence for a complex gene regulatory machinery including extensive transcript diversity generated by differential splicing. Some of the isoforms characterized contain alternative first coding exons and are expressed in different tissues at greatly differing levels, suggesting that their expression is driven and controlled by tissue-specific promoters.[16] In addition, rodless splice variants expressed in a variety of tissues[16] and a splice variant lacking 12 bp at the 3' end of exon 9[26] have been identified.

Mutant phenotypes/disease state

Defects in the expression of plectins have been reported in epidermolysis bullosa simplex (EBS)-MD, an autosomal recessive human disease characterized by severe skin blistering combined with muscular dystrophy.[26–29,32] All the mutations so far characterized lead to premature termination of translation and reduced mRNA expression. Defective expression of plectin also has been reported in EBS-Ogna,[30] an autosomal dominant skin blistering disease.

The disruption of the plectin gene via homologous recombination of targeting constructs in ES cells led to homozygous plectin (–/–) mice that died 1–3 days after birth, showing severe skin blistering and abnormalities of skeletal and heart muscle.[33]

References

1. Wiche, G., Herrmann, H., Leichtfried, F., and Pytela, R. (1982). *CSH Symp. Quant. Biol.*, **46**, 475–82.
2. Wiche, G. (1989). *CRC Crit. Rev. Biochem.*, **24**, 41–67.
3. Hieda, Y., Nishizawa, Y., Uematsu, J., and Owaribe, K. (1992). *J. Cell Biol.*, **116**, 1497–506.
4. Yang, H. -Y., Lieska, N., Goldman, A. E., and Goldman, R. D. (1985). *J. Cell Biol.*, **100**, 620–31.
5. Foisner, R. and Wiche, G. (1987). *J. Mol. Biol.*, **198**, 515–31.
6. Wiche, G., Krepler, R., Artlieb, U., Pytela, R., and Denk, H. (1983). *J. Cell Biol.*, **97**, 887–901.
7. Wiche, G., Krepler, R., Artlieb, U., Pytela, R., and Aberer, W. (1984). *Exp. Cell Res.*, **155**, 43–9.
8. Seifert, G. J., Lawson, D., and Wiche, G. (1992). *Eur. J. Cell Biol.*, **59**, 138–47.
9. Foisner, R., Bohn, W., Mannweiler, K., and Wiche, G. (1995). *J. Struct. Biol.*, **115**, 304–17.
10. Svitkina, T. M., Verkhovsky, A. B., and Borisy, G. G. (1996). *J. Cell Biol.*, **135**, 991–1007.
11. Foisner, R., Traub, P., and Wiche, G. (1991). *Proc. Natl Acad. Sci. USA*, **88**, 3812–16.
12. Herrmann, H. and Wiche, G. (1987). *J. Biol. Chem.*, **262**, 1320–5.
13. Wiche, G., Gromov, D., Donovan, A., Castañón, M. J., and Fuchs, E. (1993). *J. Cell Biol.*, **121**, 607–19.
14. Nikolic, B., Mac Nulty, E., Mir, B., and Wiche, G. (1996). *J. Cell Biol.*, **134**, 1455–67.
15. Liu, C. -g., Maercker, C., Castañón, M. J., Hauptmann, R., and Wiche, G. (1996). *Proc. Natl Acad. Sci. USA*, **93**, 4278–83.
16. Elliott, C. E., Becker, B., Oehler, S., Castañón, M. J., Hauptmann, R., and Wiche, G. (1997). *Genomics*, **42**, 115–25.
17. Herrmann, H. and Wiche, G. (1983). *J. Biol. Chem.*, **258**, 14610–18.
18. Malecz, N., Foisner, R., Stadler, C., and Wiche, G. (1996). *J. Biol. Chem.*, **271**, 8203–8.
19. Foisner, R., Malecz, N., Dressel, N., Stadler, C., and Wiche, G. (1996). *Mol. Biol. Cell*, **7**, 273–88.
20. Yaoita, E., Wiche, G., Yamamoto, T., Kawasaki, K., and Kihara, I. (1996). *Am. J. Pathol.*, **149**, 319–27.
21. Errante, L. D., Wiche, G., and Shaw, G. (1994). *J. Neurosci. Res.*, **37**, 515–28.
22. Weitzer, G. and Wiche, G. (1987). *Eur. J. Biochem.*, **169**, 41–523.
23. Eger, A., Stockinger, A., Wiche, G., and Foisner, R. (1997). *J. Cell Sci.*, **110**, 1307–16.
24. Foisner, R., Feldman, B., Sander, L., Seifert, G., Artlieb, U., and Wiche, G. (1994). *Acta Histochem.*, **96**, 421–38.
25. Wiche, G., Becker, B., Luber, K., Weitzer, G., Castañon, M. J., Hauptmann, R., *et al.* (1991). *J. Cell Biol.*, **114**, 83–99.
26. McLean, W. H. I., Pulkkinen, L., Smith, F. D. J., Rugg, E. L., Lane, E. B., Bullrich, F., *et al.* (1996). *Genes Dev.*, **10**, 1724–35.
27. Gache, Y., Chavanas, S., Lacour, J. P., Wiche, G., Owaribe, K., Meneguzzi, G., and Ortonne, J. P. (1996). *J. Clin. Invest.*, **97**, 2289–98.
28. Smith, F. J. D., Eady, R. A. J., Leigh, I. M., McMillan, J. R., Rugg, E. L., Kelsell, D. P., *et al.* (1996). *Nature Genet.*, **13**, 450–7.
29. Pulkkinen, L., Smith, F. J. D., Shimizu, H., Murata, S., Yaoita, H., Hachisuka, H., *et al.* (1996). *Hum. Mol. Genet.*, **5**, 1539–46.
30. Koss-Harnes, D., Jahnsen, F. L., Wiche, G., Søyland, E., Brandtzaeg, P., and Gedde-Dahl Jr, T. (1996). *Exp. Dermatol.*, **6**, 41–8.
31. Rezniczek, G. A., de Pereda, J. M., Reipert, S., and Wiche, G. (1998). *J. Cell Biol.*, **141**, 209–25.
32. Mellerio, J. E., Smith, F. J. D., McMillan, J. R., McLean, E. H. I., McGrath, J. A., Morrison, G. A. J. *et al.* (1997). *Brit. J. Dermatol.*, **137**, 898–906.
33. Andrä, K., Lassmann, H., Bittner, R., Shorny, S., Fässler, R., Propst, F., and Wiche, G. (1997). *Genes Dev.*, **11**, 3143–56.

Gerhard Wiche
Vienna Biocenter, Institute of Biochemistry and Molecular Cell Biology,
Vienna, Austria

Tight junction proteins

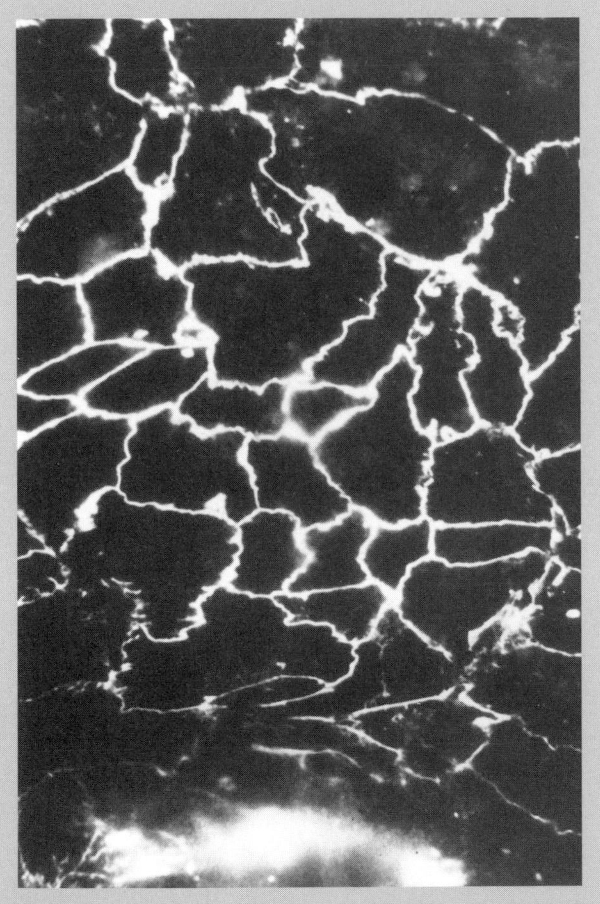

Immunofluorescence localization of cingulin in chicken kidney epithelial cell. Provided by Sandra Citi, University of Geneva, Switzerland.

Cingulin

Cingulin is a protein localized on the cytoplasmic face of vertebrate tight junctions. Its apparent molecular size is 140–160 kDa, depending on the species. The tissue distribution and subcellular localization of cingulin suggest that it is restricted to tight junction-containing polarized cells.

Protein properties

Cingulin migrates with an apparent molecular weight of 140 000 Da in mammalian and avian species[1–4] and 160 000 Da in *Xenopus laevis* extracts.[5] It can be extracted from cells and tissues without detergents, suggesting that it is a peripheral component of tight junctions.[1,2] A 108 kDa form of avian cingulin (probably a proteolytic degradation product of the 140 kDa form) was purified to homogeneity from intestinal brush border cells.[1,2] It was shown to be a heat- and ethanol-stable disulphide-linked dimer, appearing in the electron microscope as a 130 nm long and 2 nm wide flexible rod (Fig. 1), suggesting a coiled-coil structure.[1,2] Sequencing of cingulin cDNAs and secondary structure prediction suggest that in addition to a large, central coiled-coil domain that shows homology to the coiled-coil region of myosin II heavy chains,[6] cingulin dimers contain two N-terminal globular heads and two C-terminal small globular tails (unpublished observations).

Expression of cingulin has been detected in chicken polarized epithelia, stratified epithelia and some endothelia,[2] human benign and malignant epithelial neoplasias,[3] cultured cells derived from chicken embryonic hearts,[7] mouse oocytes and developing embryos,[4] and *Xenopus laevis* oocytes and developing embryos.[5] Cingulin immunoreactivity has not been detected in HUVEC cultured endothelial cells,[3] non-epithelial tissues and tumours,[3] fibroblasts and adult cardiac myocytes,[2] and invertebrates.

The localization of cingulin on the cytoplasmic face of tight junctions, at a distance of about 40 nm from the plasma membrane, has been determined by immunoelectron microscopy.[1,2] Immunofluorescent labelling of cultured epithelial cells shows junctional labelling (Fig. 2), but cingulin also appears to be localized in the nuclei in subconfluence MDCK cell cultures.[8]

Cingulin is phosphorylated in Ser residues *in vivo*[9] and phosphorylation *in vitro* occurs in the C-terminal portion of the coiled-coil domain.[10] Inhibiting Ser/Thr phosphory-

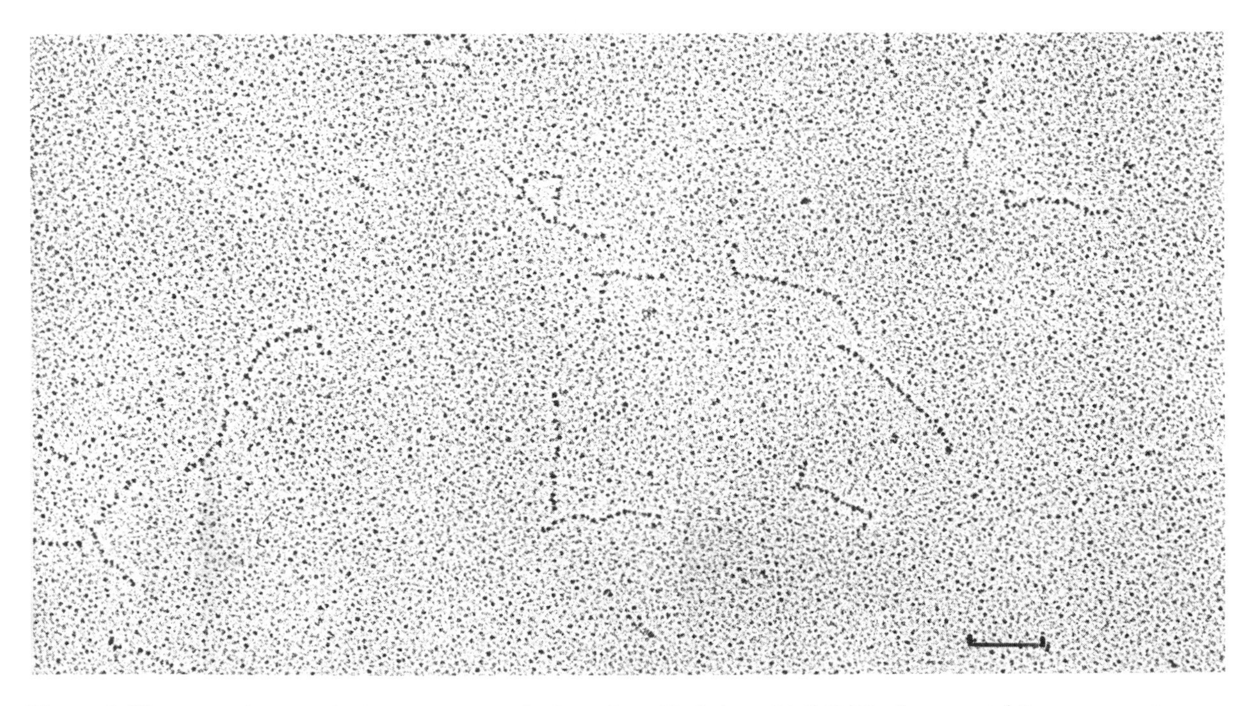

Figure 1. Electron micrograph showing rotary shadowed purified cingulin (108 kDa fragment).[1] Bar represents 50 nm.

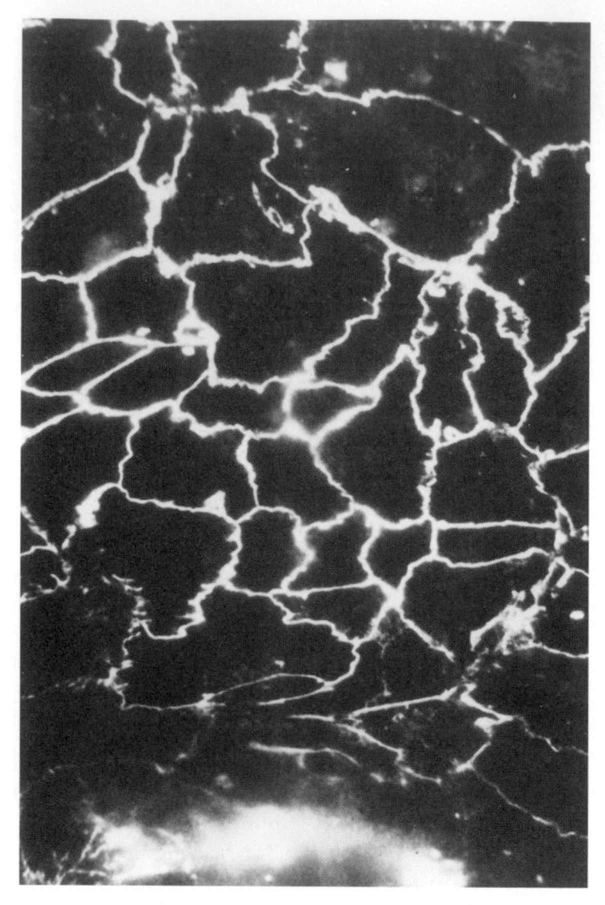

Figure 2. Immunofluorescent localization of cingulin in primary cultures of chicken kidney epithelial cells.

lation during tight junction assembly by calcium switch in cultured MDCK cells blocks the accumulation of cingulin at sites of cell–cell contact.[9]

The role of cingulin in tight junctions is not known.

Purification

Cingulin (a 108 kDa form) can be purified from chicken intestinal epithelial cells by ammonium sulphate fractionation of cellular extracts followed by gel filtration and heat treatment or ion exchange chromatography.[1,2]

Activities

Not known.

Antibodies

Monoclonal and polyclonal antibodies against native and recombinant chicken cingulin have been published.[1,5] The polyclonal antisera cross-react with mammalian and other species and can be used for immunofluorescence and immunoblotting.

Genes

Cingulin cDNA sequences have not yet been published.

Mutant phenotype/disease states

Cingulin expression persists in undifferentiated human colon adenocarcinomas, suggesting that it may be useful in the typing of epithelial-derived malignant neoplasias.[3]

Structure

Not available.

References

1. Citi, S., Sabanay, H., Jakes, R., Geiger, B., and Kendrick-Jones, J. (1988). *Nature*, **333**, 271–6.
2. Citi, S., Sabapay, H., Kendrick-Jones, J., and Geiger, B. (1989). *J. Cell Sci.*, **93**, 107–22.
3. Citi, S., Amorosi, A., Franconi, F., Giotti, A., and Zampi, G. (1991). *Am. J. Pathol.*, **138**, 781–9.
4. Fleming, T. P., Hay, M., Javed, Q., and Citi, S. (1993). *Development*, **117**, 1135–44.
5. Cardellini, P., Davanzo, G., and Citi, S. (1996). *Dev. Dynamics*, **207**, 104–13.
6. Citi, S., Kendrick-Jones, J., and Shore, D. (1990). *J. Cell Biol.*, **111**, 409a.
7. Eisenberg, C. A., and Bader, D. (1995). *Dev. Biol.*, **67**, 469–81.
8. Citi, S. (1998) In *Adhesive interactions of Cells* (ed. D.R. Garrod, A. North, and M.A.J. Chidgey), Jai Press, Greenwich, CT. (In press.)
9. Citi, S., and Denisenko, N. (1995). *J. Cell Sci.*, **108**, 2917–26.
10. Rabino, M., Denisenko, N., and Citi, S. (1993). *Mol. Biol. Cell*, **4**, 440a.

Sandra Citi
Dipartimento di Biologia,
Universita' di Padova, Italy and
Departement de Biologie Moleculaire
Université de Genève, Switzerland

Symplekin

Symplekin is a high molecular weight, membrane-associated protein specific for the tight junctions (TJ) of polarized epithelial cells, which is present as well in the nucleus of both TJ-containing and TJ-deficient cells.[1] In contrast to other known TJ-associated components, symplekin is absent from TJ of vascular endothelia. The polypeptide shares no striking homologies with any known proteins and its biological function(s) are currently unknown.

Protein properties

Symplekin, a M_r = 150 kDa polypeptide, was originally detected by a monoclonal antibody E150 at a continuous zone of cell–cell contact in cultured human colon carcinoma cells. Immunoelectron microscopy of thin sections of human colon tissue revealed that the protein is localized at the electron dense cytosolic structure underlying the *zonulae occludente.*[1] By indirect immunofluorescence, symplekin was found to be present at the tight junctions (TJ) of polarized epithelia of stomach, intestine, liver, pancreas, and mammary gland, as well as at the 'specialized junctions' of Sertoli cells in testis. Interestingly, symplekin is not detected at the TJ of vascular endothelia, including brain capillary endothelia. The only detection of the protein at non-epithelial cell junctions is its presence at a newly characterized adherens-type junction identified in an astrocytoma derived glioma cell line (unpublished results). At these cell junctions, which appear punctate or as short linear parallel arrays, ZO-1 is also detected (ZO-1 has also been reported at similar junctions in astrocytes[2]), but occludin is not.

Interestingly, symplekin mRNA and protein are detected in a broad range of cell types including those that are incapable of forming stable cell–cell contacts. Immunofluorescence studies using precautions to minimize loss and/or redistribution of soluble protein resulted in the detection of symplekin in the nucleus of cells in culture and *in situ* (see Plate 1). In cell fractionation experiments, symplekin fractionates into the nuclear pellet following hypotonic cell lysis and can be enriched in crude membrane preparations of liver bile canaliculi.

A cDNA was originally cloned from a λgt11 expression library derived from cultured cells of a human adenocarcinoma line. The cDNA encodes a slightly acidic, 1142 a.a. polypeptide with a predicted MW of 126 000 Da. Recently a larger cDNA (clone R6E1; 4.0 kb),[9] containing a start codon upstream of and in-frame with the original symplekin cDNA (clone Sym5b; 3.7 kb), was isolated from a human brain cDNA library. In addition, the human symplekin gene has been mapped to an approximately 40 kb region telomeric to gene 59 and the myotonic dystrophy (DM) gene loci on chromosome 19q13.3.[9] *In vitro* transcription and translation of the R6E1 gene results in a single dominant band that can be resolved from the faster migrating Sym5b product in low percentage polyacrylamide gels (B. Keon; unpublished results). Another human symplekin gene (4.0 kb) has been submitted to GenBank by authors of the DM expanded repeat mapping project. This cDNA contains the entire open reading frame of clone Sym5b, the 5' extension of clone R6E1, and an additional 49 bp upstream of the 5' extension that includes a termination codon. The 5' extension is therefore regarded as the full-length symplekin cDNA. Guinea pig antibodies raised against peptides corresponding to distinct regions of the cDNA-derived amino acid sequence immunoprecipitate the 150 kDa polypeptide and recognize it in immunoblots. Analyses of symplekin cDNA and the encoded amino acid sequences revealed no striking homologies to any known proteins or protein families, including the membrane-associated guanylate kinase (MAGUK)[3] family to which members of the TJ protein complex, ZO-1,[4,5] ZO-2,[6] and ZO-3[7,8] belong. The primary structure of symplekin, however, does contain some interesting features including a small region of homology to serine threonine kinases, a negatively charged serine- and proline-rich carboxyl-terminal domain, and two consensus nuclear localization signals. Symplekin is phosphorylated *in vivo* and at least one phosphorylation site has been identified by mass spectrometric analysis (unpublished results, B. Keon and P. Jedrzejewski). Any potential biological significance of these features remains to be determined.

Purification

Symplekin can be partially extracted from hypotonically lysed cells with high alkaline buffers, consistent with its peripheral membrane association. Complete recovery of the polypeptide, however, requires extraction of the cellular material with non-ionic detergents, the removal of which does not affect the protein's solubility *in vitro*. The non-ionic detergent is likely to be required to access the nuclear pool of the protein. Using polyclonal antibodies raised against synthetic peptides based on the cDNA-derived amino acid sequence of symplekin, the polypeptide can be cleared from cell lysates of both TJ-containing and TJ-deficient cells following their extraction in a modified RIPA buffer.

Activities

None known.

Antibodies

Sym-TJ-E150, a mouse monoclonal IgG1, is available from Progen Biotechnik GmbH, Heidelberg, Germany and Research Diagnostics, Flanders, NJ, USA. Also available are Sym 139 (a mouse monoclonal IgM) and various partially characterized guinea pig antisera.

Genes

(1) Human symplekin clones: Sym5b (GenBank accession no. U49240); R6E1 (no. U88726); 5C-K7 (Y10931); chromosome 19q13.3;

(2) *Caenorhabditis elegans* possible homologue (28% identity, 50% similarity): CELF25G6 (accession no. AFO22973).

Mutant phenotype/disease states

No information available.

Structure

Not available.

References

1. Keon, B. H., Schaefer, S., Kuhn, C., Grund, C., and Franke, W. W. (1996). *J. Cell Biol.*, **134**, 1003–18.
2. Howarth, A. G., Hughes, M., and Stevenson, B. R. (1992). *Am. J. Physiol.*, **262**, C461–9.
3. Anderson, J. M. (1996). *Curr. Biol.* **6**, 382–4.
4. Stevenson, B. R., Siliciano, J. D., Mooseker, M. S., and Goodenough, D. A. (1986). *J. Cell Biol.*, **103**,:755–66.
5. Anderson, J. M., Stevenson, B. R., Jesaitis, L. A., Goodenough, D. A., and Mooseker, M. S. (1988). *J. Cell Biol.*, **106**, 1141–9.
6. Jesaitis, L. A. and Goodenough, D. A. (1994). *J. Cell Biol.*, **124**, 949–61.

Brigitte H. Keon
Department of Neurobiology,
Harvard Medical School,
240 Longwood Avenue,
Boston, MA, USA

ZO-1 and ZO-2

ZO-1 and ZO-2 are peripheral membrane proteins which are associated with tight junctions in all epithelia. They form a complex which can be co-immunoprecipitated from RIPA buffers.[1] Both belong to the MAGUK (membrane-associated guanylate kinase) family of proteins, a family involved in the clustering of integral membrane proteins at synapses.[2–4] The ZO-1/ZO-2 complex has been shown to interact with occludin,[5] an integral membrane protein involved in the permeability barrier function of the tight junction. The function of the ZO-1/ZO-2 complex is not known.

Synonymous names

None.

Homologous proteins

Uniting features are summarized in Fig. 1.

Protein properties

ZO-1, a 220 kDa protein, was the first protein shown localized to the tight junction using a monoclonal antibody approach.[11,12] In addition to tight junctions, ZO-1 is found at the specialized pedicel interactions of kidney podocytes[13] and in other non-tight junction locations,[14,15] indicating that ZO-1 is not uniquely a tight junction marker. ZO-1 has been cloned from human[16] and mouse.[15] ZO-2, a 160 kDa protein,[1] has been cloned from dog[17] and from human.[18] ZO-3, a third member of the family, which co-immunoprecipitates with ZO-1 and ZO-2, has also been characterized.[31] ZO-1 and ZO-2 are members of the MAGUK family of proteins,[19] which includes the product of the *Drosophila* lethal(1)discs-large-1 (*dlg*) tumour suppressor gene,[7] PSD-95 from rat brain synapses,[6] the erythrocyte membrane protein p55,[8] and the *C. elegans* vulval induction gene *lin-2*.[9] These proteins each have an SH3 domain and a domain homologous to yeast guanylate kinase (GK). Key catalytic regions of this latter domain are poorly conserved in ZO-1, such that ZO-1 is unlikely to have GK activity.

Molecular cloning of ZO-1 has permitted the demonstration of an 80 amino acid domain, motif α, near the N-terminal boundary of a proline-rich domain which is variably present as a result of alternative RNA splicing. Isoform-specific antisera have revealed that the ZO-1α^{+} isoform is found in most junctions between epithelial cells and the ZO-1α^{-} isoform is found in both endothelial cells and between kidney podocytes and Sertoli cells of seminiferous tubules.[20] ZO-2 is also found at the tight junction and co-immunoprecipitates with ZO-1.[1,17,21] A

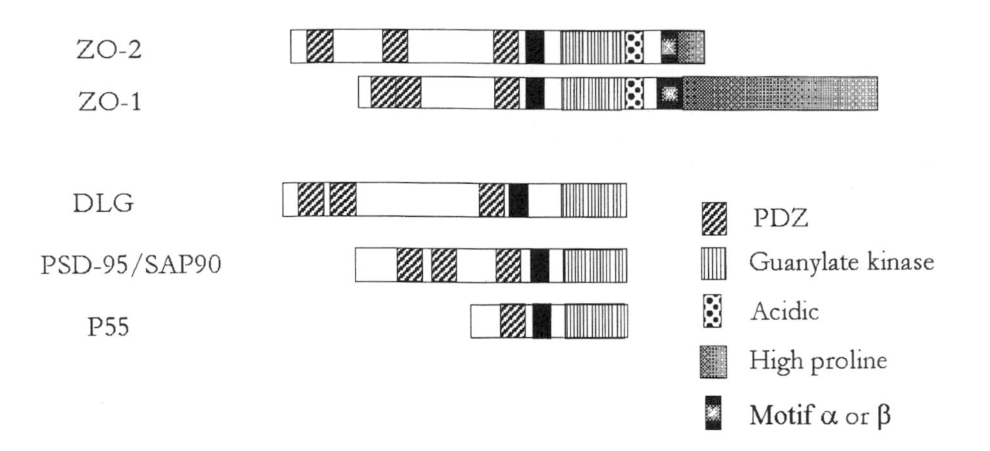

Figure 1. 1. PSD-95/SAP90, PSD-93, and SAP102;[6] 2. *Drosophila* discs-large (Dlg[7]); 3. red cell p55;[8] 4. *C. elegans* lin-2;[9] 5. CASK.[10]

comparison of some of the MAGUK family members, including ZO-1 and ZO-2 is shown in Fig. 1. Like ZO-1, ZO-2 contains at least one alternatively spliced domain, called motif β.[17]

MAGUK family members also have a novel structural motif, the PDZ domain,[22] which is shared by a rapidly expanding number of proteins. PDZ domains have been shown to bind to motifs at the absolute C termini of proteins, and to be involved in the clustering of integral membrane proteins in a number of systems. Initially, it was shown that the second PDZ domain of the MAGUK protein PSD-95 would bind to the motif S/TXV at the absolute C terminus of the NMDA receptor.[3] Using the yeast two-hybrid system, Kim *et al.*[4] showed that Shaker-type K^+ channels bind hDlg/SAP97, PSD-95, and another MAGUK called clone 5 via a C-terminal TDV sequence, and also showed that these interactions resulted in clustering of the K^+ channels. In addition to these peptides, PDZ domains have also been shown to bind to each other: for example the PDZ domain of nNOS (neuronal nitric oxide synthase) binds the second PDZ of PSD-93 and PSD-95, and can also bind to the PDZ domain of α1-syntrophin.[23] In a broadly based study using peptide libraries, it was found that PDZs in the MAGUK family selected peptides which contained the consensus motif Glu–(Ser/Thr)–Xxx–(Val/Ile) at the carboxyl terminus. In contrast, another family of PDZ domains, including those of Lin-2 and p55 selected peptides with hydrophobic or aromatic side chains at the carboxyl terminal three residues.[24] AF-6, a Ras target, has a PDZ domain. In MDCK epithelial cells, immunoelectron microscopy shows a close association between ZO-1 and AF-6 at tight junctions, ZO-1 interacts with the Ras binding domain of AF-6 and this interaction is inhibited by activated Ras. Thus, AF-6 may participate in the regulation of cell–cell contacts including tight junctions, via direct interaction with ZO-1 downstream of Ras.[32]

The binding domains in the human homologue of the discs-large protein have been studied,[25,26] providing experimental evidence that the first two of the PDZ domains bind to Shaker-type K^+ channels and to ATP while a novel domain between the SH3 and GK domains binds to the actin binding protein 4.1. These observations directly implicate this MAGUK family member as a protein linking integral membrane proteins to the cytoskeleton. Neither ZO-1 nor ZO-2 contain any known actin binding motifs, however, suggesting that an adapter protein such as 4.1 may be involved in the interactions suspected between these proteins and the cytoskeleton.

In addition to these binding functions, MAGUK proteins are likely to be involved in signalling pathways in development. In *C. elegans*, Lin-2 and Lin-7 are required for vulval induction.[9,27] Mutation of *Drosophila dlg* causes epithelial cells in the imaginal disc to lose polarity and undergo neoplastic transformation.[7] The functions of ZO proteins in epithelia and embryos are not currently known.

Purification

ZO-1[12] and ZO-2[21] are purified by electroelution from SDS–polyacrylamide gels or by immunoprecipitation following extraction of MDCK cells with RIPA buffer using antibodies covalently coupled to sepharose beads.

Activities

Not known.

Antibodies

1. ZO-1: rat monoclonal mAb 1520; available from Chemicon International, Inc., 28835 Single Oak Drive, Temecula, CA 92590, USA.
2. ZO-1: rabbit polyclonal cat No 61–7300; available from Zymed Laboratories, Inc., 458 Carlton Court, South San Francisco, CA 94080, USA.

3. ZO-2: rabbit polyclonal cat. no. 71–1400 available from Zymed Laboratories, Inc., 458 Carlton Court, South San Francisco, CA 94080, USA.

Genes

Human ZO-1: GenBank accession number L14837,[16] chromosome 15q13.[28] Mouse ZO-1: EMBL/GenBank DDBJ accession number D14340.[15] Canine ZO-2: GenBank/EBI Data Bank accession number L27152.[17] Human ZO-2: GenBank accession number L27476.[18]

Mutant phenotype/disease states

No information available.

Structure

Crystallographic structures of PDZ domains from hDlgA, human discs-large,[29] and from PSD-95[30] have been determined.

References

1. Gumbiner, B., Lowenkopf, T. and Apatira, D. (1991). *Proc. Natl Acad. Sci. USA*, **88**, 3460–4.
2. Woods, D. F. and Bryant, P. J. (1993). *Mech. Deve.*, **44**, 85–9.
3. Kornau, H. C., Schenker, L. T., Kennedy, M. B., and Seeburg, P. H. (1995). *Science*, **269**, 1737–40.
4. Kim, E., Niethammer, M., Rothschild, A., Jan, Y. N., and Sheng, M. (1995). *Nature*, **378**, 85–8.
5. Furuse, M., Itoh, M., Hirase, T., Nagafuchi, A., Yonemura, S., Tsukita, S., and Tsukita, S. (1994). *J. Cell Biol.*, **127**, 1617–26.
6. Cho, K.-O., Hunt, C. A., and Kennedy, M. B. (1992). *Neuron*, **9**, 929–42.
7. Woods, D. F. and Bryant, P. J. (1991). *Cell*, **66**, 451–64.
8. Marfatia, S. M., Lue, R. A., Branton, D., and Chishti, A. H. (1994). *J. Biol. Chem.*, **269**, 8631–4.
9. Hoskins, R., Hajnal, A. F., Harp, S. A., and Kim, S. K. (1996). *Development*, **122**, 97–111.
10. Hata, Y., Butz, S., and Südhof, T. C. (1996). *J. Neurosci.*, **16**, 2488–94.
11. Stevenson, B. R., Siliciano, J. D., Mooseker, M. S., and Goodenough, D. A. (1986). *J. Cell Biol.*, **103**, 755–66.
12. Anderson, J. M., Stevenson, B. R., Jesaitis, L. A., Goodenough, D. A., and Mooseker, M. S. (1988). *J. Cell Biol.*, **106**, 1141–9.
13. Kurihara, H., Anderson, J. M., and Farquhar, M. G. (1995). *Am. J. Physiol.*, **37**, F514–24.
14. Howarth, A. G., Hughes, M. R., and Stevenson, B. R. (1992). *Am J Physiol.*, **262**, C461–9.
15. Itoh, M., Nagafuchi, A., Yonemura, S., Kitani-Yasuda, T., Tsukita, S., and Tsukita, S. (1993). *J. Cell Biol.*, **121**, 491–502.
16. Willott, E., Balda, M. S., Fanning, A.S., Jameson, B., Van Itallie, C. and Anderson, J. M. (1993). *Proc. Natl Acad. Sci. USA*, **90**, 7834–8.
17. Beatch, M., Jesaitis, L. A., Gallin, W. J., Goodenough, D. A., and Stevenson, B. R. (1996). *J. Biol. Chem.*, **271**, 25723–6.
18. Duclos, F., Rodius, F., Wrogemann, K., Mandel, J.-L., and Koenig, M. (1994). *Hum. Mol. Genet.*, **3**, 909–14.
19. Anderson, J. M. (1996). *Curr. Biol.*, **6**, 382–4.
20. Balda, M. S. and Anderson, J. M. (1993). *Am. J. Physiol.*, **264**, C918–24.
21. Jesaitis, L. A. and Goodenough, D. A. (1994). *J. Cell Biol.*, **124**, 949–61.
22. Kennedy, M. B. (1995). *TIBS*, **20**, 350.
23. Brenman, J. E., Chao, D. S., Gee, S. H., McGee, A. W., Craven, S. E., Santillano, D. R., *et al.* (1996). *Cell*, **84**, 757–67.
24. Songyang, Z., Fanning, A. S., Fu, C., Xu, J., Marfatia, S. M., Chishti, A. H., *et al.* (1997). *Science*, **275**, 73–7.
25. Marfatia, S. M., Morais Cabral, J. H., Lin, L., Hough, C., Bryant, P. J., Stolz, L., and Chishti, A. H. (1996). *J. Cell Biol.*, **135**, 753–66.
26. Lue, R. A., Brandin, E., Chan, E. P., and Branton, D. (1996). *J. Cell Biol.*, **135**, 1125–37.
27. Simske, J. S., Kaech, S. M., Harp, S. A., and Kim, S. K. (1996). *Cell*, **85**, 195–204.
28. Mohandas, T. K., Chen, X. N., Rowe, L. B., Birkenmeier, E. H., Fanning, A. S., Anderson, J. M., and Korenberg, J. R. (1995). *Genomics*, **30**, 594–7.
29. Cabral, J. H. M., Petosa, C., Sutcliffe, M. J., Raza, S., Byron, O., Poy, F., *et al.* (1996). *Nature*, **382**, 649–52.
30. Doyle, D. A., Lee, A., Lewis, J., Kim, E., Sheng, M., and Mackinnon, R. (1996). *Cell*, **85**, 1067–76.
31. Haskins, J., Gu, L. J., Wittchen, E. S., Hibbard, J. and Stevenson, B. R. (1998) *J. Cell Biol.*, **141**, 199–208.
32. Yamamoto, T., Harada, N., Kano, K., Taya, S., Canaani, E., Matsuura, Y., Mizoguchi, A., Ide, C., and Kaibuchi, K. (1997). *J. Cell Biol.*, **139**, 785–95.

Daniel A. Goodenough
Department of Cell Biology,
Harvard Medical School,
240 Longwood Avenue,
Boston, MA 02115, USA

2

Cell adhesion and cell-cell contact proteins

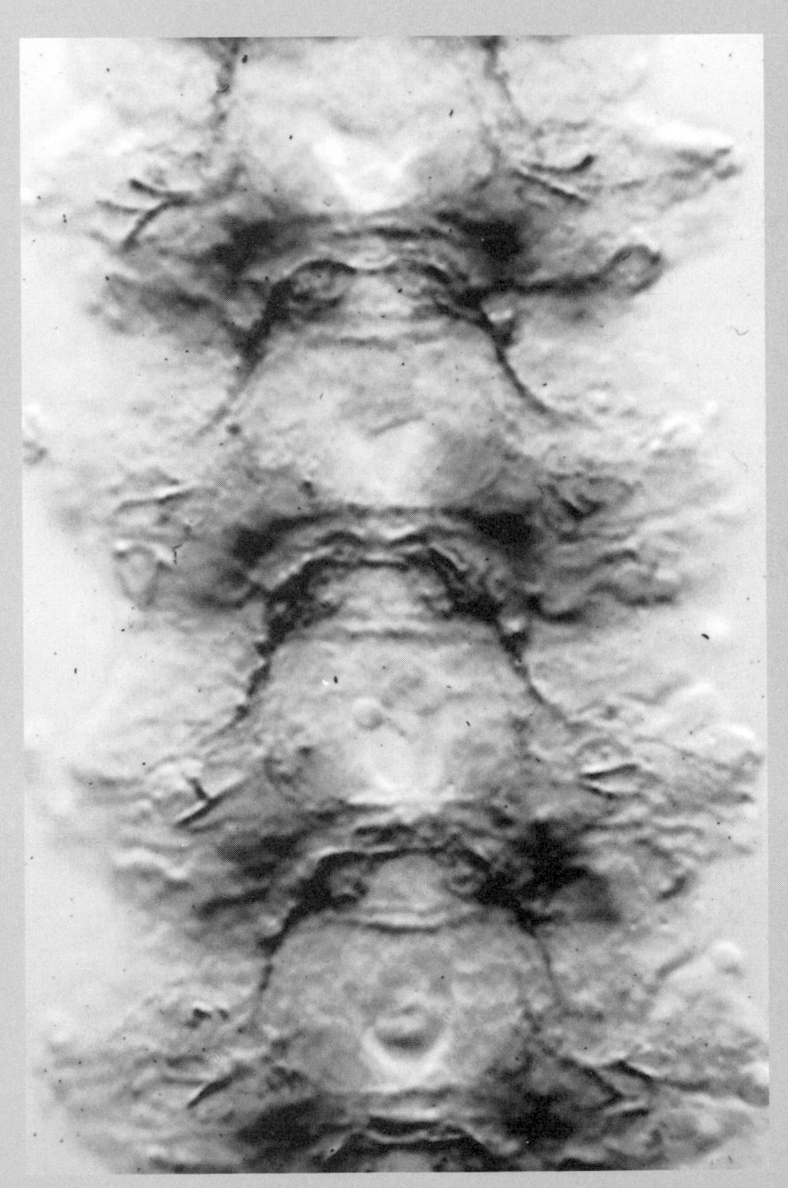

Fasciclin III is expressed on a subset of embryonic neurones and axon pathways. The figure shows four contiguous segmental neuromeres of the *Drosophila* CNS (anterior is towards the top) at hour 12 of embryonic development. At this stage fasciclin III protein is expressed on the surface of axons in five commissural fascicles and on the continuation of one of the A commissure fascicles as it turns posteriorly and laterally towards one of the two peripheral nerves, the intersegmental nerve. (Courtesy of N. Patel.)

Introduction

The chapters compiled in this book assemble the current knowledge on different families of cell adhesion molecules, including integrins, cadherins, members of the immunoglobulin superfamily, selectins, and others. These molecules mediate adhesion by binding other components which are located either on neighbouring cells or in the extracellular matrix. Cell adhesion is provided by protein–protein (in the case of integrins, cadherins and immunoglobulin superfamily members) or protein–carbohydrate interactions (in the case of selectins and certain immunoglobulin superfamily members, with the sialoadhesins). That cell adhesion is caused by the molecular interaction of proteins or carbohydrate molecules located on opposing cell surfaces is a concept that has developed in the last 30 years. Classical biological experiments performed in the first half of this century with sponges and amphibian embryos had already demonstrated that cells possess specific capacities for adhesion.[1,2] The subsequent development of cell biological and immunological techniques allowed the isolation and characterization of the responsible molecules (see ref. 3). Milestones in the molecular phase of adhesion research were the discovery of the extracellular matrix component fibronectin (in the early 1970s,[4–7]) which interacts with integral membrane proteins called the integrins (discovered in the early 1980s,[8–11]) or the discovery of the cell–cell adhesion molecules N CAM and E-cadherin (in the late 1970s and early 1980s[12–20]). Today, over 100 different cell adhesion molecules have been identified and molecularly characterized, and their function in cell adhesion has been verified in numerous laboratories. Besides these classical adhesion molecules, additional molecules that possess dual functions are discussed; these molecules mediate adhesive processes and, for instance, have protease activity (ADAMs), glycosyltransferase activity, function in the recognition of molecules in the immune system (the T-cell receptor in connection with MHC class I and II molecules), recognize chemokines and growth and differentiation factors (e.g. tyrosine kinases, Notch, netrin receptors), or form gap and tight junctions (connexins, occludin). The range of molecules discussed in the various chapters of this volume is thus comprehensive and covers many different aspects of current cell adhesion research. Here, I will emphasize aspects discussed in this book which have had a high impact in the field.

Old and new molecules

This revised edition summarizes our present knowledge on the main families of adhesion molecules which mediate primarily cell adhesion, i.e. the cadherins, integrins, the immunoglobulin superfamily members and the selectins. The chapter by Johnson *et al*. describes the classical, desmosomal, prototypical and atypical cadherins. The integrins and their subfamilies are introduced in a comprehensive review by Hemler.

Selectins and their ligands are discussed in chapters by Rosen and Vestweber. Selectins are a family of three carbohydrate-binding adhesion molecules (L-, P-, and E-selectins) which mediate the docking and rolling of leukocytes on the blood vessel wall. This initiates the entry of leukocytes into sites of inflammation and the homing of lymphocytes into lymphatic tissues. The discovery and characterization of selectins and their ligands have validated the long-held view that carbohydrates can serve as recognition determinants in eukaryotic cell interactions. Selectins contain at their N-terminus a calcium-type (C-type) lectin domain of approximately 120 amino acid residues followed by other motifs and the transmembrane domain. This lectin domain binds the selectin ligands which contain the saccharide sialyl Lewis X (sLex) or modifications of it as recognition motifs. Several of these protein ligands have now been characterized in detail, e.g. PSGL-1, ESL-1, GlyCAM-1, CD34 and MadCAM-1. Others await further molecular characterization. Lectin domains are also found on mannose-binding proteins (MBPs), which act as opsonins for a variety of microorganisms (see chapter by Sastry and Ezekowitz, this volume).

The immunoglobulin superfamily of adhesion receptors is represented by a series of articles (on N CAMs, fasciclins, TAG-1/axonin, V CAMs, MadCAM, PECAM-1 and others). The many members of the family are involved in a variety of homophilic and heterophilic adhesion processes (see chapter by Thiery, this volume). I am always astonished by the multiplicity of possible interactions of immunoglobulin superfamily members of adhesion molecules. For instance, N CAM can homodimerize, or heterodimerize with TAG-1/axonin or L1. It can also bind components of the extracellular matrix, heparin or proteoglycans, and receptor tyrosine phosphatases. Other members of this family bind integrins or laminin (see chapters by Brümmendorf). Thus, complex networks of interactions are formed by immunoglobulin-family members, which are regulated by selected proteolysis, differential gene expression, differential glycosylation and other mechanisms.

The dystroglycan–dystrophin glycoprotein complex is an interesting assembly of proteins that mediate cell adhesion and transmembrane linkage. The dystroglycan gene encodes two proteins of this complex, α- and β-dystroglycan. α-Dystroglycan is a peripheral membrane protein that functions as a cell adhesion molecule in Schwannoma cells; it interacts with laminin-2 and agrin in muscle cells. β-Dystroglycan is an integral membrane protein which associates with other integral membrane proteins, the sarcoglycans, and with the cytoskeletal protein dystrophin, which binds F-actin. Interacting molecules of dystroglycan in non-muscle tissues need to be identified. The dystroglycan–dystrophin complex functions in neuromuscular junctions and in epithelial morphogenesis. Members of the dystrolycan–dystrophin complex have been found to be mutated in various human muscle dystrophies (see chapter by Henry and Campbell, this volume).

CD44 is a glycoprotein involved in cell adhesion to the extracellular matrix component hyaluronan. Intracellularly, CD44 interacts with members of the esrin/radixin/moesin family of cytoskeletal proteins. Alternative splicing of 10 or 11 additional exons produces various CD44 isoforms, which have additional peptide sequences inserted into their extracellular domains. The various CD44 isoforms are differentially expressed in development, in adult tissues and during tumour progression. CD44 variants play a role in inflammation processes by regulating cellular interaction, and have been implicated in signal transduction and metastasis (see chapter by Pure and Hodge-Dufour, this volume).

Syndecans are major integral heparan sulphate proteoglycans which are encoded by four different genes. The heparan sulphate chains interact with a wide variety of soluble and insoluble ligands such as components of the extracellular matrix, growth factors, chemokines, or surface proteins of microbial and viral pathogens. Syndecans are adhesion molecules for various cells in culture, e.g. mammary epithelial cells, which adhere through syndecan to fibronectin, collagens, tenascin or thrombospondin. Moreover, syndecans bind fibroblast growth factor-2, an interaction which is essential for high-affinity binding to fibroblast growth factor receptor-1 (see chapter by Bernfield, this volume).

Several new adhesion molecules have come onto the stage, and several previously known molecules have been recognized as having primary importance since the last edition of this book. This new edition includes a chapter on Ep-CAM (which was long-known as a carcinoma-associated antigen, 17-1A), one on the leucine-rich repeat family of proteins, on occludin (the tight junction integral membrane protein), and one on the platelet GPIb–XI–V complex.

Knock-out of genes for cell adhesion molecules

Cell adhesion molecules frequently show characteristic spatio-temporal expression patterns during development. This regulated expression controls morphogenic events and results in adhesion or segregation of particular cellular populations. Cell adhesion molecules are bound to the plasma membranes by transmembrane domains or GPI anchors; the expression of transmembrane and GPI anchored protein isoforms can also be regulated in a developmental manner. Moreover, homotypic or heterotypic binding of adhesion molecules can be modulated during development. This new edition describes many of such developmental changes, or gives essential references for the interested reader.

In the last five years, many genes encoding cell adhesion molecules have been mutated in the mouse by the use of homologous recombination and embryonal stem cell technology. These knock-outs are described in a recent review by Hynes.[21] Together with the knowledge on spontaneous mutations in humans, a comprehensive picture of the role of the various cell adhesion molecules in development and disease is emerging.

Three classical cadherins (E-, N- and P-) as well as three of the cytoplasmically associated catenins have been mutated. Most of these mutations interfere with appropriate development and cause an embryonal lethality.[22–28] E-cadherin and α-catenin –/– embryos show the most extreme phenotypes: the polarity of blastocyst epithelium cannot be maintained and the blastocysts fail to implant.[22,23,26] Other mutations interfere with development at later stages or in a more subtle manner. Mutation of the desmosomal cadherin desmoglein 3 produces a phenotype similar to Pemphigus vulgaris.[28] These findings demonstrate that cadherins play essential roles in adhesion during early development, and show that their linkage to components of the cytoskeleton is crucial for adhesion to occur.

Integrins have been extensively analysed by mouse genetics; most genes for integrin subunits have now been mutated, and several double mutations have already been described,[21] (see also chapter by Hemler, this volume). Ablation of the integrin subunit β1, which is known to assemble with over 10 different α-subunits, causes an early embryonal lethality. Mutations of the respective α-subunits lead to different, again mostly lethal phenotypes at later stages of development, demonstrating that the different integrins have independent, non-overlapping functions. Moreover, all integrins which interact with fibronectin produce a distinct phenotype, but none is as severe as a fibronectin null-mutation. This again points to independent functions of the different integrins. Ablation of the integrins subunits located in hemidesmosomes (α6 and β4) results in detachment of skin epithelia from the basement membrane; similar mutations have been reported in humans with epidermolysis bullosa.[29–31]

It was a surprise to me that the mutation of the first immunoglobulin superfamily member of cell adhesion molecules, the widely expressed NCAM, had a rather mild phenotype; the mice were viable and fertile and showed only slight neuronal defects.[21,32] Mutation of L1, MAG, and Po had no effect or only mild neuronal defects. Apparently, the complex processes of neuronal migration, axon guidance and myelination require multiple adhesion receptors with overlapping functions.

Mutation of the selectin genes produced defects in inflammatory responses and leukocyte homing, but none had an effect on development.[21] Apparently, selectins do not play an essential role in embryonal vasculogenesis or angiogenesis, but selectively regulate interactions of leukocytes with the endothelial cell surface.

Junctional complexes

Single adhesion molecules on opposing cell surfaces may interact during the initial adhesion steps. Strong adhesion is, however, mediated by cell adhesion molecules that are assembled at high local concentrations in cell junctions. For instance, classical cadherins (E-, N-) are concentrated in adherens junctions or synapses, the cadherins, desmogleins and desmocollins are enriched in desmosomes, and certain integrins are assembled in hemidesmosomes, focal contacts or structures such as lamillipodia or filopodia. This results in cooperative interactions between the adhesion molecules and their ligands, and thereby strengthens adhesion by orders of magnitudes. Experimental evidence for oligomeric organization of selectins in the plasma membrane exists, that results in an increased ligand affinity of the lectin domain. NCAM also forms oligomers, and the aggregation rate of NCAM-containing vesicles increases strongly when the NCAM concentration is only slightly increased. Interaction with the cytoskeleton participates in such local assemblies of adhesion molecules, which again strengthen adhesion. Associations with the cytoskeleton are often mediated by additional, intermediary molecules, e.g. catenins, members of the esrin/radixin/moesin family, α-actinin or ankyrin. These linkage proteins can form prominent coats below the junction surface, for instance in desmosomes.[33] Cells may be able to remove the entire adhesion complexes from the cell surface by endocytosis, and place them back onto the surface as a whole.[34] Thus, adhesion complexes are frequently large organelle-like assemblies of both integral and peripheral membrane proteins, that associate extracellularly with the extracellular matrix or other cell adhesion molecules, and intracellularly with cytoskeletal components.

Three dimensional structures

Functionally important subdomains of adhesion molecules have recently been analysed by X-ray diffraction and NMR, which provide molecular models of the interactions. For instance, classical cadherins are presented on the cell surface as dimers, and one subdomain of a cadherin molecule interacts in an antiparallel fashion with the same subdomain in the cadherin molecule located on the adjacent cell,[35] (see chapter by Johnson *et al.*, this volume). The three-dimensional structures of immunglobulin-like domains of cell adhesion molecules (VCAM-1, NCAM and ICAM-2) have been elucidated. The folds of these domains are similar to those in the fibronectin type III repeat. Immunoglobulin-type adhesion molecules have longer β-strands in their immunoglobulin-like domains than the Ig domains of immunoglobulins, and these β-strands are directly implicated in cell adhesion. The protein surface responsible for homophilic interaction between two immunoglobulin-like domains has been defined in detail (see chapter by Thiery, this volume).

The structure of integrins has been solved by low resolution electron microscopy. The N-terminal sequences of the α and β subunits of LFA-1 form a globular region, which is followed by two stalks which cross the plasma membrane. X-ray analysis of the I-domain of the α-chain reveals a dinucleotide binding fold with a novel binding site for divalent cations (MIDAS). The seven repeats of the α-chain have been modelled as a propeller. Less is known about the structure of the β-chain, although similarities to the I-domain of the α-chain exist (see chapter by Dustin and Springer, this volume). The three-dimensional structures of the lectin and EGF domains of E-selectin have been reported. Lysine residues in the lectin domain appear to be necessary for the interaction with the ligand sLex: when lysine residues are introduced into the homologous MBP, sLex binding sites are generated. Co-crystallization established the residues of the lectin domain which interact with fucose of sLex and with Ca^{2+}, and thus produced an elegant structural explanation for the fucose and Ca^{2+} requirement of sLex binding (see chapter by Rosen, this volume).

Nuclear signalling

It has long been registered that cell adhesion molecules can signal to the nucleus. I always try to point out in my lectures that cell adhesion is as a good growth factor as EGF. In the last 5 years, the molecular basis for nuclear signalling has been identified. We now know that adhesion to the extracellular matrix via integrins activates signalling pathways which control passage through the cell cycle and apoptosis. A number of proto-oncogene products play a crucial role in these signalling events; these include members of the MAP kinase pathway, focal adhesion kinase, and numerous other signalling molecules (see chapter by Dans and Giancotti, this volume). Cytoskeletal assembly and disassembly is regulated by integrins and cadherins via members of the Rho family of small G-proteins (see chapter by Hall, this volume). Moreover, cadherin-associated proteins of the β-catenin/armadillo family are also components of the wnt/wingless signalling pathway; wnt/wingless molecules control differentiation and morphogenesis during development. How cell adhesion and wnt signalling pathways interact and influence each other is an active area of research (see chapter by Behrens, this volume). The wnt/wingless signalling pathway contains oncogene and tumour suppressor gene products, highlighting the connection between cell adhesion and oncogenesis (see also below). Progress has also been made in the investigation of signalling by immunoglobulin family members of cell adhesion molecules: NCAM and L1/NgCAM bind FGF-receptors, members of the src family of tyrosine kinases, esrin and ankyrin. Several transmembrane tyrosine kinases or phosphatases contain immunoglobulin-like domains on their extracellular domains (e.g. trk of *Drosophila*, the Eph receptor tyrosine kinases, the DLAR tyrosine phosphatases) and may transiently mediate cell adhesion (see chapter by Thiery, this volume). Selectins also provide signals, since they cause changes in cell shape, activation of integrins, tyrosine phosphorylation, and changes in Ca^{2+} fluxes. Finally, the recent finding that the receptor tyrosine kinases DDR-1 and DDR-2 bind selected types of collagens and are activated by these molecules may also influence our future thinking about the signals the extracellular matrix provides.[36,37]

Cancer and other diseases

Cell adhesion molecules play important roles in cancer and in blistering and inflammatory diseases. My laboratory became interested in cancer in the late 1980s, when we realized that a monoclonal antibody that interferes with E-cadherin-mediated cell adhesion, induced invasiveness of epithelial cells, and that invasive carcinoma cells have lost frequently E-cadherin expression.[38,39] At around the same time other laboratories found that changes in integrin function also affect the malignancy of cells.[40–42] Today we know that E-cadherin is mutated in about 50 per cent of diffuse-type gastric and lobular carcinomas,[43,44] and that various human carcinomas do not express E-cadherin, which often correlates with the invasive and metastatic potential of the tumours (see ref. 45 for a review). Interestingly, the *fat* gene of *Drosophila*, which encodes a protein with 34 cadherin units in the extracellular domain, was characterized as a tumour suppressor gene.[46] The cadherin-associated β-catenin has recently been found to induce transformation when mutated, and directly affects gene expression in colon carcinomas and melanomas by interacting with the transcription factors TCF4 or LEF-1.[47–50] Many other cell adhesion molecules have been implicated in cancer, including CEA (a member of the immunoglobulin-family), syndecan, EpCAM, CD44 and others.

Two types of severe blistering disease are due to production of autoimmune antibodies against desmosomal cadherins. In pemphigus vulgaris, antibodies against desmoglein-3 cause oral and skin blisters. In pemphigus foliaceus, antibodies against desmoglein-1 result in a epidermis that splits above the basal layer. The serum of some pemphigus patients contains antibodies against desmocollins (see chapter by Johnson *et al.*, this volume).

References

1. Wilson, H. V. (1909). *J. Exp. Zool.*, **5**, 245–58.
2. Holtfreter, J. (1939). *Arch. Exp. Zellforschung*, **23**, 169–209.
3. Beug, H., Katz, F. E., and Gerisch, G. (1973). *J. Cell Biol.*, **56**, 647–58.
4. Hynes, R.O. (1973). *Proc. Natl Acad. Sci. USA*, **70**, 3170–74.
5. Gahmberg, C. G. and Hakomori, S. I. (1973). *Proc. Natl Acad. Sci. USA*, **70**, 3329–33.
6. Ruoslahti, E., Vaheri, A., Kuusela, P., and Linder, E. (1973). *Biochim. Biophys. Acta*, **322**, 352–8.
7. Yamada, K. M. and Weston, J. A. (1974). *Proc. Natl Acad. Sci. USA*, **71**, 3492–6.
8. Tarone, G., Galetto, G., Prat, M., and Comoglio, P. M. (1982). *J. Cell Biol.*, **94**, 179–86.
9. Neff, N. T., Lowrey, C., Decker, C., Tovar, A., Damsky, C., Buck, C., and Horwitz, A. F. (1982). *J. Cell Biol.*, **95**, 654–66.
10. Pytela, R., Pierschbacher, M. D., and Ruoslahti, E. (1985). *Proc. Natl Acad. Sci. USA*, **82**, 5766–70.
11. Tamkun, J. W., DeSimone, D. W., Fonda, D., Patel, R. S., Buck, C., Horwitz, A. F., and Hynes, R. O. (1986). *Cell*, **46**, 271–82.
12. Brackenbury, R., Thiery, J. P., Rutishauser, U., and Edelman, G. M. (1977). *J. Biol. Chem.*, **252**, 6835–40.
13. Thiery, J. P., Brackenbury, R., Rutishauser, U., and Edelman, G. M. (1977). *J. Biol. Chem.*, **252**, 6841–5.
14. Hyafil, F., Babinet, C., and Jacob, F. (1981). *Cell*, **26**, 447–54.
15. Damsky, C. H., Richa, J., Solter, D., Knudsen, K., and Buck, C. A. (1983). *Cell*, **34**, 455–66.
16. Shirayoshi, Y., Okada, T. S., and Takeichi, M. (1983). *Cell*, **35**, 631–8.
17. Imhof, B. A., Vollmers, H. P., Goodman, S. L., and Birchmeier, W. (1983). *Cell*, **35**, 667–75.
18. Cunningham, B. A., Hemperly, J. J., Murray, B. A., Prediger, E. A., Brackenbury, R., and Edelman, G. M. (1987). *Science*, **236**, 799–806.
19. Nagafuchi, A., Shirayoshi, Y., Okazaki, K., Yasuda, K., and Takeichi, M. (1987). *Nature*, **329**, 341–3.
20. Ringwald, M., Schuh, R., Vestweber, D., Eistetter, H., Lottspeich, F., Engel, J., Dolz, R., Jahnig, F., Epplen, J., Mayer, S., et al (1987). *EMBO J.*, **6**, 3647–53.
21. Hynes, R. O. (1996). *Dev. Biol.*, **180**, 402–12.
22. Larue, L., Ohsugi, M., Hirchenhain, J., and Kemler, R. (1994). *Proc. Natl Acad. Sci. USA*, **91**, 8263–7.
23. Riethmacher, D., Brinkmann, V., and Birchmeier, C. (1995). *Proc. Natl Acad. Sci. USA*, **92**, 855–9.
24. Haegel, H., Larue, L., Ohsugi, M., Fedorov, L., Herrenknecht, K., and Kemler, R. (1995). *Development*, **121**, 3529–37.
25. Ruiz, P., Brinkmann, V., Ledermann, B., Behrend, M., Grund, C., Thalhammer, C., Vogel, F., Birchmeier, C., Gunthert, U., Franke, W. W., and Birchmeier, W. (1996). *J. Cell Biol.*, **135**, 215–25.
26. Torres, M., Stoykova, A., Huber, O., Chowdhury, K., Bonaldo, P., Mansouri, A., Butz, S., Kemler, R., and Gruss, P. (1997). *Proc. Natl Acad. Sci. USA*, **94**, 901–6.
27. Bierkamp, C., Mclaughlin, K. J., Schwarz, H., Huber, O., and Kemler, R. (1996). *Dev. Biol.*, **180**, 780–5.
28. Koch, P. J., Mahoney, M. G., Ishikawa, H., Pulkkinen, L., Uitto, J., Shultz, L., Murphy, G.F., Whitaker-Menezes, D., and Stanley, J.R. (1997). *J. Cell Biol.*, **137**, 1091–102.
29. van der Neut, R., Krimpenfort, P., Calafat, J., Niessen, C. M., and Sonnenberg, A. (1996). *Nature Genet.*, **13**, 366–9.
30. Georges-Labouesse, E., Messaddeq, N., Yehia, G., Cadalbert, L., Dierich, A., and Le Meur, M. (1996). *Nature Genet.*, **13**, 370–3.
31. Dowling, J., Yu, Q. C., and Fuchs, E. (1996). *J. Cell Biol.*, **134**, 559–72.
32. Cremer, H., Lange, R., Christoph, A., Plomann, M., Vopper, G., Roes, J., Brown, R., Baldwin, S., Kraemer, P., Scheff, S., Barthels, D., Rajewsky, K., and Wille, W. (1994). *Nature*, **367**, 455–9.
33. Schwarz, M. A., Owaribe, K., Kartenbeck, J., and Franke, W. W. (1990). *Ann. Rev. Cell Biol.*, **6**, 461–91.
34. Kartenbeck, J., Schmelz, M., Franke, W. W., and Geiger, B. (1991). *J. Cell Biol.*, **113**, 881–92.
35. Shapiro, L., Fannon, A. M., Kwong, P. D., Thompson, A., Lehmann, M. S., Grubel, G., Legrand, J. F., Als Nielsen, J., Colman, D. R., and Hendrickson, W. A. (1995). *Nature*, **374**, 327–37.
36. Shrivastava, A., Radziejewski, C., Campell, E., Kovac, L., McGlynn, M., Ryan, T. E., Davis, S., Goldfarb, M. P., Glass, D. J., Lemke, G., and Yancopoulos, G. D. (1997). *Mol. Cell*, **1**, 25–34.
37. Vogel, W., Gish, G. D., Alves, F., and Pawson, T. (1997). *Mol. Cell*, **1**, 13–23.
38. Behrens, J., Mareel, M. M., Van Roy, F. M., and Birchmeier, W. (1989). *J. Cell Biol.*, **108**, 2435–47.

39. Frixen, U. H., Behrens, J., Sachs, M., Eberle, G., Voss, B., Warda, A., Lochner, D., and Birchmeier, W. (1991). *J. Cell Biol.*, **113**, 173–85.
40. Giancotti, F. G. and Ruoslahti, E. (1990). *Cell*, **60**, 849–59.
41. Albelda, S. M., Mette, S. A., Elder, D. E., Stewart, R., Damjanovich, L., Herlyn, M., and Buck, C. A. (1990). *Cancer Res.* **50**, 6757–64.
42. Chan, B. M., Matsuura, N., Takada, Y., Zetter, B. R., and Hemler, M. E. (1991). *Science*, **251**, 1600–2.
43. Becker, K. F., Atkinson, M. J., Reich, U., Becker, I., Nekarda, H., Siewert, J. R., and Höfler, H. (1994). *Cancer Res*, **54**, 3845–52.
44. Berx, G., Cleton-Jansen, A.-M., Nollet, F., De Leeuw, W. J. F., van de Vijver, M. J., Cornelisse, C., and Van Roy, F. (1995). *EMBO J.*, **14**, 6107–15.
45. Birchmeier, W. and Behrens, J. (1994). *Biochim. Biophys. Acta*, **1198**, 11–26.
46. Mahoney, P. A., Weber, U., Onofrechuk, P., Biessmann, H., Bryant, P. J., and Goodman, C. S. (1991). *Cell*, **67**, 853–68.
47. Behrens, J., von Kries, J. P., Kühl, M., Bruhn, L., Wedlich, D., Grosschedl, R., and Birchmeier, W. (1996). *Nature*, *382*, 638–42.
48. Korinek, V., Barker, N., Morin, P.J., van Wichen, D., de Weger, R., Kinzler, K. W., Vogelstein, B., and Clevers, H. (1997). *Science*, **275**, 1784–7.
49. Morin, P. J., Sparks, A. B., Korinek, V., Barker, N., Clevers, H., Vogelstein, B., and Kinzler, K. W. (1997). *Science*, **275**, 1787–1789.
50. Rubinfeld, B., Robbins, P., El Gamil, M., Albert, I., Porfiri, E., and Polakis, P. (1997). *Science*, **275**, 1790–92.

Walter Birchmeier
Max Delbrück Center for Molecular Medicine,
13125 Berlin, Germany

The Ig superfamily of adhesion molecules

The immunoglobulin (Ig) superfamily comprises a large variety of glycoproteins mostly expressed at the cell surface. Molecules carrying Ig-like domains are endowed with different functions including cytoskeletal organization, endocytosis, adhesion, migration, growth control, immune recognition, viral receptors, inflammatory reactions, and tumour progression. Numerous studies have been devoted to this increasingly complex superfamily. The reader may get confused by the somewhat disparate nomenclature; some of the members have been classified as CD.

Structure

Williams and colleagues have largely contributed to defining this superfamily, particularly with respect to the structural diversity of the constant and variable Ig domains[1] (see also comprehensive reviews)[2,3] In brief, the basic Ig domain is formed by a tightly packed barrel of seven or nine antiparallel β strands arranged in two layers. Ig-like domains are usually recognized by the presence of two cysteine residues separated by 55–75 amino acids and a tryptophan residue located 10 to 15 amino acids carboxy to the first cysteine. Sequence identity between different members of the superfamily may not exceed 25–30 per cent.

As in immunoglobulins, these seven or nine domains have been classified as C1 or C2 (seven strands) or V (nine strands) despite the lack of diversity in V domains throughout the Ig superfamily. More recently, the I set was described; Ig domains belonging to this fourth set were originally found in telokin. The I set is characterized by a shortened V-like domain and can be found in the first domain of VCAM-1, NCAM, and in several Ig domains of titin.[4]

The three-dimensional structure has been established for several Ig domains including class II MHC, CD2, CD4, VCAM-1, NCAM, and ICAM-2.[5] A similar folding is evident in fibronectin III repeats although proteins containing such domains do not belong to this superfamily. The approximate size of one domain is 25 × 30 × 40 Å. The V-Ig domains of members involved in adhesion are characterized by longer β strands with shorter connecting loops in contrast to the antigen binding sites in Ig. These β strands are directly implicated in cell adhesion.[2,6] The area of interaction between two Ig domains, in homophilic interaction between cells, may be of the order of 800 Å. A similar observation has been made for the Ig-like amino terminal domain of E and N-cadherin involved in homotypic interactions.[7]

Classification

The regrouping of Ig superfamily members into distinct families and subfamilies is based on either different structural criteria or common functions. Proteins implicated in antigen recognition include TCR, CD3, MHC class I and II. Recognition, endocytosis, or transcytosis of antigen–antibody complexes are controlled by Ig and poly-Ig receptors. Numerous growth factor and cytokine receptors including PDGF, FGF, c-kit, CSF-1, IL-1 and IL-6 receptors, characterized by several Ig-like domains in their extracellular portion, can be classified as one family. Cytoplasmic proteins associated with muscle cytoskeleton such as titin and myosin light chain kinase constitute another family while some extracellular matrix proteoglycans containing

Ig-like domains may be regrouped as a separate family. Members clearly involved in adhesion have also been classified as subfamilies based on relatedness of their sequence and domain organization, but in some cases, function is also taken into consideration. Originally, these proteins were recognized as members of the Ig superfamily based on similarities with the prototype member, NCAM. NCAM related members involved with myelin assembly and maintenance include thy1, P_0, and MAG. One subfamily regroups L1/NgCAM with NrCAM/bravo, neurofascin, and CHL1 (a mouse homologue of L1 but most related to NgCAM in chicken); another subfamily includes contactin/F11, TAG-1/axonin-1, BIG-1 and BIG-2. These adhesion molecules are mostly implicated in the development of neuronal networks particularly in the control of axonal guidance and fasciculation. Although the chicken SC1/DM-Grasp/BEN is expressed in subsets of neurones during development, it is now proposed (see pp. 328–30) to be incorporated in the V-V-C2-C2-C2 subfamily which includes its homologues: ALCAM in human and neurolin in goldfish. This subfamily also comprises human MUC18, chicken HEMCAM (gycerin), and human Lu blood group (BCAM). Similarly, telencephalin expressed on soma and dendrites of neurones has now been regrouped with the ICAM subfamily based on their strong sequence identities. CCAM is a four Ig-like cell adhesion molecule expressed in a variety of cell types including epithelial cells, granulocytes, and platelets, this molecule belongs to a family including the tumour suppressor CEA and other related molecules (see pp. 150–54). Some adhesion molecules, such as PECAM-1, found in large amounts on endothelial cells and expressed to a lesser extent on leukocytes and platelets, have not been regrouped into subfamilies (see pp. 283–5). Other molecules such as sialoadhesin, a macrophage receptor, characterized by its 17 Ig domains, may represent a prototype member of sialic acid-binding proteins which serve as adhesion receptors. This family may include CD22 in B cells and CD33 in myelomonocytic cells.[8]

Evolution

Evolutionarily, the first priority for metazoans was to ensure cell–cell interactions and selective adhesion. Certainly, different molecular structures have been selected for this purpose, for instance the most primitive animals such as sponges have used large proteoglycans. The discovery, in insects, of cell adhesion molecules homologous to vertebrate Ig superfamily members led to the hypothesis that this type of folding emerged early during evolution prior to the emergence of immunoglobulins known to be restricted to the vertebrate. NCAM is highly conserved in vertebrates and homologues have been found in invertebrates. ApCAM, isolated in the marine mollusc *Aplysia californica*, is closely related to rat NCAM and to *Drosophila* fasciclin II.9 In the grasshopper, *Schistcerca americana*, and *Drosophila melanogaster*, fasciclin II is most homologous to NCAM. *Drosophila* and moth (*Manduca sexta*) neuroglians are closely related to L1/ NgCAM while an L1-like molecule has also been identified in *C. elegans*. However, *Drosophila* amalgam and fasciclin III, which are also characterized by three Ig-like domains, do not appear so far to have homologues in vertebrates,[10] none the less lachesin, discovered recently in grasshopper and Drosophila is closely related to amalgam, and is possibly homologue of the bovine opioid binding cell adhesion molecule (OBCAM) and the human poliovirus receptor.[11] These findings are strengthened by the fact that even unicellular organisms express molecules containing Ig-like domains; for example, the conjugation of the two mating types in yeast is controlled by the glycoprotein α agglutinin which contains three Ig-like domains. It was hypothesized that the Ig domain was originally encoded by two exons, a phenomenon still found in many adhesion molecules belonging to the Ig superfamily molecules and in the first Ig domain of CD4, while molecules implicated in self–non-self discrimination and in antigen recognition are made up of Ig domains encoded by one exon. Progressive diversification of cell adhesion molecules, particularly in the nervous system, may have evolved by gene duplication of adhesion molecule precursors. In addition, such precursors may have given rise to MHC molecules with a high degree of polymorphism and diversity augmented by recombination and somatic mutations in the V domains.[12] Early attempts to use Ig domains for immunity may be found in insects. Haemolin is a four Ig-like domain protein which is involved in immune response to bacterial infections in moths permitting the binding of haemocytes to bacteria. Interestingly this protein is closely related to neuroglian in *Drosophila*, sharing common intronic sequences with NCAM but is characterized by Nf-B-like consensus sequences in its promoter.[13] With the exception of ICAM-1, -3 and -4 and the telencephalin cluster, the genes encoding Ig superfamily adhesion molecules have widespread localization on chromosomes (see Table 1). A new NCAM like protein has been identified on chromosome 21.[14]

Functions

Members of the diverse subfamilies described above have been shown to serve as adhesion receptors during development, in the adult, and may also be activated or repressed in disease states. Some of the members, such as neurofascin or axonin-1/TAG-1 may have restricted spatio-temporal distributions in the nervous system thereby implicating a role in guidance and proper targeting of axons. NCAM, which appears in many tissues very early during development, is considered to be a primary CAM expressed in a dynamic pattern during primary and secondary inductive events. In cooperation with other members such as L1/NgCAM in *cis* interaction, NCAM may serve more specialized adhesion functions. Although it was originally proposed that these adhesion molecules were primarily acting through homophilic interactions, it now appears that heterophilic interactions may prevail (Fig. 1). As in other adhesion molecule superfamilies, the pattern of recognition is degenerate, with each member

Table 1

Human CAM-Ig	Chromosomal mapping
NCAM	11q23
NCAM-2	21q21
MAG	19q12–13
L1	Xq28
NrCAM	7q31
Neurofacscin	1q31–32
F11/F3 Contactin	12q11–12
Axonin-1/TAG-1	1q32.1
CD2	1p13
LFA-3	1p13
ICAM-1	19p13.2
ICAM-2	17q23–25
ICAM-3	19p13.2
ICAM-4	19p13.2
Telencephalin	19p13.2
PECAM1	17q
MUC18	11
Lu blood group	19q13.2–13.3
BCAM	
ALCAM	3q13.1–2

recognizing multiple ligands. The ligands may belong to the Ig superfamily, to the integrin family, or to a variety of glycoconjugates. The homophilic interaction may involve one or several Ig-like domains contacting each other in an antiparallel manner. Heterotypic binding may preferentially involve the amino-terminal domain. In the case of ICAM-1, the binding sites for the different ligands have been mapped very precisely; the integrin LFA-1, *Plasmodium falciparum*, and rhinoviruses all bind to overlapping but not identical sites (see p. 216). The affinity and rate constants were determined for several members using a BIAcore (Pharmacia Biotech AB, Uppsala, Sweden). These interactions have been studied extensively for CD2 and CD48.[15] The affinity constant is low, of the order of 10^4 M^{-1}, while the association rate constant is similar to those found for antibodies. However, the dissociation rate constant is very high, over 5 s^{-1}, clearly showing that these adhesion molecules mediate transient binding which facilitates rapid detachment. Weak interaction, as determined by measurement of the interaction of individual molecules, may be underestimated since these molecules may dimerize or oligomerize. NCAM was shown to form trimers and, *in vitro*, the aggregation rate was increased 30-fold when the concentration of NCAM was doubled in lipid vesicles. Recently, CD4 was shown by X-ray crystallography to dimerize through the amino-terminal domain.[16] A similar situation was found recently for N-cadherin, allowing the formation of a zipper-like stucture between interacting dimers.[17] The Ig superfamily adhesion molecules may not change their affinity as markedly as integrins, which can interconvert between a low and high affinity state through conformational changes. However, the adhesive strength and stability of the adhesive process may be controlled by the prevalence of these molecules on the surface, the type and the extent of glycosylation as well as by cis interactions. Molecules such as ICAM-1 can be induced very rapidly during inflammatory processes by a variety of cytokines. Many adhesion molecules expressed in a highly regulated manner during development may be controlled by specific transcriptional regulators including homeobox gene products.[18]

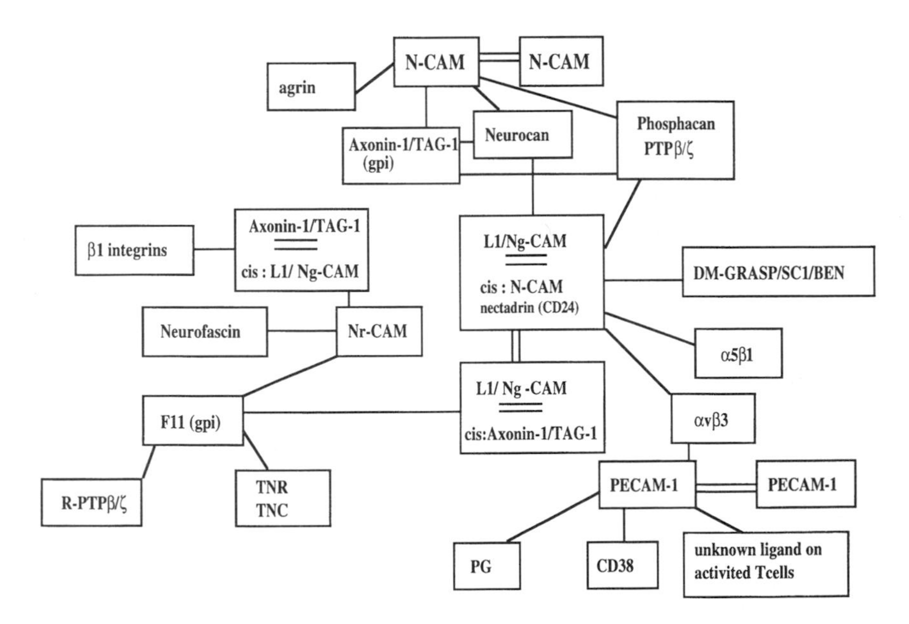

Figure 1. The diagram exemplifies the complex network of homophilic (double line) and heterophilic (single line) interactions between members of different subfamilies. *Cis* interaction can also contribute to the modulation of adhesion (double line in rectangle).

The adhesive function has been investigated in different types of *in vitro* assays using either purified ligands or expressed on cell surfaces. *In vivo*, pioneer experiments were mostly based on pertubations induced by blocking antibodies. More recently, the function of these molecules has been studied through inactivation of their respective genes. As found for many other genes, inactivation of individual CAM genes does not necessarily lead to lethality during development nor to major defects. Inactivation of the NCAM gene leads to an almost complete disappearance of the olfactory bulb while other defects have been detected in several central nervous system loci, for example, the hippocampus. Mutations in the L1 gene are responsible for a variety of neurological disorders including X-linked hydrocephalus, MASA syndrome, and spastic paraplegia type 1, which have been regrouped into the acronym CRASH. However, to date, there is no clear relationship between the type of mutation (point mutations, deletions, duplications) and the severity of the disease[10]). Inactivation of the VCAM-1 gene in embryos is lethal due to severe malformation of heart and placenta where it is abundantly expressed. A similar phenotype is obtained with the inactivation of the VCAM-1 ligand, integrin $\alpha 4\beta 1$. Inactivation of *Drosophila* homologues has also provided interesting phenotypes clearly showing the importance of these molecules in neurogenesis.

Our knowledge of the mechanisms transducing the adhesive signal has not progressed as rapidly as for adhesion molecules belonging to other superfamilies. Ankyrin is almost the only protein found to bind to the cytoplasmic domain of several members including L1/NgCAM, neurofascin, and neuroglian. It should be stressed that several members are only glycosyl phostatidylinositol (GPI)-linked to the plasma membrane. Recently it was proposed that signalling of NCAM and L1/NgCAM could also occur via *cis* interactions with the FGF receptors through a CAM homology domain on the FGF receptor.[19] Several cell adhesion molecules of the Ig superfamily may have in-built transduction machinery, as they contain either a tyrosine kinase or a tyrosine phosphatase in their cytoplasmic domain. The Dtrk receptor was one of the first molecules to be described as an adhesive protein in *Drosophila* neurogenesis.[20] Tyrosine phosphatase β binds specifically to contactin/F11 allowing the adhesive signal to be transduced directly by the ligand.[21] Several receptor tyrosine phosphatases containing three-Ig-like domains were shown to control motor axon guidance in *Drosophila*.[22,23] Perhaps the Ig-like domain localized in the extracellular domain of Eph family tyrosine kinase receptors may also be involved in the transient adhesive process preceding repulsion of axons or in segregation of two cell populations expressing either the receptor or the cognate ligand.[24] Although nothing is known about signalling mediated by semaphorins, a new class of receptors containing one Ig-like domain in the extracellular portion, it may also be considered that this Ig domain may permit transient contact, either in *trans* or in *cis*, before inducing collapse of axons and/or preventing cells to penetrate into a semaphorin rich territory.[25] Clearly, much remains to be done to decipher the mechanochemical transduction machineries controlling the adhesive properties of the CAM–Ig superfamily members. Integration of all the data provided by the three-D structural and functional studies of such a diverse group of molecules remains a challenging task. As with other superfamily members, one of the striking features is the multiplicity of ligands and the increasing role of *cis* interactions. Even more formidable will be the integration of data derived from the many studies carried out on the different adhesion systems discussed in this book to ensure a better understanding of the basic principles of morphoregulation.

References

1. Williams, A. F. and Barclay, A. N. (1988). *Ann. Rev. Immunol.* **6**, 381–405.
2. Vaughn, D. E. and Bjorkman, P. J. (1996). Neuron, 16, 261–73.
3. Chotia, C. and Jones, Y. E. (1997). *Ann. Rev. Biochem.*, **66**, 823–62.
4. Bateman, A., Jouet, M., MacFarlane, J., Du, J.-S., Kenwrick, S., and Chothia, C. (1996). *EMBO J.*, **15**, 6050–9.
5. Casasnovas, J. M., Springer, T. A., Liu, J-H., Harrison, S. C., and Wang, J.-H. (1997). *Nature*, **387**, 312–5.
6. Holness, C. L. and Simmons, D. L. (1994). *J. Cell Sci.*, **107**, 2065–70.
7. Overduin, M., Harvey, T. S., Bagby, S., Tong, K. I., Yau, P., Takeichi, M., and Ikura, M. (1995). *Science*, **267**, 386–9.
8. Crocker, P. R., Mucklow, S., Bouckson, V., McWilliam, A., Willis, A. C., Gordon, S., et al. *EMBO J.*, **13**, 4490–503.
9. Mayford, M., Barzilai, A., Keller, F., Schacher, S., and Kandel, E. R. (1992). *Science*, **256**, 638–44.
10. Hortsch, M. (1996). *Neuron*, **17**, 587–93.
11. Karlstrom, R. O., Wilder, L. P., and Bastiani, M. J. (1993). Development, 118, 509–22.
12. Edelman, G. M. (1987). Immun. Rev., 100, 11–45.
13. Lindström, I., Sun, S.-C., and Faye, I. (1995). *Eur. J. Biochem.*, **230**, 920–5.
14. Paoloni-Giacobino, A., Chen, H., and Antonarakis, S. E. (1997). *Genomics*, **43**, 43–51.
15. Van der Merwe, P. A. and Barclay, A. N. (1994) TIBS, 19, 354–9.
16. Wu, H., Kwong, P. D., and Hendrickson, W. A. (1997). *Nature*, **387**, 527–30.
17. Shapiro, L., Fannon, A. M., Kwong, P. D., Thompson, A., Lehmann, M. S., Grübel, G., et al. (1995). *Nature*, **374**, 327–37.
18. Cunningham, B. A. (1995). *Curr. Opin. Cell Biol.*, **7**, 628–33.
19. Hall, H., Walsh, F. S., and Doherty, P. (1996). *Cell Adhes. Commun.*, **3**, 441–50.
20. Pulido, D., Campuzano, S., Koda, T., Modolell, J., and Barbacid, M. (1992). *EMBO J.*, **11**, 391–404.
21. Peles, E., Nativ, M., Campbell, P. L., Sakurai, T., Martinez, R., Lev, S., et al. (1995). *Cell*, **82**, 251–260.
22. Desai, C. J., Gindhart, J. G., Goldstein, L. S. B., and Zinn, K. (1996). *Cell*, **84**, 599–609.
23. Krueger, N. X., Van Vactor, D., Wan, H. I., Gelbart, W. M., Goodman, C. S., and Saito, H. (1996). *Cell*, **84**, 611–622.
24. Nieto, M. A. (1996). Neuron, 17, 1039–48.
25. Kolodkin, A. L. (1996). *Trends Cell Biol.*, **6**, 15–21.

Jean Paul Thiery
CNRS UMR 144 and Institut Curie,
26 rue d'Ulm, 75248 Paris cedex 05,
France

Cadherin/catenin-mediated signal transduction

Cadherins and their cytoplasmic interaction partners, the catenins, build up cell adhesion complexes which regulate various developmental and pathological processes. Besides their function in cell adhesion, these components are also involved in signal transduction pathways. Current research analyses how the interplay between cell adhesion and signalling functions may regulate differentiation processes. This review will summarize the structural and functional aspects which point to a role of cadherins/catenins in signal transduction and discuss possible molecular mechanisms.

Structural aspects

Cadherins are transmembrane glycoproteins which provide strong intercellular adhesion in a Ca^{2+}-dependent manner. 'Classic' cadherins (see article by Johnson, Wheelock and Takeichi, this volume, for details and references) are composed of a large extracellular domain, which mediates homophilic cell adhesion, a transmembrane domain and a highly conserved cytoplasmic domain, which interacts with the actin cytoskeleton. Other members of the cadherin family are the desmosomal cadherins (desmogleins and desmocollins), which are associated with the keratin filament network and differ mainly in their cytoplasmic domain from classic cadherins.

The extracellular domain of classic cadherins consists of five repeated units of about 110 amino acids which are proposed to form lateral dimers on the cell surface.[1,2] By head-to-head interactions of dimers from opposing cells, a zipper-like structure is proposed to form which may co-operatively enhance cell–cell adhesion (Johnson, Wheelock and Takeichi, this volume). *In vitro* studies show that the lateral dimerization of cadherin molecules is indeed required for homophilic binding.[3] Analyses of deletion mutants have also revealed a crucial role for the cytoplasmic tail of cadherins.[4,5] Truncated cadherins lacking this domain produce only weak cell adhesion after ectopic expression in recipient cells. Moreover, these mutants no longer interact with actin filaments, indicating that association with the cytoskeleton is essential for cadherin function. The cytoplasmic domain contains two highly conserved regions which are separated by an intervening part that is more specific for various cadherin subtypes.[6] The C-terminal conserved region is the binding site for the catenins, a group of cytoplasmic proteins which establish a link to the actin cytoskeleton. β-Catenin and its relative, plakoglobin (also termed γ-catenin), directly interact with the cytoplasmic tail of cadherins but form mutually exclusive complexes.[7] They belong to the armadillo family of proteins, members of which are characterized by a central domain of 12 repeats of about 40 amino acids (so-called arm-repeats[8]). This domain was originally identified in armadillo, which is the *Drosphila* homologue of β-catenin. The structure of the central domain of β-catenin has been revealed by X-ray crystallography, which showed that the arm repeats form a superhelix.[9] The binding site for the various interaction partners of β-catenin (see below) forms a positively charged groove. Via their N-terminal domains, β-catenin and plakoglobin associate with the vinculin-related protein α-catenin which, in turn, makes contact to actin filaments.[7,10,11] The role of the conserved region at the N-terminus of the cytoplasmic domain of cadherins is not known but it might bind to the armadillo family protein $p120^{CAS}$.[12]

Cadherins and catenins form highly organized cell adhesion complexes which establish a direct connection between adjacent cells and their cytoskeletal networks. This organization is probably a prerequisite for the rigidity of cell interactions, but it may also provide a basis for intercellular signal transfer. From the signalling point of view, cadherins can be regarded as forming specialized ligand–receptor pairs which transmit information between opposing cells. By this process, neighbouring cells could co-ordinate cell junction formation and cytoskeletal architecture. It is not known how such signals are further propagated inside the cell, but it is likely that the cytoplasmic catenins are involved. β-Catenin can associate with putative regulators of the cytoskeleton: it binds to α-catenin in complexes with cadherins, and to fascin[13] and the tumour suppressor gene product APC[14,15] in cadherin-independent complexes. α-Catenin acts as a linker of cadherins to actin or actin-associated proteins;[16] fascin is involved in bundling of F-actin,[13] and APC establishes a possible link to the microtubular network,[17] but also has a role in the degradation of β-catenin (see below). Through these multiple interactions β-catenin may regulate the interplay between the cytoskeleton and cadherins.

In analogy to what is known from integrin signal transduction (see chapter by Dans and Giancotti, this volume), it is possible that cadherins directly activate other cytoplasmic signalling molecules. As a consequence of the formation of the extracellular adhesion zipper, cadherin molecules are clustered on the cell surface; this might induce co-clustering and activation of associated cytoplasmic components. Tyrosine kinases and phosphatases have been shown to interact with the cadherin–catenin complex[18] (see also below) and might be regulated by such a mechanism. Interestingly, the growth factor activated adaptor protein Shc binds to N-cadherin and might thereby feed signals into mitogenic pathways.[19]

Functional aspects

Cadherins show characteristic tissue-specific expression patterns and have been proposed to act as morphogenetic regulators during development. This chapter will summarize the *in vitro* and *in vivo* evidence which demonstrates that cadherins affect cell growth, differentiation and transformation. It will also discuss the signalling events that regulate the function of cadherins. The various cellular responses elicited by cadherins indicate that these proteins not only function as cell adhesion receptors but also have distinct roles as signalling molecules.

In vitro studies

Gene transfection approaches as well as experiments with interfering antibodies have shown that cadherins affect cellular behaviour in multiple ways. Forced expression of cadherins in mammalian cells not only increases cell–cell adhesion but also alters differentiation, motility and invasiveness of the cells.[20,21] On the other hand disturbance of cadherin function by specific antibodies leads to loss of cell–cell contacts and to increased migration of the cells and invasiveness. Furthermore, mixtures of cells expressing different cadherins efficiently sort out *in vitro*, a process which resembles sorting out of tissues *in vivo*.[22] Differentiation of cells is also dependent on cadherins: cadherins promote skeletal muscle differentiation in three-dimensional cultures[23] and direct tissue differentiation of ES cells grown as teratocarcinomas.[24] Moreover, cadherins induce specific molecular responses, such as changes in the organization of the cytoskeleton and in the expression of proteases or other cell adhesion receptors.[25] It will be important to dissect whether cadherins actively trigger signalling pathways or whether the described cellular responses are secondary consequences of alterations in adhesion and shape induced by cadherins.

In vivo studies

Direct evidence for the *in vivo* function of cadherins and catenins has been obtained through the analysis of 'knock out' animals and through overexpression studies in chimeric animals. Targeted mutation of the epithelial E-cadherin gene leads to early embryonic lethality. Homozygous mutants do not develop past the compacted blastocyst at which stage the adhesive cells of the morula dissociate.[26,27] A similar phenotype is observed after mutation of the α-catenin gene, indicating that the connection to the actin cytoskeleton normally provided by α-catenin is essential for cadherin function *in vivo*.[28] The knock-out of the neural N-cadherin gene leads to disturbances of neural structures such as neural tube and somites and failures in heart development.[29] P (placental)-cadherin mutant mice are viable but show precocious differentiation of the mammary gland, as well as hyperplasia and dysplasia of the mammary epithelium with age.[30] P-cadherin is normally expressed in myoepithelium surrounding the mammary epithelium. It is apparently needed for proper growth control signalling between the two tissues. Mutation of β-catenin affects mesoderm development at gastrulation, a phenotype which might again be due to disturbance of cell adhesion but could also result from the loss of the signalling function of β-catenin in the wnt pathway[31] (see also below). Mutation of the plakoglobin (γ-catenin) gene leads to cell adhesion defects in heart muscle development.[32,33] Taken together, the evidence from gene ablation experiments clearly supports the essential role of these components in cell adhesion, but also hints at their possible functions in signal transduction.

Expression of full-size cadherins and deletion mutants in chimeric mice has revealed further physiological functions *in vivo*. Forced expression of E-cadherin in the intestinal epithelium resulted in decreased cell migration and proliferation, and increased apoptosis.[34] Expression of a dominant-negative cadherin mutant led to the disruption of cell–cell contacts and to accelerated cell migration. In addition, neoplastic lesions formed in the villus epithelium, and the crypt cells showed increased proliferation and apoptosis.[35,36] Thus, alterations in cadherin-mediated cell adhesion have profound effects on cell migration, proliferation and apoptosis in the intestinal epithelium. Again, the question arises whether these effects are due to specific signalling functions of cadherins and their associated proteins or a mere consequence of alterations in cell adhesion. Interestingly, in chimeras expressing dominant-negative cadherins, β-catenin was displaced from the cell membrane and found in the cytoplasm.[34,36] It is possible that the free pool of β-catenin is able to activate the wnt signalling pathway by transmitting signals to the nucleus, e.g. by interacting with LEF/TCF transcription factors (see below). Transgenic expression of E-cadherin in a mouse model system for pancreas carcinoma formation prevented the progression of well-differentiated adenomas to invasive carcinomas, whereas expression of a dominant-negative form of E-cadherin promoted early invasion and metastasis.[37] Taken together with the earlier results indicating a role for E-cadherin in the prevention of metastases[38] and more recent findings of mutations of the E-cadherin gene in tumours (reviewed in ref. 39), the data suggest that cadherins generate signals which are able to counteract tumour progression.

Regulation of cadherin-mediated cell adhesion

Cell adhesion mediated by cadherins is not a static process but is highly dynamic. Cell adhesion can be modulated according to the specific requirements of the cells, for instance during cell migration and cytokinesis. Thus, cadherins are also subject to regulation by intracellular signals, a process commonly referred to as 'inside-out' signaling. Prime candidates to control cadherin function are the catenins. Cells lacking functional catenins, e.g. cancer cells with mutations in catenin genes, do not adhere properly. Adhesion can be restored by transfecting wildtype catenin cDNA.[40] There is also evidence that β-catenin, despite its essential function as a linker of α-catenin to the cadherins, can negatively regulate cadherin function.[41] Members of the Rho family of small GTPses (Rho, Rac, Cdc42), which are known regulators of the actin cytoskeleton, also affect cadherins. Blocking of endogenous Rho or Rac led to the disappearence of cadherins from adherens junctions. On the other hand, activation of these components increased the amount of detergent-insoluble, i.e. cytoskeleton-associated, E-cadherin and β-catenin.[42,43] Moreover, the guanine nucleotide exchange factor Tiam1, which specifically activates Rac, led to increased cadherin-dependent cell adhesion.[44] Taken together the data indicate that Rho family members have a critically role in the assembly of the cadherin/actin connection.

Several investigations indicate an interplay of cadherin/catenins with tyrosine kinase pathways (see. ref. 18 for a comprehensive review). Tyrosine kinases have been shown to affect drastically cell adhesion mediated by cadherins. Activation of cytoplasmic (pp60^{src}) or receptor tyrosine kinases (EGF receptor, c-Met, c-ErbB2 and others) leads to profound disturbances of epithelial cell–cell adhesion. Concomitantly, strong and reversible tyrosine phosphorylation of β-catenin is observed, and this might affect the activity of cadherins. A number of phosphatases and kinases have been shown to interact with cadherins/catenins, again pointing to a regulatory relationship between tyrosine phosphorylation and cadherin-mediated cell adhesion.[18]

Cadherins are also regulated by other cell adhesion components. For instance, expression of N-cadherin in epithelial cells leads to the down-regulation of epithelial cadherins such as E- and P-cadherin as well as the acquisition of a dedifferentiated phenotype of the transfectants.[45] Expression of EpCAM (epithelial cell adhesion molecule), which is structurally unrelated to cadherins, abolishes cadherin-mediated cell adhesion and association with the cytoskeleton.[46] Finally, cadherin function is also affected by cell–substrate adhesion molecules of the integrin family.[47] This cross-talk of cell adhesion molecules might provide a basis for the co-ordination of cell–cell as well as cell–substrate interactions.

Signalling by β-catenin/armadillo in the wnt/wingless pathway

β-Catenin not only is involved in cell adhesion but also participates in the wnt/wingless signalling pathway. Originally, the connection to wingless was detected by genetic approaches in *Drosophila*. Mutation of armadillo, the *Drosophila* homologue of β-catenin, results in a segment polarity phenotype which is also observed after mutation of wingless and other activating components of the pathway. Genetic epistasis analysis proved that armadillo is in fact a direct target of wingless signal transduction (reviewed in ref. 48). In *Xenopus* embryos, signalling through β-catenin is crucial for the determination of the body axis. Overexpression of β-catenin in ventral cells of the embryo leads to dorsalization of the derived structures and ultimately results in the formation of two-headed embryos. The same effects can be seen when activating members of the wnt pathway are expressed, and the *Xenopus* system (as well as *Drosophila* genetics) has been instrumental in identifying the hierarchy of components of the signalling cascade (reviewed in ref. 49).

The combination of genetic, developmental and biochemical studies has shown that the action of wingless in *Drosophila* and of wnts in vertebrates follows a strikingly similar pathway (summarized in refs 49 and 50; see Fig. 1). The wnt/wingless proteins are secreted factors which bind to transmembrane receptors of the frizzled family. This results in activation of the cytoplasmic phosphoprotein dishevelled by an as yet unknown mechanism. Dishevelled inhibits the activity of the serine/threonine kinase GSK3β (glycogen synthase kinase 3β; zeste-white-3/shaggy in *Drosophila*) which in its active state phosphorylates β-catenin at specific residue in its N-terminal domain. Phosphorylation of β-catenin appears to be a trigger for its ubiquitination and subsequent degradation by proteasomes.[51] Thus, when the wnt signal is present, GSK3β becomes inactivated and cytoplasmic β-catenin is stabilized. As a result, β-catenin is able to bind to transcription factors of the LEF/TCF family (pangolin/dTCF in *Drosophila*) and enters the cell nucleus.[52–55]

The association of β-catenin with LEF/TCF appears to be the essential step by which the wnt/wingless signalling pathway is connected to the regulation of gene expression. Lymphoid enhancer factor-1 (LEF-1) and T-cell factor (TCF)-1, -3, -4 were originally identified in B- and T-cells (reviewed in ref. 56). They are closely related to each other and belong to the HMG box family of transcription factors. These factors bind to target promoter sites via their HMG box at the minor groove of the DNA and cause a sharp bend in the DNA. Because of this activity, LEF-1 has been proposed to act as an architectural transcription factor which regulates gene expression by modulating higher-order nucleoprotein structures.[56,57] The binding of β-catenin to LEF-1 alters DNA bending, and this might result in changes in gene transcription.[52] When bound to LEF/TCF factors, β-catenin can also directly stimulate transcription of reporter genes through a transactivation domain present in its C-terminal part.[58] Apparently, LEF/TCF and β-catenin form bipartite transcription factors in which DNA binding is performed by the HMG box and transcriptional activation by the C-terminal domain of β-catenin. A number of investigations have shown that the LEF/TCF-β-catenin interaction is of biological relevance. LEF-1 caused axis duplication in *Xenopus*,[52] and a dominant-negative mutant of TCF-1 blocked the formation of the endogenous body axis.[53] Furthermore, in *Drosophila* the homologous HMG box factor dTCF/pangolin is genetically downstream of wingless and armadillo.[55,58,59] The complexes between LEF/TCF and β-catenin change expression of downstream genes of the wnt/wingless pathway, such as *ultrabithorax* in *Drosophila* and *siamois* in *Xenopus*.[59,60] Wnt signalling has also been analysed in *C. elegans* and shown to involve homologues of β-catenin (named WRM-1), APC (named APR-1), and TCF-LEF (named POP-1). In this system, WRM-1

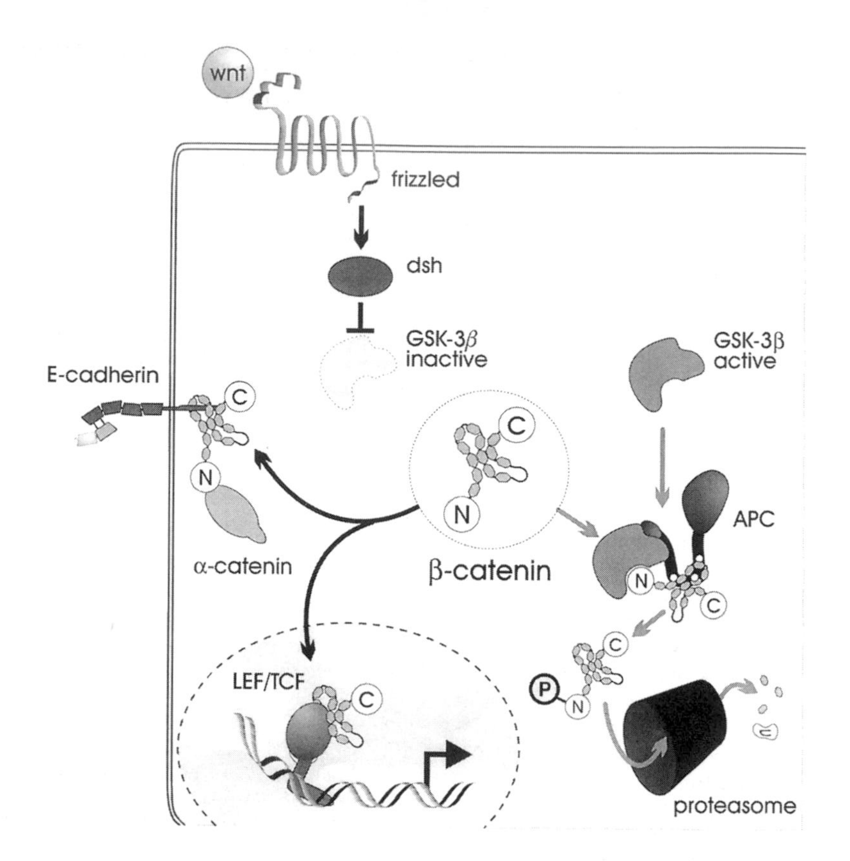

Figure 1. Model for the roles of β-catenin in cell adhesion and wnt signaling. β-Catenin (shown with the central arm repeat region flanked by N- and C-terminal domains) forms cell adhesion complexes with cadherins and α-catenin at the cell membrane. In the absence of the wnt signal, cytoplasmic levels of β-catenin are negatively regulated through interaction of β-catenin with the tumour-suppressor gene product APC and the serine/threonine kinase GSK3β. Phosphorylation of the N-terminal domain of β-catenin by GSK3β promotes ubiquitination of β-catenin and its degradation by proteasomes. Wnt signalling is initiated by binding of wnt factors to members of the frizzled family of transmembrane receptors. This leads to activation of dishevelled (dsh) which blocks GSK3β activity. As a result cytoplasmic β-catenin is stabilized and associates with LEF/TCF transcription factors. β-Catenin can alter LEF/TCF-induced DNA bending and activate transcription of wnt target genes through its C-terminal transactivation domain.

appears to co-operate with APR-1 to block rather than activate POP-1 function.[61]

The data indicate that control of cytoplasmic levels of β-catenin is essential for the regulation of the wnt pathway. A role in the degradation of β-catenin has been found for the tumour suppressor gene product APC (adenomatous polyposis coli). APC is a large cytoplasmic protein which binds to β-catenin via a central domain containing repeats of 20 amino acids.[14,15,62] APC is also found in a complex with GSK3β which phosphorylates the 20 amino acid repeat region of APC and thereby promotes its binding to β-catenin.[63] The function of APC in β-catenin degradation has been shown by transfection experiments. When introduced into APC-deficient cells, APC induces degradation of β-catenin and blocks the transcription of TCF/LEF-dependent promoters.[14,15,64] In human tumours, mutations of APC result in truncated proteins which are no longer able to down-regulate β-catenin. Moreover, in tumours containing wild-type APC, β-catenin mutations are observed that interfere with the APC controlled degradation.[65,66] These mutations alter serine and threonines at the NH_2-terminus which are putative target sites for GSK3β. Truncated versions of β-catenin lacking the NH_2-terminus are resistant to degradation and can transform fibroblasts.These results point to an essential role of β-catenin in tumour development and indicate that one of the main functions of APC is to control cytoplasmic β-catenin levels and to prevent signalling into the cell nucleus. The role of APC in normal wnt signalling during embryonal development is less clear. Genetic experiments do not support a function of a *Drosophila* APC homologue in wingless signalling,[67] and expression of APC in *Xenopus* leads to the formation of duplicated body axes, indicating that APC in this system activates rather than inhibits wnt signalling.[68] Recently, two related proteins, axin and conductin, have been implicated in the regulation of β-catenin stability.[69–72] Both proteins inhibit wnt signalling in *Xenopus* embryos,[69,70] and promote β-catenin degradation in transfected colorectal carcinoma cells.[70,71] Biochemically, axin and conductin interact with β-catenin, APC and GSK3β, indicating that they assemble multiprotein complexes which target β-catenin for degradation.[70–72]

Are the dual functions of β-catenin in cell adhesion and wnt signalling dependent on each other? Experimentally, the overexpression of the cytoplasmic domain of cadherins in *Xenopus* embryos leads to depletion of free β-catenin and to a block of the endogenous axis formation.[73] Whether this is of physiological relevance is unclear, but through the sequestration of β-catenin cadherins may antagonize the wnt signalling pathway. Tumours are frequently deficient for expression of cadherins[38] which may result in an increase of cytoplasmic β-catenin and activation of the wnt signal. As mentioned above, such a mechanism could account for the formation of neoplastic lesions in chimeric animals expressing dominant-negative cadherins.[35,36] Conversely, forced expression of cadherins promotes normalization and differentiation of tumour cells.[37,74] Wnt signalling could also act as an activator of cell adhesion under certain conditions: constitutive expression of Wnt-1 in cultured cells induces cadherin/catenin complexes and leads to increased cell–cell adhesion.[75] Thus in certain experimental systems a cross-talk between the adhesion system and the wnt signal pathway can be clearly demonstrated. However, mutational analyses show that the functions of β-catenin in cell adhesion and signalling can also be experimentally separated. For instance, in *Xenopus* embryos binding of β-catenin to cadherins is not required for double axis formation i.e. for induction of the wnt phenotype.[76] Similarly, specific mutations of armadillo have been found in *Drosophila* which affect adherens junction formation but not segment polarity, and vice versa.[77] Further functional evidence from *Drosophila* indicates that cadherin-mediated cell adhesion can be uncoupled from wingless signalling.[78] Obviously the degree of interplay of the adhesion and the signalling systems depends on the specific biological situation. Under conditions where the adhesion system is highly dynamic, i.e. in developmental processes where cadherins are down-regulated, β-catenin might be free to enter the wnt signalling pathway. In contrast, under conditions where cell adhesion is favoured and cadherins neutralize free β-catenin, wnt signalling might be blocked. In yet other situations wnt signalling may activate cell adhesion by increasing the amount of β-catenin which interacts with cadherins.

References

1. Nagar, B., Overduin, M., Ikura, M., and Rini, J. M. (1996). *Nature*, **380**, 360–4.
2. Shapiro, L., Fannon, A. M., Kwong, P. D.*et al.* (1995). *Nature*, **374**, 327–37.
3. Brieher, W. M., Yap, A. S., and Gumbiner, B. M. (1996). *J. Cell Biol.*, **135**, 487–96.
4. Nagafuchi, A. and Takeichi, M. (1988). *EMBO J.*, **7**, 3679–84.
5. Ozawa, M., Ringwald, M., and Kemler, R. (1990). *Proc. Natl Acad. Sci. USA*, **87**, 4246–50.
6. Rimm, D. L. and Morrow, J. S. (1994). *Biochem. Biophys. Res. Commun.*, **200**, 1754–61.
7. Hülsken, J., Birchmeier, W., and Behrens, J. (1994). *J. Cell Biol.*, **127**, 2061–9.
8. Peifer, M., Berg, S., and Reynolds, A. B. (1994). *Cell*, **76**, 789–91.
9. Huber, A. H., Nelson, W. J., and Weis, W. I. (1997). *Cell*, **90**, 871–82.
10. Herrenknecht, K., Ozawa, M., Eckerskorn, C., Lottspeich, F., Lenter, M., and Kemler, R. (1991). *Proc. Natl Acad. Sci. USA*, **88**, 9156–60.
11. Nagafuchi, A., Takeichi, M., and Tsukita, S. (1991). *Cell*, **65**, 849–57.
12. Reynolds, A. B., Herbert, L., Cleveland, J. L., Berg, S. T., and Gaut, J. R. (1992). *Oncogene*, **7**, 2439–45.
13. Ying, S. T., Edwards, R. A., Tubb, B., Wang, S., Bryan, J. and McCrea, P. D. (1996). *J. Cell Biol.*, **134**, 1271–81.
14. Su, L. K., Vogelstein, B., and Kinzler, K. W. (1993). *Science*, **262**, 1734–7.
15. Rubinfeld, B., Souza, B., Albert, I., Müller, O., Chamberlain, S. H., Masiarz, F. R., Munemitsu, S., and Polakis, P. (1993). *Science*, **262**, 1731–4.
16. Rimm, D. L., Koslov, E. R., Kebriaei, P., Cianci, C. D., and Morrow, J. S. (1995). *Proc. Natl Acad. Sci. USA*, **92**, 8813–7.
17. Pollack, A. L., Barth, A. I. M., Altschuler, Y., Nelson, W. J., and Mostov, K. (1997). *J. Cell Biol.*, **137**, 1651–62.
18. Daniel, J. M.and Reynolds, A. B. (1997). *BioEssays*, **19**, 883–91.
19. Xu, Y., Guo, D.-F., Davidson, M., Inagami, T., and Carpenter, G. (1997). *J. Biol. Chem.*, **272**, 13463–6.
20. Takeichi, M. (1990). *Annu. Rev. Biochem.*, **59**, 237–52.
21. Frixen, U. H., Behrens, J., Sachs, M., Eberle, G., Voss, B., Warda, A., Löchner, D., and Birchmeier, W. (1991). *J. Cell Biol.*, **113**, 173–185.
22. Nose, A., Nagafuchi, A. and Takeichi, M. (1988). *Cell*, **54**, 993–1001.
23. Redfield, A., Nieman, M. T., and Knudsen, K. A. (1997). *J. Cell Biol.*, **138**, 1323–31.
24. Larue, L., Antos, C., Butz, S., Huber, O., Delmas, V., Dominis, M., and Kemler, R. (1996). *Development*, **122**, 3185–94.
25. Hodivala, K. and Watt, F. (1994). *J. Cell Biol.*, **124**, 589–600.
26. Larue, L., Ohsugi, M., Hirchenhain, J., and Kemler, R. (1994). *Proc. Natl Acad. Sci. USA*, **91**, 8263–7.
27. Riethmacher, D., Brinkmann, V. and Birchmeier, C. (1995). *Proc. Natl Acad. Sci. USA*, **92**, 855–9.
28. Torres, M., Stoykova, A., Huber, O., Chowdhury, K., Bonaldo, P., Mansouri, A., Butz, S., Kemler, R., and Gruss, P. (1997). *Proc. Natl Acad. Sci. USA*, **94**, 901–6.
29. Radice, G., Rayburn, H., Matsumani, H., Knudsen, K.A., Takeichi, M., and Hynes R. O. (1997). *Dev. Biol.*, **181**, 64–78.
30. Radice, G. L., Ferreira-Cornwell, M. C., Robinson, S. D., Rayburn, H., Chodosh, L. A., Takeichi, M., and Hynes, R. (1997). *J. Cell Biol.*, **139**, 1025–32.
31. Haegel, H., Larue, L., Ohsugi, M., Fedorov, L., Herrenknecht, K. and Kemler, R. (1995). *Development*, **121**, 3529–37.
32. Ruiz, P. *et al.* (1996). *J. Cell Biol.*, **135**, 215–25.
33. Bierkamp, C., McLaughlin, K. J., Schwartz, H., Huber, O., and Kemler, R. (1996). *Dev. Biol.*, **180**, 780–5.
34. Hermiston, M. L., Wong, M. H. and Gordon, J. I. (1996). *Genes Dev.*, **10**, 985–96.
35. Hermiston, M. L. and Gordon, J. I. (1995). *J. Cell Biol*, **129**, 489–506.
36. Hermiston, M. L. and Gordon, J. I. (1995). *Science*, **270**, 1203–1207.
37. Perl., A.-K., Wilgenbus, P., Dahl, U., Semb, H., and Christofori, G. (1998). *Nature*, **392**, 190–93.
38. Birchmeier, W. and Behrens, J. (1994). *Biochim. Biophys. Acta*, **1198**, 11–26.
39. Guilford *et al.* (1998). *Nature*, **392**, 402–5.

40. Hirano, S., Kimoto, N., Shimoyama, Y., Hirohashi, S. and Takeichi, M. (1992). *Cell*, **70**, 293–301.
41. Nagafuchi, A., Ishihara, S., and Tsukita, S. (1994). *J. Cell Biol.*, **127**, 234–45.
42. Braga, V. M. M., Machesky, L. M., Hall, A., and Hotchin, N. A. J. (1997). *J. Cell Biol.*, **137**, 1421–31.
43. Takaishi, K., Sasaki, T., Kotani, H., Nishioka, H., and Takai, Y. (1997). *J. Cell Biol.*, **139**, 1047–59.
44. Hordijk, P. L., ten Klooster, J. P., van der Kammen, R., Michiels, F., Oomen, L. C. J. M., and Collard, J. G. (1997). *Science*, **278**, 1464–6.
45. Islam, S., Carey, T. E., Wolf, G. T., Wheelock, M. J., and Johnson, K. R. (1996). *J. Cell Biol.*, **135**, 1643–54.
46. Litvinov, S. V., Balzar, M., Winter, M. J., Bakker, H. A. M., Briaire-de Bruijn, I. H., Prins, F., Fleuren, G. J., and Warnaar, S. O. (1997). *J. Cell Biol.*, **139**, 1337–48.
47. Monier-Gavelle, F., and Duband, J.-L. (1997). *J. Cell Biol.*, **137**, 1663–81.
48. Peifer, M. (1995). *Trends Cell Biol.*, **5**, 224–9.
49. Miller, J. R. and Moon, R. T. (1996). *Genes Dev.*, **10**, 2527–39.
50. Moon, R. T., Brown, J. D., and Torres, M. (1997). *Trends Genet.*, **13**, 157–62.
51. Aberle, H., Bauer, A., Stappert, J., Kispert, A., and Kemler, R. (1997). *EMBO J.*, **16**, 3797–3804.
52. Behrens, J., von Kries, J. P., Kühl, M., Bruhn, L., Wedlich, D., Grosschedl, R., and Birchmeier, W. (1997). *Nature*, **382**, 638–42.
53. Molenaar, M., van de Wetering, M., Oosterwegel, M., Peterson-Maduro, J., Godsave, S., Korinek, V., Roose, J., Destree, O., and Clevers, H. (1996). *Cell*, **86**, 391–9.
54. Huber, O., Korn, R., McLaughlin, J. M. O., Herrmann, B. and Kemler, R. (1996). *Mech. Dev.*, **59**, 3–10.
55. Brunner, E., Peter, O., Schweizer, L., and Basler, K. (1997). *Nature*, **385**, 829–33.
56. Clevers, H., and van de Wetering, M. (1997). *Trends Genet.*, **13**, 485–9.
57. Giese, K., Kingsley, C., Kirshner, J., and Grosschedl, R. (1995). *Genes Dev.*, **9**, 995–1008.
58. van de Wetering, M., Cavallo, R., Dooijes, D., van Beest, M., van Es, J., Loureiro, J., Ypma, A., Hursh, D., Jones, T., Bejsovic, A., Peifer, M., Mortin, M., and Clevers, H. (1997). *Cell*, **88**, 789–99.
59. Riese, J., Yu, X., Munnerlyn, A., Eresh, S., Hsu, S.-C., Grosschedl, R., and Bienz, M. (1997). *Cell*, **88**, 777–87.
60. Brannon, M., Gomperts, M., Sumoy, L., Moon, R. T., and Kimelman, D. (1997). *Genes Dev.*, **11**, 2359–70.
61. Han, M. (1997). *Cell*, **90**, 581–4.
62. Polakis, P. (1997). *Biochim. Biophys. Acta*, **1332**, F127–47.
63. Rubinfeld, B., Albert, I., Porfiri, E., Fiol, C., Munemitsu, S. and Polakis, P. (1996). *Science*, **272**, 1023–6.
64. Korinek, V., Barker, N., Morin, P. J., van Wichen, D., de Weger, R., Kinzler, K. W., Vogelstein, B., and Clevers, H. (1997). *Science*, **275**, 1784–7.
65. Morin, P. J., Sparks, A. B., Korinek, V., Barker, N., Clevers, H., Vogelstein, B., and Kinzler, K. W. (1997). *Science*, **275**, 1787–90.
66. Rubinfeld, B., Robbins, P., El-Gamil, M., Albert, I., Porfiri, E., and Polakis, P. (1997). *Science*, **275**, 1790–2.
67. Hayashi, S., Rubinfeld, B., Souza, B., Polakis, P., and Levine, A. J. (1997). *Proc. Natl Acad. Sci. USA*, **94**, 242–7.
68. Vleminckx, K., Rubinfeld, B., Polakis, P., and Gumbiner, B. (1996). *J. Cell Biol.*, **136**, 411–20.
69. Zeng *et al.* (1997). *Cell*, **90**, 181–92.
70. Behrens, J. *et al.* (1988). *Science*, **280**, 596–9.
71. Hart, M. J., de los Santos, R., Albert, I. N., Rubinfeld, B., and Polakis, P. (1998). *Curr. Biol.*, **8**, 573–81.
72. Ikeda, S., Kishida, S., Yamamoto, H., Murai, H., Koyama, S., and Kikuchi, A. (1998). *EMBO J.* **17**, 1371–84.
73. Heasman, J., Crawford, A., Goldstone, K., Garner-Hamrick, P., Gumbiner, B., McCrea, P., Kintner, C., Noro, C.Y. and Wylie, C. (1994). *Cell*, **79**, 791–803.
74. Meiners, S., Brinkmann, V., Naundorf, H., and Birchmeier, W. (1998). *Oncogene*, **16**, 9–20.
75. Bradley, R.S., Cowin, P., and Brown, A.M. (1993). *J. Cell Biol.*, **123**, 1857–65.
76. Fagotto, F., Funayama, N., Glück, U., and Gumbiner, B.M. (1996). *J. Cell Biol.*, **132**, 1105–14.
77. Orsulic, S., and Peifer, M. (1996). *J. Cell Biol.*, **134**, 1283–1300.
78. Sanson, B., White, P. and Vincent, J.-P. (1996). *Nature*, **383**, 627–30.

Jürgen Behrens
Max Delbrück Center for Molecular Medicine,
Robert-Rössle-Strasse 10,
13122 Berlin, Germany

Signaling via integrins

Introduction

The integrins are a family of cell adhesion receptors, composed of an α and a β subunit. Although most integrins mediate adhesion to the extracellular matrix (ECM), some bind to soluble multivalent plasma proteins (such as fibrinogen) or other adhesion receptors (such as intracellular adhesion molecules, ICAMs), thereby mediating homotypic aggregation or heterotypic cell-to-cell adhesion, respectively (see also pp. 196–212). It has recently become clear that the physiologic role of integrins involves much more than adhesion. The initial observation that normal fibroblasts need to adhere to ECM components, such as fibronectin and vitronectin, in order to survive and proliferate *in vitro*, while transformed fibroblasts do not (i.e. they are 'anchorage-independent'), suggests that cell adhesion to the ECM generates signals that are necessary for normal cells to progress through the cell cycle but are constitutively active in neoplastic cells.[1] In accordance with this hypothesis, recent studies have revealed that a number of proto-oncogene products play a crucial role in integrin signalling.[2,3] In a number of cell types, however, adhesion to certain ECM proteins causes exit from the cell-cycle and differentiation, suggesting that certain integrins may regulate proliferation in a negative manner.[4,5] Finally, there is evidence that adhesion to an appropriate ECM is required for cell survival.[6,7] Thus, it appears that integrins transduce to the cell interior positional information encoded by the ECM so that cells proliferate and eventually differentiate in the correct environment, but undergo apoptosis if displaced. Because of these multiple functions, integrins play a major role in a variety of important physiological processes such as embryonic development, wound healing, and tumorigenesis.

There has been great interest in understanding the signalling mechanisms by which integrins affect cell behaviour. Ligation of integrins results in a number of intracellular events similar to those elicited by growth factor receptors, including tyrosine phosphorylation of intracellular proteins, activation of protein kinase C, elevation of intracellular calcium concentrations, and changes in lipid metabolism.[2,8] However, unlike most growth factor receptors, integrins possess no intrinsic tyrosine kinase activity and must therefore signal via cytoplasmic kinases, such as the Src family kinases and focal adhesion kinase (FAK). Through these tyrosine kinases, integrins activate the GTPase Ras, leading to the activation of mitogen activated protein kinase (MAPK) cascades capable of controlling immediate-early gene expression.[3] Because integrins and receptor tyrosine kinases share many common signalling molecules, it is not surprising that they act in concert to regulate cell proliferation. In addition to activating tyrosine kinase pathways leading to the activation of Ras, integrins also affect the Rho family GTPases Rac and Rho (see also pp. 136–137). These proteins have been implicated in regulating the cytoskeleton as well as cooperating with Ras to activate immediate-early genes. Finally, it is becoming increasingly clear that the integrin-dependent pathways which regulate cell cycle progression and those which regulate the cytoskeleton do not function independently, but 'cross-talk' extensively.

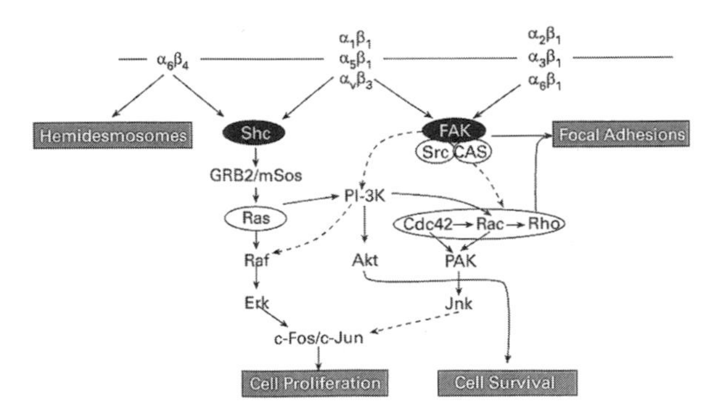

Figure 1. Major integrin-mediated signal transduction pathways. The diagram shows three classes of integrins which (left) organize hemidesmosomes and recruit Shc to activate MAP kinase cascades and induce gene expression; (middle) recruit Shc and activate FAK, which mediates focal adhesion formation and possibly gene expression via PI3K or CAS; (right) activate FAK, but not Shc. The Rho family proteins mediate both focal adhesion formation and gene expression and can be activated by Ras via PI3K or possibly by FAK via CAS. PI3K mediates adhesion-dependent cell survival via Akt and serves as a potential point of cross talk between the Shc and FAK pathways.

$\beta1$ and αv integrins

Signalling via $\beta1$ and αv subunit-containing integrins has been examined predominantly in fibroblasts and endothelial cells. Upon binding to ECM ligands, these integrins aggregate in the plane of the plasma membrane and undergo a conformational change which promotes the interaction of the β subunit cytoplasmic tail with talin and possibly α-actinin. This initial event is followed by the recruitment of other cytoskeletal proteins, such as paxillin, vinculin, and tensin, as well as signalling molecules, resulting in the nucleation of actin microfilaments. The entire process results in the formation of cell–ECM junctions called focal adhesions.[9] There is evidence suggesting that all $\beta1$ and αv integrins are linked to tyrosine kinase-dependent signalling pathways, which play a role in the formation and/or regulation of focal adhesions.[2] In addition, a subgroup of them can induce the recruitment of the adaptor proteins Shc and Grb2, thereby activating the Ras/MAP kinase cascade.[7] Finally, recent data suggest that $\beta1$ and αv integrins can also activate the tyrosine kinase Abl[10] as well as suppress the activity of integrin linked kinase (ILK).[11]

Focal adhesion kinase

FAK is a cytoplasmic tyrosine kinase which is activated and recruited to focal adhesions in response to ligation of all $\beta1$ and αv integrins. Experiments of gene targeting and dominant negative inhibition suggest that FAK regulates the assembly and/or disassembly of focal adhesions during cell migration,[12,13] but it may also influence cell survival and proliferation.[14] FAK contains a central catalytic domain flanked by an N-terminal and a C-terminal segment which are devoid of the SH2 and SH3 domains typical of other cytoplasmic tyrosine kinases. The mechanism of FAK activation is unclear, but one model proposes that the binding of its N-terminal segment (which may be autoinhibitory) to the cytoplasmic domain of the integrin β subunit exposes the catalytic domain, putting it in an open and active conformation. The aggregation of integrins due to ligand binding would then result in a concomitant oligomerization and transautophosphorylation of FAK. The major FAK autophosphorylation site (Tyr-397) finally binds the SH2 domain of Src, enabling Src to phosphorylate the FAK-associated proteins paxillin and $p130^{CAS}$, as well as other sites at the C terminus of FAK, such as Tyr-925.[8] It has been suggested that FAK Tyr-925 binds the SH2 domain of the adaptor molecule Grb2, which may activate Ras by recruiting the Ras GTP exchange factor Sos to the plasma membrane where Ras is located.[15] This mechanism, however, appears to be limited to cells which overexpress FAK or Src, such as certain cancer cells.

FAK contains proline-rich motifs which mediate its interaction with the SH3 domain-containing molecules p130^{CAS} and Graf.[8] Binding to FAK enables p130^{CAS} to be phosphorylated by FAK-bound Src. This process generates SH2 binding sites for additional adaptor molecules such as Crk and Nck.[16] While Crk and Nck can both bind effectors that can lead to Ras activation, Nck can also bind a number of putative effectors of Rho family proteins. Graf, which is a GTPase activating protein (GAP) for Rho and Cdc 42, may also link FAK to Rho family proteins.[8] Since Rho, Rac and Cdc42 are key players in actin cytoskeletal regulation, the binding of p130^{CAS} and/or Graf to FAK may serve as a possible mechanism for FAK's regulation of focal adhesion assembly and cell migration.

Finally, there is evidence suggesting that the major autophosphorylation site in FAK can also bind the SH2 domain of phosphatidylinositol (PI)-3 kinase (PI3K),[17] a key player in both protecting adherent cells from apoptosis and in stimulating cell proliferation. PI3K acts by phosphorylating the membrane lipids PI(4) phosphate (PIP) to generate PI(3,4)P2 or PI(4,5) bisphosphate (PIP_2) to generate PI(3,4,5)P3. Although it is still unclear how such lipid products affect cell behaviour, there is evidence that they can promote the recruitment to the plasma membrane and dimerization of target effectors containing pleckstrin homology (PH) domains.[18] One example of this mechanism is seen with the serine/threonine kinase Akt, which mediates PI-3K's protection of cells from apoptosis.[19] PI-3K may also participate in the activation of Ras and one of its downstream targets, Raf, as well as of Rac in response to cell adhesion.[20,21] Activation of both Raf and Rac can lead to cell proliferation through the MAP kinases ERK and c-Jun N-terminal kinase (JNK), respectively.[22] Therefore, the binding of PI3K to FAK may be another mechanism by which integrins regulate cell behavior.

Shc

While all $\beta1$ and αv integrins can activate FAK, $\alpha1\beta1$, $\alpha5\beta1$ and $\alpha v\beta3$, but not $\alpha2\beta1$, $\alpha3\beta1$ and $\alpha6\beta1$, are also able to recruit the adaptor protein shc and induce its tyrosine phosphorylation.[7] Shc contains a C-terminal SH2 domain and, at the N-terminus, another phosphotyrosine binding module named PTB domain.[23] The central portion of Shc contains two tyrosine phosphorylation sites that can bind the Grb2/Sos complex, leading to the activation of Ras.[24] The recruitment of Shc to activated integrins is mediated by the extracellular or transmembrane segment of the integrin α subunit, with the transmembrane protein caveolin possibly serving as an intermediary molecule. Caveolin is associated with certain Src family kinases[25] and may thus provide both the transmembrane adoptor and the kinase necessary for the recruitment and tyrosine phosphorylation of Shc. It is interesting that $\alpha2\beta1$, $\alpha3\beta1$ and $\alpha6\beta1$, which are unable to activate shc signaling, combine with members of the tetraspans family of transmembrane proteins and thereby with PI4K.[26] Thus, different transmembrane adapters may function to link different subgroups of integrins to distinct intracellular signalling pathways.

While FAK has not yet been proven necessary for integrin induced activation of ERK in normal untransformed cells, Shc has.[27,7] Integrin-mediated Shc signalling results in activation of ERK and c-Fos transcription and promotes cell cycle progression in response to growth factors. In contrast, ligation of integrins not linked to Shc results in exit from the cell cycle and either apoptosis or differentiation, depending on, for example, whether survival and differentiation factors are available. Remarkably, this occurs even if the cells are exposed to otherwise mitogenic concentrations of growth factors. The selective interaction of certain integrins with Shc may help to explain why the response of a particular cell to the ECM depends on which integrins it expresses and the composition of the ECM. For example, it is known that in various cell types adhesion to fibronectin promotes proliferation and inhibits differentiation, while adhesion to laminin promotes cell cycle withdrawl and differentiation. For instance, myoblasts proliferate on fibronectin (via $\alpha5\beta1$) but fuse to form myotubes on laminin (via $\alpha6\beta1$).[28] In addition, the same matrix molecule may exert different effects depending on which integrin it ligates. For example, both $\alpha5\beta1$ and $\alpha4\beta1$ bind fibronectin, but only the binding of fibronectin to $\alpha5\beta1$ results in induction of the collagenase gene in synovial fibroblasts.[29] By these and additional mechanisms, such as its potential ability to activate PI3K via Ras,[21,30] Shc is likely to play a crucial role in regulating embryonic development.

Rho family GTPases

There is increasing evidence that integrins can activate the Rho family of small GTPases. These proteins, which were originally identified as capable of regulating the cytoskeleton in response to serum components, are now considered capable as well of activating the expression of immediate-early genes.[22] Cdc42 has been implicated in the formation of filopodia, which are fine needle-like structures containing a single bundle of actin filaments and may have a sensory function at the leading edge of migrating cells and in neuronal growth cones. Rac promotes the formation of lamellipodia, which are flat protrusions supported by a criss-cross network of actin fibres. Finally, Rho promotes the organization of focal adhesions and associated actin stress fibres. Although the various Rho family members may be activated separately by upstream signals, there is evidence that they are also connected in a horizontal hierarchy, with Cdc42 activating Rac, and Rac activating Rho. This may explain why filopodia often merge to form lamellipodia, in which focal adhesions then arise. How the Rho family members are activated by cell adhesion is still unclear. They may lie downstream of FAK, as described above. Alternatively, they may lie downstream of Ras since Ras can activate PI3K, which in turn may activate Rac.[21,30]

In addition to regulating the actin cytoskeleton, the Rho family proteins can also affect gene expression and cell cycle progression by activating the MAP kinase JNK. JNK activates c-Jun, which heterodimerizes with c-Fos to bind to AP1 sites and promote transcription of genes necessary for cell cycle progression.[31] Interestingly, mutational analysis has revealed that different target effectors of Rho proteins are responsible for propagating these divergent signals. For example, certain point mutations in Rac and Cdc42 abolish cytoskeletal regulation but do not impair JNK activation. Unexpectedly, these mutations block entry into the S phase, suggesting that the cytoskeleton is linked to the control of the cell cycle.[32] Finally, Rho cooperates with the Ras–ERK pathway to induce c-Fos expression rather than acting on c-Jun as Rac and Cdc42 do.[33] Clearly Cdc42, Rac and Rho have complex properties which enable them to play key roles in both cytoskeletal formation and cell cycle progression.

Abl and integrin-linked kinase

The two kinases Abl and integrin-linked kinase (ILK) have recently been shown to be regulated by integrins. Abl is a tyrosine kinase stimulated by cell adhesion to various ECM components and is thought to have separate functions in the cytoplasm and nucleus.[10] Cytoplasmic Abl may regulate focal adhesions by phosphorylating mena. This is a protein which interacts with the focal adhesion components zyxin and vinculin and regulates actin polymerization via profilin.[34] Integrin-mediated translocation of Abl to the nucleus may be necessary for the phosphorylation of RNA polymerase II. This event may contribute to regulating the transcription of genes involved in S-phase entry.

The serine/threonine kinase, ILK, was identified by virtue of its ability to interact with the cytoplasmic domain of the integrin $\beta1$ subunit in the yeast two-hybrid system and was later shown to reside in focal adhesions.[11] Surprisingly, it appears to be inhibited rather than activated upon adhesion, yet its overexpression leads to anchorage-independent cell growth.[35] Therefore, it is unique in being a transforming kinase that is negatively regulated by cell adhesion.

Leukocyte $\beta2$ and platelet αIIb$\beta3$ integrins

The $\beta2$ integrins and the αIIb$\beta3$ integrin are restricted to the haemopoietic system and differ from other integrins because they are prominently susceptible to intracellular signals which modify their affinity and/or avidity for extracellular ligand (inside-to-outside signalling).[36]

The $\beta2$ integrins participate in the homing and extravasation of leukocytes. Although leukocytes do not have well organized focal adhesions, the $\beta2$ integrins activate signalling pathways similar to those involved in the organization of the actin cytoskeleton in response to ligation of $\beta1$ integrins in fibroblasts. Ligation of $\beta2$ integrins activates the tyrosine kinases Syk and fakB/Pyk2 and also elevates intracellular calcium via activation of PLC-γ and formation of IP3.[37–39] Moreover, the αL$\beta2$ integrin (LFA-1), which binds ICAM-1, has been shown to activate Ras through tyrosine phosphorylation of the guanine nucleotide exchange factor Vav.[40] Thus, in addition to activating intracellular signalling pathways capable of affecting the actin cytoskeleton, the $\beta2$ integrins may also regulate gene expression. For example, it is clear that in T cells they cooperate with the T cell receptor to promote activation and other changes in cell behaviour.[41,42]

The platelet-specific αIIb$\beta3$ integrin is a major fibrinogen receptor which plays a crucial role in platelet aggregation. Like $\beta1$ and αv integrins, αIIb$\beta3$ interacts intracellularly with talin and other focal adhesion components. Although αIIb$\beta3$ does not control gene expression, since platelets lack a nucleus, this integrin does stimulate intracellular signalling via tyrosine phosphorylation of Syk, FAK, and Src.[43] These events are likely to control various aspects of platelet activation.

$\alpha6\beta4$ integrin

The $\alpha6\beta4$ integrin is a laminin receptor and is expressed at high levels in various types of epithelial cells.[44] The cytoplasmic domain of $\beta4$ is very large and contains, toward the C terminus, two pairs of type III fibronectin-like repeats joined by a 142 amino acid sequence called the connecting segment. Upon $\alpha6\beta4$ binding to laminin, the unique intracellular portion of $\beta4$ interacts, by means of intermediary cytoskeletal elements, with the keratin filament system, resulting in the nucleation of hemidesmosomes. The hemidesmosomes are punctuate junctions connecting the basal cells of stratified and complex epithelia to the basement membrane. Known cytoskeletal components of hemidesmosomes include the 230 kDa bullous pemphigoid antigen 1 (BPAG1) and the 400 kDa protein HD1/plectin. Experiments of immune electron microscopy have indicated that the hemidesmosomes consist of an outer plaque containing the transmembrane proteins $\alpha6\beta4$ and 180 kDa bullous pemphigoid antigen 2 (BPAG2) and an inner plaque containing BPAG1 and HD1/plectin (see pp. 105–10). In accordance with the notion that the $\alpha6\beta4$ integrin is required for nucleation of hemidesmosomes and stable adhesion of the epidermis to the underlying basement membrane, $\beta4$ knockout mice lack hemidesmosomes and die soon after birth because of severe blistering.[45,46]

Ligation of $\alpha6\beta4$ activates an associated tyrosine kinase and causes phosphorylation of the cytoplasmic domain of $\beta4$. This phosphorylation mediates both the assembly of hemidesmosomes and the recruitment of Shc.[47] Evidence suggests that this adaptor protein links $\alpha6\beta4$ to the Ras/ERK and Rac/JNK MAP kinase cascades, thereby regulating cell proliferation. In keratinocytes, for example, ligation of $\alpha6\beta4$, but not $\alpha3\beta1$ or $\alpha2\beta1$, stimulates ERK and JNK and promotes progression through the G_1 phase of the cell cycle in response to EGF.[21] Furthermore, mice lacking the cytoplasmic domain of $\beta4$, the Shc-binding region, have defects in epithelial cell proliferation.[49]

Conclusions

Although the signalling pathways linking integrins to cytoskeletal and nuclear events are still incompletely understood, the recent rapid progress of this field has provided a plausible explanation for the multiple and often divergent effects of the ECM on cellular behaviour. Integrin stimulation of MAP kinase cascades helps to explain the ability of the ECM to regulate immediate-early gene expression and promote cell survival and proliferation. It also suggests that neoplastically transformed cells are anchorage independent because of constitutive activation of these signalling pathways. The recently uncovered signalling differences between integrins suggests that the repertoire of integrins expressed by a given cell will dictate its behaviour in response to ECMs of different composition. This notion is likely to explain the divergent effects of fibronectin and laminin in various cell types. Finally, it is evident that integrins and growth factor receptors cooperate to regulate cell behaviour. The signalling pathways activated by these two receptor systems share several components which may serve as points of convergence, divergence, and/or cross talk. In addition, some integrins may directly associate with growth factor receptors.[48] A more detailed map of integrin signalling will undoubtedly enable us better to understand cell proliferation and differentiation and begin to approach how these fundamental processes are deregulated in neoplastic cells.

References

1. Giancotti, F. G. and Mainiero, F. (1994). *Biochim. Biophys. Acta*, **1198**, 47–64.
2. Schwartz, M. A., Schaller, M. D., and Ginsberg, M. H. (1995). *Ann. Rev. Cell. Dev. Biol.*, **11**, 549–99.
3. Giancotti, F. G. (1997). *Curr. Opin. Cell Biol.* **9**, 691–700.
4. Adams, J. C. and Watt, F. M. (1993). *Development*, **117**, 1183–98.
5. Lin, C. Q. and Bissell, M. J. (1993). *FASEB J.* **7**, 737–43.
6. Ruoslahti, E. and Reed, J. C. (1994). *Cell*, **77**, 477–8.
7. Wary, K. K., Mainiero, F., Isakoff, S. J., Marcantonio, E. E., and Giancotti, F. G. (1996). *Cell*, **87**, .733–43.
8. Parsons, J. T. (1996). *Curr. Opin. Cell Biol.*, **8**,146–52.
9. Gilmore, A. P. and Burridge, K. (1996). *Structure*, **4**, 647–51.
10. Lewis, J. M., Baskaran, R., Taagepera, S., Schwartz, M. A., and Wang, Y. J. (1996). *Proc. Natl Acad. Sci. USA*, **93**, 15174–9.
11. Hannigan, G. E., Leung-Hagesteijn, C., Fitz-Gibbon, L., Coppolino, M. G., Radeva, G., Filmus, J., *et al.* (1996). *Nature*, **379**, 91–6.
12. Ilic, D., Furuta, Y., Kanazawa, S., Takeda, N., Sobue, K., Nakatsuji, N., *et al.* (1995). *Nature*, **377**, 539–44.
13. Richardson, A. and Parsons, J. T. (1996). *Nature*, **380**, 538–40.
14. Frisch, S. M., Vuori, K., Ruoslahti, E., and Chan-Hui, P. Y. (1996). *J. Cell Biol.*, **134**, 793–9.
15. Schlaepfer, D. D., Hanks, S.K., Hunter, T., and van der Geer, P. (1994). *Nature*, **372**, 786–91.
16. Schlaepfer, D. D., Broome, M. A., and Hunter, T. (1997). *Mol. Cell Biol.*, **17**,1702–13.
17. Chen, H. C., Appeddu, P. A., Isoda, H., and Guan, J. L. (1996). *J. Biol. Chem.*, **271**, 26329–34.
18. Toker, A. and Cantley, L. C. (1997). *Nature*, **387**, 673–6.
19. Franke, T. F., Kaplan, D. R., and Cantley, L. C. (1997). *Cell*, **88**, 435–7.
20. King, W. G., Mattaliano, M. D., Chan, T. O., Tsichlis, P. N. and Brugge, J. S. (1997). *Mol. Cell Biol.*, **17**, 4406–18.
21. Maniero, F., Murgia, C., Wary, K. K., Curatola, A. M., Pepe, A., Blumemberg, M., *et al.* (1997). *EMBO J.*, **16**, 2365–75.
22. Vojtek, A. B. and Cooper, J. A. (1995). *Cell*, **82**, 527–9.
23. Pawson, T. (1995). *Nature*, **373**, 573–80.
24. van der Geer, P., Wiley, S., Gish, G. D., and Pawson, T. (1996). *Curr. Biol.*, **6**, 1435–44.
25. Corley Mastick, C., Brady, M. J., and Saltiel, A. R. (1995). *J. Cell Biol.*, **129**, 1523–31.
26. Berditchevski, F., Talias, K. F., Wong, K., Carpenter, C. L., and Hemler, M. (1997). *J. Biol. Chem.*, **272**, 2595–8.
27. Lin, T. H., Aplin, A. E., Shen, Y., Chen, Q., Schaller, M., Romer, L., Aukhil, I., and Juliano, R. L. (1997). *J. Cell Biol.*, **136**, 1385–95.
28. Sastry, S. K., Lukonishok, M., Thomas, D. A., Muschler, J., and Horowitz, A. F. (1996). *J. Cell Biol*, **133**, 169–84.
29. Huhtala, P., Humphries, M. J., McCarthy, J. B., Tremble, P. M., Werb, Z., and Damsky, C. H. (1995). *J. Cell Biol.*, **129**, 867–79.
30. Rodriguez-Viciana P, Warne, P. H., Khwaja, A., Marte, B. M., Pappin, D., Das, P., *et al*, (1997). *Cell*, **89**, 457–67.
31. Karin, K. A. (1995). *J. Biol. Chem.*, **270**, 16483–6.
32. Lamarche, N., Tapon, N., Stowers, L., Burbelo, P. D., Aspenstrom, P., Bridges, T., *et al.* (1996). *Cell*, **87**, 519–29.
33. Hill, C. S., Wayne, J., and Treismann, R. (1995). *Cell*, **81**, 1159–70.
34. Gertler, F. B., Niebuhr, K., Reinhard, M., Wehland, J., and Soriano, P. (1996). *Cell*, **87**, 227–39.
35. Radeva, G., Petrocelli, T., Behrend, E., Leung-Hagesteijn, C., Filmus, J., Slingerland, J., and Dedhar, S. (1997). *J. Biol. Chem.*, **272**, 13937–44.
36. Shattil, S. J. and Ginsberg, M. H. (1997). *J. Clin. Invest.*, **100**, 1.–5.
37. Yan, S. R., Huang, M. and Berton, G. (1997). *J. Immunol.*, **158**, 1902–10.
38. Kanner, S. B. (1996). *Cell. Immunol.* **171**, 164–9.
39. Hellberg, C., Molony, L., Zheng, L., and Anderson, T. (1996). *Biochem. J.*, **317**, 403–9.
40. Zheng, L., Sjolandera, A., Eckerdal, J., and Anderson, T. (1996) *Proc. Natl Acad. Sci. USA*, **93**, 8431–6.
41. Kanner, S. B., Grosmaire, L. S., Ledbetter, J. A., and Damle, N. K. (1993). *Proc. Natl Acad. Sci. USA*, **90**, 7099–103.
42. Voss, L. M., Abraham, R. T., Rhodes, K. H., Schoom, R. A., and Leibson, P. J. (1991). *J. Clin. Immunol.*, **11**, 175–83.
43. Shattil, S. J., Ginsberg, M. H., and Brugge, J. S. (1994). *Curr. Opin. Cell Biol.*, **6**, 695–704.
44. Giancotti, F. G. (1996). *J. Cell Sci.*, **109**, 1165–72.

45. van der Neut, R., Krimpenfort, P., Calafat, J., Niessen, C. M., and Sonnenberg, A. (1996). *Nature Genet.*, **13**, 366–9.
46. Dowling, J., Yu, Q.-C., and Fuchs, E. (1996). *J. Cell Biol.*, **134**, 559–72.
47. Mainiero, F., Pepe, A., Wary, K. K., Spinardi, L., Mahammadi, M., Schlessinger, J., and Giancotti, F. G. (1995). *EMBO J.*, **14**, .4470–81.
48. Schneller, M., Vuori, K., and Ruoslahti, E. (1997). *EMBO J.*, **16**, 5600–7.
49. Murgia, C., Blaikie, P., Kim, N., Dans, M., Petrie, H. T., and Giancotti, F. G. (1998). *EMBO J.*, **17**, 3940–51.

Michael J. Dans and Filippo G. Giancotti
Cellular Biochemistry and Biophysics Program,
Memorial Sloan-Kettering Cancer Center, New York,
NY 10021, USA

ADAMs

An ADAM is a transmembrane protein containing both a disintegrin and metalloproteinase domain and therefore potentially has both cell adhesion and protease activities. The first identified ADAM is fertilin, a sperm surface protein that functions in sperm binding to and fusing with the plasma membrane of the egg. Fertilin is a heterodimer composed of two ADAM family transmembrane subunits, fertilin α and β; fertilin β contains a disintegrin domain with an identified active site that binds sperm to the egg plasma membrane, while fertilin α has a proposed, but unproven, function in promoting membrane fusion. The best characterized ADAM shown to have a metalloproteinase activity is TACE (tumour-necrosis factor-α [TNF-α] converting enzyme) which specifically cleaves the transmembrane form of TNF-α to the soluble active form.

Synonyms and homologies

Fertilin was originally termed PH-30.[1] The ADAM family of proteins has also been called MDC proteins[2] (metalloproteinase, disintegrin, cysteine-rich), cellular disintegrins,[2] or metalloproteinase-disintegrins.[3–5] ADAM family members all contain the following domains: prodomain, metalloproteinase, disintegrin, cysteine-rich, EGF-like, transmembrane, and cytoplasmic tail (Fig. 1). As of December 1997, the ADAM family has 22 members for which full length cDNA sequences have been reported[3–5] (partially reviewed in ref. 6).

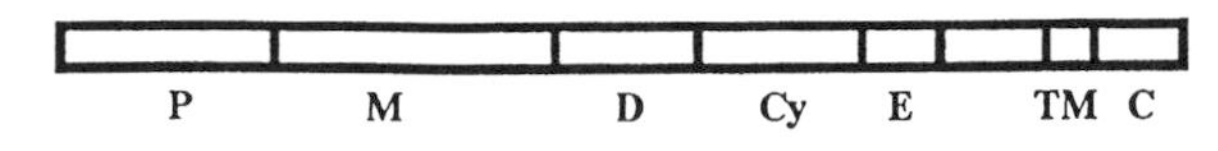

Figure 1. A diagram showing the domain organization shared by all ADAMs. The proposed fusion activity of fertilitin α is in the cystein-rich domain (Cy). The cytoplasmic tail domain (C) is of variable length and may have signalling function(s) in some ADAMs. P, prodomain; M, metalloproteinase; D, disintegrin; E, EGF-like domain; tM, transmembrane domain.

Protein properties

Although each ADAM has a disintegrin and metalloproteinase domain, for most ADAMs it is not yet known if they actually have adhesive and/or proteolytic activity. Among the 17 sequenced ADAMs, nine have the appropriate residues in the metalloproteinase active site to actually have protease activity.[3–6] Only three ADAM proteins, fertilin, TACE, and MADM, have been purified and partially characterized; the other identified ADAMs are cDNAs and the corresponding proteins have not yet been isolated.

Our further description here will be restricted to two ADAMs acting in sperm–egg adhesion/fusion, fertilin and cyritestin (ADAM 3), the latter available only as a cDNA sequence.[8]

Fertilin α and β, in guinea pig, are each made as precursors on spermatogenic cells with molecular masses of 100–110 kDa on reducing SDS–PAGE. They are proteolytically processed to a smaller form on mature sperm: mature (processed) fertilin α has a molecular weight of ~60 kDa and mature fertilin β 44 kDa on reduced SDS–PAGE.[1] N-terminal sequence analysis of the mature forms indicates that in both α and β, proteolytic cleavage occurs between the metalloproteinase and disintegrin domains, so that the mature subunits begin with the N terminus of the disintegrin domain[9] (C. Blobel; personal communication).

Several types of experiments indicate that fertilin functions in sperm–egg fusion. Two monoclonal antibodies (PH-30 and PH-1 mAbs) that recognize different epitopes on the β subunit of guinea-pig fertilin have different effects in an *in vitro* sperm–egg fusion assay. One of the mAbs, PH-30, showed a dose-dependent inhibition of sperm–egg fusion. The other mAb, PH-1, had no effect on fusion and thus served as a control, non-function blocking, antibody.

Direct evidence that the fertilin β disintegrin domain functions in sperm–egg binding leading to fusion has come from peptide inhibition studies. As in other

ADAMS, the disintegrin domain of fertilin has high sequence homology to a class of soluble snake venom peptides from pit vipers, the disintegrins.[9] The soluble snake disintegrins bind to the platelet integrin αIIbβ3 (GP IIb–IIIa). Disintegrin binding to αIIbβ3 prevents fibrinogen binding and subsequent platelet aggregation and thus promotes the spread of snake toxins from the site of the initial wound.[10] NMR structural analysis of the short snake venom disintegrins has shown that their active site region, containing the signature RGD sequence, is located at the tip of a flexible loop with the base of the loop formed by disulphide bonds.[10,11] In guinea-pig fertilin β, the sequence TDE is in the same position as RGD in the snake disintegrins. TDE-containing peptides bind to guinea-pig eggs and potently inhibit the fusion of sperm with the egg, presumably by preventing a step in sperm–egg membrane binding that leads to membrane fusion.[12] Similar results have been obtained with mouse peptides (containing the corresponding sequence QDE) from the same active site loop of the mouse fertilin β sequence. These peptides inhibit sperm fusion with eggs,[13–15] as do recombinant fragments of mouse fertilin β (J. Evans, G. Kopf, and R. Schultz; personal communication). These results collectively indicate that fertilin has an active site in the same region of the disintegrin domain as the snake disintegrins and binds sperm to the egg plasma membrane in a fashion that leads to sperm–egg fusion.

Another ADAM, cyritestin,[8] also functions along with fertilin in the process of sperm–egg adhesion/fusion. Based on the cDNA sequence, cyritestin disintegrin domain active site peptides were designed and antibodies to the disintegrin active site generated. Both active site peptides and antibodies to the active site of cyritestin strongly inhibit sperm binding and fusion. These results indicate that sperm–egg fusion may be a multistep process and the disintegrin-binding sites of different ADAMs may participate in more than one of the binding steps.[15]

It was initially proposed that guinea-pig fertilin α, which has a hydrophobic sequence with the properties of a viral fusion peptide at residues 90–110 of the mature subunit, promoted membrane fusion following sperm–egg adhesion.[9] Biochemical or genetic evidence to support this proposal has so far been limited,[16] so the function of fertilin α and the mechanism of membrane fusion remain to be elucidated.

Purification

Fertilin been purified from guinea-pig sperm using affinity chromatography on the MAb PH-30.[1] Fertilin has been purified from bovine sperm using ConA affinity chromatography followed by ion exchange chromatography.[17]

Activities

No activities have yet been studied using purified fertilin.

Antibodies

The mAbs PH-30 and PH-1, described above, bind to guinea-pig fertilin[1] and do not have species cross-reactivity. Polyclonal rabbit antiserum raised against purified guinea-pig fertilin[1] recognizes bovine fertilin.[17] Various antipeptide antibodies have been raised to mouse ADAMS, including antibodies to the disintegrin domain active sites of fertilin β, or cyritestin which have been useful in functional studies and immunofluorescence.[15] Immunofluorescence analysis has shown that fertilin is localized to the posterior head of guinea-pig sperm while both fertilin and cyritestin are localized to the equatorial region of the head of mouse sperm. As these are sperm plasma membrane regions where fusion with the egg is initiated,[18] these localizations are consistent with the proposed functions of these ADAMs.[15]

Genes

Full length cDNAs for 22 distinct ADAM family members have been reported (as of December 1997). These cDNAs have been isolated mainly from mammals but also from *Drosophila melanogaster* and *C. elegans*.[6] cDNAs for fertilin α and β and cyritestin have been cloned from various mammals. Expression of ADAM genes can show wide tissue distribution (e.g. fertilin α, ADAM 15) or be tissue specific (e.g. fertilin β and cyritestin, both testis-specific).[6] The genes for various ADAMs are found on different chromosomes in the mouse.[19]

Mutant phenotype

The only loss of function phenotype thus far reported at the organismal level is the null mutant in the *kuzbanian* gene in *Drosophila*. The absence of this ADAM leads to a loss of the correct partitioning of precursor cells into neural and non-neuronal cells in the developing central and peripheral nervous systems. In the study of TACE, TACE$^{-/-}$ T-cells were found to show an 80–90 per cent reduction in release of TNF-α from the cell surface.[4]

Structure

Not available.

Note added in proof

The following noteworthy findings have recently been made on ADAMs:

1. The crystal structure of the catalytic (metalloprotease) domain of TACE (ADAM 17) has been reported.[21]
2. The *Drosophila* gene *kuzbanian* codes for an ADAM (ADAM 10) which functions as a proteolytic activator of Notch, a surface receptor important in various developmental pathways.[22]

3. A soluble form of ADAM 12 (meltrin α) has been discovered. Embryonal rhabdomyosarcoma cells were stably transfected with a mini-gene encoding soluble ADAM 12 and injected to produce tumours in nude mice. These tumours contained ectopic muscle cells (probably of mouse origin), suggesting that ADAM 12 may be able to induce muscle differentiation.[23,24]
4. Evidence has been presented that ADAM 15 (metargidin) binds specifically to the integrin $\alpha v\beta 3$ and does not bind to nine other integrins tested.[25]
5. Mice lacking fertilin β (gene knockout of fertilin β) are normal except that males are infertile. Sperm from the mutant mice are defective in adhering to the egg plasma membrane, show a reduced rate of fusion with the egg membrane and, surprisingly, are also unable to bind to the egg zona pellucida or migrate from the uterus to the oviduct.[25]

References

1. Primakoff, P., Hyatt, H., and Tredick-Kline, J. (1987)., *J. Cell. Biol.*, **104**, 141–9.
2. Weskamp, G., Krätzschmar, J., Reid, M. S., and Blobel, C. P. (1996). *J. Cell. Biol.*, **132**, 717–26.
3. Rooke, J., Pan, D., Xu, T., and Rubin, G. M. (1996). *Science*, **273**, 1227–31.
4. Black, R. A., Rauch, C. T., Kozlosky, C. J,., Peschon, J. J., Slack, J. L., Wolfson, M. F., *et al.* (1997). *Nature*, **385**, 729–32.
5. Moss, M. L., Jin, S. -L. C., Milla, M. E., Burkhart, W., Carter, H. L., Chen, W. -J., *et al.* (1997). *Nature*, **385**, 733–6.
6. Wolfsberg, T. and White, J. M. (1996). *Dev. Biol.*, **180**, 389–401.
7. Howard, L., Lu, X., Mitchell, S., Griffiths, S., and Glynn, P. (1996). *Biochem. J.*, **317**, 45–50.
8. Heinlein, U. A. O., Wallat, S., Senleben, A., and Lemaire, L. (1994). *Dev. Growth Differ.*, **36**, 49–58.
9. Blobel, C. P., Wolfsberg, T. G., Turck, C. W., Myles, D. G., Primakoff, P., and White, J. M. (1992). *Nature*, **356**, 248–52.
10. Niewiarowski, S., Mclane, M. A., Kloczewiak, M., and Stewart, G. J. (1994). *Semin. Hematol.*, **31**, 289–300.
11. Adler, M., Lazarus, R. A., Dennis, M. S., and Wagner, G. (1991). *Science*, **53**, 445–448.
12. Myles, D. G., Kimmel, L. H., Blobel, C. P., White, J. M., and Primakoff, P. (1994). *Proc. Natl Acad. Sci. USA*, **91**, 4195–8.
13. Almeida, E. A., Huovila, A. P., Sutherland, A. E., Stephens, L. E., Calarco, P. G., Shaw, L. M., *et al.* (1995). *Cell*, **81**, 1095–104.
14. Evans, J. P., Schultz, R. M., and Kopf, G. S. (1995). *J. Cell,. Sci.*, **108**, 3267–78.
15. Yuan, R., Primakoff, P., and Myles, D. G. (1997). *J. Cell. Biol.*, **137**, 105–12.
16. Muga, A., Neugebauer, W., Hirama, T., and Surewiez, W. K. (1996). *Biochemistry*, **33**, 4444–8.
17. Waters, S. and White, J. M. (1997). *Biol. Reprod.*, **56**, 1245–54.
18. Myles, D. G. (1993). *Dev. Biol.*, **158**, 35–45.
19. Cho, C., Primakoff, P., White, J. M., and Myles, D. G. (1996). *Genomics*, **34**, 413–17.
20. Wolfsberg, T. G., Bazan, J. F., Blobel, C. P., Myles, D. G., Primakoff, P. and White, J. M. (1993). *Proc. Natl Acad. Sci. USA*, **90**, 10783–7.
21. Maskos, K. *et al.* (1998). *Proc. Natl Acad. Sci. USA*, **95**, 3408–12.
22. Pan, D. and Rubin, G. M. (1997). *Cell*, **90**, 271–80.
23. Gilpin, B. J., Loechel, F., Mattei, M.-G., Engvall, E., Albrechtsen, R., and Wewer, U. M. (1998). *J. Biol. Chem.*, **273**, 157–66.
24. Loechel, F., Gilpin, B. J., Engvall, E., Albrechtsen, R., and Wewer, U. M. (1998). *J. Biol. Chem.*, **273**, 16993–7.
25. Cho, C. *et al.* (1998). *Science*, **281**, 1857–9.

Paul Primakoff
Department of Cell Biology and Human Anatomy, University of California at Davis, Davis, CA 95616, USA

Diana G. Myles
Section of Molecular and Cellular Biology, University of California at Davis, Davis, CA 95616, USA

Cadherins

Overview of the family

Cadherins constitute a large family of transmembrane glycoproteins that play a key role in calcium-dependent cell–cell adhesion.[1] They function not only for establishing tight cell–cell associations but also for defining adhesive specificities of cells. In development, the expression of each member of the family is spatiotemporally regulated to be correlated with morphogenetic events in which adhesion or segregation of cells is involved.[2]

Members of the cadherin family can be divided into four subgroups: classic cadherins, desmosomal cadherins, protocadherins, and atypical cadherins. Proteins are designated as members of the broadly defined cadherin family if they have one or more 'cadherin repeats'. Each cadherin repeat (CR) consists of approximately 110 amino acids and contains amino acid motifs with the conserved sequences LDRE, DXNDN, and DXD.[3] Calcium is essential for cadherin function and the extracellular domain of cadherin can be divided into independently folding structural domains (CADs) which have been shown to be

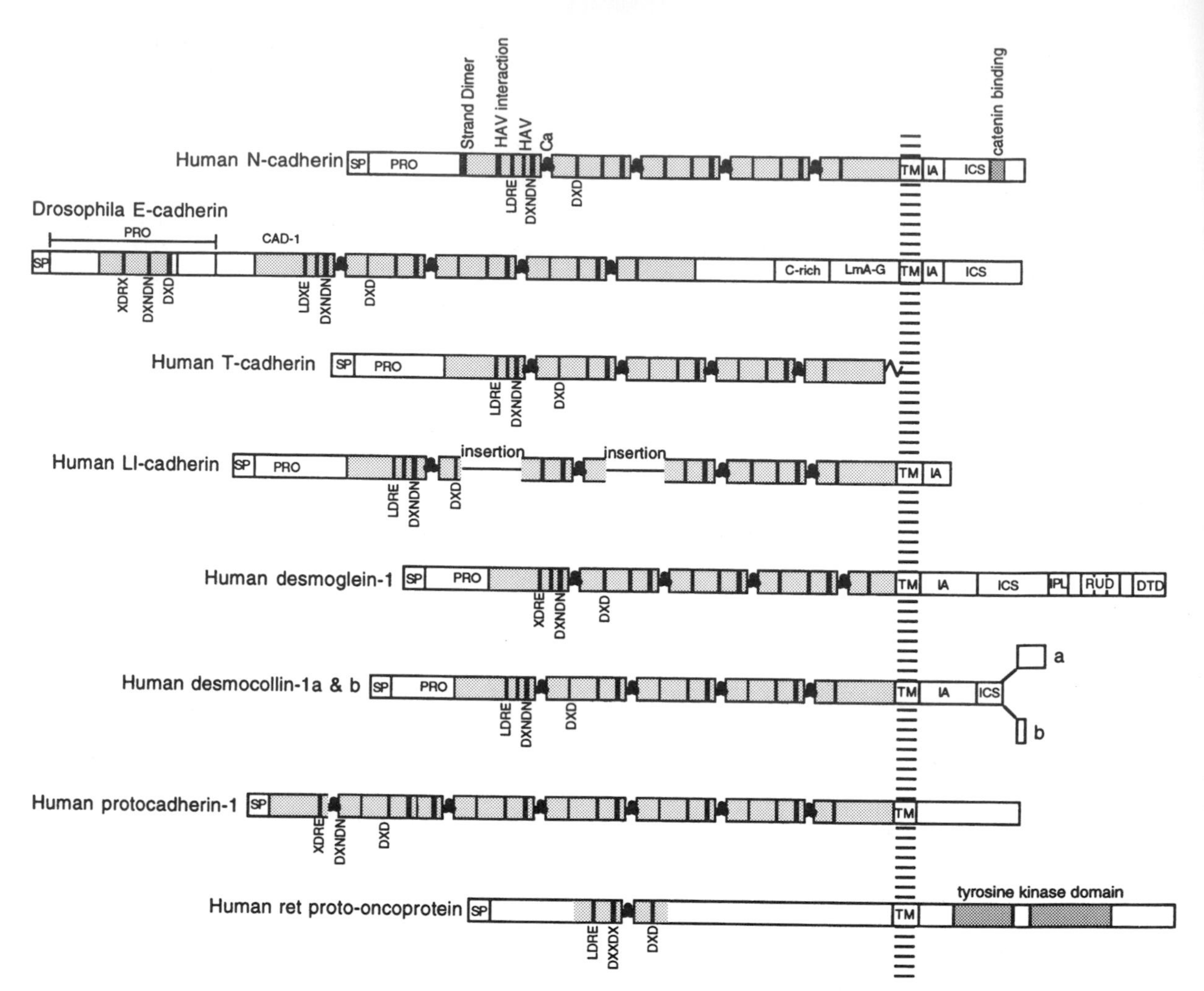

Figure 1. Comparison of the structural domains of several cadherin family members. The cadherin structural domain (CAD) is represented as a stippled box. The signal peptide (SP), prosequence (PRO), and transmembrane (TM) regions are indicated. Calcium ions (Ca) are shown between adjacent DXNDN and DXD sequences. The DXD, LDRE, and DXNDN sequences are shown for the first structural domain of each cadherin. When the amino acids do not agree with the consensus, an X has been used. The remainder of the DXD, LDRE, and DXNDN consensus sequences are represented as black bars through the CAD. The intracellular anchoring domain (IA), intracellular cadherin-specific sequence (ICS), and catenin-binding domain are indicated where appropriate. *Drosophila* E-cadherin has a cysteine-rich domain (C-rich) and a laminin-A globular repeat (LmA-G). The human desmoglein-1 cytoplasmic domain has an intracellular proline-rich linker (IPL), a repeat region (RUD), and a desmoglein-specific terminal domain (DTD). Desmocollin-1 is expressed as two splice variants, a and b. The ret proto-oncoprotein has a tyrosine kinase domain.

important in binding calcium ions. Classic cadherins have five CADs while the number of CADs in other members of the family varies widely. The most N-terminal CAD is designated CAD-1. In Fig. 1, each CAD is shown as a stippled box with three calcium ions between adjacent CADs.

Most, but not all, cadherins are type I transmembrane proteins that are associated with the cytoskeleton through interactions with cytoplasmic proteins. The classic cadherins have a highly conserved cytoplasmic domain that interacts with a group of proteins termed catenins that in turn associate with the actin cytoskeleton.[4] The desmosomal cadherins are not closely related to the classic cadherins in their cytoplasmic domains; they associate with plakoglobin and are linked to the intermediate filament cytoskeleton.[5] Some of the atypical cadherins are linked to the membrane by lipid tails or very short cytoplasmic domains and thus are not likely to be associated with catenins.[6] The cytoplasmic domains of the protocadherins are unrelated to the classic cadherins and there is no evidence that they associate with the cytoskeleton.[7]

Classic cadherins

Protein properties

A growing number of classic cadherins have been characterized at the molecular level, including E-cadherin, N-cadherin and P-cadherin. These members share amino acid homology and have a common basic structure of 110–150 kDa on SDS–PAGE.[8] Classic cadherins are composed of an extracellular domain (EC) that can be divided into five cadherin structural domains (CADs) designated CAD-1 through CAD-5, a transmembrane domain, and a highly conserved cytoplasmic domain that can be divided into an intracellular anchoring domain (IA) and an intracellular cadherin–typical sequence (ICS). CAD-1 is the most N-terminal cadherin structural domain and, in those cadherins where it has been studied, is responsible for cadherin–cadherin interactions in adjacent cells.[8] A sequence termed the cell adhesion recognition (CAR) sequence is found in this domain.[9] The CAR sequence in many classic cadherins is composed of the amino acids His–Ala–Val (HAV) and surrounding sequences.[10] Sequences on either side of the HAV tripeptide in E-cadherin and P-cadherin have been shown to confer homospecificity upon the cadherin.[11] Although cadherin adhesion is typically due to homophilic interactions between cadherin molecules on adjacent cells, it has been reported that E-cadherin expressed by epithelial cells can interact with the $\alpha_E\beta_7$ integrin on T lymphocytes.[12]

Recent structural studies have identified the sequence, also in CAD-1, that interacts with HAV.[13] These structural studies have also shown that the second amino acid in mature N-cadherin (Trp) is responsible for strand dimerization (or dimerization of cadherins from the same cell). Thus, the cadherin is probably present on the cell surface as a parallel dimer, and CAD-1 of one dimer interacts in an antiparallel fashion with CAD-1 of a cadherin dimer on an adjacent cell. Cadherins require calcium for binding activity; structural studies have shown that the calcium ions act as a bridge between adjacent domains to rigidify

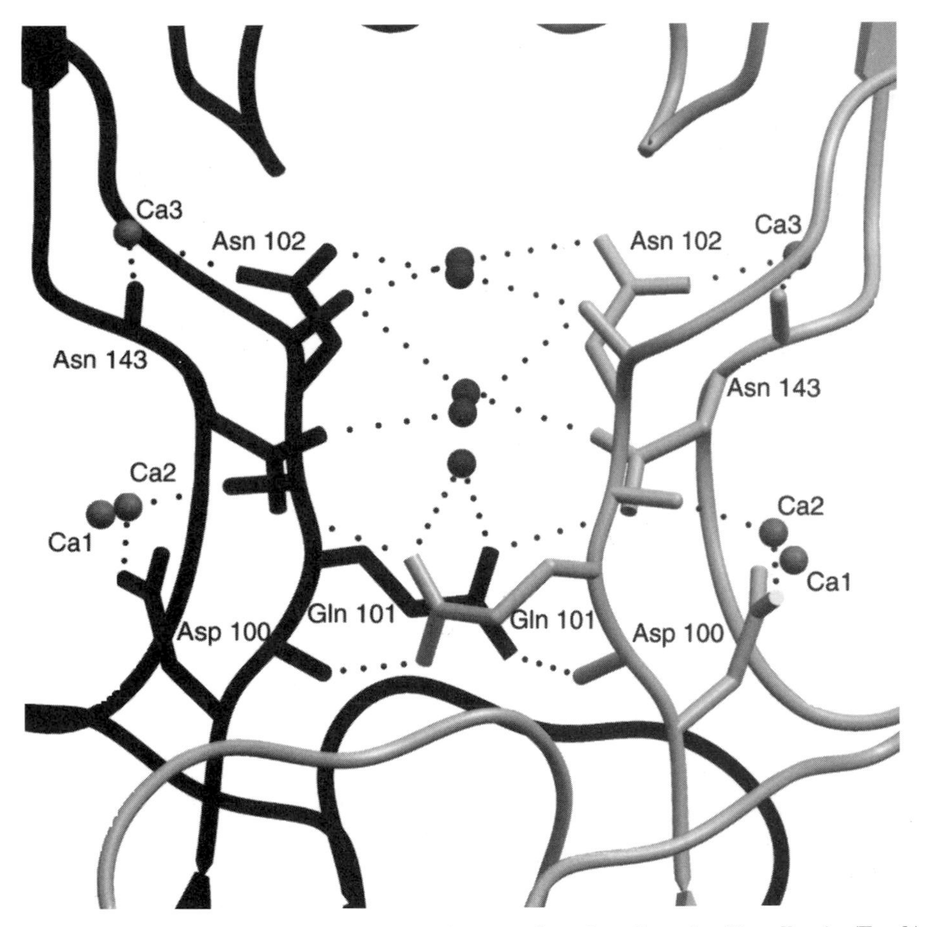

Figure 2. A fragment containing the two most N-terminal repeat domains of murine E-cadherin (Ecad1, 2) was crystallized in the presence of calcium.[16] Shown is the heart of the calcium binding sites which mediate calcium-dependent dimer interactions. The amino acids of one strand are shown in black while those of the other strand are shown in light grey. Calcium ions are labelled Ca1, Ca2 etc. and water is dark grey. (Courtesy of J. Rini and M. Ikura.)

the protein.[13–15] Fig. 2 illustrates the binding of calcium ions between CADs 1 and 2 of E-cadherin.[16]

Since classic cadherins were initially identified using antibodies that disrupted calcium-dependent adhesion and the work was done in a number of different laboratories using different systems, the most abundant members of the family were given several different names. Uvomorulin,[17] Arc-1,[18] cell–CAM 120/80,[19] and rr-1[20] are names that were given to E-cadherin from different species. ACAM[21] and N-Cal-CAM[22] are alternative names for N-cadherin. Table 1 presents the classic cadherins that have been characterized to date. The table contains all members that have been sequenced and submitted to the database. An attempt has been made to identify orthologues from different species. One invertebrate classic cadherin, *Drosophila* E-cadherin, has been identified.[23] It does not exactly fit the mould for vertebrate classic cadherins, but it does promote homotypic adhesion and does bind catenins. In addition, *Drosophila* E-cadherin localizes to the apical poles of epithelial cells, as does vertebrate E-cadherin. The structure of *Drosophila*

Table 1 Classic cadherins

Cadherin	Species	Accession number	Chromosome location	Accession number
N-cadherin	Human	X54315	18q11.2	114020
	Mouse	M31131, M22556	18–6.0	1882
	Bovine	X53615		
	Xenopus	X57675		
	Chicken	X07277		
	Zebrafish	X67648		
E-cadherin	Human	Z13009	16q22.1	192090
	Mouse	X06115	8–52.0	1881
	Xenopus	U04708		
LCAM	Chicken	M16260, J04074		
B-cadherin	Chicken	X58518		
KCAM		M81894		
EP-cadherin	*Xenopus*	X63720		
C-cadherin				
XB-cadherin	*Xenopus*	X78546		
F-cadherin	*Xenopus*	X85330		
P-cadherin	Human	X63629	16q22.1	114021
	Mouse	X06340, X68057	8–52.0	1883
	Bovine	X53614		
PB-cadherin	Rat	D83348		
R-cadherin	Human	L34059		
Cadherin-4	Mouse	D14888, X69966	2–106.0	15975
	Chicken	D14459		
(XMN-cad)	*Xenopus*	S82457		
Cadherin-5	Human	X79981	16q22.1	601120
VE-cadherin	Mouse	X83930	8–51.0	32096
Cadherin-6	Human	D31784	Chr. 5	G26765*
K-cadherin	Mouse	D82029		
	Rat	D25290		
Cadherin-6B	Chicken	D42149		
Cadherin-7	Chicken	D42150		
Cadherin-8	Human	L34060		
	Mouse	X95600		
Cadherin-10	Chicken	X95600		
Cadherin-11	Human	L34056, D21254	Chr. 16	600023
OB-cadherin	Mouse	D21253, D31963	8–51	15974
Cadherin-12	Human	L34057	5p14-p13	600562
Br-cadherin		L33477		
Cadherin-14	Human	U59325		
M-cadherin	Human	D83542	16q24.1	114019
Cadherin-15	Mouse	M74541	8–67.0	16328
DE-cadherin	*Drosophila*	D28749	Chr. 2 (57B)	
DN-cadherin	*Drosophila*	AB002397	Chr. 2 (36C/D)	

GenBank accession numbers are presented for complete sequences of classic cadherins. Alternative names for the same cadherin are given in italics. Given in italics are names that are likely to be the same cadherin in a different species. Most of the accession numbers for human chromosome locations were obtained from the Database of Mendelian Inheritance in Man. Those indicated by an asterisk are GenBank accession numbers. The accession numbers for mouse chromosome locations were obtained from Mouse Genome Informatics at The Jackson Laboratory.

E-cadherin is compared with that of vertebrate cadherins in Fig. 1.

Classic cadherins generally mediate cell–cell adhesion via homophilic interactions. The homophilic nature of the cadherin–cadherin interaction allows cells expressing different complements of cadherins to segregate from one another. Expression of the various classic cadherins is regulated both spatially and temporally during development. In addition, cadherins have been shown to delineate boundaries between subdivisions in the brain of developing mouse embryos.[24] Cadherins have thus been implicated as mediators of the complicated cell sorting events that lead to tissue formation in the developing embryo.[25] In the adult organism, E-cadherin has been identified as a suppressor of metastasis since disruption of E-cadherin function often results in an invasive carcinoma. In addition, transfection of E-cadherin into invasive cells has been shown to suppress motility and invasion.[26–28] Thus, cell–cell interactions mediated by classic cadherins are likely critical to the proper sorting of cells during development and to the maintenance of normal tissues in the adult. Fig. 3 shows boundaries in the mouse embryonic hindbrain that are defined by cadherin-6 expression.

The strong adhesive activity of the classic cadherins depends not only on extracellular protein–protein interactions but also on the cadherin being associated with the cytoskeleton.[29] The cadherin molecule has been localized to the adherens junction and in epithelial cells has been implicated in regulation of the junctional complex, including desmosomes and tight junctions.[30] The adherens junction is composed of the transmembrane cadherin which is associated with the cytoplasmic plaque proteins α-catenin, β-catenin, and plakoglobin. The catenins are in turn linked to actin filaments directly as well as through an association with α-actinin, an actin-binding protein.[29] A core region of 30 amino acids in the cytoplasmic domain of E-cadherin has been shown to be essential for interacting with the catenins[31] and is depicted in Fig. 1.

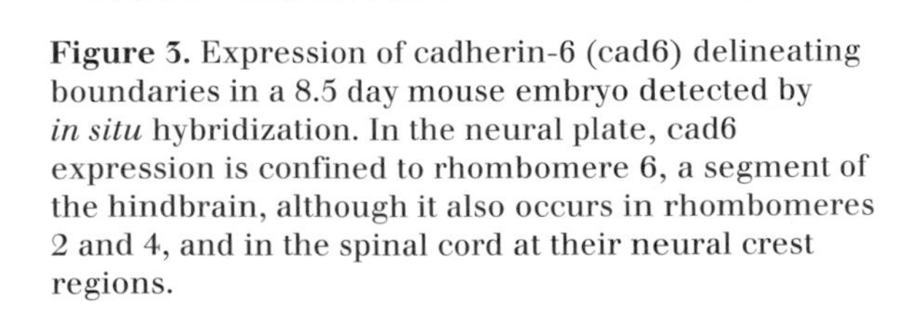

Figure 3. Expression of cadherin-6 (cad6) delineating boundaries in a 8.5 day mouse embryo detected by *in situ* hybridization. In the neural plate, cad6 expression is confined to rhombomere 6, a segment of the hindbrain, although it also occurs in rhombomeres 2 and 4, and in the spinal cord at their neural crest regions.

Purification

Classic cadherins have mainly been identified by antibodies that disrupt their activities, not by purification of the proteins. Thus, most reports of purification have made use of antibody affinity chromatography. Affinity purification has been most successfully done on extracellular fragments, often generated by proteolytic cleavage. The native cadherins have been purified as a complex with the catenins from detergent extracts of cells, again using antibody affinity chromatography. Recent identification of new cadherins has made use of molecular biological techniques and structural studies have made use of bacterially generated fragments of cadherins, rather than purification of native proteins.

Activities

Classic cadherins mediate homotypic, calcium dependent cell–cell adhesion which can be measured by aggregation assays. The classic cadherin has been shown to be the transmembrane component of the adherens junction and regulates the organization of the junctional complex in epithelial cells including desmosomes, tight junctions, and gap junctions. It has been proposed that classic cadherins are responsible for cell sorting events during embryonic development; in addition, E-cadherin mediates the establishment of cellular polarity and suppresses tumour invasion.[8,25–27]

Antibodies

Both monoclonal and polyclonal antibodies against many of the classic cadherins are available through the laboratories of individual investigators. Many of the antibodies inhibit adhesive activity of the cadherin. Although many of the original anti-cadherin antibodies are species specific, newly available commercial antibodies have been generated that recognize multiple species. A polyclonal pan-cadherin antibody against the conserved cytoplasmic domain of classic cadherins which recognizes E-cadherin, P-cadherin, and N-cadherin as well as some of the other classic cadherins is commercially available, from Sigma Immunochemicals. Transduction Laboratories, Sigma Immunochemicals, Zymed Laboratories, Becton Dickinson, and Takara all have collections of well characterized classic cadherin antibodies.

Genes

cDNA sequences have been published for most of the classic cadherins identified to date. The majority of the characterization has been done in human, mouse, chicken, and *Xenopus*. The GenBank accession numbers are included in Table 1. Chromosome mapping in the mouse and human has been done for a limited number of the classic cadherins and is also included in Table 1. Human chromosome locations were obtained from the on-line database of Mendelian Inheritance in Man which is located at http://www.ncbi.nlm.nih.gov/Omim/. Mouse chromosome locations were obtained from Mouse Genome Informatics at The Jackson Laboratory which is located at http://www.informatics.jax.org/.

Mutant phenotype/disease states

The best studied disease problem that implicates the classic cadherins is cancer.[26–28] Many studies have shown that mutations in E-cadherin result in decreased adhesive characteristics in carcinomas resulting in increased invasion and metastasis. Disruptions of cadherin genes in mice have shown that E-cadherin is critical for compaction in the early embryo[32,33] and that N-cadherin plays a vital role in heart development.[34] In *Drosophila* E-cadherin mutants, epithelial rearrangement processes are impaired.[35,36]

Structure

The CAD domain consists of seven anti-parallel β strands and is approximately 45 Å × 25 Å × 25 Å.[13] It is likely that three calcium ions act as bridges between adjacent CADs[16] (shown in Figs. 1 and 2). Studies of the crystal structure of CAD-1 of N-cadherin suggest that, in addition to the antiparallel self-association of cadherins expressed on two different cells, cadherins on the surface of the same cell self-associate in a parallel fashion to form a dimer. A tryptophan near the N terminus of CAD-1 plays a role in the formation of the strand dimer[13] (see Fig. 1). The crystal structure of CAD-1 plus CAD-2 of E-cadherin, however, showed no evidence for this tryptophan being involved in strand dimerization.[16]

Desmosomal cadherins

Protein properties

The desmosomal cadherins include desmoglein and desmocollin. There are three distinct, genotypically different desmogleins and three distinct genotypically different desmocollins.[37] Each desmocollin exists as two splice variants, a and b, which differ in the length of their C-terminal end.[38] The desmosomal cadherins are the transmembrane components of the desmosome and associate with the cytoplasmic plaque protein plakoglobin which, together with desmoplakin, presumably links the cadherin to the intermediate filament cytoskeleton. Desmogleins are slightly larger than classic cadherins and migrate at about 165 kDa on SDS–PAGE while the desmocollins migrate at approximately 116–120 kDa.

Unlike the classic cadherins, it does not appear that one cadherin is sufficient to form a desmosome. Desmosome-expressing cells always contain at least one desmoglein and one desmocollin. The desmosomal cadherins are expressed in a tissue-specific manner. Most tissues express the ubiquitous desmosomal cadherins, desmoglein-2 and desmocollin-2. However, in addition to the ubiquitous cadherins, stratifying squamous epithelial tissues also express differentiation-specific isoforms of the other desmosomal cadherins.[37,39,40]

Like the classic cadherins, the desmosomal cadherins have five extracellular cadherin structural domains (CADs), a single-pass transmembrane domain, and an intracellular anchoring domain (IA).[41] The desmosomal cadherins do not have the same cell adhesion recognition (CAR) sequence that is found in the classic cadherins. In human desmoglein the tripeptide is RAL, in human desmocollin it is FAT, and in bovine desmocollin it is

Table 2 Desmosomal cadherins

Cadherin	Species	Accession number	Chromosome location	Accession number
Desmoglein-1	Human	X56654	18q12.1	125670
	Bovine	X58466		
	Mouse	X74335	18–7.0	8905
Desmoglein-2	Human	Z26317, S64273	18q12.1–q12.2	125671
Desmoglein-3	Human	M76482	18q12.1	169615
	Mouse		18–7.0	16273
Desmocollin-1	Human	Z34522	18q12	125643
	Bovine	M67489		
	Mouse	X97986		
Desmocollin-2	Human	X56807	18q12	
	Bovine	M8190		
	Mouse	L33779	18–7.0	24091
Desmocollin-3	Human	X83929	18q12.1	600271
	Bovine	L33774		

GenBank accession numbers are presented for complete sequences of desmosomal cadherins. Most of the accession numbers for human chromosome locations were obtained from the Database of Mendelian Inheritance in Man. The accession numbers for mouse chromosome locations were obtained from Mouse Genome Informatics at The Jackson Laboratory.

YAT.[42] The desmosomal cadherins have an intracellular cadherin-typical sequence (ICS) that is shortened in the case of the b isoforms of desmocollin.[41] In addition, the cytoplasmic domain contains sequences specific to desmocollin or desmoglein.[41] Fig. 1 compares the structural organization of desmoglein-1 and desmocollin-1 with other cadherins. Table 2 presents the forms of desmoglein and desmocollin that have been characterized and submitted to GenBank.

The desmosomal cadherins are markers for differentiation in stratified epithelia, particularly in the epidermis. Desmoglein-1 and desmocollin-1 are most prominent in the upper spinous and granular layers; desmoglein-2 is predominantly localized in the basal layer, while desmocollin-2 has been detected throughout the living layers; desmoglein-3 is expressed strongly in the lower layers and becomes weaker in the upper layers while desmocollin-3, like desmocollin-2, is expressed in all living layers.[47] Interestingly, the switch in desmosomal cadherin expression in the epidermis upon differentiation is accompanied by a switch in intermediate filament (keratin) isoform expression, suggesting these may be coupled.[44]

Two severe blistering diseases have been shown to be due to production of autoimmune antibodies against the desmosomal cadherins. In pemphigus vulgaris, patients produce antibodies against desmoglein-3 and develop oral and skin blisters. In pemphigus foliaceus, the autoimmune antigen is desmoglein-1 and the patients develop a split in the epidermis just above the basal layer.[41] Despite the expression of classic cadherins in the same cells, the autoimmune antibodies are sufficient to produce severe disruption of cell–cell adhesion. Due to the clinical significance of these cadherins, chromosomal localizations have been assigned to all the human isoforms and are presented in Table 2.

Neither the precise nature of the extracellular protein–protein interactions nor the precise nature of the association of desmosomal cadherins with the intermediate filament cytoskeleton are known. The extracellular portion of desmoglein-3 has been shown to mediate weak aggregation when transfected as a chimeric molecule with the intracellular domain of a classic cadherin.[45] No studies to date have convincingly shown strong aggregation mediated by desmosomal cadherins alone. Desmoglein and the longer b isoforms of desmocollin have been shown to bind to plakoglobin.[46–48] The precise nature of the association of the cadherin–plakoglobin complex with the intermediate filament network is not known.

Purification

The desmosomal cadherins, along with the desmosomal plaque proteins, were first isolated from bovine nasal epidermis as part of intact desmosomal structures. Further characterization determined that they were the transmembrane glycoproteins of the desmosome.[49] The human homologues were identified as the antigens for the autoimmune diseases pemphigus vulgaris and pemphigus foliaceus.[41]

Activities

The desmosomal cadherins have been shown to be the transmembrane component of the desmosome; as such, they mediate the strong cell–cell interactions seen in epithelial tissues and cardiac muscle.

Antibodies

Both monoclonal and polyclonal antibodies against many of the desmosomal cadherins are available through the laboratories of individual investigators. In addition, human autoantibodies against some desmogleins and desmocollins are available from individual investigators. Commercial monoclonal antibodies against desmogleins are available through Progen or its distributor American Research Products.

Genes

cDNA sequences have been published for all three desmogleins and all three desmocollins. The majority of the characterization has been done in human and bovine with some sequences available for mouse desmosomal cadherins. The GenBank accession numbers are included in Table 2. Chromosome mapping in human has been done for most of these proteins and the locations are given in Table 2. Human chromosome locations were obtained from the on-line database of Mendelian Inheritance in Man which is located at http://www. ncbi.nlm.nih.gov/Omim/. Mouse chromosome locations were obtained from Mouse Genome Informatics at The Jackson Laboratory which is located at http://www.informatics.jax.org/.

Mutant phenotype/disease states

Desmoglein-1 is the autoantigen for the human autoimmune blistering disease pemphigus foliaceus and desmoglein-3 is the autoantigen for pemphigus vulgaris.[41] In addition, desmocollin has been shown to be recognized by serum from some pemphigus patients.[50] Disruption of the desmoglein-3 gene in mice results in a phenotype that is similar to pemphigus vulgaris patients (Peter J. Koch; personal communication). Loss of plakoglobin causes disruption of desmosomes in the heart.[51]

Structure

No structural studies have been published to date.

Atypical cadherins and protocadherins

Protein properties

In addition to the classic and desmosomal cadherins, several cadherin-related proteins have been identified.

Table 3 Protocadherins

Cadherin	Species	Accession number	Chromosome location
Protocadherin-1 *Protocadherin-42*	Human	L11370	
Protocadherin-2 *Protocadherin-43*	Human	L11373	
Protocadherin-3	Human	L34592	5q31–5q33
	Mouse	L43592	18
	Rat	L43592	
Fat	Human	X87241	4q34–4q35
	Drosophila	M80537	24D
Dachsous	*Drosophila*	L08811	21D
cdh-3	*C. elegans*	L14324, L18807	

GenBank accession numbers are presented for complete sequences and chromosome locations of protocadherins. Alternative names for the same cadherin are given in italics.

The protocadherins have varying numbers of extracellular cadherin structural domains (CADs) but their cytoplasmic domains show no significant homology to the classic cadherins. The protocadherin family appears to be a very large, divergent one and to date, no biological function has been assigned to these proteins.[52] However, protocadherins 1 and 2 show weak calcium-dependent cell aggregation when transfected into L-cells.[53] The ability of the extracellular domain to mediate aggregation can be enhanced by making a chimera with the intracellular domain of a classic cadherin, suggesting that the protocadherins are capable of homophilic interactions.[52]

Mammalian protocadherins 1–3 have six or seven CADs[53,54] while other protocadherin-like molecules such as Fat, which has been cloned from *Drosophila* and human, or Dachsous which has been cloned from *Drosophila* have 34 and 27 CADs, respectively. In addition these large molecules can have other motifs such as EGF repeats. Fat was first characterized from *Drosophila* as a tumour suppressor gene[55] and Dachsous plays a role in imaginal disc morphogenesis;[56] their roles in cell adhesion remain to be determined. The structure of protocadherin-1 is compared with that of other cadherins in Fig. 1. The characterized vertebrate and invertebrate protocadherins are presented in Table 3.

One protocadherin, cdh-3, has been characterized in *Caenorhabditis elegans*. It is expressed in epithelial cells and has been proposed to function in epithelial cell morphogenesis.[57] Disruption of the gene encoding cdh-3 affects the morphogenesis of a single cell, hyp10 in the tail. In addition to the characterized protocadherins, a number of protocadherin sequences have been identified by the *Caenorhabditis elegans* sequencing project. These are sequences of uncharacterized genes and are available in GenBank, but we have not presented them here.

Several atypical cadherins have been identified. T-cadherin lacks the cytoplasmic domain but is otherwise very similar to a classic cadherin.[58] It is anchored to the membrane by a lipid tail and is capable of mediating weak calcium dependent cell aggregation. The expression pattern of T-cadherin in the developing chicken embryo suggests that it may be involved in segregation of cell types.[59] In the human, T-cadherin, which has been called cadherin 13 and H-cadherin, was shown to have decreased expression in breast cancer suggesting that it may also be a tumour suppressor.[60]

LI-cadherin and Ksp-cadherin are similar to one another in that they have a short cytoplasmic domain and five extracellular CADs.[61] Unlike the classic cadherins, CADs 2 and 3 of LI-cadherin and Ksp-cadherin are interrupted by amino acid insertions. In addition, the cytoplasmic domain shares no homology with the classic cadherins. LI-cadherin, also known as HPT-1, has been associated with intestinal peptide transport.[62] In addition, the ret proto-oncoprotein, which has been implicated in several forms of human cancer,[63] shares limited sequence homology with cadherins.[64] The biological roles these atypical cadherins play and their functional relationship to the cadherin superfamily are yet to be elucidated. The structures of T-cadherin, LI-cadherin and the ret proto-oncoprotein are compared with other cadherins in Fig. 1. The characterized atypical cadherins are presented in Table 4.

Purification

There have been limited purification schemes developed for the atypical and protocadherins which have been mainly identified using molecular biological techniques.

Activities

Some of the atypical and protocadherins have been shown to mediate calcium-dependent cell aggregation when transfected into cadherin-negative L-cells, but as yet, no distinct biological function has been assigned to these proteins.

Antibodies

In some cases antibodies are available from individual investigators.

Table 4 Atypical cadherins

Cadherin	Species	Accession number	Chromosome location
Cadherin-13	Human	L34058, U59289	16q24
T-cadherin	Chicken	M81779, J04891	
H-cadherin			
LI-cadherin	Human	U0769, X83228	
HPT-1	Rat	X78997	
Ksp-cadherin	Rabbit	U28945	
C-ret	Human	P07949	
	Mouse	X67812	
	Chicken	912987	
	Drosophila	546972	
	Zebrafish	1711302	

GenBank accession numbers are presented for complete sequences and chromosome locations of atypical cadherins. Alternative names for the same cadherin are given in italics.

Genes

cDNA sequences have been published for all of the atypical and protocadherins discussed here. The GenBank accession numbers and chromosome locations are included in Tables 3 and 4.

Mutant phenotype/disease states

T-cadherin (also known as H-cadherin or cadherin 13) has been implicated in human breast cancer,[60] and the ret proto-oncoprotein has been implicated in numerous human cancers.[63]

Structure

No structural studies have been published to date.

References

1. Takeichi, M. (1990). *Ann. Rev. Biochem.*, **59**, 237–52.
2. Takeichi, M. (1992). *Development*, **102**, 639–55.
3. Oda, H., *et al.* (1994). *Dev. Biol.*, **165**, 716–26.
4. Wheelock, M. J., *et al.* (1996). *Curr. Topics Memb.*, **43**, 169–85.
5. Koch, P. J. and Franke, W. W. (1994). *Curr. Opin. Cell Biol.*, **6**, 682–7.
6. Pouliot, Y. (1992). *BioEssays*, **14**, 743–8.
7. Sano, K. *et al.* (1993). *EMBO J.*, **12**, 2249–56.
8. Takeichi, M. (1990). *Ann. Rev. Biochem.*, **59**, 237–52.
9. Pouliot, Y. (1992). *BioEssays*, 14, 743–8.
10. Blaschuk, O. W., *et al.* (1990). *Dev. Biol.*, **139**, 227–9.
11. Nose, A., *et al.* (1990). *Cell*, **61**, 147–55.
12. Cepek, K. L., *et al.* (1994). *Nature*, **372**, 190–3.
13. Shapiro, L., *et al.* (1995). *Nature*, **347**, 327–37.
14. Pokutta, S., *et al.* (1994). *Eur. J. Biochem.*, **223**, 1019–26.
15. Shapiro, L., *et al.* (1995). *Proc. Natl Acad. Sci. USA*, **92**, 6793–7.
16. Nager, B., *et al.* (1996). *Nature*, **380**, 360–364.
17. Ringwald, M., *et al.* (1987). *EMBO J.*, **6**, 3647–53.
18. Behrens, J., *et al.* (1985). *J. Cell Biol.*, **101**, 1307–15.
19. Damsky, C. D., *et al.* (1983). *Cell*, **34**, 455–66.
20. Gumbiner, B. and Simons, K. (1986). *J. Cell Biol.*, **102**, 457–68.
21. Volk, T. and Geiger, B. (1986). *J. Cell Biol.*, **103**, 1441–50.
22. Crittenden, S. L., *et al.* (1987). *Development*, **101**, 729–40.
23. Oda, H., *et al.* (1994). *Dev. Biol.*, **165**, 716–26.
24. Redies, C. and Takeichi, M. (1996). *Dev. Biol.*, **180**, 413–23.
25. Takeichi, M. (1991). *Science*, **251**, 1451–5.
26. Blaschuk, O., *et al.* (1995). *Endocrine*, **3**, 83–9.
27. Takeichi, M. (1993). *Curr. Opin. Cell Biol.*, **5**, 806–11.
28. Hülsken, J., *et al.* (1994). *Curr. Opin. Cell Biol.*, **6**, 711–16.
29. Wheelock, M. J., *et al.* (1996). *Curr. Topics Memb.*, **43**, 169–85.
30. Watabe, M., *et al.* (1994). *J. Cell Biol.*, **127**, 247–56.
31. Stappert, J. and Kemler, R. (1994). *Cell Adhes. Commun.*, **2**, 319–27.
32. Larue, L., *et al.* (1994). *Proc. Natl Acad. Sci. USA*, **91**, 8263–7.
33. Riethmacher, D., *et al.* (1995). *Proc. Natl Acad. Sci. USA*, **92**, 855–9.
34. Radice, G., *et al.* (1997). *Dev. Biol.* **181**, 64–78.
35. Uemura, T., *et al.* (1996). *Genes Dev.*, **10**, 659–71.
36. Tepass, U., *et al.* (1996). *Genes Dev.*, **10**, 672–85.
37. Koch, P. J. and Franke, W. W. (1994). *Curr. Opin. Cell Biol.*, **6**, 682–7.
38. Troyanovsky, S. M., *et al.* (1993). *Cell*, **72**, 561–74.
39. Nuber, U. A., *et al.* (1995). *Eur. J. Cell Biol.*, **66**, 69–74.
40. Schafer, S., *et al.* (1994). *Exp. Cell Res.*, **211**, 391–9.
41. Amagai, M. (1994). *Prog. Dermatol.*, **28**, 1–16.
42. Magee, A. I. and Buxton, R. S. (1991). *Curr. Opin. Cell Biol.*, **3**, 854–61.
43. Jensen, P. J. and Wheelock, M. J. (1996). *Cell Death Diff.*, **3**, 357–71.
44. Fuchs, E. (1995). *Ann. Rev. Cell Dev. Biol.*, **11**, 123–53.
45. Amagai, M., *et al.* (1994). *J. Invest. Dermatol.*, **102**, 402–8.
46. Wahl, J. K., *et al.* (1996). *J. Cell Sci.*, **109**, 1143–54.
47. Troyanovsky, S. M., *et al.* (1994). *Proc. Natl Acad. Sci. USA*, **91**, 10790–4.
48. Troyanovsky, S. M., *et al.* (1994). *J. Cell Biol.*, **127**, 151–60.
49. Kapprell, H.-P., *et al.* (1990). In *Morphoregulatory molecules* (ed. G.M. Edelman, B.A. Cunningham, and J.P. Thiery), pp. 285–314. John Wiley.
50. Dmochowski, M., *et al.* (1993). *J. Invest. Dermatol.*, **100**, 380–4.
51. Ruiz, P., *et al.* (1996). *J. Cell Biol.*, **135**, 215–25.
52. Suzuki, S. T. (1996). *J. Cell. Biochem.*, **61**, 531–42.
53. Sano, K., *et al.* (1993). *EMBO J.*, **12**, 2249–56.
54. Sago, H., *et al.* (1995). *Genomics*, **29**, 631–40.
55. Mahoney, P. A., *et al.* (1991). *Cell*, **67**, 853–68.

56. Clark, H. F., *et al.* (1995). *Genes Dev.*, **9**, 1530–42.
57. Pettitt, J., *et al.* (1996). *Development*, **122**, 4149–57.
58. Vestal, D. J. and Ranscht, B. (1992). *J. Cell Biol.*, **119**, 451–61.
59. Ranscht, B. (1994). *Curr. Opin. Cell Biol.*, **6**, 740–6.
60. Lee, S. W. (1996). *Nature Medi.*, **2**, 776–82.
61. Thomson, R. B., *et al.* (1995). *J. Biol. Chem.*, **270**, 17594–601.
62. Dantzig, A. H., *et al.* (1994). *Science*, **264**, 430–3.
63. van der Geer, P., *et al.* (1994). *Ann. Rev. Cell Biol.*, **10**, 251–337.
64. Iwamoto, T. (1993). *Oncogene*, **8**, 1087–91.

Keith R. Johnson and Margaret J. Wheelock
Department of Biology,
University of Toledo,
Toledo, OH 43606, USA

Masatoshi Takeichi
Department of Biophysics,
Faculty of Science, Kyoto University,
Kyoto 606–8502, Japan

CCAM and other CEA-related CAMs

CCAMs (cell–cell adhesion molecules) belong to the CEA (carcinoembryonic antigen) gene family, which contains several different cell surface proteins with cell adhesion activies.[1–4] The CCAMs are transmembrane proteins which are phylogeneticaly well conserved. The other CEA-related molecules with adhesion activities are bound to the cell surface via a glycosylphosphatidyl inositol anchor; they are expressed in the human species but not in rodents. The CEA-related adhesion molecules are multifunctional proteins with signal-regulatory properties;[3] one of the isoforms of CCAM has potent tumour suppressor activity.

Synonymous names and homologous proteins

The nomenclature of the CEA family is complex and confusing and CEA-related adhesion molecules, especially the CCAMs, are known under a large number of names.[4] CCAM was originally used for the rat proteins, whereas the murine and the human homologuous proteins are known as biliary glycoproteins (Bgp and BGP, respectively). Here, CCAM is used for the homologous proteins of all three species. For more information see Fig. 1.

Protein properties

The CEA gene family, which is a subfamily within the immunoglobulin (Ig) superfamily, contains two subgroups named the CEA subgroup and the PSG (pregnant-specific glycoproteins) subgroup, respectively.[1,4] Whereas all PSGs are secreted, the CEA subgroup contains both secreted and cell surface-bound proteins. The cell surface bound CEA-related molecules are shown in Fig. 1; cell adhesion activities have been documented for CEA, NCA, CGM6, and the CCAMs. The CCAMs, which are well conserved in rodents and man, are single-pass transmembrane proteins which presumably represent the primordial adhesion molecules of this family.[4] The other CEA-related adhesion proteins are bound to the cell surface via a glycosylphosphatidyl inositol (GPI) anchor. The GPI-linked CEA-related molecules are expressed in human but have not been found in rat or mouse.

In adult organisms the CEA-related cell surface molecules are expressed primarily in different epithelia, vessel endothelia, and haematopoietic cells, but individual molecules show distinct expression patterns (see references in ref. 3). All CEA-related cell surface molecules have an extracellular amino-terminal V-type Ig domain, which is followed by a variable number (0–6) of C2-type Ig domains. They are heavily *N*-glycosylated, but cell- and tissue-specific variations in the glycosylation patterns occur. The CCAMs are subject to alternative splicing and a large number of isoforms exist (see references in refs 3 and 4). The two major isoforms of CCAM have four Ig domains (one V-type and three C2-type domains) in the extracellular portion, but differ in their cytoplasmic parts due to differential splicing of one exon. The longer isoform (CCAM-L) has a cytoplasmic domain of 71–73 amino acid residues depending on species; that of the shorter isoform (CCAM-S) only contains 10–12 amino acids. Splice variants containing fewer C2-type Ig domains (1–2) also occur. In addition, allelic variants denoted *a* and *b* exist in rat and mouse. They differ from each other primarily in the amino-terminal Ig domain.

CCAM and the GPI-linked molecules mediate adhesion primarily by homophilic binding to molecules on adjacent cells, but heterophilic binding to produce the complexes CEA–NCA, CEA–BGP, and NCA–BGP has also been found (see references in refs 3 and 4). The amino-terminal V-type Ig domains mediate these interactions, but whereas CCAM homophilic binding results from a reciprocal binding between amino-terminal domains of opposing molecules, the amino-terminal domain of CEA binds to another, more membrane-close Ig domain of its ligand CEA partner (see references in refs 3 and 4). Human CCAM and NCA, which carry the sialyl Lewis carbohydrate epitope, can also mediate adhesion of human granulocytes to E-selectin on vessel endothelial cells (see references in refs 3 and 4).

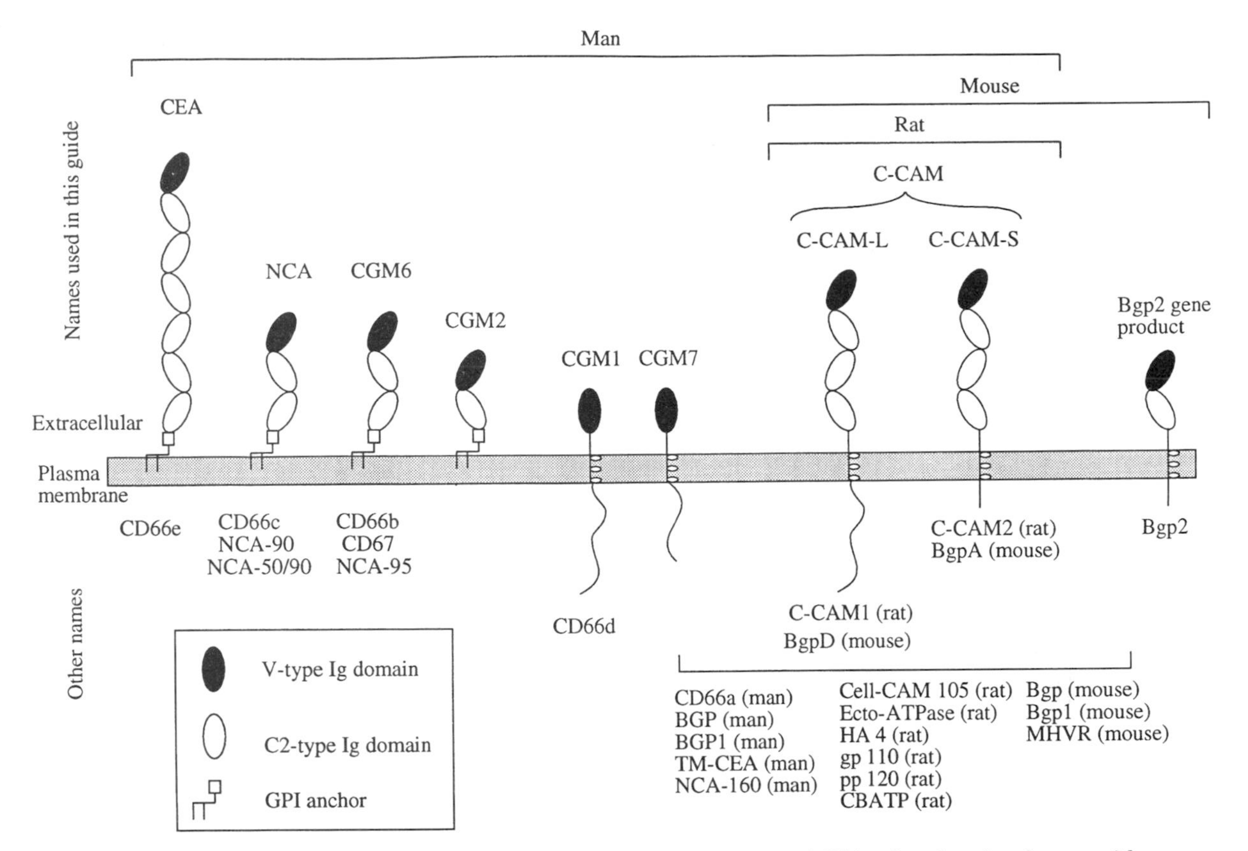

Figure 1. Cell-surface bound CEA-related molecules. The cell surface bound CEA-related molecules are either transmembrane proteins or bound to the plasma membrane via a GPI anchor. The GPI-linked molecules (CEA, NCA, CGM6, and CGM2) do not occur in rodents, nor do two transmembrane molecules, CGM1 and CGM7, but they are found in human. In contrast, the CCAMs, which are phylogenetically conserved, are expressed in rats, mice, and humans. Several of these molecules are known under a large number of names in the literature, and this confusion over nomenclature has contributed to the complexity of this field. In this guide a simplified nomenclature, shown in the upper half of the figure, is used. Some of the many other names that are in use are shown in the lower part of the figure. In addition to the two CCAM isoforms (CCAM-L and CCAM-S) shown in the figure, several alternatively spliced isoforms with fewer Ig domains occur. They have been given distinct names, and are not shown in the figure. Also CGM1 can be alternatively spliced, yielding either a soluble isoform or a transmembrane isoform with a shorter cytoplasmic domain; only the isoform with the long cytoplasmic domain is shown in the figure. BGP, biliary glycoprotein in man; Bgp, biliary glycoprotein in mouse; CBATP, canalicular bile acid transport protein; CCAM, cell–cell adhesion molecule; CD, cluster of differentiation; CEA, carcinoembryonic antigen; CGM, CEA gene family member; gp, glycoprotein; GPI, glycosylphosphatidy linositol; HA, hepatocyte antigen; Ig, immunoglobulin; L, large; MHVR, mouse hepatitis virus receptor; NCA, non-specific cross-reacting antigen; pp, phosphoprotein; S, small; TM-CEA, transmembrane CEA.

The expression of CCAMs is developmentally regulated, and the spatiotemporal expression patterns suggest important functions during embryonic development in, for example, placental development, vascularization of the central nervous system, muscle development, and development and eruption of teeth (see references in ref. 3). There are also numerous reports showing that molecules of the CEA family are associated with malignancy (see references in refs 3 and 4). It has recently been demonstrated that CCAM-L can inhibit tumour growth of a variety of different carcinoma cells (see references in ref. 3). The tumour inhibitory activity is provided by the long cytoplasmic domain of CCAM-L.

CCAMs are multifunctional proteins with signal-regulatory properties which can regulate, for example, cell and tumour growth, respiratory activity and integrin activation in granulocytes, ecto-ATPase activity, and transport of bile salts (see references in ref. 3). In the cytoplasmic domain of CCAM-L there are two tyrosines which are located in sequences resembling ITAM (immunoreceptor tyrosine-based activation motif) and ITIM (immunoreceptor tyrosine-based inhibition motif) sequences (see references in ref. 3). Both src-family kinases and the insulin receptor kinase can phosphorylate CCAM-L, which then becomes able to bind SH2 domain-containing signalling molecules such as src-family kinases and the protein tyrosine phosphatases SHP1 and SHP2 (see references in ref. 3). CCAM can also bind the calcium-regulated protein calmodulin in a calcium-dependent manner; both the cytoplasmic domains of CCAM-L and CCAM-S contain calmodulin-binding sites (see references in ref. 3). Furthermore, both of these CCAM isoforms can dimerize in the plane of the membrane in a process which is controlled by the cells. Dimerization is regulated by the calcium concentration; binding of calmodulin leads to dimer dissociation (see references in ref. 3).

The GPI-linked molecules CEA and NCA can block cellular differentiation by an unknown mechanism, when transfected into rat myoblasts (see references in refs 3 and 4). CEA-related adhesion molecules can also serve as microbial receptors. Fimbrial proteins of several strains of *Escherichia coli* and *Salmonella* bind to D-mannosyl residues on CEA, NCA, and human CCAM (see references in ref. 3), and *Neisseria gonorrhoeae* and *Neisseria meningitidis* bind to the amino-terminal Ig domain of human CCAM, NCA, CGM1, and CEA by means of their virulence-associated Opa proteins (see references in ref. 3). Binding of Opa_{52}-expressing gonococci to CCAM on human myelomonocytic cells activates a signalling cascade involving src-like protein tyrosine kinases, Rac1, PAK, and Jun-N-terminal kinase.[27] Mouse CCAM is the receptor for the mouse hepatitis virus, a murine coronavirus (see references in refs 3 and 4).

Purification

CCAM is purified from nonionic detergent extracts of cells or plasma membranes by immunoaffinity chromatography on polyclonal and monoclonal antibodies.[5] Further purification can be achieved by gel permeation chromatography and ion exchange chromatography.[5] Different domains of CCAM have succesfully been produced as bacterial fusion proteins, or with baculovirus vectors in insect cells.[6] Fusion proteins can easily be purified by affinity chromatographic procedures utilizing the specific properties of the fusion partners.

Activities

Purified rat liver CCAM can bind both to itself in a calcium-independent manner,[7] and to calmodulin in a calcium-dependent manner (see references in ref. 3). Most activities have been observed in transfection experiments. Both isoforms of CCAM, CCAM-L and CCAM-S, can mediate cell–cell adhesion by homophilic binding (see references ref. 3) CCAM-L stimulates ecto-ATPase and bile salt transport activities, it binds src-family kinases and protein tyrosine phosphatases, and inhibits tumour formation in mice of colon, prostate, bladder, and breast carcinoma cells (see references in ref. 3). CCAM-L also stimulates internalization of the insulin receptor after insulin stimulation (see references in ref. 3). Several forms of mouse CCAM function as receptors for mouse hepatitis virus (see references in refs 3 and 4). Human CCAM, NCA, CGM1, and CEA are receptors for various bacteria (see references in ref. 3). Transfection of rat myoblasts with CEA or NCA blocks biochemical and morphological differentiation (see references in refs 3 and 4).

Antibodies

Polyclonal rabbit antibodies against rat[8] and mouse[9] CCAMs are readily raised. Some monoclonal antibodies against both rat and mouse CCAM exist,[9–12] but several laboratories have experienced difficulties in raising monoclonal antibodies against these molecules. There are several commercially available polyclonal and monoclonal antibodies against CEA, NCA, CGM6, and human CCAM.[13,14] However, the different human CEA-related molecules are highly immunologically cross-reactive. Therefore, most antibodies against human CCAM also react with CEA. Only a few monoclonal antibodies specific for human CCAM have been reported.[15,16] Antibodies, specific for the cytoplasmic domains of CCAM-L and CCAM-S have been made by immunization with synthetic peptides.[17,18]

Genes

The human CEA family genes are clustered on chromosome19q31.2.[4,19] and those of the mouse are in a syntenic segment of chromosome 7.[4,20,21] The human gene family consists of 29 different genes, 12 of which belong to the CEA-subgroup.[1,4] The rat and mouse gene families, which do not contain genes for GPI-linked proteins, have fewer members. For the transmembrane CCAMs there is one gene in human (*BGP*),[22] one gene in rat (*CCAM1*)[23] but two genes in mouse (*Bgp1* and *Bgp2*)[24,25] identified so far. Complete cDNA sequences have been published for a large number of the human, as well as the rat and the mouse CEA-related molecules.

Disease states

There are numerous reports that members of the CEA gene family are associated with malignancy (see references in refs 3 and 4). Alterations in expression patterns in various tumours have been reported for CEA, NCA, CGM2, and CCAM. Generally, it seems that CEA and NCA are upregulated whereas CCAM and CGM2 are downregulated in many carcinomas. CEA has for a long time been the most widely used tumour marker in clinical medicine and CCAM has recently been demonstrated to have tumour-inhibitory properties. However, no clear picture of the functional roles of these molecules in cancer has yet emerged.

Structure

No high resolution structure has yet appeared for any of the CEA-related proteins or their domains, but recently a low resolution structure was produced for CEA by X-ray and neutron scattering.[26] This demonstrated that the seven Ig domains of CEA are arranged in an extended rigid arrangement with an overall length of 27 to 33 nm.

In an effort to unify the nomenclature, 29 different groups involved in research in the CEA field began extensive discussions at a CEA/PSG Workshop held in Estes Park Colorado in September 1997. A Nomenclature Subcommittee was formed, and a proposal was put forward at the most recent CEA/PSG Workshop held in Ratzeburg, Germany (September 1998), where the final nomenclature was adopted by a majority vote. The name of this gene family characterized in all species will now be recognized as the CEA gene family. The family is composed of two separate branches identified as the CEACAM and the PSG branches, respectively. Three orthologous genes can unambiguously be defined in the CEACAM branch in the human, mouse and rat, i.e. *CEACAM1*, *CEACAM9*, and *CEACAM10*. These genes will therefore bear the same numbers in all species in which their products have been identified. The table shows the new nomenclature for the CEACAM branch of the CEA family.

References

1. Thompson, J. A., Grunert, F. and Zimmermann W. (1991). *J. Clin. Lab. Anal.*, **5**, 344–66.
2. Öbrink, B. (1991). *Bioessays*, **13**, 227–34.
3. Öbrink, B. (1997). *Curr. Opin. Cell Biol.*, **9**, 616–26.
4. Stanners, C. P., (ed.) (1998). *Cell adhesion and communication mediated by the CEA family: basic and clinical perspectives*. Harwood Academic, Amsterdam, The Netherlands.
5. Odin, P., Tingström, A., and Öbrink, B. (1986). *Biochem. J.*, **236**, 559–68.
6. Cheung, P. H., Luo, W., Qui, Y., Zhang, X., Earley, K., Millirons, P., and Lin, S.-H. (1993). *J. Biol. Chem.*, **268**, 24303–10.
7. Tingström, A., Blikstad, I., Aurivillius, M., and Öbrink, B. (1990). *J. Cell Sci.*, **96**, 17–25.
8. Odin, P., Asplund, M., Busch, C., and Öbrink, B. (1988). *J. Histochem. Cytochem.*, **36**, 729–39.
9. Daniels, E., Letourneau, S., Turbide, C., Kuprina, N., Rudinskaya, T., Yazova, A. C., *et al.* (1996). *Dev. Dynamics*, **206**, 272–90.
10. Hixson, D. C. and McEntire, K. D. (1989). *Cancer Res.*, **49**, 6788–794.
11. Hubbard, A., Bartles, J. R., and Braiterman, L. T. (1985). *J. Cell Biol.*, **100**, 1115–25.
12 Becker, A., Neumaier, R., Park, C.S., Gossrau, R., and Reutter, W. (1985). *Eur. J. Cell Biol.*, **39**, 417–23.
13 Stocks, S. C., Ruchaud-Sparagano, M. H., Kerr, M. A., Grunert, F., Haslett, C., and Dransfield, I. (1996). *Int. J. Immunol.*, **26**, 2924–32.
14. Skubitz, K. M., Campbell, K. D., and Skubitz, A. P. N. (1996). *J. Leuk. Biol.*, **60**, 106–17.
15. Prall, F., Nollau, P., Neumaier, M., Haubeck, H. D., Drzeniek, Z., Helmchen, U., *et al.* (1996). *J. Histochem. Cytochem.*, **44**, 35–41.

Table 1 New nomenclature for the CEA family

Old Name	New Gene Name	New Protein Name
(A) Human		
Biliary glycoprotein, BGP1-CEA, TM, NCA-160, CD66a	*CEACAM1*	CEACAM1
CEA gene family member 1, CGM1, CD66d	*CEACAM3*	CEACAM3
CEA gene family member 7, CGM7	*CEACAM4*	CEACAM4
Carcinoembryonic antigen, CEA, CD66e	*CEACAM5*	CEA
Non-specific cross-reacting antigen, NCA, NCA-50/90, NCA-90, CD66c	*CEACAM6*	CEACAM6
CEA gene family member 2, CGM2	*CEACAM7*	CEACAM7
CEA gene family member 6, CGM6, NCA-95, CD66b, CD67	*CEACAM8*	CEACAM8
(B) Mouse		
Bgp1, mCEA1, mmCGM1a, MHVR	*Ceacam1*	Ceacam1
Bgp2	*Ceacam2*	Ceacam2
Cea5, mmCGM8	*Ceacam9*	Ceacam9
Cea10	*Ceacam10*	Ceacam10
(C) Rat		
Cell-CAM105, C-CAM, C-CAMn, gp110, pp120, HA4, ecto-ATPase, CBATP	*CEACAM1*	CEACAM1
rnCGM2	*CEACAM9*	CEACAM1
C-CAM4	*CEACAM10*	CEACAM10

16. Jantscheff, P. Nagel, G., Thompson, J., von Kleist, S., Embleton, M. J., Price, M. R., and Grunert, F. (1996). *J. Leuk. Biol.* **59**, 891–901.
17. Hunter, I., Sawa, H., Edlund, M., and Öbrink, B. (1996). *Biochem. J.*, **320**, 847–53.
18. Baum, O., Troll, S. and Hixson, D. C. (1996). *Biochem. Biophys. Res. Commun.*, **227**, 775–81.
19. Brandriff, B. F., Gordon, L. A., Tynan, K. T., Olsen, A. S., Mohrenweiser, H. W., Fertitta, A., *et al.* (1992). *Genomics*, **12**, 773–9.
20. Robbins, J., Robbins, P. F., Kozak, C.A., and Callahan, R. (1991). *Genomics*, **10**, 583–7.
21. Stubbs, L., Carver, E. A., Shannon, M. E., Kim, J., Geisler, J., Generoso, E. E., *et al.* (1996). *Genomics*, **35**, 499–508.
22. Barnett, T. R., Drake, L., and Pickle, W. (1993). *Mol. Cell Biol.*, **13**, 1273–82.
23. Najjar, S. M., Boisclair, Y. R., Nabih, Z. T., Philippe, N., Imai, Y., Suzuki, Y., and Ooi, G. T. (1996). *J. Biol. Chem.*, **271**, 8809–17.
24. Nédellec, P., Dveksler, G. S., Daniels, E., Turbide, C., Chow, B., Basile, A. A., *et al.* (1994). *J. Virol.*, **68**, 4525–37.
25. Nédellec, P., Turbide, C., and Beauchemin, N. (1995). *Eur. J. Biochem.*, **231**, 104–14.
26. Boehm, M. K., Mayans, M. O., Thornton, J. D., Begent, R. H., Keep, P. A., and Perkins, S. J. (1996). *J. Mol. Biol.*, **259**, 718–36.
27. Hauck, C. R., Meyer, T. F., Lang, F., and Gulbins, E. (1998). *EMBO J.* **17**, 443–54.

B. Öbrink
Department of Cell and Molecular Biology,
Medical Nobel Institute, Karolinska Institute,
PO Box 285, S-17177 Stockholm, Sweden

CD2/LFA-3

The CD2 and LFA-3 molecules are members of the immunoglobulin superfamily that interact directly to mediate adhesion of T lymphocytes expressing CD2 to diverse cells that express LFA-3.[1,2] The interaction of CD2 and LFA-3 is one of the best characterized heterophilic adhesion mechanisms.

Synonyms

CD2: T11, E rosette receptor; LFA-3: CD58.

Protein properties

CD2 is a 50 kDa M_r glycoprotein and LFA-3 is a 55–75 kDa M_r glycoprotein that migrates on SDS–PAGE as a broad smear. CD2 and LFA-3 form a heterophilic adhesion mechanism which has an important role in T lymphocyte interactions in the context of antigen recognition and T cell development.[3,4] The structural and functional relationship of these molecules makes it appropriate to discuss them both in the same chapter. CD2 was one of the first pan T cell markers in humans.[5] In fact, CD2 was used as a marker for human T cells prior to the advent of monoclonal antibodies in that the major clinical test for T cells is the 1970s and early 1980s was sheep erythrocyte rosetting.[6] This interaction is based on the interaction of human CD2 with the sheep homologueue of LFA-3, T11TS. CD2 is expressed on T cells and some B cells in rodents.[7] CD2 interacts with LFA-3 (CD58) in humans[3] and CD48 in rodents.[8,9] LFA-3 and CD48 are widely expressed glycoproteins.[8,10] Human CD2 does not interact with CD48 in a manner that supports physiological interactions.[11] CD2 plays an important role in T cell repertoire development and in mature T cell function.[10,12]

Activation of T cell through CD2

CD2 engagement with antibodies can trigger T cell activation in a T cell antigen receptor dependent manner.[13,14] In contrast to antibody cross-linking, the interaction of CD2 with its natural ligands does not directly activate T cells.[15] This discrepancy may be due to the fundamental difference between the high affinity and multivalent interaction of pairs of antibodies compared to the low affinity and transient interaction of CD2 with LFA-3 and other natural ligands.

Low affinity interaction of CD2 and LFA-3 in solution leads to a high 2D affinity in contact areas

The affinity of CD2 for LFA-3 has been determined by surface plasmon resonance.[11] The interaction of CD2 and LFA-3 has a low affinity in solution with a k_d of 15 μM at 37°C. The off-rate is estimated to be < 5^{-1}. The interaction or rat CD2 and its major ligand rat CD48 has an even lower affinity of 60–80 μM and a similar off-rate of < 5^{-1}.[16] Despite this low affinity in solution, both the human and rat adhesion mechanisms are remarkably efficient at forming bonds in cell–cell and cell–artificial bilayer contact areas. Recently, the 2D k_d for interaction of CD2 and CD58 has been determined. The 2D k_d for interaction of human CD2 and LFA-3 is on the order of 2 molecules/μm^2.[17,18] This 2D k_d indicates that physiological densities of CD2 and LFA-3, which are of the order of 100–200 molecules/μm^2, will drive efficient equilibrium binding in a self-assembled contact area. The best explanation for the high 2D affinity of the CD2/LFA-3 interaction is that the cooperative activity of many bonds aligns the interacting cell membranes with ~1 nm precision, in effect concentrating the CD2 and LFA-3 binding sites in a very small volume.[17] The fast off-rate of the CD2/LFA-3 interaction is evident in rapid bond turnover in contact areas.[18] Thus, adhesion molecules produce a situation that is not typically encountered in cell–cell signalling through soluble molecules, an extremely large number (< 10 000 easily) of transient bonds. These unique kinetics may explain why cell–cell adhesion through CD2/LFA-3 interactions does not trigger T cell activation, the individual bonds may not exist long enough to allow assembly of signalling complexes. This is an important point since activation directly through CD2 alone could be disastrous in terms of T cell selection or autoimmunity.

The nature of the CD2/LFA-3 interaction

The crystal structure of rat CD2 showed not only the expected immunoglobulin like folding arrangement of the two domains, but revealed a bonus, a dimeric unit cell in which two CD2 molecules interact through the previously defined CD48 binding interface.[19] Alan Williams had proposed earlier that the CD2/LFA-3 interaction may have evolved from a homophilic interaction of ancestral CD2 molecules.[20] The Jones *et al.* crystal structure appeared to capture this ancestral interaction at 1.8 nm resolution, including prediction of several specific charged interactions that could be modelled on to a CD2/CD48 interaction (Fig. 1). These predictions have been tested by site directed mutagenesis and the importance of the predictions is elegantly upheld.[2] The interface between CD2 and CD48 contains many charged residues that form salt bridges. Interestingly, when mutated, these charged interactions are shown to make no contribution to the affinity of interaction, although they contribute to the specificity.[59] The structure of rat CD2 was also solved by NMR. Using NMR, it was possible to focus on the environment of the CD2 residues thought to be involved in binding CD48 in the absence and presence of the ligand.[21] The changes in NMR spectra for these residues directly demonstrated that they were involved in a binding interaction with CD48. The crystal structure of CD2 also predicts that the length of the bond is 15 nm. Combining this information with the 2D k_d studies, the conclusion can be reached that the CD2/LFA-3 interaction will hold interacting membranes at a distance of 15±1 nm.

A model for CD2 cooperation with the T cell antigen receptor

The T cell antigen receptor is a unique type of signalling machine in that it is tasked with distinguishing a subtle structural difference between MHC molecules with self-peptides and the same MHC molecules with foreign peptides. The affinity of the TCR interaction with activating foreign MHC–peptide complexes is low, with k_d of 50 μM,

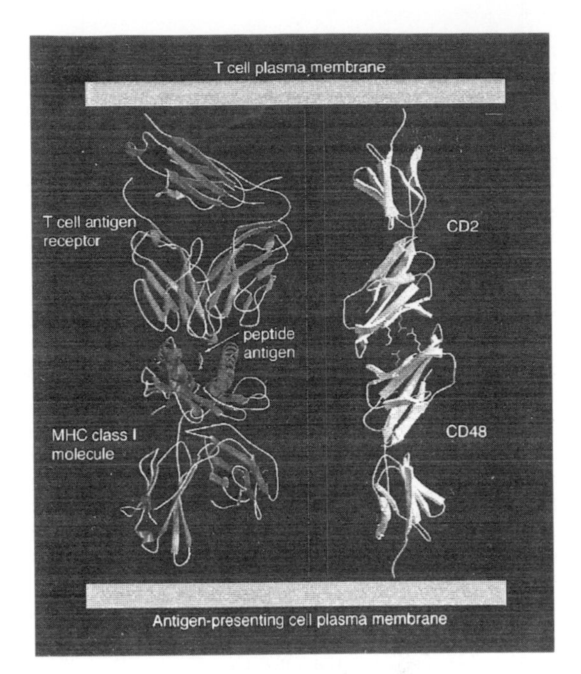

Figure 1. Model for interaction of rat CD2 and rat CD48. The T-cell receptor structure is included to demonstrate the similarity in the gap spanned for CD2/ligand and T-cell receptor/ligand interactions. (Figure provided by P. A. van der Merwe, Oxford University.)

and as few as 100 MHC peptide complexes on an antigen presenting cell (0.1 molecules/μm^2) can activate a T cell.[22,23] The TCR is also a relatively short molecule which when interacting with MHC peptide complexes, spans a gap of 15 nm (Fig. 1). The observation that CD2/LFA-3 and TCR/MHC interactions share the same bond length suggests that the numerically superior CD2/CD58 interaction may set up an alignment of the T cell and antigen presenting cell membranes in which both the CD2/LFA-3 and TCR/MHC interaction will have an optimal 2D affinity. Another T cell adhesion molecule that shares a ~15 nm bond length is the CD28/CD80 interaction.[24] The hypothesis is that T cells compensate for low antigen density on antigen presenting cells by using professional adhesion molecules to bring the membranes into alignment and increase the concentration of the TCR and MHC–peptides complexes. CD2/LFA-3, CD28/CD80, and other adhesion mechanisms may share this function. This may explain the subtle phenotypes of individual CD2 and CD28 knockout mice.[25,26] In fact, preliminary data on the phenotype of the CD2/CD28 double null mutant mouse support the idea that CD2 and CD28 have partially overlapping functions (J. Green and A. Shaw, personal communication).

CD2 mediated signalling

In addition to this extracellular function of aligning membranes, it is clear that CD2 can participate in signal transduction.[27–29] CD2 has a large cytoplasmic tail of 116 amino acids. The tail contains three proline rich regions that can interact with proteins containing SH3 domains. It is know that src family tyrosine kinases interact with these sequences.[30] It is also likely that other classes of SH3 domain-containing molecules may interact with these sequences. Functionally, the cytoplasmic tail is required for optimal activation of T cell by antigen presenting cells bearing the appropriate MHC–peptide complexes and LFA-3.[31–33] These interactions appear to be important *in vivo* also since a human CD2 transgene in mice, which does not interact with mouse ligands, has a dominant negative phenotype for positive selection of thymocytes expressing a well characterized T cell receptor as a transgene.[12] The proximity of CD2 and the TCR in the same subdomains of the contact area, determined by the shared bond length or by direct lateral interactions, may contribute to this synergistic signalling since CD2 could then recruit kinases and cytoskeletal components to the site of TCR engagement.

Regulation of CD2 mediated adhesion

As long as LFA-3 is laterally mobile, CD2 is a constitutively active adhesion mechanism.[34] When LFA-3 is immobilized artificially on a substrate, CD2 is much less effective at mediating adhesion.[32,34] Under conditions of LFA-3 immobility, T cell activation results in a dramatic increase in T-cell adhesion. The mechanism of this increase in not known but may involve modulation of CD2 lateral mobility on the cell surface.[35] In cell–cell interactions, this antigen receptor mediated adhesion strengthening is subtle. This may reflect the lateral mobility of LFA-3 on the surface of antigen presenting cells. The CD2 adhesion mechanism is indirectly regulated by the surface charge of the interacting cells. CD2 and LFA-3 are buried in the glycocalyx of the T cell and antigen presenting cells and the repulsive activity of the negatively charged sialic acid may greatly reduce the rate of CD2/CD58 encounters until a contact area is established and the membranes are brought into close range.[1] Thymocytes and activated T cells have lower surface charge than resting T cells and these changes in surface charge may modulate the activity of the CD2/LFA-3 adhesion mechanism.

Purification

CD2 can be purified from human T cells using TS2/18 affinity chromatography.[36] Rat CD2 can be purified using OX34.[16] Human LFA-3 can be purified using TS2/9.[37] Rat CD48 can be purified using OX45.[16] All of these purification strategies involve elution at pH 3, which is well tolerated by Ig domains and results in full retention of activity. It is generally accepted that these proteins have fixed natural conformations and do not exist in low and high affinity forms.

Activities

Full length CD2 activity can be measured by direct binding to LFA-3 expressing cells following iodination and adsorption of detergent with excess BSA.[38] LFA-3 can be assayed by forming protein micelles by rapid detergent removal, iodinating the micelles, and testing for binding to CD2 positive cells.[39] These assays are multivalent. The most effective presentation for adhesion is to reconstitute LFA-3 (GPI form) or CD48 into liposomes followed by formation of glass supported planar bilayers.[40] The density of adhesion molecules is evaluated by binding of iodinated mAb. As little as 20 molecules/μm^2 of LFA-3 or 100 molecules/μm^2 rat CD48 will mediate significant adhesion of the appropriate T cells.

Antibodies

The anti-human CD2 mAb TS2/18 blocks binding to LFA-3[41] and is available from ATCC (cell line) or Endogen (Woburn, MA). The anti-human CD2 mAb 6F10.3 does not block binding to LFA-3[42] and is available form Immunotech (Westbrook, ME). Anti-rat (OX34) and anti-mouse (RM2–5) mAb are available from Biosource (Camarillo, CA) and Pharmingen (San Diego, CA), respectively. Anti-human LFA-3 mAb TS2/9 blocks binding to CD2[41] and is available from ATCC or Endogen (Woburn, MA). Anti-rat CD48 mAb OX45 is available from Serotech, USA, Westbrook, ME).

Genes

cDNA sequences have been reported for human CD2[43–45] (M16445, M16336, M14362), rat CD2 [51] (XO5111), mouse CD2[47–49] (Y00023, X06143, M18934), horse CD2[50] (X69884), human LFA-3[51,52] (X06296, Y00636), rat CD48[53] (X13016), and mouse CD48[54] (X17501, X53526). The genomic organization has been determined for the human and mouse CD2 genes[55,56] (J03622, J03623, X07871–74). PDB accession codes: rat CD2: 1HNG; human CD2: 1GYA and 1HNF.

Structures

LFA-3 is homologueous to CD2 in having two Ig-like domains.[51] LFA-3 is heavily glycosylated with six N-linked glycans. LFA-3 can be GPI-anchored or may have a short transmembrane domain.[57] Both isoforms arise from the same gene by alternative splicing.[52] There is no evidence that this splicing is regulated and all cells examined have had equal ratios of both forms. CD48 exists only in a glycolipid-anchored form.[53] The GPI-anchored forms are predicted to have a relatively high lateral mobility and to associate with glycolipid enriched membrane domains.[58]

References

1. Dustin, M. L. and Springer, T. A. (1991). *Ann. Rev. Immunol.*, **9**, 27–66.
2. van der Merwe, P. A., McNamee, P. N., Davies, E. A., Barclay, A. N., and Davis, S. J. (1995). *Curr. Biol.*, **5**, 74–84.
3. Shaw, S., Luce, G. E. G., Quinones, R., Gress, R. E., Springer, T. A., and Sanders, M. E. (1986). *Nature*, **323**, 262–4.
4. Vollger, L. W., Tuck, D. T., Springer, T. A., Haynes, B. F., and Singer, K. H. (1987). *J. Immunol.*, **138**, 358–63.
5. Reinherz, E. L., Kung, P. C., Goldstein, G., and Schlossman, S. F. (1979). *Proc. Natl Acad. Sci. USA*, **76**, 4061–5.
6. Hünig, T., Mitnacht, R., Tiefenthaler, G., Kohler, C., and Miyasaka, M. (1986). *Eur. J. Immunol.*, **16**, 1615–21.
7. Nakamura, T., Takahashi, K., Fukazawa, T., *et al.* (1990). *J. Immunol.*, **145**, 3628–34.
8. Kato, K., Koyanagi, M., Okada, H., *et al.* (1992). *J. Exp. Med.*, **176**, 1241–9.
9. Brown, M. H., Cantrell, D. A., Brattsand, G., Crumpton, M. J., and Gullberg, M. (1989). *Nature*, **339**, 551–3.
10. Krensky, A. M., Sanchez-Madrid, F., Robbins, E., Nagy, J., Springer, T. A., and Burakoff, S. J. (1983). *J. Immunol.*, **131**, 611–16.
11. van der Merwe, P. A., Barclay, A. N., Mason, D. W., *et al.* (1994). *Biochemistry*, **33**, 10149–10160.
12. Melton, E., Sarner, N., Torkar, M., *et al.* (1996). *Eur. J. Immunol.*, **26**, 2952–63.
13. Bockenstedt, L. K., Goldsmith, M. A., Dustin, M., Olive, D., Springer, T. A., and Weiss, A. (1988). *J. Immunol.*, **141**, 1904–11.
14. Alcover, A., Alberini, C., Acuto, O., *et al.* (1988). *EMBO J.*, **7**, 1973–7.
15. Mentzer, S. J., Smith, B. R., Barbosa, J. A., Crimmins, M. A. V., Herrmann, S. H., and Burakoff, S. J. (1987). *J. Immunol.*, **138**, 1325–30.
16. van der Merwe, P. A., Brown, M. H., Davis, S. J., and Barclay, A. N. (1993). *EMBO J.*, **12**, 4945–54.
17. Dustin, M. L., Ferguson, L. M., Chan, P. Y., Springer, T. A., and Golan, D. (1996). *J. Cell Biol.*, **132**, 465–74.
18. Dustin, M. L. (1997). *J. Biol. Chem.*, **272**, 15782–8.
19. Jones, E. Y., Davis, S. J., Williams, A. F., Harlos, K., and Stuart, D. I. (1992). *Nature*, **360**, 232–9.
20. Williams, A. F. and Barclay, A. N. (1988). *Annu. Rev. Immunol.*, **6**, 381–405.
21. McAlister, M. S., Mott, H. R., van der Merwe, P. A., Campbell, I. D., Davis, S. J., and Driscoll, P. C. (1996). *Biochemistry*, **35**, 5982–91.
22. Matsui, K., Boniface, J. J., Steffner, P., Reay, P. A., and Davis, M. M. (1994). *Proc. Natl Acad. Sci. USA*, **91**, 12862–6.
23. Harding, C. V. and Unanue, E. R.(1990). *Nature*, **346**, 574–6.
24. Shaw, A. S. and Dustin, M. L. (1997). *Immunity*, **6**, 361–9.
25. Killeen, N., Stuart, S. G., and Littman, D. R. (1992). *EMBO J.*, **11**, 4329–36.
26. Kündig, T. M., Shahinian, A., Kawai, K., *et al.* (1996). *Immunity*, **5**, 41–52.
27. Meuer, S. C., Hussey, R. E., Fabbi, M., *et al.* (1984). *Cell*, **36**, 897–906.
28. Bernard, A., Knowles, R. W., Naito, K., *et al.* (1986). *Hum. Immunol.*, **17**, 388–405.
29. Bierer, B. E., Peterson, A., Gorga, J. C., Herrmann, S. H., and Burakoff, S. J.(1988). *J. Exp. Med.*, **168**, 1145–56.
30. Bell, G. M., Fargnoli, J., Bolen, J. B., Kish, L., and Imboden, J. B. (1996). *J. Exp. Med.*, **183**, 169–78.
31. He, Q., Beyers, A. D., Barclay, A. N., and Williams, A. F. (1988). *Cell*, **54**, 979–84.
32. Hahn, W. C. and Bierer, B. E. (1993). *J. Exp. Med.*, **178**, 1831–6.
33. Chang, H.-C., Moingeon, P., Lopez, P., Krasnow, H., Stebbins, C., and Reinherz, E. L. (1989). *J. Exp. Med.*, **169**, 2073–83.
34. Chan, P. Y., Lawrence, M. B., Dustin, M. L., Ferguson, L. M., Golan, D. E., and Springer, T. A. (1991). *J. Cell Biol.*, **115**, 245–55.
35. Liu, S. J., Hahn, W. C., Bierer, B. E., and Golan, D. E. (1995). *Biophys. J.*, **68**, 459–70.
36. Plunkett, M. L. and Springer, T. A. (1986). *J. Immunol.*, **136**, 4181–7.
37. Dustin, M. L., Sanders, M. E., Shaw, S., and Springer, T. A. (1987). *J. Exp. Med.*, **165**, 677–92.
38. Selvaraj, P., Plunkett, M. L., Dustin, M., Sanders, M. E., Shaw, S., and Springer, T. A. (1987). *Nature*, **326**, 400–3.
39. Dustin, M. L., Olive, D., and Springer, T. A. (1989). *J. Exp. Med.*, **169**, 503–17.
40. McConnell, H. M., Watts, T. H., Weis, R. M., and Brian, A. A. (1986). *Biochim. Biophys. Acta*, **864**, 95–106.
41. Sanchez-Madrid, F., Krensky, A. M., Ware, C. F., *et al.* (1982). *Proc. Natl Acad. Sci. USA*, **79**, 7489–93.
42. Olive, D., Cerdan, C., Ragueneau, M., and Mawas, C. (1987). In *Leukocyte typing III* (ed., A.J. McMichael), Vol. **1**, 148–53. Oxford University Press, Oxford.
43. Seed, B. and Aruffo, A. (1987). *Proc. Natl Acad. Sci. USA*, **84**, 3365–9.
44. Sewell, W. A., Brown, M. H., Dunne, J., Owen, M. J., and Crumpton, M. J. (1986). *Proc. Natl Acad. Sci. USA*, **83**, 8718–22.
45. Sayre, P. H., Chang, H.-C., Hussey, R. E., *et al.* (1987). *Proc. Natl Acad. Sci. USA*, **84**, 2941–5.
46. Williams, A. F., Barclay, A. N., Clark, S. J., Paterson, D. J., and Willis, A. C. (1987). *J. Exp. Med.*, **165**, 368–80.
47. Sewell, W. A., Brown, M. H., Owen, M. J., Fink, P. J., Kozak, C. A., and Crumpton, M. J. (1987). *Eur. J. Immunol.*, **17**, 1015–20.
48. Clayton, L. K., Sayre, P. H., Novotny, J., and Reinherz, E. L. (1987). *Eur. J. Immunol.*, **17**, 1367–70.
49. Yagita, H., Okumura, K., and Nakauchi, H. (1988). *J. Immunol.*, **140**, 1321–6.
50. Tavernor, A. S., Kydd, J. H., Bodian, D. L., *et al.* (1994). *Eur. J. Biochem.*, **219**, 969–76.
51. Seed, B. (1987). *Nature*, **329**, 840–2.
52. Wallner, B. P., Frey, A. Z., Tizard, R., *et al.* (1987). *J. Exp. Med.*, **166**, 923–34.
53. Killeen, N., Moessner, R., Arvieux, J., Willis, A., and Williams, A. F. (1988). *EMBO J.*, **7**, 3087–91.
54. Wong, Y. W., Williams, A. F., Kingsmore, S. F., and Seldin, M. F. (1990). *J. Exp. Med.* **171**, 2115–30.
55. Diamond, D. J., Clayton, L. K., Sayre, P. H., and Reinherz, E. L. (1988). *Proc. Natl Acad. Sci. USA*, **85**, 1615–9.
56. Lang, G., Wotton, D., Owen, M. J., *et al.* (1988). *EMBO J.*, **7**, 1675–82.
57. Dustin, M. L., Selvaraj, P., Mattaliano, R. J., and Springer, T. A. (1987). *Nature* **329**, 846–8.
58. Ishihara, A., Hou, Y., and Jacobson, K. (1987).*Proc. Natl Acad. Sci. USA*, **84**, 1290–3.
59. Davis, S. J., Davies, E. A., Tucknott, M. G., Jones, E. Y., and van der Merwe, P. A. (1998). *Proc. Natl Acad. Sci. USA*, **95**, 5490–94.

Michael L. Dustin
Department of Pathology,
Washington University School of Medicine,
St Louis, MO 63110, USA

Timothy A. Springer
Center for Blood Research and
Department of Pathology,
Harvard Medical School,
Boston, MA 02115, USA

CD44

CD44 is a family of glycoproteins generated by alternative splicing of a single gene that are expressed on many cell types of neuroectodermal and mesenchymal origin and belong to the link module superfamily. CD44 undergoes a variety of activation or differentiation-dependent, cell-type specific post-translational modifications that can affect its ligand specificity and affinity. CD44 is a principal receptor for the extracellular matrix glycosaminoglycan hyaluronan and can exhibit affinity for several extracellular matrix proteins. The known functions of CD44 indicate that it may play an important role in inflammation by virtue of its capacity to mediate cell–cell and cell–matrix interactions as well as signal transduction in leukocytes.

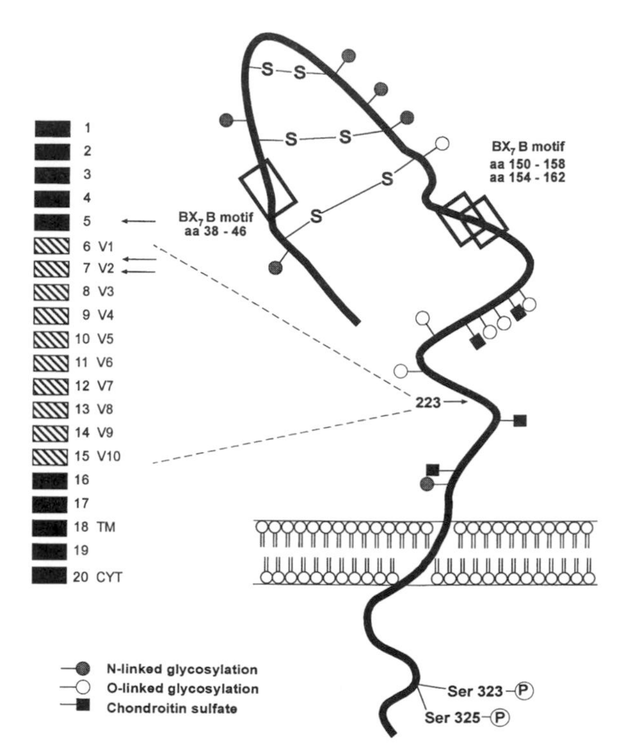

Figure 1. Schematic representation of the standard and variant forms of CD44. A schematic of CD44s (right) shows the position of the three potential HA-binding motifs (BX_7B). The insertion site for the variant peptides is indicated at amino acid 223. On the left, the exons encoding the extracellular, transmembrane, and cytoplasmic domains present in all isoforms of CD44 are indicated by the black boxes. The alternatively spliced exons that encode the variant peptides that are inserted in the membrane-proximal region of the extracellular domain are indicated by the hatched boxes. The arrows indicate the location of the potential splice donor and acceptor sites.

Synonymous names

CD44, CD44s, and CD44H are used to designate the most prevalent, standard hematopoietic, 85–90 kDa form of CD44. The variant forms of CD44 that are generated by alternative splicing are designated as CD44v followed by the number of the alternatively spliced exon encoding the variant peptide region. CD44E refers to the first alternatively spliced form described that is expressed on epithelial cells.

Also known as pgp-1 (phagocyte glycoprotein-1), blood group antigen In(Lu)p80, hyaluronate receptor, extracellular matrix receptor III, and lymphocyte homing receptor.

Homologous proteins

Functionally related HA-binding motif containing proteins include cartilage link protein, aggrecan, versican, TSG-6 and RHAMM.

Protein properties

CD44 is a family of type I transmembrane glycoproteins generated by alternative splicing of a single gene.[1] The gene for CD44 maps to chromosome 11 in humans and 2 in mice.[2] Seven exons encode for the extracellular domain of CD44s and an additional 10 or 11 exons, depending on the species, encode for peptides found in the extracellular domain of the alternatively spliced isoforms (Figs. 1 and 2).[3] The most prevalent form, CD44s, has an apparent molecular weight of 85–90 kDa. The 361 amino acid peptide core of CD44s contains none of the alternatively spliced exons of the extracellular domain and has a molecular weight of ~37 kDa with the remainder of the apparent molecular weight being contributed by both *N*- and *O*-linked carbohydrate moieties. A multitude of variant isoforms of CD44 that range in molecular weight to up to 200 kDa are generated by the inclusion of one or more alternatively spliced exons that encode for variant peptides inserted in the membrane proximal region of the extracellular domain. In addition, variations in CD44 arise from post-translational modifications including variations in the *N*- and *O*-linked carbohydrate structures and covalent

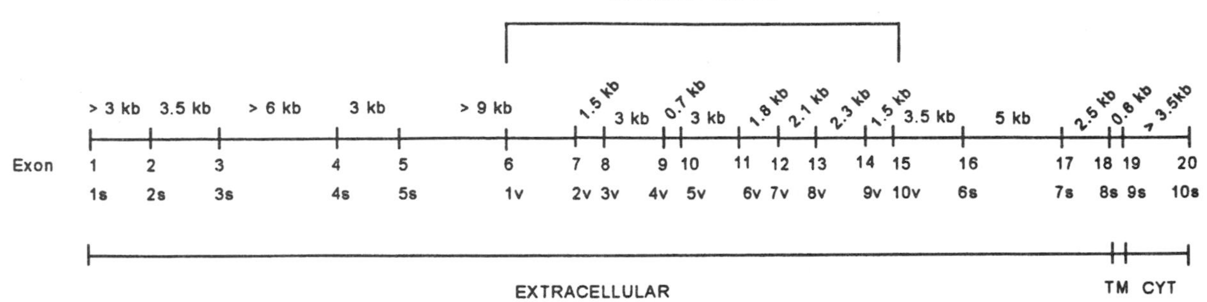

Figure 2. Genomic organization of CD44. Linear map designating the exon nomenclature and approximate lengths of introns according to Screaton *et al.*[1]

modification with chondroitin sulphate, that are both cell-type specific and vary with activation and differentiation of some cell lineages. These modifications can influence the ligand-binding affinity and/or specificity of CD44. The high sialic acid content contributes to the acidic p*I* of 4.2–5.8.

The amino terminus of the extracellular domain consists of a structural 'link module' formed by the first ~100 amino terminal residues defines members of the link module superfamily and includes cartilage link protein, TSG-6, and aggrecan.[4] This module also contains one of the three potential hyaluronan (HA) binding motifs, $B(X_7)B$ (Fig. 1).[5] The amino terminus also contains six cysteines that participate in intramolecular disulphide bridging, exhibits 80–90 per cent homology among species and contains the majority of the six potential sites of *N*-glycosylation (Fig. 1). The membrane proximal region of the extracellular domain is less well conserved (50 per cent) and contains three or four potential chondroitin sulphate attachment sites and five of the potential sites for *O*-glycosylation (Fig. 1). The transmembrane and intracellular domains of CD44 are highly conserved (80–90 per cent) among species. The intracellular domain of 72 amino acids contains five highly conserved serine residues, two (ser 325 and ser 327) of which are constitutively phosphorylated in some cell types (Fig. 1).[6] Transmembrane CD44 partitions into both the soluble and insoluble phases of detergent lysates.[6] In the case of fibroblasts the partitioning of CD44 into the detergent-insoluble fraction has been attributed to an interaction between the transmembrane domain of CD44 and plasma membrane phospholipids.[7] Other studies suggest that CD44 interacts via the cytoplasmic domain of the receptor with components of the cytoskeleton. Evidence for an association of CD44 with the cytoskeleton include colocalization with actin stress fibres in fibroblasts and co-immunoprecipitation of CD44 in a complex with members of the ezrin/radixin/moesin family of cytoskeletal proteins.[2,6] However, the importance of CD44–cytoskeletal interactions to ligand binding is not yet clear.

A soluble form of CD44 that is generated by proteolytic cleavage of the transmembrane protein is found in serum.[8] Although present at significant levels in normal serum, the levels correlate with immune activation, being elevated in autoimmune animals, during graft vs. host reactions and in tumour-bearing mice but reduced in immunodeficient animals.[8] Cleavage of CD44 can also be induced by anti-CD44 antibodies both *in vitro* and *in vivo*.[6] Finally, an alternatively spliced exon between v9 and v10 was recently described which contains stop codons and which can generate a soluble form of murine CD44.[3]

Both standard and variant isoforms of CD44 are expressed during development.[9] In the adult, CD44 is expressed on many cell types of neuroectodermal and mesenchymal origin including lymphocytes, macrophages, fibroblasts, glial cells, epithelial cells, and smooth muscle cells. Interestingly, the only cell types consistently demonstrated to lack CD44 in the adult are liver hepatocytes, kidney tubular epithelium, myocardium, and portions of the skin and testis. The variant isoforms are expressed in the adult mainly in pathologic states, on epithelial cells and certain variant epitopes (v6 and v9) are expressed transiently following stimulation of some cell types. Variant isoforms have also been detected on various tumour cell types. Upregulated expression of CD44 isoforms has been reported to correlate with tumour growth and metastatic potential of some tumour types. The mechanisms whereby CD44s or variant isoforms of CD44 may contribute to the character of some tumour types is not known but may include effects on growth regulation, matrix degradation, and angiogenesis.

Purification

CD44 can be purified from detergent extracts by affinity chromatography on wheat germ agglutinin, by ligand affinity chromatography, and with numerous monoclonal anti-CD44 antibodies. The abundance of CD44 on the surface of many cell types makes its purification relatively easy.

Activities

CD44 binds the extracellular matrix glycosaminoglycan hyaluronan.[2,6] A potentially important function of CD44 is its role in the assembly and turnover of hyaluronan-containing matrices.[10,11] CD44 also binds several protein components of extracellular matrices including collagen types I and VI,[2] fibrinogen,[12] and fibronectin[13] and may also mediate cell–cell interactions. More recently, it was reported that CD44 can bind osteopontin.[14] The affinity for some extracellular matrix components may be restricted to the 180–200 kDa chondroitin sulphate modified proteoglycan form of CD44.[2] Binding of hyaluronan to CD44-positive cells can be detected by fluorescence activated flow cytometry as well as adhesion to HA-coated surfaces or HA-containing pericellular matrices. The specificity of binding can be established with blocking anti-CD44 antibodies and based on sensitivity to hyaluronidase. Importantly, the affinity of CD44 for HA is tightly regulated. Thus, cells expressing substantial levels of CD44 may not exhibit affinity for HA, may exhibit high affinity for HA constitutively, or may be inducible by cross-linking the receptor with enhancing mAbs.[2,6] Thus, in many cell types, CD44 is required but not sufficient for HA binding. The molecular bases for regulation of CD44 adhesion function include levels of expression,[15] receptor clustering,[15] post-translational modifications including glycosylation,[16,17] serine phosphorylation of the cytoplasmic domain,[6] and potentially, interactions with the cytoskeleton.[6,15] A recent study demonstrated that cytokines may play a role in regulating CD44-mediated adhesion.[36] Thus, IL-1α, IL-1β, IL-3, GM-CSF, TNFα and LPS induced human peripheral blood monocytes to bind HA. Cytokine-, but not LPS-induced HA binding is inhibited by a neutralizing anti-TNF-α antibody.

Both CD44 standard and higher molecular weight isoforms are expressed early in ontogeny including high levels in the heart, somites and condensing limb bud mesenchyme[9] (Fig. 3). Additionally, CD44 has been noted on instructive epithelia. In the haematopoietic system, CD44 is expressed at high levels early, then wanes but is upregulated again on mature lymphoid and myeloid cells.[2] Antibodies against CD44 block the generation of lymphoid and myeloid cells in long-term bone marrow cultures and hematopoiesis *in vivo*.[2,6] The expression of CD44 on murine embryonic neurones of the developing optic chiasm inhibits embryonic retinal axon outgrowth *in vitro*.[18] However, inhibition of retinal axon growth by CD44 was insensitive to mAbs that block CD44-mediated HA binding or hyaluronidase but was partially reversed by mAb to another epitope of CD44.[18]

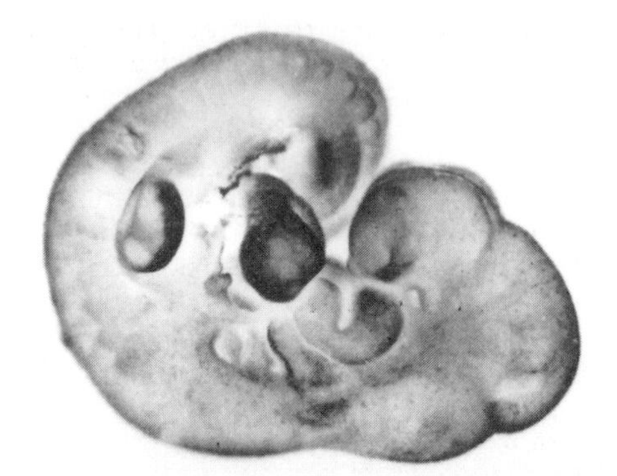

(a)

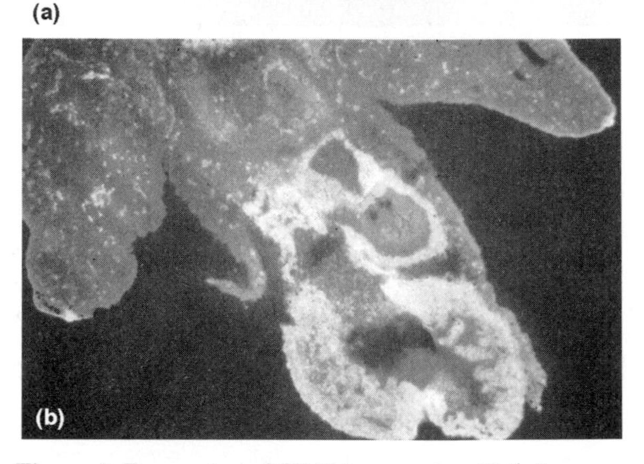

(b)

Figure 3. Expression of CD44 in a murine 10.5-day embryo. (a) Whole mount immunolocalization of CD44 in the embryonic day 10.5 embryo. Despite diffuse distribution of hyaluronan throughout the mesoderm of the embryo, there is a clear accentuation of CD44 expression within the maxillary (maxP) and mandibular processes (manP) of the first pharyngeal arches, the heart (ht), the forelimb bud (flb), and hindlimb bud (hlb). (b) Immunolocalization of CD44 expression on a cross-section through the heart and forelimb buds of a 10.5 day embryo confirms expression within apical ectodermal ridge (aer) of the limb bud and expression within the pericardium (pc) as well as the myocardium of the atrium (at) and ventricle (vt) of the developing heart. Interestingly, there is attenuation of CD44 expression by endothelial cells which have undergone epithelial:mesenchymal transformation within the endocardial cushions (edc) to begin formation of the mitral and tricuspid valves. This area is known to be particularly rich in hyaluronan. (S. Baldwin and E. Puré; unpublished observations.)

The evidence for whether or not CD44–HA interactions are involved in lymphoid trafficking by mediating lymphocyte adhesion to lymphoid high endothelial venules (HEV) has not been definitive but recent studies performed under flow rather than static conditions indicate that CD44–HA interactions can mediate lymphocyte rolling on endothelium[19] as well as rolling of lymphocytes along reticular fibres of lymphoid organ stroma that radiate from the HEV.[20] Although CD44 may contribute to normal lymphocyte circulation into lymphoid organs, it is not required.[21] CD44 may, however, be required for optimal induction of leukocyte recruitment or retention in inflammation[21] and, in fact, anti-CD44 antibodies have been used successfully to prevent tissue oedema and leukocyte infiltration in murine arthritis.[22] The proteoglycan form of CD44 may also play a role in inflammation by immobilizing cytokines such as MIP-1β.[23]

Cross-linking of CD44 with mAbs or ligation of the receptor with a low molecular weight form of HA induces production of inflammatory mediators including IL-1 and TNF, the chemokines MIP-1α, MIP-1β, RANTES, and MCP-1 and activation of the Nf-B regulatory loop.[24–26] Based on these and other data, CD44 may play an important role in regulating inflammatory and fibrotic responses. In addition, antibody cross-linking of CD44 on human T lymphocytes results in the rapid tyrosine phosphorylation of the cytoplasmic tyrosine kinases p56^{lck} and Zap-70 and can provide costimulatory signals to T cells and can induce cytotoxic responses.[2,27,28] Thus, in addition to mediating cell–cell and cell–matrix interactions, CD44 transduces signals from external stimuli leading to regulation of cellular function.

Antibodies

Many monoclonal antibodies that react with all isoforms of mouse, human, and rat CD44 have been generated and characterized. Monoclonal antibodies specific for alternatively spliced isoforms of human and, more recently, murine CD44 have been described.[29,30] In addition, antibodies to epitopes of murine CD44 that are dependent on specific post-translational modifications have also recently been generated.[30] Some of the frequently used anti-CD44 mAbs that have been characterized and their functional activities are listed in Table 1.

Genes

The sequences of the genes encoding CD44 from different species that have been deposited in the GenBank and their accession numbers are: baboon, M22452; human, M24925, M25078; mouse,M27129, M27130, M30655, J05163; hamster, M33827; canine Z27115; bovine, S63418, D12725–D12733; equine, X66862.

Mutant phenotype

A mouse strain deficient in expression of CD44 was recently generated by homologous recombination technology.[31] CD44–/– mice are born in Mendelian ratio without any obvious developmental or neurological impairments. However, CD44-deficient mice were characterized by a defect in the distribution of granulocyte-macrophage progenitor cells between bone marrow and spleen such that increased levels of CFU-GM were detected in the bone marrow while they were detected in reduced numbers in spleen. Furthermore, infection with *C. parvum* resulted in exaggerated granuloma formation in livers of CD44-deficient mice. This study also provided added evidence for a role for CD44 in tumorigenesis since SV40-transformed CD44-deficient fibroblasts

Table 1 Partial list of monoclonal anti-murine and anti-human antibodies

mAb	Name	Activity
Anti-mouse	KM201, KM81, KM114, KM703[32]	Block HA-mediated adhesion
	IM7.8.1[2]	No effect; blocks HA binding under some conditions
	IRAWB14[2]	Induces HA binding
	RAMB44[2]	Allele specific (CD44.1)
	C71/26[2]	Allele specific (CD44.2)
	H2Y2 and Lyks[30]	React with CD44v6, CD44v10 and post-translational modification-dependent determinants
Anti-human	5F12,[33] HP2/9,[34]	Block HA-mediated adhesion
	Hermes-1[2]	No effect
	Hermes-3[2]	Blocks lymphocyte binding to mucosal HEV
	F10–44–2[35]	Induces HA binding
	FW11–9–2[29]	CD44v6
	FW11–10–3[29]	CD44v4
	FW11–24–17–36[29]	CD44v9

were highly tumorigenic in nude mice and reintroduction of CD44s resulted in a dramatic inhibition in growth of these tumours.

References

1. Screaton, G. R., Bell, M. V., Jackson, F. B., Gerth, U., and Bell, J. I. (1992). *Proc. Natl Acad. Sci. USA*, **89**, 12160–4.
2. Lesley, J., Hyman, R., and Kincade, P. W. (1993). *Adv. Immunol.*, **54**, 271–335.
3. Yu, Q. and Toole, B. P. (1996). *J. Biol. Chem.*, **271**, 20603–7.
4. Kohda, D., Morton, C. J., Parkar, A. A., Hatanaka, H., Inagaki, F. M., Campbell, I. D., and Day, A. J. (1996). *Cell*, **86**, 767–75.
5. Yang, B., Yang, B. L., Savani, R. C., and Turley, E. A. (1994). *EMBO J.*, **13**, 286–96.
6. Lazaar, A. L. and Puré, E. (1995). *The Immunologist*, **3**, 19–25.
7. Perschl, A., Lesley, J., English, N., Hyman, R., and Trowbridge, I. S. (1995). *J. Cell Sci.*, **108**, 1033–41.
8. Katoh, S., McCarthy, J. B., and Kincade, P. W. (1994). *J. Immunol.*, **153**, 3440–9.
9. Wheatley, S. C., Isacke, C. M., and Crossley, P. H. (1993). *Development*, **119**, 295–306.
10. Knudson, W., Bartnik, E., and Knudson, C. B. (1993). *Proc. Natl Acad. Sci. USA*, **90**, 4003–7.
11. Underhill, C. B., Nguyen, H. A., Shizari, M., and Culty, M. (1993). *Dev. Biol.*, **155**, 324–36.
12. Henke, C. A., Roongta, U., Mickelson, D. J., Knutson, J. R., and McCarthy, J. B. (1997). *J. Clin. Invest.*,
13. Jalkanen, S. and Jalkanen, M. (1992). *J. Cell Biol.*, **116**, 817–25.
14. Weber, G. F., Ashkar, S., Glimcher, M. J., and Cantor, H. (1996). *Science*, **271**, 509–12.
15. Perschl, A., Lesley, J., English, N., Trowbridge, I., and Hyman, R. (1995). *Eur. J. Immunol.*, **25**, 495–501.
16. Katoh, S., Zheng, Z., Oritani, K., Shimozato, T., and Kincade, P. (1995). *J. Exp. Med.*, **182**, 419–29.
17. Lesley, J., English, N., Perschl, A., Gregoroff, J., and Hyman, R. (1995). *J. Exp. Med.*, **182**, 431–7.
18. Sretavan, D. W., Feng, L., Puré, E., and Reichardt, L. F. (1994). *Neuron*, **12**, 957–75.
19. DeGrendele, H. C., Estess, P., Picker, L. J., and Siegelman, M. H. (1996). *J. Exp. Med.* **183**, 1119–30.
20. Clark, R. A., Alon, R., and Springer, T. A. (1996). *J. Cell Biol.*, **134**, 1075–87.
21. Camp, R. L., Scheynius, A., Johansson, C., and Puré, E. (1993). *J. Exp. Med.*, **178**, 497–508.
22. Mikecz, K., Brennan, F. R., Kim, J. H., and Glant, T. T. (1995). *Nat Med.*, **1**, 558–63.
23. Tanaka, Y., Adams, D. H., Hubscher, S., Hirano, H., Siebenlist, U., and Shaw, S. (1993). *Nature*, **361**, 79–82.
24. Webb, D. S. A., Shimizu, Y., van Seventer, G. A., Shaw, S., and Gerrard, T. L. (1990). *Science*, **249**, 1295–7.
25. Noble, P. W., McKee, C. M., Cowman, M., and Shin, H. S. (1996). *J. Exp. Med.*, **183**, 2373–8.
26. McKee, C. M., Penno, M. B., Cowman, M., Bao, C., and Noble, P. W. (1996). *J. Clin. Invest.*, **98**, 2403–13.
27. Taher, T. E. I., Smit, L., Griffioen, A. W., Shilder-Tol, E. J. M., Borst, J., and Pals, S. T. (1996). *J. Biol. Chem.*, **271**, 2863–7.
28. Pericle, F., Sconocchia, G., Titus, J. A., and Segal, D. M. (1996). *J. Immunol.*, **157**, 4657–63.
29. Mackay, C. R., Terpe, H., Stauder, R., Marston, W. L., Stark, H., and Gunthert, U. (1994). *J. Cell Biol.*, **124**, 71–82.
30. Yamashita, Y., Hirano, H., and Hodes, R. J. (1997). *Cell. Immunol.*, **176**, 22–33.
31. Schmits, R., Filmus, J., Gerwin, N., Senaldi, G., Kiefer, F., Kundig, T., *et al.* (1997). *Blood*, **90**, 2217–33.
32. Miyake, K., Underhill, C. B., Lesley, J., and Kincade, P. W. (1990). *J. Exp. Med.*, **172**, 69–75.
33. Liao, H. X., Levesque, M. C., Patton, K., Bergamo, B., Jones, D., Moody, M. A., Telen, M. J., and Haynes, B. F. (1993). *J. Immunol.*, **151**, 6490–9.
34. de la Hera, A., Acevedo, A., Marston, W., and Sanchez-Madrid, F. (1989). *Int. Immunol.*, **1**, 598–604.
35. Dalchau, R., Kirkley, J., and Fabre, J. W. (1980). *Eur. J. Immunol.*, **10**, 737–44.
36. Levesque, M. C. and Haynes, B. F. (1996). *J. Immunol.*, **156**, 1557–65.

Ellen Puré and Jennifer Hodge-Dufour
Wistar Institute,
Philadelphia, PA, USA

Cell surface galactosyltransferase

β1,4-Galactosyltransferase I (GalTase) is one member of the glycosyltransferase family of enzymes responsible for oligosaccharide chain biosynthesis. GalTase has two distinct subcellular distributions resulting in two different biological functions. In the Golgi, GalTase catalyses the transfer of galactose from UDP-galactose to terminal *N*-acetylglucosamine residues on oligosaccharides of membrane-bound and secretory glycoproteins. GalTase is also found on the plasma membrane of many cells, where it mediates a variety of cell–cell and cell–matrix interactions by binding to appropriate glycoside ligands on adjacent cell surfaces or in the extracellular matrix.

Synonymous names

Galactosyltransferase, GalTase, GTase, GT, β1,4-GT.

Homologous proteins

GalTase is a member of the glycosyltransferase superfamily, some of which have been reported to be homologous to secreted fringe-like signalling proteins.[1] Recently, five new proteins have been identified that are homologous to GalTase (GalTII to GalTVI), thus constituting a newly defined family of β1,4-galactosyltransferases.[36–38] Whether any of these newly defined GalTs are expressed on the cell surface, as is GalTase (GalTI), is not yet known.

Protein properties

GalTase isoforms and distribution

GalTase is a 54–60 kDa type II integral membrane glycoprotein with a carboxyl-terminal catalytic domain oriented lumenally in the Golgi compartment and extracellularly on the plasma membrane (Fig. 1).[2] Most cells contain two size classes of GalTase mRNA that are produced by differential transcription initiation from a single GalTase gene.[2] The two classes of GalTase mRNA encode identical proteins except that one has an additional 13 amino acid residue in its amino-terminal cytoplasmic domain (Fig. 2). Both isoforms are found in the Golgi, where they participate in glycoprotein biosynthesis. However, a portion of the long isoform is routed to the plasma membrane, where it functions as a receptor for extracellular glycoside ligands.[3] A spermatid-specific transcript that encodes the long GalTase isoform has also been described and results from alternative polyadenylation signals.[4] A catalytically active, proteolytic fragment of GalTase is found in most bodily fluids including milk, where it interacts with α-lactalbumin to form the lactose synthetase complex.[5] Other than in milk, the function of soluble GalTase is unknown.

GalTase has been detected on the surface of virtually all cells examined, although its cell adhesion function has been studied in most detail on sperm, embryonic cells, mesenchymal cells, and neuronal cells.[6] It is unknown whether GalTase functions catalytically on the cell surface or acts in a non-enzymatic, lectin-like capacity. GalTase expression is restricted to specific plasma membrane

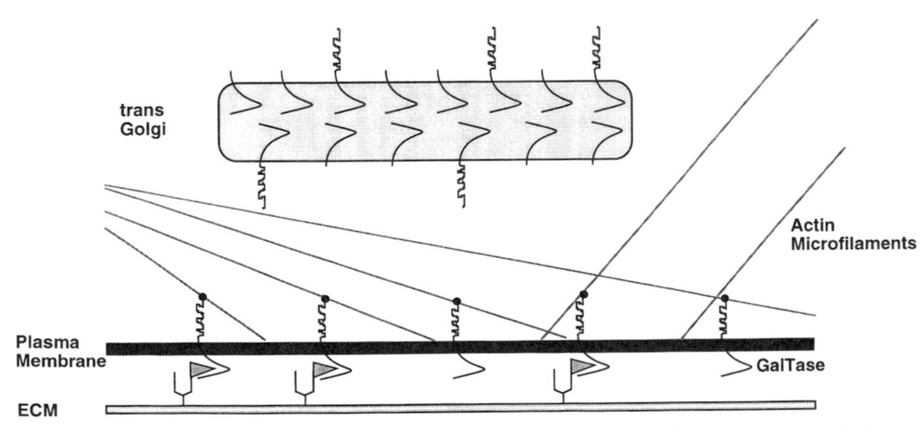

Figure 1. Cells make two different GalTase isoforms that differ in the length of their cytoplasmic domains. The shorter isoform is confined to the Golgi complex where it participates in glycoprotein biosynthesis. The longer isoform has an additional 13 amino acid cytoplasmic extension (denoted by the wavy line), and resides in both the Golgi compartment as well as on the cell surface. When expressed on the plasma membrane, GalTase associates with the cytoskeleton and signal transduction machinery enabling it to function as a receptor for extracellular glycoside ligands.

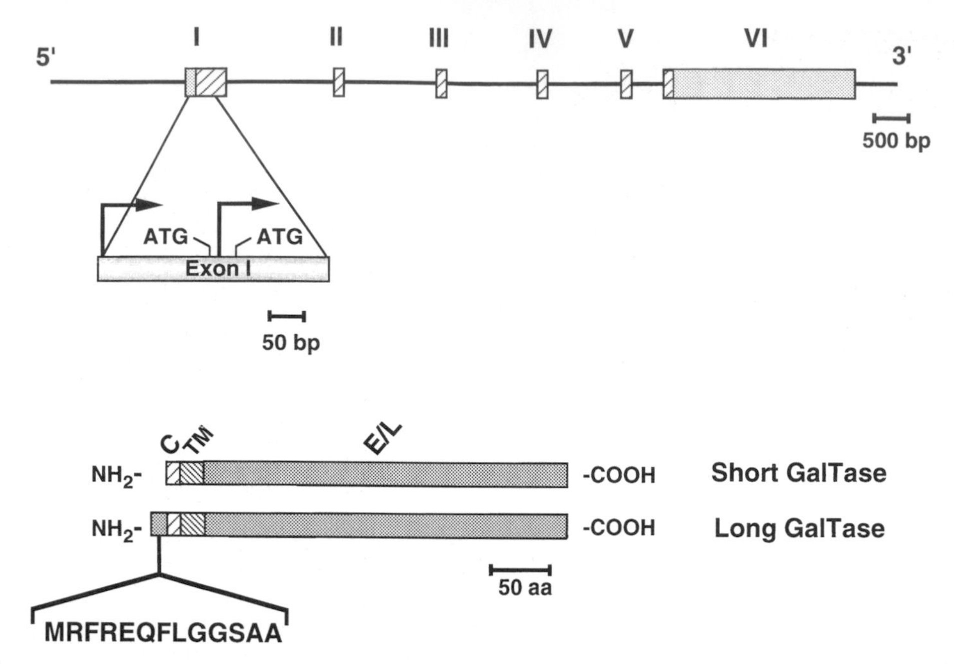

Figure 2. The murine GalTase gene has six exons, the first of which encodes two in-frame translation initiation codons. Transcription initiation upstream of the first ATG encodes the long GalTase isoform that has an amino-terminal 13 amino acid extension in the cytoplasmic domain, relative the the short GalTase isoform that is encoded by transcripts that initiate between the two ATGs. C, cytoplasmic domain; TM, transmembrane domain; E/L, extracellular/lumenally-oriented domain.

domains appropriate for its adhesive function (Fig. 3). In this regard, GalTase is localized to broad areas of intercellular contact during embryonal carcinoma cell adhesion,[7] but is restricted to the dorsal aspect of the anterior mouse sperm head, where it participates during fertilization.[8] GalTase is preferentially localized to newly formed lamellipodia[9] and neurites on mesenchymal cells and neurones, respectively, where it functions during cell spreading, migration, and neurite outgrowth by binding to *N*-linked oligosaccharides within the E8 domain of laminin.[6]

Since not all of the long GalTase isoform is found on the cell surface, some component required for GalTase transport and/or association with the cell surface must be rate-limiting. Evidence discussed below suggests that the rate-limiting elements are cytoskeletally-associated proteins.[6,10] GalTase colocalizes by indirect immunofluorescence with actin-containing microfilaments in a number of cell types. Furthermore, most surface GalTase partitions with the detergent-insoluble cytoskeleton and the degree of association is affected by the integrity of the cytoskeleton.[10]

GalTase function during fertilization

Mouse sperm are unique in that all of their GalTase is confined to the dorsal, anterior aspect of the sperm surface, where it mediates adhesion to the egg coat by binding to specific *O*-linked oligosaccharides on the ZP3 glycoprotein.[8] Following gamete binding, sperm surface GalTase is aggregated by multivalent ZP3 oligosaccharides, thus activating a pertussis toxin-sensitive G-protein complex that is associated with the GalTase cytoplasmic domain. The activated G-protein cascade induces the acrosome reaction.[11] Sperm from transgenic mice that overexpress GalTase on their surface bind more ZP3 ligand than wild-type sperm. This leads to accelerated G-protein activation and precocious acrosome reactions. Thus, overexpression of GalTase makes sperm hypersensitive to their egg ligand.[12] In contrast, sperm from GalTase-null mice are unable to bind ZP3 and fail to undergo an acrosome reaction.[13] The inability of GalTase-null sperm to bind ZP3 and undergo an acrosome reaction is due directly to a lack of GalTase on the sperm surface, and is not the secondary result of defective galactosylation during spermatogenesis.

GalTase function during cell interactions with the basal lamina

During cell migration, laminin induces GalTase expression on the cell surface, at which time it becomes associated with the cytoskeleton.[9] The cytoskeleton has a defined and saturable number of binding sites for surface GalTase, which has allowed the production of a dominant negative phenotype that interferes with surface GalTase function.[10] Similar results have been reported for the cadherins, emphasizing the importance of the cytoskeleton in regulating the function of various cell adhesion molecules.[14] Since the number of cytoskeleton-binding sites for GalTase is limiting, migration rates are inversely

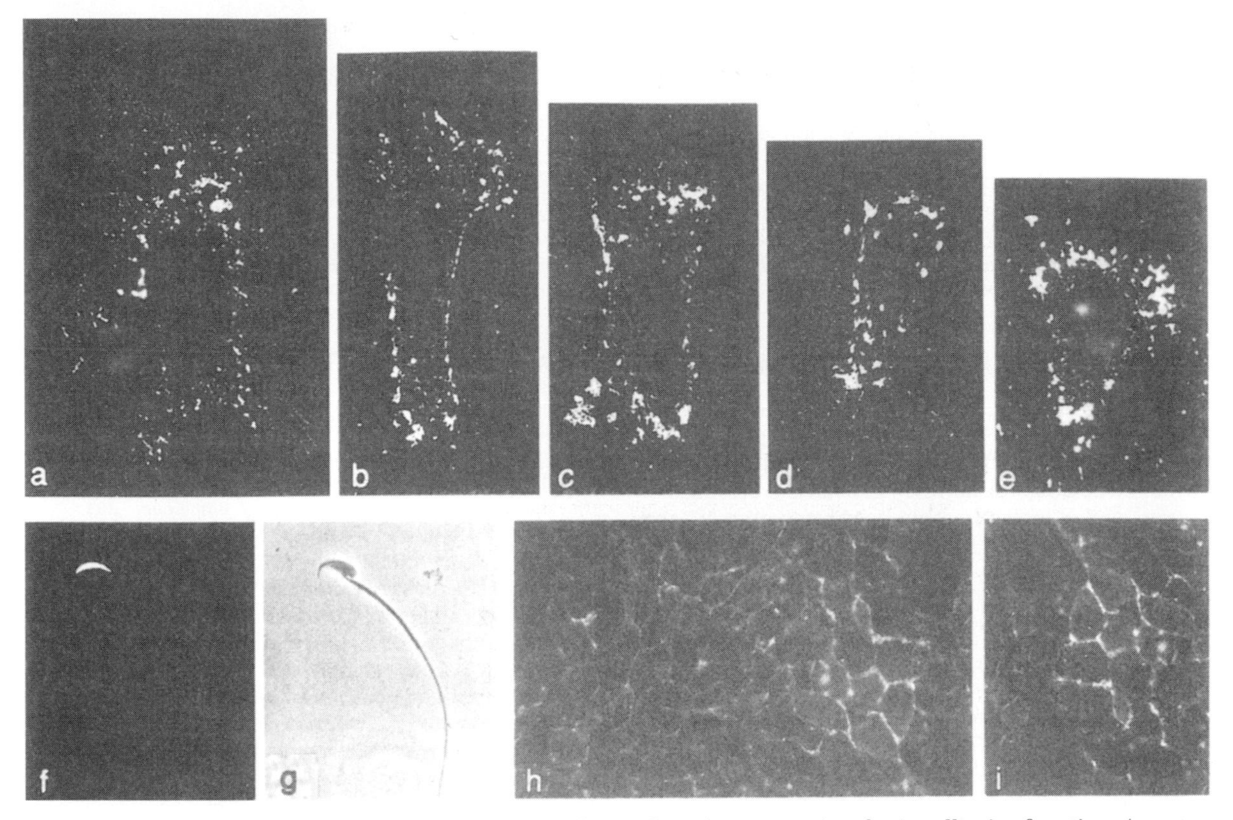

Figure 3. Surface GalTase is restricted to plasma membrane domains appropriate for its adhesive function. As mesenchymal cells polarize to initiate migration, surface GalTase becomes concentrated to the leading and trailing edges of the migrating cell (a–e). GalTase is restricted to the dorsal aspect of the anterior mouse sperm head where it mediates fertilization (f,g), and is localized to areas of intercellular contact during embryonal carcinoma cell adhesion (h,i).

related to the amount of surface GalTase expressed in stably transfected fibroblasts.[15]

Perturbing surface GalTase activity in avian embryos inhibits cell migration and leads to teratological abnormalities, suggesting that surface GalTase functions during embryonic development *in vivo*.[6,16] Overexpressing surface GalTase in transgenic mice leads to neonatal lethality, apparently by interfering with epithelial interactions with the basal lamina during branching morphogenesis. Analysis of mammary glands from these animals illustrate that overexpression of GalTase inhibits the anastomosing of epithelial cells into functional tubules.[17]

Purification

GalTase is most often purified by affinity chromatography using α-lactalbumin-conjugated columns.[18]. Additional purification is achieved by application on *N*-acetylglucosamine and/or UDP substrate affinity columns. There are many reports of GalTase purified from cell lysates and one report of GalTase purified specifically from the cell surface.[19]

Activities

GalTase is assayed by the incorporation of radiolabelled galactose, donated from UDP–galactose, to glycoside acceptor substrates terminating in *N*-acetylglucosamine. The radiolabelled galactosylated product can be isolated by high voltage paper electrophoresis, acid precipitation, ion exchange or thin layer chromatography, or can be visualized by autoradiography.[6,19] Its cell adhesion activity can be assessed by reagents that perturb enzyme activity and/or substrate binding, including appropriate antibodies, competitive substrates, α-lactalbumin, hexosaminidase digestion, and UDP–galactose.[6,8–10,16]

Antibodies

The following antibodies against GalTase have been reported: monospecific rabbit polyclonal[7] and mouse monoclonal[20] antibodies against affinity-purified bovine milk GalTase, some of which have been affinity purified on recombinant bovine fusion protein,[21] rabbit polyclonal antibodies against recombinant murine GalTase,[22] monospecific rabbit polyclonal antibodies against affinity-purified chicken[16,18] and human[23] GalTase, and rabbit polyclonal antibodies against the cytoplasmic extension unique to the long GalTase isoform.[24] Most antibodies have been used for immunolocalization and immunoprecipitation studies; the anti-recombinant antibodies have been particularly useful for inhibiting GalTase-mediated cellular interactions. Polyclonal antibodies generally cross-react with other mammalian GalTases to varying degrees, but the mouse monoclonal antibodies prepared against bovine GalTase do not cross-react with other species.

Genes

cDNAs encoding GalTase have been published for mouse (J03880, D00314), cow (J05217, M13214, M13569), human (M13701, M22921, X13223, X14085, X55415), and rat (S81025). The gene for GalTase has been characterized in mouse[25] and is located on mouse chromosome 4[26] and human chromosome 9.[27]

Mutant phenotypes/disease states

Surface GalTase activity is elevated on metastatic cells[28] and recent evidence suggests that the elevated surface expression facilitates the metastatic process. There is evidence that surface GalTase plays some undefined role in growth control, since cell proliferation can be inhibited by selectively perturbing surface GalTase activity.[29] Alterations in GalTase expression are associated with rheumatoid arthritis,[30] galactosaemia,[31] cystic fibrosis,[32] limb dysmorphogenesis,[33] and sperm and embryos that express mutant *t*-alleles.[34] Overexpression of GalTase leads to defective branching morphogenesis[17] and renders sperm hypersensitive to their egg coat ligand.[12] Eliminating GalTase produces pituitary insufficiency[35] (as a result of defective galactosylation) and selectively eliminating the long isoform renders sperm refractory to their zona ligand.[13]

Structure

Not available.

Web sites

http://www.vei.co.uk/TGN/gt_guide.htm

References

1. Yuan, Y. P., Schultz, J., Mlodzik, M., and Bork, P. (1997). *Cell*, **88**, 9–11.
2. Russo, R. N., Shaper, N. L., and Shaper, J. H. (1990). *J. Biol. Chem.*, **265**, 3324–31.
3. Evans, S. C., Youakim, A., and Shur, B. D. (1995). *BioEssays*, **17**, 261–8.
4. Shaper, N. L., Wright, W. W., and Shaper, J. H. (1990). *Proc. Natl Acad. Sci. USA*, **87**, 791–5.
5. Brodbeck, U., Denton, W. L., Tanahashi, N., and Ebner, K. E. (1967). *J. Biol. Chem.*, **242**, 1391–7.
6. Shur, B. D. (1993). *Curr. Opin. Cell Biol.*, **5**, 854–63.
7. Bayna, E. M., Shaper, J. H., and Shur, B. D. (1988). *Cell*, **53**, 145–57.
8. Miller, D. J., Macek, M. B., and Shur, B. D. (1992). *Nature*, **357**, 589–93.
9. Eckstein, D. J. and Shur, B. D. (1989). *J. Cell Biol.*, **108**, 2507–17.
10. Evans, S., Lopez, L. C., and Shur, B. D. (1993). *J. Cell Biol.*, **120**, 1045–57.
11. Gong, X., Dubois, D. H., Miller, D. J., and Shur, B. D. (1995). *Science*, **269**, 1718–21.
12. Youakim, A., Hathaway, H. J., Miller, D. J., Gong, X., and Shur, B. D. (1994). *J. Cell Biol.*, **126**, 1573–84.
13. Lu, Q.-X. and Shur, B.D. (1997). *Development*, **124**, 4121–31.
14. Kintner, C. (1992). *Cell*, **69**, 225–36.
15. Appeddu, P. A. and Shur, B. D. (1994). *Proc. Natl Acad. Sci. USA*, **91**, 2095–9.
16. Hathaway, H. and Shur, B. D. (1992). *J. Cell Biol.*, **117**, 369–82.
17. Hathaway, H. J. and Shur, B. D. (1996). *Development*, **122**, 2859–72.
18. Hathaway, H. J., Runyan, R. B., Khounlo, S., and Shur, B. D. (1991). *Glycobiology*, **1**, 211–21.
19. Shur, B. D. and Neely, C. A. (1988). *J. Biol. Chem.*, **263**, 17706–14.
20. Ulrich, J. T., Schenck, J. R., Rittenhouse, H. G., Shaper, N. L., and Shaper, J. H. (1986). *J. Biol. Chem.*, **261**, 7975–81.
21. Teasdale, R. D., D'Agostaro, G., and Gleeson, P. A. (1992). *J. Biol. Chem.*, **267**, 4084–96.
22. Nguyen, T., Hinton, D., and Shur, B. D. (1994). *J. Biol. Chem.*, **269**, 28000–9.
23. Roth, J., Lentze, M. J., and Berger, E. G. (1985). *J. Cell Biol.*, **100**, 118–25.
24. Youakim, A., Dubois, D., and Shur, B. D. (1994). *Proc. Natl Acad. Sci. USA*, **91**, 10913–17.
25. Hollis, G. F., Douglas, J. G., Shaper, N. L., Shaper, J. H., Stafford-Hollis, J. M., Evans, R. J., and Kirsch, I. R. (1989). *Biochem. Biophys. Res. Commun.*, **162**, 1069–75.
26. Shaper, N. L., Shaper, J. H., Hollis, G. F., Chang, H., Kirsch, I. R., and Zozak, C. A. (1987). *Cytogenet. Cell Genet.*, **44**, 18–21.
27. Duncan, A. M. V., McCorquodale, M. M., Morgan, C., Rutherford, T. J., Appert, H. E., and McCorquodale, D. J. (1986). *Biochem. Biophys. Res. Commun.*, **141**, 1185–8.
28. Passaniti, A. and Hart, G. W. (1990). *Cancer Res.*, **50**, 7261–71.
29. Hinton, D., Evans, S. and Shur, B. D. (1995). *Exp. Cell Res.*, **219**, 640–9.
30. Alavi, A. and Axford, J. (1995). *Adv. Exp. Med. Biol.*, **376**, 185–192.
31. Ornstein, K. S., McGuire, E. J., Berry, G. T., Roth, S., and Segal, S. (1992). *Pediatric Res.*, **31**, 508–11.
32. Rao, G. J., Spells, G., and Nadler, H. L. (1977). *Pediatric Res.*, **11**, 981–5.
33. Elmer, W. A., Pennybacker, M. F., Knudsen, T. B., and Kwasigroch, T. E. (1988). *Teratology*, **38**, 475–84.
34. Shur, B. D. and Bennett, D. (1979). *Dev. Biol.*, **71**, 243–59.
35. Lu., Q.-X., Hasty, P., and Shur, B. D. (1997). *Dev. Biol.*, **181**, 257–67.
36. Ameida, R., Amado, M., David, L., Levery, S. B., Holmes, E. H., Merkx, G., van Kessel, A. G., Rygaard, E., Hassan, H., Bennett, E., and Clausen, H. (1997). *J. Biol. Chem.*, **272**, 31979–91.
37. Sato, T., Furukawa, K., Bakker, H. van den Eijnden, D. H., and van Die, I. (1998). *Proc. Natl Acad. Sci. USA*, **95**, 472–7.
38. Lo, N.-W., Shaper, J. H., Pevsner, J., and Shaper, N. L. (1998). *Glycobiology*, **8**, 517–26.

Barry D. Shur
Department of Cell Biology,
Emory University School of Medicine,
Atlanta, GA, USA

Chemokine receptors

Chemokine receptors are seven-transmembrane-domain proteins of 350 to 368 amino acids which bind chemotactic cytokines (chemokines).[1–5] They are constitutively expressed in blood monocytes and granulocytes and induced by IL-2 in CD45RO+ T cells. The main function mediated by chemokine receptors is the trafficking of blood leukocytes to sites of inflammation and disease or to sites of haematopoiesis and primary antigen recognition. Certain chemokine receptor are essential coreceptors, together with CD4, for infection by HIV-1.[6,7]

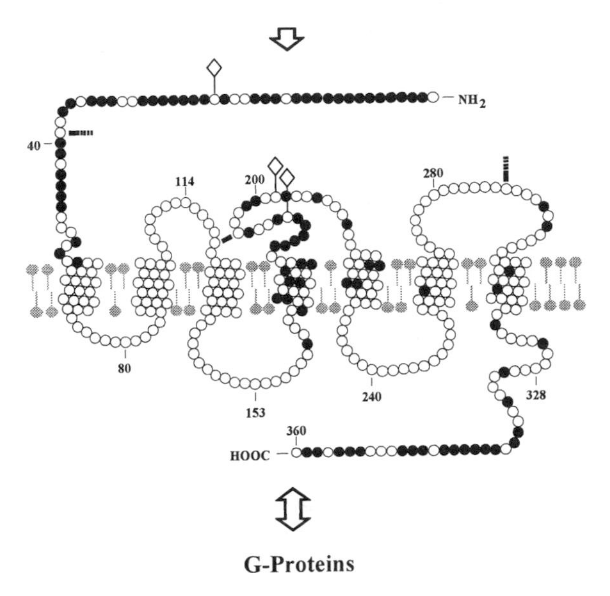

Figure 1. Schematic representation of the human IL-8 receptors, CXCR1 and CXCR2. Amino acid positions that differ between CXCR1 and CXCR2 are indicated by filled circles, diamonds designate putative *N*-linked glycosylation sites, and pinhead symbols represent lipids in the plasma membrane. Bold lines denote two potential disulphide bridges formed by Cys_{119}/Cys_{196} and Cys_{39}/Cys_{287}, respectively.

Synonymous names

The new nomenclature and synonymous names for human chemokine receptors are listed in Table 1. The first chemokine receptors to be identified are the human interleukin 8 (IL-8) receptors[8,9] and designated Cdw128 according to the nomenclature rules for human leukocyte differentiation antigens.

Protein properties

Size/structure

Chemokine receptors are members of the large family of serpentine receptors with seven transmembrane domains (TM) which couple to *B. pertussis* toxin-sensitive heterotrimeric G-proteins for signal transduction.[1–5] Figure 1 shows the schematic representation of

Table 1 Human chemokine receptors

	Receptor nomenclature		
	New	Old	Chemokine selectivity
CXC chemokines	CXCR1	IL-8R1 (type A)	IL-8
	CXCR2	IL-8R2 (type B)	IL-8, GROα, β, γ, NAP-2, ENA78, GCP-2
	CXCR3	IP10/MigR	IP10, Mig, I-TAC
	CXCR4	LESTR, HUMSTR, fusin	SDF-1
	CXCR5	BLR1	BCA-1 (BLC)
CC chemokines	CCR1	RANTES/MIP-1αR	RANTES, MIP-1α, MCP-2, MCP-3
	CCR2a/b	MCP-1RA/B	MCP-1, MCP-2, MCP-3, MCP-4
	CCR3	EotaxinR, CKR-3, CC CKR3	Eotaxin, RANTES, MCP-3, MCP-4
	CCR4	CC CKR4	TARC, MDC
	CCR5	CC CKR5	RANTES, MIP-1α, MIP-1β
	CCR6	CKR-L3, STRL33	LARC (MIP-3α)
	CCR7	EBI1, BLR2	ELC (MIP-3β), SLC
	CCR6	TER1, ChemR1, CKR-L1	I-309
	CCR10	D6	CC chemokines
CX_3C chemokine	CX_3CR1	V28	Neurotactin/fraktalkine
C chemokine	CR1	GPR5	Lymphotactin

the two human IL-8 receptors, CXCR1 and CXCR2. The 16 known human chemokine receptors are glycoproteins of 350–368 amino acids and are grouped into four classes, CXCRs, CCRs, CX_3CR and CR, depending on their selectivity for either CXC, CC, CX_3C or C chemokines (Table 1). The majority of chemokine receptors recognize either CXC or CC chemokines whereas only one CX_3C and C chemokine receptor are known. Amino acid sequence identity among CXC and CC chemokine receptors is 36–77 per cent and 46–74 per cent, respectively. Highest sequence conservation is found in the seven TM regions and in the second intracellular loops (IC2) which contain the 'DRYLAIVHA' motif. Several orphan receptors are known with undefined ligand selectivity but prominent sequence similarity to chemokine receptors.

The model for the structure of chemokine receptors proposes that the seven hydrophobic domains form membrane-spanning α-helices which may be linked by disulphide bonds and which define three extracellular and three intracellular loop regions. Studies with anti-receptor antibodies have demonstrated that the amino-terminal regions are extracellular.

Function

Chemokines are soluble proteins of 68 to 127 amino acids with four conserved cysteine residues linked by two essential disulphide bonds.[1,10–12] Based on the arrangement of the first two cysteines, human chemokines are divided into two subfamilies, CXC and CC chemokines. Two additional chemokines, lymphotactin/fractalkine (in which the N-terminal two cysteines are separated by three amino acids) and lymphotactin (which lacks two of the four conserved cysteines) are designated as CX_3C and C chemokine, respectively. Figure 2 shows the amino acid sequences and disulphide linkages of the two prototypical CXC and CC chemokines, IL-8 and monocyte chemotactic protein 1 (MCP-1); Table 2 lists the currently known human chemokines. The majority of chemokines are induced locally during inflammation by tissue cells and infiltrated leukocytes and recruit effector leukocytes from blood. Another emerging class of chemokines are produced constitutively in lymphoid organs and regulate lymphocyte trafficking at sites of haematopoiesis and lymphocyte maturation. In addition to chemotaxis, chemokine-mediated activation of leukocytes and lymphocytes results in enzyme release from intracellular stores, oxygen radical formation, shape change through cytoskeletal rearrangement, generation of lipid mediators, and induction of adhesion to endothelium or extracellular matrix proteins. Immune response-unrelated functions of chemokines have been reported in haematopoiesis,[1,2,4,5] angiogenesis,[1,2,4,5] and, as recently shown for SDF-1,[13a,b] in embryogenesis. Chemokines may also be important in angiogenesis but the receptors they bind to in tissue cells are not defined. Certain chemokine receptors, including CXCR4 and CCR5, are essential for infection of CD4-bearing blood cells by HIV-1.[6,7]

Chemokine binding

Chemokine binding is complex and involves multiple, in part overlapping (rather than single) extracellular regions and possibly sites formed by the transmembrane domains.[14,15] The extracellular amino-terminal regions are highly charged with an excess of acidic residues which is typical for chemokine receptors and shown to be essential for binding of chemokines, which themselves are basic. Of over 40 human chemokines, a total of 28 function as ligands for the 16 human chemokine receptors (Table 2).[1,3,11] Most chemokine receptors recognize more than one chemokine and many chemokines bind to more than one receptor. Chemokine–receptor dissociation constants (K_ds) range from 0.2 to 4 nM.

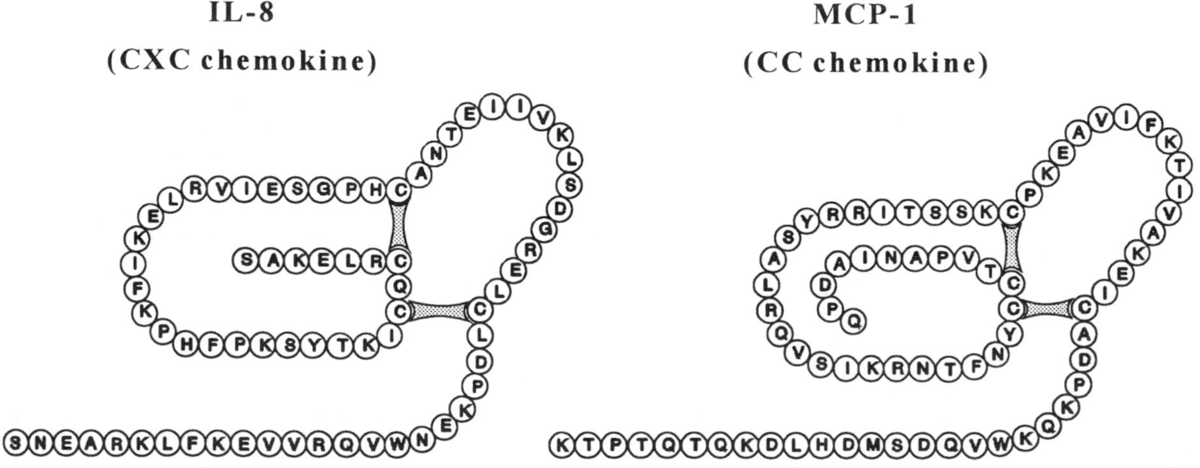

Figure 2. 'Two-loop' structures of IL-8 and MCP-1. The mature 72 amino acid and the 76 amino acid forms of IL-8 and MCP-1, respectively, are shown.

Table 2 Human chemokines with known function on leukocytes and/or lymphocytes

CXC Chemokines	IL-8	Interleukin 8
	GROα, β, γ	Growth-related protein α, β, γ
	NAP-2	Neutrophil-activating peptide 2
	ENA78	Epithelial cell-derived neutrophil-activating peptide 78
	GCP-2	Granulocyte chemotactic protein 2
	IP10	IFN-γ-inducible 10 kDa protein
	Mig	Monocyte/macrophage-activating, IFN-γ-inducible protein
	I-TAC	IFN-inducible, T cell-activating alpha chemokine
	SDF-1	Stromal cell-derived factor 1
	BCA-1	B cell-attracting chemokine 1
CC Chemokines	MCP-1, -2, -3, -4	Monocyte chemotactic protein 1, 2, 3, 4
	RANTES	Regulated on activation, normal T cell expressed and secreted
	MIP-1α, -1β	Macrophage inflammatory protein 1α, 1β
	Eotaxin	Eosinophil chemoattractant protein
	I-309	Intercrine-β glycoprotein 309
	TARC	Thymus and activation-regulated chemokine
	MDC	Macrophage-derived chemokine
	LARC	Liver and activation-regulated chemokine
	ELC	EBI1-ligand chemokine
	SLC	Secondary lymphoid-tissue chemokine
CX_3C chemokine		Neurotactin/fractalkine
C chemokine	Ltn	Lymphotactin

Cellular expression

Leukocytes and lymphocytes are the primary target cells for chemokine action which express the corresponding receptors whereas chemokine receptor expression in tissue cells is generally low or not detectable. Strong expression is found in monocytes, neutrophils, basophils, eosinophils, macrophages, dendritic cells, lymphocytes, and natural killer cells, with densities ranging from several hundreds to several thousands of receptors per cell.[1,3,11] Monocytes and granulocytes constitutively express chemokine receptors whereas low numbers or no receptors are found in circulating lymphocytes (except for the number of CXCR4 which is high in all blood cells). Notably, activated $CD45RO^+$ T lymphocytes with $CD4^+$ or $CD8^+$ phenotype are the principal lymphocyte populations which become activated by chemokines, and IL-2 treatment was shown to be required for expression of CXCR3, CCR1, CCR2, and CCR5.[16] IL-2 can be substituted in part by IL-4, IL-10, and IL-12 whereas cell activation *via* T cell receptor/CD3 complex leads to downregulation of receptor expression. CXCR3 is found selectively in activated T lymphocytes, the eotaxin receptor CCR3 is abundant in eosinophils, whereas CXCR1 and CXCR2 are highly expressed in neutrophils which are the primary target cells for IL-8 and other CXC chemokines. Except for CXCR3 (and possibly CCR6), chemokine receptors are not restricted to a single type of leukocytes or lymphocytes.[1,3,11]

Signal transduction

The α subunits of heterotrimeric G proteins that have been shown to interact with chemokine receptors are $G_{\alpha i2}$, $G_{\alpha i3}$, $G_{\alpha 14}$, $G_{\alpha 15}$ and $G_{\alpha 16}$.[17,18] Chemokine binding induces dissociation of G proteins from the receptor and GTP-bound G_{α} subunits and $\beta\gamma$ subunits activate a large range of intracellular signal transduction elements, including phospholipase isoforms (which leads to inositol triphosphate/diacylglycerol-mediated increases in the intracellular Ca^{2+} concentration, protein kinase C activation and the production of other metabolites), small GTP-binding proteins (rho, ras), phosphatidylinositol 3-phosphate kinases as well as serine/threonine and possibly tyrosine protein kinases.[19] Cellular responses to chemokines are generally transient and short in duration which may be due to rapid receptor inactivation through phosphorylation of serine/threonine residues by specific kinases and receptor ligand internalization.[1,3,11]

Other chemokine receptors

The Duffy blood group antigen DARC on erythrocytes, which serves as port of entry for the malarial parasite *Plasmodium vivax*, is a highly promiscuous receptor for chemokines but lacks signalling function.[20] Certain viruses carry genes for functional chemokine receptors, including US28 from human cytomegalovirus and ECRF3 from herpes virus saimiri, which are receptors for CC and CXC chemokines, respectively.[2,3] Of note, the chemokine receptor KSHV-GPCR, encoded by Kaposi's sarcoma-associated herpesvirus 8, is oncogenic and constitutively signals in an agonist-independent fashion.[21] In mice, receptors with the same chemokine selectivity as CXCR2–5, CCR1–3, CCR5, CCR6, CCR8, and CCR10 are known.

Antibodies

Several laboratories have overcome the difficulties, due to sequence conservation in mammals and the small size of the extracellular regions, in raising anti-receptor antibodies. Monoclonal and polyclonal antibodies to CXCR1–5, CCR1–3, and CCR5 are described.

Genes

The genes for CXC chemokine receptors CXCR1, CXCR2, and CXCR4 are on human chromosome 2q21–35 and the genes for the CC chemokine receptors CCR1–5 are on human chromosome 3p21–24.[1–5] The gene for CXCR3 is on human chromosome X. Most chemokine receptors are encoded by a single exon and many receptor genes contain additional exons which encode 5'-untranslated sequences and give rise to alternative mRNA splice products.[2,3]

References

1. Baggiolini, M., Dewald, B., and Moser B. (1997). *Ann. Rev. Immunol.*, **15**, 675–706.
2. Luster, A. D. (1998). *N. Engl. J. Med.*, **338**, 436–45.
3. Murphy, P. M. (1996). *Cytokine Growth Fact. Rev.*, **7**, 47–64.
4. Gerard, C. and Gerard, N. P. (1994). *Curr. Opin. Immunol.*, **6**, 140–5.
5. Premack, B. A. and Schall, T. J. (1996). *Nature Med.*, **2**, 1174–8.
6. D'Souza, M. P. and Harden, V. A. (1996). *Nature Med.*, **2**, 1293–1300.
7. Moore, J. P., Trkola, A., and Dragic, T. (1997). *Curr. Opin. Immunol.*, **9**, 551–62.
8. Holmes, W. E., Lee, J., Kuang, W.-J., Rice, G. C., and Wood, W. I. (1991). *Science*, **253**, 1278–80.
9. Murphy, P. M. and Tiffany, H. L. (1991). *Science*, **253**, 1280–3.
10. Baggiolini, M., Dewald, B., and Moser B. (1994). *Adv. Immunol.*, **55**, 97–179.
11. Moser, B., Loetscher, M., Piali, L., and Loetscher, P. (1998). *Intern. Rev. Immunol.*, **16**, 323–44.
12. Schall, T. J. and Bacon, K. B. (1994). *Curr. Opin. Immunol.*, **6**, 865–73.

13a. Tachibana, K., Hirota, S., Iizasa, H., Yoshida, H., Kawabata, K., Kataoka, Y., Kitamura, Y., Matsushima, K., Yoshida, N., Nishikawa, S.-I., Kishimoto, T., and Nagasawa, T. (1998). *Nature*, **393**, 591–4.

13b. Zou, Y.-R., Kottmann, A. H., Kuroda, M., Taniuchi, I., and Littman, D. R. (1998). *Nature*, **393**, 595–9.

14. Leong, S. R., Kabakoff, R. C., and Hébert, C. A. (1994). *J. Biol. Chem.*, **269**, 19343–8.
15. Ahuja, S. K., Lee, J. C., and Murphy, P. M. (1996). *J. Biol. Chem.*, **271**, 225–32.
16. Loetscher, P., Seitz, M., Baggiolini, M., and Moser, B. (1996). *J. Exp. Med.*, **184**, 569–77.
17. Wu, D., LaRosa, G. J., and Simon, M. I. (1993). *Science*, **261**, 101–3.
18. Kuang, Y. N., Wu, Y. P., Jiang, H. P., and Wu, D. Q. (1996). *J. Biol. Chem.*, **271**, 3975–8.
19. Bokoch, G. M. (1995). *Blood*, **86**, 1649–60.
20. Horuk, R., Martin, A., Hesselgesser, J., Hadley, T., Lu, Z. H., Wang, Z. X., and Peiper, S. C. (1996). *J. Leukocyte Biol.*, **59**, 29–38.
21. Bais, C., Santomasso, B., Coso, O., Arvanitakis, L., Geras Raaka, E., Gutkind, J. S., Asch, A. S., Cesarman, E., Gerhengorn, M. C., and Mesri, E. A. (1998). *Nature*, **391**, 86–9.

Bernhard Moser
Theodor-Kocher Institute,
University of Bern, Freiestrasse 1,
CH-3012 Bern, Switzerland

c-Kit receptor tyrosine kinase–kit ligand/stem cell factor

Steel factor/stem cell factor and the receptor c-kit are a growth factor–receptor pair which promotes diverse cellular responses including proliferation, survival, adhesion, migration, potentiation of secretion and maturation and differentiation during embryogenesis and in the adult animal.[1–3] The c-kit receptor fuctions prominently in gametogenesis, melanogenesis and haematopoiesis as well as in other cell systems.[1–4] c-Kit is a receptor tyrosine kinase and its cognate ligand, kit-ligand/stem cell factor (KL/SCF), is a membrane growth factor which may function in a juxtacrine fashion and/or as a soluble growth factor.[5,6]

Synonymous names

c-Kit proto-oncogene, c-kit receptor, stem cell factor receptor, kit-ligand (KL), steel factor (SF), stem cell factor (SCF), mast cell growth factor (MCGF).

Protein properties

Kit-ligand is a growth factor with a four-helical bundle topology found in certain cytokines and growth factors and is known to function as a non-covalent dimer. Two alternatively spliced KL RNA transcripts encode two cell-associated KL protein products, KL-1 and KL-2, of 248 and 220 amino acids respectively.[5,6] Both the KL-1 and KL-2 proteins contain *O*-linked and *N*-linked carbohydrate modifications. KL-1 protein is efficiently processed by proteolytic cleavage to produce soluble KL; KL-2 is also processed to form soluble KL, but not as effectively. The KL-2 protein lacks exon 6 sequences that include the major proteolytic cleavage site for the generation of the soluble KL protein from KL-1. KL-2 therefore represents a differentially more stable cell-associated form of KL[6] (Fig. 1). The phenotype of Sl mutant mice producing only a secreted form of KL (*Sl^d^*) imply that the cell membrane

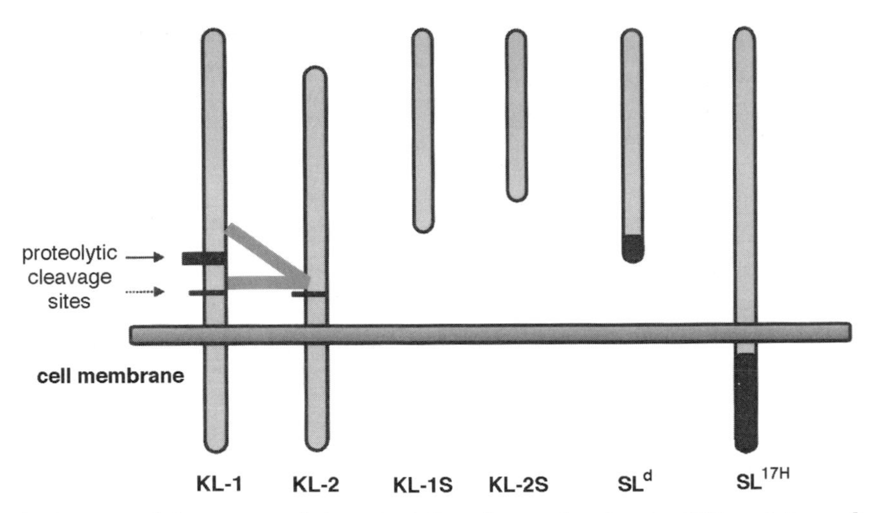

Figure 1. Schematic diagram of the structural characteristics of normal and mutant KL protein products. The normal KL proteins KL-1 and KL-2 are membrane proteins and KL-1S and KL-2S are soluble proteins produced by proteolytic cleavage of KL-1 and KL-2. Sl^{d} and Sl^{17H} are mutant proteins. The dark shaded areas in the Sl^{d} and Sl^{17H} proteins represent altered protein sequences.

form of KL and consequently juxtacrine signalling are critical for many aspects of kit function.

Kit is a receptor tyrosine kinase belonging to the PDGF receptor family.[7,8] The extracellular domain of kit consists of five immunoglobulin domains and the cytoplasmic domain contains a tyrosine kinase domain which is separated into two subdomains by the kinase insert segment consisting of 79 amino acids (Fig. 2). An isoform generated by use of an alternative splice donor site contains a four amino acid insertion (GNNK) between amino acids 513 and 514. The kit protein contains *N*-linked and *O*-linked carbohydrate side chains in the extracellular domain and the molecular weight of kit isolated from different cell types varies from 140 to 160 kDa.[9] Potential sites for tyrosine autophosphorylation and serine/threonine phosphorylation are contained in the juxtamembrane region, the kinase insert, the C-terminal kinase subdomain (major tyrosine phosphorylation site Y821), and the C-terminal domain. Tyrosine phosphates may serve as docking sites for downstream signalling molecules.[10] Known substrates of kit include PI 3-kinase, src family members, grb2, shc and cbl. Upon stimulation with KL, kit receptor molecules form dimers, the intrinsic kinase is activated and autophosphorylation sets in motion various signalling cascades. On the other hand the activated receptor is downregulated by rapid internalization and degradation.[11]

The c-kit receptor and its ligand are expressed in close cellular environments. In agreement with the cell autonomous nature of *W* mutations, c-*kit* is expressed in cellular targets of *W* and *Sl* mutations during embryogenesis and in the postnatal animal in melanogenesis, gametogenesis, and in cells of the haematopoietic system.[4,12] Expression of KL has been shown to be associated with migratory pathways of melanoblasts and germ cells, and homing sites of both germ cells and haematopoietic progenitors during embryonic development. However, c-*kit* and KL are also expressed in tissues and cell types that are not known targets of *W* and *Sl* mutations. Such sites include the neural tube, dorsal root ganglia, portions of the developing central nervous system, the olfactory epithelium, the digestive tract, the lung, and other tissues, whereas KL expression is seen in the floor plate of the neural tube, the thalamus and in the olfactory epithelium. In the adult animal c-*kit* and KL expression are prominent in the lung and in the brain, including the hippocampus and the cerebellum where c-*kit* expression is evident in basket, stellate, and Golgi inter-neurones and KL expression is evident in Purkinje neurones.

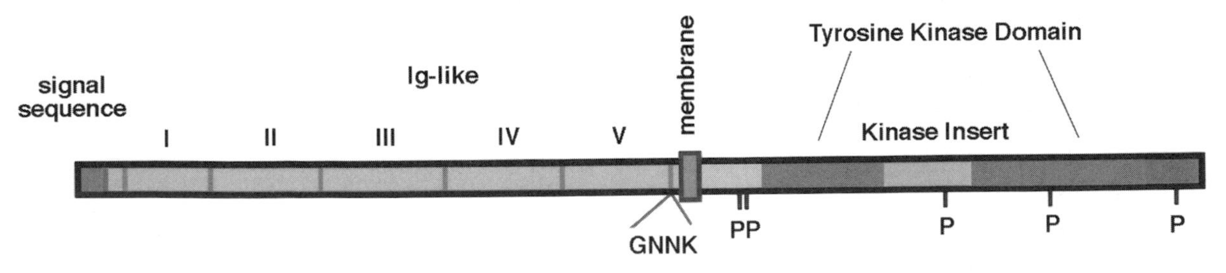

Figure 2. Schematic diagram of the kit receptor tyrosine kinase.

Purification

The soluble form of KL was initially purified from conditioned medium of Balb/c 3T3 cells and BRL-3a rat cells. Recombinant versions of KL are available from various suppliers.

Activities

In vitro, the kit kinase is assayed in immune complexes by transfer of radioactivity from [γ-^{32}P]ATP to appropriate substrates.[9] *In vivo*, kit ligand-induced autophosphorylation of the receptor may be determined by Western blotting using antiphosphotyrosine antibodies.[10]

The kit receptor mediates diverse cellular responses including cell proliferation, survival, migration, adhesion, differentiation, secretion and radio-resistance.[2,3] The adhesion mediated by kit on one hand involves tethering by means of the membrane growth factor/receptor interaction and on the other hand KL induces activation of integrin mediated adhesion (VLA-4 and VLA-5).[5,13,14] These activities provide the basis for the role KL displays in various cell lineages during embryogenesis and in the postnatal organism. Several cell systems including bone-marrow derived mast cells, primordial germ cells, and human myeloid Mo7e cells have been studied to characterize the functional activities of KL/SCF.

Antibodies

Polyclonal and monoclonal antibodies have been raised for both human and murine KL/SCF as well as for kit and many are commercially available. A useful antagonistic kit monoclonal antibody (ACK2) is available for murine kit.[15]

Genes

The KL/SCF gene maps on human chromosome 12 (12q22–24) and on mouse chromosome 10 at the steel (*Sl*) locus.[1] The mouse KL gene has been partially characterized and spans approximately 50 kb.[16] cDNA sequences have been published for the mouse (GenBank M38511, U44724, U44725), rat, pig, human, and chicken.

V-kit was identified as the oncogene of an acute transforming feline retrovirus, the HZ4-FeSV.[17] The c-*kit* gene maps on human chromsome 4 (4q11–13) in the vicinity of the PDGF receptor A chain and the flk-1 receptor genes and the mouse gene is in a syntenic segment of chromsome 5 at the white spotting (*W*) locus.[18,19] Both the human and the mouse c-*KIT* genes have been characterized and they span approximately 70 kb.[20,21] The cDNA sequences have been published for the mouse, rat, cat, human, and chicken (EMBL/GenBank Y00864, D12524, D01190, X06182, and D13225 respectively).[7,8]

Mutant phenotype/disease states

Mutations at both the *white spotting* and the *steel* locus of the mouse result in deficiencies in three cell systems: haematopoiesis, the pigmentary system and germ cells during embryogenesis and in the adult animal.[2,3,4,22,23] Many alleles are known at these loci with differing degrees of severity in the heterozygous and the homozygous state and in the different cellular targets. In haematopoiesis, during early development as well as in adult life, *W* and *Sl* mutations affect cells within the haematopoietic stem cell hierarchy, distinctive cell populations in the erythroid cell lineage and mast cells. *W* and *Sl* mutant mice therefore typically suffer from macrocytic anaemia, they lack tissue mast cells, and display an increased sensitivity to irradiation. In the pigmentary system the *W* and *Sl* mutations affect various aspects of melanogenesis including the development of melanoblasts form the neural crest as well as later aspects of melanocyte development and differentiation, causing depigmentation. In gametogenesis, *W* and *Sl* mutations affect primordial germ cells, their proliferation/survival, and/or migration, spermatogensis, and oogenesis, giving rise to differing degrees of infertility. In addition, *W* and *Sl* mutant animals develop megacolon, presumably because of a role of c-kit in development and/or function of intestinal pacemaker cells (interstitial cells of Cajal) and they have a specific spatial learning deficit because of a role for c-kit in neuronal sub-populations.

References

1. Besmer, P. (1991). *Curr. Opin. Cell Biol.*, **3**, 939–46.
2. Besmer, P. (1997). In *Colony stimulating factors: molecular biology and cellular biology* (ed. J. M. Garland, P. J. Queensberry, and D. J. Hilton), pp 369–404. Marcel Dekker, New York.
3. Galli, S. J., Zsebo, K. M., and Geissler, E. N. (1994). *Adv. Immunol.*, **55**, 1–96.
4. Besmer, P., Manova, K., Duttlinger, R., Huang, E., Packer, A. I., Gyssler, C., Bachvarova, R. (1993). *Development* (Suppl.) **1993**, 125–37.
5. Flanagan, J. G., Chan D., and Leder P. (1991). *Cell*, **64**, 1025–35.
6. Huang, E., Nocka, K. C., Buck, J. and Besmer, P. (1992). *Mol. Biol. Cell*, **3**, 349–62.
7. Yarden, Y., Kuang, W.-J., Yang-Feng, T., Coussens, L., Munemitsu, S., Dull, T. J., *et al.* (1987). *EMBO J.*, **6**, 3341–51.
8. Qiu, F., Ray, P., Brown, K., Parker, P. E., Jhanwar, S., Ruddle, F. H., and Besmer, P. (1988). *EMBO J.*, **7**, 1003–11.
9. Majumder, S., Brown, K., Qiu, F. H. and Besmer, P. (1988). *Mol. Cell. Biol.*, **8**, 4896–4903.
10. Rottapel, R., Reedijk, M., Williams, D. E., Lyman, S. D., Anderson, D. M., Pawson, T. and Bernstein, A. (1991). *Mol. Cell. Biol.*, **11**, 3043–51.
11. Yee, N. S., Hsiau, C.-W. M., Serve, H., Vosseller, K., and Besmer, P. (1994). *J. Biol. Chem.*, **269**, 31991–8.
12. Motro, B., Van der Kooy, D., Rossant, J., Reith, A., and Berstein, A. (1991). Development, **113**, 1207–21.
13. Kinashi, T. and Springer, T. A. (1994). *Blood*, **83**, 1033–8.
14. Serve, H., Yee, N. S., Stella, G., Sepp-Lorenzino, Tan, J. C., and Besmer, P. (1995). *EMBO J.*, **14**, 473–83.
15. Nishikawa, S., Kusakabe, M., Yoshinaga, K., Ogawa, M., Hayashi, S.-I., Kunisada, T., *et al.* (1991). *EMBO J.*, **10**, 2111–8.
16. Bedell, M. A., Copeland, N. G. and Jenkins, N. A. (1996). *Genetics*, **142**, 927–34.
17. Besmer, P., Murphy, J. E. George, P. C. Qiu, F. H. Bergold, P. J. Lederman, L. *et al.* (1986). *Nature*, **320**, 415–21.

18. Chabot, B., Stephenson, D. A., Chapman, V. M., Besmer, P., and Bernstein, A. (1988). *Nature*, **335**, 88–9.
19. Geissler, E. N., Ryan, M. A., and Housman, D. E. (1988). *Cell*, **55**, 185–92.
20. André, C., Martin, E., Cornu, F., Hu, W.-X., Wang, X.-P., and Galibert, F. (1992). *Oncogene*, **7**, 685–91.
21. Gokkel, E., Grossman, Z., Ramot, B., Yarden, Y., Rechavi, G., and Givol, D. (1992). *Oncogene*, **7**, 1423–9.
22. Russell, E. S. (1979). *Adv. Genet.*, **20**, 357–459.
23. Silvers, W. K. (1979). *The coat colors of mice*. Springer, New York.

Peter Besmer
Molecular Biology Program,
Sloan-Kettering Memorial Cancer Center,
New York, NY 10021, USA

Connexins

The connexins comprise a family of gap junction structural proteins. Hexameric assemblies of these proteins in the plasma membranes of adjacent cells interact to form intercellular channels. These channels are permeable to a variety of molecules smaller than 1200–2000 Da. They play an essential role in the coordination of contraction in smooth and cardiac muscle and compose electrical synapses which couple many neurones. In addition, signalling through gap junctions critically regulates processes as diverse as myocardial patterning, ovarian follicular development, and myelination of peripheral nerve.

Protein properties

Structural models based on X-ray diffraction (Fig. 1) show that gap junctions are aggregations of intercellular channels which directly connect the cytoplasms of adjacent cells. Unlike other membrane channels, intercellular channels span two plasma membranes and require the contribution of hemi-channels, called connexons, from both participating cells (Fig. 2). Two connexons interact in the extracellular space to form the complete intercellular channel. Each connexon is composed of six similar or identical proteins, which have been termed connexins.

The connexins are a multigene family of highly related proteins. DNAs encoding 14 rodent connexins have been reported.[1–12,54] Orthologues of some of these, in addition to several possibly unique connexins, have been cloned from human,[13–16] bovine,[17,18] ovine,[19] avian,[20–23] teleost,[24–26] canine[27] and amphibian[28–30] sources. In addition, at least six connexin genes have been isolated from other vertebrate species[25,30,55,56] for which rodent orthologues have not yet been identified. Thus, the total number of vertebrate connexin genes is likely to exceed 20. No connexin has thus far been isolated from an invertebrate source and a variety of evidence suggests that invertebrate gap junctions may be composed of proteins unrelated to connexins.[31] To distinguish individual connexins, a system of nomenclature based on species of origin and molecular mass predicted by cDNA analysis has been suggested.[9] For

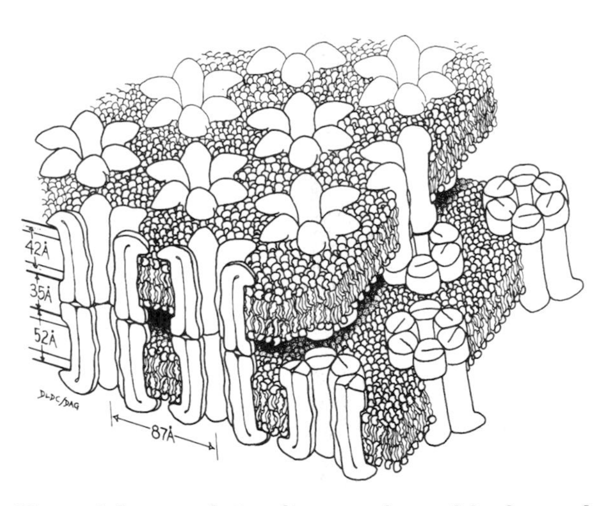

Figure 1. Low resolution diagram of one of the forms of isolated gap junctions from mouse liver, based on data from X-ray diffraction and electron microscopy. The protein subunits are arrayed in hexamers, or connexons, in each of the paired junctional membranes. (Reprinted with permission from the *Journal of Cell Biology*, 1977, **74**, 643.)

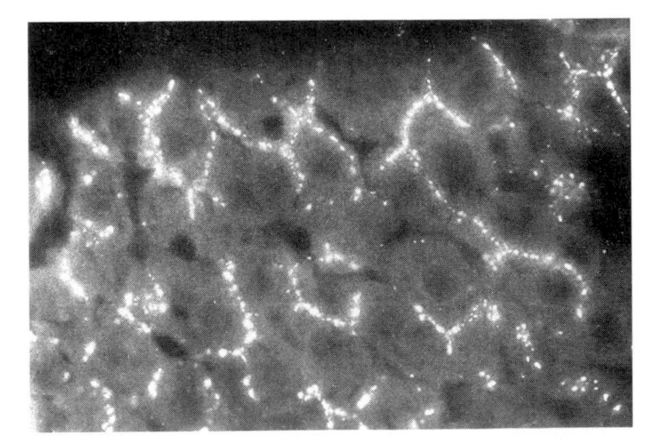

Figure 2. Immunofluorescence localization of Cx32 in a frozen section of rat liver.

example, the connexins expressed in rat hepatocytes were named rat connexin (Cx) 26 and Cx32. An alternative nomenclature termed these proteins β_2 and β_1[32], respectively.

The topological orientation of connexin proteins in the plasma membrane is depicted in Fig. 3. This model is supported by several independent lines of evidence[33,34] including high-resolution crystallographic analysis.[35] The four transmembrane, two extracellular and short N-terminal cytoplasmic domains are well conserved among all connexins, while the central (Figure 3 (a)) and C-terminal cytoplasmic (Fig. 3 (b)) domains are highly variable in sequence and in size.

The connexins are expressed in complex, overlapping patterns, partially summarized in Table 1. Specific connexins may exhibit relatively broad (e.g. Cx32, Cx43) or relatively restricted distributions (e.g. Cx50, Cx33). Many cells express multiple connexins which can often, although not always, form channels of mixed composition.[36–38] Since each connexin makes channels with distinct biophysical and regulatory properties, the incorporation of multiple connexins into a single channel could dramatically expand the range of channel properties (see Activities).

Purification

Methods for bulk purification of gap junction maculae have been reported using rat hepatocytes,[39,40] rat myocardium,[41] and bovine lens fibres.[42]

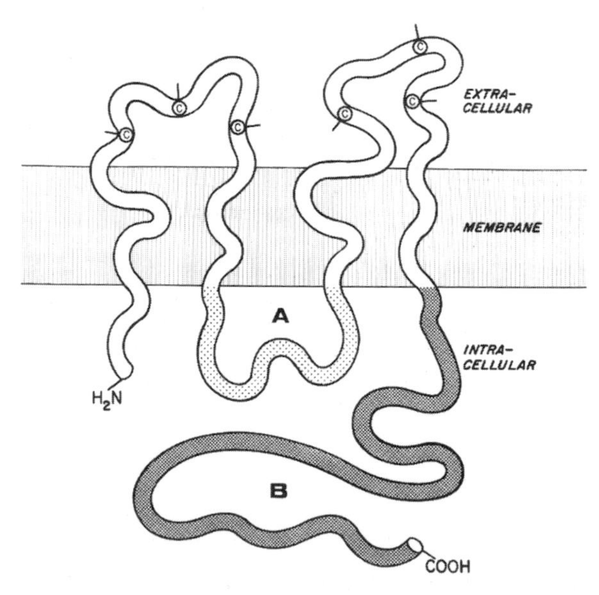

Figure 3. Structure and topology of the connexins relative to the junctional plasma membrane. Unshaded portions represent regions of connexins which are relatively conserved among connexins. Two extracellular domains each contain three invariant cysteines. Cytoplasmic domains A and B differ among connexins in both sequence and length.

Activities

Although all connexins share a high degree of similarity, there are striking differences in their functional properties which may provide insight into the significance of connexin diversity. First, it has been shown that gating of intercellular channels is critically dependent on connexin type, responding differently to phosphorylation, calcium, pH, and voltage.[33,34] For example, pp60$^{v\text{-}src}$ inhibits communication based on Cx43 by inducing tyrosine phosphorylation at residue 265 in the C-terminal cytoplasmic domain. In contrast, neither tyrosine phosphorylation nor inhibition of communication occurs with rat Cx32.

Second, a likely consequence of multiple connexin expression is the establishment of communication compartments. In many animal tissues, communication occurs within a group of cells but not between adjacent groups, even in the absence of anatomical boundaries. This could be explained by the observation that some pairs of connexins do not form functional gap junctions when expressed in adjacent cells.[34,35] For example, Cx40 and Cx43 do not interact with each other although they can form active channels with themselves and with other connexins.[43] This may have biological significance in the heart, where communication compartments isolate Purkinje fibres of the conduction system from the majority of the myocardial muscle to facilitate coordinated excitation and contraction. Since the Purkinje fibres express Cx40 while the surrounding muscle cells express predominantly Cx43, it has been suggested that connexin incompatibility constrains the fibres to couple only to each other and to a subset of myocardial cells expressing Cx40.

A third benefit of connexin diversity is the ability selectively to propagate different second messenger molecules between cells. It has been established that there are connexin-specific differences in channel permeability based on size, charge and/or conformation.[33,34] For example, cells expressing Cx26, Cx37, Cx40, Cx43 or Cx45 transferred propidium iodide (414 Da) and DAPI (314 Da) while those expressing Cx31 or Cx32 transferred only DAPI.[44] Selectivity in this size range could permit discrimination between known second messengers such as cAMP/cGMP and IP_3. Because the half-lives of these signalling molecules are short, even small differences in the permeability of intercellular channels could alter the outcome of intercellular signal transduction.

Antibodies

A variety of polyclonal and monoclonal antibodies that are specific for individual connexins have been reported; rat Cx26;[1] rat Cx32;[6] rat Cx40, rat Cx43;[45] rat Cx40[43] rat Cx46;[11] sheep MP70[46] (rodent Cx50). Antibodies for Cx26, Cx32, Cx43, and Cx45 are also available from commercial sources such as Chemicon and Zymed.

Genes

cDNA and genomic sequences are available for a large number of connexins.[1–30] Genomic analysis of rodent

Table 1 Expression pattern of connexins (from refs. 33, 34 and unpublished observations)

Connexin	Cell/organ location	Possible physiological roles
Rat Cx26 (β_2)	Hepatocytes, mammary epithelium, endometrium, cochlea	In humans, Cx26 mutations cause familial sensorimotor deafness. Mice with targeted ablation die in early embryogenesis
Mouse Cx30	Skin, brain	?
Mouse Cx30.3	Skin, preimplantation blastocyst	?
Rat Cx31	Skin, placenta, Harderian gland, preimplantation blastocyst	?
Rat Cx31.1	Skin, oesophagus, preimplantation blastocyst	?
Rat Cx32 (β_1)	Many secretory and absorptive some neurones, Schwann cells, oligodendrocytes	In humans, Cx32 mutations are associated with a common demyelinating peripheral neuropathy. Mice with null mutations display higher rates of hepatocellular carcinoma
Rat Cx33	Sertoli cells	?
Mouse Cx36	Neurones	?
Rat Cx37	Ovarian follicle, vascular	Mice with gene knockout are female sterile because oocyte development is arrested in early antral stage
Rat Cx40	Vascular endothelium, Purkinje cells of cardiac conduction system atrial myocardium, preimplantation blastocyst	Mice with gene knockout have abnormal ECG indicating conduction block
Rat Cx43 (α_1)	Smooth and cardiac muscle, fibroblasts and osteoblast, ovarian granulosa cells, lens and corneal endothelium, pancreatic β-cells, distal convoluted tubules, astrocytes, preimplantation lastocyst	Mouse knockout dies as neonate due to gross deformations of pulmonary outflow tract in the heart. In humans, point mutations may be associated with defects of laterality (visceroatrial heterotaxia)
Mouse Cx45	Ventricular myocytes, preimplantation blastocyst	?
Rat Cx46	Lens fibres, Schwann cells	Mouse knockout develops lens cataracts
Mouse Cx50	Lens fibres, corneal epithelium	Mouse knockout exhibits small lenses and cataracts

connexins indicates a simple gene structure in which a sole intron interrupts the 5′ UTR and the entire coding region is contained in the second exon.

Mutant phenotypes/disease states

Several connexin genes have been implicated in human disease states. First, mutations in Cx32 appear to cause an X-linked form of Charcot–Marie–Tooth disease, a hereditary neuropathy affecting peripheral myelination.[47] This hypothesis is consistent with the report of demyelination in Cx32 knockout mice, even though functional deficits are minor compared to the human disorder.[48] Second, several forms of hereditary non-syndromic sensorineural deafness are associated with coding region alterations in Cx26.[49] Third, a relationship between human cardiac malformation (visceroatrial heterotaxia) and mutations in Cx43 has been proposed.[50] Although controversial,[51] this idea is supported by the observation of malformations in the pulmonary outflow tract in mice with targeted ablation of Cx43.[52] Fourth, a form of congenital cataract exhibiting zonular pulverulent opacities (CZP1, OMIM 116200) has been associated with mutations in Cx50.[58] Consistent with this finding, pulverulent cataracts and a lens growth defect are observed in mice lacking the Cx50 gene.[59] Loss of Cx46, which is co-expressed with x50 at high levels in lens fibres, also causes cataracts.[60] Knockout studies suggest that Cx40 has an important functional role in synchronizing contraction of cardiac muscle. Cx40 is expressed at high levels in the His-Purkinje system and ablation causes partial conduction block.[61,62] Finally, targeted ablation of Cx37 results in female sterility due to a failure of ovarian follicular development.[53] Most oocytes fail to achieve meiotic competence and inappropriate luteinization of the follicle occurs in the absence of ovulation. The phenotype is similar to karyotypically normal spontaneous ovarian failure, a human disorder of unknown aetiology.

References

1. Zhang, J.-T. and Nicholson, B. (1989). *J. Cell Biol.*, **109**, 3391–401.
2. Dahl, E., Manthey, D., Chen, Y., Schwarz, H. J., Chang, S., Lalley, P. A., *et al.* (1996). *J. Biol. Chem.*, **271**, 17903–10.
3. Hennemann, H., Dahl, E., White, J. B., Schwarz, H. J., Lalley, P. A., Chang, S., *et al.* (1992). *J. Biol. Chem.*, **267**, 17225–33.
4. Hoh, J. H., John, S. A., and Revel, J.-P. (1991). *J. Biol. Chem.*, **266**, 6524–31.
5. Haefliger, J.-A., Bruzzone, R., Jenkins, N. A., Gilbert, D. J., Copeland, N. G., and Paul, D. L. (1992). *J. Biol. Chem.*, **267**, 2057–64.

6. Paul, D. L. (1986). *J. Cell Biol.*, **103**, 123–34.
7. Kumar, N. and Gilula, N. B. (1986). *J. Cell Biol.*, **103**, 767–76.
8. Willecke, K., Heykes, R., Dahl, E., Stutenkemper, R., Hennemann, H., Jungbluth, S., *et al.* (1991). *J. Cell Biol.*, **114**, 1049–58.
9. Beyer, E. C., Paul, D. L., and Goodenough, D. A. (1987). *J. Cell Biol.*, **105,** 2621–9.
10. Willecke, K. (1992). *Eur. J. Cell Biol.*, **57**, 51–8.
11. Paul, D. L., Swenson, K. I., Takemoto, L. J., and Goodenough, D. A. (1991). *J. Cell Biol.,* **115**, 1077–89.
12. White, T. W., Bruzzone, R., Goodenough, D. A. and Paul, D. L. (1992). *Mol. Biol. Cell,* **3**, 711–20.
13. Lee, S. W., Tomasetto, C., Paul, D. Keyomarsi, K., and Sager, R. (1992). *J. Cell Biol.*, **118**, 1213–21.
14. Bergoffen, J., Scherer, S. S., Wang, S., Oronzi Scott, M., Bone, L. J., Paul, D. L. *et al.* (1993). *Science*, **262**, 2039–42.
15. Reed, K. E, Westphale, E. M., Larson, D. M., Wang, H. Z., Veenstra, R. D., and Beyer, E. C. (1993). *J. Clin. Invest.,* **91**, 997–1004.
16. Fishman, G. I., Spray, D. C., and Leinwand, L. A. (1990). *J. Cell Biol.,* **111**, 589–98.
17. Lash, J. A., Critser, E. S., and Pressler, M. L. (1990). *J. Biol. Chem.*, **265**, 13113–17.
18. Gupta, V. K., Berthoud, V. M., Atal, N., Jarillo, J. A., Barrio, L. C., and Beyer, E. C. (1994). *Invest. Opthalmol. Vis. Sci.,* **35**, 3747–58.
19. Yang, D. I. and Louis, C. F. (1996). *Curr. Eye Res.*, 15, 307–14.
20. Musil, L. S., Beyer, E. C., and Goodenough D. A. (1990). *J. Membr. Biol.*, **116**, 163–75.
21. Beyer, E. C. (1990). *J. Biol. Chem.*, **265**, 14439–43.
22. Jiang, J.X., White, T. W., Goodnough, D. A. and Paul, D. L. (1994). *Mol. Biol. Cell,* **5**, 363–73.
23. Rup, D. M., Veenstra, R. D., Wang, H. Z., Brink, P. R. and Beyer, E. C. (1993). *J. Biol. Chem.*, **268**, 706–12.
24. Yoshizaki, G., Patino, R., and Thomas, P. (1994). *Biol. Reprod.*, **51**, 493–503.
25. O'Brien, J., al-Ubaidi, M. R., and Ripps, H. (1996). *Mol. Biol. Cell.,* **7**,: 233–243.
26. Essner, J. J., Laing, J. G., Beyer, E. C., Johnson, R. G. and Hackett, P. B. (1996). *Dev. Biol.*, **177**, 449–462.
27. Kanter, H. L., Saffitz, J. E., and Beyer, E. C. (1992). *Circ. Res.*, **70**, 438–444.
28. Ebihara, L., Beyer, E. C., Swenson, K. I., Paul, D. L., and Goodenough, D. A. (1988). *Science*, **243**, 1194–5.
29. Gimlich, R. L., Kumar, N. M., and Gilula, N. B. (1990). *J. Cell Biol.*, **110**, 597–605.
30. Yoshizaki, G. and Patino, R. (1995). *Mol. Reprod. Dev.* **42**, 7–18.
31. Barnes, T. M. (1994). *Trends Genet.*, **10**, 303–5.
32. Risek, B., Guthrie, S., Kumar, N., and Gilula, N. B. (1990). *J. Cell Biol.*, **110**, 269–282.
33. Goodenough, D. A., Goliger, J. A., and Paul, D. L. (1996). *Ann. Rev. Biochem.*, **65**, 475–502.
34. Bruzzone, R., White, T. W., and Paul, D. L. (1996). *Eur. J. Biochem.*, **238**, 1–27.
35. Unger, V. M., Kumar, N. M., Gilula, N. B., and Yeager, M. (1997). *Nat. Struct. Biol.*, **4**, 39–43.
36. Stauffer, K. A. (1995). *J. Biol. Chem.*, **270**, 6768–72.
37. Jiang, J. X. and Goodenough, D. A. (1996). *Proc. Natl Acad. Sci. USA*, **93**, 1287–91.
38. White, T. W., Paul, D. L., Goodenough, D. A. and Bruzzone, R. (1995). *Mol. Biol. Cell*, **6**, 459–70.
39. Baker, T. S., Sosinsky, G., Caspar, D. L. D., Gall, C., and Goodenough, D. A. (1985). *J. Mol. Biol.*, **184**, 81–98.
40. Hertzberg, E. L. (1984). *J. Biol. Chem.*, **259**, 9936–43.
41. Manjunath, C. K., Nicholson, B. J., Teplow, D., Hood, L., Page, E., and Revel, J.-P. (1987). *Biochem. Biophys. Res. Comm.*, **142**, 228–34.
42. Paul, D. L. and Goodenough, D. A. (1983). *J. Cell Biol.*, **96**, 625–632.
43. Beyer E. C., Kistler J., Paul D. L., and Goodenough D. A. (1988). *J. Cell Biol.*, **108**, 595–605.
44. Bruzzone, R., Haefliger, J.-A., Gimlich, R. L., and Paul, D. L. (1993). *Mol. Biol. Cell,* **4**, 7–20.
45. Kistler, J., Christie, D., and Bullivant, S. (1988). *Nature*, **331**, 721–723.
46. Elfgang, C., Eckert, R., Lichtenberg-Frate, H., Butterweck, A., Traub, O., Klein, R. A., Hulser, D. F. and Willecke, K. (1995). *J. Cell Biol.*, **129**, 805–817.
47. Bergoffen, J., Scherer, S. S., Wang, S., Oronzi Scott, M., Bone, L. J., Paul, D. L., *et al.* (1993). *Science,* **262**, 2039–42.
48. Anzini, P., Neuberg, D. H. H., Schachner, M., Nelles, E., Willecke, K., Zielasek, *et al.* (1997). *J. Neurosci.*, **17**, 4545–51.
49. Kelsell, D. P., Dunlop, J., Stevens, H. P., Lench, N. J., Liang, J. N., Parry, G., Mueller, R. F. and Leigh, I. M. (1997). *Nature*, **387**, 80–3.
50. Britz-Cunningham, S. H., Shah, M. M., Zuppan, C. W., and Fletcher, W. H. 1995. *New Engl. J. Med.*, **332**, 1323–9.
51. Penman Splitt, M., Tsai, M. Y., Burn, J., and Goodship, J. A. (1997). *Heart*, **77**, 369–370.
52. Reaume, A. G., de Sousa, P. A., Kulkarni, S., Langille, B. L., Zhu, D., Davies, T. C., *et al.* (1995). *Science*, **267**, 1831–4.
53. Simon, A. M., Goodenough, D. A., Li, E. and Paul, D. L. (1997). *Nature*, **385**, 525–9.
54. Condorelli, D. F., Parenti, R., Spinella, F., Salinaro, A. T., Belluardo, N. *et al.* (1998). *Eur. J. Neurosci.*, **10**, 1202–8.
55. O'Brien, J., Bruzzone, R., White, T. W., al-Ubaidi, M. R., and Ripps, H. (1998). *J. Neurosci.*, **18**, 7625–37.
56. Itahana, K., Tanaka, T., Morikazu, Y., Komatu, S., Ishida, N., and Takeya, T. (1998). *Endocrinology*, **139**, 320–9.
57. Yoshizaki, G. and Patiño, R. (1995). *Mol. Reprod. Devel.*, **42**, 7–18.
58. Shiels, A., Mackay, D., Ionides, A., Berry, V., Moore, A., and Bhattacharya, S. (1998). *Am. J. Hum. Genet.*, **62**, 526–32.
59. White, T., Goodenough, D. A., and Paul, D. L. (1998). *J. Cell Biol.*, (Submitted).
60. Gong, X., Li, E., Klier, G., Huang, Q., Wu, Y. *et al.* (1997). *Cell*, **91**, 833–43.
61. Simon, A. M., Goodenough, D. A., and Paul, D. L. (1998). *Curr. Biol.*, **8**, 295–8.
62. Kirchhoff, S., Nelles, E., Hagendorff, A., Krüger, O. Traub, O., and Willecke, K. (1998). *Curr. Biol.*, **8**, 299–302.
63. Phelan, P., Bacon, J. P., Davies, J. A., Stebbings, L. A., Todman, M. G. *et al.* (1998). *Trends Genet.* (In press).

■ *David L. Paul*
Department of Neurobiology,
Harvard Medical School,
Boston. MA, USA

Contact site A

The contact site A (csA) cell-surface glycoprotein is a developmentally regulated cell adhesion molecule which mediates EDTA-stable (Ca^{2+}-independent) cell-to-cell contacts during early development of Dictyostelium discoideum.

Synonymous names

gp80,[1] antigen 117[2]

Homologous proteins

CsA shows limited relationships to chick NCAM.[3]

Protein properties

The contact site A (csA) glycoprotein from *Dictyostelium discoideum* is derived from a single-copy gene and encodes a protein with a calculated molecular mass of 53 kDa.[4,5] According to the cDNA-derived sequence, the translation product consists of three regions: a large extracellular N-terminal domain that is predicted to have an extended β-sheet configuration and five potential *N*-glycosylation sites,[4,5] a region rich in proline, serine, and threonine, which resembles the hinge region of immunoglobulins, and a carboxyl-terminal stretch of hydrophobic amino acids.[4,5] The protein is co-translationally modified by *N*-linked carbohydrate residues and the carboxyl-terminal stretch (amino acids 484–514) is replaced by a ceramide-based phospholipid anchor.[6,7] In the Golgi apparatus the Pro/Ser/Thr-rich stretch of the C-terminal region close to the plasma membrane is decorated with *O*-linked carbohydrates, where *N*-linked carbohydrates also become sulphated.[8,9] With its modifications the csA protein shows an apparent molecular mass of 80 kDa.

CsA expression is strictly developmentally regulated.[4,10,11] The protein is absent from growth-phase cells and is maximally expressed at the aggregation stage, at the transition from the single-cell state to a multicellular organism.[12] The expression of csA is strongly stimulated by periodic pulses of cyclic AMP.[13,14]

The functional relationship of the various csA moieties hasbeen evaluated using biochemical and genetic approaches. The phosholipid anchor prevents internalization and guarantees long persistence of the csA protein on the cell surface.[7] ModB mutants that are defective in *O*-linked glycosylation show strongly reduced EDTA-stable adhesion, because the csA glycoprotein is proteolytically cleaved by a cell-surface bound protease.[15] A 50 kDa N-terminal fragment of csA is released into the medium of these mutants.[15] Elimination of the protease by chemical mutagenesis restores EDTA-stable cell

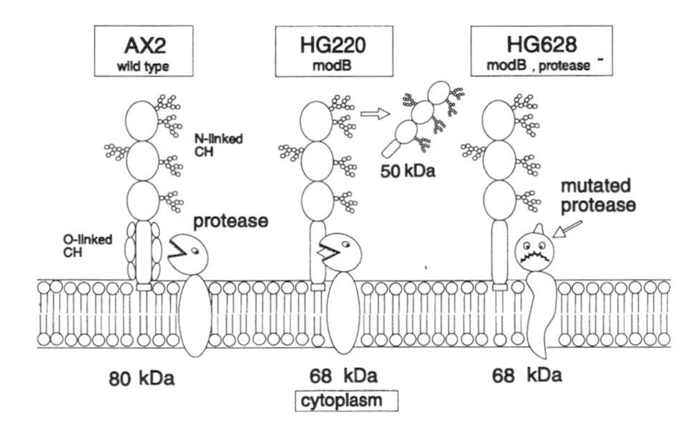

Figure 1. Scheme of carbohydrate (CH) modifications of the csA glycoprotein in the AX2 wild type and modB mutants lacking *O*-linked oligosaccharides. The normal, fully glycosylated csA has an apparent molecular mass of 80 kDa (left). Without *O*-linked sugars the molecular mass shifts to 68 kDa and the protein is proteolytically cleaved by a cell surface bound protease, giving rise to a soluble 50 kDa fragment (middle). Elimination of this protease by mutagenesis results in a protein that is stable without *O*-linked sugars (right).

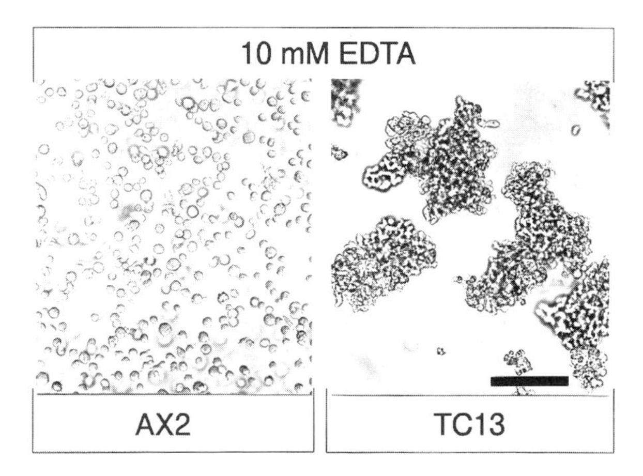

Figure 2. EDTA-stable adhesion in cell suspension induced by overexpression of the csA protein. Left: Control of growth-phase cells of the AX2 wild type in which csA is not expressed; Right: growth-phase cells of transformant TC13 which strongly expresses csA under control of a constitutive actin promoter. Bar represents 100 μm.

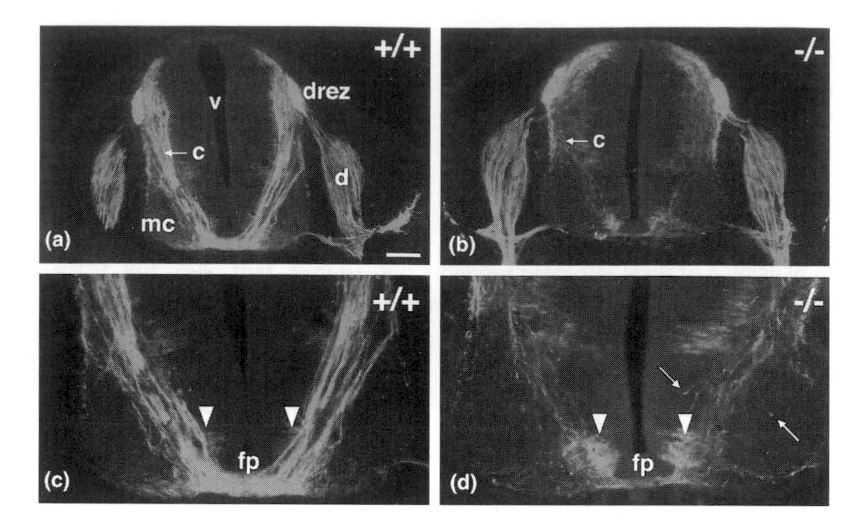

Figure 2. Defects in commissural axon projections in *Dcc*−/− embryos. Trajectories of commissural axons are visualized using an antibody to TAG-1 in sections of wild-type (a, c) and *Dcc*−/− (b, d) E1.5 embryos. In *Dcc*−/− embryos, TAG-1-positive commissural neurones are present but few axons extend into the ventral spinal cord (b) and those that do, project along aberrant trajectories (arrows in d). Projections of sensory axons and motor axons in the ventral roots appear largely normal (b). Arrowheads in (c) and (d) indicate a population of TAG-1-positive cells adjacent to the floor plate. Additional abbreviations: d, dorsal root ganglia; drez, dorsal root entry zone; mc, motor column; v, ventricle; c, commissural axons; fp, floor plate. Scale bar is 100 μm in (a), (b) and 50 μm in (c), (d). (Adapted, with permission, from Fazeli *et al.* 1996.)

Neogenin expression is also seen in the adult and embryonic nervous system and correlates primarily with the onset of neural differentiation and the onset of axon extension.[1] Like their vertebrate counterparts, *Drosophila* Frazzled and *C. elegans* UNC-40 are highly expressed on axons in the developing nervous system.[4,5]

DCC was first identified as a candidate tumour suppressor gene on human chromosome 18q.[2] This region is frequently subject to loss of heterozygosity (LOH) in colorectal tumours and there has been a great effort to elucidate its role in tumourigenesis.[11] The finding that *C. elegans unc-40* is a homologue of DCC has turned attention to a possible role for DCC and related proteins in axon guidance and cell migrations.[3–5]

UNC-40 is expressed on motile cells and pioneer neurones and acts cell autonomously to orient movement toward a source of the netrin UNC-6.[4] In cells that coexpress UNC-5, UNC-40 appears to help orient cells away from netrin/UNC-6 sources.[4,18] Double mutant analysis indicates that UNC-40 requires UNC-6 for most of its functions and vice versa.[12,18] This evidence is consistent with the possibility that UNC-40 is a receptor or a component of a receptor for netrin UNC-6. Direct evidence comes from vertebrate systems, where it has been shown that DCC binds netrin-1 with a k_d in the nanomolar range, and that antibodies to DCC can inhibit netrin-dependent outgrowth of commissural neurones into collagen gels.[3] Antibodies to DCC also inhibit the growth cones of single *Xenopus* retinal and embryonal spinal neurones from turning toward a point source of netrin-1.[19,20] The role for DCC as a netrin receptor is further supported by the finding that the phenotype of DCC−/− mice is strikingly similar to the phenotype of netrin-1 mutant mice.[13] This function appears to be conserved in *Drosophila* as well, as mutants in *frazzled* have a phenotype that is similar to that of mutants in the *Drosophila* netrin genes.[5]

Purification

Not applicable.

Activities

DCC family members have no known enzymatic activity. Rat DCC and rat neogenin, possess specific netrin-binding activity.[3] The function of *C.elegans* UNC-40 and *Drosophila* Frazzled as receptors for netrin-mediated axon guidance has been inferred from the phenotypic analysis of mutant and transgenic organisms.

Antibodies

There are numerous antibodies to the vertebrate DCC and neogenin proteins. Polyclonal antibodies to chick DCC,[8] *Xenopus* DCC,[7] and human DCC[7] are available. There is a commercially available monoclonal antibody to human DCC (AF5 mouse monoclonal, Oncogene Science) which cross-reacts with murine and rat DCC and is capable of blocking DCC function in commissural axon outgrowth assays.[3] In addition, this antibody can be used for immunoblotting, immunoprecipitation, and immunohistochemistry. A monoclonal antibody (10–22A8) is available for chick neogenin.[1] In *Drosophila*, a polyclonal antibody to the C-terminal portion of Frazzled is available.[5] Antibodies that can be used for immunoblotting, immunoprecipitation and immunohistochemistry are available for all species noted. Transgenic *C. elegans* are available that express either GFP (*Aequoria victoria* green fluorescent protein) or an UNC-40–GFP fusion protein from the *unc-40* promoter.[4]

Genes

Full length DCC cDNA clones are available from human[6] (GenBank X76132), rat[3] (U68725), mouse[9] (X85788), and *Xenopus*[10] (U10986). Full length neogenin clones are available for human,[14,15] (U61262, U72391) and chickens (U07644). A partial rat neogenin cDNA clone is also available (U68726). The full length *Drosophila frazzled* cDNA[5] (U71001, U71002) and the full length *C. elegans unc-40* cDNA[4] (Chan, Killeen, Zheng, and Culotti; unpublished results) are available. The mouse *Dcc* gene maps to chromosome 18 (45.OcM). Mouse neogenin (*Ngn*) maps to the central region of chromosome 9 in a region of homology with human chromosome 15q (D. J. Gilbert, N. G. Copeland, and N. A. Jenkins; unpublished observations). The genomic *unc-40* clone is on cosmid T0D27 and has been sequenced by the *C. elegans* genome sequencing consortium (U70618).

Mutant phenotypes

Mice

DCC-deficient mice die at birth with severe abnormalities of the nervous system.[13] Most of the major commissures of the brain and spinal cord are reduced or absent. In the spinal cord, commissural axons are impaired in their ability to extend along their initial dorsal to ventral trajectory in the dorsal spinal cord, and those that reach the ventral spinal cord become misrouted, such that few reach the normal intermediate target, the floor plate. In the brain, the corpus callosum and hippocampal commissure are completely absent, and the anterior commissure is greatly reduced. In addition, a novel commissure is seen at the hindbrain–midbrain junction. However, not all commissures are affected, as the posterior and habenular commissures appear to be intact. Finally, the pontine nuclei are absent. This phenotype is strikingly similar to that of netrin-1-deficient mice.[17]

Extensive analysis of heterozygote $DCC^{+/-}$ has failed to reveal any increased predisposition toward tumours or cancer of any kind.[13]

Drosophila

In embryos that lack Frazzled, CNS axon commissures are partly missing or thinner, with the posterior commissure more severely disrupted than the anterior commissure. In addition, abnormalities are seen in the trajectories of the ISN and SNb motor axons, particularly in their dorsal paths.[5]

C. elegans

unc-40 mutants have defects in a variety of cell and pioneer axon-growth cone migrations, most of which are oriented on the dorsoventral axis of the epidermis.[4] Affected cells include motoneurones (AS, DA, DB, DD, VD), the neurone-like excretory cell, and three mesodermal cell types (distal tip cell, head mesodermal cell, and male linker cell), plus a variety of sensory neurones with lateral cell bodies and pioneer axons that normally extend to the ventral midline. In these mutants, migrations still occur on the epidermis but are frequently misoriented.

References

1. Fearon, K. R., Nigro, J. M., Kern, S. E., Simons, J. W., Ruppert, J. M., *et al*. (1990) *Science*, **247**, 49–56.
2. Vielmetter, J., Kayyem, J. F., Roman, J. M., and Dreyer, W. J. (1994). *Cell Biol.*, **127**, 200–20.
3. Keino-Masu, K., Masu, M., Hinck, L., Leonardo, E. D., Chan, S. S.-Y., Culotti, J. G., and Tessier-Lavigne, M. (1996). *Cell*, **87**, 75–185.
4. Chan, S. S.-Y., Zheng, H., Su, M.-W., Wilk, R., Killeen, M. T., Hedgecock, E. M. and Culotti, J. G. (1996). *Cell*, **87**, 187–96.
5. Kolodziej, P. A., Timpe, L. C., Mitchell, K. J., Fried, S. R., Goodman, C. S., Jan, L. Y. and Jan, Y. N. (1996). *Cell*, **87**, 197–204.
6. Hedrick, L., Cho, K. R., Fearon, E. R., Wu, T.-C., Kinzler, K. W., and Vogelstein, B. (1994). *Genes Dev.*, **8**, 1174–83.
7. Reale, M. A., Hu, G., Zafar, A. I., Getzenberg, R. H., Levine, S. M., and Fearon, E. R. (1994). *Cancer Res.*, **54**, 4493–501.
8. Chuong, C.-M., Xiang, T.-X., Yin, E. and Widelitz, R. B. (1994). *Dev. Biol.*, **164**, 383-97.
9. Cooper, H. M., Armes, P., Britto, J., Gad, J., and Wilks, A. F. (1995). *Oncogene*, **11**, 2243–54.
10. Pierceall, W. E., Reele, M. A., Candia, A. F., Wright, C. V. E., Cho, K. R., and Fearon, E. R. (1994). *Dev. Biol.*, **166**, 654–65.
11. Cho, K. R., and Fearon, E. R. (1995) *Curr. Opin. Genet. Dev.*, **5**, 72–18.
12. Hedgecock, E. M., Culotti, J. G., and Hall D. H. (1990) *Neuron*, **2**, 61–*85*.
13. Fazeli, A., Dickinson, S. L., Hermiston, M. L., Tighe, R. V., Steen, R. G., Small, C. G., *et al*. (1997) *Nature*, **386**, 796–804.
14. Vielmetter, J., Chen, X. N., Miskevich, F., Lane, R. P., Yamakawa, K., Korenberg, J. R., and Dreyer, W. J. (1997). *Genomics*, **41**, 414–21.
15. Meyerhardt, J. A., Look, A. T., Bigner, S. H., and Fearon, E. R. (1997). *Oncogene*, **14**, 1129–36.
16. Justice, M. J., Gilbert, D. J., Kinzler, K. W., Vogelstein, B., Buchberg, A. M., Ceci, J. D., *et al*. (1992). *Genomics*, **13**, 1281–8.
17. Serafini, T., Colamarino, S. A., Leonardo, E. D., Wang, H., Beddington, R., Skarnes, W. C., and Tessier-Lavigne, M. (1996). *Cell*, **87**, 1001–14.
18. Colavita, A. and Culotti, J. G. (1998). *Dev. Biol.*, **194**, 72–85.
19. de la Torre, J. R., Höpker, V. H., Ming, G.-L., Poo, M. -M., Tessier-Lavigne, M., Hemmati-Brivanlov, A., and Holt, C. E. (1997). *Neuron*, **19**, 1211–24.
20. Ming, G.-L., Song, H.-J., Berninger, B., Holt, C. E., Tessier-Lavingne, M., and Poo, M.-M. (1997). *Neuron*, **19**, 1225–35.

E. David Leonardo and Marc Tessier-Lavigne
Howard Hughes Medical Institute and
Departments of Anatomy and of
Biochemistry and Biophysics,
University of California,
San Francisco, CA 94122, USA

Joseph G. Culotti
Division of Molecular Immunology
and Neurobiology, Samuel Lunenfeld
Research Institute, Mount Sinai Hospital,
Toronto M5G 1X5, Canada

Dystroglycan

Dystroglycan was first purified from rabbit skeletal muscle as a membrane glycoprotein component of the dystrophin–glycoprotein complex (DGC).[1] In muscle, it binds to both laminin-2 in the extracellular matrix and to dystrophin in the cytoskeleton.[2] Mutation of genes encoding a number of proteins associated with dystroglycan in the DGC leads to distinct forms of muscular dystrophy.[3] Dystroglycan is also expressed in a wide variety of developing and adult non-muscle tissues where it mediates cellular interactions with the extracellular matrix.

Synonyms

α-dystroglycan: SL 156,[4] 156-dystrophin-associated glycoprotein (DAG),[5] cranin,[6] laminin binding protein (LBP)-120.[7] β-dystroglycan: 43-DAG,[5] A3a.[8]

Protein properties

Dystroglycan is composed of two subunits.[9] The α subunit is an extracellular peripheral membrane glycoprotein which ranges in size from around 120 kDa to over 190 kDa depending on the tissue source and method of analysis.[1,6,7,10] The β subunit is an integral membrane glycoprotein that, in contrast to the α subunit, displays a conserved molecular weight of around 43 kDa in a wide range of species and tissues.[1,8] Both of these subunits are the post-translationally derived products of a single mRNA transcript encoded by a single dystroglycan gene.[9] The considerable heterogeneity in the apparent molecular weight of α-dystroglycan is most probably the result of differential carbohydrate modification.[2,6] Indeed, α-dystroglycan is extensively glycosylated and tends to migrate as a broad band in SDS–polyacrylamide gels. The nature of these modifications is only partially understood at present. There are both *N*- and *O*-linked carbohydrate moieties present on α-dystroglycan. The *N*-linked sugars are of the high mannose variety. The *O*-linked sugars are less well characterized, but are thought to be of a mucin-type with an unusual *O*-mannosyl glycosidic linkage.[11]

Dystroglycan is expressed in a broad array of adult and developing tissues and cell types.[9,31] In general, dystroglycan is localized to some, but not all, cellular domains that are in close apposition to laminin-containing extracellular matrix. Examples include the sarcolemma, the neuromuscular junction, the Schwann cell outer membrane surrounding peripheral nerve, epithelial basement membranes, smooth muscle, and the cellular constituents of Reichert's membrane in the developing rodent embryo (Fig. 1). Dystroglycan is expressed early in mouse development with abundant expression in the egg cylinder stage embryo, and in the maternal decidual tissue.

At least several binding partners for both α- and β-dystroglycan are known. Complementing dystroglycan's juxtaposed localization to basement membranes *in vivo* is the ability of α-dystroglycan to bind laminin-1[9] and laminin-2[12] *in vitro* with high affinity. As α-dystroglycan binding to these laminins appears to be mediated through the G domains at the carboxyl terminus of the laminin α chain,[7] it is likely that α-dystroglycan binds to other laminin heterotrimeric isoforms. Dystroglycan also binds to agrin,[10] an extracellular matrix proteoglycan critically involved in neuromuscular junction formation, through a conserved G domain.[13] It is not yet clear whether α-dystroglycan exhibits specific binding to a single extracellular ligand in cases where more than one of its potential ligands are co-expressed within the extracellular matrix. The apparently homogeneous primary structure of dystroglycan indicates that ligand selectivity would have to be regulated by post-translational mechanisms, such as carbohydrate modification. Some data indicates that heparan sulphate proteoglycans might mediate ligand binding specificity.[14] Alternatively, in at least one tissue, peripheral nerve, dystroglycan may be able to bind both agrin and laminin-2 through overlapping binding sites.[15]

β-dystroglycan also binds to several related intracellular ligands. In skeletal muscle, at the sarcolemma, β-dystroglycan binds to dystrophin which, in turn, binds to F-actin.[2] This interaction is mediated between the carboxyl-terminal 15 amino acids of β-dystroglycan and a larger cysteine-rich domain in dystrophin.[16,17] This same interacting region is also present in dystrophin isoforms derived from alternative promoters within in the dystrophin gene, which are expressed in non-muscle tissues, and is also conserved in utrophin, a broadly expressed dystrophin homologue. In fact, utrophin may replace dystrophin as β-dystroglycan's binding partner at the neuromuscular junction, indicating that distinct dystroglycan complexes can exist within the same cell type[18]. Interestingly, the proline-rich carboxyl terminus of β-dystroglycan also interacts with the cell signalling molecule grb-2 via its SH3 domain.[19] However, a cell signalling function for dystroglycan has yet to be characterized.

The sarcoglycans are a group of four intergal membrane proteins that are known to interact with dystroglycan in muscle. The details of their association with dystroglycan are unclear at present. The sarcoglycans do form a tight complex with dystroglycan that can only be dissociated by relatively harsh biochemical treatments.[20] One possible role for the sarcoglycans may be to stabilize the association of α-dystroglycan with the sarcolemma. Although α- and β-dystroglycan can be isolated together *in vitro* as a non-covalently associated unit,[20] this connection may be tenuous *in vivo* and could require the support of other proteins like the sarcoglycans.[21] Figure 2 summarizes the current state of understanding of dystroglycan and its binding partners in skeletal muscle, where this complex is best understood, but similar complexes are likely to exist in other tissues. Defining dystroglycan's molecular partners in non-muscle tissues is a clear target for future research.

Dystroglycan function has been inferred from several lines of evidence. First, based on both the abundance of dystroglycan in muscle and the consequences of disrupting the dystroglycan-mediated linkage between the extracellular matrix and the cytoskeleton, as occurs in several forms of muscular dystrophy, it was proposed that dystroglycan served as a mechanical linkage protecting the sarcolemma from contraction-induced shear forces.[5] Accordingly, dystroglycan behaves as a cell adhesion molecule in Schwannoma cell cultures.[22] Antibody inhibition studies have revealed important roles for dystroglycan in various developmental contexts. These show that dystroglycan function is necessary for acetylcholine receptor clustering[23] and epithelial morphogenesis.[24] However, the mechanistic details for dystroglycan's involvement in these processes have remained elusive. The phenotype of the dystroglycan-null mutant mouse offers a plausible hypothesis for these developmental effects of inhibiting dystroglycan function. The dystroglycan-null mutant mice die early in embryonic development as a result of structural and functional defects in Reichert's membrane, a laminin-rich extraembryonic basement membrane.[25] The details of the phenotype indicate that dystroglycan is required for the development of Reichert's membrane, possibly by mediating the assembly of laminin networks present in that structure. Therefore, dystroglycan may also be mediating the assembly of the extracellular matrix in developing neuromuscular synapses, epithelia and other tissues in which it is expressed. Perhaps dystroglycan acts in the assembly of extracellular matrices

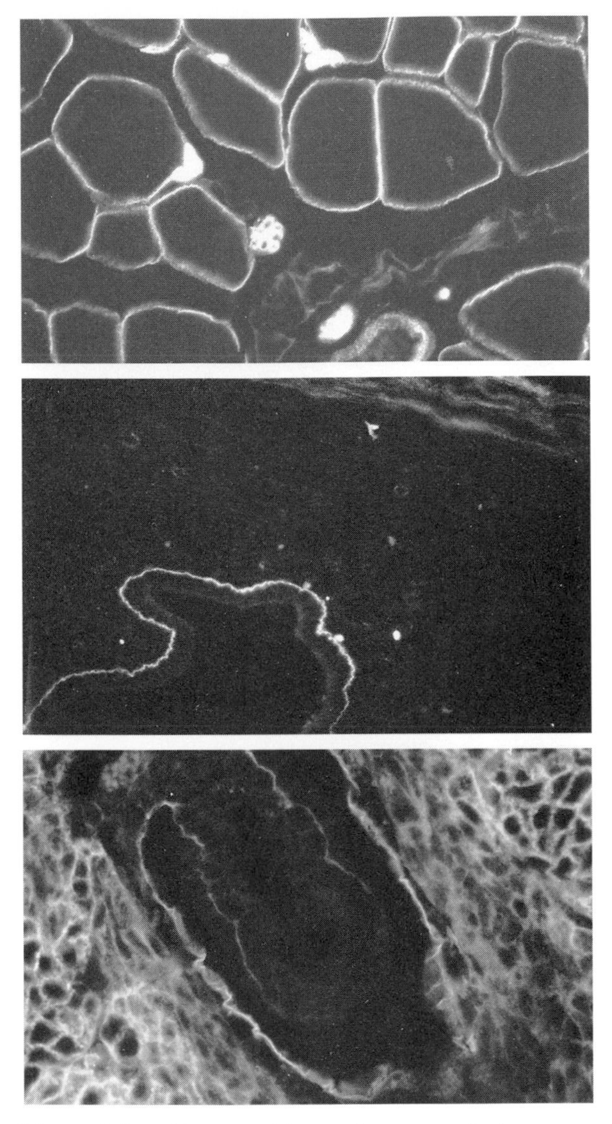

Figure 1. Localization of dystroglycan protein in various tissues. Top: section of mouse skeletal muscle showing dystroglycan localization in the sarcolemma, neuromuscular junction, peripheral nerve bundle, and the wall of a major blood vessel. Middle: section of mouse uterus showing dystroglycan localization to the lumenal epithelial basement membrane, and smooth muscle in the myometrium. Bottom: Parasaggital section through an embryonic day 6.5 mouse embryo *in utero* showing embryonic dystroglycan localization to Reichert's membrane, and a basement membrane between the visceral endoderm, and ectodermal layers, and maternal localization to the decidual cells surrounding the embryo.

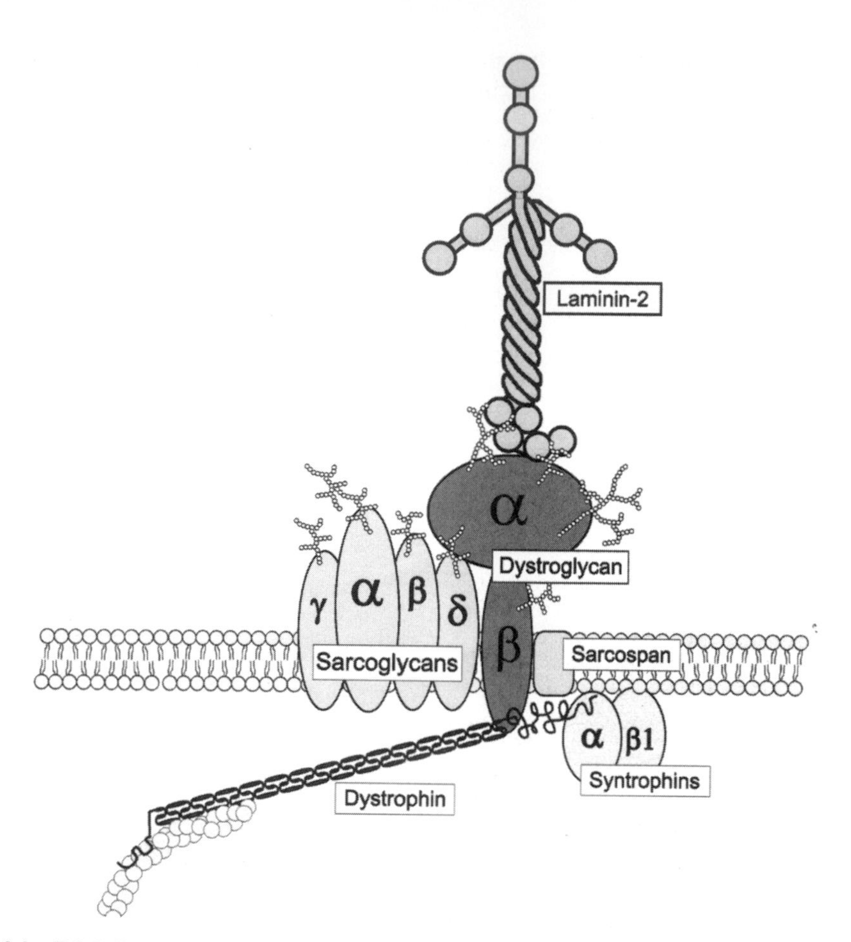

Figure 2. Model of the DGC. Dystroglycan is a central component of the DGC as it is known to link laminin-2 in the extracellular matrix, and dystrophin in the cytoskeleton. Mutations in genes encoding proteins associated with dystroglycan–dystrophin, each of the sarcoglycans, and the laminin $\alpha2$ chain lead to distinct forms of muscular dystrophy. The sarcoglycans form a subcomplex within the DGC. The relationship of sarcospan, a newly identified component of the DGC, is not yet full characterized.

during embryonic development or tissue remodeling and persists in mature tissues as a stable linkage between the cell and extracellular matrix.

Purification

α- and β- dystroglycan were first purified from skeletal muscle as a part of a glycoprotein complex which anchors dystrophin to the muscle cell membrane. This complex, the DGC, is purified from digitonin-extracted rabbit skeletal muscle microsomes by affinity chromatography on succinylated wheat germ agglutinin–sepharose followed by DEAE chromatography and a final step on a sucrose gradient.[1] A similar scheme has been used to purify dystroglycan from bovine peripheral nerve.[26] Dystroglycan was purified as an agrin receptor from the post-synaptic membranes of the electric organ from *Torpedo californica* utilizing an immunoaffinity approach.[10] α- Dystroglycan can be purified by pretreatment of skeletal muscle microsomes with high pH[14] or by laminin or lectin affinity chromatography from brain homogenates.[6,7]

Activities

α-Dystroglycan binds to laminin and agrin isoforms with affinities in the low nanomolar range as judged by blot overlay assays.[2,12] This binding is calcium-dependent, salt-sensitive, and heparin-inhibitable. β-Dystroglycan binds tightly and specifically to dystrophin in a variety of *in vitro* assays.[16,17]

Antibodies

Monoclonal antibodies specific for either α- or β-dystroglycan have been raised by several laboratories. Some of these are commercially available. Applications and cross-reactivities for these reagents vary. One monoclonal antibody, IIH6, developed in our laboratory, blocks the association between α-dystroglycan and laminin or agrin *in vitro*.[2,12] This blocking antibody also specifically inhibits neuromuscular junction formation and kidney epithelial morphogenesis *in vivo*[23,24]. Polyclonal antibodies have been raised against α-dystroglycan, β-dystroglycan, or both subunits by several laboratories. In general, the polyclonal antibodies against dystroglycan tend to be widely species cross-reactive. Currently, no polyclonal antibodies against dystroglycan are commercially available.

Genes

Only a single dystroglycan gene is known to exist in each species in which it has been identified. Both the genomic structure and coding potential of this gene are highly conserved among mammalian species.[9,27] The primary amino acid sequence encoded by the dystroglycan gene is rather well conserved between mammals and the electric ray *Torpedo*. Clones exhibiting similarity to dystroglycan have recently been discovered in invertebrate species (*Drosophila*, *C. elegans*) (S. Baumgartner; personal communication). In humans, this gene is located at 3p21.[27] In mice, it is located in a syntenic region at the distal tip of chromosome 9[28](EMBL/GenBank U48854). In both of these species, dystroglycan is organized into two exons. A single 5.8 kb mRNA species has been characterized in each of the tissues and developmental stages that have been analysed.[9,27] Full length or partial cDNA clones exist for human (GenBank L19711), rabbit (GenBank X64393), and mouse (EMBL Z34532) dystroglycan.

Mutant phenotype/disease states

Despite the fact that mutations in genes encoding a number of dystroglycan's binding partners within the DGC lead to distinct forms of muscular dystrophy in humans and amimals,[3] there is as yet no linkage between mutations in the dystroglycan gene and muscular dystrophy. Nevertheless, a common feature of muscular dystrophies involving other DGC components is a reduction or destabilization of dystroglycan in association with the sarcolemma.[1,21] A likely explanation for the lack of a null mutant form of dystroglycan linked to muscular dystrophy is provided by the phenotype of the dystroglycannull mutant mouse that dies during early embryogenesis.[25] Therefore, certain DGC components may support the muscle-specific functions of dystroglycan, but not its non-muscle roles.

Structure

Very little is known about the structure of dystroglycan at present. Electron microscopic analysis of purified preparations of α-dystroglycan suggests that it is an elongated, dumb-bell-shaped molecule.[29] Ultrastructural analysis *in situ* also argues for an extended conformation of α-dystroglycan.[30]

Web sites

www-camlab.physlog.uiowa.edu

References

1. Ervasti, J. M., Ohlendieck, K., Kahl, S. D., Gaver, M. G., and Campbell, K. P. (1990). *Nature*, **345**, 315–9.
2. Ervasti, J. M. and Campbell, K. P. (1993). *J. Cell. Biol.*, **122**, 809–23.
3. Campbell, K. P. (1995). *Cell*, **80**, 675–9.
4. Ohlendeick, K., Ervasti, J. M., Snook, J. B., and Campbell, K. P. (1991). *J. Cell. Biol.*, **112**, 135–48.
5. Ervasti, J. M. and Campbell, K. P. (1991). *Cell*, **66**, 1121–31.
6. Smalheiser, N. R. and Kim, E. (1995). *J. Biol. Chem.*, **270**, 15425–33.
7. Gee, S. H., Blacher, R. W., Douville, P. J., Provost, P. R., Yurchenco, P. D., and Carbonetto, S. (1993). *J. Biol. Chem.*, **268**, 14972–80.
8. Yoshida, M. and Ozawa, E. (1990). *J. Biochem.*, **108**, 748–56.
9. Ibraghimov-Beskrovnaya, O., Ervasti, J. M., Leveille, C. J., Slaughter, C. A., Sernett, S. W., and Campbell, K. P. (1992). *Nature*, **355**, 696–702.
10. Bowe, M. A., Deyst, K. A., Leszyk, J. D., and Fallon, J. F. (1994). *Neuron*, **12**, 1173–80.
11. Chiba, A., Matsumura, K., Yamada, H., Inazu, T., Shimizu, T., Kusunoki, S., *et al.* (1997). *J. Biol. Chem.*, **272**, 2156–62.
12. Yamada, H., Shimizu, T., Tanaka, T., Campbell, K. P., and Matsumura, K. (1994). *FEBS Lett.*, **352**, 49–53.
13. Gesemann, M., Cavalli, V., Denzer, A. J., Brancaccio, A., Schumacher, B., and Ruegg M. A. (1996). *Neuron*, **16**, 755–67.
14. Pall, E. A., Bolton, K. M., and Ervasti J. M. (1996). *J. Biol. Chem.*, **271**, 3817–21.
15. Yamada, H., Denzer, A. J., Hori, H., Tanaka, T., Anderson, L. V. B., Fujita, S., *et al.* (1996). *J. Biol. Chem.*, **271**, 23418–23.
16. Jung, D., Yang, B., Meyer, J., Chamberlain, J. S., and Campbell, K. P. (1995). *J. Biol. Chem.*, **270**, 27305–10.
17. Suzuki, A., Yoshida, M., Hayashi, K., Mizuno, Y., Hagiwara, Y., and Ozawa, E. (1994). *Eur. J. Biochem.*, **220**, 283–92.
18. Ohlendieck, K., Ervasti, J. M., Matsumura, K., Kahl, S. D., Leveille, C. J., and Campbell, K. P. (1991). *Neuron*, **7**, 499–508.
19. Yang, B., Jung, D., Motto, D., Meyer, J., Koretzky, G., Campbell, K. P. (1995). *J. Biol. Chem.*, **270**, 11711–4.
20. Yoshida, M., Suzuki, A., Yammamoto, H., Noguchi, S., Mizuno, Y., and Ozawa, E. (1994). *Eur. J. Biochem.*, **222**, 1055–61.
21. Roberds, S. L., Ervasti, J. M., Anderson, R. D., Ohlendeick, K., Kahl, S. D., Zoloto, D., and Campbell, K. P. (1993). *J. Biol. Chem.*, **268**, 11496–9.
22. Matsumura, K., Chiba, A., Yamada, H., Fukuta-Ohi, H., Fujita, S., Endo, T., *et al.* (1997). *J. Biol. Chem.*, **272**, 13904–10.
23. Fallon, J. R. and Hall, Z. W. (1994). *Trends Neurosci.*, **17**,469–73.
24. Durbeej, M., Larsson, E., Ibraghimov-Beskrovnaya, O., Roberds, S. L., Campbell, K. P., and Ekblom, P (1995). *J. Cell,. Biol.*, **130**, 79–91.
25. Williamson, R. A., Henry, M. D., Daniels, K. J., Hrstka, R. F., Lee, J. C., Sunada, Y., *et al.* (1997). *Hum. Mol. Genet.*, **6**, 831–41.

26. Yamada, H., Chiba, A., Endo, T., Kobata, A., Anderson, L. V. B., Hori, H., *et al.* (1996). *J. Neurochem.*, **66**, 1518–24.
27. Ibraghimov-Beskrovnaya, O., Milatovich, A., Ozcelik, T., Yang, B., Koepnick, K., Francke, U., and Campbell, K. P. (1993). *Hum. Mol. Genet.*, **2**, 1651–7.
28. Gorecki, D. C., Derry, J. M. J., and Barnard E. A. (1994). *Hum. Mol. Genet.*, **3**, 1589–97.
29. Brancaccio, A., Schulthess, T., Gesemann, M., and Engel, J. (1995). *FEBS Lett.*, **368**, 139–42.
30. Cullen, M. J., Walsh, J., Roberds, S. L., and Campbell, K. P. (1996). *Neuropathol. Appl. Neurobiol.*, **22**, 30–7.
31. Durbeej, M., Henry, M. D., Ferletta, M., Campbell, K. P., and Ekblom, P. (1998). *J. Histochem. Cytochem.*, **46**, 449–57.

Michael D. Henry and Kevin P. Campbell
Howard Hughes Medical Institute,
Department of Physiology and Biophysics
and Department of Neurology,
University of Iowa College of Medicine,
Iowa City, IA 52242, USA

β_3-Endonexin

β_3-endonexin is a 12.6 kDa protein that interacts selectively with the cytoplasmic domain of the β_3 integrin subunit. It may be involved in activation of the ligand-binding function of integrin $a_{IIb}\beta_3$.

Protein properties

β_3-Endonexin was identified from an EBV-transformed B-cell library in a yeast two-hybrid screen using the integrin β_3 cytoplasmic domain as a bait.[1] The cDNA sequence of β_3-endonexin encodes a 111 residue polypeptide with a molecular mass of 12.6 kDa and reveals no obvious homology or biological significance. The protein contains a number of serine and threonine residues suggesting that it may be subjected to modification by phosphorylation. Indeed, Ser-28 is in an appropriate context for phosphorylation by protein kinase A,[2] and Ser-6, Thr-24, and Thr-79 are in contexts favourable for protein kinase C.[3] RT-PCR using the original B lymphocyte cDNA library identified a β_3-endonexin-related mRNA which differs due to insertion of 93 and 50 base pairs and encodes a 170 amino acid, 19.2 kDa polypeptide that differs from β_3-endonexin due to an additional 59 amino acids at the carboxy terminus. In contrast to β_3-endonexin this longer, related polypeptide does not bind to the cytoplasmic domain of the β_3 integrin subunit, either in yeast or *in vitro*.[1]

β_3-endonexin fails to bind to other integrin cytoplasmic domains, including those of β_1, β_2 and α_{IIb}. The membrane-distal NITY motif at β_3 756–759, which has been shown to be important for integrin signalling function,[4] is a critical motif for this selective interaction of β_3-endonexin.[5]

β_3-Endonexin polypeptide could be detected in platelets and a mononuclear leukocyte fraction of blood (containing lymphocytes and monocytes) by immunoblotting with two different specific polyclonal antisera.[1] The subcellular distribution of β_3-endonexin transfected as a fusion protein with green fluorescent protein (GFP) into human endothelial cells or CHO cells showed both cytoplasmic and nuclear localization.

In a CHO cell model system, expression of β_3-endonexin as a GFP-fusion protein is associated with an increase in the affinity state of integrin $\alpha^{IIb}\beta_3$ and with fibrinogen binding and fibrinogen-dependent cell aggregation.[6]

Purification

The β_3-endonexin protein has not been purified.

Activities

β_3-endonexin has been shown to bind selectively to the integrin β_3 cytoplasmic domain *in vitro* and in yeast. Expression of a GFP/β_3-endonexin fusion protein in CHO cells increased the affinity state of integrin $\alpha^{IIb}\beta_3$ in this model system.

Antibodies

Two polyclonal antibodies and four monoclonal antibodies against β_3-endonexin have been generated against a carboxyl-terminal peptide or bacterially expressed β_3-endonexin. These antibodies are suitable for immunoblotting.[1]

Genes

Full length cDNAs for human β_3-endonexin have been isolated. The sequence data are available from EMBL/GenBank/DDBJ under accession number U37139.

Mutant phenotype/disease states

No information available.

Structure

Not available.

References

1. Shattil, S. J., O'Toole, T., Eigenthaler, M., Thon, V., Williams, M., Babior, B. M., and Ginsberg, M. H. (1995). *J. Cell Biol.*, **131**, 807–16.
2. Glass, D.B., el-Maghrabi, M.R., and Pilkis, S.J. (1986). *J. Biol. Chem.*, **261**, 2987–93.
3. Woodgett, J.R., Gould, K.L., and Hunter, T. (1986). *Eur. J. Biochem.*, **161**, 177–84.
4. Ylanne, J., Huuskonen, J., O'Toole, T. E., Ginsberg, M. H., Virtanen, I., and Gahmberg, C. G. (1995). *J. Biol. Chem.*, **270**, 9550–7.
5. Eigenthaler, M., Höfferer, L., Shattil, S. J. and Ginsberg, M. H. (1997). *J. Biol. Chem.*, **272**, 7693–8.
6. Kashiwagi, H., Schwartz, M. A., Eigenthaler, M., Davis, K. A., Ginsberg, M. H., and Shattil, S. J. (1997). *J. Cell Biol.*, **137**, 1433–43.

Martin Eigenthaler
Medizinische Universitätsklinik,
Klinische Biochemie,
97080 Würzburg, Germany

Sanford J. Shattil
The Scripps Research Institute,
Department of Vascular Biology,
La Jolla, CA 92037, USA

Ep-CAM (epithelial cell adhesion molecule)

Ep-CAM is a 40 kDa type transmembrane glycoprotein with properties of a Ca^{2+}-independent homotypic intercellular adhesion molecule. Structurally, Ep-CAM is not related to any of the major families of adhesion molecules.

Protein properties

Ep-CAM (alias KS1/4, GA733-2, EGP40 or 17-1A antigen) was identified as a human carcinoma associated antigen.[1] The full length of the molecule is 314 amino acids, of which the first 20 constitute the leader sequence. The extracelluar domain of 257 a.a. consists of two EGF-like domains (residues 27–59 and 66–135 respectively) followed by a cysteine-poor region, a transmembrane domain (22 a.a.) and a cytoplasmic domain (26 a.a.).[2,3,4] The protein is synthesized as a 34 kDa precursor. Subsequent *N*-linked glycosylation (Ep-CAM contains three potential *N*-linkage sites) produces several glycoforms that most probably differ slightly in the oligosaccharides attached. Up to three forms of MW 39–43 kDa can be detected in various cell lines, but, in some cells only one form is present.[5] A post-translational cleavage of the Ep-CAM molecule at the position Arg^{80}–Arg^{81} may occur, but the resulting fragments remain covalently associated.[6] No additional post-translational modifications of the molecule were described. The Ep-CAM molecule is very conserved among mammalian species,[7] with murine Ep-CAM being 86 per cent identical to the human molecule.[8]

Ep-CAM is predominantly expressed in human epithelial tissues (e.g. see Plate 6), although in squamous epithelia the expression is observed only during embryogenesis and in preneoplastic/malignant lesions.[9,10] In mice the tissue pattern of Ep-CAM expression is similar to that in humans; however, muEp-CAM was demonstrated to be additionally expressed in thymocytes[11] and dendritic cells.[12] No data on human Ep-CAM being expressed in non-epithelial cells are currently available.

Only one human protein of homology to Ep-CAM is known: GA733-1.[7,13] This molecule is encoded by an intronless gene, originated from a retroposition of the Ep-CAM cDNA into the germline. The GA733-1 protein is 49 per cent homologous to Ep-CAM. The molecule has in general a tissue pattern of expression complementary to Ep-CAM, being predominantly expressed in tissues where Ep-CAM expression is low (squamous epithelia), but not

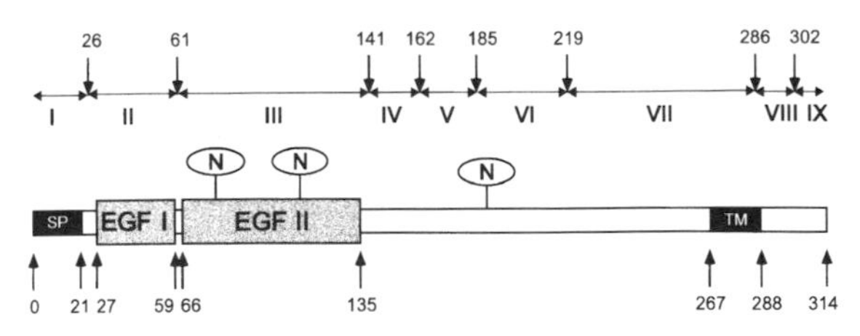

Figure 1. Domain organization of the Ep-CAM molecule. The position and size of the various domains are shown on the structural map of the Ep-CAM molecule: signal peptide (SP), EGF-like domains (I and II), the transmembrane domain (TM). The numbers below show the respective position in the polypeptide chain for the first and the last amino acid of each domain. The roman numerals (I–IX) correspond to the exons encoding the respective fragment of the molecule. The potential *N*-linked glycosylation sites (N) are shown.

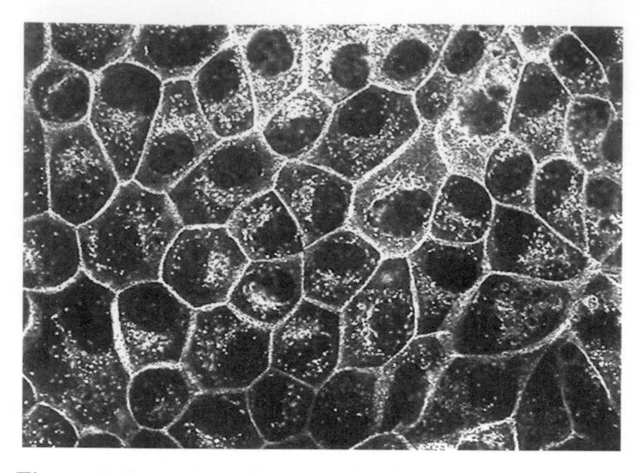

Figure 2. Detection of human Ep-CAM in transfected MDCK cells by immunofluorescent staining using mAb 323/A3. Note the substantial intracellular fraction formed by the internalized Ep-CAM molecules.

in tissues with a high level of Ep-CAM (colon).[14] GA733-1 is slightly larger than Ep-CAM (323 a.a.), and contains one additional site for *N*-linked glycosylation.[13] This 45 kDa molecule can be phosphorylated at Ser^{303} in the cytoplasmic domain[15] (this amino acid is absent in the corresponding region of Ep-CAM). The function of GA733–1 is unknown.

Purification

Ep-CAM can be purified by affinity chromatography from cell lysates. A secreted form of Ep-CAM with the deleted transmembrane domain and hexa-His tag at the carboxyl terminus was generated and expressed in a baculovirus system for protein purification in large quantities.[16]

Activities

Ep-CAM molecules form dimers, interacting collaterally, and this interaction is independent of the molecule's cytoplasmic domain. The cytoplasmic domain contains two sites of binding to α-actinin; the latter mediates interaction of Ep-CAM with the actin-based cytoskeleton. The interaction with the cytoskeleton is important for the molecule's ability to form stable adhesions.[17] In epithelial cells Ep-CAM is present strictly on the lateral domains of the cell membrane and is excluded from tight junctions, adherens junctions, and desmosomes. When introduced into cells deficient in other means of cell–cell interactions, Ep-CAM is capable of mediating homotypic Ca^{2+}-independent intercellular interactions, cell aggregation, and of preventing cell scattering.[5] The contribution of the molecule to interaction of epithelial cells is unclear.[18] Recent findings, that expression of Ep-CAM leads to a partial abrogation of E-cadherin mediated junctions, when considered together with an observed enhanced expression of the molecule in actively proliferating epithelial and carcinoma cells, suggest a role for Ep-CAM in defining the plasticity of the epithelial cell phenotype and in regulation of cell proliferation.[19]

Antibodies

Most monoclonal antibodies to human Ep-CAM recognize one of a few overlapping conformational epitopes (that can be destroyed by β-mercaptoethanol) within the EGF-like domains of the molecule.[20] One antibody to an epitope in the cysteine-poor region of Ep-CAM has been generated[20]. No function-blocking mAbs are known. Most antibodies to huEp-CAM have no cross-species reactivity. A rat monoclonal antibody to murine Ep-CAM (G8.8) exists.[21] Most antibodies can be used in immunohistochemistry on frozen sections, immunoprecipitation, and immunoblotting. Some can recognize Ep-CAM on formalin fixed sections after trypsin pretreatment. Several antibodies specific to GA733–1 exist, none of these is cross-reactive with Ep-CAM.[22]

Genes

Sequences have been published for human (M32306; M33011) and murine (M76124) cDNAs. The human Ep-CAM gene consists of nine exons and has a size over 4 kb (the exon sequences: M93029-M93036). The Ep-CAM gene (GA733-2) was mapped to chromosome 4q. The sequences for both the GA733-1 cDNA (J04152) and the gene, located at chromosome 1, have been published (X13425).

References

1. Herlyn, M., Steplewski, Z., Herlyn, D., and Koprowski, H. (1979). *Proc. Natl Acad. Sci. USA*, **76**, 1438–52.
2. Perez, M. S., and Walker, L. E. (1989). *J. Immunol.*, **142**, 3662–7.
3. Simon, B., Podolsky, D. K., Moldenhauer, G., Isselbacher, K. J., Gattoni-Celli, S., and Brand, S. J. (1990). *Proc. Natl Acad. Sci. USA*, **87**, 2755–9.
4. Szala, S., Froehlich, M., Scollon, M., Kasai, Y., Steplewski, Z., Koprowski, H., and Linnenbach, A. J. (1990). *Proc. Natl Acad. Sci. USA*, **87**, 3542–6.
5. Litvinov, S. V., Velders, M. P., Bakker, H. A., Fleuren, G. J., and Warnaar, S. O. (1994). *J. Cell Biol.*, **125**, 437–46.
6. Bjork, P., Jonsson, U., Svedberg, H., Larsson, K., Lind, P., Dillner, J., *et al.* (1993). *J. Biol. Chem.*, **268**, 24232–41.
7. Linnenbach, A. J., Seng, B. A., Wu, S., Robbins, S., Scollon, M., Pyrc, J. J., *et al.* (1993). *Mol. Cell. Biol.*, **13**, 1507–15.
8. Bergsagel, P. L., Victor-Kobrin, C., Timblin, C. R., Trepel, J., and Kuehl, W. M. (1992). *J. Immunol.*, **148**, 590–6.
9. Quak, J. J., van Dongen, G., Brakee, J. G., Hayashida, D. J., Balm, A. J., Snow, G. B., and Meijer, C. J. (1990). *Hybridoma*, **9**, 377–87.
10. Litvinov, S. V., van Driel, W., van Rhijn, C. M., Bakker, H. A. M., van Krieken, H., Fleuren, G. J., and Warnaar, S. O. (1996). *Am. J. Pathol.*, **148**, 865–75.
11. Nelson, A. J., Dumm, R. J., Peach, R., Aruffo, A., and Farr, A. J. (1996). *J. Immunol.*, **26**, 401–8.
12. Borkowski, T. A., Nelson, A. J., Farr, A. G., and Udey, M. C. (1996). *J. Immunol.*, **26**, 110–4.

13. Linnenbach, A. J., Wojcierowski, J., Wu, S., Pyrc, J. A., Ross., A. H., Dietzschold, B., *et al.* (1989). *Proc. Natl Acad. Sci. USA*, **86**, 27–31.
14. Stein, R., Basu, A., Chen, S., Shih, L. B., and Goldenberg, D. M. (1993). *Int. J. Cancer*, **55**, 938–46.
15. Basu, A., Goldenberg, D. M., and Stein, R. (1995). *Int. J. Cancer*, **62**, 472–9.
16. Strassburg, C. P., Kasai, Y., Seng, B. A., Miniou, P., Zaloudik, J., Herlyn, D., *ey al.* (1992). *Cancer Res.*, **52**, 815–21.
17. Balzar, M., Bakker, H. A. M., Briaire, I. H., Fleuren, G. J., Warnaar, S. O., and Litvinov, S.V. (1998). *Mol. Cell. Biol.*, **18**, 4833–43.
18. Litvinov, S. V., Bakker, H. A., Gourevitch, M. M., Velders, M. P., and Warnaar, S. O. (1994). *Cell Adhes. Commun.*, **2**, 417–28.
19. Litvinov, S. V., Balzar, M., Bakker, H. A. M., Briaire-de Bruijn, I. H., Prins, F., Fleuren, G. J., and Warnaar, S. O. (1997). *J. Cell Biol.* **139**, 1337–48.
20. De Leij, L., Helrich, W., Stein, R., and Mattes, M. J. (1994). *Int. J. Cancer*, (Suppl. 8), 60–3.
21. Farr A., Nelson, A., Truex, J., and Hosier, S. (1991). *J. Histochem. Cytochem.*, **39**, 645.
22. Stein, R., Basu, A., Goldenberg, D. M., Lloyd, K. O., and Mattes, M. J. (1994). *Int. J. Cancer*, (Suppl. 8), 98–102.

■ *Sergey V. Litvinov*
Department of Pathology,
Leiden University, Leiden,
The Netherlands

Fasciclin I

Fasciclin I[1,2] is a homophilic, Ca^{2+}-independent cell adhesion molecule[3] in grasshopper and *Drosophila* that is expressed on a subset of growth cones, axon pathways, and glia cells during embryonic development; it is also expressed at other times and places outside of the developing nervous system.[4] During neuronal development, fasciclin I appears to play a role in growth cone guidance and axon fasciculation.[5,6]

■ Protein properties

The fasciclin I protein was first identified in the grasshopper embryo as a result of a monoclonal antibody (mAb) screen 1; it is a membrane-associated glycoprotein with an apparent molecular weight of 70 kDa which is transitionally expressed in a segmentally repeated pattern on the surface of certain neuroepithelial cells and later by a subset of neurones which pioneer specific axonal pathways.[1,7] During *Drosophila* embryogenesis, fasciclin I is expressed on the surface of all PNS neurones and initially on all commissural axons in the CNS. Similar to the expression pattern in grasshopper embryos, *Drosophila* fasciclin I later becomes restricted to specific axonal pathways.[2,4] Fasciclin I is also expressed by many non-neuronal tissues in the fly embryo, such as the hindgut, the salivary glands, and the ovaries.

Fasciclin I cDNAs were initially isolated in grasshopper by using degenerate oligonucleotides based on the known sequences of amino acid fragments. Subsequently, *Drosophila* cDNAs were identified by a low stringency screen.[2,8] The cDNA sequences of grasshopper and *Drosophila* fasciclin I encode polypeptides of about 660 amino acids 2. Both proteins comprise four tandem domains of 150 amino acids each (see Fig. 1), identified by virtue of their weak homology to each other and by the presence of more highly conserved ~40 amino acid repeats at the end of the second, third, and fourth domains. The last carboxyl-terminal 30 amino acids of the fasciclin I protein are mainly uncharged and provide the signal for the attachment of a glycosyl phosphatidylinositol (GPI) lipid membrane anchor.[9] In *Drosophila* embryos this GPI linkage is developmentally regulated and a large fraction of the protein appears to lose its membrane anchor during embryonic development.

In both grasshopper and *Drosophila*, the *fasciclin I* gene gives rise to multiple transcripts which show different temporal and spatial regulation.[10] The *Drosophila* transcripts differ primarily in their 3′ untranslated regions and by the alternative splicing of two nine nucleotide micro-exons (encoding a SFK and a FMN amino acid sequence, respectively) at the end of the conserved amino acid repeat in the second protein domain. In *Drosophila* these micro-exons give rise to three different forms of the protein, the type I (none of the micro-exons is included), type II (containing only the first SFK-encoding micro-exon), and type III form (including both micro-exons). The first SFK-encoding micro-exon is conserved in grasshopper, whereas the second micro-exon is only 6 bp long and encodes the amino acids GF. Alternative splicing creates at least three different protein forms of grasshopper fasciclin I, corresponding to the type I and II proteins observed in *Drosophila* and a type IV protein containing only the GF motif which is encoded by the second micro-exon. The inclusion of these micro-exons appears to influence the ability of fasciclin I to mediate homophilic adhesion in an adhesion assay using *Drosophila* S2 cells. *Drosophila* type I and II fasciclin I induces strong S2 cell aggregation, whereas the *Drosophila* type III and the grasshopper type IV proteins are considerably less active in the same assay system (Seeger and Goodman; unpublished observations). A second molecule with four fasci-

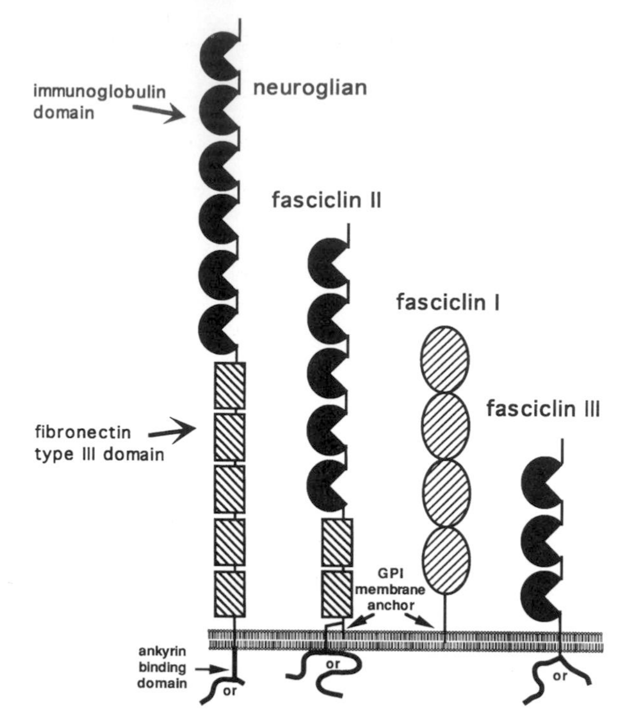

Figure 1. Schematic domain structures of the neuroglian, fasciclin II, fasciclin I, and fasciclin III glycoproteins. Three of the four glycoproteins are members of the immunoglobulin superfamily. Two of these proteins, neuroglian and fasciclin II, have multiple immunoglobulin domains followed by multiple fibronectin type III domains and belong to the L1 and the NCAM gene family respectively. The third, fasciclin III, has three, more divergent, immunoglobulin domains. Fasciclin I has a unique structure made up of four tandem protein domains and is anchored in the plasma membrane by a glycosyl phosphatidyinositol (GPI) lipid membrane anchor. Neuroglian and fasciclin III have alternatively spliced cytoplasmic domains; fasciclin II exists in at least two different protein forms, including one with a GPI membrane anchor and one transmembrane form, with a cytoplasmic domain which contains a PEST sequence motif.

clin-I-type repeats has been closed from *Drosophila* and initially characterized.[21] This molecule is dynamically expressed during embryogenesis in the blastoderm, CNS midline cells (cell bodies of midline neurones and glia and on midline axons) and trachea and was therefore named midline fasiclin.

Purification

Fasciclin I protein was originally isolated from grasshopper embryonic lysates by affinity chromatography on 3B11 mAb columns.[1,8] Wang *et al.*[11] used the same type of 3B11 mAb affinity matrix to purify fasciclin I from supernatants of transfected CHO or *Drosophila* S2 cells which secrete a soluble form of the grasshopper fasciclin I protein.[11]

Activities

Transfection of *Drosophila* S2 cells and *in vitro* cell aggregation experiments show that fasciclin I can mediate homophilic, Ca^{2+}-independent cell adhesion.[3]

Antibodies

The 3B11 mAb recognizing grasshopper fasciclin I[1] is available from the Developmental Studies Hybridoma Bank at the University of Iowa. Additional monoclonal and polyclonal antibodies in mice and rats against grasshopper fasciclin I have been reported.[1,8] Rat antisera and several mouse mAbs (e.g the 6D8 mAb used in Fig. 2) which recognize *Drosophila* fasciclin I on Western blots

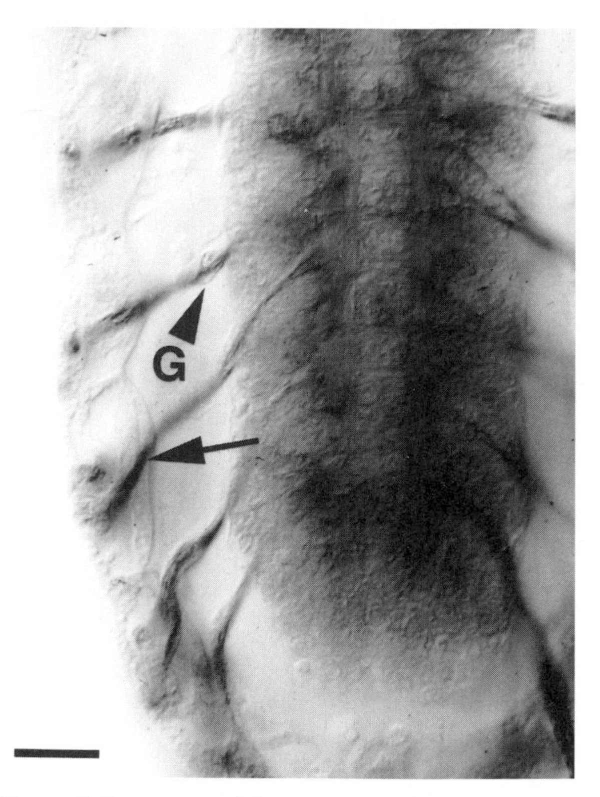

Figure 2. Expression of *Drosophila* fasciclin I in the developing embryonic nervous system. A dissected, late developmental stage 15 embryo was stained for fasciclin I protein using the mAb 6D8. Shown is a view of the posterior end (pointing down). The segmental and intersegmental nerve, which extends out to the periphery, is indicated in one segment by an arrow. They join near the periphery of the CNS at a fasciclin I-positive glial cell which is marked in one segment with an arrowhead. Scale bar: represents 25 μm.

and in immunocytochemical procedures have also been generated using fasciclin I fusion proteins.[2,9]

Genes

Full cDNA sequences for grasshopper (GenBank M20544) and *Drosophila* fasciclin I (GenBank M20545, genomic accession number M32311) have been published.[2,10] A second fasciclin I-type gene has been reported from *Drosophila*. The mRNA sequence of this molecule, called midline fasciclin, can be found in GenBank under AF038843.[21] The following molecules bear various degrees of sequence homology to the insect fasciclin I polypeptides: the secreted MPT70 antigen from *Mycobacterium tuberculosis* (GenBank D37968);[12] Algal CAM from *Volvox carteri* (GenBank X80416);[13] the gynecophoral canal protein SmGCP from *Schistosoma mansoni* (GenBank U47862);[14] a family of cell surface proteins from the sea urchin *Paracentrotus* lividus (a partial genomic DNA sequence for the *bep4* gene is available under GenBank L11925,[22,23] the TGF-β-induced βIG-H3 proteins from chick chondrocytes (GenBank AB005553) and from human adenocarcinoma cells (GenBank M77349; the mouse mRNA can be found under L19932);[15] and the osteoblast-specific factor 2 (OSF-2), a putative adhesion molecule expressed in mammalian osteoblasts (the GenBank accession number for mouse OSF-2 is D13664 and for two forms of human OSF-2, OSF-2os and OSF-2pl, D13666 and D13665, respectively).[16] An open reading frame encoding a fasciclin I-like protein has also been identified on a cosmid containing genomic DNA from *Caenorhabditis elegans* (GenBank CEF26E4).

Mutant phenotypes

The *Drosophila fasciclin I* gene maps to the third chromosome at position 89D. Mutations have been identified and generated in the *fasciclin I* gene.[17] A large (~100 kb) transposable element, called a TE element, in one stock, TE77, was found to be located in a large intron in the *fasciclin I* gene. The TE77 insert leads to a mRNA and protein null mutation in the *fasciclin I* gene, $fasI^{TE77}$. Additional *fas I* alleles were isolated by screening for revertants which had lost the TE77 P element insert together with adjacent genomic DNA. At least two such revertants, $fasI^{R3}$ and $fasI^{R40.1}$, appear to represent complete protein null mutations for *fasciclin I 17*. *Drosophila* lines with an additional copy of a functional fasciclin I gene have been generated by P element-mediated germline transformation of flies.[17]

Genetic analysis has shown that fasciclin I is involved in growth cone guidance and in the arborization of synaptic terminals. Protein null mutations in the *fasciclin I* gene are homozygous viable, display strong behavioural defects (uncoordinated motor activity), but show no gross anatomical defects in CNS morphogenesis. However, the level of fasciclin I on the presynaptic side of the larval neuromuscular junction is important for the strength of synaptic function and inversely correlates with the complexity of nerve terminal arborization.[18] Embryos doubly mutant for *fas I* and *abl*, the *Drosophila* Abelson proto-oncogene homologue, which encodes a cytoplasmic tyrosine kinase, display major defects in CNS axon pathways, particularly in the commissural tracts where the expression of these two proteins normally overlaps.[17] Whereas in *fasciclin I*-deficient larvae only a slight increase of wing sensory neurone branching was observed, in larvae doubly mutant for fasciclin I and III these neurones also exhibited some degree of axonal misrouting.[19] An anti-fasciclin I mAb was used to perturb fasciclin I protein on the surface of Ti1 pioneer neurones in the developing grasshopper limb bud by chromophore-assisted laser inactivation (CALI).[5,6] Fasciclin I-targeted CALI disrupted the fasciculation of these axons without affecting their growth or guidance. Deletions covering the *Drosophila* midline fasciclin gene similar to fasciclin I mutations display only weak defects in axonogenesis, but also strongly interact phenotypically with mutations in *abl*.[21] The gene for *Drosophila* midline fasciclin maps to the third chromosome at the cytological location 87A4-9.

Structure

Soluble fasciclin I expressed by transfected mammalian or *Drosophila* tissue culture cells exists as a monomer in solution, indicating a low affinity for the homophilic fasciclin I interaction.[20] A circular dichroism study of purified grasshopper fasciclin I suggests that its structure is primarily α-helical.[20] This is in contrast to immunoglobulin- and fibronectin type III-domain cell adhesion molecules which have a predominantly β-sheet structure.

References

1. Bastiani, M. J., Harrelson, A. L., Snow, P. M., and Goodman, C. S. (1987). *Cell*, **48**, 745–55.
2. Zinn, K., McAllister, L., and Goodman, C. S. (1988). *Cell*, **53**, 577–87.
3. Elkins, T., Hortsch, M., Bieber, A. J., Snow, P. M., and Goodman, C. S. (1990). *J. Cell Biol.*, **110**, 1825–32.
4. McAllister, L., Goodman, C. S., and Zinn, K. (1992). *Development*, **115**, 267–76.
5. Jay, D. G. and Keshishian, H. (1990). *Nature*, **348**, 548–50.
6. Diamond, P., Mallavarapu, A., Schnipper, J., Booth, J., Park, L., O'Connor, T. P., and Jay, D. G. (1993). *Neuron*, **11**, 409–21.
7. Boyan, G., Therianos, S., Williams, J. L., and Reichert, H. (1995). *Development*, **121**, 75–86.
8. Snow, P. M., Zinn, K., Harrelson, A. L., McAllister, L., Schilling, J., Bastiani, M. J., *et al.* (1988). *Proc. Natl Acad. Sci. USA*, **85**, 5291–5.
9. Hortsch, M. and Goodman, C. S. (1990). *J. Biol. Chem.*, **265**, 15104–9.
10. McAllister, L., Rehm, E. J., Goodman, G. S., and Zinn, K. (1992). *J. Neurosci.*, **12**, 895–905.
11. Wang, S.-Y. and Gudas, L. J. (1983). *Proc. Natl Acad. Sci. USA*, **80**, 5880–4.
12. Matsumoto, S., Matsuo, T., Ohara, N., Hotokezaka, H., Naito, M., Minami, J., and Yamada, T. (1995). *Scand. J. Immunol.*, **41**, 281–7.
13. Huber, O. and Sumper, M. (1994). *EMBO J.*, **13**, 4212–22.

14. Bostic, J. R. and Strand, M. (1996). *Mol. Biochem. Parasitol.*, **79**, 79–89.
15. Skonier, J., Neubauer, M., Madisen, L., Bennett, K., Plowman, G. D., and Purchio, A. F. (1992). *DNA Cell Biol.*. **11**, 511–22.
16. Takeshita, S., Kikuno, R., Tezuka, K., and Amann, E. (1993). *Biochem. J.*, **294**, 271–8.
17. Elkins, T., Zinn, K., McAllister, L., Hoffmann, F. M., and Goodman, C. S. (1990). *Cell*, **60**, 565–75.
18. Zhong, Y. and Shanley, J. (1995). *J. Neurosci.*, **15**, 6679–87.
19. Whitlock, K.E. (1993). *Development*, **117**, 1251–60.
20. Wang, W. C., Zinn, K., and Bjorkman, P. J. (1993). *J. Biol. Chem.*, **268**, 1448–55.
21. Hu, S., Sornenfeld, M., Stahl, S., and Crews, S. T. (1998). *J. Neurobiol.*, **35**, 77–93.
22. Di Carlo, M., Montana, G., and Bonura, A. (1990). *Mol. Reprod. Dev.*, **25**, 28–36.
23. Romancinc, D. P., Ghersi, G., Montana, G., Bonura, A., Perriera, S., and Di Carlo, M. (1992). *Differentiation*, **50**, 67–74.
24. Kawamoto, T., Noshiro, M., Shen, M., Nakamasu, K., Hashimoto, K., Kawashima-Ohya, Y., Gotoh, O., and Kato, Y. (1998). *Biochim. Biophys. Acta*, **1395**, 288–92.

Michael Hortsch
Department of Anatomy and Cell Biology, University of Michigan, Ann Arbor, MI 48109–0616, USA

Fasciclin II

Fasciclin II[1–3] is a homophilic, Ca^{2+}-independent cell adhesion molecule in insects[4] which is a member of the immunoglobulin gene superfamily; it shares a common ancestor with vertebrate NCAM.[4] It is expressed on a subset of longitudinal axon fascicles and other axon pathways in grasshopper and *Drosophila* embryos. The protein comes in at least three different forms which are generated by alternative splicing.[3,11] During neuronal development, fasciclin II appears to play a role in specific growth cone guidance, selective fasciculation and the establishment and maintenance of synaptic connections.

Protein properties

The fasciclin II protein was first identified in the grasshopper embryo as a result of a monoclonal antibody (mAb) screen;[1] it is a membrane glycoprotein with an apparent molecular weight of about 100 kDa which is expressed on the surface of a specific subset of axon fascicles in both the CNS and PNS during embryogenesis.[1,2,5] Fasciclin II is primarily expressed on a subset of longitudinal axon pathways, the intersegmental nerve root, and transiently on some commissural axon bundles (Fig. 1).[1,2] The protein is also expressed in a restricted fashion outside of the developing nervous system. The expression pattern of fasciclin II suggests that it might be involved in growth cone guidance. When grasshopper embryos are incubated with anti-fasciclin II antibodies before the MP1 growth cone contacts the MP1 fascicle, the MP1 growth cone stalls at the choice point; it either takes longer than normal to join the appropriate axon fascicle or sometimes follows the wrong pathway, and often extends an aberrant second growth cone across the anterior commissure.[2]

Fasciclin II cDNAs were initially isolated in grasshopper by using degenerate oligonucleotides and cDNA expression cloning[2,5] and were subsequently isolated from *Drosophila* by using PCR.[3] The cDNA open reading frame of grasshopper fasciclin II encodes a transmembrane glycoprotein of 897 amino acids; the extracellular domain contains five immunoglobulin C2-type domains and two fibronectin type III domains 2 (see Fig. 1, Fasciclin I, p. 190); the *Drosophila* homologue has a similar extracellular domain structure.[3] The deduced grasshopper and *Drosophila* fasciclin II proteins have approximately 45 per cent amino acid identity. Fasciclin II is expressed in a very similar fashion in the developing CNS of both insects.[2,3] In *Drosophila* at least three forms of the fasciclin II protein can be detected using anti-fasciclin II antibodies (Grenningloh *et al.*; unpublished results). The two longer forms with apparent molecular weights of about 110 kDa are transmembrane proteins, one of which contains a PEST degeneration sequence in its cytoplasmic domain (+PEST form), whereas the shortest form has an apparent molecular weight of 98 kDa and is linked to the membrane by a glycosyl phosphatidylinositol (GPI) lipid moiety.

Fasciclin II has a similar domain organization to vertebrate NCAM and the grasshopper protein shares 28 per cent amino acid identity over the entire extracellular domain to mouse or chicken NCAM.[6,7] Insect fasciclin II and vertebrate NCAMs probably evolved from a common ancestral molecule which predated the split in the two evolutionary lines leading to the arthropods and chordates.[4]

Purification

Fasciclin II protein was isolated from grasshopper embryonic protein extracts by affinity chromatography on 8C6 mAb columns.[1,5]

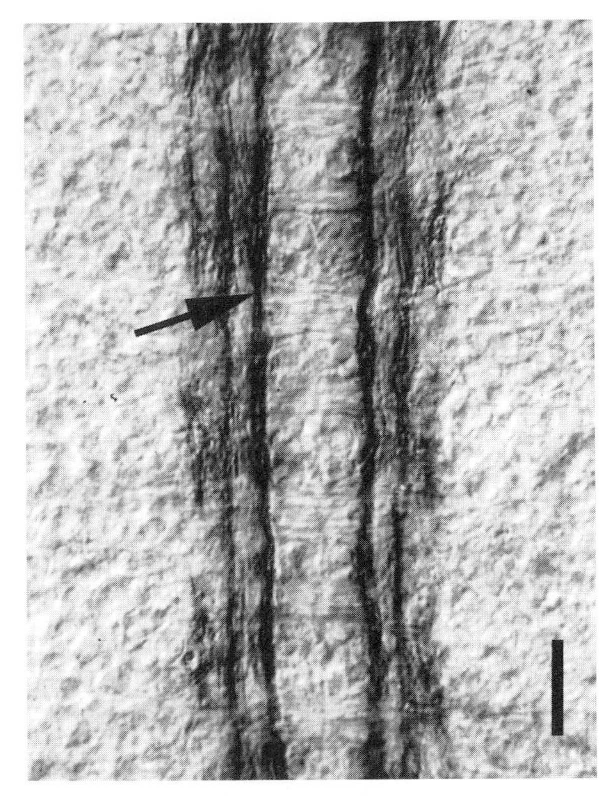

Figure 1. Expression of fasciclin II in the developing *Drosophila* ventral nerve cord (Courtesy of G. Grenningloh and C. S. Goodman). Shown is a ventral view of a developmental stage 16 *Drosophila* embryo stained with a rat anti-fasciclin II antiserum. Fasciclin II protein is expressed at high levels on three longitudinal axon bundles, the most medial being the MP1 fascicle (marked with an arrow). Scale bar: represents 10 μm.

Activities

Transfection of *Drosophila* S2 cells and *in vitro* cell aggregation experiments show that fasciclin II can mediate homophilic, Ca^{2+}-independent cell adhesion.[4]

Antibodies

The generation of monoclonal and polyclonal antibodies in mice and rats against grasshopper fasciclin II has been reported by Bastiani *et al.*[1] The 8C6 mAb which recognizes grasshopper fasciclin II and has been used for the immunoaffinity purification of the grasshopper protein 1 is available from the Developmental Studies Hybridoma Bank at the University of Iowa. Rat antisera and mouse mAbs have also been generated using different *Drosophila* fasciclin II fusion proteins.[3] One mAb, called 1D4, recognizes an intracellular epitope which is specific for the transmembrane, PEST sequence-containing protein form of *Drosophila* fasciclin II (G. Helt and C. S. Goodman, unpublished results).

Genes

A small segment of the 5′ end of the grasshopper fasciclin II mRNA sequence has been published and can be accessed under GenBank accession number J03789.[5] Two mRNA sequences encoding two different protein forms of *Drosophila* fasciclin II (a protein form with a GPI-membrane anchor and a transmembrane form with a cytoplasmic PEST sequence motif) 3 can be found under GenBank M77165 and M77166, respectively. Fasciclin II/NCAM-like molecules have been cloned from other invertebrate species including the mollusc *Aplysia californica* (GenBank M89648, M89649, and M89650)[8] and the nematode *Caenorhabditis elegans* (GenBank U49830).

Mutant phenotypes

The *Drosophila fasciclin II* gene is located on the X chromosome at position 4B1–2. Two classes of mutations have been isolated by the imprecise excision of a P element near the 5′ end of the *Drosophila fasciclin II* gene (enhancer trap line A31 9). The first class includes two viable, hypomorphic alleles, *fasII*^{e86} and *fasII*^{e76}, which express less than 50 per cent and 10 per cent, respectively, of the wild type fasciclin II protein level.[3] The second class encompasses more than 20 homozygous lethal lines, such as *fasII*^{eB112} and *fasII*^{eB78}, which show no fasciclin II protein expression.[3]

During insect embryonic nervous system development, fasciclin II expression is important for the proper development of several axonal pathways (including the vMP2, the MP1, and the FN3 pathways) which normally express fasciclin II protein. Although *fasciclin II*-mutant *Drosophila* embryos display no gross structural abnormalities of the CNS, a complete or partial defasciculation of these axonal pathways can be demonstrated using specific antibody probes.[4,10] Overexpression of fasciclin II in these CNS neurones results in the abnormal fusion of axon fascicles, aberrant pathway choices of growing axons, and sometimes in the stalling of growth cone advance.[10,11] In contrast to these findings in the *Drosophila* embryo, a study using chromophore-assisted laser inactivation (CALI) in grasshopper embryos indicates that fasciclin II protein on one particular fasciclin II-positive neurone, the Ti1 neurone, is necessary for axonal outgrowth but not for fasciculation.[12]

Fasciclin II appears to play an important role in the stabilization and growth of certain neuromuscular synapses in the *Drosophila* embryo where it is required on both the pre- and the postsynaptic membrane.[13] Before synapse formation, fasciclin II protein is evenly expressed at low levels on specific muscle cells. During synapse formation fasciclin II becomes concentrated at the synapse and disappears from the rest of the muscle. This change of fasciclin II cell surface distribution appears to be mediated by a direct interaction between the carboxy-terminus of fasciclin II and the PDZ-containing protein Disc-Large (Dlg).[19,20] In *fasciclin II*-null mutant embryos, initial synapse formation is normal, but subsequently synaptic contacts are eliminated during larval development.[13] Hypomorphic mutations in the *fasciclin II* gene display fewer synapse-bearing nerve terminal varicosities, but the level of synaptic strength is compensated for by enhanced neurotransmitter release from mutant varicosities and by enlarged synaptic contacts.[14,15] cAMP appears to regulate structural plasticity of the neuromuscular junction via fasciclin II, whereas synaptic function is under control of a parallel pathway which is mediated by CREB, the cAMP response element-binding protein.[16] Transient increases of fasciclin II levels in target and non-target muscles stabilize growth cone contacts, result in novel, functional synapses, and have a profound effect on target muscle selection.[21]

In larval imaginal discs, mutations in the *Drosophila* Abelson proto-oncogene homologue show a dominant interaction with fasciclin II mutations in the control of proneural gene expression.[17] This suggests that the *Abelson tyrosine kinase* gene product might mediate the fasciclin II-dependent induction of proneural genes in these tissues. In wing imaginal discs of *fasciclin II*-hypomorphic mutant animals, an occasional misrouting of sensory axons and the appearance of ectopic neurones on the posterior margin has been observed.[18]

References

1. Bastiani, M. J., Harrelson, A. L., Snow, P. M., and Goodman, C. S. (1987). *Cell*, **48**, 745–55.
2. Harrelson, A. L. and Goodman, C. S. (1988). *Science*, **242**, 700–8.
3. Grenningloh, G., Rehm, E. J., and Goodman, C. S. (1991). *Cell*, **67**, 45–57.
4. Grenningloh, G., Bieber, A. J., Rehm, E. J., Snow, P. M., Traquina, Z. R., Hortsch, M., *et al*. (1990). *Cold Spring Harbor Symp. Quant. Biol.*, **55**, 327–40.
5. Snow, P. M., Zinn, K., Harrelson, A. L., McAllister, L., Schilling, J., Bastiani, M. J., *et al*. (1988). *Proc. Natl Acad. Sci. USA*, **85**, 5291–5.
6. Cunningham, B. A., Hemperly, J. J., Murray, B. A., Prediger, E. A., Brackenbury, R., and Edelman, G. M. (1987). *Science*, **236**, 799–806.
7. Barthels, D., Santoni, M. J., Wille, W., Ruppert, C., Chaix, J. C., Hirsch, M. R., *et al*. (1987). *EMBO J.*, **6**, 907–14.
8. Mayford, M., Barzilai, A., Keller, F., Schacher, S., and Kandel, E. R. (1992). *Science*, **256**, 638–44.
9. Ghysen, A. and O'Kane, C. (1989). *Development*, **105**, 35–52.
10. Lin, D. M., Fetter, R. D., Kopczynski, C., Grenningloh, G., and Goodman, C. S. (1994). *Neuron*, **13**, 1055–69.
11. Lin, D. M. and Goodman, C. S. (1994). *Neuron*, **13**, 507–23.
12. Diamond, P., Mallavarapu, A., Schnipper, J., Booth, J., Park, L., O'Connor, T. P., and Jay, D. G. (1993). *Neuron*, **11**, 409–21.
13. Schuster, C. M., Davis, G. W., Fetter, R. D., and Goodman, C. S. (1996). *Neuron*, **17**, 641–54.
14. Stewart, B. A., Schuster, C. M., Goodman, C. S., and Atwood, H. L. (1996). *J. Neurosci.*, **16**, 3877–86.
15. Schuster, C. M., Davis, G. W., Fetter, R. D., and Goodman, C. S. (1996). *Neuron*, **17**, 655–67.
16. Davis, G. W., Schuster, C. M., and Goodman, C. S. (1996). *Neuron*, **17**, 669–79.
17. Garcia-Alonso, L., VanBerkum, M. F., Grenningloh, G., Schuster, C., and Goodman, C. S. (1995). *Proc. Natl Acad. Sci. USA*, **92**, 10501–5.
18. Whitlock, K. E. (1993). *Development*, **117**, 1251–60.
19. Zito, K., Fetter, R. D., Goodman, C. S., and Isacoff, E. Y. (1997). *Neuron*, **19**, 1007–16.
20. Thomas, J., Kim, E., Kuhlendahl, S., Koh, Y. H., Gundelfinger, E. D., Sheng, M., Garner, C. C., and Budnik, V. (1997). *Neuron*, **19**, 787–99.
21. Davis, G. W., Schuster, C. M., and Goodman, C. S. (1997). *Neuron*, **19**, 561–73.

Michael Hortsch
Department of Anatomy and Cell Biology, University of Michigan, Ann Arbor, MI 48109–0616, USA

Fasciclin III

Fasciclin III[1,2] is a homophilic, Ca^{2+}-independent cell adhesion molecule[2] in *Drosophila* which is a member of the immunoglobulin gene superfamily;[3] it has divergent immunoglobulin domains and is unrelated to any known vertebrate member of the immunoglobulin superfamily.[2,3] It is expressed on a subset of commissural axon fascicles and other axon pathways in the *Drosophila* embryo, as well as on a variety of other non-neuronal tissues.[1,4–6] The protein comes in at least two different forms generated by alternative splicing.[1,2,6]

Protein properties

Drosophila fasciclin III is expressed on a subset of axon fascicles in the developing CNS, in a segmentally repeated pattern in the epidermis, and in the larval eye disc. Because of this diverse pattern of expression, anti-fasciclin III monoclonal antibodies (mAbs) were isolated by several groups using a number of different antibody screens.[1,4,5] Fasciclin III has also been called the DENS antigen.[6] During neurogenesis, fasciclin III is expressed in patches in the neurogenic region. During axon out-

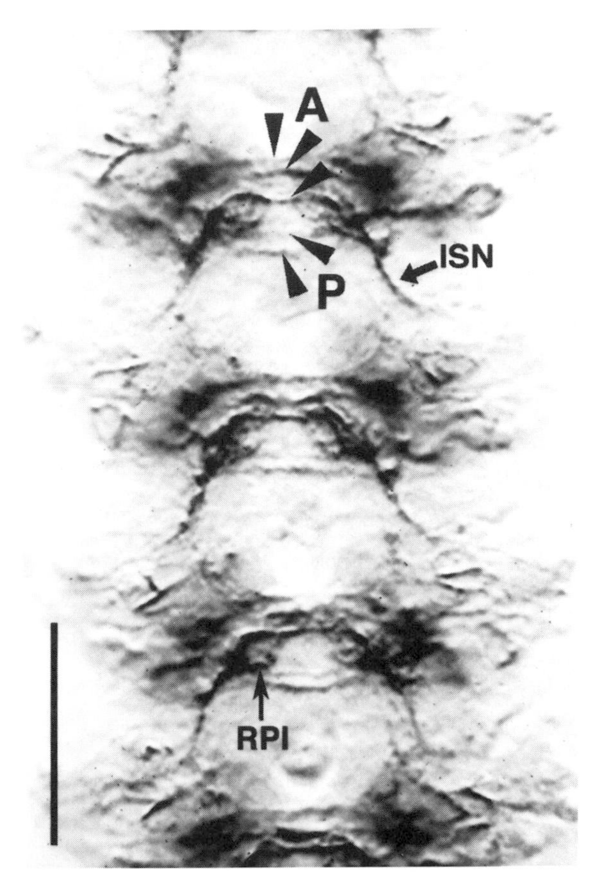

Figure 1. Fasciclin III is expressed on a subset of embryonic neurones, and axon pathways. The figure shows four contiguous segmental neuromeres of the *Drosophila* CNS (anterior is up) at h 12 of embryonic development (courtesy of N. Patel). The embryo was dissected on to a glass slide, and stained with the 7G10 mAb. At this stage fascicilin III protein is expressed on the surface of axons in five commissural fascicles (three in the anterior, A, and two in the posterior, P, commissure), and on the continuation of one of the A commissure fascicles as it turns posteriorly, and laterally (as the RP1 fascicle) towards one of the two peripheral nerves, the intersegmental nerve (ISN). Bar represents 25 μm.

growth, it is expressed on a subset of neuronal cell bodies and axons in the CNS as well as on a subset of non-neuronal (presumably glial) cells.[1] Fasciclin III is expressed at high levels on five specific commissural axon fascicles and on the RP1 fascicle as it heads towards the intersegmental nerve root (Fig. 1); it is absent from longitudinal axon pathways. During late embryogenesis, fasciclin III is expressed by a subset of motor neurones, for example the RP1 and the RP3 neurones, and their respective target muscle cells.[7,13] Outside of the developing nervous system, fasciclin III is expressed in the visceral mesoderm and the luminal surface of the salivary gland epithelium.[1,6] In the larval eye disc it is expressed on the cone cells surrounding the clusters of eight photoreceptor cells which do not express fasciclin III.[5]

Anti-fasciclin III mAbs recognize at least two different forms of the protein with apparent molecular masses of 80 kDa and 66 kDa.[1,6] The full length cDNA coding for the largest form of the fasciclin III protein (80 kDa) contains an open reading frame of 508 amino acids, 20 of which code for an amino-terminal signal sequence.[2] A transmembrane segment of 24 hydrophobic amino acids divides the mature molecule into an extracellular domain of 326 amino acids and a cytoplasmic domain of 138 amino acids. A full length cDNA that encodes the shorter form (66 kDa) of the protein reveals that this form is identical to the long form, except that it has a different cytoplasmic domain (Snow *et al.;* unpublished results).

Although initial analysis suggested that the fasciclin III protein appears unrelated to any other known protein, further analysis of its sequence reveals that the extracellular part of the molecule consists of three highly divergent immunoglobulin-type domains[3] (see Fig. 1, Fasciclin I, page 00). The third immunoglobulin domain lacks the characteristic cysteine residues. The first domain appears to be of the V type, whereas assignment of the other two Ig domains is less certain. Both are most related to the C2 type of Ig domains, although all three fasciclin III immunoglobulin domains are highly divergent from those found in other insect and vertebrate neural Ig-domain cell adhesion molecules.

Purification

Fasciclin III protein was isolated from *Drosophila* embryonic protein extracts by affinity chromatography on 2D5 or 7G10 mAb columns.[1] A similar approach was used to purify an artificial, soluble form of fasciclin III from supernatants of transfected Sf9 cells.[8]

Activities

Transfection of *Drosophila* S2 cells and *in vitro* cell aggregation experiments show that fasciclin III can mediate homophilic, Ca^{2+}-independent cell adhesion.[2]

Antibodies

The generation of monoclonal and polyclonal antibodies in mice and rats against *Drosophila* fasciclin III has been reported from several different laboratories.[1,4,5] One mAb, called 7G10, which recognizes *Drosophila* fasciclin III,[1] is available from the Developmental Studies Hybridoma Bank at the University of Iowa.

Genes

A complete cDNA sequence encoding the 80 kDa form of *Drosophila* fasciclin III has been cloned and sequenced (GenBank accession number M27813).[2]

Mutational phenotypes

The *Drosophila* fasciclin III gene maps to position 36E1 on the second chromosome.[1] A viable, protein-null mutation in the fasciclin III gene has been isolated (*fasIII*^{E25}) and has a normal embryonic nervous system.[9] However, fasciclin III appears to play a role in the development of stereotypic synaptic connections between fasciclin III-expressing motor neurones, like the RP3 neurone, and their embryonic muscle targets which also express fasciclin III.[13] Although the target recognition of muscle cells 6 and 7 by the RP3 motor neurone is normal in fasciclin III-mutant embryos, the ectoptic expression of fasciclin III protein by other embryonic muscle cells resulted in the inappropriate innervation of neighboring muscle cells by the RP3 neurone.[10,13] In the developing larva an increased branching and a misrouting of wing sensory neuronal axons was observed in fasciclin III-mutant animals.[11]

Structure

Using a sequence-to-structure algorithm, Castonguay *et al.* have proposed a structural model for the amino-terminal IG domain of *Drosophila* fasciclin III.[12]

References

1. Patel, N. H., Snow, P. M., and Goodman, C. S. (1987). *Cell*, **48**, 975–88.
2. Snow, P. M., Bieber, A. J., and Goodman, C. S. (1989). *Cell*, **59**, 313–23.
3. Grenningloh, G., Bieber, A. J., Rehm, E. J., Snow, P. M., Traquina, Z. R., Hortsch, *et al.* (1990). *Cold Spring Harbor Symp. Quant. Biol.*, **55**, 327–40.
4. Brower, D. L., Smith, R. J., and Wilcox, M. (1980). *Nature*, **285**, 403–5.
5. Zipursky, S. L., Venkatesh, T. R., Teplow, D. B., and Benzer, S. (1984). *Cell*, **36**, 15–26.
6. Gauger, A., Glicksman, M. A., Salatino, R., Condie, J. M., Schubiger, G., and Brower, D. L. (1987). *Development*, **100**, 237–44.
7. Halpern, M. E., Chiba, A., Johansen, J., and Keshishian, H. (1991). *J. Neurosci.*, **11**, 3227–38.
8. Strong, R. K., Vaughn, D. E., Bjorkman, P. J., and Snow, P. M. (1994). *J. Mol. Biol.*, **241**, 483–7.
9. Elkins, T., Zinn, K., McAllister, L., Hoffmann, F. M., and Goodman, C. S. (1990). *Cell*, **60**, 565–75.
10. Chiba, A., Snow, P., Keshishian, H., and Hotta, Y. (1995). *Nature*, **374**, 166–8.
11. Whitlock, K. E. (1993). *Development*, **117**, 1251–60.
12. Castonguay, L. A., Bryant, S. H., Snow, P. M., and Fetrow, J. S. (1995). *Protein Sci.*, **4**, 472–83.
13. Kose, H., Rose, D., Zhu, X., and Chiba, A. (1997). *Development*, **124**, 4143–52.

Michael Hortsch
Department of Anatomy, and Cell Biology, University of Michigan, Ann. Arbor, MI 48109–0616, USA.

Integrins

Overview of the family

Integrins are a family of $\alpha\beta$ heterodimers, comprising eight different β chains that associate with 16 different α chains, to give at least 22 distinct human heterodimers. Integrins primarily mediate cell adhesion. They recognize a variety of ligands including extracellular matrix proteins, cell surface proteins, and plasma proteins. Integrin-dependent adhesion controls cell growth, differentiation, gene expression, apoptosis, and motility. Deletion of genes for many of the integrin subunits results in profoundly altered development, and/or other overt phenotypes.

Protein properties

Electron microscopy studies suggest that intact integrins may have a globular head region, supported by two stalks.[1] General structural features of integrin heterodimers are shown in Fig. 1. Each α subunit has ~1000–1150 amino acids and a mature size of 140–210 kDa. All integrin α subunits have seven amino terminal repeating segments that may fold into a seven unit β-propeller

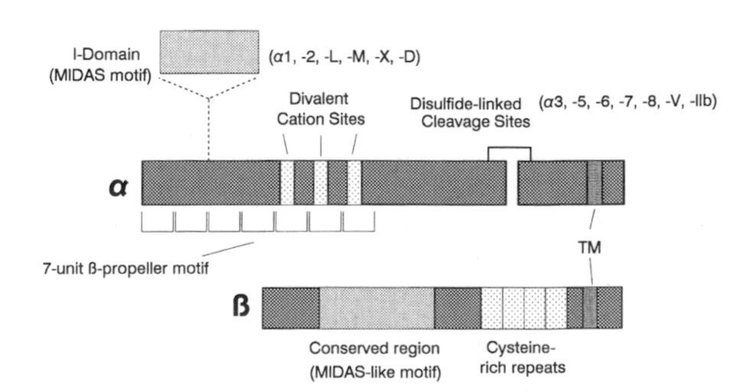

Figure 1. Schematic diagram of integrin α and β subunits. TM = transmembrane.

motif, resembling that found in G protein β subunits.[2] Some α subunits (α1, α2, αL, αM, αX, αD, αE) have an inserted 'I-domain' region (sometimes called A domain) between amino terminal repeating units 2 and 3. This I domain can function as a discrete unit, with divalent cation and ligand binding activity.[3–6] Analysis of the I-domain crystal structure revealed a 'Rossmann' folding pattern, with an unusual divalent cation coordination site at its surface.[7] This site has been called 'MIDAS', for metal ion-dependent adhesion site. One of the coordination sites of the MIDAS divalent cation may be provided by an acidic residue such as found within key regions of most integrin ligands.[8]

Many of the integrin α subunits that lack an I-domain (α3, α5, α6, α7, α8, αV, αIIb) have a disulphide-linked cleavage site towards the carboxyl end of the extracellular domain. Conversely, those with an I domain generally lack this disulphide-linked cleavage site. Furin has been identified as a candidate enzyme for mediating endoproteolytic cleavage of integrin α subunits.[9] Loci for the cleaved subunits listed above are found on human chromosomes 2, 12, and 17, most of those with I domains are on chromosomes 5 or 16 (Table 1). Standing apart are α4 and α9, which have neither I domain nor disulphide linked cleavage site; and αE, which has both an I domain and an unusual disulphide-linked cleavage site. For several α subunits that lack an I domain, a site within the third amino terminal repeat unit is critical for ligand recognition.[10,11] The cytoplasmic domains of all integrin α subunits have a conserved 'GFFKR' or closely related motif proximal to the transmembrane domain.

Integrin β subunits (with the exception of β4) are composed of ~730–800 amino acids and are ~90–130 kDa in size. Many of the 48–56 cysteine residues conserved among all β subunits are within four repeating motifs in a 'cysteine-rich' region. A conserved region in the amino-terminal half of the β subunit has some resemblance to the MIDAS motif found in I domains. However, this region in the β subunit has not been shown to function as a discrete unit and may also differ from I-domain MIDAS regions in other key respects.[12] As in I domains, β subunit MIDAS-like regions also contain amino acids essential for ligand and divalent cation binding.[13,14]

Cytoplasmic tails of integrin β1, β3, β5, β6, and β7 subunits have 'NPXY' sequences that may be important for integrin localization, endocytosis, and signalling.[15–17] Also, cytoplasmic domains of β1 can be alternatively spliced, yielding β1A, β1B, β1C, and β1D isoforms.[18–21] The β1A form is expressed ubiquitously and contains all of the sequences needed for localization into focal adhesions.[15] The β1B variant, expressed in skin and liver, fails to localize into focal adhesions because key residues have been replaced.[22] The β1C isoform may regulate cell proliferation. It correlates with a benign, non-proliferative phenotype in epithelial cells *in vivo*[23] and also is found in megakaryocytes, platelets, and erythroleukaemic cell lines.[19,24] The β1D form is found in cardiac and skeletal muscle, where it is regulated during myoblast differentiation.[20] The cytoplasmic tails of both α and β subunits are usually highly conserved between species, consistent with their having critical functions.

As a consequence of ligand binding, Mn^{2+} binding, and/or addition of EDTA, several new epitopes appear on integrin β subunits. These epitopes map to at least six distinct locations within the β1 or β3 chains see ref. 25 and references within and thus emphasize the substantial capability of integrins to change their conformations.

Purification

Purification of many integrins has been achieved by ligand affinity columns. These can be eluted using EDTA, which inhibits nearly all ligand binding function. Alternatively, integrins can be purified using immobilized monoclonal antibodies, although this approach may require harsh elution conditions. In a few cases, recombinant soluble integrins have been produced, after truncation of α and β cytoplasmic domains, but yields have generally been low.

Activities

Integrin activities typically have been demonstrated:

(1) by using inhibitory monoclonal antibodies to block specific adhesive functions;

(2) by isolation of detergent solubilized integrins on ligand affinity columns;

(3) by binding of soluble ligands to either cellular, or purified immobilized integrins;

(4) by acquisition of specific functions upon overexpression of individual subunits in transfected cell lines; and

(5) by immunostaining of cellular integrins that specifically localize to sites in contact with ligands.

The range of ligands recognized by integrins *in vitro* can be greatly expanded upon the addition of stimulatory monoclonal antibodies or manganese. However, it is sometimes difficult to determine whether these additional interactions may be meaningful *in vivo*.

The 22 distinct $\alpha\beta$ heterodimers in the integrin family mediate vertebrate cell adhesion to many different ligands, including extracellular matrix proteins, cell surface proteins, and plasma proteins (Fig. 2). Invertebrate species, such as *Drosophila*, have a more restricted repertoire of α and β subunits and may have their own specialized ligands (e.g. the *Drosophila* PS2 integrin binds tiggrin.[26]. Integrin interactions with ligand can be regulated by unknown factors in different cellular environments,[27,28] and can be rapidly altered by specific signalling pathways.[29,30] Also, integrin-dependent adhesion can be regulated by diffusion and clustering of integrins in the plasma membrane,[31,32] and by post-ligand binding mechanisms that involve cytoskeletal proteins.[33] The cytoplasmic domains of α and β subunits make both positive and negative contributions to cell adhesion.[34]

and mapped to specific chromosomal locations (Table 1). Many of the integrin α and β subunits have also been knocked out, either naturally, or by homologueous recombination in mice.[42–44] Details regarding specific integrin subunits are presented in subsequent sections.

$\alpha1\beta1$ and $\alpha2\beta1$ integrins

The $\alpha1\beta1$ and $\alpha2\beta1$ integrins are important cellular receptors for various collagens, and also can recognize laminin-1. Genes for the $\alpha1$, $\alpha2$, and $\beta1$ subunits have been cloned from multiple species. Ligand recognition activity may reside largely within the $\alpha1$ and $\alpha2$ subunit I domains. Genetic knockout of $\alpha1$ resulted in no obvious phenotype, thus indicating that $\alpha1\beta1$ is not required for normal mouse development.

Protein properties

The $\alpha1$ and $\alpha2$ subunits of integrins $\alpha1\beta1$ (VLA-1, CD49a/CD29) and $\alpha2\beta1$ (VLA-2, CD49b/CD29, Ia–IIa) are ~40 per cent identical at the amino acid level and both contain I domains (Fig. 3). Subunit sizes are 200–210 ($\alpha1$), 150–160 ($\alpha2$), and 110–130 kDa ($\beta1$). The $\alpha1\beta1$ integrin appears on several neuronal cell types,[45,46] hepatocytes,[47] smooth muscle cells,[48] microvascular endothelial cells,[49] lymphocytes late after activation,[50] and can be induced to appear on several other cell types. The $\alpha2\beta1$ integrin appears on platelets, endothelial and epithelial cells *in vivo*, and on the majority of adherent cell lines cultured *in vitro*.[51,52]

Purification

The $\alpha1\beta1$ integrin can be purified from placenta, using an anti-$\alpha1$ monoclonal affinity column,[53] and it also can be purified using laminin-1[54] or collagen affinity columns.[55] The $\alpha2\beta1$ integrin can be purified from platelets using an anti-$\alpha2$ monoclonal affinity column,[56] or collagen or laminin affinity columns.[52,57,58]

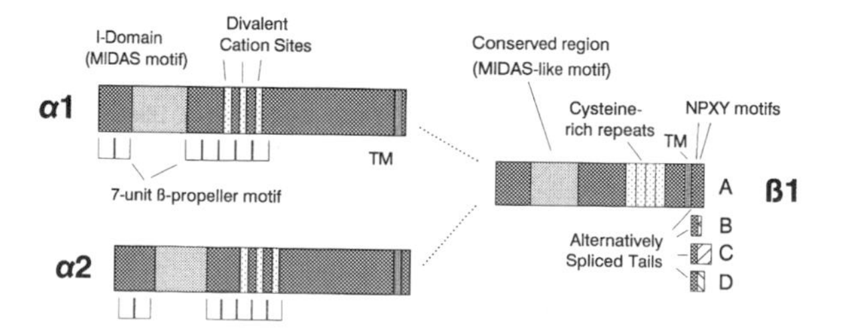

Figure 3. Schematic diagram of $\alpha1$, $\alpha2$, and $\beta1$ subunits. TM = transmembrane; A, B, C, D represent different alternatively spliced forms.

Activities

The $\alpha2\beta1$ integrin mediates cell adhesion to collagens I–V (59), while $\alpha1\beta1$ recognizes collagens I and IV.[45,46] On platelets, $\alpha2\beta1$ mediates collagen-dependent platelet aggregation.[59] In the presence of Mg^{2+}, or Mg^{2+} + Ca^{2+}, collagen bound to $\alpha1\beta1$ with a k_d of 25–30 nM and to $\alpha2\beta1$ with a k_d of 60–110 nM.[55] The triple helical structure of collagen is required for binding these integrins.[55] Notably, recombinant I domains mimic the collagen binding activities of $\alpha1\beta1$[4] and $\alpha2\beta1$[5,6] Both the $\alpha2\beta1$ and $\alpha1\beta1$ integrins also recognize laminin-1 (and possibly other laminins), but this activity is weaker than interactions with collagen and is not obvious in certain cellular environments.[52,57,60] Adhesive activities of $\alpha1\beta1$ and $\alpha2\beta1$ are supported by Mg^{2+} and Mn^{2+}, but not by Ca^{2+}. The cytoplasmic domain of $\alpha2$ may be uniquely well suited to support focal adhesion formation and collagen gel contraction.[61,62] The $\alpha2\beta1$ integrin also mediates cell attachment to a subset of echoviruses.[63]

Antibodies

Numerous monoclonal antibodies recognizing the human $\alpha1$ and $\alpha2$ subunits are available.[40,64,65] The majority of anti-$\alpha2$ antibodies have been mapped to sites within the I domain.[66]

Genes, knockouts

Cloned genes for $\alpha1$, $\alpha2$, and $\beta1$ are listed in Table 1. The integrin $\alpha1$ subunit was deleted from mice by homologous recombination, and no overt phenotype was observed. However, embryonic fibroblasts lacking $\alpha1$ were deficient in attachment to collagen IV and laminin.[47] The $\alpha2$ gene promoter has been characterized,[67] but the $\alpha2$ gene has not yet been knocked out.

Mice deficient in $\beta1$ die as embryos, shortly after implantation.[68] Hematopoietic stem cells lacking $\beta1$ readily differentiate into erythroid, myeloid and lymphoid lineages, but fail to colonize the fetal liver.[69] The $\beta1$ promoter region has been studied[70], and genomic organization of the 3′ alternatively spliced region of mouse $\beta1$ has been analysed.[21]

$\alpha3\beta1$, $\alpha6\beta1$, $\alpha7\beta1$, and $\alpha6\beta4$ integrins

The $\alpha3\beta1$, $\alpha6\beta1$, $\alpha6\beta4$, and $\alpha7\beta1$ integrins are primarily laminin receptors, although $\alpha3\beta1$ recognizes additional ligands. Genes for $\alpha3$, $\alpha6$, $\alpha7$, $\beta1$, and $\beta4$ subunits have been cloned from multiple species. The $\alpha3$, $\alpha6$, and $\alpha7$ subunits have multiple alternatively spliced forms. The $\beta4$ subunit has an unusually large cytoplasmic domain that facilitates hemidesmosome formation. Genetic knockout of $\alpha3$, $\alpha6$ or $\beta4$ results in a perinatal lethal phenotype, whereas $\alpha7$ knockout yields muscular dystrophy.

Protein properties

The $\alpha3\beta1$ (VLA-3, CD49c/CD29), $\alpha6\beta1$ (VLA-6, CD49e/CD29, Ic–Ia), $\alpha6\beta4$ (CD49e/CD104), and $\alpha7\beta1$ integrins are all primarily laminin receptors. The $\alpha3$, $\alpha6$ and $\alpha7$ subunits are more closely related to each other (35–47 per cent identity) than to any other integrin α subunits. Each is cleaved into heavy chains (~1-Ig 20 kDa) and light chains (~25–30 kDa) that are disulphide linked, typical of α subunits lacking an I domain (Fig. 4). The $\alpha6$ cleavage site was identified and mutated, yielding an uncleaved $\alpha6\beta1$ integrin that was defective in phorbol ester-stimulated adhesion to laminin.[71] Within the extracellular domain of $\alpha7$, mutually exclusive X1 and X2 sequences are found between amino terminal repeat domains III and IV.[72] The $\alpha6$ subunit also shows alternative splicing between repeat units III and IV, resulting in X1 or X1X2 variant forms.[73] Between residues 557–886, the $\alpha7$ chain undergoes ADP-ribosylation by an enzyme on the surface of skeletal muscle cells.[74]

The $\alpha3$, $\alpha6$, and $\alpha7$ subunits each show similar patterns of alternative splicing within their cytoplasmic domains. As indicated in Fig. 4, A and B forms of each have been defined. While the A and B forms are quite distinct from each other, the 3A, 6A, and 7A sequences are 47–63 per cent similar, and the 3B, 6B, and 7B sequences are 53–66 per cent similar. A relatively short α7C alternative splice form has also been identified.[75]

The $\alpha6\beta4$ integrin is unique, in that the cytoplasmic domain of $\beta4$ contains ~1000 amino acids, instead of the ~50–60 found in other β subunits. Towards the C terminus of the $\beta4$ cytoplasmic domain are two pairs of fibronectin type III repeat units separated by a variable connecting segment that possibly undergoes alternative splicing.[76] This C-terminal region is specifically required for $\alpha6\beta4$ organization into hemidesmosomes.[77] Also, $\beta4$ readily undergoes proteolytic processing within its cytoplasmic domain, causing the 200 kDa mature form to be converted to 165 and 130 kDa fragments.[78,79] The extracellular domain of $\beta4$ contains only 48 of the 56 conserved cysteines found in other integrin β subunits.

The $\alpha3\beta1$ integrin is widely expressed on nearly all tissue types and is particularly abundant on endothelial and epithelial cells in proximity to basement membranes.[80] Also it is found on nearly all rapidly growing adherent cell lines.[81] The α3B variant may be restricted to brain and heart tissue, whereas α3A is present in all tissues.[82] The $\alpha6$ subunit is widely distributed and when both $\beta1$ and $\beta4$ are present, $\alpha6$ usually shows preference for pairing with $\beta4$. The $\alpha6\beta1$ heterodimer is present on platelets, lymphocytes, monocytes, and in various neural tissues.[40,83] The $\alpha6\beta4$ integrin is a basement membrane receptor on most epithelial cells, some endothelial cells, and on Schwann cells.[84–86] Whereas $\beta1$ integrins can associate with the actin cytoskeleton in focal adhesions, the $\alpha6\beta4$ integrin is a critical component of hemidesmosomes and connects to intermediate filaments.[86] The 6A and 6B variant forms show differing distributions in a variety of epithelial and other tissues.[83] These variant cytoplasmic domains do not appear to regulate preference for association with $\beta1$ or $\beta4$.

The $\alpha7\beta1$ integrin is largely found on cardiac and skeletal muscle. The α7A isoform is found only in skeletal muscle and appearance is regulated in concert with myoblast fusion.[87] In contrast, 7B is more widespread, with expression in striated muscle, vasculature, and nervous system.[88] The 7B variant is present on a myoblast cell line,

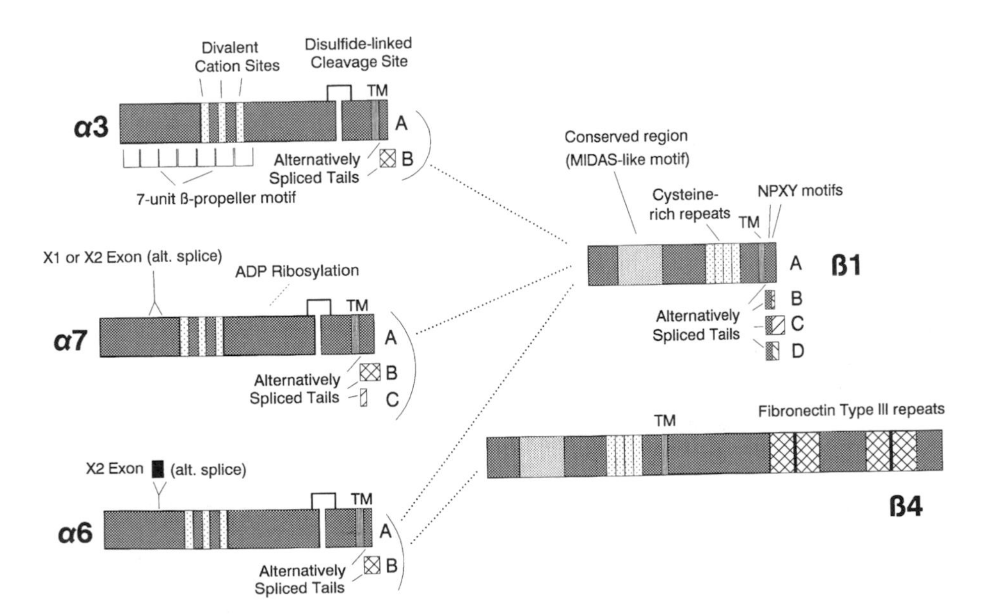

Figure 4. Schematic diagram of $\alpha3$, $\alpha6$, $\alpha7$, $\beta1$, and $\beta4$ subunits.

whereas 7A increases as myoblasts differentiate into myotubes.[89] In contrast to α7A and α7B variants that are confined to synaptic sites in adult skeletal muscle, the α7C form is present both synaptically and extrasynaptically.[90]

The $\alpha3\beta1$ and $\alpha6\beta1$ integrins associate specifically with proteins containing four transmembrane domains (TM4SF proteins), but the functional significance of this is not yet clear.[91]

Purification

The $\alpha3\beta1$ integrin can be purified from placenta, using an anti-α3 monoclonal antibody column.[53] Likewise, $\alpha6\beta1$ and/or $\alpha6\beta4$ can be isolated from platelets or placenta, using an anti-α6 monoclonal immunoaffinity column[92] and $\alpha6\beta4$ can be purified using an anti-β4 immunoaffinity column.[92] It should be feasible to purify $\alpha7\beta1$ from skeletal muscle, or muscle cell lines, using either monoclonal antibody or laminin affinity columns.

Activities

The $\alpha3\beta1$, $\alpha6\beta1$, $\alpha6\beta4$, and $\alpha7\beta1$ integrins all mediate cell adhesion and motility on laminin. $\alpha3\beta1$ interacts more strongly with human laminins 5 and 10/11. It also variably mediates cell adhesion to other laminin isoforms, and other ligands such as collagen, fibronectin, thrombospondin, and entactin. However, adhesion to these other ligands has been difficult to assess, due to dominant overlapping functions of other integrins. Also, $\alpha3\beta1$ may participate in phagocytosis of ECM molecules[93] and a role for $\alpha3\beta1$ in cell–cell adhesion has been suggested, but this has not been observed in several reports.[94,95] Integrin $\alpha6\beta1$ mediates adhesion to laminin-1, –2, –4, and –5. Also, $\alpha6\beta1$ in egg plasma membrane mediates sperm cell binding, through the counter-receptor fertilin.[96] The $\alpha7\beta1$ integrin mediates cell adhesion to laminin-1, –2/4, but not laminin-5;[89] $\alpha6\beta4$ also interacts with several laminins (laminin-5, laminin-1). Both $\alpha6\beta1$ and $\alpha6\beta4$ may also function as receptors for papillomaviruses.[97] So far, the A and B isoforms of α3, α6, and α7 appear to be interchangeable with respect to influence on integrin ligand recognition.

Antibodies

Antibodies to human α7 have recently become available,[89] and monoclonal antibodies have been prepared that distinguish between the 3A and 3B cytoplasmic domains[82] and the 6A and 6B cytoplasmic tails.[83] Also, a number of antibodies to extracellular domains of α3, α6, and β4 have been described.[40,98–100]

Genes, knockouts

Cloned genes for α3, α6, α7, β1, and β4 are listed in Table 1. Knockout of the α3 gene in mice demonstrates that $\alpha3\beta1$ plays a critical role in kidney and lung organogenesis.[101] Also, α3 is required for normal development of epidermal basement membrane, as its absence contributes to membrane blistering.[80] Knockout of the α7 gene results in a form of muscular dystrophy,[42] and naturally occurring mutations in α7 cause congenital myopathy.[297] The α6 gene promoter has been characterized,[298] and knockout of the α6 gene results in perinatal lethality, with severe blistering of skin and other epithelia.[102] This phenotype resembles epidermolysis bullosa in humans and probably results from an absence of $\alpha6\beta4$ and the absence of formation of hemidesmosomes. Consistent with this, knockout of β4 also resulted in epithelial detachment due to an absence of hemidesmosomes.[103,104] Naturally occurring deficiencies in human β4 are associated with epidermolysis bullosa with pyloric atresia.[105,106] The β4 gene is organized into 41 exons, spanning 36 kb.[107,108]

$\alpha4\beta1$, $\alpha4\beta7$, $\alpha E\beta7$, and $\alpha9\beta1$ integrins

The $\alpha4\beta1$ (VLA-4, CD49d/CD29), $\alpha4\beta7$ (LPAM-1), and $\alpha E\beta7$ (CD103/β7) integrins are found mostly on leukocytes. These integrins mediate leukocyte migration to inflammatory sites, leukocyte homing and recirculation through the gut, and lymphocyte localization into epithelial sites, respectively. Ligands include VCAM-1, fibronectin, MAdCAM-1, and E-cadherin. Knockout of α4 is embryonic lethal, whereas mice deleted for αE or β7 are viable, but show impaired immune functions. The $\alpha9\beta1$ integrin is structurally similar to $\alpha4\beta1$, but has a much broader cell and tissue distribution, and mediates cell attachment to tenascin.

Protein properties

$\alpha4\beta1$ and $\alpha9\beta1$ comprise a subfamily among β1 integrins, with the α4 and α9 subunits being closely related (~40 per cent similarity). Also, the α4 and α9 subunits differ from the others since they have neither an I domain nor a membrane proximal disulphide-linked cleavage site (Fig. 5). The α4 subunit (140–150 kDa) is variably cleaved at a site in the centre of the molecule, yielding 80 and 70 kDa fragments. However, mutation of this site did not result in overt alterations of $\alpha4\beta1$ integrin function.[109] The α9 subunit (140–150 kDa) lacks a critical dibasic motif (that is present in α4) and is not cleaved. In contrast to other α subunits, α4 has at least two unpaired cysteines, both of which can contribute to aberrant subunit migration upon SDS–PAGE.[110] Sites critical for α4 interaction with ligand have been mapped to segments within repeat units II, III, and IV, that presumably lie on the upper face of the β-propeller.[10,111] The unique cytoplasmic domain of α4 may be particularly well suited to support cell migration.[62]

The αE subunit of $\alpha E\beta7$ contains an I domain and also a unique extra domain of 55 amino-acids, on the amino terminal end of the I domain (Fig. 5). This extra domain contains a stretch of 18 consecutive charged residues, as well as an endoproteolytic cleavage site.[112] As a result

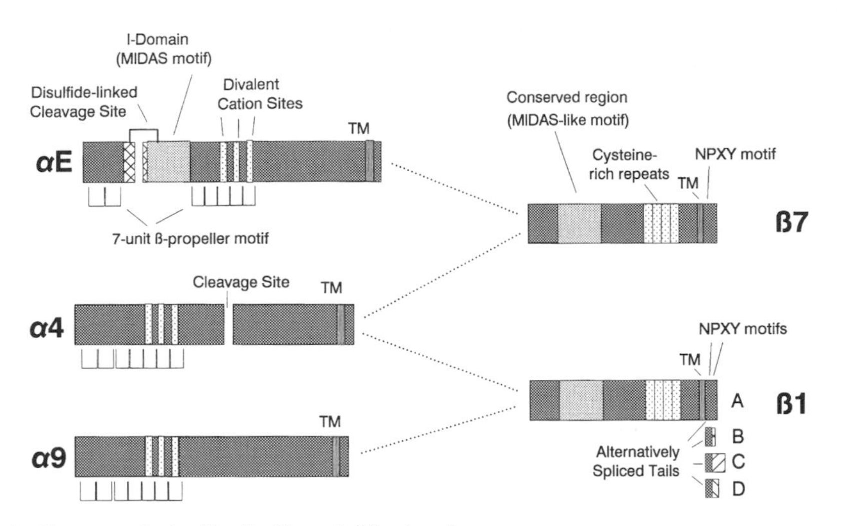

Figure 5. Schematic diagram of $\alpha4$, αE, $\alpha9$, $\beta1$, and $\beta7$ subunits.

of proteolytic cleavage, the αE subunit migrates in SDS–PAGE at 175 kDa under non-reducing conditions and as fragments of 150 and 25 kDa under reducing conditions. The αE subunit is unique in having a disulphide-linked cleavage site near the amino terminus of the molecule.[113] In primary sequence, the $\beta7$ subunit is 32–46 per cent identical to most other β subunits and, like other typical β subunits, it contains ~800 amino acids, 54 conserved extracellular cysteines, four cysteine-rich homologous repeat sequences, and a conserved NPXY motif in the cytoplasmic domain.[114,115] Like $\beta1$, $\beta7$ also contains a 'MIDAS' domain which is involved in interactions with ligands.[116]

The $\alpha4\beta1$ integrin is present on most mononuclear leukocytes, and is also found on eosinophils, basophils, and various non-haematopoietic tumour cells; it may also be on non-lymphoid tissues in the developing embryo.[117] The $\alpha4\beta7$ integrin is present on most lymphocytes resident in lymph nodes or rheumatoid synovium and is also on the gut homing subset of memory T cells and other cells.[117] αE$\beta7$ is found on > 90 per cent of intraepithelial lymphocytes and 45–50 per cent of T lymphocytes within lamina propria; it is also found on small numbers of cells in other locations.[118] The $\alpha9\beta1$ integrin is expressed on a variety of cell types *in vivo*, including airway epithelial cells, basal layers of squamous epithelium, smooth muscle, skeletal muscle, and hepatocytes.[119]

Purification

The $\alpha4\beta1$ integrin has been purified from lymphoblastoid cell lines, using affinity columns of monoclonal antibodies,[53,120] or fibronectin peptides.[121] $\alpha4\beta7$ was similarly purified using a fibronectin CS1-peptide affinity column.[122] αE$\beta7$ was purified from the spleen of a hairy cell leukaemia patient, using an anti-αE monoclonal antibody.[113] The $\alpha9\beta1$ integrin has been purified from rat liver by affinity chromatography, using sepharose conjugated with the peptide GRGDSPC.[123]

Activities

The $\alpha4\beta1$ integrin binds to sites within Ig domains 1 and 4 of VCAM-1, a ligand that appears on activated endothelium.[124] Also, it recognizes a Leu–Asp–Val motif in the fibronectin CS1 region and may facilitate a host of weaker interactions with other ligands.[117] The $\alpha4\beta7$ integrin mediates lymphocyte adhesion to MAdCAM-1,[125], a molecule expressed on the high endothelial venules of Peyer's patches and on lamina propria post-capillary venules. The $\alpha4\beta7$ integrin also may mediate adhesion to VCAM-1 and fibronectin.[122,126] Unlike many other integrins, both $\alpha4\beta1$ and $\alpha4\beta7$ have the unusual property of mediating lymphocyte tethering and rolling under shear flow.[127,128] Placental and heart defects resulting from $\alpha4$ gene deletion imply that $\alpha4$ integrins mediate additional functions involving non-haematopoietic cells.[129] The αE$\beta7$ integrin binds to E-cadherin on epithelial cells,[130] thus facilitating localization of intraepithelial lymphocytes to epithelial sites. The $\alpha9\beta1$ integrin mediates cell adhesion to tenascin[131] and, perhaps, weaker adhesion to other ligands.[123]

Antibodies

Several anti-$\alpha4$,[132–134] αE,[135,136] and $\beta7$[116] monoclonal antibodies have been characterized. Antibodies that recognize combinatorial $\alpha4\beta1$[137] and $\alpha4\beta7$[138] epitopes have been particularly useful for studying cells and tissues that have mixtures of these integrins.

Genes

Cloned genes for $\alpha4$, αE, $\alpha9$, $\beta1$, and $\beta7$ are listed in Table 1. Knockout of the $\alpha9$ gene causes mice to die within 10 days of birth.[42] Deletion of the $\alpha4$ gene results in embryonic lethality (at E10.5–E12.5), with placental and heart defects.[129] In $\alpha4$-null chimeric mice, T cell localization into Peyer's patches is impaired, as is differentiation of T and B cells in the bone marrow.[139] The human[140] and mouse[141] $\alpha 4$ gene promoter regions have been characterized,[140] and complete genomic organization for mouse $\alpha 4$ has been determined.[142] Mice lacking αE are viable, but show reduced numbers of intraepithelial lymphocytes (IEL).[42] In mice deficient in $\beta 7$, numbers of intraepithelial and lamina propria lymphocytes are greatly reduced and there is a marked reduction in size of Peyer's patches.[143] The genomic organization of $\beta 7$ has been determined.[144]

$\alpha 5\beta 1$, $\alpha 8\beta 1$, $\alpha V\beta 1$ integrins

The $\alpha 5\beta 1$ (CD49e/CD29), $\alpha 8\beta 1$ ($\alpha 8$/CD29), and $\alpha V\beta 1$ (CD51/CD29) integrins form a subgroup of $\beta 1$ integrins that primarily recognize various ligands containing Arg–Gly–Asp motifs and contain closely related α subunits. Ligands include fibronectin, vitronectin, tenascin, and osteopontin. Knockout of $\alpha 5$ is embryonic lethal, whereas kidney and vascular defects in $\alpha 8$ and αV knockout mice often allow survival until after birth.

Protein properties

The $\alpha 5$, αV, and $\alpha 8$ subunits are ~45–46 per cent identical to each other. Each α subunit undergoes a typical endoproteolytic cleavage, producing a carboxyl terminal 25–30 kDa fragment that remains disulphide-linked to a larger α chain fragment (125–140 kDa). Also, the amino terminal portions of $\alpha 5$, $\alpha 8$, and αV each contain the typical seven repeating units found in all integrin α chains (Fig. 6). Residues required for $\alpha 5\beta 1$ ligand binding have been identified within $\alpha 5$ N-terminal repeat III.[10] Also, detailed analysis of $\alpha 5\beta 1$ oligosaccharide structures has been carried out.[145]

The $\alpha 5\beta 1$ integrin is present at high levels in embryos, including developing heart, lung, and peripheral nerve tissue. Expression is more limited in adult tissues (see ref. 146 and references within). Also, $\alpha 5\beta 1$ is present on circulating T lymphocytes and monocytes.[40] The $\alpha 8\beta 1$ heterodimer in the embryo is present on neural cells, and on mesenchymal cells bordering epithelial cell sheets in many developing organs, including kidney.[147,148] In the adult it is present on smooth muscle cells, kidney mesangial cells, and lung myofibroblasts.[149] $\alpha V\beta 1$ is present on growing endothelial cells, smooth muscle cells, avian neural crest cells, and many different types of cultured adherent cell lines. Distribution in tissues has been hard to distinguish, since there is often interference from many other integrins that contain either the $\beta 1$ subunit, or the αV subunit.

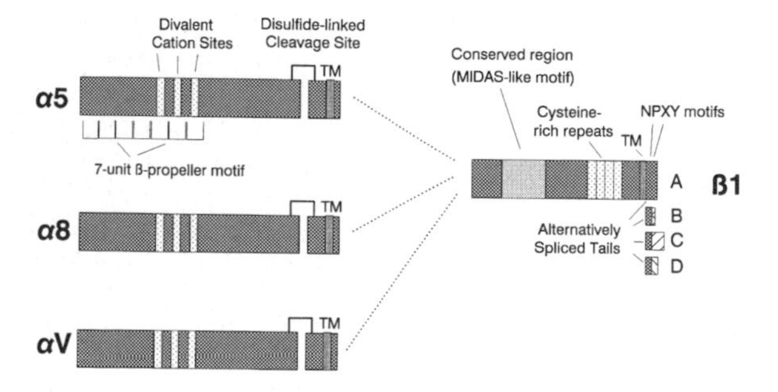

Figure 6. Schematic diagram of $\alpha 5$, $\alpha 8$, αV, and $\beta 1$.

Purification

Fibronectin affinity columns have been used to isolate the $\alpha 5\beta 1$,[150] $\alpha V\beta 1$,[151] and $\alpha 8\beta 1$[152] heterodimers. The latter has also been isolated from a human embryonic kidney cell line by affinity chromatography on vitronectin– sepharose.[152]

Activities

The $\alpha 5\beta 1$ integrin recognizes an Arg–Gly–Asp (RGD) motif in the central cell-binding domain of fibronectin.[150] Page display selection of peptides binding to the $\alpha 5\beta 1$ integrin yielded mostly peptides containing Arg–Gly–Asp, thus confirming the importance of this motif.[153] A 'synergy' site adjacent to the Arg–Gly–Asp site in fibronectin contributes to fibronectin recognition by $\alpha 5\beta 1$.[154] The $\alpha 5\beta 1$ integrin not only regulates adhesion and migration on fibronectin, but also plays a key role in the assembly of fibronectin into an extracellular matrix.[155–157]

The $\alpha 8\beta 1$ heterodimer is a receptor for fibronectin, vitronectin, and tenascin, with all activities inhibitable by Arg–Gly–Asp peptides.[152,158] Additional ligands may exist in embryonic kidney, as evidenced by staining with alkaline-phosphatase-conjugated recombinant $\alpha 8\beta 1$ protein.[148] The $\alpha V\beta 1$ integrin recognizes fibronectin, vitronectin, osteopontin, and fibrinogen,[159–161] again in an Arg–Gly–Asp inhibitable fashion. On many cells, activity of $\alpha V\beta 1$ is difficult to evaluate due to interference by the many other integrins containing αV or $\beta 1$ subunits.

Antibodies

Several anti-$\alpha 5$ monoclonal antibodies are available.[40,162] Several αV antibodies are also available,[163] but none are yet specific for $\alpha V\beta 1$. In many cases, it has been necessary to identify $\alpha V\beta 1$ biochemically (e.g. by demonstrating the presence of αV after anti-$\beta 1$ immunoprecipitation, or $\beta 1$ after anti-αV immunoprecipitation). The $\alpha 8\beta 1$ heterodimer has been studied using specific anti-$\alpha 8$ polyclonal antibodies.[149,158]

Genes, knockouts

Cloned genes for $\alpha 5$, $\alpha 8$, αV, and $\beta 1$ are listed in Table 1. The promoter region for human $\alpha 5$ has been characterized.[164] Knockout of $\alpha 5$ yields an embryonic lethal phenotype (at E9.5–11), with mesodermal and vascular defects.[146] Notably, cells derived from $\alpha 5$ deficient embryos could still assemble a fibronectin matrix, due to compensation by αV integrins.[42] Deletion of the α8 gene from mice resulted in severe kidney defects, with mice living only 1–2 days after birth.[148] Knockout of the αV gene yielded overt cerebral vascular defects and was either perinatal lethal or embryonic lethal in mice.[42]

αIIbβ3, αVβ3, αVβ5, αVβ6, and αVβ8 integrins

The αVβ3 (CD51/CD61), αIIbβ3 (IIb/IIIa, CD41/CD61), αVβ5, αVβ6, and αVβ8 integrins form another subgroup of integrins that primarily recognizes ligands containing Arg–Gly–Asp motifs. Ligands include fibrinogen, vitronectin, fibronectin, osteopontin, von Willebrand factor, and tenascin. Deletion of the αV gene yields severe defects and lethality. In contrast, absence of αIIb, β3, β5, or β6 yields mice (or humans) that are viable, but may display prolonged bleeding (αIIb, β3), or inflammatory disorders (β6).

Protein properties

The αV and αIIb subunits are 33 per cent identical, they also show similarity to α5 and α8 subunits (25–30 per cent). Both the αV and αIIb subunits undergo typical endoproteolytic cleavage, producing carboxyl-terminal fragments of ~25 kDa, that remain disulphide-linked to the larger α chain fragments of 120–125 kDa. Also, the amino terminal portions of these α subunits contain seven repeating units typical of all integrin α chains (Fig. 7).

The β3, β5 and β6 subunits are ~43–47 per cent identical, each has 56 conserved extracellular cysteines and typical ligand-binding regions partly resembling MIDAS domains. The predicted presence of MIDAS-like domains within the β3 and β5 subunits has been verified in mutagenesis and ligand-binding studies.[12] The human β8 subunit is 31–37 per cent identical to human β1-β7, but it contains only 50 of the 56 cysteines conserved in β1, β2, β3, β5, and β6. Like β1, the β3 subunit also displays multiple alternatively spliced A and B cytoplasmic domains[165] (Fig. 7). Also, the β5 tail may have differing forms,[166,167,299] and the β6 subunit contains a unique 11 amino acid extension at its carboxyl terminus.[168] The 58 amino acid cytoplasmic domain of β8 is distinct from all other β tails.[169]

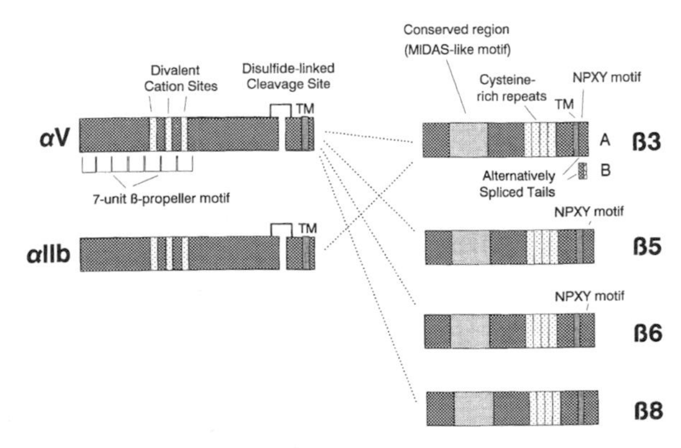

Figure 7. Schematic diagram of αV, αIIb, β5, β6, and β8.

Rotary shadowing of the αIIbβ3 integrin suggests a globular head, with two stalks.[170] Intrachain and interchain pairings of the 18 cysteines in αIIb[171] and 56 cysteines in β3[172] have been determined. Ligand-binding sites have been localized to MIDAS regions of β3 and to amino terminal regions of αIIb, including the third amino terminal repeat unit (see ref. 11 and references within). Ligand binding sites for αVβ3 have been mapped to similar locations.[173,174]

The αIIbβ3 integrin is restricted largely to platelets, whereas αVβ3 is found on endothelial cells (especially in pathologic conditions), osteoclasts, macrophages, lymphoid cells, and on many types of tumour cell lines. The integrin αVβ5 is widely distributed on most adherent cell lines and in tissues is present on epithelial, endothelial, and other cell types.[175] The αVβ6 integrin is expressed on epithelial cells during development and is associated with tumourigenesis and epithelial repair, but is almost undetectable in normal adult kidney, lung and skin.[176] Several tissues and cell lines contain mRNA for the β8 subunit, suggesting that αVβ8 expression may be moderately widespread.[169]

Purification

Using Arg–Gly–Asp peptide affinity columns, the αIIbβ3 integrin can readily be purified from platelets,[177] and αVβ3 can be purified from endothelial cells or various tumour cell lines.[178] The αVβ5 integrin has been purified using vitronectin affinity columns, whereas fibronectin-affinity chromatography was used to isolate native and recombinant soluble αVβ6.[179,180] Also, αVβ8 has been purified from a melanoma cell line using vitronectin–sepharose affinity chromatography.[181]

Activities

αIIbβ3 is primarily known as a platelet receptor for fibrinogen. This integrin becomes activated when platelets are stimulated with thrombin, ADP, collagen, or other agonists, and the subsequent fibrinogen binding leads to platelet aggregation.[182,183] The first approved anti-integrin therapeutics are antagonists of αIIbβ3 that have an anti thrombotic effect *in vivo*.[184] The αIIbβ3 integrin also binds to vitronectin and fibronectin and, in the latter case, can support fibronectin matrix assembly.[185]

The αVβ3 integrin mediates attachment to vitronectin as well as to fibronectin, fibrinogen, osteopontin, von Willebrand factor, tenascin, and a host of other proteins that display accessible Arg–Gly–Asp or related motifs. In cells lacking α5β1, αVβ3 can promote fibronectin matrix assembly.[186] The transmembrane protein CD47 (IAP) can associate with the β3 integrins αVβ3 and αIIbβ3,[187] and appears to modulate their functions.[188,189]

Integrin αVβ5 mediates strong attachment to vitronectin,[179,190] and also may function as a receptor for osteopontin[191] and sometimes, fibronectin.[175,192] While both $\alpha V\beta 3$ and $\alpha V\beta 5$ mediate cell attachment to vitronectin, the former preferentially localizes into focal contact sites,[193] there are also differences in associated signalling events.[194] Furthermore, the $\alpha V\beta 3$ and $\alpha V\beta 5$ integrins both contribute to tumour cell angiogenesis, but each may be responsive to different growth factors.[195] Both $\alpha V\beta 3$ and $\alpha V\beta 5$ also participate in adenovirus entry into host cells,[196] and $\alpha V\beta 5$ may also bind to a basic domain in the HIV Tat protein.[197] The $\alpha V\beta 6$ integrin mediates cell attachment to fibronectin[180] and tenascin.[198] In addition, $\alpha V\beta 6$ exerts a growth stimulatory effect on epithelial cells which is dependent on the sequence EKXKVDL at the carboxyl terminus of $\beta 6$.[199,200]

Like $\alpha 5\beta 1$, the $\alpha IIb\beta 3$, $\alpha V\beta 3$, $\alpha V\beta 5$, and $\alpha V\beta 6$ integrins all recognize an Arg–Gly–Asp motif within the tenth type III repeat of fibronectin. Whereas $\alpha 5\beta 1$ and $\alpha IIb\beta 3$ also interact with a separate 'synergy' site in the ninth type III repeat, the other integrins do not utilize this site.[192,201] Recombinant soluble $\alpha V\beta 8$ bound to vitronectin, but native $\alpha V\beta 8$ did not mediate cell adhesion to vitronectin unless the $\beta 8$ transmembrane and cytoplasmic domains were replaced by those of $\beta 3$.[181] $\beta 8$ integrin on chick sensory neurones mediates interactions with laminin-1, collagen IV, and fibronectin.[202]

Antibodies

Monoclonal antibodies to human αV,[178,203] $\alpha V\beta 3$,[178] $\beta 5$,[175] $\alpha V\beta 5$,[193,204] $\beta 6$,[180] $\beta 8$, and $\alpha V\beta 8$ complex[181] have been generated.

Genes, knockouts

Cloned genes for αV, αIIb, $\beta 5$, $\beta 6$, and $\beta 8$ are listed in Table 1. Naturally occurring deletion or disruption of either αIIb or $\beta 3$ causes Glanzmann thrombasthenia, a non-lethal bleeding disorder associated with defective platelet aggregation.[205,206] Complete αIIb[207] and $\beta 3$[208,209] genomic organizations have been determined. Also, the promoter regions for human[210] and avian[211] $\beta 3$, human αIIb,[212] and mouse αV[300] have been characterized. Knockout of the αV gene yielded overt cerebral vascular defects and was either perinatal lethal or embryonic lethal in mice.[42] The $\beta 5$ gene has been deleted in mice and the mice are viable and fertile, with no obvious defects.[42] Knockout of $\beta 6$ produced viable and fertile mice, but with inflammation in lungs and skin.[213] $\beta 8$ has not yet been knocked out.

$\alpha L\beta 2$, $\alpha M\beta 2$, $\alpha X\beta 2$, and $\alpha D\beta 2$ integrins

The $\alpha L\beta 2$ (LFA-1, CD11a/CD18), $\alpha M\beta 2$ (Mac-1, Mo-1, CR3, CD11b/CD18), $\alpha X\beta 2$ (p150,95, CR4, CD11c/CD18), and $\alpha D\beta 2$ (CD11d/CD18) integrins are restricted to leukocytes. These integrins mediate cell–cell adhesion and recognize counter-receptors ICAM-1, –2, or –3, iC3b, fibrinogen, and several other ligands. Absence of a functional $\beta 2$ gene results in leukocyte adhesion deficiency disease (LAD); deletion of αM causes impaired neutrophil functions.

Protein properties

The αM, αX, and αD subunits show 60–66 per cent identity to each other and are 35–36 per cent identical to αL. The cytoplasmic tails (19–53 residues) are substantially more dissimilar than the rest of these molecules. The αL (180 kDa), αM (170 kDa), αX (150 kDa), and αD (155 kDa) subunits each contain I domains (A domains), that play essential roles during ligand binding. Each also has three typical EF-hand type cation-binding sites and a cytoplasmic GFFKR motif (Fig. 8). Recombinant I domains from αM and αL have been crystallized, both have a Rossmann folding pattern, with a divalent cation coordination site (MIDAS motif) on the surface.[7,214] Whereas the αM I-domain structure varies depending on the presence of Mn^{2+} or Mg^{2+},[215] the αL I-domain structure is essentially unchanged whether Mn^{2+}, Mg^{2+}, or no divalent cation is present.[214]

The $\beta 2$ subunit is more similar to $\beta 7$ and $\beta 1$ (46 per cent) than to other β subunits. It contains 56 extracellular cysteines conserved in $\beta 1$, $\beta 3$, $\beta 5$, and $\beta 6$ and a typical four-fold repeat in a cysteine-rich region. The cytoplasmic domain contains an NPXF instead of the NPXY motif found in $\beta 1$, $\beta 3$, $\beta 5$, and $\beta 7$. Residues within the $\beta 2$ MIDAS motif are critical for ligand binding.[216]

The $\alpha L\beta 2$ integrin is present on nearly all leukocytes. In contrast, $\alpha M\beta 2$ and $\alpha X\beta 2$ are found on monocytes, macrophages, granulocytes, large granular lymphocytes and a subpopulation of immature B cells. The $\alpha X\beta 2$ heterodimer is also present on some activated lymphocytes, and is a marker for hairy cell leukaemia.[217] The $\alpha D\beta 2$ heterodimer is less obvious on peripheral blood leukocytes, but is highly expressed on tissue-compartmentalized macrophages and related cells.[218,219]

Purification

Monoclonal antibody affinity columns have been used to purify the $\alpha L\beta 2$,[220] $\alpha M\beta 2$,[221] $\alpha X\beta 2$,[221] and $\alpha D\beta 2$[219] inte-

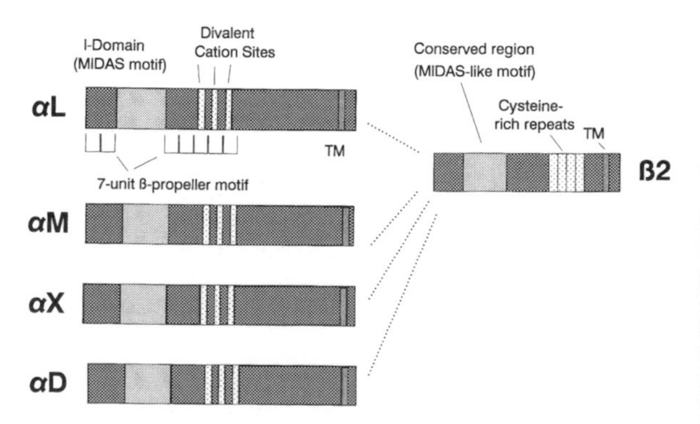

Figure 8. Schematic diagram of αL, αM, αX, αD and $\beta 2$.

grins from a leukaemic cell line, a hairy cell leukaemia spleen, and normal spleen. The αMβ2 and αXβ2 proteins have also been isolated from neutrophils and spleen cells by iC3b–Sepharose affinity chromatography.[222]

Activities

The αLβ2 integrin mediates leukocyte adhesion to cells bearing any of three counter-receptors, ICAM-1, ICAM-2, and ICAM-3, which are either inducibly or constitutively expressed on nearly all cell types.[223] Thus, the αLβ2 integrin participates in leukocyte recirculation, homing, and localization to inflammatory sites. The integrin αMβ2 binds to ICAM-1, iC3b, fibrinogen, and serum factor X, and it may also mediate cell adhesion to heparin (see ref. 224 and references within). In addition, αMβ2 has been suggested to bind to denatured proteins,[194] deoxy-oligonucleotides,[225] elastase,[226] high molecular weight kininogen,[227] and other substances. Furthermore, αMβ2 may have a lectin-like site that allows interaction with carbohydrate β-glucan structures.[228] The αXβ2 integrin binds to iC3b, iC3b opsonized particles, fibrinogen (see ref. 229 and references within), and ICAM-1.[230] The array of ligands recognized by αMβ2 and αXβ2 facilitates myeloid cell adhesion to endothelium, transmigration, chemotaxis, and phagocytosis of opsonized particles. In contrast to αMβ2 and αXβ2, the αDβ2 integrin binds preferentially to ICAM-3.[218]

Antibodies

Many monoclonal antibodies have been prepared to αL,[231] αM,[232] and αX,[229] and their epitopes have been mapped. Monoclonal antibodies are also available to the recently discovered canine[219] and human[218] αD subunit.

Genes, knockouts

Cloned genes for αL, αM, αX, αD, and β2 are listed in Table 1. Notably, αD, αL, αM, and αX all map to human chromosome 16.[233] The complete genomic sequences for αX,[234] αM[235] and β2[236] have been determined. Also, promoter regions of αL,[237] αM,[238], αX[237,239] and β2[240] have been characterized. Knockout of the αM gene has shown that αMβ2 plays a critical role in neutrophil degranulation and binding to fibrinogen,[241] also αMβ2 may accelerate phagocytosis-induced neutrophil apoptosis.[44] Absence or disruption of the β2 gene results in Leukocyte adhesion deficiency (LAD) disease in humans,[43] and a similar phenotype in mice,[242] and cows.[243] Impaired immune functions include neutrophil transendothelial migration, macrophage oxidative burst and phagocytosis, and lymphocyte proliferation.[43]

References

1. Nermut, M. V., Green, N. M., Eason, P., Yamada, S., and Yamada, K. M. (1988). *EMBO J.*, **7**, 4093–9.
2. Springer, T. A. (1997). *Proc. Natl Acad. Sci. USA*, **94**, 65–72.
3. Michishita, M., Videm, V., and Arnaout, M. A. (1993). *Cell* **72**, 857–67.
4. Calderwood, D. A., Tuckwell, D. S., Eble, J., Kuhn, K., and Humphries, M. J. (1997). *J. Biol. Chem.*, **272**, 12311–7.
5. Tuckwell, D. S., Calderwood, D. A., Green, L. J., and Humphries, M. J. (1995). *J. Cell Sci.*, **107**, 1629–37.
6. Kamata, T. and Takada, Y. (1995). *J. Biol. Chem.*, **269**, 26006–10.
7. Lee, J. -O., Rieu, P., Arnaout, M. A., and Liddington, R. (1995). *Cell*, **80**, 631–8.
8. Bergelson, J. M. and Hemler, M. E. (1995). *Curr. Biol.*, **5**, 615–7.
9. Lehmann, M., Rigot, V., Seidah, N. G., Marvaldi, J., and Lissitzky, J. C. (1996). *Biochem. J.*, **317**, 803–9.
10. Irie, A., Kamata, T., Puzon-Mclaughlin, W., and Takada, Y. (1995). *EMBO J.*, **14**, 5550–6.
11. Kamata, T., Irie, A., Tokuhira, M., and Takada, Y. (1996). *J. Biol. Chem.*, **271**, 18610–5.
12. Lin, E. C. K., Ratnikov, B. I., Tsai, P. R., Gonzalez, E. R., McDonald, S., Pelletier, A. J., and Smith, J. W. (1997). *J. Biol. Chem.*, **272**, 14236–43.
13. Takada, Y., Ylänne, J., Mandelman, D., Puzon, W., and Ginsberg, M. H. (1992). *J. Cell. Biol.*, **119**, 913–21.
14. Loftus, J. C., O'Toole, T. E., Plow, E. F., Glass, A., Frelinger, A. L., and Ginsberg, M. H. (1990). *Science*, **249**, 915–8.
15. Reszka, A. A., Hayashi, Y., and Horwitz, A. F. (1992). *J. Cell. Biol.*, **117**, 1321–30.
16. Van Nhieu, G. T., Krukonis, E. S., Reszka, A. A., Horwitz, A. F., and Isberg, R. R. (1996). *J. Biol. Chem.*, **271**, 7665–72.
17. Johansson, M. W., Larsson, E., Lüning, B., Pasquale, E. B., and Ruoslahti, E. (1994). *J. Cell. Biol.*, **126**, 1299–309.
18. Altruda, F., Cervella, P., Tarone, G., Botta, C., Balzac, F., Stefanuto, G., and Silengo, L. (1990). *Gene* **95**, 261–6.
19. Languino, L. R. and Ruoslahti, E. (1992). *J. Biol. Chem.*, **267**, 7116–20.
20. Belkin, A. M., Zhidkova, N. I., Balzac, F., Altruda, F., Tomatis, D., Maier, *et al.* (1996). *J. Cell. Biol.*, **132**, 211–26.
21. Baudoin, C., van der Flier, A., Borradori, L., and Sonnenberg, A. (1996). *Cell Adh., Commun.* **4**, 1–11.
22. Balzac, F., Belkin, A. M., Koteliansky, V. E., Balabanov, Y. V., Altruda, F., Silengo, L., and Tarone, G. (1993). *J. Cell. Biol.*, **121**, 171–8.
23. Zheng, D. Q., Fornaro, M., Bofetiado, C. J., Tallini, G., Bosari, S., and Languino, L. R. (1997). *Kidney Int.* **51**, 1434–40.
24. Meredith, J. Jr, Takada, Y., Fornaro, M., Languino, L. R., and Schwartz, M. A. (1995). *Science*, **269**, 1570–2.
25. Bazzoni, G., Shih, D. -T., Buck, C. A., and Hemler, M. E. (1995). *J. Biol. Chem.*, **270**, 25570–7.
26. Fogerty, F. J., Fessler, L. I., Bunch, T. A., Yaron, Y., Parker, C. G., Nelson, R. E., *et al.* (1994). *Development*, **120**, 1747–58.
27. Masumoto, A., and Hemler, M. E. (1993). *J. Biol. Chem.*, **268**, 228–34.
28. Faull, R. J., Kovach, N. L., Harlan, J. M., and Ginsberg, M. H. (1994). *J. Exp. Med.*, **179**, 1307–16.
29. Dustin, M. L. and Springer, T. A. (1989). *Nature*, **341**, 619–24.
30. Hughes, P. E., Renshaw, M. W., Pfaff, M., Forsyth, J., Keivens, V. M., Schwartz, M. A., and Ginsberg, M. H. (1997). *Cell*, **88**, 521–30.
31. Kucik, D. F., Dustin, M. L., Miller, J. M., and Brown, E. J. (1996). *J. Clin. Invest.*, **97**, 2139–44.
32. Yauch, R. L., Felsenfeld, D., Kraeft, S. -K., Chen, L. B., Sheetz, M., and Hemler, M. E. (1997). *J. Exp. Med.*, (In press.)
33. Peter, K., and O'Toole, T. E. (1995). *J. Exp. Med.*, **181**, 315–26.
34. Hemler, M. E., Weitzman, J. B., Pasqualini, R., Kawaguchi, S., Kassner, P. D., and Berdichevsky, F. B. (1994). In *Integrin:*

the biological problem (ed. Y. Takada), pp. 1–35.CRC Press, Ann. Arbor.
35. Miyamoto, S., Teramoto, H., Coso, O. A., Gutkind, J. S., Burbelo, P. D., Akiyama, S. K., and Yamada, K. M. (1995). *J. Cell. Biol.*, **131**, 791–805.
36. Wary, K. K., Mainiero, F., Isakoff, S. J., Marcantonio, E. E., and Giancotti, F. G. (1996). *Cell,* **87**, 733–43.
37. Schwartz, M. A., Schaller, M. D., and Ginsberg, M. H. (1995). *Ann. Rev. Cell. Dev. Biol.*, **11**, 549–99.
38. Burridge, K., and Chrzanowska-Wodnicka, M. (1996). *Ann. Rev. Cell. Dev. Biol.*, **12**, 463–519.
39. Berditchevski, F., Tolias, K. F., Wong, K., Carpenter, C. L., and Hemler, M. E. (1997). *J. Biol. Chem.*, **272**, 2595–8.
40. Hemler, M. E. (1990). *Ann. Rev. Immunol.*, **8**, 365–400.
41. (1995). *Leukocyte typing V–white cell differentiation antigens*, Oxford University Press.
42. Hynes, R. O. (1996). *Dev. Biol.*, **180**, 402–12.
43. Anderson, D. C. and Springer, T. A. (1987). *Ann. Rev. Med.*, **38**, 175–94.
44. Coxon, A., Rieu, P., Barkalow, F. J., Askari, S., Sharpe, A. H., von Andrian, U. H., *et al.* (1996). *Immunity,* **5**, 653–66.
45. Turner, D. C., Flier, L. A., and Carbonetto, S. (1989). *J. Neurosci.*, **9**, 3287–96.
46. Lein, P. J., Higgins, D., Turner, D. C., Flier, L. A., and Terranova, V. P. (1991). *J. Cell. Biol.*, **113**, 417–28.
47. Gardner, H., Kreidberg, J., Koteliansky, V., and Jaenisch, R. (1996). *Dev. Biol.*, **175**, 301–13.
48. Belkin, V. M., Belkin, A. M., and Koteliansky, V. E. (1990). *J. Cell. Biol.*, **111**, 2159–70.
49. Defilippi, P., van Hinsbergh, V., Bertolotto, A., Rossino, P., Silengo, L., and Tarone, G. (1991). *J. Cell. Biol.*, **114**, 855–63.
50. Hemler, M. E., Jacobson, J. G., Brenner, M. B., Mann, D., and Strominger, J. L. (1985). *Eur. J. Immunol.*, **15**, 502–8.
51. Zutter, M. M. and Santoro, S. A. (1990). *Am. J. Pathol.*, **137**, 113–20.
52. Elices, M. J. and Hemler, M. E. (1989). *Proc. Natl Acad. Sci. USA*, **86**, 9906–10.
53. Takada, Y., Strominger, J. L., and Hemler, M. E. (1987). *Proc. Natl Acad. Sci. USA,* **84**, 3239–43.
54. Forsberg, E., Paulsson, M., Timpl, R., and Johansson, S. (1990). *J. Biol. Chem.*, **265**, 6376–81.
55. Kern, A., Eble, J., Golbik, R., and Kühn, K. (1993). *Eur. J. Biochem.*, **215**, 151–9.
56. Takada, Y. and Hemler, M. E. (1989). *J. Cell. Biol.*, **109**, 397–407.
57. Kirchhofer, D., Languino, L. R., Ruoslahti, E., and Pierschbacher, M. D. (1990). *J. Biol. Chem.*, **265**, 615–8.
58. Santoro, S. A., Rajpara, S. M., Staatz, W. D., and Woods, V. L. (1988). *Biochem. Biophys. Res. Commun.*, **153**, 217–23.
59. Santoro, S. A. and Zutter, M. M. (1995). *Thromb. Hemostas.* **74**, 813–21.
60. Wong, L. D., Sondheim, A. B., Zachow, K. R., Reichardt, L. F., and Ignatius, M. J. (1996). *Cell Adh. Commun.* **4**, 201–21.
61. Chan, B. M. C., Kassner, P. D., Schiro, J. A., Byers, H. R., Kupper, T. S., and Hemler, M. E. (1992). *Cell,* **68**, 1051–60.
62. Kassner, P. D., Alon, R., Springer, T. A., and Hemler, M. E. (1995). *Mol. Biol. Cell*, **6**, 661–74.
63. Bergelson, J. M., St. John, N., Kawaguchi, S., Chan, M., Stubal, H., Modlin, J., and Finberg, R. W. (1993). *J. Virol.*, **67**, 6847–52.
64. Hemler, M. E., and Bodorova, J. (1995). In *Leukocyte typing V–white cell differentiation antigens* (ed. S. F. Schlossman, L. Boumsell, W. Gilks *et al.*), p. 1614. Oxford University Press.
65. Hemler, M. E., Kawaguchi, S., and Bodorova, J. (1995). In *Leukocyte typing V–white cell differentiation antigens*, (ed. S. F. Schlossman, L. Boumsell, W. Gilks *et al.*), p. 1615. Oxford University Press.
66. Bergelson, J. M., St. John, N., Kawaguchi, S., Pasqualini, R., Berdichevsky, F., Hemler, M. E., and Finberg, R. W. (1994). *Cell Adh. Commun.* **2**, 455–64.
67. Zutter, M. M., Santoro, S. A., Painter, A. S., Tsung, Y. L., and Gafford, A. (1994). *J. Biol. Chem.*, **269**, 463–9.
68. Fassler, R. and Meyer, M. (1995). *Genes Dev.,* **9**, 1896–908.
69. Hirsch, E., Iglesias, A., Potocnik, A. J., Hartmann, U., and Fassler, R. (1996). *Nature,* **380**, 171–5.
70. Hirsch, E., Balzac, F., Pastore, C., Tarone, G., and Silengo, L. (1993). *Cell Adh. Commun.* **1**, 203–12.
71. Delwel, G. O., Kuikman, I., van der Schors, R. C., de Melker, A. A., and Sonnenberg, A. (1997). *Biochem. J.*, **324**, 263–72.
72. Ziober, B. L., Vu, M. P., Waleh, N., Crawford, J., Lin, C. -S., and Kramer, R. H. (1993). *J. Biol. Chem.*, **268**, 26773–83.
73. Delwel, G. O., Kuikman, I., and Sonnenberg, A. (1995). *Cell Adh. Commun.* **3**, 143–61.
74. Zolkiewska, A. and Moss, J. (1993). *J. Biol. Chem.*, **268**, 25273–6.
75. Song, W. K., Wang, W., Sato, H., Bieiser, D. A., and Kaufman, S. J. (1993). *J. Cell Sci.*, **106**, 1139–52.
76. Tamura, R. N., Rozzo, C., Starr, L., Chambers, J., Reichardt, L. F., Cooper, H. M., and Quaranta, V. (1990). *J. Cell. Biol.*, **111**, 1593–604.
77. Spinardi, L., Ren, Y. -L., Sanders, R., and Giancotti, F. G. (1993). *Mol. Biol. Cell*, **4**, 871–84.
78. Giancotti, F. G., Stepp, M. A., Suzuki, S., Engvall, E., and Ruoslahti, E. (1992). *J. Cell. Biol.*, **118**, 951–9.
79. Potts, A. J., Croall, D. E., and Hemler, M. E. (1994). *Exp. Cell Res.*, **212**, 2–9.
80. Dipersio, C. M., Hodivala-Dilke, K. M., Jaenisch, R., and Kreidberg, J. A. (1997). *J. Cell. Biol.*, **137**, 729–42.
81. Elices, M. J., Urry, L. A., and Hemler, M. E. (1991). *J. Cell. Biol.*, **112**, 169–81.
82. de Melker, A. A., Sterk, L. M., Delwel, G. O., Fles, D. L., Daams, H., Weening, J. J., and Sonnenberg, A. (1997). *Lab. Invest.*, **76**, 547–63.
83. Hogervorst, F., Admiraal, L. G., Niessen, C., Kuikman, I., Janssen, H., Daams, H., and Sonnenberg, A. (1993). *J. Cell. Biol.*, **121**, 179–91.
84. Sonnenberg, A., Linders, C. J. T., Daams, J. H., and Kennel, S. J. (1990). *J. Cell Sci.*, **96**, 207–17.
85. Feltri, M. L., Arona, M., Scherer, S. S., and Wrabetz, L. (1997). *Gene,* **186**, 299–304.
86. Giancotti, F. G. (1996). *J. Cell Sci.*, **109**, 1165–72.
87. Collo, G., Starr, L., and Quaranta, V. (1993). *J. Biol. Chem.*, **268**, 19019–24.
88. Velling, T., Collo, G., Sorokin, L., Durbeej, M., Zhang, H., and Gullberg, D. (1996). *Dev. Dynamics,* **207**, 355–71.
89. Yao, C. C., Ziober, B. L., Sutherland, A. E., Mendrick, D. L., and Kramer, R. H. (1996). *J. Cell Sci.*, **109**, 3139–50.
90. Martin, P. T., Kaufman, S. J., Kramer, R. H., and Sanes, J. R. (1996). *Dev. Biol.*, **174**, 125–39.
91. Hemler, M. E., Mannion, B. A., and Berditchevski, F. (1996). *Biochim. Biophys. Acta*, **1287**, 67–71.
92. Hemler, M. E., Crouse, C., and Sonnenberg, A. (1989). *J. Biol. Chem.*, **264**, 6529–35.
93. Coopman, P. J., Thomas, D. M., Gehlsen, K. R., and Mueller, S. C. (1996). *Mol. Biol. Cell*, **7**, 1789–804.
94. Weitzman, J. B., Chen, A., and Hemler, M. E. (1995). *J. Cell Sci.*, **108**, 3635–44.
95. Jensen, P. J. and Wheelock, M. J. (1995). *Exp. Cell Res.*, **219**, 322–31.
96. Almeida, E. A. C., Huovila, A. -P. J., Sutherland, A. E., Stephens, L. E., Calarco, P. G., Shaw, L. M., *et al.* (1995). *Cell*, **81**, 1095–104.

97. Evander, M., Frazer, I. H., Payne, E., Qi, Y. M., Hengst, K., and McMillan, N. A. J. (1997). *J. Virol.*, **71**, 2449–56.
98. Hemler, M. E., Weitzman, J. B., and Bodorova, J. (1995). In *Leukocyte typing V–white cell differentiation antigens* (ed. S. F. Schlossman, L. Boumsell, W. Gilks, *et al.*), pp. 1616–1617. Oxford University Press.
99. Hemler, M. E., Quaranta, V., Starr, L., and Bodorova, J. (1995). In *Leukocyte typing V–white cell differenatiation antigens* (ed. S.F. Schlossman, L. Boumsell, W. Gilks, *et al.*), pp. 1619–1620. Oxford University Press.
100. Wong, D. A. and Springer, T. A. (1995). In *Leukocyte typing V–white cell differentiation antigens* (ed. S.F. Schlossman, L. Boumsell, W. Gilks, *et al.*), pp. 1667–1668. Oxford University Press.
101. Kreidberg, J. A., Donovan, M. J., Goldstein, S. L., Rennke, H., Shepherd, K., Jones, R. C., and Jaenisch, R. (1996). *Development,* **122**, 3537–47.
102. Georges-Labouesse, E. N., Messaddeq, N., Yehia, G., Cadalbert, L., Dierich, A., and Le Meur, M. (1996). *Nature Genet.*, **13**, 370–3.
103. van der Neut, R., Krimpenfort, P., Calafat, J., Carien, M., Niessen, C., and Sonnenberg, A. (1996). *Nature Genet.*, **13**, 366–9.
104. Dowling, J., Yu, Q. C., and Fuchs, E. (1996). *J. Cell. Biol.*, **134**, 559–72.
105. Gil, S. G., Brown, T. A., Ryan, M. C., and Carter, W. G. (1994). *J. Invest. Dermatol.*, **103**, 31S–8S.
106. Vidal, F., Aberdam, D., Miquell, C., Christiano, A. M., Pulkkinen, L., Uitto, J., *et al.* (1995). *Nature Genet.*, **10**, 229–34.
107. Pulkkinen, L., Kurtz, K., Xu, Y. L., Bruckner-Tuderman, L., and Uitto, J. (1997). *Lab. Invest.*, **76**, 823–33.
108. Iacovacci, S., Gagnoux-Palacios, L., Zambruno, G., Meneguzzi, G., and D'Alessio, M. (1996). *Mamm. Genome,* **109**, 1656–706.
109. Teixidó, J., Parker, C. M., Kassner, P. D., and Hemler, M. E. (1992). *J. Biol. Chem.*, **267**, 1786–91.
110. Pujades, C., Teixidó, J., Bazzoni, G., and Hemler, M. E. (1996). *Biochem. J.*, **313**, 899–908.
111. Irie, A., Kamata, T., and Takada, Y. (1997). *Proc. Natl Acad. Sci. USA*, **94**, 7198–203.
112. Shaw, S. K., Cepek, K. L., Murphy, E. A., Russell, G. J., Brenner, M. B., and Parker, C. M. (1994). *J. Biol. Chem.*, **269**, 6016–25.
113. Parker, C. M., Cepek, K., Russell, G. J., Shaw, S. K., Posnett, D., Schwarting, R., and Brenner, M. B. (1992). *Proc. Natl Acad. Sci. USA*, **89**, 1924–28.
114. Yuan, Q., Jiang, W. -M., Krissansen, G. W., and Watson, J. D. (1990). *Int. Immunol.*, **2**, 1097–108.
115. Dipersio, C. M., Jackson, D. A., and Zaret, K. S. (1991). *Mol. Cell. Biol.*, **11**, 4405–14.
116. Tidswell, M., Pachynski, R., Wu, S. W., Qiu, S. -Q., Dunham, E., Cochran, N., *et al.* (1997). *J. Immunol.*, **159**, 1497–505.
117. Lobb, R. R., and Hemler, M. E. (1994). *J. Clin. Invest.*, **94**, 1722–8.
118. Cerf-Bensussan, N., Jarry, A., Brousse, N., Lisowska-Grospierre, B., Guy-Grand, D., and Griscelli, C. (1987). *Eur. J. Immunol.*, **17**, 1279–85.
119. Palmer, E. L., Rüegg, C., Ferrando, R., Pytela, R., and Sheppard, D. (1993). *J. Cell. Biol.*, **123**, 1289–97.
120. Hemler, M. E., Huang, C., Takada, Y., Schwarz, L., Strominger, J. L., and Clabby, M. L. (1987). *J. Biol. Chem.*, **262**, 11478–85.
121. Guan, J.-L. and Hynes, R. O. (1990). *Cell,* **60**, 53–61.
122. Rüegg, C., Postigo, A., Sikorski, E. E., Butcher, E. C., Pytela, R., and Erle, D. J. (1992). *J. Cell. Biol.*, **117**, 179–89.
123. Forsberg, E., Ek, B., Engström, Å., and Johansson, S. (1994). *Exp. Cell Res.*, **213**, 183–90.
124. Hemler, M. E., Elices, M. J., Parker, C., and Takada, Y. (1990). *Immunol. Rev.*, **114**, 45–65.
125. Berlin, C., Berg, E. L., Briskin, M. J., Andrew, D. P., Kilshaw, P. J., Holzmann, B., *et al.* (1993). *Cell,* **74**, 185–95.
126. Chan, B. M. C., Elices, M. J., Murphy, E., and Hemler, M. E. (1992). *J. Biol. Chem.*, **267**, 8366–70.
127. Berlin, C., Bargatze, R. F., Campbell, J. J., von Andrian, U. H., Szabo, M. C., Hassien, S. R., *et al.* (1995). *Cell,* **80**, 413–22.
128. Alon, R., Kassner, P. D., Carr, M. W., Finger, E. B., Hemler, M. E., and Springer, T. A. (1995). *J. Cell. Biol.*, **128**, 1243–53.
129. Yang, J. T., Rayburn, H., and Hynes, R. O. (1995). *Development,* **121**, 549–60.
130. Cepek, K. L., Shaw, S. K., Parker, C. M., Russell, G. J., Morrow, J. S., Rimm, D. L., and Brenner, M. B. (1994). *Nature,* **372**, 190–3.
131. Yokosaki, Y., Palmer, E. L., Prieto, A. L., Crossin, K. L., Bourdon, M. A., Pytela, R., and Sheppard, D. (1994). *J. Biol. Chem.*, **269**, 26691–6.
132. Pulido, R., Elices, M. J., Campanero, M. R., Osborn, L., Schiffer, S., García-Pardo, A., *et al.* (1991). *J. Biol. Chem.*, **266**, 10241–5.
133. Hemler, M. E., Kassner, P. D., and Bodorova, J. (1995). In *Leukocyte typing V–white cell differentiation antigens* (ed. S.F. Schlossman, L. Boumsell, W. Gilks, *et al.*), pp. 1617–18. Oxford University Press.
134. Kamata, T., Puzon, W., and Takada, Y. (1995). *Biochem. J.*, **305**, 945–51.
135. Cepek, K. L., Wong, D. A., Brenner, M. B., and Springer, T. A. (1995). In *Leukocyte typing V–white cell differentiation antigens* (ed. S.F. Schlossman, L. Boumsell, W. Gilks *et al.*), Oxford University Press.
136. Russell, G. J., Parker, C. M., Cepek, K. L., Mandelbrot, D. A., Sood, A., Mizoguchi, E., *et al.* (1994). *Eur. J. Immunol.*, **24**, 2832–41.
137. Bednarczyk, J. L., Szabo, M. C., Wygant, J. N., Lazarovits, A. I., and McIntyre, B. W. (1994). *J. Biol. Chem.*, **269**, 8348–54.
138. Schweighoffer, T., Tanaka, Y., Tidswell, M., Erle, D. J., Horgan, K. J., Ginther-Luce, G. E., *et al.* (1993). *J. Immunol.*, **151**, 717–29.
139. Arroyo, A. G., Yang, J. T., Rayburn, H., and Hynes, R. O. (1996). *Cell,* **85**, 997–1008.
140. Rosen, G. D., Birkenmeier, T. M., and Dean, D. C. (1991). *Proc. Natl Acad. Sci. USA*, **88**, 4094–8.
141. De Meirsman, C., Schollen, E., Jaspers, M., Ongena, K., Matthijs, G., Marynen, P., and Cassiman, J. J. (1994). *DNA Cell Biol.,* **13**, 743–54.
142. De Meirsman, C., Jaspers, M., Schollen, E., and Cassiman, J. J. (1996). *DNA Cell Biol.,* **15**, 595–603.
143. Wagner, N., Löhler, J., Kunkel, E. J., Ley, K., Leung, E., Krissansen, G., *et al.* (1996). *Nature,* **382**, 366–70.
144. Jiang, W. -M., Jenkins, D., Yuan, Q., Leung, E., Choo, A. K. H., Watson, J. D., and Krissansen, G. W. (1992). *Int. Immunol.*, **4**, 1031–40.
145. Nakagawa, H., Zheng, M., Hakomori, S., Tsukamoto, Y., Kawamura, Y., and Takahashi, N. (1996). *Eur. J. Biochem.*, **237**, 76–85.
146. Yang, J. T., Rayburn, H., and Hynes, R. O. (1993). *Development,* **119**, 1093–1105.
147. Bossy, B., Bossy-Wetzel, E., and Reichardt, L. F. (1991). *EMBO J.*, **10**, 2375–85.
148. Müller, U., Wang, D., Denda, S., Meneses, J. J., Pedersen, R. A., and Reichardt, L. F. (1997). *Cell,* **88**, 603–13.

149. Schnapp, L. M., Breuss, J. M., Ramos, D. M., Sheppard, D., and Pytela, R. (1995). *J. Cell Sci.*, **108**, 537–44.
150. Pytela, R., Pierschbacher, M. D., and Ruoslahti, E. (1985). *Cell,* **40**, 191–8.
151. Vogel, B. E., Tarone, G., Giancotti, F. G., Gailit, J., and Ruoslahti, E. (1990). *J. Biol. Chem.*, **265**, 5934–7.
152. Schnapp, L. M., Hatch, N., Ramos, D. M., Klimanskaya, I. V., Sheppard, D., and Pytela, R. (1995). *J. Biol. Chem.*, **270**, 23196–202.
153. Koivunen, E., Gay, D. A., and Ruoslahti, E. (1993). *J. Biol. Chem.*, **268**, 20205–10.
154. Aota, S., Nomizu, M., and Yamada, K. M. (1994). *J. Biol. Chem.*, **269**, 24756–61.
155. Roman, J., LaChance, R. M., Broekelmann, T. J., Kennedy, C. J. R., Wayner, E. A., Carter, W. G., and McDonald, J. A. (1989). *J. Cell. Biol.*, **108**, 2529–43.
156. Akiyama, S. K., Yamada, S. S., Chen, W. -T., and Yamada, K. M. (1989). *J. Cell. Biol.*, **109**, 863–75.
157. Giancotti, F. G. and Ruoslahti, E. (1990). *Cell,* **60**, 849–59.
158. Müller, U., Bossy, B., Venstrom, K., and Reichardt, L. F. (1995). *Mol. Biol. Cell*, **6**, 433–48.
159. Bodary, S. C. and McLean, J. W. (1990). *J. Biol. Chem.*, **265**, 5938–41.
160. Liaw, L., Skinner, M. P., Raines, E. W., Ross, R., Cheresh, D. A., Schwartz, S. M., and Giachelli, C. M. (1995). *J. Clin. Invest.*, **95**, 713–24.
161. Marshall, J. F., Rutherford, D. C., McCartney, A. C., Mitjans, F., Goodman, S. L., and Hart, I. R. (1995). *J. Cell Sci.*, **108**, 1227–38.
162. Hemler, M. E., Pujades, C., and Bodorova, J. (1995). In *Leukocyte typing V–white cell differentiation antigens* (ed. S.F. Schlossman, L. Boumsell, W. Gilks, *et al.*), pp. 1618–19. Oxford University Press.
163. Wong, D. A. and Springer, T. A. (1995). In *Leukocyte typing V–white cell differentiation antigens* (ed. S.F. Schlossman, L. Boumsell, W. Gilks, *et al.*), pp. 1663–4. Oxford University Press.
164. Birkenmeier, T. M., McQuillan, J. J., Boedeker, E. D., Argraves, W. S., Ruoslahti, E., and Dean, D. C. (1991). *J. Biol. Chem.*, **266**, 20544–9.
165. Van Kuppevelt, T. H., Languino, L. R., Gailit, J. O., Suzuki, S., and Ruoslahti, E. (1989). *Proc. Natl Acad. Sci. USA*, **86**, 5415–8.
166. Ramaswamy, H. and Hemler, M. E. (1990). *EMBO J.*, **9**, 1561–8.
167. McLean, J. W., Vestal, D. J., Cheresh, D. A., and Bodary, S. C. (1990). *J. Biol. Chem.*, **265**, 17126–31.
168. Sheppard, D., Rozzo, C., Starr, L., Quaranta, V., Erle, D. J., and Pytela, R. (1990). *J. Biol. Chem.*, **265**, 11502–7.
169. Moyle, M., Napier, M. A., and McLean, J. W. (1991). *J. Biol. Chem.*, **266**, 19650–8.
170. Heino, J. and Massague, J. (1989). *J. Biol. Chem.*, **264**, 21806–11.
171. Calvete, J. J., Henschen, A., and Gonzalez-Rodriguez, J. (1989). *Biochem. J.*, **261**, 561–8.
172. Calvete, J. J., Henschen, A., and González-Rodríquez, J. (1991). *Biochem. J.*, **274**, 63–71.
173. Smith, J. W. and Cheresh, D. A. (1988). *J. Biol. Chem.*, **263**, 18726–31.
174. Smith, J. W. and Cheresh, D. A. (1990). *J. Biol. Chem.*, **265**, 2168–72.
175. Pasqualini, R., Bodorova, J., Ye, S., and Hemler, M. E. (1993). *J. Cell Sci.*, **105**, 101–111.
176. Breuss, J. M., Gallo, J., DeLisser, H. M., Klimanskaya, I. V., Folkesson, H. G., Pittet, J. F., *et al.* (1995). *J. Cell Sci.*, **108**, 2241–51.
177. Pytela, R., Pierschbacher, M. D., Ginsberg, M. H., Plow, E. F., and Ruoslahti, E. (1986). *Science*, **231**, 1559–62.
178. Cheresh, D. A. (1987). *Proc. Natl Acad. Sci. USA*, **84**, 6471–5.
179. Busk, M., Pytela, R., and Sheppard, D. (1992). *J. Biol. Chem.*, **267**, 5790–6.
180. Weinacker, A., Chen, A., Agrez, M., Cone, R. I., Nishimura, S., Wayner, E., Pytela, R., and Sheppard, D. (1994). *J. Biol. Chem.*, **269**, 6940–8.
181. Nishimura, S. L., Sheppard, D., and Pytela, R. (1994). *J. Biol. Chem.*, **269**, 28708–15.
182. Phillips, D. R., Charo, I. F., and Scarborough, R. M. (1991). *Cell,* **65**, 359–62.
183. Ginsberg, M. H., Loftus, J. C., and Plow, E. F. (1988). *Thromb. Hemostas.* **59**, 1–6.
184. Coller, B. S. (1997). *J. Clin. Invest.*, **99**, 1467–71.
185. Wu, C., Kevins, V., O'Toole, T. E., McDonald, J. A., and Ginsberg, M. H. (1995). *Cell* **83**, 715–24.
186. Wu, C., Hughes, P. E., Ginsberg, M. H., and McDonald, J. A. (1996). *Cell Adh. Commun.*, **4**, 149–58.
187. Lindberg, F. P., Gresham, H. D., Schwarz, E., and Brown, E. J. (1993). *J. Cell. Biol.*, **123**, 485–96.
188. Chung, J., Gao, A. G., and Frazier, W. A. (1997). *J. Biol. Chem.*, **272**, 14740–6.
189. Lindberg, F. P., Gresham, H. D., Reinhold, M. I., and Brown, E. J. (1996). *J. Cell. Biol.*, **134**, 1313–22.
190. Smith, J. W., Vestal, D. J., Irwin, S. V., Burke, T. A., and Cheresh, D. A. (1990). *J. Biol. Chem.*, **265**, 11008–13.
191. Hu, D. D., Lin, E. C., Kovach, N. L., Hoyer, J. R., and Smith, J. W. (1995). *J. Biol. Chem.*, **270**, 26232–8.
192. Chen, J., Maeda, T., Sekiguchi, K., and Sheppard, D. (1996). *Cell Adh. Commun.*, **4**, 237–50.
193. Wayner, E. A., Orlando, R. A., and Cheresh, D. A. (1991). *J. Cell. Biol.*, **113**, 919–29.
194. Lewis, J. M., Cheresh, D. A., and Schwartz, M. A. (1996). *J. Cell. Biol.*, **134**, 1323–32.
195. Friedlander, M., Brooks, P. C., Shaffer, R. W., Kincaid, C. M., Varner, J. A., and Cheresh, D. A. (1995). *Science*, **270**, 1500–2.
196. Wickham, T. J., Mathias, P., Cheresh, D. A., and Nemerow, G. R. (1993). *Cell,* **73**, 309–19.
197. Vogel, B. E., Lee, S. J., Hildebrand, A., Craig, W., Pierschbacher, M. D., Wong-Staal, F., and Ruoslahti, E. (1993). *J. Cell. Biol.*, **121**, 461–8.
198. Wu, C., Bauer, J. S., Juliano, R. L., and McDonald, J. A. (1993). *J. Biol. Chem.*, **268**, 21883–8.
199. Agrez, M., Chen, A., Cone, R. I., Pytela, R., and Sheppard, D. (1994). *J. Cell. Biol.*, **127**, 547–56.
200. Dixit, R. B., Chen, A., Chen, J., and Sheppard, D. (1996). *J. Biol. Chem.*, **271**, 25976–80.
201. Bowditch, R. D., Hariharan, M., Tominna, E. F., Smith, J. W., Yamada, K. M., Getzoff, E. D., and Ginsberg, M. H. (1994). *J. Biol. Chem.*, **269**, 10856–63.
202. Venstrom, K. and Reichardt, L. (1995). *Mol. Biol. Cell*, **6**, 419–31.
203. Savill, J., Dransfield, I., Hogg, N., and Haslett, C. (1990). *Nature,* **343**, 170–3.
204. Stuiver, I. and Smith, J. W. (1995). *Hybridoma,* **14**, 545–50.
205. Newman, P. J., Seligsohn, U., Lyman, S., and Coller, B. S. (1991). *Proc. Natl Acad. Sci. USA*, **88**, 3160–4.
206. Newman, P. J. (1991). *Thromb. Hemostas.,* **66**, 111–8.
207. Heidenreich, R., Eisman, R., Surrey, S., Delgrosso, K., Bennett, J. S., Schwartz, E., and Poncz, M. (1990). *Biochemistry,* **29**, 1232–44.
208. Zimrin, A. B., Gidwitz, S., Lord, S., Schwartz, E., Bennett, J. S., White, G. C., and Poncz, M. (1990). *J. Biol. Chem.*, **265**, 8590–5.

209. Lanza, F., Kieffer, N., Phillips, D. R., and Fitzgerald, L. A. (1990). *J. Biol. Chem.*, **265**, 18098–18103.
210. Villa-Garcia, M., Li, L., Riely, G., and Bray, P. F. (1994). *Blood,* **83**, 668–76.
211. Cao, X., Ross, F. P., Zhang, L., MacDonald, P. N., Chappel, J., and Teitelbaum, S. L. (1993). *J. Biol. Chem.*, **268**, 27371–380.
212. Martin, F., Prandini, M. H., Thevenon, D., Marguerie, G., and Uzan, G. (1993). *J. Biol. Chem.*, **268**, 21606–12.
213. Huang, X. Z., Wu, J. F., Cass, D., Erle, D. J., Corry, D., Young, S. G., *et al.* (1997). *J. Cell. Biol.*, **133**, 921–8.
214. Qu, A. and Leahy, D. J. (1996). *Structure,* **4**, 931–42.
215. Lee, J. -O., Bankston, L. A., Arnaout, M. A., and Liddington, R. C. (1995). *Structure,* **3**, 1333–40.
216. Bajt, M. L., Goodman, T., and McGuire, S. L. (1995). *J. Biol. Chem.*, **270**, 94–8.
217. Kishimoto, T. K., Larson, R. S., Corbi, A. L., Dustin, M. L., Staunton, D. E., and Springer, T. A. (1989). *Adv. Immunol.*, **46**, 149–82.
218. Van der Vieren, M., Trong, H. L., Wood, C. L., Moore, P. F., St. John, T., Staunton, D. E., and Gallatin, W. M. (1995). *Immunity,* **3**, 683–90.
219. Danilenko, D. M., Rossitto, P. V., Van der Vieren, M., Le Trong, H., McDonough, S. P., Affolter, V. K., and Moore, P. F. (1995). *J. Immunol.*, **155**, 35–44.
220. Larson, R. S., Corbi, A. L., Berman, L., and Springer, T. A. (1989). *J. Cell. Biol.*, **108**, 703–12.
221. Miller, L. J., Wiebe, M., and Springer, T. A. (1987). *J. Immunol.*, **138**, 2381–3.
222. Micklem, K. J. and Sim, R. B. (1985). *Biochemistry,* **231**, 233–6.
223. Springer, T. A. (1995). *Ann. Rev. Physiol.*, **57**, 827–72.
224. Diamond, M. S., Alon, R., Parkos, C. A., Quinn, M. T., and Springer, T. A. (1995). *J. Cell. Biol.*, **130**, 1473–82.
225. Benimetskaya, L., Loike, J. D., Khaled, Z., Loike, G., Silverstein, S. C., Cao, L., *et al.* (1997). *Nature Med.*, **3**, 414–20.
226. Cai, T. Q. and Wright, S. D. (1996). *J. Exp. Med.*, **184**, 1213–23.
227. Wachtfogel, Y. T., DeLa Cadena, R. A., Kunapuli, S. P., Rick, L., Miller, M., Schultze, R. L., *et al.* (1994). *J. Biol. Chem.*, **269**, 19307–12.
228. Thornton, B. P., Vetvicka, V., Pitman, M., Goldman, R. C., and Ross, G. D. (1996). *J. Immunol.*, **156**, 1235–46.
229. Bilsland, C. A., Diamond, M. S., and Springer, T. A. (1994). *J. Immunol.*, **152**, 4582–9.
230. Blackford, J., Reid, H. W., Pappin, D. J., Bowers, P. S., and Wilkinson, J. M. (1996). *Eur. J. Immunol.*, **26**, 525–31.
231. Huang, C. and Springer, T. A. (1995). *J. Biol. Chem.*, **270**, 19008–16.
232. Diamond, M. S., Garcia-Aguilar, J., Bickford, J. K., Corbi, A. L., and Springer, T. A. (1993). *J. Cell. Biol.*, **120**, 1031–43.
233. Wong, D. A., Davis, E. M., LeBeau, M., and Springer, T. A. (1997). *Gene,* **171**, 291–4.
234. Corbi, A. L., Garcia-Aguilar, J., and Springer, T. A. (1990). *J. Biol. Chem.*, **265**, 2782–8.
235. Fleming, J. C., Pahl, H. L., Gonzalez, D. A., Smith, T. F., and Tenen, D. G. (1993). *J. Immunol.*, **150**, 480–90.
236. Weitzman, J. B., Wells, C. E., Wright, A. H., Clark, P. A., and Law, S. K. A. (1991). *FEBS Lett.*, **294**, 97–103.
237. Lopez-Rodriguez, C., Nueda, A., Rubio, M., and Corbí, A. L. (1995). *Immunobiol.*, **193**, 315–21.
238. Hickstein, D. D., Baker, D. M., Gollahon, K. A., and Back, A. L. (1992). *Proc. Natl Acad. Sci. USA*, **89**, 2105–9.
239. Noti, J. D., Gordon, M., and Hall, R. E. (1992). *DNA Cell Biol.*, **11**, 123–38.
240. Agura, E. D., Howard, M., and Collins, S. J. (1992). *Blood,* **79**, 602–9.
241. Lu, H., Smith, C. W., Perrard, J., Bullard, D., Tang, L., Shappell, S. B., *et al.* (1997). *J. Clin. Invest.*, **99**, 1340–50.
242. Wilson, R. W., Ballantyne, C. M., Smith, C. W., Montgomery, C., Bradley, A., O'Brien, W. E., and Beaudet, A. L. (1993). *J. Immunol.*, **151**, 1571–8.
243. Shuster, D. E., Kehrli, M. E.,Jr., Ackermann, M. R., and Gilbert, R. O. (1992). *Proc. Natl Acad. Sci. USA*, **89**, 9225–9.
244. Briesewitz, R., Epstein, M. R., and Marcantonio, E. E. (1993). *J. Biol. Chem.*, **268**, 2989–96.
245. Ignatius, M. J., Large, T. H., Houde, M., Tawil, J. W., Barton, A., Esch, F., *et al.* (1990). *J. Cell Biol.*, **111**, 709–20.
246. Kamata, T., Puzon, W., and Takada, Y. (1994). *J. Biol. Chem.*, **269**, 9659–63.
247. Tsuji, T., Hakomori, S., and Osawa, T. (1991). *J. Biochem.*, **109**, 659–65.
248. Takada, Y., Murphy, E., Pil, P., Chen, C., Ginsberg, M. H., and Hemler, M. E. (1991). *J. Cell Biol.*, **115**, 257–66.
249. Tsuji, T., Yamamoto, F., Miura, Y., Takio, K., Titani, K., Pawar, S., *et al.* (1990). *J. Biol. Chem.*, **265**, 7016–21.
250. Takeuchi, K., Hirano, K., Tsuji, T., Osawa, T., and Irimura, T. (1994). *J. Cell. Biochem.*,
251. Takada, Y., Elices, M. J., Crouse, C., and Hemler, M. E. (1989). *EMBO J.*, **8**, 1361–8.
252. Neuhaus, H., Hu, M. C-T., Hemler, M. E., Takada, Y., Holzmann, B., and Weissman, I. L. (1991). *J. Cell Biol.*, **115**, 1149–58.
253. Argraves, W. S., Suzuki, S., Arai, H., Thompson, K., Pierschbacher, M. D., and Ruoslahti, E. (1987). *J. Cell Biol.*, **105**, 1183–90.
254. Fitzgerald, L. A., Poncz, M., Steiner, B., Rall, S. C., Jr, Bennett, J. S., and Phillips, D. R. (1987). *Biochemistry,* **26**, 8158–65.
255. Holers, V. M., Ruff, T. G., Parks, D. L., McDonald, J. A., Ballard, L. L., and Brown, E. J. (1989). *J. Exp. Med.*, **169**, 1589–1605.
256. MacLaren, L. A. and Wildeman, A. G. (1995). *Biol. Reprod.*, **53**, 153–65.
257. Hogervorst, F., Kuikman, I., Van Kessel, A. G., and Sonnenberg, A. (1991). *Eur. J. Biochem.*, **199**, 425–33.
258. de Curtis, I., Quaranta, V., Tamura, R. N., and Reichardt, L. F. (1991). *J. Cell Biol.*, **113**, 405–16.
259. Song, W. K., Wang, W., Forster, R. F., Bielser, D. A., and Kaufman, S. J. (1992). *J. Cell Biol.*, **117**, 643–57.
260. Hibi, K., Yamakwa, K., Ueda, R., Horio, Y., Murata, Y., Tamari, M., *et al.* (1994). *Oncogene*, **9**, 611–9.
261. Smith, T. J., Ducharme, L. A., Shaw, S. K., Parker, C. M., Brenner, M. B., Kilshaw, P. J., and Weis, J. H. (1994). *Immunity,* **1**, 393–403.
262. Suzuki, S., Argraves, W. S., Arai, H., Languino, L. R., Pierschbacher, M., and Ruoslahti, E. (1987). *J. Biol. Chem.*, **262**, 14080–5.
263. Wada, J., Kumar, A., Liu, Z., Ruoslahti, E., Reichardt, L., Marvaldi, J., and Kanwar, Y. S. (1996). *J. Cell Biol.*, **132**, 1161–76.
264. Bossy, B. and Reichardt, L. F. (1990). *Biochemistry,* **29**, 10191–8.
265. Poncz, M., Eisman, R., Heidenreich, R., Silver, S. M., Vilaire, G., Surrey, S., Schwartz, E., and Bennett, J. S. (1987). *J. Biol. Chem.*, **262**, 8476–82.
266. Uzan, G., Frachet, P., Lajmanovich, A., Frandini, M., Denarier, E., Duperray, A., *et al.* (1988). *Eur. J. Biochem.*, **171**, 87–93.
267. Poncz, M. and Newman, P. J. (1990). *Blood,* **75**, 1282–9.
268. Kaufmann, Y., Tseng, E., and Springer, T. A. (1992). *J. Immunol.*, **147**, 369–74.
269. Corbi, A. L., Kishimoto, T. K., Miller, L. J., and Springer, T. A. (1988). *J. Biol. Chem.*, **263**, 12403–11.

270. Arnaout, M. A., Gupta, S. K., Pierce, M. W., and Tenen, D. G. (1988). *J. Cell. Biol.*, **106**, 2153–8.
271. Pytela, R. (1988). *EMBO J.*, **7**, 1371–8.
272. Corbi, A. L., Miller, L. J., O'Connor, K., Larson, R. S., and Springer, T. A. (1987). *EMBO J.*, **6**, 4023–8.
273. Wehrli, M., DiAntonio, A., Fearnley, I. M., Smith, R. J., and Wilcox, M. (1993). *Mech. Dev.*, **43**, 21–36.
274. Bogaert, T., Brown, N., and Wilcox, M. (1988). *Cell*, **51**, 929–40.
275. Tamkun, J. W., DeSimone, D. W., Fonda, D., Patel, R. S., Buck, C., Horwitz, A. F., and Hynes, R. O. (1986). *Cell*, **46**, 271–82.
276. Tominaga, S. (1988). *FEBS Lett.*, **238**, 315–19.
277. Malek-Hedayat, S. and Rome, L. H. (1995). *Gene*, **158**, 287–90.
278. DeSimone, D. W. and Hynes, R. O. (1988). *J. Biol. Chem.*, **263**, 5333–40.
279. Kishimoto, T. K., O'Connor, K., Lee, A., Roberts, T. M., and Springer, T. A. (1987). *Cell*, **48**, 681–90.
280. Law, S. K. A., Gagnon, J., Hildreth, J. E. K., Wells, C. E., Willis, A. C., and Wong, A. J. (1987). *EMBO J.*, **6**, 915–9.
281. Wilson, R. W., O'Brien, W. E., and Beaudet, A. L. (1989). *Nucl. Acids Res.*, **17**, 5397.
282. Shuster, D. E., Bosworth, B. T., and Kehrli, M. E.,Jr. (1992). *Gene*, **114**, 267–71.
283. Bilsland, C. A. and Springer, T. A. (1994). *J. Leuk. Biol.*, **55**, 501–6.
284. Ransome, D. G., Hens, M. D., and DeSimone, D. W. (1993). *Dev. Biol.*, **160**, 265–75.
285. Fitzgerald, L. A., Steiner, B., Rall, S. C.,Jr, Lo, S., and Phillips, D. R. (1987). *J. Biol. Chem.*, **262**, 3936–9.
286. Rosa, J. P., Bray, P. F., Gayet, O., Johnston, G. I., Cook, R. G., Jackson, K. W., *et al.* (1988). *Blood*, **72**, 593–600.
287. Mimura, H., Cao, X., Ross, F. P., Chiba, M., and Teitelbaum, S. L. (1994). *Endocrinology*, **134**, 1061–6.
288. Hogervorst, F., Kuikman, I., von dem Borne, A. E. G. Kr., and Sonnenberg, A. (1990). *EMBO J.*, **9**, 765–70.
289. Suzuki, S. and Naitoh, Y. (1990). *EMBO J.*, **9**, 757–63.
290. Kennel, S. J., Foote, L. J., Cimino, L., Rizzo, M. G., Chang, L. Y., and Sacchi, A. (1993). *Gene*, **130**, 209–16.
291. Erle, D. J., Rüegg, C., Sheppard, D., and Pytela, R. (1991). *J. Biol. Chem.*, **266**, 11009–16.
292. Yuan, Q., Jiang, W. -M., Leung, E., Hollander, D., Watson, J. D., and Krissansen, G. W. (1992). *J. Biol. Chem.*, **267**, 7352–8.
293. Hu, M. C-T., Crowe, D. T., Weissman, I. L., and Holzmann, B. (1992). *Proc. Natl Acad. Sci. USA*, **89**, 8254–8.
294. MacKrell, A. J., Blumberg, B., Haynes, S. R., and Fessler, J. H. (1988). *Proc. Natl Acad. Sci. USA*, **85**, 2633–7.
295. Gettner, S. N., Kenyon, C., and Reichardt, L. F. (1995). *J. Cell. Biol.*, **129**, 1127–41.
296. Marsden, M. and Burke, R. D. (1997). *Dev. Biol.*, **181**, 234–45.
297. Hayashi, Y. K., Chou, F. L., Engvall, E., Ogawa, M., Matsuda, C., Hirabayashi, S., Yokochi, K., Ziober, B. L., Kramer, R. H., Kaufman, S. J., Ozawa, E., Goto, Y., Nonaka, I., Tsukahara, T., Wang, J. Z., Hoffman, E. P., and Arahata, K. (1998). *Nature Genet.*, **19**, 94–7.
298. Lin, C. S., Chen, Y., Huynh, T., and Kramer, R. (1997). *DNA Cell Biol.*, **17**, 929–37.
299. Zhang, H., Tan, S. M., and Lu, J. (1998). *Biochem. J.*, **331**, 631–7.
300. Kambe, M., Miyamoto, Y., and Hayashi, M. (1998). *Biochim. Biophys. Acta*, **1395**, 209–19.
301. Obata, H., Hayashi, K., Nishida, W., Momiyama, T., Uchida, A., and Sobue, K. (1997). *J. Biol. Chem.*, **272**, 26643–51.
302. Edelman, J. M., Chan, B. M., Uniyal, S., Onodera, H., Wang, D. Z., John, N. F., Damjanovich, L., Latzer, D. B., Finberg, R. W., and Bergelson, J. M. (1994). *Cell Adhes. Commun.*, **2**, 131–43.
303. Whittaker, C. A. and DeSimone, D. W. (1993). *Development*, **117**, 1239–49.
304. Joos, T. O., Whittaker, C. A., Meng, F., DeSimone, D. W., Gnau, V., and Hausen, P. (1995). *Mech Dev.*, **50**, 187–99.
305. Hierck, B. P., Thorsteinsdottir, S., Niessen, C. M., Freund, E., Iperen, L. V., Feyen, A., Hogervorst, F., Poelmann, R. E., Mummery, C. L., and Sonnenberg, A. (1993). *Cell Adhes. Commun.*, **1**, 33–53.
306. Dena, S., Muller, U., Crossin, K. L., Erickson, H. P., and Reichardt, L. F. (1998). *Biochemistry*, **37**, 5464–74.
307. Joos, T. O., Reintsch, W. E., Brinker, A., Klein, C., and Hausen, P. (1998). *Int. J. Dev. Biol.*, **42**, 171–9.
308. Stark, K. A., Yee, G. H., Roote, C. E., Williams, E. L., Zusman, S. and Hynes, R. O. (1997). *Development*, **124**, 4583–94.
309. Grotewiel, M. S., Beck, C. D. O., Wu, K. H., Zhu, X. R., and Davis, R. L. (1998). *Nature*, **391**, 455–60.
310. Pancer, Z., Kruse, M., Muller, I. and Muller, W. E. (1997). *Mol. Biol. Evol.*, **14**, 391–8.
311. Brower, D. L., Brower, S. M., Hayward, D. C., and Ball, E. E. (1997). *Proc. Natl Acad. Sci. USA*, **94**, 9182–7.
312. Holmblad, T., Thornqvist, P. O., Soderhall, K., and Johansson, M. W. (1997). *J. Exp. Zool.*, **277**, 255–61.

Martin E. Hemler
Dana-Farber Cancer Institute,
44 Binney Street, Boston,
MA 02115, USA

Integrin-associated protein

Integrin-associated protein (IAP) is a ubiquitously expressed plasma membrane protein which is named for its physical and functional association with the integrin $\alpha V\beta 3$. Antibodies to IAP inhibit $\alpha V\beta 3$-mediated activation of chemotaxis, respiratory burst, and phagocytosis by polymorphonuclear leukocytes (PMN), $\alpha V\beta 3$-initiated increase in intracytoplasmic calcium concentration in endothelial cells, and binding of vitronectin-coated particles to $\alpha V\beta 3$ on several different cell types. Cells from IAP-deficient mice appear to have the same defects in $\alpha V\beta 3$ function as are induced by anti-IAP antibodies.

Synonymous names

CD47,[1,2] OA-3[3] The most closely related proteins are from the A38L open reading frames from the Vaccinia and Variola pox viruses. The function of the A38L proteins is unknown. Vaccinia with A38L deleted make smaller plaques *in vitro*, but have no alteration in virulence *in vivo*.[4]

Protein properties

IAP was first identified as the target of antibodies made against a copurifying molecule in preparations of human placental $\alpha V\beta 3$. Subsequent work showed that IAP could be coimmunoprecipitated with $\alpha V\beta 3$ from a variety of cells. IAP homologues have been identified in many vertebrate, but not invertebrate species. IAP is a transmembrane member of the immunoglobulin family. It has three basic domains: a single extracellular immunoglobulin domain, a multiply membrane-spanning domain, and an alternatively spliced carboxyl-terminal cytoplasmic tail (Fig. 1).[3,5,6] The amino-terminal immunoglobulin domain is in the Ig V family. Its closest relatives other than the pox virus A38L protein are the Ig V domains from cartilage link protein, CD4, and myelin P_0. The Ig V domain is approximately 60 per cent carbohydrate by weight and has five potential *N*-linked glycosylation sites. All commercially available monoclonal antibodies to IAP recognize the Ig domain. The multiply membrane-spanning domain is thought to span the membrane five times, but transmembrane segments have not been experimentally mapped. The carboxyl terminus of the molecule has four alternatively spliced forms which arise from a single promiscuous splice donor at the 5′ end of the exon encoding the 3′ untranslated region of the mRNA. The alternative splicing leads to cytoplasmic tails of IAP which vary between 3 and 36 amino acids in length. The amino acid sequences of the alternatively spliced cytoplasmic tails are conserved between murine and human IAP.

IAP is ubiquitously expressed.[7] There is less IAP on ovarian and renal tubular epithelium and on skeletal muscle than on lymphocytes, erythrocytes, endothelial cells, and fibroblasts, but all tissues contain IAP.[6,7] The alternatively spliced forms show cell-type specific expression, with a 16 amino acid cytoplasmic tail the most common form in bone marrow-derived cells, and the 36 amino acid cytoplasmic tail the most common form in neurones. IAP is identical to the cluster determinant CD47.[1,8] CD47 was defined as an erythrocyte membrane protein which was poorly expressed on Rh-null erythrocytes.[9] However, CD47 is expressed at normal levels on other cells from individuals with Rh-null erythrocytes, and the mechanism and significance of decreased expression of IAP on Rh-null erythrocytes is unknown. Although CD47 expression is reduced on Rh-null erythrocytes, the Rh-null mutation does not map to CD47.[10,11] IAP also is identical to OA-3, the antigen recognized by the OVTL-3 antibody prepared against ovarian carcinoma antigens.[3,12,13]

The IAP Ig domain is the site for interaction with $\alpha V\beta 3$.[14] The $\alpha V\beta 3$/IAP interaction is believed to occur within the plasma membrane of a single cell rather than between two different cells. The binding site for IAP on the integrin has not been mapped. The only other known ligand for IAP is thrombospondin.[15] Thrombospondin binds to the IAP Ig V domain. The IAP binding site in thrombospondin has been mapped to a carboxyl-terminal domain peptide sequence FYVVM, which is well conserved among the different thrombospondin isoforms.

Purification

The conventional protocol for purification of IAP is antibody affinity chromatography, using human placenta as starting material.[16] Other cell sources have been used for antibody affinity purification as well.[2,17]

Activities

Most studies on IAP biology have used antibodies to probe function on intact cells. A variety of $\alpha V\beta 3$-dependent functions are inhibited by anti-IAP (Fig. 2). For example, anti-IAP inhibits PMN activation by Arg–Gly–Asp-containing extracellular matrix proteins;[16,18] PMN chemotaxis toward entactin and other Arg–Gly–Asp-containing proteins and peptides;[19] and binding of vitronecrin-coated particles by a variety of cells.[14,20,21] Anti-IAP also inhibits $\alpha V\beta 3$-dependent increase in intracytoplasmic calcium concentration in endothelial cells,[22] C32

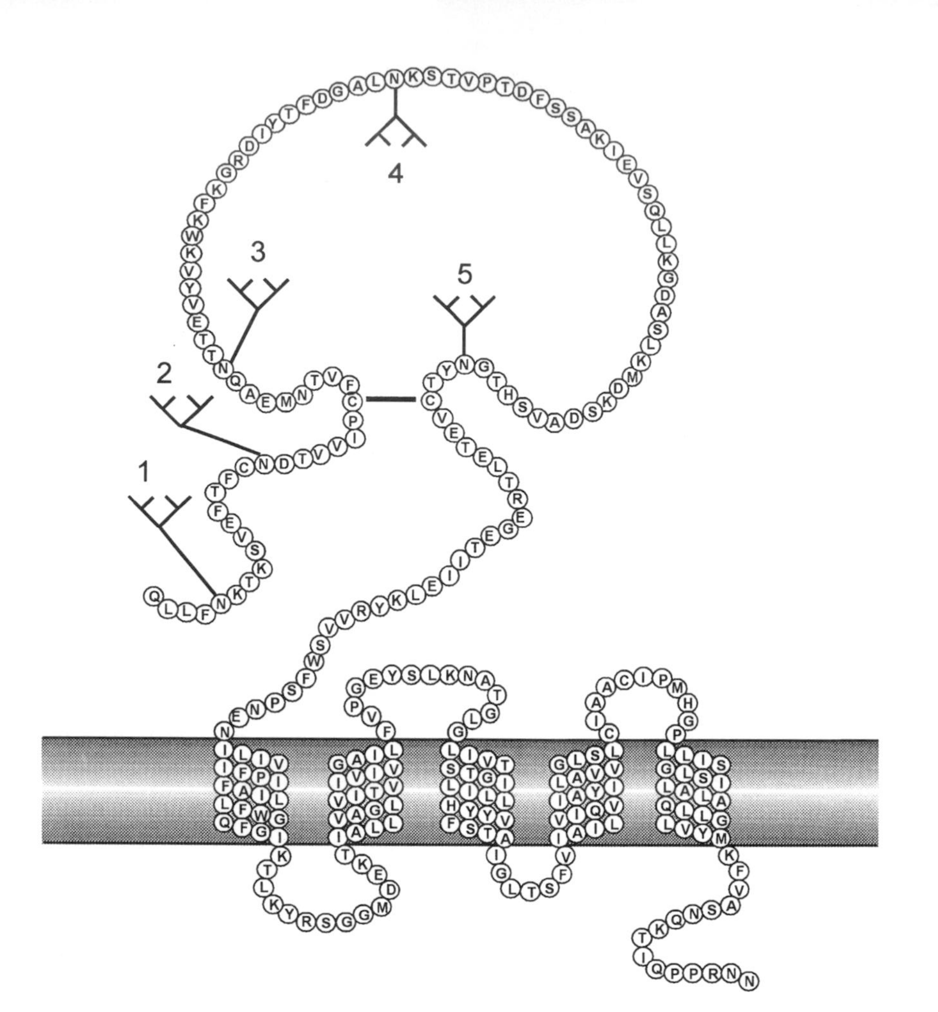

Figure 1. A structural model of IAP. IAP has three functional domains: an extracellular Ig V domain, a multiply membrane-spanning domain, and a short carboxyl-terminal cytoplasmic tail. The Ig V domain has five potential *N*-linked glycosylation sites, numbered 1–5 in this diagram. The carboxyl-terminal tail shown is form 2, the most commonly expressed form of the IAP cytoplasmic tail.[6]

melanoma cell spreading on vitronectin,[23] and $\alpha V\beta 3$-dependent integrin cross-talk in K562 cells.[24] IAP is expressed on some cells, such as erythrocytes and some lymphocytes, which have minimal or undetectable $\alpha V\beta 3$, suggesting it may have roles in addition to association with this integrin. Anti-IAP inhibits transendothelial and transepithelial migration[17,25] and chemotaxis toward thrombospondin peptides,[15] effects that are not clearly associated with $\alpha V\beta 3$. When associated with an antigen presenting cell, anti-IAP will enhance the response of $\alpha V\beta 3$-negative lymphocytes to suboptimal concentrations of antigen.[26] Thrombospondin binding to IAP can activate platelets.[27] IAP mRNA is increased during memory formation in rats, and antisense IAP oligonucleotides inhibit memory formation.[32]

In an IAP-negative ovarian carcinoma cell line, transfection of IAP will reconstitute binding of vitronectin-coated particles by both $\alpha v\beta 3$ and $\alpha v\beta 5$.[14] There are no described assays for functions of purified IAP.

Antibodies

A large number of anti-human IAP and anti-CD47 mAb exist. Both inhibitory antibodies and antibodies with equal affinity, which do not block function, have been described.[16,18,28] Anti-CD47 (BRIC125, BRIC126, CIKM1[29]) is available from many commercial sources for monoclonal antibodies, but these reagents generally have not been tested for function-blocking activity. An anti-murine IAP mAb (mIAP301[31]) is available from Pharmingen. A cell line, B6H12, which secretes a function-blocking antibody that is useful for immunoprecipitation, Western blotting, and fluorescence flow cytometry is available through the American Type Culture Collection.

Genes

cDNAs for human, murine, and rat IAP have been described.[5,30] The GenBank accession numbers are Z25521

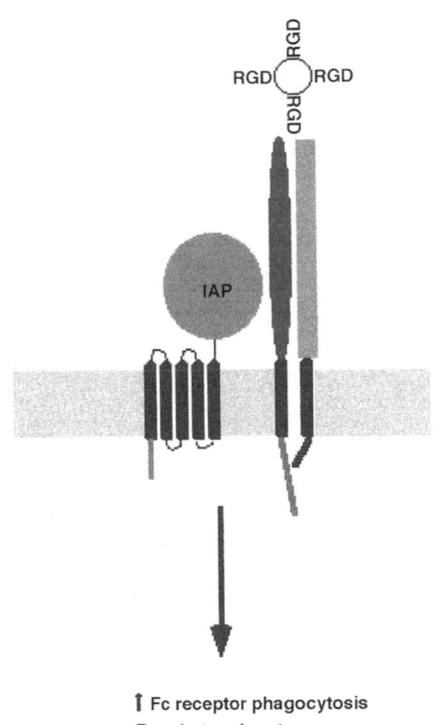

Figure 2. A model for the IAP/$\alpha V\beta 3$ complex. IAP, and $\alpha V\beta 3$ associate in a lateral interaction within the plasma membrane. The IAP Ig V domain is sufficient for this interaction.[14] The IAP/$\alpha V\beta 3$ complex is required for the binding of RGD-coated particles to the integrin and the complex regulates cell activation as measured by increase in intracytoplasmic calcium concentration, enhanced phagocytosis, respiratory burst, and activation of a serine/threonine kinase, perhaps protein kinase C.[20]

(human); Z25524 (mouse); and D87659 (rat). There is also a rat EST (accession number C06656) similar to IAP. The human IAP gene has been mapped to chromosome 3 q13.1–13.2.[1] The murine gene has been mapped to a syntenic region of chromosome 16.

Mutant phenotype/disease state

An IAP-deficient mouse line has been created through homologous recombination.[31] PMN from these mice show similar activation defects *in vitro* as IAP-expressing PMN treated with anti-IAP. *In vivo,* IAP-deficient PMN migrate more slowly into a site of infection and the IAP-deficient mice show increased susceptibility to *E. coli* infection.

Structure

Not available.

Web sites

Display has information on IAP reagents available through our laboratories.
http://www.id.wustl.edu/~lindberg/ has information on IAP reagents available through our laboratories.

References

1. Lindberg, F. P., Lublin, D. M., Telen, M. J., *et al.*(1994). *J. Biol. Chem.*, **269**, 1567–70.
2. Mawby, W. J., Holmes, C. H., Anstee, D. J., Spring, F. A., and Tanner, M. J. A. (1994). *Biochem. J.*, **304**, 525–30.
3. Campbell, I. G., Freemont, P. S., Foulkes, W., and Trowsdale, J. (1992). *Cancer Res.*, **52**, 5416–20.
4. Parkinson, J. E., Sanderson, C. M., and Smith, G. L. (1995). *Virology,* **214**, 177–88.
5. Lindberg, F. P., Gresham, H. D., Schwarz, E., and Brown, E. J. (1993). *J. Cell. Biol.*, **123**, 485–96.
6. Reinhold, M. I., Lindberg, F. P., Plas, D., Reynolds, S., Peters, M. G., and Brown, E. J. (1995). *J. Cell Sci.*, **108**, 3419–25.
7. Hadam, M. R. (1989). In *Leukocyte typing IV: white cell differentiation antigens* (ed. W. Knapp, B. Dorken, W. R. Gilks, E. P. Rieber, R. E. Schmidt, H. Stein, and A. E. G. K. von dem Borne) pp. 658–60. Oxford University Press.
8. Anstee, D. J., and Tanner, M. J. (1993). *Bailliere's Clin. Haematol.*, **6**, 401–22.
9. Spring, F. A., Judson, P. A., and Anstee, D. J. (1989). In *Leukocyte typing IV: white cell differentiation antigens* (ed W. Knapp, B. Dorken, W. R. Gilks, E.P. Rieber, R. E. Schmidt, H. Stein and A. E. G. K. von dem Borne) pp. 660–1. Oxford University Press.
10. Cherif-Zahar, B., Raynal, V., Gane, P., *et al.* (1996). *Nature Genet.*, **12**, 168–73.
11. Cartron, J. P. (1994). *Blood Rev.,.* **8**, 199–212.
12. Poels, G., Peters, D., van Megen, Y., *et al.* (1986). *JNCI,.* **76**, 781–7.
13. Boerman, O. C., van Niekerk, C., Makkink, K., Hanselaar, T. G. J. M., Kenemans, P., and Poels, L. G. (1991). *Int. J. Gynecol. Path.*, **10**, 15–25.
14. Lindberg, F. P., Gresham, H. D., Reinhold, M. I., and Brown, E. J. (1996). *J. Cell. Biol.*, **134**, 1313–22.
15. Gao, A. -G., Lindberg, F. P., Finn, M. B., Blystone, S. D., Brown, E. J., and Frazier, W. A. (1996). *J. Biol. Chem.*, **271**, 21–4.
16. Brown, E. J., Hooper, L., Ho, T., and Gresham, H. D. (1990). *J. Cell. Biol.*, **111**, 2785–94.
17. Parkos, C. A., Colgan, S. P., Liang, T. W., *et al.* (1996). *J. Cell. Biol.*, **132**, 437–50.
18. Zhou, M. -J., and Brown, E. J. (1993). *J. Exp. Med.*, **178**, 1165–74.
19. Senior, R. M., Gresham, H. D., Griffin, G. L., Brown, E. J., and Chung, A. E. (1992). *J. Clin. Invest.*, **90**, 2251–7.
20. Brown, E. J., and Gresham, H. D. (1993). In *Structure, function, and regulation of molecules involved in leukocyte adhesion* (ed. P. E. Lipsky, R. Rothlein, T. K. Kishimoto, R. B. Faanes, and C. W. Smith), Vol. 1, pp. 78–91 (Springer, New York.)
21. Brown, E. J. (1993) In *Cell adhesion molecules* (ed. M. E. Hemler, and E. Mihich), pp. 105–26 Plenum, New York.

22. Schwartz, M. A., Brown, E. J., and Fazeli, B. (1993). *J. Biol. Chem.*, **268**, 19931–4.
23. Gao, A. G., Lindberg, F. P., Dimitry, J. M., Brown, E. J., and Frazier, W. A. (1996). *J. Cell. Biol.*, **135**, 533–44.
24. Blystone, S. D., Graham, I. L., Lindberg, F. P., and Brown, E. J. (1994). *J. Cell. Biol.*, **127**, 1129–37.
25. Cooper, D., Lindberg, F. P., Gamble, J. R., Brown, E. J., and Vadas, M. A. (1995). *Proc. Natl Acad. Sci. USA*, **92**, 3978–82.
26. Reinhold, M. I., Lindberg, F. P., Kersh, G. J., Allen, P. M., and Brown, E. J. *J. Exp. Med.*, **185**, 1–11.
27. Dorahy, D. J., Thorne, R. F., Fecondo, J. V., and Burns, G. F. (1997). *J. Biol. Chem.*, **272**, 1323–30.
28. Rosales, C., Gresham, H. D., and Brown, E. J. (1992). *J. Immunol.*, **149**, 2759–64.
29. Avent, N., Judson, P. A., Parsons, S. F., *et al.* (1988). *Biochem. J.*, **251**, 499–505.
30. Nishiyama, Y., Tanaka, T., Naitoh, H., *et al.* (1997). *Jpn J. Cancer Res.*, **88**, 120–8.
31. Lindberg, F. P., Bullard, D. C., Caver, T. E., Gresham, H. D., Beaudet, A. L., and Brown, E. J. (1996). *Science*, **274**, 795–8.
32. Huang, A., Wang, H. L., Tang, Y. P., and Lee, E. H. Y. (1998). *J. Neurosci.*, **18**, 4305–13.

Eric J. Brown and Frederik P. Lindberg
Division of Infectious Diseases, Washington University School of Medicine, St Louis, MO 63110, USA

Intercellular adhesion molecules (ICAMs)

Intercellular adhesion molecules are a subset of the immunoglobulin superfamily that share the ability to interact with $\beta2$ integrins. There are now five members of this family of which three have been characterized in many studies dating back to the identification of ICAM-1 as an LFA-1 ligand in 1986. This chapter will describe the unique structural and functional features of each ICAM and a general discussion of the LFA-1 binding surface in ICAMs which is defined by the recent crystal structure of ICAM-2.

ICAM-1 (CD54)

Protein properties

ICAM-1 is a cellular ligand for the leukocyte integrins LFA-1 (integrin $\alpha L\beta2$) and Mac-1 (integrin $\alpha M\beta2$) through distinct binding sites. ICAM-1 is a type I transmembrane glycoprotein of 85 000–110 000 Da with a polypeptide of 55 000 Da[1] composed of five tandem Ig-like domains, a transmembrane domain, and a short cytoplasmic tail[2] that binds α-actinin[3] (Fig. 1). LFA-1 binds to domain 1 of ICAM-1,[4] while Mac-1 binds to domain 3 of ICAM-1 in a manner that is partially obstructed by a natural *N*-linked oligosaccharide.[5] ICAM-1 is also a receptor for the major group of rhinoviruses (cold viruses) and is the receptor for *Plasmodium falciparum* (malaria) infected erythrocytes.[6–8] LFA-1, rhinovirus, and *Plasmodium falciparum* bind to overlapping, but non-identical parts of ICAM-1 domain 1.[4,9] Rotary shadowing electron microscopy indicates that there is a bend between domains 3 and 4.[4,10] Human and mouse ICAM-1 form dimers on the cell surface that depend on the transmembrane domains and normal presentation of domain 4.[11,12] (unpublished observations). These dimers have a profound advantage for binding purified LFA-1, but the advantage of dimerization has been more difficult to assess in cell–cell adhesion.[11]

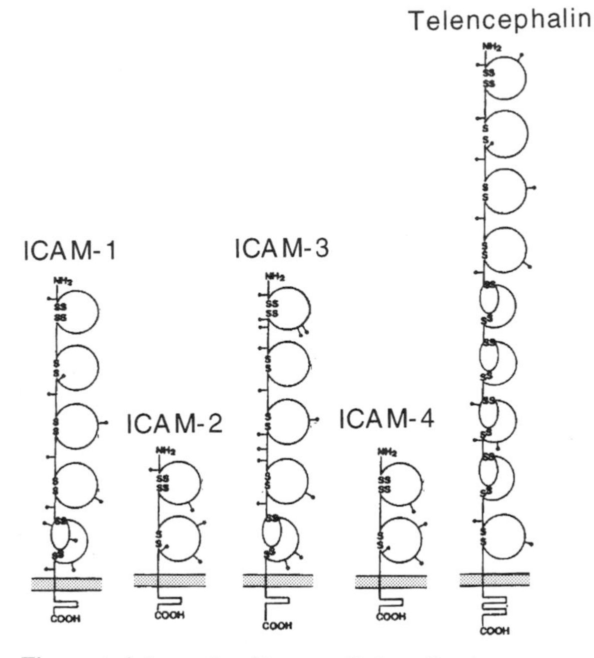

Figure 1. Schematic of interacullular adhesion molecules. The Ig-like domains are indicated as disulphide-linked loops. The glycosylation sites are indicated as lollipops. Glycosylation patterns are for mouse ICAM-1, ICAM-2 and telencephalin, and human ICAM-3 and ICAM-4.

ICAM-1 is expressed on endothelial cells, some epithelial cells, and is present in an intracellular pool in monocytes.[1,13,14] Activation of lymphocytes results in an increase in ICAM-1 on T and B lymphocytes.[1,15] Treatment of fibroblasts and endothelial cells with inflammatory mediator such as interleukin 1, tumour necrosis factor, and lipopolysaccharide results in a dramatic upregulation of ICAM-1 on the surface of these cells beginning at 1–2 h and peaking at 24–48 h.[1,16,17] Treatment of fibroblasts, endothelial cells, and epithelial cells with the immunological mediator interferon-γ also results in a profound induction of ICAM-1 expression.[1,16–18] ICAM-1 expression is very high in reactive lymphoid tissues.[1] Therefore, ICAM-1 is normally restricted in expression, but becomes an abundant molecule in tissues undergoing immune or inflammatory reactions. Transcriptional control of ICAM-1 expression involves STAT1 for interferon γ signalling[58] and NF-κB and c-Rel for tumour necrosis factor signalling.[59]

Site-directed mutagenesis on ICAM-1 revealed that the most important single residue is a glutamic acid at position 34.[4] ICAM-1 is heavily glycosylated, with eight *N*-linked glycans, but none on the first domain. By several criteria, ICAM-1 is the most potent ligand for LFA-1-mediated adhesion.[19]

Two ICAM-1 deficient mouse strains have been generated.[20,21] The ICAM-1 gene is organized so that each domain is encoded by a single exon such that exon 2 encodes domain 1, and so on.[22] ICAM-1-mutant mice with disruption of exon 4 and exon 5 were generated independently. These mice have significant defects in leukocyte migration *in vivo* and activation *in vitro*. Interestingly, the mice are resistant to endotoxin-induced shock.[21] This suggests that ICAM-1 is an important ligand for pathological effects of systemic leukocyte activation. These mice were also the vehicle through which alternative splicing of ICAM-1 was discovered.[23] While ICAM-1 expression is severely decreased or undetectable in many tissues of mutant mice, both mice have ICAM-1 (domain 1) expression in the thymus and in multiple tissues following lipopolysaccharide injection. When the source of this ICAM-1 expression was investigated it was revealed that alternatively spliced forms of ICAM-1 were generated that lacked domains 2, 3, and 4 in different combinations. The function of the alternatively spliced forms of ICAM-1 in normal animals is not known. It is likely that distance spanned by a bond between interacting cells may be important for the organization of cell–cell contact areas.[24] Therefore, the spliced forms of ICAM-1 may serve to shorten the distance between cells interacting through LFA-1/ICAM-1 and may alter the utility of the LFA-1/ICAM-1 interaction.

Recently, two groups solved the structure of human ICAM-1 domains 1 and 2.[60,61] Both structures reveal the flat LFA-1-binding surfaces first described for ICAM-2 (see below and Plate 4). The crystal structure of the longer ICAM-1 fragment shows an ICAM-1 dimer with the LFA-1 binding interfaces directed outward and a small ICAM-1/ICAM-1 interface in domain 1.[60] This dimer might reflect part of the interaction that leads to ICAM-1 dimerization at the cell surface.[11,12] The small size of the interface suggests that the ICAM-1 dimer may 'breathe' rapidly, alternately presenting single and paired conformations of domain 1 that may each be preferentially bound by different receptors.

Other ligands for ICAM-1

It has been reported that ICAM-1 binds to CD43, fibrinogen, and heparan sulphate proteoglycans.[25–27] These binding sites in ICAM-1 have not been mapped and their physiological significance is not known.

Purification

ICAM-1 can be immunoaffinity purified from B lymphoblastoid cell detergent lysates using RR1/1 or 84H10. Elution at pH 12.5 (RR1/1)[28] or pH 3 (RR1/1 or 84H10)[29] yields active material.

Activity

Highly purified ICAM-1 adsorbs efficiently to untreated polystyrene surfaces. Cell adhesion is cation, energy and activation dependent.[28,30]

Antibodies

The anti-human ICAM-1 domain 1 adhesion blocking (for LFA-1) mAb 84H10 is available from Immunotech (Westbrook, ME). The anti-human ICAM-1 domain 2 adhesion blocking (for LFA-1 and Mac-1) mAb R6.5 is available from Genesys Biotech (The Woodlands, TX). Mouse ICAM-1 is recognized by YN1/1,[31] available from ATCC. The anti-mouse ICAM-1 mAb KAT-1 is available from Accurate Chemical and Scientific (Westbury, NY). All of the anti-ICAM-1 antibodies can be used in Western blotting on non-reduced samples only.

Genes

ICAM-1 cDNA have been obtained for human[2,32,33] (J03132, X06990, M24283), mouse[34–36] (X16624, M31585, X52264, X15372), dog37 (L31625), rat[38] (D00913), and chimpanzee (M86848). The gene organization has been determined for the human[39] (X59286–88) and mouse[22] (M90546–51). ICAM-1 domain 1+2 protein structure coordinates are available: PDB accession code 1IAM.

ICAM-2 (CD102)

Protein properties

ICAM-2 was identified by expression cloning from an endothelial cell library.[40] ICAM-2 is a glycoprotein of 60 000 Da with a polypeptide chain of 28 393 Da composed of two Ig-domains that are homologous to the two most amino terminal domains of ICAM-1, a transmembrane domain, and a cytoplasmic tail that binds ezrin[41] (Fig. 1). ICAM-2 is abundantly expressed on endothelial cells and at lower levels on leukocytes, lymphocytes, and platelets.[42] ICAM-2 is not inducible by cytokine treatment of endothelial cells or other cells that have been examined. Functionally, ICAM-2 appears to be a less potent LFA-1 ligand than ICAM-1.[16,19] Binding to ICAM-2 may shorten the distance between membranes compared to binding ICAM-1 and may allow LFA-1 to cooperate more closely with smaller adhesion molecules such as CD2/LFA-3. The function of ICAM-2 *in vitro* is best demonstrated when combinations of ICAM-1 and –3 antibodies are used to block processes such as T cell activation by mitogen or allogeneic lymphocytes or HIV-induced syncitia formation, thus revealing a clear ICAM-2-dependent component.[19,43]

ICAM-2 is the first ICAM for which a crystal structure has been solved.[44] The crystal structure shows that the glutamic acid residue in ICAM-2 corresponding to Glu-34 in ICAM-1, Glu-37, is on a flat surface on the CFG face of the first Ig-like domain. This was a surprise since the key binding residues for other integrins, the arginine–glycine–aspartic acid (RGD) sequence, were always on loops that projected away from the immunoglobulin like (VCAM) or fibronectin type III (fibronectin cell binding domain) repeat.[44] This is also the first look at a binding surface for an I-domain integrin and suggests a different type of binding interface compared to non-I-domain integrins. Another interesting feature of the ICAM-2 structure is that the six N-linked glycans are visualized and they form a tripod-like array of three oligosaccharides on the bottom of the second Ig-like domain that may maintain ICAM-2 in a configuration most accessible to LFA-1 on another cell. This positioning may partly compensate for the small size of ICAM-2.

Antibodies

An adhesion-blocking mAb to human ICAM-2 (CBRIC2.2[45]) is available from Biosource International (Camarillo, CA).

Genes

ICAM-2 cDNA for human[40] (X15606) and mouse[46] (X65493, S46669) have been isolated and expressed. The human (M32331) and murine (X65490, X65491) ICAM-2 genes have been characterized. ICAM-2 protein structure coordinates are available: PDB accession code 1ZXQ.

ICAM-3 (ICAM-R, CD50)

Protein properties

ICAM-3 was identified by monoclonal antibodies that block adhesion that is LFA-1 dependent, but resistant to anti-ICAM-1 and anti-ICAM-2 in combination. ICAM-3 appears to be the least active ligand for LFA-1, but is a high affinity ligand for $\alpha d\beta 2$, the forth member of the $\beta 2$ integrin family. ICAM-3 is a glycoprotein of 124 000 Da composed of five Ig-like domains, a transmembrane domain, and a short cytoplasmic tail; unlike ICAM-1, there is no bend visible by rotary shadowing electron microscopy.[47–49] There is also no evidence that ICAM-3 forms dimers (unpublished observations). ICAM-3 is strongly expressed on non-activated leukocytes and expression decreases on activation of lymphocytes. Therefore, ICAM-3 is the major LFA-1 ligand on resting lymphocytes. ICAM-3 is not expressed on non-haematopoietic cells except some forms of cancer.[50] The combination of anti-ICAM-1, ICAM-2, and ICAM-3 is as effective as anti-LFA-1 in *in vitro* lymphocyte activation.[19] Therefore, with ICAM-3, it was initially thought that all LFA-1 ligand activities had been defined.

Site-directed mutagenesis on the first domain of ICAM-3 identified a similar LFA-1-binding site as was defined in ICAM-1 and ICAM-2.[49,51] A critical glutamic acid residue surrounded by important hydrophobic and charged residues was identified. ICAM-3 is heavily glycosylated, including glycans on domain 1. The ICAM-3-binding site on the I domain of LFA-1 appears to be distinct from the binding site for ICAM-1.[52]

Antibodies

ICR3.3 is available from Ancell (Bayport, ME) and CBRIC3.1[19] is available from Research Diagnostics (Flanders, NJ).

Genes

ICAM-3 cDNAs have been cloned in human[47,48] (S50015, X69711, S49855, X69819) and cow [53] (L41844).

Newest ICAMs: ICAM-4 (LW blood group antigen) and telencephalin

Protein properties

Recently, the sequencing of proteins related to ICAMs has revealed two new LFA-1 ligands that were subsequently confirmed to bind LFA-1 in functional assays. ICAM-4 (LW) is a 60 000 Da glycoprotein that carries the LW blood group antigen.[54] ICAM-4 has two Ig-domains, a single transmembrane domain, and a short cytoplasmic tail and in this respect is most similar to ICAM-2. The critical residues for binding to LFA-1 are conserved and this binding has recently been demonstrated.[55] The physiological significance of an ICAM on erythrocytes is not known although the expression of ICAM-4 on highly proliferative haematopoietic precursors may provide these cells with an additional handle for immune surveillance to detect carcinogenic events.

Telencephalin is a 130 000 Da glycoprotein consisting of nine Ig-like domains, a transmembrane domain, and a short cytoplasmic tail.[56] Telencephalin is expressed on the soma and dendritic projections, but not axons, of neurones in the central nervous system. The first five Ig-like domains are 50 per cent similar to ICAM-1 and ICAM-3. Soluble recombinant telencephalin and telencephalin expressed on L-cells mediated adhesion of LFA-1-expressing cells in a divalent cation, energy, and activation dependent manner that was blocked by anti-LFA-1 mAb.[57] Telencephalin is well positioned in the central nervous system to play a role in leukocyte/neurone interactions.

Antibodies

ICAM-4 and telencephalin are recognized by polyclonal antisera.

Genes

Human ICAM-4 cDNAs have been cloned and expressed[54,56] (X93093). Telencephalin cDNAs have been obtained for human[56] (AA421394), rabbit[56] (L13199) and mouse (MMU06483) and the human form has been expressed.[57]

References

1. Dustin, M. L., Rothlein, R., Bhan, A. K., Dinarello, C. A., and Springer, T. A. (1986). *J. Immunol.*, **137**, 245–54.
2. Staunton, D. E., Marlin, S. D., Stratowa, C., Dustin, M. L., and Springer, T. A. (1988). *Cell*, **52**, 925–33.
3. Carpén, O., Staunton, D. E., and Springer, T. A. (1992). *J. Cell. Biol.*, **118**, 1223–34.
4. Staunton, D. E., Dustin, M. L., Erickson, H. P., and Springer, T. A. (1990). *Cell*, 243–54.
5. Diamond, M. S., Staunton, D. E., Marlin, S. D., and Springer, T. A. (1991). *Cell*, **65**, 961–71.
6. Staunton, D. E., Merluzzi, V. J., Rothlein, R., Barton, R., Marlin, S. D., and Springer, T. A. (1989). *Cell*, **56**, 849–53.
7. Cunningham, B. C., Jhurani, P., Ng, P., and Wells, J. A. (1989). *Science*, **243**, 1330–6.
8. Berendt, A. R., Simmons, D. L., Tansey, J., Newbold, C. I., and Marsh, K. (1989). *Nature*, **341**, 57–9.
9. Berendt, A. R., McDowall, A., Craig, A. G., *et al.* (1992). *Cell*, **68**, 71–81.
10. Kirchhausen, T., Staunton, D. E., and Springer, T. A. (1995). *J. Leuk. Biol.* **53**, 342–6.
11. Miller, J. M., Knorr, R., Ferrone, M., Houdei, R., Carron, C., and Dustin, M. L. (1995). *J. Exp. Med.*, **182**, 1231–41.
12. Reilly, P. L., Woska, J. R., Jeanfavre, D. D., McNally, E., Rothlein, R., and Bormann, B. J. (1995). *J. Immunol.*, **155**, 529–32.
13. Dustin, M. L. and Springer, T. A. (1991). *Ann. Rev. Immunol.*, **9**, 27–66.
14. Dougherty, G. J., Murdoch, S., and Hogg, N. (1988). *Eur. J. Immunol.*, **18**, 35–9.
15. Clark, E. A., Ledbetter, J. A., Holly, R. C., Dinndorf, P. A., and Shu, G. (1986). *Hum. Immunol.*, **16**, 100–13.
16. Dustin, M. L. and Springer, T. A. (1988). *J. Cell. Biol.*, **107**, 321–31.
17. Pober, J. S., Gimbrone Jr., M. A., Lapierre, L. A., *et al.* (1986). *J. Immunol.*, **137**, 1893–6.
18. Dustin, M. L., Singer, K. H., Tuck, D. T., and Springer, T. A. (1988). *J. Exp. Med.*, **167**, 1323–40.
19. de Fougerolles, A. R., Qin, X., and Springer, T. A. (1994). *J. Exp. Med.*, **179**, 619–629.
20. Sligh, J. E., Ballantyne, C. M., Rich, S. S., *et al.* (1993). *Proc. Natl Acad. Sci. USA*, **90**, 8529–33.
21. Xu, H., Gonzalo, J. A., St. Pierre, Y., *et al.* (1994). *J. Exp. Med.*, **180**, 95–109.
22. Ballantyne, C. M., Sligh, J. E. Jr., Dai, X. Y., and Beaudet, A. L. (1992). *Genomics*, **14**, 1076–80.
23. King, P. D., Sandberg, E. T., Selvakumar, A., Fang, P., Beaudet, A. L., and Dupont, B. (1995). *J. Immunol.*, **154**, 6080–93.
24. Shaw, A. S. and Dustin, M. L. (1997). *Immunity*, **6**, 361–9.
25. Rosenstein, Y., Park, J. K., Hahn, F. S., Rosen, F. S., Beirer, B. E., and Burakoff, S. J. (1995). *Nature*, **354**, 233–5.
26. Languino, L. R., Duperray, A., Joganic, K. J., Fornaro, M., Thornton, G. B., and Altieri, D. C. (1995). *Proc. Natl Acad. Sci. USA*, **92**, 1505–9.
27. McCourt, P. A. G., Ek, B., Forsberg, N., and Gustafson, S. (1994). *J. Biol. Chem.*, **269**, 30081–4.
28. Marlin, S. D. and Springer, T. A. (1987). *Cell*, **51**, 813–9.
29. Makgoba, M. W., Sanders, M. E., Ginther Luce, G. E., *et al.* (1988). *Eur. J. Immunol.*, **18**, 637–40.
30. Dustin, M. L. and Springer, T. A. (1989). *Nature*, **341**, 619–24.
31. Takei, F. (1985). *J. Immunol.*, **134**, 1403–7.
32. Simmons, D., Makgoba, M. W., and Seed, B. (1988). *Nature*, **331**, 624–7.
33. Tomassini, J. E., Graham, D., DeWitt, C. M., Lineberger, D. W., Rodkey, J. A., and Colonno, R. J. (1989). *Proc. Natl Acad. Sci*, **86**, 4907–11.
34. Horley, K. J., Carpenito, C., Baker, B., and Takei, F. (1989). *EMBO J.*, **8**, 2889–96.
35. Siu, G., Hedrick, S. M., and Brian, A. A. (1989). *J. Immunol.*, **143**, 3813–20.
36. Ballantyne, C. M., O'Brien, W. E., and Beaudet, A. L. (1989). *Nucl. Acids Res.*, **17**, 5853.
37. Manning, A. M., Lu, H. F., Kukielka, G. L., *et al.* (1995). *Gene*, **156**, 291–5.
38. Kita, Y., Takashi, T., Iigo, Y., Tamatani, T., Miyasaka, M., and Horiuchi, T. (1992). *Biochim. Biophys. Acta*, **1131**, 108–10.
39. Voraberger, G., Schafer, R., and Stratowa, C. (1991). *J. Immunol.*, **147**, 2777–86.
40. Staunton, D. E., Dustin, M. L., and Springer, T. A. (1989). *Nature*, **339**, 61–4.
41. Helander, T. S., Carpén, O., Turunen, O., Kovanen, P. E., Vaheri, A., and Timonen, T. (1996). *Nature*, **382**, 256–68.
42. de Fougerolles, A. R., Stacker, S. A., Schwarting, R., and Springer, T. A. (1991). *J. Exp. Med.*, **174**, 253–67.
43. Butini, L., de Fougerolles, A. R., Vaccarezza, M., *et al.* (1997). *Eur. J. Immunol.*, **24**, 2191–5.
44. Casasnovas, J. M., Springer, T. A., Liu, J., Harrison, S. C., and Wang, J.(1997). *Nature*, **387**, 312–5.
45. de Fougerolles, A. R. Schwarting, R., and Springer, T. A. (1990). *Characterization of intercellular adhesion molecule 2 (ICAM-2) reveals the presence of a novel ligand for lymphocyte function antigen 1 (LFA-1)* (Unpublished.)
46. Xu, H., Tong, I. L., de Fougerolles, A. R., and Springer, T. A. (1992). *J. Immunol.*, **149**, 2650–5.
47. Fawcett, J., Holness, C. L., Needham, L. A., *et al.* (1992). *Nature*, **360**, 481–4.
48. Vazeux, R., Hoffman, P. A., Tomita, J. K., *et al.* (1992). *Nature*, **360**, 485–8.
49. Sadhu, C., Lipsky, B., Erickson, H. P., *et al.* (1994). *Cell Adh. Commun.*, **2**, 429–40.
50. Cordell, J. L., Pulford, K., Turley, H., *et al.* (1994). *J. Clin. Pathol.*, **47**, 143–7.

51. Holness, C. L., Bates, P. A., Little, A. J., *et al.* (1995). *J. Biol. Chem.*, **270**, 877–84.
52. van Kooyk, Y., Binnerts, M. E., Edwards, C. P., *et al.* (1996). *J. Exp. Med.*, **183**, 1247–52.
53. Lee, E. K., Kehrli, M. E. Jr., Dietz, A. B., Bosworth, B. T., and Reinhardt, T. A. (1996). *Gene*, **174**, 311–3.
54. Bailly, P., Hermand, P., Callebaut, I., *et al.* (1994). *Proc. Natl Acad. Sci. USA*, **91**, 5306–10.
55. Bailly, P., Tontti, E., Hermand, P., Cartron, J. P., and Gahmberg, C. G. (1995). *Eur. J. Immunol.*, **25**, 3316–20.
56. Yoshihari, Y., Oka, S., Nemoto, Y., *et al.* (1994). *Neuron*, **12**, 541–53.
57. Tian, L., Yoshihara, Y., Mizuno, T., Mori, K., and Gahmberg, C. G. (1997). *J. Immunol.*, **158**, 928–36.
58. Walter, M. J., Look, D. C., Tidwell, R. M., Roswit, W. T., and Holtzman, M. J. (1997). *J. Biol. Chem.*, **272**, 28582–9.
59. Aoudjit, F., Brochu, N., Belanger, B., Stratowa, C., Hiscott, J., and Audette, M. (1997). *Cell Growth Differ.*, **8**, 335–42.
60. Casassnovas, J. M., Stehle, T., Liu, J. H., Wang, J. H., and Springer, T. A. (1998). *Proc. Natl Acad. Sci. USA*, **95**, 4134–9.
61. Bella, J., Kolatkar, P. R., Marlor, C. W., Greve, J. M. and Rossmann, M. G. (1998). *Proc. Natl Acad. Sci. USA*, **95**, 4140–5.

Michael L. Dustin
Department of Pathology,
Washington University School of Medicine,
St Louis, MO 63110, USA

Timothy A. Springer
Center for Blood Research and
Department of Pathology,
Harvard Medical School,
Boston, MA 02115, USA.

Glypicans

The glypican-related integral membrane proteoglycans (GRIPs)[1] or glypicans compose a family of heparan sulphate proteoglycans that are linked to the cell surface via glycosyl phosphatidylinositol (GPI). Their core proteins have similar general structures. In vertebrates the glypican family comprises at least five distinct members. Each member shows a specific expression pattern. The glypicans are thought to function as coreceptors for growth factors and adhesion molecules and to play a role in the control of cell division and patterning during development.[2,3]

Synonomous/homologous proteins

All glypicans share a conserved amino acid sequence motif that includes a characteristic pattern of 14 cysteine residues. Additional similarities include the overall sizes of the proteins (approximately 60 kDa), the presence of N-terminal and C-terminal signal peptide-like sequences (the first predicted to be involved in the membrane translocation of the nascent polypeptides, the second in the temporary membrane anchorage and subsequent glypiation of the proteins), and the presence of glycosaminoglycan attachment consensus sequences close to the C termini of the proteins. All these features together represent the hallmarks of a glypican.

Protein properties

Glypican-1

Glypican-1, the first member of this family to be identified as a heparan sulphate proteoglycan, was originally purified from cultured human lung fibroblasts and the corresponding cDNA was initially cloned from the same source.[4] Many cultured cells, of epithelial, endothelial, mesothelial, and mesenchymal origin do, however, express glypican-1.[5] The glypican-1 core protein is encoded as a polypeptide of 558 amino acids with a predicted M_r of 61.6 kDa. The mature glypican-1 core protein carries *N*-linked oligosaccharide and up to three heparan sulphate chains. The proteoglycan is polydisperse in size and migrates as a broad smear with an apparent M_r of approximately 200 kDa in SDS–PAGE. After treatment with heparitinase, the glypican-1 core protein migrates as a discrete band, with an apparent molecular mass of

Table 1 Summary of the glypican nomenclature

Family name	Trivial name	Human gene (location)
Glypican-1	Glypican	*GPC1* (2q35–37)
Glypican-2	Cerebroglycan	*GPC2*
Glypican-3	OCI-5	*GPC3* (Xq26)
Glypican-4	K-glypican	*GPC4* (Xq26)
Glypican-5	–	*GPC5* (13q32)

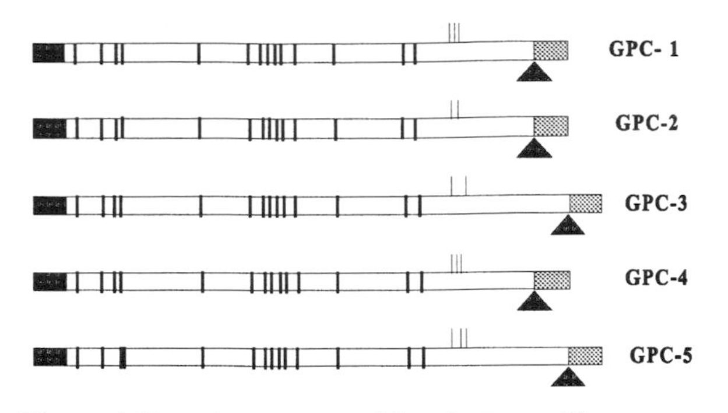

Figure 1. Domain structure of the glypicans. The N-terminal signal peptides are indicated in black, the C-terminal signal peptides for GPI-anchorage in grey. Presumptive GPI-anchoring sites are indicated by black triangles. The positions of potential glycosaminoglycan attachment sites are indicated by the lines above the bars. Cysteine residues are indicated by the lines in the bars.

64–68 kDa. Glypican-1 is rapidly and quantitatively shed to the culture medium, by confluent fibroblasts and by several other cultured cells. The shed and membrane associated forms have similar N termini but the core protein of the form shed by fibroblasts is distinctly smaller (3 kDa) than the core protein of the membrane-associated form and lacks the hydrophobic properties of the latter.[4] Glypican-1 from Schwann cells, in contrast, appears to be shed in a more intact form (presumably through cleavage of the GPI anchor by a GPI-specific phospholipase).[6]

The glypican-1 mRNA is expressed in several tissues: specific regions of the fetal and adult brain,[7] fetal and adult lung,[4] adult and fetal heart,[8] striated muscle,[6] vascular endothelium,[11] Schwann cells[6] and several cell lines.[5] *In situ* hybridization studies in rodents show that glypican-1 is highly expressed in the brain and skeletal system. Expression is also observed in skeletal and smooth muscle, epidermis, and in developing tubules and glomeruli of the kidney.[22] Immunocytochemical data show that in neonatal peripheral nerve, brain, heart and striated muscle, the glypican-1 protein is present at regions of the cell surface that are in contact with the basal membrane.[6] Basolateral cell surface expressions have also been observed in several polarised epithelia, which seems at odds with the rule that GPI-anchored proteins are exclusively expressed at the apical side of polarized cells. Mutagenesis studies, however, have indicated that the apical sorting of glypican-1 is inversely correlated with its heparan sulphate substitution.[10] Although the glypican-1 protein is normally present at the cell surface, a significant portion of glypican immunoreactivity can be found in cell nuclei of the rat central nervous tissue and in C6 glioma cells. Glypican-1 contains a functional nuclear localization signal and the levels of glypican-1 in the nucleus change during the cell cycle, suggesting a potential role in the process of cell division.[23]

Glypican-2

Glypican-2, originally called cerebroglycan, was cloned from a neonatal rat brain cDNA library. It has a predicted M_r of 58.6 kDa and contains five potential heparan sulphate attachment sites. Glypican-2 is only expressed in the developing nervous system. *In situ* hybridization experiments suggest that the glypican-2 mRNA is transiently expressed by immature neurones. The mRNA appears around the time of final mitosis and disappears after cell migration and axon outgrowth have been completed. Glypican-2, therefore, may have a function related to the motile behaviour of developing neurones.[16] The glypican-2 protein is predominantly expressed on axon tracts throughout the developing brain and spinal cord. The protein can also be found on neuronal growth cones where it may regulate the growth or guidance of axons.[24]

Glypican-3

Glypican-3, originally called OCI-5, was isolated as a developmentally regulated transcript from rat intestine.[13] Subsequent experiments identified it as a GPI-anchored heparan sulphate proteoglycan of the glypican family, with a core protein M_r of 69 kDa.[14] The glypican-3 transcript is expressed in various fetal tisues (kidney, liver and lung), but it is down-regulated in most adult tissues, except in the lung.[14,15] Mutations in the glypican-3 gene in humans cause the Simpson–Golabi–Behmel syndrome, an X-linked condition characterized by pre- and postnatal overgrowth with visceral and skeletal abnormalities.[15] This implies a possible role for glypican-3 in the control of cell division and patterning during embryonic development. It has been suggested that glypican-3 may form a complex with insulin-like growth factor-2 (IGF2) and modulate the action of this growth factor.[15] The formation of a complex between glypican-3 and IGF2 has not be shown by others, but an interaction betwen bFGF and the heparan sulfate chains of glypican-3 has been shown.[25] Glypican-3 also interacts with tissue factor pathway inhibitor and this interaction appears to be independent of the heparan sulfate chains.[26] Glypican-3 induces apoptosis in a cell-line specific manner. This apoptosis-inducing activity requires the anchoring to the cell membrane but not the glycosaminoglycan chains. Interestingly, glypican-3-transfected cells can be rescued from cell death by IGF2.[27]

Glypican-4

Glypican-4, originally called K-glypican, has a predicted molecular mass of 57.5 kDa. The glypican-4 mRNA is highly expressed in the developing mouse kidney and brain, moderately in liver and lung, and at low levels in adult brain and spleen. As revealed by *in situ* hybridization, the major sites of expression in the developing embryo are the tubular epithelial cells in the kidney and proliferating neuro-epithelial cells in the brain. Other tissues that express glypican-4 mRNA during embryonic development are the developing adrenal gland and the smooth muscle cells of large blood vessels.[17]

Glypican-5

Glypican-5 is the most recently discovered member of the glypican family. Its protein core has an apparent M_r of 65 kDa. Glypican-5 is more homologous to glypican-3 (40 per cent) than to the other glypicans (homology of only 20 per cent). It is expressed in fetal brain, lung, and liver. In adult tissues the mRNA appears to be found only in the brain.[18] *In situ* hybridizations show that in murine embryos the glypican-5 transcript is expressed in restricted regions of the developing central nervous system, limb and kidney in a pattern consistent with a role in the control of cell growth or differentiation.[28]

A *Drosophila* glypican, *dally*, appears to be involved in the signalling activities of *Decapentaplegic* (*Dpp*), a *Drosophila* member of the transforming growth factor-β (TGF-β)/bone morphogenetic protein (BMP) superfamily of growth factors. Ectopic expression of *dally* can alter the patterning activity of *Dpp* suggesting a role in the modulation of the *Dpp* signal.[29]

Purification

The shed form of glypican can be purified from conditioned culture media by CsCl density gradient ultracentrifugation, ion exchange chromatography and gel filtration, or by immunoaffinity chromatography using anti-glypican antibodies.[4] Membrane-associated glypicans have been purified from non-ionic detergent extracts of the cells by ion exchange chromatography,[7,16–18] intercalation into liposomes, and immunoaffinity chromatography.[4]

Activities

The cell surface proteoglycan fraction from human umbilical vein endothelial cells that binds with high affinity to antithrombin III is specifically enriched in glypican-1, implying a potential role for glypican-1 in the anticoagulant properties of the vascular wall.[11] Glypican-1 also appears to interact with thrombin.[12] As purified from rat PC-12 cells, glypican-1 binds to the amyloid precursor protein and inhibits APP-induced neurite outgrowth,[19] an intriguing observation since glypican-1 is also found in neuritic type plaques from Alzheimer's disease patients. Glypican-1 potentially plays a role as coreceptor in fibroblast growth factor signalling, by stimulating the binding of bFGF to FGFR-1 and subsequent signal transduction.[9] As isolated from human lung fibroblasts, glypican-1 binds also with high affinity to type I collagen fibres and to fibronectin. Glypican-1 also binds to laminin *in vitro*,[6] suggesting a potential general role in cell–extracellular matrix interactions.

All the above interactions depend on the heparan sulphate moieties of the glypicans. The glypican-3 core protein, however, forms a complex with insulin-like growth factor 2 and might therefore modulate IGF2 action.[15]

No specific activities have been assigned to glypican-2, glypican-4 and glypican-5.

Glypican-3, like glypican-1, is able to interact with bFGF through its heparan sulfate chains.[25] The core protein of glypican-3 is able to interact with tissue factor pathway inhibitor, independently of its heparan sulfate chains.[26] Glypican-3 is able to induce apoptosis in certain cell lines and this requires the anchoring of the protein to the cell surface, but not the heparan sulfate chains.[27]

Antibodies

Two monoclonal antibodies have been isolated that react with the core protein of human glypican-1.[4] Both can be used for immunoblotting, immunoprecipitation, and immunohistochemistry. Monoclonal antibody S1 reacts only with the unreduced core protein. Monoclonal antibody F81–1G11 reacts with both the reduced and unreduced forms of the core protein.

Monoclonal antibodies to rat glypican-2 have been described.[24]

Polyclonal antibodies against rat glypican-1,[7] rat glypican-2,[16] and human glypican-3[15] have been described.

Genes and chromosomal mapping

A human glypican-1 cDNA[4] (mRNA of 3.8 kb, GenBank accession number X54232), a rat glypican-1 cDNA[7] (mRNA of 3.7 kb, GenBank L34067) and an avian glypican-1 cDNA[20] (mRNA of 5.0 kb, GenBank L29089) have been cloned. In general, there is a high degree of conservation between the human and rodent forms of the glypicans (approximately 90 per cent at the amino acid level).

cDNA sequences for a rat glypican-2 cDNA[16] (mRNA of 2.7 kb, GenBank L20468), a human glypican-3 cDNA[15] (mRNA of 2.5 kb, GenBank Z37987), a rat glypican-3 cDNA[13] (mRNA of 2.6 kb, GenBank M22400), a mouse glypican-4 cDNA[17] (mRNA of 3.4 kb, GenBank X83577), and a human glypican-5 cDNA[18] (mRNA of 3 kb, GenBank U66033) are also available.

A *Drosophila* glypican, *dally*, has been identified[3] (mRNA of 4.4 kb, GenBank U31985).

The human *GPC1* gene has been mapped to chromosome 2q35–37,[21] the human *GPC3* gene to Xq26,[15] and the human *GPC5* gene to 13q23.[18]

The rat glypican-1 gene, consisting of eight exons, and its promoter have been cloned.[30] The complete glypican-3 gene has also been cloned. It consists of eight exons and spans more than 500 kb.[31] The complete glypican-4 gene has been cloned. The *GPC4* gene flanks *GPC3* on chromosome Xq26 with both genes oriented in a tandem array. The *GPC4* gene consists of eight exons.[32]

Mutant phenotype/disease states

The *dally (division abnormally delayed*) locus was identified by a genetic screening for *Drosophila* mutants affected in cell division patterning in the developing central nervous system. The *dally* mutations affect viability and produce morphological defects in several adult tissues, including the eyes, antennae, wings and genitalia. The *dally* cDNA reveals an open reading frame of 626 amino acids which shows significant sequence homology to the vertebrate glypicans.[3]

Mutations in *GPC3* have been shown to cause the Simpson–Golabi–Behmel syndrome.[15] This condition is characterized by pre- and postnatal overgrowth.

Additional clinical findings include congenital heart defects, hypoplastic kidneys, supernumerary nipples, cleft palate, hernias, dysplastic kidneys, polydactyly and syndactyly.

The patients are at high risk for the development of embryonal tumours, mostly Wilms' tumour and neuroblastoma. The mortality rate is approximately 50 per cent in affected males.

One or more Simpson–Golabi–Behmel syndrome families have double deletions of both *GPC3* and *GPC4* from the glypican gene cluster on chromosome Xq26. These double deletions might in part explain the phenotypic variability observed in this syndrome.[32]

Structure

Not available.

References

1. David, G. (1993). *FASEB J.* **7**, 1023–30.
2. Weksberg, R., Squire, J. A., and Templeton, D. M. (1996). *Nature Genet.*, **12**, 225–7.
3. Nakato, H., Futch, T. A., and Selleck, S. B. (1995). *Development*, **121**, 3687–702.
4. David, G., Lories, V., Decock, B., Marynen, P., Cassiman, J. J., and van den Berghe, H. (1990). *J. Cell,. Biol.*, **111**, 3165–76.
5. Lories, V., Cassiman, J. J., Van den Berghe, H. David, G. (1992). *J. Biol. Chem.*, **267**, 1116–22.
6. Carey, D. J., Stahl, R. C., Asundi, V. K., and Tucker, B. (1993). *Exp. Cell Res.*, **208**, 10–18.
7. Litwack, D. E., Stipp, C. S., Kumbasar, A., Lander, A. D. (1994). *J. Neurosci.*, **14**, 3713–24.
8. Asundi, V. K., Keister, B. F., Stahl, R. C., and Carey, D. J. (1997). *Exp. Cell Res.*, **230**, 145–53.
9. Steinfeld, R., Van Den Berghe, H., and David, G. (1996). *J. Cell. Biol.*, **133**, 405–16.
10. Mertens, G., Van der Schueren, B., van den Berghe, H., and David, G. (1996). *J. Cell. Biol.*, **132**, 487–97.
11. Mertens, G., Cassiman, J. -J., Van den Berghe, H., Vermylen, J., and David, G. (1992). *J. Biol. Chem.*, **267**, 20435–43.
12. Bar-Shavit, R., Maoz, M., Ginzburg, Y., Vlodavsky, I. (1996). *J. Cell. Biochem.* **61**, 278–91.
13. Filmus, J., Church, J. G., and Buick, R. N. (1988). *Mol. Cell. Biol.*, **8**, 4243–9.
14. Filmus, J., Shi, W., Wong, Z. M., and Wong, M. J. (1995). *Biochem. J.*, **311**, 561–5.
15. Pilia, G., Hughes-Benzie, R. M., MacKenzie, A., Babayan, P., Chen, E. Y., Huber, R., Neri, G., Cao, A., Forabosco, A., and Schlessinger, D. (1996). *Nature Genet.*, **12**, 241–7.
16. Stipp, C. S., Litwack, E. D., and Lander, A. D. (1994). *J. Cell. Biol.*, **124**, 149–60.
17. Watanabe, K., Yamada, H., and Yamaguchi, Y. (1995). *J. Cell. Biol.*, **130**: 1207–18.
18. Veugelers, M., Vermeesch, J., Reekmans, G., Steinfeld, R., Marynen, P., David,G. (1997). *Genomics*, **40**, 24–30.
19. Williamson, T. G., Mok, S. S., Henry, A., Cappai, R., Lander, A. D., Nurcombe, V., Beyreuther, K., Masters, C. L., and Small, D. H. (1996). *J. Biol. Chem.*, **271**, 31215–21.
20. Niu, S., Antin, P. B., Akimoto K., Morkin, E. (1996). *Dev. Dyn.* **207**, 25–34.
21. Vermeesch, J. R., Mertens, G., David, G., and Marynen, P. (1995). *Genomics*, **25**, 327–9.
22. Litwack, E. D., Ivins, J. K., Kumbasar, A., Paine-Saunders, S., Stipp, C. S., and Lander, A. D. (1998). *Dev. Dyn.*, **211**, 72–87.
23. Liang, Y., Haring, M., Roughley, P. J., Margolis, R. K., and Margolis, R. U. (1997). *J. Cell Biol.*, **139**, 851–64.
24. Ivins, J. K., Litwack, E. D., Kumbasar, A., Stipp, C. S., and Lander, A. D. (1997). *Dev. Biol.*, **184**, 320–32.
25. Song, H. H., Shi, W., and Filmus, J. (1997). *J. Biol. Chem.*, **272**, 7574–7.
26. Mast, A. E., Higuchi, D. A., Huang, Z. F., Warshawsky, I., Schwartz, A. L., and Broze, G. J., Jr (1997). *Biochem J.*, **327**, 577–83.
27. Gonzalez, A. D., Kaya, M., Shi, W., Song, H., Testa, J. R., Penn, L. Z., and Filmus, J. (1998). *J. Cell Biol.*, **141**, 1407–14.
28. Saunders, S., Paine-Saunders, S., and Lander, A. D. (1997). *Dev. Biol.*, **190**, 78–93.
29. Jackson, S. M., Nakato, H., Sugiura, M., Jannuzi, A., Oakes, R., Kaluza, V., Golden, C., and Selleck, S. B. (1997). *Development*, **124**, 4113–20.
30. Asundi, V. K., Keister, B. F., and Carey, D. J. (1998). *Gene*, **206**, 255–61.
31. Huber, R., Crisponi, L., Mazzarella, R., Chen, C. N., Su, Y., Shizuya, H., Chen, E. Y., Cao, A., and Pilia, G. (1997). *Genomics*, **45**, 48–58.
32. Veugelers, M., Vermeesch, J., Watanabe, K., Yamaguchi, Y., Marynen, P., and David, G. (1998). *Genomics*, (In press).

Mark Veugelers and Guido David
Laboratory for Glycobiology,
and Developmental Genetics,
Center for Human Genetics,
University of Leuven and Flanders
Interuniversity Institute for Biotechnology,
Campus Gasthuisberg O and N,
Herestraat 49, 3000 Leuven,
Belgium

Limbic system-associated membrane protein

The limbic system-associated membrane protein (LAMP)[1] is a 64–68 kDa, neurone-specific glycoprotein expressed on the surface of somata and proximal dendrites of mature neurones and axons and growth cones of developing neurones in cortical and subcortical regions of the limbic system in the brain. LAMP is a member of the Ig superfamily of cell adhesion molecules and functions to mediate outgrowth and targeting of neurones in limbic areas of the developing central nervous system (CNS).

Synonymous names

None.

Protein properties

The limbic system associated membrane protein (LAMP)[1] is a 64–68 kDa glycoprotein that is expressed by a subset of neurones in the central nervous system of all vertebrate species screened thus far (bird, fish, amphibian, rodents, rabbit, cat, New and Old World monkey, human). LAMP is specifically expressed in the CNS, with no evidence of protein expression by peripheral neurones or cells in other organ systems.[1,2] LAMP is not expressed by non-neuronal cells of the nervous system, including astrocytes, oligodendrocytes, microglia, and Schwann cells. LAMP is anchored to the membrane via a glycosylphosphatidylinositol linkage. The *Lamp* cDNA encodes a 338 amino acid polypeptide with six putative *N*-linked glycosylation sites and three immunoglobulin domains (Fig. 1) with homology to the cell adhesion molecules of the immunoglobulin superfamily.[3] Carbohydrate linkages do not contain the HNK-1 epitope nor any sialic acid. The human homologue was cloned by RT-PCR and its amino acid sequence shows 99 per cent identity to the rat LAMP with only four amino acids substitutions, two of which are conservative, indicating a very strong phylogenetic conservation of protein structure and associated function.[4] The expression of *Lamp* transcripts from early stages of development[5] through adulthood[6] shows close correlation with the distribution pattern of the protein in the developing[2,7] and adult rat limbic structures.[1] LAMP is initially expressed on somata, dendrites, and axons, including growth cones. During the period of synapse formation, LAMP expression is eliminated on axons, expressed only postsynaptically on somata and dendrites (Fig. 2). In each brain nucleus or cortical area that expresses LAMP, both projection and interneurones express the protein (Fig. 3). LAMP is an early marker of limbic cortical regions of the CNS.[8]

Functional and biochemical studies indicate that LAMP, through homophilic interactions, selectively promotes neurite outgrowth of LAMP+ neurones comprising limbic pathways.[3,9] In addition, antibody perturbation studies investigating hippocampal connections *in vitro* and *in vivo* show that LAMP is required for normal targeting of limbic axons during development to attain proper circuit organization.[3,10]

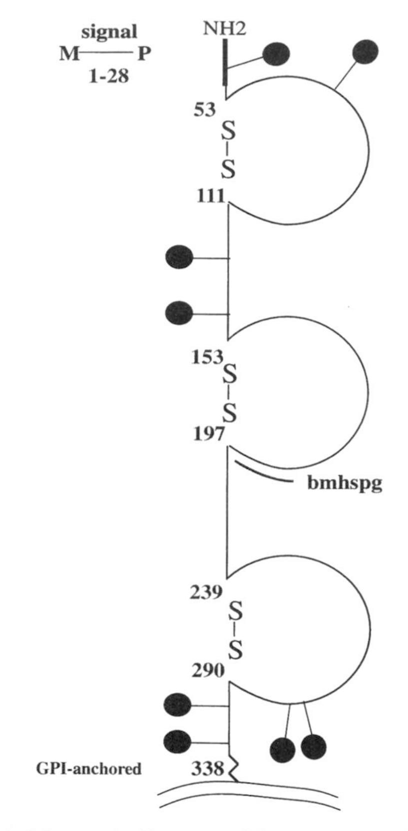

Figure 1. Schematic diagram of the structure of LAMP. Lollipop structures depict eight putative *N*-glycosylation sites. Disulphide linkages are shown to illustrate the location of the three Ig loops. The signal peptide occupies the first 28 amino acids. A region in the second Ig loop contains a sequence highly homologous to a region of the basement membrane heparan sulphate proteoglycan (bmhspg).

Purification

Native LAMP can be affinity purified using a mouse monoclonal antibody. Starting material from mammalian brain or substructures, such as the hippocampus, has been used. Standard membrane preparations should be

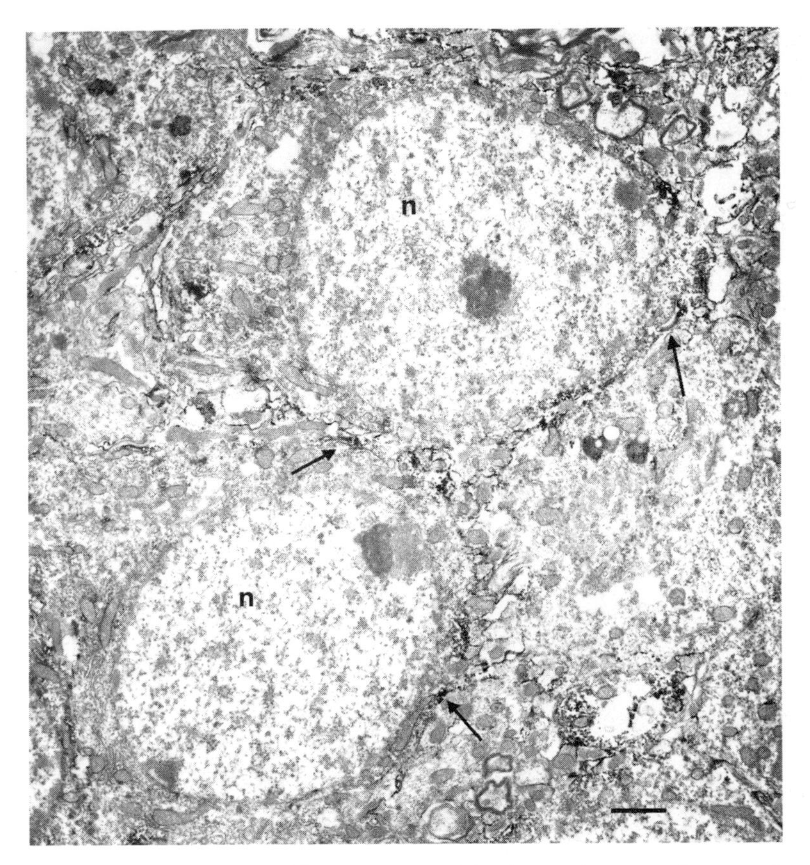

Figure 2. Electron micrograph illustrating LAMP immunoreactivity (arrows) on membranes of neuronal cell bodies (n) in the amygdala, a limbic region of the CNS. Scale bar represents 1 μm.

used, with a strategy that begins with solubilization of the membrane proteins in 2–4 per cent CHAPS.[2] The monoclonal antibody is covalently coupled to protein A–Sepharose (Pharmacia) using 20 mM dimethyl pimelimidate in 0.2 M triethanolamine. The CHAPS-solubilized material is passed over the affinity column, washed in high salt then low salt buffers, followed by a wash in 0.5 per cent deoxycholate and elution of LAMP in a high pH buffer (50 mM diethylamine, 0.5 per cent sodium deoxycholate, pH 11.5). Fractions are immediately neutralized with 0.5 M NaH_2PO_4.

Functions

Affinity purified LAMP, when attached to a tissue culture substrate, enhances neurite outgrowth specifically of $LAMP^+$ neurones,[8] as measured by determining neurite lengths. $LAMP^+$ neurones grown on a substrate of CHO cells transfected with the full length *lamp* cDNA also exhibit enhanced neurite outgrowth.[3] Both effects are blocked by the anti-LAMP monoclonal antibody at concentrations of 1–5 μg/ml. Flow cytometry studies show that LAMP can bind homophilically; fluorescent biobeads coated with purified native LAMP will self-aggregate.[9] In addition, the LAMP-coated beads bind specifically to *lamp*-transfected CHO cells.[3] The adhesion activity is Ca^{2+} independent and blocked by anti-LAMP monoclonal antibodies. LAMP is induced in progenitor cells in the CNS by activation of the EGF receptor in combination with collagen extracellular matrix.[11]

Antibodies

A mouse IgG_{2a} monoclonal antibody, isolated in a fusion in which hippocampal membranes were used as the immunogen,[1] specifically cross-reacts with LAMP native and recombinant proteins. LAMP does not cross-react with the two most closely related Ig superfamily members, neurotrimmin and OBCAM (unpublished data). A rabbit serum antibody cross-reacts with the native and recombinant LAMP protein. Cross-reactivity has not been checked against neurotrimmin or OBCAM. Both antibodies are suitable for affinity purification, immunoprecipitation, and light microscopic and ultrastructural immunocytochemistry. Fixation solutions that work best for preservation of tissue and antigenicity include 4 per cent paraformaldehyde, 0.1 per cent glutaraldehyde or 2 per cent paraformaldehyde, 3.75 per cent acrolein. Both solutions are prepared in physiological phosphate buffer, pH 7.2–7.4.

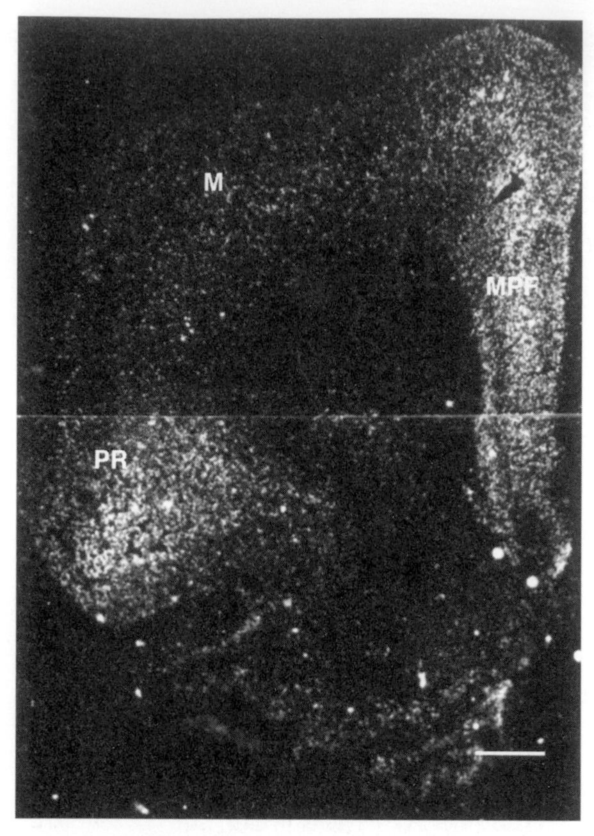

Figure 3. Dark-field photomicrograph showing the expression pattern of the *lamp* transcript by *in situ* hybridization. Heaviest labelling, represented by the white grains, is present in limbic regions of the rat frontal lobe, including medial prefrontal (MPF) and perirhinal (PR) cortex. Note the absence of labelling in the motor cortex (M). Scale bar represents 1 mm.

Genes

cDNA clones

The rat *lamp* cDNA was cloned[3] from a rat λgt11 hippocampal library using degenerate oligonucleotide probes generated from N-terminal peptide sequence (VRSVDFNRGTDNITVRQGDTA). The full length rat cDNA (GenBank U31554) encodes a 338 amino acid peptide with consensus sequences for three Ig loops, six *N*-glycosylation sites, and a GPI anchor. Northern blot analysis documents two transcripts of 8.0 and 1.6 kb. There is no conclusive evidence for alternative splicing. Southern blot analysis documents that LAMP is encoded by a single gene. The homologue of rat *lamp* expressed in the human CNS was cloned[4] by RT-PCR using human cerebral cortex poly(A)$^+$ RNA and a combination of oligo primers derived from the rat *lamp* nucleotide sequence. The nucleotide sequence (GenBank U41901) shows 94 per cent homology to rat and 99 per cent at the amino acid level (four substitutions).

Genomic clones

A mouse 129/Rej genomic library constructed in the λFIX II vector was screened with random-primed probes derived from the first 453 base pairs of the rat *Lamp* cDNA. Isolated genomic clones were restriction mapped and clone m*Lamp*-λ-1c isolated for chromosomal location analysis. The clone contains the partial sequence of the *Lamp* mRNA including the 5′ UTR leader sequence, the complete first exon including the translation initiator (first methionine codon), the coding region for the signal peptide and 24 amino acids of the N terminus, 3.0 kb of 5′ upstream sequences containing a putative TATA box and several consensus sequences for interaction with transcription factors, and approximately 6.0 kb of the first intron. Chromosomal analysis using FISH[12] documents location of the rat *lamp* gene on chromosome 16, in a region corresponding to band 16B5. The human *lamp* gene is located on chromosome 3 region q13.2–q21.

Mutant phenotype/disease states

No information available.

Structure

Not available.

Web site

http://bigmouth.neurobio.pitt.edu/levitt.htm

References

1. Levitt, P. (1984). *Science*, **223**, 299–301.
2. Zacco, A., Cooper, V., Chantler, P. D., Fisher-Hyland, S., Horton, H. L. and Levitt, P. (1990). *J. Neurosci.*, **10**, 73–90.
3. Pimenta, A. F., Zhukareva, V., Barbe, M. F., Reinoso, B. S., Grimley, C., Henzel, W., *et al.* (1995). *Neuron*, **15**, 287–97.
4. Pimenta, A. F., Fischer, I., and Levitt, P. (1996). *Gene*, **170**, 189–95.
5. Pimenta, A. F., Reinoso, B. S., and Levitt, P. (1996). *J. Comp. Neurol.* **375**, 289–302.
6. Reinoso, B. S., Pimenta, A. F., and Levitt, P. (1996). *J. Comp. Neurol.* **375**, 274–88.
7. Horton, H. L. and Levitt, P. (1988). *J. Neurosci.*, **8**, 4653–61.
8. Barbe, M. F. and Levitt, P. (1991). *J. Neurosci.*, **11**, 519–33.
9. Zhukareva, V. and Levitt, P. (1995). *Development*, **121**, 1161–72.
10. Keller, F., Rimvall, K., Barbe, M. F., and Levitt, P. (1989). *Neuron*, **3**, 551–61.
11. Ferri, R. T. and Levitt, P. (1995). *Development*, **121**, 1151–60.
12. Pimenta, A. F., Tsui, L. C, Heng, H. H. Q., and Levitt, P. (1998). *Genomics*, **49**, 472–4.

Pat Levitt and Aurea Pimenta
Department of Neurobiology,
University of Pittsburgh School of Medicine,
Pittsburgh, PA 15261, USA

Leucine-rich repeat family

A family of proteins containing leucine-rich repeats (LRRs) share a consensus motif consisting of an ordered spacing of leucine or other alphiphatic residues. The individual proteins represent a diverse group of molecules with varying functions but most often involved in protein–protein interaction. The consensus LRR motif is 24 residues in length and varies in copy number from 1 to 30 among the individual members of the family. LRRs represent a basic motif utilized by many proteins to generate a structural architecture necessary to support function.

Protein properties

Leucine-rich repeats (LRRs) were first recognized as a 24 residue motif present as eight tandem copies in a human serum protein of unknown function, termed leucine-rich α_2-glycoprotein.[1] Since their initial identification, LRRs have been found in a growing number of functionally unrelated proteins with the most common unifying functional attribute of the family being an involvement in protein–protein interaction. Indeed, many of the members of the LRR family participate in signal transduction as membrane receptors or cytoplasmic signalling molecules.

The consensus motif for LRRs consists predominantly of leucines at positions 2, 5, 7, 12, 15, 20, and 23 within a 24 residue repeat (Fig. 1). Substitutions at these positions, when they occur, are most often represented by other aliphatic residues, such as isoleucine or alanine. Another common feature of most LRRs is an asparagine at position 10. The best characterized examples of LRRs are the 15 found within porcine ribonuclease inhibitor, a protein composed of only LRRs and whose crystal structure has been solved.[2] This structure reveals the LRR motif forming a short loop composed of an α helix on one side and a β sheet on the opposite side. The repeating nature of the LRR generates an unusual horseshoe-shaped structure. However, it should be noted that the LRRs of ribonuclease inhibitor are either 28 residues (A-type repeat) or 29 residues (B-type repeat) in length and, as such, do not completely fit the consensus motif originally identified within the leucine-rich α_2-glycoprotein. Nevertheless, the well-characterized α–β structural unit of ribonuclease inhibitor has led to speculation that the role of LRRs is to promote a nonglobular shape within proteins, generating increased surface area for protein–protein interaction.[3]

The LRR is a distinct motif from the 'leucine zipper' commonly found in DNA-binding transcription factors. Based on the crystal structure of ribonuclease inhibitor, several striking structural differences exists between the two motifs. First, leucine zippers contain a periodic spacing of leucine residues within a coiled-coil motif that aligns leucine residues along a hydrophobic interface that interacts with a similar coiled-coil structure from another region of the leucine zipper. The net effect of a leucine zipper is dimerization, generating a DNA-binding motif within the protein. In contrast, the leucines within ribonuclease inhibitor do not align to form a hydrophobic interface and many LRRs contain proline residues that tend to prevent the formation of any structure resembling the prototypic coiled coil of leucine zippers.

Activities

The functional diversity of proteins containing LRRs is exemplified by primary sequence comparisons using a sequence alignment programs such as BLAST.[4] This program is available through the Internet at http://www.ncbi.nlm.nih.gov and allows an updated

a)

```
    LHLSENLLYTFSLATLMPYT     36-55
RLTQLNLDRCELTKL**QVDGTLP     56-77
VLGTLDLSHNQLQSLP*LLGQTLP     78-100
ALTVLDVSFNRLTSLPLGALRGLG     101-124
ELQELYLKGNELKTLPPGLLTPTP     125-148
KLEKLSLANNNLTELPAGLLNGLE     149-172
NLDTLLLQENSLYTI              173-187
```

b)

```
RTRHLLLANNSLQSVPPGAFDHLP     36-59
```

c)

```
xLxxLxLxxNxLxxLxxxxLxxLx
1                      24
```

Figure 1. A family of proteins containing leucine-rich repeats (LRRs) share a consensus motif consisting of an ordered spacing of leucine or other alphiphatic residues.[5] Shown are (a) seven adjacent LRRs from the α subunit of the platelet glycoprotein Ib–IX–V receptor; (b) the single LRR present in the glycoprotein IX subunit of the same receptor; and (c) a consensus LRR motif. Amino acid numbering refers to residue number within the mature polypeptide (GenBank accession numbers M22403, and M80478). Conserved residues within different repeats are shaded. The importance of LRRs within the GP Ib–IX–V complex is illustrated by a number of mutations associated with the Bernard–Soulier syndrome (normal residues mutant in GP Ib, and GP IX are highlighted as black-boxed residues).

search of GenBank databases. Thus, a current listing of proteins containing LRRs can be obtained through the Internet using the LRR frameworks from one of the prototypic member of the family, such as leucine-rich α_2-glycoprotein[1] or members of the platelet glycoprotein Ib–IX–V complex (pp. 287–288, this volume). A search performed in June 1998 identified more than 100 proteins containing the LRR motif, emphasizing the expanding nature of the family (40 members were identified by Kobe and Deisenhofer in October 1994[5]). A listing of members of the LRR family also highlights the seemingly unrelated functions of these proteins.

Members of the LRR family perform very diverse functions throughout the cell. First, there are a number of *Drosophila* proteins containing LRRs (toll, toll-related proteins, slit, connectin, chaoptin, flightless-1, and 18-wheeler) critical for normal development of the fly. Other proteins in the family are involved in adhesive interactions with other proteins (biglycan, decorin, fibromodulin, and the platelet adhesive receptor complex, glycoprotein Ib–IX–V). In the case of the platelet glycoprotein Ib–IX–V complex, four different gene products each contain from 1 to 15 copies of a LRR (see platelet GP Ib–IX–V complex, pp. 287–288, this volume). Their importance to the normal structure/function of platelets is exemplified by autosomal recessive mutations within individual subunits, resulting in abormal platelet morphology and a bleeding tendency (the Bernard–Soulier syndrome).[6]

Genes

The gene structure coding for LRRs is quite variable with examples of each repeat within a single gene having a similar exon/intron junction (luteinizing hormone receptor gene[7]), as well as examples where the tandem copies of LRRs are encoded by a single exon (genes of the platelet glycoprotein Ib–IX–V receptor complex [8–11]). This overall genomic diversity suggests that the LRR family may have evolved through convergent schemes owing to structural requirements met through an ordered arrangement of leucine residues.

References

1. Takahashi, N., Takahashi, Y., and Putnam, F. W. (1985). *Proc. Natl Acad. Sci. USA*, **82**, 1906–10.
2. Kobe, B. and Deisenhofer, J. (1993). *Nature*, **366**, 751–6.
3. Kobe, B. and Deisenhofer, J. (1995). *Nature*, **374**, 183–6.
4. Altschul, S. F., Gish, W., Miller, W., Myers, E. W., and Lipman, D. J. (1990). *J. Mol. Biol.*, **215**, 403–10.
5. Kobe, B. and Deisenhofer, J. (1994). *Trends Biochem. Sci.* **19**, 415–21.
6. Ware, J., Russell, S. R., Marchese, P., Murata, M., Mazzucato, M., De Marco, L., and Ruggeri, Z. M. (1993). *J. Clin. Invest.*, **92**, 1213–20.
7. Koo, Y. B., Ji, I., Slaughter, R. G., and Ji, T. H. (1991). *Endocrinology*, **128**, 2297–308.
8. Wenger, R. H., Kieffer, N., Wicki, A. N., and Clemetson, K. J. (1988). *Biochem. Biophys. Res. Commun.*, **156**, 389–95.
9. Hickey, M. J. and Roth, G. J. (1993). *J. Biol. Chem.*, **268**, 3438–43.
10. Yagi, M., Edelhoff, S., Disteche, C. M., and Roth, G. J. (1994). *J. Biol. Chem.*, **269**, 17424–7.
11. Lanza, F., Morales, M., La Salle, C., Cazenave, J., Clemetson, K. J., Shimomura, T., and Phillips, D. R. (1993). *J. Biol. Chem.*, **268**, 20801–7.

Jerry Ware and Zaverio M. Ruggeri
Roon Research Center for Arteriosclerosis, and Thrombosis, Departments of Molecular, and Experimental Medicine, and of Vascular Biology, The Scripps Research Institute, La Jolla, California, USA

LFA-1

LFA-1 (lymphocyte function associated-1) is an adhesion receptor on leukocytes that is a member of the integrin family.[1] LFA-1 was originally identified by monoclonal antibodies that inhibit T lymphocyte mediated antigen-specific-killing.[2–4] LFA-1 participates in a wide variety of cell adhesion interactions of leukocytes by binding intercellular adhesion molecules[5,6] and provides an important model for regulation of adhesion molecule function through changes in receptor activity.[7]

Synonyms

CD11a/CD18, integrin $\alpha L\beta 2$.

Protein properties

The α subunit of LFA-1 (CD11a, integrin αL) is a type 1 transmembrane glycoprotein of 180 000 Da. The β

subunit of LFA-1 (CD18, integrin β2) is a type 1 transmembrane glycoprotein of 95 000 Da.[8] These subunits become non-covalently associated in the endoplasmic reticulum and only then are transported to the cell surface (Fig. 1).[9,10] LFA-1 shares β2 with three other leukocyte adhesion molecules.[9,11,12] LFA-1 is expressed on all mature leukocytes and committed haematopoietic progenitor cells, but is not expressed on the most primitive haematopoietic stem cells.[13] LFA-1 functions in leukocyte adhesion by binding to five known members of the intercellular adhesion molecule subfamily of the immunoglobulin superfamily in a divalent cation- and energy-dependent manner.[14,15]

LFA-1 is important for leukocyte interactions with endothelial cells, which is required for extravasation. The interactions of leukocytes with endothelial cells is a multi-step process requiring tethering or rolling adhesion, triggering through pertussis toxin sensitive signalling pathways, and firm adhesion.[6] The best established role of LFA-1 in this process is in the generation of firm adhesion following activation by a pertussis toxin-sensitive Gαi-linked receptor, such as chemokine receptors.[16] LFA-1/ICAM-1 interaction is also the most potent adhesion pathway for leukocyte extravasation *in vitro*.[17] Recent *in vivo* studies suggest that LFA-1/ICAM-1 interaction may also contribute to the initial rolling adhesion and earlier studies support the concept that LFA-1 synergizes with L-selectin in mediating initial attachment of naïve lymphocytes to specialized high endothelial venules of lymph nodes.[18,19] LFA-1 also plays an important role in mature lymphocyte functions including T-cell mediated killing and T cell-B-cell collaboration to produce antibodies.[20–22]

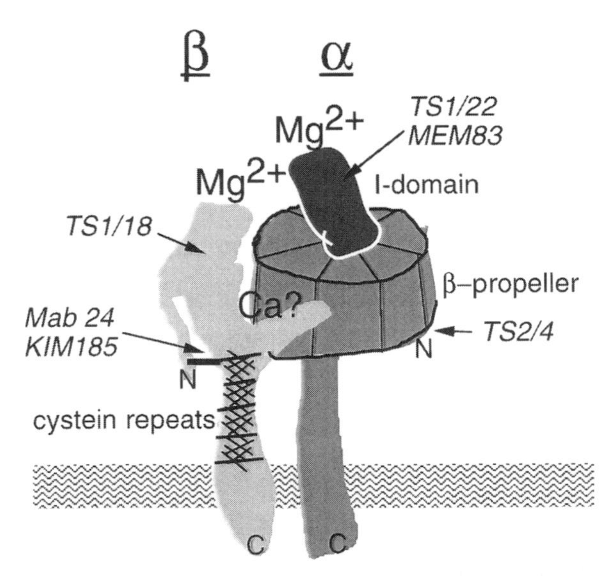

Figure 1. Schematic of LFA-1. The approximate locations of important monoclonal antibody and divalent cation binding are indicated.

LFA-1 functions in both adhesion and locomotion of lymphocytes.[5] This property may be important for LFA-1's function in extravasation. It also may play a important role in the ability of leukocytes to navigate inflamed tissues where ICAM-1 expression on cells is abundant. T lymphocyte antigen receptor engagement delivers a stop signal that suspends lymphocyte locomotion, thus enabling cell–cell communication over minutes to hours.[23]

LFA-1 function is tightly regulated on leukocytes. The regulation of LFA-1 function has been best characterized on lymphocytes, where LFA-1 is the major β2 integrin. LFA-1 is virtually inactive on resting lymphocytes as measured by adhesion to substrates coated with purified ICAM-1.[7] LFA-1 activity is rapidly stimulated by treatment of T and B lymphocytes with phorbol esters; LFA-1 activity is transiently increased by cross-linking antigen receptors on T lymphocytes.[7,24] This phenomenon is known as inside-out signalling based on the concept that cytoplasmic signals change adhesion outside the cell. Changes in LFA-1 activity do not require changes in surface expression. It is now well established that many integrins, including LFA-1, exist in two affinity states. The low affinity form of LFA-1 binds ICAM-1 with a k_d of ~100 μM, whereas the high affinity form has a k_d of ~130 nM.[25,26] Interestingly, the high affinity form of LFA-1 is not directly stimulated by T cell activation, but also requires ligand binding, leading to an induced fit model for LFA-1 affinity regulation.[27] Therefore, a critical parameter in LFA-1 regulation is access to ligand. One mechanism for controlling encounters between membrane proteins is to regulate lateral diffusion. LFA-1 on non-activated lymphocytes is laterally immobile and thus is unlikely to encounter ICAM-1 in cell–cell contacts.[28] This immobilization requires the actin cytoskeleton and is reversed by low concentrations of cytochalasin D (0.1–0.3 μg/ml, 15 min at 37°C). When lymphocytes are activated, LFA-1 lateral mobility increases 10-fold, leading to increased LFA-1/ICAM-1 encounters.[28] After ligand binding, LFA-1 is converted to the high affinity form and this form of LFA-1 may then provide an outside-in signal to organize the actin cytoskeleton to nucleate LFA-1 clusters.[29] Thus, the actin cytoskeleton appears to have a dual role in regulating LFA-1 function.

Purification

LFA-1 has been purified from human T and B cells.[30] While several mAb have been tried in different laboratories, to date only TS2/4 has been used successfully with elution at pH 11.5 in the presence of Mg^{2+}. This form of LFA-1 has a high affinity for ICAM-1 and appears to be a good model for high affinity LFA-1.[26,31] There are no published *in vitro* models for the low affinity form of LFA-1.

Activity

Purified LFA-1 can be adsorbed to plastic substrates by detergent dilution.[7] Cell adhesion or ICAM-1 dimers binding to the substrates can then be measured. Soluble ICAM-1 dimers are formed most easily by preparing chimeric ICAM-1 with the C2 and C3 domains of human IgG1 (ICAM-1 immunohesin).[32]

Antibodies

MAb to LFA-1 can be divided into several categories. Binding sites of representative antibodies are indicated in Fig. 1. The most abundant mAb, represented by TS1/22, recognize the I domain of the α subunit and block adhesion to all ligands.[33] TS1/22 recognizes the dissociated α subunit and the individually expressed I domain and can be used for Western blotting. TS1/22 is available from ATCC and from Endogen (Woburn, MA). One antibody to the I domain, MEM83, activates LFA-1 binding to ICAM-1, but blocks binding to ICAM-3.[34] MEM83 is available from Accurate Chemical and Scientific Co (Westbury, NY). Another useful antibody to LFA-1 is TS2/4 which recognizes the amino terminus in a conformation-dependent manner.[30] TS2/4 does not block function and can be used to purify LFA-1 with simultaneous conversion of the LFA-1 to the high affinity form. TS2/4 is available as a cell line from ATCC. Finally, NKI-L16 recognizes a Ca^{2+}-dependent conformation of LFA-1 and activates LFA-1-dependent adhesion through an unknown mechanism, possibly involving LFA-1 clustering .[35] The most abundant mAb to the β subunit recognize the highly conserved region in a conformation-dependent manner. A representative of this group is TS1/18.[36] TS1/18 is available from ATCC. A group of interesting β subunit mAb recognize the less conserved sequence between the highly conserved region and the cysteine-rich repeats. KIM185 and KIM127 are activating mAb that increase binding of LFA-1 to all ligands by decreasing the activation energy for conversion of $\beta 2$ integrins from the low to high affinity forms in response to ligands.[37] Both KIM185 and KIM127 can be used for Western blotting. MAb 24 recognizes the high affinity form of LFA-1 selectively in several different contexts.[38] Since the high affinity form of LFA-1 is physiologically generated through a ligand-induced conformational change, mAb 24 has also been characterized as recognizing a ligand induced binding site (LIBS).[27] A blocking antibody specific for mouse LFA-1 (M17/4) is available from ATCC or from Pharmingen (San Diego, CA).

cDNA and genes

GenBank accession numbers in parentheses. The cDNAs have been cloned for human LFA-1 αL[39] (Y00796), human $\beta 2$[1,40] (M15395, Y00057, M19545), mouse αL[41] (M60778), mouse $\beta 2$[42] (X14951), cow $\beta 2$[43] (M81233), pig $\beta 2$[44] (U13941), and chicken $\beta 2$[45] (X71786). The organization of the human $\beta 2$ gene has been determined[46] (X64071–83, X63924–6, X63835).

Mutant phenotypes/disease states

Leukocyte adhesion deficiency (LAD) is a genetic disease based on absence of the $\beta 2$ subunit of LFA-1 and the other members of the subfamily.[47] LAD is characterized by recurrent life-threatening bacterial infections and absence of pus formation. The primary defect is in leukocyte extravasation and migration. The LFA-1 and Mac-1 knockout mice both have subtle defects in leukocyte extravasation.[22,48] This has been taken to mean that LFA-1 and Mac-1 are redundant in normal leukocyte extravasation. The LFA-1 knockout mouse also had a profound defect in rejection of immunogenic tumours.[21,22]

Structure

LFA-1 and other integrins are large molecules that have only begun to yield to structural analysis. Low resolution electron microscopy-based structures of integrins show a globular domain with two stalks that attach the globular domain to the membrane, the entire structure is about 20 nm tall. The α subunit contains seven repeats interrupted by a 200 amino acid inserted domain (I domain). The I domain is inserted in the same position in the α subunits of the other leukocyte integrins and three other integrin α subunits, but is absent from most integrin α subunits. There is strong evidence that the major ICAM-1 binding site is in the I domain. Recently, the I domain has been crystallized and a number of high resolution structures of different forms have been described. The LFA-1 I domain assumes a dinucleotide binding fold with a novel divalent cation binding site at the end opposite the site of attachment to the integrin.[49] This site has been referred to as a MIDAS or metal ion dependent adhesion site.[50] It has been proposed that this bound cation (Mg^{2+}) is utilized directly in ligand binding. While no other part of LFA-1 has been crystallized, the seven repeats in the α subunit have been modelled. Contrary to previous models in which these repeats form independent domains, the repeats can effectively be modelled as forming a single structure known as a β propeller (Fig. 2).[51] The β propeller structure is also predicted to bind four cations. The folding of the β propeller region requires the β subunit. The structure of the β subunit is less understood. The amino terminal third of the $\beta 2$ subunit contains a highly conserved region that has been suggested to have similarity to an I domain.[50] Mutagenesis data make a strong case that this region of LFA-1 contains a MIDAS motif like the I domain.[52] However, unlike the I domain, which folds independently, the highly conserved region of the β subunit is absolutely dependent on the α subunit for folding, based on studies with dozens of conformation-sensitive monoclonal antibodies.[36] The membrane proximal segment of the β subunit is made up of four cysteine-rich repeats that are highly conserved among all integrin β subunits and are likely to form part of a rigid stalk connecting the β globular portion of the integrin to

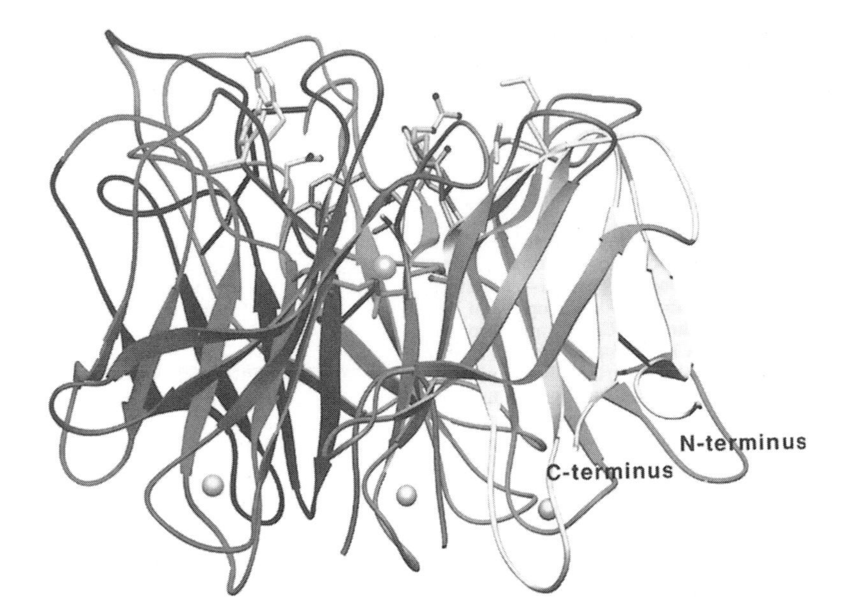

Figure 2. Model of the β-propeller domain of the integrin $\alpha 4$ subunit. The seven FG-GAP repeats (residues 1–452) are predicted to fold up into a β-propeller domain with seven anti-parallel β-sheets. The ribbon diagram shows each sheet in a different tint. Three Ca^{2+} ions shown as spheres are bound to loops on the bottom of the domain. Ligands are predicted to bind to the upper surface. From T. A. Springer (1997) *Proc. Natl Acad. Sci. USA*, **94**, 65–72.

the membrane. Many of the disulfide bonds in the α and β subunit have been determined for the integrin platelet glycoprotein IIbIIIa (αIIbβ3).[53,54] Many of the disulfides are conserved and are likely to form similarly in LFA-1.

The LFA-1 I domain has been characterized functionally in terms of ICAM-1 binding.[55] The I domain binds ICAM-1 with a k_d of ~150 μM. This is similar to the k_d for binding of ICAM-1 to the low affinity form of LFA-1.[25] These data suggest that the I domain interaction with ICAM-1 provides the low affinity interaction of intact LFA-1 with ICAM-1. The interaction of an isolated I domain with ICAM-1 mediates rolling adhesion in flow.[55] This indicates an off-rate of > 1 s^{-1} and strain resistance of the bond. The ability of the I domain to interact transiently with ICAM-1 may account for the role of ICAM-1 in regulation of leukocyte rolling velocities in mice.[18]

A protein that interacts with the cytoplasmic tail of β2 is implicated in regulation of LFA-1 avidity. The cytoplasmic tail of the β subunit of LFA-1 is critical for LFA-1 function; when the β2 tail is truncated, but a putative salt bridge between the α and β subunit cytoplasmic tails is left intact,[56] the molecule is inactive with respect to inside-out signalling.[57] Furthermore, the single chain transmembrane glycoprotein CD4 with the cytoplasmic domain of β2 acts as a dominant negative inhibitor of LFA-1 function in transfected T and B cells.[63] This suggests that there are limiting factors that interact with the β2 cytoplasmic domain to augment LFA-1 avidity. Recently, a protein referred to as cytohesin-1 or B2–1 and a related protein have been shown to bind the β2 cytoplasmic domain and to activate LFA-1 on a T cell line.[58,59] This protein binds the β2 cytoplasmic tail through a sequence with homology to yeast Sec7. This domain turns out to have binding activity for low molecular weight G proteins including ARF-1, a factor involved in protein secretion.[60] Cytohesin-1 also has a pleckstrin homology domain which has been shown to bind phosphatidylinositol 3,4,5-trisphosphate.[61] It is possible that this protein links activation of phosphatidylinositol 3-kinase, a critical signal for integrin activation,[62] to the ligand-induced association of the integrin with the actin cytoskeleton. The protein structure coordinates for the LFA-1 I domain with different cations bound in the MIDAS site are available: PDB accession codes 1ZOP, 1ZOO, 1ZON, and 1LFA.

References

1. Kishimoto, T. K., O'Connor, K., Lee, A., Roberts, T. M., and Springer, T. A. (1987). *Cell*, **48**, 681–90.
2. Maio, M., Tessitori, G., Pinto, A., Temponi, M., Colombatti, A., and Ferrone, S. (1989). *J. Immunol.*, **143**, 181–8.
3. Pierres, M., Goridis, C., and Golstein, P. (1982). *Eur. J. Immunol.*, **12**, 60–9.
4. Sanchez-Madrid, F., Krensky, A. M., Ware, C. F., *et al.* (1982). *Proc. Natl Acad. Sci. USA*, **79**, 7489–93.
5. Dustin, M. L. and Springer, T. A. (1991). *Ann. Rev. Immunol.*, **9**, 27–66.
6. Springer, T. A. (1995). *Ann. Rev. Physiol.* **57**, 827–72.
7. Dustin, M. L. and Springer, T. A. (1989). *Nature*, **341**, 619–24.

8. Sanchez-Madrid, F., Simon, P., Thompson, S., and Springer, T. A. (1983). *J. Exp. Med.*, **158**, 586–602.
9. Sanchez-Madrid, F., Nagy, J., Robbins, E., Simon, P., and Springer, T. A. (1983). *J. Exp. Med.*, **158**, 1785–803.
10. Kishimoto, T. K., Hollander, N., Roberts, T. M., Anderson, D. C., and Springer, T. A. (1987). *Cell*, **50**, 193–202.
11. Springer, T. A., Thompson, W. S., Miller, L. J., Schmalstieg, F. C., and Anderson, D. C. 1984). *J. Exp. Med.*, **160**, 1901–18.
12. Van der Vieren, M., Trong, H. L., Wood, C. L., *et al.* (1995). *Immunity,* **3**, 683–90.
13. Torensma, R., Raymakers, R. A., van Kooyk, Y., and Figdor, C. G. (1996). *Blood*, **87**, 4120–8.
14. Marlin, S. D. and Springer, T. A. (1987). *Cell*, **51**, 813–9.
15. de Fougerolles, A. R., Qin, X., and Springer, T. A. (1994). *J. Exp. Med.*, **179**, 619–29.
16. Lawrence, M. B. and Springer, T. A. (1991). *Cell*, **65**, 859–73.
17. Kavanaugh, A. F., Lightfoot, E., Lipsky, P. E., and Oppenheimer-Marks, N. (1991). *J. Immunol.*, **146**, 4149–56.
18. Kunkel, E. J., Jung, U., Bullard, D. C., *et al.* (1996). *J. Exp. Med.*, **183**, 57–65.
19. Hamann, A., Westrich, D. J., Duijevstijn, A., *et al.* (1988). *J. Immunol.*, **140**, 693–9.
20. Krensky, A. M., Sanchez-Madrid, F., Robbins, E., Nagy, J., Springer, T. A., and Burakoff, S. J. (1983). *J. Immunol.*, **131**, 611–6.
21. Schmits, R., Kündig, T. M., Baker, D. M., *et al.* (1996). *J. Exp. Med.*, **183**, 1415–26.
22. Shier, P., Otulaksowski, G., Ngo, K., *et al.* (1996). *J. Immunol.*, **157**, 5375–86.
23. Dustin, M. L., Bromely, S. K., Kan, Z., Peterson, D. A., and Unanue, E. R. (1997). *Proc. Natl Acad. Sci. USA*, **94**, 3909–13.
24. van Kooyk, Y., van de Wiel-van Kemenade, P., Weder, P., Kuijpers, T. W., and Figdor, C. G. (1989). *Nature*, **342**, 811–3.
25. Lollo, B. A., Chan, K. W. H., Hanson, E. M., Moy, V. T., and Brian, A. A. (1993). *J. Biol. Chem.*, **268**, 21693–700.
26. Woska, J. R., Morelock, M. M., Jeanfavre, D. D., and Bormann, B. J. (1996). *J. Immunol.*, **156**, 4680–5.
27. Cabanas, C. and Hogg, N. (1993). *Proc. Natl Acad. Sci. USA*, **90**, 5838–42.
28. Kucik, D. F., Dustin, M. L., Miller, J. M., and Brown, E. J. (1996). *J. Clin. Invest.*, **97**, 2139–44.
29. Lub, M., van Kooyk, Y., van Vliet, S. J., and Figdor, C. G. (1997). *Mol. Biol. Cell,* **8**, 341–51.
30. Dustin, M. L., Carpén, O., and Springer, T. A. (1992). *J. Immunol.*, **148**, 2654–63.
31. Miller, J. M., Knorr, R., Ferrone, M., Houdei, R., Carron, C., and Dustin, M. L. (1995). *J. Exp. Med.*, **182**, 1231–41.
32. Staunton, D. E., Ockenhouse, C. F., and Springer, T. A. (1992). *J. Exp. Med.*, **176**, 1471–6.
33. Huang, C. and Springer, T. A. (1995). *J. Biol. Chem.*, **270**, 19008–16.
34. Bazil, V., Stefanova, I., Higert, I., Kristofova, H., Vanek, S., and Horejsi, V. (1990). *Folia Biol. (Prague)* **36**, 41–50.
35. van Kooyk, Y., Weder, P., Heije, K., and Figdor, C. G. (1994). *J. Cell. Biol.*, **124**, 1061–70.
36. Huang, C., Lu, C., and Springer, T. A. (1997). *Proc. Natl Acad. Sci. USA*, **94**, 3156–61.
37. Andrew, D., Shock, A., Ball, E., Ortlepp, S., Bell, J., and Robinson, M. (1993). *Eur. J. Immunol.*, **23**, 2217–22.
38. Dransfield, I. and Hogg, N. (1989). *EMBO J.*, **8**, 3759–65.
39. Larson, R. S., Corbi, A. L., Berman, L., and Springer, T. A. (1989). *J. Cell. Biol.*, **108**, 703–12.
40. Law, S. K. A., Gagnon, J., Hildreth, J. E. K., Wells, C. E., Willis, A. C., and Wong, A. J. (1987). *EMBO J.*, **6**, 915–9.
41. Kaufmann, Y., Tseng, E., and Springer, T. A. (1991). *J. Immunol.*, **147**, 369–74.
42. Wilson, R., O'Brien, W., and Beaudet, A. (1989). *Nucleic Acids Res.*, **17**, 5397.
43. Schuster, D. E., Bosworth, B. T., and Kehrli, M. E. Jr. (1992). *Gene,* **114**, 267–71.
44. Lee, J. K., Schook, L. B., and Rutherford, M. S. (1994). *Xenotransplantation,* **3**, 222–30.
45. Bilsland, C. A. and Springer, T. A. (1994). *J. Leuk. Biol.* **55**, 501–6.
46. Weitzman, J. B., Wells, C. E., Wright, A. H., Clark, P. A., and Law, S. K. (1991). *FEBS Lett.*, **294**, 97–103.
47. Anderson, D. C., Kishimoto, T. K., and Smith, C. W. (1995). In *The metabolic, and molecular basis of inherited disease* (ed. C. R. Scriver, A. L. Beaudet, W. S. Sly, and D. Valle), Vol. **7**, pp. 3955–94. (McGraw-Hill, New York).
48. Lu, H., Smith, C. W., Perrard, J., *et al.* (1997). *J. Clin. Invest.*, **99**, 1340–50.
49. Qu, A. and Leahy, D. J. (1996). *Structure,* **4**, 931–42.
50. Jones, E. Y., Harlos, K., Bottomley, M. J., *et al.* (1995). *Nature*, **373**, 539–44.
51. Springer, T. A. (1997). *Proc. Natl Acad. Sci. USA*, **94**, 65–72.
52. Goodman, T. G. and Bajt, M. L. (1996). *J. Biol. Chem.*, **271**, 23729–36.
53. Calvete, J. J., Henschen, A., and González-Rodríguez, J. (1991). *Biochem. J.*, **274**, 63–71.
54. Calvete, J. J., Henschen, A., and Gonzalez-Rodriguez, J. (1989). *Biochem. J.*, **261**, 561–8.
55. Knorr, R. and Dustin, M. L. *J. Exp. Med.* (In press.)
56. Hughes, P. E., O'Toole, T. E., Ylanne, Y., Shattil, S. J., and Ginsberg, M. H. (1995). *J. Biol. Chem.*, **270**, 12411–7.
57. Hibbs, M. L., Xu, H., Stacker, S. A., and Springer, T. A. (1991). *Science*, **251**, 1611–3.
58. Kolanus, W., Nagel, W., Schiller, B., *et al.* (1996). *Cell*, **86**, 233–42.
59. Liu, L. and Pohajdak, B. (1992). *Biochim. Biophys. Acta*, **1132**, 75–8.
60. Chardin, P., Paris, S., Jackson CL., Antonny, B., *et al.* (1996). *Nature*, **284**, 481–4.
61. Klarlund, J. K., Guilherme, A., Holik, J. J., Virbasius, J. V., Chawla, A., and Czech, M. P. (1997). *Science*, **275**, 1927–30.
62. Zell, T., Hunt, S. W., Mobley, J. L., Finkelstein, L. D., and Shimizu, Y. (1996). *J. Immunol.*, **156**, 883–6.
63. Rey-Ladino, J. A., Pyszniak, A. M., and Takei, F. (1998). *J. Immunol.*, **160**, 3494–501.

Michael L. Dustin
Department of Pathology,
Washington University School of Medicine,
St Louis, MO 63110, USA

Timothy A. Springer
Center for Blood Research and Department of Pathology, Harvard Medical School,
Boston, MA 02115, USA.

MAdCAM-1

MAdCAM-1[1–3] is a cell-surface adhesion molecule predominantly expressed on high endothelial venules (HEV) of gut-associated lymphoid tissues and on venules at chronically inflamed sites. It is a vascular addressin, serving to direct and increase leukocyte traffic to these tissues. The multidomain MAdCAM-1 molecule possesses two (human) or three (mouse) Ig domains, where the first domain supports binding to the lymphocyte homing receptor LPAM-1 (integrin $\alpha 4\beta 7$). A membrane-proximal mucin-like domain is a scaffold for *O*-linked glycosylation, and is predicted to bind L-selectin, enabling MAdCAM-1 to participate in both initial L-selectin-dependent lymphocyte tethering and rolling, as well as subsequent firm cell adhesion.

Protein properties

The primary structure of mature human MAdCAM-1 (38 340 kDa), predicted from cDNAs isolated from brain,[1] and mesenteric lymph node,[2] comprises two extracellular immunoglobulin (Ig) domains sharing most homology with VCAM and ICAM, followed by a mucin-like domain rich in serine–threonine residues, a transmembrane domain, and a 43 amino acid residue cytoplasmic domain (Fig. 1). The N-terminal half of the mucin domain, termed the major mucin domain, is formed from eight tandem repeats of the MUC2-related consensus sequence DTTSPEP/SP. Alternatively spliced variants have been identified lacking parts of the second Ig domain and all

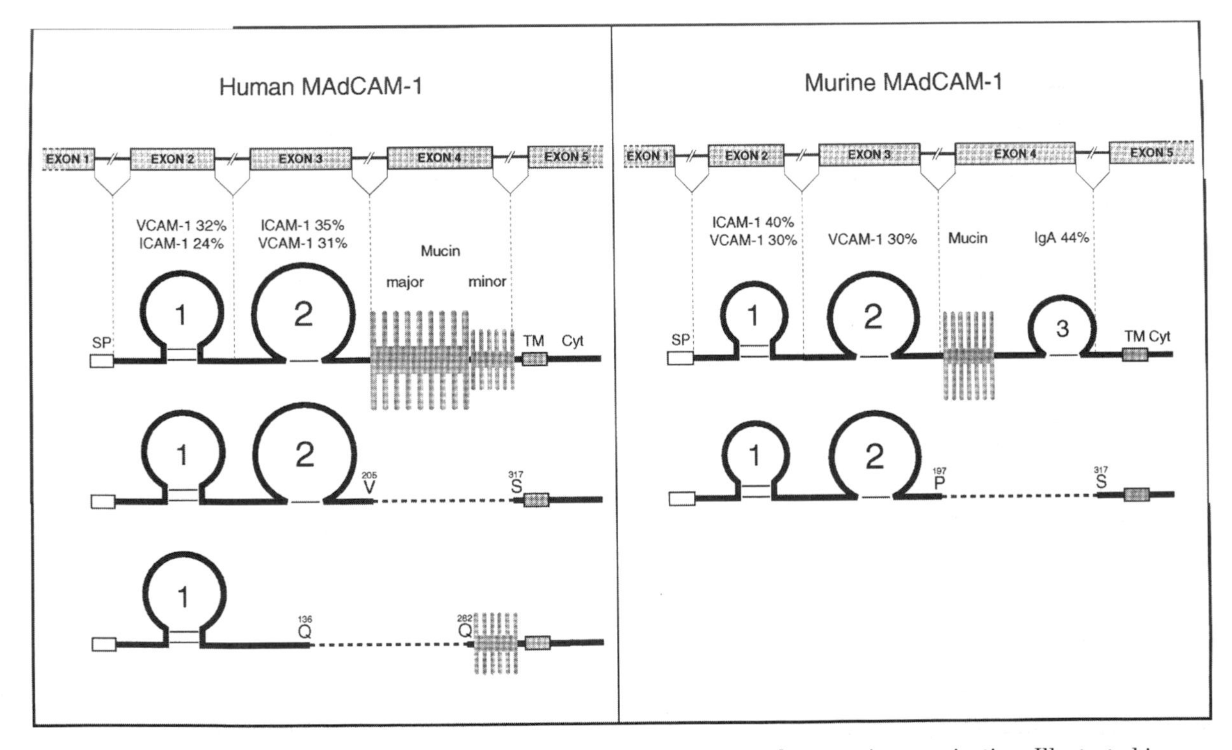

Figure 1. Structures of human and mouse MAdCAM-1 and correlation with genomic organization. Illustrated is a schematic diagram depicting the modular structures of human and mouse MAdCAM-1 and their correlation with the exon–intron organization of the respective genes. Alternatively spliced human mRNA transcripts have been identified lacking parts of the second Ig domain (exon 3) and all or parts of the major mucin domain (exon 4); a mucin domain-deficient form of mouse MAdCAM-1 has also been identified. Mouse MAdCAM-1 contains an additional IgA-related third Ig domain, whereas the mucin domain in human MAdCAM-1 has been extended. The relationships (percentage amino acid identity) of the Ig domains to other related Ig CAMs are indicated above each Ig loop. SP, signal peptide; TM, transmembrane segment; Cyt, cytoplasmic domain.

or parts of the major mucin domain, indicating that the functions of MAdCAM-1 are regulated by extensive modifications to its multidomain structure.[1,4,5] There is a lack of conservation across species of the C-terminal half of MAdCAM-1. Thus the mouse[3] and rat[25] MAdCAM-1 homologues have additional IgA-like and MHC Class I-like third Ig domains, respectively, and lack the tandem repeats of the major mucin domain. Nevertheless the N-terminal Ig domain that alone can mediate receptor binding and the second Ig domain which supports this interaction[6] are highly conserved between humans and rodents (59 and 65 per cent, respectively).

Transcripts encoding human and mouse MAdCAM-1 are predominantly expressed in the small intestine, mesenteric lymph nodes, colon, and spleen; and very weakly expressed in human pancreas and brain.[1,3] Murine MAdCAM-1 protein is detectable on endothelia in Peyer's patches, mesenteric lymph nodes, lamina propria of the small and large intestine, the lactating mammary gland,[7,8] as well as on sinus-lining cells surrounding the periarteriolar lymphocyte sheath and follicle areas of the spleen.[9] It is also displayed on follicular dendritic cells in Peyer's patches at sites associated with microenvironmental homing decisions, and on FDC in peripheral lymph nodes after primary immunization with antigen.[26] In disease states it is upregulated on HEV-like vessels in the chronically inflamed pancreas of the non-obese diabetic mouse,[10] on endothelia in chronic relapsing experimental autoimmune encephalomyelitis (EAE),[11] on venules in the genital tract of mice infected with *Chlamydia trachomatis*,[27] and on lamina propria venules in inflammatory bowel disease.[2] It appears not to be restricted exclusively to endothelia, as evidenced by expression on choroid plexus epithelial cells in the brains of EAE mice.[12]

The upregulation of MAdCAM-1 on chronically inflamed endothelia, especially HEV, is in accord with the notion that this vascular addressin has a particular role to play in recruiting pathogenic lymphocytes and non-specific leukocytes during the progression of chronic inflammatory disease. It interacts with the lymphocyte homing receptor LPAM-1 ($\alpha 4\beta 7$),[13,14] belonging to the $\beta 7$ integrin family which plays a critical role in forming and probably retaining the gut-associated lymphoid system.[15] A key LPAM-1 binding motif is composed of the three linear residues LDT within the C–D loop in the first Ig domain.[28] Antibodies to either the integrin $\alpha 4\beta 7$ complex or MAdCAM-1 can inhibit the homing of gut-seeking lymphocytes.[16] The receptor-binding site for MAdCAM-1 overlaps the binding sites for the other LPAM-1 ligands, namely VCAM-1 and the CS-1 domain of fibronectin.[17] LPAM-1 is not always the exclusive integrin receptor for MAdCAM-1 since a MAdCAM-1–Fc chimera binds to the related integrin VLA-4 ($\alpha 4\beta 1$) expressed on activated monocytes;[18] whereas VLA-4 on T and B cells appears unable to bind to MAdCAM-1.[13,19] MAdCAM-1 purified from mesenteric lymph nodes is able to support the rolling of lymphocytes under shear by binding leukocyte L-selectin, in a fashion similar to the selectin-dependent rolling of neutrophils which precedes leukocyte extravasation.[20] The selectin-binding carbohydrate determinants that decorate the mucin-like domain of MAdCAM-1 are probably generated by cell-type-specific glycosyltransferases, since MAdCAM-1 isolated from transfected lymphoid cells[20] is unable to support lymphocyte attachment to L-selectin.

Purification

MAdCAM-1 has been purified from the lysates of mouse mesenteric lymph nodes and Peyer's patches by immunoaffinity chromatography and can support both lymphocyte adhesion,[13] and L-selectin-mediated lymphocyte rolling.[20] Such CAMs are now more commonly expressed as Fc fusion proteins which retain receptor-binding properties[14] (Fig. 2).

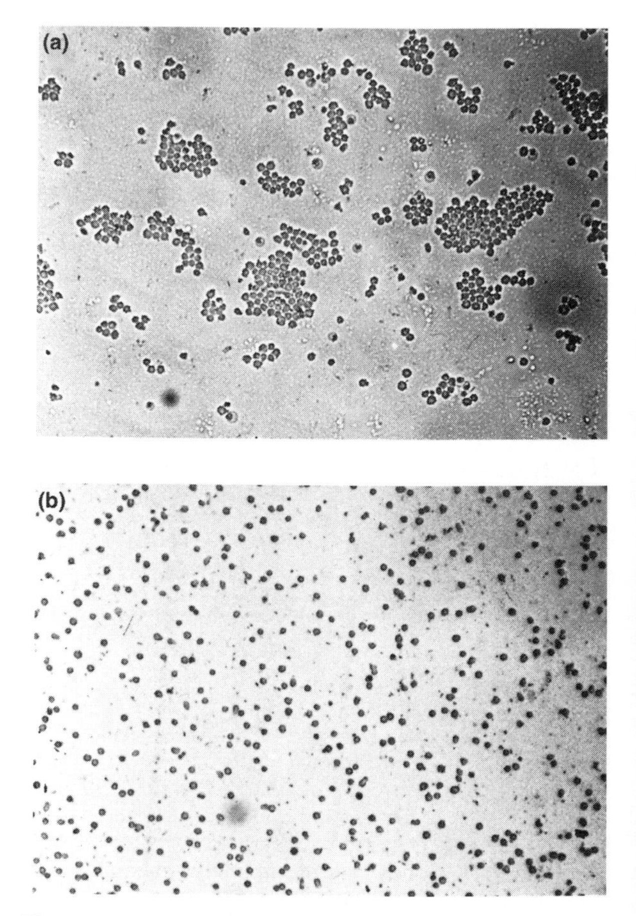

Figure 2. A recombinant MAdCAM-1–Fc chimera retains its receptor binding function. Uniform protein A microspheres (0.28 μm diameter) coated with MAdCAM-1–Fc cause spontaneous aggregation of $\alpha 4\beta 7^+$ TK-1 T cells (a), whereas cells incubated with control human IgG1-coated microspheres remain disaggregated (b).

Activities

MAdCAM-1 promotes the adhesion of T and B cells, monocytes/macrophages, and potentially eosinophils, basophils, and differentiated mast cells to the vascular endothelium. It is predicted to mediate both the L-selectin-dependent and -independent lymphocyte tethering and rolling that precedes firm adhesion by binding LPAM-1, which is highly concentrated on microvillus sites of initial cell contact.[21] Its potential roles in endothelial-mediated antigen presentation, T cell costimulation, cellular differentiation, and signalling are largely unexplored. In accord with its expression on FDC, MAdCAM-1 efficiently costimulates the proliferation of $CD4^+$ T cells exclusively via LPAM-1, and can both synergize with and induce hyperresponsiveness to the classical costimulator B7-2 (CD86) (Lehnert *et al.*, submitted).

Antibodies

A number of monoclonal antibodies against both murine[22] (MECA-367, MECA-89) and human[23] (8C1–12) MAdCAM-1 have been characterized. A subset of MAdCAM-1 from Peyer's patches is recocognized by the MECA-79 antibody believed to recognize post-translational carbohydrate modifications of the peripheral lymph node addressin (PNAd) and other glycoproteins.[19] Polyclonal antibodies raised against human MAdCAM-1 do not cross-react with a panel of other human Ig CAMs.

Genes

A single MAdCAM-1 gene in humans is located at 19p13.3, in relatively close proximity to the related ICAM-1 and ICAM-3 genes (p13.2–p13.3).[5] The mouse gene is located on a homologous region of chromosome 10.[4] MAdCAM-1 cDNAs have been cloned and sequenced from human[1,2] (GenBank U82483 and U43628) and mouse[3] (GenBank L21203), rat[25] (GenBank D87840), and a partial deduced sequence is available for the macaque monkey.[2] Sequences of genomic clones available for human[5] (GenBank U80013/14/15/16) and mouse[4] (GenBank U14552 and U14729) are in accord with the proposed structures of mouse and human MAdCAM-1. Thus the human MAdCAM-1 gene contains five exons where the signal peptide, two Ig domains, and mucin domain are each encoded by separate exons (Fig. 1). The transmembrane domain, cytoplasmic domain, and 3'-untranslated region are encoded together on exon 5. The mouse gene has a similar organization, except that a third Ig domain is encoded together with the mucin domain on exon 4. Gene promoter sequences are also available (human:[5] GenBank U80012; mouse:[4] GenBank U14552), the most significant feature of which is tandem nuclear factor-B (Nf–B) sites that are critical for induction of mouse MAdCAM-1 expression by TNF-α.[24]

References

1. Leung, E., Greene, J., Ni, J., Raymond, R. G., Lehnert, K., Langley, R., and Krissansen, G. W. (1996). *Immunol. Cell Biol.*, **74**, 490–6.
2. Shyjan, A. M., Bertagnolli, M., Kenney, C. J. and Briskin, M. J. (1996). *J. Immunol.*, **156**, 2851–7.
3. Briskin, M. J., McEvoy, L. M. and Butcher, E. C. (1993). *Nature*, **363**, 461–3.
4. Sampaio, S. O., Li, X., Takeuchi, M., Mei, C., Francke, U., Butcher, E. C. and Briskin, M. J. (1995) *J. Immunol.*, **155**, 2477–86.
5. Leung, E., Berg, R. W., Langley, R., Greene, J., Raymond, L. A., Augustus, M., *et al.* (1997). *Immunogenet.*, **46**, 111–9.
6. Briskin, M. J., Rott, L. and Butcher, E. C. (1996). *J. Immunol.*, **156**, 719–26.
7. Berg, E. L., Picker, L. J., Robinson, M. K., Streeter, P. R. and Butcher, E. C. (1991). In *Cellular and molecular mechanisms of inflammation*, (ed. C. G. Cochrane and M. A. Grimbrone), Vol. **2**, pp. 111–29. Academic Press, New York.
8. Picker, L. J. and Butcher, E. C. (1992). *Ann. Rev. Immunol.*, **10**, 561–91.
9. Kraal, G., Schornagel, K., Streeter, P. R., Holzmann, B. and Butcher, E. C. (1995). *Am. J. Pathol.*, **147**, 763–71.
10. Hanninen, A., Taylor, C., Streeter, P. R., Stark, L. S., Sarte, J. M., Shizuru, J. A., *et al.* (1993). *J. Clin. Invest.*, **92**, 2509–15.
11. O'Neill, J. K., Butter, C., Baker, D., Gschmeissner, S. E., Kraal, G., Butcher, E. C. and Turk, J. L. (1991). *Immunology*, **72**, 520–5.
12. Steffen, B. J., Brier, G., Butcher, E. C., Schulz, M. and Engelhardt, B. (1996). *Am. J. Pathol.* **148**, 1819–38.
13. Berlin, C., Berg, E. L., Briskin, M. J., Andrew, D. P., Kilshaw, P. J., Holzmann, B., *et al.* (1993). *Cell*, **74**, 185–95.
14. Yang, Y., Sammar, M., Harrison, J. E. B., Lehnert, K., Print, C. G., Leung, E., *et al.* (1995). *Scand. J. Immunol.*, **42**, 235–47.
15. Wagner, N., Lohler, J., Kunkel, E,J., Ley, K., Leung, E., Krissansen, G., *et al.* (1996). *Nature*, **382**, 366–70.
16. Hamann, A., Andrew, D. P., Jablonski-Westrich, D., Holzmann, B. and Butcher, E. C. (1994). *J. Immunol.*, **152**, 3282–93.
17. Andrew, D. P., Berlin, C., Honda, S., Yoshino, T., Hamann, A., Holzmann, B., *et al.* (1994). *J. Immunol.*, **153**, 3847–61.
18. Yang, Y., Harrison, J. E. B., Print, C. G., Lehnert, K., Sammar, M., Lazarovits, A. and Krissansen, G. W. (1996). *Immunol. Cell Biol.*, **74**, 383–93.
19. Erle, D. J., Briskin, M. J., Butcher, E. C., Garcia-Pardo, A., Lazarovits, A. I. and Tidswell, M. (1994). *J. Immunol.*, **153**, 517–28.
20. Berg, E. L., McEvoy, L. M., Berlin, C., Bargatze, R. F. and Butcher, E. C. (1993). *Nature*, **366**, 695–8.
21. Berlin, C., Bargatze, R. F., Campbell, J. J., Von Andrian, U. H., Szabo, M. C., Hasslen, S. R., *et al.* (1995). *Cell*, **80**, 413–22.
22. Streeter, P. R., Berg, E. L., Rouse, B. T. N., Bargatze, R. F. and Butcher, E. C. (1988). *Nature*, **331**, 41–6.
23. Salmi, M., Rajala, P. and Jalkanen, S. (1997). *J. Clin. Invest.*, **99**, 2165–72.
24. Takeuchi, M. and Baichwal, V. R. (1995). *Proc. Natl Acad. Sci. USA*, **92**, 3561–5.
25. Iizuka, T., Koike, R., Miyasaka, N., Miyasaka, M., and Watanabe, T. (1998). *Biochem. Biophys. Acta*, **1395**, 266–70.
26. Szabo, M. C., Butcher, E., and McEvoy, L. M. (1997). *J. Immunol.*, **158**, 5584–8.
27. Kelly, K. A. and Rank, R. G. (1997). *Infect. Immun.*, **65**, 5198–208.

28. Viney, J. L., Jones, S., Chiu, H. H., Lagrimus, B., Renz, M. E., Presta, L. G., Jackson, D., Hillan, K. J., Lew, S., and Fong, S. (1996). *J. Immunol.*, **157**, 2488–97.

Geoffrey W. Krissansen, Euphemia Leung, and Klaus Lehnert
Department of Molecular Medicine, School of Medicine and Health Sciences, University of Auckland, Auckland, New Zealand

Mannose binding proteins (MBP)

Mannose binding proteins (MBP) are hepatocyte-derived serum proteins that have been isolated from the liver and serum of a number of mammalian species.[1] MBP is synthesized as a 28–32 kDa polypeptide, which forms trimers. The trimers oligomerize. The highest order multimer is a hexamer of trimers. MBP has a cysteine-rich N terminus, a collagen domain, and a neck or trimerization domain followed by a carboxyl-terminal carbohydrate-recognition domain (CRD). MBP is a member of the collectin (collagenous lectin) family of the C-type (calcium-dependent) animal lectin superfamily.[2] MBP may be considered as an ante-antibody. MBP binds to a wide array of microorganisms. It is able to act as a direct opsonin, like antibody. Alternatively, MBP initiates the classical or the alternative complement pathway activation. The predominant biological role for MBP is as a pattern recognition molecule in first line of host defence.

Synonymous names

Mannose-binding lectin, mannan-binding lectin, mannan-binding protein, Man/GlcNAc-specific lectin, core-specific lectin, ra-reactive factor (RaRF).

Homologous proteins

Surfactant protein-A (SP-A), surfactant protein-D (SP-D), conglutinin (BK), and collectin-43 (CL-43) are other members of the collectin family having a carboxyl-terminal CRD attached to N_2-terminal collagen-like tails.

Protein properties

Mannose-binding proteins have been isolated from human, rabbit, chicken, rat, mice, and cow serum and/or liver, where they exist as multimers of a ~28–32 kDa polypeptide (Fig. 1). Recent studies indicate that MBP exists in serum in a calcium-mediated complex with two MBP- associated serine proteases,[3] called MASP-1 and MASP-2. Two homologueous forms of MBP, designated MBP-A and MBP-C, have been characterized by cDNA isolation from rats, mice and rhesus monkey liver.[4–7] Humans

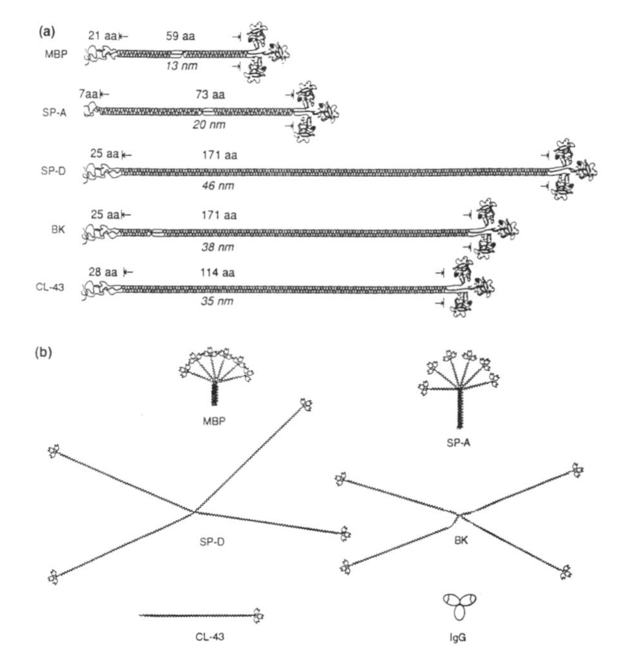

Figure 1. Collectin structures. The structures in (a) have been drawn according to the dimensions of a collagen type I structure, with a triple helix of 1.5 nm diameter, and a pitch of 0.91 nm per triplet. The carbohydrate recognition domains (CRDs) are illustrated by the struture determined by X-ray crystallography. The diameter of the three heads together appears to be 5–6 nm. The quarternary structures are indicated in (b) with IgG included to show the relative size of the molecules. aa, amino acid. (Reproduced, with permission, from Holmskov *et al.* (1994). *Immunol. Today*, 68: 67–73.

and chimpanzees appear to have a single form of MBP, which has features of rodent MBP-A and MBP-C.[7] Human MBP has overall structural similarity to C1q (Fig. 2). In rodents, MBP-A appears to be the predominant serum form and assembles into oligomers of ~650–750 kDa. Rodent MBP-C is believed to form multimers of ~200 kDa. Studies on the biosynthesis of MBP in hepatocytes and a

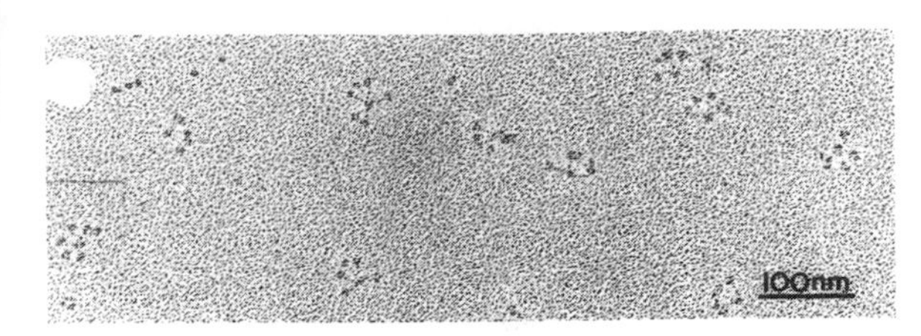

Figure 2. Electron micrograph showing human MBP. (Picture courtesy of Dr K. Reid.)

hepatoma cell line indicate that rat MBP-C undergoes post-translational modifications – hydroxylation of proline and lysine followed by glycosylation of hydroxylysine in the Golgi complex– which are necessary for the secretion of this soluble protein.[8] MBP-C is secreted slowly from rat hepatoma cells with a half-time for secretion of >4 h.[8] Levels of MBP in human sera vary widely from 0.07 to 6.4 μg/ml (mean 1.7 μg/ml).[9] MBP-A levels in mice appear to be higher, with an average of 50 μg/ml. The human serum MBP level appears to be elevated 1.5- to 2-fold as a part of the acute phase response. MBP levels have also been shown to be associated with mutations in the promoter region of the human MBP gene.[10] Low levels are observed in serum of individuals with B, C, and D allelic forms (see below).

Human MBP has been shown to function directly as a opsonin.[11] MBP-A from rodents as well as human MBP have been shown to activate the complement pathway after binding to a mannan-rich surface. Rodent MBP-A and MBP-C show ~52 per cent identity at the amino acid level.[4,5] Rat MBP-A and MBP-C have been shown to differ in their affinity for various monosaccharides. Both MBP-A and MBP-C have high affinity for mannose. Rat MBP-A appears to have higher affinity for fucose and *N*-acetylglucosamine and glucose compared to rat MBP-C. Furthermore, both rat MBP-A and MBP-C appear to have low affinity for galactose. Most of the carbohydrate structures recognized by MBP appear to be well represented on the surface glycoproteins of many microorganisms. MBP binds to cells via a collectin receptor. A ~60–65 kDa protein similar or identical to calreticulin was first reported to bind human MBP, C1q, SP-A, and BK and was termed the 'collectin' receptor.[12] However, calreticulin appears to be secreted by neutrophils activated by FMLP and PMA[13] and is therefore a collectin-binding protein rather than a receptor. A 120 kDa transmembrane protein has been identified as a collectin receptor. Isolation of the cDNA that encodes this molecule indicates that the encoded protein, called $C1qR_p$, has an extracellular N_2-terminal C-type CRD, five EGF-like domains, a transmembrane domain, and a short cytoplasmic tail.[14]

Purification

MBPs have been traditionally purified by affinity chromatography on a mannan/D-Mannose column followed by elution with 10 mM EDTA, taking advantage of calcium-dependence for CRD activity.[4,15] However, it has been recognized that this approach with human serum, invariably copurifies serum amyloid protein (SAA), which binds to pyruvate acetal carbohydrate on the agarose backbone of the column in a calcium-dependent manner. SAA can be removed by first passing the serum through an underivatized Sepharose column.[16] Isolation of MBP by mannan column chromatography by elution with mannose instead of EDTA allows the recovery of a MBP + MASP complex.[17] However, immunoglobulins have to removed from this preparation by additional steps. Mouse MBP have also been purified by batch adsorption on to the Ra chemotype of *Salmonella typhimurium.*[6]

Activities

MBP binds carbohydrates and its concentration can be measured either by a mannan-binding assay or a polyclonal/specific monoclonal antibody capture sandwich assay using a mouse monoclonal antibody–HRP conjugate for detection.[9,18] No international reference standard has as yet been developed for MBP. MBP can activate classical (by associating with C1r and C1s),[19] alternative (directly cleave C3 via MASP),[20] and a novel pathway of complement activation, involving two newly characterized serine proteases, MASP-1 and MASP-2, which are presumed to be activated sequentially upon binding of carbohydrate by MBP, leading to the cleavage of the C4 and C2 components of the classical pathway of complement activation.[1] It is currently believed that the MBP–MASP pathway is the most physiologically significant pathway of complement activation by MBP[3] and that MASP activity and complex formation is regulated by α_2 macroglobulin.[21] In addition to enhancing of phagocytosis of microorganisms due to deposition of iC3b on the surface of microorganisms, bearing appropriate sugars, by activation of the complement pathway, MBP devoid of MASP could also enhance phagocytosis via cross-linking of the monocyte cell surface receptor $C1qR_{p.}$[14] Allelic forms of MBP with a mutation in the collagen domain (Gly-54 to Asp-54) are unable to activate complement either via association with $C1r_2/C1s_2$ or MASP,[18,22] but can act as an opsonin *in vitro*.

Antibodies

Polyclonal antibodies against human MBP and rat, mouse, bovine, and chicken MBP have been developed.[16] Monoclonal antibodies against human MBP have also been developed.[9,18] Polyclonal rabbit antibodies against rat MBP-C cross-react with human, rhesus, cynomolgus, and chimpanzee MBP.[7] Mouse monoclonal antibody clone 3 against human MBP[18] cross-reacts with rhesus MBP-C (unpublished data). Currently, no commercial sources of human or rodent MBP reagents are known. Several types of immobilized rabbit MBP columns are available commercially for purification of bovine, goat, sheep, rat, and human IgM (Pierce, Rockford, IL, USA)

Genes

Rat (accession number M14103), human (X15422), mouse [MBP-A: S42292 (CBA/J), D11441 (Balb/c) and MBP-C: S42294 (CBA/J), D11440 (Balb/c)] and rhesus MBP-A and MBP-C (L43911 and L439112) cDNAs have been completely characterized. The human MBP gene (gene symbol: *MBL*) (X15954 to X15957), rat MBP-A gene (M14104 and 14105), rat MBP-B pseudogene (M14106), and mouse MBP-A and MBP-C genes (*Mbl1* and *Mbl2*: U09006 to U09017) have also been characterized. The MBP genes are organized in four protein coding exons (Fig. 3). The first exon encodes a signal sequence, an N-terminal region, and the first six or seven Gly–X–Y repeats of the collagen-like region. The second exon encodes the rest of the collagen domain while the third exon encodes the so called neck, linker, or trimerization domain. The C-terminal CRD is always encoded in a single, last exon. The human MBP gene has been localized to chromosome 10q21–23, to a region where other collectin genes, SP-AI (*SFTP1*), SP-AII, SP-Aψ (pseudogene), and SP-D (*SFTP4*), have also been localized.[23,24] The mouse MBP-A gene (*Mbl1*) has been localized to chromosome 14 to a region syntenic to human chromosome 10, where the human MBP gene has also been localized. The mouse MBP-C gene (*Mbl2*) has been localized to chromosome 19 to a region syntenic to the human chromosome 10q23–25 region, suggesting the possibility that the two ancestral MBP genes were linked.[25]

Alternative splicing in the 5′ end of a rodent MBP-C mRNA transcript either by inclusion or exclusion of an additional 80 bp (rat MBP-C) or 45 bp (mouse MBP-C) from an intron preceding the second 5′ untranslated exon has been observed. The significance of this phenomenon is not clear as the region included in the alternative transcript is a part of the 5′ untranslated region.[26,27]

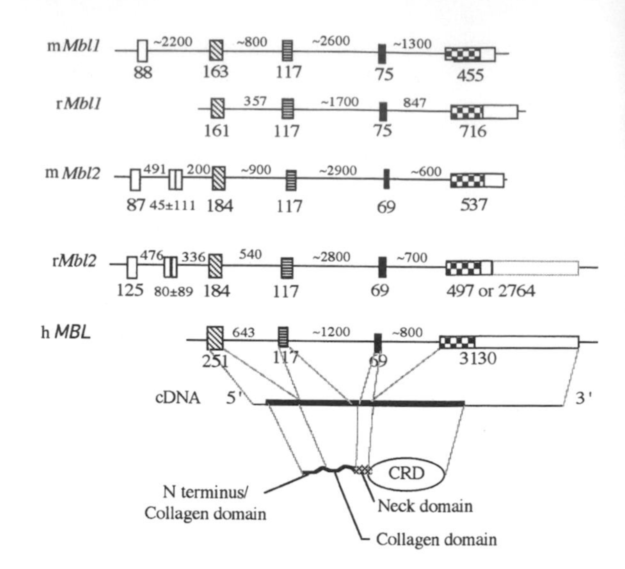

Figure 3. The relationship between the domains of MBP, and organization of the genes. The numbers below the boxes indicate exon size while the numbers above the thin lines indicate size of introns in base pairs. Open boxes at the 5′, and 3′ end indicate untranslated exons or regions of exon. The regions encoded by various exons are as follows: cross-hatched box, signal peptide/N_2 terminus/collagen-like domain; striped box, collagen-like domain; filled box, neck domain; checkered box, carbohydrate recognition domain. m, mouse; r, rat; h, human; *Mbl1*, MBP-A; *Mbl2*, MBP-C. (Adapted, with permission, from Sastry *et al.*[27]).

Alleles of MBP and association with disease states

The human MBP gene is found in the population as four distinct allelic forms, called A, B, C, and D. The A allele is predominant and is considered Wild-type. The B, C, and D alleles have point mutations in the first exon of the MBP gene in codons 54, 57, and 52, respectively, leading to changes in amino acids in the collagen-like domain of the protein. The A allele encodes for Gly in the region corresponding to codons 54 and 57 while Arg is encoded by codon 52. The B, C and D alleles have Asp, Glu and Cys instead of Gly, Gly, and Arg, respectively, at these amino acid positions. The frequencies of the three collagen-like domain mutations vary in different ethnic populations studied (Table 1).

Table 1 Frequencies in various populations of three mutations in exon 1 of the human MBP gene

	Observed frequency of the mutant gene		
Population studied	Codon 52	Codon 54	Codon 57
UK (Caucasian)	0.06	0.16	0.00
Denmark.	0.05	0.13	0.02
Hong Kong Chinese	0.01	0.11	0.00
Greenland Eskimo	0.00	0.13	0.00
Papua New Guinea.	0.00	0.07	0.00
Vanuatu (SW Pacific)	0.00	0.01	0.00
The Gambia	0.02	0.00	0.29
Kenya	0.05	0.03	0.23
Xhosa (S. Africa)	0.00	0.00	0.27
San Bushmen (Namibia)	0.00	0.03	0.07

Reproduced, with permission, from Turner.[1]

MBP was first identified as an opsonin *in vitro*.[11] MBP bound to *Salmonella montevideo* enhanced uptake and killing of the opsonized bacteria by phagocytes. An association between low levels of serum MBP, mutant genotype, and susceptibility to recurrent infection was first reported by Summerfield and colleagues.[29] Individuals homozygous for mutant alleles of MBP apper to be at an increased risk for acquiring certain bacterial, fungal, and viral infections and may also have an increased predisposition to develop autoimmune disorders like the systemic lupus erythematosus.[29] Due to the high prevalence and persistence of mutant MBP alleles in the population, it is believed that low MBP levels may protect individuals from excessive activation of complement by the MBP pathway by microrganisms with appropriate sugars or in an autoimmune disease like rheumatoid arthritis, where elevated levels of MBP binding the agalactosyl form of IgG are believed to contribute to the pathology of synovial joint inflammation.[29]

Transgenic mice expressing wild-type human MBP have been useful in providing evidence regarding their role in first-line host defence in an experimental model of *Candida albicans* infection.[30] Human MBP serum levels were found to have decreased by more than 50 per cent in the first hour following injection. Similarly, endogenous mouse MBP-A levels also decreased by 25 per cent in the first hour following injection and were elevated by at least two-fold by 72 h, confirming it to be an acute-phase protein.[30]

Structure

The three-dimensional crystal structures of truncated rat MBP-A and MBP-C and human MBP (corresponding to neck and CRD regions) complexed with carbohydrates have been elucidated (Figs 4 and Plate 15). Five amino acids; Glu-185, Asn-187, Glu-193, Asn-205, and Asp-206 were shown to involved in binding to both Ca^{2+} and 3- and 4-OH groups of mannose in the X-ray crystal structure of rat MBP-A CRD at 1.7 Å resolution.[31] Rat MBP-C X-ray crystal studies revealed a similar structure, with five amino acids in the CRD (Glu-190, Asn-192, Glu-198, Asn-110 and Asp-211) involved in binding both Ca^{2+} and α-methyl-mannose. Interestingly, the orientation of mannose was rotated 180° relative to the orientation observed in rat MBP-A.[32] Crystals of the human and rat MBP neck + CRD have revealed that the neck region forms an α-helical coiled coil that orientates the individual CRDs, which were found to be spaced 45 Å or 53 Å apart in human and rat, respectively. This arrangement of the CRDs favours multivalent, high-affinity binding of oligosaccharides on the surfaces of microorganisms.[33]

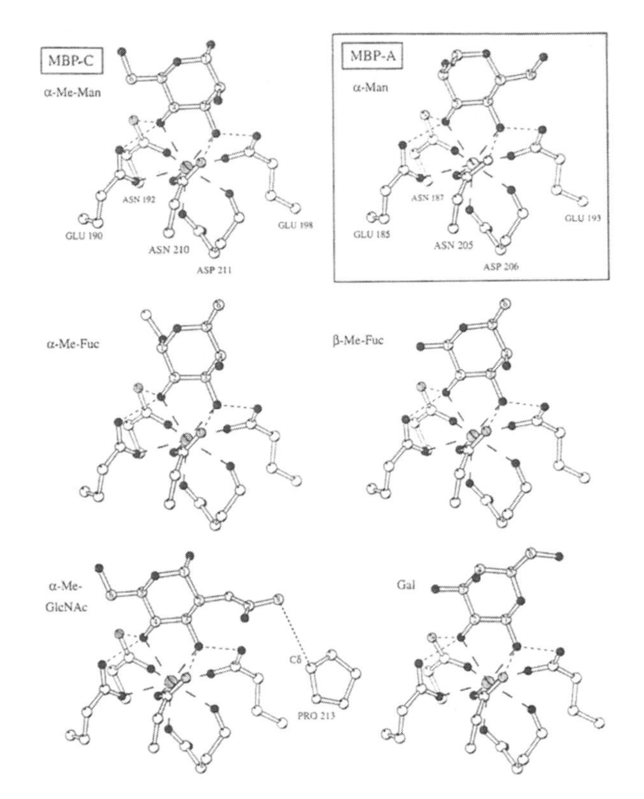

Figure 4. The structures of five different sugars bound to rat MBP-C are shown along with the structure of the terminal mannose residue bound to rat MBP-A. The remaining portion of the oligosaccharide bound to rat MBP-A has been omitted for clarity. The ligands have been superimposed, and are presented in a common view. Carbon, nitrogen, and oxygen atoms are shown as *white*, *grey*, and *black* spheres, respectively. Ca^{2+} 2 is shown as a *larger grey* sphere. Carbon atoms of the bound sugars are numbered. Broken lines with *long-dashes* denote coordination bonds with Ca^{2+}, with *medium dashes* denote hydrogen bonds, and with *short dashes* denote van der Waals' contacts. Note that the methyl aglycon is visible only in α-Me-Fuc, and in one of the two copies of α-Me-Man (not shown). (Reproduced, with permission, from Ng *et al.*[32])

References

1. Turner, M. W. (1996). *Immunol. Today*, **17**, 541–9.
2. Drickamer, K. (1988). *J. Biol. Chem.*, **263**, 9557–60.
3. Theil, S., Vorup-Jensen, T., Stover, C., Schwaeble, W., Laursen, S. B., Poulsen, K., *et al.* (1997). *Nature*, **386**, 506–10.
4. Drickamer, K., Dordal, M. S., and Reynolds, L. (1986). *J. Biol. Chem.*, **261**, 6878–87.
5. Sastry, K., Zahedi, K., Lelias, J. -M., Whitehead, A. S., and Ezekowitz, R. A. B. (1991). *J. Immunol.*, **147**, 692–7.
6. Kuge, S., Ihara, S., Watanabe, E., Watanabe, M., Takishima, K., Suga, T., Mamiya, G., and Kawakami, M. (1992). *Biochemistry*, **31**, 6943–50.
7. Mogues, T., Ota, T., Tauber, A. I., and Sastry, K. N. (1996). *Glycobiology*, **6**, 543–50.
8. Colley, K. J. and Baenziger, J. U. (1987). *J. Biol. Chem.*, **262**, 10296–303.
9. Terai, I., Kobayashi, K., Fujita, T., and Hagiwara, K. (1993). *Biochem. Med. Metbol. Biol.* **50**, 111–19.
10. Madsen, H. O., Garred, P., Thiel, S., Kurtzhals, J. A. L., Lamm, L. U., Ryder, L. P., and Svejgaard, A. (1995). *J. Immunol.*, **155**, 3013–20.

11. Kuhlman, M., Joiner, K., and Ezekowitz, R. A. B. (1989). *J. Exp. Med.*, **169**, 1733–45.
12. Malhotra, R., Thiel, S., Reid, K. B., and Sim, R. B. (1990). *J. Exp. Med.*, **172**, 955–9.
13. Eggleton, P., Lieu, T. S., Zappi, E. G., Sastry, K., Coburn, J., Zaner, K. S., *et al.* (1994). *Clin. Immunol., Immunopathol.*, **72**, 405–9.
14. Nepomuceno, R. R., Henschen-Edman, A. H., Burgess, W. H., and Tenner, A. J. (1997). *Immunity*, **6**, 119–29.
15. Kawasaki, T., Etoh, R., and Yamashina, I. (1978). *Biochem. Biophys. Res. Commun.*, **81**, 1018–24.
16. Colley, K. J., Beranek, M. C., and Baenziger, J. U. (1988). *Biochem. J.*, **256**, 61–8.
17. Matsushita, M. and Fujita, T. (1992). *J. Exp. Med.*, **176**, 1497–502.
18. Super, M., Gillies, S. D., Foley, S., Sastry, K., Schweinle, J.-E., Silverman, V. J., and Ezekowitz, R. A. B. (1992). *Nature Genet.*, **2**, 50–5.
19. Ohta, M., Okada, M., Yamashina, I., and Kawasaki, T. (1990). *J. Biol. Chem.*, **265**, 1980–4.
20. Matsushita, M. and Fujita, T. (1995). *Immunobiol.*, **194**, 443–8.
21. Terai, I., Kobayashi, K., Matsushita, M., Fujita, T., and Matsuno, K. (1995). *Int. Immunol.*, **7**, 1579–84.
22. Matsushita, M., Ezekowitz, R. A. B., and Fujita, T. (1995). *Biochem. J.*, **311**, 1021–3.
23. Sastry, K., Herman, G. A., Day, L., Deignan, E., Bruns, G., Morton, C. C., and Ezekowitz, R. A. B. (1989). *J. Exp. Med.*, **170**, 1175–89.
24. Kolble, K., Lu, J., Mole, S. E., Kalutz, S., and Reid, K. B. M. (1993). *Genomics*, **17**, 294–8.
25. White, R. A., Dowler, L. L., Adkinson, L. R., Ezekowitz, R. A. B., and Sastry, K. N. (1994). *Mamm. Genome*, **5**, 807–9.
26. Wada, M., Itoh, N., Ohta, M., and Kawasaki, T. (1992). *J. Biochem.*, **111**, 66–73.
27. Sastry, R., Wang, J. -S., Brown, D. C., Ezekowitz, R. A. B., Tauber, A. I., and Sastry, K. N. (1995). *Mamm. Genome*, **6**, 103–10.
28. Sumiya, M., Super, M., Tabona, P., Levinsky, R. J., Arai, T., Turner, M. W., and Summerfield, J. A. (1991). *Lancet*, **337**, 1669–70.
29. Sumiya, M. and Summerfield, J. A. (1996). *Quart. J. Med.* **89**, 723–6.
30. Tabona, P., Mellor, A., and Summerfield, J. A. (1995). *Immunol.*, **85**, 153–9.
31. Weis, W. I., Drickamer, K., and Hendrickson, W. A. (1992). *Nature*, **360**, 127–34.
32. Ng, K. K. -S., Drickamer, K., and Weiss, W. I. (1996). *J. Biol. Chem.*, **271**, 663–74.
33. Sheriff, S., Cheng, C. Y., and Ezekowitz, R. A. B. (1994). *Nature Struct. Biol.*, **1**, 789–94.
34. Satow Y, Cohen, G. H., Padian, E. A., Davies, D. R. (1986). *J. Mol. Biol.*, **190**,593–604.
35. Kaulis, P. J. (1991). *J. Appl. Crystallogr.* **24**, 946–50.

Kedarnath N. Sastry
Department of Pathology,
Boston University School of Medicine,
Boston, MA, USA

R. Alan B. Ezekowitz
Laboratory of Developmental Immunology,
Department of Pediatrics, Massachusetts
General Hospital, Harvard Medical School,
Boston, MA, USA

MHC class I and II

Major histocompatibility complex (MHC) class I and II proteins are highly polymorphic and structurally related membrane glycoproteins which play an important role in the antigen-specific immune response. An MHC class I molecule is a trimer of the class I heavy chain, the light chain, β_2-microglobulin (β_2m), and a short antigenic peptide of 8–10 a.a. which binds within two α helices on the surface of the heavy chain. An MHC class II molecule is likewise a heterotrimer of α chain, β chain, and antigenic peptide bound in a similar manner. During an immune response, the structural entity recognized by the antigen-specific receptor on T cells is formed by the composite surface of MHC and bound peptide.

Synonymous names

HLA (human) = H2 (mouse).

Homologous proteins

Class I-like genes (class Ib)

These encode proteins which have similarity with class I MHC, yet are less polymorphic and have more restricted tissue-specific expression patterns.[1] These include, in humans, HLA-E, F, G, H, and J. Of these only HLA-G has been shown to have an antigen-presenting ability and its expression pattern in the placenta suggests a role in modulation of the immune system during pregnancy.

In mice, over 40 class I-like genes are found in three clusters: H2-T, Q, and M.[1,2] Included among these are H2-Qa-1, H2-Qa-2, and H2-M3, that have each been shown to bind and present peptides to T cells. Particularly interesting is the observation that M3 preferentially binds *N*-formylated peptides, found in bacteria and some mitochondrial proteins.

CD1 proteins

These class I-like molecules, encoded outside the MHC region, have recently been shown to present microbial lipid and glycolipid antigens to T cells.[3]

FcRn

Structurally, FcRn is very similar to MHC class I, yet with no peptide-binding groove.[4] This protein is a pH-sensitive IgG-binding protein involved in transporting antibodies from mother to young and may also serve to regulate serum IgG catabolism.[5,6]

Class II-like genes

HLA-DM, which plays an important role in peptide loading on to class II molecules, is related to class II MHC, but, apparently, does not itself present peptides on the cell surface.[7]

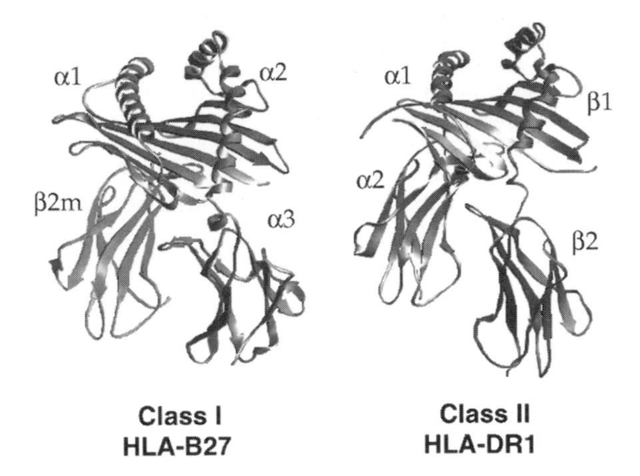

Figure 1. Comparison of crystal structures of MHC class I, and class II. These proteins are shown in a similar orientation, with the peptide binding helices at the top. The transmembrane, and cytoplasmic domain are not shown, as they were removed for crystallization. Thanks to Don Wiley, and Larry Stern for help obtaining this figure.

Protein properties

MHC class I heavy chain is a type-I membrane glycoprotein of approximately 45 kDa which engages in a non-covalent interaction with the soluble secretory protein β_2m (12 kDa). As first revealed in the crystal structure of HLA-A2,[8] the extracellular (lumenal) portion of MHC class I is formed by the interaction of three structural domains, $\alpha1$–$\alpha3$. The $\alpha3$ domain and β_2m interact to form a base that supports a β sheet platform of eight strands, four each contributed by $\alpha1$ and $\alpha2$. The $\alpha1$ and $\alpha2$ domains also have regions of helical structure which form an extended groove into which binds the antigenic peptide of, on average, nine amino acids (Fig. 1, left). Generally between one (human) and three (mouse) *N*-linked glycans are present on the heavy chain. Glycosylation is required for surface expression of some but not all class I molecules.

The two MHC class II chains assemble into a structure nearly indistinguishable in folding pattern from class I[9] (Fig. 1, right), although the protein backbone and subunit composition is quite different. One notable difference is that the peptide-binding helices in class II are open at the ends. This allows the binding of longer peptides, which may extend from the groove. The β sheet platform and the two α helices, which are both specified by the heavy chain polypeptide for class I, are found on separate polypeptides for class II. These are the α chain (33–35 kDa) and the β chain (25–30 kDa). The lumenal/extracellular domains of these chains are divided into two domains of 90–100 amino acids, a transmembrane segment, and a cytoplasmic tail of about 12–15 amino acids. There is a direct correspondence between the $\alpha1/\alpha2$ domains of class II and the $\alpha1/\beta_2$m of class I (Fig. 2); likewise the $\beta1/\beta2$ domains of class II correspond with the $\alpha2/\alpha3$ domains of class I.

The assembly of the mature heterotrimers of MHC molecules is complex and has been reviewed.[7,10,11] As discussed in these reviews, there is a major difference between the intracellular pathways taken by class I MHC products, compared with class II products as they travel to the cell surface. The upshot of these differences in intracellular trafficking is that class I molecules, generally, bind peptides derived from cytosolic proteins, whereas class II molecules bind peptides derived from extracelluar proteins, which have entered the endosomal pathway.

Peptides that are destined for class I MHC are typically derived from the action of cytosolic proteases which include the proteasome complex.[12] These peptides are then transported into the endoplasmic reticulum (ER) via the TAP transporter, where they bind to empty class I molecules. This process probably involves accessory pro-

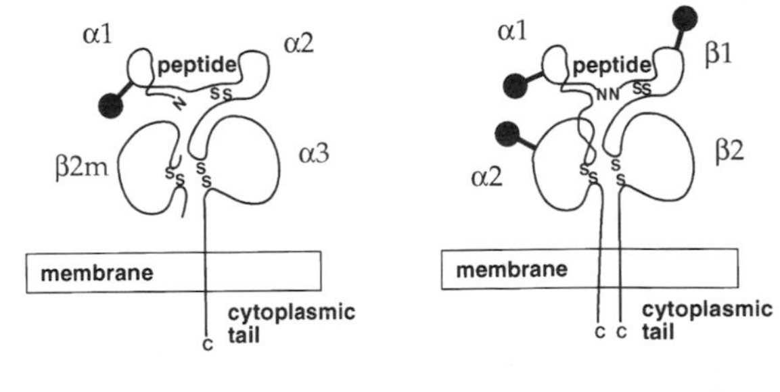

Figure 2. Schematic diagram of the MHC class I, and class II protein backbones, showing the relative locations of the domains, peptide binding site, carbohydrate attachments, orientation in membrane (N/C termini), and disulphide bond locations (SS).

teins such as calnexin and tapasin.[13] Once β_2m and peptide are bound, quality control systems in the ER allow the transport of the mature class I MHC to the Golgi complex and cell surface.

The situation for class II peptide loading is different.[7] A third polypeptide, invariant chain (Ii), binds to the $\alpha\beta$ class II chains in the ER, effectively preventing peptides from binding in this compartment. The Ii chain then ferries the complex to a specialized endosomal compartment where Ii is proteolytically cleaved. Peptides found in this acidic compartment are then able to access the groove. The removal of Ii peptides from the class II chains is catalysed by HLA-DM (H2-M in mice).

In general, MHC class I products are expressed constitutively on most nucleated cells of the body, while class II is normally expressed on professional antigen-presenting cells such as B cells, macrophages, and dendritic cells. Levels of expression of both class I and class II are sensitive to cytokines such as IFN-γ.

Antibodies

A large number of monoclonal and polyclonal antibodies against class I and class II proteins have been raised by different groups of researchers. For class I proteins, the cytoplasmic domain has been used to generate antisera reacting with both folded and unfolded heavy chains; other sera raised in rabbits preferentially react with free, unfolded class I heavy chains. Monoclonal antibodies which react with β_2m light chain as well as folded class I heavy chains bound to light chain are also available. Several monoclonal antibodies have differential specificities for different class I genes. Antibodies reactive against the free class I heavy chains are useful in Western blotting and immunohistochemistry, and most of these antibodies work very well in immunoprecipitation, or for affinity purification. Both one-dimensional (IEF) and two-dimensional electrophoretic methods (2D SDS–PAGE) are routinely used to resolve the different allelic forms of MHC products.

Anti-class II reagents likewise include monoclonal and polyclonal antibodies with varied binding specificities. Reagents are available which react with peptide-loaded class II as well as only free α and β chains. Antibodies which react with either the lumenal or cytoplasmic region of invariant chain have been made.

Genes

The human MHC is located on chromosome 6 and the genes encoding the classical class I and II products are shown schematically in Fig. 3. The non-classical class I-like genes in the human are generally clustered near the HLA-A locus. Not shown in the figure is the class III region, which lies between the class I and II regions. The class III region in humans contains genes with immune function, such as complement components, hsp70, and TNF-α and -β genes, as well as other genes with no

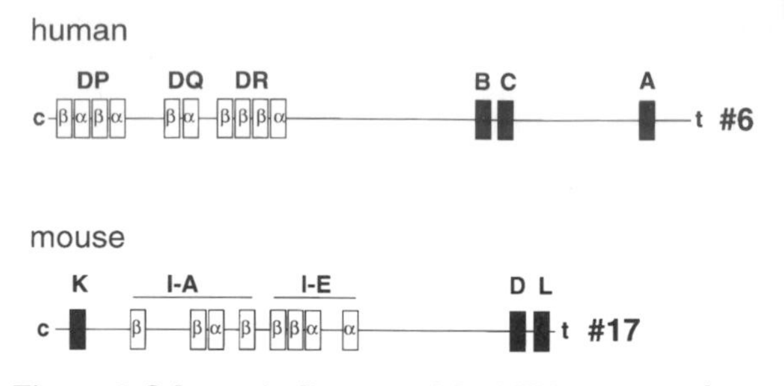

Figure 3. Schematic diagram of the MHC genes as they are found on the chromosomes in humans, and mice. The size of the human MHC is about 3500 kb, while the mouse MHC region is about 10 kb in size. The class II genes are indicated as open boxes (α or β), while the class I genes are shown as filled boxes.

obvious immune function. In mice, on chromosome 17, the class I genes H2-K, D, and L, are interrupted by the class II region, as shown. The class II α and β chains are encoded by separate genes grouped in clusters. The class I light chain, β_2m, is encoded outside the MHC complex on chromosome 15 in humans and chromosome 2 in mice. It is appropriate to note here that MHC genes are present in all vertebrate groups down to sharks.[14] A comprehensive review of the MHC was published by J. Trowsdale.[15]

Mutant phenotypes/disease states

Mice with knockouts of the class I light chain, β_2m,[16,17] have a dramatic decrease in the CD4–/CD8+ subset of T cells and greater susceptibility to tumours. Mice with interrupted class II biosynthesis, which include class II knockouts,[18] invariant chain knockouts,[19,20] and H2-M knockouts[21] likewise have compromised immune function, with major perturbations in the CD4+/CD8– subset of T cells.

Diseases that are in some way influenced by MHC products include autoimmunity, allergy, and cancer.[22–24] Certain haplotypes in humans have an increased association with diseases including type I diabetes, myasthenia gravis, lupus, and reactive arthritis. Tumour aggressiveness correlates with loss of MHC class I expression in many types of cancer.[24,25] The importance of MHC class I proteins in viral pathogenesis is clear, as certain viruses, notably those of the herpesviridae, encode products which specifically interfere with class I biosynthesis.[26,27]

Structures

X-ray crystal structures of MHC proteins may be found in the Brookhaven structural protein database (PDB). This web site is at http://www.pdb.bnl.gov/

As of June 1998, there are over 30 different crystal structures related to MHC molecules, including the revolutionary 1987 class I structure published by Pamela Bjorkman and colleagues (1HLA), class II structures

(1DLH, 1A6A), and the impressive co-crystal structure of the TcR/MHC/peptide complex (2CKB). Also within the structure database are structures of several class Ib proteins, including rat FcRn (1FRT, 3FRU) and mouse CD1 (1CD1).

Web sites

These links are accurate as of July 1998 and, due to the volatility of the web, may change without notice.

Ploegh lab pages

http://www.hms.harvard.edu/pathol/ploegh/

The Ploegh group studies antigen processing and presentation in MHC molecules. In the past several years they have been investigating viral proteins that interfere with these processes, attempting to define the molecular mechanism of the interference.

Stuart Schreiber's group

http://slsiris2.harvard.edu/home/structures/hlalig.html
This page is an image of a modified peptide-like antigen bound in MHC class I.

The Dinah Singer lab

Studies of MHC gene transcriptional regulation; web site may be found at
http://www.nci.nih.gov/intra/EIB/singerd.htm

Mouse gene knockouts

A list may be found at
http://vega.crbm.cnrs-mop.fr/bioscience/knockout/indxlef.htm

References

1. Melian, A., Beckman, E. M., Porcelli, S. A., and Brenner, M. B. (1996). *Curr. Opin. Immunol.*, **8**, 82–8.
2. Shawar, S. M., Vyas, J. M., Rodgers, J. R., and Rich, R. R. (1994). *Ann. Rev. Immunol.*, **12**, 839–80.
3. Porcelli, S., Morita, C. T., and Brenner, M. B. (1992). *Nature*, **360**, 593–7.
4. Burmeister, W. P., Gastinel, L. N., Simister, N. E., Blum, M. L., and Bjorkman, P. J. (1994). *Nature*, **372**, 336–43.
5. Story, C. M., Mikulska, J. E., and Simister, N. E. (1994). *J. Exp. Med.*, **180**, 2377–81.
6. Israel, E. J., Wilsker, D. F., Hayes, K. C., Schoenfeld, D., and Simister, N. E. (1996). *Immunology*, **89**, 573–8.
7. Wolf, P. R. and Ploegh, H. L. (1995). *Ann. Rev. Cell. Dev. Biol.*, **11**, 267–306.
8. Bjorkman, P. J., Saper, M. A., Samraoui, B., Bennett, W. S., Strominger, J. L., and Wiley, D. C. (1987). *Nature*, **329**, 506–512.
9. Brown, J. H., Jardetzky, T. S., Gorga, J. C., Stern, L. J., Urban, R. G., Strominger, J. L., and Wiley, D. C. (1993). *Nature*, **364**, 33–9.
10. Heemels, M. T. and Ploegh, H. (1995). *Annu. Rev. Biochem.*, **64**, 463–91.
11. York, I. A. and Rock, K. L. (1996). *Ann. Rev. Immunol.*, **14**, 369–96.
12. Rock, K. L., Gramm, C., Rothstein, L., Clark, K., Stein, R., Dick, L., Hwang, D., and Goldberg, A. L. (1994). *Cell*, **78**, 761–71.
13. Solheim, J. C., Carreno, B. M., and Hansen, T. H. (1997). *J. Immunol.*, **158**, 541–3.
14. Salter-Cid, L. and Flajnik, M. F. (1995). *Crit. Rev. Immunol.*, **15**, 31–75.
15. Trowsdale, J. (1995). *Immunogenetics*, **41**, 1–17.
16. Koller, B. H., Marrack, P., Kappler, J. W., and Smithies, O. (1990). *Science*, **248**, 1227–30.
17. Zijlstra, M., Bix, M., Simister, N. E., Loring, J. M., Raulet, D. H., and Jaenisch, R. (1990). *Nature*, **344**, 742–6.
18. Grusby, M. J. and Glimcher, L. H. (1995). *Ann. Rev. Immunol.*, **13**, 417–35.
19. Viville, S., Neefjes, J., Lotteau, V., Dierich, A., Lemeur, M., Ploegh, H., Benoist, C., and Mathis, D. (1993). *Cell*, **72**, 635–48.
20. Bikoff, E. K., Huang, L. Y., Episkopou, V., van Meerwijk, J., Germain, R. N., and Robertson, E. J. (1993). *J. Exp. Med.*, **177**, 1699–712.
21. Wolf, P. R. and Ploegh, H. L. (1995). *Nature*, **376**, 464–5.
22. Feltkamp, T. E., Khan, M. A., and Lopez de Castro, J. A. (1996). *Immunol. Today*, **17**, 5–7.
23. Corzo, D., Salazar, M., Granja, C. B., and Yunis, E. J. (1995). *Exp. Clin. Immunogenet.*, **12**, 156–70.
24. Ferrone, S. and Marincola, F. M. (1995). *Immunol. Today*, **16**, 487–94.
25. Garrido, F., Ruiz-Cabello, F., Cabrera, T., Perez-Villar, J. J., Lopez-Botet, M., Duggan-Keen, M., and Stern, P. L. (1997). *Immunol. Today*, **18**, 89–95.
26. Beersma, M. F., Bijlmakers, M. J., and Ploegh, H. L. (1993). *J. Immunol.*, **151**, 4455–64.
27. Hill, A., Jugovic, P., York, I., Russ, G., Bennink, J., Yewdell, J., Ploegh, H., and Johnson, D. (1995). *Nature*, **3755**, 411–15.

Craig M. Story and Hidde L. Ploegh
Department of Pathology,
Harvard Medical School,
200 Longwood Avenue,
Boston, MA 02115, USA

Myelin sheath proteins

Overview

The myelin sheath of vertebrates is the newest of nature's inventions in the nervous system. Compact myelin is a multi-layered, flat sheet of plasma membrane generated by myelinating cells (oligodendrocytes) in the central nervous system and Schwann cells in the peripheral nervous system. The primary function of the sheath is to permit saltatory conduction of the nerve impulse; the major effect of this is that an axon can conduct action potentials with extreme rapidity while maintaining axonal diameters of small calibre. In other words, in the absence of the myelin sheath, axons would need to have to have significantly larger diameters in order to sustain rapid conduction. Thus, the development of the myelin sheath has maximized the integrative abilities of the nervous system while keeping brain volume to a minimum.

The proteins and lipids which compose the myelin sheath in both nervous system subdivisions have been well characterized, and their functional roles in establishing the myelin sheath superstructure have been in some cases determined. During the process of myelin formation, the myelinating cell establishes contact with an axon and then with its own plasma membrane as it begins to enwrap the axon. After several turns of loosely wrapped plasma membrane have been generated, the cytoplasm is removed and the plasma membrane layers become closely apposed, or 'compacted'. Complete compaction gives the hallmark ultrastructure of a myelinated nerve fibre, with a tightly wrapped myelin membrane surrounding an axon.

The myelin sheath, whether central or peripheral, is organized into discrete subdomains (see Fig. 1). Both types of myelin sheath have a highly specialized architecture which consists of 'islands' of compact myelin bordered by a highly convoluted but continuous cytoplasmic channel network. Particularly in Schwann cell derived myelin, channels are found in register across adjacent lamellae such that one channel loop abuts another all along its length. While membrane adhesion within the compact myelin subdomains is mediated by protein zero (P_0), the myelin basic proteins (MBPs) in the peripheral nervous system, and certain proteolipid proteins (PLPs) and MBPs in the central nervous system, at least in the peripheral nervous system the cytoplasmic channel membranes adhere to one another via adhesive interactions (cadherins and integrins) which are likely to be common elements in all cells of epithelial derivation.

■ *David R. Colman*
Brookdale Center for Developmental and Molecular Biology, The Mount Sinai School of Medicine, One Gustave Levy Place, New York, NY 10029, USA

■ *Marie T. Filbin*
Department of Biology, Hunter College, 695 Park Avenue, New York, NY 10021, USA

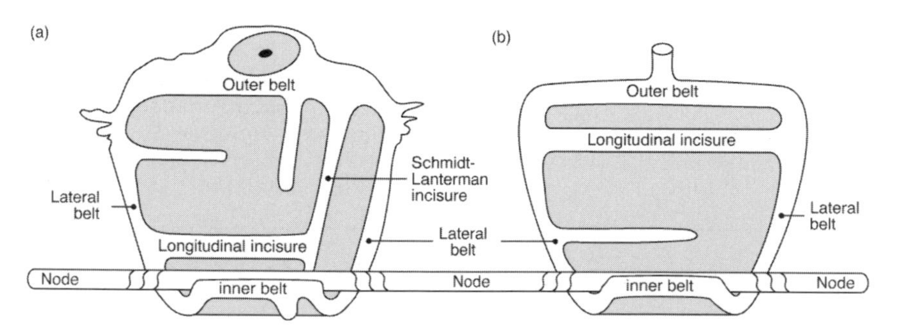

Figure 1. In the peripheral nervous system, each Schwann cell myelinates a single internodal axon segment. In the central nervous system, one oligodendrocyte may myelinate several internodes via numerous cytoplasmic 'arms'. In both myelin sheath types, compact myelin (shaded) is bordered by cytoplasmic channels which are continuous with one another.

Myelin associated glycoprotein (MAG)

MAG is a transmembrane glycoprotein of the immunoglobulin (Ig) superfamily expressed by oligodendrocytes and Schwann cells. It has five Ig domains that mediate interactions between myelinating glia and axons during the development of the myelin sheath. MAG is also a myelin-associated growth inhibitory protein that blocks axon regeneration in the mature central nervous system (CNS).

Synonymous names

1B236.[1]

Homologous proteins

CD22, CD23, sialoadhesin and Schwann cell myelin protein.[2]

Protein properties

MAG runs as a broad band of about 100 kDa on SDS–PAGE. N-linked carbohydrates contribute 30 per cent of its weight, and the protein backbone is either 67 or 72 kDa (S-MAG, L-MAG) due to alternative splicing of the cytoplasmic domain (Fig. 1). The high percentage of acidic amino acids and the content of sialic acid and sulphate residues contribute to a pI in the range of 3–4.5. MAG is also palmitoylated. Other members of the sialoadhesion cell surface adhesion protein family contain variable numbers of Ig domains and a single transmembrane segment, and bind sialylated glycans.

The cytoplasmic domain of both forms of MAG can be phosphorylated on tyrosine, serine and threonine, both *in vivo* and *in vitro*.[3,4] MAG has potential protein kinase C (PKC) phosphorylation sites,[5] and PKC can phosphorylate both L-MAG and S-MAG.[4] During myelination, L-MAG expression precedes that of S-MAG, with S-MAG predominant in the adult CNS (1 per cent of myelin protein). In the peripheral nervous system (PNS), MAG is a minor myelin protein (0.1 per cent of myelin protein).[4,6] The kinase Fyn associates with the cytoplasmic domain of L-MAG, and Fyn is activated at the early stages of myelination.[7] Moreover Fyn deficient mice exhibit hypo- myelination. The localization of MAG at the interface between myelin sheaths and axons, together with the above data, implicate MAG in the process of myelination.

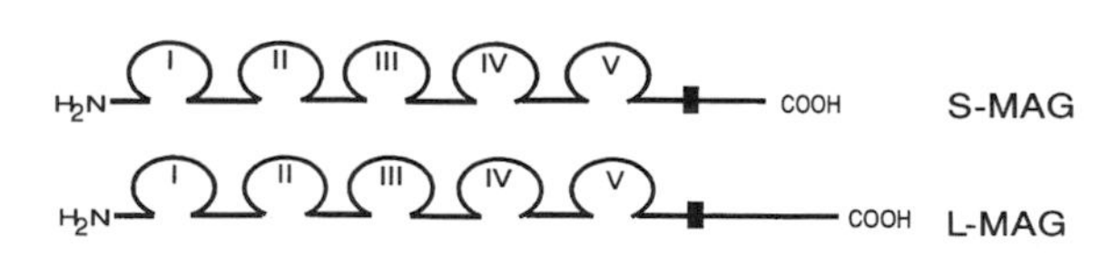

Figure 1. Schematic representation of MAG. MAG has five Ig domains, a transmembrane domain (rectangle), and two splice variants of the cytoplasmic domain.

By immunocytochemistry, MAG staining is seen throughout the wraps of loose myelin during development of the myelin sheath; after compaction MAG is restricted to the first wrap of myelin i.e. the periaxonal space, the Schmidt–Lanterman incisures, the paranodal loops, and the inner and outer mesaxons.[8,9] While cell culture experiments suggest that MAG is important in the process of myelination,[10] axons in the CNS and PNS of MAG-null mutant mice are myelinated. In younger mice lacking MAG there are subtle morphological abnormalities in the periaxonal region and cytoplasmic collar.[11,12] Older MAG null mutant mice develop a neuropathy of axonal and myelin degeneration,[13] therefore implicating MAG in the maintenance of myelin.

Recently, MAG has been identified as one of the growth-inhibitory proteins present in CNS myelin.[14,15] MAG induces growth cone collapse when growth-cones encounter MAG-coated beads, but soluble MAG is not reported to induce collapse, suggesting that MAG must be anchored as substrate to exert its growth-inhibitory activity.[11] MAG substrates inhibit neurite outgrowth from a wide variety of CNS neurones. Dorsal root ganglion (DRG) neurones undergo a developmental switch in their response to MAG since MAG promotes neurite growth of embryonic neurones, but inhibits MAG growth of postnatal neurones.[16] Studies of regeneration of injured axons in MAG-null mutant mice demonstrate that, *in vivo,* MAG has growth-inhibitory activity; regeneration is improved in the absence of MAG.[11,17]

Purification

MAG was first identified as the major glycoprotein of CNS myelin by [^{3}H]fucose labelling. It can be purified from myelin by extraction with lithium diiodosalicylate and phenol, or by DEAE chromatography followed by immunoaffinity chromatography[14] with monoclonal antibody 513.

Activities

Carbohydrate binding

MAG binds with high specificity to cell-surface carbohydrates which contain the terminal α2,3-*N*-acetylneuraminic acid–galactose moiety.[18] Recombinant MAG–Fc fusions have been used to demonstrate that the lectin activity is contained within the three N-terminal Ig domains.[2] Inhibition of this lectin activity, either by removal of cell-surface sialic acid or incubation with soluble sialic-acid- bearing sugars, blocks both the stimulatory and inhibitory activities of MAG. For some neurones trypsin treatment blocks MAG activities, while other studies have demonstrated MAG binding to neuronal gangliosides.[16] The identity of the sialoglycoprotein target on neurones is unknown.

MAG mediated cell–cell interactions

Several studies have demonstrated that MAG mediates cell–cell adhesive interactions. MAG antibodies inhibit both homophilic oligodendrocyte–oligodendrocyte and heterophilic oligodendrocyte–neurone aggregation.[19] Recombinant MAG mediates cell-surface binding when incorporated into liposomes or expressed in cultured fibroblasts.[20,21] These interactions appear to involve the lectin activity of MAG, as removal of sialic acid from the surface of cultured neurones inhibits both MAG binding and associated neurite outgrowth effects.[16] The functional consequence of MAG interaction with neuronal growth cones is an inhibition of neurite outgrowth (Fig. 2).[11,14,15]

Genes

MAG cDNAs have been cloned from rat (N14871,[5] X05301,[22] M16800[1]), human (M29273[23]) and mouse (M31811[24]). The rat MAG gene consists of 13 exons over ~16 kB; including exon 12, which has a stop codon that terminates the p67 protein. Alternative RNA splicing deletes exon 12 to produce the larger p72 MAG isoform. The MAG locus has been mapped to human chromosome 19 q12–13, and to mouse chromosome 7.[25,26]

Antibodies

The 513 antibody (Boehringer Manniheim) recognizes a three-dimensional epitope near the hinge region of native MAG, and 513 antibody blocks adhesive interactions between native MAG and neurones.[27] The GenS3 and B11F7 monoclonal antibodies have been mapped to amino acids 167–177, and 375–388, respectively, and both epitopes are only accessible in denatured MAG.[28] An important carbohydrate epitope, HNK-1, is highly expressed in MAG and is present on N-linked carbohydrates in Ig domain 4 or 5.[29] Various other monoclonal and polyclonal antibodies against MAG exist.

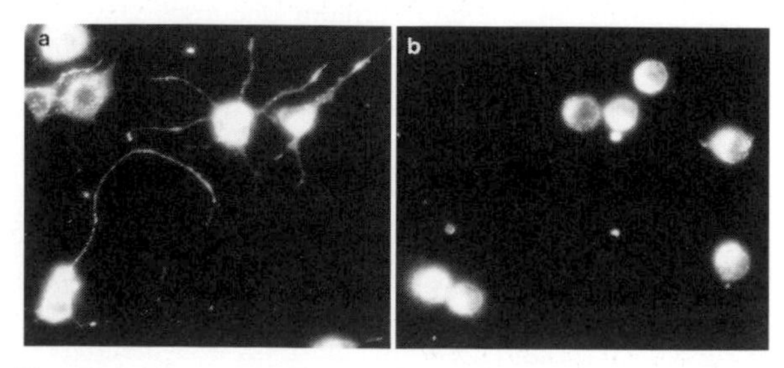

Figure 2. Fluorescently labelled NG108 neuroblastoma cells grow long neurites on control poly-L-lysine substrates (a), but on MAG substrates the cells remain rounded and do not extend neurites (b).

Structure

X-ray crystallographic and NMR structures are not available. Analytical ultracentrifugation analysis suggests that the MAG ectodomain is in a folded, non-linear conformation.[30] This conformation has been supported by electron microscope analysis of rotary-shadowed MAG extracellular domain, where MAG appears to be rod-shaped and may be bent around the third Ig domain to bring the first and fifth Ig domains in close apposition.[27]

References

1. Lai, C., Brow, M. A. Nave, K. -A. Noronha, A. B. Quarles, R. H. Bloom, F. E. Milner, R. J., and Sutcliffe, J. G. (1987). *Proc. Natl Acad. Sci. USA*, **84**, 4337–41.
2. Kelm, S., A. *et al.* (1994). *Curr. Biol.*, **4**, 965–72.
3. Yim, S. H., Toda, K. Goda, S. and Quarles, R.H. (1995). *J. Mol. Neurosci.* **6**, 63–74.
4. Eichberg, J. and Iyer, S. (1996). *Neurochem Res.*, **21**, 527–35.
5. Arquint, M., Roder, J. C. Chia, L. -S. Down, J. Wilkinson, D. Bayley, H. Braun, P. and Dunn, R. J. (1987). *Proc. Natl Acad. Sci. USA*, (1987). **84**: 600–604.
6. Pedraza, L., Spagnol, G. Latov, N. and Salzer, J. L. (1995). *J. Neurosci. Res.*, **40**, 716–27.
7. Umemori, H., Sato, S. Yagi, T. Aizawa, S. and Yamamoto, T. (1994). *Nature*, **367**, 572–6.
8. Sternberger, N. H., Quarles, R. H. Itoyama, Y. and Webster, H. D. (1979). *Proc. Natl Acad. Sci. USA*, (1979). **76**, 1510–14.
9. Martini, R. and Schachner, M. (1986). *J. Cell Biol.*, **103**, 2439–48.
10. Owens, G. C. and R. P. Bunge, R. P. (1991). *Neuron*, **7**, 565–75.
11. Li, M., Shibata, A. Li, C. Braun, P. E. McKerracher, L. Roder, J. Kater, S. B. and David, S. (1996). *J. Neurosci. Res.*, **46**, 404–14.
12. Montag, D., *et al.* (1994). *Neuron*, **13**, 229–46.
13. Fruttiger, M., Montag, D. Schachner, M. and Martini, R. (1995). *Eur. J. Neurosci.*, **7**, 511–5.
14. McKerracher, L., David, S. Jackson, D. L. Kottis, V. Dunn, R. J. and Braun, P. E. (1994). *Neuron*, **13**, 805–11.
15. Mukhopadhyay, G., Doherty, P. Walsh, F. S. Crocker, P. R. and Filbin, M. T. (1994). *Neuron*, **13**, 757–67.
16. DeBellard, M. Tang, E., S. Mukhopadhyay, G. Shen, Y. J. and Filbin, M. T. (1996). *Mol. Cell. Neurosci.*, **7**, 89–101.
17. Schafer, M., Fruttiger, M. Montag, D. Schachner, M. and Martini, R. (1996). *Neuron*, **16**, 1107–13.
18. Collins, B. E., Yang, L. J. Mukhopadhyay, G. Filbin, M. T. Kiso, M. Hasegawal, A. and Schnaar, R. L. (1997). *J. Biol. Chem.*, **272**, 1248–55.
19. Poltorak, M., Sadoul, R. Keilhauer, G. Landa, C. Fahrig, T. and Schachner, M. (1987). *J. Cell Biol.*, **105**, 1893–9.
20. Johnson, P. W., Abramow-Newerly, W. Seilheimer, B. Sadoul, R. Tropak, M. B. Arquint, M. Dunn, R. J. Schachner, M. and Roder, J. C. (1989). *Neuron*, **3**, 377–85.
21. Afar, D. E. H., Marius, R. M. Salzer, J. L. Stanners, C. P. Braun, P. E. and Bell, J. C. (1991). *J. Neurosci. Res.*, **29**, 429–36.
22. Salzer, J. L., Holmes, W. P. and Colman, D. R. (1987). *J. Cell Biol.*, **104**, 957–965.
23. Sato, S., Fujita, N. Kurihara, T. Kuwano, R. Sakimura, K. Takahashi, Y. and Miyatake, T. (1989). *Biochem. Biophys. Res. Commun.* **163**, 1473–80.

24. Nakano, R., Fujita, N. Sato, S. Inuzuka, T. Sakimura, K. Ishiguro, H. Mishina, M. and Miyatake, T. (1991). *Biochem. Biophys. Res. Commun.*, **178**, 282–90.
25. Barton, D. E., Arquint, M. Roder, J. Dunn, R. and Francke, U. (1987). *Genomics*, (1987). **1**, 107–112.
26. Garcia, E., *et al.* (1995). *Genomics*, **27**, 52–66.
27. Meyer-Franke, A., Tropak, M. B. Roder, J. C. Fischer, P. Beyreuther, K. Probstmeier, R. and Schachner, M. (1995). *J. Neurosci. Res.* **41**, 311–23.
28. Tropak, M. B. and Roder, J. C. (1994). *J. Neurochem.* **62**, 854–62.
29. Fahrig, T., Probstmeier, R. Spiess, E. Meyer-Franke, A. Kirchhoff, F. Drescher, B. and Schachner, M. (1993). *Eur. J. Neurosci.* **5**, 1118–26.
30. Attia, J., Hicks, L. Oikawa, K. Kay, C. M. and Dunn, R. J. (1993). *J. Neurochem.*, **61**, 718–26.

L. McKerracher
Department of Pathology,
University of Montreal,
Montreal, Canada.

R. J. Dunn
Center for Research in Neuroscience,
Montreal General Hospital Institute,
Montreal, Canada H3G 1A4.

Myelin basic proteins (MBPs)

The myelin basic proteins (MBPs) are structural proteins of the myelin membrane in the nervous system. A striking property of the major MBP isoforms is their ability to 'seal' the cytoplasmic leaflets of the myelin membrane bilayer to form the 'major dense line'. The most direct insight into the function of MBP comes from the study of a recessive neurological mutant, the *shiverer* mouse, in which a major deletion of the MBP gene occurs. In this mutant, CNS myelin is uncompacted, demonstrating that MBPs play a critical role in maintaining the compact structure of the myelin sheath.

Despite extensive primary sequence homology, the individual MBPs exhibit different intracellular localization, so that the isoforms may not be functionally equivalent in myelinating cells. While the major MBPs are intracellular, membrane binding adhesion molecules, some of the less abundant isoforms that are expressed early in development localize to the nucleus and so may have regulatory effects on the myelination program (Fig. 1).

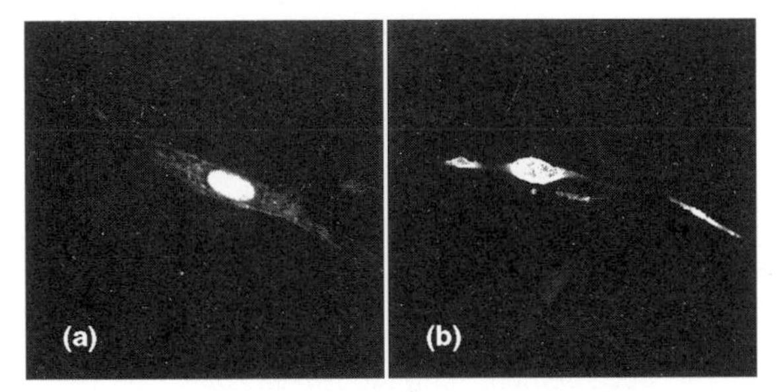

Figure 1. Intracellular distribution of MBP–GFP fusion proteins in Schwann cells. MBP–GFP constructs were obtained by subcloning the sequences corresponding to either 14 kDa MBP or 21 kDa MBP into the multiple cloning site of the mammalian expression vector pEGFP (Clontech). Optical sections through the equatorial plane of nuclei of cells transiently expressing 21MBP-GFP (a) or 14MBP-GFP (b) are shown.

Protein properties

The MBPs are a family of closely related polypeptide isoforms. As the name 'basic protein' implies, all of these isoforms have numerous positively charged amino acids which are relatively uniformly distributed throughout the molecule. The MBPs are the product of a single gene composed of seven exons spanning 32 kb. Alternative splicing of exons 2, 5 and 6 generates the major isoforms.[1–3] In mouse, developmental regulation of the expression of these isoforms has been observed.[4] The isoforms containing exon II are more abundant early in myelinogenesis, whereas in compact myelin the isoforms lacking this exon are most abundant. Isoforms containing exon II appear to have different properties and perhaps different functions in myelination.[5]

MBP as an intracellular adhesion molecule

The MBPs are highly enriched in purified myelin and can be immunochemically localized to the major dense line, which is the close apposition of the cytoplasmic faces of the myelin membrane. The MBPs have a key role in the molecular architecture of CNS myelin. Thus, in the *shiverer* mutant mouse, in which the MBP gene is functionally deleted, the major dense line fails to form and some oligodendrocyte cytoplasm remains there, indicating that MBPs play a critical role in maintaining the compact structure of the myelin sheath. Once compacted, the myelin membrane is highly stable.

The association of MBPs with membranes is primarily electrostatic, by virtue of its high positive charge. Thus, MBPs are easily extracted in soluble form under conditions of low pH or high ionic strength. The binding properties of the 18.5 kDa MBP isoform in solution have been extensively studied. This molecule is highly reactive with lipids and membranes, binding preferentially to acidic lipids. The binding of lipids to MBP is probably followed by partial dehydration of the bilayer surface.[6] This loss of water might contribute to the close apposition ('compaction') of the two cytoplasmic aspects of the myelin membrane (which in mature compact myelin are 30–40 Å apart). Further, MBP is capable of inducing membrane fusion in the presence of Ca^{2+}.[7]

The high reactivity of MBP with membranes presents a challenge to the protein synthesis machinery of myelin-producing cells. The 14 kDa and 18.5 Kda MBP isoforms expressed in cultured *shiverer* oligodendrocytes show a plasma membrane distribution. Moreover, a perinuclear concentration of MBP immunoreactivity is never seen in cultured normal oligodendrocytes. Thus, myelinating cells must have a mechanism to directly target the MBP molecules to the forming myelin sheath. This hypothesis was confirmed by expression of the cDNAs encoding 14 and 18.5 kDa isoforms in non-glial cells. The transfected isoforms had a distribution that was predominantly perinuclear. This distribution would be expected if translation ensued shortly after export of mRNA from the nucleus.

Intracellular movement of MBP mRNAs

Clues to how the MBPs are targeted were first obtained in subcellular fractionation experiments. Polysomes purified from myelin fractions were shown to be enriched in MBP mRNA,[8] suggesting that myelin-producing cells could transport MBP mRNA to the site of myelin synthesis. This observation has been confirmed by *in situ* hybridization. Early in development, prior to myelin compaction, MBP mRNA appears to be distributed in oligodendrocyte cell bodies. Later, the hybridization pattern becomes diffusely distributed in regions enriched in myelin sheaths.[9–11] The movement of MBP mRNA along oligodendrocyte processes has been visualized *in vitro*.[12] Recently, transport and localization elements have been defined in MBP mRNAs.[13]

The precise mechanism by which MBP mRNA is transported remains to be elucidated. It could involve recognition of a sequence within the mRNA or a sequence in the nascent MBP polypeptides. The transport is likely to be facilitated by association of the mRNA transport complex with a cytoskeletal element, such as microtubules.[14]

Purification

MBPs can be purified from CNS myelin.[15,16] The MBPs are released and solubilized from acetone-delipidated myelin by dilute acid, and purified by ion-exchange chromatography.

Antibodies

A monoclonal antibody against MBP which can be used for immunolabelling of brain cryosections as well as of cells in culture is available from Boehringer Mannheim.

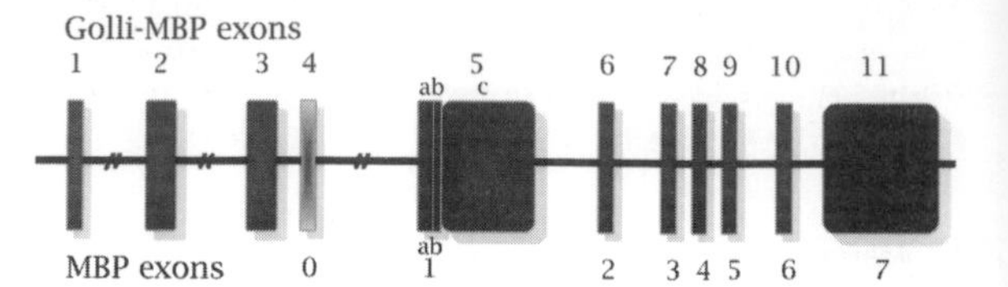
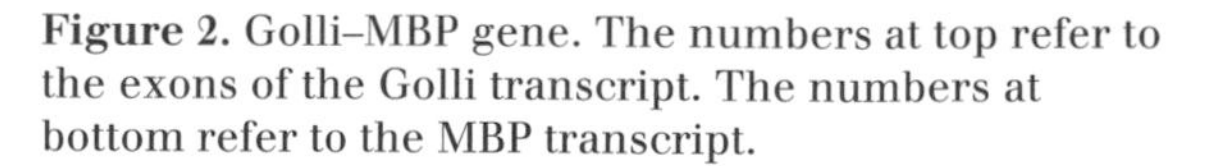

Figure 2. Golli–MBP gene. The numbers at top refer to the exons of the Golli transcript. The numbers at bottom refer to the MBP transcript.

This antibody recognizes MBP from human, monkey, rabbit, bovine and rat.

Genes

A large gene (Fig. 2) which includes the MBP gene has been characterized.[17,18] This gene is over 100 kb in length and has been designated Golli (for 'genes expressed in oligodendrocyte lineage')–MBP. The Golli–MBP gene contains at least 11 exons. In mouse, the Golli–MBP gene produces two families of mRNAs from different transcription start sites that generates either MBPs or Golli proteins. Some of the Golli transcripts include sequences that are also contained in MBP transcripts and could be translated into polypeptides with immunological cross-reactivity with MBP. Others are antigenically unrelated to MBP. Transcripts produced from the classical MBP promoter are confined to myelin-forming cells. On the other hand, transcripts produced from the Golli promoter, which lies 70 kb upstream, are present in neurones as well as oligodendrocytes. These two major promoters of the Golli–MBP gene are under independent developmental regulation.[19]

Mutant phenotypes/disease states

Mutations

The spontaneous *shiverer* mutation of the mouse (*shi*) has been mapped to chromosome 18. The genetic defect is a large deletion in the MBP gene which encompasses exons 3–7 (Golli–MBP exons 7–11).[20,21] *Shiverer* mice are unable to synthesize any MBP isoform. The mutation results in a visible 'shivering' and convulsive phenotype. The intact MBP gene has been re-introduced into *shi* mice in rescue experiments, with a subsequent reduction in the clinical signs of the neurological disorder cause by the mutation.[22]

The mutation *myelin deficient* (shi^{mld}) is allelic to *shi*. The molecular defect of shi^{mld} is a duplication–inversion of the entire MBP gene.[23] Transcription results in antisense MBP mRNA. The shi^{mld} phenotype is milder; there are myelin-forming MBP-positive oligodendrocytes which have apparently eliminated the abnormal gene.

MBPs and autoimmunity

Injection of MBP into susceptible strains of rodents produces an autoimmune disease, termed experimental allergic encephalomyelitis (EAE), which has been studied as a model for multiple sclerosis. In pioneering work in this area, EAE was induced by injection of total brain homogenates. Later, it was found that MBP was the main encephalitogenic agent in these homogenates.

Structure

The conformation of MBP varies depending on the environment. In aqueous solutions it is different from that in organic solvents or when interacting with detergents and lipids. Thus, the structural features of MBP obtained from those systems may not apply to MBP in myelin, where the protein is located at the interface of a solid and an aqueous phase.

References

1. de Ferra, F. *et al.* (1995). *Cell*, **43**, 721–7.
2. Takahashi, A., Teplow, D. B., Prusiner, S. B., and Hood, L. (1985). *Cell*, **42**, 139–44.
3. Newman, S., Kitamura, K., and Campagnoni, A. T. (1987). *Proc. Natl Acad. Sci. USA*, **84**, 886–90.
4. Barbarese, E., Carson, J. H., and Braun, P. E. (1978). *J. Neurochem.*, **31**, 779–82.
5. Pedraza, L., Fidler, L., Staugaitis, S. M., and Colman, D. R., (1997). *Neuron*, **18**, 579–89.
6. Sankaram, M. B., Brophy, P. J., and Marsh, D. (1989). *Biochemistry*, **28**, 9699–707.
7. Monferran, C. G., Maggio, B., Roth, G. A., Cumar, F. A., and Caputto, R. (1979). *Biochim. Biophys. Acta*, **553**, 417–23.
8. Colman, D. R., Kreibich, G., Frey, A. B., and Sabatini, D. D. (1982). *J. Cell Biol.*, **95**, 598–608.
9. Jordan, C., Friedrich, V., and Dubois-Dalcq, M. (1989). *Neurosci.*, **9**, 248–57.
10. Trapp, B. D. *et al.* (1987). *Proc. Natl Acad. Sci. USA*, **84**, 7773–7.
11. Verity, A. N., and Campagnoi, A. T. (1988). *J. Neurosci. Res.*, **21**, 238–48.
12. Ainger, K. *et al.* (1993). *J. Cell Biol.*, **123**, 431–41.
13. Ainger, K. *et al.* (1997). *J. Cell Biol.*, **138**, 1077–87.
14. Brophy, P. J., Boccaccio, G. L., and Colman, D. R. (1993). *Trends Neurosci.*, **16**, 515–21.
15. Eylar, E. H., Salk, J., Beveridge, G. C., and Brown, L. V. (1969). *Arch. Biochem. Biophys.*, **132**, 34–48.
16. Martenson, R. E., Deibler, G. E., and Kies, M. W. (1970). *Biochim. Biophys. Acta*, **200**, 353–62.
17. Campagnoni, A. T. *et al.* (1993). *J. Biol. Chem.*,**268**, 4930–38.
18. Pribyl, T. M. *et al.* (1993). *Proc. Natl Acad. Sci. USA*, **90**, 10695–9.
19. Landry, C. F. *et al* (1996). *J. Neurosci.*, **16**, 2452–62.
20. Roach, A., Boylan, K. Horvath, S., Prusiner, S. B., and Hood, L. E. (1983). *Cell*, **34**, 799–806.
21. Roach, A., Takahashi, N., Pravtcheva, D., Ruddle, F., and Hood, L. (1985). *Cell*, **42**, 149–55.
22. Readhead, C. *et al.* (1987). *Cell*, **48**, 703–12.
23. Popko, B., Puckett, C., and Hood, L. (1988). *Neuron*, **1**, 221–5.

Liliana Pedraza
Brookdale Center for Developmental and Molecular Biology, The Mount Sinai School of Medicine, One Gustave Levy Place, New York, NY 10029, USA.

Myelin oligodendrocyte glycoprotein (MOG)

The myelin oligodendrocyte glycoprotein (MOG) is expressed specifically in the central nervous system of mammals during myelinogenesis. The *Mog* gene is located in the major histocompatibility complex (MHC) in mice and humans and encodes a protein that is a member of the immunoglobulin (Ig) superfamily with a single variable Ig-like domain. MOG shares homology with several other proteins whose genes also map within MHC regions. While in mice only one transcript is generated from the *Mog* gene, six alternatively spliced mRNAs have been identified in humans and one of these contains an exon not present in the mouse genome.

Synonymous name

M2.

Protein properties

Myelin oligodendrocyte glycoprotein (MOG) is an integral membrane protein with a 27 amino acid cleaved signal sequence that is expressed specifically by oligodendrocytes in the central nervous system (CNS).[1] The *Mog* gene probably arose recently relative to the evolution of the CNS myelin membrane because this protein has been identified in mammals but not in birds or fish which nonetheless synthesize compact myelin.[2] The overall level of amino acid conservation of MOG is at least 93 per cent between the human protein and the mouse, rat or bovine proteins. On SDS–polyacrylamide gels, MOG runs predominantly as a doublet with an M_r of 26 and 28 kDa (a dimer band runs at 54 kDa) that is reduced to a single band of 25 kDa by *N*-glycanase treatment.[3] The mature apoprotein in rodents comprises 218 amino acids with a molecular mass of 25.0 kDa and is *N*-glycosylated on a single site at Asp31 in the extracellular domain of the mature polypeptide.[1] The extracellular domain of MOG contains a single variable region immunoglobulin-like (Ig-like) domain of 75 amino acids, placing the protein in the superfamily (IgSF) of adhesion and recognition proteins and thereby indicating potential functions of this protein.[1] Also included in the mature polypeptide are two hydrophobic stretches of amino acids with sufficient length to serve as transmembrane domains. However, only the most N-terminal of these domains appears to span the bilayer and the resulting type I membrane protein topology is identical to that of other transmembrane IgSF members.[4] Nonetheless, the presence of more than one hydrophobic domain in an IgSF molecule is very unusual, being shared by only one other family member, ovarian antigen 3/integrin-associated protein.[5]

In primary cell cultures derived from CNS tissue, MOG is expressed only in differentiated oligodendrocytes and subsequent to the expression of other markers of oligodendrocyte differentiation.[6,7] *In vivo* MOG expression follows the temporal and spatial expression patterns of the other major myelin-specific genes such as MBP and PLP, first appearing in white matter tracts perinatally in the medulla and moving in a caudorostral gradient.[7,8] At the cellular level, MOG is absent from the bulk of the compact myelin sheath and is found in the outermost lamellae as well as the oligodendrocyte plasma membrane.[9] In these locations, the protein is exposed to the extracellular environment and, significantly, the immune system. When expressed in transfected fibroblasts, MOG is principally found in the plasma membrane (Fig. 1).

Purification

MOG is a relatively insoluble glycoprotein and as such is most easily purified from mammalian white matter or purified myelin in the presence of detergents.[3,10] Recombinant MOG has been purified from *E. coli*.[30]

Activities

While the function of MOG is currently unknown, this protein shares considerable amino acid identity in its Ig domain with several other proteins[11,12] (Fig. 2) that lie within major histocompatibility complex (MHC) regions; these include butyrophilin, which is an IgSF member that is expressed in lactating mammary glands; two recently identified butyrophilin-like proteins, called BT2.1 and BT3.2/B7–3 (GenBank accessions U90142 and U90144, respectively), which have unknown tissue distributions; and the polymorphic B-G antigens, which form a cluster of closely related genes in the MHC of chickens with overlapping but distinct tissue expression patterns.[13,14] Interestingly, MOG, butyrophilin, BT2.1 and BT3.2 all map to chromosome 6p21.3–22 in humans, suggesting that they may also be part of a gene cluster, perhaps with similar functions but distinct expression patterns (e.g. MOG expression in the brain and butyrophilin expression in lactating mammary tissue). These proteins have been included in the B7 (CD80) subfamily of IgSF proteins[12] and, although published data regarding the co-receptor functions of B7–1 and B7–2 have shed little light on the role that MOG plays in myelin, the homology between these proteins as well as the *in vivo* modification of a proportion of MOG apoprotein molecules with the L2/HNK-1 epitope[15] has fuelled speculation that MOG may act as a co-receptor for T-cell activation or play a role in antigen presentation to these cells in the CNS.[16]

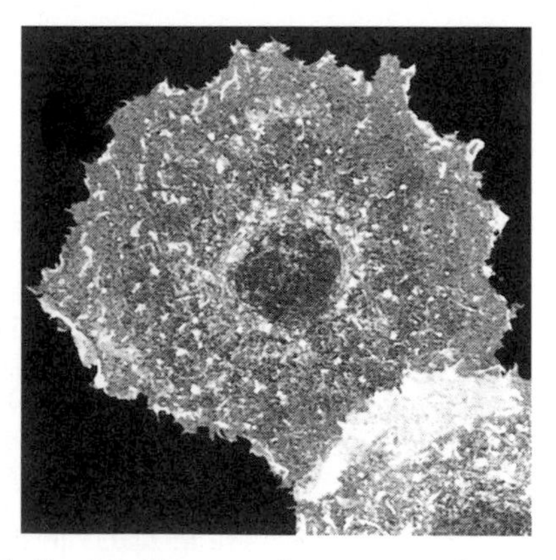

Figure 1. Confocal immunofluorescence image of two transfected COS-7 cells expressing the rat *Mog* cDNA under transcriptional control of the cytomegalovirus immediate early promoter. MOG protein, detected with the 8–18C5 monoclonal antibody, is abundant at the cell surface.

Antibodies

Monoclonal and polyclonal antibodies have been generated against MOG.[2,4,17]

Genes

The proximal promoter region of the mouse *Mog* gene has been analysed *in vitro* and a number of potential

```
                                  ⇓
HuMOG     AGQF.VIGPR.PIRALVGDE.ELPCRISP.K.AT.VEVGWYR...S..V.LYRNGKD.D
HuButyro  SAPF.VIGPP.PILAVVGED.ELPCRLSP.A.AE.LELRWFR...S..V.VHRDGRE.E
HuBT2.1   SAHF.VVGPT.PILATVGEN.TLRCHLSP.K.AE.MEVRWFR...S..V.VYKGGRE.T
HuBT3.2   SAQF.VLGPS.PILAMVGED.DLPCHLFP.M.AE.MELKWVS...R..V.VYADGKE.E
ChB-G     SAQL.VVAPS.RVTAIVGQD.VLRCHLCP.K.AW.LDIRWIL...S..V.HYQNG--.D
          +    *+ *   + * **    * *++ *   *  +++ *        *     *

                                                 ⇓
HuMOG     .DQA.EYRGRT.LLKDAIG.GKVTLRI.NVRFSDEG.FTCFFRDH..QEEAAMELKVED
HuButyro  .EQM.EYRGRA.LVQDGIA.GRVALRI.GVRVSDDG.YTCFFRED..YEEALVHLKVAA
HuBT2.1   .EQM.EYRGRT.FVSKDIS.GSVALVI.NITAQGNG.YRCYFQEG..YDEAILHLVVAE
HuBT3.2   .RQS.PYRGRT.ILRDGIT.GKAALRI.NVTASDSG.YLCYFQDG..YEKALVELKVAA
ChB-G     .GQM.GYKGRT.LLRDGLY.GNLDLRI.AVSTSDSG.YSCAVQDG..YADAVVDLEVSD
            *   *+**+  +   +  *   * *  +     * + *  ++      * + * *
```

Figure 2. The immunoglobulin-like domains of several proteins have been aligned to demonstrate their extensive homologies. The cysteine residues that participate in disulphide bonds are indicated by the downward-pointing arrows. Identical (*) or conserved (+) amino acids in four or more of the proteins at each position are shown. Non-conserved amino acids are indicated by dots, and gaps in the sequences by dashes.

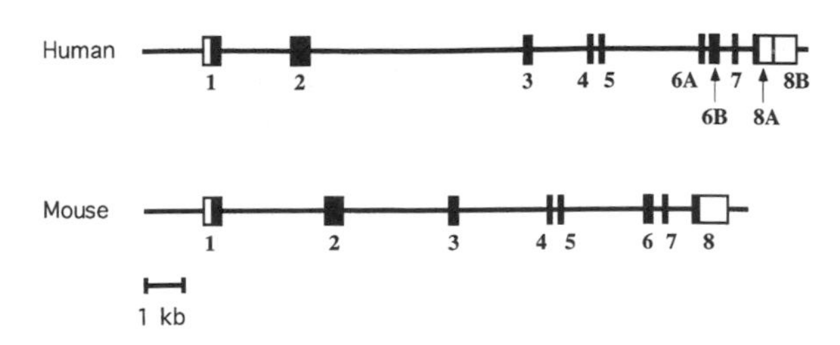

Figure 3. The structures of the mouse, and human *MOG* genes. While the overall layout of these genes is similar, the human gene contains an exon not present in mice (exon 6A), and exon 8 can be differentially spliced to yield two distinct C termini for the encoded protein isoforms.

recognition sites for DNA-binding proteins were identified, including several that are found in promoters of other myelin genes.[18] The structure of the *Mog* gene has been characterized in mouse and human (Fig. 3). The 5′ splice sites for each intron are positioned in the coding region in similar fashion to most IgSF members,[19] i.e. splicing occurs between the first and second nucleotides of the final codon specified in exons 1–7. Exon 8 in mice contains a long 3′ untranslated region which is a feature typical of many myelin genes. In humans, an additional exon has been identified at the 3′ end of the gene.[20] Surprisingly, the human *MOG* gene gives rise to at least six alternatively spliced isoforms, including a secreted ectodomain, while only a single mRNA has been found in rodents.[21] The human and mouse genes have been physically mapped to 6p21.3–22 and the syntenic region of chromosome 17 within the M region of the MHC, respectively.[22] The cDNAs encoding MOG from several species have been sequenced including: bovine L21757 (GenBank); U64564–71, U18798–801, U18803, U18840, U18843 (GenBank); mouse U64572 (GenBank) and rat M99485 (GenBank). Genomic sequences are available for human Z48051 (EMBL) and mouse L29498–503 (GenBank).

Mutant phenotypes/disease states

The MOG protein was originally identified as an autoantigen from guinea-pig myelin and was subsequently found in the CNS of many mammals. Although this protein is, quantitatively, a minor myelin protein, immunization of guinea pigs with whole brain homogenate or purified myelin emulsified in adjuvant nonetheless elicits a strong immune response against MOG leading to complement fixing antibody-mediated CNS demyelination and a T-cell response against oligodendrocytes.[23] The selective oligodendrocyte killing and resulting pathology associated with this form of experimental allergic encephalomyelitis (EAE) may bear greater similarity to demyelinating lesions in the CNS of patients with multiple sclerosis (MS) than has been observed for forms of EAE induced by other myelin proteins such as MBP and PLP.[24,25] Furthermore, protocols that induce EAE in rats by adoptive transfer of MBP-reactive T-cells in combination with injections of MOG antibody generate persistent demyelinating plaques specifically in the CNS and have attracted considerable attention as useful models of MS.[24–26] Attesting to the relevance of MOG as an autoantigen, several groups have shown that a proportion of MS patients have high titres of anti-MOG antibodies in the cerebrospinal fluid and harbour MOG-reactive T-cells in the blood.[27–29]

Structure

The structure is not available.

Web sites

Detailed summaries of the *Mog* gene and disease phenotypes associated with mutations can be obtained from the 'Online Mendelian inheritance of man' site at http://www3.ncbi.nlm.gov/Omim/.

References

1. Gardinier, M. V., Amiguet, P., Linington, C., and Matthieu, J. M. (1992). *J. Neurosci., Res.*, **33**, 177–87.
2. Birling, M. C., Roussel, G., Nussbaum, F., and Nussbaum, J. L. (1993). *Neurochem. Res.*, **18**, 937–45.
3. Amiguet, P., Gardinier, M. V., Zanetta, J. P., and Matthieu, J. M. (1992). *J. Neurochem.*, **58**, 1676–82.
4. Kroepfl, J. F., Vase, L. R., Charron, A. J., Linington, C., and Gardinier, M. V. (1996). *Neurochem.*, **67**, 2219–62.
5. Campbell, I, G., Freemont, P. S., Foulkes, W., and Trowsdale, J. (1992). *Cancer Res.*, **52**, 5416–10.
6. Scolding, N. J. *et al.* (1989). *J. Neuroimmunol.*, **22**, 169–76.
7. Solly, S. K. *et al.* (1996). *Glia*, **18**, 39–48.
8. Pham-Dinh, D. *et al.* (1993). *Proc. Natl Acad. Sci. USA*, **90**, 7990–4.
9. Brunner, C., Lassmann, H., Waehneldt, T. V., Matthieu, J. M., and Livington, C. (1989). *J. Neurochem.*, **52**, 296–304.
10. Abo, S., *et al.* (1993). *Biochem. Mol. Biol. Int.*, **30**, 945–58.
11. Gow, A. (1997). *J. Neurosci. Res.*, **50**, 659–64.
12. Linsley, P. S., Peach, R., Gladstone, P., and Bajorath, J. (1994). *Protein Sci.*, **3**, 1341–3.
13. Salomonsen, J. *et al.* (1991). *Proc. Natl Acad. Sci. USA*, **88**, 1359–63.
14. Bikle, D. D., Munson, S., and Komuves, L. (1996). *J. Biol. Chem.*, **271**, 9075–83.
15. Burger, D., Steck, A. J., Bernard, C. C., and Kerlero de Rosbo, N. (1993). *J. Neurochem.*, **61**, 1822–7.
16. Steinman, L. (1993). *Proc. Natl Acad. Sci. USA*, **90**, 7912–4.
17. Piddlesden, S. J., Lassmann, H., Zimprich, F., Morgan, B. P., and Linington, C. (1993). *Am. J. Pathol.*, **143**, 555–64.
18. Solly, S. K., Duabas, P., Monge, M., Dautigny, A., and Zalc, B. (1997). *J. Neurochem.*, **68**, 1705–11.
19. Sharp, P. A. (1980). *Cell*, **23**, 643–6.
20. Pham-Dinh, D., Della Gaspera, B., Kerlero de Rosbo, N., and Dautigny, A. (1995). *Genomics*, **29**, 345–52.
21. Ballenthin, P. A. and Gardinier, M. V, (1996). *J. Neurosci. Res.*, **46**, 271–81.
22. Pham-Dinh, D. *et al.* (1995). *Immunogenetics*, **42**, 386–91.
23. Lebar, R., Vincent, C., and Boubennec, D. F. -L. (1979). *J. Neurochem.*, **32**, 1451–60.
24. Bernard, C. C. *et al.* (1997). *J. Mol. Med.*, **75**, 77–88.

25. Wekerle, H., Kojima, K., Lannes-Vieira, J., Lassmann, H., and Linington, C. (1994). *Ann. Neurol.*, **36** (Suppl.), S47–53.
26. Linington, C., Engelhardt, B., Kapocs, G., and Lassman, H. (1992). *J. Neuroimmunol.*, **40**, 219–24.
27. Sun, J. *et al.* (1991). *J. Immunol.*, **146**, 1490–5.
28. Xiao, B. G., Linington, C., and Link, H. (1991). *J. Neuroimmunol.*, **31**, 91–6.
29. Ben-Nun, A., *et al.* (1996). *J. Neurol.*, **243**, S14–22.
30. Bettadapura, J. *et al.* (1998). *J. Neurochem.*, **70**, 1593–9.

■ *Alexander Gow*
Brookdale Center for Developmental and Molecular Biology, Box 1126,
The Mount Sinai School of Medicine,
One Gustave Levy Place,
New York, NY 10029, USA.

Peripheral myelin protein 22 (PMP22)

The PMP22/*gas3* gene encodes an axonally regulated Schwann cell protein that is assembled into peripheral nervous myelin and that is also expressed in several non-neural tissues. PMP22 is involved in regulating myelin stability and cell growth. Point mutations or an altered number of copies of this gene have been linked to inherited peripheral neuropathies in rodents and humans.

Synonymous names

PASII,[1] CD25,[2] Sr13,[3] gas3[4] and PMP22.[5]

Protein properties

Peripheral myelin protein 22 (PMP22) is a transmembrane glycoprotein highly expressed by myelinating Schwann cells of the peripheral nervous system (PNS). PMP22 was cloned by differential hybridization of cDNA libraries generated from injured versus non-injured rat sciatic nerve.[2,3] It is homologous to the growth-arrest-specific gene *gas3*, which is expressed by fibroblasts when their proliferation is inhibited following serum starvation or contact inhibition.[4] Conversely, it is down-regulated when these cells are induced to proliferate. The expression of the PMP22 gene is differentially regulated during development, nerve injury and nerve regeneration.[5,6] The mRNA is abundant in intact sciatic nerve, but after injury it quickly decreases in the degenerating segment distal to the lesion. During regeneration PMP22 transcript levels resume control values indicating that its expression is axonally regulated. The pattern of PMP22 expression in the PNS during development and nerve injury is very similar to that of other myelin genes such as protein zero (P_0) and myelin basic protein (MBP).[2,5,6] PMP22 is also expressed in motor neurones,[7] brain, intestine, lung and heart,[2,3,8] and is widely distributed during embryonic mouse development.[9] The PMP22 gene is regulated by two independent promoters that are located upstream of two alternative 5′-non-coding exons (exons 1A and 1B).[10] Thus, two alternatively spliced PMP22 mRNAs can arise, which encode the same protein. The expression of these transcripts is regulated in a tissue-specific fashion. The activation of the distal promoter is coupled to myelination, whereas the more proximal is preferentially activated in all non-neural tissues.[10,11]

Electron and confocal microscopy of immunohistochemistry experiments have localized PMP22 mainly into peripheral compact myelin,[5,12,13] where it comprises 2–5 per cent of the total protein content.[14] Based on sequence analysis and biochemical studies, PMP22 consists of 160 amino acids and it is inserted into the plasma membrane via an N-terminal non-cleaved signal sequence. The first extracellular domain carries an N-linked glycosylation at an asparagine (Asn41). On SDS–polyacrylamide gels, glycosylation increases the relative molecular mass of the native protein from 18 kDa to 22 kDa.[2,3,4] As predicted by computer analysis and experimentally demonstrated by topology studies, PMP22 consists of four transmembrane domains and two extracellular hydrophilic regions; the N and C termini are exposed to the cytoplasm[15] (Fig. 1).

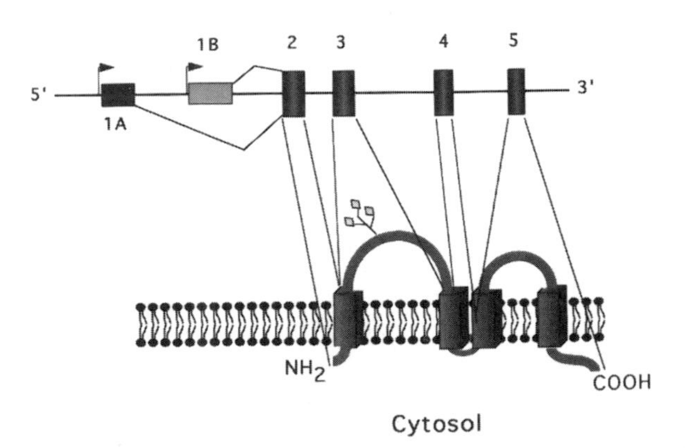

Figure 1. Schematic representation showing the structure of the PMP22 gene deduced from the human, and the mouse sequence, and the topology of the protein within the plasma membrane. Lines indicate the domains encoded by each exon. A glycosylation site is present in the first extracellular domain.

PMP22-related epithelial membrane proteins 1, 2 and 3 (Emp1, EMP2 and EMP3)[16,17] have been identified, which are putative four-transmembrane-domain molecules, which share approximately 40 per cent amino acid identity with PMP22, and have a molecular mass of 18 kDa. Their mRNA transcripts are found in most tissues with an expression pattern partially overlapping with that of PMP22, but with differences in relative expression levels. In contrast to PMP22 and EMP1, EMP2 and EMP3 expression has been also detected in liver. No functional studies on EMP proteins have been reported.

In vivo and *in vitro* studies have established that PMP22 is involved in the maintenance of myelin thickness and stability in the PNS and in regulating Schwann cell proliferation and differentiation. Furthermore the phenotypes obtained by point mutations or an altered dosage of the gene imply that function of PMP22 depends on both the molecular structure and the level of expression.

Purification

The protein can be immunoprecipitated from Schwann cell cultures[14] and peripheral nerve myelin (D'Urso D., unpublished), but preparative purification has not been reported.

Activities

Beside its specialized function in PNS myelin formation and maintenance, PMP22 also seems to play a more general role as a regulator of cell proliferation. It has been reported that PMP22 acts as a negative modulator of Schwann cell growth. Rat Schwann cells engineered to overexpress PMP22 are delayed in their transition from G_0/G_1 to S phase of the cell cycle.[18] Furthermore, overexpression of *gas3*/PMP22 cDNA induced by plasmid microinjection[8] or retroviral gene transfer[19] into NIH-3T3 fibroblasts leads to an apoptotic-like phenotype. The biological significance of this is still under debate.

Antibodies

Both polyclonal[5,8,15,19] and monoclonal[13] antibodies raised against synthetic peptides contained in the extracellular domains and in the C terminus of PMP22 have been described. These antibodies show cross-reactivity with human, rat and mouse. They are suitable for immunoblotting, immunohistochemistry and immunoprecipitation. No commercial antibodies are available.

Genes

PMP22 sequences have been published for human (EMBL accession numbers: D11428, M94048, L03203, X65968 and S61788), mouse (EMBL accession numbers: M32240 and Z38110) and rat (EMBL accession numbers: X62431, M69139 and S55427), and they show almost 90 per cent homology. The PMP22 gene is mapped to human chromosome 17 (region p11.2-p12) and mouse chromosome 11. The entire human PMP22 gene spans about 40 kb. It contains four coding exons (exons 2–5) and two alternative exons, 1A and 1B, located in the 5'-untranslated region and regulating the expression of two alternative transcripts (CD25 and SR13, each approximately 1.8 kb)[11] which encode the same protein. Exons 2 and 3 encode the first transmembrane region and the first extracellular loop, respectively. Exon 4 corresponds to the second transmembrane sequence and half of the third, and exon 5 covers the remaining part of the third, the second extracellular loop, the fourth transmembrane domain and the 3'-untranslated region.[10] The mouse PMP22 gene has the same structure (Bosse, F., unpublished observation).

Mutant phenotype/disease states

Duplication,[20,21] deletion[22] and point mutations[23–26] of the PMP22 gene have been identified and associated with various hereditary motor and sensory neuropathies

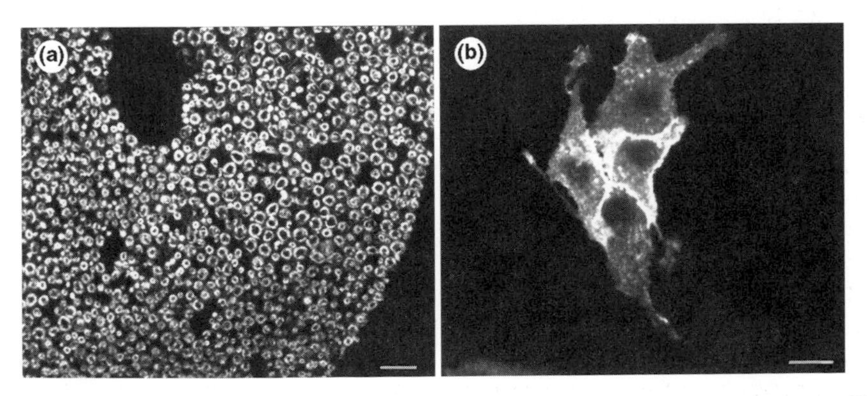

Figure 2. PMP22 localization. (a) Immunofluorescence staining of PMP22 in adult rat sciatic nerve. PMP22 is localized in compact myelin. (b) PMP22 expression in transfected HeLa cells. Immunostaining shows the protein distributed in the cytoplasmic compartments, and at the plasma membrane. Scale bars represent 20 μm.

in humans (Charcot–Marie–Tooth type 1A (CMT1A), Dejerine-Sottas syndrome (DSS), hereditary neuropathy with liability to pressure palsy (HNPP)) and mice (*Trembler* and *TremblerJ*). In particular, the majority of CMT1A patients have a duplication of a 1.5 megabase region of chromosome 17 containing the PMP22 gene, the disease is also associated with several PMP22 mutations (L16P, S79C, T118M and L147R). Other point mutations have been found in patients affected by severe DSS (H12Q, M69K and S72L). Deletion of one copy of the gene, resulting from a reciprocal deletion of the CMT1A duplication, or a 2 bp null mutation) is linked to HNPP. The mouse mutants *Trembler* and its allele *Trembler J*, which display a severe peripheral neuropathy, carry non-conservative missense mutations, G150D and L16P respectively (see ref. 27 for a review). PMP22-deficient mice, or mice with only one functional copy of PMP22, show delayed onset of myelination, tomacula formation at a young age and then severe demyelination.[28] Transgenic rats[29] or mice[30] which carry additional copies of the PMP22 gene present peripheral hypomyelination and a phenotype similar to that described in CMT1A patients.

Structure

To date, no information on X-ray crystallographic or NMR-derived structure is available.

References

1. Kitamura, K., Suzuki, M., and Uyemura, K. (1976). *Biochim. Biophys. Acta*, **455**, 806–16.
2. Spreyer, P., Kuhn, G., Hanemann, C. O., Gillen, C., Schaal, H., Kuhn, R., Lemke, G., and Müller, H. W. (1991). *EMBO J.*, **10**, 3661–8.
3. Welcher, A. A., Suter, U., De Leon, M., Snipes, G. J., and Shooter, E. M. (1991). *Proc. Natl Acad. Sci. USA*, **88**, 7195–9.
4. Manfioletti, G., Ruaro, M. E., Del Sal, G., Philipson, L., and Schneider, C. (1990). *Mol. Cell. Biol.*, **10**, 2924–30.
5. Snipes, G., Suter, U., Welcher, A. A., and Shooter, E. (1992). *J. Cell Biol.*, **117**, 225–38.
6. Kuhn, G., Lie, A., Wilms, S., and Müller, H. W. (1993). *Glia*, **8**, 256–64.
7. Parmantier, E., Cabon, F., Braun, C., D'Urso, D., Müller, H. W., and Zalc, B. (1995). *Eur. J. Neurosci.*, **7**, 1080–88.
8. Fabbretti, E., Edomi, P., Brancolini, C., and Schneider, C. (1995). *Genes Dev.*, **9**, 1846–56.
9. Baechner, D., Liehr, T., Hameister, H., Altenberger, H., Grehl, H., Suter, U., and Rautenstrauss, B. (1995). *J. Neurosci., Res.*, **42**, 733–41.
10. Suter, U., Snipes, G. J., Schoener-Scott, R., Welcher, A. A., Pareek, S., Lupski, J. R., Murphy, R. A., Shooter, E. M., and Patel, P. I. (1994). *J. Biol. Chem.*, **269**, 25795–806.
11. Bosse, F., Zoidl, G., Gillen, C. P., Wilms, S., Kuhn, G., and Müller, H. W. (1994). *J. Neurosci. Res.*, **37**, 529–37.
12. Haney, C., Snipes G. J., Shooter, E. M., Suter, U., Garcia, C., Griffin, J. W., and Trapp, B. D. (1996). *J. Neuropathol. Exp. Neurol.*, **55**, 290–99.
13. D'Urso, D., Schmalenbach, C., Zoidl, G., Prior, R., and Müller, H. W. (1996). *J. Neurosci. Res.*, **48**, 31–42.
14. Pareek, S., Suter, U., Snipes, G. J., Welcher, A. A., Shooter, E. M., and Murphy, R. A. (1993). *J. Biol. Chem.*, **268**, 10372–9.
15. D'Urso, D. and Müller, H. W. (1997). *J. Neurosci. Res.*, **49**, 551–62.
16. Taylor, V., Welcher, A. A., Amigen EST Programm, and Suter, U. (1995). *J. Biol. Chem.*, **270**, 28824–33.
17. Taylor, V. and Suter, U. (1996). *Gene*, **175**, 115–20.
18. Zoidl, G., Blass-Kampmann, S., D'Urso, D., Schmalenbach, C., and Müller, H. W. (1995). *EMBO J.*, **14**, 1122–8.
19. Zoidl, G., D'Urso, D., Blass-Kampmann, S., Schmalenbach, C., Kuhn, R., and Müller, H. W. (1997). *Cell Tissue Res.*, **287**, 459–70.
20. Lupski, J. R. *et al.* (1991). *Cell*, **66**, 219–32.
21. Matsunami, N. *et al.* (1992). *Nature Genet.*, **1**, 176–9.
22. Chance, P. F. *et al.* (1993). *Cell*, **72**, 143–51.
23. Roa, B. R. (1993). *New Engl. J. Med.*, **329**, 96–101.
24. Valentijn, L. J., Bass, F., Woltermann, R. A., Hoogendijk, J. E., van den Bosch, N. H. A., Zorn, I., Gabreesl-Festen, A. A. W. M., de Visser, M., and Bolhuis, P. A. (1993). *Nature Genet.*, **2**, 288–91.
25. Suter, U., Welcher, A. A., Ozcelik, T., Snipes, G. J., Kosaras, B., Francke, U., Billings-Gagliardi, S., Sidman, R. L., and Shooter, E. M. (1992). *Nature*, **356**, 241–4.
26. Suter, U., Moskow, J. J., Welcher, A. A., Snipes, G. J., Kosaras, B., Sidman, R. L., Buchberg, A. M., and Shooter, E. M. (1992). *Proc. Natl Acad. Sci. USA*, **89**, 4382–6.
27. Suter, U., and Snipes, G. J. (1995). *Annu. Rev. Neurosci.*, **18**, 45–75.
28. Adlkofer, K., Martini, R., Aguzzi, A., Zielasek, J., Toyka, K. V., and Suter, U. (1995). *Nature Genet.*, **11**, 274–80.
29. Sereda, M. *et al.* (1996). *Neuron*, **16**, 1049–60.
30. Magyar, J. P., Martini, R., Ruelicke, T., Aguzzi, A., Adlkofer, K., Dembic, Z., Zielasek, J., Toyka, K. V., and Suter, U. (1996). *J. Neurosci.*, **16**, 5351–80.

Donatella D'Urso
Neurologische Klinik, Neurochemisches Labor, University of Dusseldorf, Moorenstrasse 5, 40225 Dusseldorf, Germany

Protein zero (P_0)

Protein zero (P_0) is the major structural protein of peripheral nerve myelin. It has been shown to mediate membrane adhesion in the spiral wraps of the myelin sheath. The importance of P_0 in the formation and maintenance of PNS myelin has been demonstrated in several ways. In P_0 gene knock-out experiments, severe hypomyelination has been observed.[1] Although Schwann cells in these animals can contact and enwrap axons, membrane compaction is severely impaired. There are two disease states associated with mutations in the expressed protein. Instances of both Charcot–Marie–Tooth (CMT) type I and Dejerine–Sottas syndrome (DSS), diseases which affect myelin structure, have been attributed to mutations in the human *MPZ* gene.[2,3]

Protein properties

P_0 has a 124-residue extracellular domain that shows sequence similarity to immunoglobulins and has a single N-linked glycosylation site,[4,5] a single transmembrane domain, and a highly basic 69-residue cytoplasmic domain. The intracellular membrane apposition of peripheral myelin is thought to be mediated by the cytoplasmic domain,[6] which may interact directly with lipid headgroups through electrostatic interactions.[6,7]

Activity

P_0 is a homophilic adhesion membrane molecule. Surfaces coated with the P_0 extracellular domain have been used as substrates to which cells expressing P_0 adhere.[8] When non-adherent cells are transfected with P_0 cDNA they rapidly aggregate,[9,10] and the membrane contacts of these cells are very close and regular, like the intraperiod line of PNS myelin.[11] It is now generally accepted that P_0 through the homophilic adhesion capacity of its extracellular domain, holds together adjacent membrane wraps of peripheral nerve myelin.

Genes

cDNAs for *P_0* have been cloned in many animals including mouse, rat, chicken, human, dogfish and trout. The gene encoding the P_0 protein is called *MPZ* in humans[12] (GenBank accession U10017) and *P_0* in other animals.[5]

Structure

The crystal structure of the P_0 extracellular domain has been determined.[13] The arrangement of molecules in the crystal indicates that this domain may exist on the membrane surface as a tetramer (Plate 16) that can link to other tetramers from the opposing membrane, thus forming an adhesive network. The structure also suggests that the P_0 extracellular domain mediates adhesion through the direct interaction of apical tryptophan side-chains with the opposing membrane, in addition to homophilic protein–protein interactions.

References

1. Giese, K. P., Martini, R., Lemke, G., Soriano, P., and Schachner, M. (1992). *Cell*, **71**, 565–76.
2. Patel, P. I. and Lupski, J. R. (1994). *Trends Genet.*, **10**, 128–33.
3. Warner, L. E. *et al.* (1996). *Neuron*, **17**, 451–60.
4. Lai, C., Brow, M. A., Nave, K., Noronha, A. B., Quarles, R. H., Bloom, F. E., Milner, R. J., and Sutcliffe, J. G. (1987). *Proc. Natl Acad. Sci. USA*, **84**, 4337–41.
5. Lemke, G., Lamar, E., and Patterson, J. (1988). *Neuron*, **1**, 73–83.
6. Lemke, G. and Axel, R. (1985). *Cell*, **40**, 501–8.
7. Ding, Y. and Brunden, K. R. (1994). *J. Biol. Chem.*, **269**, 10764–770.
8. Griffith, L. S., Schmitz, B., and Schachner, M. (1992). *J. Neurosci. Res.*, **33**, 639–48.
9. Filbin, M. T., Walsh, F. S., Trapp, B. D, Pizzy, J. A., and Tennekoon, G. . (1990). *Nature*, **344**, 871–2.
10. Schneider-Schaulies, J., Von Brunn, A., and Schachner, M. (1990). *J. Neurosci. Res.* **27**, 286–97.
11. D'Urso, D., Brophy, P. J., Staugaitis, S. M., Gillespie, C. S., Frey, A. B., Stempak, J. G., and Colman, D. R. (1990). *Neuron*, **4**, 449–60.
12. Hayasaka, K., Himoro, M., Wang, Y., Takata, M., Minoshima, S., Shimizu, N., Miura, M., Uyemura, K., and Takata, G. (1993). *Genomics*, **17**, 755–8.
13. Shapiro, L., Doyle, J. P., Henseley, P., Colman, D. R., and Hendrickson, W. A. (1996). *Neuron*, **17**, 435–49.

Lawrence Shapiro
Program in Structural Biology,
Department of Physiology,
The Mount Sinai School of Medicine,
One Gustave Levy Place,
New York, NY 10029, USA.

David R. Colman
Brookdale Center for Developmental
and Molecular Biology, The Mount Sinai
School of Medicine, One Gustave Levy Place,
New York, NY 10029, USA.

Proteolipid proteins (lipophilins)

The lipophilins (proteolipid proteins, PLPs) are a small group of very hydrophobic integral membrane proteins that arose at least 440 million years ago. Most family members are bona fide glycoproteins or have consensus *N*-glycosylation sequences in extracellular domains. The physical properties and function of the archival member of this family, encoded by the PLP gene, have been studied in greatest detail and appear to play a structural role in stabilizing the extracellular membrane surfaces of central nervous system (CNS) compact myelin from amphibians to humans. Other family members are expressed inside and outside of the CNS and may well have functions distinct from that of PLP.

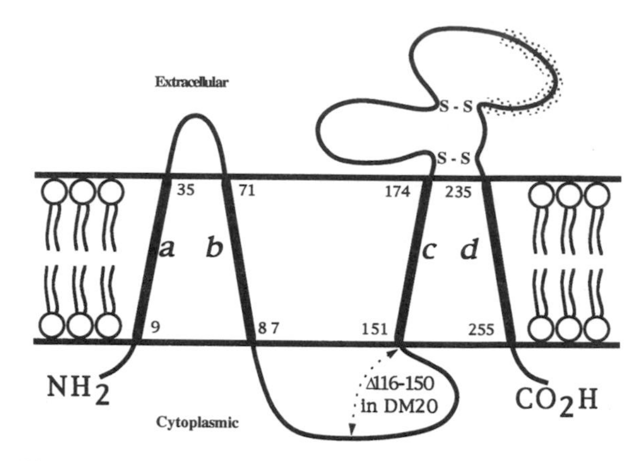

Figure 1. Schematic representation of the topology of DM-20/PLP in the lipid bilayer. Other lipophilins are expected to adopt the same topology.[9] DM-20 and PLP possess four transmembrane domains with both N and C termini at the cytoplasmic surfaces of the membrane.[3] Disulphide bridges in the second extracellular domain are shown; and the position of the motif illustrated in Fig. 2 is indicated by the stippled box. DM-20 and PLP differ only by the absence or presence, respectively, of a 35 amino acid peptide in the cytoplasmic loop. This PLP-specific peptide is composed largely of polar and charged amino acids.

Family members

Proteolipid protein (PLP), Folch–Pi–Lees proteolipid (FLPL), P7 proteolipid apoprotein, N2, lipophilin.
DM-20: Intermediate proteins, Agrawal protein.
DMα, DMβ, DMγ.
M6a, M6b, M6, EMA.

Protein properties

The lipophilins are integral membrane proteins that share at least 50 per cent overall amino acid similarity. The transmembrane domains are the most highly conserved with over 80 per cent amino acid similarity in the uncleaved membrane-embedded sequence at the N terminus. Four cysteine residues in the second extracellular domain of all family members are conserved and have been shown for DM-20 and PLP to form disulphide bridges[1] (Fig. 1).

The physical properties of the DM-20 and PLP proteins are aptly portrayed by their solubility in chloroform:methanol (2:1 v/v), although a 'water-soluble' (probably micellar) form of these proteins can be prepared from the organic phase. Roughly half of the amino acids in these proteins are non-polar and, in addition, at least two cytoplasmically-exposed cysteine residues are thioacylated.[1] The molecular masses and other characteristics of lipophilin family members are summarized in Table 1. Note that the apparent molecular masses of

Table 1 Physical properties of the lipophilin family members

Gene	No. of amino acids[1]	Mol. mass (kDa)	pI[2]	M_r (kDa)[3]	N×S/T consensus sites	Glycoproteins	species
DMα	245	26.9	4.7	27	yes	yes	*Squalus*
DMβ	278	31.2	5.0	nd	yes	nd	*Squalus*
DMγ	246	26.8	6.2	nd	yes	nd	*Squalus*
M6a	278	31.1	5.0	35	yes	yes	*Mus*
M6b	288	31.2	5.1	35	yes	yes	*Mus*
DM-20	242	26.3	9.0	20.5	no	no	*Mus*
PLP	277	30.0	9.4	24.8	no	no	*Mus*

[1]The initiation methionine has been included.
[2]Calculated from the coding regions of each gene.
[3]Determined by SDS–PAGE.
nd, not determined.

DM-20 and PLP determined from SDS–PAGE[2] are significantly smaller than those calculated from the cDNAs, probably due to increased detergent binding compared with most proteins. PLP and DM-20 possess four transmembrane domains,[3] the first being an uncleaved signal anchor sequence, and both the N and C termini are located at the cytoplasmic membrane surfaces (Fig. 1). Other lipophilin family members are likely to share this topology which has been inferred from the demonstration of *N*-glycosylation on asparagine residues in these proteins.[4] Quaternary structures have not been defined for any of the lipophilins but oligomerization of DM-20 and PLP as heteromeric or homomeric complexes is indicated from analytical ultracentrifugation data[5,6] and the co-expression of these proteins in transfected fibroblasts.[7,8]

The first lipophilin family member arose at least 440 million years ago in marine vertebrates.[9] In elasmobranchs (sharks and rays), contemporaries of the most primitive animals in which lipophilins have been identified, three related genes are present and it is currently thought that two of these, DMα and DMγ, are ancestral forms of the DM-20/PLP and M6b genes in mammals, respectively, while the third gene, DMβ, is probably the homologue of M6a (Fig. 2). The primordial lipophilin gene remains unidentified in early marine vertebrates; however, the human genome project has provided compelling evidence that this gene is the ancestor of M6b/DMγ. A 200 kb BAC contig from the human chromosome Xp22 AC003037 (gb) spans a large portion of the M6b gene including six exons that are present in the published cDNA sequence of human M6b.[33] The 5′ end of the final exon of this gene encodes the C-terminus of M6b which has high homology to exon 7 of DM-20/PLP, as well as a conserved intron/exon boundary. Surprisingly, the 3′ end of this M6b exon contains a consensus splice acceptor site preceding an open reading frame that is highly homologous to published M6a/DMβ cDNA sequences.[9,18] Moreover, EST clones indicate that the last exon of M6b, at least in mice, is alternatively spliced to encode C-termini that are homologous to either M6a S65737 (gb) or DM-20/PLP AA163802 (gb). In similar fashion, the 5′ end of the M6b gene contains exons that encode both DM-20/PLP-like and M6a-like N-termini; the first identifiable exon in the M6b contig encodes an M6a-like peptide but the published human M6b cDNA[33] is homologous to DM-20/PLP at the 5′ end. This portion of the cDNA is not present in the M6b contig and is, therefore, at least 100 kb upstream. Together, these data explain the divergence in the published amino acid sequences deduced for the mouse and human M6b proteins[18,33] and suggest that DMγ has been duplicated twice during evolution; the first event likely gave rise to DMβ and the second to DMα (Figs. 2 and 3). Two additional alternatively spliced exons are also present in the human M6b contig identified from EST clones (GenBank accession numbers AA327738 and AA379915) and RT-PCR experiments (A. G., unpublished observations).

DMα and DMγ share homology at the N-terminal regions with the glutamate (GluR1) and acetylcholine (nACHRα) receptors, suggesting that these proteins may stem from a common primitive channel protein(s) that was co-opted for specific purposes in different cell types.[9] In this regard, the lipophilins can be divided into two groups on the basis of the presence or absence of a short motif,[10] with unknown function, in the second extracellular domain (Fig. 3). In the CNS, family members in group A are expressed predominantly in glial cells in white matter regions[9,10] (although M6b also appears postnatally in neurones) while M6a and DMβ in group B are expressed in grey matter (M6a is localized to growth cones in cultured neurones).[11,12] DMβ mRNA localization in the CNS of elasmobranchs has not been determined; however in *Xenopus* this gene is expressed in grey matter.[12] In addition to their expression in oligodendrocytes, DM-20 and PLP are present in Schwann cells, thymus and, at very low levels, in skin fibroblasts; DM-20 is also expressed in ventricular cells during early development and in heart.[13–17] The M6 genes are also expressed in choroid plexus and outside of the nervous system in kidney proximal tubules.[18]

The PLP protein can be viewed as the most recent lipophilin family member arising from the insertion of

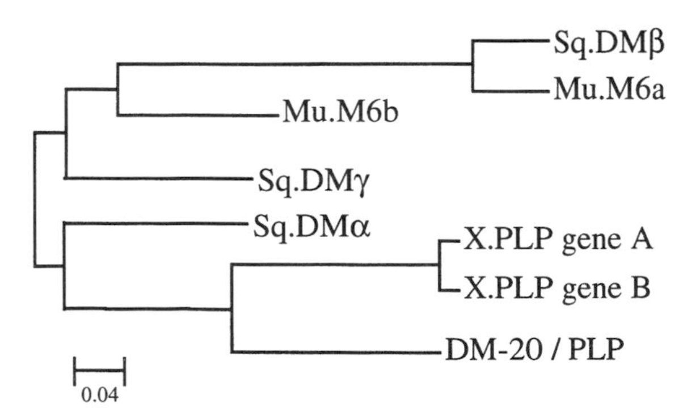

Figure 2. Dendrogram of the lipophilin family. DM-20/PLP, mammalian, avian and reptilian species; Mu., mouse; Sq., *Squalus acanthias*; X., *Xenopus*.

GROUP A	DM-20/PLP/M6b		PWNAFPGK
	DMα/DMγ		PWNASPGK
Consensus*		C X_{16-17} C X_9	*motif* X C X_7 C
GROUP B	DMβ		VPIHEQKT
	M6a		VTIGEEKK
Consensus*		C X_{11} C X_8	*motif* X C X_7 C

Figure 3. The lipophilins can be divided into two groups according to the presence (group A) or absence (group B) of a short peptide motif in the second extracellular domain.[10] Also present in this external domain of all lipophilins are four conserved cysteine residues that have been shown for DM-20/PLP to participate in disulphide bridges (Fig. 1). *The consensus sequence begins at codon 183 of human PLP and shows conserved cysteines in the second extracellular domain of lipophilin family members.

22. Rosenbluth, J., Stoffel, W., and Schiff, R. (1996). *J. Comp. Neurol.*, **371**, 336–44.
23. Lees, M. B. and Bizzozero, O. A (1992). In *Myelin: biology and chemistry* (ed. R. E. Martenson), pp. 237–255. CRC Press, Boca Raton.
24. Diaz, R. S., Regueiro, P., Monreal, J., and Tandler, C. J. (1991). *J. Neurosci. Res.*, **29**, 114–20.
25. Skalidis, G., Trifilieff, E., and Liu, B. (1986). *J. Neurochem.*, **46**, 297–99.
26. Yamamura, T., Konola, J. T., Wekerle, H., and Lees, M. B. (1991). *J. Neurochem.*, **57**, 1671–80.
27. Hudson, L., Friedrich, V. L. J., Behar, T., Dubois-Dalcq, M., and Lazzarini, R. A. (1989). *J. Cell Biol.*, **109**, 717–27.
28. Berndt, J. A., Kim, J. G., and Hudson, L. D. (1992). *J. Biol. Chem.*, 267, 14730–7.
29. Ikenaka, K. and Kagawa, T. (1995). *Dev. Neurosci.*, **17**, 127–36.
30. Hodes, M., Pratt, V., and Dlouhy, S. (1993). *Dev. Neurosci.*, **15**, 383–94.
31. Nave, K.-A. and Boespflug-Tanguy, O. (1996). *Neuroscientist*, **2**, 33–43.
32. Gow, A., Southwood, C. M. and Lazzarini, R. A. (1998). *J. Cell Biol.*, **140**, 925–34.

Alexander Gow
Brookdale Center for Developmental and Molecular Biology, Box 1126,
The Mount Sinai School of Medicine,
One Gustave Levy Place,
New York, NY 10029, USA

Neural cell adhesion molecule (NCAM)

NCAM is the first neural member of the immunoglobulin superfamily (IgSF) which had been isolated and characterized.[1] It mediates cell–cell interactions by homophilic binding. Multiple isoforms of NCAM are known, which differ with respect to the mode of cell membrane anchorage or the presence of short sequence motifs. NCAM is post-translationally modified by a particular carbohydrate structure, the polysialic acid (PSA) which modulates homophilic binding. Expression of PSA is developmentally regulated. NCAM is not restricted to the nervous system and it is expressed in the developing and the adult brain.

Synonymous names

NCAM is also referred to as CD56, leu-19 or 5B4-CAM in mammals.[2]

Related proteins

The mammalian protein NCAM/OCAM[51–53] is about 46% identical with mammalian NCAM. The invertebrate molecule ApCAM[3], isolated from the marine snail *Aplysia californica,* as well as the *Drosophila* protein fasciclin II,[4] are structurally and functionally[5] related to NCAM. Another related molecule is LeechCAM, from the leech, *Hirudo medicinalis.*[54]

Protein properties

Different isoforms of NCAM are known which include species of 180 kDa, 140 kDa, and 120 kDa.[1,2,6–8] Both large forms have a transmembrane domain and differ in the lengths of their cytoplasmic extensions. The small form is anchored by glycosyl phosphatidylinositol (GPI). The membrane-anchored isoforms have five immunoglobulin- (Ig-) like domains and two fibronectin type III- (FnIII-) like domains. A secreted form is also known which lacks the carboxyl-proximal FnIII-like domain (Fig. 1). The homophilic binding activity of NCAM is modulated by the post-translational attachment of PSA which is almost exclusively found on the NCAM molecule.[9,10] PSA is attached by polysialyltransferases[11,12,55] to Asn-459 and Asn-430 in the fifth Ig-like domain.[13] Expression of PSA is temporally and spatially correlated with dynamic cellular interactions, for instance cell migration[56] and neurite outgrowth in neurohistogenesis.[8,10] PSA transferase is Ca^{2+}-dependent[57] which may allow a dynamic regulation of NCAM polysialylation in subsets of neurones[57] and glial cells.[56]

Like other multidomain members of the IgSF, NCAM undergoes multiple extracellular interactions.[14] The most prominent activity is its homophilic binding. It resides in the Ig-like region but its mechanistic details are controversial.[15] Other membrane glycoproteins interacting with NCAM are the IgSF members axonin-1/TAG-1 and L1/NgCAM. Furthermore, NCAM binds to extracellular matrix components. A heparin-binding region is located in the second Ig-like domain. Neurocan, a neural proteoglycan, and phosphacan, an isoform of receptor protein tyrosine phosphatase β/ζ, have also been shown to interact with NCAM.[16] Recently, NCAM was found to bind to agrin, a heparan sulphate proteoglycan implicated in synaptogenesis.[17]

The cytoplasmic domain of NCAM may interact with brain spectrin and activate second messenger systems in

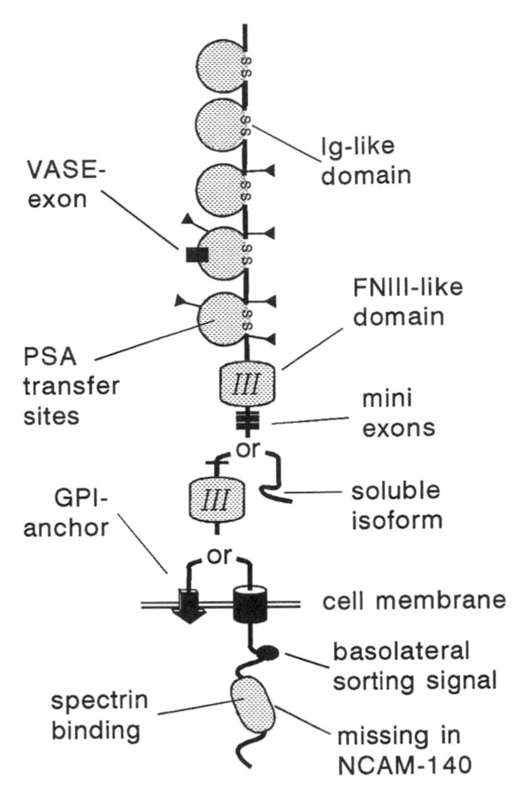

Figure 1. Domain model of NCAM. The amino terminus is at the top, and putative conserved *N*-linked glycosylation sites are marked by arrowheads.

neurones.[2,8] The cytoplasmic domain of NCAM-140 contains a sequence motif which directs basolateral sorting in Madin–Darby canine kidney cells.[18] Extracellular triggering of NCAM-140 induces the formation of a complex[19] with the src-related tyrosine kinase p59^{fyn} and the focal adhesion kinase p125^{fak}.

Purification

Studies describing the isolation of NCAM by immunoaffinity chromatography from non-ionic detergent extracts of brain plasma membranes of different species, including chick,[20] have been cited in a recent review.[2]

Activities

A multitude of *in vitro* assay systems suggest that NCAM is involved in the induction of neurite outgrowth,[21] cell adhesion, and cell migration.[2] *In vivo,* NCAM-specific antibodies interfere with axon fasciculation and nerve sprouting,[22,23] cause disordered axonal growth,[24,25] and lead to projection errors in the plexus region of the limb bud.[26] Antibodies also interfere with cell migration.[27,28] Enzymatic removal of PSA from NCAM *in vivo* interferes with axon fasciculation,[29] axon branching,[30] and cell migration.[31] NCAM modulates astrocyte proliferation[58] by signalling pathways which are distinct from those which regulate neurite outgrowth.[21]

In the adult brain, NCAM has been implicated in synaptic plasticity.[32,33,59,65] Antibodies interfere with memory consolidation in chick[34] and rat[35] models and with long-term potentiation in the hippocampus.[36,37] Enzymatic removal of PSA from NCAM interferes with the acquisition and retention of spatial memory[38] and prevents induction of long-term potentiation and depression in the hippocampus.[39]

NCAM is also involved in cell–cell interactions outside the nervous system,[2] for instance in myogenesis[40] and seems to influence the metastatic potential of tumour cells, as exemplified by glioma cells.[41]

Antibodies

Studies describing polyclonal and monoclonal antibodies to human, mouse, rat, chick, and *Xenopus* NCAM have been reviewed recently.[2] Antibodies are also available from the *Developmental Studies Hybridoma Bank* (Johns Hopkins University, Iowa City, USA; http://www.uiowa.edu/~dshbwww/). Commercial suppliers include Sigma (St Louis, USA), Santa Cruz Biotechnology (Santa Cruz, USA), Dianova (Hamburg, Germany), and Boehringer Mannheim (Mannheim, Germany).

Genes

Complete NCAM cDNA sequences have been reported for human,[42] mouse,[43] rat,[44] bovine,[45] chicken[1] and *Xenopus*[46] NCAM. Multiple isoforms of NCAM are known, which may be secreted, may have a GPI anchor, or may have cytoplasmic domains.[1,2,6,7] Alternative splicing of the VASE exon in the fourth Ig-like domain is developmentally regulated and modulates the neurite outgrowth promoting activity.[2,32] Several small alternatively spliced exons are also found between the FnIII-like domains.[6,7] The mouse NCAM promoter is regulated in part by synaptic activty[60] and binds the Pax-6 transcription factor.[61]

Mutant phenotype

Transgenic mice lacking NCAM show defects in granule cell migration[47] and axon fasciculation[62] in the olfactory system. They show abnormal axonal growth in the hippocampal mossy fibre pathway[48] and changes in circadian rhythmicity.[63] Another study revealed that NCAM-deficient mice show impaired long-term potentiation[39] and have deficits in spatial learning whereas activity and motor abilities appear normal.[49] Abnormalities in social behaviour have also been observed.[64]

Structure

The three-dimensional structure of the first domain was determined by NMR spectroscopy and has been categorized into the intermediate set (I set) of Ig-like domains.[50]

References

1. Cunningham, B. A., Hemperly, J. J., Murray, B. A., Prediger, E. A., Brackenbury, R. and Edelman, G. M. (1987). *Science*, **236**, 799–806.
2. Brümmendorf, T. and Rathjen, F. G. (1995). *Protein Profile*, **2**, 963–1108.
3. Mayford, M., Barzilai, A., Keller, F., Schacher, S. and Kandel, E. R. (1992). *Science*, **256**, 638–44.
4. Harrelson, A. L. and Goodman, C. S. (1988). *Science*, **242**, 700–8.
5. Martin, K. C. and Kandel, E. R. (1996). *Neuron*, **17**, 567–70.
6. Walsh, F. S. and Doherty, P. (1991). *Sem. Neurosci.*, **3**, 271–84.
7. Goridis, C. and Brunet, J. F. (1992). *Sem. Cell Biol.*, **3**, 189–97.
8. Rutishauser, U. (1993). In *Guidebook to the extracellular matrix and adhesion proteins*, (ed. T. Kreis and R. Vale), pp. 158–9. Oxford University Press.
9. Rougon, G. (1993). *Eur. J. Cell. Biol.*, **61**, 197–207.
10. Rutishauser, U. and Landmesser, L. (1996). *Trends Neurosci.*, **19**, 422–7.
11. Eckhardt, M., Mühlenhoff, M., Bethe, A., Koopman, J., Frosch, M. and Gerardy-Schahn, R. (1995). *Nature*, **373**, 715–8.
12. Nakayama, J., Fukuda, M. N., Fredette, B., Ranscht, B. and Fukuda, M. (1995). *Proc. Natl Acad. Sci. USA*, **92**, 7031–5.
13. Nelson, R. W., Bates, P. A., and Rutishauser, U. (1995). *J. Biol. Chem.*, **270**, 17171–9.
14. Brümmendorf, T. and Rathjen, F. G. (1996). *Curr. Opin. Neurobiol.*, **6**, 584–93.
15. Kiselyov, V. V., Berezin, V., Maar, T. E., Soroka, V., Edvardsen, K., Schousboe, A., and Bock, E. (1997). *J. Biol. Chem.*, **272**, 10125–34.
16. Margolis, R. K., Rauch, U., Maurel, P., and Margolis, R. U. (1996). *Perspect. Dev. Neurobiol,*. **3**, 273–90.
17. Tsen, G., Halfter, W., Kroger, S., and Cole, G. J. (1995). *J. Biol. Chem.*, **270**, 3392–9.
18. Le Gall, A. H., Powell, S. K., Yeaman, C. A., and Rodriguez Boulan, E. (1997). *J. Biol. Chem.*, **272**, 4559–67.
19. Beggs, H. E., Baragona, S. C., Hemperly, J. J., and Maness, P. F. (1997). *J. Biol. Chem.*, **272**, 8310–9.
20. Hoffman, S., Sorkin, B. C., White, P. C., Brackenbury, R., Mailhammer, R., Rutishauser, U., *et al.* (1982). *J. Biol. Chem.*, **257**, 7720–9.
21. Doherty, P. and Walsh, F. S. (1996). *Mol. Cell. Neurosci.*, **8**, 99–111.
22. Landmesser, L., Dahm, L., Schultz, K., and Rutishauser, U. (1988). *Dev. Biol.*, **130**, 645–70.
23. Booth, C. M., Kemplay, S. K., and Brown, M. C. (1990). *Neuroscience*, **35**, 85–91.
24. Thanos, S., Bonhoeffer, F., and Rutishauser, U. (1984). *Proc. Natl Acad. Sci. USA*, **81**, 1906–10.
25. Fraser, S. E., Carhart, M. S., Murray, B. A., Chuong, C. M., and Edelman, G. M. (1988). *Dev. Biol.*, **129**, 217–30.
26. Tang, J., Rutishauser, U., and Landmesser, L. (1994). *Neuron*, **13**, 405–14.
27. Schwanzel Fukuda, M., Reinhard, G. R., Abraham, S., Crossin, K. L., Edelman, G. M., and Pfaff, D. W. (1994). *J. Comp. Neurol.*, **342**, 174–85.
28. Bronner Fraser, M., Wolf, J. J., and Murray, B. A. (1992). *Dev. Biol.*, **153**, 291–301.
29. Landmesser, L., Dahm, L., Tang, J. C., and Rutishauser, U. (1990). *Neuron*, **4**, 655–67.
30. Daston, M. M., Bastmeyer, M., Rutishauser, U., and O'Leary, D. D. (1996). *J. Neurosci.*, **16**, 5488–97.
31. Ono, K., Tomasiewicz, H., Magnuson, T., and Rutishauser, U. (1994). *Neuron*, **13**, 595–609.
32. Doherty, P., Fazeli, M. S., and Walsh, F. S. (1995). *J. Neurobiol.*, **26**, 437–46.
33. Fields, R. D. and Itoh, K. (1996). *Trends Neurosci.*, **19**, 473–80.
34. Scholey, A. B., Rose, S. P., Zamani, M. R., Bock, E., and Schachner, M. (1993). *Neuroscience*, **55**, 499–509.
35. Doyle, E., Nolan, P. M., Bell, R., and Regan, C. M. (1992). *J. Neurochem.*, **59**, 1570–3.
36. Ronn, L. C., Bock, E., Linnemann, D., and Jahnsen, H. (1995). *Brain Res.*, **677**, 145–51.
37. Lüthi, A., Laurent, J. P., Figurov, A., Müller, D., and Schachner, M. (1994). *Nature*, **372**, 777–9.
38. Becker, C. G., Artola, A., Gerardy Schahn, R., Becker, T., Welzl, H., and Schachner, M. (1996). *J. Neurosci. Res.* **45**, 143–52.
39. Müller, D., Wang, C., Skibo, G., Toni, N., Cremer, H., Calaora, V., *et al.* (1996). *Neuron*, **17**, 413–22.
40. Peck, D. and Walsh, F. S. (1993). *J. Cell. Biol.*, **123**, 1587–95.
41. Edvardsen, K., Pedersen, P. H., Bjerkvig, R., Hermann, G. G., Zeuthen, J., Laerum, O. D., *et al.* (1994). *Int. J. Cancer,* **58**, 116–22.
42. Barton, C. H., Dickson, G., Gower, H. J., Rowett, L. H., Putt, W., Elsom, V., *et al.* (1988). *Development,* **104**, 165–73.
43. Barthels, D., Santoni, M. J., Wille, W., Ruppert, C., Chaix, J. C., Hirsch, M. R., *et al.* (1987). *EMBO J.* **6**, 907–14.
44. Small, S. J., Shull, G. E., Santoni, M. J., and Akeson, R. (1987). *J. Cell Biol.*, **105**, 2335–45.
45. Lipkin, V. M., Khramtsov, N. V., Andreeva, S. G., Moshnyakov, M. V., Petukhova, G. V., Rakitina, T. V., *et al.* (1989). *FEBS Lett.*, **254**, 69–73..
46. Krieg, P. A., Sakaguchi, D. S., and Kintner, C. R. (1989). *Nucl. Acids Res.*, **17**, 10321–35.
47. Tomasiewicz, H., Ono, K., Yee, D., Thompson, C., Goridis, C., Rutishauser, U., and Magnuson, T. (1993). *Neuron,* **11**, 1163–74.
48. Cremer, H., Chazal, G., Goridis, C., and Represa, A. (1997). *Mol. Cell Neurosci.*, **8**, 323–35.
49. Cremer, H., Lange, R., Christoph, A., Plomann, M., Vopper, G., Roes, J., *et al.* (1994). *Nature*, **367**, 455–9.
50. Thomsen, N. K., Soroka, V., Jensen, P. H., Berezin, V., Kiselyov, V. V., Bock, E., and Poulsen, F. M. (1996). *Nature Struct. Biol.*, **3**, 581–5.
51. Yoshihara, Y., Kawasaki, M., Tamada, A., Fujita, H., Hayashi, H., Kagamiyama, H., and Mori, K. (1997). *J. Neurosci.*, **17**, 5830–42.
52. Alenius, M. and Bohm, S. (1997). *J. Biol. Chem.*, **272**, 26083–6.
53. Paoloni Giacobino, A., Chen, H., and Antonarakis, S. E. (1997). *Genomics*, **43**, 43–51.
54. Huang, Y. Q., Jellies, J., Johansen, K. M., and Johansen, J. (1997). *J. Cell Biol.*, **138**, 143–57.
55. Angata, K., Nakayama, J., Fredette, B., Chong, K., Ranscht, B., and Fukuda, M. (1997). *J. Biol. Chem.*, **272**, 7182–90.

56. Wang, C., Pralong, W. F., Schulz, M. F., Rougon, G., Aubry, J. M., Pagliusi, S., Robert, A., and Kiss, J. Z. (1996). *J. Cell Biol.*, **135**, 1565–81.
57. Bruses, J. L. and Rutishauser, U. (1998). *J. Cell Biol.*, **140**, 1177–86.
58. Krushel, L. A., Tai, M. H., Cunningham, B. A., Edelman, G. M., and Crossin, K. L. (1998). *Proc. Natl Acad. Sci. USA*, **95**, 2592–6.
59. Ronn, L. C., Pedersen, N., Jahnsen, H., Berezin, V., and Bock, E. (1997). *Adv. Exp. Med. Biol.*, **429**, 305–22.
60. Holst, B. D., Vanderklish, P. W., Krushel, L. A., Zhou, W., Langdon, R. B., McWhirter, J. R., Edelman, G. M., and Crossin, K. L. (1998). *Proc. Natl Acad. Sci. USA*, **95**, 2597–602.
61. Holst, B. D., Wang, Y., Jones, F. S., and Edelman, G. M. (1997). *Proc. Natl Acad. Sci. USA*, **94**, 1465–70.
62. Treloar, H., Tomasiewicz, H., Magnuson, T., and Key, B. (1997). *J. Neurobiol.*, **32**, 643–58.
63. Shen, H., Watanabe, M., Tomasiewicz, H., Rutishauser, U., Magnuson, T., and Glass, J. D. (1997). *J. Neurosci.*, **17**, 5221–9.
64. Stork, O., Welzl, H., Cremer, H., and Schachner, M. (1997). *Eur. J. Neurosci.*, **9**, 1117–25.
65. Schachner, M. (1997). *Curr. Opin. Cell Biol.*, **9**, 627–34.

Thomas Brümmendorf
Max-Delbrück-Center for Molecular Medicine,
Robert-Rössle Strasse 10,
13122 Berlin, Germany

Neural cell recognition molecule F11 (contactin)

F11 is a neural member of the immunoglobulin superfamily (IgSF) involved in neurite outgrowth and fasciculation. It is a glycosyl phosphatidylinositol (GPI)-anchored cell surface glycoprotein expressed in axon-rich regions of the developing central and peripheral nervous system. The molecule shows multiple interactions with other membrane proteins and with components of the extracellular matrix. F11 is also expressed in the adult brain.

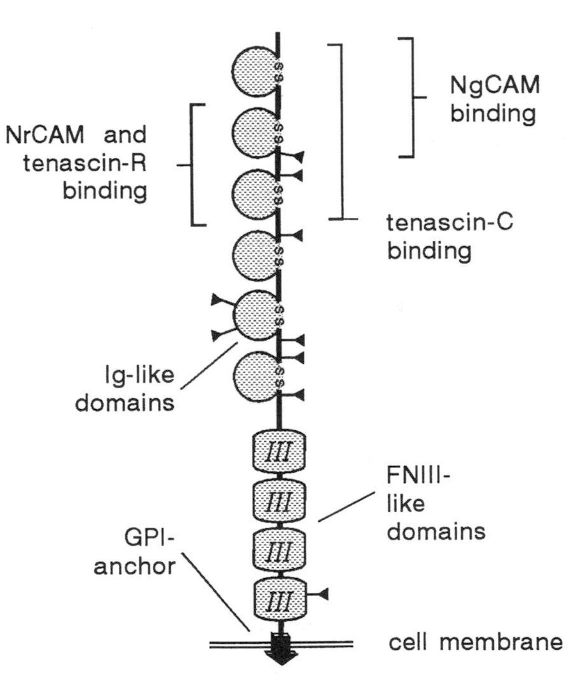

Figure 1. Domain model of F11/contactin, and mapping of ligand binding sites. The amino terminus is at the top, and putative conserved *N*-linked glycosylation sites are marked with arrowheads.

Synonymous names

F11 is also referred to as F3 or contactin.[1]

Protein properties

The F11 molecule is a 130 kDa cell surface glycoprotein containing complex type and high mannose/hybrid type *N*-linked carbohydrates, the HNK-1/L2 carbohydrate epitope,[2,3] and a glycosyl phosphatidylinositol (GPI) membrane anchor.[4] It has six immunoglobulin- (Ig-) like domains and four fibronectin type III domains (Fig. 1).[5–12] This arrangement of domains is also found in the structurally related molecules axonin-1/TAG-1, BIG-1, BIG-2, NB-2 and NB-3 which form a neural subgroup of the IgSF.[13,14,30]

F11 is expressed in axon-rich regions (Fig. 2) of the developing and adult nervous system.[1,2,5,7] The molecule is also found at synapses in the cerebellum, either pre- or postsynaptically.[15]

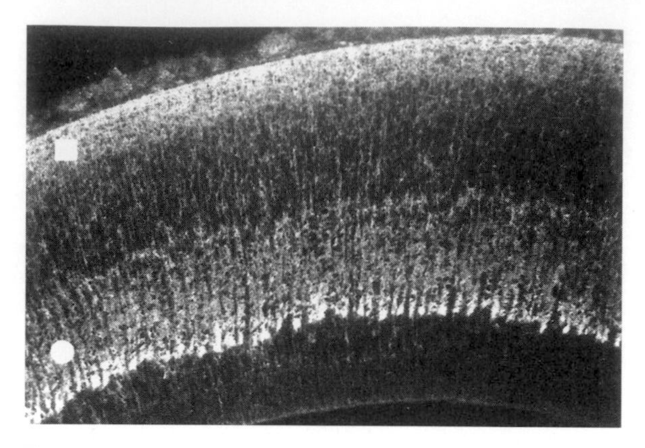

Figure 2. Expression of F11 on incoming retinal ganglion cell axons (white square), and tectofugal fibres (white point) of the developing chick optic tectum. A horizontal section (ventricular zone at the bottom) stained with F11-specific antibodies in an immunofluorescence protocol is shown.

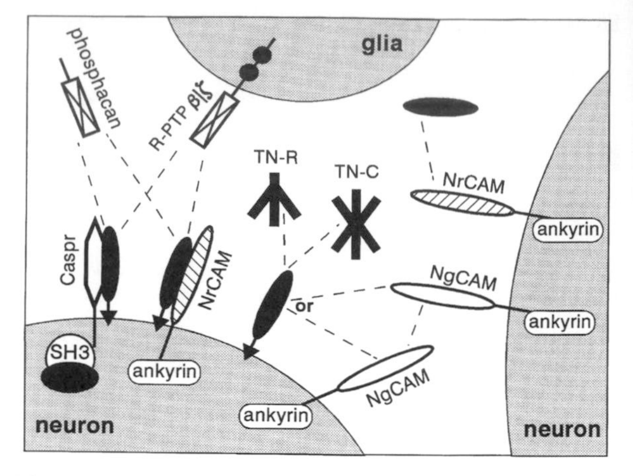

Figure 3. Interactions of F11/contactin with molecules on neuronal, and glial cell surfaces, and in the brain extracellular matrix. GPI-linked F11 is depicted as a black ellipse with an arrow, and free F11 as a black ellipse. The topology of the F11–NgCAM binding is unclear at present. R-PTP: receptor protein tyrosine phosphatase, TN: tenascin.

In line with its supposed roles in cell–cell interactions, a variety of different binding assays have revealed that the molecule interacts with several other proteins.[16] F11 binds to the neural IgSF molecules NrCAM[17] and NgCAM[18] and to the glial receptor protein tyrosine phosphatase (R-PTP) β/ζ.[11] It also interacts with Caspr/Paranodin, a transmembrane molecule at paranodal regions of central and peripheral nervous system axons.[19,32,33] Caspr/Paranodin has a proline-rich cytoplasmic sequence which binds SH3 domains of signalling molecules.[19] In addition, F11 interacts with the extracellular matrix components tenascin (TN)-R[18] and TN-C [20] (Fig. 3). Homophilic binding of F11 has not been demonstrated so far.

Domain deletion mutagenesis of the F11 molecule has revealed that the four N-terminal Ig-like domains are sufficient to bind the ligands NgCAM,[18] NrCAM,[17] TN-R[18] and TN-C[20]. The domains 1 and 2 are most important for NgCAM binding[18] while domains 2 and 3 are important for interactions with NrCAM[17] or with TN-R[18] (Fig. 1).

Purification

F11 can be purified from detergent extracts or from phosphatidylinositol-specific phospholipase C (PI-PLC) supernatants of brain plasma membranes by immunoaffinity chromatography. Studies describing its purification, for instance from human,[8,9] mouse,[7] or chick[2] brains, have been listed in a recent review.[1]

Activities

In vitro antibody perturbation experiments suggest that F11 is involved in axon fasciculation[2] and in elongation of axons on pre-existing axons.[21] F11 isolated from brain promotes neurite outgrowth.[17,18,22] F11 expressed in eukaryotic cells may activate[23] or inhibit neurite outgrowth, depending on the neurones examined.[24] F11-induced outgrowth of chick tectal neurones is mediated by interaction with the axonal IgSF member NrCAM[17] which most probably binds in a *cis* orientation[25] and with R-PTP β/ζ which interacts in *trans*[11] (Fig. 3). The outgrowth promoting activity of F11 is enhanced by binding to the extracellular matrix component TN-R[26] and is modulated by integrins.[34] The F11–TN-R interaction has been implicated in a serine/threonine kinase-mediated[35] avoidance response of mouse cerebellar neurones.[27,28]

Antibodies

Studies describing polyclonal and monoclonal antibodies to F11 isolated from different species, including human,[8,9] mouse,[7] and chick,[2,5,18] have been reviewed recently.[1]

Genes

Complete cDNA sequences have been reported for human,[8,9] mouse,[7] rat,[10,11] chick,[5,6] and *Xenopus*.[12] The exon/intron structure of the F11 gene suggests evolution via exon shuffling and each domain is encoded by two exons.[29] Only two isoforms have been reported which differ with respect to a short stretch of 11 amino acids near the amino terminus.[8] The human F11 gene[8] is located in region q11–q12 of chromosome 12 and the promoter has been characterized in the mouse.[36]

Mutant phenotype/disease states

No information available.

Structure

Not available.

References

1. Brümmendorf, T. and Rathjen, F. G. (1995). *Protein Profile*, **2**, 963–1108.
2. Rathjen, F. G., Wolff, J. M., Frank, R., Bonhoeffer, F. and Rutishauser, U. (1987). *J. Cell Biol.*, **104**, 343–53.
3. Rathjen, F. G. (1993). In *Guidebook to the extracellular matrix and adhesion proteins* (ed. T. Kreis, and R. Vale), pp. 139–40. Oxford University Press.
4. Wolff, J. M., Brümmendorf, T., and Rathjen, F. G. (1989). *Biochem. Biophys. Res. Commun.*, **161**, 931–38.
5. Ranscht, B. and Dours, M. T. (1988). *J. Cell Biol.*, **107**, 1561–73.
6. Brümmendorf, T., Wolff, J. M., Frank, R. and Rathjen, F. G. (1989). *Neuron.*, **2**, 1351–61.
7. Gennarini, G., Cibelli, G., Rougon, G., Mattei, M. G. and Goridis, C. (1989). *J. Cell Biol.*, **109**, 775–88.
8. Berglund, E. O. and Ranscht, B. (1994). *Genomics*, **21**, 571–82.
9. Reid, R. A., Bronson, D. D., Young, K. M. and Hemperly, J. J. (1994). *Brain Res. Mol. Brain Res.*, **21**, 1–8.
10. Hosoya, H., Shimazaki, K., Kobayashi, S., Takahashi, H., Shirasawa, T., Takenawa, T. and Watanabe, K. (1995). *Neurosci. Lett.*, **186**, 83–6.
11. Peles, E., Nativ, M., Campbell, P. L., Sakurai, T., Martinez, R., Lev, S., *et al.* (1995). *Cell*, **82**, 251–60.
12. Nagata, S., Fujita, N., Takeuchi, K. and Watanabe, K. (1996). *Zool. Sci.*, **13**, 813–20.
13. Sonderegger, P. and Rathjen, F. G. (1992). *J. Cell. Biol.*, **119**, 1387–94.
14. Yoshihara, Y., Kawasaki, M., Tamada, A., Nagata, S., Kagamiyama, H. and Mori, K. (1995). *J. Neurobiol.*, **28**, 51–69.
15. Faivre-Sarrailh, C., Gennarini, G., Goridis, C., and Rougon, G. (1992). *J. Neurosci.*, **12**, 257–67.
16. Brümmendorf, T., and Rathjen, F. G. (1996). *Curr. Opin. Neurobiol.*, **6**, 584–93.
17. Morales, G., Hubert, M., Brümmendorf, T., Treubert, U., Tarnok, A., Schwarz, U., and Rathjen, F. G. (1993). *Neuron*, **11**, 1113–22.
18. Brümmendorf, T., Hubert, M., Treubert, U., Leuschner, R., Tarnok, A., and Rathjen, F. G. (1993). *Neuron*, **10**, 711–27.
19. Peles, E., Nativ, M., Lustig, M., Grumet, M., Schilling, J., Martinez, R., *et al.* (1997). *EMBO J.*, **16**, 978–88.
20. Zisch, A. H., D'Alessandri, L., Ranscht, B., Falchetto, R., Winterhalter, K. H., and Vaughan, L. (1992). *J. Cell Biol.*, **119**, 203–13.
21. Chang, S., Rathjen, F. G., and Raper, J. A. (1987). *J. Cell Biol.*, **104**, 355–62.
22. Durbec, P., Gennarini, G., Goridis, C., and Rougon, G. (1992). *J. Cell Biol.*, **117**, 877–87.
23. Gennarini, G., Durbec, P., Boned, A., Rougon, G., and Goridis, C. (1991). *Neuron*, **6**, 595–606.
24. Buttiglione, M., Revest, J. M., Rougon, G., and Faivre Sarrailh, C. (1996). *Mol. Cell Neurosci.*, **8**, 53–69.
25. Sakurai, T., Lustig, M., Nativ, M., Hemperly, J. J., Schlessinger, J., Peles, E., and Grumet, M. (1997). *J. Cell Biol.*, **136**, 907–18.
26. Nörenberg, U., Hubert, M., Brümmendorf, T., Tarnok, A., and Rathjen, F. G. (1995). *J. Cell Biol.*, **130**, 473–84.
27. Pesheva, P., Gennarini, G., Goridis, C., and Schachner, M. (1993). *Neuron*, **10**, 69–82.
28. Xiao, Z. C., Taylor, J., Montag, D., Rougon, G., and Schachner, M. (1996). *Eur. J. Neurosci.*, **8**, 766–82.
29. Plagge, A., and Brümmendorf, T. (1997). *Gene*, **192**, 215–25.
30. Ogawa, J., Kaneko, H., Masuda, T., Nagata, S., Hosoya, H., and Watanabe, K. (1996). *Neurosci. Lett.*, **218**, 173–6.
31. Peles, E., Schlessinger, J., and Grumet, M. (1998). *Trends Biochem. Sci.*, **23**, 121–4.
32. Einheber, S., Zanazzi, G., Ching, W., Scherer, S., Milner, T. A., Peles, E., and Salzer, J. L. (1997). *J. Cell Biol.*, **139**, 1495–506.
33. Menegoz, M., Gaspar, P., Le Bert, M., Galvez, T., Burgaya, F., Palfrey, C., Ezan, P., Arnos, F., and Girault, J. A. (1997). *Neuron*, **19**, 319–31.
34. Treubert, U. and Brümmendorf, T. (1998). *J. Neurosci.*, **18**, 1795–805.
35. Xiao, Z. C., Hillenbrand, R., Schachner, M., Thermes, S., Rougon, G., and Gomez, S. (1997). *J. Neurosci. Res.*, **49**, 698–709.
36. Cangiano, G., Ambrosini, M., Patruno, A., Tino, A., Buttiglione, M., and Gennarini, G. (1997). *Brain Res. Mol. Brain Res.*, **48**, 279–90.

Thomas Brümmendorf
Max-Delbrück-Center for Molecular Medicine,
Robert-Rössle Strasse 10,
13122 Berlin, Germany

Neural cell recognition molecule L1

L1 is a transmembrane member of the immunoglobulin superfamily (IgSF) implicated in human hereditary diseases affecting brain development. It has a complex biology and is involved in many processes involving cell–cell interactions, including cell migration, neurite fasciculation and elongation, myelination, and growth cone morphology. The protein is expressed in axon tracts of the developing and adult nervous system but is also found on non-neural cells. L1 has a highly conserved, ankyrin binding, cytoplasmic domain. The molecule binds to a variety of other membrane proteins, including those of the IgSF and integrins as well as to components of the extracellular matrix.

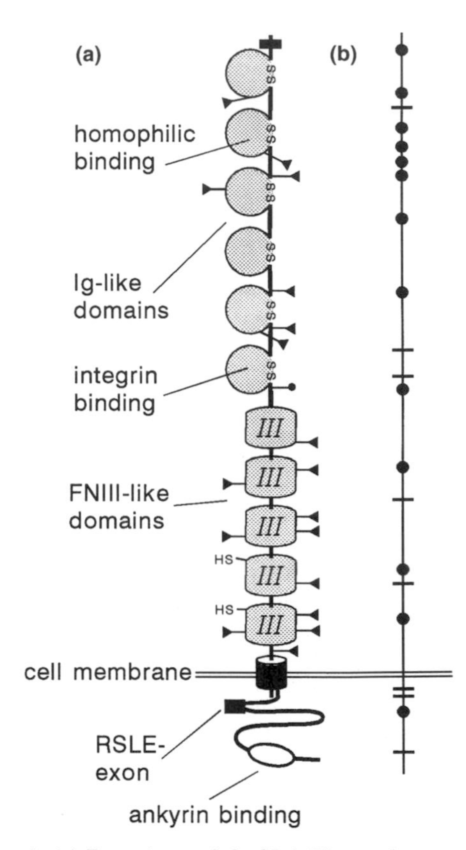

Figure 1. (a) Domain model of L1. The amino terminus is at the top, and putative conserved *N*-linked glycosylation sites are marked by arrowheads. (b) Positions of mutations in L1 associated with human hereditary brain malformations. Mis sense mutations are depicted as dots, and mutations leading to premature termination of the polypeptide chain (nonsense mutations, frameshifts, and splicing errors) are indicated as short horizontal lines.

Synonymous names

In mammals, L1 is also referred to as L1-CAM, 5G3 antigen, NILE or 69A1 antigen.[1]

Related proteins

In non-mammalian vertebrates, putative candidates for species homologues are NgCAM (also termed G4 or 8D9 antigen) in chicken,[2,3] L1 in pufferfish,[53] L1.1 in zebrafish,[4] and E587 antigen in goldfish.[5] L1 is distantly related to neuroglian in *Drosophila*[6] and to the protein tractin in *Hirudo medicinalis*.[44]

Protein properties

The L1 glycoprotein is a 200 kDa cell surface molecule with the HNK-1/L2 carbohydrate epitope.[7] The extracellular part of the molecule is composed of six immunoglobulin-like domains and five fibronectin type III domains[8–13] (Fig. 1(a)). This domain arrangement has also been found in other members of the L1-subgroup[3] of the IgSF, like CHL1,[14] NrCAM, and neurofascin.[1,15]

In the nervous system, L1 is expressed by subpopulations of postmitotic neurones but also by glial cells.[7] L1 is also found on some cells outside the nervous system,[1] including melanoma cells[16] and those of haematopoietic origin.[17,18]

L1 shows a complex interaction pattern and binds to a variety of cell surface and extracellular matrix proteins.[7,19,45] In different assay systems, L1 and its chicken relative NgCAM have been shown to interact in a homophilic manner, to bind to the IgSF cell surface molecule axonin-1/TAG-1 and to the extracellular matrix protein laminin. Additionally, NgCAM has been shown to interact with the IgSF members F11(contactin) and DM-grasp and with the proteoglycans phosphacan and neurocan.[20] L1 undergoes *cis* interactions with NCAM and nectadrin/CD24 and has been documented to interact with the integrins $\alpha_5\beta_1$ and $\alpha_v\beta_3$ which recognize an RGD peptide sequence within the sixth immunoglobulin-like domain.[16,17]

The highly conserved cytoplasmic domain of L1[7,21,46] and NgCAM[2,22] can be phosphorylated. L1 is linked to the cytoskeleton via ankyrin,[23] which may be modulated by tyrosine phosphorylation.[47] Signalling and cytoskeletal interactions[47,48] of L1 have been reviewed recently.[49]

Purification

Studies describing the isolation of L1/NgCAM by immunoaffinity chromatography from nonionic detergent extracts of brain plasma membranes of different species, including human,[9,24] mouse,[7] or chick,[25] have been cited in a recent review.[1]

Activities

Different *in vitro* assay systems, making use of antibody perturbation, isolated protein or transfected fibroblasts suggest that L1[7,50] and NgCAM[2,51] are involved in axon fasciculation, induction of neurite outgrowth by different types of neurones and cell adhesion.[1] In the context of neurite outgrowth, the FGF receptor[26,52] and src family kinases[27] have been implicated in signal transduction. *In vivo,* NgCAM-specific antibodies interfere with nerve branching in the chick[28] and E587-specific antibodies disrupt the fascicle pattern of goldfish retinal ganglion cell axons.[29]

Increasing evidence suggests that L1 may not only be involved in early brain development, but also in activity-dependent plasticity in late development and in synaptic plasticity in the adult brain.[30] Expression of L1 in dorsal root ganglia is modulated by specific frequencies of neural impulses.[31] L1-specific antibodies interfere with hippocampal long-term potentiation[32] and intracranial injection of antibodies interferes with memory consolidation in a chick model.[33]

Outside the nervous system, on the basis of *in vitro* assays, L1 binding to integrins is involved in cell adhesion,[17] cell spreading, and motility.[16]

Antibodies

Studies describing polyclonal and monoclonal antibodies have been cited in a recent review[1], including those specific for human[9,24] and mouse[7] L1 or chicken NgCAM.[25] Antibodies are also available from the Developmental Studies Hybridoma Bank (John Hopkins University, Iowa City, USA; http://www.uiowa.edu/~dshbwww/). Commercial suppliers for L1 specific antibodies include Santa Cruz Biotechnology (Santa Cruz, USA), Transduction Laboratories (Lexington, USA) and Boehringer Mannheim (Mannheim, Germany).

Genes

Full length cDNA sequences have been reported for human[9,10,13], mouse[8] and rat[11,12] L1. Sequences of putative species homologues are also available, including chick NgCAM,[34] goldfish E587-antigen,[5] L1 in pufferfish,[53] L1.1 in zebrafish[4] and *Drosophila* neuroglian.[6] The mouse and human L1 gene is located on the X chromosome[35] and the human gene has been mapped to the Xq28 region near the telomere of the long arm.[35,36] Isoforms differing with respect to an RSLE motif in the cytoplasmic region[10,12,13] and a short brain-specific segment near the amino terminus[18] have been described. The gene organization of L1 is well conserved between human and pufferfish[53] and neuronal expression is regulated by a neural-restrictive silencer element.[54]

Mutant phenotype/disease states

Mutations in the L1 gene have been implicated in human genetic disorders affecting embryonal brain development.[15,37,55] The clinical spectrum is relatively broad and includes corpus callosum hypoplasia, mental retardation, adducted thumbs, spastic paraplegia, and hydrocephalus.[38–42] The mutations are distributed along the entire length of the polypeptide (Fig. 1(b)). A significant correlation was observed between the likelihood of severe hydrocephalus and possessing a mutation in the extracellular regions.[56] Up-to date lists of L1 mutations are maintained at http://www.uwcm.ac.uk/uwcm/mg and http://dnalab-www.uia.ac.be/dnalab/l1. L1 mutations have also been found in patients with Hirschsprung's disease.[57]

L1-deficient mice showed abnormalities in nervous system development which resemble some of the aspects of the human diseases, for instance malformations of the corticospinal tract[58] and enlarged ventricles.[59] Neurones of L1-deficient mice did not extend neurites on immobilized L1 and in the peripheral nervous system, axon–Schwann cell interactions are impaired.[59]

Structure

The outline structure which describes the major secondary structures and residue solvent accessibility has been derived by comparison with related proteins whose three-dimensional structures had been resolved.[43]

References

1. Brümmendorf, T. and Rathjen, F. G. (1995). *Protein Profile,*. **2**, 963–1108.
2. Grumet, M. (1992). *J. Neurosci. Res.*, **31**, 1–13.
3. Sonderegger, P. and Rathjen, F. G. (1992). *J. Cell Biol.*, **119**, 1387–94.
4. Tongiorgi, E., Bernhardt, R. R. and Schachner, M. (1995). *J. Neurosci. Res.*, **42**, 547–61.
5. Giordano, S., Laessing, U., Ankerhold, R., Lottspeich, F. and Stuermer, C. A. O. (1997). *J. Comp. Neurol.*, **377**, 286–97.
6. Bieber, A. J., Snow, P. M., Hortsch, M., Patel, N. H., Jacobs, J. R., Traquina, Z. R., *et al.* (1989). *Cell*, **59**, 447–60.
7. Schachner, M. (1993). In *Guidebook to the extracellular matrix and adhesion molecules*, (ed. T. Kreis and R. Vale), pp.147–8. Oxford University Press.
8. Moos, M., Tacke, R., Scherer, H., Teplow, D., Früh, K. and Schachner, M. (1988). *Nature*, **334**, 701–3.
9. Hlavin, M. L. and Lemmon, V. (1991). *Genomics*,. **11**, 416–23.
10. Kobayashi, M., Miura, M., Asou, H. and Uyemura, K. (1991). *Biochim. Biophys. Acta*, **1090**, 238–40.
11. Prince, J. T., Alberti, L., Healy, P. A., Nauman, S. J. and Stallcup, W. B. (1991). *J. Neurosci. Res.*, **30**, 567–81.

12. Miura, M., Kobayashi, M., Asou, H. and Uyemura, K. (1991). *FEBS Lett.*, **289**, 91–5.
13. Reid, R. A. and Hemperly, J. J. (1992). *J. Mol. Neurosci.*, **3**, 127–35.
14. Holm, J., Hillenbrand, R., Steuber, V., Bartsch, U., Moos, M., Lubbert, H., *et al.* (1996). *Eur. J. Neurosci.*, **8**, 1613–29.
15. Hortsch, M. (1996). *Neuron*, **17**, 587–93.
16. Montgomery, A. M. P., Becker, J. C., Siu, C. H., Lemmon, V. P., Cheresh, D. A., Pancook, J. D., *et al.* (1996). *J. Cell Biol.*, **132**, 475–85.
17. Ruppert, M., Aigner, S., Hubbe, M., Yagita, H. and Altevogt, P. (1995). *J. Cell Biol.*, **131**, 1881–91.
18. Jouet, M., Rosenthal, A. and Kenwrick, S. (1995). *Brain Res. Mol. Brain Res.*, **30**, 378–80.
19. Brümmendorf, T. and Rathjen, F. G. (1996). *Curr. Opin. Neurobiol.*, **6**, 584–93.
20. Margolis, R. K., Rauch, U., Maurel, P. and Margolis, R. U. (1996). *Perspect. Dev. Neurobiol.*, **3**, 273–90.
21. Wong, E. V., Schaefer, A. W., Landreth, G. and Lemmon, V. (1996). *J. Biol. Chem.*, **271**, 18217–223.
22. Kunz, S., Ziegler, U., Kunz, B. and Sonderegger, P. (1996). *J. Cell Biol.*, **135**, 253–67.
23. Davis, J. Q. and Bennett, V. (1994). *J. Biol. Chem.*, **269**, 27163–6.
24. Wolff, J. M., Frank, R., Mujoo, K., Spiro, R. C., Reisfeld, R. A. and Rathjen, F. G. (1988). *J. Biol. Chem.*, **263**, 11943–7.
25. Rathjen, F. G., Wolff, J. M., Frank, R., Bonhoeffer, F., and Rutishauser, U. (1987). *J. Cell Biol.*, **104**, 343–53.
26. Doherty, P., and Walsh, F. S. (1996). *Mol. Cell. Neurosci,*. **8**, 99–111.
27. Maness, P. F., Beggs, H. E., Klinz, S. G., and Morse, W. R. (1996). *Perspect. Dev. Neurobiol.*, **4**, 169–81.
28. Landmesser, L., Dahm, L., Schultz, K., and Rutishauser, U. (1988). *Dev. Biol.*, **130**, 645–70.
29. Bastmeyer, M., Ott, H., Leppert, C. A., and Stuermer, C. A. (1995). *J. Cell Biol.*, **130**, 969–76.
30. Fields, R. D., and Itoh, K. (1996). *Trends Neurosci.*, **19**, 473–80.
31. Itoh, K., Stevens, B., Schachner, M., and Fields, R. D. (1995). *Science*,. **270**, 1369–72.
32. Lüthi, A., Laurent, J. P., Figurov, A., Muller, D., and Schachner, M. (1994). *Nature*, **372**, 777–9.
33. Scholey, A. B., Mileusnic, R., Schachner, M., and Rose, S. P. R. (1995). *Learning Memory,* **2**, 17–25.
34. Burgoon, M. P., Grumet, M., Mauro, V., Edelman, G. M., and Cunningham, B. A. (1991). *J. Cell Biol.* **112**, 1017–29.
35. Djabali, M., Mattei, M. G., Nguyen, C., Roux, D., Demengeot, J., Denizot, F., *et al.* (1990). *Genomics*, **7**, 587–93.
36. Dietrich, A., Korn, B., and Poustka, A. (1992). *Mamm. Genome*, **3**, 168–72.
37. Wong, E. V., Kenwrick, S., Willems, P. J., and Lemmon, V. (1995). *Trends Neurosci.*, **18**, 168–72.
38. Fransen, E., Vits, L., Van Camp, G., and Willems, P. J. (1996). *Am. J. Med. Genet.*, **64**, 73–77.
39. Kenwrick, S., Jouet, M., and Donnai, D. (1996). *J. Med. Genet.*, **33**, 59–65.
40. Gu, S. M., Orth, U., Veske, A., Enders, H., Klunder, K., Schlosser, M., *et al.* (1996). *J. Med. Genet.*, **33**, 103–6.
41. Ruiz, J. C., Cuppens, H., Legius, E., Fryns, J. P., Glover, T., Marynen, P., and Cassiman, J. J. (1995). *J. Med. Genet.*, **32**, 549–52.
42. Schrander Stumpel, C., Howeler, C., Jones, M., Sommer, A., Stevens, C., Tinschert, S., *et al.* (1995). *Am. J. Med. Genet.*, **57**, 107–16.
43. Bateman, A., Jouet, M., MacFarlane, J., Du, J. S., Kenwrick, S., and Chothia, C. (1996). *EMBO J.*, **15**, 6050–9.
44. Huang, Y. Q., Jellies, J., Johansen, K. M., and Johansen, J. (1997). *J. Cell Biol.*, **138**, 143–57.
45. Kadmon, G. and Altevogt, P. (1997). *Differentiation*, **61**, 143–50.
46. Zisch, A. H., Stallcup, W. B., Chong, L. D., Dahlin Huppe, K., Voshol, J., Schachner, M., and Pasquale, E. B. (1997). *J. Neurosci. Res.*, **47**, 655–65.
47. Garver, T. D., Ren, Q., Tuvia, S., and Bennett, V. (1997). *J. Cell Biol.*, **137**, 703–14.
48. Dahlin Huppe, K., Berglund, E. O., Ranscht, B., and Stallcup, W. B. (1997). *Mol. Cell. Neurosci.*, **9**, 144–56.
49. Kamiguchi, H. and Lemmon, V. (1997). *J. Neurosci. Res.*, **49**, 1–8.
50. Burden Gulley, S. M., Pendergast, M., and Lemmon, V. (1997). *Cell Tissue Res.*, **290**, 415–22.
51. Sonderegger, P. (1987). *Cell Tissue Res.*, **290**, 429–39.
52. Saffell, J. L., Williams, E. J., Mason, I. J., Walsh, F. S., and Doherty, P. (1997). *Neuron*, **18**, 1–11.
53. Coutelle, O., Nyakatura, G., Taudien, S., Elgar, G., Brenner, S., Platzer, M., Drescher, B., Jouet, M., Kenwrick, S., and Rosenthal, A. (1998). *Gene*, **208**, 7–15.
54. Kallunki, P., Edelman, G., and Jones, F. S. (1997). *J. Cell Biol.*, **138**, 1343–54.
55. Brümmendorf, T., Kenwrick, S., and Rathjen, F. G. (1998). *Curr. Opin. Neurobiol.*, **8**, 87–97.
56. Yamasaki, M., Thompson, P., and Lemmon. V. (1997). *Neuropediatrics*, **28**, 175–8.
57. Okamoto, N., Wada, Y., and Goto, M. (1997). *J. Med. Genet.*, **34**, 670–71.
58. Cohen, N. R., Taylor, J. S. H., Scott, L. B., Guillery, R. W., Soriano, P., and Furley, A. J. (1997). *Curr. Biol.*, **8**, 26–33.
59. Dahme, M., Bartsch, U., Martini, R., Anliker, B., Schachner, M., and Mantei, N. (1997). *Nature Genet.*, **17**, 346–9.

Thomas Brümmendorf
Max-Delbrück-Center for Molecular Medicine,
Robert-Rössle Strasse 10,
13122 Berlin, Germany

Neurofascin

Neurofascin belongs to the L1 subgroup of the immunoglobulin superfamily (IgSF) and is implicated in neurite formation as suggested by its association with developing fibre tracts, interference of neurofascin-specific antibodies with axonal fasciculation, and the neurite outgrowth promoting activity of purified neurofascin.[1-3] Extracellular interactions may be linked to cytoskeletal rearrangements by binding of neurofascin to ankyrin.[4]

Synonymous names

According to cytoskeletal associations, rat neurofascin was previously termed ankyrin-binding glycoprotein (ABGP).[4]

Homologous proteins

Structurally related L1/NgCAM, CHL1 and NrCAM form, with neurofascin, the L1 subgroup within the IgSF[2,5-9] that share as common features six immunoglobulin-like domains, five fibronectin type III (FNIII)-like repeats, a transmembrane domain, and a highly conserved cytoplasmic domain that binds to ankyrin.[10]

Protein properties

Chick embryonic neurofascin comprises two molecular mass forms of 160 kDa and 185 kDa.[1] At least 50 different isoforms are formed in brain by alternative splicing: three major sequences are differentially expressed including the third and the fifth FNIII-like repeat and the PAT domain which is located between the fourth and fifth FNIII-like repeat.[11] In addition, four minor alternatively spliced sequences have been found at the NH_2-terminus, between the second and third Ig-like domain, at the junction between the Ig-like and FNIII-like domains and in the cytoplasmic domain. One major isoform expressed at embryonal day 6 is replaced by five different major isoforms at embryonal day 16 (Fig. 1). Neurofascin polypeptides may be modified by the addition of *O*-linked and *N*-linked carbohydrates.[2]

Neurofascin is expressed in the central and the peripheral nervous system.[1,3,12] At early stages of development, neurofascin is expressed in the optic fibre layer of the developing retina while in the optic tectum, neurofascin is concentrated in the tectobulbar tract.[1,13] In the developing spinal cord, strong neurofascin expression is found on circumferential axons within the floor plate and on the distal part of circumferential axons that have already crossed the floor plate.[14] Neurofascin isoforms derived from alternatively spliced exons are differentially expressed as shown for an isoform including the fifth FNIII-like repeat which is located at the nodes of Ranvier and an isoform containing the third FNIII-like repeat which is expressed on unmyelinated axons.[12]

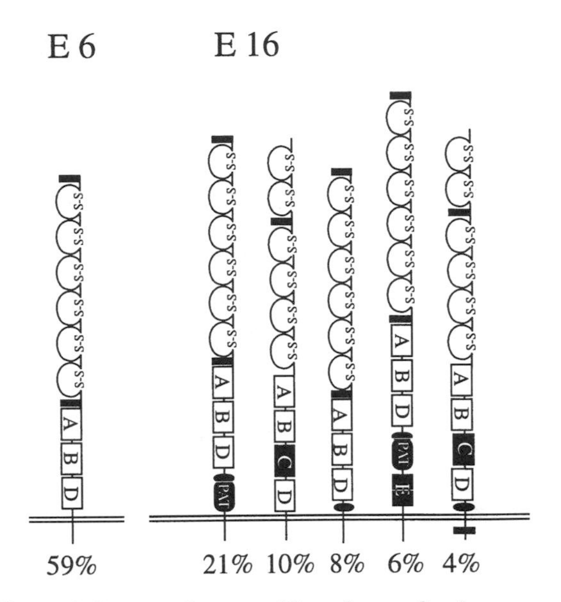

Figure 1. Structural composition of neurofascin isoforms. The neurofascin gene encodes six Ig-like domains, five FNIII-like repeats (A–E), the PAT domain which is located between the FNIII-like repeats D, and E, a transmembrane domain, and a cytoplasmic domain. Alternatively spliced sequences are shown in black. Major neurofascin isoforms (frequency > 4 per cent) are shown with their relative frequency at either embryonal day 6 (E6) or 16 (E16).

Purification

Neurofascin can be purified from detergent extracts of brain membranes by immunoaffinity chromatography.[1] Expression of neurofascin cDNA by eukaryotic vectors and affinity purification of fusion proteins have been described.[3]

Activities

Fab fragments of neurofascin-specific polyclonal antibodies inhibit fasciculation of retinal axons, implying that neurofascin participates in axon–axon interactions.[1] Immobilized neurofascin promotes neurite outgrowth of primary tectal cells by interactions of its Ig-like domains with NrCAM.[3] Ankyrin, expressed as a bacterial fusion protein, binds neurofascin at the cytoplasmic domain.[4] This interaction is regulated by phosphorylation of a tyrosine residue within the cytoplasmic domain of neurofascin.[15]

Antibodies

Monoclonal as well as polyclonal antibodies to neurofascin have been described.[1]

Genes

Chick genomic sequences encompassing 33 exons spread over 72 kb have been cloned and partially sequenced.[11] Neurofascin was mapped on murine chromosome 1 in a region that corresponds to human chromosome 1q31–32.[16]

Mutant phenotype/disease states

No information available.

Structure

Not available.

References

1. Rathjen, F. G., Wolff, J. M., Chang, S., Bonhoeffer, F., and Raper, J. A. (1987). *Cell*, **51**, 841–9.
2. Volkmer, H., Hassel, B., Wolff, J. M., Frank, R., and Rathjen, F. G. (1992). *J. Cell Biol.*, **118**, 149–61.
3. Volkmer, H., Leuschner, R., Zacharias, U., and Rathjen, F. G. (1996). *J. Cell Biol.*, **135**, 1059–69.
4. Davis, J. Q., McLaughlin, T., and Bennett, V. (1993). *J. Cell Biol.*, **121**, 121–33.
5. Holm, J., Hillenbrand, R., Steuber, V., Bartsch, U., Moos, M., Lubbert, H., *et al.* (1996). *Eur. J. Neurosci.*, **8**, 1613–29.
6. Grumet, M., Mauro, V., Burgoon, M. P., Edelman, G. M., and Cunningham, B. A. (1991). *J. Cell Biol.*, **113**, 1399–412.
7. Burgoon, M. P., Grumet, M., Mauro, V., Edelman, G. M., and Cunningham, B. A. (1991). *J. Cell Biol.*, **112**, 1017–29.
8. Moos, M., Tacke, R., Scherer, H., Teplow, D., Fruh, K., and Schachner, M. (1988). *Nature*, **334**, 701–3.
9. Kayyem, J. F., Roman, J. M., de la Rosa, E. J., Schwarz, U., and Dreyer, W. J. (1992). *J. Cell Biol.*, **118**, 1259–70.
10. Davis, J. Q. and Bennett, V. (1994). *J. Biol. Chem.*, **269**, 27163–6.
11. Hassel, B., Rathjen, F. G., and Volkmer, H. (1998). *J. Biol. Chem.*, **272**, 28742–9.
12. Davis, J. Q., Lambert, S., and Bennett, V. (1996). *J. Cell Biol.*, **135**, 1355–67.
13. Kröger, S. and Schwarz, U. (1990). *J. Neurosci.*, **10**, 3118–34.
14. Shiga, T. and Oppenheim, R. W. (1991). *J. Comp. Neurol.* **310**, 234–52.
15. Garver, T. D., Ren, Q., Tuvia, S., and Bennett, V. (1997). *J. Cell Biol.*, **137**, 703–14.
16. Burmeister, M., Ren, Q., Makris, G. J., Samson, D., and Bennett, V. (1996). *Mamm. Genome*, **7**, 558–9.

Hansjürgen Volkmer
Max-Delbrück-Center for Molecular Medicine,
Robert-Rössle Strasse 10,
13122 Berlin, Germany

Neuroglian

Neuroglians are homophilic, Ca^{2+}-independent cell adhesion molecules in insects which are members of the L1 family of immunoglobulin domain neural cell adhesion molecules.[1] The *Drosophila* and the *Manduca sexta* neuroglian proteins come in two different forms generated by alternative splicing; the long form is specific for neurones, whereas the short form is widely expressed outside of the developing nervous system. The grasshopper gene appears to generate only one form of the protein.

Protein properties

Insect neuroglians are transmembrane glycoproteins. The extracellular region consists of six immunoglobulin C2-type domains and five fibronectin type III domains (see Fig. 1, Fasciclin I, p. 190). In the number and the arrangement of their extracellular domains, they resemble vertebrate representatives of the L1 gene family, such as L1-CAMs, Nr CAMs, and neurofascins. Neuroglians and vertebrate L1 family members display low levels of amino acid identity (below 35 per cent) extending throughout most of the molecule. The exceptions are two short segments in the cytoplasmic domain which are highly conserved throughout the L1 family. The second conserved segment of 36 amino acid residues comprises a binding site for the cytoskeletal linker protein ankyrin.[2]

The *neuroglian* gene in *Drosophila* generates at least two different protein products by tissue-specific alternative splicing: a short (more abundant) form (nrg^{167};167 kDa) and a long (less abundant) form (nrg^{180};180 kDa).[3] The nrg^{180} protein is restricted to the surface of neurones in the CNS and neurones and some support cells in the PNS; in contrast, the nrg167 form is expressed on a wide range of other cells and tissues (Fig. 1).[3] The gene product of the *elav* locus (embryonic lethal abnormal visual system), an RNA binding protein, mediates the generation of the neurone-specific, long protein form (nrg180)-encoding neuroglian splice product in *Drosophila* embryos.[4] The two neuroglian protein forms are identical in their extracellular domains, the transmembrane segment, and the first 68 amino acids of their cytoplasmic domains, including the ankyrin-binding domain.[2] The more abundant nrg167 protein form continues for another 17 amino acids and the neurone-specific nrg180 protein form extends for another 62 amino acids. These extra 62 amino acid residues have a unique amino acid composition. Glycine, serine, alanine, and proline constitute 77 per cent of these 62 extra residues.

Invertebrate neuroglian homologues have been cloned in grasshopper (Grenningloh and Rehm; unpublished results), the leech *Hirudo medicinalis* (this molecule is also called tractin)[3] and in the moth *Manduca sexta*.[5] An L1/neuroglian-like molecule has also been identified in the nematode *Caenorhabditis elegans* (see GenBank accession number U50067). Whereas the *Drosophila* and the *Manduca sexta neuroglian* genes both generate two different protein forms with different cytoplasmic domains, the grasshopper gene appears to generate only one protein form as detected on Western blots (Grenningloh *et al.*, unpublished results). This single form in grasshopper is expressed throughout many tissues of the embryo, similar to the short form of the *Drosophila* protein.

Purification

Both *Drosophila* neuroglian protein forms can be copurified by affinity chromatography using the anti-*Drosophila* neuroglian monoclonal antibody (mAb) 1B7.[6]

Activities

The transfection and expression of both *Drosophila* neuroglian protein forms in *Drosophila* S2 cells demonstrated that both mediate homophilic, Ca^{2+}-independent cell aggregation.[7] In constrast to vertebrate L1 family members, no extracellular heterophilic binding partner has been identified so far for any of the insect neuroglians. Although *Drosophila* neuroglian interacts most strongly with itself in S2 cell aggregation assays, it dis-

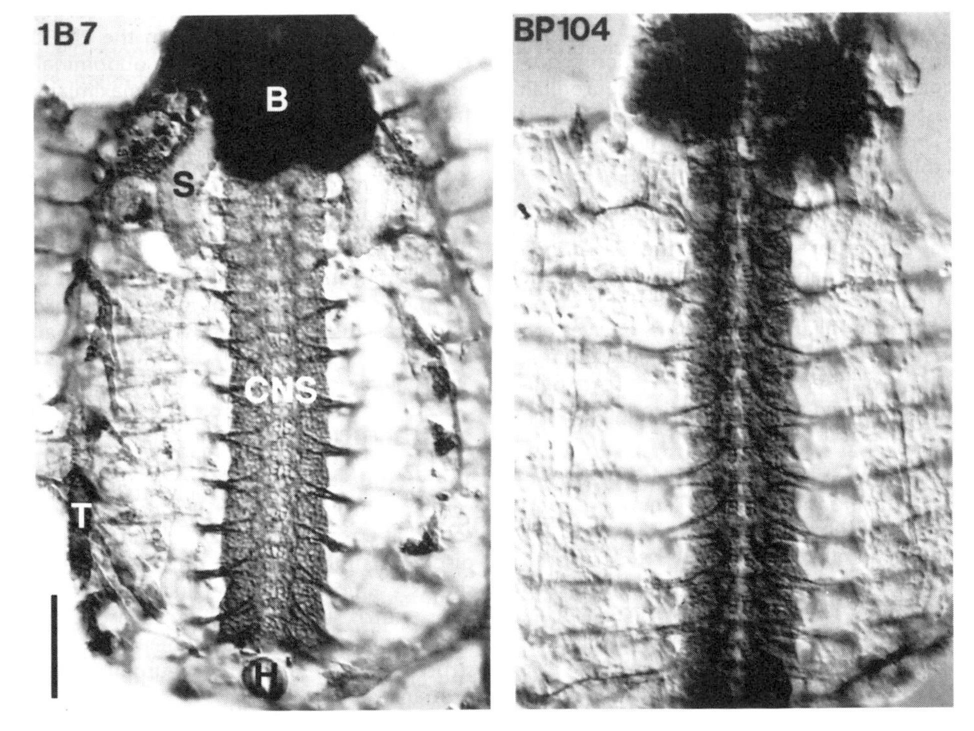

Figure 1. Expression of neuroglian protein in the *Drosophila* embryo. Shown are 12 h *Drosophila* embryos dissected on to glass slides, and stained with either anti-*Drosophila* neuroglian mAb 1B7 (left) or mAb BP-104 (right). MAb BP-104 specifically recognizes the neuroneal nrg180 protein, whereas mAb 1B7 recognizes an epitope common to both protein forms. Neuroglian protein is detected by mAb 1B7 in the brain (B), central, and peripheral nervous system (CNS), salivary gland (S), trachea (T), and the hindgut (H). In contrast, mAb BP-104 stains only the CNS, and PNS, and none of these other tissues. Bar represents 50 μm (left); 38 μm (right).

since authentic *Drosophila* AChE does not behave as an adhesive molecule in transfected S2 cells.[12]

Neurotactin expression at points of cell–cell contact supports the hypothesis that this protein acts as a cell recognition/adhesion molecule. Indeed, cell transfection experiments indicate that cells expressing neurotactin may bind to a heterologous ligand (soluble or membrane bound) expressed by primary *Drosophila* embryonic cells.[12] Neurotactin is only detected during cell proliferation and differentiation in the embryonic and larval–pupal stages and it is found mainly in neural tissue (Fig. 1). Non-neuronal expression is detected in certain mesodermal derivatives and some epidermal cells of imaginal discs (Fig. 2).

Purification

Neurotactin is routinely purified from embryo lysates by antibody immunoprecipitation.[1]

Figure 1. Immunoperoxidase staining of a section showing neurotactin expression in the developing imaginal central nervous system within the larval brain of *Drosophila*.

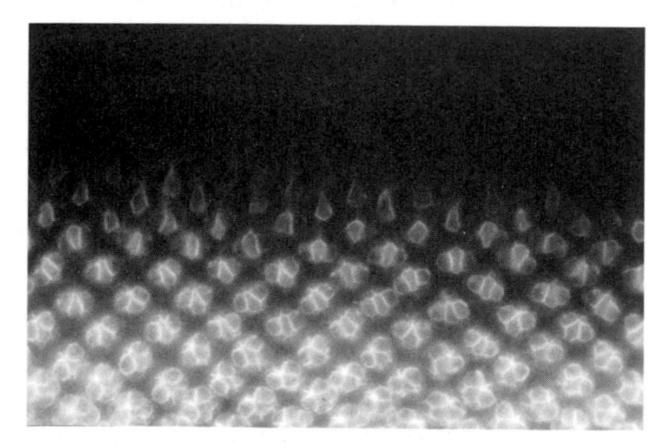

Figure 2. Immunofluorescence staining of neurotactin in the clusters of photoreceptor cells within the eye imaginal disc in the *Drosophila* third instar larva.

Activities

Neurotactin mediates heterophilic cell adhesion. In a cellular binding assay or aggregation assays *Drosophila* S2 cells transfected with *nrt* cDNA are able to bind to embryonic cells[3] or to self-aggregate in the presence of a crude soluble fraction containing the ligand activity.[12]

Antibodies

Monoclonal antibodies have been described.[1,4] Several others have been produced against the extracellular and the cytoplasmic domain. Polyclonal antibodies have been raised against the full length molecule (Barthalay and Piovant; unpublished results).

Genes and mutants

The sequence of two *nrt* cDNAs is available (EMBL data library X5387 and X54999).[2,4] The gene maps to 73C1–2 on the third chromosome. Loss-of-function mutations are viable and fertile and produce mild neural phenotypes.[13] Strong synergistic phenotypes affecting axon growth and guidance are found in embryo double mutants for *nrt* and several other cell adhesion and signalling genes.[13] Neurotactin lacking the cytoplasmic domain behaves as a dominant negative molecule and its overexpression and/or ectopic expression in wild-type embryos also leads to defects in axogenesis.[10]

Structure

A tridimensional model deduced from the crystal structure of *Torpedo* AChE has been constructed for neurotactin and suggests that its extracellular domain consists of two subdomains organized around a gorge: an N-terminal region whose tridimensional structure is almost identical to that of *Torpedo* AChE and a less conserved C-terminal region.[12] Aggregation assays which involve truncated molecules lacking this latter region have demonstrated that the two subdomains are structurally independent and that the ligand-binding site is in the N-terminal region.

References

1. Piovant, M. and Léna, P. (1988). *Development*, **103**, 145–56.
2. De la Escalera, S., Bockamp, E. O., Moya, F., Piovant, M., and Jiménez, F. (1990). *EMBO J.*, **9**, 3593–601.
3. Barthalay, Y., Hipeau-Jacquotte, R., de la Escalera, S., Jiménez, F., and Piovant, M. (1990). *EMBO J.*, **9**, 3603–9.
4. Hortsch, M., Patel, N. H., Bieber, A. J., Traquina, Z. R., and Goodman, C. S. (1990). *Development*, **110**, 1327–40.
5. Olson, P. F., Fessler, L. I., Nelson, R. E., Sterne, R. E., Campbell, A. G., and Fessler, J. H. (1990). *EMBO J.*, **9**, 1219–27.
6. Auld,V. J., Fetter, R. D., Broadie, K., and Goodman, C. S. (1995). *Cell*, **81**, 757–67.
7. Merken, L., Simons, M. J., Swillens, S., Massaer, M., and Vassart, G. (1985). *Nature*, **316**, 647–51.

8. Ichtchenko, K., Hata, Y., Nguyen, T., Ullrich, B., Missler, M., Moomaw, C., and Sudhof, T. C. (1995). *Cell*, **81**, 435–43.
9. Haruna, M.,. Hayashi, K., Yano, H., Takeuchi, O, and Sobue, K. (1993). *Biochem. Biophys. Res. Commun.*, **197**, 145–53.
10. Darboux, I., Hipeau-Jacquotte, R., Fremion, F., Singer, J., Gorde, S., and Piovant, M. (1997). (Submitted.)
11. Arno, H., J. Müller, and E. Wieschaus (1996). *J. Cell. Biol.*, **134**, 149–63.
12. Darboux, I., Barthalay, Y., Piovant, M., and Hipeau-Jacquotte, R. (1996). *EMBO J.*, **15**, 4835–43.
13. Speicher, S., García-Alonso, L., Carmena, A., Martín-Bermudo, M. D., de la Escalera, S., and Jiménez, F. (1998). *Neuron*, **20**, 221–33.

Fernando Jiménez
Centro de Biología Molecular, CSIC-UAM, Madrid, Spain

Michel Piovant
Laboratoire de Génétique et Biologie Cellulaire, CNRS, Marseille, France

Notch/delta/serrate

Notch, *Delta*, and *Serrate* are genes that were first identified in *Drosophila melanogaster*. All three encode cell surface molecules and are members of the notch signalling pathway,[1,2] which is conserved from worms to humans. The pathway's central element is the notch receptor, whose activity is regulated by the membrane-bound ligands delta and serrate. Genetic and molecular analyses have revealed the existence of intracellular members of the pathway that include cytoplasmic and nuclear effectors. Notch signalling has a broad and fundamental developmental action in that it seems to control the progression of non-terminally differentiated precursor cells to more differentiated states. Experimentally, activation of the pathway in precursor cells inhibits their differentiation. Notch signalling thus appears to define a fundamental and general cell-fate controlling mechanism that is essential for the differentiation of tissues throughout development.

Synonymous names

None.

Homologous proteins[2]

- Lin-12: notch homologue in *C. elegans*;
- Glp-1: notch homologue in *C. elegans*;
- Lag-2: delta/serrate homologue in *C. elegans*;
- Apx-1: delta/serrate homologue in *C. elegans*;
- TAN-1: notch homologue in humans (also named human notch 1);
- Jagged: serrate homologue in vertebrates.

Other homologues for all three proteins are found in *Xenopus*, zebrafish, chicken, mouse, rat, and humans. They are not given novel names other than notch 2, X[*enopus*]-delta-1, jagged 1, and so on. Whereas *Drosophila* has only one notch gene, the mammalian genomes harbour four. Presently two jagged (serrate homologue) and two delta proteins have been identified in mammals, but it is likely that as many as four genes for each ligand may exist.

Protein properties

Notch is an approximately 300 kDa protein (2703 amino acids) with a single transmembrane domain.[1] In addition to the signal peptide near its N terminus, the extracellular domain of notch consists of 36 tandem epidermal growth factor (EGF)-like repeats followed by three notch/Lin-12 repeats (Fig. 1). In its intracellular domain there are six cdc10/ankyrin repeats followed by an opa repeat and a PEST sequence. Notch is detected in a wide range of tissues throughout development in both vertebrates and invertebrates.[3,4] In flies, notch-expressing cells represent non-terminally differentiated cell populations and, in fact, there is a conspicuous absence of notch expression in terminally differentiated tissues. This is not always the case in mammals where, for example, postmitotic neurones in the retina have been shown to express notch.[5] Immunocytochemical analysis has identified antigens both on the cell surface and in the cytoplasm. In mammalian cells, immunoreactive material has also been seen in the nucleus.[5,6] In the polarized epithelium of a *Drosophila* imaginal disc, notch is confined apically to the adherens junctions (Fig. 2).[3] Although the regulation of notch is not understood, certain studies have suggested the existence of autoregulatory loops.[7,8] The intracellular domain of notch harbours nuclear localization signals, but it remains unclear whether cleavage of this domain followed by translocation to the nucleus is part of notch signalling. On the other hand, recent data have shown that notch is cleaved in the extracellular domain in the *trans* Golgi network and reaches the surface as a het-

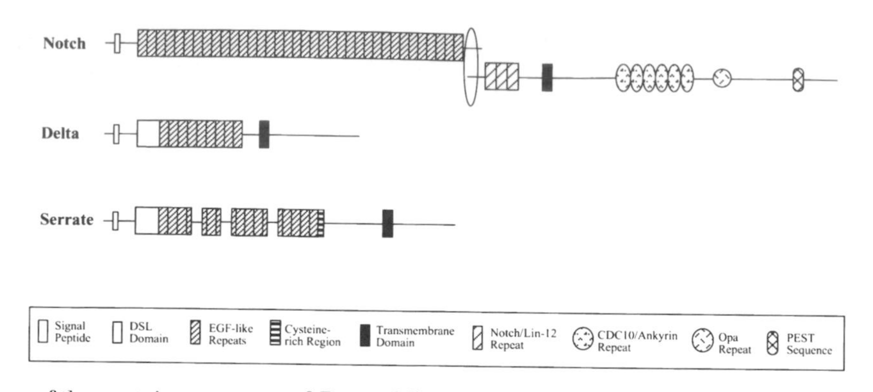

Figure 1. Diagram of the protein structures of *Drosophila* notch, delta, and serrate. The various sequence motifs are indicated by shaded boxes or circles, and defined at the bottom of the figure.

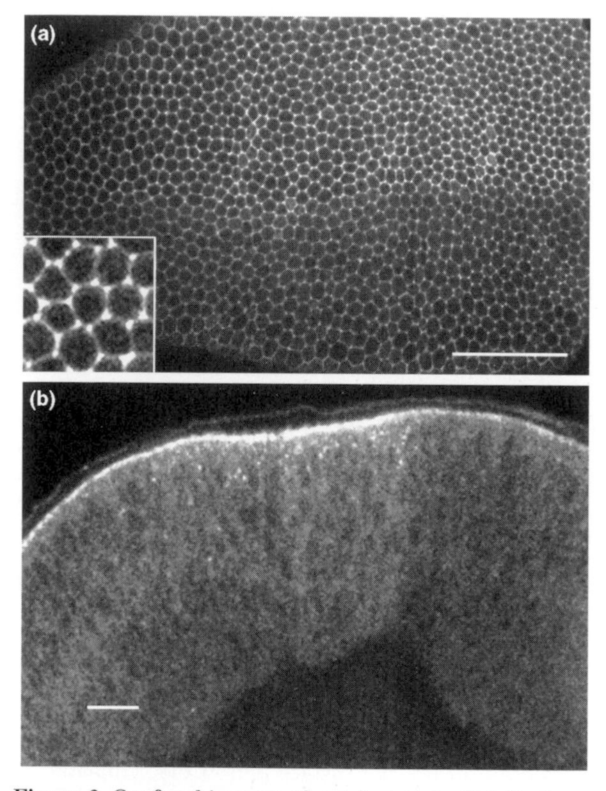

Figure 2. Confocal images of notch protein distribution in *Drosophila* tissues.[3] (a) Tangential section showing notch expression in a blastoderm stage embryo. Notch protein is associated with the cell surface, and increased expression is found just below the apical surface of each cell. Also note (inset) the bright points at the apices where three cells come into contact. (b) Notch expression in a wing imaginal disc. Note that it is polarized to the apical surface. Electron microscopy has shown that notch is localized in the adherens junctions (R. Fehon, and S. A.-T.; unpublished observations).

erodimer.[9] The evidence so far suggests – but does not yet prove – that the extracellular part is attached to the rest of the protein via disulphide bridges.

The molecular masses of delta and serrate are approximately 100 kDa (832 amino acids) and 160 kDa (1404 amino acids) respectively.[1] Both proteins have tandem EGF-like repeats in the extracellular region, a single transmembrane domain, and an intracellular domain with no known sequence motifs (Fig. 1). They share homology in the extracellular domain and both contain a DSL (delta, serrate, lag-2) motif, the signature motif for all notch ligand homologues. Although no homology is shared between the intracellular domains of delta and serrate, deletions of these domains result in dominant-negative mutations. These ligand forms, as well as forms that secrete the extracellular domain, act as antagonists of notch signalling in *Drosophila* and *Xenopus*.[10,11] Little is known about the transcriptional regulation of delta and serrate. Similar to notch, delta protein is detected in many tissues in *Drosophila*.[12] At the cellular level, the expression of notch and delta generally overlaps, with the exception of pupal wing and eye discs, where the two expression patterns appear to be complementary. The vertebrate homologues of delta and serrate (jagged) have been found to have distinct but overlapping patterns in *Xenopus*, chicken, and rat.[2] Subcellularly, delta is seen primarily in vesicles located close to the apical surface. In some tissues, such as the embryo, delta is initially found on the cell surface before being internalized, and it has been shown that shibire (*Drosophila* dynamin) plays a role in this internalization.[13] Serrate is found in *Drosophila* embryos and wing imaginal discs on the apical cell surface, as well as in vesicles.[14] A recent study[15] suggests that *serrate* gene expression responds to the juxtaposition of cells expressing fringe, a protein important for wing boundary determination. Even though internalization of these ligands has been documented, it is unclear whether this is linked to signalling rather than to the clearing of the cell surface.

Cell aggregation assays have revealed that delta and serrate bind directly via the DSL domains, to notch expressed on neighbouring cells. For aggregation, the two EGF repeats 11 and 12 of notch were shown to be both necessary and sufficient. Functionally, at least in some circumstances, the two ligands are interchangeable. For example, phenotypes associated with delta mutants can be rescued by the ectopic expression of serrate. Nevertheless, it seems that apart from the distinct expression patterns, other qualitative differences such as interactions with different effectors may separate the function of the two ligands. The studies reported so far have identified one downstream effector (suppressor of hairless in *Drosophila* and its mammalian homologue CBF1). Genetic analyses in flies and mammalian cell culture studies suggest the existence of additional downstream effectors. It remains to be seen whether the DSL-containing family of ligands are the only ligands for notch, or whether distinct ligands are capable of triggering distinct downstream events.

Purification

No purification of full-length notch, delta, or serrate has been reported.

Activities

The biochemical nature of notch signalling is not understood. However, molecular interaction between many of the components has been documented. An aggregation assay revealing the ability of delta and serrate to bind notch has been mentioned above. In addition, adhesive properties between cells that express delta have also been documented. Finally, colocalization assays have raised the possibility of interactions between notch and delta expressed on the surface of the same cell.[16]

Truncations of the extracellular domain result in the constitutive activation of the receptor judged by two criteria. Expression of such mutant forms are associated with gain of function phenotypes and the ectopic induction of *m*δ, a gene whose expression is known to depend on supressor of hairless-dependent notch signalling. Mδ is an HLH nuclear protein and a member of the Enhancer of split gene complex (the vertebrate homologue is *HES-1*), a locus harbouring several HLH transcriptional modulators thought to be involved in controlling the activity of downstream targets such as some proneural genes.[11,17–19]

Physical interactions have been documented between the intracellular part of notch and

(1) deltex, a cytoplasmic positive regulator of notch;

(2) suppressor of hairless, a transcription factor and downstream effector of notch;

(3) dishevelled, a cytoplasmic protein involved in wingless signalling;[20] and

(4) numb, a protein involved in neural differentiation.[21]

The activities of all three proteins can also be measured using phenotypic analyses and comparing them to known phenotypes. This approach has been used in transgenics and vertebrate cell cultures.[22,23] It is noted that the minigene of notch only partly rescues the phenotypes associated with loss of function notch mutations.

Antibodies

Monoclonal and polyclonal antibodies against the extracellular and intracellular epitopes of *Drosophila* notch have been reported.[3] They generally work well for immunoprecipitation, immunofluorescence, and Western analysis. Some do display low cross-reactivity with the vertebrate notch homologues. Antibodies against the extracellular domains of *Drosophila* delta and serrate have also been generated.[3,14] Both antibodies work for immunofluorescence and Western analysis but only delta antibodies work for immunoprecipitations.

Both polyclonal and monoclonal antibodies have been generated against the intracellular domains of human notch 1 and human notch 2.[6,24] They have been used in Western analysis and immunofluorescence.

Genes

All three *Drosophila* genes have been precisely mapped to the polytene chromosomes.[25] Both full length cDNA and genomic DNA (40 kb) are available for *Drosophila* notch.[3,26] Full length cDNAs are available for delta and serrate.[27] Several truncated forms of the cDNAs are also available.[11,27,28] Notch homologous genes from *C. elegans*,[29,30] the Australian sheep blowfly,[31] zebrafish,[32] *Xenopus*,[10,33] chick,[34] mouse,[35,36] rat,[4,22,37] and human[24,38,39] have been reported.

Four notch paralogues – notch 1, 2, 3, 4 – have been described in vertebrates. This is consistent with the notion that two duplication events occurred during the evolution from flies to vertebrates and it follows the general rule that single genes in flies are represented by several paralogues in vertebrates. It is currently known that in vertebrates, two ligands are serrate-like (jagged 1, 2) and one is delta-like. It is expected that more vertebrate ligands exist.

It is noteworthy that in the two *C. elegans* notch-like molecules, Lin-12 and Glp-1, and in the two ligands, Apx-1 and Lag-2, the overall structure is significantly different, while the biological activities and the structural motifs carried by those molecules are similar to the fly or vertebrate counterparts (see Fig. 3).

Mutant phenotype/disease states

Consistent with the fundamental and general developmental action of notch, malfunctions in this signalling pathway lead to a very broad spectrum of mutant phenotypes. Loss-of-function notch and delta lead to similar embryonic phenotypes in flies, whereas loss-of-function serrate leads to pupal lethality. Moreover, *notch*, *delta* and *serrate* mutations display phenotypes in wings, eyes, bristles, oocytes, muscles, and so on. There seems to be a

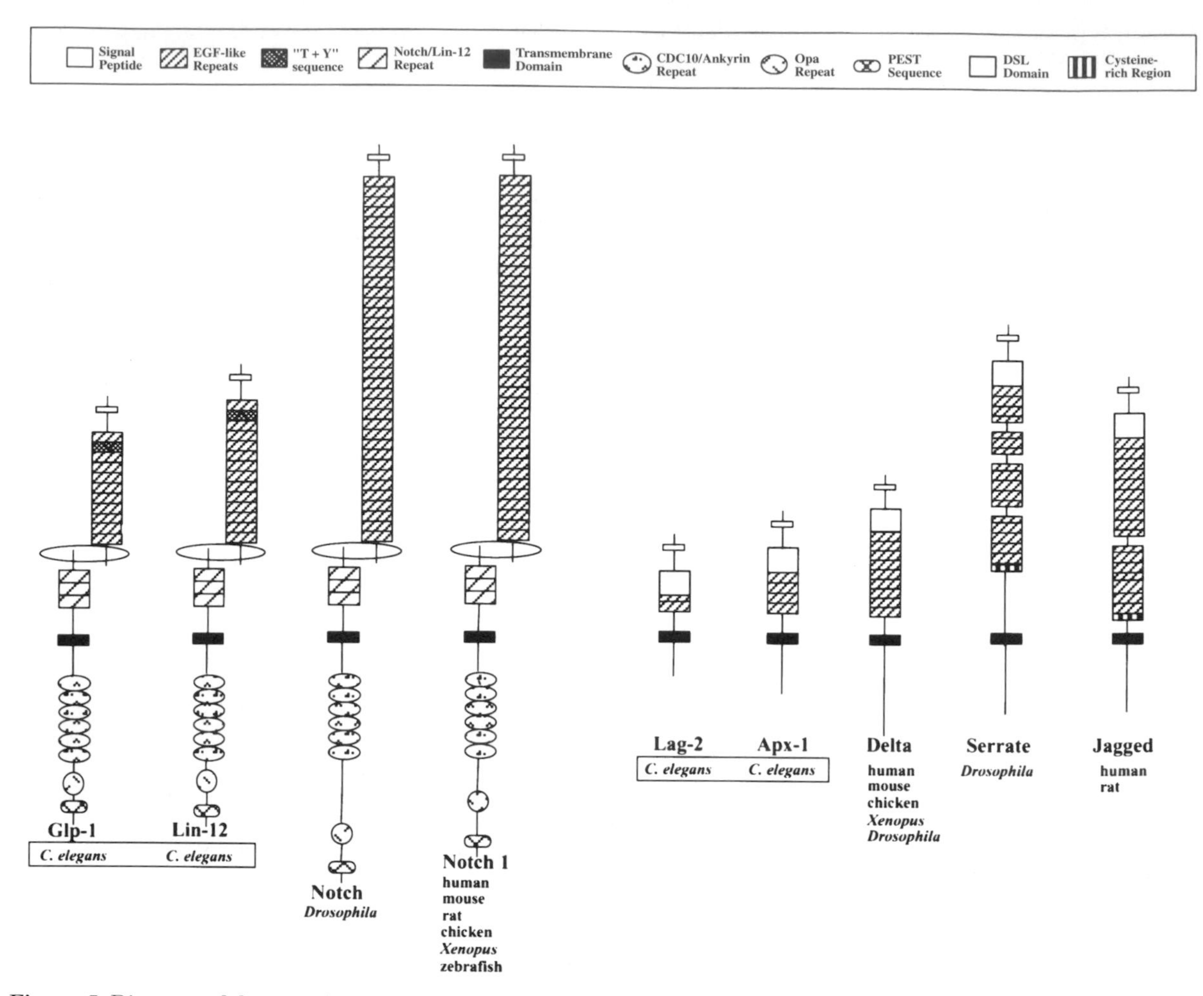

Figure 3. Diagram of the protein structures of *Drosophila* notch, delta, and serrate, and their homologues. Note that the overall structure of the two *C. elegans* notch-like molecules (Glp-1, and Lin-12), and the two ligands (Lag-2, and Apx-1) is significantly different, although the biological activities, and the structural motifs carried by those molecules are similar to the fly or vertebrate counterparts.

common feature in the diverse phenotypes associated with notch signalling. The abnormal regulation of notch in precursor cells leads to abnormal acquisition of cell fates. In *C. elegans*, mutations in *glp-1*, *lin-12*, *apx-1*, and *lag-2* result in defects in early embryogenesis, in anchor cell/ventral uterine cell decision, and in germline induction by the distal tip cell. Mouse knockouts in notch 1 lead to early lethality.

Pathologies associated with notch signalling malfunction have been reported for three out of the four vertebrate *notch* genes. Mutations leading to the constitutive activation of the human notch 1 receptor (originally named *TAN1* for translocation associated notch 1) cause T-cell lymphoblastic leukaemias.[38] The human syndrome CADASIL, associated with late onset strokes and dementia, has been associated with mutations in notch 3.[40] Finally, MTV insertions in notch 4 cause mammary tumours in mice.[41]

Mutations in human jagged 1 have been shown to be associated with the Alagille syndrome, an autosomal dominant disorder characterized by pleiotropic developmental defects, including the abnormal development of liver, heart, skeleton, eye, face, and kidney.[42,43]

Structure

Not available.

References

1. Artavanis-Tsakonas, S., Delidakis, C. and Fehon, R. G. (1991). *Ann. Rev. Cell. Biol.*, **7**, 427–52.
2. Artavanis-Tsakonas, S., Matsuno, K. and Fortini, M. E. (1995). *Science*, **268**, 225–32.

3. Fehon, R. G., Johansen, K., and Artavanis-Tsakonas, S. (1991). *J. Cell Biol.*, **113**, 657–69.
4. Weinmaster, G., Roberts, V. J., and Lemke, G. (1991). *Development*, **113**, 199–205.
5. Ahmad, I., Zagouras, P., and Artavanis-Tsakonas, S. (1995). *Mech. Deve.*, **53**, 73–85.
6. Zagouras, P., Stifani, S., Blaumueller, C. M., Carcangiu, M. L., and Artavanis-Tsakonas, S. (1995). *Proc. Natl Acad. Sci. USA*, **92**, 6414–18.
7. Heitzler, P. and Simpson, P. (1991). *Cell*, **64**, 1083–92.
8. Wilkinson, H. A., Fitzgerald, K., and Greenwald, I. (1994). *Cell*, **79**, 1187–98.
9. Blaumueller, C. M., Qi, H., Zagouras, P., and Artavanis-Tsakonas, S. (1997). *Cell*, (Submitted.)
10. Chitnis, A., Henrique, D., Lewis, J., Ish-Horowicz, D., and Kintner, C. (1995). *Nature*, **375**, 761–6.
11. Sun, X. and Artavanis-Tsakonas, S. (1996). *Development*, **122**, 2465–74.
12. Kooh, P. J., Fehon, R. G., and Muskavitch, M. A. T. (1993). *Development*, **117**, 431–40.
13. Parks, A. L., Turner, F. R., and Muskavitch, M. A. T. (1995). *Mech. Deve.*, **50**, 201–16.
14. Thomas, U., Speicher, S. A., and Knust, E. (1991). *Development*, **111**, 749–61.
15. Kim, J., Irvine, K. D., and Carroll, S. B. (1995). *Cell*, **82**, 795–802.
16. Fehon, R. G., Kooh, P. J., Rebay, I., Regan, C. L., Xu, T., Muskavitch, M. A., and Artavanis-Tsakonas, S. (1990). *Cell*, **61**, 523–34.
17. Jennings, B., Preiss, A., Delidakis, C., and Bray, S. (1994). *Development*, **120**, 3537–48.
18. Jarriault, S., Brou, C., Logeat, F., Schroeter, E. H., Kopan, R., and Israel, A. (1995). *Nature*, **377**, 355–8.
19. Wettstein, D. A., Turner, D. L., and Kintner, C. (1997). *Development*, **124**, 693–702.
20. Axelrod, J. D., Matsuno, K., Artavanis-Tsakonas, S., and Perrimon, N. (1996). *Science*, **271**, 1826–32.
21. Zhong, W., Feder, J. N., Jiang, M. M., Jan, L. Y., and Jan, Y. N. (1996). *Neuron*, **17**, 43–53.
22. Lindsell, C. E., Shawber, C. J., Boulter, J., and Weinmaster, G. (1995). *Cell*, **80**, 909–17.
23. Nye, J. S., Kopan, R., and Axel, R. (1994). *Development*, **120**, 2421–30.
24. Aster, J., Pear, W., Hasserjian, R., Erba, H., Davi, F., Luo, B., *et al*. (1994). *Cold Spring Harbor Symp. Quant. Biol.*, **59**, 125–36.
25. Lindsley, D. L. and Zimm, G. G. (1992). The genome of *Drosophila melanogaster*. Academic Press, New York.
26. Ramos, R. G. P., Grimwade, B. G., Wharton, K. A., Scottgale, T. N., and Artavanis-Tsakonas, S. (1989). *Genetics*, **123**, 337–48.
27. Rebay, I., Fehon, R. G., and Artavanis-Tsakonas, S. (1993). *Cell*, **74**, 319–29.
28. Rebay, I., Fleming, R. J., Fehon, R. G., Cherbas, L., Cherbas, P., and Artavanis-Tsakonas, S. (1991). *Cell*, **67**, 687–99.
29. Fitzgerald, K. and Greenwald, I. (1995). *Development*, **121**, 4275–82.
30. Lambie, E. J. and Kimble, J. (1991). *Development*, **112**, 231–40.
31. Chen, Z., Newsome, T., McKenzie, J. A., and Batterham, P. (1998). *Insect Biochem. Mole. Biol.*. (Submitted.)
32. Bierkamp, C. and Campos-Ortega, J. A. (1993). *Mech. Dev.*, **43**, 87–100.
33. Coffman, C. R., Skoglund, P., Harris, W. A., and Kintner, C. R. (1993). *Cell*, **73**, 659–71.
34. Henrique, D., Adam, J., Myat, A., Chitnis, A., Lewis, J., and Ish-Horowicz, D. (1995). *Nature*, **375**, 787–90.
35. Bettenhausen, B., de Angelis, M. H., Simon, D., Guenet, J. -L., and Gossler, A. (1995). *Development*, **121**, 2407–18.
36. Franco del Amo, F., Gendron-Maguire, M., Swiatek, P. J., Jenkins, N. A., Copeland, N. G., and Gridley, T. (1993). *Genomics*, **15**, 259–64.
37. Weinmaster, G., Roberts, V. J., and Lemke, G. (1992). *Development*, **116**, 931–41.
38. Ellisen, L. W., Bird, J., West, D. C., Soreng, A. L., Reynolds, T. C., Smith, S. D., and Sklar, J. (1991). *Cell*, **66**, 649–61.
39. Larsson, C., Lardelli, M., White, I., and Lendahl, U. (1994). *Genomics*, **24**, 253–58.
40. Joutel, A., Corpechot, C., Ducros, A., Vahedi, K., Chabriat, H., Mouton, P., *et al*. (1996). *Nature*, **383**, 707–10.
41. Robbins, J., Blondel, B. J., Gallahan, D., and Callahan, R. (1992). *J. Virol.*, **66**, 2594–9.
42. Oda, T. *et al*., (1997). *Nature Genet.*, **16**, 235–42.
43. Li, L. *et al*., (1997). *Nature Genet.*, **16**, 243–51.

■ *Xin Sun, Laurent Caron, and Spyros Artavanis-Tsakonas*

Howard Hughes Medical Institute, and Department of Cell Biology, and Biology, Boyer Center for Molecular Medicine, Yale University, New Haven, CT 06536–0812, USA.

NgCAM-related cell adhesion molecule (NrCAM)

NrCAM is a member of the L1 subgroup of the immunoglobulin superfamily (IgSF) which is expressed on neurones and axon tracts of the central and peripheral nervous system. NrCAM undergoes distinct interactions with other members of the IgSF and is involved in the induction of neurite outgrowth. Multiple isoforms of NrCAM are known, which differ with respect to the presence of a fibronectin type III-like domain and small interdomain segments.

Synonymous names

NrCAM is also known as bravo antigen.[1]

Protein properties

NrCAM[1–4] and the related proteins L1/NgCAM, CHL1,[5] and neurofascin[6] are members of the L1 subgroup[7,8] of neural IgSF members composed of six immunoglobulin-like domains, five fibronectin type III-like repeats, a transmembrane domain, and a highly conserved cytoplasmic domain (Fig. 1(a)).[9] It is a cell surface glycoprotein of about 140 kDa.[10] NrCAM is expressed on neurones of the central and peripheral nervous system and is predominantly found in axon tracts.[1–3,10–12] Expression in the developing spinal cord is found in the floor plate where interactions with axonin-1 expressed by commissural axons contribute to the guidance into the floor plate.[13,22] Axonin-1–NrCAM interactions also account for contact formations between axons and peripheral glia.[14] Binding of NrCAM to the neural IgSF molecules F11[15,23] or neurofascin[16] induces neurite outgrowth of chick tectal cells. Most probably, the former interaction occurs in a *cis* configuration within the same cell membrane.[17,24] NrCAM may cooperate with NgCAM to promote neurite outgrowth of retinal cells.[18]

Purification

NrCAM is involved in multiple molecular interactions[25] and has been implicated in axon guidance at the spinal cord floor plate.[26] It can be isolated by immunoaffinity chromatography using monoclonal antibodies from nonionic detergent extracts of brain plasma membranes.[2,10,18]

Activities

NrCAM is involved in multiple molecular interactions[25] and has been implicated in axon guidance at the spinal

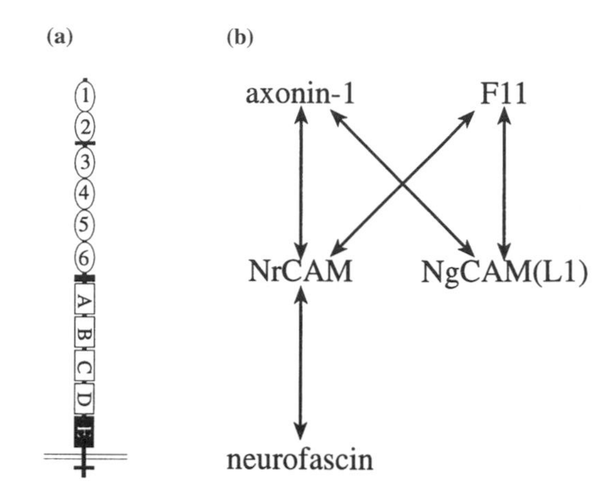

Figure 1. (a) Structure of NrCAM. NrCAM is composed of six Ig-like domains (1–6) followed by five FNIII-like domains (A–E), a transmembrane domain, and a cytoplasmic domain at the C terminus. Alternatively spliced sequences are shown in black. (b) Schematic representation of NrCAM interactions with other members of the immunoglobulin superfamily.

cord floor plate.[26] It undergoes heterophilic interactions with F11,[15] neurofascin,[16] and axonin-1[14] via its extracellular domains as analysed in binding assays based on microspheres coated with purified molecules (Fig. 1(b)). A homophilic Ca^{2+}-dependent interaction has also been shown.[19] The intracellular domain is presumably linked to the cytoskeleton via ankyrin as suggested by affinity chromatography.[20]

Antibodies

Polyclonal and monoclonal antibodies to chicken NrCAM have been described.[2,10]

Genes

Complete cDNA sequences of chicken,[1,2] rat,[3] and human NrCAM[4,27] have been reported. NrCAM isoforms are generated by alternative splicing of the fifth FnIII-like repeat and three minor sequences located between the second and third Ig-like domain, between the Ig-like and FnIII-like repeats, and in the cytoplasmic domain. The murine NrCAM gene[21] has been mapped to chromosome 12 while human NrCAM[4] is located at chromosome 7q31.

Disease states/mutant phenotype

No information available.

Structure

Not available.

References

1. Kayyem, J. F., Roman, J. M., de la Rosa, E. J., Schwarz, U., and Dreyer, W. J. (1992). *J. Cell Biol.*, **118**, 1259–70.
2. Grumet, M., Mauro, V., Burgoon, M. P., Edelman, G. M., and Cunningham, B. A. (1991). *J. Cell Biol.*, **113**, 1399–412.
3. Davis, J. Q., Lambert, S., and Bennett, V. (1996). *J. Cell Biol.*, **135**, 1355–67.
4. Lane, R. P., Chen, X. N., Yamakawa, K., Vielmetter, J., Korenberg, J. R., and Dreyer, W. J. (1996). *Genomics*,. **35**, 456–65.
5. Holm, J., Hillenbrand, R., Steuber, V., Bartsch, U., Moos, M., Lubbert, H., *et al.* (1996). *Eur. J. Neurosci.*, **8**, 1613–29.
6. Volkmer, H., Hassel, B., Wolff, J. M., Frank, R., and Rathjen, F. G. (1992). *J. Cell Biol.*, **118**, 149–61.
7. Sonderegger, P. and Rathjen, F. G. (1992). *J. Cell Biol.*, **119**, 1387–94.
8. Brümmendorf, T. and Rathjen, F. G. (1995). *Protein Profile*, **2**, 963–1108.
9. Grumet, M. and Sakurai, T. (1996). *Sem. Neurosci.*, **8**, 379–89.
10. de la Rosa, E. J., Kayyem, J. F., Roman, J. M., Stierhof, Y. D., Dreyer, W. J., and Schwarz, U. (1990). *J. Cell Biol.*, **111**, 3087–96.
11. Denburg, J. L., Caldwell, R. T., and Marner, J. M. (1995). *J. Comp. Neurol.*, **354**, 533–50.
12. Krushel, L. A., Prieto, A. L., Cunningham, B. A., and Edelman, G. M. (1993). *Neuroscience*, **53**, 797–812.
13. Stoeckli, E. T. and Landmesser, L. (1995). *Neuron*, **14**, 1165–79.
14. Suter, D. M., Pollerberg, G. E., Buchstaller, A., Giger, R. J., Dreyer, W. J., and Sonderegger, P. (1995). *J. Cell Biol.*, **131**, 1067–81.
15. Morales, G., Hubert, M., Brümmendorf, T., Treubert, U., Tarnok, A., Schwarz, U., and Rathjen, F. G. (1993). *Neuron*, **11**, 1113–22.
16. Volkmer, H., Leuschner, R., Zacharias, U., and Rathjen, F. G. (1996). *J. Cell Biol.*, **135**, 1059–69.
17. Sakurai, T., Lustig, M., Nativ, M., Hemperly, J. J., Schlessinger, J., Peles, E., and Grumet, M. (1997). *J. Cell Biol.*, **136**, 907–18.
18. Morales, G., Sanchez Puelles, J. M., Schwarz, U., and de la Rosa, E. J. (1996). *Eur. J. Neurosci.*, **8**, 1098–1105.
19. Mauro, V. P., Krushel, L. A., Cunningham, B. A., and Edelman, G. M. (1992). *J. Cell Biol.*, **119**, 191–202.
20. Davis, J. Q. and Bennett, V. (1994). *J. Biol. Chem.*, **269**, 27163–6.
21. Burmeister, M., Ren, Q., Makris, G. J., Samson, D., and Bennett, V. (1996). *Mamm. Genome*, **7**, 558–9.
22. Stoeckli, E. T., Sonderegger, P., Pollerberg, G. E., and Landmesser, L. T. (1997). *Neuron*, **18**, 209–21.
23. Treubert, U. and Brümmendorf, T. (1998). *J. Neurosci.*, **18**, 1795–805.
24. Peles, E., Schlessinger, J. and Grumet, M. (1998). *Trends Biochem. Sci.*, **23**, 121–4.
25. Grumet, M. (1997). *Cell Tissue Res.*, **290**, 423–8.
26. Stoeckli, E. T. and Landmesser, L. T. (1998). *Curr. Opin. Neurobiol.*, **8**, 73–9.
27. Wang, B., Williams, H., Du, J. S., Terrett, J. and Kenwrick, S. (1998). *Mol. Cell. Neurosci.*, **10**, 287–95.

Hansjürgen Volkmer and Thomas Brümmendorf
Max-Delbrück-Center for Molecular Medicine,
Robert-Rössle Strasse 10,
13122 Berlin, Germany

Occludin

Occludin is an integral membrane protein with four transmembrane domains which constitutes the tight junction strand. Tight junctions play dual roles, barrier and fence functions, in both of which occludin is directly involved. Occludin also functions as a membrane-binding partner for tight junction-associated peripheral membrane proteins such as ZO-1 and ZO-2.

Protein properties

In epithelial and endothelial cells, the tight junction (TJ) seals cells to create a primary barrier to the diffusion of solutes across the cell sheet; it also functions as a boundary between the apical and basolateral membrane domains to maintain polarization.[1] In thin-section electron microscopy, TJ appear as a series of discrete sites of apparent fusion, involving the outer leaflet of the plasma membranes of adjacent cells.[2] In freeze-fracture electron microscopy, these junctions appear as a set of continuous, anastomosing intramembrane strands or fibrils in the P face with complementary grooves in the E face.[3]

Occludin is an integral membrane protein that is localized at TJ. Occludin, with a molecular mass of ~65 kDa, was first isolated from the chick liver using monoclonal antibody production[4] and its mammalian homologues were also identified.[5] The amino acid sequences of human, murine, and canine occludins are very closely related (~90 per cent identity), whereas they diverge con-

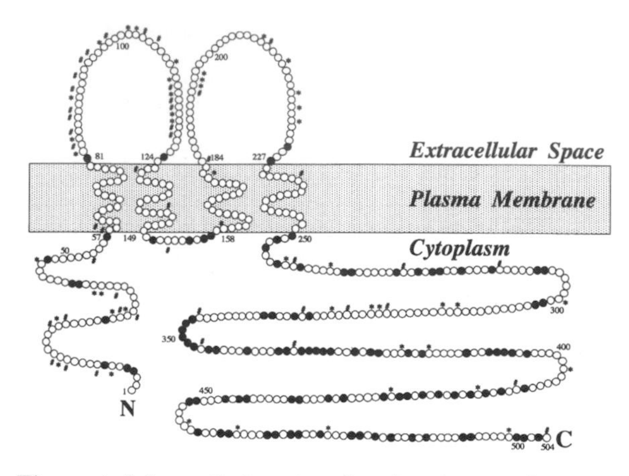

Figure 1. Schematic drawing showing the membrane-folding model for chicken occludin. Filled circles, charged amino acids; #, glycine;.*, tyrosine.

siderably from those of the chicken. Regardless of this inter-species diversity in the occludin sequence, the basic structure appears to have been conserved during phylogenetic evolution (Fig. 1). Mammalian as well as chicken occludin comprises four transmembrane domains, a long carboxyl-terminal cytoplasmic domain, a short amino-terminal cytoplasmic domain, two extracellular loops, and one intracellular turn. Characteristic conserved aspects are the high content of tyrosine and glycine residues in the first extracellular loop (~60 per cent), which may be directly involved in cell adhesion, and the α-helical coiled-coil structure in the long carboxyl-terminal cytoplasmic domain.

In most simple epithelial cells, occludin is colocalized precisely with ZO-1, a TJ-associated peripheral membrane protein, at the most apical part of their lateral membranes.[4,6] At the electron microscopic level, occludin signals are detected directly over the points of membrane contact in TJ in ultrathin sectional images[4] or over the TJ strand itself in freeze-fracture replica images[6–8] (Fig. 2). Northern blotting identifies occludin mRNA in various mouse tissues.[6] In testis, kidney, liver, lung, and brain, an intense 2.9 kb band and two additional faint bands of 3.6 and 2 kb are detected, but these are undetectable in skeletal muscle, spleen, and heart. Anti-mouse occludin mAb and pAb recognize several bands around 60 kDa in mouse cultured epithelial cells, the smallest of which is the most intense. This multi-banding pattern of occludin is also observed in other species such as chicken, pig, dog, and human. By contrast, occludin is barely detected in cultured fibroblasts either at the protein or MRNA level.

Our knowledge of occludin in endothelial cells is still fragmentary. Occludin is highly expressed and concentrated at cell–cell borders of brain endothelial cells, where TJ are highly developed. By contrast, the occludin signal is very weak or undetectable in endothelial cells in non-neuronal tissues, where relatively poor TJ are detected. Brain endothelial cells express a large amount of occludin in culture, in good agreement with this immunofluorescence observation, whereas only trace amounts of occludin are detected in cultured aorta endothelial cells.

Figure 2. Localization of occludin on a freeze-fracture replica of chicken liver. ×200 000.

Activities

When occludin is overexpressed in insect cells by recombinant baculovirus infection, peculiar occludin-enriched multilamellar structures accumulate in the cytoplasm.[8] Each lamella is transformed from intracellular membranous cisternae with a completely collapsed luminal space and in each lamella, the outer leaflets of opposing membranes appear to be fused, with no gaps, as in TJ. These findings suggest that occludin can obliterate the extracellular space at TJ and that occludin is directly involved in the barrier function of TJ. In good agreement, under certain conditions, occludin shows a Ca^{2+}-independent cell adhesion activity in fibroblast transfectants expressing exogenous occludin.

When full length occludin is overexpressed in epithelial cells, the barrier function is upregulated with a concomitant increase of the number of TJ strands.[9] On the other hand, the expression of carboxyl-terminally truncated occludin in epithelial cells leads to a several-fold increase in paracellular flux of small molecular weight tracers and to destruction of the fence function of TJ to maintain a fluorescent lipid in a specifically labelled cell surface domain.[10] Furthermore, a synthetic peptide corresponding to the second extracellular domain of chicken occludin perturbs the TJ permeability barrier in epithelial cells.[11] Taken together, it is safe to say that occludin is directly involved in barrier as well as fence functions of TJ.

Little is known about the regulation mechanism of occludin functions. ZO-1 directly binds to the carboxyl-terminal 150 amino acids of occludin[12] and occludin is heavily phosphorylated at serine and threonine residues when it is incorporated into TJ.[13] Interactions with the actin-based cytoskeleton through ZO-1 and post-translational modifications, such as phosphorylation, may be important in the regulation of occludin functions.

Antibodies

A number of monoclonal and polyclonal antibodies against chicken occludin have been characterized. Most of them are specific to chicken occludin,[4] but some cross-react with mammalian[10] and/or *Xenopus*[11] occludin homologues. Monoclonal antibodies specific for mammalian occludin are now available.[5,6]

Genes

Occludin cDNA has been cloned and sequenced from chicken[4] (EMBL/GenBank/DDBJ D21837), rat-kangaroo[5] (EMBL/GenBank/DDBJ U49183), mouse[5] (EMBL/GenBank/DDBJ U49185), dog[5] (EMBL/GenBank/DDBJ U49221), and human[5] (EMBL/GenBank/DDBJ U49184).

References

1. Gumbiner, B. (1993). *J.Cell Biol.*, **123**, 1631–3.
2. Farquhar, M.G. and Palade, G.E. (1963). *J. Cell Biol.*, **17**, 375–409.
3. Staehelin, L.A. (1973). *J. Cell Sci.*, **13**, 763–86.
4. Furuse, M., Hirase, T., Itoh, M, Nagafuchi, A., Yonemura, S., Tsukita, S., and Tsukita, S. (1993). *J. Cell Biol.*, **123**, 1777–88.
5. Ando-Akatsuka,Y., Saitou, M., Hirase, T., Kishi, M., Sakakibara, A., Itoh, M., *et al.* (1996). *J. Cell Biol.*, **133**, 43–7.
6. Saitou, M., Ando-Akatsuka, Y., Itoh, M., Furuse, M., Inazawa, J., Fujumoto, K., and Tsukita, S. (1997). *Eur. J. Cell Biol.* (In press.)
7. Fujimoto, K. (1995). *J. Cell Sci.*, **108**, 3443–9.
8. Furuse, M., Fujimoto, K., Sato, N., Hirase, T., Tsukita, S., and Tsukita, S. (1996). *J. Cell Sci.*, **109**, 429–35.
9. McCarthy, K. M., Skare, I. B., Stankewich, M. C., Furuse, M., Tsukita, S., Rogers, R. A., *et al.* (1996). *J. Cell Sci.*, **109**, 2287–98.
10. Balda, M. S., Whitney, J. A., Flores, C., González, S., Cereijido, M. and Matter, K. (1996). *J. Cell Biol.*, **134**, 1031–49.
11. Wong, V. and Gumbiner, B. M. (1997). *J. Cell Biol.*, **136**, 399–409.
12. Furuse, M., Itoh, M., Hirase, T., Nagafuchi, A., Yonemura, S., Tsukita, S., and Tsukita, S. (1994). *J. Cell Biol.*, **127**, 1617–26.
13. Sakakibara, A., Saitou, M., Ando-Akatsuka, Y., Furuse, M., and Tsukita, S. (1997). *J. Cell Biol.* (In press.)

Shoichiro Tsukita
Department of Cell Biology,
Kyoto University Faculty of Medicine,
Yoshida-Konoe, Sakyo-ku,
Kyoto 606, Japan

PECAM-1/CD31

PECAM-1/CD31 is emerging as an important vascular cell adhesion molecule of the immunoglobulin gene superfamily. It is able to mediate cell–cell adhesion, as well as transduce intracellular signals that upregulate the function of integrins. The regulation of PECAM-1-mediated adhesion or signalling phenomena is still not fully understood but is very likely to involve activational events such as phosphorylation of its cytoplasmic domain. Data now indicate that this molecule may play important roles in the recruitment of leukocytes at inflammatory sites, T-cell immune responses, angiogenesis, and cardiovascular development.

Protein properties

Platelet endothelial cell adhesion molecule or PECAM-I/CD31 (reviewed in refs 1 and 2) is a single-membrane spanning glycoprotein with a molecular mass of 130 kDa that varies slightly among different cell types, presumably due to differences in glycosylation. A distinguishing feature of PECAM-1 is its expression on a number of vascular-associated cells. It is present in large amounts on cultured endothelial cells ~10^6 copies per cell) where it concentrates at cell–cell junctions[3] (Fig. 1). *In situ*, PECAM-1 is constitutively expressed on continuous endothelium of all vessel types and has thus provided a useful immunostaining marker of blood vessels. It has been identified on most leukocytes (at up to 50 000 copies/cell), bone marrow stem cells, and transformed cell lines of the myeloid and megakaryocytic lineage. Lastly, PECAM-1 is expressed on platelets, although in relatively small amounts (~5000 copies per platelet).

The mature form of human PECAM-1 (Fig. 2) consists of a large extracellular domain of 574 amino acids, a single membrane-spanning region of 19 hydrophobic residues, and a cytoplasmic tail of 118 amino acids.[4] The extracellular region is organized into six Ig-like homology domains. Of these, domains 1 and 2 appear to be directly involved in mediating ligand binding.[5–8] The nine potential extra-

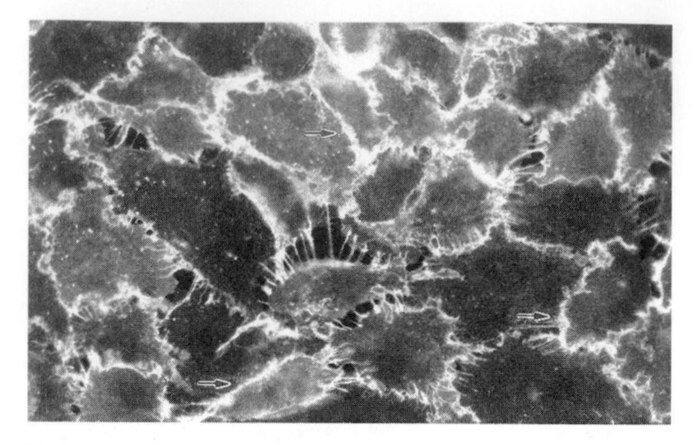

Figure 1. Localization of PECAM-1 at the intercellular junctions of endothelial cells. Immunofluorescence staining of human umbilical vein endothelial cells stained with an anti-PECAM-1 monoclonal antibody. Adjacent endothelial cells show the characteristic concentration of PECAM-1 at intercellular borders (arrows).

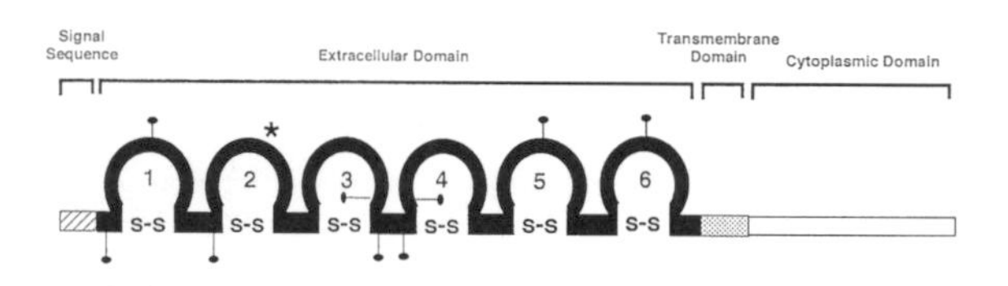

Figure 2. Protein structure and genomic organization of PECAM-1. PECAM-1 protein comprises six disulphide-bonded (S–S) domains with Ig C2 homology. Filled circles represent potential sites of *N*-linked glycosylation and the asterisk indicates the position of a heparin-binding sequence in domain 2. (Adapted from ref. 1 with permission.)

cellular sites for *N*-linked glycosylation are consistent with the finding that carbohydrate constitutes 40 per cent of the mass of the mature protein. Finally, in the cytoplasmic domain are several serine and tyrosine residues that are phosphorylated following cell activation.[9–12]

The ligand interactions of PECAM-1 are complex. The molecule has been reported to interact with itself (homophilic adhesion)[6–8,13–15] and with an ever growing number of heterophilic ligands including the integrin $\alpha V\beta 3$,[6,16] an unidentified proteoglycan,[5,6,14] human CD38,[17] and an unidentified ligand on activated T cells.[18] In addition to mediating cell–cell adhesion, a variety of studies have established that engagement of surface PECAM-1 may transduce intracellular signals that activate the function of integrins on leukocytes and platelets.[2] This ability to activate integrin function may be an intrinsic property of PECAM-1, as engagement of human PECAM-1 transfected into COS cells led to upregulation of the adhesive function of the endogenous primate $\beta 1$ integrin.[13] The regulation of ligand binding and the mechanism(s) by which engagement of PECAM-1 triggers downstream signalling events are currently being defined. It does appear, however, that phosphorylation of tyrosine residues in exons 13 and 14 may be important to these activities.[1,2,11] Recently, it has also been demonstrated that antibody binding of PECAM-1 on endothelial cells and cellular transfectants expressing PECAM-1 induces a slow but sustained increase in intracellular calcium accompanied by the secretion of prostacyclin.[19]

Purification

Significant amounts of human PECAM-1 can be purified from outdated human platelets by affinity chromatography using anti-PECAM-1 monoclonal antibodies.[2]

Activities

An expanding number of functions have been defined or proposed for PECAM-1.[1,2] Data indicate that PECAM-1 may play important roles in the recruitment of leukocytes (particularly neutrophils) at inflammatory sites,[20] angiogenesis,[21] cardiovascular development,[22] and T-cell mediated immune responses.[16,23]

Antibodies

A number of monoclonal antibodies against human[24] and murine PECAM-1[25] have been characterized. Antibodies that bind to the first, second, and sixth Ig-like domains appear to have the greatest effect in blocking or activating PECAM-1 dependent function *in vitro* and *in vivo.*

Genes

The gene for human PECAM-1 is located on the long arm of chromosome 17 and encompasses more than 65 kb of DNA, making it the largest gene for a cell adhesion molecule of the Ig gene superfamily reported to date. It is complex, consisting of 16 exons separated by introns ranging in size from 86 to 12×10^3 bp (EMBL/GenBank L34631–L34657).[26] The promoter for PECAM-1 is GC-rich, devoid of TATA and CAAT elements, contains several closely spaced transcription initiation sites, and possesses consensus binding sequences for a number of transcriptional regulatory elements (EMBL/GenBank X96848 and 96849).[27] Several alternatively spliced forms that differ within either the transmembrane or cytoplasmic domains have been identified but their functional significance is unknown.[1] PECAM-1 has been cloned and sequenced from human (EMBL/GenBank M28526)[4] and mouse (EMBL/GenBank L06039) cDNAs.[22,28]

Mutant phenotype/disease states

There is one report that polymorphisms of the PECAM-1 gene may be important in patents receiving allogenic bone marrow transplants, as the risk of graft-versus-host disease increases when the donor and recipient are not identical with respect to PECAM-1 alleles.[29] However, this has not been reproduced by others.[30]

References

1. DeLisser, H. M., Baldwin, H. S., and Albelda, S.M. (1997). *Trends Cardiovasc. Med.*, **7**, 203–10.
2. Newman, P. J. (1997). *J. Clin. Invest.*, **99**, 3–8.
3. Albelda, S. M., Oliver, P. D., Romer, L. H., and Buck, C. A. (1990). *J. Cell Biol.*, **110**, 1227–37.
4. Newman, P. J., Berndt, M. C., Gorski, J., White, G. C., II, Lyman, S., Paddock, C., and Muller, W. A. (1990). *Science*, **247**, 1219–22.
5. DeLisser, H. M., Yan, H., Newman, P. J., Muller, W. A., Buck, C. A., and Albelda, S. M. (1993). *J. Biol. Chem.*, **268**, 16037–46.
6. Piali, L., Hammel, P., Uherek, C., Bachmann, F., Gisler, R. H., Dunon, D., and Imhof, B. A. (1995). *J. Cell Biol.*, **130**, 451–60.
7. Sun, J., Williams, J., Yan, H., Amin, K. M., Albelda, S. M., and DeLisser, H. M. (1996). *J. Biol. Chem.*, **271**, 18561–70.
8. Sun, Q., DeLisser, H. M., Zukowski, M. M., Paddock, C., Albelda, S. M., and Newman, P. J. (1996). *J. Biol. Chem.*, **271**, 11090–8.
9. Newman, P. J., Hillery, C. A., Albrecht, R., Parise, L. V., Berndt, M. C., Mazurov, A. V., *et al.* (1992). *J. Cell Biol.*, **119**, 239–46.
10. Zehnder, J. L., Hirai, K., Shatsky, M., McGregor, J. L., Levitt, L. J., and Leung, L. L. K. (1992). *J. Biol. Chem.*, **267**, 5243–9.
11. Jackson, D. E., Ward, C. M., Wang, R., and Newman, P. J. (1997). *J. Biol. Chem.*, **272**, 6986–93.
12. Lu, T. T., Yan, L. G., and Madri, J. A. (1996). *Proc. Natl Acad. Sci. USA*, **93**, 11808–13.
13. Fawcett, J., Buckley, C., Holness, C. L., Bird, I. N., Spragg, J. H., Saunders, J., *et al.* (1995). *J. Cell Biol.*, **128**, 1229–41.
14. Watt, S. M., Williamson, J., Genevier, H., Fawcett, J., Simmons, D. L., Hatzfeld, A., *et al.* (1993). *Blood*, **82**, 2649–63.
15. Lastres, P., Almendro, N., Bellón, T., López-Guerrero, J.A., Eritja, R., and Bernabéu, C. (1994). *J. Immunol.*, **153**, 4206–18.
16. Buckley, C. D., Doyonnas, R., Newton, J. P., Blystone, S. D., Brown, E. J., Watt, S. M., and Simmons, D. L. (1996). *J. Cell Sci.*, **109**, 437–45.
17. Deaglio, S., Dianzani, U., Horenstein, A. L., Fernandez, J. E., van Kooten, C., Bragardo, M., *et al.* (1996). *J. Immunol.*, **156**, 727–34.
18. Prager, E., Sunder-Plassmann, R., Hansmann, C., Koch, C., Holter, W., Knapp,W., and Stockinger, H. (1996). *J. Exp. Med.*, **184**, 41–50.
19. Garubhagavatula, I., Amrani, Y., Pratico, D., Ruberg, F. L., Albelda, S. M., and Panettieri, R. (1998). *J. Clin. Invest.*, **101**, 212–22.
20. Vaporciyan, A. A., DeLisser, H. M., Yan, H., Mendiguren, I. I., Thom, S. R., Jones, M. L., *et al.* (1993). *Science*, **262**, 1580–2.
21. DeLisser, H. M., Christofidou-Solomidou, M., Strieter, R. M., Burdick, M. D., Robinson, C. S., Wexler, R. S. *et al.* (1997). *Am. J. Pathol.*, **151**, 671–7.
22. Baldwin, H. S., Shen, H. M., Yan, H., DeLisser, H. M., Chung, A., Mickanin, C., *et al.* (1994). *Development*, **120**, 2539–53.
23. Zehnder, J. L., Shatsky, M., Leung, L. L. K., Butcher, E. C., McGregor, J. L., and Levitt, L. J. (1995). *Blood*, **85**, 1282–8.
24. Yan, H., Pilewski, J. M., Zhang, Q., DeLisser, H. M., Romer L., and Albelda, S. M. (1995). *Cell Adh. Commun.*, **3**, 45–66.
25. Yan, H., Baldwin, H. S., Sun, J., Buck, C. A., Albelda, S. M., and DeLisser, H. M. (1995). *J. Biol. Chem.*, **270**, 23672–80.
26. Kirschbaum, N. E., Gumina, R. J., and Newman, P. J. (1994). *Blood*, **84**, 4028–37.
27. Almendro, N., Bellón, T., Rius, C., Lastres, P., Langa, C., Corbi, A., and Bernabéu, C. (1996). *J. Immunol.*, **157**, 5411–21.
28. Xie, Y., and Muller, W. A. (1993). *Proc. Natl Acad. Sci. USA*, **90**, 5569–73.
29. Behar, E., Chao, N. J., Hiraki, D. D., Krishnaswamy, S., Brown, B. W., Zehnder, J. L., and Grumet, F. C. (1996). *New Engl. J. Med.*, **334**, 286–91.
30. Nichols, W. C., Antin, J. H., Lunetta, K. L., Terry, V. H., Hertel, C. E., Wheatley, M. A., *et al.* (1996). *Blood*, **88**, 4429–34.

Horace M. DeLisser and Steven M. Albelda
Pulmonary and Critical Care Division,
Department of Medicine,
University of Pennsylvania
Medical Center,
Philadelphia, PA 19104–4283, USA

PH-20

PH-20 is a sperm-specific surface protein which has two essential functions in fertilization: it has a hyaluronidase activity which enables sperm to pass through the cumulus cell layer surrounding the egg and it has an activity in adhesion when acrosome-reacted sperm bind to the zona pellucida of the egg (secondary binding). It is present on both the plasma membrane and secretory granule (acrosomal) membrane and is anchored in the membrane by covalent linkage to glycosylphosphatidyl inositol.

Protein properties

The PH-20 protein has been found on the surface of sperm from all mammals tested (guinea pig, mouse, monkey and human) and is sperm specific.[1] The protein is bifunctional: it has a hyaluronidase activity which digests hyaluronic acid in the sperm's path through the extracellular matrix of the cumulus cells surrounding the egg.[2] Unrelated to the hyaluronidase activity is a second PH-20 activity required by sperm, after they acrosome react, to bind to the *zona pellucida* of the egg (secondary sperm–zona binding).[3] The activity of PH-20 that makes it required for secondary binding is unknown.[1] The N-terminal domain of PH-20 (residues 1–311) contains the hyaluronidase activity while the activity of the C-terminal domain (residues 312–468) is undefined (Fig. 1).[3]

In guinea pigs where it has been most extensively studied, the PH-20 protein is synthesized as a single chain, M_r 66 kDa on reducing SDS–PAGE.[4,5] Based on cDNA sequence, N-terminal sequencing, and N-glycanase digestion, mature guinea pig PH-20 is 468 amino acids long, contains six *N*-linked glycosylations sites of which five or six are probably used, and has twelve cysteines, eight of which are clustered near the C terminus (Fig. 1). The cDNA sequence indicates that the N-terminal domain of PH-20 has significant sequence homology with bee hyaluronidase[6] whereas the C-terminal domain is a novel sequence. Biochemical analysis of the protein as well as sequencing of various cDNAs, genomic Southern blots, and Northern blots, have thus far revealed only a single membrane protein structure, one mRNA and one gene, indicating that PH-20 on the plasma and acrosomal membranes may be structurally identical.[4,5]

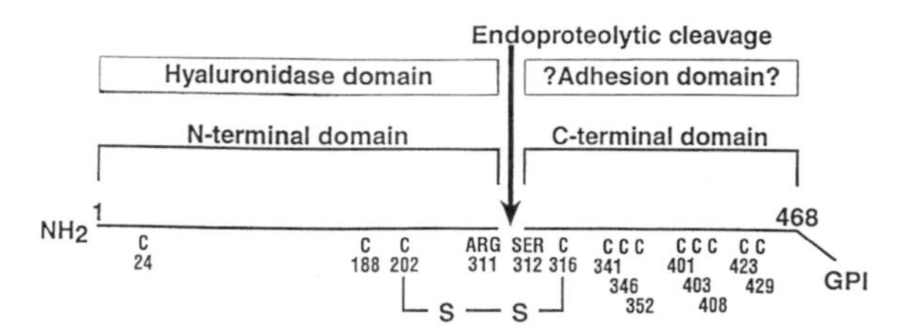

Figure 1. A diagram showing the key structural features of PH-20. In guinea pig sperm, prior to acrosome reaction, PH-20 is a single chain. After acrosome reaction, the proteolytic cleavage between Arg-311 and Ser-312 forms a two subunit protein with (one or more) interchain disulphide bonds. Residues 1–307 have sequence homology with bee venom hyaluronidase.

PH-20 can initially be detected by immunofluoresence in round spermatids, first on the acrosomal membrane and subsequently on the plasma membrane.[7] PH-20 undergoes post-translational modifications later in the life of the sperm, during final differentiation of sperm in the epididymis, and during fertilization after sperm exocytosis (acrosome reaction). The initial 66 kDa form is reduced to a size of 64 kDa in sperm that have completed differentiation in the epididymis (cauda epididymal sperm).[8] The second processing step is an endoproteolytic cleavage that generates two fragments held together by disulphide bond(s), so that when the protein is run on reducing SDS–PAGE it separates into two bands at 41–48 kDa and 27 kDa.[4] Although this second processing step may begin while sperm are traversing the epididymis, the bulk of this endoproteolytic cleavage occurs at the time of the acrosome reaction.[4,8]

The localization of PH-20 to specific surface domains, its level of surface expression, and its ability to diffuse in the membrane are developmentally regulated. In the guinea pig, PH-20 is found on the whole plasma membrane of testicular sperm.[7,8] On cauda epididymal sperm, before acrosome reaction, PH-20 is localized to the posterior head domain.[9] After the acrosome reaction, which results in insertion of the inner acrosomal membrane into the plasma membrane, PH-20 migrates out of the posterior head into the inner acrosomal membrane. There, the plasma membrane PH-20 population joins the other population of PH-20 that is associated with the inner acrosomal membrane and therefore is revealed by the acrosome reaction. The revealing of the second population of PH-20 increases the level of PH-20 surface expression 2.5-fold, a change that may regulate the avidity of sperm–zona adhesion.[9,10]

PH-20 on the plasma membrane has highly restricted diffusion early in development and free diffusion at the end of development when it functions in sperm–zona adhesion. The measured diffusion coefficients vary over a 250 fold range: 1.9×10^{-11} cm^2/s (testicular sperm); 1.8×10^{-10} cm^2/s (cauda epididymal sperm before acrosome reaction); 4.9×10^{-8} cm^2/s (cauda epididymal sperm after acrosome reaction).[11,12]

Purification

PH-20 is purified routinely from octylglucoside extracts of acrosome-intact or acrosome-reacted cauda epididymal sperm by affinity chromatography using an anti-PH-20 monoclonal antibody (mAb PH-22) coupled to Sepharose.[4] PH-20 has also been purified from mixtures of spermatogenic cells and testicular sperm using the same protocol.[7]

Activities

The hyaluronidase activity of purified sperm or recombinant PH-20 has been measured in standard turbidimetric enzyme assays.[2] The precise activity of PH-20 in sperm–zona adhesion remains to be established.[1] Acrosome-reacted sperm adhesion to the zona pellucida is blocked by mAbs to certain epitopes of PH-20 and by specific removal of PH-20 from the sperm surface with phosphatidyl inositol specific phospholipase C (PI-PLC).[1]

Antibodies

Three anti-guinea-pig PH-20 mAbs have been described[13] and two rabbit polyclonal antisera.[5]

Genes

Because of the interest in using PH-20 as an antigen in a contraceptive vaccine for wild animal (pests) and humans,[14] PH-20 cDNA has been cloned and sequenced from a large variety of mammals. These include guinea pig, mouse, rat, fox, rabbit, monkey and human. Mouse and human PH-20 genes have been assigned chromosomal locations.[15,16]

Mutant phenotype

No information available.

Structure

Not available.

References

1. Myles, D. G. and Primakoff, P. (1997). *Biol. Reprod.*, **56**, 320–7.
2. Lin, Y., Mahan, K., Lathrop, W. F., Myles, D. G., and Primakoff, P. (1994). *J. Cell Biol.*, **125**, 1157–63.
3. Hunnicutt, G., Primakoff, P., and Myles, D. G. (1996) *Biol. Reprod.*, **55**, 80–6.
4. Primakoff, P., Cowan, A. E., Hyatt, H., Tredick-Kline, J., and Myles, D. G. (1988). *Biol. Reprod.*, **38**, 921–34.
5. Lathrop, W. F., Carmichael, E. P., Myles, D. G., and Primakoff, P. (1990). *J. Cell Biol.*, **111**, 2939–49.
6. Gmachl, M. and Kreil, G. (1993). *Proc. Natl Acad. Sci. USA*, **90**, 3569–73.
7. Phelps, B. and Myles, D. G. (1987). *Dev. Biol.*, **123**, 63–72.
8. Phelps, B. M., Koppel, D. E., Primakoff, P., and Myles, D. G. (1990). *J. Cell Biol.*, **111**, 1839–47.
9. Myles, D. G. and Primakoff, P. (1984). *J. Cell Biol.*, **99**, 1634–41.
10. Cowan, A. E., Primakoff, P., and Myles, D. G. (1986). *J. Cell Biol.*, **103**, 1289–97.
11. Phelps, B. M., Primakoff, P., Koppel, D. E M., Loew, G., and Myles, D. G. (1988). *Science*, **240**, 1780–2.
12. Cowan, A. E., Myles, D. G., and Koppel, D. E. (1987). *J. Cell. Biol.*, **104**, 917–23.
13. Primakoff, P., Hyatt, H., and Myles, D. G. (1985). *J. Cell. Biol.*, **101**, 2239–2244.
14. Primakoff, P. (1994). *Am. J. Reprod. Immunol.*, **31**, 208–10.
15. Deng, X., Moran, J., Copeland, N. G., Gilbert, D. J., Jenkins, N. A., Primakoff, P., and Martin-DeLeon, P. A. (1997). *Mamm. Genome*, **8**, 94–7.
16. Jones, M. H., Davey, P. M., Aplin, H., and Affara, N. A. (1995). *Genomics*, **29**, 796–800.
17. Primakoff, P. and Myles, D. G. (1983). *Dev. Biol,*. **98**, 417–28.

Paul Primakoff
Department of Cell Biology and Human Anatomy, University of California at Davis, Davis, CA 95616, USA

Diana G. Myles
Section of Molecular and Cellular Biology, University of California at Davis, Davis, CA 95616, USA

Platelet GP Ib–IX–V complex

The GP Ib-IX-V complex is a heterooligomeric platelet membrane receptor essential for normal platelet adhesion and thrombus formation at sites of vascular injury. It functions by interacting with the adhesive protein, von Willebrand factor (vWF) and also has a binding site for the platelet agonist, α-thrombin. Congenital deficiency of the GP Ib–IX–V complex results in the haemorrhagic disorder known as Bernard–Soulier syndrome.

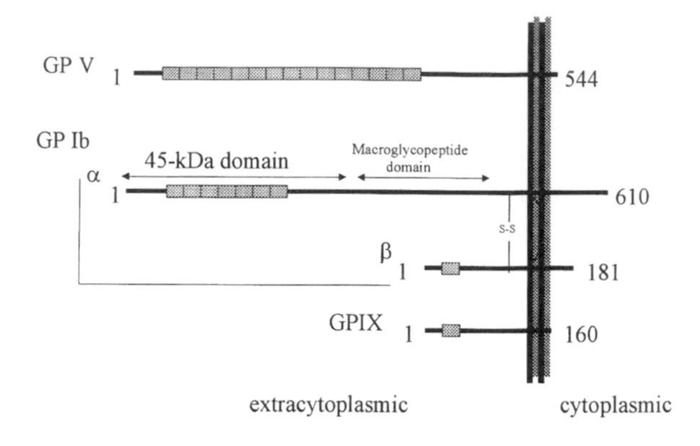

Figure 1. A schematic representation of the GP Ib–IX–V complex is depicted with the amino terminus of each mature polypeptide indicated as residue 1, and the carboxyl-terminal residue indicated as 610, 181, 160, and 544 for GP Ibα, GP Ibβ, GP IX, and GP V, respectively. The 24 residue leucine-rich repeat is represented as a shaded box within the linear sequence of each polypeptide, present as single copies within GP Ibβ, and GP IX, and multiple copies within GP Ibα and GP V. The α and β subunits of GP Ib are linked via an interchain disulphide bond between cysteine residues just beyond the extracytoplasmic side of the transmembrane domain within each subunit. The 45 kDa domain of GP Ibα has been extensively characterized as containing binding sites for von Willebrand factor, and α-thrombin. A polymorphic macroglycopeptide domain of GP Ibα links the GP Ib 45 kDa domain, and transmembrane domain. The non-covalent bonds linking GP Ib, GP IX, and GpV have not been characterized. Reprinted from ref. 1 with permission of the publisher (F. K. Schattauer, Stuttgart and New York).

Protein properties

The glycoprotein (GP) Ib–IX–V complex[1] is composed of four distinct gene products (Fig. 1) the disulphide-linked α and β subunits of GP Ib (also designated CD42b and CD42c) and non-covalently associated GP IX (CD42a) and GP V (CD42d). Electron microscopy of GP Ibα demonstrates an elongated structure with a longitudinal axis of 59.6 nm and a globular N-terminal domain.[2] GP Ib has an apparent molecular mass of ~170 kDa (α chain, ~140 kDa; β chain, ~24 kDa); GP IX of ~18 kDa, and GP V of ~82 kDa. All four proteins have single transmembrane and cytoplasmic domains. Glycocalicin is a soluble fragment of GP Ibα corresponding essentially to its extracytoplasmic domain.[3] The GP Ib–IX–V complex is synthesized in megakaryocytes and expressed on the platelet surface in 16 000 to 30 000 copies per cell, second in density only to the GP IIb–IIIa complex (integrin $\alpha_{IIb}\beta_3$). GP Ibα can appear on the platelet membrane in the absence of the other subunits,[4] but concomitant synthesis of GP Ibβ, GP IX, and GP V results in more efficient surface expression of the complex.[5–7]

The GP Ibα mRNA codes for a 16 residue signal peptide and a mature 610 residue polypeptide[8]. GP Ibα can be divided into four distinct domains: the N-terminal domain (~45 kDa), residues 1–293, containing the binding sites for vWF and α-thrombin; the carbohydrate-rich macroglycopeptide domain (~84 kDa), residues 294–485; the transmembrane domain, residues 486–514; and the intracytoplasmic domain, residues 515–610. The N-terminal domain contains two *N*-linked and one *O*-linked carbohydrate chains and seven copies of a 24 residue leucine-rich sequence, found in a variety of proteins displaying diverse functions (see Leucine-rich repeat family, pp. 227–228). The N-terminal domain of GP Ibα contains three disulphide loops and one unpaired cysteine.[9] The macroglycopeptide domain is very rich in *O*-linked carbohydrate (more than half of the molecular mass) and is responsible for GP Ibα heterogeneity owing to a polymorphic 13 residue repeat containing signals for *O*-linked glycosylation.[10,11] The cytoplasmic domain of GP Ibα interacts with acting-binding protein (filamin) and phospholipase A_2 (δ isoform 14–3–3 protein).[12,13]

The GP Ibβ mRNA codes for a 25 residue signal peptide and a mature 181 residue polypeptide.[14] The extracytoplasmic domain of GP Ib is composed of 122 residues with one *N*-linked glycosylation site and one leucine-rich repeat. It also contains nine cysteine residues, forming three intrachain loops, two cysteines linked to cysteine residues in GP Ibα, and one unpaired cytoplasmic cysteine which may be linked to membrane lipids. The transmembrane domain of GP Ibβ is composed of 25 residues and the intracytoplasmic domain of 34 residues.

The GP IX mRNA codes for a 16 residue signal peptide and a mature 160 residue polypeptide.[15] The extracytoplasmic domain contains one *N*-linked glycosylation site and one copy of the leucine-rich motif; it also contains eight cysteine residues, probably arranged in four intrachain loops. The 20 residue transmembrane domain contains one cysteine residue, presumably involved in a thioester bond with a membrane lipid,[16] and precedes a short, six residue intracytoplasmic tail. The physicochemical nature of the non-covalent association between GP Ib and GP IX is not known.

The GP V cDNA codes for a 16 residue signal peptide and mature 544 residue polypeptide,[17] containing 15 tandem copies of the leucine-rich repetitive motif, eight *N*-linked glycosylation sites, a 24 residue transmembrane domain, and a 16 residue cytoplasmic domain. GP V associates with GP Ib–IX in a 1:2 ratio (GP V:GP Ib–IX).[18] It contains a thrombin cleavage site releasing a 69 kDa fragment representing most of the extracellular domain of the protein. The expression of GP V increases surface expression of the GP Ib–IX–V complex, resulting in an increased binding capacity for vWF by the platelet.[6]

Purification

The GP Ib–IX–V complex is purified from detergent extracts of platelet membranes by lectin (wheatgerm agglutinin) affinity chromatography followed by immunoaffinity chromatography.[19] The α and β chains of GP Ib are separated by molecular sieve chromatography following disulphide reduction. The soluble extracytoplasmic domain of GP Ibα (glycocalicin) is isolated following activation of an endogenous Ca^{2+}-dependent protease (calpain).[3]

Activities

The GP Ib–IX–V complex mediates the initiation and propagation of platelet responses to vascular injury by anchoring platelets to subendothelial surfaces or damaged endothelium.[20] Specific interaction with surface-bound vWF may transmit intracellular signals, causing propagation of the initial response through platelet activation. These functions require ligand-recognition sites, located in the N-terminal 45 kDa domain of GP Ibα. Binding to vWF tethers platelets to exposed thrombogenic surfaces in a wide range of flow conditions, but is the exclusive receptor/ligand interaction that is efficient under conditions of high wall shear rate.[21] Platelets tethered to the vascular wall continue to move with reduced velocity until the platelet integrin $\alpha_{IIb}\beta_3$ receptor becomes activated, resulting in irreversible platelet adhesion, a process dependent upon a second receptor-binding domain within vWF. It is not clear whether α-thrombin binding to GP Ibα is coupled to signal generation inside the platelet.

Antibodies

Monoclonal antibodies recognizing the N terminus of GP Ibα have been characterized in detail and can be used to to study the structure–function relationships of the two distinct binding sites for vWF and α-thrombin.[22] Several anti-GP Ib and anti-GP IX monoclonal antibodies are commerically available (DAKO, Carpinteria, CA; Becton-Dickinson, Bedford, MA).

Genes

The genes for the components of the GP Ib–IX–V complex each has a relatively simple structure with the majority of protein-encoding sequence present within a single exon. Each gene of the complex is present as a single copy in the haploid genome and, to date, no pseudogenes have been identified. Human GP Ibα and GP Ibβ genes have been assigned to chromosomes 17p12-ter and 22q11.2, respectively; GPIX and GPV map to chromosome 3, 3q21 and 3q29, respectively. The nucleotide sequence of each gene is available from GenBank with accession numbers M22403, Z23091, U07983, and M80478 for GP Ib, GP V, GP Ib, and GP IX respectively.

The expression of a GP Ib–IX–V complex is restricted to megakaryocytes. GP Ibα mRNA has been reported in endothelial cells along with a variant form of the GP Ibβ transcript.[23,24] The variant GP Ibβ transcript results from a more 5′ gene containing an imperfect polyadenylation signal sequence that utilizes the downstream polyadenylation site within the platelet GP Ibβ gene.[24]

Mutant phenotypes/disease states

Mutations within the GP Ib–IX–V complex cause the Bernard–Soulier syndrome (BSS) characterized by (i) a lack of von Willebrand factor binding to platelets and (ii) giant platelets.[25] The latter phenotype suggests a linkage between the GP Ib–IX–V complex and normal platelet morphology. The BSS is an autosomal-recessive disease and classical genetic lesions have been described in the GP Ibα, GP Ibβ and GP IX genes. Heterozygous mutations within a single subunit can impair surface expression of the entire complex, resulting in the BSS phenotype. Dominant mutations within GP Ibα can also result in a receptor with increased affinity for vWF, a condition termed pseudo- or platelet-type von Willebrand disease.[26]

Structure

Not available.

References

1. Ware, J. (1998). *Thromb. Haemost.*, **79**, 466–78.
2. Fox, J. E. B., Aggerbeck, L. P. and Berndt, M. C. (1988). *J. Biol. Chem.*, **263**, 4882–90.
3. Titani, K., Takio, K., Handa, M., and Ruggeri, Z. M. (1987). *Proc. Natl Acad. Sci. USA*, **84**, 5610–5614.
4. Meyer, S., Kresbach, G., Haring, P., Schumpp-Vonach, B., Clemetson, K. J., Hadvary, P., and Steiner, B. (1993). *J. Biol. Chem.*, **268**, 20555–62.
5. Lopez, J. A., Leung, B., Reynolds, C. C., Li, C. Q., and Fox, J. E. B. (1992). *J. Biol. Chem.*, **267**, 12851–9.

6. Calverley, D. C., Yagi, M., Stray, S. M., and Roth, G. J. (1995). *Blood*, **86**, 1361–7.
7. Meyer, S. C. and Fox, J. E. (1995). *J. Biol. Chem.*, **270**, 14693–9.
8. Lopez, J. A., Chung, D. W., Fujikawa, K., Hagen, F. S., Papayannopoulou, T., and Roth, G. J. (1987). *Proc. Natl Acad. Sci. USA*, **84**, 5615–9.
9. Hess, D., Schaller, J., Rickli, E. E., and Clemetson, K. J. (1991). *Eur. J. Biochem.*, **199**, 389–93.
10. Lopez, J. A., Ludwig, E. H., and McCarthy, B. J. (1992). *J. Biol. Chem.*, **267**, 10055–61.
11. Ishida, F., Furihata, K., Ishida, K., Yan, J., Kitano, K., Kiyosawa, K., *et al.* (1995). *Blood*, **86**, 1357–60.
12. Andrews, R. K. and Fox, J. E. (1992). *J. Biol. Chem.*, **267**, 18605–11.
13. Du, X., Harris, S. J., Tetaz, T. J., Ginsberg, M. H., and Berndt, M. C. (1994). *J. Biol. Chem.*, **269**, 18287–90.
14. Lopez, J. A., Chung, D. W., Fujikawa, K., Hagen, F. S., Davie, E. W., and Roth, G. J. (1988). *Proc. Natl Acad. Sci. USA*, **85**, 2135–9.
15. Hickey, M. J., Deaven, L. L., and Roth, G. J. (1990). *FEBS Lett.*, **274**, 189–92.
16. Muszbek, L. and Laposata, M. (1989). *J. Biol. Chem.*, **264**, 9716–19.
17. Hickey, M. J., Hagen, F. S., Yagi, M., and Roth, G. J. (1993). *Proc. Natl Acad. Sci. USA*, **90**, 8327–31.
18. Modderman, P. W., Admiraal, L. G., Sonnenberg, A., and von dem Borne, A. E. G. K. (1992). *J. Biol. Chem.*, **267**, 364–9.
19. Canfield, V. A., Ozols, J., Nugent, D., and Roth, G. J. (1987). *Biochem. Biophys. Res. Commun.*, **147**, 526–34.
20. Ruggeri, Z. M. (1997). *J. Clin. Invest.*, **99**, 559–64.
21. Savage, B., Saldivar, E., and Ruggeri, Z. M. (1996). *Cell*, **84**, 289–97.
22. Handa, M., Titani, K., Holland, L. Z., Roberts, J. R., and Ruggeri, Z. M. (1986). *J. Biol. Chem.*, **261**, 12579–85.
23. Konkle, B. A., Shapiro, S. S., Asch, A. S., and Nachman, R. L. (1990). *J. Biol. Chem.*, **265**, 19833–8.
24. Zieger, B., Hashimoto, Y., and Ware, J. (1997). *J. Clin. Invest.*, **99**, 520–5.
25. Lopez, J. A. (1998). *Blood*, **91**, 4397–418.
26. Miller, J. L. (1996). *Thromb. Haemost.*, **75**, 865–9.

Jerry Ware and Zaverio M. Ruggeri
Roon Research Center for Arteriosclerosis, and Thrombosis, Departments of Molecular, and Experimental Medicine and of Vascular Biology, The Scripps Research Institute, La Jolla, California, USA

Selectins

The selectins are a family of cell–cell adhesion proteins, which are involved in the dynamic interactions of leukocytes with the vascular endothelium. In contrast to the integrins, cadherins, and members of the immunoglobulin superfamily (exclusive of the sialoadhesins), whose functions rely on protein–protein interactions, the selectins mediate adhesive interactions through recognition of specific carbohydrate determinants presented on a discrete set of macromolecular ligands.[1] The selectins have dramatically and unequivocally validated the long-held view that carbohydrates serve as recognition determinants in eukaryotic cellular interactions.

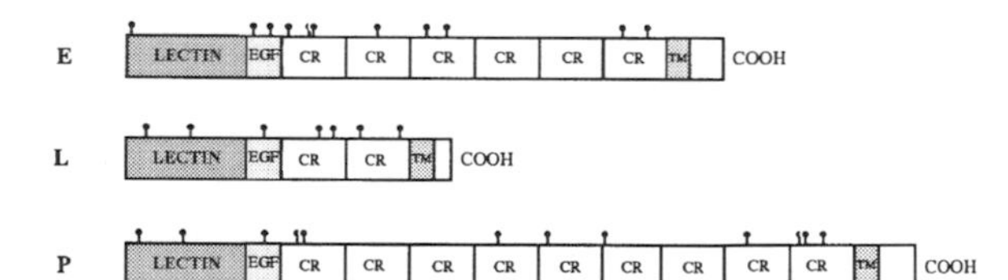

Figure 1. The selectin family. The organization of the protein motifs for the human selectins is shown (E-, L-, and P-selectin). Potential sites for N-glycosylation are denoted by small pinhead symbols.

Overview of the family

The selectins consist of three members (Fig. 1), which have various designations (see below).[2–7] L-selectin is the leukocyte selectin, E-selectin and P-selectin are the endothelial selectins, and P-selectin is also expressed in platelets. These cell surface-associated glycoproteins are type I transmembrane proteins (N_{out}/C_{in}), which exhibit a common tandem organization of protein motifs: an amino terminal calcium-type (C-type) lectin domain of ~120 residues, an EGF motif of ~36 residues, a series of contiguous (2–9) short consensus repeats (SCRs) of ~63 residues, a single-pass transmembrane domain, and a short cytosolic tail (17–35 residues). C-type lectin domains are found in a variety of animal proteins, many of which manifest calcium-dependent lectin activity.[8] So-called complement-regulatory (SCR) domains are sometimes associated with binding of complement factors C3b or C4b. However, no such activity has as yet been reported for any of the selectins. Comparison of the lectin and EGF domains of the three selectins within the same species reveals homologies of ~50 per cent, which are substantially higher than the similarities with comparable domains in other proteins (20–40 per cent). When the same domains are compared between species, the homology is generally highest for the lectin (~72 per cent) and transmembrane domains (~80 per cent). Each of the modular extracellular domains of the selectins, as well as the transmembrane domain, is encoded by single exons within the genes. Reflecting the likely evolutionary origin of these proteins from a primordial gene, the genes for the three proteins are clustered (< 340 kb) on chromosome 1 in both human and mouse. A more detailed description of each selectin follows. References in some cases are to primary sources but to limit the total number of citations, a restricted number of recent reviews are cited.

E-selectin

E-selectin or ELAM-1, one of the two endothelial selectins, is biosynthetically induced on cultured endothelial cells after several h of stimulation with inflammatory mediators such as IL-1, TNF, and bacterial lipopolysaccharide.[2,9] Through recognition of specific counter-receptors ('ligands') on appropriate partner leukocytes, E-selectin mediates the tethering and rolling of the leukocytes on activated endothelium. It functions, in cooperation with the other selectins and with leukocyte integrins, in the recruitment of leukocytes from the blood into inflammatory sites. A number of recent reviews describe the cascade of steps involved in inflammatory leukocyte trafficking and lymphocyte homing.[10–12]

Synonymous names

CD62e, ELAM-1.

Protein properties

The biosynthetic induction of E-selectin on HUVEC (human umbilical vein endothelial cells) is at the transcriptional level and involves the cytokine-induced activation and nuclear translocation of Nf-κB.[2,9,13] In response to IL-1 or TNF, the protein is rapidly induced (~60 min), but its cell surface expression declines after 3–6 h, reflecting downregulation of gene activation, the short-half life of its mRNA, and its internalization from the cell surface and degradation in lysosomes. The molecular mass of E-selectin on SDS gels is 115 kDa. The difference between this value and protein mass predicted from the cDNA (64 kDa) is accounted for by *N*-glycosylation. E-selectin is induced to associate with the cytoskeleton within endothelial cells upon leukocyte adhesion.[14]

All of the selectins, by virtue of their C-type lectin domains, are capable of binding the carbohydrate determinant known as sialyl Lewis^x (sLex),15 Siaα2,3Galβ1,4[Fucα1,3]GlcNAc–and related structures. Fucosylation of GlcNAc and 3' sialylation of Gal (or alternatively its 3' sulphation) are essential for this binding.[16–18] Recently, direct measurements via NMR establish that the affinity of sLex for E-selectin is low (K_D = 0.72 mM) and the interaction is characterized by a fast off-rate (k_{off} = 164 s^{-1}).[19] Parallel determinations for P-selectin and L-selectin yield values (K_D, k_{off}) of 3.9 mM, 1080 s^{-1} and 7.8 mM, 522 s^{-1}, respectively. More complex sLex-containing structures, identified within *N*-linked carbohydrate chains from myeloid cells, have submicromolar affinities for E-selectin.[20] The macromolecular ligands for E-selectin demonstrate dramatically enhanced affinities, which may be due to a number of factors: multivalent presentation of epitopes, further complexity of the epitopes, and/or precise presentation of different epitopes.[5,21] PSGL-1 and ESL-1 represent two myeloid cell ligands for E-selectin (see Selectin ligands, pp. 298–303).[22,23] The HECA 452 mAb, which reacts with an sLex-related epitope, identifies a set of E-selectin ligands (CLA) on a subset of memory T cells.[12,24]

E-selectin, expressed on cells or coated on plastic wells as a recombinant protein, is capable of mediating adhesion of appropriate leukocyte populations under static conditions. Of more physiological relevance is the ability of E-selectin to support the tethering and rolling of leukocytes in a parallel plate flow chamber, which closely recapitulates the shear conditions of blood flow in vessels.[10,25] Myeloid cells and a small population of lymphocytes have been shown to roll on E-selectin, establishing the potential of these cells to utilize E-selectin for initial deceleration in the recruitment process.

In vivo evidence for an actual trafficking function of E-selectin derives from numerous studies showing the induction of E-selectin on blood vessels at sites of inflammation. Frequently, it is seen in postcapillary venules at sites of neutrophil infiltration.[26,27] However, its expression is also seen on blood vessels at chronic sites of inflammation where mononuclear cells predominate, for example, in multiple skin lesions.[24] Supporting these histological findings, inhibition of E-selectin function with antibodies, Ig chimeras of E-selectin, or with sLex-based carbohydrates ameliorates inflammatory reactions in a number of animal models.[26,27]

Purification

Generally, soluble recombinant forms or transfected cells are used for study.

Antibodies

The adhesion blocking mAb H18/7 maps to the lectin domain of E-selectin whereas the non-adhesion blocking H4/18 requires the first three SCRs for binding.[28] Two comprehensive listings of commercially sources of antibodies are *Linscott's Directory of Immunological and Biological Reagents*, (9th ed) 1996 and MSRS Catalog, *Manufacturer's specifications and reference synopsis, primary antibodies*, (3rd ed) (Robert V. Weiner, Aerie Corp.).

Genes

cDNAs for E-selectin have been sequenced from human, mouse, rat, rabbit, pig, cow, and dog.

Mutant phenotype/disease states

Like the other two selectin knockouts, E-selectin-null mice are viable and exhibit no developmental abnormalities.[29] Neutrophil trafficking is not affected in a number of inflammatory models in the absence of E-selectin. However, when P-selectin function is blocked in the null mice by a mAb, neutrophil emigration is blunted in models of peritonitis and delayed type hypersensitivity (DTH), whereas there is no effect of the mAb in wild-type mice. In further support of functional redundancy between the endothelial selectins, mice that are lacking both E- and P-selectin exhibit profound deficiencies in leukocyte rolling, leading to a marked leukocytosis and a susceptibility to bacterial infections.[30,31] The phenotype is much more severe than in P-selectin-null mice (see below). In a human inherited disease, known as leukocyte deficiency II, functional ligands for E-selectin and P-selectin are lacking on neutrophils, due to a generalized fucosylation defect.[32] Predictably, neutrophil rolling is impaired in these individuals with leukocytosis and recurrent bacterial infections as the clinical consequences.

Structure

A crystal structure of the lectin and EGF domains of E-selectin has been reported.[33] There is no obvious binding cleft in the lectin domain to accomodate oligosaccharides such as sLex. Based on mutagenesis studies,[4,34] two lysine residues within a stretch of three continuous lysines (111–113 in the lectin domain) are necessary for sLex binding. Grafting of this three-lysine segment into the homologous region of the mannose-binding protein (MBP) confers sLex binding.[35] A cocrystal of the modified MBP with bound sLex confirms the involvement of this lysine stretch in binding to Gal in sLex, but surprisingly there is no direct interaction between sialic acid and the protein (Fig. 2).[36] The co-crystal establishes the residues, which, together with fucose, ligate calcium, thus providing an elegant structural explanation for both the calcium and fucose requirements for sLex binding.

L-selectin

L-selectin was discovered as a 'lymph node homing receptor' involved in the initial interactions of lymphocytes with high endothelial venules (HEV) of lymph nodes, a step necessary for the recruitment of lymphocytes into this lymphoid organ.[37] Subsequently, the protein was shown to be broadly distributed on all classes of leukocytes in the blood.[12,38] It is now known to function in many instances of leukocyte–endothelial interactions and leukocyte–leukocyte interactions that occur in inflammatory trafficking.[7] Recent evidence points to a diverse range of signal transduction functions for L-selectin.

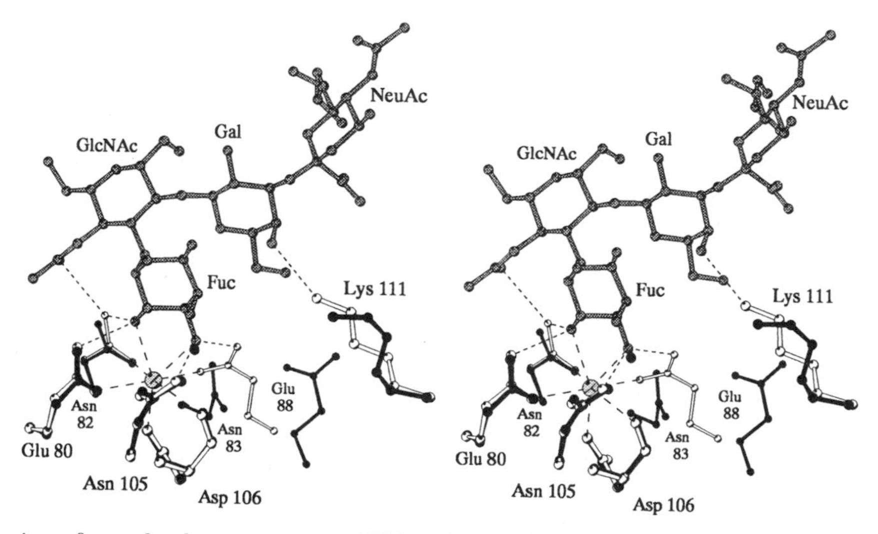

Figure 2. Stereo view of complex between mutant MBP, and sLex. Unligated E-selectin (black) is superimposed on the complex between sLex (grey), and the mutant MBP (white) in which residues 211–213 have been changed to the Lys–Lys–Lys sequence in E-selectin (111–113). Hydrogen bonds are indicated with short dashed lines, and coordination bonds by long dashed lines. The structure of the MBP mutant does not change appreciably upon sLex binding. There is a high degree of overlap between E-selectin, and the MBP mutant in the region of sugar binding. Lys-211 makes a hydrogen bond with 4-OH of Gal. The sialic acid (NeuAc) does not contact the protein. The carboxylate moiety of sialic acid comes closest to Lys-211 (4.5 Å) but not close enough to form a salt bridge. The 2-OH, and 3-OH positions of fucose form coordination bonds with calcium (larger round sphere), and form four hydrogen bonds with amino acids that also coordinate calcium. (Taken, with permission from Ng, and Weis, *Biochemistry*, 36, 979–988 (1997). The figure was kindly provided by Kenneth Ng, and Bill Weis of Stanford University.)

Synonymous names

MEL-14 antigen, gp90^{MEL}, homing receptor (HR), LAM-1, LECAM-1, Leu 8, TQ-1.

Protein properties

In the blood, L-selectin is present on almost all monocytes, neutrophils and eosinophils and a majority of B cells and naïve T lymphocytes.[6,39] It is restricted to a subset of NK cells and memory T cells. Developmentally, L-selectin is found on early myeloid cells where it is maintained through maturation. For B cells, however, L-selectin appears at a relatively late stage of maturation. L-selectin is concentrated on microvillous projections, a distribution which depends on its cytoplasmic tail and an intact cytoskeleton.[7,40]

A substantial proportion (> 40 per cent) of the mass of L-selectin is due to *N*-linked glycosylation (seven potential N-linked sites).[42] The amount of N-linked glycosylation varies with the leukocyte class.[7] The apparent mass of L-selectin (SDS–PAGE) is 74 kDa and 95 kDa on human neutrophils and lymphocytes, respectively. Neutrophil L-selectin (human) appears to be modified by sLex and may serve as one of the ligands for E-selectin.[42] Before its cloning, mouse L-selectin was reported to be covalently modified by ubiquitin,[41] but this claim has not been substantiated.

L-selectin is shed from the cell surface of leukocytes in response to a variety of activating stimuli including phorbol esters.[43] Shedding results from a metalloproteinase which cleaves the receptor near its membrane-spanning region.[6,44] The interaction of calmodulin with the cytoplasmic tail of L-selectin is proposed to regulate the shedding process.[110] Shed L-selectin, which retains its ligand-binding activity, is detected in plasma at a level of micrograms per ml.[45] The functional utility of this cell surface modulation and the high levels of soluble L-selectin in the blood is not understood.

L-selectin is clearly indispensable for the homing of lymphocytes to lymph nodes during the process of lymphocyte recirculation, as initially deduced in the mouse with the classic MEL-14 mAb.[37] Reflecting its broad distribution on blood-borne leukocytes, L-selectin also participates in many instances of leukocyte–endothelial interactions during inflammatory reactions. In a large number of acute models of inflammation[27] and one chronic disease model to date[46] (diabetes in NOD mice), administration of antagonists to L-selectin (mAbs or soluble L-selectin) blunts the disease process. Recently, it has become recognized that L-selectin is capable of mediating leukocyte–leukocyte interactions under shear.[47,48] This mode of interaction provides a mechanism to amplify leukocyte recruitment through binding of secondary leukocytes to those that have already become associated with the endothelium. Thus, some of the anti-inflammatory effects of L-selectin antagonists may be due to blockade of these secondary clusters. Although evidence for the involvement of L-selectin in chronic inflammation is limited, a significant role is strongly suspected, based on the induction of HEV-like vessels in many chronic sites and the expression of L-selectin ligands on these vessels.[5,49]

L-selectin, like the other two selectins, mediates the tethering and rolling of leukocytes.[50] In parallel plate flow chambers, L-selectin-mediated neutrophil rolling is faster than that observed with either of the endothelial selectins, possibly because of faster k_{on} and k_{off} values for its ligand interactions.[51] The microvillous distribution of L-selectin is essential for its tethering activity.[7,52] Surprisingly, L-selectin-mediated rolling does not take place at low shear stresses.[53] This threshold requirement may help to limit leukocyte–endothelial and leukocyte–leukocyte interactions under conditions of stasis.

The application of intravital microscopy to experimental animals confirms the ability of L-selectin to mediate tethering and rolling of leukocytes at inflammatory sites and along HEV in secondary lymphoid organs.[54,55] L-selectin-mediated rolling in the absence of contributions from the endothelial selectins is fast, consistent with the *in vitro* findings.

A number of HEV-associated ligands for L-selectin have been identified at the molecular level:[65,112] GlyCAM-1, CD34, MAdCAM-1, and, most recently, podocalyxin.

The binding of L-selectin to its physiological ligands on HEV requires sulphation in addition to sialylation and fucosylation.[56–58] A detailed analysis of the *O*-glycans of one these ligands (GlyCAM-1) reveals the presence of core-2 based structures, capped by groups containing sLex modified with sulphate esters on position 6 of Gal and/or position 6 of GlcNAc.[59] The importance of the GlcNAc 6-sulphate is supported by several studies,[60,61,107] whereas the contribution of Gal-6-sulphate requires further study.[61–63] L-selectin also binds to a number of sulphated or phosphorylated compounds with no apparent relationship to sLex.[21] Perhaps the recognition of these compounds occurs through a second site within L-selectin, which is distinct from the sLex-binding site.[64,65] As described below, there is very strong evidence for a two-site mechanism in P-selectin.

Like many other adhesion receptors, L-selectin is capable of signal transduction. Integrin activation, cell shape changes, tyrosine phosphorylation, ras activation, and calcium fluxes are all documented to occur within leukocytes after ligation of L-selectin by antibodies, carbohydrate mimetics, or actual physiological ligands.[7,66–70,108]

Purification

Chemical quantities of L-selectin can be purified to homogeneity from detergent lysates of mouse spleen by immunoaffinity chromatography (0.1 μg per spleen).[71] Typically, soluble recombinant chimeras of the receptor or transfected lymphoid cells (pre-B cells) have been used for most studies.

Antibodies

MEL-14 mAb, which recognizes the mouse version of L-selectin, was the first described function-blocking antibody.[37] Ly-22 mAb recognizes L-selectin in certain strains of mice.[72] Two extensive series of mAb (DREGs and LAMs) against human L-selectin have been generated and thoroughly characterized.[73,74,109] For many of the adhesion-blocking antibodies, epitopes are within the lectin domain, but some react with sequences within the EGF domain (e.g. Ly-22).[7] One highly novel antibody, EL-246, is adhesion-blocking for both E- and L-selectin and maps to their SCR regions.[75] As mentioned above, listings of commercial sources for antibodies are available.

Genes

cDNAs for L-selectin have been sequenced from human, orangutan, chimpanzee, rhesus monkey, baboon, rabbit, rat, mouse, and cow.

Mutant phenotype/disease states

An L-selectin-null mouse confirms the indispensable participation of L-selectin in lymphocyte homing to lymph nodes.[76] Homing to a gut-associated lymphoid organ is markedly reduced in these animals as well. With respect to neutrophils, L-selectin-deficient mice show normal rolling immediately after surgical trauma but dramatically reduced rolling within 30 min of injury.[55] The residual rolling at early time points is attributable to P-selectin. The knockout animals also show reduced inflammatory trafficking in models of peritonitis and DTH.[6,77] Analysis of single knockouts (L, P) and the double knockout (EP) reveal significant synergy between the endothelial selectins and L-selectin.

It has not been feasible to draw conclusions about L-selectin from the LAD II patients. However, mice null for fucosyltransferase VII, an enzyme that can elaborate sLex determinants, show the absence of functional-ligands for L-selectin in the HEV of lymph nodes and markedly reduced lymphocyte homing.[59]

Structure

No crystal structure for L-selectin is as yet available. Based on the structure of E-selectin and energy minimization determinations, a model of L-selectin has been proposed.[78]

P-selectin

P-selectin is stored in platelets and endothelial cells in granules. It can be elicited rapidly to the cell surface where it functions in platelet–leukocyte, platelet–endothelial, and leukocyte–endothelial interactions.[2,3] A large body of evidence establishes that P-selectin is essential for inflammatory trafficking, both in acute and chronic settings. Roles for P-selectin in thrombosis and haemostasis are also indicated.[79]

Synonymous names

CD62p, GMP-140, PADGEM.

Protein properties

P-selectin is found in the membranes of Weibel–Palade bodies of endothelial cells and the α granules of platelets.[80] It is mobilized from these intracellular pools to the cell surface within min by a variety of agonists, including thrombin, histamine, and complement fragments.[3] *In vitro*, P-selectin persists on the surface for only a few min before it is internalized and sorted to lysosomes for degradation.[81] Sequences within the cytoplasmic tail are responsible for the intracellular sorting patterns of this receptor.[81,82] A soluble form of P-selectin lacking the transmembrane segment is present at low levels in the blood and is probably the product of an alternatively spliced transcript.[3,83] Like E-selectin, the synthesis of new P-selectin can be induced at the transcriptional level by cytokines such as IL-1 and TNF-α.[84]

Human P-selectin has an apparent mass of 140 kDa and a predicted protein core of 86 kDa.[83] Its cytoplasmic tail is acylated with a fatty acid and the extracellular domain has 12 potential sites for *N*-linked glycosylation.[3] The cytoplasmic tail of P-selectin shows a complex pattern of phosphorylation involving tyrosine, serine, threonine, and, quite novelly, histidine, in response to cell activation.[85,86] The extracellular domain of P-selectin, as visualized by rotary shadowing, is a rigid rod of 48 nm.[87] Interestingly, intact P-selectin, containing a membrane-spanning domain, forms dimers in the presence of detergent, consistent with the possibility of an oligomeric organization in the plasma membrane.[87] For other C-type lectins, dramatic increases in ligand affinity are achieved by clustering of the lectin domains.[88]

P-selectin is capable of binding to carbohydrate-based ligands on myeloid cells, activated B-cells, and a subpopulation of memory T cells.[89] In accord with the activities of the other selectins, P-selectin can mediate the tethering and rolling of leukocytes in the parallel plate flow chamber.[10,90] A variety of *in vivo* experiments, employing knockout mice and antagonists, establish that P-selectin is responsible for the earliest rolling behaviour of neutrophils in an inflammatory response.[27,54,79] Subsequently, L-selectin and E-selectin participate and synergize with P-selectin.[30,31,54]

The functions of platelet P-selectin are beginning to emerge. P-selectin on activated platelets helps to recruit leukocytes into thrombi.[91] Platelet binding to monocytes, via P-selectin, leads to the induction of tissue factor, which initiates coagulation.[92] The mild deficiencies of haemostasis observed in mice null for P-selectin may reflect the loss of this signalling response.[93] Activated platelets can mediate the rolling of lymphocytes on lymph node HEV.[94] The platelet performs a bridging function using its P-selectin molecules to bind to the lymphocyte on one side and HEV-associated ligands on the other side. More work is required to reveal the physiologic or pathophysiologic significance of this novel homing mechanism, in which L-selectin ligands appear to be co-opted by P-selectin. Platelet binding to certain tumour cells can occur through P-selectin.[95,96] This interaction could contribute to the formation of tumour/platelet emboli in the blood, which are thought to be important in the metastases of tumour cells.

Analysis of the *O*-glycans of PSGL-1, the best characterized myeloid ligand for P-selectin, reveals the presence of a novel core-2 type based structure, containing a trifucosylated, polylactosamine structure that is capped by sLex.[97] Whether this entire structure (11 saccharide units) or only part of it constitutes the optimal recognition epitope for P-selectin remains to be determined. P-selectin binding to PSGL-1 requires not only carbohydrate-based determinants but also an amino-terminal peptide with 1–3 tyrosine sulfation modifications.[98,99] The available evidence strongly argues for a two-site model for ligand recognition: a calcium-dependent lectin site which recognizes sLex-like structures and a site for anionic peptides, which stabilizes the first interaction.[65,100] An attractive notion is that the second site is within the EGF domain, which would explain the considerable influence this domain exerts in ligand specificity and the particularly high evolutionary conservation of this domain in P-selectin.[7,101] The existence of a second binding site might explain the ability of a variety of anionic carbohydrates to be recognized by P-selectin and to block its adhesive function.[21] It is noteworthy that the major ligands for both L-selectin and P-selectin require sulphation,[65] on tyrosines in some cases and on carbohydrates in other cases.

Purification

Chemical quantities of native P-selectin (membrane form) can be purified from platelets by immunoaffinity chromatography.[87] Most studies employ fibroblast transfectants or soluble recombinant forms of P-selectin.

Antibodies

Polyclonal antibodies and mAbs against P-selectin have been widely used to study this receptor. See above for sources of antibody listings.

Genes

cDNAs for P-selectin have been sequenced from human, sheep, cow, pig, rabbit, rat, and mouse.

Mutant phenotype/disease states

There are now several studies in which P-selectin-null mice have been used to investigate the recruitment of mononuclear cells during later stages of inflammatory responses. The advantage of using null mice over the administration of selectin antagonists is that one does not have to sustain the inhibitor level over extended periods. This issue is particularly pertinent when the antagonist is a small molecular weight sugar, which typically would have a short half-life in the blood. An early indication that P-selectin is relevant to chronic responses was the finding that this receptor is expressed on the cell surface of venules in rheumatoid arthritic synovium and can support the *in vitro* attachment of monocytes.[102] P-selectin-null mice demonstrate a significant reduction in late monocyte entry into inflamed peritoneum and T-cell recruitment into DTH sites of the skin.[103,104] Furthermore, the absence of P-selectin delays the formation of fatty streaks in mice that are prone to develop atherosclerotic lesions.[111]

Structure

The E-selectin crystal model has guided mutagenesis and chemical modification studies of P-selectin. Mutations that affect sugar binding and specificity have been described.[4,105,106] One of the striking observations of the E-selectin crystal structure is the very limited interaction between the lectin and EGF domains.[33] In view of the evidence that the EGF domain of P-selectin strongly influences ligand-binding specificity,[7] this observation has been taken as further evidence that the EGF module does not just serve a structural support role for the lectin domain but may interact directly with ligands (e.g. recognition of tyrosine-sulphated region in PSGL-1).

Web site

A comprehensive web site devoted to the selectins, and their ligands has been initiated by Professor Bruce Macher, and his students at California State University/San Francisco. The URL is http://lewis.sfsu.edu/selectins.html.

References

1. Crocker, P. R. and Feizi, T. (1996). *Curr. Opin. Struct. Biol.*, **6**, 679–91.
2. Bevilacqua, M. P. and Nelson, R. M. (1993). *J. Clin. Invest.*, **91**, 379–87.
3. McEver, R. P. (1994). *Curr. Opin. Immunol.*, **6**, 75–84.
4. Lasky, L. A. (1995). *Ann. Rev. Biochem.*, **64**, 113–39.
5. Rosen, S. D. and Bertozzi, C. R. (1994). *Curr. Opin. Cell Biol.*, **6**, 663–73.
6. Tedder, T. F., Steeber, D. A., Chen, A., and Engel, P. (1995). *FASEB J.*, **9**, 866–73.
7. Kansas, G. S. (1996). *Blood*, **88**, 3259–87.

8. Drickamer, K. (1988). *J. Biol. Chem.*, **263**, 9557–60.
9. Bevilacqua, M. P., Spengeling, S., Gimbrone, M. A., Jr., and Seed, B. (1989). *Science*, **243**, 1160–5.
10. Springer, T. A. (1995). *Ann. Rev. Physiol.*, **57**, 827–72.
11. Imhof, B. A. and Dunon, D. (1995). *Adv. Immunol.*, **58**, 345–416.
12. Butcher, E. C. and Picker, L. J. (1996). *Science*, **272**, 60–6.
13. Read, M. A., Whitley, M. Z., Williams, A. J., and Collins, T. (1994). *J. Exp. Med.*, **179**, 503–12.
14. Yoshida, M., Westlin, W. F., Wang, N., Ingber, D. E., Rosenzweig, A., Resnick, N., and Gimbrone, M. A., Jr. (1996). *J. Cell Biol.*, **133**, 445–55.
15. Foxall, C., Watson, S. R., Dowbenko, D., Fennie, C., Lasky, L. A., Kiso, M., *et al.* (1992). *J. Cell Biol.*, **117**, 895–902.
16. Lowe, J. B., Stoolman, L. M., Nair, R. P., Larsen, R. D., Berhend, T. L., and Marks, R. M. (1990). *Cell*, **63**, 475–84.
17. Imai, Y., Lasky, L. A., and Rosen, S. D. (1992). *Glycobiology*, **2**, 373–81.
18. Feizi, T. (1993). *Curr. Opin. Struct. Biol.*, **3**, 701–10.
19. Poppe, L., Brown, G. S., Philo, J. S., Nikrad, P. V., and Shah, B. H. (1997). *J. Am. Chem. Soc.*, **119**, 1727–36.
20. Patel, T. P., Goelz, S. E., Lobb, R. R., and Parekh, R. B. (1994). *Biochemistry*, **33**, 14815–24.
21. Varki, A. (1994). *Proc. Natl Acad. Sci. USA*, **91**, 7390–7.
22. Sako, D. *et al.* (1993). *Cell*, **75**, 1179–86.
23. Steegmaler, M., Levinovitz, A., Isenmann, S., Borges, E., Lenter, M., Kocher, H. P., *et al.* (1995). *Nature*, **373**, 615–20.
24. Picker, L. J., Kishimoto, T. K., Smith, C. W., Warnock, R. A., and Butcher, E. C. (1991). *Nature*, **349**, 796–9.
25. Lawrence, M. B. and Springer, T. A. (1993). *J. Immunol.*, **151**, 6338–46.
26. Mulligan, M. S., Varani, J., Dame, M. K., Lane, C. L., Smith, C. W., Anderson, D. C., and Ward, P. A. (1991). *J. Clin. Invest.* **88**, 1396–1406.
27. Albelda, S. M., Smith, C. W., and Ward, P. A. (1994). *FASEB J.*, **8**, 504–12.
28. Walz, G., Aruffo, A., Kolanus, W., Bevilacqua, M., and Seed, B. (1990). *Science*, **250**, 1132–5.
29. Labow, M. A. *et al.* (1994). *Immunity*, **1**, 709–20.
30. Frenette, P. S., Mayadas, T. N., Rayburn, H., Hynes, R. O., and Wagner, D. D. (1996). *Cell*, **84**, 563–74.
31. Bullard, D. C., Kunkel, E. J., Kubo, H., Hicks, M. J., Lorenzo, I., Doyle, N. A., *et al.* (1996). *J. Exp. Med.*, **183**, 2329–36.
32. Etzioni, A., Frydman, M., Pollack, S., Avidor, I., Phillips, M. L., Paulson, J. C., and Gershoni, B. R. (1992). *New Engl. J. Med.*, **327**, 1789–92.
33. Graves, B. J. *et al.* (1994). *Nature*, **367**, 532–8.
34. Erbe, D. V. *et al.* (1992). *J. Cell Biol.*, **119**, 215–27.
35. Blanck, O., Iobst, S. T., Gabel, C., and Drickamer, K. (1996). *J. Biol. Chem.*, **271**, 7289–92.
36. Ng, K. K. S. and Weis, W. I. (1997). *Biochemistry*, **36**, 979–88.
37. Gallatin, W., Weissman, I., and Butcher, E. (1983). *Nature*, **304**, 30–4.
38. Lewinsohn, D. M., Bargatze, R. F., and Butcher, E. C. (1987). *J. Immunol.*, **138**, 4313–21.
39. Tedder, T. F., Penta, A. C., Levine, H. B., and Freedman, A. S. (1990). *J. Immunol.*, **144**, 532–40.
40. Pavalko, F. M., Walker, D. M., Graham, L., Goheen, M., Doerschuk, C. M., and Kansas, G. S. (1995). *J. Cell. Biol.*, **129**, 1155–64.
41. Siegelman, M., Bond, M. W., Gallatin, W. M., St. John, T., Smith, H. T., Fried, V. A., and Weissman, I. L. (1986). *Science*, **231**, 823–9.
42. Zöllner, O., Lenter, M. C., Blanks, J. E., Borges, E., Steegmaier, M., Zerwes, H. G., and Vestweber, D. (1997). *J. Cell Biol.*, **136**, 707–16.
43. Kishimoto, T. K., Jutila, M. A., Berg, E. L., and Butcher, E. C. (1989). *Science*, **245**, 1238–41.
44. Kahn, J., Ingraham, R. H., Shirley, F., Magaki, G. I., and Kishimoto, T. K. (1994). *J. Cell Biol.*, **125**, 461–70.
45. Schleiffenbaum, B., Spertini, O., and Tedder, T. F. (1992). *J. Cell Biol.*, **119**, 229–38.
46. Yang, X. -D., Karin, N., Tisch, R., Steinman, L., and McDevitt, H. O. (1993). *Proc. Natl Acad. Sci. USA*, **90**, 10494–8.
47. Walcheck, B., Moore, K. L., McEver, R. P., and Kishimoto, T. K. (1996). *J. Clin. Invest.*, **98**, 1081–7.
48. Alon, R., Fuhlbrigge, R. C., Finger, E. B., and Springer, T. A. (1996). *J. Cell. Biol.*, **135**, 849–65.
49. Michie, S. A., Streeter, P. R., Bolt, P. A., Butcher, E. C., and Picker, L. J. (1993). *Am. J. Path.*, **143**, 1688–98.
50. Lawrence, M. B., Berg, E. L., Butcher, E. C., and Springer, T. A. (1995). *Eur. J. Immunol.*, **25**, 1025–31.
51. Puri, K. D., Finger, E. B., and Springer, T. A. (1997). *J. Immunol.*, **158**, 405–13.
52. Ley, K., Tedder, T. F., and Kansas, G. S. (1993). *Blood*, **82**, 1632–8.
53. Finger, E. B., Puri, K. D., Alon, R., Lawrence, M. B., Von Andrian, U. H., and Springer, T. A. (1996). *Nature*, **379**, 266–9.
54. Ley, K. and Tedder, T. F. (1995). *J. Immunol.*, **155**, 525–8.
55. Bargatze, R. F., Jutila, M. A., and Butcher, E. B. (1995). *Immunity*, **3**, 99–108.
56. Imai, Y., Lasky, L. A., and Rosen, S. D. (1993). *Nature*, **361**, 555–7.
57. Rosen, S. D., Singer, M. S., Yednock, T. A., and Stoolman, L. M. (1985). *Science*, **228**, 1005–7.
58. Maly, P. *et al.* (1996). *Cell*, **86**, 643–53.
59. Hemmerich, S., Leffler, H., and Rosen, S. D. (1995). *J. Biol. Chem.*, **270**, 12035–47.
60. Scudder, P. R., Shailubhai, K., Duffin, K. L., Streeter, P. R., and Jacob, G. S. (1994). *Glycobiology*, **4**, 929–33.
61. Saunders, W. J., Katsumoto, T. R., Bertozzi, C. R., Rosen, S. D., and Kiessling, L. L. (1996). *Biochemistry*, **35**, 14862–7.
62. Tsuboi, S., Isogai, Y., Hada, N., King, J. K., Hindsgaul, O., and Fukuda, M. (1996). *J. Biol. Chem.*, **271**, 27213–6.
63. Koenig, A., Jain, R., Vig, R., Norgard-Sumnicht, K. E., Matta, K. L., and Varki, A. (1997). *Glycobiology*, **7**, 79–93.
64. Malhotra, R., Taylor, N., and Bird, M. I. (1996). *Biochem. J.*, **314**, 297–303.
65. Rosen, S. D. and Bertozzi, C. B. (1996). *Curr. Biol.*, **6**, 261–4.
66. Simon, S. I., Burns, A. R., Taylor, A. D., Gopalan, P. K., Lynam, E. B., Sklar, L. A., and Smith, C. W. (1995). *J. Immunol.*, **155**, 1502–14.
67. Hwang, S. T., Singer, M. S., Giblin, P. A., Yednock, T. A., Bacon, K. B., Simon, S. I., and Rosen, S. D. (1996). *J. Exp. Med.*, **184**, 1343–8.
68. Harris, H. and Miyasaka, M. (1995). *Immunology*, **84**, 47–54.
69. Waddell, T. K., Fialkow, L., Chan, C. K., Kishimoto, T. K., and Downey, G. P. (1995). *J. Biol. Chem.*, **270**, 15403–11.
70. Brenner, B. *et al.* (1996). *Proc. Natl Acad. Sci. USA*, **93**, 15376–81.
71. Lasky, L. A., Singer, M. S., Yednock, T. A., Dowbenko, D., Fennie, C., Rodriguez, H., *et al.* (1989). *Cell*, **56**, 1045–55.
72. Siegelman, M. H., Cheng, I. C., Weissman, I. L., and Wakeland, E. K. (1990). *Cell*, **61**, 611–22.
73. Kishimoto, T. K., Jutila, M. A., and Butcher, E. C. (1990). *Proc. Natl Acad. Sci. USA*, **87**, 2244–8.

74. Spertini, O., Kansas, G. S., Reimann, K. A., Mackay, C. R., and Tedder, T. F. (1991). *J. Immunol.*, **147**, 942–9.
75. Jutila, M. A., Watts, G., Walcheck, B., and Kansas, G. S. (1992). *J. Exp. Med.*, **175**, 1565–73.
76. Arbonés, M. L., Ord, D. C., Ley, K., Ratech, H., Maynard-Curry, C., Otten, G., *et al.* (1994). *Immunity*, **1**, 247–60.
77. Tedder, T. F., Steeber, D. A., and Pizcueta, P. (1995). *J. Exp. Med.*, **181**, 2259–2264.
78. Bajorath, J. and Aruffo, A. (1995). *Biochem. Biophys. Res. Commun.*, **216**, 1018–23.
79. Frenette, P. S. and Wagner, D. D. (1996). *New Engl. J. Med.*, **335**, 43–5.
80. McEver, R. P., Beckstead, J. H., Moore, K. L., Marshall-Carlson, L., and Bainton, D. F. (1989). *J. Clin. Invest.*, **84**, 92–9.
81. Green, S. A., Setiadi, H., McEver, R. P., and Kelly, R. B. (1994). *J. Cell Biol.*, **124**, 435–48.
82. Koedam, J. A., Cramer, E. M., Briend, E., Furie, B., Furie, B. C., and Wagner, D. D. (1992). *J. Cell Biol.*, **116**, 617–25.
83. Johnston, G. I., Cook, R. G., and McEver, R. P. (1989). *Cell*, **56**, 1033–44.
84. Weller, A., Isenmann, S., and Vestweber, D. (1992). *J. Biol. Chem.*, **267**, 15176–83.
85. Crovello, C. S., Furie, B. C., and Furie, B. (1993). *J. Biol. Chem.*, **268**, 14590–3.
86. Crovello, C. S., Furie, B. C., and Furie, B. (1995). *Cell*, **82**, 279–86.
87. Ushiyama, S., Laue, T. M., Moore, K. L., Erickson, H. P., and McEver, R. P. (1993). *J. Biol. Chem.*, **268**, 15229–37.
88. Drickamer, K. (1995). *Nature Struct. Biol.*, **2**, 437–9.
89. McEver, R. P., Moore, K. L., and Cummings, R. D. (1995). *J. Biol. Chem.*, **270**, 11025–8.
90. Lawrence, M. B. and Springer, T. A. (1991). *Cell*, **65**, 859–73.
91. Palabrica, T., Lobb, R., Furie, B. C., Aronovitz, M., Benjamin, C., Hsu, Y. M., Sajer, S. A., and Furie, B. (1992). *Nature*, **359**, 848–51.
92. Celi, A., Pellegrini, G., Lorenzet, R., De Blasi, A., Ready, N., Furie, B. C., and Furie, B. (1994). *Proc. Natl Acad. Sci. USA.*, **91**, 8767–71.
93. Subramaniam, M., Frenette, P. S., Saffaripour, S., Johnson, R. C., Hynes, R. O., and Wagner, D. D. (1996). *Blood*, **87**, 1238–42.
94. Diacovo, T. G., Puri, K. D., Warnock, R. A., Springer, T. A., and von Andrian, U. H. (1996). *Science*, **273**, 252–5.
95. Stone, J. P. and Wagner, D. D. (1993). *J. Clin. Invest.*, **92**, 804–13.
96. Mannori, G., Crottet, P., Cecconi, O., Hanasaki, K., Aruffo, A., Nelson, R. M., *et al.* (1995). *Cancer Res.*, **55**, 4425–31.
97. Wilkins, P. P., McEver, R. P., and Cummings, R. D. (1996). *J. Biol. Chem.*, **271**, 18732–42.
98. Sako, D., Comess, K. M., Barone, K. M., Camphausen, R. T., Cumming, D. A., and Shaw, G. D. (1995). *Cell*, **83**, 323–31.
99. Pouyani, T. and Seed, B. (1995). *Cell*, **83**, 333–43.
100. Varki, A. (1997). *J. Clin. Invest.*, **99**, 158–62.
101. Kansas, G. S., Saunders, K. B., Ley, K., Zakrzewicz, A., Gibson, R. M., Furie, B. C., *et al.* (1994). *J. Cell Biol.*, **124**, 609–18.
102. Grober, J. S., Bowen, B. L., Ebling, H., Athey, B., Thompson, C. B., Fox, D. A., and Stoolman, L. M. (1993). *J. Clin. Invest.*, **91**, 2609–19.
103. Johnson, R. C., Mayadas, T. N., Frenette, P. S., Mebius, R. E., Subramaniam, M., Lacasce, A., *et al.* (1995). *Blood*, **86**, 1106–14.
104. Subramaniam, M., Saffaripour, S., Watson, S. R., Mayadas, T. N., Hynes, R. O., and Wagner, D. D. (1995). *J. Exp. Med.*, **181**, 2277–82.
105. Erbe, D. V., Watson, S. R., Presta, L. G., Wolitzky, B. A., Foxall, C., Brandley, B. K., and Lasky, L. A. (1993). *J. Cell Biol.*, **120**, 1227–35.
106. Bajorath, J., Hollenbaugh, D., King, G., Harte, W., Eustice, D. C., Darveau, R. P., and Aruffo, A. (1994). *Biochemistry*, **33**, 1332–9.
107. Mitsuoka, C., Sawada-Kasugai, M., Ando-Furui, K., Izawa, M., Nakanishi, H., Ishida, H., Kiso, M., and Kannagi, R. (1998). *J. Biol. Chem.*, **273**, 11225–33.
108. Giblin, P. A., Hwang, S. T., Katsumoto, T. R., and Rosen S. D. (1997). *J. Immunol.*, **159**, 3498–507.
109. Steeber, D. A., Engel, P., Miller, A. S., Sheetz, M. P., and Tedder, T. (1997). *J. Immunol.*, **159**, 952–63.
110. Kahn, J., Walcheck, B., Migaki, G. I., Jutila, M. A., and Kishimoto, T. K. (1998). *Cell*, **92**, 809–18.
111. Johnson, R. C., Chapman, S. M., Oong, Z. M., Ordovos, J. M., Mayados, T. N., Herz, J., Hynes, R. O., Schaefer, E. J., and Wagner, D. L. (1997). *J. Clin. Invest.*, **99**, 1037–43.
112. Sassetti, C., Tangemann, K., Singer, M. S., Kershaw, D. B., and Rosen, S. D. (1998). *J. Exp. Med.*, **187**, 1965–75.

Steven Rosen
Department of Anatomy, and Program in Immunology, University of California, San Francisco, CA 94143–0452, USA

Selectin ligands

Overview

The selectins are a family of three carbohydrate-binding cell adhesion molecules (L-, E-, and P-selectin) which mediate the docking of leukocytes to the blood vessel wall and the rolling of these cells along the endothelial cell surface. These adhesion phenomena initiate the entry of leukocytes into sites of inflammation as well as the homing of lymphocytes into lymphatic tissue. Blocking selectin function with antibodies or oligosaccharides has proved to be beneficial in various animal models of inflammation. This has raised much interest in the identification of the physiological ligands of the selectins. Carbohydrate structures such as the tetrasaccharide sialyl Lewis^x (sLex) NeuAcα2→3Galβ1→4(Fucα1 →3)GlcNAc or stereoisomers of this compound are known to bind to all three selectins. Several glycoprotein ligands of the selectins have been identified (Fig. 1). Only four of them (PSGL-1, ESL-1, GlyCAM-1, and CD34) have been defined by direct affinity isolation from cell detergent extracts with selectin affinity probes and have been characterized on the sequence level. These four ligands will be described in detail here. The vascular addressin MAdCAM-1, of which subpopulation can bind to L-selectin, is described in a separate chapter.

Other glycoprotein ligands are, for L-selectin: Sgp200 from mouse lymph nodes, calcium-dependent heparin-like ligands on non-lymphoid endothelial cells, several sulphated glycoproteins in rat lymph nodes; for P-selectin: CD24 and its mouse homologue heat stable antigen, a 160 kDa glycoprotein on mouse and human neutrophils; for E-selectin: a 250 kDa ligand on bovine γ/δ T cells, members of the NCA family, a subpopulation of sLex-carrying β2 integrins, sLex-carrying LAMP-1 lysosomal protein on the surface of carcinoma cells, not yet characterized glycoproteins on human T-lymphocytes carrying the sLex-like carbohydrate epitope defined by the mAb HECA452. These glycoprotein ligands will not be described here but have been discussed in recent reviews.[1,2] Furthermore, L-selectin has been described as a carbohydrate-presenting ligand on human neutrophils (but not on mouse neutrophils) for E-selectin,[3] which can even be affinity isolated as a major glycoprotein ligand from neutrophil detergent extracts. Since L-selectin is discussed in this book in the selectin chapter, it will not be further decribed here.

While several of the glycoprotein ligands described in this chapter bind very selectively and can be affinity isolated with a selectin affinity probe from cellular detergent extracts in single-step purification protocols, at present PSGL-1 is still the only ligand which has been shown to be important for leukocyte recruitment *in vivo*.

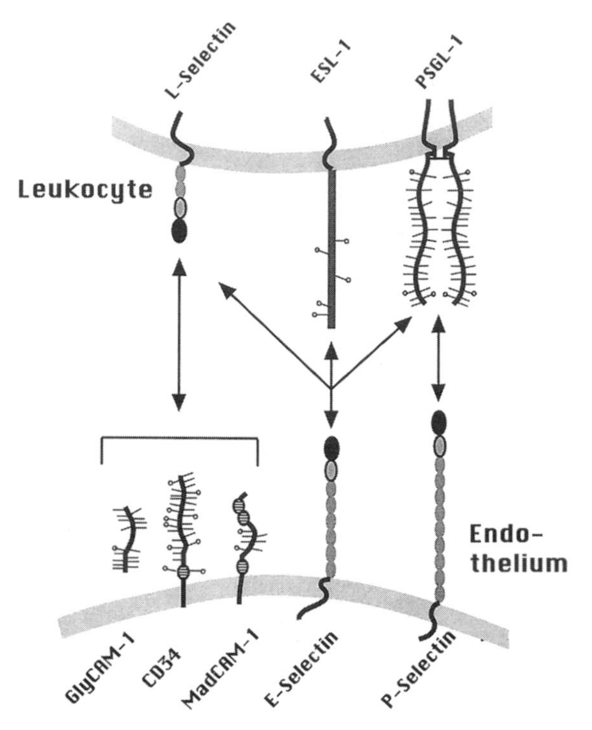

Figure 1. Major glycoprotein ligands of the selectins. The ligands GlyCAM-1, CD34, ESL-1, and PSGL-1 can be isolated directly and purified from cell-detergent extracts using the respective selectin as affinity probe. MAdCAM-1 functions as a vascular addressin, mediating lymphocyte recirculation by binding to the lymphocyte integrin $\alpha_{4\beta7}$. A subpopulation of MAdCAM-1 molecules carries post-translational modifications, which allow binding of L-selectin. Antibodies against MAdCAM-1 block lymphocyte migration into mucosa associated lymphatic tissue (probably via blocking of its function as an integrin ligand). Antibodies against mouse PSGL-1 block lymphocyte migration into inflamed skin[17] as well as neutrophil migration into inflamed peritoneum.[18]

P-selectin-glycoprotein ligand-1 (PSGL-1)

PSGL-1 was originally identified by expression cloning as a ligand of human P-selectin[4] and was also identified by affinity isolation from human myeloid cells with a human P-selectin probe[5] and from mouse myeloid cells with a mouse P-selectin–Ig fusion protein.[6] PSGL-1 also binds to L-selectin[7] and to E-selectin.[6,8]

Homologous proteins

PSGL-1 is a sialomucin, mouse and human homologues have been cloned.

Protein properties

PSGL-1 is a 110 kDa glycoprotein which forms a disulphide linked homodimer (220 kDa). The cDNA of human PSGL-1 codes for an open reading frame of 402 amino acids, with an 18 amino acid signal sequence, followed by a 23 amino acid propeptide. The mature form of PSGL-1 starts at amino acid 42 and has a single putative transmembrane region between amino acids 309 and 333. PSGL-1 classifies as a sialomucin, which are glycoproteins that carry large clusters of *O*-linked carbohydrate side chains, that are rich in sialic acid. The structures of Ser/Thr linked *O*-glycans on human PSGL-1 were analysed recently.[9] Unexpectedly, only a minority of the O-glycans were $\alpha(1,3)$-fucosylated and occurred as two major species containing the sLe^x Antigen (Fig. 2). Beside carbohydrate side chains a peptide region at the N terminus of mature PSGL-1 is involved in the binding to P-selectin. Cleavage of the first 10 amino acids of mature human PSGL-1 abolishes binding to P-selectin,[10] while adhesion-blocking antibodies recognize an epitope either within the first 14 amino acids or the first 19 amino acids of the mature forms of human[11] or mouse PSGL-1,[12] respectively. Sulfation of the tyrosine residues in this peptide segment is known to be essential for the binding to P-selectin,[13,14] but not for the binding to E-selectin.[8] PSGL-1 has been shown to bind to all three selectins.

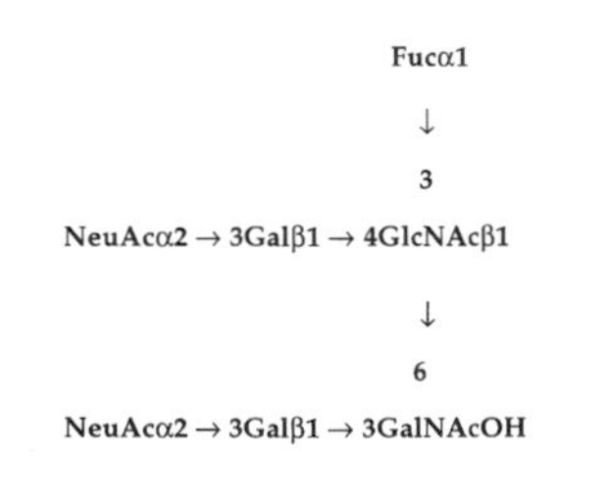

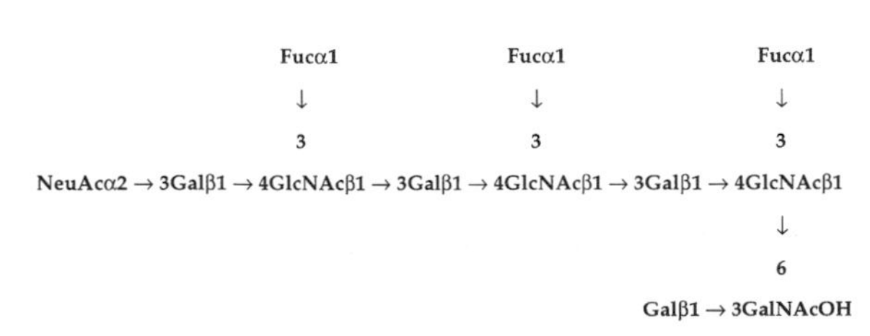

Figure 2. Two major species of $\alpha(1,3)$-fucosylated *O*-glycans of human PSGL-1 containing the sialyl Lewis antigen.[9] The two depicted carbohydrate species represent only a minority of the *O*-glycans that were removed from human PSGL-1. Interestingly, CD43 analysed from the same cellular source lacked these fucosylated glycans. (For details see Wilkins *et al.*[9])

PSGL-1 is the major P-selectin ligand on human neutrophils and on several subsets of activated human lymphocytes, as was demonstrated in *in vitro* cell adhesion assays.[15] Rolling of human leukocytes perfused into rat postcapillary venules was demonstrated to be blocked by a mAb against human PSGL-1.[16] Furthermore, mouse PSGL-1 has been demonstrated to be important for the extravasation of mouse T helper 1 lymphocytes into inflamed skin[17] and of neutrophils into inflamed peritoneum of the mouse.[18]

PSGL-1 is found on myeloid cells, most peripheral lymphocytes, on dendritic cells, and also on a few non haematopoietic cells such as epithelium of the Fallopian tube as well as on microvascular endothelium in some pathologic tissues.[19] On myeloid cells, PSGL-1 is found on the tips of microvillous processes.[15]

Purification7

Using a soluble form of human P-selectin immobilized on an Emphaze matrix, PSGL-1 can be affinity purified from membranes of human neutrophils.[20]

Activities

PSGL-1 binds to all three selectins. The affinity of PSGL-1 for immobilized human P-selectin is 50-fold stronger than for human E-selectin.[20]

Antibodies

Monoclonal antibody (mAb) PL-1 blocks human PSGL-1-mediated cell adhesion and binds to an epitope between amino acid 49 and 62 (amino acids 42 forms the N terminus of mature human PSGL-1).[11] The mAb PL-2 has no adhesion-blocking activity and binds to a region between amino acids 188 and 325.[11] The adhesion-blocking mAb 2PH1 against mouse PSGL-1 binds to an epitope between amino acids 42 and 60 (as for human PSGL-1, amino acid 42 forms the N terminus of mouse PSGL-1).[18]

Genes

Full length cDNAs for human[4] and mouse[12] PSGL-1 have been published.

Structure

No information about X-ray structure is available. The structures of Ser/Thr linked *O*-glycans on human PSGL-1 have been analysed.[9]

E-selectin ligand-1 (ESL-1)

ESL-1 was originally identified on mouse neutrophils and mouse myeloid cell lines by affinity isolation with a mouse E-selectin–Ig fusion protein.[21,22] Antibodies against ESL-1 can partially inhibit the binding of mouse myeloid cells to E-selectin.

Homologous proteins

The following strongly homologous proteins were found in different species by unrelated approaches: chicken CFR,[23] rat MG160,[24] a human homologue of rat MG160.[25]

Protein properties

ESL-1 is a glycoprotein of 150 kDa apparent molecular mass, consisting of a 1114 amino acid extracellular domain, a 21 amino acid transmembrane region, and a short 13 amino acid cytoplasmic tail (relative molecular mass of 131 kDa). It was identified on mouse myeloid cells by affinity isolation with an antibody-like fusion protein containing the lectin, EGF, and the first two short consensus repeats of mouse E-selectin fused to the Fc part of human IgG1. In contrast to sialomucin-like selectin ligands, which require *O*-linked carbohydrates for selectin binding, ESL-1 requires *N*-linked carbohydrates for binding. ESL-1 contains five potential *N*-glycosylation sites and carries hardly any *O*-linked carbohydrate side chains. Modification with sialic acid and α (1,3)-linked fucose is essential for binding to E-selectin. ESL-1 does not bind to P-selectin.

Affinity purified, polyclonal antibodies against ESL-1 can partially block the binding of myeloid cells to immobilized E-selectin–Ig and to activated, E-selectin-expressing endothelial cells.[22] Subcellular localization by immunogold scanning electron microscopy has revealed that the cell surface located ESL-1 on lymphoma cells is clearly enriched on the surface of microvilli[26] (Fig. 3). In contrast to PSGL-1 and L-selectin, it is not concentrated on the tips of these microvilli. In addition to its expression at the cell surface, ESL-1 is strongly expressed in the Golgi, as was analysed by indirect immunofluorescence of permeabilized leukocytes.[26]

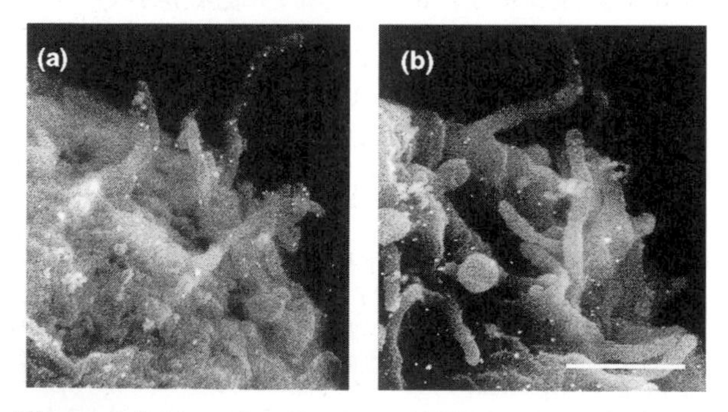

Figure 3. Immunolocalization of ESL-1 on microvilli of K46 lymphoma cells by scanning electron microscopy. Immunogold staining (the bright white dots) is shown for affinity purified anti-ESL-1 antibodies (A) and for a monoclonal antibody against B220 (B). (For details see Steegmaier *et al.*[26]) (printed by permission of the Company of Biologists Ltd).

Expression of the ESL-1 glycoprotein is not restricted to myeloid or lymphoid cells, but it was also found on more sessile cells such as fibroblasts, colon carcinomas, endotheliomas, and various tertocarcinomas and is probably very broadly distributed. However, only on myeloid cells and certain types of activated lymphoid cells is ESL-1 glycosylated in a functionally active form which binds to E-selectin. ESL-1 is a preferential target for certain α (1,3)-fucosyltransferases (such as FTVII and FTIV) in CHO cells.[27]

Two strongly homologous proteins were found in chicken and rat by unrelated approaches: a chicken protein called CFR was found by affinity isolation with an FGF-affinity matrix. It binds to FGF-1, -2, -4, and -7, but has no structural similarity to the classical types of FGF receptors. Potential signals which could be transmitted by this putative receptor have not yet been described. The amino acid sequence of mouse ESL-1 is 94 per cent identical to CFR except for a 70 amino acid domain at the N terminus, replacing the first 35 amino acids of the N terminus of CFR. The biological function of CFR has not yet been revealed. A rat protein called MG160 was identified with the help of a mAb which stained the Golgi of rat brain neurones.[28] Its amino acid sequence is 98 per cent identical to that of mouse ESL-1.[24] MG160 has been suggested to mediate not yet further defined functions in the Golgi apparatus.

Purification

Mouse ESL-1 can be affinity purified from mouse myeloid cells by a single step purification procedure using immobilized, recombinant E-selectin–Ig as affinity probe.[21,22]

Activities

ESL-1 from myeloid cells binds specifically to E-selectin, but not to P-selectin or L-selectin.

Antibodies

Polyclonal rabbit antibodies have been raised against a mouse ESL–IgG fusion protein containing the complete extracellular domain of ESL–1.[22] The mouse mAb 10A8 recognizes rat MG160,[28] but not the mouse homologue.

Genes

Full length cDNAs have been cloned for mouse ESL-1,[22] chicken CFR,[23] rat MG160,[24] and the human homologue of rat MG160.[25]

GlyCAM-1

GlyCAM-1 (glycosylation-dependent cell adhesion molecule) is a mouse glycoprotein of 50 kDa apparent molecular mass, which is secreted by high endothelial cells in lymph nodes. It was identified by affinity isolation with an antibody like fusion protein of L-selectin.[29,30]

Synonymous names

Originally named Sgp50.

Homologous proteins

GlyCAM-1 is a sialomucin; mouse and rat homologues have been cloned.

Protein properties

GlyCAM-1 is secreted by endothelial cells of high endothelial venules in mouse lymph nodes. Although its sequence comprises 151 amino acids, GlyCAM-1 separates in SDS–PAGE at an apparent molecular mass of 50 kDa. GlyCAM-1 is a sialomucin carrying two large clusters of *O*-linked carbohydrate side chains, interrupted by a non-glycosylated short peptide region. This region allowed the production of polyclonal antibodies against glycosylation-independent epitopes of GlyCAM-1. Sialylation[29] and sulphation (on carbohydrates)[31] of GlyCAM-1 are essential for binding to L-selectin, while fucosylation is suggested from inhibitor studies. Analysis of the capping structures of GlyCAM-1 revealed 6′-sulphated sialyl Lewis X (Siaα2→3(SO_4–6)Galβ1→4(Fucα1→3) GlcNAc) as a major capping structure including all three critical elements for L-selectin-binding.[32]

Since GlyCAM-1 is a secreted protein, it is probably not involved in adhesion mechanisms. However, binding of GlyCAM-1 to L-selectin on lymphocytes can trigger the activation of β2 integrins, arguing for a possible role of GlyCAM-1 in signalling via L-selectin.[33]

The L-selectin-binding form of GlyCAM-1 is specifically expressed by endothelial cells of high endothelial venules in lymph nodes and its expression is regulated by afferent lymphatic flow.[34] A non-binding form of the glycoprotein is secreted by mammary epithelial cells.[35]

Purification

GlyCAM-1 is found at 1.3–1.6 μg/ml in mouse serum, from which it can be efficiently purified.[36]

Activities

GlyCAM-1 binds to all three selectins.[36]

Antibodies

Rabbit polyclonal antibodies (called CAM02) have been raised against an internal peptide of GlyCAM-1 (CKEPSIFREELISKD).[37]

Genes

The cDNA sequences of mouse[30] and rat[38] GlyCAM-1 and the genomic sequence of mouse GlyCAM-1 were published.[39]

Mutant phenotype/disease states

Mice deficient for a functional gene for GlyCAM-1 have been generated at Genentech Inc.

CD34

CD34 is a 90 kDa sialomucin which is expressed on endothelial cells throughout the organism and on certain haematopoietic precursor cells. A specially glycosylated form of CD34 on endothelial cells in high endothelial venules of lymph nodes can be selectively isolated with antibody like fusion proteins of L-selectin.[29,40]

Synonymous names

Originally named as Sgp90.

Homologous proteins

CD34 is a sialomucin; mouse and human homologues have been cloned.

Protein properties

With the help of an L-selectin–Ig fusion protein, a 90 kDa sulphated glycoprotein was affinity isolated from detergent extracts of cultured, $^{35}SO_4$-labelled mouse lymph nodes.[29] Cloning of this protein allowed it to be identified as the mouse homologue of human CD34. Like GlyCAM-1, it needs to be sialylated and sulphated (and probably fucosylated) for binding to L-selectin. CD34 is expressed throughout the endothelial cells of the vasculature and on haematopoietic precursor cells,[41] but it only seems to be correctly glycosylated for L-selectin recognition in high endothelial venules of lymph nodes. Besides expression in the vascular system, CD34 is strongly expressed in a number of embryonic fibroblast cell lines and in brain.

CD34 is a cell surface glycoprotein of 105–120 kDa apparent molecular mass with a single transmembrane region. The cDNA predicts a 382 (mouse) or 373 (human) amino acid protein. The extracellular part of the protein contains a 140 amino acid N-terminal domain, 40 per cent of whose residues are serine or threonine, potential attachment sites for *O*-linked carbohydrates. Proximal to the membrane there is a cysteine-rich extracellular segment of about 70 amino acids. The intracellular part is 73 amino acids long and shows 90 per cent sequence identity between human and mouse, while the extracellular part of the human and mouse protein sequences only show 43 per cent amino acid identity.

The mAb MECA79, which blocks binding of lymphocytes to high endothelial venules of peripheral lymph nodes,[42] defines a sulphate-dependent carbohydrate epitope which is found on a panel of glycoproteins made by HEV endothelial cells. One of these proteins is CD34. MECA79 affinity purified glycoproteins from human tonsil, immobilized in laminar flow chambers, were found to support rolling of lymphocytes. CD34 was reported to be an important component of these antigens.[43] However, genetic disruption of CD34 expression in mice does not give an obvious loss of lymphocyte trafficking to lymph nodes.[44]

Purification

The lymph node derived, specifically glycosylated form of CD34 that is able to bind to L-selectin has been purified with L-selectin–Ig as an affinity probe.[40]

Activities

CD34 from lymphatic high endothelial venules binds to L-selectin and to E-selectin.

Antibodies

A rabbit antiserum against mouse CD34 has been raised against a recombinant form of this antigen.[40] Thirty-three mAbs against human CD34 were described on CD workshop V.[45]

Genes

The cDNAs for human[46] and mouse[47] CD34 have been published.

Mutant phenotype/disease states

Gene-deficient mice have been generated, they show haematopoietic defects but have no defects in lymphocyte trafficking.[44]

Podacalyxin-like protein (PLCP)

Podocalyxin, originally identified in rat, was first described as a sialoprotein present on the foot processes of podocytes in kidney glomeruli[48] and on vascular endothelium at some sites. Its structural organization is similar to that of CD34. It was shown recently that the podocalyxin-like protein is a 160 kDa glycoprotein present in HEV of human secondary lymphoid organs. PCLP is one of four major protein bands of the human peripheral node addressins (PNAd), which are glycoproteins purified from human tonsils by the anti-carbohydrate antibody MECA79[42] in immunoblots.[49] PCLP binds to L-selectin and supports rolling of lymphocytes under physiological flow conditions *in vitro*.[49]

Synonymous names

Podocalyxin was named in the rat, but not cloned.[48] The rabbit[50] and the human[51] homologues have been called podocalyxin-like proteins; they share all biochemical characteristics and the tissue distribution with podocalyxin, but antibodies do not cross-react with rat podocalyxin. A more distant relative has been identified in chicken and named thrombomucin.[52]

Purification

To obtain L-selectin-binding PCLP, the antigen can be purified from human tonsils by a two-step affinity isolation procedure[49] using as first affinity matrix the anti-PCLP mAb 3D3 coupled to protein A–Sepharose and as a second matrix the mAb MECA79 coupled to Sepharose. This antibody defines a carbohydrate epitope which the L-selectin-binding peripheral lymph node addressins (PNAd) have in common.[42]

Activities

PCLP purified from human tonsils can support the rolling of human lymphocytes under flow when coated onto one surface of a parallel plate laminar flow chamber.[49] In the kidney the strongly negatively charged podacalyxin is thought to function as an anti-adhesion molecule which maintains the filtration slits in the glomeruli between podocyte foot processes via charge repulsion.[48]

Antibodies

Mouse mAbs 5F7 and 4B3[53] have been raised against rabbit podocalyxin-like protein, and 3D3 (IgG), 2A4 (IgM), and 4F10 (IgM)[54] against human PCLP. The mouse mAb MEP-21 recognizes chicken thrombomucin.[55]

Genes

cDNAs have been cloned for rabbit[53] and human[54] PCLP and for the chicken homologue thrombomucin.[55]

Mutant phenotype/disease states

Gene-deficient mice are not yet available.

Structure

PCLP is similar in structure to CD34 in that both consist of a large N-terminal mucin-like domain followed by a disulfide-containing (and presumably globular) domain, a transmembrane domain and a cytoplasmic tail. No information about X-ray is structure available.

Note

MAdCAM-1, a selection ligand, is discussed in the chapter by Krissansen *et al*.

References

1. Vestweber, D. (1997). In *Advances in vascular biology* (ed. M. A. Vadas and J. Harlan), p. 225. Harwood Academic, Amsterdam.
2. Vestweber, D. (1996). *J. Cell Biochem.*, **61**, 585–91.
3. Zöllner, O., Lenter, M. C., Blanks, J. E., Borges, E., Steegmaier, M., Zerwes, H. -G., and Vestweber, D. (1997). *J. Cell Biol.*, **136**, 707–15.
4. Sako, D., Chang, X. J., Barone, K. M., Vachino, G., White, H. M., Shaw, G., *et al.* (1993). *Cell*, **75**, 1179–86.
5. Moore, K. L., Stults, N. L., Diaz, S., Smith, D. F., Cummings, R. D., Varki, A., and McEver, R. P. (1992). *J. Cell Biol.*, **118**, 445–56.
6. Lenter, M., Levinovitz, A., Isenmann, S., and Vestweber, D. (1994). *J. Cell Biol.*, **125**, 471–81.
7. Spertini, O., Cordey, A. -S., Monai, N., Giuffre, L., and Schapira, M. (1996). *J. Cell Biol.*, **135**, 523–31.
8. Li, F., Wilkens, P. P., Crawley, S. Weinstein, J. Cummings, R. D., and McEver, R. P. (1996). *J. Biol. Chem.*, **271**, 3255–64.
9. Wilkins, P. P., McEver, R. P., and Cummings, R. D. (1996). *J. Biol. Chem.*, **271**, 18732–42.
10. De Luca, M. Dunlop, L. C., Andrews, R. K., Flannnery, J. V., Jr., Ettling, R., Cumming, D. A., *et al.* (1995). *J. Biol. Chem.*, **270**, 26734–7.
11. Li, F., Erickson, H. P., James, J. A., Moore, K. L. Cummings, R. D., and McEver, R. P. (1996). *J. Biol. Chem.*, **271**, 6342–8.
12. Yang, J., Galipeau, J., Kozak, C. A., Furie, B. C., and Furie, B. (1996). *Blood*, **87**, 4176–86.
13. Pouyani, T. and Seed, B. (1995). *Cell*, **83**, 333–43.
14. Sako, D., Comess, K. M., Barone, K. M., Camphausen, R. T., Cummings, D. A., and Shaw, G. D. (1995). *Cell*, **83**, 323–31.
15. Moore, K. L., Patel, K. D., Bruehl, R. E., Li, L., Johnson, D. A., Lichenstein, H. S., *et al.* (1995). *J. Cell Biol.*, **128**, 661–7.
16. Norman, K. E., Moore, K. L., McEver, R. P., and Ley, K. (1995). *Blood*, **86**, 4417–21.
17. Borges, E. Tietz, W., Steegmaier, M., Moll, T., Hallmann, R., Hamann, A., and Vestweber, D. (1997). *J. Exp. Med.*, **185**, 573–8.
18. Borges, E., Eytner, R., Moll, T., Steegmaier, M., Matthew, A., Campell, P., *et al.* (1997). *Blood*, **90**, 1934–42.
19. Laszik, Z., Jansen, P. J., Cummings, R. D., Tedder, T. F., McEver, R. P., and Moore, K. L. (1996). *Blood*, **88**, 3010–21.
20. Moore, K. L., Eaton, S. F., Lyons, D. E., Lichenstein, H. S. Cummings, R. D., and McEver, R. P. (1994). *J. Biol. Chem.*, **269**, 23318–27.
21. Levinovitz, A., Mühlhoff, J., Isenmann, S., and Vestweber, D. (1993). *J. Cell Biol.*, **121**, 449–59.
22. Steegmaier, M., Levinovitz, A., Isenmann, S., Borges, E., Lenter, M., Kocher, H. P., *et al.* (1995). *Nature*, **373**, 615–20.
23. Burrus, L. W., Zuber, M. E., Lueddecke, B. A., and Olwin, B. B. (1992). *Mol. Cell. Biol.*, **12**, 5600–9.
24. Gonatas, J. O., Murelatos, Z., Stieber, A., Lane, W. S., Brosius, J., and Gonatas, N. K. (1995). *J. Cell. Sci*, **108**, 457–67.
25. Mourelatos, Z., Gonatas, J. O., Cinato, E., and Gonatas, N. K. (1996). *DNA Cell Biol.*, **15**, 1121–8.
26. Steegmaier, M., Borges, E., Berger, J., Schwarz, H., and Vestweber, D. J. (1997). *J. Cell Sci.*, **110**, 687–94.
27. Zöllner, O. and Vestweber, D. (1996). *J. Biol. Chem.*, **271**, 33002–8.
28. Gonatas, J. O., Mezitis, S. G., Stieber, A., Fleischer, B., and Gonatas, N. K. (1989). *J. Biol. Chem.*, **264**, 646–53.
29. Imai, Y., Singer, M. S., Fennie, C., Lasky, L. A., and Rosen, S. D. (1991). *J. Cell Biol.*, **113**, 1213–21.
30. Lasky, L. A., Singer, M. S., Dowbenko, D., Imai, Y., Henzel, W. J., Grimley, C. *et al.* (1992). *Cell*, **69**, 927–38.
31. Imai, Y., Lasky, L. A., and Rosen, S. D. (1993). *Nature*, **361**, 555–7.
32. Hemmerich, S. and Rosen, S. D. (1994). *Biochemistry*, **33**, 4830–5.
33. Hwang, S. T., Singer, M. S., Giblin, P. A., Yendock, T. A., Bacon, K. B. Simon, S. I., and Rosen, S. D. (1996). *J. Exp. Med.*, **184**, 1343–8.
34. Mebius, R. E., Dowbnk, D., Williams, A., Fennie, C., Lasky, L. A., and Watson, S. R. (1993). *J. Immunol.*, **151**, 6769–76.
35. Dowbenko, D., Kikuta, A., Fennie, C., Gillett, N., and Lasky, L. A. (1993). *J. Clin. Invest.*, **92**, 952–60.
36. Singer, M. S. and Rosen, S. D. (1996). *J. Immunol. Meth.* **196**, 153–61.
37. Hemmerich, S., Bertozzi, C. R., Leffler, H., and Rosen, S. D. (1994). *Biochemistry,* **33**, 4820–9.
38. Dowbenko, D., Watson, S. R., and Lasky, L. A. (1993). *J. Biol. Chem.*, **268**, 14399–403.
39. Dowbenko, D., Andalibi, A., Young, P. E., Lusis, A. J., and Lasky, L. A. (1993). *J. Biol. Chem.*, **268**, 4525–9.
40. Baumhueter, S., Singer, M. S., Henzel, W. Hemmerich, S., Renz, M., Rosen, S. D., and Lasky, L. A. (1993). *Science*, **262**, 436–8.
41. Baumhueter, S., Dybdal, N., Kyle, C., and Lasky, L. A. (1994). *Blood*, **84**, 2554–65.
42. Streeter, P. R., Rouse, B. T. N., and Butcher, E. C. (1988). *J. Cell Biol.*, **107**, 1853–62.
43. Puri, K. D., Finger, E. B., Gaudernack, G., and Springer, T. A. (1995). *J. Cell Biol.*, **131**, 261–70.
44. Cheng, J., Baumhueter, S., Cacalano, G., Carver-Moore, K., Thibodeaux, R., Thomas, H. E., *et al.* (1996). *Blood*, **87**, 479–90.
45. Gaudernack, G., and Egeland, T. (1995). In *Leukocyte typing V: white cell differentiation antigens* (ed. S. Schlossman), Oxford University Press, New York.
46. Simmons, D. L., Satterthwaite, A. B., Tenen, D. G., and Seed, B. (1992). *J. Immunol.*, **148**, 267–71.
47. Brown, J., Greaves, M. F., and Molgaard, H. V. (1991). *Int.. Immunol.*, **3**, 175–84.
48. Kerjaschki, D., Sharkey, D. J., and Farquhar, M. G. (1984). *J. Cell Biol.*, **98**, 1591–6.
49. Sassetti, C., Tangemann, K., Singer, M. S., Kershaw, D. B., and Rosen, S. D. (1998). *J. Exp. Med.*, **187**, 1965–75.
50. Kershaw, D. B., Thomas, P. E., Wharram, B. L., Goyal, M., Wiggins, J. E., Whiteside, C. I., and Wiggins, R. C. (1995). *J. Biol. Chem.*, **270**, 29439–46.
51. Kershaw, D. B., Beck, S. G., Wharram, B. L., Wiggins, J. E., Goyal, M., Thomas, P. E., and Wiggins, R. C. (1997). *J. Biol. Chem.*, **272**, 15708–14.
52. McMagny, K. M., Petterson, I., Rosi, F., Flamme, I., Shevchenko, A., Mann, M., and Graf, T. (1997). *J. Cell Biol.*, **138**, 1395–407.

Dietmar Vestweber
Institute of Cell Biology, ZMBE,
University of Münster, Technologiehof,
Mendelstrasse 11, D-48149 Münster,
Germany

Sevenless and bride of sevenless

Neuronal induction of the R7 photoreceptor cell in the compound eye of *Drosophila* depends on a signal from the neighbouring R8 neurone. The bride of sevenless (boss) ligand, a transmembrane protein with seven membrane-spanning segments, transmits this signal by activating the sevenless receptor tyrosine kinase on the surface of the R7 precursor cell.

Synonymous names

None.

Homologous proteins

Boss and sevenless have been cloned from *Drosophila melanogaster* and *D. virilis* and are highly conserved between these two species.[1–4] Outside of Diptera, no homologues of boss have yet been identified.

Sevenless is a member of the family of receptor tyrosine kinases.[2] Within this family only the mammalian receptor protein c-ros shares similarity with sevenless outside of the kinase domain. The two receptors exhibit low levels of conservation in their exceptionally large extracellular domains, particularly the organization of the eight fibronectin type III domains.[5]

Protein properties

The compound eye of *Drosophila* arises from a monolayered epithelium of undifferentiated cells. These cells organize themselves into about 800 ommatidia that comprise the adult eye through interactions between cells in the epithelium.[6] The best understood interaction is the induction of the R7 photoreceptor cell by the neighbouring R8 cell.[7] This inductive signal is mediated by two transmembrane proteins: the sevenless receptor tyrosine kinase and the boss transmembrane ligand.

The sevenless receptor is synthesized as a 280 kDa precursor[8,9] that is processed into a 220 kDa extracellular subunit and a 58 kDa subunit that contains the transmembrane and kinase domains (Fig. 1). Two 280 kDa and two 60 kDa subunits assemble by non-covalent interactions into the heterotetrameric sevenless receptor.[10]

The most striking feature of the Boss protein is the presence of seven membrane-spanning segments (Fig. 1). This transmembrane domain is preceded by a signal sequence and an extracellular domain of 498 amino acids. The cytoplasmic tail of boss is 115 amino acids 1. While the overall topology of boss is reminiscent of G protein-coupled receptors or the recently identified family of wnt receptors,[11] there is no sequence homology.

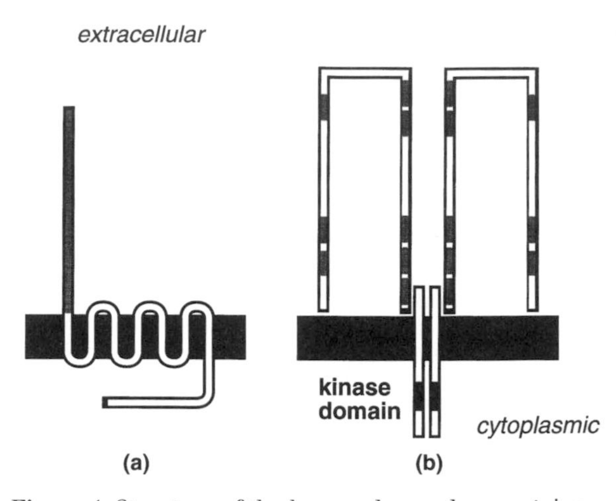

Figure 1. Structure of the boss and sevenless proteins. (a) The boss protein consists of a large extracellular domain of 498 amino acids, a transmembrane domain with seven membrane-spanning segments, and a cytoplasmic tail of 123 amino acids.[1] A recombinant secreted form of the extracellular domain (grey box) appears to bind to sevenless *in vivo*, but acts as an antagonist,[15] suggesting a requirement for the transmembrane domain in sevenless activation. Such a requirement is also supported by a comparison of the boss proteins between *Drosophila melanogaster*, and *D. virilis*: while the extracellular domain is only 71 per cent conserved, the transmembrane domain, and the cytoplasmic tail exhibit more than 90 per cent identity between these two species. (b) The sevenless receptor tyrosine kinase is unusual in its size of more than 2500 amino acids. The role of the eight fibronectin type III repeats (grey boxes) that are conserved between sevenless, and the mammalian receptor c-ros 5 has not been established. A functional kinase domain is required for sevenless activity.[8]

The initial suggestion that the boss gene encodes the ligand for the sevenless receptor was based on genetic data. Mutations in boss cause the same phenotype as those in sevenless: the transformation of the R7 photoreceptor neurone into a non-neuronal cone cell.[7,12] However, genetic mosaic studies indicated that sevenless is required in R7 precursor cells[7] while boss is required in R8 cells, suggesting that Boss either acts as a ligand or a regulator of the ligand for sevenless.[13]

More direct evidence for this interaction was obtained from adhesion assays with *Drosophila* S2 cells. Untransfected S2 cells do not adhere to each other, but cell lines expressing either boss or sevenless exhibit specific adhesion when they are mixed together.[14] The specificity of this binding was confirmed by its inhibition using either anti-boss or anti-sevenless antibodies[14] or a recombinant secreted form of the extracellular domain of boss.[15] Additional evidence for the direct interaction between boss and sevenless came from a surprising finding. The entire Boss transmembrane protein is internalized into multivesicular bodies in R7 cells.[14,16] Boss internalization is dependent on its interaction with sevenless receptor because in protein-null sevenless mutants, boss is no longer internalized.[14,16]

The boss ligand and its receptor sevenless cooperate in a developmental process that induces a single R7 cell in each ommatidium. This cell is recruited from a group of cells that all have the potential to become R7 cells, the R7 equivalence group.[17–19] However, the two proteins have quite different roles in this process. The sevenless receptor tyrosine kinase is the switch that triggers neuronal development in the R7 cell. The principal mechanism that relays the information from the activated receptor to the nucleus appears to be the ras/raf/MAP kinase pathway.[7,19–25]

The dynamic expression pattern of sevenless (Fig. 2) suggested that its regulation also played a crucial role.[26,27] However, ubiquitous expression of the sevenless receptor had no effect on the specification of R7 cell fates,[28,29] indicating that a protein other than sevenless is critical for the spatial control of R7 development. This protein turns out to be boss. The boss transmembrane ligand is specifically expressed on the central R8 cell in each ommatidium (Fig. 2). As a consequence, its access is restricted only to the R7 precursor cell among the cells of the R7 equivalence group. This restricted localization is critical for proper ommatidial development: ubiquitous expression of boss results in the activation of the sevenless pathway in the remaining cells of the R7 equivalence group. As a result, cells that normally develop as non-neuronal cone cells can be transformed into R7 neurones.[18]

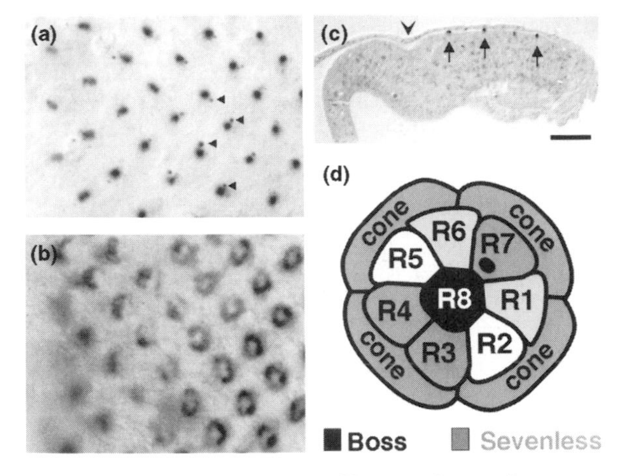

Figure 2. Expression pattern of boss and sevenless. (a) The boss protein localizes to the apical region of R8 photoreceptor cells in the developing eye imaginal disc (large dots of staining). The arrowheads indicate the boss protein internalized into MVBs in R7 cells.[14,16] (b) Sevenless expression is not restricted to the R7 precursor cell;[26,27] its expression pattern is complex, and tightly regulated. In (a), and (b) the left border of the panel corresponds to the morphogenetic furrow, an indentation in the apical surface of the eye disc that demarcates the initiation of cellular differentiation in the eye disc. The morphogenetic furrow is labelled by an arrowhead in (c), which shows a cross section of an eye imaginal disc stained with anti-boss antibodies.[18] Note the restriction of boss localization (arrows) to the apical surface, similar to the apical localization for sevenless.[26,27] (d) Summary of the expression pattern of boss, and sevenless in the eye disc. The scale bar in (c) corresponds to 4 μm in (a) and (b) and 25 μm in (c).

Purification

No purification of the full-length sevenless or boss proteins has been reported. A recombinant secreted form of the extracellular domain of boss of 483 amino acids has been purified to apparent homogeneity from Schneider S2 cell supernatants by peanut agglutinin affinity chromatography. This soluble protein acted as an antagonist of the sevenless receptor.[15]

Activities

Binding of boss to sevenless has been demonstrated by their ability to specifically promote cell adhesion in tissue culture cells[14] and the activation of tyrosine phosphorylation of the sevenless receptor upon binding of boss containing membranes.[15] Tyrosine phosphorylation of sevenless enhances binding of the DRK/sos complex[23,24] which appears to be the initial step in the activation of the ras/raf/MAP kinase pathway downstream of sevenless.[7]

Antibodies

Several antibodies against the extracellular and intracellular domains of both boss and sevenless have been raised using peptides as well as fusion proteins.[10,14,16,24,27,29]

Genes

The sevenless protein is encoded by a 8 kb mRNA that is composed of 10 exons (GenBank accession number J03158). The 3.2 kb mRNA that encodes the bride of sevenless protein is composed of six exons (GenBank accession number L08133).

Mutant phenotype

The sevenless and boss mutations were both initially identified based on the transformation of the R7 photoreceptor neurone into a cone cell.[7,12] No other phenotypes have been identified in these mutants.

Web sites

Flybase is an excellent resource for all *Drosophila* genes. The following URL can be used to access Flybase: http://flybase.bio.indiana.edu/

References

1. Hart, A. C., Krämer, H., Van Vactor, D. L. J., Paidhungat, M., and Zipursky, S. L. (1990). *Genes Dev.*, **4**, 1835–47.
2. Hafen, E., Basler, K., Edstroem, J. E., and Rubin, G. M. (1987). *Science*, **236**, 55–63.
3. Michael, W. M., Bowtell, D. D., and Rubin, G. M. (1990). *Proc. Natl Acad. Sci. USA*, **87**, 5351–3.
4. Hart, A. C., Harrison, S. D., Van Vactor, D. L. J., Rubin, G. M., and Zipursky, S. L. (1993). *Proc. Natl Acad. Sci. USA*,. **90**, 5047–51.
5. Birchmeier, C., Sonnenberg, E., Weidner, K. M., and Walter, B. (1993). *BioEssays*, **15**, 183–90.
6. Wolff, T., and Ready, D. F. (1993) In *The development of Drosophila melanogaster*. (ed. M. Bate and A. Martinez-Arias), pp. 1277–325. Cold Spring Harbor Laboratory Press, Cold Spring Harbor.
7. Zipursky, S. L. and Rubin, G. M. (1994). *Ann. Rev. Neurosci.*, **17**, 373–97.
8. Basler, K. and Hafen, E. (1988). *Cell*, **54**, 299–311.
9. Bowtell, D. D., Simon, M. A., and Rubin, G. M. (1988). *Genes Dev.*, **2**, 620–34.
10. Simon, M. A., Bowtell, D. D., and Rubin, G. M. (1989). *Proc. Natl Acad. Sci. USA*,. **86**, 8333–7.
11. Bhanot, P., *et al.* (1996). *Nature,* **382**, 225–30.
12. Tomlinson, A. and Ready, D. F. (1986). *Science*, **231**, 400–2.
13. Reinke, R. and Zipursky, S. L. (1988). *Cell*, **55**, 321–30.
14. Krämer, H., Cagan, R. L., and Zipursky, S. L. (1991). *Nature*, **352**, 207–12.
15. Hart, A. C., Krämer, H., and Zipursky, S. L. (1993). *Nature*, **361**, 732–6.
16. Cagan, R. L., Krämer, H., Hart, A. C., and Zipursky, S. L. (1992). *Cell,* **69**, 393–9.
17. Basler, K., Christen, B., and Hafen, E. (1991). *Cell*, **64**, 1069–81.
18. Van Vactor, D. L. J., Cagan, R. L., Krämer, H., and Zipursky, S. L. (1991). *Cell,* **67**, 1145–55.
19. Fortini, M. E., Simon, M. A., and Rubin, G. M. (1992). *Nature*, **355**, 559–61.
20. Simon, M. A., Bowtell, D. D., Dodson, G. S., Laverty, T. R., and Rubin, G. M. (1991). *Cell*, **67**, 701–16.
21. Dickson, B., Sprenger, F., Morrison, D., and Hafen, E. (1992). *Nature*, **360**, 600–3.
22. Brunner, D., *et al.* (1994). *Cell*, **76**, 875–88.
23. Olivier, J. P., *et al.* (1993). *Cell,* **73**, 179–91.
24. Simon, M. A., Dodson, G. S., and Rubin, G. M. (1993). *Cell*, **73**, 169–77.
25. Rogge, R., Cagan, R., Majumdar, A., Dulaney, T., and Banerjee, U. (1992). *Proc. Natl Acad. Sci. USA*,. **89**, 5271–5.
26. Tomlinson, A., Bowtell, D. D., Hafen, E., and Rubin, G. M. (1987). *Cell*, **51**, 143–50.
27. Banerjee, U., Renfranz, P. J., Hinton, D. R., Rabin, B. A., and Benzer, S. (1987). *Cell,* **51**, 151–8.
28. Bowtell, D. D., Simon, M. A., and Rubin, G. M. (1989). *Cell*, **56**, 931–6.
29. Basler, K. and Hafen, E. (1989). *Science*, **243**, 931–4.

Helmut Krämer
Department of Cell Biology, and Neuroscience,
University of Texas Southwestern
Medical Center at Dallas,
5323 Harry Hines Blvd,
Dallas, TX 75235–9111, USA

Syndecans

Syndecans are integral membrane heparan sulphate proteoglycans (HSPG) that are the major source of heparan sulfate (HS) at cell surfaces. Syndecan expression is highly regulated during embryogenesis, wound repair, and neoplastic transformation. These proteoglycans associate extracellularly via their HS chains with a wide variety of soluble and insoluble cellular effectors, acting as receptors and co-receptors, and intracellularly with the actin cytoskeleton. Syndecan extracellular domains undergo regulated shedding from the cell surface, a process which instantly changes their function to soluble effectors.

Synonymous names

- Syndecan-1, syndecan, B-B4, CD138;
- Syndecan-2, fibroglycan;
- Syndecan-3, N-syndecan;
- Syndecan-4, ryudocan, amphiglycan.

Homologous proteins

No proteins show significant homology, but other cell surface proteins that contain covalently linked heparan sulphate (HS) are the glypicans, a family of lipid-linked HSPGs, a variant of CD44, and betaglycan, the TGF-β type III receptor.

Protein properties

Syndecans are the major source of cell surface heparan sulphate

These are transmembrane proteoglycan products of a gene family of four distinct genes in mammals (Fig. 1).

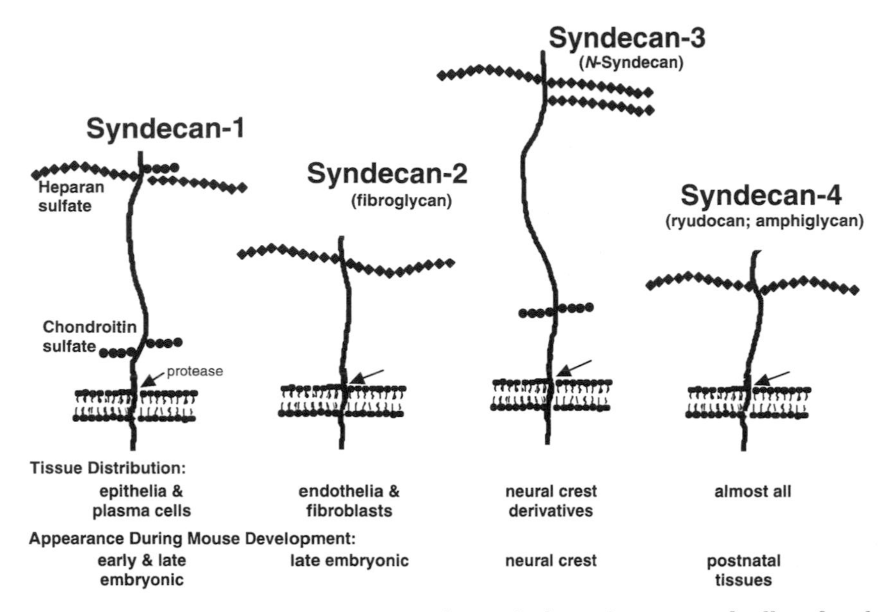

Figure 1. The syndecan family of transmembrane proteoglycans is the major source of cell surface heparan sulphate (HS). The four mammalian syndecans vary in the size of their extracellular domains (ectodomains). Each is a single membrane-spanning protein that bears HS chains distal from the plasma membrane, near the N terminus. Syndecan-1, and –3 can also contain chondroitin sulphate chains and, where examined, these are near the plasma membrane. Each has a putative protease-susceptible site adjacent to the plasma membrane. The cytoplasmic domains are similar in size, and mostly conserved in sequence. Syndecan expression is highly regulated during development, and in tissue distribution. Nearly every adherent cell expresses at least one of the syndecans on its surface.

Each of the four mammalian syndecan core proteins is a single membrane-spanning protein with an apparently extended extracellular domain (ectodomain) that bears HS chains near its N terminus, distal from the plasma membrane. Each syndecan core protein has a similar domain structure; an N-terminal signal sequence, an ectodomain that includes repetitive Ser–Gly sequences where serine residues serve as attachment sites for HS or chondroitin sulphate glycosaminoglycan (GAG) chains, a conserved site adjacent to the plasma membrane that is susceptible to trypsin-like proteases, a hydrophobic transmembrane domain, and a short (28–34 amino acids) cytoplasmic domain with serine and tyrosine residues at conserved positions. Although the extracellular domains are mostly non-homologous, the transmembrane and cytoplasmic domains are highly homologous (ca. 90 per cent sequence identity) among family members and across species, suggesting that these regions interact with evolutionarily stable membrane-associated and intracellular components, such as membrane and cytoskeletal proteins.[1] Their chromosomal localization, exon organization, and sequence relationships with *Drosophila* syndecan suggest that the syndecan gene family arose by gene duplication and divergent evolution from a single ancestral gene, and that syndecan-1 and –3, and –2 and –4 comprise subfamilies.[2]

The syndecan core proteins behave anomalously on SDS–PAGE, possibly due to their ability to form multimers that resist SDS treatment. The calculated M_r for the mature core proteins of mammalian syndecan-1, –2, –3, –4 is 30.6, 20.2, 44.6, and 19.5 kDa, while they migrate at ca. 80, 48, 120, and 35 kDa respectively. The calculated M_r for the mature core protein of *Xenopus* syndecan-2 (XS-2) is 20.9 kDa and of *Drosophila* syndecan is 39 kDa, while their respective ectodomains migrate at 46 kDa and 90 kDa.

The GAG chains on the syndecans are predominantly HS although syndecan-1 and –3 can also contain chondroitin sulphate. HS is a highly negatively charged GAG of variable composition which interacts extensively with proteins.[3] The HS chains on syndecan-1 are compoed of highly sulphated (heparin-like) domains containing N-sulphated regions of ca. 10–30 disaccharides that alternate with larger low-sulphated regions.[4] While the size and macroscopic structure of the HS chains reproducibly vary in syndecan-1 from different cell types, the process responsible for the consistency of this differentiated characteristic is unknown.[3]

Syndecans bind to a diverse group of ligands via the HS chains on their extracellular domains

Ligands include extracellular matrix components, growth factors, growth factor binding proteins, chemokines, proteases, antiproteases, cell adhesion molecules, lipid carrier proteins, lipolytic enzymes, and surface proteins of several microbial pathogens.[1,3,5] Where studied, these ligands bind under physiological conditions and with affinities ranging from 1 to 100 nM. The peptide sequences that bind HS are generally rich in arginine and lysine but no HS binding consensus sequence is known.[3,6] The binding enables syndecans to serve as receptors or co-receptors for a wide variety of soluble and insoluble extracellular ligands (cf. Fig. 2; ref. 36). Following binding, a soluble or viral ligand may be internalized (e.g. lipoprotein lipase, low density lipoprotein, cytomegalovirus). In other instances, the ligand binds to a syndecan prior to or simultaneously with its binding to a signal transducing receptor (e.g. FGF growth factor family members, fibronectin).

While most studies have been done with syndecan-1, the receptor and co-receptor roles are likely to be shared among the syndecan family members.[7] Consistent with its role as a matrix co-receptor, syndecan-1 binds mammary epithelial cells to fibronectin, fibrillar collagens, tenascin and thrombospondin, and cells of the B lineage to type I collagen. Syndecan-4 acts as a co-receptor with integrins during the assembly of microfilaments and focal adhesions involved in cell spreading.[8] Syndecans also act as co-receptors for a variety of heparin-binding growth factors. For example, interaction with cell surface syndecans facilitates the binding of FGF-2 to FGFR-1, activating the signalling receptor by formation of a ternary complex at the cell surface. Each of the syndecan ectodomains can be cleaved proteolytically from the cell surface in a process known as shedding. The shedding of syndecan-1 and -4 can be accelerated by growth factors.[9] Shedding converts the cell surface syndecan into a soluble effector. The soluble ectodomain retains all its HS and thus can bind the same ligands as the cell surface syndecans, enabling it to function as a potent inhibitor or activator depending on the macroscopic structure of the HS chains and the nature of the ligand.[10,32]

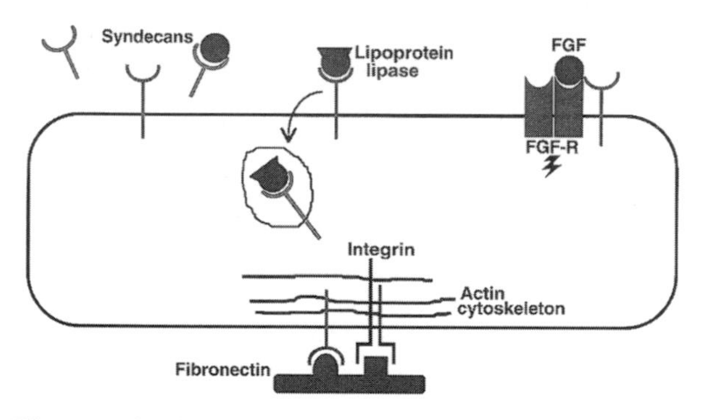

Figure 2. Syndecans can act as cell surface receptors, and co-receptors and as soluble effectors. Via their heparan sulphate (HS) chains, syndecans bind a wide variety of extracellular ligands at affinites of 1–100 nM. Following binding, some soluble ligands are endocytosed (e.g. lipoprotein lipase). Some soluble ligands, such as FGFs, are bound as a ternary complex of a receptor tyrosine kinase (FGFR) in concert with a syndecan which accelerates receptor activation. Insoluble ligands, such as fibronectin, bind simultaneously to both a syndecan, and an integrin, leading to immobilization of the receptor complex, reorganization of the actin cytoskeleton and cell spreading. Receptor-activated cleavage of the core protein by a zinc-dependent metalloproteinase instantly changes the function of the ectodomain from a cell surface receptor or co-receptor to a soluble effector. Because they can interact via their HS chains with the same ligands as cell surface syndecans, the shed syndecan ectodomains can function as inhibitors or activators depending on the macroscopic structure of the HS chains and the nature of the ligand.

Syndecan expression is regulated during morphogenesis, wound repair, and neoplastic transformation

In mouse embryos, syndecan-1 appears earliest and is at cell surfaces in the morula; expression thereafter coincides with the cells fated to become the embryo proper. During epithelial–mesenchymal interactions, syndecan-1 is transiently induced on mesenchymal condensations and is transiently lost from epithelia when the cells change shape; thus, its expression follows morphogenetic rather than histologic boundaries. Syndecan-2 is expressed later during organogenesis and predominates in mesenchymal cells.[11] Syndecan-3 arises at a similar time, primarily in limb bud and neural crest-derived cranial mesenchyme and in the neural crest; high levels are restricted to the early postnatal nervous system.[12] Syndecan-4 is the latest to appear, near the end of embryogenesis, and is nearly ubiquitously distributed.[13] All adherent cells studied express syndecans, but the relative abundance of each depends on the cell type;[14] syndecan-1 is expressed predominantly by plasma cells and epithelia, where it polarizes to basolateral surfaces, and is absent from most terminally differentiated cells; syndecan-2 is prominent in endothelia and hepatocytes; syndecan-3 is primarily in neural tissues; and syndecan-4 is expressed in most tissues. Although this cell-type specific distribution implies that these proteoglycans have distinct functions, the single *Drosophila* syndecan is expressed in tissues analogous to those of the four mammalian syndecans.[15]

Syndecan-1 and –4 are induced and lost in response to tissue injury.[16] Skin wounds are the best studied example. Keratinocytes migrating into the wound from the margins transiently lose cell surface syndecan-1. Concomitantly, syndecan-1 increases on the endothelial cells and syndecan-4 increases on the fibroblasts that form the wound granulation tissue. The inductions are, in part, due to the action of neutrophil-derived antimicrobial peptides.[16] The ectodomains of syndecan-1 and -4 are shed into the wound being repaired,[9] where they maintain the balance between proteinases and anti-proteinases.[37] Expression returns to normal upon re-epithelialization of the wound.

Reduced syndecan-1 on epithelial cell surfaces correlates with the acquisition of malignant characteristics and with carcinomatous invasion and metastasis.[17] Depletion of syndecan-1 from mouse mammary epithelial cells by antisense constructs causes the cells to change shape, lose cell–cell contacts, invade collagen gels, and acquire anchorage-independent growth.[18] Similarly, syndecan-1 is reduced by androgen treatment of hormone-responsive mouse mammary tumour cells, correlating with changes in cell shape and loss of growth control. Overexpression of cell surface syndecan-1 in transformed epithelia or in Schwann cells, which express little syndecan-1, causes the cells to become 'epithelial-like' in shape.

Syndecan expression is regulated at multiple levels

The transcriptional activity of the syndecan-1 gene is regulated by growth factors and retinoic acid.[19,20] Syndecan-1 mRNA transcription can be activated by WT-1, the product of a tumour suppressor gene which is essential for normal urogenital development, and by action of FGF-2 on a specific enhancer complex in mesenchymal cells.[21] Changes in syndecan-1 transcript levels correlate with changes in proteoglycan expression in condensing dental mesenchyme, in keratinocytes and granulation tissue of healing skin wounds, and during muscle differentiation. However, despite reduced cell surface syndecan-1, syndecan-1 mRNA levels are not changed in transformed cells, suggesting that surface expression can also be post-transcriptionally controlled. Post-transcriptional controls have also been detected during B-lymphocyte differentiation and peritoneal macrophage activation. Post-translational regulation of syndecan-1 glycanation results in HS chains of varying size and fine structure on different cell types.[4]

Syndecan ectodomains can be shed from cell surfaces in a regulated manner. All four mammalian and the single *Drosophila* syndecan ectodomain are shed by cultured cells,[14,22,23] and shedding has been detected *in vivo* during formation of the murine ectoderm. The shed proteoglycan corresponds to the ectodomain, although the precise site of proteolytic cleavage is unknown. Shedding is induced by stress (e.g. harvesting), by activation of at least two distinct receptor classes (e.g. EGF, thrombin), and by agents that activate second messengers (e.g. phorbol esters) or inhibit protein tyrosine phosphatases (e.g. pervanadate).[9,24] Cleavage of syndecan ectodomains from the cell surface is mediated by a TIMP-3-sensitive metalloproteinase(s), although the precise nature of the enzyme involved is unknown.[38] Upon shedding, the function of syndecans instantly changes from a cell surface receptor or co-receptor to a soluble effector. Because soluble and cell surface syndecans compete for the same ligands, the soluble syndecan ectodomain may function as a paracrine effector.

The syndecan core proteins place the HS chains in specific plasma membrane domains, for example syndecan-1 at basal epithelial cell surfaces and syndecan-4 in fibroblast focal adhesions. Antibody ligation of the syndecan-1 ectodomain core protein causes the actin cytoskeleton to reorganize and the cell to spread, even when the GAG chains have been removed and the cytoplasmic domain has been deleted, suggesting that the transmembrane or extracellular domains of the core protein are involved in mediating cell spreading.[25] The syndecan-4 core protein, by itself, can serve as an adhesive substrate for cells, an activity that localizes to a specific core protein region of the syndecan-4 extracellular domain.[26] Each of the syndecans forms multimers, a process which is reported to involve the extracellular and transmembrane domains,[1] the variable region of the cytoplasmic domain of syndecan-4,[27] and a region adjacent to the plasma membrane and the transmembrane domain of syndecan-3.[28] This multimerization is non-covalent, but the multimers resist SDS treatment. The juxtamembrane region of each syndecan core protein is involved in the regulated proteolytic cleavage that leads to shedding of the ectodomains.[9]

The syndecan core proteins may interact physically with their partner when functioning as a co-receptor, for example with integrins or FGF receptors.[10,29,39] A specific region of the syndecan-4 cytoplasmic domain, the central variable region, interacts directly with protein kinase Cα, potentiating its activity and localizing this cytoplasmic enzyme to focal adhesions.[27] This interaction results from multimerization of syndecan-4 via the central-variable region, a sequence which is unique to syndecan-4.[29] A highly conserved region of the syndecan cytoplasmic domains is the C-terminal EFYA sequence. This sequence fits precisely into the binding pocket of one class of PDZ domain proteins, a large group of cytoplasmic proteins that bind the C termini of widely disparate membrane proteins. Syndecan, a PDZ protein that binds to syndecan cytoplasmic domains, could be involved in the receptor and co-receptor functions of these proteoglycans.[40]

Purification

The ectodomain shed from cell surfaces is purified from the conditioned media of cultured cells by sequential anion exchange chromatography, isopycnic centrifugation in CsCl, and immunoaffinity chromatography.[22] Medium from NMuMG cells yields ~10 μg per litre of pure syndecan-1 ectodomain. The intact proteoglycan is hydrophobic, and thus is either extracted and purified as described above in the presence of detergent, or the proteoglycan is first purified from the extract by incorporation into lipid vesicles.[30]

Activities

The activity of cell surface syndecan is assessed by inference from the action of the ligand, for example syndecan-4 is active in fibroblasts if focal contacts and actin stress fibres form on a fibronectin substratum,[8] or syndecans are active if cell proliferation or differentiation results from the addition of FGF-2.[10,41]

Shedding instantly changes the function of the cell surface proteoglycan from a receptor or co-receptor to a soluble effector (Fig. 2). The soluble syndecan-1 ectodomain inhibits cell proliferation when added to cultured transformed cells,[31] inhibits heparin-mediated FGF-2 mitogenicity[32] and HB-EGF mitogenicity[42] when added to suspension cells containing the relevant receptor, and binds to neutrophil-derived proteases, reducing the affinity of their physiologial inhibitors.[37]

Antibodies

Monoclonal antibodies specific to syndecan ectodomains include antibody 281–2 against mouse syndecan-1,[22] one against human syndecan-1 (available from Serotec); 10H4 against human syndecan-2 (cross-reacts with mouse); 1C7 against human syndecan-3;[11] and 5G9 and 8C7 against human syndecan-4.[33] Polyclonal antisera against ectodomains include MSE-2, MSE-3, and MSE-4, against their respective recombinant mouse syndecan,[14] anti-N-syndecan against recombinant rat syndecan-3,[12] and HSE-1 against recombinant human syndecan-1.[33] Antibodies specific for syndecan cytoplasmic domains include S7C and S1CD against peptides corresponding to the C-terminal 7 or 10 amino acids of syndecan-1, respectively[24] and SCD-4 against a 13 amino acid peptide unique to the syndecan-4 cytoplasmic domain.[9]

Genes

The syndecans are an ancient gene family that apparently arose during the Metazoan radiation because their genes are conserved and organized similarly in nematodes, arthropods, and chordates. Because Metazoans evolved in large part due to their ability to establish epithelia that generate an intracellular compartment segregated from the outside world, syndecan-1, the syndecan that predominates in epithelia, may be the most primitive of the vertebrate syndecans. It is also most closely related to the syndecans of *Drosophila* and *C. elegans* where the family is represented by only a single gene[23] (GenBank accession number Z69646). The syndecan genes are dispersed throughout the mouse genome. *Synd1* maps on proximal mouse chromosome 12 (syntenic with human chromosome 2p23–24), *Synd2* on proximal chromosome 15 (human 8q23), *Synd3* on distal chromosome 4 (human 1p32–p36), and *Synd4* on mouse chromosome 2 (human 20q12–q13). The genomic organization of the mammalian syndecan genes is similar to that of the *Drosophila* syndecan: there are five exons and each encodes a discrete functional domain.[12,19] Genomic clones are available for mouse syndecan-1,[19] rat syndecan-3,[12] chick syndecan-4,[13] and mouse syndecan-4. Complete cDNA sequences have been published for syndecan-1 (human, hamster, rat, mouse), syndecan-2 (rat, mouse), syndecan-3 (chicken, rat, mouse), syndecan-4 (human, chicken, rat, mouse), for the *Xenopus* homologue of syndecan-2, and for the single *Drosophila* syndecan.

Mutant phenotype

A P-element induced mutation in the *Drosophila* syndecan gene results in larval lethality, indicating that syndecan is essential for *Drosophila* viability.[15] Targeted deletion of syndecan-1 in mice does not appear to affect viability, normal embryonic development, or fertility, indicating that its early and widespread embryonic expression does not reflect an essential function of syndecan-1.[35] However, syndecan-1 does appear to have an essential role in cutaneous wound repair.[35]

References

1. Bernfield, M., Kokenyesi, R., Kato, M., Hinkes, M. T., Spring, J., Gallo, R. L., and Lose, E. J. (1992). *Ann. Rev. Cell. Biol.*, **8**, 365–98.
2. Bernfield, M., Hinkes, M. T., and Gallo, R. L. (1993). *Development,* (Suppl.) 205–12.
3. Salmivirta, M., Lidholt, K., and Lindahl, U. (1996). *FASEB J.* **10**, 1270–9.
4. Kato, M., Wang, H., Bernfield, M., Gallagher, J. T., and Turnbull, J. E. (1994). *J. Biol. Chem.*, **269**, 1881–980.
5. Salmivirta, M. and Jalkanen, M. (1995). *Experientia*, **51**, 863–72.
6. Silbert, J. L., Bernfield, M., and Kokenyesi, R. (1995). In *Glycoproteins* (ed. J. Montreuil, H. Schachter, and J. F. G. Vliegenthart), pp. 1–31. Elsevier, Amsterdam.
7. Steinfeld, R., Van Den Berghe, H., and David, G. (1996). *J. Cell. Biol.*, **133**, 405–16.
8. Couchman, J. R. and Woods, A. (1996). *J. Cell Biochem.*, **61**, 578–84.
9. Subramanian, S. V., Fitzgerald, M. L., and Bernfield, M. (1997). *J. Biol. Chem.*, **272**, 14713–20.
10. Schlessinger, J., Lax, I., and Lemmon, M. (1995). *Cell*, **83**, 357–60.
11. David, G. (1993). *FASEB J.* **7**, 1023–30.
12. Carey, D. J., Conner, K., Asundi, V. K., O'Mahoney, D. J., Stahl, R. C., Showalter, L., *et al.* (1997). *J. Biol. Chem.*, **272**, 2873–79.
13. Baciu, P. C. and Goetinck, P. F. (1994). *Mol. Biol. Cell*, **6**, 1503–13.
14. Kim, C. W., Goldberger, O. A., Gallo, R. L., and Bernfield, M. (1994). *Mol. Biol. Cell*, **5**, 797–805.
15. Paine-Saunders, S., Spring, J., Bernfield, M., and Hynes, R. (1997). (Unpublished.)
16. Gallo, R. L., Ono, M., Povsic, T., Page, C., Eriksson, E., Klagsbrun, M., and Bernfield, M. (1994)., *Proc. Natl Acad. Sci. USA*, **91**, 11035–9.
17. Inki, P. and Jalkanen, M. (1996). *Ann. Med.*, **28**, 63–7.
18. Kato, M., Saunders, S., Nguyen, H., and Bernfield, M. (1995). *Mol. Biol. Cell*, **6**, 559–76.
19. Vihinen, T., Auvinen, P., Alanen-Kurki, L., and Jalkanen, M. (1993). *J. Biol. Chem.*, **268**, 17261–9.
20. Larraín, J., Cizmeci-Smith, G., Troncoso, V., Stahl, R. C., Carey, D. J., and Brandan, E. (1997). *J. Biol. Chem.*, **272**, 18418–24.
21. Jaakkola, P., Vihinen, T., Määttä, A., and Jalkanen, M. (1997). *Mol. Cell. Biol.*, **17**, 3210–19.
22. Jalkanen, M., Rapraeger, A., Saunders, S., and Bernfield, M. (1987). *J. Cell Biol.*, **105**, 3087–96.
23. Spring, J., Paine-Saunders, S. E., Hynes, R. O., and Bernfield, M. (1994). *Proc. Natl Acad. Sci. USA*, **91**, 3334–8.
24. Reiland, J., Ott, V. L., Lebakken, C. S., Yeaman, C., McCarthy, J., and Rapraeger, A. C. (1996). *Biochem. J.*, **319**, 39–47.

25. Lebakken, C. S. and Rapraeger, A. C. (1996). *J. Cell Biol.*, **132**, 1209–21.
26. McFall, A. J. and Rapraeger, A. C. (1997). *J. Biol. Chem.*, **272**, 12901–4.
27. Oh, E. S., Woods, A., and Couchman, J. R. (1997) *J. Biol. Chem.*, **272**, 11805–11.
28. Asundi, V. K. and D. J. Carey (1995). *J. Biol. Chem.*, **270**, 26404–10.
29. Oh, E. S., Woods, A., and Couchman, J. R. (1997) *J. Biol. Chem.*, **272**, 8133–6.
30. Lories,V., DeBoeck, H., David, G., Cassiman, J. J., and Van den Berghe, H. (1987). *J. Biol. Chem.*, **262**, 854–9.
31. Mali, M., Andtfolk, H., Miettinen, H. M., and Jalkanen, M. (1994). *J. Biol. Chem.*, **269**, 27795–8.
32. Kato, M., Wang, H., Kainulainen, V., Fitzgerald, M. L., Ornitz, D., Ledbetter, S., and Bernfield, M. (1998). *Nature Med.*, **4**, 691–7.
33. Gallo, R., Kim, C., Kokenyesi, R., Adzick, N. S., and Bernfield, M. (1996). *J. Invest Derm.*, **107**, 676–83.
34. Spring, J., Goldberger, O. A., Jenkins, N. A., Gilbert, D. J., Copeland, N. G., and Bernfield, M. (1994). *Genomics*, **21**, 415–8.
35. Hinkes, M. T., Gibson, H., and Bernfield, M. (Unpublished).
36. Carey, D. J. (1997). *Biochem. J.*, **327**, 1–16.
37. Kainulainen, V., Wang, H., Schick, C., and Bernfield, M. (1998). *J. Biol. Chem.*, **273**, 11563–9.
38. Fitzgerald, M. L., Murphy, G., and Bernfield, M. (Unpublished).
39. Woods, A. and Couchman, J. R. (1998). *Trends Cell Biol.*, **8**, 189–92.
40. Grootjans, J. J., Zimmermann, P., Reekmans, G., Smets, S., Degeest, G., Durr, J., and David, G. (1997). *Proc. Natl Acad. Sci. USA*, **94**, 13683–8.
41. Olwin, B. B. and Rapraeger, A. (1992). *J. Cell Biol.*, **118**, 631–9.
42. Wang, H. and Bernfield, M. (Unpublished).

Marilyn L. Fitzgerald and Merton Bernfield
Department of Pediatrics,
The Children's Hospital,
Harvard Medical School, Boston,
MA, USA

TAG-1/axonin-1

TAG-1/axonin-1 is a glycosyl phosphatidylinositol (GPI)-linked, 135 kDa glycoprotein expressed transiently on the surfaces of subsets of neurones especially during migration and axonogenesis.[1–3] Comprising six immunoglobulin-like (Ig) domains and four fibronectin type III (FnIII) repeats,[4,5] it is a member of a family of Ig/FnIII cell adhesion molecules expressed on developing neural cells.[6] *In vitro*, in addition to binding homophilically, it can interact with a number of partners, both in *cis* and in *trans*, and mediates axon outgrowth promoting and cell adhesion events. Perturbation of axonin-1 function *in vivo* disrupts the guidance of commissural axons across the floor plate of the developing spinal cord.[7,8]

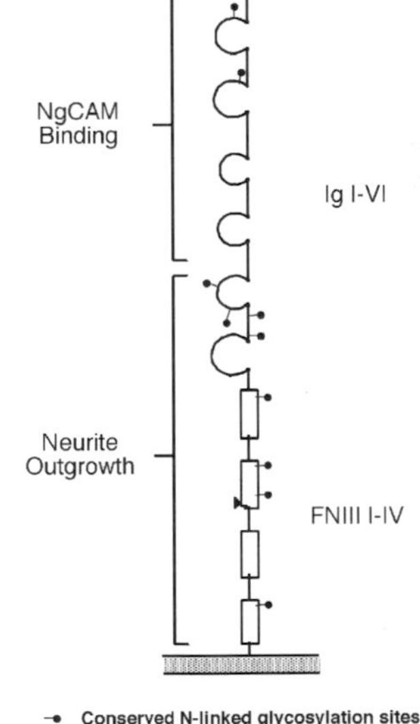

Figure 1.

Synonyms

- SNAP mouse;[9]
- TAG-1 rat,[4] zebrafish;[40]
- Axonin-1 chicken;[10,11]
- SC2 chicken;[12]
- TAX-1 human.[13,14]

Homologues

TAG-1/axonin-1 is a member of the Ig/FnIII family of vertebrate neural cell adhesion molecules comprising at least nine members that are related at the primary sequence and domain organization levels (Fig. 1).[6] The closest relatives of TAG-1/axonin-1 are F3/F11/contactin, BIG-1/PANG, and BIG-2, which are ~50 per cent identical to each other at the amino acid level.[15] Although the

Ig/FnIII family is represented in *Drosophila*, for example neuroglian is an L1-like molecule,[16] TAG-1/axonin-1-like molecules have not be found (Grenningloh, Furley, and Goodman, unpublished observations).

Protein properties

The primary structure of TAG-1/axonin-1, deduced by cDNA cloning and direct protein sequencing,[4,5] indicates the presence of six immunoglobulin-like domains (C2 type) and four fibronectin type III domains, preceded by a membrane insertion leader sequence and followed by a hydrophobic tail that directs attachment to the cell surface via a GPI anchor. In addition to the cell surface GPI-linked form of the protein a substantial fraction of TAG-1/axonin-1 is released from neurones.[10,11,17,18] It is probable that the two forms of the protein arise through differential post-translational processing of the same precursor protein.[4] This is likely to be due to cleavage of the GPI linkage of the nascent protein by an endogenous phospholipase D[19] which probably occurs before the protein reaches the cell surface.[10] Further post-translational modification occurs in the form of specific glycosylation.[20]

Intracellular and tissue distribution

Light microscopic[3] and ultrastructural[2] studies of surface bound TAG-1 indicate that the protein is present on neuronal cell bodies, axons, and growth cones, although this distribution appears to vary according to the developmental stage of the neurone (see also below). Where TAG-1 is present in axon fascicles its localization is punctate and appears at points of membrane contact with a regular periodicity.[2] Recent studies using adenoviral vectors to express axonin-1 in neurones in culture demonstrate that newly synthesized protein is transported directly to the growth cone and inserted in the growth cone membrane rather than the axolemma of the axon shaft.[21]

Expression of TAG-1/axonin-1 in development begins as the first axons are extended in the early neural tube[3] and is reiterated on many different, though not all, sets of neurones throughout the nervous system at various stages of development and into the adult[22–25] (see especially ref. 1 for a comprehensive study). Its expression is transient on many axons[1,3,26] with changes in its expression coinciding with changes in axon trajectory, suggesting a role in pathfinding[3] (see also below). The expression of TAG-1/axonin-1 in the adult hippocampus raises the possibility that it may also be involved in the maintenance of synaptic connections.[1,15]

Functional properties and intermolecular interactions

In common with other members of this family, TAG-1/axonin-1 can cause cell–cell adhesion by cross-linking cell membranes through homophilic binding.[27,28] Moreover, purified TAG-1/axonin-1 protein promotes axon extension when presented as a substrate *in vitro*.[4,17] However, the ability of TAG-1/axonin-1 to promote neurite outgrowth does not depend on the homophilic binding of neuronal TAG-1/axonin-1 to the TAG-1/axonin-1 substrate, but rather on heterophilic interactions, the best characterized of which is that with the Ig/FnIII members, L1 (with TAG-1 in rodents) and NgCAM (with axonin-1 in chick), presumed species homologues.[27,29,30] Blockade of neuronal NgCAM/L1 with anti-NgCAM/L1 antibodies prevented axon extension on the TAG-1/axonin-1 substrate. This led to the suggestion that NgCAM and L1 are neuronal receptors for axonin-1 and TAG-1 respectively, a hypothesis supported by the observation that synthetic microspheres coated with purified NgCAM adhered to similar microspheres coated with axonin-1.[29] Paradoxically, however, cells expressing TAG-1 on their surface failed to form aggregates with L1 expressing cells[31]. Subsequently, it has become clear that the physical binding of TAG-1/axonin-1 to L1/NgCAM occurs not between cells in *trans*, but rather on the same cell surface, that is, in *cis*.[30,32] Thus, while both TAG-1/axonin-1 and L1/NgCAM can function separately to cause cell adhesion by homophilic binding, axon extension on either substrate requires, in addition, the formation of a complex between the two molecules on the same cell surface. It seems likely that the formation of this complex generates the intracellular signals necessary to initiate the complex process of axon elongation; interestingly it has recently been shown that clustering of NgCAM with axonin-1 leads to an increase in the phosphorylation of NgCAM by a casein kinase II-related activity.[33] By contrast, when axonin-1 is not complexed with NgCAM, the src-family kinase, fyn, is found associated with axonin-1.[33]

A number of other heterophilic interactions of TAG-1/axonin-1 have been documented, including β1 integrin;[27] NrCAM;[34] NCAM, neurocan and phosphacan (PTPβ/ζ),[35] and F3.[41]

Relation of structure to function

The binding of NgCAM to axonin-1 has been analysed in detail by studying the interaction of purified NgCAM immobilized on synthetic microspheres with cell lines transfected with mutated versions of the axonin-1 gene.[36] Surprisingly, full length axonin-1 expressed at the cell surface was found to interact poorly with NgCAM-bearing microspheres in contrast to the strong interaction found when both proteins were presented on microspheres.[29] However, a strong NgCAM-binding activity was uncovered when Ig domains 5 or 6 were deleted from transfection constructs, suggesting that, on the cell surface, axonin-1 assumes a tertiary conformation in which the NgCAM-binding site is masked; deletion of Ig domains 5 or 6, or covalent coupling to microspheres, appears to expose this site. Two further pieces of evidence support this interpretation: first, further deletion experiments locate the NgCAM-binding site within the first four Ig domains of axonin-1, which appear to form a

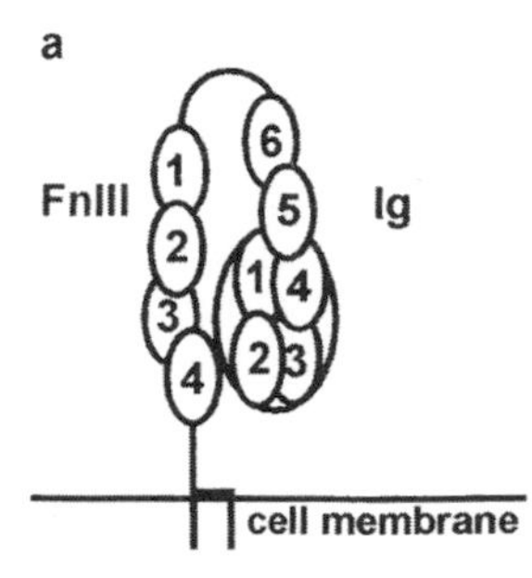

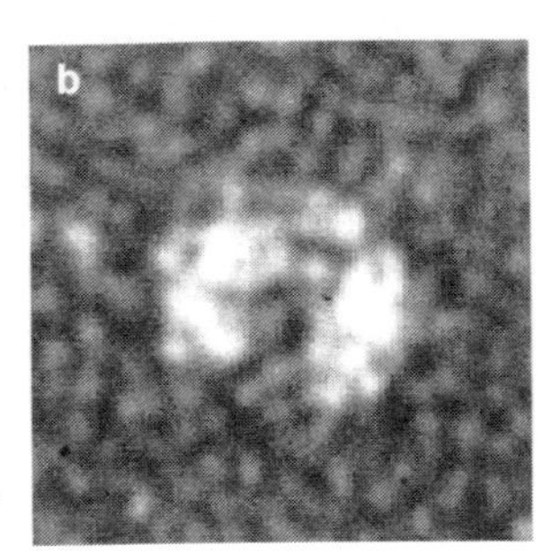

Figure 2.

discrete unit since deletion of any of the four domains disrupts NgCAM binding; second, electron microscopic studies reveal that, rather than the rod-like structure suggested in Fig. 1, purified axonin-1 assumes a horseshoe-like appearance in which the molecule folds back on itself (Fig. 2(b)). These observations have led to the model in Fig 2(a) in which the first four Ig domains form a conglomerate and fold back to associate with the FnIII domains. This would restrict access of the NgCAM binding site to *trans* interactions, but none the less allow *cis* interactions with NgCAM on the same membrane (see also above).

Biological roles

Of the interactions described above, the physiological significance of the interaction of axonin-1 with NrCAM has been best elucidated. Commissural axons of the developing spinal cord express TAG-1/axonin-1 on their surface as they encounter the midline of the spinal cord, the floor plate.[3,8] In contrast, NrCAM is expressed at low levels throughout the neuroepithelium, but is notably concentrated on the non-neuronal cells of the floor plate.[8] Using both antibodies to NrCAM and axonin-1, and purified axonin-1 protein, Stoeckli and Landmesser have shown that interference with axonin-1 function on the commissural neurones, or with NrCAM on the floor plate, leads to the failure of commissural axons to cross the floor plate, which instead turn prematurely.[8] These studies have been extended to show that NrCAM/axonin-1 function is necessary to mask inhibitory properties of the floor plate that would otherwise prevent commissural axons from crossing.[7]

Purification

TAG-1 has been purified from both embryonic rat brain lysates[4] and from medium conditioned by cell lines expressing transfected TAG-1 genes[27] by immunoaffinity chromatography using monoclonal anti-TAG-1 antibodies. Axonin-1 is abundant in vitreous fluid from embryonic chick eyes and can be purified by a series of chromatographic steps.[10] Typically, lectin affinity chromatography on concanavalin A–sepharose is followed by ion exchange (DEAE), hydrophobic interaction chromatography (phenyl-substituted polymer), and finally gel filtration (e.g. Superose 12) chromatography.

Activities

The neurite outgrowth activity of TAG-1/axonin-1 is usually measured by immobilizing purified protein on to tissue culture plastic (e.g. ref. 32). Binding activities have been measured either by coating synthetic microspheres with purified protein (e.g.ref. 29), or transfecting non-adherent cell lines (e.g. ref. 27). See also discussion above.

Antibodies

A substantial number of both monoclonal and polyclonal antibodies to rat TAG-1 and chick axonin-1 have been described, some of which also cross-react with murine TAG-1.[3,9,36] The anti-rat TAG-1 monoclonal 1C12 is available from the Developmental Studies Hybridoma Bank, University of Iowa http://www.uiowa.edu/~dshbwww/.

Genes

cDNA

Full length cDNAs are available for rat TAG-1;[4] chick axonin-1;[5] and human TAX-1.[13,14]

Gene structures

The structural organization of the chicken,[37] human,[38]and mouse (Furley, Kozlov, and Sonderegger; unpublished results) genes have been studied; each comprises 23 exons spread over ~40 kb. Characteristically the first exon contains only 5′ untranslated sequences.

Chromosomal locations

Human TAX-1 maps to 1q32.1;[14,39] mouse TAG-1 has been mapped to chromosome 1 (N. Jenkins, personal communication.)

Mutant phenotype/disease states

Although chromosomal mapping of human TAX-1 localizes the gene in a region associated with microcephaly and Van der Woude syndrome, there is no evidence to suggest that mutations in TAX-1 are causal in these diseases. Mutations in the mouse TAG-1 gene are yet to be described.

Structure

See Fig. 2.

References

1. Wolfer, D. P., Henehan, B. A., Stoeckli, E. T., Sonderegger, P., and Lipp, H. P. (1994). *J. Comp. Neurol.*, **345**, 1–32.
2. Yamamoto, M., Hassinger, L., and Crandell, J. E. (1990). *J. Neurocytol.*, **19**, 619–27.
3. Dodd, J., Morton, S. B., Karagogeos, D., Yamamoto, M., and Jessell, T. M. (1988). *Neuron*, **1**, 105–16.
4. Furley, A. J., Morton, S. B., Manalo, D., Karagogeos, D., Dodd, J., and Jessell, T. M. (1990). *Cell*, **61**, 157–70.
5. Zuellig, R. A., *et al.* (1992). *Eur. J. Biochem.* **204**, 453–63.
6. Sonderegger, P. and Rathjen, F. G. (1992). *J. Cell Biol.*, **119**, 1387–94.
7. Stoeckli, E. T., Sonderegger, P., Pollerberg, E. G., and Landmesser, L. T. (1997). *Neuron*, **18**, 209–22.
8. Stoeckli, E. T. and Landmesser, L. T. (1995). *Neuron*, **14**, 1165–79.
9. Yamamoto, M., Boyer, A. M., Crandall, J. E., Edwards, M., and Tanaka, H. (1986). *J. Neurosci.*, **6**, 3576–94.
10. Ruegg, M. A., Stoeckli, E. T., Kuhn, T. B., Heller, M., Zuellig, R., and Sonderegger, P. (1989). *EMBO J.*, **8**, 55–63.
11. Stoeckli, E. T., Lemkin, P. F., Kuhn, T. B., Ruegg, M. A., Heller, M., and Sonderegger, P. (1989). *Eur. J. Biochem.*, **180**, 249–58.
12. Sakurai, T., Shiga, T., Shirai, T., Tanaka, H., and Grumet, M. (1994). *Brain Res. Dev. Brain Res.*, **83**, 99–108.
13. Hasler, T. H., Rader, C., Stoeckli, E. T., Zuellig, R. A., and Sonderegger, P. (1993). *Eur J. Biochem.*, **211**, 329–39.
14. Tsiotra, P. C., Karagogeos, D., Theodorakis, K., Michaelidis, T. M., Modi, W. S., Furley, A. J., *et al.* (1993). *Genomics*, **18**, 562–7.
15. Yoshihara, Y., Kawasaki, M., Tamada, A., Nagata, S., Kagamiyama, H., and Mori, K. (1995). *J. Neurobiol.*, **28**, 51–69.
16. Bieber, A. J., Snow, P. M., Hortsch, M., Patel, N. H., Jacobs, J. R., Traquina, Z. R., *et al.* (1989). *Cell*, **59**, 447–60.
17. Stoeckli, E. T., Kuhn, T. B., Duc, C. O., Ruegg, M. A., and Sonderegger, P. (1991). *J. Cell Biol.*, **112**, 449–55.
18. Karagogeos, D., Morton, S. B., Casano, F., Dodd, J., and Jessell, T. M. (1991). *Development*, **112**, 51–67.
19. Lierheimer, R., Kunz, B., Vogt, L., Savoca, R., Brodbeck, U., and Sonderegger, P. (1997). *Eur. J. Biochem.*, **243**, 502–10.
20. Denzinger, T., Savoca, R., Sonderegger, P., and Przybylski, M. (1997). *Biochim. Biophys. Acta* (In press.)
21. Vogt, L. *et al.* (1996). *Curr. Biol.*, **6**, 1153–8.
22. Rager, G., Morino, P., Schnitzer, J., and Sonderegger, P. (1996). *J. Comp. Neurol*, **365**, 594–609.
23. Redies, C., Arndt, K., and Ast, M. (1997). *J. Comp. Neurol.*, **381**, 230–52.
24. Honig, M. G. and Kueter, J. (1995). *Dev. Biol.*, **167**, 563–83.
25. Halfter, W., Yip, Y., and Yip, J. W. (1994). *Dev. Brain Res.*, **78**, 87–101.
26. Morino, P., Buchstaller, A., Giger, R., Sonderegger, P., and Rager, G. (1996). *Dev. Brain Res.*, **91**, 252–9.
27. Felsenfeld, D. P., Hynes, M. A., Skoler, K. M., Furley, A. J., and Jessell, T. M. (1994). *Neuron*, **12**, 675–90.
28. Rader, C., Stoeckli, E. T., Ziegler, U., Osterwalder, T., Kunz, B., and Sonderegger, P. (1993). *Eur J. Biochem.*, **215**, 133–41.
29. Kuhn, T. B., Stoeckli, E. T., Condrau, M. A., Rathjen, F. G., and Sonderegger, P. (1991). *J. Cell Biol.*, **115**, 1113–26.
30. Stoeckli, E. T., Ziegler, U., Bleiker, A. J., Groscurth, P., and Sonderegger, P. (1996). *Dev. Biol.*, **177**, 15–29.
31. Felsenfeld, D., Hynes, M. A., Furley, A. J., and Jessell, T. M. (1992). *J. Cell Biochem.*, **16F**, 148.
32. Buchstaller, A., Kunz, S., Berger, P., Kunz, B., Ziegler, U., Rader, C., and Sonderegger, P. (1996). *J. Cell Biol.*, **135**, 1593–607.
33. Kunz, S., Ziegler, U., Kunz, B., and Sonderegger, P. (1996). *J. Cell Biol.*, **135**, 253–67.
34. Suter, D. M., Pollerberg, G. E., Buchstaller, A., Giger, R. J., Dreyer, W. J., and Sonderegger, P. (1995). *J. Cell Biol.*, **131**, 1067–81.
35. Milev, P., Maurel, P., Haring, M., Margolis, R. K., and Margolis, R. U. (1996). *J. Biol. Chem.*, **271**, 15716–23.
36. Rader, C., Kunz, B., Lierheimer, R., Giger, R. J., Berger, P., Tittmann, P., Gross, H., and Sonderegger, P. (1996). *EMBO J.*, **15**, 2056–68.
37. Giger, R. J., Vogt, L., Zuellig, R. A., Rader, C., Henehan Beatty, A., Wolfer, D. P., and Sonderegger, P. (1995). *Eur. J. Biochem.*, **227**, 617–28.
38. Kozlov, S. V., Giger, R. J., Hasler, T. A., Korvatska, E., Schorderet, D. F., and Sonderegger, P. (1995). *Genomics*, **30**, 141–8.
39. Kenwrick, S., Leversha, M., Rooke, L., Hasler, T., and Sonderegger, P. (1993). *Hum. Mol. Genet.*, **2**, 1461–2.
40. Warren, J. T., Chandrasekhar, A., Kanki, J. P., Rangarajan, R., Furley, A. J., and Kuwada, J. W. (1998). *Mech. Dev.*, (In press).
41. Buttiglione, M., Revest, J.-M., Pavlou, O., Karagogeos, D., Furley, A., Rougon, G., and Faivre-Sarrailh, C. (1998). *J. Cell Biol.*, (In press).

Andrew J. W. Furley
Developmental Genetics Programme,
Department of Biomedical Science,
University of Sheffield,
Sheffield S10 2TN, UK.

Peter Sonderegger
Department of Biochemistry,
University of Zurich, Winterthurerstrasse 190,
CH-8057 Zurich, Switzerland.

TCR/CD3 complexes and the CD4 and CD8 co-receptors

T cells recognize antigen by means of a clonotypic membrane-bound T-cell receptor (TCR). Two distinct forms of TCRs, TCR$\alpha\beta$ and TCR$\gamma\delta$, define the $\alpha\beta$ and $\gamma\delta$ T cell lineages. Both TCRs transduce their signals via components of the CD3 complex which is non-covalently associated with the TCR. TCR$\alpha\beta$ does not interact with intact protein, but requires degradation of proteins into short peptides. A TCR$\alpha\beta$ recognizes the composite surface of a single specific peptide and the highly polymorphic, membrane-distal α helices of a class I or class II protein of the major histocompatibility complex (MHC). Selective expression of the CD4 or CD8 co-receptors defines two major sublineages of $\alpha\beta$ T cells. CD4 and CD8 assist in the recognition of peptide/MHC by binding to non-polymorphic, membrane-proximal domains of class II and class I MHC molecules, respectively. T cells of the $\gamma\delta$ lineage are divided into several sublineages, differing by their usage of variable gene repertoire, homing, antigen recognition, and cytokine production. Beside playing a role in specific detection of antigen and signalling, the TCR, CD3, CD4, and CD8 chains are pivotal in controlling survival, growth, and differentiation of T cells.

Protein properties

TCR$\alpha\beta$ recognizes small peptide fragments bound to MHC molecules. The recognition of the specific MHC/peptide complex is mediated by the clonotypic, heterodimeric $\alpha\beta$ TCR. Surface expression of TCR/CD3 complexes requires the assembly of TCR$\alpha\beta$ dimers with the signal transducing modules, collectively called the CD3 complex. The exact stoichiometry and three-dimensional organization of the TCR/CD3 complex is not known (see below). Experiments addressing this issue are still controversial, especially with respect to the presence of one or two TCR heterodimers in a single TCR/CD3 complex. Directly linked to this problem is the role of the TCR in adhesion. Biophysical studies with soluble TCR and MHC/peptide complexes revealed a low affinity of the order of 10^{-5} M. However the off rate was found to be 10–40 fold lower in the presence of chemically cross-linked dimeric MHC molecules. Thus, dimeric MHC molecule, or possibly bivalent TCR$\alpha\beta$, can stabilize the MHC/TCR complex by increasing the avidity of the interaction.

Genes encoding the T cell antigen receptors

The genes encoding TCRα, β, γ, and δ chains are similar in primary sequence, gene organization, modes of rearrangement, and function.[1] Like the immunoglobulin (Ig) loci, the TCR genes are divided into arrays of non-functional gene segments scattered over large tracts of chromosomal DNA. The tremendous diversity of TCRs is created by somatic rearrangement of DNA, known as V(D)J recombination. This process, which is catalysed by the recombination activating gene (RAG) products (RAG-1 and RAG-2) and components of the DNA double-strand repair machinery, takes place in precursor lymphocytes and joins variable (V), diversity (D) and joining (J) gene segments, thereby forming a unique V-domain coding exon, adjacent to a constant (C) region.[2] Conserved heptamer and nonamer sequences spaced by 12 or 23 nucleotides (recognition sequences), which flank each germline V, D, and J segment mediate this site specific reaction. Additional diversification is generated in the junctional regions by a potential loss and/or addition of nucleotides at the coding junction. Since formation of an open reading frame is a stochastic event, lymphocytes often contain non-functionally rearranged antigen-receptor alleles.[2] Precursor lymphocytes that do not manage to produce a functional TCR die.

Developmental stages controlled by the pre-TCRα, TCRα, β, and CD3, CD4, and CD8 chains

Intrathymic differentiation of $\alpha\beta$ thymocytes is strictly controlled by the sequential appearance of the TCRβ and TCRα chains (Fig. 1). In V(D)J-recombination deficient (severe combined immunodeficient, SCID and RAG-mutant) mice, thymocyte development is arrested at the $CD3^{vl}$ (vl, very low) $CD4^-$, $CD8^-$, $CD25^+$ (IL-2Rα chain) pro-T cell stage. Further development into $CD4^+$, $CD8^+$, $CD25^-$ pre-T cells is controlled by the pre-TCR/CD3 receptor complex.[3] The pre-TCR consists of a conventional TCRβ chain and the invariant pre-TCR α chain (pTα).[4] The pre-TCR requires CD3γ,[5] ϵ,[6] and ζ[7–9] but not CD3δ[10–12] to induce efficiently the developmental transition from pro-T cells to pre-T cells.[12] The pre-TCR/CD3 complex controls the number of $CD4^+/CD8^+$ thymocytes, induces TCRα rearrangement and blocks further rearrangments at the TCRβ locus, that is TCRβ allelic exclusion.[3,12] Subsequent expression of an appropriate TCRα chain allows assembly and surface expression of low levels of the TCR$\alpha\beta$/CD3 complex, indicative of immature $CD4^+$, $CD8^+$, that is double-positive (DP) thymocytes.[13] Further development of the $\alpha\beta$ T-cell lineage requires the presence of CD3δ[10] in the TCR/CD3 complex and MHC ligands. The kind of interaction between the TCR and MHC/peptide ligands deter-

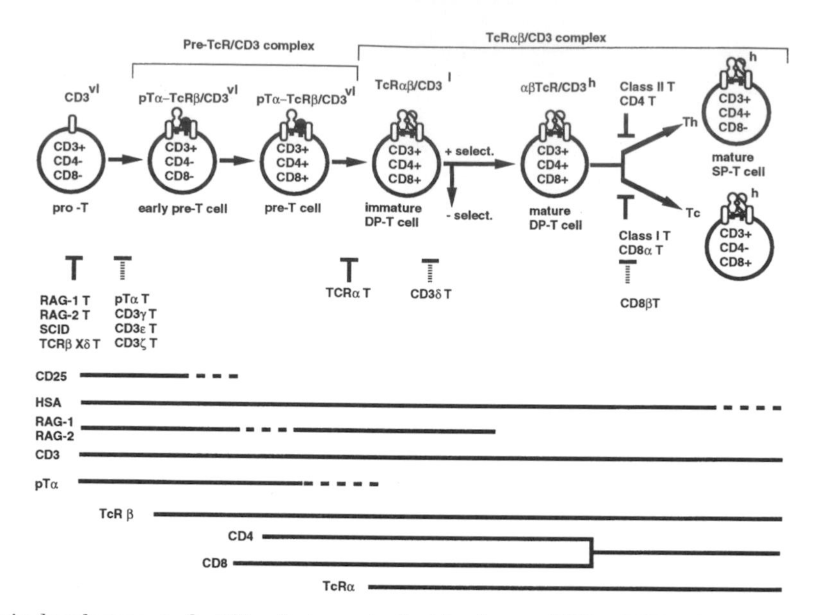

Figure 1. Intrathymic development of $\alpha\beta$ T cells is controlled by the pre-TCRα, TCRα, TCRβ, CD3, CD4, and CD8 chains. The developmental stage that is affected in mice lacking either of these chains is indicated. A 'T' indicates inactivation of the gene by gene targeting. HSA (heat-stable antigen), CD24; see also text.

mines whether $\alpha\beta$ thymocytes will be positively and negatively selected.[13] Negative selection involves clonal elimination or anergy. Positive selection into mature DP thymocytes is characterized by high surface levels of the TCR$\alpha\beta$/CD3 complex. The transcription of RAG-1 and RAG-2 is turned off at this stage. The final differentiation into MHC class II-restricted CD4$^+$, CD8$^-$ helper T cells or MHC class I-restricted CD4$^-$, CD8$^+$ cytotoxic T cells is characterized by a selective shut down of CD8 or CD4 transcription, respectively.[13]

Stoichiometry and composition of the TCR/CD3 complex

Expression of normal TCR$\alpha\beta$/CD3 levels at the cell surface requires the presence of disulphide-linked TCR$\alpha\beta$, the non-covalently linked CD3$\gamma\epsilon$, and CD3$\delta\epsilon$, and the disulphide-linked $\zeta\zeta$ homodimer or, less frequently, a CD3$\zeta\eta$ heterodimer.[14] The three-dimensional organization of the TCR/CD3 complex is not known, but present experimental data on the composition of the TCR/CD3 complex favour either of the following two models:[15]

- a monovalent octameric TCR$\alpha\beta$ CD3$\gamma\delta\epsilon_2\zeta\zeta$ complex;
- a bivalent decameric (TCR$\alpha\beta$)$_2$ CD3$\gamma\delta\epsilon_2\zeta\zeta$ complex.

Physical characteristics of the pTα, TCR, CD3, CD4 and CD8 chains

The physical characteristics of the pTα, TCRα, -β, -γ, -δ, CD3γ, -δ, -ϵ, ζ, η, CD4, and CD8 chains are summarized in Table 1. All chains are type I transmembrane proteins and belong to the Ig superfamily. The components within the pre-TCR and TCR/CD3 complex harbour conserved charged residues in the transmembrane region (Table 1). These highly conserved residues are expected to form neutralizing salt bridges within the hydrophobic transmembrane region of the TCR/CD3 complex, leading to a facilitated assembly and stabilization. In addition, these residues target degradation of incomplete TCR/CD3 complexes.[14]

The crystal structures have been determined for a variable TCRα domain,[16] a TCRβ chain,[17] a TCRβ chain bound to superantigen,[18] TCR$\alpha\beta$ (human and mouse) bound to their specific peptide/MHC ligands,[19–20] CD4,[22–24] CD8$\alpha\alpha$,[25] and the CD8$\alpha\alpha$ homodimer bound to a MHC class I complex.[26] The crystal structures of the TCR$\gamma\delta$, pTα, CD3, and CD8β chains are not known.

The $\alpha\beta$ TCR

The crystal structures of TCR$\alpha\beta$ heterodimers[19–21] revealed a basic antigen-binding fragment (Fab)-like structure. Specific recognition of a peptide complexed within the polymorphic α helices of MHC molecules by TCR$\alpha\beta$ occurs through variable loops, the complementarity-determining regions CDRα 1,2,3 and CDRβ 1,2,3, which are located on the membrane-distal surface of the TCR structure.[19–21] The TCR$\alpha\beta$ combining site is relatively flat and fits diagonally across the MHC peptide-binding site.

The $\gamma\delta$ TCR

Based on sequence homology, a Fab-like structure is expected for the TCR$\gamma\delta$. With a few noted exceptions, TCR$\gamma\delta$ dependent antigen recognition is non-MHC restricted. The TCR$\gamma\delta$ can recognize non-peptidic molecules associated with microorganisms and stressed cells. Recognition of these antigens involves the antigen recep-

Table 1 Physical characteristics of the TCR/CD3 components and the coreceptors CD4 and CD8

Chain	Synonyms: human/ mouse	Knockout mice (ref.)	Relative mass (kDa) of glycosylated forms		Function	Crystal structure (ref.)	N-linked carbo-hydrates	ITAM motifs in Cy	Protein–domains N → C terminus	Charged residues in TM	Remarks
			Human	Mouse							
pTα	gp33	4	33	33	development of $\alpha\beta$ T cells		2	0	LP–C–CP–TM–Cy	2+	Cy of pTα contains two potential PKC phosphorylation sites
TCRα		32, 33	45–60	45–55	development of $\alpha\beta$ T cells and antigen recognition	Vα (16) TCRβ (17, 18) TCR$\alpha\beta$(19, 20)	5	0	LP–V–C–CP–TM–Cy	2+	Cy of TCRα contributes to PKC-mediated down regulation of TCP/CD3
TCRβ		33	40–50	40–55	development of $\alpha\beta$ T cells and antigen recognition		3	0	LP–V–C–C P –TM–Cy	1+	Cy of TCRβ contains a di-lysine ER retention motif
TCRγ		–	45–60	45–60	development of $\gamma\delta$ T cells and antigen recognition		4	0	LP–V–C–CP–TM–Cy	1+	
TCRδ		34	40–60	40–60	development of $\gamma\delta$ T cells and antigen recognition		2	0	LP–V–C–CP–TM–Cy	2+	
CD3γ	T3	5	25–28	21	signal transduction		0	1	LP–C–TM–Cy	1–	Cy of CD3γ and CD3δ contain a di-leucine motif involved in TCR down regulation. The CD3γ,δ and ε genes are syntenic
CD3δ	T3	11	20	28			0	1	LP–C–TM–Cy	1–	
CD3ε	T3	6	20	25			0	1	LP–C–TM–Cy	1–	
ζ		7–9	16	16			0	3	LP–EP–TM–Cy	1–	EP of ζ; and η comprise only nine amino acids

Table 1 (Cont'd)

Chain	Synonyms: human/ mouse	Knockout mice (ref.)	Relative mass (kDa) of glycosylated forms		Function	Crystal structure (ref.)	N-linked carbo-hydrates	ITAM motifs in Cy	Protein–domains N → C terminus	Charged residues in TM	Remarks
			Human	Mouse							
η		7–9		21			0	2	LP–V–C–C–C–TM–Cy	1–	η arises by altemative splicing of the ζ pre-mRNA
CD4	Leu-3/T4 L3T4	35	55	55	signal transduction and coreceptor	2	0		LP–EP–TM–Cy	No	Cy of CD4 and CD8α but not CD8β have a binding motif for the *src*-related
CD8α	Leu-2fT8 Lyt 2	36	32–34	34–38	signal transduction and coreceptor	(22–24)	1–3	0	LP–V–H–TM–Cy	No	kinase *lck*. CD8 can exist as CD8$\alpha\alpha$ or CD8$\alpha\beta$ dimer.
CD8β	Leu-2/T8 Lyt 3	37–39	32–34	30		CD8VαVα 1 (25, 26)	1–3	0	LP–V–H–TM–Cy	No	CD8β requires CD8α for surface expression

LP, leader peptide; V , Ig-like variable domain; C , Ig-like constant domain; H, hinge region; EP, extracellular peptide; TM, transmembrane domain; Cy, cytoplasmic tail; ITAM, immunoreceptor tyrosine-based activation motif; PKC, protein kinase C.

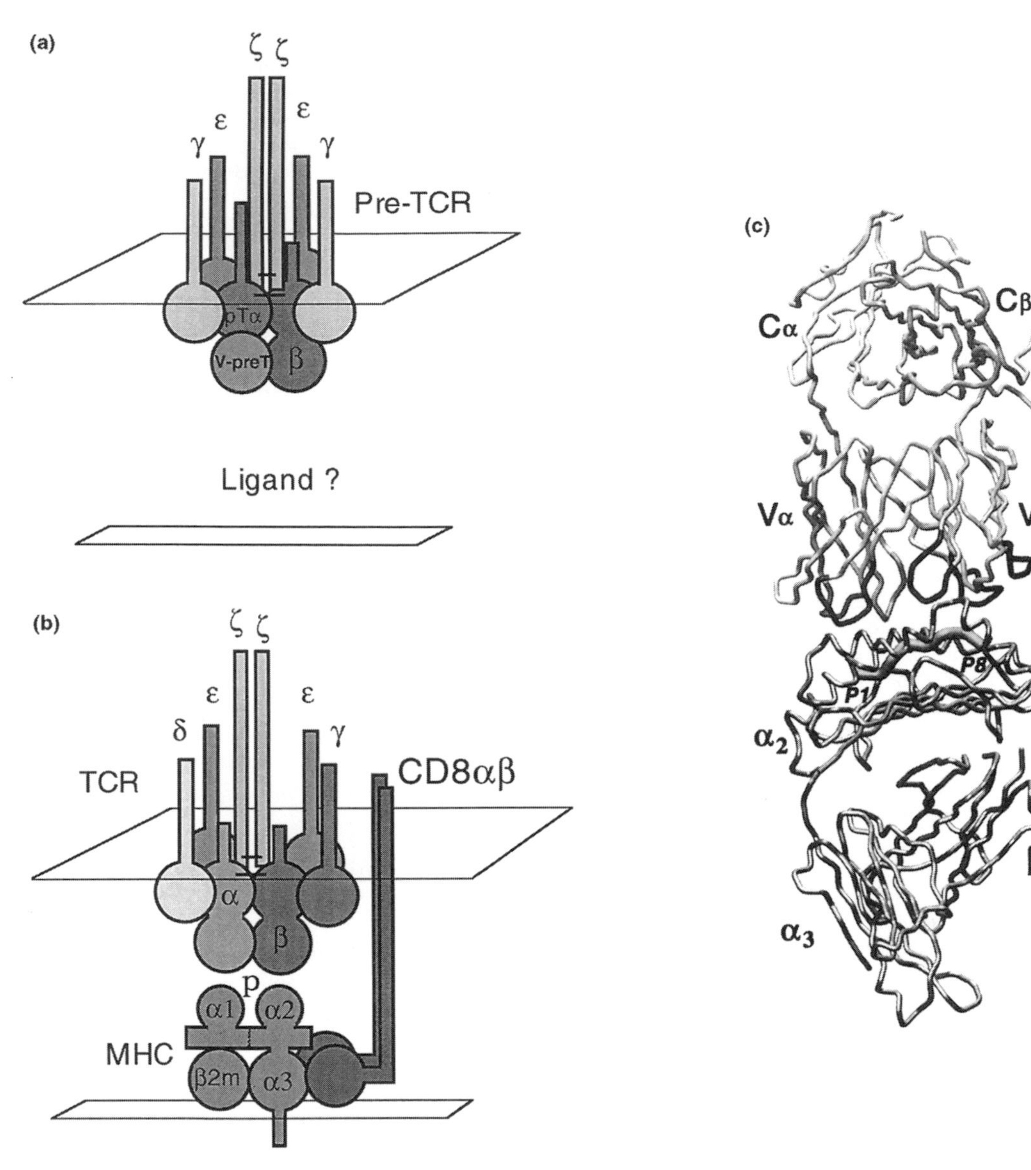

Figure 2. Cartoon of (a) the pre-TCR/CD3 complex, and (b) the TCR/CD3/CD8 complex interacting simultaneously with a MHC/peptide ligand. A ligand of the pre-TCR is not known or does not exist. The exact composition, and organization of the (pre)TCR/CD3 complex is not known. The presence of a V-pre-T chain, analogous to the V-pre-B in the pre-B cell receptor[12] is still hypothetical. (c) Crystal structure of a TCR$\alpha\beta$ interacting with its MHC/peptide ligand. The TCR recognizes a composite surface of peptide (P1–P8), and the α1, and α2 domains of an MHC class I molecule (From Ian Wilson *et al.*, with permission; see also references 16–20).

tor but does not require functional antigen processing machinery or MHC molecules on target cells.[27] $\gamma\delta$ T cells consist of several subsets with distinct receptor repertoires and homing properties.[28] Unlike lymphoid $\gamma\delta$ T cells and $\alpha\beta$ T cells, activated epithelial $\gamma\delta$ T cells secrete the chemotactic cytokine lymphotactin as well as the epithelial-cell-specific fibroblast growth factor (FGF)-7. These and other observations imply a specific function of epithelial $\gamma\delta$ T cells in attracting and activating other lymphocytes and controlling tissue integrity.[29]

The CD4 and CD8 co-receptors

The CD4 and CD8 surface markers define two major peripheral T-cell subsets, the helper/inducer and the cytotoxic subsets, which comprise about two thirds and one third of peripheral $\alpha\beta$ T cells, respectively. CD4 and CD8 are co-expressed on most thymocytes (Fig. 1). In human and rat, CD4 is also found on members of the macrophage lineage and in the mouse at low levels on very early thymocyte immigrants. Unlike CD4, which occurs as a monomer on T helper cells, CD8 occurs as a disulphide-linked heterodimer of two unrelated proteins α and β on cytotoxic T cells, but is found as a disulphide-linked CD8$\alpha\alpha$ homodimer on extrathymic intraepithelial T cells as well as on natural killer (NK) cells.

The extracelluar portion of CD4 comprises four Ig-like domains (D1–D4). A hinge-like junction is found between the two rigid portions D1–D2 and D3–D4.[22–24] The crystal structure of D1–D4 indicates dimerization through D4 domains. CD4 dimers can also be observed at high protein concentration in solution and might be relevant for signal transduction.[24]

The extracellular portion of CD8α and CD8β consists of a NH2-terminal Ig-like variable domain and a 'hinge' region of 50 and 30 amino acid residues, respectively, which are extensively modified by *O*-linked glycosylation.[25,26,29] The cytoplasmic domains of CD4 and CD8α, but not CD8β, contain a cysteine-based motif (CXCP) which is responsible for non-covalent association with the protein tyrosine kinase p56^{lck}.[29] This interaction provides the molecular basis for co-signalling by the CD4 and CD8 co-receptors with the TCR/CD3 complex.

Homologous sites on class I and class II MHC molecules have been found to interact with the membrane distal Ig-like variable domain of CD8 and CD4, respectively. Exposed loops in the α3 domain of class I and β2 domain of class II MHC molecules are the primary binding sites for CD8 and CD4, respectively.[29] These interactions provide the molecular basis of co-receptor recruitment in predominantly class I restricted CD8+ cytotoxic T cells and class II restricted CD4+ T helper/inducer cells.

One CD8$\alpha\alpha$ homodimer (and probably also the CD8$\alpha\beta$ heterodimer) binds one MHC class I complex by clamping the exposed loop of α3 between the CDR-like loops of both CD8 subunits, thereby interfacing also with the α2 domain of class I and β_2-microglobulin.[26] In a similar fashion CD4 dimers might interact with the non-polymorphic α2 and β2 domains of the MHC class II molecules. Residues 35–46 within the membrane-distal CD4 D1 domain are essential for binding MHC class II proteins and also encompass the binding site for the envelope glycoproptein gp120 of the human immunodeficiency virus.[30] In addition, amino acid residues that map to lateral surfaces of D1 and to the N-terminal part of D2 as well as to the D1–D2 interdomain groove affect binding to MHC class II.

The structural information on CD4 and CD8 are in good agreement with results suggesting that TCR and CD4/CD8 increase the avidity of association between a T cell and an antigen-presenting cell or target cell by binding simultaneously to MHC molecules, respectively. TCR and CD4/CD8 are considered to be integral part of a multimolecular complex that controls antigen recognition and signalling by $\alpha\beta$ T cells.

Purification

Components of the TCR/CD3 complex as well as CD4 and CD8 can be purified by cell lysis in different detergents and subsequent immunoprecipitation(s) with specific monoclonal antibodies. In contrast to ionic detergents, certain mild non-ionic detergents like digitonin or Brij preserve the non-covalent protein–protein interactions within the TCR/CD3 complex, which has allowed the isolation and characterisation of intact TCR/CD3 complexes by using a single mAb (see ref. 14 and references therein).

Activities

The TCR/CD3 complex is the critical element for clonotypic activation of $\alpha\beta$ and $\gamma\delta$ T cells. The intracellular signalling cascades that are triggered upon stimulation of TCR/CD3 and CD4/CD8 co-receptors have been reviewed in detail.[31]

Antibodies

TCR$\alpha\beta$-, $\gamma\delta$-, CD3ϵ-, CD4-, and CD8-specific monoclonal antibodies can be purchased from Pharmingen, Southern Biotechnology Associates, Caltag, Serotec, Chemicon, Zymed laboratories, Becton Dickinson, Boehringer-Mannheim, Gibco BRL, Molecular Probes, and other companies. In addition, a large number of polyclonal sera have been raised against these chains. These reagents are useful in affinity puri-fication, immuno-flow cytometry, immunoprecipitation, immunoblotting, immunohistochemistry, and T-cell stimulation.

Genes, transgenic mice, and gene-targeted mice

Many TCR genes with known specificity for peptide/MHC have been cloned from MHC class I and class II restricted T cell clones and have been used to generate TCR$\alpha\beta$ transgenic mice. Idiotype-specific monoclonal antibodies against these TCRs allows the specific detection of transgenic T cells *in vivo*. These systems are useful in studying questions related to tolerance, survival, autoimmune disease, and tumour surveillance. Mouse strains lacking functional pTα, CD3ϵ, -γ, and -δ, ζ, CD4, CD8α, CD8β, and TCR-α, -β or -δ but not yet TCRγ have been generated by gene targeting (Table 1) and are also valuable in studying these processes.

References

1. Davis, M. M. and Bjorkman, P. J. (1988). *Nature,* **334**, 395–402.
2. Gellert, M. (1996). *Genes Cells*, **1**, 269–75.
3. Fehling, H. J. and von Boehmer, H. (1997). *Curr. Opin. Immunol.*, **9**, 263–75.
4. Fehling, H. J., Krotkova, A., Saint-Ruf, C., and von Boehmer. H. (1995). *Nature,* **375**, 795–8.
5. Haks, M. C., Krimpenfort, P., Borst, J., and Kruisbeck, A. M. (1998). *EMBO J.*, **17**, 1871–82.
6. Malissen, M., Gillet, A., Ardouin, L., Bouvier, G, Trucy, J., Ferrier, P., *et al.* (1995). *EMBO J.*, **14**, 4641–53.
7. Love, P. E., Shores, E. W., Johnson, M. D., Tremblay, M. L. Lee, E. J., Grinberg, A., *et al.* (1993). *Science*, **261**, 918–21.

8. Malissen, M., Gillet, A., Rocha, B., Trucy, J., Vivier, E., Boyer, C., *et al.* (1993). *EMBO J.*, **12**, 4347–55.
9. Ohno, H., Aoe, T., Taki, S., Kitamura, D., Ishida, Y., Rajewsky, K., and Saito, T. (1993). *EMBO J.*, **12**, 4357–66.
10. Jacobs, H., Vandeputte, D., Tolkamp, L., De Vries, E, Borst, J., and Berns, A. (1994). *Eur. J. Immunol.*, **24**, 934–9.
11. Dave, V. P., Cao, Z., Browne, C., Alarcon, B., Fernandez-Miguel, G., Lafaille, J., *et al.* (1997). *EMBO J.*, **16**, 1360–70.
12. Borst, J., Jacobs, H., and Brouns, G. (1996). *Curr. Opin. Immunol.* **8**, 181–90.
13. Kisielow, P. and von Boehmer, H. (1995). *Adv. Immunol.*, **58**, 87–209.
14. Klausner, R. D., Lippincott-Schwartz, J., and Bonnofacino, J. S. (1990). *Ann. Rev. Cell Biol.*, **6**, 403–31.
15. Jacobs, H. (1997). *Immunol. Today*, **18**, 565–9.
16. Fields, B. A., Ober, B., Malchiodi, E. L., Lebedeva, M. I., Braden, B. C., Ysern, X., *et al.* (1995). *Science*, **270**, 1821–4.
17. Bentley, G. A., Boulot, G., Karjalainen, K., and Mariuzza, R. A. (1995). *Science*, **267**, 1984–7.
18. Fields, B. A., Malchiodi, E. L., Li, H., Ysern, X., Stauffacher, C. V., Schlievert, P. M., *et al.* (1996). *Nature*, **384**, 188–92.
19. Garcia, K. C., Degano, M., Stanfield, R. L., Brunmark, A, Jackson, M. R., Peterson, P. A., *et al.* (1996). *Science*, **274**, 209–19.
20. Garboczi, D. N., Ghosh, P., Utz, U., Fan, Q. R., Biddison, W. E., and Wiley, D. C. (1996). *Nature*, **384**, 134–41.
21. Bjorkman, P. J. (1997). *Cell*, **89**, 167–70.
22 Wang, J., Yan, Y., Garrett, T. P. J., Liu, J., Rodgers, D. W., Garlick, R. L., *et al.* (1990). *Nature*, **348**, 411–8.
23. Brady, R. L., Dodson, E. J., Dodson, G. G., Lange, G., Davis, S. J., Williams, A. F., and Barclay, A. N. (1993). *Science*, **260**, 979–83.
24. Wu, H., Kwong, P. D., and Hendrickson, W. A. (1997). *Nature*, **387**, 527–30.
25. Leahy, D. J., Axel, R., and Hendrickson, W. A. (1992). *Cell*, **68**, 1145–62.
26. Gao, G. F., Tormo, J., Gerth, U. C., Wyer, J. R., McMichael, A. J., Stuart, D. I., *et al.* (1997). *Nature*, **387**, 630–4.
27. Boismenu, R., and Havran, W. L. (1997). *Curr. Opin. Immunol.* **9**, 57–63.
28. Haas, W., Pereira, P., and Tonegawa, S. (1993). *Ann. Rev. Immunol.*, **11**, 637–85.
29. Leahy, D. J. (1995). *FASEB J.*, 17–25.
30. Ryu, S-E., Kwong, P. D., Truneh, A., Porter, T. G., Arthos, J., Rosenberg, M *et al.* (1990). *Nature*, **348**, 419–26.
31. Alberola-Ila, J., Takaki, S., Kerner, J. D., and Perlmutter, R. M. (1997). *Ann. Rev. Immunol.*, **15**, 125–54.
32. Philpott, K. L., Viney, J. L., Kay, G., Rastan, S., Gardiner, E. M., Chae, S *et al.* (1992). *Science*, **256**, 1448–52.
33 Mombaerts, P., Clarke, A. R., Rudnicki, M. A., Iacomini, J., Itohara, S., Lafaille, J. J., *et al.* (1992). *Nature*, **360**, 225–31.
34. Itohara, S., Mombaerts, P., Lafaille, J. J., Iacomini, J., Nelson A., Clarke, A. R., *et al.* (1993). *Cell*, **72**, 337–348.
35. Rahemtulla, A., Fung-Leung, W. -P., Schilham, S. R., Kündig, T. M., Sambhara, S. R., Narendran, A., *et al.* (1991). *Nature*, **353**, 180–4.
36. Fung-Leung, W. -P., Schillham, M. W., Rahemtulla, A., Kündig, T. M., Vollenweider, M., Potter, J., *et al.* (1991). *Cell*, **65**, 443–9.
37. Nakayama, K., Nakayama, K., Negishi, I., Kuida, K., Louie, M., Kanagawa, O., *et al.* (1994). *Science*, 1131–3.
38. Fung-Leung, W. -P., Kündig, T. M., Ngo, K., Panakos, J., DeSouza Hitzler, J., *et al.* (1994). *J. Exp. Med.*, **180**, 959–67.
39. Crooks, M. E. C. and Littman, D. R. (1994). *Immunity*, **1**, 277–85.

Heinz Jacobs
Basel Institute for Immunology,
Grenzacherstrasse 487,
CH-4005 Basel, Switzerland

Tetraspans

The tetraspans are homologous surface proteins with four transmembrane domains which are found in most cell types. Although their exact function is still unknown, a major feature of these molecules is their ability to associate together and with several other surface molecules such as β1 integrins or HLA antigens. Some of them have been shown to modulate adhesion and cell migration and to deliver activation signals. Because the tetraspans are engaged in large molecular complexes (the tetraspan network), it is possible that these effects are mediated through associated molecules.

Synonyms

Transmembrane 4 superfamily (TM4SF), tetraspanins.

Protein properties

The molecules of the tetraspan superfamily are 200–300 amino acids in length (Table 1) and are characterized by four hydrophobic regions of sufficient length to span the membrane bilayer. The N and C termini are predicted to form short cytoplasmic domains (5–14 amino acids). The two extracellular loops are unequal, the small domain having 20–27 amino acids and the larger domain, which is the most variable region, 75–130 amino acids.[1]

A major group of tetraspans can be distinguished based on sequence homology and gene structure when available. It includes several leukocyte antigens (although their expression is not restricted to leukocytes): CD9, CD37, CD53, CD63, CD81 (TAPA-1), CD82, CD151 (PETA-3/SFA-1), and non-CD molecules such as Co 029, A15/TALLA-1 and NAG-2. The parasitic worm *Schistosoma*

Table 1

	MW (kDa)	Length (a.a.)	GenBank number	Gene	Chromosome
CD9	24–26	228	M38690	Y	12p13/m6
CD81 (TAPA-1)	26	236	M33680	Y	11p15.5/m7
CO-029	27–34	237	M35252	N	NK
CD151/PETA-3	27	253	U14650	Y	11p15.5
CD82 (R2/IA4)	50–80	267	X53795	Y	11p12/m2
CD53	32–40	219	M37033	Y	1p12–p31/m3
CD63 (ME491)	30–60	237	M58485	Y	12q12–12q14/m10–18
CD37	40–50	281	X14046	Y	19p13–q13.4/m7
A15/TALLA-1	38–45	244	D29808	N	Xq11
NAG-2	28–35	238	AF022813	N	NK
TI1/UPIb	28	260	Z29378	N	3q13.3–q21
UPIa	27	258	Z29475	N	NK
SAS	NK	210	U01160	Y	12q13–14
Peripherin/RDS	39	345	J02884	Y	6p
Rom-1	33	351	L07894	Y	11q13
il-TMP	22–40	202	U31449	N	NK
L6	24	202	M90657	Y	3q21–25
Lbl	NK	208	U49081		
Sm23 and Sj23	23	218	M34453		

Y, characterized; N, uncharacterized; NK, not known.

antigens Sm23 and Sj23 are closely related to these molecules. These molecules have an identity score (determined by Clustal) ranging from 25 to 45 per cent to each other. SAS, a molecule amplified in sarcomas, is more distant. Another group comprises the bladder epithelium molecules UPIa and UPIb (TI1) which have 39 per cent identity to each other and 14 to 23 per cent identity to the proteins of the first group. Other molecules have been suggested to belong to the tetraspan family, namely the photoreceptor membrane proteins rom-1 and peripherin/RDS (35 per cent identity to each other) and il-TMP and L6 (50 per cent identity to each other) but these molecules have little homology with the molecules of the first two groups. Interestingly, a *Drosophila* protein involved in synapse formation, late bloomer (lbl), belongs to the tetraspan family.[2]

Most of these molecules are *N*-glycosylated in the large extracellular loop, except CD9, which is glycosylated in the short extracellular domain, and CD81 which is not. This explains the pattern of apparent molecular weights of these molecules (Table 1). Four cysteines are conserved in the large cellular loop; they are involved in the correct folding of the molecules since mAbs do not recognize the tetraspans in reducing conditions.

Results from the workshops on human differentiation antigens suggest that all nucleated cells express several tetraspans. Some molecules, such as CD81 and CD82, are widely expressed, whereas others have a restricted pattern of expression. CD37 is specific for the B lymphoid cell lineage whereas CD53 is strictly expressed by leukocytes. While these proteins have usually been identified on the plasma membrane, two of them, CD9 and CD63, are also present in internal granules. CD63 is the best characterized and is readily expressed on the cell surface after platelet, endothelial cell, or neutrophil activation.

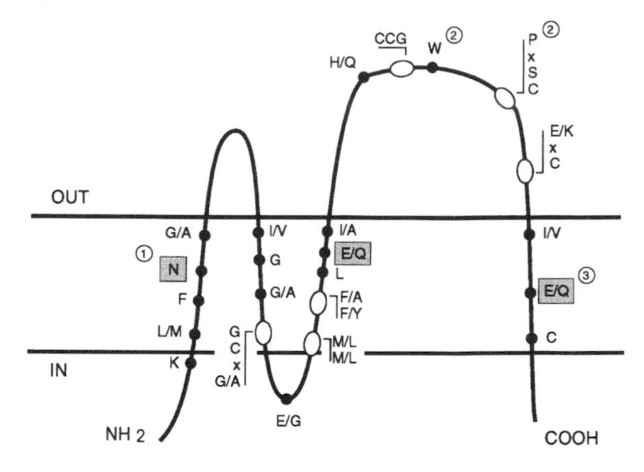

Figure 1. Schematic drawing of the tetraspans. The amino acids that are conserved in most of the human members of the first group (one or two choices) are indicated. The polar amino acids in the transmembrane domains are shaded. x; any amino acid. (1) not present in CD63, or TALLA-1; (2) not in CD9, CD81, or CO-029; (3) not present in CD9.

One of the most striking features of the tetraspans is their ability to aggregate and to associate with other identical molecules, including several $\beta1$ integrins (in particular $\alpha3\beta1$, $\alpha4\beta1$, and $\alpha6\beta1$) HLA class I and II.[3–7] These complexes are of high molecular weight as determined by cross-linking and continuous sucrose gradient analyses,[4] and they contain several tetraspans, including multiple copies of one given tetraspan.[5] The non-tetraspan molecules associated with tetraspans can be found in the same complexes as demonstrated for $\beta1$ integrins and HLA-DR antigens.[5] When a tetraspan is expressed after transfection, it is apparently included into the pre-exist-

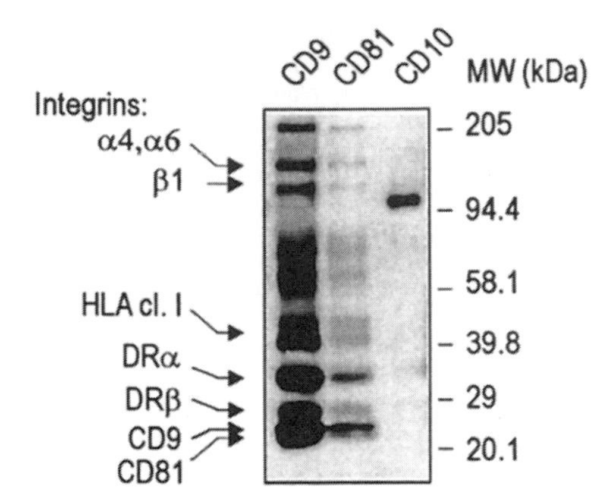

Figure 2. Immunoprecipitation performed after biotin labelling of the CD9-transfected B lymphoid cell line Raji, and cell lysis in 1 per cent Brij 97. CD9, and CD81 mAbs, but not the control CD10 mAb, coprecipitate the same molecules. Identified bands are indicated on the left.

ing complexes.[5] Finally, these complexes contain a large fraction of these molecules as determined by colocalization or cocapping studies.[4,8] These features, together with the fact that mAbs directed to tetraspans and associated molecules produce similar effects, suggest their existence on the cell surface in macromolecular complexes, possibly as a molecular network (the tetraspan network or tetraspan web).[5,7]

Various studies suggest a role in adhesion, migration, and costimulation. Because the tetraspans are associated with diverse surface molecules, part of these functions may be mediated by these partner molecules. Thus the modulation of cell adhesion and migration induced by tetraspans could be mediated by β1 integrins.[8–13] MAbs against CD81, originally described for their antiproliferative effects, can induce the phosphorylation of several proteins in B cells[14] and potentiate the activation through surface immunoglobulins.[15] Because CD81 is associated with CD19, a signal-transducing molecule, on B cells, it is possible that the activity of CD81 is at least partially mediated by this latter molecule.[15] CD9, CD81, and CD82 engagements deliver costimulatory signals for CD3 activation of T cells.[16–18] In these cells, CD81 and CD82 are associated with CD4 or CD8.[19] Cross-linking of mAbs against CD53 induces cytoplasmic calcium fluxes in monocytes and B cells.[20] MAbs against CD9 are also known to activate platelets, by a mechanism dependent on the recruitment of the platelet Fc receptor (FcγRII/CD32).[21] Finally, CD53 and CD63 have been shown to be associated with a tyrosine phosphatase activity in rat cells[22] and CD63 and CD81 with PI4 kinase.[23]

Since no extracellular ligand has been identified and because the tetraspans interfere with important cellular functions such as signal transduction, adhesion, and migration, it is possible that their function is the formation of the tetraspan network. The role of this network could be to organize the cell surface, thus coupling different surface molecules to various cellular functions. A role as a molecular chaperone is also possible, since the CD9 antigen also associates in renal carcinoma cells with the transmembrane precursor of the growth factor heparin-binding EGF (HB-EGF), in a way that upregulates the juxtacrine activity of this growth factor, as well as its activity as a receptor for diphtheria toxin.[24]

Purification

No protocols available.

Activity

No information available.

Antibodies

Most of the tetraspans, except SAS, are recognized by monoclonal antibodies that recognize the large extracellular loop.

Genes

See Table 1.

Disease states

The expression of CD9, CD63, and CD82 is inversely correlated with metastasis in different cancers and transfection of these molecules in tumour cell lines has been shown to reduce metastasis.[9,25,26] CD9 mAbs inhibit infection by FIV and CDV (canine distemper virus) but CD9 does not directly bind the virus, suggesting a role as a viral co-receptor.[27,28] It is also a co-receptor for diphtheria toxin.[24] Uroplakin I mediates anchorage of of *E. coli* to the urothelial surface.[29] Retinitis pigmentosa can be caused by mutation in rom-1 or peripherin.[30]

References

1. Wright, M. D. and Tomlinson, M. G. (1994). *Immunol. Today*, **15**, 588–94.
2. Kopczynski, C. C., Davis, G. W., and Goodman, C. S. (1996). *Science*, **271**, 1867–70.
3. Angelisova, P., Hilgert, I., and Horejsi, V. (1994). *Immunogenetics*, **39**, 249–56.
4. Berditchevski, F., Zutter, M. M., and Hemler, M. E. (1996). *Mol. Biol. Cell*, **7**, 193–207.
5. Rubinstein, E., Le Naour, F., Lagaudrière, C., Billard, M., Conjeaud, H., and Boucheix, C. (1996). *Eur. J. Immunol.*, **26**, 2657–65.
6. Lagaudriere-Gesbert, C., LebelBinay, S., Wiertz, E, Ploegh, H. L., Fradelizi, D., and Conjeaud, H. (1997). *J. Immunol.*, **158**, 2790–7.
7. Szöllosi, J., Horejsi, V., Bene, L., Angelisova, P., and Damjanovgich, S. (1996). *J. Immunol.*, **157**, 2939–46.
8. Rubinstein, E., Le Naour, F., Billard, M., Prenant, M., and Boucheix, C. (1994). *Eur. J. Immunol.*, **24**, 3005–13.

9. Ikeyama, S., Koyama, M., Yamaoko, M., Sasada, R., and Miyake, M. (1993). *J. Exp. Med.*, **177**, 1231–7.
10. Masellis-Smith, A. and Shaw, A. R. E. (1994). *J. Immunol.*, **152**, 2768–77.
11. Shaw, A. R. E., Domanska, A., Mak, A., Gilchrist, A., Dobler, K., Visser, L., *et al.* (1995). *J. Biol. Chem.*, **270**, 24092–9.
12. Behr, S. and Schriever, F. (1995). *J. Exp. Med.*, **182**, 1191–9.
13. Radford, K. J., Thorne, R. F., and Hersey, P. (1997). *J. Immunol.*, **158**, 3353–8.
14. Schick, M. R., Nguyen, V. Q., and Levy, S. (1993). *J. Immunol.*, **151**, 1918–25.
15. Tedder, T. F., Zhou, L. J., and Engel, P. (1994). *Immunol. Today*, **15**, 437–42.
16. Lebel-Binay, S., Lagaudrière, C., Fradelizi, D., and Conjeaud, H. (1995). *J. Immunol.*, **155**, 101–10.
17. Tai, X. G., Yashiro, Y., Abe, R., Toyooka, K., Wood, C. R., Morris, J., *et al.* (1996). *J. Exp. Med.*, **184**, 753–8.
18. Todd, S. C., Lipps, S. G., Crisa, L., Salomon, D. R., and Tsoukas, C. D. (1996). *J. Exp. Med.*, **184**, 2055–60.
19. Imai, T., Kakizaki, M., Nishimura, M., and Yoshie, O. (1995). *J. Immunol.*, **155**, 1229–39.
20. Rasmussen, A. M., Blomhoff, H. K., Stokke, T., Horejsi, V., and Smeland, E. B. (1994). *J. Immunol.*, **153**, 4997–5007.
21. Rubinstein, E., Boucheix, C., Worthington, R. E., and Carroll, R. C. (1995). *Sem. Thromb. Hemost.*, **21**, 10–22.
22. Carmo, A. M. and Wright, M. D. (1995). *Eur. J. Immunol.*, **25**, 2090–5.
23. Berditchevski, F., Tolias, K., Wong, K., Carpenter, C., and Hemler, M. (1997). *J. Biol. Chem.*, **272**, 2595–8.
24. Iwamoto, R., Higashiyama, S., Mitamura, T., Taniguchi, N., Klagsbrun, M., and Mekada, E. (1994). *EMBO J.*, **13**, 2322–30.
25. Radford, K. J., Mallesch, J., and Hersey, P. (1995). *Int. J. Cancer*, **62**, 631–5.
26. Dong, J. T., Lamb, P. W., Rinker-Schaeffer, C. W., Vukanovic, J., Ichikawa, T., Isaacs, J. T., and Barrett, J. C. (1995). *Science*, **268**, 884–6.
27. Willett, B. J., Hosie, M. J., Jarrett, O., and Neil, J. C. (1994). *Immunology*, **81**, 228–33.
28. Löffler, S., Lottspeich, F., Lanza, F., Azorsa, D. O., ter Meulen, V., and Schneider-Schaulies, J. (1997). *J. Virol.*, **71**, 42–9.
29. Wu, X. R., Sun, T. T., and Medina, J. J. (1996). *Proc. Natl Acad. Sci. USA*, **93**, 9630–5.
30. Bascom, R. A., Manara, S., Collins, L., Molday, R. S., Kalnins, V. I., and McInnes, R. R. (1992). *Neuron*, **8**, 1171–84.

Eric Rubinstein and Claude Boucheix
INSERM U268, Hôpital Paul Brousse,
94807 Villejuif Cedex, France

UNC-5 family

The UNC-5 family of proteins defines a subfamily of the immunoglobulin (Ig) superfamily characterized by two immunoglobulin-like (Ig) domains followed by two thrombospondin type-1 (Tsp1) repeats in the extracellular domain.[1] These proteins are putative netrin receptors involved in specifying migration away from netrin sources (i.e. repulsion).[1–6]

Homologous proteins

The UNC-5 family consists of *C. elegans* UNC-5[1] and three vertebrate homologues, UNC5H1, UNC5H2 and UNC5H3.[4,6] All UNC-5 proteins share a similar domain structure.

Protein properties

Members of the UNC-5 family have predicted molecular masses of approximately 110 kDa and share common structural features. They are defined as a subfamily of the Ig superfamily due to the unique combination of two immunoglobulin-like domains followed by two thrombospondin type-1 repeats in the extracellular domain. These proteins are presumably glycosylated based on the presence of one *N*-linked glycosylation site in the second Ig domain but the post-translational modification(s) of the UNC-5s has not been defined.

The UNC-5s contain a single predicted transmembrane domain. The vertebrate UNC-5s all have an apparent amino-terminal signal sequence for secretion. In contrast, the predicted protein product of *C. elegans unc-5* does not have an obvious amino terminal signal sequence. There is possible alternative splicing of the *unc-5* mRNA that replaces the first predicted coding exon (exon 2) with an alternative exon (exon 1). This alternative exon

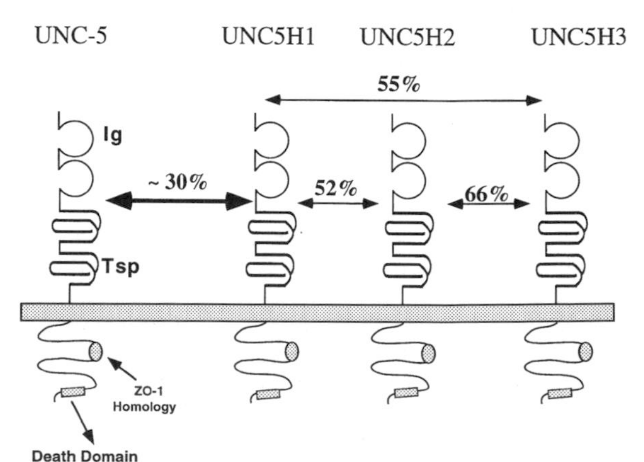

Figure 1.

would encode a different amino terminus which also lacks an obvious signal sequence for secretion.[1]

The intracellular domain of these proteins is large and does not contain any known catalytic domains. However, the cytoplasmic domain does possess two small regions of homology to known molecules. One region of homology is to zona occludens-1 (ZO-1), a protein that localizes to adherens junctions and is implicated in junction formation.[7] The second region is a death domain at the carboxyl termini of these proteins. Death domains are protein interaction domains found in a number of receptors including the low affinity NGF receptor (p75NGF-R) and the tumour necrosis factor receptor (TNFR).[8] In addition, there is a sequence in *C. elegans* UNC-5 that weakly resembles known SH3 domains, but does not contain the conserved tryptophan normally found near the middle of these modules. How these different domains contribute to the function of UNC-5 family members is not known.

Unc-5 was originally identified in screens of *C. elegans* locomotor mutants defective for guidance of commissural and motor axons along the dorsoventral body axis.[2] These screens, as well as an independent effort to identify factors secreted from floor plate cells that attract spinal commissural axons to the ventral midline in vertebrates, led to the identification of the netrin family of laminin-related secreted proteins, which include the *C. elegans* netrin UNC-6.[9,10] Several lines of evidence strongly suggest that UNC-5 is a cell-autonomous receptor or component of a receptor complex that mediates chemorepulsive responses to UNC-6.[1] First, mosaic analysis indicates that UNC-5 function is required within migrating cells and growth cones to confer normal guidance properties to these migrations.[1] Consistent with this, the expression of a haemagglutinin (HA)-tagged version of *unc*-5 rescues most defects of a null mutant and is detected on the rescued cells and axons (M. Killeen and J. Culotti; unpublished observations). Second, double mutant analysis indicates that the dorsal guidance function of UNC-5 requires UNC-6 and vice versa.[2] Finally, the expression of UNC-5 is not only necessary, but also sufficient, to confer responses to UNC-6, as demonstrated by the finding that ectopic expression of UNC-5 in some neurones redirects their axons away from an unc-6 source, presumably through chemorepulsion, in an UNC-6-dependent manner.[3] The ability of UNC-5 to steer growth cones was used as the basis for a suppressor screen to identify additional genes involved in *unc-5*-mediated guidance.[11] Eight genes were identified, including *unc-40*. This result is consistent with the hypothesis that repulsion, at least under some circumstances, requires the expression of both UNC-5 and UNC-40 (DCC).

Two of the vertebrate homologues of *Unc-5, Unc5h1* and *Unc5h2,* were isolated on the basis of their homology to the *C. elegans* gene.[4] The mouse *Unc5h3^{rcm}*, was independently discovered as the product of the mouse *rostral cerebellar malformation (rcm)* gene.[5,6] The vertebrate homologues have all been shown to bind netrin-1 with k_d values in the nanomolar range, consistent with the effective dose for the axon outgrowth-promoting effects of netrin-1.[4] Northern blot analysis reveals that these three genes are expressed both during development and in the adult organism.[4,6] *In situ* hybrization analysis of the developing embryo reveals complex patterns of expression of the three vertebrate *unc-5* homologues in a variety of neuronal and non-neuronal tissues, in several cases in overlapping patterns.[4,12] One example of the overlapping expression of these genes is in the developing cerebellum, where all three vertebrate *unc-5* homologues are expressed.[4,6] This is striking in light of the phenotype of *rcm*-mutant mice which display ataxia.[6] Histological analysis of these mice suggests a role for UNC5H3 in cell migration because of the presence of a variety of defects in cerebellar development including the misrouting of migrating granule cells.[6]

Purification

No methods available.

Activities

The UNC-5 protein has no known enzymatic activity. It is believed to function as a chemorepellent receptor for netrin ligands. This assignment of function is based on phenotypic analysis of *C. elegans* mutations in the *unc-5* and *unc-6* genes and on the fact that the vertebrate homologues of UNC-5 bind netrin-1 with k_d values in the nanomolar range, consistent with the effective dose for the axon outgrowth-promoting effects of netrin-1.[4]

Protein detection

Transgenic *C. elegans* are available that either express an HA-tagged version of UNC-5 that can be detected by indirect immunofluorescence, or that express GFP (*Aequoria victoria* green fluorescent protein) from the *unc-5* regulatory region in the same set of cells that express the HA-tagged protein (M.-W. Su, M. Killeen, and J. Culotti; unpublished observations). Rabbit polyclonal antisera raised against peptides corresponding to sequences in the extracellular domain of UNC5H1 and UNC5H2 are available.[4] These antibodies recognize the protein expressed on the surface of tissue culture cells transfected with the corresponding gene.

Genes

A full length *C. elegans unc-5* cDNA has been cloned and sequenced (GenBank accession number S47134).[1] The corresponding gene on cosmid C37G6 has been entirely sequenced by the *C. elegans* genome sequencing consortium (EMBL: S47168.1). Full length cDNAs are also available for *Unc5h1*,[4] *Unc5h2*,[4] and *Unc5h3^{rcm}*.[6] The GenBank accession numbers are U87305, U87306 and U72634, respectively. *Unc5h1* maps to the central region of mouse chromosome 13 in a region of homology with human chromosome 5q (D. J. Gilbert, N. G. Copeland, and N. A. Jenkins; unpublished observations). *Unc5h2* maps to the central region of mouse chromosome 10 in a region of homology with human chromosome 10q and 22q (D. J. Gilbert, N. G. Copeland, and N. A. Jenkins; unpublished observations). *Unc5h3^{rcm}* maps to mouse chromosome 3 in a region of homology with human chromosome 4q21–24.[5]

Mutant phenotypes/disease states

In *C. elegans*, *unc-5* mutants have defects in a variety of cell and pioneer axon growth cone migrations, all of which are dorsally oriented migrations that occur on the basal surface of the epidermis.[2] Affected cells include motoneurones (AS, DA, DB, DD, VD), the neurone-like excretory cell, and three mesodermal cell types (distal tip cell, head mesodermal cell, and male linker cell). Mice homozygous for the spontaneous *rostral cerebellar malformation* mutation (*Unc5h3^{rcm}*) exhibit laminar structure abnormalities in lateral regions of the rostral cerebellum.[5] Both *Unc5h3^{rcm}* and another transgenic insertion allele (*Unc5h3$^{rcm-TgN(Ucp)1.23Kz}$*) exhibit cerebellar and midbrain defects with fewer and smaller cerebellar folia and ectopic cerebellar Purkinje and granule cells present in the midbrain.

References

1. Leung-Hagesteijn, C., Spence, A. M., Stern, B. D., Zhou, Y., Su, M.-W., Hedgecock, E. M., and Culotti, J. G. (1992). *Cell*, **71**, 289–99.
2. Hedgecock, E. M., Culotti, J. G., and Hall, D. H. (1990). *Neuron*, **2**, 61–85.
3. Hamelin, M., Zhou, Y., Su, M.-W., Scott, I. M., and Culotti, J. G. (1993). *Nature*, **364**, 327–30.
4. Leonardo, E. D., Hinck, L., Masu, M., Keino-Masu, K., Ackerman, S. L., and Tessier-Lavigne, M. (1997). *Nature*, **386**, 833–8.
5. Lane, P. W., Bronson, R. T., and Spencer, C. A. (1992). *J. Hered.*, **83**, 315–8.
6. Ackerman, S. L., Kozak, L. P., Przyborski, S. A., Rund, L. A., Boyer, B. B., and Knowles, B. B. (1997). *Nature*, **386**, 838–42.
7. Willot, E., Balda, M. S., Fanning, A. S., Jameson, C., Van Itallie, C., and Anderson, J. M. (1993). *Proc. Natl Acad. Sci. USA*, **90**, 7834–8.
8. Hofman, K. and Tschopp, (1995). *FEBS Lett.*, **371**, 321–3.
9. Ishii, N., Wadsworth, W. G., Stern, B. D., Culotti, J. G., and Hedgecock, E. M. (1992). *Neuron*, **9**, 873–81.
10. Serafini, T., Kennedy, T. E., Galko, M. J., Mirzayan, C., Jessell, T. M., and Tessier-Lavigne, M. (1994). *Cell*, **78**, 409–24.
11. Colavita, A. and Culotti, J. G. (1998). *Dev. Biol.*, **194**, 72–85.
12. Przyborski, S., Knowles, B. B., and Ackerman, S. L. (1998). *Development*, **125**, 41–50.

Joseph G. Culotti
Division of Molecular Immunology and Neurobiology Samuel Lunenfeld Research Institute, Mount Sinai Hospital, Toronto M5G 1X5, Canada

Lindsay Hinck
Department of Biology, University of California, Santa Cruz, CA, USA

Marc Tessier-Lavigne
Howard Hughes Medical Institute and Department of Anatomy, University of California, San Francisco, CA, USA

VCAM-1

VCAM-1 (vascular cell adhesion molecule-1) is an immunoglobulin (Ig) superfamily adhesion molecule induced on the surface of endothelial cells at sites of inflammation. It binds the integrin VLA-4 ($\alpha4\beta1$) present on all leukocytes except neutrophils, recruiting these cells from the bloodstream to sites of infection and/or inflammation.

Protein properties

VCAM-1 (also called INCAM-110) was originally cloned by a functional assay from a cDNA library of IL-1 stimulated human umbilical vein endothelial cells.[1] It binds a variety of lymphoid and monocytic cell lines[1–3] as well as certain tumour cell lines,[4] via the integrin VLA-4[5]. More recent studies show that VCAM-1 can bind to the closely related integrin $\alpha4\beta7$,[6,7] which is expressed on a number of leukocyte subsets.[7]

Two forms of VCAM-1 mRNA are present in stimulated HUVECs; the originally reported sequence predicts six Ig homologous domains of the H or C2 type,[8] while the more abundant seven-domain form[9–11] is identical except for an additional domain in the middle, probably encoded by an exon that is occasionally skipped during splicing to generate the minor six-domain form. The protein has a classic N-terminal signal sequence that is cleaved to generate the mature protein, and a predicted hydrophobic transmembrane region followed by a short (19 amino acid) cytoplasmic tail. The six-domain form is

~95 kDa while the seven-domain form is ~110 kDa. The six-domain form has six predicted *N*-linked glycosylation sites, while the seven-domain form has seven. Both six- and seven-domain forms of the protein bind VLA-4-bearing cells when expressed on the surface of *cos*7 cells. Rabbit VCAM-1 can be expressed as an eight-domain protein,[12] and in mouse as a glycosyl phosphatidylinositol-linked three-domain form.[12]

Cultured endothelial cells can be stimulated to produce VCAM-1 by the inflammatory cytokines IL-1 and TNF[1], bacterial endotoxin (LPS),[2,4] and the T-cell stimulatory cytokine IL-4.[13,14] In response to TNF or IL-1, VCAM-1 mRNA can be detected at 1–2 h after addition of cytokine, is present at maximal levels by 2.5 h, and is sustained at substantial levels for at least 72 h in the continued presence of cytokine.[1,11] Binding activity for the T-cell line Jurkat[1] and for anti-VCAM-1 mAb 4B9[2] was found to reach maximal levels 4–6 h after stimulation and to be sustained at substantial levels for at least 48 h.

In vivo, VCAM-1 has been implicated in a number of physiological and pathological processes, including leukocyte extravasation, production and maturation of B lymphocytes, and perhaps metastasis of certain solid tumours. In support of its role in recruitment of leukocytes to sites of inflammation, wound healing, and/or infection, VCAM-1 is found on endothelial cells in a variety of inflamed tissues, on lymphoid dendritic cells, synovial linking cells, some tissue macrophages, and reactive mesothelial cells.[15] It is required for adhesion to endothelial cells of CD18 lymphocytes from LAD (leukocyte adhesion deficiency) patients, who show nearly normal lymphocyte extravasation but deficient neutrophil extravasation due to lack of CD11/CD18-dependent adhesion of neutrophils to endothelium.[16] VCAM-1 is present on endothelium of atherosclerotic plaques in a rabbit model system.[17] VCAM-1 has also been implicated in adhesion of human B cells to lymphoid germinal centres in the spleen.[18] In murine bone marrow, a VCAM-like antigen is present on stromal cells; mAbs to this antigen interfere with B lymphocyte formation in long-term bone marrow culture. Finally, several melanoma cell lines express VLA-4 and bind to VCAM-1, suggesting that this adhesion pathway could play a role in metastasis of the solid tumours from which these lines are derived.[4]

Purification

Several purified forms of VCAM-1 have been reported. A recombinant soluble form of VCAM-1, in which the seven extracellular domains are retained, but the cytoplasmic tail and transmembrane region have been removed, has been described.[19] VCAM-1 is proteolytically cleaved from the surfaces of cells and circulates at significant levels in plasma and has been purified from human plasma or serum[20] as the seven-domain form. VCAM–Ig fusion proteins containing the first two or three Ig domains (encompassing a VLA-4 binding site) have also been described.[21]

Activities

VCAM-1 binds VLA-4-bearing cells, such as the lymphoblastoid and myeloid cell lines RAMOS, Jurkat, HL60, and U937, and peripheral blood mononuclear leukocytes. This binding can be measured by a simple adhesion assay, either to VCAM-1-expressing cells or to immobilized recombinant forms of VCAM-1.[1,19] VCAM-1 also binds the integrin $\alpha4\beta7$.[6,7] The interaction of VCAM-1 with both $\alpha4$ integrins can be monitored directly using a novel solution phase assay which measures the direct binding of VCAM–Ig to the cell surface.[6,21] The interaction of biotinylated VCAM-1 with immobilized, partially purified VLA-4 integrin has also been described.[22]

Antibodies

Monoclonal antibodies binding to human,[2,4,14] macaque,[11] rabbit,[17] mouse,[7] and rat[23] VCAM-1 have been described; most can block adhesion to VLA-4-bearing cells, and some have been used successfully *in vivo* to evaluate the role of VCAM-1 in inflammatory disease models.[7,23]

Genes

cDNA sequences have been published for the human six-domain form[1] (GenBank M30257) and seven-domain form[9–11] (GenBank M60335).

Mutant phenotype/disease states

Naturally occurring mutants of VCAM-1 have not been described. However, mice lacking VCAM-1 through gene targeting have been described.[24] Major defects have been noted in the formation of the placenta and the coronary vessels due to lack of VCAM-1 expression in the allantois and myocardium, respectively, and leading to embryonic lethality in almost all cases.

Structure

Rotary shadowing electron microscopy of recombinant seven-domain VCAM-1 indicates an extended, slightly bent rod-like structure.[25] The first two domains of human VCAM-1, containing a VLA-4 binding site, have recently been crystallized and a high resolution structure of this portion of VCAM-1 obtained.[26,27] The molecule has an extended loop in the first domain believed to interact with the integrin counter-ligand (Plate 9).

References

1. Osborn, L., Hession, C., Tizard, R., Vassallo, C., Luhowskyj, S., Chi-Rosso, G., and Lobb, R. (1989). *Cell,* **59**, 1203–11.
2. Carlos, T. M., Schwartz, B. R., Kovach, N. L., Yee, E., Russo, M., Osborn, L. *et al.* (1990). *Blood,* **76**, 965–70.

3. Rice, G. E., Munro, J. M., and Bevilacqua, M. P. (1990). *J. Exp. Med.*, **171**, 1369–1374.
4. Rice, G. E. and Bevilacqua M. P. (1989). *Science*, **246**, 1303–6.
5. Elices, M. J., Osborn, L., Takada, Y., Crouse, C., Luhowskyj, S., Hemler, M. E., and Lobb, R. R. (1990). *Cell*, **60**, 577–84.
6. Lobb, R. R., Antognetti, G., Pepinsky, R. B., Burkly, L. C., Leone, D. R., and Whitty, A. (1995). *Cell Adhes. Commun.*, **3**, 385–97.
7. Lobb, R. R. and Hemler, M. E. (1994). *J. Clin. Invest.*, **94**, 1722–8.
8. Hunkapillar, T. and Hood, L. (1989). *Adv. Immunol.*, **44**, 1–63.
9. Polte, T., Newman, W., and Venkat Gopal, T. (1990). *Nucl. Acids Res.*, **18**, 5901.
10. Cybulsky, M. I., Fries, J. W. U., Williams, A. J., Sultan, P., Davis, V. M., Gimbrone Jr, M. A., and Collins, T. (1990). *Am. J. Pathol.*, **138**, 815–20.
11. Hession, C., Tizard, R., Vassallo, C., Schiffer, S. G., Goff, D., Moy, P., *et al.* (1991). *J. Biol. Chem.*, **266**, 6682–5.
12. Moy, P., Lobb, R., Tizard, R., Olson, D., and Hession, C. (1993). *J. Biol. Chem.*, **268**, 8835–41.
13. Masinovsky, B., Urdal, D., and Gallatin, W. M. (1990). *J. Immunol.*, **145**, 2886–95.
14. Thornhill, M. H., Wellcome, S. M., Mahlouz, D. L., Lanchbury, J. S. S., Kyan- aung, U., and Haskard, D. O. (1991). *J. Immunol.*, **145**, 592–8.
15. Rice, G. E., Munro, J. M., Corless, C., and Bevilacqua, M. P. (1990). *Am. J. Pathol.*, **138**, 385–93.
16. Schwartz, B. R., Wayner, E. A., Carlos, T. M., Ochs, H. D., and Harlan, J. M. (1990). *J. Clin. Invest.*, **85**, 2019–22.
17. Cybulsky, M. I. and Gimbrone, M. A. (1991). *Science*, **251**, 788–91.
18. Freedman, A. S., Munro, J. M., Rice, G. E., Bevilacqua, M. P., Morimoto, C., McIntyre, B. W., *et al.* (1990). *Science*, **249**, 1030–3.
19. Lobb, R., Chi-Rosso, G., Leone, D., Rosa, M., Newman, B., Luhowskyj, S *et al.* (1991). *Biochem. Biophys. Res. Commun.*, **178**, 1498–504.
20. Gearing, A. J. H. and Newman, W. (1993). *Immunol. Today*, **14**, 506–12.
21. Jakubowski, A., Rosa, M. D., Bixler, S., Lobb, R., and Burkly, L. C. (1995). *Cell Adhes. Commun.* **3**, 131–42.
22. Makarem, R., Newham, P., Askari, J. A., Green, L. J., Clements, J., Edwards, M., *et al.* (1994). *J. Biol. Chem.*, **269**, 4005–11.
23. May, M. J., Entwistle, G., Humphries, M. J., and Ager, A. (1993). *J. Cell Sci.*, **106**, 109–19.
24. Curtner, G. C., Davis, V., Li, H., McCoy, M. J., Sharpe, A., and Cybulsky, M. I. (1995). *Genes Dev.*, **9**, 1–14.
25. Osborn, L., Vassallo, C., Browning, B. G., Tizard, R., Haskard, D. O., Benjamin, C. D., *et al.* (1994). *J. Cell Biol.*, **124**, 601–8.
26. Jones, E. Y., Harlos, K., Bottomley, M. J., Robinson, R. C., Driscoll, P. C., Edwards, R. M., *et al.* (1995). *Nature*, **373**, 539–44.
27. Wang, J.-H., Pepinsky, R. B., Stehle, T., Liu, J.-H., Karpusas, M., Browning, B., and Osborn, L. (1995). *Proc. Natl Acad. Sci. USA*, **92**, 5714–18.

Roy R. Lobb and Laurelee Osborn
Biogen, Inc., Cambridge, MA, USA

V-V-C2-C2-C2 CAMs

The V-V-C2-C2-C2 subfamily of the immunoglobulin superfamily comprises cell adhesion molecules that share structural characteristics, sequence homology, and, in part, tissue distribution and function. The first member of this family to be characterized structurally was the human melanoma cell adhesion molecule MUC18.[1] Additional members now include gicerin,[2] HEMCAM[3] haematopoietic MUC18 related cell adhesion molecule), ALCAM[4] activated leukocyte cell adhesion molecule), DM-grasp/SC1/BEN,[5–7] neurolin,[8] BCAM[9] basal cell adhesion molecule), and the Lutheran blood group antigen.[10]

Synonyms

The gene encoding MUC18[11] has been named MCAM (melanoma cell adhesion molecule) and the molecule was recently designated CD146. It is also known in the literature as Mel-CAM and S-Endo. Human ALCAM (CD166) has been identified in the chicken as DM-grasp,[5] SC1,[6] and BEN,[7] and in the rat as KG-CAM.[12] Neurolin, identified in the goldfish and zebrafish,[8] appears to be the ALCAM homologue in these species. Gicerin and HEMCAM, both identified in the chicken, are identical in most of their sequence and are likely to be splice variants of the same gene.[3] BCAM and the Lutheran blood group antigen are splice variants of a single human gene.[13]

Protein properties

All of these molecules are cell surface glycoproteins and consist of five Ig-like extracellular domains (two N-terminal V type domains followed by three C2 type domains), a transmembrane region, and a relatively short cytoplasmic tail (Fig. 1). All contain sites for *N*-linked glycosylation and the native molecules are glycosylated and range between 78 kDa and 113 kDa in apparent molecular mass. The members of this family are more similar to each other in sequence than to other members of the Ig superfamily and in addition to their Ig domain organization, share a conserved peptide sequence located between the second and third Ig-like domains (Fig. 1).

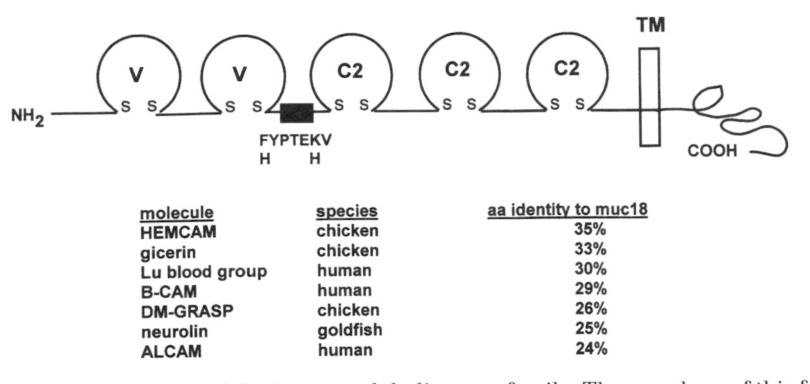

molecule	species	aa identity to muc18
HEMCAM	chicken	35%
gicerin	chicken	33%
Lu blood group	human	30%
B-CAM	human	29%
DM-GRASP	chicken	26%
neurolin	goldfish	25%
ALCAM	human	24%

Figure 1. The V-V-C2-C2-C2 subfamily of the immunoglobulin superfamily. The members of this family share a conserved seven amino acid peptide sequence between the second, and third Ig like domains. The comparison of the members to human MUC18 was made with the FastA program, and represents the percentage amino acid identity over 520 amino acids. TM, putative transmembrane region.

The proteins comprising this subfamily all appear to mediate intercellular adhesion and most have been shown to function as homophilic as well as heterophilic cell adhesion molecules. Human ALCAM is a ligand of the leukocyte molecule CD6[4] and the chicken homologue has been shown to be a ligand for the neural cell adhesion molecule NgCAM.[14] Gicerin binds to neurite outgrowth factor[2] and MUC18 binds to an undefined heterophilic ligand expressed by melanoma cells.[15]

Although each molecule demonstrates its own unique tissue distribution, a striking number of characteristics are shared. Most of these molecules are expressed in the placenta[10,16] and in the developing nervous and haematopoietic systems.[2,3,10] Both gicerin and the ALCAM homologues in the chicken and goldfish have been shown in *in vitro* studies to promote neurite extension and are thought to play a role in axon guidance.[2,5–8] All of these molecules may therefore play important roles in the development of different organ systems. In the adult organism, expression of these molecules can be associated with malignant transformation and with cell activation. The expression of MUC18,[17] BCAM,[18] and gicerin[19] is associated with malignant transformation and tumour progression in particular cell lineages. For MUC18, expression of this molecule by human melanoma cells has been shown to be associated with increased tumour growth and metastasis formation in animal models.[20,21] Although undetectable on resting lymphocytes, ALCAM,[4] gicerin/HEMCAM,[3] and MUC18[22] are expressed by populations of activated leukocytes. In addition, expression of HEMCAM, MUC18, and BCAM is observed on vascular smooth muscle and endothelium.[3,15,18]

Purification

These cell surface molecules are isolated from cell lysates using affinity chromatography with polyclonal or monoclonal antibody columns.

Activities

Homophilic and/or heterophilic cell adhesion has been demonstrated for most of these molecules by quantitating the aggregation of cDNA-transfected cell lines.

Antibodies

Monoclonal antibodies directed against each of these molecules have been described.

Genes

To date only the genomic organization of human and mouse MUC18 has been described[11] human, X68264–X68271; mouse, X74627, X74628). Based on mapping of EST clones, human MUC18 is located on chromosome 11. The human gene for BCAM and Lutheran blood group antigen has been localized to chromosome 19q13.2–13.3;[13] and the gene for human ALCAM has been localized to chromosome 3q13.1–2.[4]

cDNA sequences are available for human ALCAM (L38608), the chicken homologue (SC1, S63276; DM-grasp, M76678; BEN, X64301), the mouse homologue (L25274), the goldfish homologue (L25056), and the zebrafish homologue (L25057, L25273). Two human MUC18 cDNA sequences differing in the 3′ untranslated region have been identified (M29277, M28882), as have a variety of variants of the chicken gicerin cDNA showing differences in the cytoplasmic and/or in the extracellular region (D38559, D49849, Y08856, Y08855,Y08854). Human BCAM cDNA (X80026) and the Lutheran antigen cDNA (X83425) differ only in the length of the cytoplasmic regions.

Mutant phenotype/disease states

No information available.

Structure

Not available.

References

1. Lehmann, J. M., Riethmüller, G., and Johnson, J. P. (1989). *Proc. Natl Acad. Sci. USA*, **86**, 9891–5.
2. Taira, E., Takaha, N., Taniura, H., Kim, C. H., and Miki, N. (1994). *Neuron*, **12**, 861–72.
3. Vainio, O., Dunon, D., Aissi, F., Dangy, J. P., McNagny, K. M., and Imhof, B. A. (1996). *J. Cell Biol.*, **135**, 1655–68.
4. Bowen, M. A., Patel, D. D., Li, X., Modrell, B., Malacko, A. R., Wang, H., *et al.* (1995). *J. Exp. Med.*, **181**, 2213–20.
5. Burns, F. R., Kannen, S., Guy, L., Raper, J. A., Kamholz, J., and Chang, S. (1991). *Neuron*, **7**, 209–20.
6. Tanaka, H., Matsui, T., Agata, A., Tomura, M., Kubota, I., McFarland, K. C., *et al.* (1991). *Neuron*, **7**, 535–45.
7. Pourquie, O., Corbel, C., LeCaer, J. P., Rossier, J., and Le Douarin, N. M. (1992). *Proc. Natl Acad. Sci. USA*, **89**, 5261–5.
8. Laessing, U., Giordano, S., Stecher, B., Lottspeich, F., and Stuermer, C. A. O. (1994). *Differentiation*, **56**, 21–29.
9. Campbell, I. G., Foulkes, W. D., Senger, G., Trowsdale, J., Garin-Chesa, P., and Rettig, W. J. (1994). *Cancer Res*, **54**, 5761–5.
10. Parsons, S. F., Mallinson, G., Holmes, C. H., Houlihan, J. M., Simpson, K. L., Mawby, W. J., *et al.* (1995). *Proc. Natl Acad. Sci. USA*, **92**, 5496–500.
11. Sers, C., Kirsch, K., Rothbächer, U., Riethmüller, G., and Johnson, J. P. (1993). *Proc. Natl Acad. Sci. USA*, **90**, 8514–18.
12. Peduzzi, J. D., Irwin, M. H., and Geisert, E. E. (1994). *Brain Res.*, **640**, 296–307.
13. Rahuel,C., Le Van Kim, C., Mattei, M. G., Cartron, J. P., and Colin, Y. (1996). *Blood*, **88**, 1865–72.
14. DeBernardo, A. P. and Chang, S. (1996). *J. Cell Biol.*, **133**, 657–66.
15. Johnson, J. P., Rummel, M. M., Rothbächer, U., and Sers, C. (1996). *Curr. Topics Microbiol. Immunol.*, **213/I**, 95–105.
16. Shih, I. M. and Kurman, R. J. (1996)., *Lab. Invest.*, **75**, 377–88.
17. Lehmann, J. M., Holzmann, B., Breitbart, E. W., Schmiegelow, P., Riethmüller, G., and Johnson, J. P. (1987). *Cancer Res.*, **47**, 841–5.
18. Garin-Chesa, P., Sanz-Moncasi, M. P., Campbell, I. G., and Rettig, W. J. (1994). *Int. J. Oncol.*, **5**, 1261–6.
19. Takaha, E., Taira, E., Taniura, H., Nagino, T., Tsukamoto, Y., Matsumoto, T., *et al.* (1995). *Differentiation*, **58**, 313–20.
20. Xie, S., Luca, M., Huang, S., Gutman, M., Reich, R., Johnson, J. P., and Bar-Eli, M. (1997). *Cancer Res.*, **57**, 2295–303.
21. Bani, M. R., Rak, J., Adachi, D., Wiltshire, R., Trent, J. M., Kerbel, R. S., and Ben-David,Y. (1996). *Cancer Res.*, **56**, 3075–86.
22. Pickl, W. F., Majdic, O., Fischer, G. F., Petzelbauer, P., Fae, I., Waclavicek, M., *et al.* (1997). *J. Immunol.*, **158**, 2107–15.

Judith P. Johnson
Institute for Immunology,
Goethestrasse 31, 80336 Munich,
Germany

Yeast sexual agglutinins

In *Saccharomyces cerevisiae* α-agglutinin and a-agglutinin are complementary cell adhesion glycoproteins that cause aggregation of cells during conjugation.[1] Cells of α mating type express α-agglutinin, a cell wall-bound member of the immunoglobulin superfamily. α-Agglutinin binds with high affinity to a-agglutinin expressed by cells of **a** mating type. a-agglutinin is composed of two subunits, one a small glycoprotein that binds α-agglutinin, the other a large proteoglycan-like subunit mediating attachment to the cell wall.

Synonyms

α-agglutinin: α-agglutination substance[2] a-agglutinin: a-agglutination substance.[3]

Homologues

α-Agglutinin has functional analogues in other yeasts, the best characterized being the 21-factor of *Hansenula wingei*.[1,4] Related sequences and genes have been reported in *Candida albicans*, and some of these gene products mediate yeast adhesion to epithelial cells.[5,27,28] a-Agglutinin is structurally and functionally related to mating agglutinins in other yeasts, the best characterized being 5-factor of *Hansenula wingei*.[1,6]

Protein properties

In the yeasts, cell–cell binding during mating is mediated by complementary cell wall glycoproteins called agglutinins.[1] The agglutinins have been investigated in several species that have two mating types. The *S. cerevisiae* agglutinins have been well characterized genetically and biochemically and are described here.

The agglutinins are constitutively expressed on haploid cells, α-agglutinin on α cells, and a-agglutinin on **a** cells, although many **a** strains have very low constitutive a-agglutinin levels. Exposure of the cells to the sex pheromone secreted by cells of the opposite mating type induces increased expression of the agglutinin mRNAs

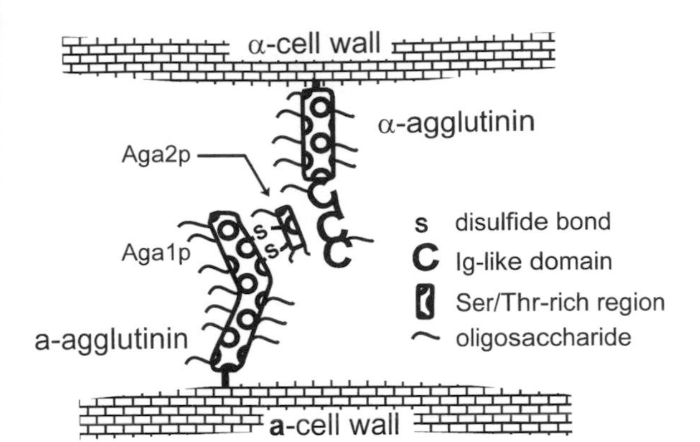

Figure 1. Schematic drawing of the *S. cerevisiae* agglutinins. The agglutinins are shown covalently bound to the cell walls of mating type **a** and α cells. α-Agglutinin (top) has, starting at the N terminus, three immunoglobulin-like domains, a Ser/Thr-rich stalk region, and a cell wall anchorage derived from a GPI anchor. a-Agglutinin (bottom) has two subunits, both of which are Ser/Thr-rich and glycosylated. The small binding subunit is encoded by the *AGA2* gene. The proteoglycan-like cell wall anchorage subunit is encoded by *AGA1*.

and a rise in cell surface agglutinin levels to 2–5 × 10^4 molecules per cell.[7–9] The pheromones act through well-defined signal transduction cascades.[10]

α-Agglutinin is the product of the *AGα1* gene, which encodes a 650 residue protein.[7,11] There is an N-terminal secretion signal, followed by three tandem immunoglobulin-like domains (domain I, residues 20–104; domain II, 105–199, and domain III, 200–324).[12] Residues in domain III that identify the ligand-binding site have been localized to the end of the β barrel opposite that corresponding to the CDR in immunoglobulins.[12–14] At least one other immunoglobulin-like domain participates in binding, and is required for activity.[15] Residues 325–627 of α-agglutinin probably constitute an extended and rigid 'stalk', with 40 per cent Ser and Thr residues and numerous *N*- and *O*-glycosylation sites.[12,16] This stalk holds the binding region of the Ig-like domain away from the surface of the cell wall.[15,16] The 23 C-terminal residues are a signal for addition of a glycosyl phosphatidylinositol (GPI) anchor. The anchor is added to the nascent α-agglutinin in the ER and remains until the protein is transported to the cell surface as a membrane-associated glycoprotein.[12,17] The anchor is then processed by an uncharacterized mechanism that results in loss of membrane attachment, addition of β1,6-glucan, and linkage into the cell wall at its outer surface.[18,19] Expression of α-agglutinin is specific to cells of α mating type, with most strains showing 1.5- to 2-fold increase in cell surface concentrations in response to exposure to the sex pheromone **a**-factor.[7,9,11]

a-Agglutinin is the product of the *AGA1* and *AGA2* genes.[14,20,21] *AGA2* encodes an 87 residue protein that contains an 18 residue N-terminal secretion signal, and the mature 69 residue Aga2p glycopeptide contains the binding determinants for α-agglutinin.[14,16] Glycosylation at 10–20 Ser and Thr residues is important for activity, but a non-glycosylated synthetic peptide of the C-terminal 10 residues has been reported to bind at low affinity.[16] Aga2p is expressed only in **a** cells, and transcription and cell surface levels are increased in response to pheromone.[20]

The *AGA1* product is a 725 residue Ser/Thr-rich protein (45 per cent) that is highly *O*-glycosylated. It has an N-terminal secretion signal and a C-terminal GPI addition signal. Aga1p attaches to the a-agglutinin binding subunit Aga2p through a pair of interchain disulphide bonds,[16] and also attaches to the **a** cell surface, presumably by a GPI-derived glucan link. Surprisingly, Aga1p is expressed in **a** and α cells and is pheromone induced in both mating types.[20]

The agglutinins bind to each other in a complex manner, and there is evidence for at least two binding states, weak and strong.[14,22] Binding of ^{125}I-labelled agglutinin shows that the strong binding state has a K_A of about 10^9 l/mol, with a slow association rate. No other activities of the proteins are known and strains carrying null mutations in the agglutinin genes mate on solid media with efficiencies 25–100 per cent of wild type. In contrast, wild-type alleles of these genes are required for efficient mating in liquid cultures, suggesting that the agglutinins are necessary to maintain intimate contact between potential **a** and α mating partners under these conditions.[11,20,21]

Purifications

α-Agglutinin has been purified from whole cell extracts (probably precursor forms) and from glucanase extracts of cell walls by conventional means, including ultrafiltration, partial deglycosylation, ion exchange chromatography, and HPLC.[7,23] An active form of α-agglutinin truncated after the immunoglobulin-like domains can be purified from medium after expression from a plasmid.[15] Affinity methods with **a** cells or a-agglutinin have not worked, probably because of the slow association of the components. Affinity chromatography has been successful with *H. wingei* agglutinins which show stronger interactions.[6] Aga2p, the a-agglutinin binding subunit, has also been purified by ion exchange and gel permeation.[24] Aga1p has only been isolated as fragments in association with Aga2p, these forms have molecular weights in the millions and carbohydrate content of 90–98 per cent.[1,21] The *H. wingei* homologue is of similar size and composition, with every Ser and Thr residue *O*-glycosylated.[1,6]

Activities

Binding activities are the only known function for these proteins. They are assayed by their ability to promote or inhibit cell–cell interaction in agglutination assays, either by visual inspection of agglutinates formed in the presence of serial dilutions of agglutinins,[24] or by light scattering in a colorimeter.[2,25]

Antibodies

High affinity polyclonal antibodies to α-agglutinin and to its Ig-like region have been produced.[7,9,11] Polyclonal antibodies to Aga2p have also been reported.[16] There are no monoclonals or commercial sources currently available.

Genes

AGα1 (also called *SAG1)* is YJR004C in the yeast genome sequence (chromosome X and its GenBank accession number is M28164). *AGA1* is YNR044W in the yeast genome sequence (chromosome XIV, accession M60590). *AGA2* is YGL032C in the yeast genome sequence (chromosome VII, accession X62877).

Mutant phenotypes

Mutations in any of the agglutinin genes lower mating efficiencies by 10^4–10^5 -fold in matings in liquid media, but have only minor effects in matings on the surface of agar or on nitrocellulose filters. There is no further loss of mating efficiency or additional phenotype for matings between pairs of non-agglutinable cells.[20]

Structure

No X-ray or NMR structures have been published. There is a homology model of Ig-like domain III of α-agglutinin based on sequence similarity, amino acid accessibility, and site-specific mutagenesis.[26]

References

1. Lipke, P. N. and Kurjan, J. (1992). *Microbiol. Rev.*, **56**, 180–94.
2. Yanagishima, N. (1984). In *Cellular interactions. Encyclopedia of plant physiology*, Series N, Vol. 17 (ed. H. F. Linskens and J. Haslop-Harrison), Springer.
3. Yamaguchi, M., Yoshida, K., and Yanagishima, N. (1984). *Arch. Microbiol.*, **140**, 113–19.
4. Burke, D., Mendonca-Previato, L., and Ballou, C. E. (1980). *Proc. Natl Acad. Sci. USA*, **77**, 318–22.
5. Hoyer, L. L., Scherer, S., Schatzman, A. R., and Livi, G. P. (1995). *Mol. Microbiol.* **15**, 39–54.
6. Yen, P. H. and Ballou, C. E. (1974). *Biochemistry*, **13**, 2428–37.
7. Hauser, K. and Tanner, W. (1989). *FEBS Lett.*, **255**, 290–4.
8. Terrance, K. and Lipke, P. N. (1987). *J. Bacteriol.*, **169**, 4811–5.
9. Wojciechowicz, D. and Lipke, P. (1989). *Biochem. Biophys. Res. Commun.*, **161**, 45–51.
10. Kurjan, J. (1993). *Ann. Rev. Genet.*, **27**, 147–79.
11. Lipke, P., Wojciechowicz, D., and Kurjan, J. (1989). *Mol. Cell. Biol.*, **9**, 3155–66.
12. Wojciechowicz, D., Lu, C. -F., Kurjan, J., and Lipke, P. N. (1993). *Mol. Cell. Biol.*, **13**, 2554–63.
13. de Nobel, H., Lipke, P. N., and Kurjan, J. (1996). *Cell. Mol. Biol.*, **7**, 143–53.
14. Cappellaro, C., Hauser, K., Watzele, M., Watzele, G., Gruber, C., and Tanner, W. (1991). *EMBO J.*, **10**, 4081–8.
15. Chen, M. -H., Shen, Z. -M., Bobin,S., Kahn, P. C., and Lipke, P. N. (1995). *J. Biol. Chem.*, **270**, 26168–77.
16. Cappellaro, C., Baldermann, C., Rachel, R., and Tanner, W. (1994). *EMBO J.*, **13**, 4737–44.
17. Lu, C. -F., Kurjan, J., and Lipke, P. N. (1994). *Mol. Cell. Biol.*, **14**, 4825–33.
18. Lu, C. -F., Montijn, R. C., Brown, J. L., Klis, F., Kurjan, J., Bussey, H., and Lipke, P. N. (1995). *J. Cell Biol.*, **128**, 333–40.
19. Kapteyn, J. C., Montijn, R. C., Vink, E., de la Cruz, J., Llobell, A., Douwes, J. E., *et al.* (1996). *Glycobiology*, **6**, 337–45.
20. De Nobel, H., Pike, J., Lipke, P. N., and Kurjan, J. (1995). *Mol. Gen. Genet.*, **247**, 409–15.
21. Roy, A., Lu, C. -F., Marykwas, D., Lipke, P. N., and Kurjan, J. (1991). *Mol. Cell. Biol.*, **11**, 4196–206.
22. Lipke, P., Terrance, K., and Wu, Y. -S. (1987). *J. Bacteriol.*, **169**, 483–8.
23. Terrance, K., Heller, P., Wu, Y. -S., and Lipke, P. (1987). *J. Bacteriol.*, **169**, 475–82.
24. Watzele, M., Klis, F., and Tanner, W. (1988). *EMBO J.*, **7**, 1483–8.
25. Terrance, K. and Lipke, P. N. (1981). *J. Bacteriol.* **148**, 889–96.
26. Lipke P. N., Chen, M. -H., de Nobel, H., Kurjan, J., and Kahn, P. C. (1995). *Protein Sci.*, **4**, 2168–78.
27. Gaur, N. K. and Klotz, S. A. (1997). *Infect. Immun.*, **65**, 5289–94.
28. Fu, Y., Reig, G., Fonzi, W. A., Belanger, P. H., Edwards, J. E., Jr, and Filler, S. G. (1998). *Infect. Immun.*, **66**, 1783–6.

Peter N. Lipke
Department of Biology, Hunter College of City University of New York,
New York, NY 10021, USA

Janet Kurjan
Department of Microbiology and Molecular Genetics and the Vermont Cancer Center, University of Vermont,
Burlington, VT 05405, USA

3

Extracellular matrix molecules

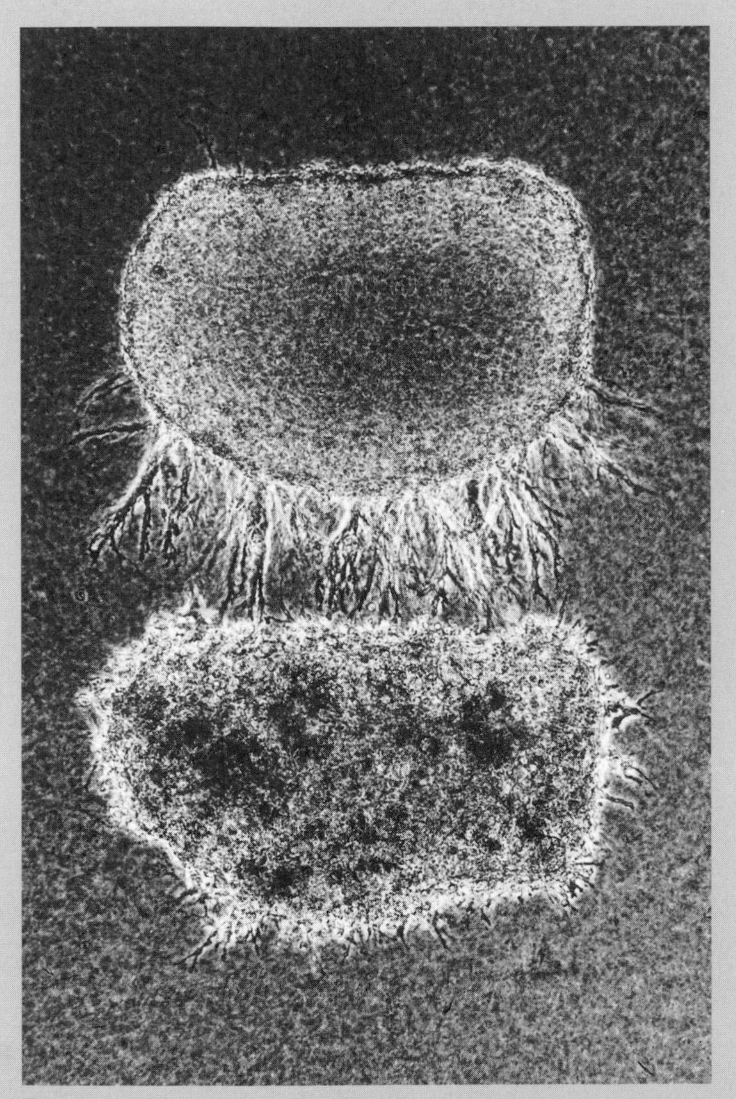

Chemotropic action of netrin-1 on spinal commissural axons. An aggregate of COS cells secreting recombinant netrin-1 (bottom) elicits directional outgrowth of commissural axons from an explant of rat dorsal spinal cord (top) embedded in a three dimensional collagen matrix. (Reproduced from Kennedy *et al.* 1994. *Cell*, **78**, 425–35.)

Introduction

The extracellular matrix regulates the development and function of almost all cells in metazoan organisms. Our understanding of extracellular matrix functions has been enhanced in recent years by discoveries of novel extracellular matrix constituents, mapping of sites crucial for interactions of extracellular matrix constituents with each other and with cells, characterization of proteases and protease inhibitors which regulate extracellular matrix assembly and turnover, and identification of novel receptors and signalling mechanisms that mediate cellular responses in extracellular matrix constituents.

The development and normal functioning of all cell types in an organism depend upon interactions with molecules in their environments. The major classes of molecules that regulate cellular development and function include growth and differentiation factors, cell adhesion molecules, and components of the extracellular matrix (ECM). The extracellular matrix is evolutionarily ancient, present in all phyla of metazoan organisms.[1] The extracellular matrix functions to provide order in extracellular space and has been shown to have many functions associated with establishment, separation, and maintenance of differentiated tissues and organs. For example, the matrix provides a substrate for organization of cells which adhere to it, such as epithelia. Matrix components have been shown to be essential for differentiation of many cell types, such as epithelial keratinocytes and mammary epithelial cells.[2] Matrix components include ligands that activate intracellular signalling pathways within cells, thereby regulating cell proliferation, survival, and differentiation.[3–7] Matrix components also interact with growth factors, chemotropic agents and other soluble factors that regulate cell proliferation, differentiation, and migration.[8,9] The ECM thereby helps regulate the spatial and temporal properties of signals conveyed by these molecules. The space between cells provided by the ECM has been shown to permit migration of cells and movements of growth cones.[10,11] The ordered structure made possible by the existence of the ECM makes it possible to establish gradients of molecules that direct the movements of these cells and growth cones.[12] Glycosaminoglycans in the ECM are particularly important in regulating the size and properties of the extracellular spaces in organs and embryos, in controlling macromolecular transport across the basal lamina permeation barriers, and for creating permissive environments for cell migration. The importance and mechanisms of action of these extended, highly charged, sugar chains have been confirmed by mutational analyses and other perturbations.[13,14] Finally, the ECM contributes to the mechanical integrity, rigidity, and elasticity of skin, the vasculature, tendons, lungs, and other organs. This has been confirmed by analyses of phenotypes of mice and humans with alterations in collagen or laminin genes.[15,16]

This chapter focuses on the extracellular matrix, its receptors, and their functions. More extended reviews on these topics include those of Venstrom and Reichardt,[11] Schwartz *et al.*,[3] Hynes,[16] Ruoslahti,[17] and Uitto and Pulkkinen.[18] A brief review of the salient properties of proteoglycans is included in this volume (Protweoglycans, pp. 351–356.). An overview of proteases, protease inhibitors, and their actions is also included later in this book (chapters by Sternlicht and Werb, pp. 505–563).

Composition and function of the extracellular matrix

The extracellular matrix was originally defined morphologically as extracellular material visible as fibrils or sheets in the electron microscope. It is now defined more broadly to include essentially all secreted molecules that are immobilized outside cells. In the brain, for example, a morphologically visible extracellular matrix is not detected, even though many of the molecules are present that are constituents of visible extracellular matrices in other tissues.[10,11]

Major constituents identified initially in extracellular matrices included collagens, non-collagenous glycoproteins, and proteoglycans. Each of these major classes of molecules appears to be present in all metazoa and must therefore be evolutionarily ancient. Evolutionary pressures have also, however, resulted in substantial diversification, including generation of new members of these families, within individual phyla. As examples of this diversification, specialized collagens essential for formation of the cuticle have been characterized in nematodes and for attachment in mussels.[19] Multiple noncollagenous glycoproteins, unique to vertebrates, have been identified that are associated with platelets or serum and regulate coagulation and vascular repair. Additional ECM glycoproteins, such as reelin, have evolved to permit novel developmental processes, such as laminar organization of the cerebellum and cortical regions in the vertebrate brain.[20] Within vertebrates, there is tremendous diversity in these molecules. The number of distinct collagens in humans is now greater than 18 and this does not include the multiple isoforms generated by expression of homologous genes or alternative splicing (e.g. collagen IV). There is similar diversity in adhesive glycoproteins. For example, laminin trimers can be assembled using one each of the products of five α, five β and three γ subunit genes with differential splicing and proteolysis contributing further to the diversity of structure and function present in the assembled trimers[21,22] (Burgeson, personal communication). Multiple sets of genes also encode isoforms of tenascin and thrombospondin. Differential splicing generates further diversity in the thrombospondin, tenascin, fibronectin, and many other ECM macromolecules. Many of the more recently discovered molecules or isoforms have very restricted and developmentally regulated distributions in developing embryos.[20,23]

Available evidence indicates that this diversity in molecules of the vertebrate ECM has important functional and developmental consequences. Over the past five years, a combination of human and murine genetics has provided several convincing examples of the roles of ECM proteins, their receptors, and cytoskeletal linker proteins in vertebrate development and function. Similar studies in *Drosophila* and *C. elegans* have further illuminated our understanding of the roles of these molecules. I will review briefly here studies on the roles of ECM in maintaining integrity of the epithelium and skeletal musculature, function of the kidney, and development of the nervous system.

Epithelial health is essential for viability and has been shown to be compromised by the absence of ECM proteins in the epithelial basal lamina of ECM proteins which strengthen the adhesive interaction between dermal and epidermal layers, of epithelial cell receptors mediating basal lamina formation or cellular interactions, or of cytoskeletal proteins within epithelial cells which form junctional complexes with the underlying basal lamina. Of particular interest is the expression of a specific epithelial laminin and a specific collagen isoform which strengthen interactions between the epithelium and underlying dermis. Mutations in any of the the three chains of laminin-5, whose expression is restricted to the epithelial basal lamina ($\alpha3$, β 3 or $\gamma2$) result in a severe and often lethal epithelial disorder named epidermolysis bullosa.[15] Mutations in collagen VII, whose fibres connect the ECM of the dermis with the basal lamina at the epithelial–dermal interface also result in a form of epidermolysis bullosa. A targeted mutation in the integrin $\alpha3$ gene, resulting in loss of the laminin receptor $\alpha3\beta1$, prevents assembly of a continuous basal lamina, causing comparatively mild skin blistering in newborn animals.[16] Probably, blistering would become more severe with age, but the animals do not survive because of deficits in kidney and lung function. Either murine or human mutations in the integrin $\alpha6\beta4$, a laminin receptor which is an essential constituent of hemidesmosomes, or human mutations in BPAG2/collagen XVII, an epithelial transmembrane linker protein, result in shear-mediated loss of epidermal adhesion to the underlying basal lamina and dermis with lethal consequences.[15,16] Similarly, mutations in either BPAG1 or plectin, two proteins in the cytoplasm of the epithelial cells which are associated with hemidesmosomes cause forms of epidermolysis bullosa in humans and severe skin blistering in mice. In summary, the basal lamina and ECM constituents that strengthen interactions with adjacent tissues are essential to maintain epithelial integrity. Analysis, primarily of humans with genetic disorders, has illuminated the roles of proteins mediating a transcellular linkage from the interior of the dermis through the basal lamina to the cytoskeleton within keratinocytes.

The basal lamina is similarly essential to maintain viability and function of skeletal muscle. Mutations reducing or eliminating expression of laminin $\alpha2$ (merosin), a subunit of laminin-2 ($\alpha2\beta1\gamma1$), which is the major laminin isoform in the basal lamina surrounding skeletal myotubes, result in murine and human muscular dystrophies.[24,25] Other human muscular dystrophies have been shown to be caused by mutations in subunits of the transmembrane dystroglycan/adhalin complex, which functions as a receptor for laminin-2. In mice, targeted ablation of the integrin $\alpha7$, causing absence of the laminin receptor $\alpha7\beta1$, has also been shown to induce a dystrophic phenotype. Finally, the cytoskeletal linker protein dystrophin received its name because genetic studies showed that mutations in its gene resulted in Duchenne's muscular dystrophy in humans. Subsequently a dystrophic mouse was shown to have a mutation in the same gene.[16,25] Thus analyses, largely of humans with genetic disorders, again document the importance of a basal lamina, transmembrane receptors, and proteins mediating receptor linkage to the cytoskeleton.

Alport syndrome, a progressive glomerulonephritis caused by deficits in serum filtration within the kidney glomerulus, has long been associated with mutations in the specific collagen IV isoforms expressed in the glomerular basal lamina.[26] Human mutations have now been identified in each of the subunits [$\alpha3$(IV), $\alpha4$(IV) and $\alpha5$(IV)] which assemble to form the specific isoform of collagen IV within the glomerulus. In each case, the glomerular collagen IV[$\alpha3$(IV) $\alpha4$(IV), $\alpha5$(IV)] is replaced with a network of a homologous collagen IV [$\alpha1(IV)^2$ $\alpha2$(IV)], but this is not sufficient to prevent disease. Murine models of Alport syndrome have been generated by targeted mutagenesis of collagen IV subunits.[27,28] Targeted mutagenesis has also been used to demonstrate that specific isoforms of laminin must be present in the kidney. Mutation of the laminin $\beta2$ subunit (s-laminin) results in progressive nephritis despite replacement of this subunit with the laminin $\beta1$ subunit.[29] It seems likely that genetic deletion of any of the basal lamina constituents within the glomerular basement membrane will result in nephritis, provided they do not cause earlier lethal phenotypes.

Studies on formation and regeneration of vertebrate neuromuscular junctions provide a particularly elegant system to illustrate the more dramatic functions of extracellular matrix. Conclusive experiments performed two decades ago demonstrated that the extracellular matrix at the neuromuscular junction contains all the information needed to induce synaptic differentiation by both axons and skeletal myotubes *in vivo*.[5,6] The basal lamina at the neuromuscular junction, which covers approximately 0.1 per cent of the myotube surface, is distinct molecularly from the non-synaptic basal lamina which covers the remainder of the myotube. The synaptic basal lamina contains a collagen IV trimer assembled from the $\alpha3$(IV) and $\alpha4$(IV) chains, while the non-synaptic basal lamina contains a collagen IV assembled from the $\alpha1$(IV) and $\alpha2$(IV) subunits. Similarly, the laminin isoform in the synapse is assembled primarily from the $\alpha5$, $\beta2$ (s-laminin), and $\gamma1$ chains, while laminin in the non-synaptic myotube basal lamina is assembled from the $\alpha2$ (merosin), $\beta1$, and $\gamma1$ chains. Additional molecules, such as isoforms of acetylcholinesterase containing collagen-like tails, are also localized specifically in the synaptic basal lamina. Inhibition of function of this esterase dramatically perturbs synaptic transmission by prolonging action of the major transmitter, acetylcholine.[30]

Of particular interest, the ECM glycoprotein, agrin, is specifically localized at the neuromuscular junction and has been shown to be essential for synapse formation.[5,6] Agrin, one of the few extracellular matrix molecules discovered based upon its biological activity, was detected as a factor in biochemical preparations of synaptic ECM which induces clustering of acetylcholine receptors (AChRs) on skeletal myotubes in cell culture. Clustering of this receptor is one of the most dramatic events in neuromuscular synapse formation and regeneration *in vivo*. Purified agrin has been shown to induce clustering of synaptic molecules *in vitro*, such as ECM-associated acetylcholinesterase and membrane-associated AChRs. A specific exon, inserted by differential splicing in neurones, but not muscle, must be present in agrin for it to have significant biological activity in these assays. As predicted for an agent that induces differentiation in skeletal myotubes, agrin containing this exon is synthesized in motor neurones and is transported to their terminals in skeletal muscles where it is inserted into the synaptic basal lamina. Analysis of a mouse lacking the specific exon required for activity in AChR clustering assays has shown that agrin is essential for normal synapse formation.[31] Few or no functional neuromuscular junctions are present in mutant animals, which die shortly before birth. Few clusters of AChRs are present on skeletal myotubes and these are dramatically reduced in size compared to those in normal litter mates. Many clusters are not associates with nerve terminals. Other synaptic molecules normally clustered in the muscle membrane, submembranous cytoplasm, or basal lamina, such as the tyrosine kinase receptor ErbB, the transmembrane proteoglycan dystroglycan, the laminin $\beta2$ subunit, and the cytoplasmic AChR-associated protein rapsyn remain associated with the minute aneural and subneural AChR clusters in the mutant myotubes, indicating that nerve agrin is not essential for assembly of this postsynaptic complex.

Early studies of agrin signalling revealed a potential role for tyrosine kinases.[5,6] Agrin was shown to induce phosphorylation on tyrosine on the AChR β subunit. Treatments that inhibited receptor aggregation were shown to prevent tyrosine phosphorylation. Results thus suggested that the agrin receptor regulates a tyrosine protein kinase or phosphatase that in turn regulates receptor clustering. More recent work has confirmed these earlier observations and demonstrated that the transmembrane tyrosine kinase MuSK plays an essential role in agrin signalling.[32] MuSK is concentrated on the postsynaptic side of neuromuscular junctions and is rapidly phosphorylated by application of agrin. Cultured myoblasts lacking MuSK are completely unresponsive to agrin. Mice lacking MuSK exhibit severe deficits in neuromuscular junction formation.[33] Postsynaptic differentiation is completely absent. Normal levels of AChR are present, but are not clustered at any location. Similarly, other constituents of the synaptic basal lamina, postsynaptic membrane and submembranous space are not clustered. While MuSK is an essential element in the agrin signalling cascade, it does not appear to bind agrin directly. Instead, it appears to be part of a receptor complex with an unidentified agrin-binding subunit named MASC. The transmembrane proteoglycan dystroglycan has also been shown to bind agrin (and other ECM constituents), but agrin binding to dystroglycan is not essential for agrin-mediated signalling.[5,6] There is thus convincing evidence that the ECM constituent agrin activates a transmembrane tyrosine kinase which in turn induces AChR clustering and other steps in postsynaptic differentiation. The real picture is more complicated because both the extracellular domain of MuSK and the cytoplasmic AChR-associated protein rapsyn have been shown to be essential for this signalling pathway, possibly because they coordinately form a scaffold for assembly of synaptic components.[34] Intriguingly, agrin and MuSK mutants exhibit equally striking deficiencies in presynaptic differentiation, indicating that this signalling pathway induces retrograde signals in muscle which affect the behaviour of nerve terminals. Thus, a single ECM constituent has been shown to be important for the reciprocal cellular inductive interactions that result in formation of a synapse.

Their striking localizations made it seem likely that one or more of the synapse-specific collagen IV or laminin subunits, all of which are synthesized by skeletal myotubes, regulate synapse formation by the terminals of motor neurones. Analysis of a laminin $\beta2$ mutant mouse has revealed mild deficits in synaptic transmission and presynaptic nerve terminal differentiation.[5,6] Mild deficits have also been reported in differentiation of the basal lamina and junctional folds. The mutant phenotype suggests that laminin $\beta2$ is one, but not the only retrograde signal responsible for directing nerve growth and differentiation. In particular, the triamino acid sequence – leucine, arginine, glutamate – which is present in laminin $\beta2$ and several additional ECM glycoproteins present at the synapse may be a motor neurone-selective site that acts to stop axonal outgrowth at synaptic sites,[35] although definitive evidence challenging this role for this triamino acid sequence within native isoforms of laminin has recently been published.[36]

The above results are important because they demonstrate that the extracellular matrix contains, in at least one instance, all the essential information needed to form a complex structure, the synapse. Heterogeneity in molecular composition of the ECM is extensive and may explain the unique developmental and functional properties of the synaptic ECM. Individual ECM molecules present at the synapse have dramatic inductive effects, on both neurones and myotubes, that are distinct from simply promoting cell adhesion. Specific signals, including tyrosine phosphorylation, are responsible. With appropriate assays, it seems likely that other matrix proteins will prove to have equally important and specific effects on cell differentiation and function in other organs formed by inductive reciprocal interactions between cells.

Mutational analysis has not been as productive in analysing functions of all ECM proteins. Despite the high concentration of vitronectin in serum, a mouse lacking this protein is viable without discernible deficiencies.[37] Thrombospondin-1, a constituent of platelet α granules, is clearly important in regulating cell proliferation and differentiation *in vitro*.[38] Many distinct classes of molecules appear to mediate cellular interactions with this protein, suggestive of multiple distinct functions. Despite this, a mouse lacking this protein is viable.[16] As there are four members of the thrombospondin gene family, functional redundancy could account for this surprising phenotype.

Tenascins are another family of related ECM proteins – tenascin-C (also named cytotactin and hexabranchion), tenascin-R (also named restrictin, janusin, or J1–160/180), tenascin-X (MHC-tenascin), and tenascin-Y where specific distribution patterns suggest specific functions for individual family members.[39] Tenascin subunits assemble into multimers, linked by disulphide bonds. Domains present in each subunit include epidermal growth factor-like domains, fibronectin type III-like domains, and a C-terminal fibrinogen-like domain. Additional diversity in tenascin-like proteins is generated by differential splicing of several exons encoding fibronectin type III repeats. Initially, tenascin-C appeared to be particularly interesting because of its distribution along pathways of migrating cells, such as neural crest and cerebellar granule cells, and because antibody injections indicated that it promoted the migration of these same cells during embryogenesis. In addition, tenascin-C and tenascin-X have been shown to contain both adhesive and 'antiadhesive' domains which have striking effects on cell spreading and motility *in vitro*. Despite these provocative data, it has proved surprisingly difficult to observe a developmental deficit in tenascin-C mutant mice.[40] Recently, though, tenascin-X deficiency has been associated with a connective tissue disorder named Ehlers–Danlos syndrome, perhaps rescuing the reputation of this family of intriguing molecules.[41]

While definitive functions for some ECM proteins have been difficult to discern using genetics, positional cloning or more classical genetics has revealed extremely interesting functions for novel, putatively ECM-associated proteins. As one example, genetics has shown that the product encoded by the murine reeler locus, the brain ECM protein reelin, has crucial roles in regulating migration of cells during development of cerebellum, hippocampus, and cerebral cortex.[20] Fibronectin-deficient mice do not have an obvious deficit in gastrulation as would have been predicted from antibody-blocking studies in amphibian and avian embryos, but do exhibit clear deficits in somitogenesis and angiogenesis.[16] A tenascin-related protein named tenm, identified by homology in *Drosophila*, has been shown to function as a pair-rule gene.[42] A genetic screen in *C. elegans* resulted in identification of the *unc-6* gene and its product, a netrin.[43,44] Netrins are laminin-related proteins which bind heparin avidly and function as chemoattractants and chemorepellants, guiding migrating cells and growth cones in *C. elegans*, *Drosophila*, and vertebrates.[12]

Until recently, proteoglycans were poorly characterized and categorized according to the nature of their glycosaminoglycan side chains (see Proteoglycans, pp. 351–356). Recently, the core proteins of several proteoglycans have been cloned and sequenced. The analysis of these sequences has clarified a previously confusing literature and has shown that this class of molecules is best considered as a group of diverse glycoproteins with functions mediated both by their protein cores and by their carbohydrate side chains. The domains found in these molecules include putative hyaluronic acid-binding domains, Ca^{2+}-dependent lectin (sugar-binding) domains, leucine-rich repeats, epidermal growth factor (EGF) repeats, and immunoglobulin-like domains. Often many of these domains are found in the same core protein. The very large chondroitin sulphate proteoglycan named versican, for example, contains a functional hyaluronic acid-binding domain, two EGF repeats, a lectin-like domain, and a complement-regulatory protein-like domain. The basement membrane proteoglycan perlecan contains LDL receptor-ligand binding domains, laminin domain III-like regions, Ig domains, laminin G domains, and EGF repeats. The nature of these domains suggests an adhesive or mitogenic function for these proteoglycans.

Unfortunately, equivalent progress has not been made in determining the sequences and heterogeneity of glycosaminoglycan side chains associated with individual core proteins, even though these chains are crucial for normal proteoglycan function. We have known for decades that carbohydrate structure and sequence is important in regulating associations of glycosaminoglycans with proteins. In particular, interactions of heparin with thrombin have been extensively studied.[45] It has also been appreciated that tremendous diversity can potentially be encoded by carbohydrate side chains, particularly in glycosaminoglycans where these chains are modified by *N*- and *O*-sulphation and *N*-acetylation. Sequence determination remains difficult, however, making it virtually impossible to analyse the complete structures of individual proteoglycans. In principle, the fidelity of chain sequence and modification must limit the specificity of information encoded by glycosaminoglycan chains expressed on individual proteoglycans. Cloning and expression of the genes encoding the enzymes that catalyse glycosaminoglycan synthesis may facilitate studies on fidelity. Methods of analysing the interactions between glycosaminoglycans and proteins are also limited, but new, more sensitive methods are being developed that should help rectify this deficiency (see Proteoglycans, pp. 351–356).

Comparatively slow progress in proteoglycan biochemistry has been frustrating because it now seems that individual proteoglycans have quite important biological functions. Heparan sulphate proteoglycans, such as syndecan, can function as receptors, binding cells to ECM glycoproteins.[46, 47] Proteoglycans can also modulate the binding of other receptors. A chondroitin sulphate proteoglycan with weak affinity for fibronectin, for example, has been shown to reduce binding of fibronectin to both cells and the purified fibronectin receptor (the integrin $\alpha5\beta1$), probably by steric hindrance.[48] Finally, as will be discussed below, proteoglycans bind and modulate the activities of many growth and differentiation factors. In individual examples, these interactions are mediated by both glycosaminoglycan and protein moieties of individual proteoglycans.[49]

In an additional, important class of interactions, the extracellular matrix is now known to bind to several growth factors, including fibroblast growth factors (FGF), wnt proteins, transforming growth factor-β (TGF-β), granulocyte-macrophage colony stimulating factor (GM-CSF), platelet-derived growth factor, and interleukin-3.[49] It seems likely that the matrix captures most of these growth factors, as they are released by cells, restricting their diffusion and range of action. Many of these growth factors contain extremely basic domains able to interact with glycosaminoglycans. Differential splicing of such a domain in PDGF-A regulates its association with the cell surface and ECM. Interactions of TGF-β with the proteoglycan decorin inhibits its activity. Interactions with a second, cell surface-associated proteoglycan, β-glycan, appears to modulate subsequent binding of TGF-β to signal-transducing receptors.[50,51] Interactions between basic FGF and heparan sulphate chains of proteoglycans have been shown to protect it from proteolysis and to regulate its diffusion.[52] Most dramatically, the activation of the FGF receptor by basic FGF requires binding of the ligand to heparan sulphate.[53] Thus interactions with the extracellular matrix localize and regulate the activities of many growth factors. Moreover, studies using both genetics and antibodies have provided strong evidence that interactions between proteoglycans and wnt and FGF family members are crucial for normal morphogenesis *in vivo*.[13,54,55]

New classes of molecules have also been detected in the extracellular matrix in recent years. These include isoforms of several cell adhesion molecules, including NCAM, myelin-associated glycoprotein, and L1;[56,57] members of the semaphorin family which have been shown to function in axon guidance,[12] and the netrins, which function as chemotropic attractants and repellants for both migrating cells and growth cones.[12] Finally, several matrix constituents have been shown to bind to proteases or protease inhibitors, again regulating the diffusion and activities of these important enzymes.[58] Proteases and their inhibitors have reciprocal actions on the ECM[59,60] (see chapters by Sternlicht and Werb, pp. 505–563), regulating assembly of ECM constituents, such as the collagens and epithelial laminins, accessibility of otherwise cryptic binding sites for cells and other proteins within these ECM proteins, and turnover of ECM proteins and their receptors.[59,60]

Domains in ECM constituents

Many extracellular matrix glycoproteins are unusually large molecules with extended conformations spanning distances of several hundred nanometres. Analyses of cDNA and genomic sequences indicate that they can be considered as protein chimeras, containing domains assembled by exon shuffling.[61] Combinatorial assembly of different domains has been a versatile mechanism for generating ECM molecules that serve specialized functions (e.g. ref. 19 and this volume, pp. 344–51). These include distinct structural modules, such as immunoglobulin, fibronectin type III, epidermal growth factor, and numerous other domains. The names suggest specific functions, and indeed multiple functional domains can be identified in most extracellular matrix proteins. These include binding sites for cells, other extracellular matrix glycoproteins, proteoglycans, glycosaminoglycans, glycolipids, growth factors, and proteases or protease inhibitors. In many cases, it has been possible to localize specific functions, such as cell binding, to specific domains. In several instances recognition domains have been shown to consist, at least in part, of peptide sequences as short as 3–10 amino acids. One of these sequences, arginine–glycine–aspartate (RGD), functions as a cell attachment site in several different extracellular matrix glycoproteins, including fibronectin, thrombospondin, fibrinogen, von Willebrand's factor, and vitronectin.

Once a specific function for a domain in one protein has been demonstrated, it has seemed possible that similar domains have similar functions wherever they are found. By this criterion, most of the domain structures in sequenced extracellular matrix proteins are potential mediators of protein–protein interactions or cell adhesion. As one example, the fibronectin type III repeat, one of which in fibronectin contains the sequence RGDS as a major cell attachment site, is found in many extracellular matrix proteins and cell adhesion molecules, sometimes in multiple copies. Currently, at least four of 18 such domains in fibronectin, one of seven in tenascin, and one of six in thrombospondin have been shown to mediate cell attachment.[62] It seems likely that more of these domains will prove to have adhesive functions when tested with appropriate cells. Adding to potential diversity, exons encoding fibronectin type III domains are differentially spliced in isoforms of RNAs encoding both fibronectin and tenascin.[62,63] One of the differentially spliced type III domains in fibronectin has been shown to contain an important cell adhesion site. In contrast, a differentially spliced type III repeat in tenascin interacts with cells, but disrupts adhesive interactions promoted by other ECM glycoproteins.[64]

A second prominent domain implicated in cell adhesion, the immunoglobulin domain, exists in several copies in perlecan, a major basal lamina heparan sulphate proteoglycan.[65] Similar domains are functionally important structures in virtually all Ca^{2+}-independent adhesion molecules, such as NCAM, the ICAMs and other neural or immune system-associated adhesion molecules. Many of these cell adhesion molecules also contain fibronectin type III repeats. Thus, neither class of domain distinguishes cell membrane-associated from extracellular matrix-associated adhesive molecules.

Domains homologous to epidermal growth factor, named EGF repeats, are found in many extracellular matrix glycoproteins, and proteoglycans including laminin, entactin, thrombospondin, tenascin, aggrecan, and versican. In many proteins, EGF repeats have been shown to mediate protein–protein interactions, making it seem likely that many of these repeats function similarly in extracellular matrix constituents. Particular interest has been directed to the possibility that some repeats may function similarly to EGF, that is, interact with receptors that initiate intracellular signalling cascades. Consistent with this possibility, tenascin contains an anti-adhesive site localized to its EGF repeat region which inhibits cell flattening and motility on more adhesive substrates.[63] Thrombospondin, SPARC, and laminin also have similar antiadhesive activities.[64,66] These results are difficult to explain without postulating effects on intracellular second messenger systems.

In addition to effects on adhesion, many matrix proteins contain EGF-like domains that may more directly regulate cell proliferation. Thrombospondin, tenascin, and laminin promote proliferation of smooth muscle cells, fibroblasts, and Schwann cells, respectively.[67] In the case of soluble laminin, the mitogenic activity has been localized within a fragment containing multiple EGF repeats which does not contain the major cell attachment site. A variant cell line lacking EGF receptors does not respond to soluble laminin, suggesting that the EGF receptor may mediate some of these mitogenic actions. Despite experimental efforts, though, there is no biochemical evidence that laminin can interact directly with this receptor.

Receptors for ECM constituents

Many classes of molecules are believed to function as receptors for extracellular matrix constituents. These include transmembrane glycoproteins, proteoglycans, gangliosides and other glycolipids.

Virtually all ECM glycoproteins have been shown to interact with integrins, a large family of receptor heterodimers to be discussed below. Integrin receptor function is required in most cell adhesion assays, perhaps because integrins can interact with and reorganize the cytoskeleton, strengthening weak primary interactions.[3,4] It also seems clear, though, that additional cell surface molecules interact with the ECM. As examples, CD44 is a hyaluronic acid-binding receptor present on many cells which has been shown to associate with both ECM and cell surface-associated ligands (cf. ref. 68). The syndecan family consists of four intrinsic membrane-associated proteoglycans that can mediate cellular interactions with fibronectin, thrombospondin, and several collagens.[46,47] Dystroglycan is a transmembrane constituent of the dystroglycan complex which has been shown to bind dystrophin inside the cell and laminin or agrin outside the cell.[25] The galactase-specific lectin CBP35 binds laminin with high affinity.[69] A cell surface enzyme, galactosyltransferase, also mediates motility responses of neural crest cells and neuronal growth cones to laminin.[70] In addition, gangliosides and other glycolipids, differentially expressed macromolecules present on the surfaces of all cells, have been shown to interact directly with ECM glycoproteins, such as laminin, fibronectin, and thrombospondin.[71] They can also interact with and modulate integrin functions directly.[72]

Integrins are non-covalently associated heterodimeric glycoprotein complexes, expressed on the surfaces of most cell types, that function as receptors for essentially all of the major extracellular matrix glycoproteins.[3,4] Integrins have been identified in nematodes, arthropods, and Porifera and therefore appear to be present in all metazoan phyla.[73] Each heterodimer consists of one α chain non-covalently associated with a β subunit. Multiple genes encode families of both α and β subunits. At last count, cDNAs encoding 15 distinct α subunits and eight distinct β subunits have been characterized in vertebrates. Multiple integrins also are present in *C. elegans* and *Drosophila*, but total numbers are almost certainly fewer than those in vertebrates. Differential splicing, particularly of exons encoding cytoplasmic domains of both α and β subunits, further extends the diversity of both vertebrate and invertebrate integrin subunits. Some of these splice variants have been shown to affect ligand binding outside of the cell and cytoskeletal associations inside the cell. The ligand-binding regions of integrin α subunits have been modelled as a β propeller; the ligand-binding regions of integrin β subunits have been predicted to adopt a structure similar to that of the I domain in Mac-1, whose crystal structure has been solved.[74,75] Reversible associations between the ligand-binding surfaces of both subunits, modelled using the structure of a trimeric G protein, have been proposed to regulate ligand access and integrin activity.[75]

Integrins bind virtually all major constituents of the extracellular matrix. Integrin ligands in the extracellular matrix include several collagens, fibronectin, laminin, osteopontin, tenascin, thrombospondin, vitronectin, von Willebrand's factor, and fibrinogen. Even agrin has recently been shown to interact with one integrin (Sanes; personal communication). A second set of integrin ligands consists of cell surface-associated members of the Ig superfamily. These include ICAM-1, ICAM-2, and VCAM-1, each of which has been implicated in lymphocyte homing. Integrin-mediated adhesion to VCAM-1 has also been shown to be important in mediating fusion to primary skeletal myotubes of secondary myoblasts in culture, but not *in vivo*,[76] and in heart development during embryogenesis.[16] It seems likely that additional members of this class of integrin ligands will be discovered, particularly in tissues such as the nervous system, where cell to cell interactions appear more prominent than cell–matrix interactions. Integrins have been shown also to interact with disintegrins, some of which are cell-surface associated. In particular, evidence indicates that, as a crucial step in sperm–egg fusion, the integrin $\alpha6\beta1$ on the surface of unfertilized eggs binds to a sperm protein named fertilin which contains a disintegrin domain.[77] Integrins also interact with a number of proteins expected to regulate and localize proteolysis. The lymphocyte integrin αmac-1$\beta2$ has been shown to interact with the complement factor C3bi, the coagulation factor X, and human leukocyte elastase, each of which leads to local generation of proteases.[78,79] The integrin $\alpha v\beta3$ has been shown to form a binding site for matrix metalloproteinase MMP-2.[80] Several integrins have also been shown to interact with the urokinase-type plasminogen activator receptor (uPAR).[81] uPAR binding to activated integrins was shown to inhibit native integrin function, but to promote cell adhesion via uPAR to a different ECM ligand, vitronectin. These interactions seem likely to be important for regulation of cellular migratory and invasive behaviours.[82] As major cell surface proteins, integrins have also been exploited by pathogenic bacteria and viruses for which they function inadvertently as receptors or coreceptors.

In addition to binding extracellular ligands, integrins also transmit signals to cells. One major type of signalling is almost certainly mediated through the cytoskeleton. For example, binding by many integrins promotes formation of focal contacts in which integrins associate with the termini of F-actin filaments. Based on protein–protein interactions detected *in vitro* and colocalization studies *in vivo*, talin, α-actinin, and vinculin are believed to link integrins with the cytoskeleton.[4] The cytoplasmic domains of integrin β subunits mediate these associations which are also regulated by ligand binding via the cytoplasmic domains of associated integrin α subunits. The amino acid sequence present in the β subunit cytoplasmic domain determines whether it associates with focal contacts (e.g. $\beta1$, $\beta3$), hemidesmosomes (e.g. $\beta4$), or fails to associate with these structures (e.g. $\beta5$, $\beta7$). Obviously, the different associations of different integrin receptors have important functional consequences. In epithelia, for example, the two major laminin receptors, $\alpha3\beta1$ and $\alpha6\beta4$, have non-redundant functions which reflect their associations with focal adhesions and hemidesmosomes, respectively.[16]

Formation of focal contacts and hemidesmosomes is regulated by intracellular signals which promote or inhibit integrin clustering and cytoskeletal associations.[4,83] Formation of focal adhesions has been shown to be driven by activation of the small monomeric G protein rho and activation of myosin light chain kinase. Activation of a variety of tyrosine kinases appears to inhibit their formation. Formation of hemidesmosomes following ligand binding and clustering of the integrin $\alpha6\beta4$ has similarly been shown to be regulated by tyrosine kinase activity.[83]

Clustering and ligand engagement of integrins also catalyse recruitment of a variety of signalling molecules to focal adhesions and hemidesmosomes.[4,83] For example, ligand engagement and clustering of $\alpha6\beta4$ leads to tyrosine phosphorylation of the $\beta4$ cytoplasmic domain, followed by recruitment of shc and grb-2, and activation of ras.[84] Ligand binding and clustering of $\beta1$ integrins induces assembly of a signalling complex consisting of cytoskeletal components plus shc, grb-2, the pp125FAK tyrosine kinase, and the EGF, PDGF, and FGF receptors.[85] A number of signal mediating proteins, including pp125FAK, calreticulin and integrin linked kinase have been identified which appear to associate with cytoplasmic domains of certain integrins.[86,87] Some, but not all $\beta1$ integrins, are associated with calveoli and therefore with the numerous signalling proteins concentrated in these membrane structures.[7]

Downstream consequences of ligand engagement and clustering of integrins depend upon both the integrin involved and upon cell-specific factors.[3,4] Particularly relevant for understanding integrin synergy with growth factor receptors are observations indicating that integrins can activate a phosphatidylinositol 4-phosphate 5-kinase which exists in a complex with the small G protein rho.[88] By activating this kinase, integrins increase the concentration of phosphatidylinositol 4,5-bisphosphate. This compound is a substrate for PLC-$\gamma1$, which is activated by many receptor tyrosine kinases. The result is generation of diacylglycerol and inositol-trisphosphate, which increase cytoplasmic calcium and kinase-C activity. Phosphatidylinositides have also been shown to interact with many proteins which regulate F-actin assembly and disassembly. In addition, they promote focal adhesion formation by exposing cryptic binding sites in vinculin for talin and actin.[89] Activated rho also helps drive integrin clustering and focal adhesion formation by activating myosin-driven contractility.[4] Ligand engagement or clustering of integrins has been shown to activate ras and many downstream effectors of ras in many cells.[4] Perhaps the most dramatic downstream consequence of disrupting integrin function is induction of apoptosis, which has been observed in many types of cultured cells and has been shown seriously to impede angiogenesis *in vivo*.[90] During the next few years, it seems virtually certain that investigators will focus not only on understanding mechanisms by which integrins regulate intracellular signalling pathways, but also on mechanisms by which the other receptors described above affect these same processes.

Intracellular metabolism can also alter the structure and function of extracellular domains of integrins. Physiological stimuli, modulators of second messenger systems, and ligand binding to other non-integrin receptors regulate appearance of conformation-dependent epitopes and ligand-binding activity.[3] Inside-out signalling requires the cytoplasmic domains of both integrin subunits and appears to reflect changes in the interactions of these cytoplasmic domains with each other. Mutations in the cytoplasmic domains can result in constitutively active or inactive integrins. Cytoplasmic proteins have been identified in recent years which are strong candidates to regulate integrin activation.[91] Cycling of integrins between active and inactive states appears essential for their normal function.[92] In at least some conditions, signals activated by integrin engagement negatively regulate integrin activity, thereby promoting integrin cycling. Both protein kinase C and MAP kinase activation have been shown to suppress integrin activation.[93] Intracellular metabolism also regulates surface concentration of some integrins. In neutrophils, large fractions of the integrins $\alpha mac\beta 2$ and $\alpha gp150\beta 2$ are sequestered in intracellular vesicles.[94] Physiological stimuli induce exocytosis, increasing integrin concentrations on the cell surface. Integrins have been shown to cycle rapidly between intracellular and surface compartments in fibroblasts, being preferentially inserted at leading edges of motile cells.[95] Regulation and targeting of integrin insertion may be one mechanism by which growth factors, chemotropic agents, and other molecules affect cell motility.

Conclusions and perspectives

Recent progress has increased our appreciation of the diversity and importance of the ECM in regulating embryonic development and physiological functioning of the adult organism. Discoveries in other phyla have documented the evolutionarily ancient ancestry of the major constituents of the extracellular matrix and the central role that the ECM plays in all metazoans. A major feature of recent progress has been an explosion in the number and categories of constituents of the ECM. Many of these appear to have evolved after separation of the major phyla in response to particular evolutionary pressures, permitting organisms to function in specialized environments.[19] Exon shuffling appears to have been essential for evolution of these novel ECM proteins within individual groups of organisms. Focusing on traditional constituents of the ECM, new collagens, adhesive glycoproteins and proteoglycans are discovered each year in vertebrates. The rate of discovery in invertebrate phyla has been slower, but still significant. The new ECM constituents include completely new proteins such as agrin and reelin, new members of previously identified families, such as novel laminin subunits, and growing arrays of isoforms generated by differential splicing and other post-transcriptional mechanisms. It seems safe to predict that large numbers of similar discoveries will be made in the next few years. The new additions seem likely to have more restricted distributions. As illustrated by agrin and reelin, though, many will probably have quite important functions.

Impressive but uneven progress has been made in the past few years in characterization of proteoglycans. Molecular biology has increased our appreciation of the structures and functions of the protein moieties of these macromolecules. Our understanding of carbohydrate structure and function remains deficient. New methods of carbohydrate analysis seem necessary to pursue this topic more efficiently.

New classes of molecules, including growth factors, chemotropic factors, and cell adhesion molecules, have been localized to the extracellular matrix in the past few years. It seems safe to predict that the numbers of such molecules localized there will continue to increase. A major function of the ECM is clearly to modulate the diffusion, activity, stability, and other properties of these classes of molecules.[49]

The classical constituents of the ECM tend to be very large molecules with extended conformations, assembled by exon shuffling. Even if one postulates that all important matrix constituents will soon be identified, one will still have to identify the functions of the individual modules in these proteins. Fibronectin, for example, has 32 spatially distinct modules. Other ECM proteins also have large numbers of domains. Each of these seems capable in principle of mediating interactions with cells or other macromolecules. Increasing numbers of these modules in proteins such as fibronectin or thrombospondin have been shown to have specific functions. A high degree of evolutionary conservation between species suggests that many regions of characterized ECM glycoproteins, such as the laminins, have important functions.[21,62] Analysis of mutants in invertebrates has resulted in discovery of novel matrix proteins and receptors.[44,63] Although a daunting prospect, a genetic approach in vertebrates seems essential to provide a comprehensive insight into the functions of individual ECM domains.

The discovery of integrins as prominent receptors for the extracellular matrix should not obscure the importance of other classes of receptors, such as dystroglycan, or prevent recognition of our ignorance about other aspects of cell–matrix interactions. While there have been significant advances, comparatively little is known about how adhesion affects cytoskeletal structure and cytoplasmic signalling systems. We know little more about mechanisms by which cytoplasmic signals regulate integrin structure and function. Finally, an abundance of evidence indicates that there must be additional classes of receptors. Integrins are not strong candidates to mediate the 'antiadhesive' effects of tenascin, thrombospondin, and other matrix molecules. It remains to be seen whether known macromolecules that bind the ECM, such as CD44, proteoglycans, or gangliosides, can mediate such signals. As described above, agrin clearly induces postsynaptic differentiation at the neuromuscular junction by activating a novel transmembrane tyrosine kinase. It seems very likely that agrin will not prove to be the sole ECM protein acting via this mechanism and therefore that novel receptors, less easily detected in cell adhesion assays, remain to be discovered.

References

1. Gerhart, J. and Kirschner, M. (1997). Blackwell Science, Malden MA.
2. Lin, C. Q. and Bissell, M. J. (1993). *FASEB J.*, **7**, 737–43.
3. Schwartz, M. A., Schaller, M. D., and Ginsberg, M. H. (1995). *Ann. Rev. Cell Dev. Biol.*, **11**, 549–99.
4. Burridge, K. and Chrzanowska-Wodnicka, M. (1996). *Ann. Rev. Cell Dev. Biol.*, **12**, 463–518.
5. Kleiman, R. J. and Reichardt, L. F. (1996). *Cell*, **85**, 461–4.
6. Sanes, J. R. (1997). *Curr. Opin. Neurobiol.*, **7**, 93–100.
7. Wary, K. K., Mainiero, F., Isakoff, S. J., Marcantonio, E. E., and Giancotti, F. G. (1996). *Cell*, **87**, 733–43.

8. Lopez-Casillas, F., Payne, H. M. Andres, J. L., and Massague, J. (1994). *J. Cell Biol.*, **124**, 557–68.
9. Kaname, S. and Ruoslahti, E. (1996). *Biochem. J.*, **315**, 815–820.
10. Reichardt, L. F. and Tomaselli, K. J. (1991). *Ann. Rev. Neurosci.*, **14**, 531–70.
11. Venstrom, K. A. and Reichardt, L. F. (1993). *FASEB J.*, **7**, 996–1003.
12. Tessier-Lavigne, M. and Goodman, C. S. (1996). *Science*, **274**, 1123–33.
13. Dealy, C. N., Seghatoleslami, M. R., Ferrari, D., and Kosher, R. A. (1997). *Dev. Biol.*, **184**, 343–50.
14. Watanabe, H. Kimata, K., Line, S., Strong, D., Gao, L. Y., Kozak. C. A., and Yamada, Y. (1994). *Nature Genet.*, **7**, 154–7.
15. Uitto, J., Burgeson, R. E., Christiano, A. M., and Moshell, A. N. (1996). *J. Invest. Dermatol.*, **107**, 787–8.
16. Hynes, R. O. (1996). *Dev. Biol.*, **180**, 402–12.
17. Ruoslahti, E. (1996). *Ann. Rev. Cell Dev. Biol.*, **12**, 697–715.
18. Uitto, J. and Pulkkinen. L. (1996). *Mol. Biol. Reports*, **23**, 35–46.
19. Engel, J. (1997). *Science*, **277**, 1785–6.
20. D'Arcangelo, G., Miao, G. G., Chen, S. C., Soares, H. D., Morgan, J. I., and Curran, T. (1995). *Nature*, **374**, 719–23.
21. Burgeson, R. E., Chiquet, M., Deutzmann, R., Ekblom, P., Engel, J., Kleinman, H., *et al.* (1994). *Matrix Biol.*, **14**, 209–11.
22. Gianelli, G., Falk-Marzillier, J., Shiraldi, O., Stetler-Stevenson, W. G., and Quaranta, V. (1997). *Science*, **277**, 225–8.
23. Lentz, S. I., Miner, J. H., Sanes, J. R., and Snider, W. D. (1997). *J. Comp. Neurol.*, **378**, 547–61.
24. Xu, H., We, X. R., Wewer, U. M., Engvall, E. (1994). *Nature Genet.*, **8**, 297–302.
25. Straub, V. and Campbell, K. P. (1997). *Curr. Opin. Neurol.*, **10**, 168–75.
26. Flinter, F. (1997). *J. Med. Genet.*, **34**, 326–30.
27. Cosgrove, D., Meehan, D. T., Grunkemeyer, J. A., Kornak, J. M., Sayers, R., Hunter, W. J., and Samuelson, G. C. (1996). *Genes Dev.*, **10**, 2981–92.
28. Miner, J. H. and Sanes, J. R. (1996). *J. Cell Biol.*, **135**, 1403–13.
29. Noakes, P. G., Miner, J. H., Gautam, M., Cunningham, J. M., Sanes, J. R., and Merlie, J. P. (1995). *Nature Genet.*, **10**, 400–6.
30. Hall, Z. W., and Sanes, J. R. (1993). *Cell*, **72**/*Neuron*, **10** (Suppl.), 99–121.
31. Gautam, M., Noakes, P. G., Moscoso, L., Rupp, F., Scheller, R. H., Merlie, J. P., and Sanes, J. R. (1996). *Cell*, **85**, 525–35.
32. Glass, D. J., Bowen, D. C., Stitt, T. N., Radziejewski, C., Bruno, J., Ryan, T. E., *et al.* (1996). *Cell*, **85**, 513–24.
33. DeChiara, T. M., Bowen, D. C., Valenzuela, D. M., Simmons, M. V., Poueymirou, W. T., Thomas, S., *et al.* (1996). *Cell*, **85**, 501–12.
34. Apel, E. D., Glass, D. J., Moscoso, L. M., Yancopoulos, G. D., and Sanes, J. R. (1997). *Neuron*, **18**, 623–35.
35. Hunter, D. D., Cashman, N., Morris-Valero, R., Bulock, J. W., Adams, S. P., and Sanes, J. R. (1991). *J. Neurosci.*, **11**, 3960–71.
36. Brandenberger, R., Kammerer, R. A., Engel, J., and Chiquet, M. (1996). *J. Cell Biol.*, **135**, 1583–92.
37. Zheng, X., Saunders. T. L., Camper, S. A., Samuelson, L. C., and Gkinsburg, D. (1995). *Proc. Natl Acad. Sci. USA*, **92**, 12426–30.
38. Lawler, J., Duquette, M., Urry, L., McHenry, K., and Smith, T. F. (1993). *J. Mol. Evol.*, **36**, 509–16.
39. Hagios, C., Koch, M., Spring, J., Chiquet, M., and Chiquet-Ehrismann, R. (1996). *J. Cell Biol.*, **134**, 1499–512.
40. Saga, Y., Yagi, T., Ikawa, Y., Sakakura, T., and Aizawa, S. (1992). *Genes Dev.*, **6**, 1821–31.
41. Burch, G. H., Gong, Y., Liu, W., Dettman, R. W., Curry, C. J., Smith, L., *et al.* (1997). *Nature Genet.*, **17**, 104–8.
42. Baumgartner, S., Martin, D., Hagios, C., and Chiquet-Ehrismann, R. (1994). *EMBO J.*, **13**, 3728–40.
43. Ishii, N., Wadsworth, W. G., Stern, B. D., Culotti, J. G., and Hedgecock, E. M. (1992). *Neuron*, **9**, 873–81.
44. Wadsworth, W. G., Bhatt, H., and Hedgecock, E. M. (1996). *Neuron*, **16**, 35–46.
45. Bray, B., Lane, D. A., Freyssinet, J. M., Pejler, G., and Lindahl, U. (1989). *Biochem. J.*, **262**, 225–32.
46. Salmivirta, M. and Jalkanen, M. (1995). *Experientia*, **51**, 863–72.
47. Couchman, J. R. and Woods, A. (1996). *J. Cell Biochem.*, **61**, 578–84.
48. Hautanen, A., Gailit, J., Mann, D. M., and Ruoslahti, E. (1989). *J. Biol. Chem.*, **264**, 1437–42.
49. Ruoslahti, E. and Yamaguchi, Y. (1991). *Cell*, **64**, 867–69.
50. Andres, J. L., Stanley, D., Cheifetz, S., and Massague, J. (1989). *J. Cell Biol.*, **109**, 3137–45.
51. Wang, X.-F., Lin, H. Y., Ng-Eaton, E., Downward, J. Lodish, H. F., and Weinberg, R. A. (1991). *Cell*, **67**, 797–805.
52. Saksela, O. and Rifkin, D. B. (1990). *J. Cell Biol.*, **110**, 767–75.
53. Yavon, A., Klagsbrun, M., Esko, J. D., Leder, P., and Ornitz, D. M. (1991). *Cell*, **64**, 841–8.
54. Haerry, T. E., Heslip, T. R., Marsh, J. L., and O'Connor, M. B. (1997). *Development*, **124**, 3055–64.
55. Kispert, A., Vainio, S., Shen, L., Rowitch, D. H., and McMahon, A. P. (1996). *Development*, **122**, 3627–37.
56. Sanes, J. R., Schachner, M. and Covault, J. (1986). *J. Cell Biol.*, **102**, 420–31.
57. Fahrig, T., Landa, C. Pesheva, P. Kuhn, K. and Schachner, M. (1987). *EMBO J.*, **6**, 2875–83.
58. Tomasini, B. R. and Mosher, D. F. (1991). *Prog. Hemostasis Thrombosis*, **10**, 269–305.
59. Chin, J. R. and Werb, Z. (1997). *Development*, **124**, 1519–30.
60. Shingleton, W. D., Hodges,D. J., Brick, P., and Cawston, T. E. (1996). *Biochem. Cell Biol.*, **74**, 759–75.
61. Chothia, C. and Jones, E. Y. (1997). *Ann. Rev. Biochem.*, **66**, 823–62.
62. Hynes, R. O. (1990). *Fibronectins*, Springer, New York.
63. Spring, J., Beck, K. Chiquet-Ehrismann, R. (1989). *Cell*, **59**, 325–34.
64. Murphy-Ullrich, J. E., Lightner, V. A. Aukhil, I. Yan, Y. Z. Erickson, H. P. and Hook, M. (1991). *J. Cell Biol.*, **115**, 1127–36.
65. Kallunki, P. and Tryggvason, K. (1992). *J. Cell Biol.*, **116**, 559–71.
66. Calof, A. L. and Lander, A. D. (1991). *J. Cell Biol.*, **115**, 779–94.
67. Engel, J. (1989). *FEBS Lett.*, **251**, 1–7.
68. St. John, T., Meyer, J. Idzerda, R. and Gallatin, W. M. (1990). *Cell*, **60**, 45–52.
69. Woo, H.-J., Shaw, L. M. Messier, J. M. and Mercurio, A. M. (1990). *J. Biol. Chem.*, **265**, 7097–9.
70. Begovac, P. C. and Shur, B. D. (1990). *J. Cell Biol.*, **110**, 461–70.
71. Roberts, D. D., Rao, C. N. Magnani, J. L. Spoitalnik, S. L. Liotta, L. A. and Ginsburg, V. (1985). *Proc. Natl Acad. Sci. USA*, **82**, 1306–11.
72. Santoro, S. A. (1989). *Blood*, **73**, 484–9.
73. Muller, W. E. (1997). *Cell Tissue Res.*, **289**, 383–95.
74. Springer, T. A. (1997). *Proc. Natl Acad. Sci. USA*, **94**, 65–72.
75. Lee, J.-O., Rieu, P., Arnaout, M. A., and Liddington, M. R. (1995). *Cell*, **80**, 631–38.
76. Yang, J. T., Rando, T. A., Mohler, W. A., Rayburn, H., Blau, H. M., and Hynes, R. O. (1996). *J. Cell Biol.*, **135**, 829–35.

77. Myles, D. G. and Primakoff, P. (1997). *Biol. Reprod*, **56**, 320–27.
78. Altieri, D. C., Etingen, O. R. Fair, D. S. Brunck, T. K. Geltosky, J. E. Hajjar, D. P., and Edgington, T. S. (1991). *Science*, **254**, 1200–2.
79. Cai, T. Q. and Wright, S. D. (1996). *J. Exer. Med.*, **184**, 1213–23.
80. Brooks, F. P. C., Stromblad, S., Sanders, L. C., von Schalscha, T. L., Aimes, R. T., Stetler-Stevenson, W. G., *et al.* (1996). *Cell*, **85**, 683–93.
81. Wei, Y., Lukashev, M., Simon, D. I., Bodary, S. C., Rosenberg, S., Doyle, M. V., and Chapman, H. A. (1996). *Science*, **273**, 1551–5.
82. Preissner, K. T., May, A. E., Wohn, K. D., Germer, M., and Kanse, S. M. (1997). *Thrombosis, Haemostasis*, **78**, 88–95.
83. Mainiero, F., Pepe, A., Yeon, M., Ren, Y., and Giancotti, F. G. (1996). *J. Cell Biol.*, **134**, 241–53.
84. Mainiero, F., Pepe, A., Yeon, M., Ren, Y., and Giancotti, F. G. (1997). *EMBO J.*, **16**, 22365–75.
85. Miyamoto, S., Teramoto, H., Gutkind, J. S., and Yamada, K. M. (1996). *J. Cell Biol.*, **135**, 1633–42.
86. Hannigan, G. E., Leung-Hagesteijn, C. Y., Fitz-Gibbon, L., Coppolino, M. G., Radeva, G., Filmus, J., *et al.* (1996). *Nature*, **379**, 91–6.
87. Coppolino, M. G., Woodside, M. J., Demaurex, N., Grinstein, S., St-Arnaud, R., and Dedhar, S. (1997). *Nature*, **386**, 843–7.
88. Chong, L. D., Traynor-Kaplan, A., Bokoch, G. M., and Schwartz, M. A. (1994). *Cell*, **79**, 507–13.
89. Gilmore, A. P. and Burridge, K. (1996). *Nature*, **381**, 531–5.
90. Friedlander, M., Brooks, P. C., Shaffer, R. W., Kincaid, C. M., Varner, J. A., and Cheresh, D. A. (1995). *Science*, **270**, 1500–2.
91. Kashiwagi, H., Schwartz, M. A., Eigenthaler, M., Davis, K. A., Ginsberg, M. H., and Shattil, S. J. (1997). *J. Cell Biol.*, **137**, 1433–43.
92. Huttenlocher, A., Ginsberg, M. H., and Horwitz, A. F. (1996). *J. Cell Biol.*, **134**, 1551–62.
93. Hughes, P. E., Renshaw, M. W., Pfass, M., Forsyth, J., Keivens, V. M., Schwartz, M. A., and Ginsberg, M. H. (1997). *Cell*, **88**, 521–30.
94. Bainton, D. F., Miller, L. J. Kishimoto, T. K. and Springer, R. A. (1987). *J. Exp. Med.*, **166**, 1641–53.
95. Bretscher, M. S. (1989). *EMBO J.*, **8**, 1341–8.

Louis F. Reichardt
Department of Physiology and the Howard Hughes Medical Institute, University of California at San Francisco, San Francisco, CA 94143, USA

Modules in ECM and adhesion molecules

Proteins of the extracellular matrix (ECM) and adhesion molecules are composed of many different domains arranged along the same polypeptide chain. A domain is defined as a spatially distinct and independently folding unit. The domains of the ECM and adhesion molecules are called modules because they occur in many different proteins and are repeated in the same protein. A family of modules is defined by homology but variations in sequence in a family may be large and in case of only 15 to 25 per cent identity unambiguous decisions on the existence of homology can often not be reached.[1] It is, moreover, often difficult to define the borders between modules from sequence information alone. Elucidation of the three-dimensional structure greatly helps to solve these problems.

To date 70 to 100 families of modules have been recognized in ECM and adhesion proteins (Tables 1 and 2). During the past decade for a large number of modules the three-dimensional structure was solved by X-ray crystallography or by NMR spectroscopy. This was usually achieved with domains that were recombinantly produced in bacterial or mammalian expression systems and it was found that the correct domain borders are of crucial importance for proper folding. In turn the existence of a defined three-dimensional structure is the best proof for an independently folding domain although other biophysical or biochemical studies may also be supportive. Most importantly, structure elucidations revealed structure–function relationships of modules and many functional assignments were based on them. In many cases the three-dimensional structures have been solved only for a single member of the family. This defines the general fold in a module family but since small details in the structure are decisive for function it will be necessary to solve the structure of individual domains. It was also possible to solve the structures of complexes of modules with their ligands. In addition, pairs or larger arrays of modules were structurally resolved defining interactions between neighbouring modules and cooperations between binding sites located at different modules.

In this chapter, modules are listed applying the nomenclature introduced by Bairoch and Bork.[2–4] A more complete listing than in Tables 1 and 2 can be found in the review by Bork *et al.*[4] Modules whose structures are known at atomic resolution are indicated and examples of structure–function relationships are discussed. Finally a brief outline of the evolution of modular proteins is given.

A list of modules

In Tables 1 and 2 the two-letter nomenclature of modules introduced by Bairoch and Bork[2–4] is used. This nomenclature is also used in the SwissProt and Prosite databases and its future use may avoid many confusions caused by

Table 1 Globular modules

Name	Abbreviation	Size [Cys]	3D structure	Examples
Anaphylatoxin domain	AT [ANATO]	70 [6]	YesYesYe	Fibulin 1 and 2, netrin
Collagen I C-terminal domain	C1	250 [6]	No	Collagen I, II, III, V, XI
Collagen IV C-terminal domain	C4 [COL4C]	110 [6]	No	Collagen IV
Cadherin extracellular domain	CA [CADHE]	110 [0–2]	Yes	E-cadherin, desmoglein, desmocollin
C-type lectin domain, endostatin domain in collagen XVIII[45]	CL [CLECT]	130 [4–6]	Yes	Selectin, mannose-binding protein, aggrecan
Complement control protein/short consensus repeat/Sushi domain	CP [CCP]	70 [4]	Yes	C1r, C1s, selectin, aggrecan
CUB/C1s/C1r domain	CU [CUB]	110 [2–4]	No	C1r, C1s, sea urchin U-EGF, bone morphogenetic protein 1
Extracellular calcium-binding domain	EC	150 [4]	Yes[7]	BM40/SPARC/osteonectin, testican, SC1/hevin
Epidermal growth factor-like domain	EG [EGF]	40 [6]	Yes	thrombospondin, agrin, tenascin, selectin
Fibronectin type I domain	F1 [FnI]	40 [4]	Yes	Fibronectin, tissue plasminogen activator
Fibronectin type II domain	F2 [FnII]	60 [4]	Yes	Fibronectin, factor XII, collagenase IV
Fibronectin type III domain	F3 [FnIII]	90	Yes	Fibronectin, collagen XII, tenascin, L1-CAM
Fibrinogen γ C-terminal domain	FG [FBG]	250 [4]	Yes[8]	Fibrinogen, ficolin, tenascin
Follistatin/Kazal type inhibitor domain	FS [FOLLI]	80 [10]	Yes[9]	Agrin, SPARC/BM40 osteonectin, testican
γ-Carboxyglutamate domain	GA [GLA]	60 [2]	Yes	Prothrombin, clotting factors, osteocalcin
Kunitz-type inhibitor domain	KU [KUNIT]	60 [4–6]	Yes	Collagen VI *a*3 chain, *C. elegans* YN81, amyloid
Haemopexin domain	HX [HEMOP]	60 [0–2]	Yes	Vitronectin, haemopexin, neuramidase
Immunoglobulin domain	IG [IGSF]	100 [0–6]	Yes	L1CAM, CD2
Low density lipoprotein (LDL) receptor class A domain	LA [LDRA]	40 [6]	Yes	LDL receptor, perlecan, PKD1
LDL receptor YWTD domain	LY [LDRY]	50	No	LDL receptor, nidogen, EGF precursor, apolipoprotein E receptor
Laminin N-terminal domain VI	LN [LAMNT]	250 [6–10]	No	Laminin unc-6
Laminin EGF-like domain	LE [LAMEG]	50 [8]	Yes[10]	Laminin, agrin, perlecan
Laminin domain IV	L4 [LAMD4]	190 [8]	No	Laminin, perlecan
Laminin G domain, homology with pentraxin and TN domain (?)[46]	LG [LAMG]	190 [0–4]	No but see 46	Laminin, perlecan, neurexin, agrin
Link hyaluronate binding domain	LK [LINK]	100 [4]	Yes[11]	Link protein, CD44, aggrecan
Agrin/perlecan/enterokinase domain	SE [SEA]	120	No	Agrin, perlecan
Somatomedin B domain	SO [SOMAB]	40 [8]	No	Vitronectin, plasma protein 11
Scavenger receptor Cys-rich domain	SR [SRCR]	110 [6]	No	Scavenger receptor, MAC-2 binding protein
Thrombospondin type 1/ properdin domain	T1 [TSP1]	60 [4–6]	No	Thrombospondin-1, -2, -3 properdin
Thrombospondin type 3 calcium-binding domain	T3 [TSP3]	30	No	Thrombospondin-1, -2, -3, -4 cartilage oligomeric matrix protein/COMP

Table 1 (Cont'd)

Name	Abbreviation	Size [Cys]	3D structure	Examples
Thrombospondin C-terminal domain	TC	220	No	TGF-*b*-binding protein Thrombospondin –1, –2, –3, –4 COMP
Thyroglubulin type I domain	TY [THYGI]	50 [6, 8]	No	Nidogen, ascidian nidogen-like protein
Von Willebrand factor type A domain	VA [VWFA]	200 [0–2]	Yes	von Willebrand factor, collagen VI, cartilage matrix protein/matrilin
Von Willebrand factor type B domain	VB [VWFB]	30 [8]	No	von Willebrand factor, mucin MUC2
Von Willebrand factor type D domain	VD [VWFD	350 [28–32]	No	von Willebrand factor, mucin MUC2

Table 2 Repeat structures

Name	Abbreviation	3D structure	Examples
Coiled-coil structure	CC	Yes[12, 13]	Laminin, tenascin, cartilage oligomeric matrix protein (COMP)
Collagen-like structure	CO	Yes[14]	Triple helical regions in collagens, C1q, collectins, ficolin
Leucine rich repeat domain	LR [LRR]	Yes[15]	Biglycan, fibromodulin, decorin oligodendrocyte-myellin glycoprotein, polycystin
Serine/threonine-rich domain	ST	No	Aggrecan, mucins, sialophorin

the large and growing number of modules. The tables also contain frequently used full names and the longer abbreviations introduced by Bairoch and Bork as well as example proteins in which the modules occur. Some preliminary two-letter abbreviations were added for modules not listed by Bairoch and Bork. The approximate size of each module in number of amino acids is indicated together with the typical number of Cys residues (in brackets). It should be noted that most of the modules are defined by sequence homology which becomes uncertain when sequence identities are in the grey light zone of 15–25 per cent.[1] Modules have been found with different abundances. Excluding species redundancies, more than 1000 immunoglobulin-like domains (IG domains), about 600 epidermal growth factor (EGF)-like domains and about 400 fibronectin type III (F3)-like domains have been found. Occurrences of some domains are very small.[1–10] All frequencies may change with the discovery and sequencing of more proteins.

Relatedness of three-dimensional structures often suggests a common phylogenic origin even when this is not clear from sequence comparison. For example, the X-ray structure of the N-terminal N-cadherin module[5] suggests that cadherin (CA) modules belong to the immunoglobulin family.[6] The availability of high-resolution structures by X-ray or NMR is indicated in Tables 1 and 2. In Fig. 1 cartoon diagrams are presented of the three-dimensional structure of several important modules. Figure 1 also shows examples in which several modules are linked in the same polypeptide chain or by non-covalent interactions. More structures and references to literature can be found in the review by Bork *et al.*[4] Some interesting structure–function relationships will be discussed below.

Repeat structures

In globular modules (Table 1) the sequence does not contain repeat motifs but extended and often linear structures are frequently formed by repeating sequences. Collagen triple helices require Gly–Xaa–Yaa repeats in which Xaa and Yaa stand for any amino acid whereas Yaa is frequently 4-hydroxyproline.[20] Coiled-coil structures are characterized by a heptad repeat 'a b c d e f g' in which residues in positions a and d are predominantly hydrophobic.[21] Frequently repeat structures vary in length. For example, triple-helix forming collagen sequences may range from 25 tripeptide repeats in C1q to 330 in interstitial collagens[22] and can reach 2600 in annelid collagens.[22,23] Coiled-coil heptad repeats range from four in tenascin to 90 in laminin.[24] It is difficult to define modules of defined size although from a frequently occurring exon size of 54 base pairs in interstitial collagens a minimum collagen unit of six Gly–Xaa–Yaa repeats was suggested (reviewed in ref. 22). The short exons are combined to much longer triple helices and exons comprising very long collagen sequences were also found. The repeats are often interrupted by short irregularities. Examples are collagen IV with about 20 small interruptions of the Gly–Xaa–Yaa repeat[22,25] and three-

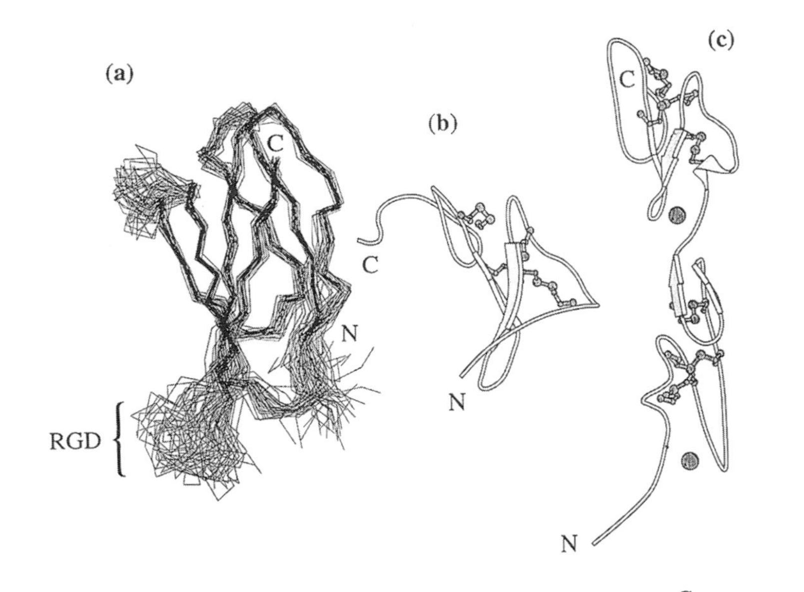

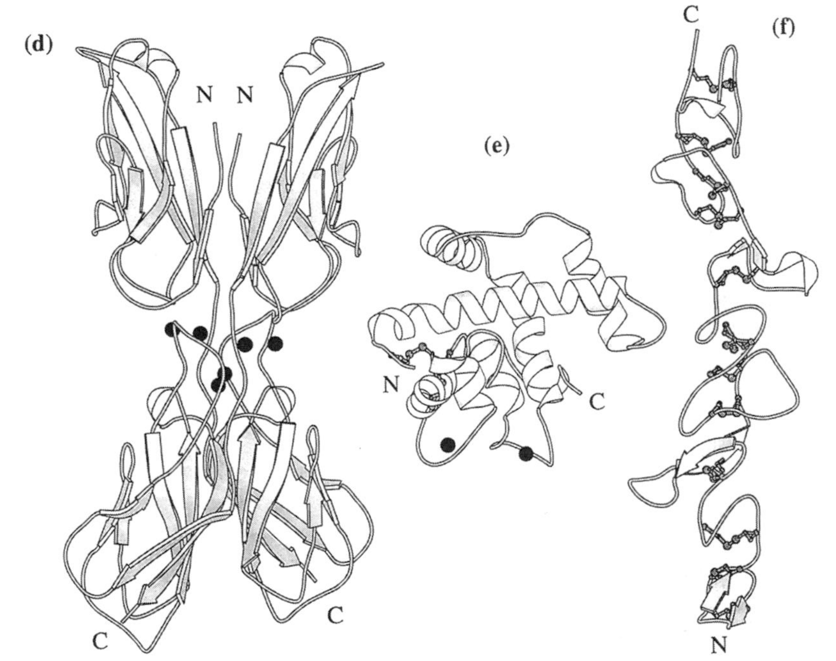

Figure 1. See overleaf for caption.

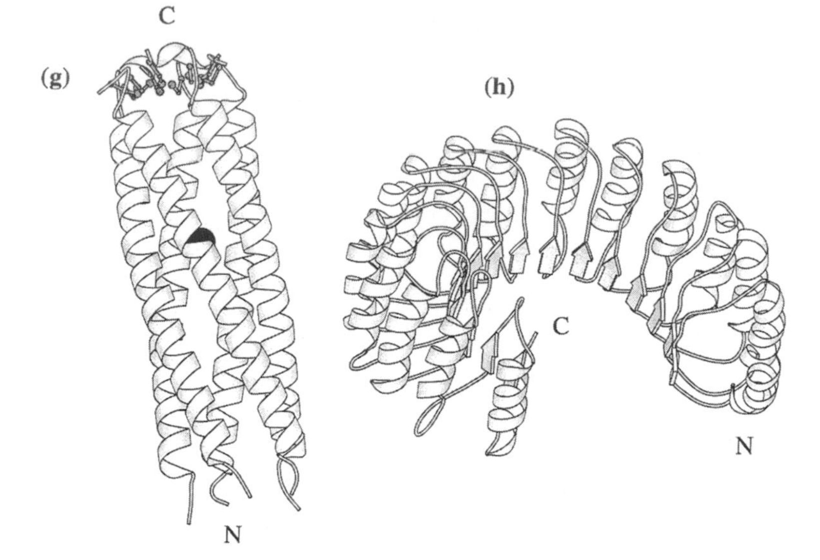

Figure 1. Cartoon diagrams of the three-dimensional structures of isolated and linked modules. Either the polypeptide backbone (a) or molscript presentations[40] are shown in which bands with arrows indicate β structures, coils α helices, and lines undefined secondary structure elements. Disulphide bridges are indicated by ball and stick representations. (a) NMR structure of the tenth type III module (cell-binding domain) of fibronectin (F3) containing an RGD sequence. Thirty-six structures were calculated independently on the basis of the MR distance constraints. Good superposition of these structures, for example in the central β core, indicates highly defined structures whereas non-matching, for example in the loop regions, indicates flexibility and co-existence of many different conformations. Note that the RGD target sequence for integrin $\alpha5\beta1$ is located in a flexible loop region. (Original data by Main *et al.*, 1992.[16]) (b) NMR structure of the epidermal growth factor (EG) module. It contains only 54 residues and is stabilized by three disulphide bonds. (Original data from Cooke *et al.*, 1987.[17]) (c) NMR structure of a pair of calcium binding EG modules in the rod-like structure of fibrillin. Ca^{2+} ions are indicated by spheres. (Original data by Downing *et al.*, 1996.[18] (d) Crystal structure of a dimer of pairs of the first and second CA modules in E-cadherin. Three Ca^{2+} ions per pair (spheres) connect the domains 1 and 2 by binding to their linking region. (Original data by Nagar *et al.*, 1996.[19]). (e) Crystal structure of the extracellular calcium binding module (EC) of BM40/SPARC/osteonectin. A calmodulin-like arrangement of two EF hand motifs with Ca^{2+} ions bound (spheres) is seen in the lower half of the domain. (Original data from Hohenester *et al.*, 1996.[7]) (f) Crystal structure of three laminin EGF-like module (LE) with four disulphide bonds each in the short arm (γ chain) of laminin 1. The binding site of nidogen resides in the central LE domain. (Original data from Stetefeld *et al.*, 1995.[10]) (g) Crystal structure of the five-stranded coiled-coil oligomerization domain in cartilage oligomeric matrix protein COMP. This domain forms a pore which is hydrophobic with the exception of five glutamines to which a chloride ion is bound (sphere). Original data from Malashkevich *et al.*, 1996.[13]) (h) Crystal structure of the leucine rich repeat domain which was solved in ribonuclease but also occurs in decorin and other ECM proteins. (For original data see Kolbe and Deisenhofer 1993.[15]) Coordinates were collected from the Brookhaven database with the following accession numbers: (a) 1 TTF; (b)1 EGF; (c) 1 EMN; (d) 1 EDH; (e) 1 SRA; (h) 1 DTJ. (Coordinates of (f) by courtesy of Dr J. Stetefeld.)

stranded coiled-coil structures in which 'stutters', links, and 'stammers' can be distinguished as discontinuities of the heptad repeats.[26]

An important feature of collagen triple helices and coiled-coil structures (Fig. 1(g)) is their formation from several chains (three for collagens, two to five for coiled-coil structures). Single chains are not stable and the definition of a module as an autonomously folding unit is not applicable to isolated chain segments. Both types of structures are frequently formed from different chains. For example, when in the absence of α and γ chains laminin β chains are expressed in a cellular system, a coiled-coil domain cannot be formed.[27]

A different type of repeat structure is the serine- and threonine rich regions in mucins and proteoglycans.[28,29] For example approximately 20 residue long repeats of the type TAP $(Xaa)_5$ (S, T) $(Xaa)_2$ P $(Xaa)_{7\text{–}8}$ are found in aggrecan and in mucin 2 (MUC1) but many other repeats exist in other mucins.[29] The serine and threonine serve as attachment sites of *O*-glycosylation. The number of repeats is highly variable in different proteins and electron microscopy suggests worm-like extended structures.[28] In other repeat structures stabilization occurs by interactions between repeats in the same chain. An example of which the crystal structure is known is ribonuclease[15] in which leucine repeats combine to an extended structure (Fig. 1(h)). Similar structures with variable numbers of repeats have been proposed for decorin, fibromodulin, and other proteins.

Structure–function relationships

Modules belonging to the same phylogenetic family may possess very different functions. Apparently during evolution the same general structure was adapted to different functions by modifications of binding sites and other features. This is not surprising in view of the rather large sequence variations within a module family.

An example for a different function of the same module is the EG domain with mitogenic activity compared to the calcium binding EG domain without growth factor activity (Fig. 1(b, c)). The first type of EG module is found in EGF, TGF-α, and other growth factors, whereas the second type is present in a large number of ECM proteins including fibrillin and fibulin. A distinction between the two functional types of EG modules was reached on the basis of their three-dimensional structures.[18,30] In particular it was possible to define the residues involved in binding of calcium ions. It should be noted that for the many hundred EG modules in different proteins, functional assignments are still missing. Elucidation of functions in multifunctional proteins is a time-consuming and difficult task and was performed only for a small fraction of modules. Of course it may not be excluded that some of the many-fold repeated modules in a protein may only act as spacer elements.

Another example of a module with different functions is the IG module. As the variable domain in antibodies it recognizes a wide range of antigens but in the Fc part of antibodies it serves to connect two subunits. A third functional type mediates homophilic association in adhesion molecules like NCAM or L1-CAM.[31,32] RGD sequence regions in IG domains may also mediate interaction of this type of module with integrins. An example is the sixth IG domain in L1-CAM.[32] Many other functions have been assigned to IG modules in different other proteins and structure–function interpretations have been given on the basis of the many X-ray and NMR structures of different IG modules.[4,31] An important example is the structure of the CA modules[5,19] (Fig. 1(d)) from which details of the homophilic interactions of cadherins and the involvement of calcium ions were derived.

The cell-binding domain in fibronectin (tenth F3 domain in the sequence)[26] (Fig. 1(a)) contains an RGD sequence which was the first recognized integrin-binding motif of this type.[33] A synergism with a second integrin binding site on the ninth F3 domain was understood on a structural basis by exploring the X-ray structure of a row of the seventh to tenth domains.[34] These and many other studies with pairs or rows of fibronectin domains (reviewed in ref. 35) revealed interactions between neighbouring domains which may bring binding sites at different domains in favourable orientations.

The fourth LE domain in the γ chain of laminins binds specifically to nidogen and its three-dimensional structure and interaction sites were elucidated by X-ray crystallography and mutational analysis[10] (Fig. 1(f)). In trimers of LE domains a very tight interaction between adjacent domains was observed, giving rise to the rod-like assembly of domains in the short arms of laminin. The nidogen-binding activity of the particular domain is unique in laminin and the other, about 35, LE domains in this protein do not exhibit such a function. In a similar way the cell-binding F3 domain in fibronectin sticks out from many other F3 domains without integrin-binding function in this and other ECM proteins.

For the reasons summarized above it is certainly not possible to predict a function on the basis of sequence homology alone. Such predictions are about as valid as predictions of the profession of a daughter or son in a human family based on the parents' professions. Additional criteria must be considered such as recognition of binding motifs in the sequence, comparison of three-dimensional structure, or docking experiments with putative ligands. In most cases direct functional assays with isolated domains will be needed. For functional considerations a possible cooperation with other domains in the same protein may also be important. Therefore functional assignments by mutations in target domains of the intact molecule are often of critical importance.

From a recognition of binding sites in a module's structure important functional assignments can be made. Some binding sites can even be recognized in the sequence. An important example is the EF hand motif[36,37] which suggests Ca^{2+} binding. EF hand motifs occur in pairs or higher multiples. In the calcium-binding domain of SPARC/BM40/osteonectin only a single EF hand motif was detected in the sequence and the second unusual one was only found in the crystal structure[7] (Fig. 1(e)). This protein may also serve as an example of a structure in which two adjacent different modules (EC and FS) were solved by X-ray crystallography.[9] It should be noted that EF hand structures do not provide the only possible calcium-binding sites. Calcium-binding EGF-like domains were already mentioned and other examples are the CL domains and the binding of Ca^{2+} between CA domains (for reviews see refs 37, 38).

Some functions of the most abundant repeat structures, namely collagen triple helices and coiled-coil structures, are suggestive. Both structures interlink protein subunits in ECM modules and are therefore sometimes called oligomerization domains. Examples are the collectins in which a cluster of three CA domains is formed by assembly of the collagenous parts of the subunits in a triple helix.[38] In the asialoglycoprotein receptor a three stranded coiled-coil structure serves the same function. For the lectins, the created multivalency thus is essential for the recognition of bacterial cell walls and glycoproteins.[38] The functional reasons for the trimeric states of thrombospondins 1 and 2 and the pentameric state of thrombospondins 3, 4 and cartilage oligomeric matrix protein (COMP) are not clear yet. A possible reason is the binding of all-*trans* retinol, vitamin D and other hydrophobic ligands to the large hydrophobic channel in the five-stranded coiled-coil domain of COMP (Y. Guo, D. Bozic and J. Engel; unpublished results). In all of these cases subunits are oligomerized by short coiled-coil structures (Fig. 1(g)). In laminins and fibrinogen three different subunits are connected by coiled-coil regions with a specificity for heteroassociation. Heteroassembly is important for secretion and probably for other functions (reviewed in ref. 39).

Collagen triple helices and coiled-coil structures may also serve as rod-like spacer elements between functional domains. The rod-like structure of collagen triple helices and the charge distribution at the triple helix surface play an important role in fibril formation by lateral association.[22] More sophisticated functions have also been observed and explained on a structural level including the binding of integrins to a collagen triple helix.[41]

Evolution of domains and modular proteins

During evolution the creation of a stable protein domain was a difficult task. Out of the very many possible arrangements of 20 amino acid residues in a sequence ($20^{50} = 1.12 \times 10^{65}$ for a domain of 50 residues) only very few have been realized. It was estimated that only about 1000 protein families exist[42] and therefore an even smaller number of archetype domains may be considered as precursors for the present domains. New domains within a family resulted from gene duplications followed by base substitution.[1] This, and biological selection, led to the creation of domains of new functions within a domain family. In the case of modular proteins a domain family is defined as a module. As discussed in the preceding section each module comprises domains with the same basic three-dimensional structure but with sequence and structure variations and possibly different functions.

Unequal crossing over between sister chromosomes can readily extend tandemly repeated genes into larger series. Duplications, deletions, inversions, conversion, slippages, and translocation of DNA segments can also arise as the result of erratic rejoining of fragments. Gene rearrangements are, however, very rare events and therefore the wide distribution of modules in modular proteins is probably caused by faster mechanisms. Clearly modular proteins are rather modern proteins[1] which developed together with the evolution of multicellular organisms and after the divergence of plants, fungi, and animals about 1000 million years ago. There is evidence that introns are of recent origin and intronic recombinations may have led to exon shuffling in modern proteins (reviewed in ref. 43). An indication among others for the importance of exon shuffling are the equal phases (0, 1, or 2) of introns demarcating both the 5′ and 3′ boundary of many modules which exhibit a wide distribution. Unequal phases would shift the reading frame and would lead to a loss of protein information downstream of the recombination point.[42] There are, however, many exceptions and frequently exons do not correspond to modules. Also, domain shuffling occurred in organisms which lack introns. Only in part may these deviations be explained by secondary events during evolution, for example by later removal or insertion of introns. It is now generally agreed[43,44] that the widespread distribution of modules which we find today in ECM and adhesion proteins was caused by a number of different mechanisms during evolution.

References

1. Doolittle, R. F. (1992). *Protein Sci.*, **1**, 192–200.
2. Bairoch, A. (1994). Nomenclature of extracellular domains. Swiss Prot. protein sequence data bank, release 30. 0.
3. Bork, P. and Bairoch, A. (1995). *TIBS* **20**, poster CO2 'Extracellular protein modules: a proposed nomenclature'.
4. Bork, P., Downing, A. K., Kieffer, B., and Campbell, I. D. (1996). *Quart. Rev. Biophys.*, **29**, 119–67.
5. Shapiro, L., Fannon, A. M., Kwong, P. D., Thompson, A. Lehmann, M. S., Grübel, G., *et al.* (1995). *Nature*, **374**, 327–37.
6. Shapiro, L., Kwong, P. D., Fannon, A. M., Colman, D. R., and Hendrickson, W. A (1995). *Proc. Natl Acad. Sci. USA*, **92**, 6793–6797.
7. Hohenester, E., Maurer, P., Hohenadl, C., Timpl, R., Jansonius, J. N., and Engel, J. (1996). *Nature Struct. Biol.*, **3**, 67–73.
8. Yee, V. C., Pratt, K., Côté, H. C. F., Le Trong, I., Chung, D. W., Davie, E. W., *et al.* (1996). *Structure*, **5**, 125–38.
9. Hohenester, E., Maurer, P., and Timpl, R. (1997). *EMBO J.*, **16**, 3778–86.
10. Stetefeld, J., Mayer, U., Timpl, R., and Huber R. (1996). *J. Mol. Biol.*, **257**, 644–57.
11. Kohda, D., Morton, C. J., Parkar, A. A., Hatanaka, H., Inagaki, F. M., Campbell, I. D., and Day, A. J. (1996). *Cell*, **86**, 767–75.
12. Harbury, P. B., Uhang, T., Kim, P. S., and Alber, T. (1993). *Science*, **262**, 1401–7.
13. Malashkevich, V. N., Kammerer, R. A., Efimov, V., Schulthess, T., and Engel, J. (1996). *Science*, **274**, 761–5.
14. Bella, J., Eaton, M., Brodsky, B., and Berman, H. M. (1994). *Science*, **266**, 75–81.
15. Kobe, B. and Deisenhofer, J. (1995). *Curr. Opin. Struct. Biol.*, **5**, 409–416.
16. Main, A. L., Baron, M., Harvey, T. S., Boyd, J., and Campbell, I. D. (1992). *Cell*, **71**, 671–8.
17. Cooke, R. M., Wilkinson, A. J., Baron, M., Pastore, A., Tappin, M. J., Campbell, I. D., *et al.* (1987). *Nature*, **327**, 339–41.
18. Downing, A. K., Knott, V., Werner, J. M., Cardy, C. M. Campbell, I. D., and Handford, P. A. *Cell*, **85**, 597–605.
19. Nagar, B., Overduin, J., Ikura, M., and Rini, J. M. (1996). *Nature*, **380**, 360–4.
20. Brodsky, B. and Ramshaw, J. (1997). *Matrix Biol.*, **15**, 545–54.
21. Cohen, C. and Parry, D. A. D. (1990). *Protein*, **7**, 1–15.
22. Bateman, J. F., Lamandé, S. R., and Ramshaw, J. A. M. (1996). *Extracellular matrix* (ed. Wayne D. Comper), Vol. 2, pp. 22–67. Harwood.
23. Gaill, F., Wiedemann, H., Mann, K., Kühn, K., Timpl, R., and Engel, J. (1991). *J. Mol. Biol.*, **115**, 1159–69.
24. Kammerer, R. A. (1997). *Matrix Biol.*, **15**, 555–65.
25. Kühn, K. (1995). *Matrix Biol.*, **14**, 439–45.
26. Brown, J. H., Cohen, C., and Parry, D. A. (1996). *Protein*, **26**, 134–45.

27. Pikkarainen, T., Schulthess, T., Engel, J., and Tryggvason K. (1992). *Eur. J. Biochem.*, **209**, 571–82.
28. Paulsson, M., Mörgelin, M. Wiedemann, H., Beardmore-Gray, M., Dunham, D., Hardingham, T. E., *et al.* (1987). *Biochem. J.*, **245**, 763–72.
29. Gendler, S. J. and Spicer, A. P. (1995). *Ann. Rev. Physiol.*, **57**, 607–34.
30. Williams, M. J. and Campbell, I. D. (1994). *Meth. Enzymol.*, **245**, 451–69.
31. Jones, E. Y. (1996). *Curr. Opin. Cell Biol.*, **8**, 602–8.
32. Hortsch, M. (1996). *Neuron*, **17**, 587–93.
33. Ruoslahti, E. (1988). *Annu. Rev. Biochem.*, **57**, 375–413.
34. Leahy, D. J., Aukhil, I., and Erickson, H. P. (1996). *Cell*, **84**, 155–64.
35. Potts, J. R. and Campbell, I. D. (1996). *Matrix Biol.*, **15**, 313–20.
36. Kretsinger, R. H. (1996). *Nature Struct. Biol.*, **3**, 12–15.
37. Maurer, P., Hohenester, E., and Engel, J. (1996). *Curr. Opin. Cell Biol.*, **8**, 609–17.
38. Kishore, U., Eggleton, P., and Reid, K. B. M. (1997. *Matrix Biol.*, **15**, 383–592.
39. Maurer, P. and Engel, J. (1996). *The laminins* (ed. P. Ekblom and R. Timpl), pp. 27–49. Harwood,
40. Kraulis, P. (1991). *J. Appl. Cryst.*, **24**, 946–50.
41. Eble, J. A., Ries, A., Lichy, A., Mann, K., Stanton, H., Gabrilovic, J. *et al.* (1996). *J. Biol. Chem.*, **271**, 30964–70.
42. Clothia, C. (1992). *Nature*, **357**, 543–4.
43. Patthy, L. (1996). *Matrix Biol.*, **15**, 301–10.
44. Bork, P. (1996). *Matrix Biol.*, **15**, 311–2.
45. Hohenester, E., Sasak, T., Olsen, B. R., and Timpl, R. (1998). *EMBO J.*, **17**, 1656–64.
46. Beckmann, G., Hanke, J., Bork, P., and Reich, J. G. (1998). *J. Mol. Biol.*, **275**, 725–30.

■ *Jürgen Engel and Damir Bozic*
Department of Biophysical Chemistry,
Biozentrum University Basel,
Klingelbergstrasse 70, CH-4056 Basel,
Switzerland.

Proteoglycans

Proteoglycans are a set of ubiquitous proteins found on cell sufaces, within intracellular vesicles, and incorporated into extracellular matrices. Unlike most proteins, which are grouped into families on the basis of amino acid similarities alone, the proteoglycans are defined by a common type of post-translational modification: the glycosaminoglycan (GAG). GAGs are polysaccharides that are quite differentfrom the *N*- and O-linked oligosaccharides-found on most cell-surface and secreted proteins, and GAGs strongly influence the structure and molecular interactions of the proteins to which they are attached.

The proteoglycans are a diverse set of macromolecules. Their protein components (usually referred to as protein 'cores') can consist of polypeptide chains as small as 10 kDa or as large as 400 000 kDa; they can be soluble or insoluble; they can be membrane-spanning, lipid-tailed, or secreted; they can carry only a single GAG chain, or well over a hundred. Why then are proteoglycans grouped together as a 'class' of molecules? One practical reason is that GAGs tend to dominate the biochemical behaviours of the molecules to which they are attached, giving proteoglycans similar fractionation properties. Among other things, GAGs are responsible for the extraordinarily poor resolution of proteoglycans by SDS–PAGE, a feature that long hindered the identification of individual proteoglycan species. A theoretical, and more important, reason for treating proteoglycans as a class, is the conviction among proteoglycan researchers that GAGs mediate many of the functions of these molecules. This conviction stems both from an appreciation of the unique biophysical properties of GAGs, and from an increasing awareness of the striking number of interesting cell-surface, secreted, and extracellular matrix proteins that bind GAGs. Great strides have been made in the past 5–7 years in elucidating the functions of proteoglycans and the roles played by GAG chains. None the less, most of what proteoglycans are believed to do is still somewhat speculative, and no single unifying theory explains why this unique class of molecules is so abundant, and has been so well conserved by evolution in animals from nematodes to man.

This chapter provides a brief introduction to the world of proteoglycans: their structure, the nomenclature used to describe them, the molecules with which they interact, and the physiological functions in which they apparently participate. The names of molecules that are the subject of individual chapters in this book are shown in **bold**. Literature citations are provided for additional information that is not contained in these chapters.

■ Glycosaminoglycan structure

Given that GAGs are among the largest of protein-bound polysaccharide moieties, one might expect GAGs to exhibit an almost indecipherable degree of complexity. Yet their structures are surprisingly easy to understand: GAGs are linear polymers; there are no branches such as are found in *N*-linked oligosaccharides. Except for the short linkage region by which GAGs are attached to the serine residues of protein cores, each GAG is synthesized from only two monosaccharides, strung together in strictly alternating fashion. There are only three such disaccharide repeat units that can be polymerized on to proteins in this fashion, giving rise to three basic 'parent polymers' from which all protein-bound GAGs are fashioned. Following polymerization of these simple chains (the lengths of which are variable), certain types of enzymatic modifications are carried out on the sugars themselves.

Thus, early in their biosynthetic histories, proteoglycans start out with simple, structurally homogeneous GAG chains. It is the subsequent modification of these chains, therefore, that generates most of the complexity in GAGs. This is not because the *number* of modifications that can be made is very large, but rather because modifications are made only *sporadically* throughout any chain. Thus, while there are only a handful of different modifications–namely addition of *O*-sulphate groups at various positions, replacement of glucosamine *N*-acetyl groups by *N*-sulphate, and an isomerization step that involves the fifth carbon in D-glucuronic acid and converts that sugar into its epimer L-iduronic acid – each disaccharide in a GAG may contain any (or none) of these modifications. As a result, GAGs may be spoken of as having linear sequence–the units of which are disaccharides, each of which has been modified in a particular way. Theoretically, such sequences can encode considerable information – just as the sequences in DNA do – provided that cellular mechanisms for 'reading' that information exist. Yet, before researchers can determine whether the sequence information in GAGs is biologically meaningful, they must themselves be able to read that information. Unfortunately, just as little was known about the meaning of sequences in nucleic acids before the advent of DNA sequencing and chemical synthesis, little is currently known about the meaning of GAG sequences, which can only be sequenced and synthesized by slow and cumbersome techniques. Despite these limitations, an early and remarkable success in this area was the identification of the antithrombin-binding sequence found on heparin (see below). More recently, sequencing efforts have also identified GAG structures that bind with high affinity to several other ligands. It seems likely that, as technical advances continue to be made, much additional progress in this area will be seen.

Glycosaminoglycan nomenclature

During the early years of evolution of GAG nomenclature, GAG composition and biosynthesis were not well understood. Accordingly, readily observable biochemical features of GAGs such as the disaccharide repeat unit, susceptibility to digestion by certain enzymes or chemicals, and overall charge, form the traditional criteria by which GAGs are classified. As one might expect, information about the sequences in which sugar modifications occur is generally not reflected in GAG nomenclature.

The GAGs *heparin* and *heparan sulphate* are both derived from the parent polymer [D-glucuronic acid β (1→4) D-*N*-acetyl glucosamine α (1→4)]$_n$, the disaccharides of which are then subjected to any or all of five different chemical modifications (epimerization of glucuronate to iduronate, *N*-deacetylation/*N*-sulphation, and *O*-sulphation at position 2 of iduronate and positions 3 and/or 6 of glucosamine). What distinguishes heparin is how heavily it is modified compared with heparan sulphate (it is an unfortunate historical accident that the less modified, and therefore the less sulphated species, was the one to acquire the term 'sulphate' in its name). The purely quantitative nature of the distinction between heparin and heparan sulphate is nicely illustrated by comparing the actions of two GAG-degrading bacterial enzymes, heparinase I and heparitinase (heparinase IIII) on these GAGs. Heparinase I cleaves only at disaccharides containing certain modifications; heparitinase makes the identical cleavage, but only at disaccharides that lack particular modifications. The result is that both enzymes degrade both heparin and heparan sulphate, but heparinase cleaves heparin into small fragments while heparitinase does not, and heparitinase cleaves heparan sulphate into small fragments while heparinase does not. It should be added that, while heparan sulphate is found in virtually all tissues and can be attached to a variety of different core proteins, heparin is now known to be derived from a single proteoglycan made by a single cell type, the mast cell. Under the circumstances, it might be more logical to rename heparin 'mast cell heparan sulphate', although such a change in nomenclature seems unlikely to happen.

Chondroitin sulphate and *dermatan sulphate* are also GAGs that derive from the same starting polymer [D-glucuronic acid β (1→3) D-*N*-acetyl galactosamine β (1→4)]$_n$. Both may be modified by sulphation at position 2 of the uronic acid and positions 4 or 6 of the amino sugar. *N*-sulphation does not occur. The distinction between chondroitin sulphate and dermatan sulphate involves epimerization of glucuronic acid to iduronic acid; if this modification is found at high frequency, the GAG is called dermatan sulphate; if not, it is called chondroitin sulphate. Since it is not necessary that epimerization occur at every dissacharide position for a GAG to be called dermatan sulphate, it follows that stretches of sequence within dermatan sulphate can be identical to those in chondroitin sulphate. As with heparin and heparan sulphate, bacterial enzymes can be used to examine the differences between chondroitin and dermatan sulphates. The enzyme chondroitinase AC cleaves only at glucuronate-containing dissacharides, while the enzyme chondroitinase ABC cleaves at glucuronate or iduronate; accordingly, dermatan sulphates (at least those that contain extensive conversion of glucuronate to iduronate) will be relatively resistant to chondroitinase AC, but not ABC. The rather unique 'ABC' terminology used with chondroitinases reflects an older nomenclature in which chondroitin sulphate A referred to chondroitin sulphate in which 4-*O*-sulfation of galactosamine is prevalent (also called chondroitin 4-sulphate); chondroitin sulphate B to chondroitin sulphate in which epimerization of glucuronic acid to iduronic acid is prevalent (i.e. dermatan sulphate), and chondroitin sulphate C to chondroitin sulphate in which 6-*O*-sulfation of galactosamine is prevalent (also called chondroitin 6-sulphate).

Keratan sulphate differs from its parent polymer, [D-galactose β (1→4) D-*N*-acetyl glucosamine β (1→3)]$_n$, only by *O*-sulfation at position 6 of glucosamine and/or position 6 of galactose. Keratan sulphate is unique in that it can be synthesized not only as an *O*-linked sugar attached to serine (as are other GAGs), but also as an *N*-linked sugar. This latter structure comes about because the parent polymer for keratan sulphate is simply a repetition of the lactosamine disaccharide that is found on many *N*-linked oligosaccharide structures. Indeed, many cells are known to produce long polylactosamine chains (also known as lactosaminoglycans) at the ends of *N*-linked structures. Addition of sulphate moieties to these chains renders them structurally identical to keratan sulphate synthesized by the *O*-linked route.

Hyaluronan (hyaluronic acid) differs from other GAGs in that it is not synthesized attached to protein (so it is not a part of proteoglycans), and it is not further modified from the structure of its parent polymer, [D-glucuronic acid β (1→3) D-*N*-acetyl glucosamine β (1→4)]$_n$. Hyaluronan polymers can be extremely long, with molecular masses in the millions.

Proteoglycan structure and nomenclature

Before the era of molecular cloning, proteoglycans were named by the type of GAG chain found on them, the tissue source from which they had been isolated (e.g. 'the basement membrane heparan sulphate proteoglycan'), and/or distinguishing biochemical features (e.g. 'the large aggregating chondroitin sulphate proteoglycan of cartilage'). With the cloning and sequencing of the cDNAs of proteoglycan core proteins, these molecules were given one-word names that typically end in the suffix '-can' or '-glycan'. Although the names represented a much needed improvement, the fact that they no longer convey information about GAGs is somewhat of a drawback. Because different cell types are known to attach different GAGs to the same protein cores (or sometimes attach no GAGs at all), one name can now potentially be used to describe structurally very different molecules.

As the following chapters show, several proteoglycan core proteins can be classified into families on the basis of core protein homology, that is, amino acid sequence similarity and the presence of specific structural motifs. For example, the small, interstitial matrix proteoglycans, **decorin**, **biglycan**, **fibromodulin**, and lumican[1] have stretches of leucine-rich repeats similar to those found in several non-proteoglycans including leucine-rich glycoprotein and *Drosophila* chaoptin. The small leucine-rich proteoglycans also share the ability to bind to and inhibit the fibrillogenesis of collagen. Another family consists of the large chondroitin sulphate proteoglycans **aggrecan**, **versican**, neurocan,[2] and brevican.[3] These molecules contain related hyaluronan-binding domains as well as domains similar to those found in the **selectin** family of lymphocyte homing receptors. When associated with large hyaluronan molecules, these proteoglycans form supramolecular aggregates that are important structural elements of the extracellular matrix of cartilage and brain, as well as the pericellular matrix of cultured fibroblasts.

The major cell surface proteoglycans belong to two gene families: the **syndecans** and the **glypicans**. Syndecans are small transmembrane proteins that possess a short but highly conserved cytoplasmic domain, a conserved transmembrane domain, and an ectodomain that is rich in serine, threonine, and proline, and carries GAG chains of either the heparan sulphate type, or a combination of heparan sulphate and chondroitin sulphate chains. Four syndecans have been identified in mammals (now called syndecans 1–4,[4]) and one in *Drosophila*.[5] Glypicans are *gpi*-anchored proteins, the major part of which consists of a structural motif in which a pattern of 14 cysteines is absolutely conserved. The significance of this motif is unknown, especially as it appears that the GAGs found on glypicans – usually heparan sulphate – are not attached to this part of these molecules. The first four mammalian glypicans to be identified were given the name glypican cerebroglycan,[6] OCI-5,[7] and K-glypican,[8] but a numerical nomenclature (gypican-1, -2, -3, and -4, respectively, plus the recently cloned glypican-5[9,10] is gradually being adopted. At least one glypican is present in *Drosophila*, the product of the *dally* gene.[11]

In addition to these proteoglycan families, other extracellular matrix proteoglycans include **perlecan** (the major heparan sulphate proteoglycan of basement membranes), bamacan (a chondroitin sulphate proteoglycan of basement membranes[12]), **agrin** (a heparan sulphate proteoglycan of certain basement membranes and synapses), phosphacan (a chondroitin sulphate/keratan sulphate proteoglycan of the brain extracellular matrix,[13] and testican.[14] Other cell surface proteoglycans include **CD44** (which contains a hyaluronan-binding domain structurally related to those found in the aggrecan family), NG2[15] and neuroglycan C,[16] which are transmembrane chondroitin sulphate proteoglycans, betaglycan (which binds TGF-βs,[17] and appican (which is a GAG-containing variant of the amyloid precursor protein of Alzheimer's disease.[18]

Functions of proteoglycans

The functions of proteoglycans are only just beginning to become clear, but much progress has been made in the past five years. For the sake of discussion it is convenient to divide known functions into those that can be mediated by core proteins without the participation of GAG chains, and those that can be mediated by GAG chains without the participation of core proteins. In reality, this distinction may not always be clear cut, since some functions may involve the participation of both core protein and GAG, at least under some conditions.

A particularly good example of core protein-mediated function is the ability of the small leucine-rich extracellular matrix proteoglycans to regulate **collagen** fibrillogenesis *in vitro*, which also appears to be an essential function of at least one of the proteoglycans (**decorin**) *in vivo*.[19] The betaglycan core protein binds TGF-βs with rather high affinity, and alters the ability of different TGF-β isoforms to interact with signalling receptors.[17] The binding of proteoglycans of the **aggrecan** family, as well as **CD44**, to **hyaluronan** also represents a core protein-mediated interaction.

GAG-dependent functions can be loosely subdivided into two classes: the biophysical and the biochemical. The former refers to functions that depend on the unique biophysical properties of GAGs – the ability to fill space, bind and organize water molecules, and repel negatively charged molecules. The large quantities of chondroitin sulphate and keratan sulphate found on **aggrecan**, for example, are thought to play an important role in the hydration of cartilage. In contrast, the heparan sulphate on the kidney glomerular basement membrane proteoglycan is thought to play a role in filtration, impeding the passage of anionic serum proteins into the urine.[20]

The other functions of GAGs are those that are mediated by specific binding to proteins. Some of the many known GAG–protein interactions are outlined in Table 1. For some of these proteins, all that is known is that they bind to GAG affinity columns (e.g. heparin–agarose) under physiological conditions of salt and pH. For others, affinity constants have been measure(K_ds tend to be in the range of 10^{-6} to 10^{-9} M) and for still others, direct evidence for interaction with proteoglycans has been obtained (e.g. copurification, proteoglycan-dependent binding to cells, etc.). Only in a few cases have clear physiological functions been associated with GAG–protein interactions, but those cases that have been well studied include the following.

1. *Regulation of the activity of proteases and antiproteases.* The best understood GAG-mediated function has been worked out in the proteolytic cascade underlying blood coagulation. The protease inhibitor antithrombin III (ATIII) binds tightly to heparin and certain heparan sulphates; several of the proteases that are inactivated by ATIII (e.g. thrombin, factor IXa, factor XIa) also bind these GAGs. In the absence of GAGs, the kinetics with which ATIII inactivates these proteases are very slow; in the presence of appropriate GAGs, these reactions are accelerated by as much as 2000-fold. The explanation of this phenomenon is two-fold:[31] first, most GAG chains are sufficiently long that both protease and protease inhibitor can bind to the same chain; as a result of being confined to the same limited space, the likelihood of the two proteins then binding to each other is increased enormously. Second, heparin appears also to have an effect on protein conformation which contributes to

Table 1. Proteins that bind glycosaminoglycans under physiological conditions.[21–30] (Molecules with entries in this book are shown in bold type.)

Extracellular matrix proteins
Laminins
Fibronectins
Collagens
Thrombospondin
Tenascins
Vitronectin
vonWillebrand factor
Netrins
Semaphorins/collapsins
Acetylcholinesterase
Amyloid fibrils (many types)

Growth factors and growth factor-binding proteins
Fibroblast growth factors (FGFs)
Platelet-derived growth factor (PDGF)
Vascular endothelial growth factor (VEGF)
Transforming growth factor-betas (TGF-βs)
Bone morphogenetic proteins (BMPs)
Interferon-γ
Wnts/wingless
Interleukin 3 (IL-3)
Granulocyte/macrophage colony stimulating factor (GM-CSF)
Hepatocyte growth factor/scatter factor (HGF)
Heparin-binding epidermal growth factor (HB-EGF)
Neuregulins
Amphiregulin
Sonic hedgehog
Chemokines (interleukin 8, MIP-Iβ, most others)
Noggin
Follistatin
Insulin-like growth factor binding proteins (IGF-BPs) 3 and 5

Proteases and antiproteases
Antithrombin III
Heparin cofactor II
Protease nexin-1
Plasminogen activator inhibitor-1 (PAI-1)
Thrombin
Elastase
Cathepsin G
Factor IXa
Factor XIa
Plasminogen activators (uPA, tPA)
Mast cell proteases APP
Various complement components

Cell adhesion molecules
NCAM
L1
Myelin-associated glycoprotein
PECAM-1
Selectins

Others
Apolipoproteins B, E
Lipoprotein lipase
Triglyceride lipases
Angiogenin
Lactoferrin
Various viruses (herpes, HIV)

improving ATIII's binding kinetics. These potent effects of GAGs account for all of the anticoagulant effects of heparin *in vivo.* Furthermore, endogenous heparan sulphate of vascular endothelial cells appears to play a critical role in producing a non-thrombogenic surface along blood vessel walls.[32]

2. *Regulation of cellular responses to growth factors.* Some of the most exciting information on the functions of proteoglycans has come from the study of cellular responses to polypeptide growth factors. Not only do numerous growth factors and cytokines bind heparin and heparan sulphate (see Table 1), but some of these factors require GAGs for biological activity. The first clear demonstration of this phenomenon came from studies of fibroblast growth factors (FGFs). For many of the biological effects of FGF-1 and -2, and even for high affinity binding of these molecules to their tyrosine kinase receptors, heparan sulphate must be present on the cell surface or in the adjacent extracellular matrix, or it must be provided in the form of exogenously added heparan sulphate, heparin, or heparan sulphate proteoglycans.[33–36] Several models have been proposed to explain this requirement. Apparently heparan sulphates bind FGF receptors, potentially forming a trimeric complex with both ligands and receptors.[37,38] Heparan sulphates can also drive dimerization of FGFs, which can promote receptor dimerization and signalling.[39] While some GAG effects may be mediated through promotion of FGF dimerization, others are apparently not.[40] In addition to the FGFs, several other classes of growth factors appear to show a strong dependence on GAGs for function – these include heparin-binding epidermal growth factor (HB-EGF; and probably several other members of the neuregulin family, cf. refs 24, 41, 42, 43, 44, hepatocyte growth factor,[43,44] and *wingless,* a member of the wnt family.[28] These observations probably only scratch the surface of GAG–growth factor interactions *in vivo.* For example, recent observations on the phenotypes that result from mutations in glypican-3 (in man) and dally (in *Drosophila*) strongly suggest that these cell surface proteoglycans of the glypican family play widespread roles in cellular growth control.[11,45]

3. *Regulation of cell–cell and cell–matrix interactions.* The ability of GAGs to bind both extracellular matrix and cell adhesion molecules (Table 1) has long raised the possibility that PGs participate in cell–cell and cell–matrix adhesion. Consistent with this view, cells deficient in GAG biosynthesis show specific deficits in attachment to extracellular matrix substrata.[46] Furthermore, expression of exogenous syndecans in lymphocytic cell lines that do not nominally produce proteoglycans can both cause cell–cell aggregation and inhibit the ability of the cells to invade collagen gels.[47,48]

4. *Regulation of extracellular matrix assembly and structure.* Most large, multidomain extracellular matrix proteins contain at least one GAG-binding site, and in some cases several. Most of these proteins bind GAGs of the heparin/heparan sulphate class, but some, such as **tenascin C** and **thrombospondin-1**, also interact strongly with chondroitin sulphates. Evidence from studies of the *nanomelic* mutant in the chick indicate that disruption of the expression of aggrecan dramatically disrupt skeletal development.[49]

5. *Immobilization of diffusible molecules.* Although difficult to demonstrate *in vivo* it is very likely that one of the major functions of the GAG chains of proteoglycans is to immobilize otherwise diffusible molecules on appropriate cell surfaces or in extracellular matrices. This can have the effect of sequestering a molecule away from potential ligands (but possibly storing it for later use), or capturing a molecule and retaining it at a location where it is more likely to interact with potential ligands. It has been proposed, for example, that extracellular matrix heparan sulphate both stores and releases FGFs.[50] In the developing brain, extracellular matrix chondroitin sulphate appears to play a role in immobilizing axon guidance cues.[51] Hepatocytes and endothelial cells are also thought to use their cell surface heparan sulphate to immobilize both lipoproteins (via interactions with apolipoproteins B and E) and the enzymes that act on those lipoproteins, such as lipoprotein lipase and hepatic triglyceride lipase (e.g. ref. 52).

How important are GAG sequences?

Proteins that bind GAGs tend to interact with contiguous stretches of 6–14 monosaccharides.[25] Given the number of covalent modifications that are possible on each disaccharide unit, the number of GAG sequences with which proteins can potentially interact is very large, especially for GAGs of the heparin/heparan sulphate family (which can exhibit any of five different modifications per disaccharide, alone or in various combinations). Studies on antithrombin III have demonstrated that proteins can be extremely selective for some GAG sequences.[53,54] This protease inhibitor binds with high affinity only to heparin or heparan sulphate species that contain at least one pentasaccharide in which a strictly specified pattern of modifications (including the relatively uncommon 3-*O*-sulfation of glucosamine) is present. Absence of this pentasaccharide reduces GAG affinity approximately 1000-fold, and accounts for the very different ability of different tissue heparan sulphates to stimulate antithrombin activity. To date, no other molecules have been shown to exhibit such strong sequence-dependent selectivity for naturally occurring GAG species (although 10–20-fold differences have been reported,[55] but studies on the binding of growth factors to GAG fragments suggest that additional examples of strong sequence-selective binding *in vivo* will eventually be established (e.g. refs 56–58).

Summary

The proteoglycans are a large, interesting, and important class of molecules. Great progress has been made in the past decade in identifying the structures of both the core proteins and GAG chains of these molecules. This work has led to an increasing number of investigations into the physiological functions of proteoglycans, investigations that have already demonstrated important roles for these molecules in areas as diverse as development, tissue organization, growth control, and haemostasis.

References

1. Iozzo, R. (1997). *Crit. Rev. Biochem. Mol. Biol.*, **32**, 141–74.
2. Margolis, R. K., Rauch, U., Maurel, P., and Margolis, R. U. (1996). *Perspect. Dev. Neurobiol.*, **3**, 273–90.
3. Yamaguchi, Y. (1996). *Perspect. Dev. Neurobiol.*, **3**, 307–17.
4. Bernfield, M., Kokenyesi, R., Kato, M., Kinkes, M. T., Spring, J., Gallo, R. L., and Lose, E. J. (1992). *Ann. Rev. Cell Biol.*, **8**, 365–93.
5. Spring, J., Paine-Saunders, S., Hynes, R. O., and Bernfield, M. (1994). *Proc. Natl Acad. Sci. USA*, **91**, 3334–8.
6. Stipp, C. S., Litwack, E. D., and Lander, A. D. (1994). *J. Cell Biol.*, **124**, 149–60.
7. Filmus, J., Shi, W., Wong, Z. M., and Wong, M. J. (1995) *Biochem. J.*, **311**, 561–5.
8. Watanabe, K., Yamada, H., and Yanaguchi, Y. (1995). *J. Cell Biol.*, **130**, 1207–18.
9. Saunders, S., Paine-Saunders, S., and Lander, A. D. (1997). *Dev. Biol.*, **190**, 78–93.
10. Veugelers, M., Vermeesch, J., Reekmansm, G., Steinfeld, R., Marynen, P., and David, G. (1997). *Genomics*, **40**, 24–30.
11. Nakato, H., Futch, T. A., and Selleck, S. B. (1995). *Development*, **121**, 3687.
12. Wu, R. R. and Couchman, J. R. (1997). *J. Cell Biol.*, **136**, 433–44.
13. Maurel, P., Rauch, U., Flad, M., Margolis, R. K., and Margolis, R. U. (1994). *Proc. Natl Acad. Sci. USA*, **81**, 2512–16.
14. Alliel, P. M., Perin, J. P., Jolles, P., and Bonnet, F. J. (1993). *Eur. J. Biochem.*, **214**, 347–50.
15. Levine, J. M. and Nishiyama, A. (1996). *Perspect. Dev. Neurobiol.*, **3**, 245–59.
16. Watanabe, E., Maeda, N., Matsui, F., Kushima, Y., Noda, M., and Oohira, A. (1995). *J. Biol. Chem.*, **270**, 26876–82.
17. López-Casillas, F., Wrana, J. L., and Massagué, J. (1993). *Cell*, **73**, 1435–44.
18. Pangalos, M. N., Shioi, J., Efthimiopoulos, S., Wu, A., and Robakis, N. K. (1996). *Neurodegeneration*, **5**, 445–51.
19. Danielson, K., Baribault, H., Holmes, D., Graham, H., Kadler, K., and Iozzo, R. (1997). *J. Cell Biol.*, **136**, 729–43.
20. Kanwar, Y., Linker, A., and Farquhar, M. (1980). *J. Cell Biol.*, **86**, 688–93.
21. Booth, B., Boes, M., Andress, D., Dake, B., Kiefer, M., Maack, C., *et al.* (1995). *Growth Regul.*, **5**, 1–17.
22. Bumcrot, D., Takada, R., and McMahon, A. (1995). *Mol. Cell. Biol.*, **15**, 2294–303.
23. Jackson, R. L., Busch, S. J., and Cardin, A. D. (1991). *Physiol. Rev.*, **71**, 481–539.
24. Johnson, G. R. and Wong, L. (1994) *J. Biol. Chem.*, **269**, 27149–54.
25. Lander, A. D. (1994). *Chem. Biol.*, **1**, 73–8.
26. Nakamura, T., Sugino, K., Titani, K., and Sugino, H. (1991). *J. Biol. Chem.*, **266**, 19432–7.
27. Nelson, R., Cecconi, O., Roberts, W., Aruffo, A., Linhardt, R., and Bevilacqua, M. (1993). *Blood*, **82**, 3253–8.
28. Reichsman, F., Smith, L., and Cumberledge, S. (1996). *J. Cell Biol.*, **135**, 819–27.
29. Ruppert, R., Hoffmann, E., and Sebald, W. (1996). *Eur. J. Biochem.*, **237**, 295–302.
30. Witt, D. and Lander, A. (1994). *Curr. Biol.*, **4**, 394–400.
31. Olson, S. and Björk, I. (1992). In *'Heparin and related polysaccharides'* (ed. D. Lane, I. Björk, and U. Lindahl), pp. 155–66. Plenum Press, New York.
32. Marcum, J. A., Reilly, C. F., and Rosenberg, R. D. (1987). In *'Biology of proteoglycans'* (ed. T.N.W. and R.P. Mecham), pp. 301–38. Academic Press, Orlando.
33. Aviezer, D., Levy, E., Safran, M., Svahn, C., Buddecke, E., Schmidt, A., *et al.* (1994). *J. Biol. Chem.*, **269**, 114–21.
34. Ornitz, D. M., Yayon, A., Flanagan, J. G., Svahn, C. M., Levi, E., and Leder, P. (1992). *Mol. Cell Biol.*, **12**, 240–7.
35. Rapraeger, A., Krufka, A., and Olwin, B. B. (1991). *Science*, **252**, 1705–8.
36. Yayon, A., Klagsbrun, M., Esko, J.D., Leder, P., and Ornitz, D. M. (1991). *Cell*, **64**, 841–8.
37. Kan, M., Wang, F., To, T. B., Gabriel, J. L., and McKeehan, W. L. (1996). *J. Biol. Chem.*, **271**, 26143–8.
38. Kan, M., Wang, F., Xu, J., Crabb, J. W., Hou, J., and McKeehan, W. L. (1993). *Science*, **259**, 1918–21.
39. Ventkataraman, G., Sasisekharan, V., Herr, A. B., Ornitz, D. M., Waksman, G., Cooney, C. L., *et al.* (1996). *Proc. Natl Acad. Sci. USA*, **93**, 845–50.
40. Krufka, A., Guimond, S., and Rapraeger, A. (1996). *Biochemistry*, 11131–41.
41. Aviezer, D. and Yayon, A. (1994). *Proc. Natl Acad. Sci. USA*, **91**, 12173–7.
42. Sudhalter, J., Whitehouse, L., Rusche, J. R., Marchionni, M. A., and Mahanthappa, N. K. (1996). *Glia*, **17**, 28–38.
43. Schwall, R., Chang, L., Godowski, P., Kahn, D., Hillan, K., Bauer, K., and Zioncheck, T. (1996). *J. Cell Biol.*, **133**, 709–18.
44. Zioncheck, T. F., Richardson, L., Liu, J., Chang, L., King, K. L., Bennett, G. L., *et al.* (1995). *J. Cell Biol.*, **270**, 16871–8.
45. Pilia, G., Hughes-Benzie, R. M., MacKenzie, A., Baybayan, P., Chen, E. Y., Huber, R., *et al.* (1996). *Nature Genet.*, **12**, 241–47.
46. LeBaron, R. G., Esko, J. D., Woods, A., Johansson, S., and Höök, M. (1988). *J. Cell Biol.*, **106**, 945–52.
47. Liebersbach, B. F. and Sanderson, R. D. (1994). *J. Biol. Chem.*, **269**, 20013–19.
48. Stanley, M., Liebersbach, B., Liu, W., Anhalt, D., and Sanderson, R. (1995). *J. Biol. Chem.*, **270**, 5077–83.
49. Li, L. H., Schwartz, N. B., and Vertel, B. M. (1993). *J. Biol. Chem.*, **268**, 23504–11.
50. Moscatelli, D. (1992). *J. Biol. Chem.*, **267**, 25803–09.
51. Emerling, D. E. and Lander, A. D. (1997). *Neuron*, **17**, 1089–100.
52. Lookene, A., Savonene, R., and Olivecrona, G. (1997). *Biochemistry*, **36**, 5267–75.
53. Lam, L. H., Silbert, J. E., and Rosenberg, R. D. (1976). *Biochem. Biophys. Res. Commun.*, **69**, 570–7.
54. Lindahl, U., Thunberg, L., Backsrom, G., Risenfeld, J., Nordling, K., and Bjork, I. (1984). *J. Cell Biol.*, **259**, 12368–76.
55. San Antonio, J. D., Slover, J., Lawler, J., Karnovsky, M. J., and Lander, A. D. (1993). *Biochemistry*, **32**, 4746–55.
56. Guimond, S., Maccarana, M., Olwin, G. G., Lindahl, U., and Rapraeger, A. C. (1993). *J. Biol. Chem.*, **268**, 23906–14.
57. Lortat-Jacob, H., Turnbull, J., and Grimaud, J. (1995). *Biochem. J.*, **310**, 497–505.
58. Turnbull, J. E., Fernig, D. G., Ke, Y., Wilkinson, M. C., and Gallagher, J. T. (1992). *J. Biol. Chem.*, **267**, 10337–41.

Arthur D. Lander
Department of Developmental and Cell Biology, University of California, Irvine, Irvine, CA 92697–2275, USA

ECM molecules

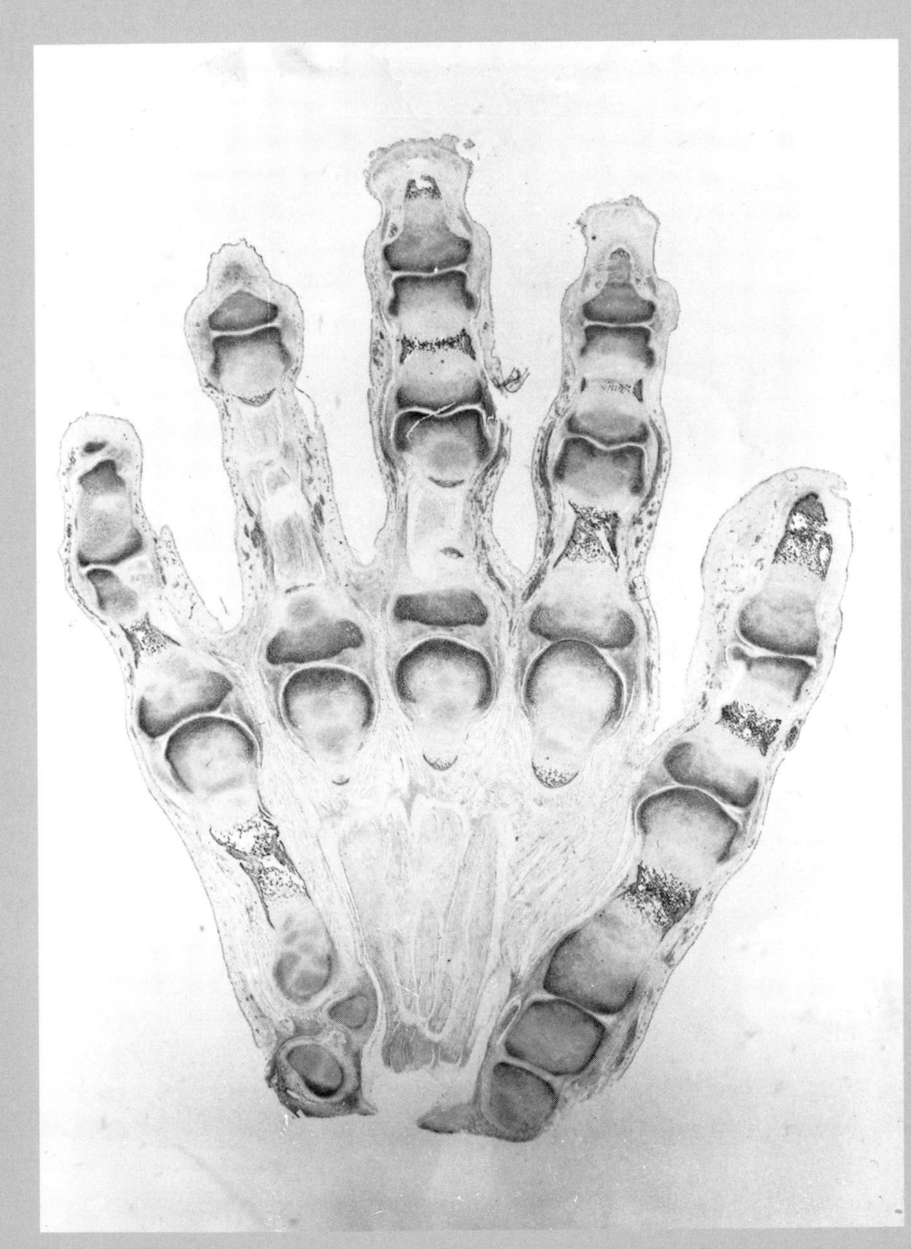

Immunolocalization of biglycan in a hand of a 15-week old fetus. Biglycan is located in the epidermis, vasculature, type I collagen-containing bone and in the very cellular articular ends of each bone rudiment. (Courtesy of Larry Fisher, National Institutes of Health.)

Aggrecan

Aggrecan[1] is the core protein of the large aggregating keratan-sulphate/chondroitin-sulphate proteoglycan[2,3] found in cartilaginous tissues. An extracellular matrix structural molecule, aggrecan localizes in cartilage as a stable ternary complex with hyaluronic acid and link protein, and provides a strongly hydrated space-filling gel due to the large number of polyanionic glycosaminoglycan chains covalently attached to the protein core.

Synonymous names

Pg-H, CSPG, PGLA. Aggrecan is a member of a family of proteoglycans including versican, neurocan, and brevican.[4] These members share the common features of conserved globular domains G1 and G3 at the N and C termini respectively, with a central glycosaminoglycan-bearing domain. The G1 domain is also related to link protein, and a number of other hyaluronan-binding proteins including CD44 and TSG-6.[5] The G3 domain contains the same structural motifs found in the selectin[6] family. This family of proteoglycans has been designated hyalectins,[4] or lecticans.[6,7]

Protein properties

Aggrecan is a monomeric polypeptide of molecular weight 220–250 kDa, as deduced from cDNA sequencing, with variations due to species-specific sequence differences, and to alternative exon splicing.[1] The mature proteoglycan monomer bears several types of covalent substituents attached to the central extended part of the core protein: glycosaminoglycans, including 20–80 keratan sulphate chains and approximately 100 chondroitin sulphate chains; as well as *N*- and *O*-linked oligosaccharides. The mature substituted molecule is about 2500 kDa.[2] The glycosaminoglycan chains are attached to specific serine residues which precede glycine residues; these attachment sites occur in groups of repeated sequences in the central core protein. The human aggrecan gene has a unique highly conserved set of these repeated sequences, which give rise to a variable number of tandem repeat (VNTR) polymorphism in the chondroitin-sulphate attachment region, producing alleles with different numbers of serine–glycine repeat sequences in different individuals.[8] Preceding the serine–glycine repeated sequences is another repeated sequence region, rich in proline, glutamic acid, and serine, where keratan sulphate attachment sites are clustered, possibly in hexameric repeats. The central glycosaminoglycan-bearing region is flanked by complex, disulphide-containing globular structures, G1 and G2 at the N terminus, and G3 at the C terminus. G1 and G2 both contain tandem repeats of homologous 100 amino acid sequences. These repeated domains, link B motifs or PTRs, are members of the C-type lectin superfamily[6] and have been shown to bind hyaluronan. The G1 domain has an additional motif, the A domain, which is also found in link protein; this is an immunoglobulin-fold motif, and mediates aggrecan–link protein interaction. The G3 domain is composed of three distinct structural motifs: one or two EGF-like sequences (depending on species), a C-type lectin domain, and a sequence related to complement regulatory protein (CRP). The lectin-like domain appears to be present in all forms of the molecule, while the EGF and CRP-like domains are alternatively spliced. While aggrecan expression is largely confined to cartilage, it has been reported in calvaria,[9] notochord,[10] and brain.[11]

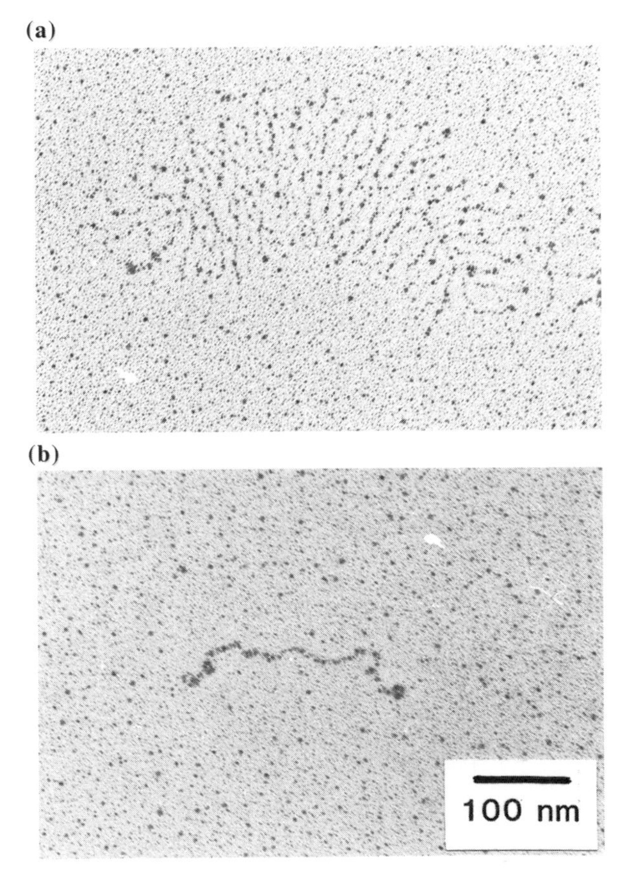

Figure 1. Rotary shadow electron micrograph of aggrecan proteoglycan monomer (a) and the aggrecan core protein (b). (Pictures provided by Drs M. Morgelin and J. Engel.)

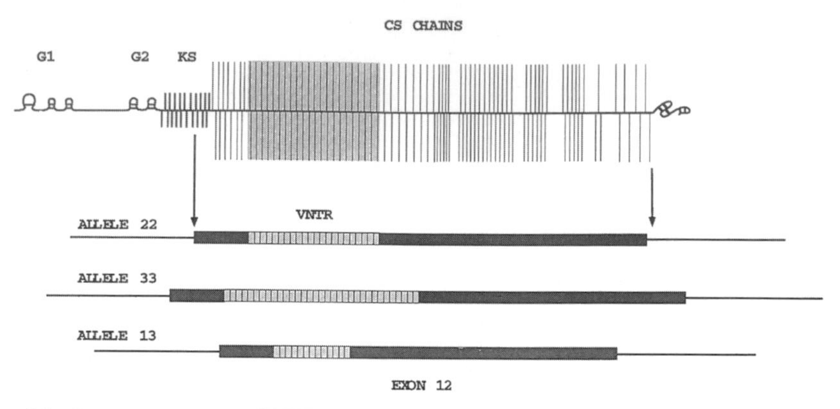

Figure 2. Diagram of the human aggrecan VNTR polymorphism. The top of the figure represents the human aggrecan monomer with structural domains indicated. The short vertical lines represent keratan sulphate (KS) chains and the long vertical lines, chondroitin sulphate (CS) chains. The portion of the CS domain corresponding to the polymorphic region is shaded. Below are pictured three of the different polymorphic alleles with 22, 33, and 13 repeats of the VNTR region; the repeats are shown as light shaded boxes within the heavy dark bar representing exon 12. Vertical arrows indicate the portion of the aggrecan core protein encoded by exon 12. (Reprinted with permission.[7])

Purification

Aggrecan has generally been purified as the proteoglycan, utilizing its high buoyant density in CsCl conferred by the numerous negative charges of the attached glycosaminoglycans.[12] Cartilage is dispersed in 4 M guanidine–HCl to reversibly denature the hyaluronic-acid-binding structure, and the extracts are then centrifuged in CsCl in 4 M guanidine, to sediment the monomer, or after reassociation in 0.5 M guanidine to sediment the aggregate with hyaluronic acid and link protein. Anion exchange and molecular sieve chromatography have also been applied, exploiting the high charge and large hydrodynamic volume of the proteoglycan.[13]

Activities

Aggrecan is a structural molecule, providing a high concentration of fixed charges in the extracellular space, which create a hyperosmotic reversibly deformable gel suitable for cyclical load bearing. The G1 domain binds irreversibly under native conditions to both the unbranched polymer hyaluronic acid and link protein, creating a stable aggregate of up to 100 monomers arrayed along the strand. Functions for G2 and the various forms of G3 have not been determined, although the lectin-like sequence of G3 can bind to fucose and galactose,[14] and the corresponding G3 domain of versican binds to tenascin-R.[15]

Antibodies

Numerous polyclonal and monoclonal immunological reagents have been prepared to both the protein and carbohydrate portions of the proteoglycan,[16] and have been applied in studies of its biosynthesis and tissue distribution.[11] One monoclonal antibody to a core protein epitope (1C-6) is available from the Developmental Hybridoma Bank (Johns Hopkins University, Baltimore, MD). Neoepitope antibodies have been used to analyse proteolytic breakdown products of aggrecan.[17]

Genes

Full length cDNAs have been reported for the rat (GenBank J03485); human (GenBank M55172); mouse (GenBank L07049); chicken (GenBank L21913); and bovine (GenBank U76615) species of aggrecan. The genomic structures have been determined for rat,[18] mouse,[19] human,[20] and chicken.[21] All species share similar intron/exon boundaries and exon numbers (18 or 19 exons, depending on the presence of the EGF1 exon) and all show the unusual feature of a single large exon encoding the glycosaminoglycan attachment region. Large first introns separate the untranslated first exons from the second exons which encode the signal peptides. The proximal promoter sequences of the rat, human, and mouse genes are similar. The human gene has been localized to 15q26.1,[22] and the mouse gene to chromosome 7 in close linkage to the *Fes* gene.[23]

Mutant phenotype/disease states

Two animal mutants have been mapped to the aggrecan locus, the nanomelic chicken[24] and the *cmd* mouse.[25] Both result in premature termination codons, near the C terminus for the chicken and near the N terminus for the mouse. Both are homozygous lethal dwarfisms, with reduced cartilage volume and low levels of aggrecan mRNA. As one of the major structural proteins of cartilage, aggrecan has been widely investigated for involvement in genetic diseases of skeletal growth, but no definitive link has been established. Research is also ongoing on the possible role of this molecule in degenerative joint conditions, including efforts to characterize the proteolytic enzymes responsible for degrading aggrecan in arthritic joints.

References

1. Doege, K. *et al.* (1991). *J. Biol. Chem.*, **266**, 894–902.
2. Wight, T. N., Heinegard, D. K., and Hascall, V. C. (1991). In *Cell biology of extracellular matrix* (ed. E. Hay), pp. 45–78. Plenum, New York.
3. Muir, H. (1995). *Bioessays*, **17**, 1039–48.
4. Iozzo, R. V. and Murdoch, A. D. (1996). *FASEB J.*, **19**, 598–614.
5. Kohda, D., Morton, C. J., Parkar, A. A., Hatanaka, H., Inagaki, F. M., Campbell, I. D., and Day, A. (1996). *Cell*, **86**, 767–75.
6 Brisset, N. C. and Perkins, S. J. (1996). *FEBS Lett.*, **388**, 211–16.
7. Ruoslahti, E. (1996). *Glycobiology*, **6**, 489–92.
8. Doege, K. J., Coulter, S. N., Meek, L. M., Maslen, K., and Wood, J. G. (1997). *J. Biol. Chem.*, **272**, 13974–9.
9. Wong, M., Lawton, T., Goetinck, P. F., Kuhn, J. L., Goldstein, S. A., and Bonadio, J. (1992). *J. Biol. Chem.*, **267**, 5592–8.
10. Stirpe, N. S. and Goetinck, P. F. (1989). *Development*, **107**, 22–33.
11. Schwartz, N. B, Domowicz, M., Krueger, R. C. Jr, Li, H., and Mangoura, D. (1996). *Persp. Dev. Neurobiol.*, **3**, 291–306.
12. Hascall, V. and Sajdera, S. (1989). *J. Biol. Chem.*, **244**, 2384–95.
13. Heinegard, D. and Sommarin, Y. (1987). *Meth. Enzymol.*, **144**, 319–72.
14. Halberg, D., Proulx, G., Doege, K., Yamada, Y., and Drickamer, K. (1988). *J. Biol. Chem.*, **263**, 9486–90.
15. Aspberg, A., Binkert, C., and Ruoslahti, E. (1995). *Proc. Natl Acad. Sci. USA*, **92**, 10590–4.
16. Caterson, B., Christner, J. E., Baker, J. R., and Couchman, J. R. (1985). *Fed. Proc.*, **44**, 386–93.
17. Hughes, C., Caterson, B., Fosang, A. J., Roughley, P. J., and Mort, J. S. (1995). *Biochem. J.*, **305**, 799–804.
18. Doege, K. J., Garrison, K., Coulter, S. N., and Yamada, Y. (1994). *J. Biol. Chem.*, **269**, 29232–40.
19. Watanabe, H., Gao, L., Sugiyama, S., Doege, K., Kimata, K., and Yamada, Y. (1995). *Biochem. J.* **308**, 433–40.
20. Valhmu, W. B., Palmer, G. D., Rivers, P. A., Ebara, S., Cheng, J.-F., Fischer, S., and Ratcliffe, A. (1995). *Biochem. J.* **309**, 535–42.
21. Li, H. and Shwartz, N. B. (1995). *J. Mol. Evol.*, **41**, 878–85.
22. Korenberg, J. R., Chen, X. N., Doege, K., Grover, J., and Roughley, P. J. (1993). *Genomics*, **16**, 546–8.
23. Walcz, E., Deak, F., Erhardt, P., Coulter, S. N., Fulop, C., Horvath, P. *et al.* (1994). *Genomics*, **22**, 364–71.
24. Li, H., Schwartz, N. B., and Vertel, B. M. (1993). *J. Biol. Chem.*, **268**, 23504–11.
25. Watanabe, H., Kimata, K., Line, S., Strong, D., Gao, L., Kozak, C. A., and Yamada, Y. (1994). *Nature Genet.*, **7**, 154–7.

Kurt J. Doege
Shriners Hospital Research, Department of Cell and Developmental Biology, Oregon Health Sciences University, Portland, OR, USA

Agrin

Agrin[1–3] was originally identified as a basal lamina-derived protein that triggers differentiation of postsynaptic specializations in vertebrate skeletal muscle myotubes in cell culture. Agrin is now known to be a member of a family of extracellular matrix proteins that are expressed in a variety of cell types and arise by alternative splicing from a single agrin gene.[4–5] Specific agrin isoforms are synthesized in the cell bodies of motor neurones in the vertebrate central nervous system and transported along their axons to skeletal muscles.[2] There, agrin is released from the axon terminal to bind to the myofibre basal lamina in the synaptic cleft and to an as yet unidentified agrin receptor in the myofibre plasma membrane, triggering the formation of the postsynaptic apparatus.

Synonymous names

None.

Homologous proteins

Agrin has several repeated structural motifs that are homologous to domains in follistatin, laminin, or EGF;[6] no proteins have been described that share an overall structural similarity.

Protein properties

Agrin was originally isolated from basal lamina-containing extracts of the electric organ of the marine ray *Torpedo californica*.[1] Such extracts contain 150 kDa and 95 kDa forms of agrin, which may be proteolytic fragments of a larger protein cleaved either during extraction or as a result of *in situ* post-translational modification. Agrin has been cloned in the marine ray, chick, and rat.[7–9] The molecular weight for the protein backbone deduced from the full length chick clone is 225 kDa; the sequence contains several sites for *N*-linked glycosylation and GAG chain attachment.[6] In homogenates isolated from various

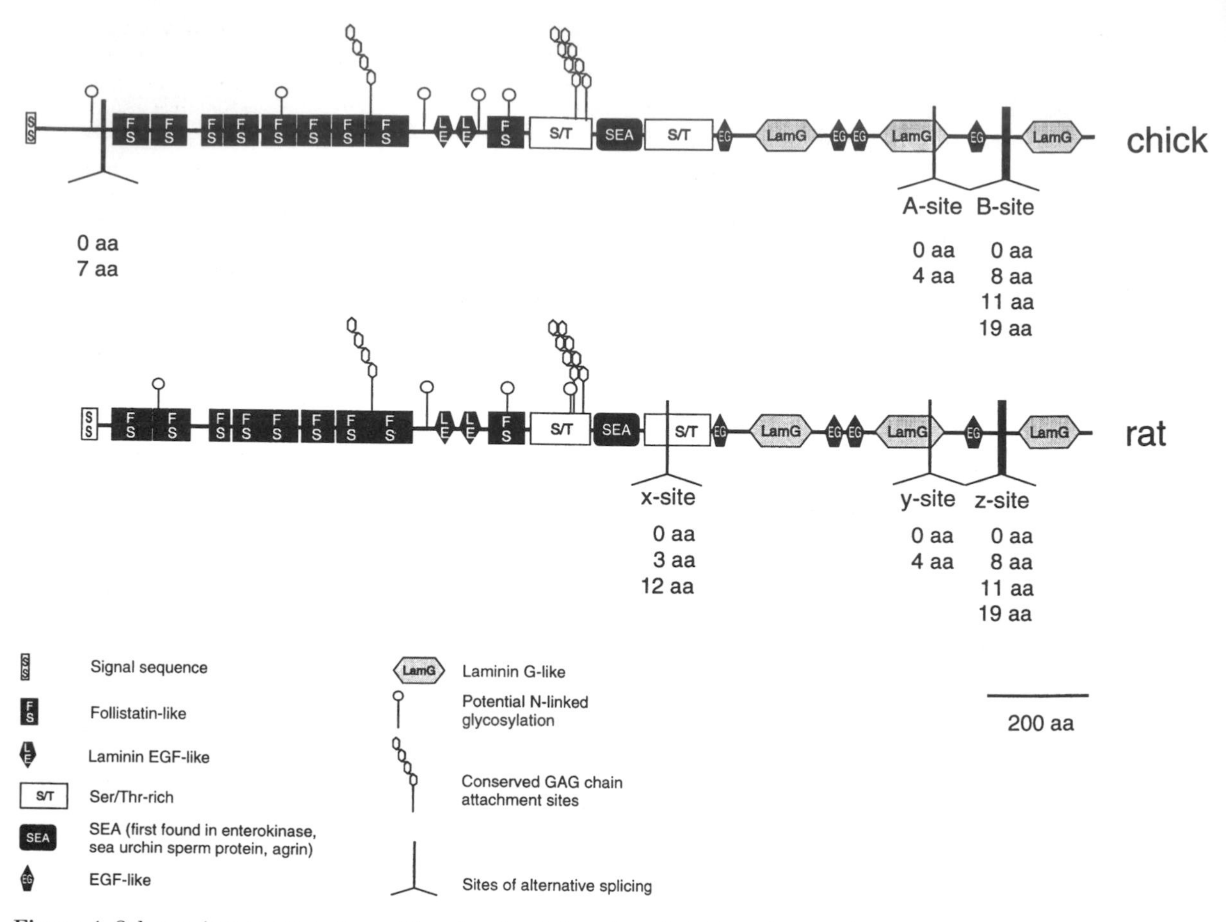

Figure 1. Schematic representation of the structural domains of chick and rat agrin.

chick tissues, including brain, spinal cord, muscle, vitreous humour, and kidney, agrin immunoreactivity has been found to be associated with proteins of ~90 and 150 kDa and with a 400–600 kDa heparan sulphate proteoglycan.[6,10]

Three alternatively spliced sites have been identified in agrin cDNA giving rise to several agrin isoforms.[4–6] Isoforms that contain inserts at two of these sites, denoted as A and B in chick and y and z in rodent, are found exclusively in the nervous system. They are expressed by many types of neurones in the central nervous system (CNS) and in ganglia of the peripheral nervous system (PNS).[11] For example, isoforms having A(y) and B(z) inserts are synthesized by motor neurones, transported down their axons, released from the axon terminals, and induce differentiation of postsynaptic specializations on myofibres.[2] Isoforms lacking the B(z) insert are synthesized in many tissues, including CNS, PNS, and muscle. In optic nerve (a CNS tract) and periperhal nerves, which contain predominately glial or Schwann cell and endothelial cell mRNA, only forms lacking B(z) inserts have been found. Such observations suggest that isoforms having the B(z) insert are expressed only by neurones, whereas non-neural cells in the nervous system and elsewhere express only isoforms lacking B(z) inserts. Agrin isoforms lacking B(z) inserts that are expressed in epithelial cells, endothelial cells, and muscle cells become associated with the extracellular matrix, as does neural agrin at the neuromuscular junction. The function of isoforms lacking the B(z) insert is not known; for example, in muscle they are neither necessary nor sufficient for agrin-induced postsynaptic differentiation.[12,13]

Agrin was originally identified by its ability to induce the formation of specializations on myotubes in culture that contain high concentrations of acetylcholine receptors (AChRs).[1,14]

Several lines of evidence suggest that agrin-induced AChR aggregation is mediated by an intracellular signalling cascade involving protein tyrosine phosphorylation.[15–18] For example, sites of AChR aggregation label with antiphosphotyrosine antibodies and inhibitors of protein tyrosine phosphorylation block AChR aggregation. Within 10 min of adding agrin to myotube cultures there is an increase in tyrosine phosphorylation of MuSK, a muscle specific protein homologous to the receptor tyrosine kinase family of proteins.[19] MuSK is required for agrin-induced postsynaptic differentiation; in MuSK-deficient mutant mice both nerve and muscle cells appear

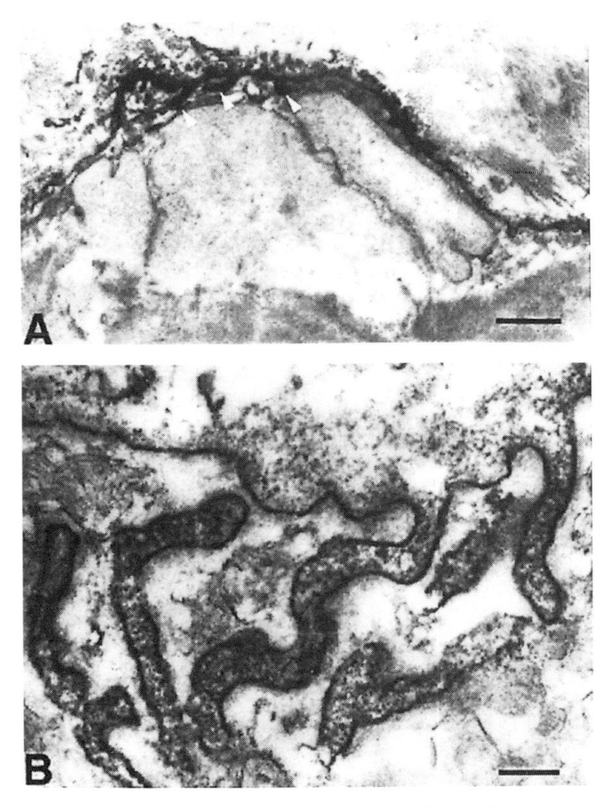

Figure 2. Myonuclei are aggregated and the plasma membrane is infolded at sites of postsynaptic protein aggregation induced by injection of Y4Z8 agrin cDNA into the extrajunctional region of denervated muscles of adult rats. Sites of postsynaptic protein aggregation are marked by electron-dense AChE immunostain. (A) At a site of postsynaptic protein aggregation there are three nuclear profiles, the plasma membrane is infolded (arrowheads), and the muscle fibre's surface bulges. (B) Infoldings at a site of postsynaptic protein aggregation are deep and filled by basal lamina. Scale bar: A, 1 μm; B, 200 nm. Reproduced from ref. 40.

to develop normally, but neuromuscular junctions fail to form.[20] Cultured myotubes derived from such MuSK knockouts fail to respond to exogenous agrin. Thus, MuSK plays an essential role in the agrin-induced intracellular signalling cascade that leads to postsynaptic differentiation, perhaps as a component of the muscle receptor for agrin, although MuSK does not appear to bind agrin directly.[19] Agrin isoforms have been shown to bind to α-dystroglycan, but such binding is not required for AChR phosphorylation or aggregation.[21,22]

Within 30 min of adding agrin to myotubes in culture there is an increase in tyrosine phosphorylation of the AChR β subunit.[15] AChR tyrosine phosphorylation appears to play a role in the formation of AChR aggregates, perhaps by strengthening the interaction of AChRs with the cytoskeleton.[17,23]

If neural agrin cDNA is injected into the extrajunctional region of a denervated soleus muscle of the adult rat, it is taken up by some of the muscle fibres.[24] They express the protein and secrete it into the extracellular space, where it acts on agrin receptors in the plasma membrane of the same and nearby muscle fibres. In response, the fibres form on their surface aggregates of extracellular, plasma membrane, and cytoplasmic proteins, including AChRs, AChE, laminin $\beta 2$ (s-laminin), utrophin, and rapsyn, as found at agrin-induced specializations on cultured myotubes. In addition, such specializations also include MuSK, neuregulin, erbB2, and erbB3. Most AChRs that accumulate initially have an $\alpha_2\beta\gamma\delta$ subunit composition, as found in the postsynaptic apparatus at embryonic neuromuscular junctions. However, if muscle fibres are made electrically active, such embryonic AChRs are entirely replaced by $\alpha_2\beta\varepsilon\delta$ AChRs, as are found at normal adult neuromuscular junctions.[25] Moreover, the myofiber plasma membrane at agrin-induced specializations has narrow and deep infoldings, the adjacent cytoplasm has an accumulation of myonuclei, and there is apparently a local increase in expression of the AChR subunit, all of which are characteristics of the normal adult postsynaptic apparatus.[24,26] Such findings lead to the conclusion that, in adult muscles, agrin released by the motor axon terminal is sufficient to induce not only protein aggregates, which mark the earliest stage of postsynaptic apparatus formation, but also changes in cell surface conformation, organelle distribution, and protein expression that appear as the postsynaptic apparatus reaches maturity.

Purification

Agrin has been purified from a basal lamina-enriched fraction of *T. californica* electric organ insoluble in isotonic saline and detergent at pH 7.5.[1] It is extracted from this fraction by treatment with 0.2 M bicarbonate buffer at pH 9.0 Subsequent purification involves affinity chromatography using Cibacron blue 3GA-agarose and immunoaffinity chromatography using anti-agrin mAbs such as mAb 5B1. Repeated extraction of the insoluble fraction with bicarbonate buffer can yield up to 10 times the amount of activity obtained in the first extraction. Recombinant ray, chick, and rat agrin have been expressed in a variety of cell lines.

Activities

AChR-aggregating activity is routinely assayed on chick myotube cultures by fluorescence microscopy after labelling AChRs with rhodamine-conjugated α-bungarotoxin.[14] Agrin causes a 3–20-fold increase in the number of AChR aggregates. It has little or no effect on the myotube size, total number of AChRs on the myotube surface, or rate of AChR synthesis or degradation. The AChR-aggregating activity is dose dependent and due, at least in part, to the lateral migration of AChRs present in the plasma membrane at the time agrin is applied. The increase in number of AChR aggregates is seen within 1–2 h after addition of agrin and reaches a plateau by 24 h. Other components of the postsynaptic apparatus accumulate together with AChRs, including acetylcholinesterase (AChE, membrane and collagen-tailed forms), butyrylcholinesterase, a synapse-specific heparan sulphate proteoglycan, the 43 kDa AChR-associated protein rapsyn, and laminin.[2,3] Homologous recombination experiments have demonstrated that rapsyn is necessary for the formation of agrin-induced synaptic specializations in developing muscle.[27] Rapsyn may act to link AChRs and other components of the postsynaptic apparatus to each other and to the cytoskeleton.[28,29]

Antibodies

Monoclonal antibodies (mAbs) against *T. californica* agrin stain basal lamina in the synaptic cleft at neuromuscular junctions of ray, frog, and chick skeletal muscle.[2] The antibodies also stain the basal lamina of capillaries in the CNS. In chick, the extrasynaptic basal lamina of slow muscle fibres stains, as does the basal lamina of smooth muscle fibres of blood vessels in skeletal muscles, the basal lamina of Schwann cells in nerves, the glomerular basal lamina in kidney and the myofibre basal lamina of cardiac muscle.[2] Golgi apparatus in motor neurones of ray, frog, and chick stain with anti-ray agrin mAbs.[2]

mAbs developed against *T. californica* agrin immunoprecipitate all isoforms of agrin, including isoforms in extracts from the CNS of ray, frog, and chick that have AChR/AChE-aggregating activity and inactive isoforms in extracts from tissues such as chick kidney and embryonic heart.[2,10] Antiserum directed against ray agrin inhibits motor neurone-induced AChR aggregation on myotubes in chick motor neurone–myotube cocultures.[12] mAbs and sera, some of which are isoform specific, have been generated against rat agrin;[30] several anti-rat agrin mAbs are available from StressGen. Antiserum specific for neuronal isoforms (i.e. those having z inserts in rat) intensely stain neuromuscular junctions. Antiserum that recognizes all isoforms stains in addition extrajunctional regions of the muscle fibre surface.

Genes

Analysis of chick and rat genomic DNA indicates that all agrin isoforms arise by alternative splicing of a single agrin gene.[5,8] The gene (symbol *AGRN*/*agrn*) has been assigned to chromosome 1 region pter-p32 in human and to mouse chromosome 4.

Mutant phenotype/disease states

Agrin-deficient mutant mice have been generated by homologous recombination.[13] In skeletal muscles of such mice, nerve and muscle appear to develop normally, but few neuromuscular junctions form. As a result, the animals are paralysed and are stillborn. On the other hand, when myoblasts from mutant embryos are allowed to form myotubes in culture, they respond normally to exogenous agrin. Thus, in the absence of agrin, muscle cells have the potential to form synaptic specializations but motor axons are unable to trigger postsynaptic differentiation, and synaptic transmission fails to occur. No other defects have been detected in agrin-deficient mice.

Structure

Not available.

References

1. Nitkin, R. M., Smith, M. A., Magill, C., Fallon, J. R., Yao, Y.-M. M., Wallace, B. G., and McMahan, U. J. (1987). *J. Cell Biol.*, **105**, 2471–8.
2. McMahan, U. J. (1990). *Cold Spring Harbor Symp. Quant. Biol.*, **50**, 407–18.
3. Bowe, M. A. and Fallon, J. R. (1995). *Ann. Rev. Neurosci.*, **18**, 443–62.
4. Ruegg, M. A., Tsim, K. W. K., Horton, S. E., Kröger, S., Escher, G., Gensch, E. M., and McMahan, U. J. (1992). *Neuron*, **8**, 691–9.
5. Rupp, F., Ozçelik, T., Linial, M., Peterson, K., Francke, U., and Scheller, R. (1992). *J. Neurosci.*, **12**, 3535–44.
6. Denzer, A. J., Gesemann, M., Schumacher, B., and Ruegg, M. A. (1995). *J. Cell. Biol.*, **131**, 1547–60.
7. Smith, M. A., Magill-Solc, C., Rupp, F., Yao, Y. -M. M., Schilling, J. W., Snow, P., and McMahan, U. J. (1992). *Mol. Cell. Neurosci.*, **3**, 406–17.
8. Tsim, K. W. K., Ruegg, M. A., Escher, G., Kröger, S., and McMahan, U. J. (1992). *Neuron*, **8**, 677–89.
9. Rupp, F., Payan, D. G., Magill-Solc, C., Cowan, D. M., and Scheller, R. H. (1991). *Cell*, **67**, 909–16.
10. Godfrey, E. W. (1991). *Exp. Cell Res.*, **195**, 99–105.
11. McMahan, U. J., Horton, S. E., Werle, M. J., Honig, L. S., Kröger, S., Ruegg, M. A., and Escher, G. (1992). *Curr. Opin. Cell Biol.*, **4**, 869–74.
12. Reist, N. E., Werle, M. J., and McMahan, U. J. (1992). *Neuron*, **8**, 677–89.
13. Gautam, M., Noakes, P. G., Moscoso, L., Rupp, F., Scheller, R. H., Merlie, J. P., and Sanes, J. R. (1996). *Cell*, **85**, 525–35.
14. Godfrey, E. W., Nitkin, R. M., Wallace, B. G., Rubin, L. L., and McMahan, U. J. (1984). *J. Cell Biol.*, **99**, 615–27.
15. Wallace, B. G., Qu, Z., and Huganir, R. L. (1991). *Neuron*, **6**, 869–78.
16. Wallace, B. G. (1994). *J. Cell Biol.*, **125**, 661–8.
17. Meier, T., Perez, G. M., and Wallace, B. G. (1995). *J. Cell Biol.*, **131**, 441–51.
18. Ferns, M., Deiner, M., and Hall, Z. (1996). *J. Cell Biol.*, **132**, 937–44.
19. Glass, D. J., Bowen, D. C., Stitt, T. N., Radziejewski, C., Bruno, J, Ryan, T. E., *et al.* (1996). *Cell*, **85**, 513–23.
20. DeChiara, T. M., Bowen, D. C., Valenzuela, D. M., Simmons, M. V., Poueymirou, W. T., Thomas, S., *et al.* (1996). *Cell*, **85**, 501–12.
21. Gesemann, M., Cavalli, V., Denzer, A. J., Brancaccio, A., Schumacher, B., and Ruegg, M. A. (1996). *Neuron*, **16**, 755–67.
22. Meier, T., Gesemann, M., Cavalli, V., Ruegg, M. A., and Wallace, B. G. (1996). *EMBO J.*, **15**, 2625–31.
23. Wallace, B. G. (1995). *J. Cell Biol.*, **128**, 1121–9.

24. Cohen, I., Rimer, M., Lømo, T., and McMahan, U. J. (1997). *Mol. Cell, Neurosci.*, **9**, 237–53.
25. Rimer, M., Mathiesen, I., Lømo, T., and McMahan, U. J. (1997). *Mol. Cell. Neurosci.*, **9**, 254–63.
26. Jones, G., Meier, T., Lichtsteiner, M., Witzmann, V., Sakmann, B., and Brenner, H. R. (1997). *Proc. Natl Acad. Sci. USA*, **94**, 2645–59.
27. Gautam, M., Noakes, P. G., Mudd, J., Nichol, M., Chu, G. C., Sanes, J. R., and Merlie, J. P. (1995). *Nature*, **377**, 232–6.
28. Froehner, S. C. (1993). *Ann. Rev. Neurosci.*, **16**, 347–68.
29. Apel, E. D., Glass, D. J., Moscoso, L. M., Yancopoulos, G. D., and Sanes, J. R. (1997). *Neuron*, **18**, 623–35.
30. Hoch, W., Campanelli, J. T., Harrison, S., and Scheller, R. H. (1994). *EMBO J.*, **13**, 2814–21.

■ *Bruce G. Wallace*
Department of Physiology and Biophysics, University of Colorado Health Sciences Center, Denver, CO, USA

■ *U. J. McMahan*
Department of Neurobiology, Stanford University School of Medicine, Stanford, CA, USA

Biglycan (BGN)

Biglycan (BGN) is a small, generally cell surface or pericellular proteoglycan composed of a ~38 kDa core protein modified with several *N*-linked oligosaccharides and, as its name implies, two chondroitin (bone) or dermatan (most soft tissues) sulphate glycosaminoglycan chains. It is member of a large superfamily of proteins that contain various number of tandem, ~25 amino acid repeats characterized by ordered spacing of hydrophobic amino acids, particularly leucine. The small, leucine-rich proteoglycans make up a discrete subfamily characterized by two highly conserved cysteine loops that flank the tandem repeats. Many functions have recently been proposed for biglycan, but its apparent ability to bind members of the TGF-β superfamily is one most often cited.

Synonymous names

Biglycan has several synonymous names generally reflecting its relative electrophoretic position on SDS–PAGE or time of elution from various purification columns. The names include PG-1, PG-I, DS-PGI, PG-S1, and DS-I.

Protein properties

Biglycan is a member of a growing family of small proteglycans whose unifying characteristics are two highly conserved cysteine loops flanking 5–10 tandem repeats with BGN having 10.[1] Each repeat is nominally ~25 amino acids in length and is based on the pattern $LxxLxLxxNxLx_{(12-14)}$. For BGN, the two glycosaminoglycan (GAG) chains are chondroitin sulphate in bone matrix and dermatan sulphate in most soft tissues. Other members of this family include decorin (DCN), fibromodulin, lumican, epiphycan, keratocan, and PG-LB (known as DSPG3 in human) (for a review see ref. 2). The biglycan sequence from a number of mammalian species has been reported, including human,[1] cow,[3,4] mouse,[5] and rat.[6] Curiously, no avian sequences have yet been reported for this highly conserved proteoglycan. Using human BGN as the model, biglycan has 368 amino acids (~41 700 Da) including 19 in the leader sequence and 18 more in the amino terminus that are often removed and are therefore considered to be a propeptide region.[1] Because the propeptide appears to remain on the proteoglycan in tissues such as the epidermis where little or no extracellular matrix accumulates, our working hypothesis is that the prepeptide may be used to bring the BGN to the cell surface and to hold it in place. According to this hypothesis, the propeptide is removed whenever the BGN is released into the surrounding matrix. In tissues such cartilage and bone, the propeptide is generally removed and the proteoglycan is found in the extracellular matrix. The 'mature' core protein (lacking the prepeptide), made by removing the disaccharide repeats of the GAG chains with chondroitinase ABC, is typically a single band of M_r ~45 000 on SDS–PAGE.[7] This core protein contains the two GAG chain linkage regions (on amino acids 42 and 47 as numbered with the starting Met=1), two *N*-linked oligosaccharides (amino acids 270 and 311) and possibly one or more *O*-linked oligosaccharide chains at unknown locations.[1]

The human biglycan gene was mapped to the X chromosome at Xq27-ter[1] and then to within 700 kb of the DXS52 marker.[8] The mouse gene, *Bgn*, has been mapped to X29.3, approximately 50 kb distal to *DXPas8*.[9] Both the human[10] and the mouse[11] genes contain dinucleotide repeats that may be useful for linkage analyses.[12,13] The ~8 kb human gene has been cloned within a single Lambda Fix phage and consists of 8 exons, the first of which does not contain any coding sequences.[10] The organization of the BGN gene is identical to that of the human decorin (DCN) gene[14] strongly suggesting that they are the direct result of gene duplication followed by divergent evolution. The human BGN gene is subject to X-inactivation but it is transcribed like an X-Y homologous gene, suggesting that transacting elements control-

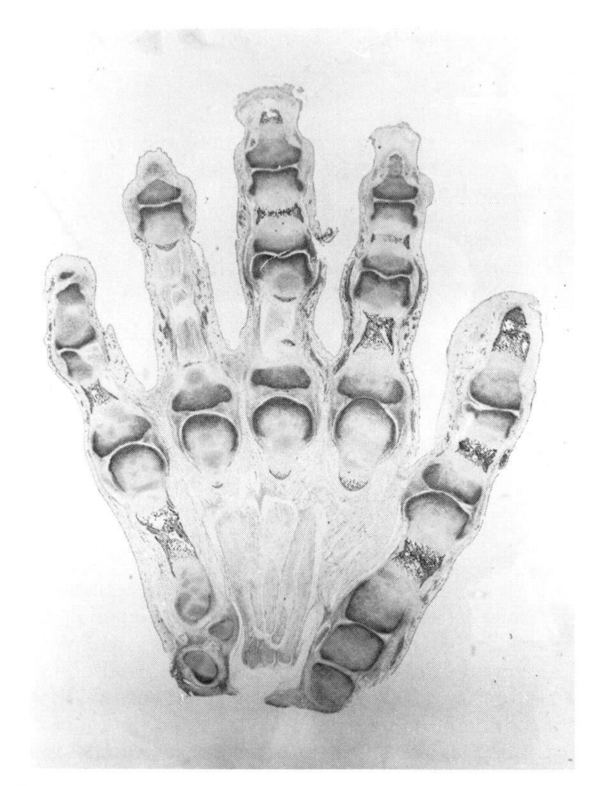

Figure 1. Immunolocalization of human biglycan in a hand of a 15-week fetus using antiserum LF-15. BGN is located in the epidermis, vasculature, type I collagen-containing bone, and in the very cellular articular ends of each bone rudiment.[16] This figure should be contrasted with a serial section stained for the related proteoglycan, decorin, on p. 409. Notice that the two are often mutually exclusive in their distribution. (See ref.16 for more details.)

ling the expression of the gene are on both the X and Y chromosomes.[15]

Biglycan has been localized to a number of tissues, often to areas distinctly different from that of decorin. In fetal humans, BGN has been observed by both immunochemistry and *in situ* experiments in kidney endothelia and collecting tubules, endocardium, and some myocardial fibres in the heart, endothelia, intima and media of the aorta, the epidermis of the skin, but generally not the dermis (except in endothelial cells of the vasculature), myofibre of skeletal muscles and all mineralized bone rudiments. In the developing cartilage of the skeleton, the BGN was seen most strongly in the highly cellular areas that give rise to chondrocytes, the growing 'caps' in articular cartilage of the appendicular elements and the inner portions of the vertebral rudiments.[16]

Purification

Chondroitin sulphate-containing BGN can be purified from fetal or young bone by a series of extraction procedures and protein chromatography.[7] Bone is milled into a fine powder, extracted with denaturing buffers to remove blood and cellular proteins, and the residue extracted with demineralizing buffer. Standard molecular sieve and ion exchange chromatography in denaturing buffers are performed. In our hands, reverse phase chromatography using standard organic solvents results in large losses. Dermatan sulphate-containing BGN can be isolated in good yield from articular cartilage using similar procedures as well as a reverse phase column and a detergent gradient.[17] We have had little luck generating soluble recombinant BGN in *E. coli* but a vaccinia-based recombinant method has been reported using UMR106 and HT-1080 as host cells resulting in ~10 mg of BGN per billion cells per day.[18]

Activities

While the function of BGN has not been unambiguously assigned, several studies have been published that suggest a wide range of possible functions. BGN has been shown to bind to type I[19] and type V[18] collagen, TGF-β,[20] and the complement protein Clq.[18] BGN has been shown selectively to increase interleukin 7-dependent proliferation of pre-B cells[21] as well as increase the survival of brain neocortical neurones *in vitro*.[22] Localization of this proteoglycan to the tips and edges of the lamellipodia of migrating endothelial cells has suggested that it may be involved in the control of cell migration.[23]

Antibodies

No monoclonal antibodies for biglycan are currently listed in ATCC's Hybridoma Data Bank (http://www.atcc.org/hdb/hdb.html). Limited amounts of the following rabbit (polyclonal) antisera are available to colleagues for research purposes only. Any use must comply completely with local and NIH's guidelines for patient care and confidentiality.

Genes

Full length BGN cDNA for human (clone P16, GenBank accession number J04599)[1] and mouse (clone 3, GenBank accession number L20276) are available from our laboratory for experimental use only. The human BGN gene in Lambda Fix DNA is also available in small quantities. Any use must comply completely with local and NIH's guidelines for patient care and confidentiality.

Mutant phenotype/disease states

No known disease has been attributed to changes in BGN. Dr Marian Young's laboratory at the National Institute of Dental Research, NIH has successfully produced a viable BGN knockout mouse, but its phenotype has not yet been published.

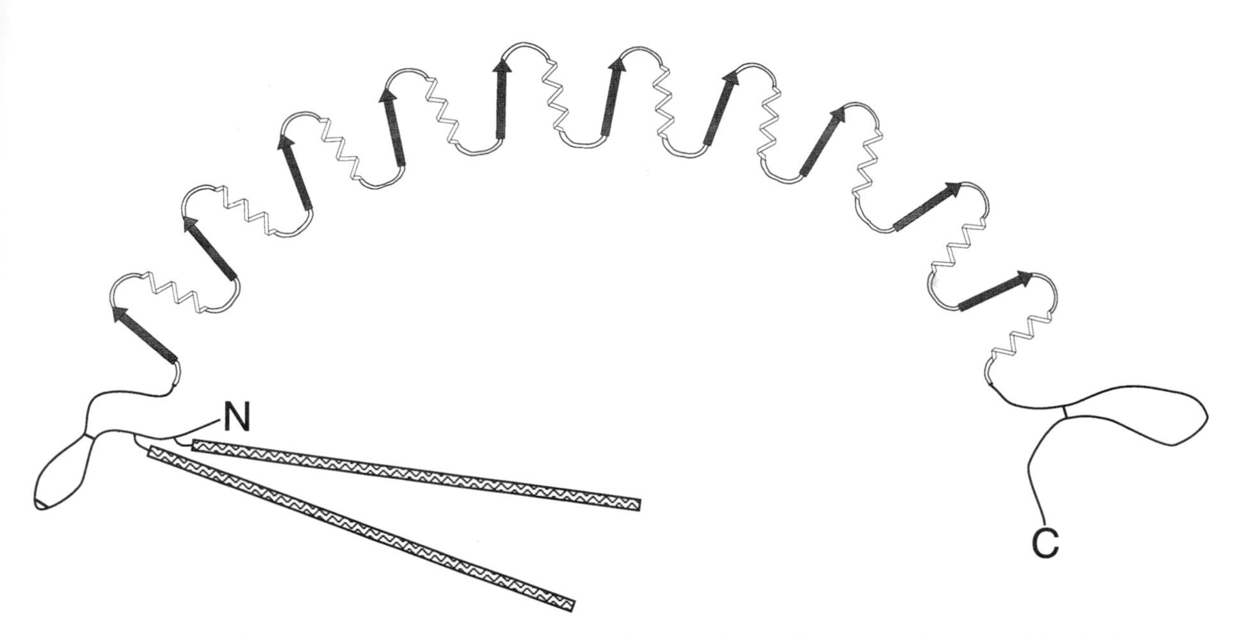

Figure 2. Diagram of the structure of biglycan based loosely on the bent-coil structure determined for the leucine-rich repeat structure of the porcine ribonuclease inhibitor.[24] The arrows represent short β sheets and the two straight rods near the N terminus represent the two glycosaminoglycans that give biglycan its name. BGN is thought to contain three disulphide bonds, represented by short connecting lines. (Drawn by Dr Andrew Hinck, NIDR, NIH.)

Table 1

Gene product	Antiserum	Antigen	Known species
Human BGN	LF-15	GVLDPDSVTPTYSA-(BSA)	H, M
Human BGN	LF-51	GVLDPDSVTPTYSA-(BSA)	H, M
Human BGN	LF-112	GVLDPDSVTPTYSA-(BSA)	H, M
Human BGN	LF-121	Recombinant BGN (w/propeptide)	H, M
Human BGN propeptide	LF-104	LPFEQRGFWDFTLDDC-(LPH)	H, M, R, Mou
Human BGN propeptide	LF-105	LPFEQRGFWDFTLDDC-(CSA)	Same as LF-104?
Bovine BGN	LF-96	LPDLDSLPPTYSC-(LPH)	Only cow tested
Bovine BGN	LF-97	LPDLDSLPPTYSC-(CSA)	Only cow tested
Mouse BGN	LP-106	VPDLDSVTPTFSAMC-(LPH)	R, Mou tested
Mouse BGN	LP-107	VPDLDSVTPTFSAMC-(CSA)	R, Mou tested

All antisera are whole rabbit sera.
H, human; M, monkey; R, rat; Mou, mouse; LPH, horseshow crab haemocyanin; CSA, chicken serum albumin.

Structure

The three-dimensional structure of BGN has not been determined. By analogy study on a porcine ribonuclease inhibitor (another protein with similar leucine-rich repeats),[24] we expect the structure of BGN to be dominated by bent-coil structure. In this hypothetical structure, each of the 10 repeats forms a single turn of the coil with each turn slightly angled to produce a structure that is somewhat horseshoe-like in appearance. Unfortunately, the ribonuclease inhibitor does not use the conserved cysteine clusters found in BGN and many other leucine-rich repeat proteins, so we can not infer the structure of the BGN protein outside of the central repeats.

References

1. Fisher, L. W., Termine, J. D., and.Young, M. F. (1989). *J. Biol. Chem.*, **264**, 4571–6.
2. Iozzo, R. V. and Murdoch, A. D. (1996). *FASEB J.*, **10**, 598–614.
3. Neame, P. J., Choi, H. U., and Rosenberg, L. C. (1989). *J. Biol. Chem.*, **264**, 8653–61.
4. Marcum, J. A., Torok, M., and Evans, S. (1993). *Biochim. Biophys. Acta*, **1173**, 81–4.
5. Just, W. (1993). Submission to GenBank, Accession No. L20276.
6. Dreher, K. L., Asundi, V., Matsura, D., and Cowan, K. (1990) *Eur. J. Cell Biol.*, **53**, 296–304.
7. Fisher, L. W., Hawkins, G. R., Tuross, N., and Termine, J. D. (1987). *J. Biol. Chem.*, **262**, 9702–9.
8. Heiss, N. S., Rogner, U. C., Kioschis, P., Korn, B., and Poustka, A. (1996). *Genome Res.*, **6**, 478–91.

9. Chatterjee, A., Faust, C. J., and Herman, G. E. (1993). *Mamm. Genome*, **4**, 33–6.
10. Fisher, L. W., Heegaard, A. M., Vetter, U., Vogel, W., Just, W., Termine, J. D., and Young, M. F. (1991). *J. Biol. Chem.*, **266**, 14371–7.
11. Wegrowski, Y, Pillarisetti, J., Danielson, K. G., Suzuki, S., and Iozzo, R. V. (1995). *Genomics*, **30**, 8–17.
12. Just, J., Rau, W., Muller, R., Geerkens, C., and Vogel, W. (1994). *Hum. Mol. Genet.*, **12**, 2268.
13. Rau, W., Just, W., Vetter, U., and Vogel, W. (1994). *Mamm. Genome*, **6**, 395–6.
14. Fisher, L.W. (1993). *Dermatan sulphate proteoglycans: chemistry, biology and chemical pathology* (ed. J. Scott), pp. 103–114. Portland Press, London.
15. Geerkens, C., Vetter, U., Just, W., Fedarko, N. S., Fisher, L. W., Young, M. F., *et al.* (1995). *Human Genet.*, **96**, 44–52.
16. Bianco, P., Fisher, L. W., Young, M. F., Termine, J. D., and Gehron Robey, P. (1990). *J. Histochem. Cytochem.*, **38**, 1549–63.
17. Choi, H. U., Johnson, T. L., Subhash, P., Tang, L.-H., Rosenberg, L. C., and Neame, P. J. (1989). *J. Biol. Chem.*, **264**, 2876–84.
18. Hocking, A. M., Stugnell, R. A., Ramamurthy, P., and McQuillan D. J. (1996), *J. Biol. Chem.*, **271**, 19571–7.
19. Pogany, G., Hemandez, D. J., and Vogel, K. G. (1994). *Arch. Biochem. Biophys.*, **313**, 102–11.
20. Yamaguchi, Y., Mann, D. M., and Ruoslahti, E. (1990). *Nature*, **346**, 281–4.
21. Oritani, K. and Kincade, P. W. (1996). *J. Cell Biol.*, **134**, 771–82.
22. Koops, A., Kappler, J., Junghans, U., Kuhn, G., Kresse, H., and Muller, H. W. (1996). *Brain Res. Mol. Brain Res.*, **41**, 65–73.
23. Kinsella, M. G., Tsoi, C. K., Javelainen, H. T., and Wight, T. N. (1997). *J. Biol. Chem.*, **272**, 318–25.
24. Kobe, B. and Deisenhofer, J. (1995). *Nature*, **374**, 183–206.
25. Fisher, L. W., Stubbs, J. T. III, and Young, M. F. (1995). *Acta Orthop. Scand.* (Suppl.) **266**, 66–70.

Larry W. Fisher
Craniofacial and Skeletal Diseases Branch, NIDR, NIH, Room 228, Building 30, Bethesda, MD 20892, USA

Bone sialoprotein (BSP)

Bone sialoprotein (BSP) is a phosphorylated and sulphated glycoprotein that is associated with most normal and many pathological mineralized matrices. It is a small (~M_r 75 000 Da) integrin-binding protein that supports cell attachment *in vitro* through both RGD-dependent and RGD-independent mechanisms and has a high affinity for hydroxyapatite It is a member of a family of acidic, integrin-binding sialoproteins which also includes osteopontin, dentin matrix protein (DMP1), and, perhaps, the dentin sialophosphoprotein (DSPP).

Synonymous names

BSP is the original name[1] but it was also known for a short time as BSP-II.[2] Care must be taken in the literature because of some confusion between BSP and osteopontin which was known originally as BSP-I,[2] as well as an occasional misnaming of the α2HS glycoprotein (or fetuin) as BSP.[3] Because the mouse genome already had used the *Bsp* locus name for another gene, the gene name for the BSP product is *IBSP* (*Ibsp* for mouse) which stands for integrin-binding sialoprotein.

Protein properties

Bone sialoprotein (BSP) constitutes about 10–15 per cent of the non-collagenous proteins found in the mineralized compartment of young bone.[1] Immunolocalization and *in situ* hybridization studies have shown BSP to be made not only by osteoblasts but also by the living cells embedded within bone, the osteocytes, as well as the multinucleated cells that resorb bone, the osteoclasts.[4] The areas richest in BSP are the 'cement lines' or collagen-poor matrix found between areas of new bone, whether that be between the cartilage anlage and bone in development or between old bone and new bone during turnover. Outside of bone, BSP has been found in three other mineralized tissues, dentin,[1] cementum,[5] and calcifying cartilage of the growth plate.[4] BSP expression is usually limited to these skeletal elements. Trophoblasts of the developing placenta, however, express high levels of BSP.[6] While this tissue is not usually considered to be a mineralized tissue, late term human placentas are well known to have hydroxyapatite crystals associated with the ageing trophoblasts.

Recently there have been reports that cancer tumours that are known to form mineralized foci also express BSP. Breast cancer tumours that have a tendency to metastasize to bone often have microcalcifications that show up on mammograms and now have been shown to express high levels of BSP.[6] Indeed, the survival rate of patients with positive lymph nodes but whose tumour biopsies exhibited no expression of BSP was higher than for those patients with BSP-positive tumours but no lymph node involvement.[7] Lung cancers with microcalcifications are also BSP positive and tend to metastagize to bone.[8] Recently two other cancers that have a high propensity to metastasize to bone, prostate[34] and thyroid[35] have also

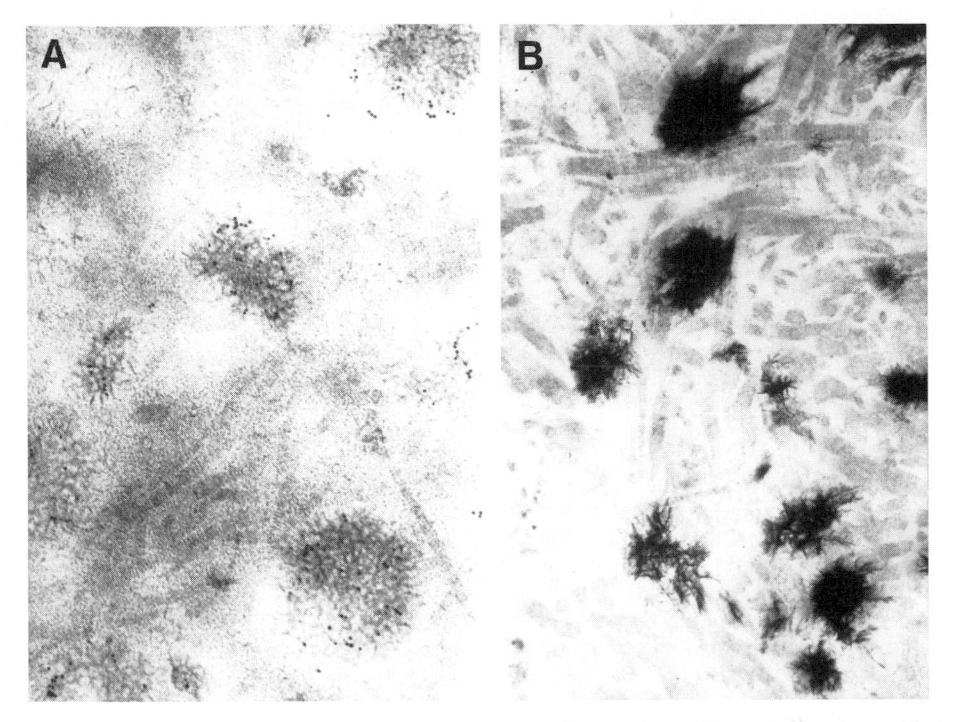

Figure 1. EM localization of BSP (A) and apatite crystals (B) in similar sections of growing rat bone. Notice the immunogold particles (BSP) associated with the electron-dense matrix between the collagen fibrils. Similar areas fixed to retain the mineral shows the earliest apatite crystals to be located in the same structures. Such stippled, electron dense structures present even at the cell membrane have been shown to contain BSP,[11] suggesting that BSP is present prior to mineral rather than BSP (with its high affinity for apatite, 2×10^{-9} M) simply binding to apatite crystals after mineralization has taken place. Antiserum LF-6 was used in this experiment. See ref. 11 for more details.

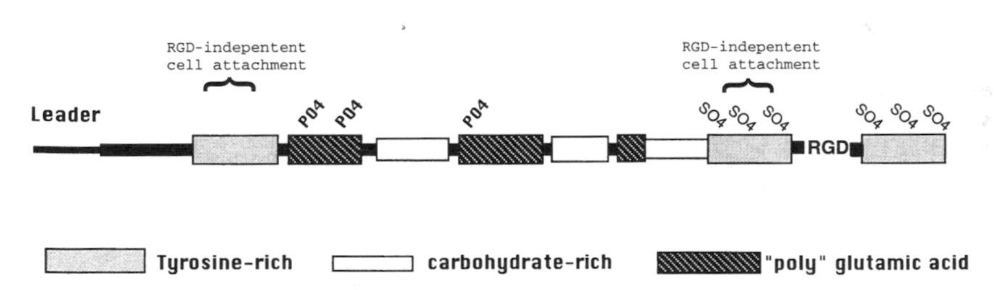

Figure 2. Diagram showing the various domains of mammalian BSP. Note that there are no disulphide bonds in BSP. The high content of hydrophilic amino acids suggests that the protein is likely to be an extended rod. PO_4 indicate likely phosphorylation sites, others are possible. SO_4 indicate a few of the many tyrosine sulphation sites found flanking the RGD domain. The more amino-terminal tyrosine-rich domain can not be sulphated. Glutamic acid-rich domains and carbohydrate-rich domains are also indicated. RGD represents the integrin-binding tripeptide. Two RGD-independent, tyrosine-rich cell attachment domains are also labelled.

been shown to often express high levels of BSP while their corresponding normal tissues do not.

All of the above localization data plus the fact that this highly acidic protein has a high affinity for apatite[9] leads quite naturally to a hypothesis that BSP may nucleate hydroxyapatite crystals *in vivo*. BSP has been shown to nucleate such crystals *in vitro*.[10] Furthermore, EM localization of BSP in developing bone has shown that electron-dense aggregates that are rich in BSP are secreted into the matrix and are associated with the earliest mineral crystals.[11] While it seems likely that any cell type that secretes large amounts of BSP over an extended period of time may eventually cause mineral to form (including placentas and tumours) it is not clear that this is the primary function of BSP. The other property of BSP and the other members of the family (osteopontin and DMP1) is the ability to support cell attachment *in vitro* through its integrin-binding tripeptide, RGD. This region has been shown to be the likely binding site to the vitronectin receptor, the $\alpha V\beta 3$ integrin.[12] BSP also has two RGD-independent cell attachment domains that have recently been shown to be two tyrosine-rich domains.[9] It is reasonable to hypothesize that cells with BSP bound to their integrins may change their own behaviour (cell shape, ability to migrate, etc.).

Using human BSP as a model[13] the protein is first made as 317 amino acid, 35 000 Da protein. The 16 amino acid leader peptide is removed during synthesis. BSP has no disulphide bonds and it is nearly uniformly hydrophilic along its length suggesting that the protein is likely to be an extended rod in solution. There are three regions particularly rich in glutamic acids residues ('polyglutamic acid domains') that have long been thought to govern the high affinity of this protein for hydroxyapatite. Recent work with recombinant fragments, however, shows that BSP's ability to bind strongly to apatite is found throughout its entire length.[9] Human BSP contains four consensus sequences for *N*-linked oligosaccharides, three of which are conserved for all mammalian species known to date. These *N*-linked and the many *O*-linked oligosaccharides make up approximately 50 per cent of the mass of BSP as it is secreted into the bone matrix.[1] (Curiously, in the rabbit, the BSP is a keratan sulphate proteoglycan.[14]) Tyrosine sulphation and serine/threonine phosphorylation make up the remainder of the known post-translational modifications. There are three tyrosine-rich domains in BSP, the last two of which flank the RGD domain and these two are subject to sulphation. The presence or absence of the sulphate groups did not appear to change the ability of fibroblasts to attach in a simple *in vitro* assay.[15]

The cDNA sequences for rat,[16] human,[13] mouse,[17] cow,[18] hamster,[19] and chicken[20] BSP have been published. The human[21] and chicken[22] genes have also been published. The human *IBSP* gene maps very close to two other members of this family, within 340 kb of *SPP1* (osteopontin) and within 150 kb of *DMP1* with the order being cen–*DMP1*–*IBSP*–*SPP1*-tel on chromosome 4.[23,24] Mouse *Ibsp* is on the homologueous region of chromosome 5 at 56.0.[17]

Purification

BSP can easily be purified from developing bone by the use of standard biochemical techniques,[1] although this approach results in protein that has been subject to denaturants. A rat osteosarcoma cell line UMR-106-BSP can produce mg/l-amounts of BSP under serum-free conditions. BSP from the media of these cells can easily be purified to > 95 per cent purity using non-denaturing conditions.[25] The UMR-106-BSP cell line makes what is probably an over-sulphated form of BSP (compared to bone-derived BSP) but the amount of sulphate groups can be lowered to as little as 5 per cent by the use of low sulphate media and chlorate.[15]

Activities

BSP has no unambiguously assigned *in vivo* activity. It binds with high affinity to hydroxyapatite crystals,[9] collagen, preferably to the α_2(I) chain[26] and to cell surface receptors including integrins.[12] It can nucleate apatite crystals *in vitro*[10] and probably causes mineralized foci in various pathologies.[4,6,8] BSP has been shown to increase, in a dose-dependent manner, osteoclast resorption in pit assays.[27] It has recently been used as a marker of

(1) breast cancers more likely to metastasize to bone,[7]
(2) relative severity of (untreated) multiple myeloma,[28]
(3) bone turnover, predominantly reflecting resorption aspects,[28] and
(4) increased joint destruction by determining BSP levels in synovial fluid.[29]

Curiously, *Staphylococcus aureus* cells isolated from patients suffering from bone infections, bound to BSP whereas *S. aureus* from other infections did not bind to BSP.[30] BSP can support cell attachment in an RGD-dependent[12] and RGD-independent[15] manner. The latter property has been determined to reside within the first two of three tyrosine-rich domains.[9] In chicken BSP, the second tyrosine-rich domain has been replaced by two additional RGD domains.[20]

Table 1[33]

Gene product	Antiserum	Antigen	Known species
Human BSP	LF-6	Human bone BSP	H, M, D, R, Mou
Human BSP	LF-83	YESENGEPRGDNYRAYED-(LPH)	H, M, D, R, Mou, C
Human BSP	LF-84	YESENGEPRGDNYRAYED-(LPH)	H, M, D, R, Mou, C
Human BSP	LF-100	AIQLPKKAGDIC-(LPH)	H, M, D, R
Human BSP	LF-101	AIQLPKKAGDIC-(CSA)	H, M, D, R, P, B
Human BSP	LF-119	Recombinant RGD domain (a.a. 257–317)	H, M, D, C
Human BSP	LF-120	Recombinant Fragment 1 (a.a. 129–281)	H, M, D, P, R
Human BSP	LF-125	Recombinant (a.a. 36–61)	H, M, P, R, S
Rat BSP	LF-87	From UMR-106 media	H, R, Mou
Rat BSP	LF-90	From chlorate-UMR-106 media	R tested

All antisera are whole rabbit sera.
H, human; M, monkey; D, dog; R, rat; Mou, mouse; C, chicken; P, pig; S, sheep; B, bovine; LPH, horseshoe crab haemocyanin; CSA, chicken serum albumin.

Antibodies

One monoclonal antibody for rat BSP (1014714) is currently listed in ATCC's Hybridoma Data Bank (http://www.atcc.org/hdb/hdb.html). Limited amounts of the following rabbit (polyclonal) antisera are available to colleagues for research purposes only. Any use must comply completely with local and NIH guidelines for patient care and confidentiality.

Genes

The *IBSP* genes for human[21] and chicken[22] have been reported. The human gene consists of one 5' untranslated exon followed by five small coding exons. The last exon is the largest and contains the integrin-binding RGD domain. Some work on the rat promoter region has been reported.[31] cCDNA for human (plasmid B6–5g, GenBank accession number J05213) and mouse (plasmid mBSP1, GenBank accession number L20232) are available from our laboratory for experimental use only. Any use must comply with local and NIH guidelines for patient care and confidentiality.

Mutant phenotype/disease states

No know disease has been attributed to changes in IBSP. The autosomal dominant disorder of dentin formation, dentinogenesis imperfecta type II (DGII), was mapped to the *IBSP* region with no recombination but sequencing of the exons from the patients revealed no disease-specific mutations[24] A knockout mouse for BSP has been reported in a preliminary study.[32] The mice are smaller, have smaller marrow spaces, smaller secondary ossification sites, and wider articular cartilage.

Structure

The three-dimensional structure for complete BSP has not been elucidated. The large amount of carbohydrates probably precludes an X-ray diffraction solution and currently at ~75 000 Da, it is too large to solve by NMR. BSP has no disulphide bonds and its high content of hydrophilic and acidic amino acids suggests that it is likely to be an extended structure. The ~60 amino acid, carbohydrate-free carboxy-terminal domain containing the integrin-binding RGD was made rich in ^{15}N by recombinant technology, analysed by NMR, and found to be a rapidly flexing or random coil.[9]

References

1. Fisher, L. W., Whitson, S. W., Avioli, L. V., and Termine, J. D. (1983). *J. Biol. Chem.*, 258, 12723–27.
2. Franzen, A. and Heinegard, D. (1985). *Biochem. J.*, **232**, 715–24.
3. Ohnishi, T., Arakaki, N., Nakamura, O., Hirono, S., and Daikuhara, Y. (1991). *J. Biol. Chem.*, **266**, 14636–45.
4. Bianco, P., Fisher, L. W., Young, M. F., Termine, J. D., and Gehron Robey, P. (1991). *Calcif. Tissue Int.*, **49**, 421–6.
5. MacNeil, R. L., Sheng, N., Strayhorn, C., Fisher, L. W., and Somerman, M. J. (1994). *Bone Min. Res.*, **9**, 1597–606.
6. Bellahcène, A., Merville, M. P., and Castrovovo, V. (1994). *Cancer Res.*, 2823–6.
7. Bellahcène, A., Menard, S., Bufalino, R., Moreau, L., and Castronovo, V. (1996). *Int. J. Cancer*, **22**, 350–3.
8. Bellahcène, A., Maloujahmoum, N., Fisher, L. W., Pasorino, H., Tagliabue, E., Menard, S., and Castronovo, V. (1997). *Calcif. Tissue Int.*, **61**, 183–8.
9. Stubbs, J. T. III, Mintz, K. P., Eanes, E. D., Torchia, D. A., and Fisher, L. W. (1997). *J. Bone Min. Res.*, **12**, 1210–22.
10. Hunter, G. K., Hauschka, P. V., Poole, A. R., Rosenberg, L. C., and Goldberg, H. A. (1996). *Biochem. J.* **317**, 59–64.
11. Bianco, P., Riminucci, Silvestrini, G., Bonucci, E., Termine, J. D., Fisher, L. W., and Gehron Robey, P. (1993). *J. Histochem. Cytochem.*, **41**, 193–203.
12. Oldberg, A., Franzen, A., Heinegard, D., Pierschbacher, M., and Ruoslahti, E. (1988). *J. Biol. Chem.*, **263**, 19433–6.
13. Fisher, L. W., McBride, O. W., Termine, J. D., and Young, M. F. (1990). *J. Biol. Chem.*, **265**, 2347–51.
14. Kinne, R. W. and Fisher, L. W. (1987). *J. Biol. Chem.*, **262**, 10206–11.
15. Mintz, K. P., Fisher, L. W., Grzesik, W. J., Hascall, V. C., and Midura, R. J. (1994). *J. Biol. Chem.*, **269**, 4845–52.
16. Oldberg, A., Franzen, A., and Heinegard, D. (1988). *J. Biol. Chem.*, **263**, 19430–2.
17. Young, M. F., Ibaraki, K., Kerr, J. M., Lyu, M. S., and Kozak, C. A. (1994). *Mamm. Genome*, **5**, 108–11.
18. Chenu, C., Ibaraki, K., Gehron Robey, P., Delmas, P. D., and Young, M. F. (1994). *J. Bone Miner. Res.*, **9**, 417–21.
19. Sasaguri, K. and Chen, L. (1996). Submission to GenBank Accession number U658890.
20. Yang, R.,Gotoh, Y., Moore, M. A., Rafidi, K., and Gerstenfeld, L. C. (1995). *J. Bone Min. Res.*, **10**, 632–40.
21. Kerr, J. M., Fisher, L. W., Termine, J. D., Wang, M. G., McBride, O. W., and Young, M. F. (1993). *Genomics*, **17**, 408–15.
22. Yang, R. and Gerstenfeld, L. C. (1997). *J. Cell. Biochem.*, **64**, 77–93.
23. Aplin, H. M., Hirst, K. L., Crosby, A. H., and Dixon, M. J. (1995). *Genomics*, **30**, 347–9.
24. Crosby, A. H., Lyu, M. S., Lin, K., McBride, O. W., Kerr, J. M., Aplin, H. M., *et al.* (1996). *Genome*, **7**, 149–51.
25. Mintz, K. P., Midura, R. J., and Fisher, L. W. (1994). *J. Tiss. Culture Meth.*, **16**, 205–9.
26. Fujisawa, R., Nodasaka, Y., and Kuboki, Y. (1995). *Calcif. Tissue Int.*, **56**, 140–4.
27. Raynal, C., Delmas, P. D., and Chenu, C. (1996). *Endocrinology*, **137**, 2347–54.
28. Seibel, M. J., Woitge, H. W., Percherstofer, M., Karmatschek, M., Horn, E., Ludwig, H., *et al.* (1996). *J. Clin. Endocrinol. Metab.*, **81**, 3289–94.
29. Saxne, T., Zunino, L., and Heinegard, D. (1995). *Arthritis Rheum.*, **38**, 82–90.
30. Yacoub, A., Lindahl, P., Rubin, K., Wendel, M., Heinegard, D., and Ryden, C. (1994). *Eur. J. Biochem.*, **15**, 919–25.
31. Ogata, Y., Yamauchi, M., Kin, R. H., Li, J. J., Freedman, L. P., and Sodek, J. (1995). *Eur. J. Biochem.*, **230**, 183–92.
32. Aubin, J. E., Gupta, A. K., Zirngbl, R., and Rossant, J. (1996). *J. Bone Min. Res.*, **11**, (Suppl. 1), S102.
33. Fisher, L. W., Stubbs III, J. T., and Young, M. F. (1995). *Acta Orthop. Scand.*, (Suppl. 266), **66**, 66–70.

34. Waltregny, D., Bellahcène, A., Van Riet, I., Fisher, L. W., Young, M. F., Fernandez, P., Dewe, W., de Leval, J., and Castronovo, V. (1998). *J. Natl Cancer Inst.*, **90**, 1000–8.
35. Bellahcène, A., Albert, V., Pollina, L., Basolo, F., Fisher, L. W., and Castronovo, V. (1998). *Thyroid*, (In press).

Larry W. Fisher
Craniofacial and Skeletal Diseases Branch, NIDR, NIH, Room 228, Building 30, Bethesda, MD 20892, USA

Cartilage matrix protein

Cartilage matrix protein (CMP) is a major non-collagenous protein in the matrix of cartilage of various organs.[1,2] It exists as a disulphide-bonded homotrimer of 148 kDa. In the growth plate during endochondral bone formation, the CMP gene is transcribed specifically by chondrocytes of the zone of maturation. The translation product is distributed in both the zones of maturation and hypertrophy, the two post-mitotic regions of an epiphyseal growth plate. Thus, CMP is a marker for post-mitotic chondrocytes.[3] In primary chondrocyte cultures, CMP forms a filamentous network that consists of both type II collagen dependent[4] and independent filaments.[5] CMP also interacts with aggrecans.[6,7]

Synonymous names

148 kDa cartilage protein,[2] matrilin-1.[22]

Homologous proteins

Based on a structure that consists of A domains, EGF-like domains, and C-terminal potential oligomerization domains, matrilin-2 (ref. 22) and matrilin-3 (refs 23 and 24) are homologous to CMP (matrilin-1). Based on the A domains, CMP also has homology to a number of serum, matrix, and cell surface proteins.[8] These are von Willebrand factor, complement factors B and C2, collagen types VI, VII, XII, and XIV, undulin, the α chains of the integrins VLA-1, VLA-2, LFA-1, Mac-1, P150/95 (in the integrins the A domains are referred to as I domains), and a *Caenorhabditis elegans* protein involved in muscle attachment as well as in malaria thrombospondin-related anonymous protein, dihydropyridine-sensitive calcium channel, and inter-α-trypsin inhibitor.

Protein properties

CMP exists as a disulphide-bonded homotrimer in the matrix of cartilage. The amino acid sequence of the CMP monomer has been deduced from chicken,[9,10] human,[11] and mouse[12] cDNA and genomic DNA sequences. Each monomer consists of two CMP-A domains which are separated by an EGF-like domain. A heptad repeat-containing[13] tail makes up the C-terminal domain of the protein. The mature form of a CMP monomer contains 12 cysteine residues. Two of these are in.each of the CMP-A domains, six in the EGF-like domain, and two in the heptad repeat-containing tail. A mutational analysis of CMP-trimer formation[14] indicates that the heptad repeats are necessary for the initiation of CMP trimerization and that the two cysteines in the heptad repeat-containing tail are both necessary and sufficient to form intermolecular disulphide bonds. The two cysteines within a CMP-A domain form an intra-domain disulphide bond. The heptad repeats are sufficient for the formation of an α-helical coiled-coil structure[15] and for maintaining the trimeric state of extracted CMP after reduction.[13]

The molecular weight of the intact protein is 148 kDa as determined by sedimentation equilibrium centrifugation.[1] In electrophoretic analyses on SDS–PAGE, CMP migrates as a 200 kDa trimer. Upon reduction of the disulphide bonds, the protein behaves as a single subunit of 54 kDa. In electron microscopic images,[13] CMP is seen as three ellipsoid subunits with approximate values of 7.6 nm for the longer axis and 5.6 nm for the shorter axis. The average diameter of whole molecules is 18 nm.

There is a single copy of the CMP gene in the genome of chicken[9,10] and human.[11] The human CMP gene has been mapped to chromosome 1p35.[11] Both the chicken and human genes consist of eight exons and seven introns. The RNA splice junctions of the seventh intron (intron G) of the chicken and human CMP gene do not conform to consensus splice sequences, suggesting a novel type of splicing mechanism in cartilage. The relationship between the structure of CMP and the CPM gene in the chicken is shown in Fig. 1. A similar organization exists for the human gene. Each of the CMP-A domains is encoded by two exons whereas the EGF-like domain is encoded by a single exon. The exon–intron junction within the CMP-A domains is at a different position within the coding regions of each of the two domains.[10] The exonic composition of the A domains of factor B, p150/95, and vWF show no distinct pattern. This domain in factor B is encoded by five exons and the A3 domain of von Willebrand factor and p150/95 are each encoded by four exons. Domains A1 and A2 of von Willebrand factor are encoded by one single large exon.

In the matrix deposited in primary chondrocyte cultures, CMP interacts with collagen fibrils.[4] The collagen binding of CMP has been localized to the CMP-A domains

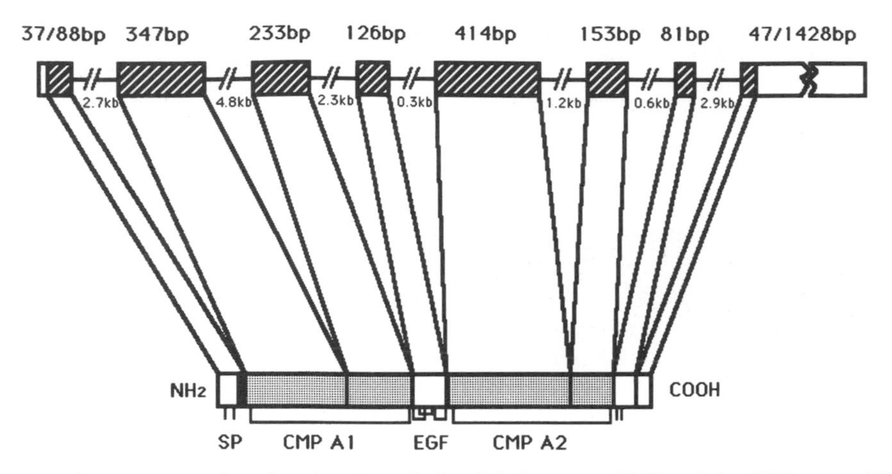

Figure 1. Diagrammatic representation showing the relationship between CMP and the CMP gene. SP, putative signal peptide; CMP-AL and CMP-A2, homologous repeats; EGF, EGF-like domain. The carboxyl-terminal tail contains heptad repeats.

and it has been suggested that CMP may be involved in collagen fibrillogenesis and in the organization of the matrix of cartilage.[16] CMP also interacts with aggrecan[6] and such interaction can involve covalent cross-linking that is associated with tissue maturation.[7]

The structural requirements for type II collagen-independent CMP filament formation have been analysed by expressing recombinant wild-type CMP and two mutant forms in chick primary cell cultures using a retrovirus expression system.[5] In chondrocytes, the wild-type virally encoded CMP is able to form disulphide-bonded trimers and to assemble into filaments. Filaments also form with CMP in which the eleventh and twelfth cysteines located at the beginning of the coiled-coil tail domain of the mature CMP (Cys-455 and Cys-457) were mutagenized to prevent interchain disulphide bond formation. Thus intermolecular disulphide bonds are not necessary for the assembly of CMP into filaments. Both the wild-type and the double-cysteine mutant also form filaments in fibroblasts indicating that chondrocyte-specific factors are not required for filament formation. A truncated form of CMP which consists only of the CMP-A2 domain and the tail domain can form trimers but fails to form filaments. Thus, the deleted CMP-A1 domain and/or the EGF domain are necessary for filament assembly but not for trimer formation. Furthermore, the expression of the virally encoded truncated CMP (mini-CMP) in chondrocyte culture acts in a dominant negative fashion and disrupts endogenous CMP filament formation. Together these data suggest a role for CMP in cartilage matrix assembly by forming filamentous networks which require participation and coordination of individual domains of CMP.[5]

The temporal expression of the gene for CMP is independent of the expression of type II collagen, link protein, and aggrecan core protein in the developing limb bud of the chick embryo. The CMP gene is also expressed in the notochord and in the early somites of the chick embryo.[17] The protein has been reported to be present in tracheal, nasal septum, xiphisternal, auricular, and epiphyseal cartilage but absent in extracts of articular cartilage, the anulus fibrosus, and the nucleus pulposus of the intervertebral disc.[2] In the growth plate during endochondral bone formation the gene is transcribed specifically by chondrocytes of the zone of maturation[3] (Fig. 2). A similar pattern of expression is evident in the human growth plate.[18] The translation product is distributed in both the zones of maturation and hypertrophy, the two post-mitotic regions of an epiphyseal growth plate. Thus, CMP is a marker for post-mitotic chondrocytes.

Purification

CMP can be extracted from cartilage using dissociative solvents (4 M guanidine–HCl) followed by CsCl density gradient centrifugation, gel chromatography and precipitation from low ionic strength solutions. Native CMP can also be extracted under non-denaturing conditions using EDTA-containing buffers.[13]

Activities

The precise function of CMP is not known. CMP binds to collagen[4,16] and aggrecan[6,7] and can also form collagen-independent fibrils.[5] Together these data suggest a role for CMP in cartilage matrix assembly.

Antibodies

Rabbit polyclonal antisera against bovine CMP have been generated.[1] Monoclonal antibodies (mAb) against chicken CMP have been used, as have rabbit polyclonal antibodies (pAb) against a synthetic peptide covering a sequence of the CMP2 domain.[4] In immunofluorescence tests, mAb 1H1 recognizes CMP from chicken, mouse, and human whereas mAb 1G1 recognizes only chicken CMP.[19]

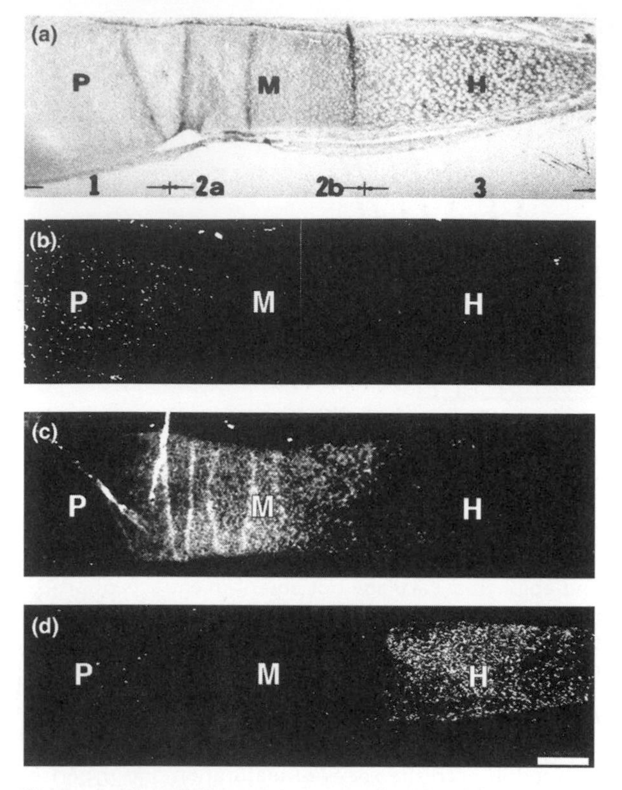

Figure 2. The CMP gene is transcribed by chondrocytes of the zone of maturation in the growth plate. Micrographs of sections of a tibiotarsus of a 15 day chick embryo. (a) Bright-field histochemical staining. P and 1, zone of proliferation; M and 2, zone of maturation; 2a, first half of the mature zone; 2b second half of the mature zone. H and 3, zone of hypertrophy. (b) Imunofluorescence micrograph with mAb against incorporated BrdU. (c) Dark-field micrograph of *in situ* hybridization with cDNA for CMP. Only the chondrocytes in the zone of maturation are positive. (d) Dark-field micrograph of *in situ* hybridization with cDNA for type X collagen. Only the chondrocytes in the zone of hypertrophy are positive. The bar represents 400 μm.

Genes

The complete nucleotide sequences are available for chicken[10] (EMBL X12346–X12354), human[11] (EMBL J05666 and J05667), and mouse[12] (EMBL U35035) CMP (matrilin-1). Nucleotide sequences for mouse matrilin-2 (EMBL U69262 and U69263), and for human (EMBL Y13341), mouse (EMBL Y10521), chicken (EMBL 001047 and 000055) matrilin-3 are available.

Mutant phenotype/disease states

CMP has been excluded as mutations in a number of heritable chondrodysplasias or as a cause of generalized osteoarthritis.[20,21]

References

1. Paulmon, M. and Heinegard. D. (1981). *Biochem. J.*, **197**, 367–75.
2. Paulsson, M. and Heinegard, D. (1982) *Biochem J.*, **207**, 207–13.
3. Chen, Q., Johnson, D. M., Haudenschild, D. R., and Goetinek, P. F. (1995). *Dev. Biol.*, **172**, 293–306.
4. Winterbottom, N., Tondravi, M. M., Harrington, T. L., Klier, G., Vertel, B., and Goetinck, P. F. (1992). *Dev. Dynamics*, **193**, 266–76.
5. Chen, Q., Johnson, D. M., Haudenschild, D. R., Tondravi, M. M., and Goetinck, P. F. (1995). *Mol. Biol. Cell*, **6**, 1743–53.
6. Paulsson, M. and Heinegard, D. (1973). *Biochem. J.*, **183**, 539–45.
7. Hauser, N., Paulsson, M., Heinegard, D., and Morgelin, M. (1996). *J. Biol. Chem.*, **271**, 32247–52.
8. Colombatti, A., Bonaldo, P., and Doliana, R. (1993). *Matrix*, **13**, 297–306.
9. Argraves, W. S., Deák, F., Sparks, K. J., Kiss, I., and Goetinck, P. F. (1987). *Proc. Natl Acad. Sci. USA*, **841**, 464–8.
10. Kiss, I., Deák, F., Holloway, R. G., Delius, H., Mebust, K. A., Frimberger, E., *et al.* (1989). *J. Biol. Chem.*, **264**, 8126–34.
11. Jenkins, R. N., Osborne-Lawrence, S. L., Sinclair, A. K., Eddy Jr, R. L., Byers, M. G., Shows, T. B., and Duby, A. D. (1990). *J. Biol. Chem.*, **265**, 19625–32.
12. Aszodi, A., Hauser, N., Studer, D., Paulsson, M., Hiripi, L., and Bosze, Z. (1996). *Eur. J. Biochem.*, **236**, 970–7.
13. Hauser, N. and Paulsson, M. (1994). *J. Biol. Chem.*, **2369**, 25747-53.
14. Haudenschild, D. R., Tondravi, M. M., Hofer, U., Chen, Q., and Goetinck, P. F. (1995). *J. Biol. Chem.*, **270**, 23150–4.
15. Beck, K., Gambee, J. E., Bohan, C. A., and Bachinger, H. P. (1996). *J. Mol. Biol.*, **256**, 909–23.
16. Tondravi, M. M., Winterbottom, N., Haudenschild, D. R., and Goetinck, P. F. (1993). In *Limb development and regeneration*, (ed. J. Falon, P. F. Boetinck, R. O. Kelley, and D. Stocum), pp. 515–22. Wiley-Liss, New York.
17. Stirpe, and Goetinck, P.F. (1989). *Development*, **107**, 23–33.
18. Mundlos, S. and Zabel. B. (1994). *Dev. Dynamics*, **199**, 241–52.
19. Chen, Q., Johnson, D. M., Haudenschild, D. R., and Goetinck, P. F. (1996). *Ann. NY Acad. Sci.*, **785**, 238–40.
20. Loughlin, J., Irven, C., Fergusson, C., and Sykes, B. (1994). *Br. J. Rheumatol.*, **33**, 1103–6.
21. Loughhn, J., Irven, C., and Sykes, B. (1994). *Hum. Genet.*, **94**, 698–700.
22. Deák, F., Piecha, D., Bachrati, C., Paulsson, M., and Kiss, I. (1997). *J. Biol. Chem.*, **272**, 9268–74.
23. Wagener, R., Kobbe, B., and Paulsson, M. (1997). *FEBS Lett.*, **413**, 129–34.
24. Belluoccio, D. and Trueb, B. (1997). *FEBS Lett.*, **415**, 212–16.

Paul F. Goetinck
Cutaneous Biology Research Center,
Massachusetts General Hospital,
Harvard Medical School,
Building 149, 13th Street,
Charlestown, MA 02129, USA

Cat-301 proteoglycan

The Cat-301 proteoglycan is one member of a family of high molecular weight chondroitin sulphate proteoglycans (CSPGs) distributed on the extracellular surface of subsets of neurones in the mature mammalian central nervous system. These neuronal cell surface CSPGs are first expressed late in development, at a time that correlates with the end of the period of developmental synaptic plasticity. Perturbation of the normal pattern of neuronal activity during the early postnatal period attenuates expression of the proteins. The function of the Cat-301 CSPG has not yet been established definitively, but the pattern of its expression and regulation suggest a role in stabilizing mature synaptic structure.

Synonymous names

None.

Homologous proteins

Neuronal cell surface CSPGs identified by antibodies Cat-315 and Cat-316.[1]

Protein properties

The Cat-301 CSPG was originally identified by its expression (detected immunocytochemically with monoclonal antibody Cat-301) on the surface of subsets of neurones in several areas of the CNS in many mammalian species.[2] Expression of the Cat-301 CSPG is largely restricted to grey matter in the CNS. A characteristic subset of neurones is recognized by monoclonal antibody Cat-301 in each area of the CNS examined to date.[3–5] In the monkey visual system the antibody demarcates a series of functionally related neurones through several orders of subcortical and cortical processing.[3] New antibodies generated to brain proteoglycans have demonstrated that the Cat-301 CSPG is one member of a larger family of neuronal cell surface CSPGs that differ from one another in the composition of both protein core and carbohydrate modification.[1]

The developmental regulation of the Cat-301 CSPG is one of its most intriguing properties. In every system examined, the surface-associated Cat-301 staining is first observed relatively late in development, at a time that correlates with the end of the period of synaptic plasticity.[6,7] Perturbation of normal patterns of neuronal activ-

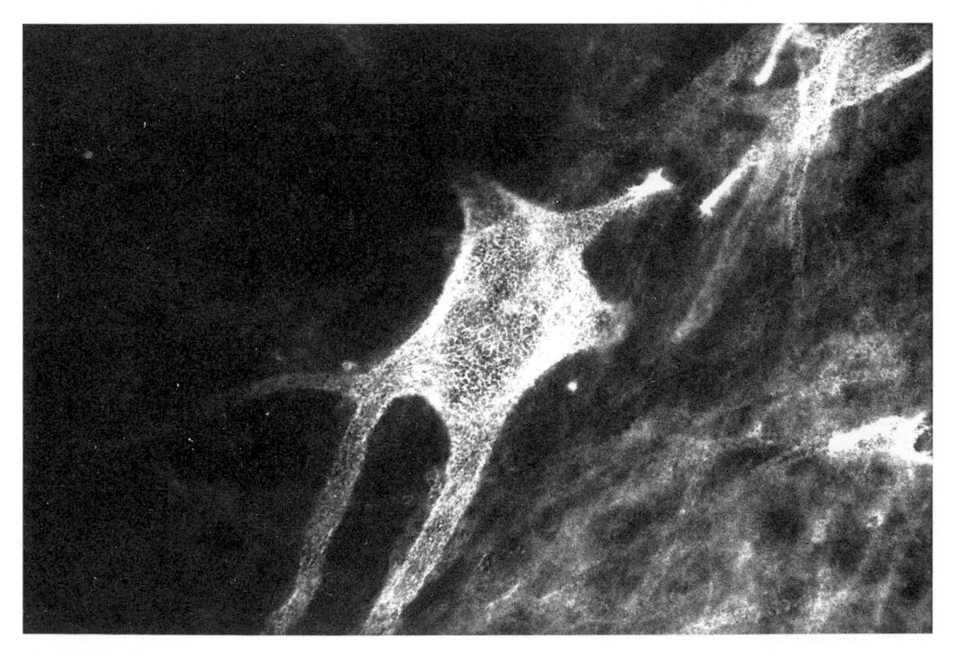

Figure 1. The Cat-301 CSPG forms an irregular lattice-work over the surface of antibody-positive neurones. The small, round areas devoid of Cat-301 staining are the sites of synapses.

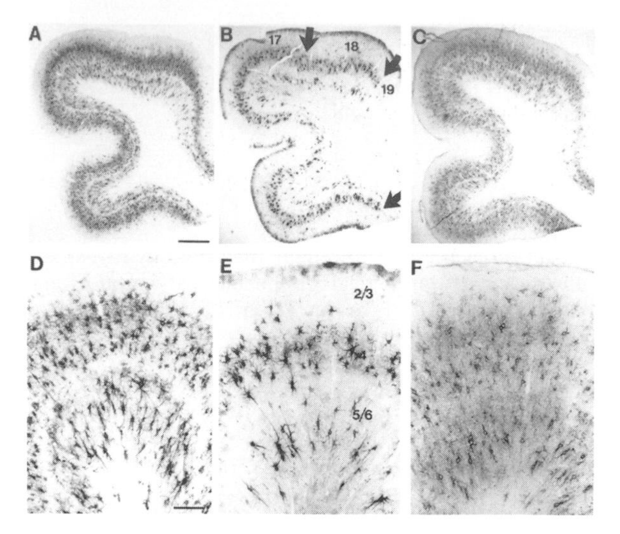

Figure 2. Cat-301, Cat-315, and Cat-316 recognize different sets of neurones in the adult cat visual cortex. (A, D) Cat-301-positive neurones in areas 17, 18, and 19 are distributed in a dense band in layer 4, and in a less dense band in layers 5/6. Layers 2/3 contain the smallest number of antibody-positive neurones. (B, E) Cat-315 recognizes the smallest subset of neurones of the three antibodies. Cat-315-immunoreactive neurones are found primarily in areas 17, and 18 of cat visual cortex, with the number of Cat-315 positive neurones dropping off sharply at the border between areas 18, and 19. Immunoreactive neurones are most dense in layer 4, and also found in layers 5/6, with few immunoreactive neurones in layers 2/3. In addition to neuronal staining, Cat-315 also stains sub-pial astrocytes. (C, F) Cat-316-immunoreactive neurones are found in areas 17, 18, and 19, with an approximately equal distribution in layers 2–6. Areal borders for panels A–C are as shown in (B); laminar assignments for panels D–F are as shown in (E). Scale bars: A–C, 1 mm; D–F, 200 μm. (From Lander *et al.*, 1997[1] with permission.).

ity during the early postnatal period (the critical period) by a variety of pharmacological, surgical and behavioural manipulations produces a marked and irreversible reduction in levels of the Cat-301 CSPG.[5–10] Perturbation of activity in adult animals has no effect on Cat-301 CSPG expression. The Cat-301 CSPG is, thus, the first protein described with a pattern of expression that parallels neuronal maturation during critical periods in CNS development. The more recently identified members of the Cat-301 family also show activity-dependent regulation of expression in the cat visual cortex.[1]

On immunoblots, the Cat-301 CSPG appears as a 650 000 Dalton polydisperse band. This band was originally identified in brain extracts.[11] Digestion by chondroitinase ABC reduces the size to approximately 550 000 Da while digestion with keratanase has no effect. Furthermore, antibodies to keratan sulphate indicate that the Cat-301 proteoglycan lacks modification by keratan sulphate. A second monoclonal antibody, Cat-304, raised using an immunosuppression protocol,[9,11,12] recognizes an independent epitope on the Cat-301 CSPG.[11]

The Cat-301 proteoglycan is antigenically related to the large, aggregating chondroitin sulphate proteoglycan from cartilage, aggrecan.[13,14] Both Cat-301 and Cat-304 recognize an antigen in cartilaginous tissues and also recognize purified aggrecan. On immunoblots purified aggrecan digested with chondroitinase and keratanase has the same apparent molecular weight as the chondroitinase-treated brain Cat-301 CSPG. Preliminary studies indicate that neither the Cat-301 nor the Cat-304 epitope is located in the G1 domain (the hyaluronic acid binding region) of aggrecan. The lack of modification of the brain-derived Cat-301 CSPG by keratan sulphate indicates that it is not identical to aggrecan. Moreover, animals that are heterozygous for a deletion in the aggrecan gene[15] show no reduction in Cat-301 (or other neuronal CSPG) expression (Lander and Hockfield, unpublished results).

Purification

The Cat-301 proteoglycan has been purified[11,13] from guanidine–HCl extracts of cat cortex using CsCl gradients under associative conditions for CSPGs and hyaluronic acid[14] followed by immunoaffinity chromatography using the Cat-301 antibody immobilized on Affigel-HZ.

Alternatively, the Cat-301 proteoglycan can be purified using standard purification procedures for the large chondroitin sulphate proteoglycan from cartilage, aggrecan.[16]

Activities

The biological role of the Cat-301 proteoglycan is not definitively known at present, but the temporal features of its expression[5–10] suggest a role in the stabilization of mature synaptic structure in the mammalian CNS.

Antibodies

Monoclonal antibody Cat-301 recognizes the brain CSPG in many mammalian species, including cat, hamster, mouse, rat, monkey and human.[2,3,6,7] Monoclonal antibody Cat-304 recognizes a different epitope on the Cat-301 CSPG[9,11] expressed in cat and cow. Both antibodies appear to be directed to peptide, rather than carbohydrate, epitopes. Both also recognize aggrecan purified from cat and cow cartilage.[13] Monoclonal antibody Cat-315 also recognizes the protein core of an overlapping set of neuronal cell surface CSPGs;[1] monoclonal antibody Cat-316 recognizes a carbohydrate epitope on a different, overlapping set of CSPGs.[1]

Genes

None identified at present.

References

1. Lander, C., Kind, P. Maleski, M., and Hockfield, S. (1997). *J. Neurosci.*, **17**, 1928–39.
2. Hockfield, S. and McKay, R. (1983). *Proc. Natl Acad. Sci. USA*, **80**, 5758–61.
3. DeYoe, E. A., Hockfield, S. Garren, H., and Van Essen, D. (1990). *Visual Neurosci.* **5**, 67–81.
4. Sahin, M. and Hockfield, S. (1990). *J. Comp. Neurol.*, **301**, 575–84.
5. Sur, M., Frost, D., and Hockfield, S. (1988). *J. Neurosci.*, **8**, 874–82.
6. Hockfield, S., Kalb, R. G. Zaremba, S., and Fryer, H. J. (1990). *Cold Spring Harbor Symp. Quant. Biol.* **55**, 505–14.
7. Kalb, R. G. and Hockfield, S. (1988). *J. Neurosci.*, **8**, 2350–60.
8. Kalb, R. G. and Hockfield, S. (1992). *Brain Res. Rev.*, **17**, 283–9.
9. Guimaraes, A., Zaremba, S., and Hockfield, S. (1990). *J. Neurosci.*, **10**, 3014–24.
10. Kalb, R. G. and Hockfield, S. (1990). *Science*, **250**, 294–6.
11. Zaremba, S., Guimaraes, A. Kalb, R. G., and Hockfield, S. (1989). *Neuron* **2**, 1207–19.
12. Hockfield, S. (1987). *Science*, **237**, 67–70.
13. Fryer, H. J., Kelly, G. M. Molinaro, L., and Hockfield, S. (1992). *J. Biol. Chem.*, **267**, 9874–83.
14. Heinegard, D. and Paulsson, M. (1984). In *Extracellular matrix biochemistry*, (ed. K. A. Piez and A. H. Reddi), pp. 277–328. Elsevier, New York.
15. Watanabe, H., Kimata, K. Line, S. Strong, D. Gao, L. Y. Kozak, C. A., and Yamada, Y. (1994). *Nature Genet.*, **7**, 154–7.
16. Hascall, V. C. and Kimura, J. H. (1982). *Meth. Enzymol.*, **82**, 769–800.

Susan Hockfield
Section of Neurobiology,
Yale University School of Medicine,
New Haven, CT, USA

Cholinesterases

The cholinesterases are a family of enzymes of common α/β hydrolase-fold structure whose only well established physiologic function is the efficient hydrolysis of acetylcholine. Acetylcholinesterase (AChE – E.C. 3.1.1.7) may be distinguished from butyrylcholinesterase (BuChE – E.C. 3.1.1.8) by specificity for acetylcholine over butyrylcholine hydrolysis. AChE is typically synthesized in nerve, muscle, and certain haematopoietic cells. In excitable tissues, its synthesis is regulated by tissue-specific development and it is localized in synapses at the extracellular face of nerve and muscle. BuChE is synthesized in liver and secreted into plasma.

Protein properties

The cholinesterases exist in multiple molecular forms which may be distinguished by their subunit associations and hydrodynamic properties (Fig. 1).[1,2] The catalytic subunits associate with a lipid linked or a collagen-like structural subunit to form distinct heteromeric species. The collagen-containing species consist of tetramers of catalytic subunits, each of which is disulphide linked to a single subunit strand of a triple-helical, collagen-like unit.[1–6] The dimensional asymmetry imparted by the filamentous collagen unit leads to the designation of asymmetric or A forms with the subscript specifying the number of attached catalytic subunits (Fig. 2). The collagen-containing subunit has non-collagenous sequences at its N and C termini[7] and associates with acidic basal laminar components within the synapse;[8,9] it has been cloned and its structure analysed.[7,10] A proline-rich attachment domain found near the N terminus is responsible for the association of the catalytic and structural subunits.[11] The lipid linked subunit is approximately 20 kDa in mass,[12] contains covalently attached fatty acids and tethers the enzyme at the outside surface of the cell. Only the hydrophilic catalytic subunits forming tetramers appear to be involved in these heteromeric associations.

The homomeric forms typically exist as dimers and tetramers, occasionally monomeric species are also found. Their hydrodynamic properties have led to the classification of globular or G forms and they may be subdivided into hydrophilic or amphiphilic G forms. The amphiphilic character arises from either the intrinsic amphipathic sequence or cleavage of a hydrophobic peptide from the C terminus and attachment of a glycophospholipid to the C terminal amino acid. The latter post-translational modification also localizes the enzyme at the outer surface of the cell membrane.[13] To date, only hydrophilic (globular).and asymmetric forms of BuChE have been identified.[1] The glycophospholipid linked forms of AChE are found in excitable tissues of lower vertebrates, whereas in mammals they appear largely restricted to the haematopoietic system.

Purification

The cholinesterases are routinely purified by affinity chromatography using a conjugated ligand that is inhibitory to the enzyme. Differential extraction in buffer, non-ionic detergent, and high ionic strength buffer yields the hydrophilic and amphiphilic, and asymmetric forms of the enzyme, respectively.[1,2]

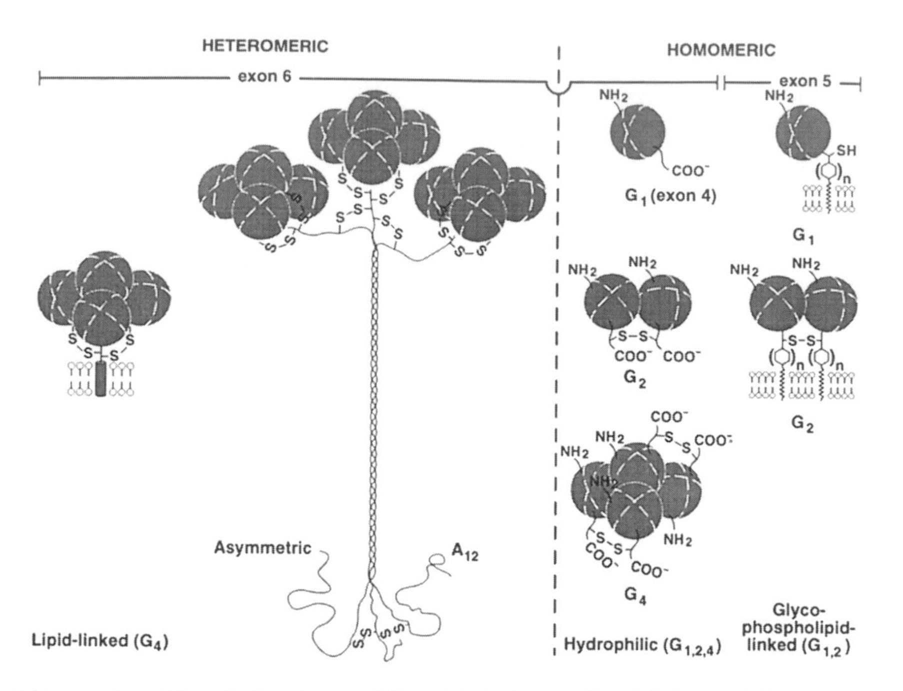

Figure 1. Molecular species of the cholinesterases[2] G and A designate the globular and the asymmetric forms of the enzyme, respectively; the subscript denotes the number of subunits. The catalytic subunits of AChE are encoded by a single gene; the C termini of most of the molecular forms are encoded by alternatively spliced exons; the glycophospholipid-linked form whose signal sequence is encoded by exon 5; a self-associating form, containing an amphipathic helix and, a cysteine for linkage and encoded by exon 6; and a monomeric soluble form whose C terminus is encoded by retaining the intron between exons 4 and 5.

Figure 2. Electron micrograph of the asymmetric form of acetylcholinesterase.[4]

Activities

Catalytic activity is typically measured by acetylthiocholine hydrolysis detecting liberated thiocholine.[1,14] Alternatively, pH stat or radioenzymatic assays are suitable for measuring acetylcholine hydrolysis. The turnover number of AChE (8.2×10^5 min^{-1}) is the fastest of the serine hydrolases. Rates for BuChE are typically 10–25 per cent of this value using butyrylthiocholine as a substrate. BuChE shows a wider substrate specificity than AChE and may have a role in the detoxification of various plant esters that are ingested.[15]

Antibodies

Both polyclonal and monoclonal antibodies have been prepared to intact cholinesterases and their structural subunits from electric fish and mammalian sources.[16,17] Some unusual carbohydrate epitopes have been detected.[18] Antibodies selective for a particular AChE species have been used to localize different forms of the cholinesterases in intact tissue.[19]

Genes

Distinct genes encode AChE and BuChE giving gene products with ~50 per cent overall amino acid identity.[1,2,20–22] The chromosomal loci of the *ACHE* and *BCHE* genes in humans are 7q22 and 3q26-ter. The open reading frame of the mammalian *ACHE* gene is encoded in four exons.[21] Alternative splicing at the 3' end of the open reading frame gives rise to three distinct C-terminal sequences which affect the state of of oligomerization of the enzyme, attachment to structural subunits, and hydrophobicity of the carboxyl-terminal sequence (Fig. 1).[21] These factors influence the cellular disposition and membrane attachment of the enzyme, but not its intrinsic catalytic parameters. To date, cDNAs for *Torpedo, Drosophila,* mouse, human, rat, rabbit, *C. elegans,* mosquito, amphioxus, snake venom (*Bungarus fasciatus*), and *Electrophorus* AChEs have been cloned.[23,24] The cholinesterases show no global sequence identity with the serine hydrolases of the trypsin or subtilisin families, but are homologous with other serine hydrolases that function as carboxylesterases, phospholipases, lipases, lyases, acyltransferases, and dehalogenases. Found within the family are thyroglobulin and several proteins which function in forming heterologous cell contacts such as glutactin, neurotactin, gliotactin, and neuroligin. Homology with the latter proteins suggests that the cholinesterases may also have non-catalytic functions. All of these proteins have a common disulphide-bonding arrangement.

Structure

Crystal structures of *Torpedo*[25] and mouse AChE[26] and some of their complexes have been determined and form a basis for structural analysis of related proteins.

References

1. Massoulié, J., Pezzementi, L., Bon, S., Krejci, E., and Valletta, F. M. (1993). *Prog. Neurobiol.*, **41**, 31–91.
2. Taylor, P. and Radic', Z., (1994). *Ann. Rev. Pharmacol. Toxicol.*, **34**, 281–320.
3. Lwebuga-Mukasa, J., Lappi, S., and Taylor, P. (1995). *Biochemistry*, **15**, 1425–35.
4. Cartaud, J., Bon, S., and Massoulié, J. (1976). *Brain Res.*, **88**, 127–30.
5. Anglister, L. and Silman, I. (1978). *J. Mol. Biol.*, **125**, 293–311.
6. Rosenberry, T. L. and Richardson, J. M. (1978). *Biochemistry*, **16**, 3550–8.
7. Krejci, E., Coussen, F., Duval, N., Chatel, J. L., Legay, C., Paype, M., *et al.* (1991) *EMBO J.*, **10**, 1285–93.
8. McMahan, U. J., Sanes, J. R., and Marshall, L. M. (1978). *Nature*, **271**, 172–4.
9. Brandan, E., Maldonado, M., Garrido, J., and Inestrosa, N. C. (1985). *J. Cell Biol.*, **101**, 985–92.
10. Krejci, E., Thomine, S., Boschetti, N., Legay, C., Sketelj, J., and Massoulié, J. (1997) *J. Biol. Chem.*, **272**, 22840–47.
11. Bon, S., Coussen, F., and Massoulié, J. (1997). *J. Biol. Chem.*, **272**, 3016–21.
12. Inestrosa, N. C., Roberts, W. L., Marshall, T. L., and Rosenberry, T. L. (1987). *J. Biol. Chem.*, **262**, 4441–4.
13. Roberts, W. L., Kim, B. H., and Rosenberry, T. L. (1987). *Proc. Natl Acad. Sci. USA*, **84**, 7817–21.
14. Rosenberry, T. L. (1975). *Adv. Enzymol.*, **43**, 103–218.
15. Soreq, H., Ehrlich, G., Schwarz, M., Loewenstein, Y., Glock, D., and Zakut, H. (1994). *Biomed. Pharmaco. Therapy*, **48**, 253–9.
16. Brimijoin, S. (1986). *Int. Rev. Neurobiol.*, **28**, 363–410.
17. Doctor, B. P., Camp, S., Gentry, M. K., Taylor, S. S. and Taylor, P. (1983). *Proc. Natl Acad. Sci. USA*, **80**, 5767–71.
18. Bon, S., Meflah, K., Musset, F., Grassi, J., and Massoulié, J. (1987). *J. Neurochem.*, **49**, 1720–31.
19. Abramson, S. N., Ellisman, M. N., Deerinck, T. J., Maulet, Y., Gentry, M. K., Doctor, B. P., and Taylor, P. (1989). *J. Cell Biol.*, **108**, 2301–11.
20. Schumacher, M., Camp, S., Maulet, Y., Newton, M., MacPhee-Quigley, K., Taylor, S. S., *et al.* (1986). *Nature*, **319**, 407–9.
21. Li, Y., Camp, S., and Taylor, P. (1993). *J Biol. Chem.*, **268**, 5790–7.
22. Arpagaus, M., Kott, M., Vatsis, K. P., Barrels, C. F., LaDu, B. N., and Lockridge, O. (1990). *Biochemistry*, **29**, 124–31.
23. Ichtchenko, K., Nguyen, T., and Sudhof, T. C. (1996). *J. Biol. Chem.*, **271**, 2676–82.
24. Cousin, X., Hotelier, T., Giles, K., Lievin, P., Toutant, J. P. and Chatonnet, A. (1997). *Nucl. Acids Res.* **25**, 143–6.
25. Sussman, J. L., Harel, M., Frolow, F., Oefner, C., Goldman, A., Toker, L., and Silman, I. (1992). *Science*, **253**, 872–9.
26. Bourne, Y., Taylor, P., and Marchot, P. (1995). *Cell*, **83**, 503–12.

International conferences on cholinesterases are held every three years and the proceedings of these meetings offer a review of the field from diverse perspectives (see Doctor, B. P., Quinn, D. M., Gentry, M. K., and Taylor, P., eds (1998) *Proceedings of the Sixth International Conference on Cholinesterases and Related Proteins.* Plenum, New York, in press).

Palmer Taylor
Department of Pharmacology 0636,
University of California, San Diego,
La Jolla, CA 92093, USA

Collagens

Overview of the family

The collagens constitute a superfamily of extracellular matrix proteins with a structural role as their primary function. Based on the exon structure of their genes as well as the configuration of the sequence domains of the proteins, they can be divided into several families or groups. Within each family, several homologous genes encode polypeptides that have domains with similar sequences. All collagenous proteins have domains with a triple-helical conformation. Such domains are formed by three subunits (α chains), each containing a $(\text{Gly–X–Y})_n$ repetitive sequence motif.

The presence of domains with a triple-helical molecular conformation (Fig. 1) provides collagens with regions of rigid, rod-like molecular structures.[1,2] In fibrillar collagens and short chain collagens each collagen molecule (after complete proteolytic removal of amino and carboxyl propeptides) contains only one such domain which accounts for almost the entire length of the molecule. In other collagens, such as FACIT collagens, basement membrane collagens, multiplexins, and collagens with transmembrane domains – MACITs, several short triple-helical domains are separated by non-triple-helical sequences.

Within triple-helical domains, each α chain is coiled into an extended left-handed polyproline II helix and three α chains are in turn twisted into a right-handed superhelix. The high resolution crystal structure of a triple-helical collagen-like peptide shows that the triple helix is surrounded by a cylinder of hydration; an extensive network of hydrogen bonds between water molecules and peptide acceptor groups stabilizes the structure.[2] Residues of 4-hydroxyproline in the Y-position of the $(\text{Gly–X–Y})_n$ repeat sequence play a critical role in the hydrogen-bonded structure. The post-translational hydroxylation of collagen polypeptides therefore causes a significant increase in the thermal denaturation (melting) temperature of collagen triple helices. The triple-helical conformation requires a close packing of every third residue in each α chain along the triple-helical axis and only glycyl residues can be accommodated in this position. This explains why collagen mutations in which such triple-helical glycyl residues are replaced by residues with more bulky side chains can cause severe abnormalities. Even replacement with an alanine residue can result in a local untwisting of the triple helix and an alteration in the characteristic hydrogen bonding pattern.[2]

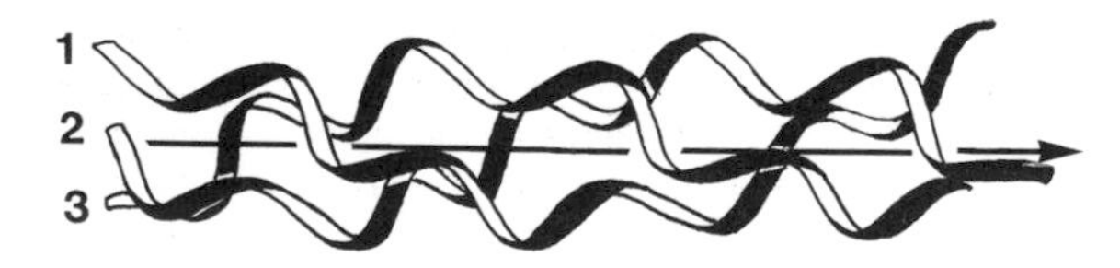

Figure 1. Molecular structure of the triple-helical conformation; three left-handed helices form a right-handed superhelix.

Collagenous proteins usually form supramolecular aggregates (fibrils, filaments, or networks), either alone or in conjunction with other extracellular matrix components. Their major function is to contribute to the structural integrity of the extracellular matrix, or to help anchor cells to the matrix. Some of the non-fibrillar collagen types appear also to have important regulatory functions.

Based on detailed analyses of the exon structures of genes that encode collagenous proteins, a comparison of protein domains, and functional considerations, the collagen superfamily can be divided into several subfamilies, as follows.

1. *Fibrillar collagens.* This group includes types I, II, III, V, and XI collagen, with molecules forming banded (cross-striated) fibrils in various tissues.
2. *FACIT collagens.* These include types IX, XII, XIV, XVI, and XIX collagens, with molecules that are associated with fibrils formed by fibrillar collagens.
3. *Short chain collagens.* These include types VIII and X collagen, with short, dumb-bell shaped molecules that form part of unique networks in basement membrane regions (type VIII) and hypertrophic cartilage (type X).
4. *Basement membrane collagens.* These include several different molecules collectively known as type IV collagens. They represent the major collagenous components of basement membranes.
5. *Multiplexins.* These are molecules with multiple short triple-helical domains that are found mostly in basement membrane regions. Types XV and XVIII collagen currently belong to this group.
6. *Collagens with transmembrane domains – MACITs.* Types XIII and XVII collagen are cell-surface molecules with multiple extracellular triple-helical domains, connected to a cytoplasmic region by a transmembrane segment. Their orientation (with the carboxyl end in the extracellular space) is similar to that of other cell surface molecules with triple-helical domains, such as the type I macrophage scavenger receptor,[3] possibly

Table 1 Chromosomal location of collagen genes

Gene locus	Chain designation	Chromosomal location in humans	Refs
Fibrillar collagens			
COL1A1	α1(I)	17q21.3–q22	17
COL1A2	α2(I)	17q21.3–q22	18
COL2A1	α1(II)	12q13–q14	19, 20
COL3A1	α1(III)	2q24.3–q3117	19
COL5A1	α1(V)	9q34	21
COL5A2	α2(V)	2q24.3–q31	22
COL5A3	α3(V)		
COL11A1	α1(XI)	1p21	23
COL11A2	α2(XI)	6p212	24
FACIT collagens			
COL9A1	α1(IX)	6q12–q14	25, 26
COL9A2	α2(IX)	1p32.3–p33	27
COL9A3	α3(IX)	20q13.3	28
COL12A1	α1(XII)	6q12–q14	29
COL14A1	α1(XIV)	8q23	30
COL16A1	α1(XVI)	1p34–p35	31
COL19A1	α1(XIX)	6q12–q14	29
Short chain collagens			
COL8A1	α1(VIII)	3q11.1–q13.2	32
COL8A2	α2(VIII)	1p32.3–p34.3	33
COL10A1	α1(X)	6q21–q22	34
Basement membrane collagens			
COL4A1	α1(IV)	13q33–q34	35–37
COL4A2	α2(IV)	13q33–q34	36, 38, 39
COL4A3	α3(IV)	2q36–q37	40
COL4A4	α4(IV)	2q35–2q37.1	41, 42
COL4A5	α5(IV)	Xq22	43
COL4A6	α6(IV)	Xq22	44, 45
Multiplexins			
COL15A1	α1(XV)	9q21–q22	46
COL18A1	α1(XVIII)	21q22.3	47
Collagens with transmembrane domains-MACITs			
COL13A1	α1(XIII)	10q22	48
COL17A1	α1(XVII)	10q24.3	49
Other collagens			
COL6A1	α1(VI)	21q22.3	50–52
COL6A2	α2(VI)	21q22.3	50, 51
COL6A3	α3(VI)	2q37	50, 53
COL7A1	α1(VII)	3p21.1–p21.3	54, 55

the B-chain of the C1q complex,[4,5] and the macrophage protein MARCO.[6] The last three molecules have not traditionally been included among the collagens, but given their triple-helical domains[7] they can, with good justification, be described as collagen types. In fact, several proteins with triple-helical domains are not included among the collagens, most likely because they were not discovered in a 'collagen laboratory'. These include the collectins[8,9] (lung surfactant protein A, lung surfactant protein D, bovine conglutinin, collectin-43, and mannose binding protein), ficolins,[10,11] hibernation proteins,[12] the asymmetric form of acetylcholinesterase,[13,14] the subunits of C1q,[15] and a component of the inner ear.[16]

7. *Other collagens*. This group includes molecules (types VI and VII) that form specialized structures in a variety of tissues (e.g. microfibrils for type VI, anchoring fibrils for type VII).

Most collagen genes have now been cloned and their chromosomal locations determined. The recently described collagen types were found through molecular cloning; additional collagenous molecules are likely to be discovered through analyses of ESTs and genome sequences. The very large number of sequence entries in the GenBank/EMBL data bank makes it impractical to list the appropriate accession numbers in the pages that follow. The reader is instead referred to original articles for specific information. The chromosomal locations of the human genes are given in Table 1. Useful databases for all collagens are Online Mendelian Inheritance in Man (OMIM) which can be accessed at http://www3.ncbi.nlm.nih.gov/Omim/ and allied resources and links.

References

1. Rich, A. and Crick, F. H. C. (1961). *J. Mol. Biol.*, **3**, 483–506.
2. Bella, J., Eaton, M., Brodsky, B., and Berman, H. M. (1994). *Science*, **266**, 75–81.
3. Resnick, D., Chatterton, J. E., Schwartz, K., Slayter, H., and Krieger, M. (1996). *J. Biol. Chem.*, **271**, 26924–30.
4. Kaul, M. and Loos, M. (1993). *Behring Inst. Mitt.*, **93**, 171–9.
5. Pihlajaniemi, T. and Rehn, M. (1995). *Prog. Nucleic Acid. Res. Mol. Biol.*, **50**, 225–62.
6. Elomaa, O., Kangas, M., Sahlberg, C., Tuukkanen, J., Sormunen, R., Liakka, A., *et al.* (1995). *Cell*, **80**, 603–9.
7. Brodsky, B. and Shah, N. K. (1995). *FASEB J.* **9**, 1537–46.
8. Hoppe, H. J. and Reid, K. B. (1994). *Protein Sci.*, **3**, 1143–58.
9. Holmskov, U. and Jensenius, J. C. (1993). *Behring. Inst. Mitt.*, **93**, 224–35.
10. Ichijo, H., Hellman, U., Wernstedt, C., Gonez, L. J., Claesson-Welsh, L., Heldin, C. H., and Miyazono, K. (1993). *J. Biol. Chem.*, **268**, 14505–13.
11. Ohashi, T. and Erickson, H. P. (1997). *J. Biol. Chem.*, **272**, 14220–6.
12. Takamatsu, N., Ohba, K., Kondo, J., Kondo, N., and Shiba, T. (1993). *Mol. Cell Biol.*, **13**, 1516–21.
13. Rosenberry, T. L. and Richardson, J. M. (1977). *Biochemistry*, **16**, 3550–8.
14. Bon, S., Coussen, F., and Massoulie, J. (1997). *J. Biol. Chem.*, **272**, 3016–21.
15. Reid, K. B. (1979). *Biochem. J.*, **179**, 367–71.
16. Davis, J. G., Oberholtzer, J. C., Burns, F. R., and Greene, M. I. (1995). *Science*, **267**, 1031–4.
17. Huerre, C., Junien, C., Weil, D., Chu, M. L., Morabito, M., Van Cong, N., *et al.* (1982). *Proc. Natl Acad. Sci. USA*, **79**, 6627–30.
18. Junien, C., Weil, D., Myers, J. C., Van Cong, N., Chu, M. L., Foubert, C., *et al.* (1982). *Am. J. Hum. Genet.*, **34**, 381–7.
19. Huerre-Jeanpierre, C., Mattei, M. G., Weil, D., Grzeschik, K. H., Chu, M. L., Sangiorgi, F. O., *et al.* (1986). *Am. J. Hum. Genet.*, **38**, 26–37.
20. Takahashi, E., Hori, T., P., O. C., Leppert, M., and White, R. (1990). *Hum. Genet.*, **86**, 14–16.
21. Greenspan, D. S., Byers, M. G., Eddy, R. L., Cheng, W., Jani-Sait, S., and Shows, T. B. (1992). *Genomics*, **12**, 836–7.
22. Emanuel, B. S., Cannizzaro, L. A., Seyer, J. M., and Myers, J. C. (1985). *Proc. Natl Acad. Sci. USA*, **82**, 3385–9.
23. Henry, I., Bernheim, A., Bernard, M., van der Rest, M., Kimura, T., Jeanpierre, C., *et al.* (1988). *Genomics*, **3**, 87–90.
24. Kimura, T., Cheah, K. S., Chan, S. D., Lui, V. C., Mattei, M. G., van der Rest, M., *et al.* (1989). *J. Biol. Chem.*, **264**, 13910–6.
25. Kimura, T., Mattei, M. G., Stevens, J. W., Goldring, M. B., Ninomiya, Y., and Olsen, B. R. (1989). *Eur. J. Biochem.*, **179**, 71–8.
26. Warman, M. L., Tiller, G. E., Polumbo, P. A., Seldin, M. F., Rochelle, J. M., Knoll, J. H., *et al.* (1993). *Genomics*, **17**, 694–8.
27. Warman, M. L., McCarthy, M. T., Perala, M., Vuorio, E., Knoll, J. H., McDaniels, C. N., *et al.* (1994). *Genomics*, **23**, 158–62.
28. Brewton, R. G., Wood, B. M., Ren, Z. X., Gong, Y., Tiller, G. E., Warman, M. L., *et al.* (1995). *Genomics*, **30**, 329–36.
29. Gerecke, D. R., Olson, P. F., Koch, M., Knoll, J. H., Taylor, R., Hudson, D. L., *et al.* (1997). *Genomics*, **41**, 236–42.
30. Schnittger, S., Herbst, H., Schuppan, D., Dannenberg, C., Bauer, M., and Fonatsch, C. (1995). *Cytogenet. Cell, Genet.*, **68**, 233–4.
31. Pan, T. C., Zhang, R. Z., Mattei, M. G., Timpl, R., and Chu, M. L. (1992). *Proc. Natl Acad. Sci. USA*, **89**, 6565–9.
32. Muragaki, Y., Mattei, M. G., Yamaguchi, N., Olsen, B. R., and Ninomiya, Y. (1991). *Eur. J. Biochem.*, **197**, 615–22.
33. Muragaki, Y., Jacenko, O., Apte, S., Mattei, M. G., Ninomiya, Y., and Olsen, B. R. (1991). *J. Biol. Chem.*, **266**, 7721–7.
34. Apte, S., Mattei, M. G., and Olsen, B. R. (1991). *FEBS Lett.*, **282**, 393–6.
35. Emanuel, B. S., Sellinger, B. T., Gudas, L. J., and Myers, J. C. (1986). *Am J. Hum. Genet.*, **38**, 38–44.
36. Griffin, C. A., Emanuel, B. S., Hansen, J. R., Cavenee, W. K., and Myers, J. C. (1987). *Proc. Natl Acad. Sci. USA*, **84**, 512–6.
37. Bowcock, A. M., Hebert, J. M., Christiano, A. M., Wijsman, E., Cavalli-Sforza, L. L., and Boyd, C. D. (1987). *Cytogenet. Cell, Genet.*, **45**, 234–6.
38. Solomon, E., Hall, V., and Kurkinen, M. (1987). *Ann. Hum. Genet.*, **51**, 125–7.
39. Killen, P. D., Francomano, C. A., Yamada, Y., Modi, W. S., and O'Brien, S. J. (1987). *Hum. Genet.*, **77**, 318–24.
40. Morrison, K. E., Mariyama, M., Yang-Feng, T. L., and Reeders, S. T. (1991). *Am J. Hum. Genet.*, **49**, 545–54.
41. Mariyama, M., Zheng, K., Yang-Feng, T. L., and Reeders, S. T. (1992). *Genomics*, **13**, 809–13.
42. Kamagata, Y., Mattei, M. G., and Ninomiya, Y. (1992). *J. Biol. Chem.*, **267**, 23753–8.
43. Hostikka, S. L., Eddy, R. L., Byers, M. G., Hoyhtya, M., Shows, T. B., and Tryggvason, K. (1990). *Proc. Natl Acad. Sci. USA*, **87**, 1606–10.
44. Zhou, J., Mochizuki, T., Smeets, H., Antignac, C., Laurila, P., de Paepe, A., *et al.* (1993). *Science*, **261**, 1167–9.
45. Oohashi, T., Sugimoto, M., Mattei, M. G., and Ninomiya, Y. (1994). *J. Biol. Chem.*, **269**, 7520–6.
46. Huebner, K., Cannizzaro, L. A., Jabs, E. W., Kivirikko, S., Manzone, H., Pihlajaniemi, T., and Myers, J. C. (1992). *Genomics*, **14**, 220–4.
47. Oh, S. P., Warman, M. L., Seldin, M. F., Cheng, S. D., Knoll, J. H., Timmons, S., and Olsen, B. R. (1994). *Genomics*, **19**, 494–9.
48. Shows, T. B., Tikka, L., Byers, M. G., Eddy, R. L., Haley, L. L., Henry, W. M., *et al.* (1989). *Genomics*, **5**, 128–33.
49. Li, K. H., Sawamura, D., Giudice, G. J., Diaz, L. A., Mattei, M. G., Chu, M. L., and Uitto, J. (1991). *J. Biol. Chem.*, **266**, 24064–9.
50. Weil, D., Mattei, M. G., Passage, E., C., N. G. V., Pribula-Conway, D., Mann, K., Deutzmann, R., *et al.* (1988). *Am. J. Hum. Genet.*, **42**, 435–45.
51. Francomano, C. A., Cutting, G. R., McCormick, M. K., Chu, M. L., Timpl, R., Hong, H. K., and Antonarakis, S. E. (1991). *Hum. Genet.* **87**, 162–6.
52. Delabar, J. M., Chettouh, Z., Rahmani, Z., Theophile, D., Blouin, J. L., Bono, R., *et al.* (1992). *Genomics*, **13**, 887–9.
53. Speer, M. C., Tandan, R., Rao, P. N., Fries, T., Stajich, J. M., Bolhuis, P. A., *et al.* (1996). *Hum. Mol. Genet.*, **5**, 1043–6.
54. Parente, M. G., Chung, L. C., Ryynanen, J., Woodley, D. T., Wynn, K. C., Bauer, E. A., *et al.* (1991). *Proc. Natl Acad. Sci. USA*, **88**, 6931–5.
55. Greenspan, D. S., Byers, M. G., Eddy, R. L., Hoffman, G. G., and Shows, T. B. (1993). *Cytogenet. Cell Genet.* **62**, 35–6.

Bjorn Reino Olsen
Department of Cell Biology, Harvard Medical School, and Harvard-Forsyth Department of Oral Biology, Harvard School of Dental Medicine, Boston, MA, USA

Yoshifumi Ninomiya
Department of Molecular Biology and Biochemistry, Okayama University Medical School, 2-5-1 Shikata-cho, Okayama 700, Japan

Fibrillar collagens

Collagen types I, II, III, V, and XI participate in the formation of fibrils with molecules packed in quarter-staggered arrays. Encoded by homologous, multiexon genes, they evolved to provide multicellular organisms, from sponges to humans, with supramolecular scaffolds for mechanical support and the proper environment for cellular migration, attachment and differentiation. Fibrillar collagens are synthesized as precursors, procollagens, that are proteolytically processed to collagen in the extracellular space.

Fibrillar collagens include five different molecular types (I, II, III, V, and XI) containing polypeptide subunits (α chains), encoded by nine distinct genes.[1–6] The molecules are either homotrimers with α chains of the same kind (types II and III) or heterotrimers with two or three different α chains (types I, V, and XI). Some α chains participate in the formation of more than one collagen type. For example, the product of the *COL2A1* gene, α1(II), forms homotrimeric type II collagen molecules and participates, as α3(XI), in the formation of heterotrimeric type XI collagen molecules. Also, α1(XI) chains appear both in heterotrimeric type XI molecules in cartilage as well as in a bone variant of type V collagen, where it replaces the α1(V) chain.[7] Finally, α2(V) chains are found both in heterotrimeric type V molecules and in the vitreous form of type XI.[8] Because of their great similarity and ability to form mixed heterotrimers, types V and XI collagen are frequently referred to as type V/XI collagen.[9]

Each fibrillar collagen α chain contains over 300 repeats of the triplet sequence -Gly–X–Y-, flanked by short non-triplet-containing sequences, telopeptides, at each end. About 50 per cent of the prolyl residues in the Y positions of the triplet domain are post-translationally converted to 4-hydroxyproline by the enzyme prolyl 4-hydroxylase (EC 1.14.11.2) located in the rough endoplasmic reticulum.[10] The active enzyme is a tetramer of two non-identical subunits, α and β. The β subunit is the enzyme protein disulphide isomerase, and the tetramer prolyl 4-hydroxylase has disulphide isomerase activity.[11] In addition to prolyl hydroxylation, some lysyl residues in the Y position are hydroxylated by lysyl hydroxylase,[12] and the sequential action of galactosyl hydroxylysyl transferase and glucosylgalactosyl hydroxylysyl transferase adds mono- and disaccharides to some hydroxylysyl residues.[13]

The α chains of fibrillar collagens are synthesized as pro-α chains, with amino (N) and carboxyl (C) propeptides flanking the central (Gly–X–Y)$_n$-containing domain (Fig. 1). Folding of a trimeric C-propeptide domain is the first step in intracellular assembly of homo- or heterotrimeric procollagen molecules; the chain composition of collagen molecules is therefore determined by the specificity by which C-propeptides of various procollagens interact. The folding of the triple-helical domain proceeds from the carboxyl end towards the amino end of the trimeric molecule in a zipper-like fashion and with a rate that is limited by *cis–trans* isomerization of peptidyl prolyl bonds.[4] Since prolyl and lysyl hydroxylases do not work on triple-helical substrates,[10,12] and the thermal stability of the triple helix depends on the level of hydroxylation of prolyl residues, the folding of the triple helix limits the degree of post-translational hydroxylation to what is needed for a stable triple helix at 37°C.[5] Substitutions of glycine residues in the Gly–X–Y repeats lower the stability, may lead to overmodification, and a decrease in the rate of secretion from cells resulting in intracellular retention and degradation.[14]

During extracellular processing of procollagen to collagen, the propeptides are removed from the major collagen triple-helical domain by specific endoproteinases.[13] Of great interest is the recent finding that the C proteinase is identical to BMP-1, the mammalian homologue of the *Drosophila* tolloid gene product.[15] In *Drosophila* the tolloid protease is involved in processing the precursor of the BMP-2/4-like product of the *decapentaplegic* (*dpp*) gene; BMP-1 may likewise activate latent forms of BMPs or other members of the TGF-β family of molecules. Controlled cleavage of the N-propeptide domain plays a role in fibrillogenesis and is thought to be important for regulation of fibril diameters.[4,16,17]

Purification and recombinant synthesis

To purify fibrillar collagens from tissues, pepsin has been commonly used to dissociate collagen triple-helical domains (these are pepsin resistant) from other extracellular matrix molecules. Repeated differential salt precipitations at neutral pH as well as in acid conditions are

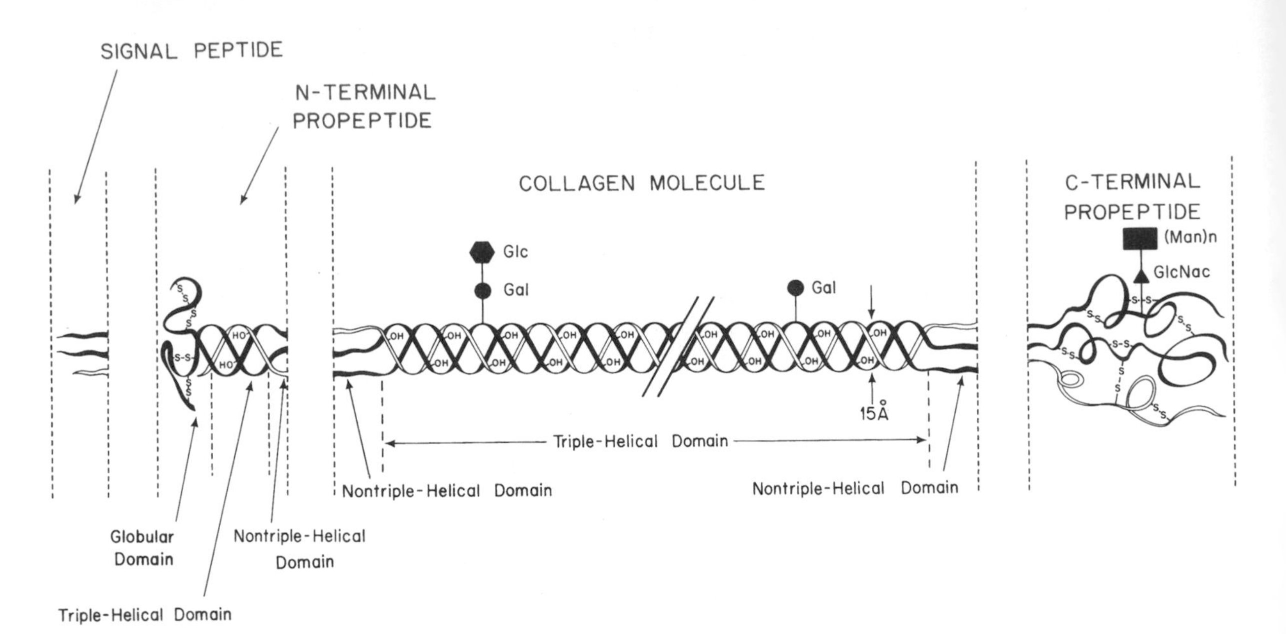

Figure 1. Diagram of the domains of a fibrillar procollagen molecule (type I). (From Olsen 1991.[13])

then used to purify each fibrillar collagen type.[18] Types II, V, and XI collagen require higher salt concentrations than types III and I to precipitate.[19] Fibrillar procollagen molecules have been purified from media of cultured cells.[20,21] Addition of protease inhibitors and avoidance of acidic pH prevent the action of endogenous proteolytic enzymes that remove the propeptides, resulting in the isolation of intact precursor molecules. Purified collagens are available from a number of commercial sources. A good listing of many suppliers is provided in the BioSupplyNet *Source Book*, available at http://www.biosupplynet.com. Recently, recombinant fibrillar procollagens have been synthesized in mammalian cells as well as in insect cells transfected with genes encoding the α and β subunits of prolyl 4-hydroxylase to ensure proper hydroxylation of prolyl residues.[22–25] A fragment of human type III collagen has been produced in *Saccharomyces cerevisiae*.[72]

Antibodies

Polyclonal and monoclonal antibodies against all fibrillar collagens from a number of animal species are available. Some of these antibodies are directed against epitopes in the propeptide domains of the procollagens;[20] others are directed against epitopes in the triple-helical domain.[26,27] Several monoclonal antibodies have been used for epitope mapping in conjunction with rotary shadowing and electron microscopy.[27,28] A variety of antibodies are available from commercial sources; a good listing of suppliers can be found in the BioSupplyNet Source Book.

Activities

The triple-helical products of procollagen processing, fibrillar collagens, polymerize to form fibrils that serve as stabilizing scaffolds in extracellular matrices.[29] Within the fibrils, the 300 nm long rod-like molecules overlap with their ends by about 30 nm and are arranged in quarter-staggered arrays. The fibrils therefore have a periodic structure. Each period is 67 nm long and consists of a 'hole' zone with more loosely packed molecules and an overlap zone with more densely packed molecules. These zones can easily be visualized by negative staining and electron microscopy. When fibrils are positively stained, a periodic cross-striation pattern is observed, reflecting the distribution of clusters of charged amino acid residues along the collagen molecules.[29] Cell differentiation and migration during development are influenced by fibrillar collagens, and collagens interact with cells through integrin receptors on cell surfaces.

Collagen fibrils usually contain more than one type of collagen,[30] and such heterotypic fibrils are arranged in different patterns in different tissues; parallel fibril bundles in tendon, criss-crossing layers in cornea, and spiral arrangements in lamellar bone (Fig. 2). Heterotypic fibrils containing types I, III, and/or V collagens are expressed in a number of tissues of mesenchymal origin such as skin, tendon, ligaments, and bone, whereas fibrils with types II and XI are found predominantly in hyaline cartilage and the vitreous body of the eye. It is believed that the presence of small amounts of collagens V and XI within the fibrils limits fibril diameters due to steric hindrance, based on the incomplete removal of N-propeptides from types V and XI mole-

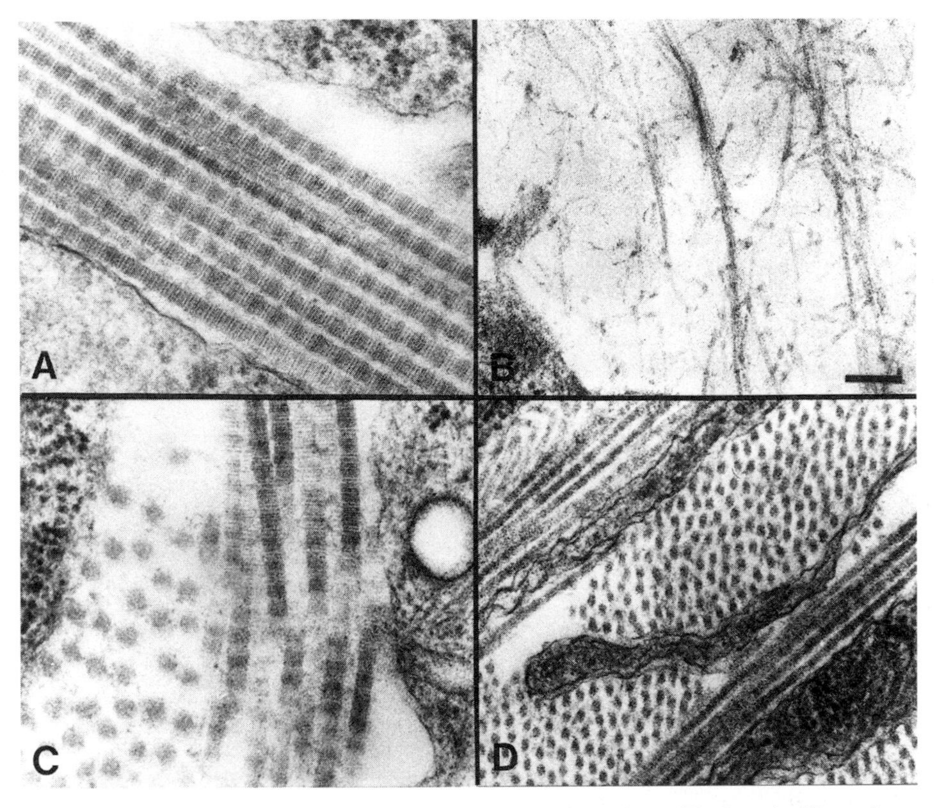

Figure 2. Collagen fibrils in 14-day chick embryo tendon (A), sternal cartilage (B), dermis (C), and corneal stroma (D). (Courtesy of Dr David Birk.)

cules.[17,30,31] Thus, the diameter of heterotypic collagen fibrils depends on the ratio between collagens V and I (or XI and II); the higher the ratio, the thinner the fibrils (Plate 11). Fibril properties are also dependent on interactions with FACIT collagens and small proteoglycans (decorin, biglycan, fibromodulin).[4,32–39] The ability of the C propeptide to serve as a ligand for the integrin $\alpha2\beta1$ may play a role in regulating fibrillogenesis at the cell surface.[40,41] Binding of fibrillar collagen to fibronectin may play a role in assembly of fibronectin fibres.[42]

Genes

The chicken $\alpha2$(I) collagen gene was the first fibrillar collagen gene to be isolated and completely characterized.[43,44] Since then, cDNAs and genomic clones have been isolated for almost all fibrillar collagen genes, from a number of species. The number and the size of exons are similar in the various fibrillar collagen genes.[2,3,45] It is likely that many of the triple-helical domain exons evolved by repeated duplications of an exon unit of 54 base pairs. Alternative splicing generates transcripts encoding fibrillar procollagens with different N-propeptide domains in the *COL2A1*, *COL11A1*, and *COL11A2* genes.[46–52] Since the number of entries in the GenBank/EMBL data bank is very large, readers are referred to original publications for accession numbers of specific sequences.

Mutant phenotypes/disease states

Mutations in mice

Homozygous *Mov*13 mice, carrying an insertion of proviral sequences in a transcriptional enhancer within the first intron of *Col1a1*, are developmentally arrested between days 11 and 12 of gestation due to a block of transcription of the gene in fibroblasts.[53] Heterozygous *Mov13* animals survive to adulthood and serve as models for the mild dominant form of osteogenesis imperfecta.[54] A frame-shift mutation in the C-propeptide coding domain of the *Col1a2* gene in the *oim* mouse also results in an osteogenesis imperfecta-like phenotype.[55] Disproportionate micromelia (*Dmm*) in mice is caused by a three-nucleotide deletion in the C-propeptide coding region of *Col2a1*,[56] while autosomal recessive chondrodysplasia (*cho*) is caused by a frame-shift mutation in *Col11a1* leading to loss of synthesis of $\alpha1$(XI) collagen chains.[57]

Transgenic mice expressing dominant-negative mutant constructs have been generated for *Col1a1*, *Col2a1*, and Col5a2.[58–63] Mice with overexpression of wild-type gene or carrying inactivated ('knock out') alleles have also been described for several fibrillar collagen genes.[64–66]

Human diseases

Mutations in *COL1A1* and *COL1A2*, the genes encoding the α chains of type I procollagen, account for the majority of cases with osteogenesis imperfecta and for certain types of the Ehlers–Danlos syndrome.[14,67–68] Mutations in *COL3A1* cause Ehlers–Danlos syndrome type III and type IV. Mutations in *COL5A1* have been described in patients with Ehlers–Danlos syndrome type I and type II.[69] Mutations in *COL2A1* cause a spectrum of chondrodysplasias, including achondrogenesis II, hypochondrogenesis, spondyloepiphyseal dysplasia, and Kniest and Stickler syndromes.[14]

Websites

A mutation database for the *COL1A1*, *COL1A2*, and *COL3A1* genes has been described[70] and is accessible at http://www.le.ac.uk/genetics/collagen/collagen.html. A review of almost 300 mutations in *COL1A1*, *COL2A1*, and *COL3A1*, as well as other collagen genes has been published.[71] The OMIM database (www3.ncbi.nlm.nih.gov/Omim/) is an excellent resource for all fibrillar collagens and their associated inherited diseases.

References

1. Miller, E. J. (1985). *Ann. N Y Acad. Sci.*, **460**, 1–13.
2. Vuorio, E. and de Crombrugghe, B. (1990). *Ann. Rev. Biochem.*, **59**, 837–72.
3. Jacenko, O., Olsen, B., and LuValle, P. (1991). In *Critical reviews in eukaryotic gene expression* (ed. G. S. Stein, J. L. Stein, and J. B. Lian), pp. 327–53. CRC Press,
4. Olsen, B. R. (1995). *Curr. Opin. Cell Biol.*, **7**, 720–7.
5. Prockop, D. J. and Kivirikko, K. I. (1995). *Ann. Rev. Biochem.*, **64**, 403–34.
6. Kadler, K. (1995). *Protein Profile*, **2**, 491–619.
7. Niyibizi, C. and Eyre, D. R. (1994). *Eur. J. Biochem.*, **224**, 943–50.
8. Mayne, R., Brewton, R. G., Mayne, P. M., and Baker, J. R. (1993). *J. Biol. Chem.*, **268**, 9381–6.
9. Fichard, A., Kleman, J. P., and Ruggiero, F. (1995). *Matrix Biol.*, **14**, 515–31.
10. Kivirikko, K. I., Myllyla, R., and Pihlajaniemi, T. (1989). *FASEB J.* **3**, 1609–17.
11. Koivu, J., Myllyla, R., Helaakoski, T., Pihlajaniemi, T., Tasanen, K., and Kivirikko, K. I. (1987). *J. Biol. Chem.*, **262**, 6447–9.
12. Kellokumpu, S., Sormunen, R., Heikkinen, J., and Myllyla, R. (1994). *J. Biol. Chem.*, **269**, 30524–9.
13. Olsen, B. R. (1991). In *Cell biology of extracellular matrix* (ed. E. D. Hay), pp. 177–220. Plenum,
14. Mundlos, S. and Olsen, B. R. (1997). *FASEB J.* **11**, 227–33.
15. Kessler, E., Takahara, K., Biniaminov, L., Brusel, M., and Greenspan, D. S. (1996). *Science*, **271**, 360–2.
16. Fleischmajer, R., Perlish, J. S., Burgeson, R. E., Shaikh-Bahai, F., and Timpl, R. (1990). *Ann. N Y Acad. Sci.*, **580**, 161–75.
17. Linsenmayer, T. F., Gibney, E., Igoe, F., Gordon, M. K., Fitch, J. M., Fessler, L. I., and Birk, D. E. (1993). *J. Cell Biol.*, **121**, 1181–9.
18. Trelstad, R. L. (1982). In *Immunochemistry of the extracellular matrix* (ed. H. Furthmayr), pp. 31–41. CRC Press,
19. Mayne, R., van de Rest, M., Bruckner, P., and Schmid, T. M. (1995). In *Extracellular matrix. a practical approach* (ed. M. A. Haralson and J. R. Hassell), pp. 73–98. IRL Press,
20. Pesciotta, D. M., Curran, S. F., and Olsen, B. R. (1982). In *Immunochemistry of the extracellular matrix* (ed. H. Furthmayr), pp. 91–109. CRC Press,
21. Peterkofsky, B., Haralson, M. A., J., DiMari, S. J., and Miller, E. J. (1995). In *Extracellular matrix. a practical approach* (ed. M. A. Haralson and J. R. Hassell), pp. 31–72. IRL Press, Oxford.
22. Ala-Kokko, L., Hyland, J., Smith, C., Kivirikko, K. I., Jimenez, S. A., and Prockop, D. J. (1991). *J. Biol. Chem.*, **266**, 14175–8.
23. Tomita, M., Ohkura, N., Ito, M., Kato, T., Royce, P. M., and Kitajima, T. (1995). *Biochem. J.*, **312**, 847–53.
24. Lamberg, A., Helaakoski, T., Myllyharju, J., Peltonen, S., Notbohm, H., Pihlajaniemi, T., and Kivirikko, K. I. (1996). *J. Biol. Chem.*, **271**, 11988–995.
25. Myllyharju, J., Lamberg, A., Notbohm, H., Fietzek, P. P., Pihlajaniemi, T., and Kivirikko, K. I. (1997). *J. Biol. Chem.*, **272**, 21824–30.
26. Linsenmayer, T. F. (1991). In *Cell biology of extracellular matrix* (ed. E. D. Hay), pp. 7–44. Plenum Press,
27. Linsenmayer, T. F., Fitch, J. M., Schmid, T. M., Birk, D. E., Bruns, R. R., and Mayne, R. (1989). In *Collagen—molecular biology* (ed. B. R. Olsen and M. E. Nimni), pp. 141–170. CRC Press,
28. Mayne, R., Wiedemann, H., Irwin, M. H., Sanderson, R. D., Fitch, J. M., Linsenmayer, T. F., and Kühn, K. (1984). *J. Cell Biol.*, **98**, 1637–44.
29. Kühn, K. (1987). In *Structure and function of collagen types* (ed. R. Mayne, and R. E. Burgeson), pp. 1–42. Academic Press,
30. Linsenmayer, T. F., Fitch, J. M., and Birk, D. E. (1990). *Ann. N Y Acad. Sci.*, **580**, 143–160.
31. Marchant, J. K., Hahn, R. A., Linsenmayer, T. F., and Birk, D. E. (1996). *J. Cell Biol.*, **135**, 1415–26.
32. Shaw, L. M. and Olsen, B. R. (1991). *Trends Biochem. Sci.* **16**, 191–4.
33. Olsen, B. R., Winterhalter, K. H., and Gordon, M. K. (1995). *Trends Glycosci. Glycotechnol.*, **7**, 115–27.
34. Nishiyama, T., McDonough, A. M., M., Bruns, R. R., and Burgeson, R. E. (1994). *J. Biol. Chem.*, **269**, 28193–9.
35. Scott, J. E. (1996). *Biochemistry*, **35**, 8795–9.
36. Hocking, A. M., Strugnell, R. A., Ramamurthy, P., and McQuillan, D. J. (1996). *J. Biol. Chem.*, **271**, 19571–7.
37. Font, B., Eichenberger, D., Rosenberg, L. M., and van der Rest, M. (1996). *Matrix Biol.*, **15**, 341–348.
38. Iozzo, R. V. (1997). *Crit. Rev. Biochem. Mol. Biol.*, **32**, 141–74.
39. Danielson, K. G., Baribault, H., Holmes, D. F., Graham, H., Kadler, K. E., and Iozzo, R. V. (1997). *J. Cell Biol.*, **136**, 729–43.
40. Weston, S. A., Hulmes, D. J., Mould, A. P., Watson, R. B., and Humphries, M. J. (1994). *J. Biol. Chem.*, **269**, 20982–6.
41. Davies, D., Tuckwell, D. S., Calderwood, D. A., Weston, S. A., Takigawa, M., and Humphries, M. J. (1997). *Eur. J. Biochem.*, **246**, 274–82.
42. Dzamba, B. J., Wu, H., Jaenisch, R., and Peters, D. M. (1993). *J. Cell Biol.*, **121**, 1165–72.
43. Ohkubo, H., Vogeli, G., Mudryj, M., Avvedimento, V. E., Sullivan, M., Pastan, I., and de Crombrugghe, B. (1980). *Proc. Natl Acad. Sci. USA*, **77**, 7059–63.
44. Boedtker, H., Finer, M., and Aho, S. (1985). *Ann. N Y Acad. Sci.*, **460**, 85–116.
45. Sandell, L. J. and Boyd, C. D. (1990). In *Extracellular matrix genes* (ed. L. J. Sandell and C. D. Boyd), pp. 1–56. Academic Press,
46. Ryan, M. C. and Sandell, L. J. (1990). *J. Biol. Chem.*, **265**, 10334–9.
47. Sandell, L. J., Morris, N., Robbins, J. R., and Goldring, M. B. (1991). *J. Cell Biol.*, **114**, 1307–19.

48. Yoshioka, H., Inoguchi, K., Khaleduzzaman, M., Ninomiya, Y., Andrikopoulos, K., and Ramirez, F. (1995). *Genomics*, **28**, 337–40.
49. Tsumaki, N. and Kimura, T. (1995). *J. Biol. Chem.*, **270**, 2372–8.
50. Zhidkova, N. I., Justice, S. K., and Mayne, R. (1995). *J. Biol. Chem.*, **270**, 9486–93.
51. Lui, V. C. H., Ng, L. J., Sat, E. W. Y., Nicholls, J., and Cheah, K. S. E. (1996). *J. Biol. Chem.*, **271**, 16945–51.
52. Lui, V. C., Ng, L. J., Sat, E. W., and Cheah, K. S. (1996). *Genomics*, **32**, 401–12.
53. Schnieke, A., Harbers, K., and Jaenisch, R. (1983). *Nature*, **304**, 315–20.
54. Bonadio, J., Saunders, T. L., Tsai, E., Goldstein, S. A., Morris-Wiman, J., Brinkley, L., *et al.* (1990). *Proc. Natl Acad. Sci. USA*, **87**, 7145–9.
55. Chipman, S. D., Sweet, H. O., McBride, D. J., Jr., Davisson, M. T., Marks, S. C., Jr., Shuldiner, A. R., *et al.* (1993). *Proc. Natl Acad. Sci. USA*, **90**, 1701–5.
56. Pace, J. M., Li, Y., Wardell, B. B., Teuscher, C., Taylor, B. A., Seegmiller, R. E., and Olsen, B. R. (1997). *Dev. Dynamics.*, **208**, 25–33.
57. Li, Y., Lacerda, D. A., Warman, M. L., Beier, D. R., Yoshioka, H., Ninomiya, Y., *et al.* (1995). *Cell*, **80**, 423–30.
58. Pereira, R. F., Hume, E. L., Halford, K. W., and Prockop, D. J. (1995). *J. Bone Miner. Res.*, **10**, 1837–43.
59. Vandenberg, P., Khillan, J. S., Prockop, D. J., Helminen, H., Kontusaari, S., and Ala-Kokko, L. (1991). *Proc. Natl Acad. Sci. USA*, **88**, 7640–4.
60. Metsaranta, M., Garofalo, S., Decker, G., Rintala, M., de Crombrugghe, B., and Vuorio, E. (1992). *J. Cell Biol.*, **118**, 203–12.
61. Rintala, M., Metsaranta, M., Garofalo, S., de Crombrugghe, B., Vuorio, E., and Ronning, O. (1993). *J. Craniofac. Genet. Dev. Biol.*, **13**, 137–46.
62. Andrikopoulos, K., Liu, X., Keene, D. R., Jaenisch, R., and Ramirez, F. (1995). *Nature Genet.*, **9**, 31–6.
63. Liu, X., Wu, H., Byrne, M., Jeffrey, J., Krane, S., and Jaenisch, R. (1995). *J. Cell Biol.*, **130**, 227–37.
64. Garofalo, S., Metsaranta, M., Ellard, J., Smith, C., Horton, W., Vuorio, E., and de Crombrugghe, B. (1993). *Proc. Natl Acad. Sci. USA*, **90**, 3825–9.
65. Li, S. W., Prockop, D. J., Helminen, H., Fassler, R., Lapvetelainen, T., Kiraly, K., *et al.* (1995). *Genes Dev.*, **9**, 2821–30.
66. Liu, X., Wu, H., Byrne, M., Krane, S., and Jaenisch, R. (1997). *Proc. Natl Acad. Sci. USA*, **94**, 1852–6.
67. Byers, P. H. (1990). *Trends Genet.*, **6**, 293–300.
68. Kivirikko, K. I. (1993). Ann. Med. **25**, 113–26.
69. De Paepe, A., Nuytinck, L., Hausser, I., Anton-Lamprecht, I., and Naeyaert, J. M. (1997). *Am. J. Hum. Genet.* **60**, 547–54.
70. Dalgleish, R. (1997). *Nucl. Acids Res.* **25**, 181–7.
71. Kuivaniemi, H., Tromp, G., and Prockop, D. J. (1997). *Hum. Mutat.*, **9**, 300–15.
72. Vaughn, P. R., Galanis, M., Richards, K. M., Tebb, T. A., Ramshaw, J. A., and Werkmeister, J. A. (1998). *DNA Cell Biol.*, **17**, 511–18.

Bjorn Reino Olsen
Department of Cell Biology, Harvard Medical School, and Harvard-Forsyth Department of Oral Biology, Harvard School of Dental Medicine, Boston, MA, USA

Yoshifumi Ninomiya
Department of Molecular Biology and Biochemistry, Okayama University Medical School, 2-5-1 Shikata-cho, Okayama 700, Japan

FACIT collagens

FACIT collagens are a group of proteins that may serve as molecular bridges between fibrillar collagens and other extracellular matrix components. Their structure is strikingly different from that of other collagens in that their molecules contain two, three, or more relatively short triple-helical domains connected by non-triple-helical sequences. For some FACIT collagens, utilization of alternative promoters and alternative splicing give rise to different transcripts that are expressed in tissue-specific and time-dependent manners during embryonic development.

The FACIT (fibril associated collagens with interrupted triple helices) group of collagens includes at least five types of molecules, IX, XII, XIV, XVI, and XIX, composed of seven distinct polypeptide chains.[1,2] cDNA clones encoding an additional chain are currently being characterized (M. Gordon; personal communication) so it is likely that the group will prove to contain even more types of collagens. The domain structure of the first molecule of this group to be described, type IX collagen, was predicted by cloning and sequencing of a cDNA encoding the chicken α1(IX) chain.[3] Type IX molecules are heterotrimers consisting of α1(IX), α2(IX), and α3(IX) chains.[4,5] As shown in Fig. 1, they contain three triple-helical (COL) domains interrupted by non-triple-helical (NC) regions. Most of the NC domains contain cysteinyl residues forming disulphide bridges between subunits. One of the subunits, α2(IX), serves as a proteoglycan core protein and contains a glycosaminoglycan side chain attached to a seryl residue in the NC3 domain.[6–10] Before the structure of collagen IX was established by cDNA cloning/sequencing, it was, in fact, isolated as a proteo-

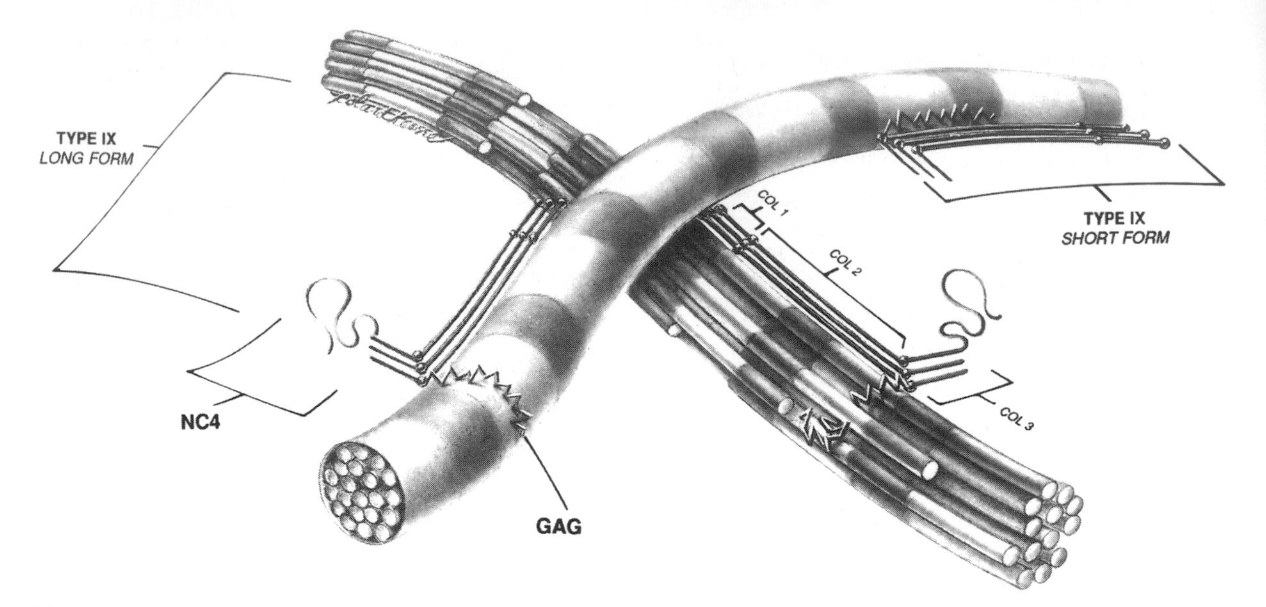

Figure 1. Diagram of type II collagen-containing fibrils with type IX collagen molecules on the surface. (From Jacenko *et al.* 1991.[83])

glycan called PG-Lt from chicken cartilage.[11,12] In cartilage, the glycosylation of the α2(IX) chain is incomplete and the glycosaminoglycan side chain is relatively short.[13,14] In the chicken vitreous body of the eye, however, the side chain is much longer and here type IX collagen may function primarily as a proteoglycan core protein.[15,16]

Type IX collagen molecules, expressed in hyaline cartilage, are associated with the surface of collagen fibrils such that two of the triple-helical domains are located at or close to the fibril surface, while an N-terminal globular domain is located in the perifibrillar space at the tip of a triple-helical arm (Fig. 2).[17] The type IX collagen molecules have an antiparallel orientation relative to the collagen II molecules within the fibrils. This has been deduced from the positions of covalent, hydroxypyridinium cross-links between the two types of molecules.[18,19] There are also covalent cross-links between different type IX collagen molecules, suggesting that type IX molecules on the surface of one fibril may be cross-linked to molecules on the surface of another fibril at points of intersection.[19,20] Immunofluorescence, *in situ* hybridization, and biochemical studies have shown that this FACIT collagen is also present in embryonic chick cornea and in the vitreous body.[21] Different tissues contain different forms of type IX molecules. These forms are translation products of two distinct mRNAs generated by alternative transcription of the α1(IX) collagen gene.[22] In chondrocytes, the majority of the α1(IX) transcripts are synthesized from an upstream transcription start site leading to the formation of mRNA that codes for a polypeptide with an N-terminal, globular domain.[23–25] In embryonic chick cornea,[22,26,27] the vitreous body,[15,16,27] neural retina,[28,29] perinotochordal matrix,[30,31] and early limb buds,[32] the majority of the transcripts are synthesized from a downstream, alternative start site, leading to the formation of mRNA encoding an α1(IX) with an alternative signal peptide sequence and lacking the N-terminal globular domain.

Types XII and XIV collagen are homologous, but distinct, homotrimeric molecules.[33–41] Their subunits contain two triple-helical (COL) domains separated by an NC region of more than 40 amino acid residues, a relatively short (<100 amino acid residues) non-triple-helical C-terminal region, and a very large (>1500 amino acid residues) N-terminal non-triple-helical domain[2] (Fig. 3). Within the native molecules the COL domains of the three subunits form a triple-helical tail attached to a central globule from which three non-triple-helical arms or finger-like structures project.[40–42] The central globule and the arms are composed of the N-terminal NC domains. For both types XII and XIV collagen, alternative splicing of primary transcripts generates molecular diversity. Two major molecular forms of type XII collagen differ in the lengths of the N-terminal NC (NC3) domains.[2] In form XIIA, the NC3 domains contain 18 fibronectin type III repeats and four von Willebrand factor A-like domains.[34] In the shorter form XIIB there are 10 fibronectin type III repeats and two von Willebrand factor A-like domains[35] (Fig. 3). Both forms contain identical signal peptides and are encoded by mRNAs with identical 5′ untranslated sequences. For form XIIB, two variants generated by alternative splicing at the 3′ end of the primary transcript have been described.[43] In variant XIIB1 the carboxyl non-triple-helical domain NC1 is 74 amino acid residues long and contains an acidic region followed by a basic region. In variant XIIB2 the NC1 domain is much shorter (only 19 residues) and lacks these features.[43]

In type XIV collagen the NC3 domain is somewhat smaller than in collagen XIIB. Splice variations affecting the 5′ untranslated region of type XIV collagen mRNA, the N-terminal fibronectin type III repeat as well as the 3′ region have been described.[38,39,44] Undulin, a protein isolated from human placenta, has been shown to be encoded by the type XIV collagen gene and represents one of these variants.[44,45] The initially isolated undulin was a protein composed of only von Willebrand factor A-like domains and fibronectin type III repeats and no collagen sequences, but this was probably caused by proteolysis during isolation of the protein.[46]

Types XII and XIV collagen are found in most dense connective tissues. There is considerable overlap between their tissue distributions, but there are also some differences.[47–49] In bovine skin, type XII collagen is particularly concentrated in the papillary dermis, while type XIV collagen is present in the reticular dermis.[40] In periosteum, type XIV collagen appears restricted to the outer fibrous layer while type XII collagen is expressed both in this layer and in the innermost layer of osteogenic cells.[50] Antibodies to both collagens show labelling along type I-containing fibrils.[51]

The structure of types XVI and XIX collagens has been deduced from cDNA sequences. Both molecules contain multiple triple-helical and non-triple-helical domains. Type XVI collagen contains 10 triple-helical domains interspersed with 11 short non-triple-helical sequences[52,53] (Fig. 3). The C-terminal triple-helical domain (COL1) shows structural homology with the COL1 domains of types IX, XII, and XIV collagens. The non-triple-helical domains contain multiple cysteine residues that are arranged in a pattern similar to that found in cuticle collagens of *C. elegans*.[54] Type XVI collagen cDNAs were initially isolated from human fibroblast[52] and placental cDNA libraries,[53] but subsequent studies have shown a wide range of expressing tissues.[55] Type XIX collagen was originally discovered through cDNA cloning with RNA from the human rhabdomyosarcoma cell line RD (CCL136).[56] The predicted polypeptide was found to contain five triple-helical (COL) domains, interspersed with and flanked by six non-triple-helical (NC) domains (Fig. 3). The coding region is relatively small in comparison to the size of the transcript due to a long 3′-UTR (5 kb). The $\alpha1$(XIX) gene is located on human chromosome 6q12–q13, syntenic to the $\alpha1$(IX) and $\alpha1$(XII) genes.[57] In mouse embryos the $\alpha1$(XIX) gene is transcribed in many organs but only a few adult tissues such as brain, eye and testis appear to express this collagen.[58]

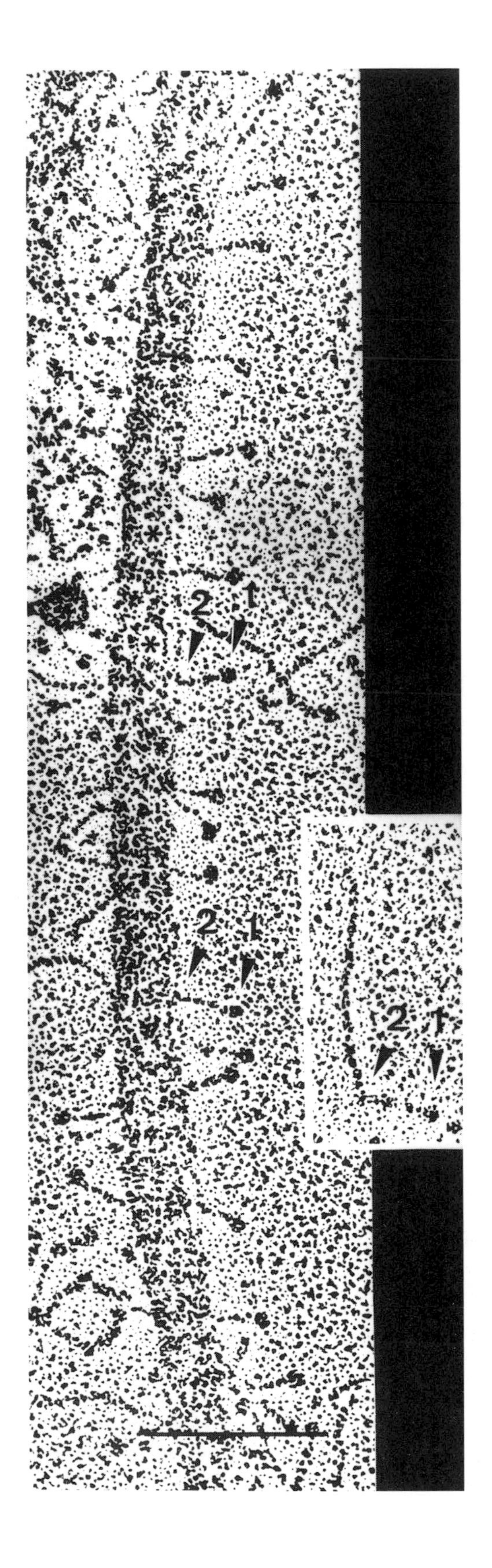

Figure 2. Rotary shadowing micrograph of type II collagen fibril with type IX molecules on the surface. (From Vaughan *et al.* 1988.[17])

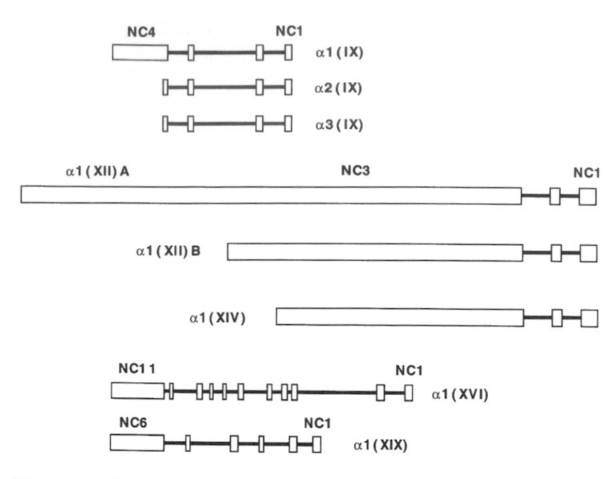

Figure 3. Diagram showing the domain structures of the members of the FACIT group of proteins. The domains are counted from the C terminus. Non-triple-helical domains are shown as open rectangles; triple-helical domains as solid lines.

Purification and recombinant synthesis

Type IX collagen can be purified from the medium of chondrocyte cultures or from cartilage tissue extracts.[59] Triple-helical fragments of the molecule have been purified from pepsin extracts of cartilage.[60] Types XII and XIV collagens have been purified from neutral salt extracts of skin and tendons[40] and as triple-helical fragments by pepsin extraction.[36] Type XVI collagen has not been purified as a protein from tissues, but a 160–210 kDa protein was detected by polyclonal antibodies in Western blots, consistent with the predicted structure from cDNA.[53,55] A recombinant α1(XIX) peptide was produced in *E. coli*.[58]

Antibodies

A large number of antibodies against FACIT collagens are available. Polyclonal antibodies against synthetic peptides deduced from nucleotide sequences[7] and polyclonal[20] as well as monoclonal antibodies[61–63] against protein fragments have been described for type IX collagen. A monoclonal antibody against a synthetic peptide derived from cDNA sequences recognizes the chicken α1(XII) chain by Western blotting, and has been used for immunohistochemical studies.[47] Monoclonal antibodies against bovine type XII (TL-A) and type XIV (TL-B) collagen are also available.[40,51] Polyclonal antibodies have been made against a synthetic peptide and a recombinant fragment of type XVI collagen.[55]

Polyclonal antibody against a recombinant α1(XIX) peptide was raised and used for Western blotting.[58]

Activities

Type IX collagen molecules are arranged in a periodic fashion along heterotypic collagen II/XI fibrils.[17] Covalent lysine-derived hydroxypyridinium cross-links between types IX and II molecules, as well as between collagen IX molecules, stabilize the fibril association.[18,19] This arrangement suggests that type IX collagen may serve as a molecular bridge between fibrils as well as between fibrils and other extracellular matrix constituents. Based on the colocalization of types XII and XIV with collagen I-containing fibrils in tissues, and the partial sequence similarity between types IX, XII, and XIV collagen, it is thought that types XII and XIV collagen associate with type I collagen fibrils in a similar fashion to type IX with type II fibrils.[2,64] Although type XIV collagen molecules did not bind to type I collagen in experiments with isolated matrix molecules,[65] it has been shown that type XII collagen can become incorporated into type I collagen fibrils when it is present during fibril formation; removal of the triple-helical domains of type XII reduced its ability to polymerize with type I collagen.[66] Both types XII and XIV collagen can bind to the dermatan sulphate chains of the fibril-associated proteoglycan decorin.[65,67] The two FACIT molecules may therefore interact with fibrils both directly and indirectly and may be important in keeping fibrils together in bundles, or alternatively, in preventing fibril fusion during tissue morphogenesis. Addition of types XII and XIV collagen to gels of type I collagen promoted fibroblast-induced gel contraction.[68] The effect was lost upon denaturation of the proteins, but was not reduced when the triple-helical domains were digested with bacterial collagenase. Since type XIIA carry glycosaminoglycan side chains while XIIB does not,[41] it is possible that cells can regulate the hydrophilic properties of perifibrillar compartments by controlling the expression of the two major splice variants of type XII collagen.

Genes

cDNA sequences are available for chicken α1(IX), α2(IX), and α3(IX),[3,69–71] mouse α1(IX), α2(IX),[72,73] and human α1(IX), α2(IX) and α3(IX).[25,74,75] Partial cDNA sequences from rat,[76] bovine,[69] and dog (GenBank L77390) α1(IX) are also reported. Genomic clones and sequences are available for the chicken α1(IX) and α2(IX) genes[23,69,77] as well as the mouse and human α1(IX) and α2(IX) genes.[9,24,73,75,78] Genomic and cDNA sequences are also available for α1(XII), α1(XIV), α1(XVI) and α1(XIX) collagens from several species.[33,34,38,39,52,53,56–58]

Mutant phenotype/disease states

Evidence for a role of type IX collagen in maintaining long-term stability of cartilage comes from studies of transgenic mice and genetic abnormalities in humans. In transgenic mice expressing an α1(IX) transgene with an in-frame deletion in the central triple-helical domain (COL2)[80] and in homozygous mice with inactivated α1(IX) alleles,[78] articular cartilage developed degenerative changes resembling those of human osteoarthritis. In humans, demonstration of linkage between the *COL9A2* locus and multiple epiphyseal dysplasia 2 (EDM2) (OMIM

600204)[81] was followed by confirmation of linkage in a second family with EDM2 and identification of a splice site mutation in *COL9A2* causing exon skipping and deletion of 12 amino acid residues in the COL3 domain of the $\alpha2$(IX) collagen chain.[82] Affected individuals in the two families develop stiffness and pain in knees during childhood and adolescence. X-rays of knees reveal flattened, irregular epiphyses and gradually appearing osteoarthritis. Because of the heterotrimeric structure of type IX collagen molecules, one would expect mutations in *COL9A1* and *COL9A3* to also cause multiple epiphyseal dysplasia with early onset osteoarthritis.

References

1. Shaw, L. M. and Olsen, B. R. (1991). *Trends Biochem. Sci.*, **16**, 191–4.
2. Olsen, B. R., Winterhalter, K. H., and Gordon, M. K. (1995). *Trends Glycosci. Glycotechnol.*, **7**, 115–27.
3. Ninomiya, Y. and Olsen, B. R. (1984). *Proc. Natl Acad. Sci. USA*, **81**, 3014–18.
4. van der Rest, M., Mayne, R., Ninomiya, Y., Seidah, N. G., Chretien, M., and Olsen, B. R. (1985). *J. Biol. Chem.*, **260**, 220–5.
5. Olsen, B. (1997). *Int. J. Biochem. Cell Biol.*, **29**, 555–8.
6. Bruckner, P., Vaughan, L., and Winterhalter, K. H. (1985). *Proc. Natl Acad. Sci. USA*, **82**, 2608–12.
7. Konomi, H., Seyer, J. M., Ninomiya, Y., and Olsen, B. R. (1986). *J. Biol. Chem.*, **261**, 6742–6.
8. Huber, S., van der Rest, M., Bruckner, P., Rodriguez, E., Winterhalter, K. H., and Vaughan, L. (1986). *J. Biol. Chem.*, **261**, 5965–8.
9. McCormick, D., van der Rest, M., Goodship, J., Lozano, G., Ninomiya, Y., and Olsen, B. R. (1987). *Proc. Natl Acad. Sci. USA*, **84**, 4044–8.
10. Huber, S., Winterhalter, K. H., and Vaughan, L. (1988). *J. Biol. Chem.*, **263**, 752–6.
11. Noro, A., Kimata, K., Oike, Y., Shinomura, T., Maeda, N., Yano, S., *et al.* (1983). *J. Biol. Chem.*, **258**, 9323–31.
12. Vaughan, L., Winterhalter, K. H., and Bruckner, P. (1985). *J. Biol. Chem.*, **260**, 4758–63.
13. Ayad, S., Marriott, A., Brierley, V. H., and Grant, M. E. (1991). *Biochem. J.*, **278**, 441–5.
14. Yada, T., Arai, M., Suzuki, S., and Kimata, K. (1992). *J. Biol. Chem.*, **267**, 9391–7.
15. Brewton, R. G., Wright, D. W., and Mayne, R. (1991). *J. Biol. Chem.*, **266**, 4752–7.
16. Yada, T., Suzuki, S., Kobayashi, K., Kobayashi, M., Hoshino, T., Horie, K., and Kimata, K. (1990). *J. Biol. Chem.*, **265**, 6992–9.
17. Vaughan, L., Mendler, M., Huber, S., Bruckner, P., Winterhalter, K. H., Irwin, M. I., and Mayne, R. (1988). *J. Cell Biol.*, **106**, 991–7.
18. Wu, J. J., Woods, P. E., and Eyre, D. R. (1992). *J. Biol. Chem.*, **267**, 23007–14.
19. Diab, M., Wu, J. J., and Eyre, D. R. (1996). *Biochem. J.*, **314**, 327–32.
20. Muller-Glauser, W., Humbel, B., Glatt, M., Strauli, P., Winterhalter, K. H., and Bruckner, P. (1986). *J. Cell Biol.*, **102**, 1931–9.
21. Gordon, M. K. and Olsen, B. R. (1990). *Curr. Opin. Cell Biol.*, **2**, 833–8.
22. Nishimura, I., Muragaki, Y., and Olsen, B. R. (1989). *J. Biol. Chem.*, **264**, 20033–41.
23. Vasios, G., Nishimura, I., Konomi, H., van der Rest, M., Ninomiya, Y., and Olsen, B. R. (1988). *J. Biol. Chem.*, **263**, 2324–9.
24. Muragaki, Y., Nishimura, I., Henney, A., Ninomiya, Y., and Olsen, B. R. (1990). *Proc. Natl Acad. Sci. USA*, **87**, 2400–4.
25. Muragaki, Y., Kimura, T., Ninomiya, Y., and Olsen, B. R. (1990). *Eur. J. Biochem.*, **192**, 703–8.
26. Svoboda, K. K., Nishimura, I., Sugrue, S. P., Ninomiya, Y., and Olsen, B. R. (1988). *Proc. Natl Acad. Sci. USA*, **85**, 7496–500.
27. Fitch, J. M., Mentzer, A., Mayne, R., and Linsenmayer, T. F. (1988). *Dev. Biol.*, **128**, 396–405.
28. Linsenmayer, T. F., Gibney, E., Gordon, M. K., Marchant, J. K., Hayashi, M., and Fitch, J. M. (1990). *Invest. Ophthalmol. Vis. Sci.*, **31**, 1271–6.
29. Liu, C. Y., Olsen, B. R., and Kao, W. W. (1993). *Dev. Dynamics*, **198**, 150–7.
30. Swiderski, R. E. and Solursh, M. (1992). *Dev. Dynamics*, **194**, 118–27.
31. Hayashi, M., Hayashi, K., Iyama, K., Trelstad, R. L., Linsenmayer, T. F., and Mayne, R. (1992). *Dev. Dynamics*, **194**, 169–76.
32. Swiderski, R. E. and Solursh, M. (1992). *Development*, **115**, 169–79.
33. Gordon, M., Gerecke, D., and Olsen, B. R. (1987). *Proc. Natl Acad. Sci. USA*, **84**, 6040–4.
34. Yamagata, M., Yamada, K. M., Yamada, S. S., Shinomura, T., Tanaka, H., Nishida, Y., *et al.* (1991). *J. Cell Biol.*, **115**, 209–21.
35. Gordon, M. K., Gerecke, D. R., Dublet, B., van der Rest, M., and Olsen, B.R. (1989). *J. Biol. Chem.*, **264**, 19772–8.
36. Dublet, B. and van der Rest, M. (1991). *J. Biol. Chem.*, **266**, 6853–8.
37. Gordon, M. K., Castagnola, P., Dublet, B., Linsenmayer, T. F., Van der Rest, M., Mayne, R., and Olsen, B. R. (1991). *Eur. J. Biochem.*, **201**, 333–8.
38. Walchli, C., Trueb, J., Kessler, B., Winterhalter, K. H., and Trueb, B. (1993). *Eur. J. Biochem.*, **212**, 483–90.
39. Gerecke, D. R., Foley, J. W., Castagnola, P., Gennari, M., Dublet, B., Cancedda, R., *et al.* (1993). *J. Biol. Chem.*, **268**, 12177–84.
40. Lunstrum, G. P., Morris, N. P., McDonough, A. M., Keene, D. R., and Burgeson, R. E. (1991). *J. Cell Biol.*, **113**, 963–9.
41. Koch, M., Bernascone, C., and Chiquet, M. (1992). *Eur. J. Biochem.*, **207**, 847–56.
42. Dublet, B., Oh, S., Sugrue, S. P., Gordon, M. K., Gerecke, D. R., Olsen, B. R., and van der Rest, M. (1989). *J. Biol. Chem.*, **264**, 13150–6.
43. Kania, A. M., Reichenberger, E. J., Karimbux, N. Y., Oh, S. P., Olsen, B. R., and Nishimura, I. (1998). (In press.)
44. Trueb, J. and Trueb, B. (1992). *Eur. J. Biochem.*, **207**, 549–57.
45. Schuppan, D., Cantaluppi, M. C., Becker, J., Veit, A., Bunte, T., Troyer, D., *et al* (1990). *J. Biol. Chem.*, **265**, 8823–32.
46. Just, M., Herbst, H., Hummel, M., Durkop, H., Tripier, D., Stein, H., and Schuppan, D. (1991). *J. Biol. Chem.*, **266**, 17326–32.
47. Sugrue, S. P., Gordon, M. K., Seyer, J., Dublet, B., van der Rest, M., and Olsen, B. R. (1989). *J. Cell Biol.*, **109**, 939–45.
48. Castagnola, P., Tavella, S., Gerecke, D. R., Dublet, B., Gordon, M. K., Seyer, J., *et al.* (1992). *Eur. J. Cell Biol.*, **59**, 340–7.
49. Oh, S. P., Griffith, C. M., Hay, E. C., and Olsen, B. R. (1993). *Dev. Dynamics*, **196**, 37–46.
50. Walchli, C., Koch, M., Chiquet, M., Odermatt, B. F., and Trueb, B. (1994). *J. Cell Sci.*, **107**, 669–81.
51. Keene, D. R., Lunstrum, G. P., Morris, N. P., Stoddard, D. W., and Burgeson, R. E. (1991). *J. Cell Biol.*, **113**, 971–8.
52. Pan, T. C., Zhang, R. Z., Mattei, M. G., Timpl, R., and Chu, M. L. (1992). *Proc. Natl Acad. Sci. USA*, **89**, 6565–9.

53. Yamaguchi, N., Kimura, S., McBride, O. W., Hori, H., Yamada, Y., Kanamori, T., *et al.* (1992). *J. Biochem.*, **112**, 856–63.
54. Kramer, J. M. (1994). *FASEB J.* **8**, 329–36.
55. Lai, C. H. and Chu, M. L. (1996). *Tissue Cell*, **28**, 155–64.
56. Myers, J. C., Yang, H., A., D. I. J., Presente, A., Miller, M. K., and Dion, A. S. (1994). *J. Biol. Chem.*, **269**, 18549–57.
57. Gerecke, D. R., Olson, P. F., Koch, M., Knoll, J. H., Taylor, R., Hudson, D. L., *et al.* (1997). *Genomics*, **41**, 236–42.
58. Sumiyoshi, H., Inoguchi, K., Khaleduzzaman, M., Ninomiya, Y., and Yoshioka, H. (1997). *J. Biol. Chem.*, **272**, 17104–111.
59. Mayne, R., van de Rest, M., Bruckner, P., and Schmid, T. M. (1995). In *Extracellular matrix. A practical approach* (ed. M. A. Haralson and J. R. Hassell), pp. 73–98. IRL Press, Oxford.
60. van der Rest, M. and Mayne, R. (1987). In *Structure, and function of collagen types* (ed. R. Mayne and R. E. Burgeson), pp. 195–221. Academic Press.
61. Irwin, M. H., Silvers, S. H., and Mayne, R. (1985). *J. Cell Biol.*, **101**, 814–23.
62. Ye, X. J., Terato, K., Nakatani, H., Cremer, M. A., and Yoo, T. J. (1991). *J. Histochem. Cytochem.*, **39**, 265–71.
63. Warman, M., Kimura, T., Muragaki, Y., Castagnola, P., Tamei, H., Iwata, K., and Olsen, B. R. (1993). *Matrix*, **13**, 149–56.
64. Olsen, B. R. (1995). *Curr. Opin. Cell Biol.*, **7**, 720–7.
65. Brown, J. C., Mann, K., Wiedemann, H., and Timpl, R. (1993). *J. Cell Biol.*, **120**, 557–67.
66. Koch, M., Bohrmann, B., Matthison, M., Hagios, C., Trueb, B., and Chiquet, M. (1995). *J. Cell Biol.*, **130**, 1005–14.
67. Font, B., Aubert-Foucher, E., Goldschmidt, D., Eichenberger, D., and van der Rest, M. (1993). *J. Biol. Chem.*, **268**, 25015–18.
68. Nishiyama, T., McDonough, A. M., Bruns, R. R., and Burgeson, R. E. (1994). *J. Biol. Chem.*, **269**, 28193–99.
69. Ninomiya, Y., Castagnola, P., Gerecke, D., Gordon, M., Jacenko, O., LuValle, P., *et al.* (1990). In *Extracellular matrix genes* (ed. L. J. Sandell and C. D. Boyd), pp. 79–114. Academic Press.
70. Brewton, R. G., Ouspenskaia, M. V., van der Rest, M., and Mayne, R. (1992). *Eur. J. Biochem.*, **205**, 443–9.
71. Har-el, R., Sharma, Y. D., Aguilera, A., Ueyama, N., Wu, J. J., Eyre, D. R., *et al.* (1992). *J. Biol. Chem.*, **267**, 10070–6.
72. Rokos, I., Muragaki, Y., Warman, M., and Olsen, B. R. (1994). *Matrix Biol.*, **14**, 1–8.
73. Perala, M., Elima, K., Metsaranta, M., Rosati, R., de Crombrugghe, B., and Vuorio, E. (1994). *J. Biol. Chem.*, **269**, 5064–71.
74. Perala, M., Hanninen, M., Hastbacka, J., Elima, K., and Vuorio, E. (1993). *FEBS Lett.*, **319**, 177–80.
75. Brewton, R. G., Wood, B. M., Ren, Z. X., Gong, Y., Tiller, G. E., Warman, M. L., *et al.* (1995). *Genomics*, **30**, 329–36.
76. Kimura, T., Mattei, M. G., Stevens, J. W., Goldring, M. B., Ninomiya, Y., and Olsen, B. R. (1989). *Eur. J. Biochem.*, **179**, 71–8.
77. Lozano, G., Ninomiya, Y., Thompson, H., and Olsen, B. R. (1985). *Proc. Natl Acad. Sci. USA*, **82**, 4050–4.
78. Faessler, R., Schnegelsberg, P. N. J., Dausman, J., Muragaki, Y., Shinya, T., McCarthy, M. T., *et al.* (1994). *Proc. Natl Acad. Sci. USA*, **91**, 5070–4.
79. Warman, M. L., Tiller, G. E., Polumbo, P. A., Seldin, M. F., Rochelle, J. M., Knoll, J. H., *et al.* (1993). *Genomics*, **17**, 694–8.
80. Nakata, K., Ono, K., Miyazaki, J., Olsen, B. R., Muragaki, Y., Adachi, E., *et al.* (1993). *Proc. Natl Acad. Sci. USA*, **90**, 2870–4.
81. Briggs, M. D., Choi, H.-C., Warman, M. L., Loughlin, J. A., Sykes, B. C., Irven, C. M. M., *et al.* (1994). *Am. J. Hum. Genet.*, **55**, 678–84.
82. Muragaki, Y., Mariman, E. C., van Beersum, S. E., Perala, M., van Mourik, J. B., Warman, M. L., *et al.* (1996). *Nature Genet.*, **12**, 103–5.
83. Jacenko, O., Olsen, B., and LuValle, P. (1991). In *Critical reviews in eukaryotic gene expression* (ed. G. S. Stein, J. L. Stein, and J. B. Lian), pp. 327–53. CRC Press.

Bjorn Reino Olsen
Department of Cell Biology,
Harvard Medical School, and Harvard-Forsyth
Department of Oral Biology,
Harvard School of Dental Medicine,
Boston, MA, USA

Yoshifumi Ninomiya
Department of Molecular Biology
and Biochemistry, Okayama
University Medical School, 2-5-1 Shikata-cho,
Okayama 700, Japan

Short chain collagens

Types VIII and X collagen, composed of the three chains α1(VIII), α2(VIII), and α1(X), form the subgroup of short chain collagens, so named because their subunits are short (only about 60 kDa) as compared with fibrillar collagen chains. Despite similarities in domain structure, amino acid sequences, and genomic exon configurations, the two types show very different temporal and spatial expression. Given the similarity in exon structure it is likely that the three genes evolved by duplication of a common precursor gene.

Type VIII collagen was originally identified as a product of bovine aortic and rabbit corneal endothelial cells, but is also synthesized by non-vascular cells.[1] The molecule is probably a heterotrimer composed of α1(VIII) and α2(VIII) chains in a ratio of two to one,[2] but the existence of homotrimeric molecules composed entirely of α1(VIII) or α2(VIII) chains cannot be ruled out.[3]

Type X collagen is a specific product of hypertrophic chondrocytes and is a useful marker for chondrocyte maturation to hypertrophy.[4] Except for the avian eggshell,[5] it does not appear to be expressed in other tissues outside hypertrophic cartilage. The molecule is a homotrimer of α1(X) chains and has a domain structure that is similar to that of type VIII collagen; a central triple-helical (COL1) domain of 50 kDa is flanked by N-terminal (NC2) and C-terminal (NC1) non-triple-helical domains.[6] Both type VIII and type X molecules appear as 130 nm long rods with knobs at both ends by electron microscopy after rotary shadowing.[5,7] The COL1 and NC1 domains of both types are encoded by one large exon, whereas the NC2 domain is encoded by a small additional exon.[7–9] Additional exons (one for α1(X) or two for α1(VIII)) encode the 5' untranslated portion of the mRNA. This exon configuration is in stark contrast to the multiexon structure of most other collagen genes.

Despite the similarities, a distinct tissue distribution has been found for the two short chain collagens: type X is restricted to hypertrophic cartilage,[4] whereas type VIII is distributed in various tissues including Descemet's membrane, vascular subendothelial matrices, heart, liver, kidney, perichondrium, and lung, as well as several malignant tumours including astrocytoma, Ewing's sarcoma, and hepatocellular carcinoma.[1,10,11]

In Descemet's membrane, type VIII collagen molecules represent major components of a hexagonal lattice structure,[12] with type VIII molecules most probably linked by interactions involving the non-triple-helical end regions (Fig. 1). Type X collagen molecules may form the same kind of polymer in hypertrophic cartilage,[13,14] and colocalization with a proteoglycan epitope[15] suggests a complex with proteoglycans. The expression of type X collagen is regulated primarily at a transcriptional level.[16,17] Evidence from 'knockout' and overexpression studies suggests that chondrocyte hypertrophy and type X collagen expression is negatively regulated by PTHrP and its receptor in growth plates.[18]

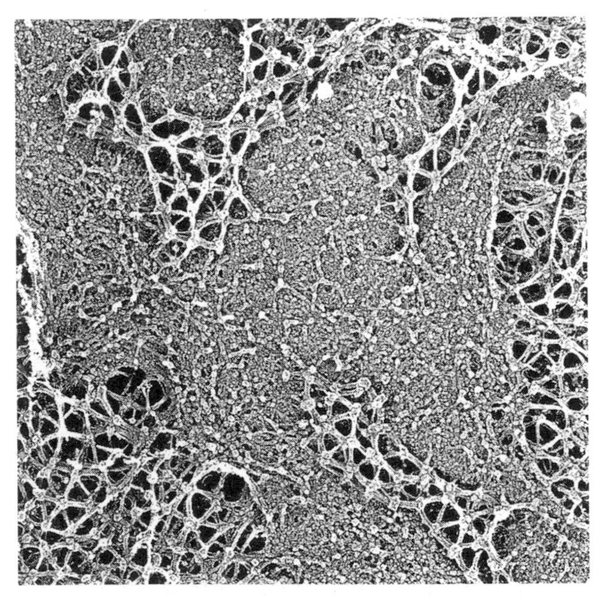

Figure 1. Hexagonal network structures formed in cultures of bovine corneal endothelial cells. The backbone of the network is composed of type VIII collagen molecules. (From Sawada, H. *et al.* 1984.[46])

Purification and recombinant synthesis

The triple-helical domain of type VIII collagen can be purified from Descemet's membrane by pepsin extraction.[1] The digested material can be precipitated with NaCl at neutral pH, and purified further by chromatography through agarose and by reverse phase HPLC.[1] Intact type VIII collagen can be recovered from the medium of cultured endothelial cells.[1] Type X collagen can be isolated intact from the medium of chicken hypertrophic chondrocytes kept in long-term culture or as a triple-helical fragment by pepsin extraction of hypertrophic cartilage.[4,19] Site-directed mutagenesis of human type X collagen has been used to determine the role of each domain in molecular assembly and secretion.[20] Three mutants with mutations in the C-terminal NC1 domain that are similar to those found in patients with Schmid type metaphyseal chondrodysplasia (see below), were unable to assemble into homotrimers *in vitro* or *in vivo* and were not secreted from cells.[20] In-frame deletions within the triple-helical domain did not prevent molecular assembly and secretion of pepsin-resistant triple-helical molecules.[20]

Antibodies

Polyclonal antibodies against bovine type VIII collagen have been used for expression studies and immunoblots.[11,21,22] Monoclonal antibodies against the bovine α1(VIII) chain have been used for immunoelectron microscopy to demonstrate that the backbone within Descemet's membrane is composed of type VIII collagen.[12] These antibodies are commercially available. Polyclonal antibodies against sheep type VIII collagen have also been produced.[10] A conformation-dependent monoclonal antibody, X-AC9, against chicken type X collagen has been used extensively for investigations of the tissue distribution, ultrastructure, and thermal stability[4]. Several other antibodies against chicken, mouse, bovine, and human type X collagens are available.[15,23–25]

Activities

Type VIII collagen is the major constituent of the hexagonal lattice observed in Descemet's membrane, as demonstrated by immunoelectron microscopy.[12] It is possible that the general function of type VIII collagen is to provide an open, porous structure that can withstand compressive force. Type X collagen in hypertrophic cartilage may play a similar role by providing a scaffold to prevent local collapse as the hypertrophic cartilage matrix is removed during endochondral ossification.[13,26,27] Several observations suggest a positive and/or negative role for type X collagen in the calcification of hypertrophic cartilage.[4]

Genes

The primary structures of the α1(VIII) and α2(VIII) chains are strikingly similar to that of α1(X) collagen. cDNA and genomic DNAs encoding rabbit, human, and mouse α1(VIII) and α2(VIII) chains have been isolated and characterized.[3,6,7,28] The human α1(VIII) and α2(VIII) genes are located on chromosomes 3 and 1, respectively.[3,28] The chicken type X gene was the first to be isolated among the short chain collagen genes.[8,28] The bovine,[29] mouse,[30–32] and human[32–34] type X genes were subsequently sequenced and characterized. The *COL10A1* gene has been localized to human chromosome 6q21–q22;[35] the mouse gene is on chromosome 10.[33]

Mutant phenotype/disease states

Mice carrying a transgene encoding an α1(X) collagen chain with an in-frame deletion in the triple-helical domain developed skeletal abnormalities within 2–3 weeks after birth.[36] Histology showed a decrease in the width of the zone of hypertrophic chondrocytes in growth plates, decreased bone formation in the metaphyses of long bones, and bone marrow abnormalities.[36] Craniofacial abnormalities were also noted.[37] Although an initial study of type X collagen null mice reported no phenotypic abnormalities,[38] subsequent studies[24] of mice that are homozygous for inactivated *Col10a1* alleles show distinct growth plate abnormalities. Recent analyses (O. Jacenko, personal communication) indicate that these are similar to those seen in the type X collagen transgenics. These findings in mice are consistent with the demonstration that Schmid metaphyseal chondrodysplasia (OMIM 156500), an autosomal dominant disorder in humans with short stature and growth plate abnormalities, is caused by mutations in the *COL10A1* gene. Since the first discovery of a frame-shift-causing deletion in the C-terminal NC1 domain,[39] a large number of mutations in the NC1 domain of type X collagen have been found in patients with Schmid metaphyseal chondrodysplasia.[40–44] Except for one report describing mis-sense mutations in the N-terminal NC2 domain,[45] all mutations have been in the NC1 domain. Studies on the molecular assembly and secretion of mutant polypeptides[20] support the initial hypothesis[39] that Schmid metaphyseal chondrodysplasia is caused by haploinsufficiency for type X collagen.

References

1. Sage, H. and Bornstein, P. (1995). In *Extracellular matrix. A practical approach* (ed. M. A. Haralson and J. R. Hassell), pp. 131–160. IRL Press, Oxford.
2. Mann, K., Jander, R., Korsching, E., Kuhn, K., and Rauterberg, J. (1990). *FEBS Lett.*, **273**, 168–72.
3. Muragaki, Y., Jacenko, O., Apte, S., Mattei, M. -G., Ninomiya, Y., and Olsen, B. R. (1991). *J. Biol. Chem.*, **266**, 7721–7.
4. Schmid, T. M., Cole, A. A., Chen, Q., Bonen, D. K., Luchene, L., and Linsenmayer, T. F. (1994). In *Extracellular matrix assembly and structure* (ed. P. D. Yurchenco, D. E. Birk, and R. P. Mecham), pp171–206. Academic Press.
5. Fernandez, M. S., Araya, M., and Arias, J. L. (1997). *Matrix Biol.*, **16**, 13–20.
6. Yamaguchi, N., Benya, P. D., van der Rest, M., and Ninomiya, Y. (1989). *J. Biol. Chem.*, **264**, 16022–9.
7. Yamaguchi, N., Mayne, R., and Ninomiya, Y. (1991). *J. Biol. Chem.*, **266**, 4508–13.
8. Ninomiya, Y., Gordon, M., van der Rest, M., Schmid, T., Linsenmayer, T., and Olsen, B. R. (1986). *J. Biol. Chem.*, **261**, 5041–50.
9. LuValle, P., Ninomiya, Y., Rosenblum, N. D., and Olsen, B. R. (1988). *J. Biol. Chem.*, **263**, 18378–85.
10. Kittelberger, R., Davis, P. F., Flyn, D. W., and Greenhill, N. S. (1990). *Connect. Tiss. Res.*, **24**, 303–18.
11. Sage, H. and Iruela-Arispe, M. -L. (1990). *Ann. NY Acad. Sci.*, **580**, 17–31.
12. Sawada, H., Konomi, H., and Hirokawa, K. (1990). *J. Cell Biol.*, **110**, 219–27.
13. Gordon, M. K. and Olsen, B. R. (1990). *Curr. Opin. Cell Biol.*, **2**, 833–8.
14. Kwan, A. P., Cummings, C. E., Chapman, J. A., and Grant, M. E. (1991). *J. Cell Biol.*, **114**, 597–604.
15. Gibson, G., Lin, D. L., Francki, K., Caterson, B., and Foster, B. (1996). *Bone*, **19**, 307–15.
16. LuValle, P., Hayashi, M., and Olsen, B. R. (1989). *Dev. Biol.*, **133**, 613–616.
17. Beier, F., Vornehm, S., Poschl, E., von der Mark, K., and Lammi, M. J. (1997). *J. Cell Biochem.*, **66**, 210–18.
18. Vortkamp, A., Lee, K., Lanske, B., Segre, G. V., Kronenberg, H. M., and Tabin, C. J. (1996). *Science*, **273**, 613–22.
19. Mayne, R., van der Rest, M., Bruckner, P., and Schmid, T. M. (1995). In *Extracellular matrix. A practical approach* (ed. M. A. Haralson and J. R. Hassell), pp. 73–97. IRL Press.
20. Chan, D., Weng, Y. M., Hocking, A. M., Golub, S., McQuillan, D. J., and Bateman, J. F. (1996). *J. Biol. Chem.*, **271**, 13566–72.
21. MacBeath, J. R., Kielty, C. M., and Shuttleworth, C. A. (1996). *Biochem. J.*, **319**, 993–8.
22. Korsching, E. and Rauterberg, J. (1995). *J. Immunol. Meth.*, **188**, 51–62.
23. Summers, T. A., Irwin, M. H., Mayne, R., and Balian, G. (1988). *J. Biol. Chem.*, **263**, 581–7.
24. Kwan, K. M., Pang, M. K. M., Zhou, S., Cowan, S. K., Kong, R. Y. C., Pfordte, T., *et al.* (1997). *J. Cell Biol.*, **136**, 459–71.
25. Girkontaite, I., Frischholz, S., Lammi, P., Wagner, K., Swoboda, B., Aigner, T., and Von der Mark, K. (1996). *Matrix Biol.*, **15**, 231–8.
26. Jacenko, O., Olsen, B. R., and LuValle, P. (1991).In *Critical reviews on eukaryotic gene expression* (ed. G. S. Stein and J. B. Lian), pp. 327–53. CRC Press.
27. Olsen, B. R. (1991). In *Articular cartilage and osteoarthritis* (ed. K. Kuettner and R. Schleyerbach), pp. 151–166. Raven Press.
28. Muragaki, Y., Mattei, M. -G., Yamaguchi, N., Olsen, B. R., and Ninomiya, Y. (1991). *Eur. J. Biochem.*, **197**, 615–22.
29. Thomas, J. T., Kwan, A. P. L., Grant, M. E., and Boot-Handford, R. P. (1991). *Biochem. J.*, **273**, 141–8.
30. Apte, S. S. and Olsen, B. R. (1993). *Matrix*, **13**, 165–79.
31. Elima, K., Eerola, I., Rosati, R., Metsaranta, M., Garofalo, S., Perala, M., *et al.* (1993). *Biochem. J.*, **189**, 247–53.
32. Beier, F., Eerola, I., Vuorio, E., LuValle, P., Reichenberger, E., Bertling, W., *et al.* (1996). *Matrix Biol.*, **15**, 415–22.
33. Apte, S. S., Seldin, M. F., Hayashi, M., and Olsen B. R. (1992). *Eur. J. Biochem.*, **206**, 217–24.

34. Reichenberger, E., Beier, F., LuValle, P., Olsen, B. R., von der Mark, K., and Bertling, W. M. (1992). *FEBS Lett.*, **311**, 305–10.
35. Apte, S., Mattei, M. -G., and Olsen, B. R. (1991). *FEBS Lett.*, **282**, 393–6.
36. Jacenko, O., LuValle, P. A., and Olsen, B. R. (1993). *Nature*, **365**, 56–61.
37. Chung, K. S., Jacenko, O., Boyle, P., Olsen, B. R., and Nishimura, I. (1997). *Dev. Dynamics*, **208**, 544–52.
38. Rosati, R., Horan, G. S., Pinero, G. J., Garofalo, S., Keene, D. R., Horton, W. A., *et al.* (1994). *Nature Genet.*, **8**, 129–35.
39. Warman, M. L., Abbott, M., Apte, S. S., Hefferone, T., Mslntosh, I., Cohn, D. H., *et al.* (1993). *Nature Genet.*, **5**, 79–82.
40. Mclntosh, I., Abbott, M., Warman, M. L., Olsen, B. R., and Francomano, C. A. (1994). *Hum. Mol. Genet.*, **3**, 303–7.
41. Wallis, G. A., Rash, B., Sweetman, W. A., and Boot-Handford, R. P. (1994). *Am. J. Hum. Genet.*, **54**, 169–78.
42. Mclntosh, I., Abbott, M., and Francomano, C. A. (1995). *Hum. Mutat.*, **5**, 121–5.
43. Bonaventure, J., Chaminade, F., and Maroteaux, P. (1995). *Hum. Genet.*, **96**, 58–64.
44. Wallis, G. A., Rash, B., Sykes, B., Bonaventure, J., Maroteaux, P., Zabel, B., *et al.* (1996). *J. Med. Genet.*, **33**, 450–7.
45. Ikegawa, S., Nakamura, K., Nagano, A., Haga, N., and Nakamura, Y. (1997). *Hum. Mutat.*, **9**, 131–5.
46. Sawada, H., Konomi, H., and Nagai, Y. (1984). *Eur. J. Cell Biol.*, **35**, 226–34.

Bjorn Reino Olsen
Department of Cell Biology,
Harvard Medical School, and
Harvard-Forsyth Department of Oral Biology,
Harvard School of Dental Medicine,
Boston, MA, USA

Yoshifumi Ninomiya
Department of Molecular Biology
and Biochemistry, Okayama
University Medical School, 2-5-1 Shikata-cho,
Okayama 700, Japan

Basement membrane collagens

Type IV collagen is the major collagenous component of basement membranes, forming a network structure with which other basement membrane components (laminin, nidogen, heparan sulphate proteoglycan) interact. Six distinct genes are identified as belonging to the type IV collagen gene family. They form three pairs of genes on three different chromosomes; within each pair the genes are arranged head-to-head and regulated by a bidirectional promoter.

Collagen molecules, composed of two α1(IV) and one α2(IV) chain, have long been recognized as a major component of basement membranes.[1–3] Each of the two chains is about 1700 amino acid residues long and contains at least three distinct domains: an N-terminal cysteine-rich (7S) domain, a central triple-helical domain, and a C-terminal non-triple-helical domain (NC1).[1] Type IV molecules assemble into a network which is quite different from the banded fibrils formed by fibrillar collagen types (Fig. 1). Within the network, separate molecules are covalently cross-linked within laterally associated 7S domains and associated by end-to-end interactions through their NC1 domains.[1,2,4] Lateral associations between the triple-helical domains also contribute to the network structure[1,2] (Fig. 1).

While type IV collagen molecules, composed of α1(IV) and α2(IV) chains, are broadly expressed, molecules containing combinations of four additional chains, α3(IV)–α6(IV), are important components of specialized basement membranes.[4–9] In the kidney glomerular basement membrane, molecules of α1(IV), α2(IV) are replaced by α3(IV), α4(IV), and α5(IV) chains as development proceeds.[10] The α6(IV) chain is present in epidermal basement membranes, around smooth muscle cells and adipocytes, and in Bowman's capsule and renal distal tubules, but absent from glomerular basement membranes.[11] The precise chain composition of triple-helical molecules assembled from the α3(IV)–α6(IV) chains is not entirely clear, but it is believed that α3(IV) and α4(IV) chains form heterotrimeric molecules. Also, analyses of bovine seminiferous tubule basement membranes have established a structural linkage between α3(IV) and α5(IV) chains.[12] α3(IV)/α4(IV) molecules and molecules containing α5(IV) chains may therefore be components of the same network. This helps to explain the observation that glomerular basement membranes from patients with Alport syndrome caused by mutations in α5(IV) (see below) are defective in α3(IV) and α4(IV).[13,14]

Purification and recombinant synthesis

Fragments of type IV collagen can be extracted from basement membranes with pepsin (resulting in triple-helical fragments) or with bacterial collagenase (resulting in non-triple-helical domains).[1,15] Intact type IV collagen composed of α1(IV) and α2(IV) chains can be isolated by acetic acid extraction of murine EHS-tumour tissue,[1] and is commercially available. Pepsinized material is also commercially available. The BioSupplyNet *Source Book* contains a good listing of suppliers. The NC1 domains of the human α1(IV)–α5(IV) chains have been synthesized in *E. coli*[16] and in insect cells.[17]

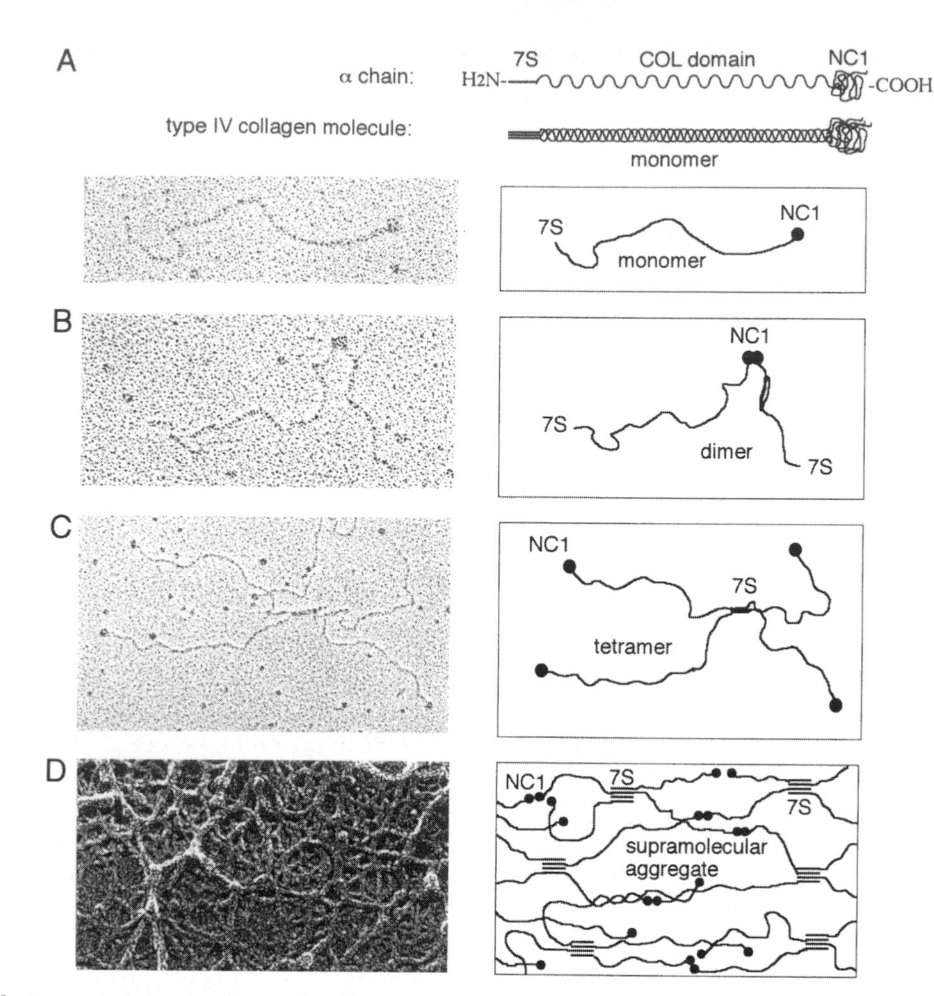

Figure 1. Electron micrographs of type IV collagen monomer, dimer, tetramer, and supramolecular aggregate after rotary shadowing, and schematic illustrations of the structure, and supramolecular assembly of type IV collagen. (a) Each α chain contributes to the 400 nm long triple-helical (COL) domain. This contains a number of interruptions in the Gly–X–Y-repeat sequence (not shown). A globular non-triple-helical domain (NC1) is located at the C-terminal end. The 7S domain is at the N-terminal end. Three α chains form a triple-helical molecule. The triple-helical molecules are the building blocks (monomers) of the basement membrane meshwork. Monomers associate into dimers that are stabilized by disulphide bonds between NC1 domains (b) or tetramers that are stabilized by disulphide bonds between the N termini (c). The supramolecular network is formed by assembly of dimers, and tetramers, and strengthened by lateral associations between molecules (d). Other basement membrane components such as laminin, proteoglycans, and nidogen are incorporated into the type IV collagen meshwork. (Courtesy of Dr Eijiro Adachi, School of Medicine, Kitazato University.)

Antibodies

A variety of antibodies are available.[1,10,18,19] These include antibodies against the 7S and NC1 domains, as well as antibodies against pepsin fragments.[1] Both polyclonal and monoclonal antibodies against type IV collagen are commercially available from several sources. FITC–anti-α5(IV) and Texas red–anti-α2(IV) antibodies for diagnosis of Alport syndrome are commercially available; a good listing of suppliers can be found in the BioSupplyNet *Source Book*.

Activities

Type IV collagen can interact with cells indirectly through laminin. Strong binding of type IV collagen to laminin is mediated by nidogen/entactin,[20,21] a glycoprotein of about 150 kDa which binds tightly to laminin[22,23] and has binding sites also for type IV collagen and cells.[3] In addition, direct low affinity interaction between laminin and type IV collagen is possible.[2,24] Type IV collagen also binds to heparin and heparan sulphate proteoglycan[2,25–27] and heparin can inhibit type IV collagen polymerization.[25]

Many cell types adhere to type IV collagen,[1,28] and peptides from within type IV sequences can inhibit this adhesion[29]. A major cell binding site in α1(IV)/α2(IV) heterotrimers is localized about 100 nm from the N terminus of the molecule; this triple-helical binding site interacts with $\alpha1\beta1$ and $\alpha1\beta2$ integrins on cells.[4] Recombinant fibulin-2 has a weak affinity for type IV collagen, but binding of nidogen to immobilized fibulin-2 allowed the formation of ternary complexes with collagen IV.[30]

Genes

Six distinct type IV collagen genes have been identified. These are organized in three sets *COL4A1/COL4A2*, *COL4A3/COL4A4*, and *COL4A5/COL4A6* which in humans are localized on three different chromosomes, 13, 2, and X, respectively (Fig. 2).[4] Within each set the genes are arranged head-to-head and their expression is regulated by bidirectional promoters between the genes. The 5' ends of the genes overlap; the transcription start sites are separated only by 130 bp in human[31,32] and mouse[33,34] α1(IV) and α2(IV) genes. The transcriptional regulation of *COL4A1/COL4A2* is well characterized.[35,36] Transcription of *COL4A6* seems to be controlled by two alternative promoters.[37] The complete primary structures of mouse and human α1(IV) and α2(IV) chains have been deduced from cDNA sequences, and mouse and human genomic clones have been extensively characterized.[38–41] The human α3(IV) and α4(IV) genes, located on chromosome 2, have also been well characterized.[42–44] The primary structure of the human α5(IV) and α6(IV) chains has been established by sequencing of cDNAs[8,9] and genomic clones.[37,45,46]

Type IV collagen genes have been characterized in several invertebrates, such as *Drosophila*,[47] *Caenorhabditis elegans*,[48–51] *Ascaris suum*,[52] and sea urchin, *Strongylocentrotus purpuratus*.[53,54] The protein encoded by the α1(IV) collagen gene in *Drosophila* is quite similar to the vertebrate type IV collagen chains, but the gene has fewer exons and is smaller than the corresponding vertebrate gene.[47] In *C. elegans* the *clb-1* and *clb-2* genes are homologous to the vertebrate α1(IV) and α2(IV) collagen genes;[48] however, these genes are located on separate chromosomes. Interestingly, mutations in the α1(IV) gene in *C. elegans* result in temperature-sensitive lethality during late embryogenesis.[49]

Mutant phenotype/disease states

Homozygous, *Col4a3*-null mice show a phenotype that is similar to Alport syndrome in humans.[55,56] Decreased glomerular filtration leads to uraemia, changes in the glomerular basement membrane causes proteinuria, and glomerulonephritis develops. Histological and molecular analyses indicate that the absence of α3(IV) chains causes loss of α4(IV) and α5(IV) chains from the glomerular basement membrane, and leads to increased levels of type VI collagen and perlecan, as well as retention of α1(IV) and α2(IV) chains.[55,56] Canine X-linked hereditary nephritis is an animal model of human X-linked Alport syndrome, and has been shown to be caused by a premature stop codon in the α5(IV) collagen chain.[57]

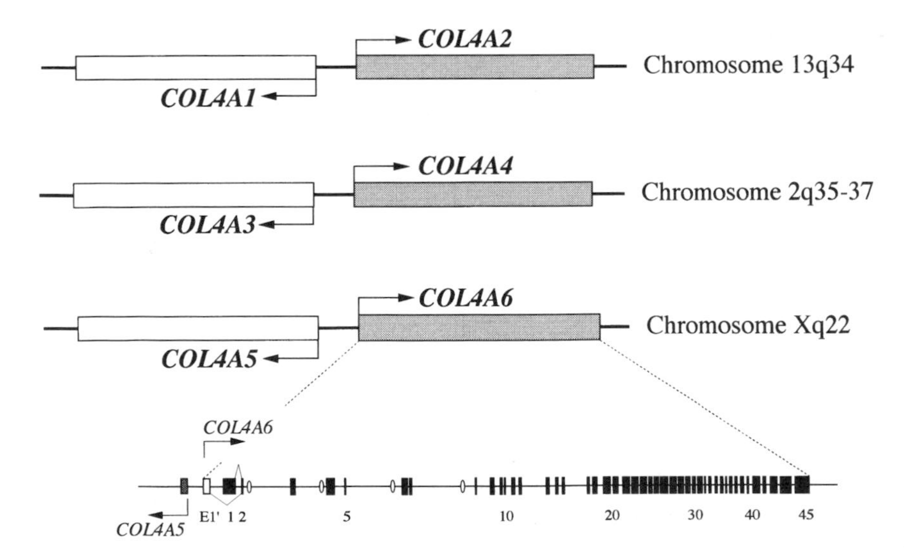

Figure 2. Illustration of the organization, and chromosomal locations of the human type IV collagen genes. The genes coding for the six type IV collagen chains are located in pairs in a head-to-head manner on three different chromosomes. The genes are depicted as rectangles, and the flanking regions by horizontal lines. For the *COL4A6* gene, the locations of exons are indicated by vertical bars, and introns by horizontal lines. The exons are numbered from the 5′ end of the gene. Introns of unknown sizes are indicated by ellipsoids. Note that the first two exons, E1′ and 1, of the gene are alternatively utilized, and spliced to exon 2 as indicated by free lines.

Three different human diseases directly involve type IV collagen genes or their translation products. Goodpasture syndrome (OMIM 233450), an autoimmune disorder causing progressive glomerulonephritis and pulmonary haemorrhage, is caused by antibodies that bind to an antigen (the Goodpasture antigen) in basement membranes of kidney glomeruli and lung alveoli.[58] The Goodpasture antigen is the NC1 domain of α3(IV) collagen chains.[59,60] Dimers of α3(IV) NC1 domains, isolated from bovine kidney, can induce an autoimmune response in rabbits similar to Goodpasture syndrome.[61] Mutations in *COL4A5*, located on the X chromosome, have been demonstrated in more than 200 cases of X-linked Alport familial nephritis (OMIM 301050).[62] In cases of autosomal recessive Alport syndrome (OMIM 203780), mutations have been identified in the α3(IV) and α4(IV) genes.[63] In rare cases of diffuse leiomyomatosis associated with Alport syndrome, large deletions involving the α5(IV) and α6(IV) genes have been found[64]. Autosomal dominant benign familial haematuria (OMIM 141200), characterized by thinning of the glomerular basement membrane and normal renal function, has been linked to the *COL4A3/COL4A4* locus and shown to be caused by a mutation in *COL4A4*.[65]

References

1. Glanville, R. W. (1987). In *Structure and function of collagen types* (ed. R. Mayne and R. E. Burgeson), pp. 43–79. Academic Press,
2. Yurchenco, P. D. and O'Rear, J. J. (1994). *Methods Enzymol.*, **145**, 489–518.
3. Timpl, R. (1996). *Curr. Opin. Cell Biol.*, **8**, 618–24.
4. Kuhn, K. (1994). *Matrix Biol.*, **14**, 439–45.
5. Hudson, B. G., Reeders S. T., and Tryggvason, K. (1993). *J. Biol. Chem.*, **268**, 26033–6.
6. Butkovski, R. J., Langeveld, J. P. M., Wieslander, J., Hamilton, J., and Hudson, B. G. (1987). *J. Biol. Chem.*, **262**, 7874–7.
7. Saus, J., Wieslander, J., Langeveld, J. P. M., Quinons, S., and Hudson, B. G. (1988). *J. Biol. Chem.*, **263**, 13374–80.
8. Hostikka, S. L., Eddy, R. L., Byers, M. G., Hoyhtya, M., Shows, T. B., and Tryggvason, K. (1990). *Proc. Natl Acad. Sci. USA*, **87**, 1606–10.
9. Oohashi, T., Sugimoto, M., Mattei, M. -G., and Ninomiya, Y. (1994). *J. Biol. Chem.*, **269**, 7520–6.
10. Miner, J. H., and Sanes, J. R. (1994). *J. Cell Biol.*, **127**, 879–91.
11. Ninomiya, Y., Kagawa, M., Iyama, K. -I., Naito, I., Kishiro, Y., Seyer, J. M., *et al.* (1995). *J. Cell Biol.*, **130**, 1219–29.
12. Kahsai, T. Z., Enders, G. C., Gunwar, S., Brunmark, C., Wieslander, J., Kalluri, R., *et al.* (1997). *J. Biol. Chem.*, **272**, 17023–32.
13. Peissel, B., Geng, L., Kalluri, R., Kashtan, C., Rennke, H. G., Gallo, G. R., *et al.* (1995). *J. Clin. Invest.*, **96**, 1948–57.
14. Hino, S., Takemura, T., Sado, Y., Kagawa, M., Oohashi, T., Ninomiya, Y., and Yoshioka, K. (1996). *Pediatr. Nephrol.*, **10**, 742–4.
15. Gunwar, S. and Hudson, B. G. (1995). In *Extracellular matrix macromolecules. A practical Approach* (ed. M. A. Haralson and J. R. Hassell), pp. 99–107. IRL Press, Oxford.
16. Neilson, E. G., Kalluri, R., Sun, M. J., Gunwar, S., Danoff, T., Mariyama, M., *et al.* (1993). *J. Biol. Chem.*, **268**, 8402–5.
17. Turner, N., Forstova, J., Rees, A., Pusey, C. D., and Mason, P. J. (1994). *J. Biol. Chem.*, **269**, 17141–5.
18. Sado, Y., Kagawa, M., Kishiro, Y., Sugihara, K., Naito, I., Seyer, J. M., *et al.* (1995). *Histochem. Cell Biol.*, **104**, 267–75.
19. Saus, J., Wieslander, J., Langeveld, J. P., Quinones, S., and Hudson, B. G. (1988). *J. Biol. Chem.*, **263**, 13374–80.
20. Timpl, R., Dziadek, M., Fujiwara, S., Nowack, H., and Wick, G. (1983). *Eur. J. Biochem.*, **137**, 455–65.
21. Carlin, B., Jaffe, R., Bender, B., and Chung, A. E. (1981). *J. Biol. Chem.*, **256**, 5209–14.
22. Paulsson, M., Aumailley, M., Deutzman R., Timpl, R., Beck, K., and Engel, J. (1987). *Eur. J. Biochem.*, **166**, 11–19.
23. Poschl, E., Mayer, U., Stetefeld, J., Baumgartner, R., Holak, T. A., Huber, R., and Timpl, R. (1996). *EMBO J.*, **15**, 5154–9.
24. Charonis, A. S., Tsilibary, E. C., Yurchenko, P. D., and Furthmayr, H. (1985). *J. Cell Biol.*, **100**, 1848–53.
25. Tsilibary, E. C., Koliakos, G. G., Charonis, A. S., Vogel, A. M., Reger, L. A., and Furcht, L. T. (1988). *J. Biol. Chem.*, **263**, 19112–8.
26. Koliakos, G. G., Louzi-Kolialos, K., Furcht, L. T., Reger, L. A., and Tsilibary, E. C. (1989). *J. Biol. Chem.*, **264**, 2313–23.
27. Fujiwara, S., Wiedeman, H., Timpl, R., Lustig, A., and Engel, J. (1984). *Eur. J. Biochem.*, **143**, 145–57.
28. Aumailley, M. and Timpl, R. (1986). *J. Cell Biol.*, **103**, 1569–75.
29. Tsilibary, E. C., Reger, L. A., Vogel, A. M., Koliakos, G. G., Anderson, S. S., Charonis, A. S., *et al.* (1990). *J. Cell Biol.*, **111**, 1583–91.
30. Sasaki, T., Gohring, W., Pan, T. C., Chu, M. L., and Timpl, R. (1995). *J. Mol. Biol.*, **254**, 892–9.
31. Poschl, E., Pollner, R., and Kuhn, K. (1988). *EMBO J.*, **7**, 2687–95.
32. Soininen, R., Houtari, M., Hostikka, S. L., Prockop, D. J., and Tryggvason, K. (1988). *J. Biol. Chem.*, **263**, 17217–20.
33. Burbelo, P. D., Martin, G. R., and Yamada, Y. (1988). *Proc. Natl Acad. Sci. USA*, **85**, 9679–82.
34. Kaytes, P., Wood, L., Theriault, N., Kurkinen, M., and Vogeli, G. (1988). *J. Biol. Chem.*, **263**, 19274–7.
35. Fischer, G., Schmidt, C., Optitz, J., Cully, Z., Kuhn, K., and Poschl, E. (1993). *Biochem. J.*, **292**, 687–95.
36. Genersch, E., Eckerskorn, C., Lottspeich, F., Herzog, C., Kuhn, K., and Poschl, E. (1995). *EMBO J.*, **14**, 791–800.
37. Sugimoto, M., Oohashi, T., and Ninomiya, Y. (1994). *Proc. Natl Acad. Sci. USA*, **91**, 11679–83.
38. Vuorio, E. and deCrombrugghe, B. (1990). *Annu. Rev. Biochem.*, **59**, 837–72.
39. Jacenko, O., Olsen, B. R., and LuValle, P. (1991). In *Critical reviews on eukaryotic gene expression* (ed. G. S. Stein, J. L. Stein, and J. B. Lian), pp. 327–353. CRC Press,
40. Sandell, L. J. and Boyd, C. D. (1990). In *Extracellular matrix genes* (ed. L. J. Sandell and C. D. Boyd), pp. 1–56. Academic Press,
41. Blumberg, B. and Kurkinen, M. (1990). In *Extracellular matrix genes* (ed. L. J. Sandell and C. D. Boyd), pp. 115–35. Academic Press,
42. Quinones, S., Bernal, D., Carcia-Sogo, M., Elena, S. F., and Saus, J. (1992). *J. Biol. Chem.*, **267**, 19780–4.
43. Sugimoto, M., Oohashi, T., Yoshioka, H., Matsuo, N., and Ninomiya, Y. (1993). *FEBS Lett.*, **330**, 122–8.
44. Heikkila, P. and Soininen, R. (1996). In *Molecular pathology and genetics of Alport syndrome* (ed. K. Tryggvason), pp. 105–129. Karger,
45. Barker, D., Hostikka, S. L., Chou, J., Chow, L. T., Olifant, A. R., Gerken, S. C., *et al.* (1990). *Science*, **248**, 1224–7.
46. Oohashi T., Ueki Y., Sugimoto M., and Ninomiya, Y. (1995). *J. Biol. Chem.*, **270**, 26863–7.
47. Blumberg, B., MacKrell, A. J., and Fessler, J. H. (1988). *J. Biol. Chem.*, **263**, 18328–37.

48. Guo, X. D. and Kramer, J. M. (1989). *J. Biol. Chem.*, **264**, 17574–82.
49. Guo, X. D, Johnson, J. J., and Kramer, J. M (1991). *Nature*, **349**, 707–9.
50. Sibley, M. H., Johnson, J. J., Mello, C. C., and Kramer, J. M. (1993). *J. Cell Biol.*, **123**, 255–64.
51. Sibley, M. H., Graham, P. L., von Mende, N., and Kramer, J. M. (1994). *EMBO J.*, **13**, 3278–85.
52. Pettitt, J. and Kingston, I. B. (1991). *J. Biol. Chem.*, **266**, 16149–56.
53. Venkatesan, M., de Pablo, F., Vogeli, G., and Simpson, R. T. (1986). *Proc. Natl Acad. Sci. USA*, **83**, 3351–5.
54. Exposito, J. Y., D'Alessio, M., Di Liberto, M., and Ramirez, F. (1993). *J. Biol. Chem.*, **268**, 5249–54.
55. Miner, J. H. and Sanes, J. R. (1996). *J. Cell Biol.*, **135**, 1403–13.
56. Cosgrove, D., Meehan, D. T., Grunkemeyer, J. A., Kornak, J. M., Sayers, R., Hunter, W. J., and Samuelson, G. C. (1996). *Genes Dev.*, **10**, 2981–92.
57. Zheng, K., Thorner, P. S., Marrano, P., Baumal, R., and McInnes, R. R. (1994). *Proc. Natl Acad. Sci. USA*, **91**, 3989–93.
58. Hudson, B. G., Kalluri, R., Gunwar, S., Noelken, M. E., Mariyama, M., and Reeders, S. T. (1993). *Kidney Int.*, **43**, 135–9.
59. Kalluri, R., Gunwar, S., Reeders, S. T., Morrison, K. C., Mariyama, M., Ebner, K. E., *et al.* (1991). *J. Biol. Chem.*, **266**, 24018–24.
60. Kalluri, R., Sun, M. J., Hudson, B. G., and Neilson, E. G. (1996). *J. Biol. Chem.*, **271**, 9062–8.
61. Kalluri, R., Gattone, V. H., 2nd., Noelken, M. E., and Hudson, B. G. (1994). *Proc. Natl Acad. Sci. USA*, **91**, 6201–5.
62. Tryggvason, K. (ed.) (1996). *Molecular pathology and genetics of Alport syndrome*, pp. 1–204. Karger, Basel.
63. Mochizuki, T., Lemmink, H. H., Mariyama, M., Antignac, C., Gubler, M. C., Pirson, Y., *et al.* (1994). *Nature Genet.*, **8**, 77–82.
64. Zhou, J., Mochizuki, T., Smeets, H., Antignac, C., Laurila, P., Paepe, A., and Tryggvason, K. (1993). *Science*, **261**, 1167–9.
65. Lemmink, H. H., Nillesen, W. N., Mochizuki, T., Schroder, C. H., Brenner, H. G., van Oost, B. A., *et al.* (1996). *J. Clin. Invest.*, **98**, 1114–18.

■ *Bjorn Reino Olsen*
Department of Cell Biology,
Harvard Medical School, and
Harvard-Forsyth Department of Oral Biology,
Harvard School of Dental Medicine,
Boston, MA, USA

■ *Yoshifumi Ninomiya*
Department of Molecular Biology
and Biochemistry, Okayama
University Medical School,
2-5-1 Shikata-cho, Okayama 700,
Japan

Multiplexins

The non-fibrillar collagens type XV and type XVIII are broadly expressed in many tissues, but are present at particularly high levels in internal organs. They contain multiple short triple-helical domains, separated and flanked by non-triple-helical regions. Based on a considerable degree of similarity in some of their structural domains, they are classified as members of a novel subfamily of collagens called multiplexins.

The two members of this class of proteins, types XV and XVIII collagen, have been given the name multiplexins[1] because they both contain multiple-triple-helix domains with interruptions. They share considerable homology at the amino acid level, but are sufficiently different to rule out the possibility that they could form mixed heterotrimers like types V and XI fibrillar collagens.

First isolated by cross-hybridization during screening of a placental cDNA library,[2] type XV collagen has now been completely characterized at the nucleotide level.[3,4] α1(XV) collagen chains contain nine triple-helical (COL) domains that are separated and flanked by non-triple-helical (NC) regions (Fig. 1). The N-terminal region (NC10*) consists of 530 amino acid residues and is almost as large as the triple-helical region; the C-terminal non-triple-helical region (NC1*) is somewhat smaller (256 amino acid residues).

The α1(XVIII) collagen chain contains 10 triple-helical domains, separated and flanked by non-triple-helical sequences[1,5] (Fig. 1). A comparison with α1(XV) shows a striking similarity in size between the six most C-terminal triple-helical domains of the two collagens.[1,6] Also, at the amino acid level there is over 60 per cent identity between the carboxyl half of the 315 residue NC1 domain of α1(XVIII) and the corresponding portion of α1(XV).[7] Both collagens contain four cysteinyl residues in this region and may therefore have a similar tertiary structure. Another region of homology is a 200 residue sequence at the amino end of the short variant (see

* For type IV basement membrane collagens, FACIT collagens, and short-chain collagens the numbering of triple-helical and non-triple-helical domains starts by counting from the C terminus of the molecule. In keeping with this tradition, this numbering system has been used also for types XV[3] and XVIII[1] collagen and is followed here. Numbering the domains from the N terminus has been suggested for α1(XV) and α1(XVIII) collagens,[5,7] but serves only to create confusion and should be avoided.

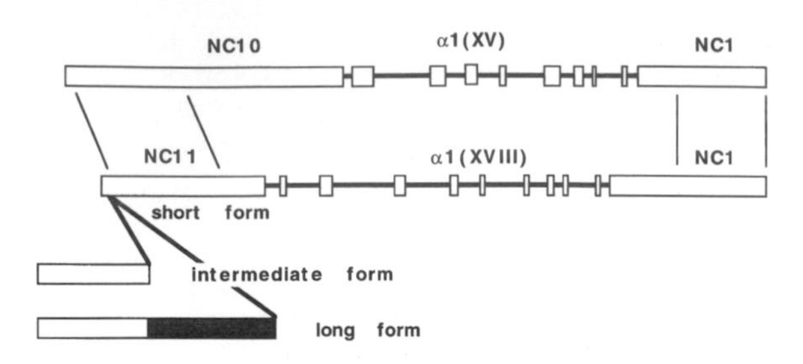

Figure 1. Diagram showing the domain structures of types XV and XVIII collagen chains. Non-triple-helical domains are indicated by rectangles; triple-helical domains are indicated by a solid line. Thin vertical lines between the two chains delineate regions of homology in the NC10/NC11, and NC1 regions of the two chains. Three variant transcripts give rise to three different α1(XVIII) chains with different NC11 domains. These are indicated as the short, intermediate, and long forms. The frizzled-like region in the long form is indicated by the stippled rectangle.

below) of α1(XVIII) and the corresponding region of α1(XV).[3] This sequence is 45 per cent identical between the two collagens and is homologous with a region of thrombospondin-1, the fibrillar procollagens V and XI, and members of the FACIT group.[7]

Northern blot analyses show that types XV and XVIII collagens are expressed in several major internal organs and in several cell types, including fibroblasts and endothelial cells.[1,3,5,8] There is considerable overlap between the expression patterns of the two collagens, but also distinct differences. For example, while both transcripts are found in the kidney, α1(XV) transcripts are low in lung and liver while those of α1(XVIII) are very high, particularly in liver. Immunohistochemical studies demonstrate a wide distribution of the two collagens, with a particular concentration in basement membrane regions.[9,10]

Purification and recombinant synthesis

A portion of the NC1 domain of type XV collagen has been expressed as a recombinant protein in bacteria and used for generation of specific polyclonal antisera.[14] Fragments of type XVIII collagen have been produced both in bacteria and in insect cells.[15]

Antibodies

Antibodies have been generated against both types XV and XVIII collagens and used for Western blotting and immunohistochemical studies.[2,9,10,14]

Activities

The supramolecular assemblies and functions of multiplexins have not been characterized. Of considerable interest, however, is the finding that a 20 kDa angiogenesis inhibitor from a murine haemangioendothelioma, called endostatin, represents a fragment of the carboxyl region of the NC1 domain of α1(XVIII) collagen chains.[15] This portion of type XVIII collagen, produced as a recombinant peptide and injected into mice, causes nearly complete suppression of tumour-induced angiogenesis and tumour growth.[15]

Genes

Genomic and cDNA sequences for mouse and human α1(XV) and α1(XVIII) collagens are available.[1–6,9,12–14,16] Of interest is that the α1(XVIII) collagen gene contains two alternative promoters and that transcripts from one of these promoters can be alternatively spliced. This gives rise to three alternative α1(XVIII) transcripts that encode α1(XVIII) collagen chains with very different N-terminal (NC11) non-triple-helical domains[9,12] (Fig. 1). The shortest variant, transcribed from the most 5′ promoter, contains mostly the thrombospondin-1 homology region. An intermediate-sized variant, transcribed from the most 3′ promoter, contains a different signal peptide and an additional region of about 200 residues that is rich in acidic amino acid residues. The longest variant, also transcribed from the most 3′ promoter, contains in addition a 250 residue region inserted between the acidic domain and the thrombospondin-1 homology region. This inserted region contains 10 cysteinyl residues and shows a striking similarity to the extracellular ligand-binding domain of frizzled receptors, with a frizzled-like distribution of the cysteines.[9,12]

References

1. Oh, S. P., Kamagata, Y., Muragaki, Y., Timmons, S., Ooshima, A., and Olsen, B. R. (1994). *Proc. Natl Acad. Sci. USA*, **91**, 4229–33.
2. Myers, J. C., Kivirikko, S., Gordon, M. K., Dion, A. S., and Pihlajaniemi, T. (1992). *Proc. Natl Acad. Sci. USA*, **89**, 10144–8.
3. Muragaki, Y., Abe, N., Ninomiya, Y., Olsen, B. R., and Ooshima, A. (1994). *J. Biol. Chem.*, **269**, 4042–6.
4. Kivirikko, S., Heinamaki, P., Rehn, M., Honkanen, N., Myers, J. C., and Pihlajaniemi, T. (1994). *J. Biol. Chem.*, **269**, 4773–9.
5. Rehn, M. and Pihlajaniemi, T. (1994). *Proc. Natl Acad. Sci. USA*, **91**, 4234–8.
6. Rehn, M., Hintikka, E., and Pihlajaniemi, T. (1994). *J. Biol. Chem.*, **269**, 13929–35.
7. Pihlajaniemi, T. and Rehn, M. (1995). *Prog. Nucl. Acids Res. Mol. Biol.*, **50**, 225–62.
8. Kivirikko, S., Saarela, J., Myers, J. C., Autio-Harmainen, H., and Pihlajaniemi, T. (1995). *Am. J. Pathol.*, **147**, 1500–9.
9. Muragaki, Y., Timmons, S., Griffith, C. M., Oh, S. P., Fadel, B., Quertermous, T., and Olsen, B. R. (1995). *Proc. Natl Acad. Sci. USA*, **92**, 8763–7.
10. Hagg, P. M., Hagg, P. O., Peltonen, S., Autio-Harmainen, H., and Pihlajaniemi, T. (1997). *Am. J. Pathol.*, **150**, 2075–86.
11. Oh, S. P., Warman, M. L., Seldin, M. F., Cheng, S. D., Knoll, J. H., Timmons, S., and Olsen, B. R. (1994). *Genomics*, **19**, 494–9.

12. Rehn, M., and Pihlajaniemi, T. (1995). *J. Biol. Chem.*, **270**, 4705–11.
13. Rehn, M., Hintikka, E., and Pihlajaniemi, T. (1996). *Genomics*, **32**, 436–46.
14. Myers, J. C., Dion, A. S., Abraham, V., and Amenta, P. S. (1996). *Cell Tissue Res.*, **286**, 493–505.
15. O'Reilly, M. S., Boehm, T., Shing, Y., Fukai, N., Vasios, G., Lane, W. S., *et al.* (1997). *Cell*, **88**, 277–85.
16. Hägg, P. M., Muona, A., Liétard, J., Kivirikko, S., and Pihlajaniemi, T. (1998). *J. Biol. Chem.*, **273**, 17824–31.

Bjorn Reino Olsen
Department of Cell Biology, Harvard Medical School, and Harvard-Forsyth Department of Oral Biology, Harvard School of Dental Medicine, Boston, MA, USA

Yoshifumi Ninomiya
Department of Molecular Biology and Biochemistry, Okayama University Medical School, 2-5-1 Shikata-cho, Okayama 700, Japan

Collagens with transmembrane domains – MACITs

Types XIII and XVII collagen are cell surface molecules with multiple extracellular triple-helical domains, connected to a cytoplasmic region by a transmembrane segment. They represent a new class of cellular adhesion molecules by which cells are connected to extracellular matrix. Type XIII collagen is expressed on the surface of fibroblasts, while type XVII collagen is a component of hemidesmosomes in epithelial cells.

Although the overall structures of types XIII and XVII collagen are quite different, it is reasonable to include them in a separate group of collagenous proteins, based on their membrane association. In analogy with the term FACIT for fibril-associated collagens, the type XIII/XVII group has therefore been designated the MACIT (membrane-associated collagens with interrupted triple-helices) group.[1] As discussed in the overview of the collagen superfamily (pp. 380–382), one can also include the macrophage scavenger receptors[2] and MARCO[3] in this group of proteins (Fig. 1).

Type XIII collagen, initially identified by cross-hybridization during screening of a human cDNA library with a mouse type IV collagen probe,[4] is encoded by a gene that gives rise to a number of transcripts by alternative splicing.[5] These transcripts encode a polypeptide chain with three triple-helical domains, separated by non-triple-helical regions. In the different splice variants the length of the N- and C-terminal triple-helical domains varies considerably. It is believed that type XIII molecules are homotrimers; how the synthesis of the different variants can be reconciled with trimerization and the proper folding of triple-helical domains is not clear.[1] Type XIII collagen is widely expressed in human tissues and cell lines. Western blots of extracts of human HT-1080 fibrosarcoma cells show the presence of bands of expected size (about 67 and 54 kDa),[4] and these antibodies show localization at discrete sites along the cell surface (Fig. 2). By *in situ* hybridization, α1(XIII) transcripts have been found in epidermis and hair follicles, muscle, intestinal wall, cartilage, and bone.[6] In placenta, stromal cells of the villi, endothelial cells of developing capillaries, and cells of the cytotrophoblastic columns are all positive for type XIII collagen transcripts.[7]

Type XVII collagen is a component of hemidesmosomes and represents the autoantigen BPAG2, causing an acquired blistering skin disease, bullous pemphigoid.[8,9] Sequencing of chicken, mouse, and human α1(XVII) collagen cDNA shows that it contains a large cytoplasmic N-terminal domain (almost 500 amino acid residues) with an extracellular triple-helical region consisting of eight heptad repeats, likely to form a coiled-coil trimer, and 15 short triple-helical domains separated by non-triple-helical regions.[10–13] Rotary shadowing of affinity-purified type XVII collagen isolated from bovine mammary gland epithelial cells showed a structure composed of a globular head, a central rod, and a flexible tail.[14] It is likely that the globular domain is the cytoplasmic region, the central rod is the heptad repeat region, and the flexible tail represents the interrupted triple-helical domain. Immunoelectron microscopy shows that type XVII collagen is a hemidesmosomal component with the extracellular domains localized in the anchoring filaments between the cell surface and the lamina densa of the underlying basement membrane.[15]

Purification and recombinant synthesis

Type XIII collagen has not been isolated as a protein from tissues, but type XVII has been isolated from primary human keratinocytes, HaCaT cells, and bovine corneal

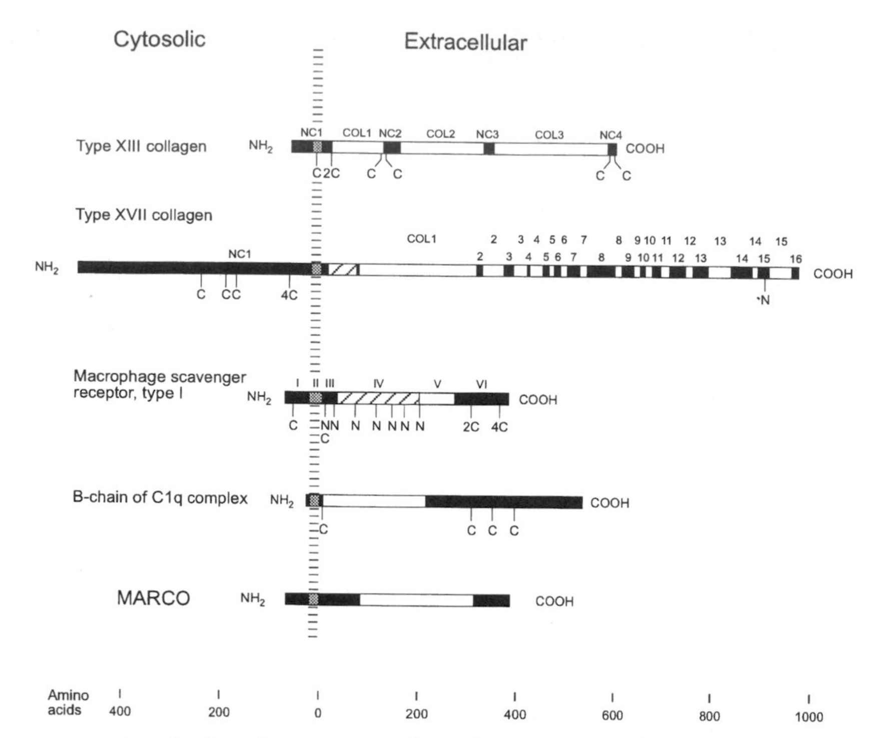

Figure 1. Diagrams comparing the domain structures of membrane-associated polypeptides containing collagenous sequences. The numbering of non-triple-helical, and triple-helical domains is shown above the corresponding polypeptide. Filled rectangles indicate non-triple-helical domains; open rectangles indicate triple-helical domains. The plasma membrane is indicated by the vertical dashed line. C, cysteine residue; N, potential *N*-glycosylation site. The scale below is given in amino acid residues, counted from the transmembrane domain. (Modified from Pihlajaniemi and Rehn (1995),[1] courtesy of Dr T. Pihlajaniemi.)

epithelial cells.[16–18] Mouse Balb/K keratinocytes were transfected with a full length type XVII collagen cDNA and shown to assemble as a triple-helical homotrimer.[17] A portion of the extracellular domain of type XVII collagen has been expressed as a recombinant protein in insect cells.[15]

Antibodies

Antipeptide antibodies are being used to study the expression and cellular localization of type XIII collagen.[1] About half the sera from patients with bullous pemphigoid and most sera from patients with herpes gestationis contain autoantibodies against type XVII collagen. Monoclonal antibodies recognizing both the extracellular and intracellular domains are available and have been used for immunofluorescence, Western blotting, and immunoelectron microscopy.[16]

Activities

Type XIII collagen is expressed at focal adhesion sites in cultured fibroblasts and may therefore represent a matrix-binding anchoring molecule at such sites (Fig. 2). Type XVII is part of the multiprotein hemidesmosome complex that mediates adhesion of epithelial cells to the underlying basement membrane.[19] Transfection experiments with various mutant cDNAs suggest that the localization of type XVII collagen in the hemidesmosome is mediated by the cytoplasmic domain and requires interaction with sequences in the cytoplasmic domain of the β4 integrin subunit.[20]

Genes

cDNA and genomic clones for mouse and human α1(XIII) collagen are available.[21–24] Alternative splicing gives rise to multiple transcripts of 2.5–2.8 kb.[5,25] For type XVII collagen cloning of chicken, mouse, and human cDNAs have been reported.[10–13] The entire human *COL17A1* gene has also been characterized.[26]

Mutant phenotype/disease states

No mutations in type XIII collagen are known. In contrast, several mutations in COL17A1 have been demonstrated in patients with generalized atrophic benign epidermolysis bullosa (OMIM 226650).[27–29] This is a rare non-lethal

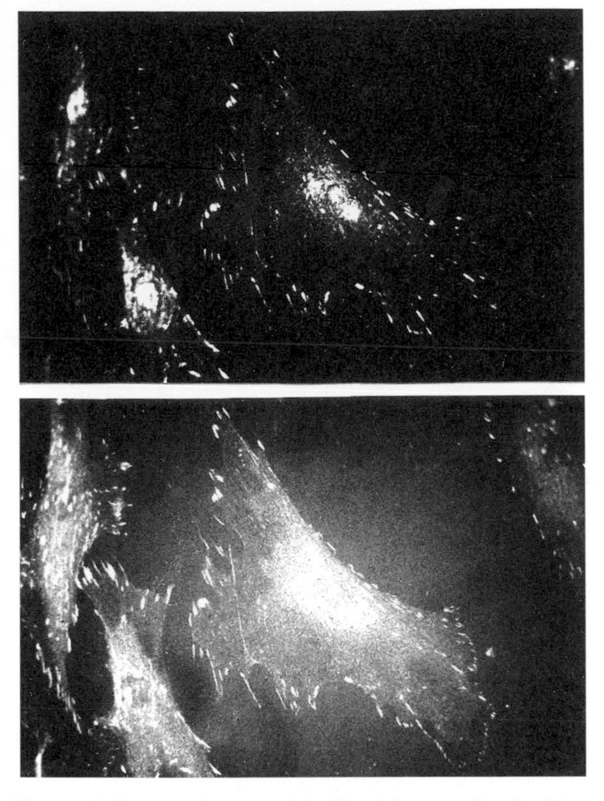

Figure 2. Primary human skin fibroblasts stained with rabbit polyclonal antibodies against type XIII collagen (top), and with monoclonal antibody against vinculin (bottom).

variant of junctional epidermolysis bullosa, usually inherited as an autosomal recessive disorder, that can be caused by mutations in the $\beta3$ chain of laminin-5[30] in addition to mutations in $\alpha1$(XVII) collagen. Most of the type XVII collagen mutations described have resulted in premature termination codons in both alleles within the largest triple-helical subdomain. In a Finnish family, the proband was a compound heterozygote, with one allele containing a 5 bp deletion and the other a nonsense mutation.[26] Homozygosity for a mis-sense mutation in type XVII collagen has also been demonstrated in a patient with the localisata variant of junctional epidermolysis bullosa.[31] Detailed and updated information is available through the OMIM database.

References

1. Pihlajaniemi, T. and Rehn, M. (1995). *Prog. Nucl. Acid. Res. Mol. Biol.*, **50**, 225–62.
2. Kodama, T., Freeman, M., Rohrer, L., Zabrecky, J., Matsudaira, P., and Krieger, M. (1990). *Nature*, **343**, 531–5.
3. Elomaa, O., Kangas, M., Sahlberg, C., Tuukkanen, J., Sormunen, R., Liakka, A., *et al.* (1995). *Cell*, **80**, 603–9.
4. Pihlajaniemi, T., Myllyla, R., Seyer, J., Kurkinen, M., and Prockop, D. J. (1987). *Proc. Natl Acad. Sci. USA*, **84**, 940–944.
5. Peltonen, S., Rehn, M., and Pihlajaniemi, T. (1997). *DNA Cell Biol.*, **16**, 227–34.
6. Sandberg, M., Tamminen, M., Hirvonen, H., Vuorio, E., and Pihlajaniemi, T. (1989). *J. Cell Biol.*, **109**, 1371–9.
7. Juvonen, M., Pihlajaniemi, T., and Autio-Harmainen, H. (1993). *Lab. Invest.*, **69**, 541–51.
8. Diaz, L. A., Ratrie, H. D., Saunders, W. S., Futamura, S., Squiquera, H. L., Anhalt, G. J., and Giudice, G. J. (1990). *J. Clin. Invest.*, **86**, 1088–94.
9. Giudice, G. J., Squiquera, H. L., Elias, P. M., and Diaz, L. A. (1991). *J. Clin. Invest.*, **87**, 734–8.
10. Gordon, M. K., Fitch, J. M., Foley, J. W., Gerecke, D. R., Linsenmayer, C., Birk, D. E., and Linsenmayer, T. F. (1997). *Invest. Ophthalmol. Vis. Sci.*, **38**, 153–66.
11. Li, K., Tamai, K., Tan, E. M., and Uitto, J. (1993). *J. Biol. Chem.*, **268**, 8825–34.
12. Giudice, G. J., Emery, D. J., and Diaz, L. A. (1992). *J. Invest. Dermatol.*, **99**, 243–50.
13. Hopkinson, S. B., Riddelle, K. S., and Jones, J. C. (1992). *J. Invest. Dermatol.*, **99**, 264–70.
14. Hirako, Y., Usukura, J., Nishizawa, Y., and Owaribe, K. (1996). *J. Biol. Chem.*, **271**, 13739–45.
15. Masunaga, T., Shimizu, H., Yee, C., Borradori, L., Lazarova, Z., Nishikawa, T., and Yancey, K. B. (1997). *J. Invest. Dermatol.*, **109**, 200–6.
16. Nishizawa, Y., Uematsu, J., and Owaribe, K. (1993). *J. Biochem.*, **113**, 493–501.
17. Limardo, M., Arffman, A., Aho, S., and Uitoo, J. (1996). *J. Invest. Dermatol.*, **106**, 860.
18. Schumann, H. and Bruckner-Tuderman, L. (1996). *J. Invest. Dermatol.*, **106**, 821.
19. Borradori, L. and Sonnenberg, A. (1996). *Curr. Opin. Cell Biol.*, **8**, 647–56.
20. Borradori, L., Koch, P. J., Niessen, C. M., Erkeland, S., van Leusden, M. R., and Sonnenberg, A. (1997). *J. Cell. Biol.*, **136**, 1333–47.
21. Rehn, M. and Pihlajaniemi, T. (1993). *Matrix Coll. Rel. Res.*, **13**, 12.
22. Pihlajaniemi, T. and Tamminen, M. (1990). *J. Biol. Chem.*, **265**, 16922–8.
23. Tikka, L., Pihlajaniemi, T., Henttu, P., Prockop, D. J., and Tryggvason, K. (1988). *Proc. Natl Acad. Sci. USA*, **85**, 7491–5.
24. Tikka, L., Elomaa, O., Pihlajaniemi, T., and Tryggvason, K. (1991). *J. Biol. Chem.*, **266**, 17713–19.
25. Juvonen, M., Sandberg, M., and Pihlajaniemi, T. (1992). *J. Biol. Chem.*, **267**, 24700–7.
26. Gatalica, B., Pulkkinen, L., Li, K., Kuokkanen, K., Ryynanen, M., McGrath, J. A., and Uitto, J. (1997). *Am. J. Hum. Genet.*, **60**, 352–65.
27. McGrath, J. A., Gatalica, B., Christiano, A. M., Li, K., Owaribe, K., McMillan, J. R., *et al.* (1995). *Nature Genet.*, **11**, 83–6.
28. McGrath, J. A., Darling, T., Gatalica, B., Pohla-Gubo, G., Hintner, H., Christiano, A. M., *et al.* (1996). *J. Invest. Dermatol.*, **106**, 771–4.
29. McGrath, J. A., Gatalica, B., Li, K., Dunnill, M. G., McMillan, J. R., Christiano, A. M., *et al.* (1996). *Am. J. Pathol.*, **148**, 1787–96.
30. McGrath, J. A., Pulkkinen, L., Christiano, A. M., Leigh, I. M., Eady, R. A., and Uitto, J. (1995). *J. Invest. Dermatol.*, **104**, 467–74.
31. Schumann, H., Hammami-Hauasli, N., Pulkkinen, L., Mauviel, A., Kuster, W., Luthi, U., *et al.* (1997). *Am. J. Hum. Genet.*, **60**, 1344–53.

Bjorn Reino Olsen
Department of Cell Biology, Harvard Medical School, and Harvard-Forsyth Department of Oral Biology, Harvard School of Dental Medicine, Boston, MA, USA

Yoshifumi Ninomiya
Department of Molecular Biology and Biochemistry, Okayama University Medical School, 2-5-1 Shikata-cho, Okayama 700, Japan

Other collagens

This is a heterogeneous group of proteins that on a genetic basis do not belong to one of the defined collagen families. They are discussed here as a group only for practical reasons. As the human genome project moves forward it is possible that identification of additional collagen genes will allow classification of the two collagens discussed below as members of their own distinct families.

Type VI collagen

Type VI collagen is broadly expressed in different tissues as the major component of beaded microfibrils.[1] Each type VI molecule appears in the electron microscope as a 105 nm long triple-helical rod flanked by two globular domains,[2] and contains three different polypeptide subunits α1(VI), α2(VI), and α3(VI). The three chains have apparent molecular masses of about 140, 130 and 250–350 kDa respectively.[1–3]

These heterotrimeric type VI molecules form disulphide bonded dimers and tetramers. The tetramers associate end-to-end and generate microfibrils, which have a characteristic periodicity of 100 nm.[4–8] The complete primary structures of the α1(VI), α2(VI), and the α3(VI) chains have been determined from amino acid and cDNA sequencing.[9–19] The chains contain a central, relatively short triple-helical domain of 335–336 amino acid residues. All three chains contain a C-terminal non-triple-helical domain composed of two repeats of a 200 residue long segment that is homologous to the A domains of von Willebrand factor. The α3(VI) chain contains in addition a proline-rich region showing homology with domains in salivary proteins, a fibronectin type 3 repeat-like domain, and a domain that is similar to a region found in serine protease inhibitors of the Kunitz type.[14,15] In the N-terminal region of the α1(VI) and α2(VI) chains there is a single 200 residue long von Willebrand factor A homology domain, while the α3(VI) chain contains up to nine such repeats in this region. Binding sites for type I collagen have been ascribed to the von Willebrand factor A region,[20] and it is possible that the homologous domains in type VI collagen have collagen binding properties as well. It is also possible that type VI collagen has a cell adhesion function;[21] several Arg–Gly–Asp sequences are found in the primary sequence of the type VI collagen subunits and experiments with neural crest cells suggest that regions in the N- and C-terminal globular domains play a role in cell adhesion and migration.[22]

Alternative splicing of exons in the 5′ region of the α3(VI) collagen gene leads to the formation of several transcripts encoding polypeptides with N-terminal globular domains of different size.[18,23–26] Splice variations in the 3′ region of the α2(VI) gene affecting the structure of the C-terminal globular domain have also been described.[10,17,27]

Purification and recombinant synthesis

The triple-helical portion of type VI collagen can be obtained by differential precipitation with NaCl from pepsin digests of various tissues in acetic acid or formic acid. Further purification can be accomplished by reprecipitation through dialysis against 0.02 M Na_2HPO_4, followed by ion exchange or molecular sieve chromatography.[11,28] Intact type VI collagen can be purified by ion exchange and molecular sieve chromatography of guanidine or urea extracts of tissues or cell cultures.[1,29] Procedures for isolating intact type VI collagen-containing microfilaments have also been described.[30]

The C-terminal Kunitz-type domain of α3(VI) chains has been generated as a recombinant protein and used for structural studies.[31] A large portion of the N-terminal globular domain of α3(VI) has also been synthesized as a recombinant protein.[32] All three type VI collagen α chains have been expressed as recombinant proteins in murine NIH/3T3 cells and shown to assemble into monomers, dimers, and tetramers.[33]

Antibodies

Several polyclonal and monoclonal antibodies are available against type VI collagen.[1,34–38] They have been used for detecting type VI chains or degradation products by immunoblotting, immunoprecipitation, and immunohistochemistry.[1] Monoclonal antibodies have been used for epitope mapping by rotary shadowing electron microscopy.[1,39] Anti-type VI collagen antibodies are available from several commercial sources. See BioSupplyNet *Source Book* for suppliers.

Activities

Type VI collagen molecules assemble into disulphide bonded polymers that form beaded microfibrils.[1,40,41] The microfibrils frequently aggregate further laterally into cross-banded fibres, referred to as Luse bodies, fusiform bodies, or zebra collagen.[1,42,43] Type VI collagen binds to hyaluronan;[44] binding sites for heparin and hyaluronan have been identified within the N-terminal globular domain of α3(VI) chains.[32] Type VI collagen also binds to the membrane-associated chondroitin sulphate proteoglycan NG2[45–47] and interacts with the microfibril-associated glycoprotein-1.[48]

Genes

cDNAs encoding type VI collagen chains in humans, chicken, and mouse have been isolated and sequenced.[19,49,50] The α2(VI) gene generates transcripts that are alternatively spliced at the 3′ end, giving rise to several mRNA variants.[3] Several variants are also generated by alternative splicing in the 5′ region of α3(VI) transcripts. The human *COL6A1* and *COL6A2* genes are organized in a head to tail arrangement on chromosome 21q22.3,[51] and both genes have been characterized and compared with the corresponding chicken genes.[52–54]

Mutant phenotype/disease states

Jobsis *et al.* (1996)[55] demonstrated linkage to the *COL6A1/COL6A2* locus on chromosome 21q22.3 in nine kindreds with the Bethlem form of autosomal dominant myopathy with contractures (OMIM 158810). A mis-sense mutation involving a glycine residue in the triple-helical domain was found in *COL6A1* in one family and in *COL6A2* in two other families.[55] Analysis of a large French Canadian family showed linkage to the *COL6A3* locus on chromosome 2q37.[56] Pan *et al.* have described a missense mutation in *COL6A3*.[93]

Structure

The crystal structure of the Kunitz-type domain in the C-terminal region of α3(VI) chains has been determined at a 1.6 Å resolution,[57] and the solution structure and backbone dynamics of the domain has been analysed by NMR.[58]

Type VII collagen

Type VII collagen is the major collagenous component of anchoring fibrils associated with the basement mem-

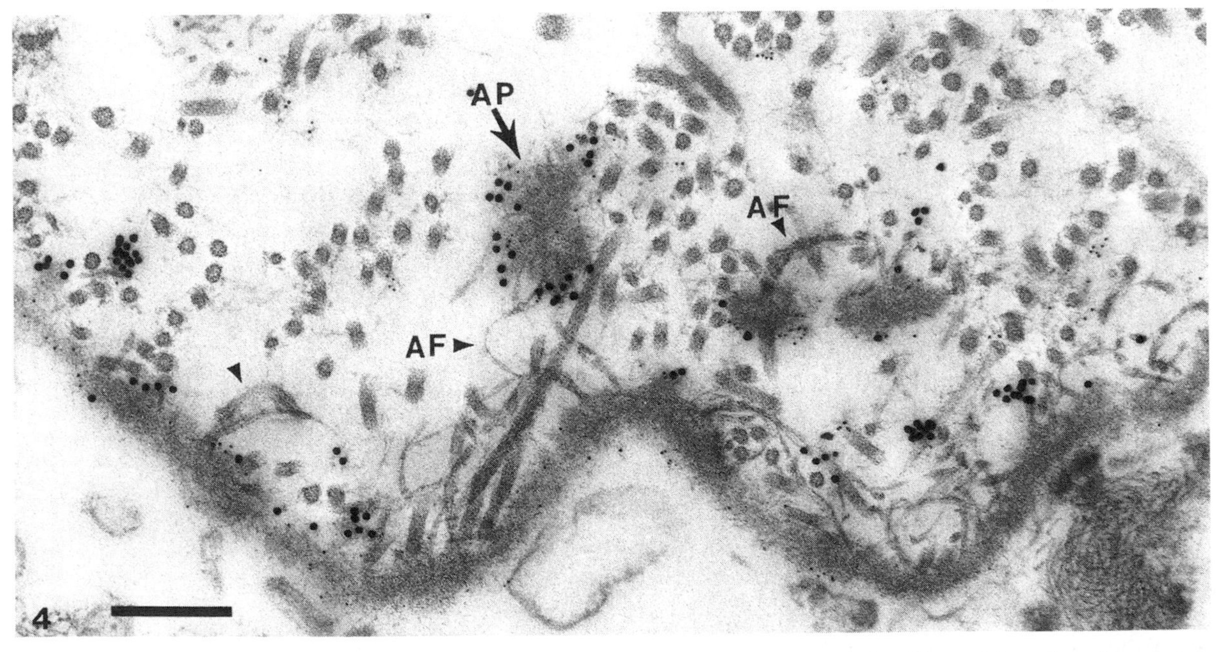

Figure 1. Ultrastructural immunolocalization of type VII collagen within the dermal–epidermal junction of neonatal human foreskin with gold-conjugated antibodies. AF, anchoring fibrils; AP, anchoring plaques. (From Keene *et al.* 1987.[62])

branes under stratified squamous epithelia.[59–61] The fibrils originate from the lamina densa and extend into the upper papillary dermis of skin where they insert into so-called anchoring plaques[62] (Fig. 1). Anchoring fibrils also connect anchoring plaques. Type VII collagen molecules are homotrimers containing a triple-helical domain that is about 50 per cent longer than the triple helix of fibrillar collagens. This domain is flanked by relatively large N- and C-terminal non-triple-helical domains, of molecular masses 150 and 30 kDa, respectively.[60,61] The 30 kDa domain is proteolytically cleaved extracellularly, and the processed molecules form antiparallel dimers, through a C-terminal overlap region. (On the basis of the initial protein studies, the large globular domain was erroneously identified as the C-terminal domain;[63] molecular cloning later showed that the large globular domain was at the N terminus.[64]) Lateral aggregation of such dimers leads to the formation of the centro-symmetrically banded anchoring fibrils.[59]

Keratinocytes are the cells of origin for type VII collagen in skin,[65] and proteolytic processing of the C-terminal globular domain precedes assembly of anchoring fibrils.[66] The N-terminal globular domain has a modular structure,[67] including nine fibronectin type III-like repeats and a von Willebrand factor A-like module.[68] The C-terminal globular domain contains eight cysteines; six of these are contained within a module that is similar to the Kunitz-type module in the C-terminal region of α3(VI) collagen chains.[69]

Purification and recombinant synthesis

The triple-helical domain of type VII collagen can be solubilized by pepsin extraction of human skin or amnion. Purification is by differential salt precipitation with NaCl, followed by ion exchange chromatography and HPLC.[59] The intact, biosynthetic form of type VII collagen has been purified from the media of KB cells (derived from a human oral basal cell carcinoma) and WISH cells (derived from amniotic epithelial cells).[59] Recombinant fusion proteins have been used for epitope mapping and detection of autoantibodies in patient sera.[70–72]

Activities

Type VII collagen molecules form the anchoring fibrils in skin, chorioamnion, oral mucosa, cornea, and the uterine cervix.[59] Laminin-5,[73] a component of anchoring filaments and a ligand for the integrin $\alpha6\beta4$ within hemidesmosomes, binds to the N-terminal globular domain of type VII collagen.[74] Interactions of the N-terminal domain with fibronectin and type I collagen,[76] and between the type VII collagen triple-helical region and fibronectin,[76] are also likely.

Antibodies

A number of polyclonal and monoclonal antibodies against type VII collagen are available.[59–79] Autoantibodies from patients with acquired epidermolysis bullosa have been shown to react with specific epitopes in α1(VII) collagen chains.[70–72,80]

Genes

Screening of a cDNA expression library with autoantibodies against type VII collagen from a patient with acquired epidermolysis bullosa resulted in the first isolation of human α1(VII) collagen cDNA.[81] This led to the isolation of cDNAs covering the entire mRNA,[68,69] and characterization of the entire human *COL7A1* gene.[82] The mouse cDNA and *Col7a1* gene has also been characterized.[83,84]

Mutant phenotype/disease states

The epidermolysis bullosa group of inherited blistering diseases in humans is classified into simplex, junctional, and dystrophic forms. The simplex forms are caused by mutations in keratins 5 and 14,[85–87] the junctional forms are caused by mutations in laminin-5,[88] and the dystrophic forms are the consequences of mutations in type VII collagen.[88] The mutations in *COL7A1* range from premature termination codons resulting in severe, mutilating recessive dystrophic epidermolysis bullosa of the Hallopeau–Siemens type (OMIM 226600)[89,90] to glycine substitutions in the triple-helical region of α1(VII) collagen resulting in clinically less severe, dominant or recessive, dystrophic epidermolysis bullosa.[91] A clinical variant of dominant dystrophic epidermolysis bullosa called the Bart syndrome (OMIM 132000) is caused by a glycine-substitution mutation in α1(VII) collagen.[92] An updated listing of all mutations in type VII collagen can be found in the OMIM database (OMIM 120120 collagen).

References

1. Timpl, R. and Engel, J. (1987). In *Structure and function of collagen types* (ed. R. Mayne and R. E. Burgeson), pp. 105–43. Academic Press,
2. Furthmayr, H., Wiedemann, H., Timpl, R., Odermatt, E., and Engel, J. (1983). *Biochem. J.*, **211**, 303–11.
3. Trueb, B. and Winterhalter, K. H. (1986). *EMBO J.*, **5**, 2815–19.
4. Bruns, R. R. (1984). *J. Ultrastruct. Res.*, **89**, 136–45.
5. Bruns, R. R., Press, W., Engvall, E., Timpl, R., and Gross, J. (1986). *J. Cell Biol.*, **103**, 393–404.
6. Linsenmayer, T. F., Bruns, R. R., Mentzer, A., and Mayne, R. (1986). *Dev. Biol.*, **118**, 425–31.
7. Keene, D. R., Engvall, E., and Glanville, R. W. (1988). *J. Cell Biol.*, **107**, 1995–2006.
8. Kuo, H. J., Keene, D. R., and Glanville, R. W. (1989). *Biochemistry*, **28**, 3757–62.
9. Weil, D., Mattei, M. G., Passage, E., C., N. G. V., Pribula-Conway, D., Mann, K., Deutzmann, R., Timpl, R., and Chu, M. L. (1988). *Am. J. Hum. Genet.*, **42**, 435–45.
10. Chu, M. L., Pan, T. C., Conway, D., Kuo, H. J., Glanville, R. W., Timpl, R., *et al.* (1989). *EMBO J.*, **8**, 1939–46.
11. Bonaldo, P., Russo, V., Bucciotti, F., Bressan, G. M., and Colombatti, A. (1989). *J. Biol. Chem.*, **264**, 5575–80.

12. Koller, E., Winterhalter, K. H., and Trueb, B. (1989). *EMBO J.*, **8**, 1073–7.
13. Trueb, B., Schaeren-Wiemers, N., Schreier, T., and Winterhalter, K. H. (1989). *J. Biol. Chem.*, **264**, 136–40.
14. Bonaldo, P. and Colombatti, A. (1989). *J. Biol. Chem.*, **264**, 20235–9.
15. Chu, M. L., Zhang, R. Z., Pan, T. C., Stokes, D., Conway, D., Kuo, H. J., *et al.* (1990). *EMBO J.*, **9**, 385–93.
16. Bonaldo, P., Russo, V., Bucciotti, F., Doliana, R., and Colombatti, A. (1990). *Biochemistry*, **29**, 1245–54.
17. Saitta, B., Stokes, D. G., Vissing, H., Timpl, R., and Chu, M. L. (1990). *J. Biol. Chem.*, **265**, 6473–80.
18. Doliana, R., Bonaldo, P., and Colombatti, A. (1990). *J. Cell Biol.*, **111**, 2197–205.
19. Ibrahimi, A., Bertrand, B., Bardon, S., Amri, E. Z., Grimaldi, P., Ailhaud, G., and Dani, C. (1993). *Biochem. J.*, **289**, 141–7.
20. Pareti, F. I., Niiya, K., McPherson, J. M., and Ruggeri, Z. M. (1987). *J. Biol. Chem.*, **262**, 13835–41.
21. Aumailley, M., Specks, U., and Timpl, R. (1991). *Biochem. Soc. Trans.*, **19**, 843–7.
22. Perris, R., Kuo, H. J., Glanville, R. W., Leibold, S., and Bronner-Fraser, M. (1993). *Exp. Cell Res.*, **209**, 103–17.
23. Stokes, D. G., Saitta, B., Timpl, R., and Chu, M. L. (1991). *J. Biol. Chem.*, **266**, 8626–33.
24. Zanussi, S., Doliana, R., Segat, D., Bonaldo, P., and Colombatti, A. (1992). *J. Biol. Chem.*, **267**, 24082–9.
25. Colombatti, A., Ainger, K., and Colizzi, F. (1989). *Matrix*, **9**, 177–85.
26. Kielty, C. M., Boot-Handford, R. P., Ayad, S., Shuttleworth, C. A., and Grant, M. E. (1990). *Biochem. J.*, **272**, 787–95.
27. Walchli, C., Marcionelli, R., Odermatt, B. F., Peltonen, J., Vuorio, E., and Trueb, B. (1996). *J. Cell. Biochem.*, **63**, 207–20.
28. Trueb, B., Schreier, T., Bruckner, P., and Winterhalter, K. H. (1987). *Eur. J. Biochem.*, **166**, 699–703.
29. Schreier, T., Winterhalter, K. H., and Trueb, B. (1987). *FEBS Lett.*, **213**, 319–23.
30. Kuo, H. J., Keene, D. R., and Glanville, R. W. (1995). *Eur. J. Biochem.*, **232**, 364–72.
31. Mayer, U., Poschl, E., Nischt, R., Specks, U., Pan, T. C., Chu, M. L., and Timpl, R. (1994). *Eur. J. Biochem.*, **225**, 573–80.
32. Specks, U., Mayer, U., Nischt, R., Spissinger, T., Mann, K., Timpl, R., *et al.* (1992). *EMBO J.*, **11**, 4281–90.
33. Colombatti, A., Mucignat, M. T., and Bonaldo, P. (1995). *J. Biol. Chem.*, **270**, 13105–11.
34. Hessle, H. and Engvall, E. (1984). *J. Biol. Chem.*, **259**, 3955–61.
35. Trueb, B. and Bornstein, P. (1984). *J. Biol. Chem.*, **259**, 8597–604.
36. Schuppan, D., Ruhlmann, T., and Hahn, E. G. (1985). *Anal. Biochem.*, **149**, 238–47.
37. Werkmeister, J. A., Tebb, T. A., White, J. F., and Ramshaw, J. A. (1993). *J. Histochem. Cytochem.*, **41**, 1701–6.
38. Sawada, H. and Yazama, F. (1994). *Biol. Reprod.*, **50**, 702–10.
39. Linsenmayer, T. F., Mentzer, A., Irwin, M. H., Waldrep, N. K., and Mayne, R. (1986). *Exp. Cell Res.*, **165**, 518–29.
40. Engvall, E., Hessle, H., and Klier, G. (1986). *J. Cell Biol.*, **102**, 703–10.
41. Engel, J., Furthmayr, H., Odermatt, E., von der Mark, H., Aumailley, M., Fleischmajer, R., and Timpl, R. (1985). *Ann. NY Acad. Sci.*, **460**, 25–37.
42. Garron, L. K., Feeney, M. L., Hogan, M. J., and McEwen, W. K. (1958). *Am. J. Ophthalmol.*, **46**, 27–35.
43. Luse, S. A. (1960). *Neurology*, **10**, 881–905.
44. Kielty, C. M., Whittaker, S. P., Grant, M. E., and Shuttleworth, C. A. (1992). *J. Cell Biol.*, **118**, 979–90.
45. Burg, M. A., Tillet, E., Timpl, R., and Stallcup, W. B. (1996). *J. Biol. Chem.*, **271**, 26110–16.
46. Nishiyama, A. and Stallcup, W. B. (1993). *Mol. Biol. Cell*, **4**, 1097–108.
47. Burg, M. A., Nichiyama, A., and Stallcup, W. B. (1997). *Exp. Cell Res.*, **235**, 254–64.
48. Finnis, M. L. and Gibson, M. A. (1997). *J. Biol. Chem.*, **272**, 22817–23.
49. Bonaldo, P., Piccolo, S., Marvulli, D., Volpin, D., Marigo, V., and Bressan, G. M. (1993). *Matrix*, **13**, 223–33.
50. Zhang, R. Z., Pan, T. C., Timpl, R., and Chu, M. L. (1993). *Biochem. J.* **291**, 787–92.
51. Heiskanen, M., Saitta, B., Palotie, A., and Chu, M. L. (1995). *Genomics*, **29**, 801–3.
52. Trikka, D., Davis, T., Lapenta, V., Brahe, C., and Kessling, A. M. (1997). *Mamm. Genome*, **8**, 342–345.
53. Hayman, A. R., Koppel, J., and Trueb, B. (1991). *Eur. J. Biochem.*, **197**, 177–84.
54. Walchli, C., Koller, E., Trueb, J., and Trueb, B. (1992). *Eur. J. Biochem.*, **205**, 583–9.
55. Jobsis, G. J., Keizers, H., Vreijling, J. P., de Visser, M., Speer, M. C., Wolterman, R. A., *et al.* (1996). *Nature Genet.*, **14**, 113–15.
56. Speer, M. C., Tandan, R., Rao, P. N., Fries, T., Stajich, J. M., Bolhuis, P. A., *et al.* (1996). *Hum. Mol. Genet.*, **5**, 1043–6.
57. Arnoux, B., Merigeau, K., Saludjian, P., Norris, F., Norris, K., Bjorn, S., *et al.* (1995). *J. Mol. Biol.*, **246**, 609–17.
58. Sorensen, M. D., Bjorn, S., Norris, K., Olsen, O., Petersen, L., James, T. L., and Led, J. J. (1997). *Biochemistry*, **36**, 10439–50.
59. Burgeson, R. E. (1987). In *Structure and function of collagen types* (ed. R. Mayne and R. E. Burgeson), pp. 145–72. Academic Press,
60. Burgeson, R. E. (1993). *J. Invest. Dermatol.*, **101**, 252–5.
61. Uitto, J. and Pulkkinen, L. (1996). *Mol. Biol. Rep.*, **23**, 35–46.
62. Keene, D. R., Sakai, L. Y., Lunstrum, G. P., Morris, N. P., and Burgeson, R. E. (1987). *J. Cell Biol.*, **104**, 611–21.
63. Lunstrum, G. P., Sakai, L. Y., Keene, D. R., Morris, N. P., and Burgeson, R. E. (1986). *J. Biol. Chem.*, **261**, 9042–8.
64. Christiano, A. M., Rosenbaum, L. M., Chung-Honet, L. C., Parente, M. G., Woodley, D. T., Pan, T. C., *et al.* (1992). *Hum. Mol. Genet.*, **1**, 475–81.
65. Regauer, S., Seiler, G. R., Barrandon, Y., Easley, K. W., and Compton, C. C. (1990). *J. Cell Biol.*, **111**, 2109–15.
66. Bruckner-Tuderman, L., Nilssen, O., Zimmermann, D. R., Dours-Zimmermann, M. T., Kalinke, D. U., Gedde-Dahl, T., Jr, and Winberg, J. O. (1995). *J. Cell Biol.*, **131**, 551–9.
67. Bork, P. (1992). *FEBS Lett.*, **307**, 49–54.
68. Christiano, A. M., Greenspan, D. S., Lee, S., and Uitto, J. (1994). *J. Biol. Chem.*, **269**, 20256–62.
69. Greenspan, D. S. (1993). *Hum. Mol. Genet.*, **2**, 273–8.
70. Lapiere, J. C., Woodley, D. T., Parente, M. G., Iwasaki, T., Wynn, K. C., Christiano, A. M., and Uitto, J. (1993). *J. Clin. Invest.*, **92**, 1831–9.
71. Tanaka, T., Furukawa, F., and Imamura, S. (1994). *J. Invest. Dermatol.*, **102**, 706–9.
72. Chen, M., Chan, L. S., Cai, X., A., O. T. E., Sample, J. C., and Woodley, D. T. (1997). *J. Invest. Dermatol.*, **108**, 68–72.
73. Burgeson, R. E., Chiquet, M., Deutzmann, R., Ekblom, P., Engel, J., Kleinman, H., *et al.* (1994). *Matrix Biol.*, **14**, 209–11.
74. Rousselle, P., Keene, D. R., Ruggiero, F., Champliaud, M. F., Rest, M., and Burgeson, R. E. (1997). *J. Cell Biol.*, **138**, 719–28.
75. Chen, M., Marinkovich, M. P., Veis, A., Cai, X., Rao, C. N., A., O. T. E., and Woodley, D. T. (1997). *J. Biol. Chem.*, **272**, 14516–22.
76. Lapiere, J. C., Chen, J. D., Iwasaki, T., Hu, L., Uitto, J., and Woodley, D. T. (1994). *J. Invest. Dermatol.*, **103**, 637–41.
77. Lapiere, J. C., Hu, L., Iwasaki, T., Chan, L. S., Peavey, C., and Woodley, D. T. (1994). *J. Dermatol. Sci.*, **8**, 145–50.

78. Tanaka, T., Matsuyoshi, N., Furukawa, F., and Imamura, S. (1994). *Dermatology*, **1**, 42–5.
79. Tanaka, T., Takahashi, K., Furukawa, F., and Imamura, S. (1994). *Br. J. Dermatol.*, **131**, 472–6.
80. Gammon, W. R., Murrell, D. F., Jenison, M. W., Padilla, K. M., Prisayanh, P. S., Jones, D. A., *et al.* (1993). *J. Invest. Dermatol.*, **100**, 618–22.
81. Parente, M. G., Chung, L. C., Ryynanen, J., Woodley, D. T., Wynn, K. C., Bauer, E. A., *et al.* (1991). *Proc. Natl Acad. Sci. USA*, **88**, 6931–5.
82. Christiano, A. M., Hoffman, G. G., Chung-Honet, L. C., Lee, S., Cheng, W., Uitto, J., and Greenspan, D. S. (1994). *Genomics*, **21**, 169–79.
83. Li, K., Christiano, A. M., Copeland, N. G., Gilbert, D. J., Chu, M. L., Jenkins, N. A., and Uitto, J. (1993). *Genomics*, **16**, 733–9.
84. Kivirkko, S., Li, K., Christiano, A. M., and Uitto, J. (1996). *J. Invest. Dermatol.*, **106**, 1300–6.
85. Bonifas, J. M., Rothman, A. L., and Epstein, E. H., Jr. (1991). *Science*, **254**, 1202–5.
86. Chen, H., Bonifas, J. M., Matsumura, K., Ikeda, S., Leyden, W. A., and Epstein, E. H., Jr. (1995). *J. Invest. Dermatol.*, **105**, 629–32.
87. Chan, Y. M., Yu, Q. C., Fine, J. D., and Fuchs, E. (1993). *Proc. Natl Acad. Sci. USA*, **90**, 7414–18.
88. Uitto, J., Pulkkinen, L., and Christiano, A. M. (1994). *J. Invest. Dermatol.*, **103**, 39S–46S.
89. Hilal, L., Rochat, A., Duquesnoy, P., Blanchet-Bardon, C., Wechsler, J., Martin, N., *et al.* (1993). *Nature Genet.*, **5**, 287–93.
90. Christiano, A. M., Anhalt, G., Gibbons, S., Bauer, E. A., and Uitto, J. (1994). *Genomics*, **21**, 160–8.
91. Christiano, A. M., McGrath, J. A., Tan, K. C., and Uitto, J. (1996). *Am. J. Hum. Genet.*, **58**, 671–681.
92. Christiano, A. M., Bart, B. J., Epstein, E. H., Jr., and Uitto, J. (1996). *J. Invest. Dermatol.*, **106**, 778–80.
93. Pan, T. C., Zhang, R. Z., Pericat-Vance, M. A., Tandan, R., Fries, T., Stajich, J. M., Viles, K., Vance, J. M., Chu, M. L., and Speer, M. C. (1998). *Human Mol. Genet.*, **7**, 807–12.

Bjorn Reino Olsen
Department of Cell Biology, Harvard Medical School, and Harvard-Forsyth Department of Oral Biology, Harvard School of Dental Medicine, Boston, MA, USA

Yoshifumi Ninomiya
Department of Molecular Biology, and Biochemistry, Okayama University Medical School, 2-5-1 Shikata-cho, Okayama 700, Japan

Decorin

Decorin (DCN) is a small proteoglycan composed of a ~38 kDa core protein usually modified with a single chondroitin sulphate (bone) or dermatan sulphate (most soft tissues) glycosaminoglycan chain and two or three *N*-linked oligosaccharides. DCN is virtually ubiquitous in the matrices of various connective tissues, being found bound to or 'decorating' the collagen fibrils. The protein portion is composed of 10 tandem repeats of ~25 amino acids characteristically rich in ordered leucines with the repeats being flanked by two cysteine disulphide loops. These tandem repeats are found a wide variety of closely related small proteoglycans including: biglycan (BGN), fibromodulin, lumican, epiphycan, keratocan, and PG-Lb. The most commonly cited functions of DCN are its roles in collagen fibril assembly (and stabilization) as well as its ability to bind to TGF-β.

Synonymous names

Decorin has several synonymous names, most reflecting its relative position on SDS–PAGE or time of elution from various purification columns. The names include PG40, PG-2, PG-II, PG-S2, CS-PGII, and DS-PGII.

Protein properties

Decorin is a member of a growing family of small proteoglycans whose unifying characteristics are two highly conserved cysteine loops flanking 5 to 10 tandem repeats. Each repeat is nominally ~25 amino acids in length and is based on the pattern $LxxLxLxxNxLx_{(12–14)}$. For DCN there are 10 repeats and the single glycosaminoglycan (GAG) chain is chondroitin sulphate in bone matrix and dermatan sulphate in most soft tissues. Other members of this family include biglycan, fibromodulin, lumican, epiphycan, keratocan, and PG-Lb (known as DSPG3 in human) (for a review see ref. 1). The DCN sequences from a number of species have been reported, including human,[2] cow,[3] mouse,[4] rat,[5] rabbit,[6] and chicken.[7] Curiously, the chicken form can have two GAG chains and these chains appear to be attached to a GlySer sites rather than the apparently universal mammalian Ser–Gly.[7] Using human DCN as the model, decorin has 359 amino acids (~39 700 Da) including 17 in the leader sequence and 14 more in the amino terminus that are often removed and are therefore considered to be a propeptide region.[2] The 'mature' core protein (lacking the propeptide), made by removing the disaccharide

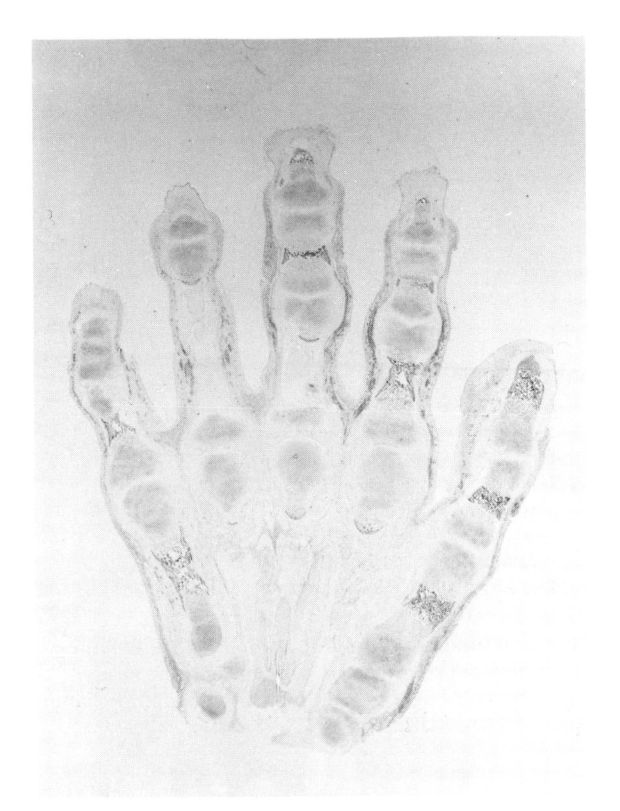

Figure 1. Immunolocalization of humin decorin in a section of a hand from a 15-week fetus using antiserum LF-30. Dark areas represent the presence of decorin. DCN is localized to all connective tissues, including dermis, cartilage, and type I collagen-containing bone.[16] Compare this with a serial section stained for a closely related proteoglycan, biglycan, on p. 366. Notice that the two are often mutually exclusive in their distribution. See ref. 16 for more details.

repeats of the GAG chain with chondroitinase ABC, is typically a doublet band of M_r ~45 and 47 kDa on SDS–PAGE.[8] This core protein contains the GAG chain linkage region (on amino acid 34, as numbered with the starting Met as 1) and two or three *N*-linked oligosaccharides on amino acids 211, 262, and 303 (leading to the doublet band.[9]).

The human DCN gene was mapped to human chromosome 12 at either q21.3[10] or q23[11] and location 55.0 on mouse chromosome 10.[12] The human gene has a complex dinucleotide repeat polymorphism that may be useful for genetic studies.[13] The human DCN gene has been cloned within three non-overlapping Lambda Fix clones making the gene at least 25 kb in size.[10] The DCN gene has eight exons with the seven protein-encoding exons matching completely with the homogeneous exons of the human biglycan gene, BGN,[14] strongly suggesting that these two genes were at one time a single gene.[15] Interestingly, DCN has two different, non-translated exon 1 thereby suggesting that the transcription of this gene is under the control of two promoters.[11]

Decorin is found wherever type I, II, or III collagen fibrils are found. This includes not only the major extracellular matrices such as skin and skeleton but also all of the finer support matrices around and within organs of the body.[16] DCN has been localized to the gap regions, near the d and e bands, on the surface of type I collagen fibrils.[17]

Purification

Chondroitin sulphate-containing DCN can be purified from fetal or young bone by a series of extraction procedures and protein chromatography.[8] Bone is milled into a fine powder, extracted with denaturing buffers to remove blood and cellular proteins, and the residue extracted with demineralizing buffer. Standard molecular sieve and ion exchange chromatography in denaturing buffers are performed. In our hands, reverse phase chromatography using standard organic solvents results in large losses. Dermatan sulphate-containing DCN can be isolated in good yield from articular cartilage using similar procedures as well as a reverse phase column and a detergent gradient.[18] Recombinant DCN with appropriate binding activity has been reported using a maltose-binding fusion protein.[19] For post-translationally modified DCN, a vaccinia-based recombinant method has been reported using UMR106 and HT-1080 as host cells resulting in ~30 mg of DCN per billion cells per day.[20]

Activities

Decorin has been reported to change the kinetics and final shape of type I collagen fibrils *in vitro*.[21] Indeed, the knockout mouse has fragile skin with unusual collagen fibril morphology.[22] DCN has also been shown to bind to TGF-β.[23] This naturally leads to an interesting hypothesis that DCN on the surface of the matrix fibrils may bind TGF-β or other members of its superfamily and release these powerful bioactive molecules when the matrix is disrupted in specific ways. Presumably other bioactive proteins may similarly be bound to matrix components. Cells sensing the increase or decrease in the levels of these different active proteins may use such mechanisms to monitor the health of the matrix within its purview. The TGF-β binding property has been proposed to be used in protecting against scarring in kidney diseases.[24] Alternatively, another report suggests that decorin induces growth suppression by up-regulating p21, an inhibitor of cyclin-dependent kinases.[25] Decorin also binds to fibronectin and this may explain its propensity to block adhesion *in vitro*.[26] DCN has been reported to induce matrix metalloproteinase collagenase (MMP-1) in synovial fibroblasts adhering to vitronectin. This activity was though to be independent of the TGF-β effects.[27]

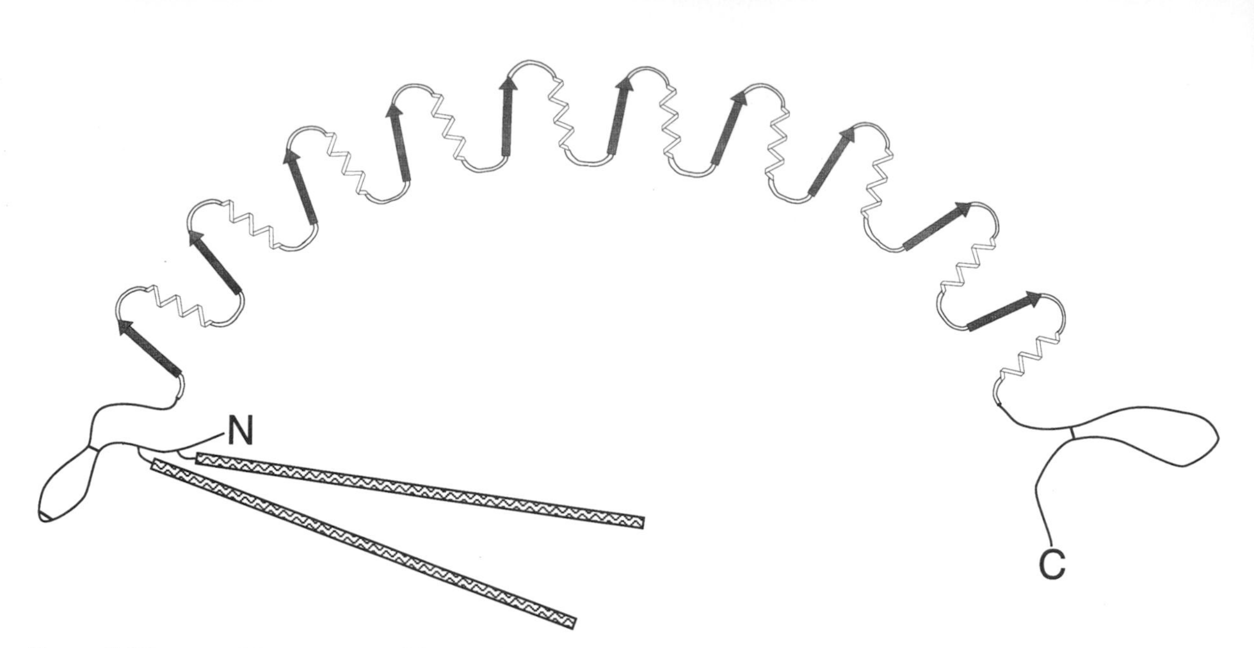

Figure 2. Diagram of the structure of decorin based loosely on the bent-coil structure determined for the leucine-rich repeat structure of the porcine ribonuclease inhibitor.[24] The arrows represent short β sheets and the straight rod near the N terminus represents the glycosaminoglycan chain. DCN is thought to contain three disulphide bonds, represented by short connecting lines. (Drawn by Dr Andrew Hinck, NIDR, NIH.).

Table 1

Gene product	Antiserum	Antigen	Known species
Human DCN	LF-30	GIGPEVPDDRDF-(KLH)	H, M
Human DCN	LF-136	GIGPEVPDDRDF-(KLH)	H, M
Human DCN	LF-122	Recombinant DCN (w/propeptide)	H, M
Human DCN Propeptide	LF-110	QVSWAGPFQQRGLFDC-(LPH)	Only H tested
Human DCN Propeptide	LF-111	QVSWAGPFQQRGLFDC-(CSA)	Only H tested
Bovine DCN	LF-94	IGPEEHFPEVPEC-(LPH)	Only cow tested
Bovine DCN	LF-95	IGPEEHFPEVPEC-(CSA)	Only cow tested
Mouse DCN	LF-113	IIPYDPDNPLISMC-(LPH)	R, Mou tested
Mouse DCN	LF-114	IIPYDPDNPLISMC-(CSA)	R, Mou tested

All antisera are whole rabbit sera.
H, human; M, monkey; R, rat; Mou, mouse; LPH, horseshoe crab haemocyanin; CSA, chicken serum albumin; KLH, keyhole limpet haemocyanin.

Antibodies

No monoclonal antibodies for decorin are currently listed in ATCC's Hybridoma Data Bank (http://www.atcc.org/hdb/hdb.html). Limited amounts of the following rabbit (polyclonal) antisera are available to colleagues for research purposes only. Any use must comply completely with local and NIHs guidelines for patient care and confidentiality.

Mutant phenotype/disease states

There have been no specific human diseases yet unambiguously ascribed to mutations in DCN. There is one report of two osteogenesis imperfecta patients, both with the same gly-415/ser mutation in the α1(I) chain of collagen, in which the patient with the more severe phenotype had little or no decorin production in fibroblasts while fibroblasts from the other patient produced normal amounts of DCN.[28] This suggest that changes in the expression of decorin may have phenotypic consequences in some tissues. The mouse DCN knockout mouse is reported to have fragile skin with coarser and irregular collagen fibrils.[22]

Structure

The three-dimensional structure of DCN has not been determined. By analogy to the X-ray diffraction study on a porcine ribonuclease inhibitor (another protein with similar leucine-rich repeats),[29] we expect the structure DCN to be dominated by bent-coil structure. In this hypothetical structure each of the 10 repeats forms a single turn of the coil with each turn slightly angled to produce a structure that is somewhat horseshoe-like in appearance. Unfortunately, the ribonuclease inhibitor does not use the conserved cysteine clusters found in DCN and many other leucine-rich repeat proteins, so we cannot infer the structure of the DCN protein outside of the central repeats.

References

1. Iozzo, R. V. and Murdoch, A. D. (1996). *FASEB J.*, **10**, 598–614.
2. Krusius, T. and Ruoslahti, E. (1986). *Proc. Natl Acad. Sci. USA*, **83**, 7683–7.
3. Day, A. A., McQuillan, C. I., Termine, J. D., and Young, M. F. (1987). *Biochem. J.*, **248**, 801–5.
4. Suzuki, S. (1990). Submission to GenBank, No X53929.
5. Asundi, V. K. and Dreher, K. L. (1992). *Eur. J. Cell Biol.*, **59**, 314–21.
6. Zhan, Q., Burrows, R., and Cintron, C. (1995). *Invest. Ophthalmol. Vis. Sci.*, **36**, 206–15.
7. Li, W., Vergnes, J. P., Cornuet, P. K., and Hassell, J. R. (1992). *Arch. Biochem. Biophys.*, **296**, 190–7.
8. Fisher, L. W., Hawkins, G. R., Tuross, N., and Termine, J. D (1987). *J. Biol. Chem.*, **262**, 9702–9.
9. Glossl, J., Beck, M., and Kresse, H. (1984). *J. Biol. Chem.*, **259**, 14144–50.
10 Vetter, U., Vogel, W., Just, W., Young, M. F., and Fisher, L. W. *Genomics*, **15**, 146–60.
11. Danielson, K. G., Fazzio, A., Cohen, I., Cannizzaro, L. A., Eichstetter, I., and Iozzo, R. V. *Genomics*, **15**, 146–60.
12 Scholzen, T., Solursh, M., Suzuki. S., Reiter, R., Morgan, J. L., Buchberg, A. M., *et al.* (1994). *J. Biol. Chem.*, **269**, 2870–81.
13. Briggs, M. D. and Cohn, D. H. (1993). *Hum. Mol. Genet.*, **2**, 1087.
14. Fisher, L. W., Heegaard, A. M., Vetter, U., Vogel, W., Just, W., Termine, J. D., and Young, M. F. (1991). *J. Biol. Chem.*, **266**, 14371–7.
15 Fisher, L. W (1993). *Dermatan sulphate proteoglycans; Chemistry, biology and chemical pathology*. (ed. J. Scott), pp 103–14. Portland Press, London.
16. Bianco, P, Fisher, L. W., Young, M. F., Termine, J. D., and Gehron Robey, P (1990). *J. Histochem. Cytochem.*, **38**, 1549–63.
17. Pringle, G. A. and Dodd, C. M. (1990). *J. Histochem. Cytochem.* **38**, 1405–11.
18. Choi, H. U., Johnson, T. L., Subhash, P., Tang, L. -H., Rosenberg, L. C., and Neame, P. J. (1989). *J. Biol. Chem.*, **264**, 2876–84.
19. Hering, T. M., Kollar, J., Huynh, T. D., and Varelas, J. B. (1996). *Anal. Biochem.*, **15**, 98–108.
20. Ramamurthy, P., Hocking, A. M., and McQuillan, D. J. (1996). *J. Biol. Chem.*, **271**, 19578–84.
21. Vogel, K. G. and Trotter, J. A. (1987). *Collagen Relat. Res.*, **7**, 105–14.
22. Danielson, K. G., Baribault, H., Holmes, D. F., Graham, H., Kadler, K. E., and Iozzo, R. V. (1997). *J. Cell Biol.*, **136**, 729–43.
23. Yamaguchi, Y., Mann, D. M., and Ruoslahti, E. (1990). *Nature*, **346**, 281–4.
24. Border, W. A., Noble, N. A., Yamamoto, T., Harper, J. R., Yamaguchi, Y., Pierschbacher, M. D., and Ruoslahti, E. (1992). *Nature*, **360**, 361–4.
25. De Luca, A., Santra, M., Baldi, A., Giordano, A., and Iozzo, R. V. (1996). *J. Biol. Chem.*, **271**, 18961–5.
26. Schmidt, G., Hausser, H., and Kresse H. (1991). *Biochem. J.*, **280**, 411–4.
27. Huttenlocher, A., Werb, Z., Tremble, P., Huhtala, P., Rosenberg, L., and Damsky, C. H. (1996). *Matrix Biol.*, **15**, 239–50.
28. Dyne, K. M., Valli, M., Forlino, A., Mottes, M., Kresse, H., and Cetta, G. (1996). *Am. J. Med. Genet.*, **63**, 161–6.
29. Kobe, B. and Deisenhofer, J. (1995). *Nature*, 374, 183–6.
30. Fisher, L. W., Stubbs III, J. T., and Young, M. F. (1995). *Acta Orthop. Scand.* (Suppl. 266), **66**, 66–70.

Larry W. Fisher
Craniofacial and Skeletal Diseases Branch,
NIDR, NIH, Room 228, Building 30,
Bethesda, MD 20892, USA

Egg zona pellucida glycoproteins

Mammalian eggs are surrounded by a relatively thick (~2–25 μm) extracellular coat, the zona pellucida (ZP), which consists of three glycoproteins, called ZP1–3.[1–3] These glycoproteins are organized, through non-covalent bonds, into an extensive network of interconnected filaments that exhibit a ~150 Å structural repeat. Free-swimming sperm bind in a relatively species-specific manner to the ZP by recognizing ZP3, the sperm receptor.[4–6] Bound sperm then undergo the acrosome reaction (exocytosis), bind to ZP2, penetrate through the ZP, and fuse with egg plasma membrane (fertilization).[3,7] Following fertilization, the ZP undergoes structural and functional changes as part of the secondary (slow) block to polyspermy. The various functions of the ZP can be accounted for fully by the properties of ZP1–3 before and after fertilization.

Protein properties

ZP glycoproteins vary considerably in size among different mammalian species and some of the variability is due

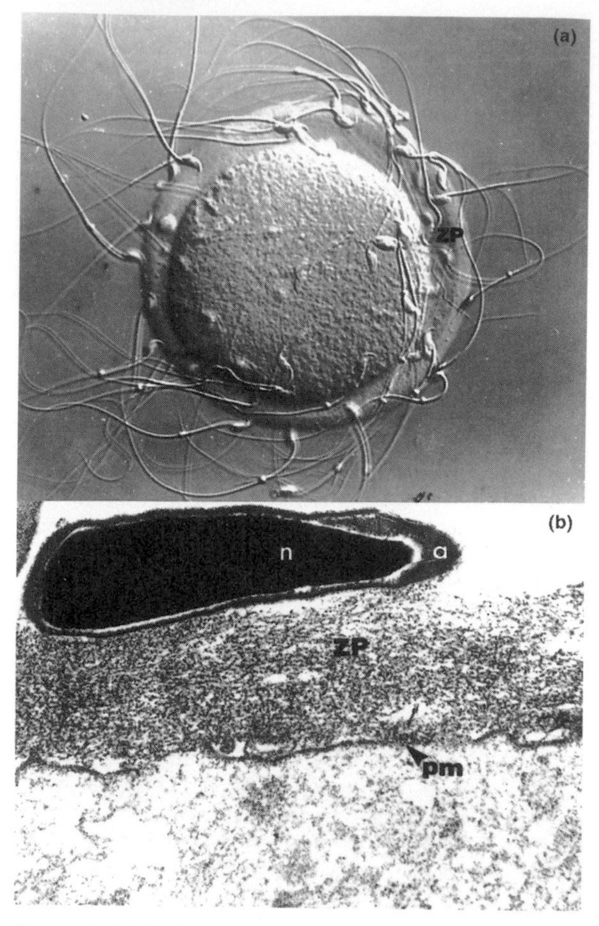

Figure 1. (a) Light micrograph of mouse sperm bound by their heads to the zona pellucida of an unfertilized mouse egg *in vitro*. The micrograph was taken using Nomarski differential interference contrast microscopy. zp, zona pellucida. (b) Transmission electron micrograph of a thin section through an acrosome-intact mouse sperm bound by plasma membrane overlying its head to the zona pellucida of an unfertilized mouse egg *in vitro*. Note the fibrillar (filamentous) nature of the egg zona pellucida. n, sperm nucleus; a, sperm acrosome; zp, zona pellucida of an unfertilized egg; pm, plasma membrane of an unfertilized egg.

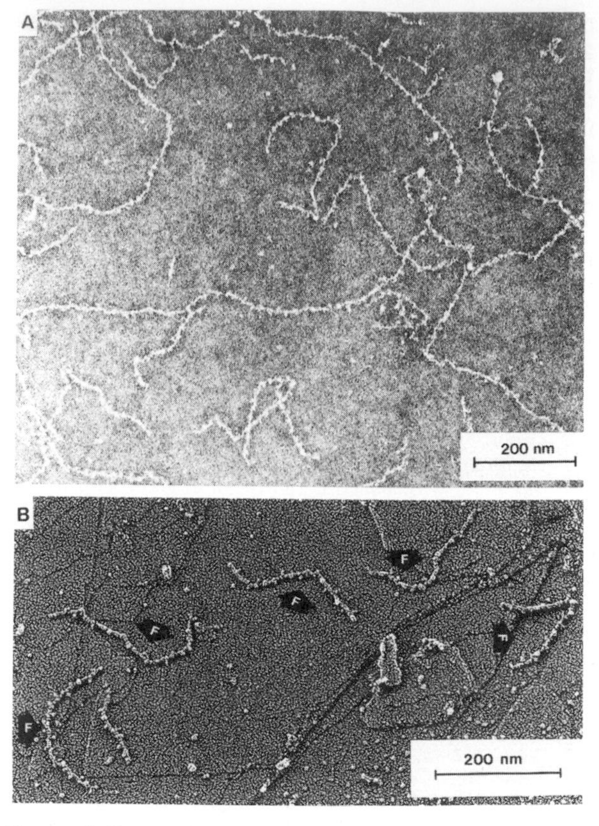

Figure 2. Transmission electron micrographs of mouse egg zona pellucida filaments. (A) Enzyme-solubilized zona pellucida preparation adsorbed to a substrate-coated grid and negatively stained. (B) Enzyme-solubilized zona pellucida preparation freeze-dried and unidirectionally shadowed. F, filaments. (Courtesy of Dr John Heuser, Washington University School of Medicine, St. Louis, MO.)

to different extents of glycosylation of the polypeptides.[1] Mouse ZP1–3 are approximately 200, 120, and 83 kDa, respectively.[8] Each glycoprotein consists of a highly conserved polypeptide and variable amounts of *N*- and *O*-linked oligosaccharides. ZP polypeptides are heterogeneously glycosylated and both sulphate and sialic acid are present on the oligosaccharides. Polypeptides of mouse ZP1–3 consist of 603 (~68 kDa), 679 (~77 kDa), and 402 (~44 kDa) amino acids, respectively.[9–11] ZP2 and ZP3 are monomers; ZP1 consists of two identical polypeptides interconnected by disulphides. ZP2 and ZP3 are present in the mouse ZP in approximately equimolar amounts; ZP1 is a relatively minor component.

Mouse ZP1–3 genes are single copy genes located on chromosomes 19, 7, and 5, respectively.[11,12] Human ZP2 and ZP3 genes are located on chromosome 7 in a region homologous to mouse chromosome 5. Overall, the organization of each of the ZP glycoprotein genes is similar in mammalian species as diverse as mice and humans.[13] The primary structures of ZP1–3 polypeptides are also quite similar (~40–70 per cent identical) in mice and humans. In addition to mice and humans, ZP glycoproteins have been cloned from hamsters, rats, rabbits, pigs, cats, dogs, marmosets, as well as other mammals (see Genes). There is evidence that fish and amphibian egg vitelline envelope glycoproteins and mammalian egg ZP glycoproteins are structurally related to each other.[6]

Mouse ZP glycoprotein genes are expressed exclusively by growing oocytes during the latter stages of oogenesis.[14] For example, as little as 153 nucleotides of mouse ZP3 5′-flanking sequence can target expression of a reporter gene exclusively to growing oocytes in transgenic mice. Mouse ZP1–3 genes are expressed concomitantly and ZP1–3 glycoproteins are synthesized and secreted concomitantly during oocyte growth. Over 2–3 weeks, approximately 3.5 ng of glycoprotein is deposited in the mouse egg ZP. Expression of ZP glycoprotein genes is terminated at about the time of ovulation when the oocyte chromosomes condense.

During oogenesis, ZP1–3 are assembled into an extensive network of long, cross-linked filaments that make up the ZP.[15] Filaments of the mouse ZP are polymers of ZP2-ZP3 dimers (i.e. $[ZP2\text{-}ZP3]_n$), with a dimer located every 150 Å or so, and are cross-linked by ZP1 to form a very porous lattice approximately 6.4 μm in width. In this manner, tens of millions of copies of ZP2 and ZP3 are located on the external surface of the ZP. Non-covalent bonds between ZP1–3 maintain the organization of the extracellular coat. Therefore, ZP1–3 are structural glycoproteins.

In mice, free-swimming, acrosome-intact sperm recognize and bind to specific *O*-linked oligosaccharides located on the C-terminal portion of ZP3 (encoded by exon 7).[5,6,16–21] ZP3 apparently consists of an N-terminal domain and a C-terminal domain ('receptor domain'), separated from each other by a relatively short, immunoglobulin-like 'hinge' region. Each acrosome-intact sperm head possesses tens of thousands of sites that are capable of binding to ZP3. Once bound to ZP3 by plasma membrane overlying their head, sperm undergo the acrosome reaction (exocytosis), lose plasma membrane and outer acrosomal membrane, and remain associated with the ZP by binding to ZP2 by their inner acrosomal membrane. Thus, during fertilization ZP3 serves as a primary sperm receptor and acrosome reaction inducer, while ZP2 serves as a secondary sperm receptor.

Shortly after fertilization, both ZP3 and ZP2 are inactivated as sperm receptors by cortical granule enzymes released into the ZP.[3] Consequently, free-swimming sperm are not able to bind to the ZP of fertilized eggs.[16] In addition, the ZP undergoes structural changes after fertilization which make it a tougher extracellular coat.[3,22] The ZP surrounds the embryo throughout preimplantation development and is shed just prior to implantation of the blastocyst in the uterus.

Purification

ZP glycoproteins routinely are purified to homogeneity from excised ovaries. For example, mouse ovaries are homogenized, fractionated on a Percoll gradient, and the band of ZP isolated and fractionated by HPLC on a size-exclusion column to yield ZP1–3.[23] Each ovary yields approximately 250 ng each of ZP2 and ZP3, and much smaller amounts of ZP1.

Activities

ZP3 routinely is assayed for sperm receptor and acrosome reaction-inducing activities *in vitro*.[16–20] The former is a competition assay in which purified ZP3 competes for binding sites on the sperm head and activity is measured as inhibition of sperm binding to ovulated eggs. The results are compared to sperm binding in culture medium alone (i.e. no ZP3). The assay for induction of the acrosome reaction involves incubation of sperm in the presence of ZP3 and assessing the extent of the reaction by light microscopy of stained (Coomassie blue, G-250) preparations. The results are compared to sperm samples incubated in culture medium alone and in the presence of calcium ionophore (A23187), an acrosome reaction inducer.

Antibodies

ZP glycoproteins are highly immunogenic. Polyclonal and monoclonal antibodies have been raised against total ZP glycoproteins, purified ZP2 and ZP3, as well as against synthetic peptides derived from ZP2 and ZP3 polypeptide sequences.[24–27] Polyclonal antibodies raised in rabbits and chickens against purified mouse ZP1 cross-react with both ZP2 and ZP3.

Genes

There are many reports describing the cloning (cDNA and/or genomic) of ZP glycoprotein genes from different mammals. Among these are (GenBank/EMBL designations) mouse ZP1 (U24227–U24230), mouse ZP2 (M90366), mouse ZP3 (X14376), hamster ZP3 (M63629, M68924), pig ZP2 (D45064), pig ZP3 (D45065, L22169), rabbit ZP glycoprotein (M58160, L12167), cat ZP2 (D45067), cat ZP3 (D45068), dog ZP2 (D45069), dog ZP3 (D45070), human ZP2 (M90366), and human ZP3 (M35109).

Mutant phenotype

Targeted mutagenesis of mouse *ZP3* has been carried out in mouse embryonic stem (ES) cells and heterozygous ($mZP3^{+/-}$) and homozygous ($mZP3^{-/-}$) mutant mice have been obtained.[28–30] Eggs from heterozygous mutant female mice have a ZP, although it is approximately one-half the width (~2.7 μm) of the ZP of eggs from wild-type ($mZP3^{+/+}$) mice (~6.2 μm), and these females are fertile. Eggs from homozygous mutant female mice do not have a ZP and these females are infertile. Targeted mutagenesis of *ZP3* has no effect on the phenotype of male mice.

References

1. Wassarman, P. M. (1988). *Annu. Rev. Biochem.*, **57**, 415–42.
2. Dietl, J. ed. (1989). *The mammalian egg coat.* Springer, New York.
3. Yanagimachi, R. (1994). In *The physiology of reproduction*, Vol. 1, (ed. E. Knobil and J. D. Neill), pp. 189–318. Raven, New York.
4. Wassarman, P. M. (1990). *Development*, **108**, 1–18.
5. Wassarman, P. M. and Litscher, E. S. (1995). *Curr. Topics Dev. Biol.*, **30**, 1–19.

6. Wassarman, P. M. (1995). *Curr. Opin. Cell Biol.*, **7**, 658–64.
7. Snell, W. J. and White, J. M. (1996). *Cell*, **85**, 629–37.
8. Bleil, J. D. and Wassarman, P. M. (1980). *Dev. Biol.*, **76**, 185–202.
9. Kinloch, R. A., Roller, R. J., Fimiani, C. M., Wassarman, D. A., and Wassarman, P. M. (1988). *Proc. Natl Acad. Sci. USA*, **85**, 6409–13.
10. Liang, L., Chamow, S. M., and Dean, J. (1990). *Mol. Cell. Biol.*, **10**, 1507–15.
11. Epifano, O., Liang, L., and Dean, J. (1995). *J. Biol. Chem.*, **270**, 27254–8.
12. Lunsford, R. D., Jenkins, N. A., Kozak, C. A., Liang, L., Silan, C. M., Copeland, N. G., and Dean, J. (1990). *Genomics*, **6**, 184–7.
13. Castle, P. E. and Dean, J. (1996). *J. Reprod. Fertil.* **50** (Suppl.) 1–8.
14. Kinloch, R., Lira, S., Mortillo, S., Roller, R., Schickler, M., and Wassarman, P. M. (1993). In *Molecular basis of morphogenesis* (ed. M. Bernfield), pp. 19–33. Wiley-Liss, New York.
15. Wassarman, P. M. and Mortillo, S. (1991). *Int. Rev. Cytol.*, **130**, 85–110.
16. Bleil, J. D. and Wassarman, P. M. (1980). *Cell*, **20**, 873–82.
17. Florman, H. M. and Wassarman, P. M. (1985). *Cell*, **41**, 313–24.
18. Rosiere, T. K. and Wassarman, P. M. (1992). *Dev. Biol.*, **154**, 309–17.
19. Kinloch, R. A., Sakai, Y., and Wassarman, P. M. (1995). *Proc. Natl Acad. Sci. USA*, **92**, 263–7.
20. Litscher, E. S. and Wassarman, P. M. (1996). *Biochemistry*, **35**, 3980–6.
21. Chen, J., Litscher, E. S., and Wassarman, P. M. (1998). *Proc. Natl Acad. Sci. USA*, **95**, 6193–7.
22. Moller, C. C. and Wassarman, P. M. (1989). *Dev. Biol.*, **132**, 103–112.
23. Bleil, J. D. and Wassarman, P. M. (1986). *J. Cell Biol.*, **102**, 1363–71.
24. Greve, J. M., Salzmann, G. S., Roller, R. J., and Wassarman, P. M. (1982). *Cell*, **31**, 749–59.
25. East, I. J. and Dean, J. (1984). *J. Cell Biol.*, **98**, 795–800.
26. East, I. J., Gulyas, B. J., and Dean, J. (1985). *Dev. Biol.*, **109**, 268–73.
27. Liu, C., Litscher, E. S., and Wassarman, P. M. (1995). *Mol. Biol. Cell*, **6**, 577–85.
28. Liu, C., Litscher, E. S., Mortillo, S., Sakai, Y., Kinloch, R. A., Stewart, C. L., and Wassarman, P. M. (1996). *Proc. Natl Acad. Sci. USA*, **93**, 5431–6.
29. Rankin, T., Familiari, M., Lee, E., Ginsberg, A., Dwyer, N., Blanchette- Mackie, J., *et al.* (1996). *Development*, **122**, 2903–10.
30. Wassarman, P. M., Qi, H., and Litscher, E. S. (1997). *Proc. Roy. Soc.*, B **264**, 323–8.

Paul M. Wassarman
Department of Cell Biology and Anatomy,
Mount Sinai School of Medicine,
New York, NY 10029–6574, USA

Elastin

Elastin is of particular importance to the structural integrity and function of tissues in which reversible extensibility or deformability are crucial, such as the major arterial vessels, the lung and the skin. The physiological importance of elastin lies in its unique elastomeric properties, which arise from its ability to form a covalently cross-linked polymer and its unusual amino acid composition. The primary sequence of elastin reveals a repetitious domain structure, consisting of hydrophobic and cross-linking sequences which provides important clues as to the function of biologically active regions of the protein.

Protein properties

Elastin is secreted from the cell as a soluble protein of molecular weight approximately 70 000 Da, called tropoelastin. It has a low content of acidic amino acids and is correspondingly rich in hydrophobic residues, particularly glycine, alanine, valine, and proline. There is no methionine, tryptophan, histidine, or hydroxylysine, although small amounts of hydroxyproline are present. The tropoelastin molecule consists, for the most part, of alternating hydrophobic and cross-linking domains (Fig. 1). Of tropoelastin's 40 lysine residues, approximately 35 will serve as cross-links. The domains in elastin that contain the cross-linking residues are of two types, with the most prevalent consisting of lysine pairs located in alanine-rich sequences. In the second type, proline or other hydrophobic residues are found between and around the lysines instead of the usual alanines. All but one of these proline-containing cross-linking regions are found in the amino one-third of the molecule.

The hydrophobic domains of elastin are thought to provide important contributions to the protein's ability to undergo elastic recoil. These domains are enriched in glycine and bulky hydrophobic amino acids and most probably adopt β-turn structures.[1] The unusually high hydrophobicity of these domains has led to models of elasticity that take into account the possibility that non-polar side chains are exposed to water when elastin is stretched. Recoil then occurs when the non-polar groups reaggregate and expel the water after the distending force is removed.[2]

Elastic fibres can be recognized in the light microscope by characteristic staining reactions, and in the electron microscope by typical ultrastructural appearances (see

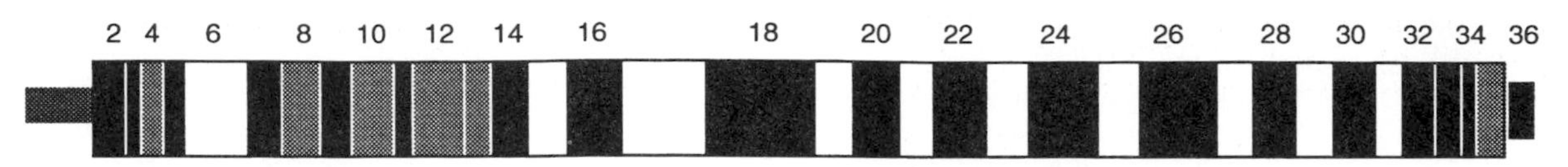

Figure 1. Diagram showing repeating domain structure of elastin. Each domain is encoded by an individual exon (numbered). Lysine-containing cross-link domains are of two types: those rich in alanine (white), and a second group usually containing proline residues (grey). Hydrophobic domains which, in many cases, contain repeating sequences, are indicated in black.

Fig. 2). Histochemical stains that selectively bind elastic fibres have enabled the visualization of elastic structure in many different tissues. It is clear from these studies that elastin is distributed as interconnected fibres in three morphologically distinct forms. In elastic ligaments, lung and skin, the fibres are small, rope-like, and variable in length. In major arteries, such as the aorta, elastic fibres form concentric sheets or lamellae (Fig. 2(a)), while in elastic cartilage a three-dimensional honeycomb arrangement of very large anastomosing fibres is apparent. These differing and complex structures are thought to arise as a consequence of the strength and direction of forces put upon the tissue.

The elastic fibre is a complex structure that contains elastin, microfibrillar proteins, lysyl oxidase, and, perhaps, proteoglycans. Elastin is the predominant protein of mature elastic fibres and endows the fibre with the characteristic property of elastic recoil.[3] Elastic fibres first appear in fetal development as aggregates of 10–12 nm microfibrils arranged in parallel array, often occupying infoldings of the cell membrane. Elastin then is deposited as small clumps of amorphous material within these bundles of microfibrils, which subsequently coalesce to form true elastic fibres (Figs. 2(b,c)). The relative proportion of microfibrils to elastin declines with increasing age, with adult elastic fibres having only a sparse peripheral mantle of microfibrillar material. Microfibrils are thought to align tropoelastin molecules in precise register so that cross-linking regions are juxtaposed. Cross-linking is initiated by the action of lysyl oxidase, which catalyses the oxidative deamination of lysine to allysine. This appears to be the only enzymatic step involved in elastin cross-linking. The subsequent formation of elastin cross-links, including the formation of the tetrafunctional amino acid isomers desmosine and isodesmosine, probably occurs as a series of spontaneous condensation reactions.[4] Once cross-linked, insoluble elastin is stable under normal physiological conditions, lasting for the life of the organism.

Purification

Elastin is isolated from elastic tissues by removing all other connective tissue and cellular components by

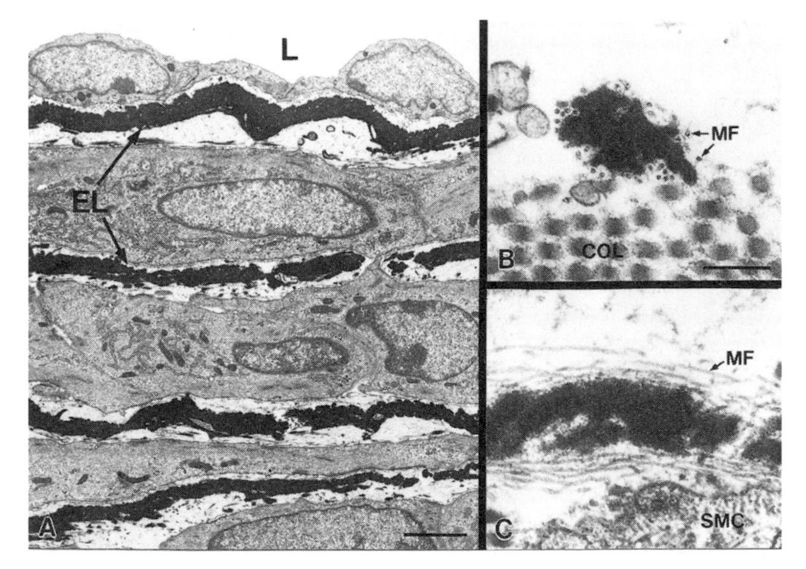

Figure 2. Electron micrographs of elastic fibres in a 5 day old mouse aorta. (A) In the developing aorta, elastic fibres fuse to form concentric elastic lamellae (EL) interposed between smooth muscle cells. Bar represents 30 μm. (B, C) At higher magnification, elastic fibres are seen to consist of two components, black amorphous elastin, and 10 nm microfibrils (MF). Note that the microfibrils have a tubular appearance when seen in cross-section (B). Panel C shows a longitudinal section. Bar represents 0.2 μm. L, lumen; COL, collagen; SMC, smooth muscle cell. (Courtesy of Elaine C. Davis.)

denaturation or degradation. The most successful purification procedures include subjecting the tissue to successive 1 h periods of autoclaving until no further protein appears in the supernatant,[5] or exposure to 0.1 N NaOH at 95°C for 45 min.[6] Elastin is the insoluble residue left behind after these treatments. Alternative purification methods include treatment of tissues with a combination of proteases, chaotropic and reducing agents, and cyanogen bromide.[7] Isolation of tropoelastin is much more difficult since the protein can only be extracted from tissues in which the cross-linking pathway is interrupted.[7,8]

Activites

Elastin is characterized by a high degree of elasticity, including the ability to deform to large extensions with small forces. The precise physicochemical properties that account for elastin's rubber-like characteristics, however, have not been fully characterized. Hydrophobic sequences within the elastin molecule have been shown to be chemotactic for both interstitial and inflammatory cells.[9,10]

Antibodies

Purified, mature elastin is a weak antigen and antisera of relatively low titre are obtained when the insoluble protein or solubilized peptides from the insoluble protein are used as antigens. Antibodies to tropoelastin have higher titres than antibodies to insoluble elastin and tend to show a broader range of cross-species reactivity. Both polyclonal and monoclonal antibodies to elastin have been described. In many cases, antigenicity of elastic fibre components can be enhanced by treating tissue sections or fixed cell preparations with 6 M guanidinium hydrochloride in 20 mM Tris (pH 8.0) containing 50 mM dithiothreitol, for 15 min. After washing with 20 mM Tris (pH 8.0), the samples are treated with 100 mM iodoacetamide in the dark for 15 min, washed, and stained using normal protocols. Commercially available antibodies to elastin and elastic fibre proteins are available from Sigma Immunochemical web site: http://www.sigma.sial.com and Elastin Products Company web site: http://www.elastin.com.

Genes

Full length cDNAs for human (GenBank M24782), bovine (GenBank J02717), chicken (GenBank M15889), rat (GenBank M60647, J05292), and mouse (U08210) elastins have been published. The mRNA for elastin is approximately 3.5 kb long of which 2.5 kb encodes the protein. The 3' end of the mRNA has been found to contain a large untranslated region which is highly conserved between species.[11] Analysis of human, bovine, and rat cDNAs have demonstrated alternative splicing of the primary transcripts.[11] S1 nuclease mapping of bovine nuchal ligament mRNA indicates that the majority of exons are spliced infrequently and that splicing in some tissues may be under developmental control.[12] The functional significance of alternative splicing is not known.

The human gene has been localized to chromosome 7q11.1–21.1 and all indications are that it exists as a single copy in all species.[13] In the human genome, the elastin gene contains an unusually large number of repetitive elements consisting of AluI repeats and large stretches of alternating purines or pyrimidines. A striking feature of the elastin gene is the small size of the translated exons (27–186 bp), which are interspersed in large expanses of introns. The intron to coding ratio is about 19:1. Another important characteristic is that coding sequences corresponding to hydrophobic and cross-link domains are found in separate exons that alternate in the gene.[11]

Mutant phenotype/disease states

Intragenic deletions, translocations, and complete deletion of the elastin gene have recently been linked to the disease supravalvular aortic stenosis.[14] In addition, deletion of one elastin allele has been implicated in the pathogenesis of Williams syndrome.[15,16] Other diseases associated with abnormalities in elastin production include forms of cutis laxa, Buschke–Ollendorff syndrome, and Hutchinson–Gilford progeria.[17]

Web sites

A web site devoted to elastin, and the elastic fibre can be found at http://ef.wustl.edu.

References

1. Mecham, R. P. and Davis, E. C. (1994). In *Extracellular matrix assembly and structure*. (eds. P. D. Yurchenko, D. E. Birk, and R. P. Mecham), pp. 281–314. Academic Press, San Diego.
2. Gosline, J. M. and Rosenbloom, J. (1984). In *Extracellular matrix biochemistry* (eds. K. A. Piez and A. H. Reddi), pp. 191–228. Elsevier, New York.
3. Partridge, S. M. (1962). *Adv. Prot. Chem.*, **17**, 227–302.
4. Reiser, K., McCormick, R. J., and Rucker, R. B. (1992). *FASEB J.*, **6**, 2439–49.
5. Partridge, S. M., Davis, H. F., and Adair, G. S. (1955). *Biochem. J.*, **61**, 11–21.
6. Lansing, A. I., Roberts, E., Ramasarma, G. B., Rosenthal, T. B., and Alex, M. (1951). *Proc. Soc. Exp. Biol. Med.*, **76**, 714–17.
7. Soskel, N. T., Wolt, T. B., and Sandberg, L. B. (1987). *Meth. Enzymol.*, **144**, 196–214.
8. Prosser, I. W., Whitehouse, L. A., Parks, W. C., Stahle-Bäckdahl, M., Hinek, A., Park, P. W., and Mecham, R. P. (1991). *Connect. Tiss. Res.*, **25**, 265–79.
9. Senior, R. M., Griffin, G. L., and Mecham, R. P. (1980). *J. Clin. Invest.*, **66**, 859–62.
10. Senior, R. M., Griffin, G. L., and Mecham, R. P. (1982). *J. Clin. Invest.*, **70**, 614–18.
11. Indik, Z., Yeh, H., Ornstein-Goldstein, N., and Rosenbloom, J. (1990). In *Genes for extracellular matrix proteins* (eds.

L. Sandell and C. Boyd), pp. 221–50. Academic Press, New York.
12. Parks, W. C. and Deak, S. B. (1990). *Am. J. Respir. Cell, Mol. Biol.*, **2**, 399–406.
13. Fazio, M. J., Mattei, M. G., Passage, E., Chu, M. -L., Black, D., Solomon, E., *et al.* (1991). *Am. J. Hum. Genet.*, **48**, 696–703.
14. Keating, M. (1995). *Circulation*, **92**, 142–7.
15. Ewart, A. K., Morris, C. A., Atkinson, D., Jin, W., Sternes, K., Spallone, P., *et al.* (1993). *Nature Genet.*, **5**, 11–16.
16. Joyce, C. A., Zorich, B., Pike, S. J., Barber, J. C. K., and Dennis, N. R. (1996). *J. Med. Genet.*, **33**, 986–92.
17. Davidson, J. M., Zang, M. C., Zoia, O., and Giro, M. G. (1995). *Ciba Found. Symp.*, **192**, 81–94.

Robert Mecham
Washington University School of Medicine,
Department of Cell Biology and Physiology,
Box 8228, 660 South Euclid Avenue,
St. Louis, MO 63110, USA

Fibrinogen/fibrin

Fibrinogen is a soluble plasma protein which after cleavage by α-thrombin, a highly specific trypsin-like enzyme generated at sites of vascular lesion, is converted to fibrin monomer molecules. The latter, in turn, self-associate to form an insoluble homopolymeric structure, the fibrin clot. Moreover, fibrinogen binds to platelets, contributing to the formation of the platelet thrombus, as well as to endothelial cells and leukocytes, thus participating in inflammatory responses. The congenital deficiency of fibrinogen results in a bleeding disorder, while increased plasma levels are associated with heightened arterial thrombotic risk.

Synonymous names

None.

Homologous proteins

The three genes coding for the fibrinogen Aα, Bβ, and γ chains have arisen from a common ancestral gene through a series of duplication events initiated perhaps almost a billion years ago.[1] The first duplication led to Aα chains evolving independently from Bβ–γ chains, while the latter two diverged more recently. This explains why Aα chains show the least homology between species, as well as to the other two chains within species, while Bβ and γ chains show about 35 per cent homology when compared within species, and even greater homology when each of them is compared between species. A recent structure-guided sequence alignment of putative fibrinogen-related proteins, based on the structure of the crystallized 30 kDa globular C terminus of the human fibrinogen γ chain, has included T-lymphocyte proteins; ficolins and invertebrate proteins like the scabrous protein of *Drosophila melanogaster*; tenascins and homologous proteins.[2] Moreover, a putative fibrinogen-like molecule in the invertebrate sea cucumber shows homologies with the C terminal regions of vertebrate fibrinogen Bβ and γ chains.

Protein properties

Fibrinogen is composed of three pairs of non identical chains, Aα, Bβ, and γ.[1,3] In the predominant form circulating in blood, Aα contains 610 amino acid residues and has an apparent molecular mass of 67 kDa; Bβ has 461 residues and 56 kDa mass; and γ has 411 residues and 47 kDa mass. The molecule, with a total mass of 340 kDa, appears as a trinodular rod when visualized by electron microscopy, approximately 48 nm long and 7 nm wide. The central nodular E domain consists of the N-terminal regions of the three polypeptides. The globular ends of the molecule form the D domain, containing the C termini of Bβ and γ chains. D and E domains are connected by the coiled-coil regions of the three chains, while the Aα C terminus appears as a flexible appendix of the oblong molecule (Fig. 1). Mammalian fibrinogen is synthesized in hepatocytes. Although synthesis has also been observed in the megakaryocytes of some species, it may not occur in the human counterpart.[4] Each of the three fibrinogen chains is encoded by a separate gene independently transcribed and translated, but coordinated expression may be controlled by homologous 5' untranslated regions preceding each coding sequence.[5] Variations in the length of the fibrinogen chains are known to occur. The Aα cDNA codes for 15 additional carboxyl terminal residues, bringing the total for the mature polypeptide to 625, but this form has never been isolated from plasma presumably because of proteolytic cleavage. A variant form of the γ chain, γ', is 16 residues longer (427 versus 411) as a result of alternative mRNA splicing. The γ' fibrinogen variant is as effective as γ fibrinogen in clotting, but has decreased affinity for platelets.[6] This may explain why both forms are present in plasma (approximately 10 per cent of the total contains γ'), but only γ chain fibrinogen is present in

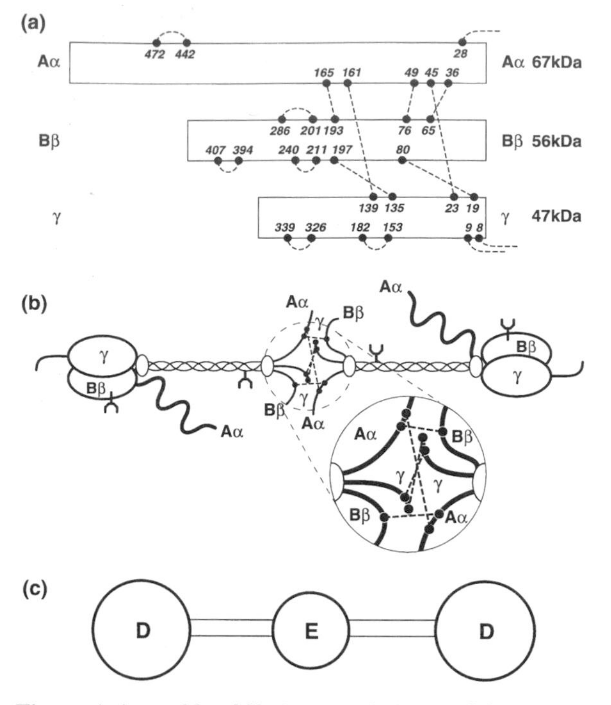

Figure 1. Assembly of fibrinogen chains, and domain structure. (a) Identification of Cys residues involved in intrachain, and interchain disulphide bonds in a fibrinogen half molecule. The three Cys residues thought to be involved in symmetric bonds linking two half molecules in mature fibrinogen are also indicated. (b) The amino terminal domain, containing all the interchain disulphide bridges, is connected to the carboxyl terminal domains, containing the intrachain disulphide bridges, by the coiled-coil regions. Note the Aα carboxyl terminus appearing as an appendix extending away from the globular carboxyl termini of the Bβ, and γ chains. The two glycosylation sites in each half molecule are indicated by forked symbols. The magnified depiction of the amino terminal ends of the three pairs of chains shows the disulphide bonds linking two half fibrinogen molecules together. The bond between Aα, and Bβ chains involves Cys-36, and Cys-65, respectively (shown in (a) as linking two chains in the same half molecule), and has recently been indicated as necessary for correct assembly of mature fibrinogen.[1,3] Whether this latter bond is present in all fibrinogen molecules remains to be established. (c). Schematic representation of the trinodular structure of fibrinogen as it can be visualized by electron microscopy. The E domain corresponds to the N termini of the three pairs of chains, while each D domain contains the C terminus of one Bβ, and γ chain, and the random coil region near the C terminus of the Aα chain. The flexible C terminus of the Aα chain is not shown in this type of representation. (Modified from refs 1 and 30.)

platelets.[7] A third variant form of the γ chain, similar to γ' but lacking the last four C-terminal residues, has also been reported.[8] A fibrinogen variant of considerably greater molecular mass than the prevalent species, 420 versus 340 kDa, can be found in plasma, and results from differential splicing of a sixth exon in the Aα gene transcript leading to a mRNA containing the coding sequence for 236 additional amino acids.[9] In this variant fibrinogen, the Aα chain (Aα_E) has an apparent molecular mass of 110 kDa.

The three fibrinogen chains are synthesized on different polysomes. The cells have surplus pools of Aα and γ chains, as well as Aα–γ complexes, but no free Bβ chains. Thus, the rate limiting step for human fibrinogen assembly appears to be the synthesis of the B chain, but species differences are known to occur in this regard. Chain assembly begins with the attachment of Aα and γ chains to newly synthesized Bβ chains. The Aα–Bβ and Bβ–γ complexes thus formed are released from polysomes, and in the lumen of the rough endoplasmic reticulum react with the respective missing chain to form half fibrinogen molecules (Aα–Bβ–γ). Subsequent combination of two half molecules leads to formation of complete fibrinogen. This complex process of assembly of three different chains into a homodimeric molecule occurs through interchain disulphide bonding. There are a total of 29 disulphide bridges in fibrinogen (Fig. 1), 11 of which, all interchain, are present in the region formed by the N-terminal ends of the three chains, the so called N-terminal disulphide knot (*N*-DSK). Twelve disulphide bonds are intrachain, one in each Aα, three in each Bβ and two in each γ chain. The disulphide connections holding the two half molecules together are still a matter of debate. A well accepted model assigned this function to symmetrical bonds between Aα Cys-28 and γ Cys-8 and Cys-9, but it now appears that a bond between Aα Cys-36 in a half molecule and Bβ Cys-65 in the other may be necessary for final assembly of fibrinogen[1,3] (Fig. 1). Glycosylation, phosphorylation, and sulfation occur in the Golgi apparatus before secretion. One *N*-linked carbohydrate chain exists on each of the Bβ (Asn-364) and γ (Asn-52) chains (Fig. 1).

Fibrinogen has complex biological roles in haemostasis, wound repair and inflammation. The active serine protease, α-thrombin, cleaves an Arg–Gly bond in the N terminus of the Aα chain, releasing the 16-residue fibrinopeptide A (FPA). The exposed end of the α chain (so designated after release of FPA) interacts with sites in the γ chain carboxyl terminus and the α chain carboxyl terminus of neighbouring molecules, initiating the non-covalent end-to-end and lateral assembly of the two-molecule thick fibrin protofibrils (Fig. 2). Subsequently, a second thrombin cleavage releases a 14-residue peptide (FPB) from the N terminus of the Bβ chain, and the exposed new β N terminus reinforces lateral aggregation, resulting in the formation of thick fibrin fibres. Ca^{2+}-binding sites in both the E and D domains are important in polymerization. Each fibrinogen molecule contains

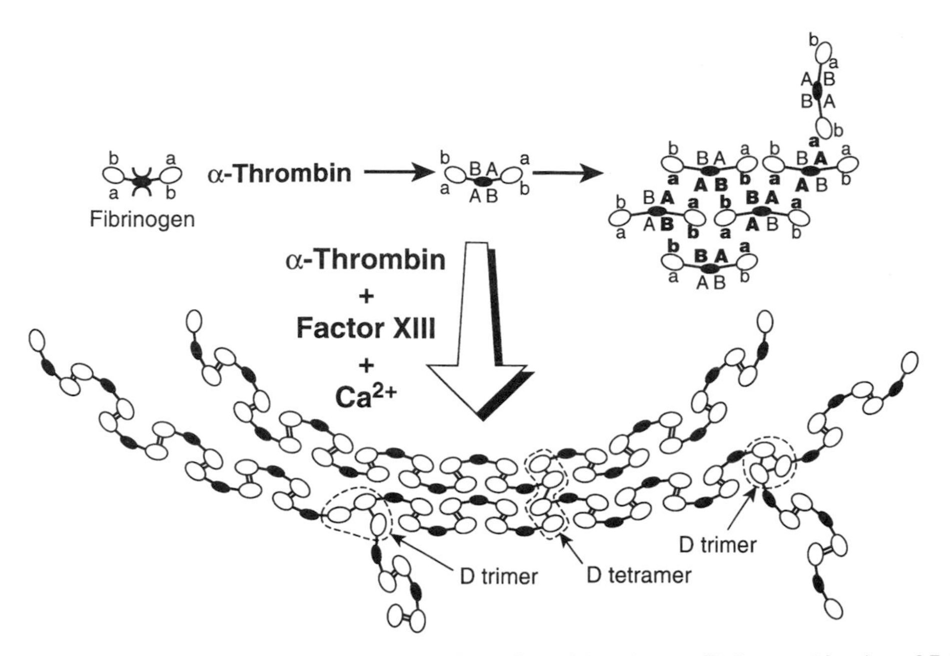

Figure 2. The process of fibrin formation. Top: cleavage by α-thrombin releases fibrinopeptides A, and B from the amino terminal ends of Aα, and Bβ chains, respectively, thus uncovering two polymerization sites (designated here as a and b). Complementary sites (designated a, and b, respectively) are present in the D domains. Site a is primarily localized in the γ chain carboxyl terminal end, but other structures in the D domain may contribute to its function. Site b may be localized in the carboxyl-terminal domain of the Aα chain, rather than in the D domain, and may be important for the branching of fibres. Bottom: a specific cleavage by α-thrombin also activates the transglutaminase, Factor XIII, that catalyses the calcium-dependent formation of lysyl–glutaminyl covalent bonds within the assembled fibrin fibres. Intermolecular cross-linking involving γ, and α chains renders fibrin insoluble in urea, and other solvents that can dissolve non-cross-linked fibrin, a reflection of the greater mechanical stability of cross-linked fibrin.

three high affinity and 10 or more low affinity Ca^{2+}-binding sites; the latter correspond to sialic acid residues. After assembly of the fibrin polymer, the transglutaminase, factor XIII, stabilizes the structure by cross-linking covalently the α and γ chains[10] (Fig. 2). This step is essential for normal haemostasis and wound repair. Plasmin, a serine protease, proteolytically degrades fibrin, as well as fibrinogen, releasing degradation products (FDPs) that may have a variety of possible pathophysiological effects. Clot remodelling through fibrinolysis is also essential for the regulation of normal haemostasis.

In addition to the role in clotting, fibrinogen binds to a specific platelet receptor, the integrin αIIbβ3 (glycoprotein IIb–IIIa complex), and performs an important, although not unique, role in mediating platelet adhesion[11] and aggregation[12] on to thrombogenic surfaces. Fibrinogen also interacts with other integrin receptors, αvβ3[13] and α5β1[14] on endothelial cells, αMβ2 on leukocytes,[15] as well as non-integrin receptors, ICAM-1 on endothelial cells,[16] participating in vascular remodelling and inflammation through mechanisms that remain to be elucidated fully with respect to their pathophysiological significance. The fibrinogen sites involved in binding to different cellular receptors have been identified with some precision. The two related β3 integrins bind selectively to distinct regions of the Aα and γ chains: αvβ3 to the RGDS sequence at Aα 572–575,[13] to which α5β1 has also been shown to bind,[14] and αIIbβ3 to the carboxyl terminus of the γ chain.[13,17] The binding to αMβ2 and ICAM-1 is mediated by distinct sequences in the γ chain, involving residues 190–202[15] and 117–133,[18] respectively.

Purification

Fibrinogen is purified from plasma using differential precipitation methods with either glycine[19] or polyethylene glycol.[20] During the purification procedure, care must be taken to avoid proteolytic degradation, particularly cleavage of the C terminus of the Aα chain.

Activities

The clotting of fibrinogen is evaluated directly by addition of thrombin or by inducing thrombin generation in plasma and measuring the time needed for the appearance of an insoluble gel.[21] The stabilization of the fibrin network by factor XIII is tested by evaluating the resistance of fibrin clots to solubilization by urea.[21] Fibrinogen interaction with platelets is measured by direct binding assays[22] and is reflected in agonist-induced aggregation.[12] Fibrinogen interaction with endothelial cells and leukocytes is measured by direct binding assays[13–15] as well as in models of leukocyte adhesion to endothelial cell monolayers.[16] The effects of plasmin on fibrin and fibrinogen are evaluated by measuring the time to clot lysis or the generation of specific degradation products.[21]

Antibodies

Numerous polyclonal and monoclonal antibodies against fibrinogen, its plasmic degradation products, and both FPA and FPB are available commercially. Monoclonal anti-peptide antibodies have been characterized which react specifically with the platelet adhesion sites.[13,23] Certain monoclonal antibodies react only with surface-bound, not soluble fibrinogen, as in the case of fibrinogen bound to its platelet receptor.[24] Other antibodies bind to neoantigens on fibrin and not to fibrinogen.[25,26]

Genes

Each of the fibrinogen chains is encoded by single copies of three distinct genes, located in close proximity within a 50 kb span on the long arm of chromosome 4, region q23–q32. The GenBank accession numbers are J00128 for the human fibrinogen Aα chain mRNA, M64983 for the Bβ chain, and M10014 for the γ and γ' chain genes. The complete coding sequence for the extended Aα_E variant has accession number M58569. Sequences are also deposited for the Aα and Aα_E chains of chicken and the γ chain of lamprey.

A mouse model of afibrinogenaemia, generated by gene targeting, has produced the expected phenotype characterized by bleeding, in most instances non fatal, but has also revealed an unexpected as well as striking failure of pregnancy.[27] Moreover, the endogenous mouse γ chain has been selectively modified by gene targeting to express a mutant protein lacking the last five carboxyl terminal residues.[28] As expected, these mice had normal clotting but impaired platelet function, confirming the location of the αIIbβ3 binding site determined for human fibrinogen.[13,17]

Mutant phenotypes/disease states

Quantitative defects of the fibrinogen molecule (afibrinogenaemia or hypofibrinogenaemia) result in an abnormal bleeding tendency,[29] while qualitative defects (dysfibrinogenaemia) may cause either bleeding or hypercoagulability.[1] Different forms of dysfibrinogenaemia, caused by distinct mutations, may result in abnormalities of fibrinopeptide release, fibrin polymerization, fibrin crosslinking, or fibrin degradation by plasmin.

Structure

The crystal structure of a 30 kDa carboxyl terminal fragment from the γ chain of human fibrinogen was reported first, and coordinates for the three crystal forms obtained have been deposited in the Protein Data Bank with identifiers 1FIB, 1FIC, and 1FID, respectively.[2] More recently, the complete structure of human fibrinogen fragment D and its crosslinked counterpart from fibrin has also been published.[31] These studies have provided the first detailed view of an important fibrin polymerization domain.

References

1. Blomback, B. (1996). *Thromb. Res.*, **83**, 1–75.
2. Yee, V. C., Pratt, K. P., Cote, H. C. F., Le Trong, I., Chung, D. W., Davie, E. W., *et al.* (1997). *Structure*, **5**, 125–38.
3. Doolittle, R. F., Everse, S. J., and Spraggon, G. (1996). *FASEB J.*, **10**, 1464–70.
4. Louache, F., Debili, J., Cramer, E., Breton-Gorius, J., and Vainchenker, W. (1991). *Blood*, **77**, 311–16.
5. Fowlkes, D. M., Mullis, N. T., Comeau, C. M., and Crabtree, G. R. (1984). *Proc. Natl Acad. Sci. USA*, **81**, 2313–16.
6. Harfenist, E. J., Packham, M. A., and Mustard, J. F. (1984). *Blood*, **64**, 1163–8.
7. Francis, C. W., Nachman, R. L., and Marder, V. J. (1984). *Thromb. Haemost.*, **51**, 84–8.
8. Francis, C. W., Muller, E., Henschen, A., Simpson, P. J., and Marder, V. J. (1988). *Proc. Natl Acad. Sci. USA*, **85**, 3358–62.
9. Fu, Y. and Grieninger, G. (1994). *Proc. Natl Acad. Sci. USA*, **91**, 2625–8.
10. Doolittle, R. F. (1984). *Annu. Rev. Biochem.*, **53**, 195–229.
11. Savage, B., Saldivar, E., and Ruggeri, Z. M. (1996). *Cell*, **84**, 289–97.
12. Ikeda, Y., Handa, M., Kawano, K., Kamata, T., Murata, M., Araki, Y., *et al.* (1991). *J. Clin. Invest.*, **87**, 1234–40.
13. Cheresh, D. A., Berliner, A. S., Vicente, V., and Ruggeri, Z. M. (1989). *Cell*, **58**, 945–53.
14. Suehiro, K., Gailit, J., and Plow, E. F. (1997). *J. Biol. Chem.*, **272**, 5360–6.
15. Altieri, D. C., Plescia, J., and Plow, E. F. (1993). *J. Biol. Chem.*, **268**, 1847–53.
16. Languino, L. R., Plescia, J., Duperray, A., Brian, A. A., Plow, E. F., Geltosky, J. E., and Altieri, D. C. (1993). *Cell*, **73**, 1423–34.
17. Kloczewiak, M., Timmons, S., Lukas, T., and Hawiger, J. (1984). *Biochemistry*, **23**, 1767–74.
18. Altieri, D. C., Duperray, A., Plescia, J., Thornton, G. B., and Languino, L. R. (1995). *J. Biol. Chem.*, **270**, 696–9.
19. Kazal, L. A., Amsel, S., Miller, O. P., and Tocantins, L. M. (1963). *Proc. Soc. Exp. Biol. Med.*, **113**, 989–94.
20. Masri, M. A., Masri, S. A., and Boyd, N. D. (1983). *Thromb. Haemost.*, **49**, 116–9.
21. Mammen, E. F. (1993). In *Hematology clinical and laboratory practice* (ed. R. L. Bick, J. M. Bennett, R. K. Brynes, M. J. Cline, L. Kass, G. Murano, *et al*), pp. 1421–33. Mosby-Year Book, St. Louis.
22. Marguerie, G. A., Plow, E. F., and Edgington, T. S. (1979). *J. Biol. Chem.*, **254**, 5357–63.
23. Savage, B., Bottini, E., and Ruggeri, Z. M. (1995). *J. Biol. Chem.*, **270**, 28812–17.
24. Ugarova, T. P., Budzynski, A. Z., Shattil, S. J., Ruggeri, Z. M., Ginsberg, M. H., and Plow, E. F. (1993). *J. Biol. Chem.*, **268**, 21080–7.

25. Scheefers-Borchel, U., Muller-Berghaus, G., Fuhge, P., Eberle, R., and Heimburger, N. (1985). *Proc. Natl Acad. Sci. USA*, **82**, 7091–5.
26. Schielen, W. J., Voskuilen, M., Tesser, G. I., and Nieuwenhuizen, W. (1989). *Proc. Natl Acad. Sci. USA*, **86**, 8951–4.
27. Suh, T. T., Holmback, K., Jensen, N. J., Daugherty, C. C., Small, K., Simon, D. I., *et al.* (1995). *Genes Dev.*, **9**, 2020–38.
28. Holmback, K., Danton, M. J. S., Suh, T. T., Daugherty, C. C., and Degen, J. L. (1996). *EMBO J.*, **15**, 5760–71.
29. al-Mondhiry, H. and Ehmann, W. C. (1994). *Am. J. Hematol.*, **46**, 343–7.
30. Dang, C. V., Bell, W. R., and Shuman, M. (1989). *Am. J. Med.*, **87**, 567–76.
31. Spraggon, G., Everse, S. J., and Doolittle, R. F. (1997). *Nature*, **389**, 455–62.

Zaverio M. Ruggeri
The Scripps Research Institute,
10550 North Torrey Pines Road,
La Jolla, California 92037, USA

Fibromodulin

Fibromodulin is a keratan sulphate proteoglycan present in many types of connective tissues, such as cartilage and tendon. Fibromodulin belongs to a family of structurally related leucine rich repeat (LRR) extracellular matrix proteoglycans and glycoproteins such as decorin, biglycan, lumican chondroadherin, PRELP, and epiphycan.[1] Fibromodulin binds to collagen and affects collagen fibrillogenesis.

Synonomous names

59 kDa cartilage protein.

Protein properties

The fibromodulin protein backbone consists of 357 amino acid residues (42 kDa) which can be divided into three structural domains.[2] The N-terminal domain has four cysteine residues of which two are involved in an intrachain disulphide bond. This region of the protein also contains five to seven closely spaced tyrosine sulphate residues.[3] The central domain, which constitutes 60 per cent of the protein, consists of ten repeats of some 25 amino acid residues. This central repeat domain, with preferentially leucine residues in conserved positions, is most homologous to similar repeats in lumican[4] but also to repeats in other LRR-proteins.[1] The C-terminal domain contains two cysteine residues, which form an intrachain disulphide bond.

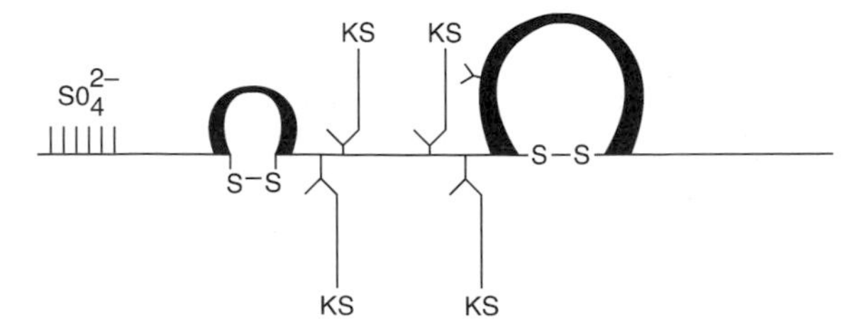

Figure 1. Schematic illustration of fibromodulin. Keratan sulphate chains (KS) and tyrosine sulphate residues (SO_4^{2-}) are indicated.

Fibromodulin from cartilage, tendon, and sclera contains asparagine-linked keratan sulphate chains.[3,5] Four of the five potential *N*-glycosylation sites in fibromodulin from bovine cartilage are substituted with keratan sulphate chains.[5]

Purification

Fibromodulin can be purified from cartilage using CsCl gradient centrifugation and ion exchange chromatography.[6]

Activities

Fibromodulin binds to type I and II collagen with a K_d of 35 nM.[7] The protein also delays the collagen fibrillation *in vitro* and causes the formation of thinner fibrils. Also mice lacking fibromodulin have abnormal collagen fibrils in tendons and cartilage (Svensson, Reinholdt, Fässler, Heinegård, and Oldberg; unpublished data). Fibromodulin and decorin knockout mice have similar phenotypes.[8]

Antibodies

Antibodies raised against the bovine protein cross-react with human and mouse fibromodulin.[6]

Genes

The human gene is composed of three exons[9] and is present as a single copy on chromosome 1q32.[10] Also, the lumican gene is composed of three exons and have similar exon/intron junctions as the fibromodulin gene, indicting a close relationship between the two proteins.[11] The bovine fibromodulin cDNA has the EMBL Data Bank number X16483.

Mutant phenotype

Mice lacking fibromodulin have abnormal collagen fibrils in tendon and cartilage (Svensson, Reinholt, Fässler, Heinegård, and Oldberg, unpublished data).

Structure

Not available.

References

1. Iozzo, R. V. and Murdoch, A. D. (1996). *FASEB J.*, **10**, 598–614.
2. Oldberg, Å, Antonssaon, P., Lindblom, K., and Heinegård, D. (1989). *EMBO. J.*, **8**, 2601–6.
3. Antonsson, P., Heinegård, D., and Oldberg, Å. (1991). *J. Biol. Chem.*, **266**, 16859–61.
4. Blochberger, T. C., Vergnes, J-P., Hempel, J., and Hassel, J. R. (1992). *J. Biol. Chem.*, **267**, 347–52.
5. Plaas, A. H. K., Neame, P. G., Nivens, C. M., and Reiss, L. (1990). *J. Biol. Chem.*, **265**, 20634–40.
6. Heinegård, D., Larsson, T., Sommarin, Y., Franzen, A., Paulsson, M., and Hedbom, E. (1986). *J. Biol. Chem.*, **261**, 13866–72.
7. Hedbom, E. and Heinegård, D. (1989). *J. Biol. Chem.*, **264**, 6898–905.
8. Danielson, K. G., Baribault, H., Holmes, D. F., Graham, H., Kadler, K.E., and Iozo, R. V. (1997). *J. Cell Biol.*, **136**, 729–43.
9. Antonsson, P., Heinegård, D., and Oldberg, Å. (1993). *Biochim. Biophys. Acta*, **1174**, 204–6.
10. Plaas, A. H. K., Neame, P. J., Nivens, C. M., and Reiss, L. (1990). *J. Biol. Chem.*, **265**, 20634–40.
11. Grover, J., Chen, X.-N., Korenberg, J. R., and Roughley, P. J. (1995). *J. Biol. Chem.*, **270**, 21942–9.

Åke Oldberg
Department of Cell/Molecular Biology, PO Box 94,
University of Lund, S22100 Lund, Sweden.

Fibronectins

Fibronectins[1,2] are high molecular weight glycoproteins found in many extracellular matrices and in blood plasma. They promote cell adhesion and affect cell morphology, migration and differentiation and cytoskeletal organization. Each subunit is made up of a series of repeating units which in turn form structural and functional domains specialized for binding tó cell surface integrin receptors or other extracellular matrix molecules. Different fibronectin isoforms arise by alternative splicing of the transcript of a single gene. Fibronectins have been found in all major groups of vertebrates but there are no convincing reports of invertebrate fibronectins.

Protein properties

The subunits of fibronectins vary in size between approximately 235 000 and 270 000 Da plus carbohydrate (see below). Each subunit is made up largely of repeating modules of three types (I, II, III). There are 12 type I repeats, each around 45 amino acids long, clustered in three groups (see Fig. 1), two adjacent type II repeats, each 60 amino acids long, and 15–17 type III repeats, each about 90 amino acids long. Type I and II repeats each contain two disulphide bonds whereas type III repeats lack disulphide bonds. There are two free cysteines per subunit. All three types of fibronectin repeat have been identified in other proteins. Fibronectin type III repeats are particularly widespread, being found in many ECM proteins, cell adhesion proteins, and cell surface receptors (kinases and phosphatases and cytokine receptors, see refs 3 and 4). Each repeat is encoded by one or two exons with introns precisely separating repeats. The initial secreted form of fibronectins is a dimer of two subunits held together by a pair of disulphide bonds near their C termini. Dimeric fibronectins are soluble molecules but, in extracellular matrix fibrils (Fig. 2), fibronectins are further disulphide-bonded into high molecular weight polymers.[1,2]

The variations in subunit size arise from alternative splicing of three segments; two type III repeats (EIIIA or ED.A and EIIIB or ED.B) and a third non-homologous segment known as V or IIICS (see Fig. 1). These three alternative splices occur in mammals, birds, and amphibians, although the precise details of splicing of the V region vary among species. A wide variety of cell types and tissues express fibronectins. The splicing patterns are cell-type specific and regulated during development and physiological processes.[1,2,5,6]

Each subunit carries 5–7 asparagine-linked complex carbohydrate side chains and one or two *O*-linked chains. Fibronectins are typically about 5 per cent carbohydrate but, in some tissues, higher levels of glycosylation occur, usually through further elaboration of the *N*-linked side

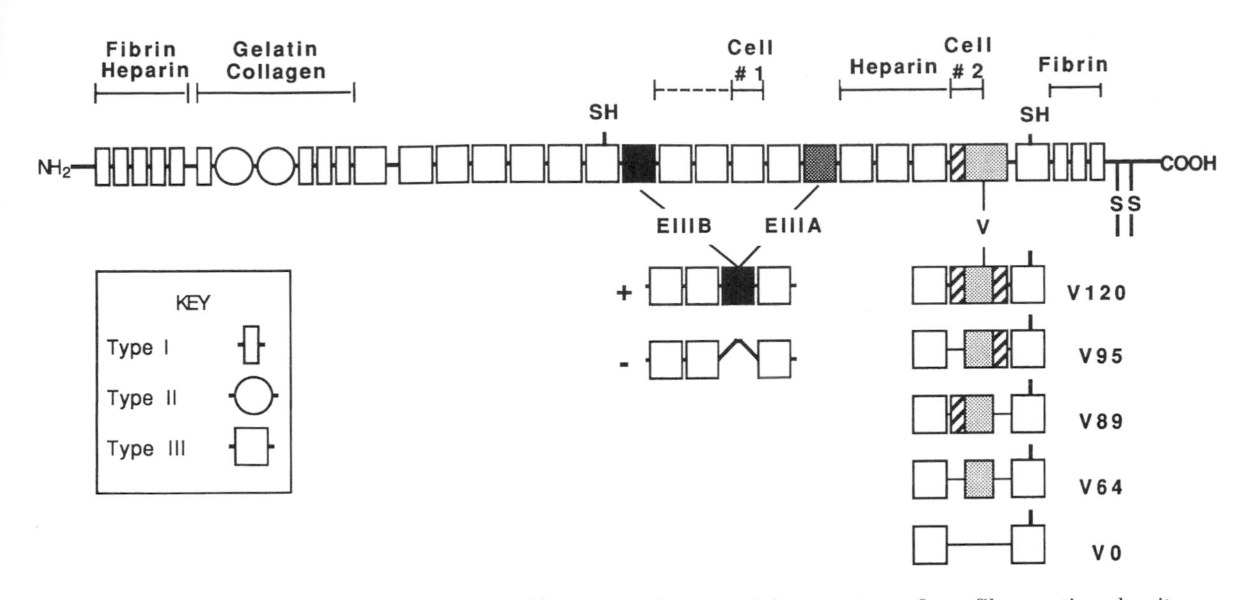

Figure 1. Structure and variants of fibronectins.[1,2] Diagram shows modular structure of one fibronectin subunit composed of three types of repeat. At three positions (EIIIB, EIIIA, and V) alternative splicing produces variations in structure.[5,6] The splicing of the V region is that seen in humans where the V region varies from 0 to 120 amino acids by inclusion or omission of three segments. Binding sites for other molecules and for cells are marked. The two cell-binding sites are recognized by different integrin receptors; #1, comprising three residues, RGD, is recognized by $\alpha 5\beta 1$ integrin while #2 which comprises EILDV is recognized by $\alpha 4\beta 1$ integrin; the efficacy of cell binding site #1 is enhanced by so-called synergy sequences N terminal to it.[28,29]

chains. Other post-translational modifications include tyrosine sulfation of the V region and serine/threonine phosphorylation of several sites near the C terminus. The functions of these post-translational modifications are unknown except that glycosylation protects the protein from proteolysis.[1,2]

The repeating modules of fibronectin subunits fold independently with 20–35 per cent β structure and no α helix. The molecule as a whole is extended and is best characterized as a tightly packed string of beads which is flexible. In solution these strings of beads are partially folded, giving a Stokes radius of 10–11 nm and sedimentation coefficient, $S_{20,w}$ of 13–15. At extremes of salt or pH or when assembled in extracellular matrix fibrils the subunits unfold into elongated forms 2–3 nm in diameter; each subunit is 60–70 nm long (Fig. 2).

Extended polypeptide segments in certain parts of the molecule are highly susceptible to proteolysis, which generates a series of protease-resistant domains, each comprising several of the repeating modules (Fig. 1). These domains contain a variety of binding sites for other molecules, including collagens, fibrin, heparin/heparan sulphate, and cell surface receptors (integrins).[1,2,7–9]

Fibronectins are widely expressed in embryos and adults especially in regions of active morphogenesis, cell migration, and inflammation. Tumor cells show reduced levels of fibronectin and levels in plasma fall in various forms of trauma. In contrast, fibronectin levels are elevated during wound healing and fibrosis and the pattern of alternative splicing is often altered in pathological situations.[1,2,5,6]

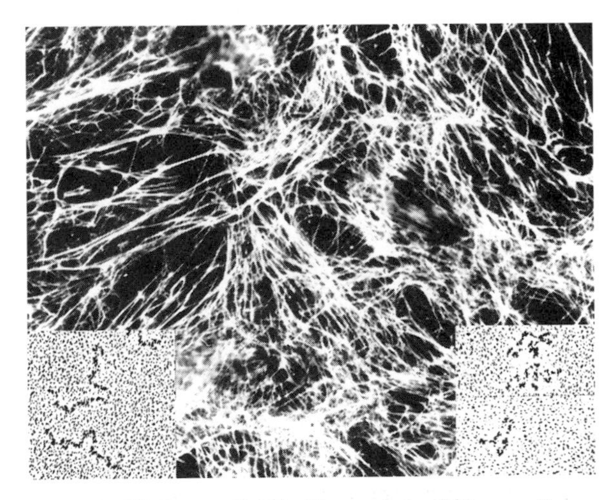

Figure 2. (Main panel) Fibrillar matrix of fibronectin surrounding cells of a confluent layer of fibroblasts as revealed by immunofluorescence. Insets: Fibronectin molecules in extended (high salt) form (left) and compact (low salt) form (right) as revealed by rotary shadowing electron microscopy.[30]

Purification

Fibronectins are most commonly purified[1,2,10] by their affinity for denatured collagen (gelatin). Fibronectins bind firmly and specifically to gelatin–Sepharose and can be eluted by chaotropes or low pH.

Activities

Fibronectins promote the adhesion and spreading of many cell types by binding to several different integrin receptors.[7,8,9] Fibronectin–integrin interactions also promote cell migration and assembly of actin microfilament bundles.[1,2] The integrin receptors act to assemble cytoskeletal structures and also trigger signal transduction cascades within the cells.

Antibodies

Although relatively conserved in sequence among species, fibronectins are highly immunogenic. Polyclonal and monoclonal antibodies are readily raised and many are available commercially. The polyclonal sera tend to be widely species cross-reactive, while the monoclonals tend to be species-specific.

Genes

There is a single fibronectin gene in species where this question has been carefully studied. The human gene is at 2q32–36 and the mouse gene is in a syntenic segment of chromosome 1.[11] Complete cDNA sequences have been published for human (GenBank X02761, A14133, X027621),[12] rat (GenBank X15906)[13,14] and *Xenopus laevis* (GenBank M77820)[15] and partial sequences for mouse (GenBank M18194), chicken (GenBank M21554/5, X06533), rabbit, cow, and an amphibian, *P. waltii* (X66813).

A complete amino acid sequence is available for bovine plasma fibronectin.[16] Genomic clones for human (X07717/8),[17] rat (GenBank X05831/2/3/4),[13,14] and chicken[18,19] are available.

Mutant phenotype

The development of mouse strains with null mutations in the fibronectin gene[20,21] demonstrates conclusively the importance of fibronectin for completion of development, since homozygous null embryos die early in gestation without forming somites or notochord and with defects in their vasculature. However, somewhat surprisingly, gastrulation initiates normally and extensive mesodermal migration occurs. This was unexpected given earlier results on injection of anti-fibronectin antibodies into amphibian or avian embryos. These knockout mice are available from Jackson Laboratories.

Structures

Three-dimensional structures have been reported for the tenth type III repeat, containing the RGD site for integrin binding[22,23] and for a set of four repeats ending with the tenth.[24] The type III repeats comprise a sandwich of two antiparallel β sheets apposed via hydrophobic surfaces. This structure is similar to an immunoglobulin domain, although the chain topology is different and there is no detectable sequence homology. Thus, FnIII domains and Ig domains appear to represent an example of convergent evolution to similar stable structures. The RGD sequence in the tenth FnIII repeat of fibronectin is in an extended and flexible loop between two β strands and is readily accessible for binding by the multiple integrins which recognize this sequence in fibronectin. The synergy site in the adjacent repeat (which enhances cell adhesion) is on the same side of the molecule and in an excellent position to bind to the same integrin recognizing the RGD.[24]

NMR structures have also been reported for type I[25,26] and type II[27] repeats.

References

1. Hynes, R. O. (1990). *Fibronectins*. Springer, New York.
2. Mosher, D. F., ed. (1989). *Fibronectin*. Academic Press, New York.
3. Bork, P. and Doolittle, R. F. (1992). *Proc. Natl Acad. Sci. USA*, **89**, 8990–4.
4. Bazan, J. F. (1990). *Proc. Natl Acad. Sci. USA*, **87**, 6934–8.
5. Ffrench-Constant, C. (1995). *Exp. Cell Res.*, **221**, 261–71.
6. Kornblihtt, A. R., Pesce, C. G., Alonso, C. R., Cramer, P., Srebrow, A., Werbajh, S., and Muro, A. F. (1996). *FASEB J.*, **10**, 248–57.
7. Hynes, R. O. (1992). *Cell*, **69**, 11–25.
8. Clark, E. A. and Brugge, J. S. (1995). *Science*, **268**, 233–9.
9. Schwartz, M. A., Schaller, M. D., and Ginsberg, M. H. (1995). *Ann. Rev. Cell Dev. Biol.*, **11**, 549–99.
10. Ruoslahti, E., Hayman, E. G., Pierschbacher, M., and Engvall, E. (1982). *Methods Enzymol.*, **82A**, 803–1.
11. Schurr, E., Skamene, E., Forget, A., and Gros, P. (1989). *J. Immunol.*, **142**, 4507–13.
12. Kornblihtt, A. R., Umezawa, K., Vibe-Pedersen, K., and Baralle, F. E. (1985). *EMBO J.*, **4**, 1755–9.
13. Patel, R. S., Odermatt, E., Schwarzbauer, J. E., and Hynes, R. O. (1987). *EMBO J.*, **6**, 2565–72.
14. Schwarzbauer, J. E., Patel, R. S., Fonda, D., and Hynes, R. O. (1987). *EMBO J.*, **6**, 2673–80.
15. DeSimone, D. W., Norton, P. A., and Hynes, R. O. (1992). *Dev. Biol.*, **149**, 357–69.
16. Skorstengaard, K., Jensen, M. S., Sahl, P., Petersen, T. E., and Magnusson, S. (1986). *Eur. J. Biochem.*, **161**, 441–53.
17. Paolella, G., Henchcliffe, C., Sebastio, G., and Baralle, F. E. (1988). *Nucleic Acids Res.*, **16**, 3547–57.
18. Hirano, H., Yamada, Y., Sullivan M., deCrombugghe, B., Pastan, I., and Yamada, K. M. (1983). *Proc. Natl Acad. Sci. USA*, **80**, 46–50.
19. Kubomura, S., Obara, M., Karasaki, Y., Taniguchi, H., Gotoh, S., Tsuda, T., *et al.* (1987). *Biochim. Biophys. Acta*, **910**, 171–81.
20. George, E. L., Georges, E. N., Patel-King, R. S., Rayburn, H., and Hynes, R. O. (1993). *Development*, **119**, 1079–91.

21. Georges-Labouesse, E. N., Rayburn, H., and Hynes, R. O. (1996). *Dev. Dynamics*, **207**, 145–56.
22. Main, A. L., Harvey, T. S., Baron, M., Boyd, J., and Campbell, I. D. (1992). *Cell*, **71**, 671–8.
23. Dickinson, C. D., Veerapandian, B., Dai, X. P., Hamlin, R. C., Xuong, N. H., Ruoslahti, E., and Ely, K. R. (1994). *J. Mol. Biol.*, **236**, 1079–92.
24. Leahy, D. J., Aukhil, A., and Erickson, H. P. (1996). *Cell*, **84**, 155–64.
25. Baron, M., Norman, D., Willis, A., and Campbell, I. D. (1990). *Nature*, **345**, 642–6.
26. Williams, M. J., Phan, I., Harvey, T. S., Rostagno, A., Gold, L. I., and Campbell, I. D. (1994). *J. Mol. Biol.*, **235**, 1302–11.
27. Constantine, K. L., Ramesh, V., Bányai, L., Trexler, M., Patthy, L., and Llinás, M. (1991). *Biochemistry*, **30**, 1663–72.
28. Yamada, K. M. (1991). *J. Biol. Chem.*, **266**, 12809–12.
29. Ruoslahti, E. (1996). *Ann. Rev. Cell Dev. Biol.*, **12**, 697–715.
30. Erickson, H. P. and Carrell, N. A. (1983). *J. Biol. Chem.*, **258**, 14539–44.

Richard Hynes
Howard Hughes Medical Institute and
Center for Cancer Research,
Department of Biology, Massachusetts
Institute of Technology, Boston, MA, USA

Fibulins

The fibulins are an emerging family of proteins that presently contain three members, fibulin-1, -2 and -3.[1–6] All three proteins contain structurally similar modules that include repeated epidermal growth factor (EGF)-like modules and a carboxy-terminal fibulin-type module (Fig. 1). Fibulin-1 and -2 share a repeated cysteine-containing module that has similarity to the disulphide-bonded loop structure of the complement anaphylatoxins C3a, C4a, and C5a. Fibulin-2 possesses two amino-terminal modules, one cysteine-free the other containing cysteine, that are not present in either fibulin-1 or -3. In each of the fibulins, the four EGF-like modules that precede the last EGF-like module each contains a potential asparagine/aspartic acid hydroxylation sequence. While nothing is known about the function of fibulin-3, evidence is building to suggest that fibulins 1 and 2 may

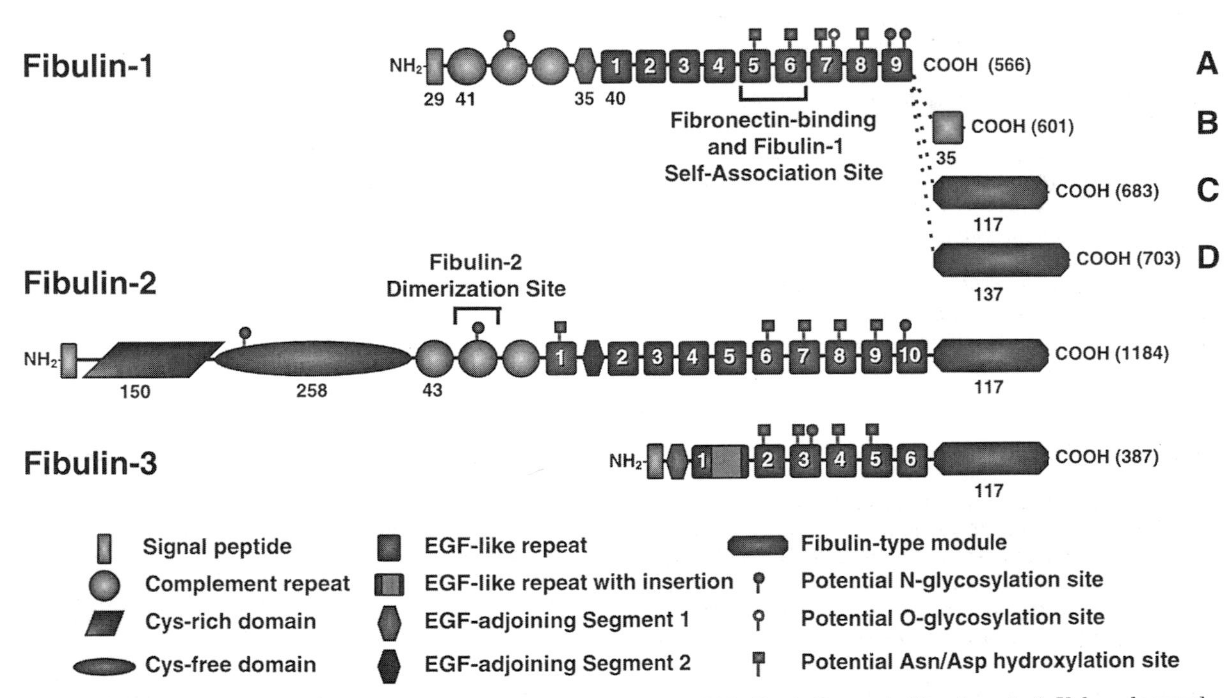

Figure 1. Schematic diagram depicting the domain organization of fibulin-1 (forms A–D), -2, and -3. Values located below modules indicate the number of amino acid residues in the selected domain. Values in parentheses indicate the number of amino acid residues in the human precursors for each protein.

function to promote the formation and/or stabilization of supramolecular structures such as elastic fibres and basement membranes. Both proteins have been found in association with these extracellular matrix structures[4,7–12] probably owing to their ability to interact with constituents such as fibronectin, laminins, fibrillin, and nidogen.[9,13–19]

Fibulin-1

Fibulin-1 is a Ca^{2+}-binding extracellular matrix and plasma glycoprotein that was the first member of the fibulin gene family to be isolated. Interspecies homologues of fibulin-1 have been described in human, mouse, chicken and nematode.

Synonymous names

BM-90.

Homologous proteins

Fibulin-1 is a member of a family of proteins that presently contains three members, fibulin-1, -2, and -3 (fibulin-3 is also known as S1–5 protein[5])).[2,4,6]

Protein properties

Alternative splicing of human fibulin-1 pre-mRNA results in four fibulin-1 transcripts that differ at their 3′ ends.[2,6,14] The C and D transcripts of fibulin-1 are widely expressed and represent the predominant fibulin-1 isoforms, whereas fibulin-1A and B are highly restricted in their pattern of expression, with the placenta being the only known site of their expression.[6] The alternatively spliced transcripts encode polypeptides (designated fibulin-1A–D) that have M_r of 58 670, 62 561, 74 463, and 77 274 D, respectively (values apply to human isoforms). Human placental fibulin-1 is glycosylated, having approximately three *N*-linked carbohydrate chains that add ~4–5 kDa to its molecular weight.[2] Using laser desorption mass spectroscopy, the conditions of which are generally considered unfavorable for preservation of non-covalent intermolecular interactions, a value of 78 842 D was determined for the molecular weight of human placental fibulin-1C monomer.[20] Rotary shadowing of fibulin-1 preparations shows dumb-bell-shaped molecules[18] that fit a model of globular amino- and carboxyl-terminal domains separated by an extended stretch of repeated EGF-like modules.

Fibulin-1 has been shown to bind the extracellular matrix proteins fibronectin, nidogen, and laminin,[13,14] and the coagulation protein fibrinogen.[20] In addition, fibulin-1 is capable of self-association.[13,14] It has been determined that fibulin-1 binds to fibronectin within the type III repeats 13–14,[13] to a site within the amino-terminal G1–G2 domains of nidogen,[14] to a site contained within the E3 fragment of Engelbreth–Holm–Swarm (EHS) tumour laminin,[14,16] a fragment derived from the carboxyl terminus of the α chain, and to a site within the carboxyl-terminal region of the fibrinogen $B\beta$ chain.[20] Fibulin-1 also binds calcium[1,3] and a calcium-binding site(s) within fibulin-1 has been mapped to EGF-like modules 5–9. Four of the nine EGF-like modules of fibulin-1 (EGF-like modules 5–8) contain the consensus sequence for post-translational hydroxylation of asparagine; such hydroxylation has been associated with calcium-binding EGF-like modules in a number of proteins.[21,22] Amino acid composition analysis indicates that fibulin-1 contains three moles of β-hydroxyasparagine per mole of protein.[23] A fibulin-1 self-association site has been localized to EGF-like modules 5 and 6 (amino acid residues 356–440).[23] Evidence from sedimentation equilibrium experiments (Tran and Argraves, unpublished results) indicates that fibulin-1 exists as a dimer presumably through a non-covalent homotypic interaction involving the EGF-like module 5–6 self-association site (Fig. 2). This pair of EGF-like modules was also found to contain the binding site for fibronectin.[23]

Fibulin-1 is widely expressed in tissues of the embryo and adult. It has been detected in some embryonic epithelial basement membranes[7,11] and has marked expression at sites of epithelial–mesenchymal transition such as endocardial cushions, developing myotomes, and neural crest.[7,11,12,24,25] Since cellular migration is associated with these processes, it has been speculated that fibulin-1 plays a role in the regulation of cell movement. In adult tissues, fibulin-1 is a widely expressed connective tissue constituent predominantly associated with matrix fibres

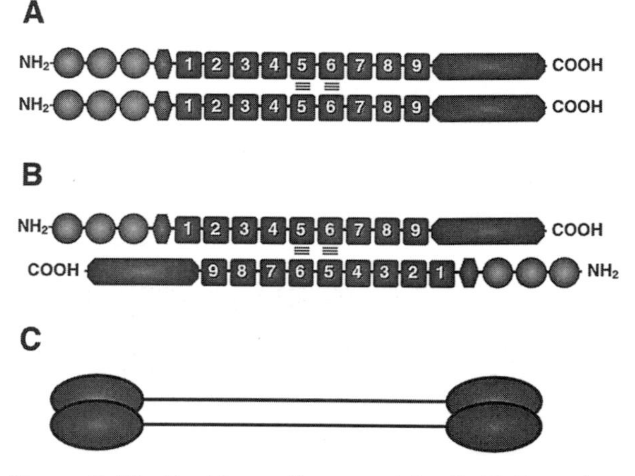

Figure 2. Fibulin-1 homodimer models. (A, B) Domain models of two fibulin-1 chains arranged in either a parallel (A) or antiparallel (B) configuration. The two subunits are depicted as non-covalently joined through an interaction between EGF-like modules 5 and 6.[23] (c) A ball and stick version of a homodimer model that resembles 33 nm long dumb-bell shaped figures observed by electron microscopy of fibulin-1 after rotary shadowing.[18]

in tissues such as the dermis, lung, placenta, intimal and medial layers of blood vessels, and meningeal tissues of the brain.[8] Some fibulin-1-containing matrix fibres contain elastin, with fibulin-1 located within the amorphous core of the fibres.[8] Fibulin-1 is also present in the plasma of both mouse and human at concentrations of 30–50 μg/ml,[2,3] and recent evidence suggests that it may have a role in haemostasis and thrombosis.[20,26]

Purification

Multiple methods have been described for the purification of fibulin-1. Fibulin-1 can be purified from detergent extracts of human placenta using immunoabsorption on mouse monoclonal fibulin-1 IgG–sepharose[2,15] or by affinity chromatography on a synthetic peptide corresponding to the cytoplasmic domain of the integrin β1 subunit coupled to sepharose.[1] In addition, fibulin-1 can be isolated from NaCl/EDTA extracts of EHS tumour by a combination of gel filtration and anion exchange chromatography techniques.[3] Cell lines stably transfected with fibulin-1 expression constructs have been generated, and recombinant fibulin-1 protein (human and mouse forms) isolated by immunoabsorption[6] or by gel filtration/anion exchange chromatography.[18]

Antibodies

Rabbit antisera have been prepared against human placental fibulin-1[1] and recombinant mouse fibulin-1C.[18] A number of mouse monoclonal antibodies to human fibulin-1 have been generated.[2,8]

Genes

The GenBank accession numbers for mRNA sequences encoding human fibulin-1A, fibulin-1B, fibulin-1C, and fibulin-1D are X53741, X53742, X53743, and U01244, respectively. The GenBank accession numbers for mRNA sequences encoding mouse fibulin-1C and fibulin-1D are X70853 and X70854, respectively. The GenBank accession numbers for chicken fibulin-1C and fibrulin-1D are AF051400 and AF051399. The GenBank accession number for *C. elegans* fibulin-1D is AF070477. The chromosomal location of the fibulin-1 gene (*FBLN1*) has been mapped to a single site on the long arm of human chromosome 22 (22q13.3) and to the E–F band of mouse chromosome 15.[27,28]

Structure

Information on the X-ray crystallographic or NMR structure of fibulin-1 is not available.

Fibulin-2

Fibulin-2 is a 175 kDa extracellular matrix protein that forms a disulphide-bonded homodimer in which the subunits are arranged in an antiparallel manner (Figs 1 and 3).[29] Fibulin-2 has been found in association with fibroblast-derived fibronectin fibrils[30] and some fibrillin-containing elastic microfibrils.[9]

Protein properties

The 175 kDa fibulin-2 polypeptide contains, in an amino- to carboxyl-terminal direction, a novel cysteine-free module, a novel cysteine-containing module, three anaphylatoxin-like domains, eleven EGF-like domains, and a globular fibulin-type carboxyl-terminal module (Fig. 1).[4] Cysteine residues within the second of the three anaphylatoxin-like modules (Cys-574) located centrally in the polypeptide are responsible for the formation of disulphide bonds that covalently link fibulin-2 subunits to form a homodimer (Fig. 3).[29]

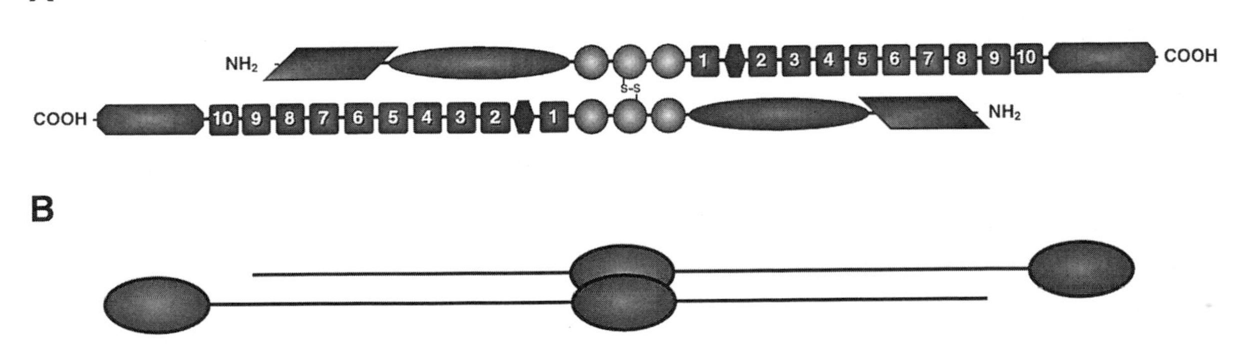

Figure 3. Fibulin-2 homodimer model. (a) The domain model of two fibulin-2 monomers arranged in an antiparallel manner as proposed by Sasaki *et al.*[29] The two subunits are covalently joined by a disulphide bridge located within the second anaphylatoxin module of each chain. (b) A ball and stick version of the homodimer model that resembles the 40–45 nm figures observed by electron microscopy of fibulin-2 after rotary shadowing.[29]

Fibulin-2, like fibulin-1, appears to be multifunctional by virtue of its ability to interact with multiple extracellular matrix proteins. For example, it has been shown to bind fibronectin and nidogen.[17] The fibronectin interaction with fibulin-2 is calcium-dependent, while the nidogen interaction is only partially inhibited by divalent metal chelators. Fibulin-2 displays low affinity for collagen type IV, perlecan, and the amino-terminal domain of the $\alpha 3$ chain of collagen type VI, but little or no binding activity for fibulin-1, vitronectin, or several other types of collagen.[17] Recently, fibulin-2 has been found to bind to the amino-terminal region of fibrillin-1, amino acid residues 45–450.[9] This binding is calcium-dependent and of high affinity (K_d 56 nM), and presumably accounts for the association of fibulin-2 with a subset of microfibrils, including ones found in the skin, perichondrium, elastic intima of blood vessels, and kidney glomerulus.[9] Fibulin-2 also binds to laminin-1 ($\alpha_1\beta_1\ \gamma_1$) through a region in the short arm of the α_1 chain (residues 654–665), and binds to laminin-5 (kalinin/nicein, epiligrin, $\alpha_3\beta_3\gamma_2$) through a region in the short arm of the γ_2 chain (residues 199–207).[19] Based on these findings, fibulin-2 (like fibulin-1) has been speculated to function as a bridge between laminin-1 and laminin-5 and other extracellular matrix proteins, thus providing a linkage between the basement membrane and the underlying stroma.

Fibulin-2 has a broad pattern of expression in tissues of the embryo and adult. It is present in the intercellular matrix of loose mesenchyme and in some basement membranes. It is also present in blood, but at very low levels, 1000-fold lower levels than fibulin-1.[4] During embryonic development, fibulin-2 expression is particularly pronounced during heart morphogenesis, initial stages of cartilage formation, and at several sites of epithelial–mesenchymal interaction such as those that give rise to hair follicles and teeth.[10,11] The expression of fibulin-2 and –1 overlap in certain tissues such as in the developing heart (e.g. endocardium and myocardium).[10] While both proteins are expressed in tissues of the nervous system, differences exist in their individual expression patterns, with fibulin-2 prominent in the neuroepithelium, spinal ganglia, and peripheral nerves and fibulin-1 in the leptomeningeal anlage, basement membranes of the neuroepithelium, and perineurium of the peripheral nerves.[10]

Purification

Cell lines stably transfected with mouse fibulin-2 expression constructs have been generated and recombinant fibulin-2 isolated from the conditioned culture medium by gel filtration/anion exchange chromatography.[4]

Antibodies

Rabbit antiserum has been prepared against recombinant mouse fibulin-2[4] and shown to be cross-reactive with the human homologue.[10]

Genes

The GenBank accession number for the mRNA sequence encoding human fibulin-2 is X82494[31] and mouse homologue is X75285.[4] The human fibulin-2 gene (FBLN2) maps to chromosome 3p24–25 and to the band D–E of mouse chromosome 6.[31]

Structure

Information on the X-ray crystallographic or NMR structure of fibulin-2 is not available.

Fibulin-3

Synonymous names

S1–5 protein.

Protein properties

Fibulin-3 is a recent addition to the family, which was identified through comparative sequence analysis[6] using the 137 amino acid carboxyl-terminal segment of human fibulin-1D. The analysis revealed that the fibulin-1D segment was not only similar to the carboxyl-terminal regions of fibulin-1C and fibulin-2, but also to the carboxyl-terminal region of a protein referred to as S1–5 protein.[5,32] In addition to having a homologous carboxy-terminal segment, S1–5 protein also contains repeated EGF-like modules arranged amino-terminally to the carboxy-terminal segment, and an EGF-adjoining segment similar to that present in fibulin-1.

The human S1–5 cDNA was originally isolated from a subtractively enriched cDNA library established from a subject with Werner syndrome.[5,32] Transcripts encoding S1–5 protein are rather widely expressed in human adult tissues except in the brain and peripheral leukocytes.[33] S1–5 mRNA expression is elevated in fibroblasts derived from subjects with Werner syndrome of premature ageing and senescent normal diploid fibroblasts.[5] It remains to be established if fibulin-3 is an extracellular matrix protein, as are its other family members.

Purification

No methods have been described for the purification of the fibulin-3/S1–5 protein.

Antibodies

Antibodies to fibulin-3/S1–5 are not presently available.

Genes

The GenBank accession number for the mRNA sequence encoding S1–5/fibulin-3 is U03877. The human S1–5/

fibulin-3 gene maps to chromosome 2p16, a position that excludes it as a candidate for Werner syndrome.[33] The fibulin-3 gene spans approximately 18 kb of genomic DNA and consists of 12 exons.[33]

References

1. Argraves, W. S., Dickerson, K., Burgess, W. H., and Ruoslahti, E. (1989). *Cell*, **58**, 23–9.
2. Argraves, W. S., Tran, H., Burgess, W. H., and Dickerson, K. (1990). *J. Cell. Biol.*, **111**, 3155–64.
3. Kluge, M., Mann, K., Dziadek, M., and Timpl, R. (1990). *Eur J. Biochem.*, **193**, 651–9.
4. Pan, T. C., Sasaki, T., Zhang, R. Z., Fassler, R., Timpl, R., and Chu, M. L. (1993). *J. Cell Biol.*, **123**, 1269–77.
5. Lecka-Czernik, B., Lumpkin, C. K., Jr., and Goldstein, S. (1995). *Mol. Cell. Biol.*, **15**, 120–8.
6. Tran, H., Mattei, M., Godyna, S., and Argraves, W. S. (1997). *Matrix Biol.*, **15**, 479–93.
7. Spence, S. G., Argraves, W. S., Walters, L., Hungerford, J. E., and Little, C. D. (1992). *Dev. Biol.*, **151**, 473–84.
8. Roark, E. F., Keene, D. R., Haudenschild, C. C., Godyna, S., Little, C. D., and Argraves, W. S. (1995). *J. Histochem. Cytochem.*, **43**, 401–11.
9. Reinhardt, D. P., Sasaki, T., Dzamba, B. J., Keene, D. R., Chu, M. L., Gohring, W., Timpl, R., and Sakai, L. Y. (1996). *J. Biol. Chem.*, **271**, 19489–96.
10. Miosge, N., Gotz, W., Sasaki, T., Chu, M. L., Timpl, R., and Herken, R. (1996). *Histochem. J.* **28**, 109–16.
11. Zhang, H. Y., Timpl, R., Sasaki, T., Chu, M. L., and Ekblom, P. (1996). *Dev. Dynamics*, **205**, 348–64.
12. Zhang, H. Y., Kluge, M., Timpl, R., Chu, M. L., and Ekblom, P. (1993). *Differentiation*, **52**, 211–20.
13. Balbona, K., Tran, H., Godyna, S., Ingham, K. C., Strickland, D. K., and Argraves, W. S. (1992). *J. Biol. Chem.*, **267**, 20120–5.
14. Pan, T. C., Kluge, M., Zhang, R. Z., Mayer, U., Timpl, R., and Chu, M. L. (1993). *Eur J. Biochem.*, **215**, 733–40.
15. Godyna, S., Mann, D. M., and Argraves, W. S. (1994). *Matrix Biol.*, **14**, 467–77.
16. Brown, J. C., Wiedemann, H., and Timpl, R. (1994). *J. Cell Sci.*, **107**, 329–38.
17. Sasaki, T., Gohring, W., Pan, T. C., Chu, M. L., and Timpl, R. (1995). *J. Mol. Biol.*, **254**, 892–9.
18. Sasaki, T., Kostka, G., Gohring, W., Wiedemann, H., Mann, K., Chu, M. L., and Timpl, R. (1995). *J. Mol. Biol.*, **245**, 241–50.
19. Utani, A., Nomizu, M., and Yamada, Y. (1997). *J. Biol. Chem.*, **272**, 2814–20.
20. Tran, H., Tanaka, A., Litvinovich, S. V., Medved, L. V., Haudenschild, C. C., and Argraves, W. S. (1995). *J. Biol. Chem.*, **270**, 19458–64.
21. Stenflo, J., Ohlin, A. K., Owen, W. G., and Schneider, W. J. (1988). *J. Biol. Chem.*, **263**, 21–4.
22. Sugo, T., Bjork, I., Holmgren, A., and Stenflo, J. (1984). *J. Biol. Chem.*, **259**, 5705–10.
23. Tran, H., VanDusen, W. J., and Argraves, W. S. (1997). *J. Biol. Chem.*, **272**, 22600–6.
24. Zhang, H. Y., Chu, M. L., Pan, T. C., Sasaki, T., Timpl, R., and Ekblom, P. (1995). *Dev. Biol.*, **167**, 18–26.
25. Bouchey, D., Argraves, W. S., and Little, C. D. (1996). *Anat Rec.*, **244**, 540–51.
26. Godyna, S., Diaz-Ricart, M., and Argraves, W. S. (1996). *Blood*, **88**, 2569–77.
27. Korenberg, J. R., Chen, X. N., Tran, H., and Argraves, W. S. (1995). *Cytogenet. Cell Genet.*, **68**, 192–3.
28. Mattei, M. G., Pan, T. C., Zhang, R. Z., Timpl, R., and Chu, M. L. (1994). *Genomics*, **22**, 437–8.
29. Sasaki, T., Mann, K., Wiedemann, H., Gohring, W., Lustig, A., Engel, J., *et al.* (1997). *EMBO J.*, **16**, 3035–43.
30. Sasaki, T., Wiedemann, H., Matzner, M., Chu, M., and Timpl, R. (1996). *J. Cell Sci.*, **109**, 2895–904.
31. Zhang, R. Z., Pan, T. C., Zhang, Z. Y., Mattei, M. G., Timpl, R., and Chu, M. L. (1994). *Genomics*, **22**, 425–30.
32. Lecka-Czernik, B., Moerman, E. J., Jones, R. A., and Goldstein, S. (1996). *Exp. Gerontol.*, **31**, 159–74.
33. Ikegawa, S., Toda, T., Okui, K., and Nakamura, Y. (1996). *Genomics*, **35**, 590–2.

W. Scott Argraves
Department of Cell Biology and Anatomy,
Medical University of South Carolina,
Charleston, SC 29425–2204, USA

Hyaluronan, hyaluronan synthases, and hyaluronan-binding proteins

Hyaluronan (HA) is a high molecular weight, highly anionic polysaccharide composed of repeating disaccharides of β1,4-glucuronate β1,3-*N*-acetylglucosamine which is synthesized by a family of plasma membrane proteins, the HA synthases. HA is a widespread component of extracellular matrices, where it plays a structural role, and of cell surfaces, where it influences cell behaviour. Important to both types of function are its unique physicochemical properties and its interactions with HA-binding proteins (HABPs). HA-binding proteoglycans and link protein contribute to the structure of extracellular matrices via ternary interactions with HA; the cell surface-associated HABPs, RHAMM and CD44, act as HA receptors which mediate the effects of HA on cell behaviour.

Synonyms

For hyaluronan: hyaluronic acid, hyaluronate.

Hyaluronan

Hyaluronan (HA) is a widely distributed extracellular and cell surface polysaccharide of very high molecular weight, usually several million, that is found in most adult and embryonic tissues. Due to intramolecular charge repulsion, hydrophilic and hydrophobic self-interactions and molecular entanglement, HA forms an expanded, stiffened, coiled and intertwined network that encompasses an enormous volume of immobilized water.[1,33,34] Consequent on these properties are its high viscosity and its dramatic effects on molecular exclusion, flow resistance, tissue osmosis, lubrication, and hydration, all of which are likely to be important in HA-rich matrices such as are found in synovium, umbilical cord, dermis, subcutaneous tissue, and so on. Reduction of HA concentrations in these tissues would result in osmotic imbalances, tissue adhesions, or decreased matrix hydration.[1] In cartilage, the major function of HA is its contribution to matrix integrity via ternary and multivalent interaction with link protein and the proteoglycan, aggrecan; similar interactions involving other proteoglycans, such as versican, contribute to the structure of other tissues.[2,3]

HA is especially enriched in pericellular matrices surrounding migrating and proliferating cells during embryonic development, tissue repair, and tumourigenesis.[1,4,5] HA influences proliferation, migration, and adhesion of cells within such matrices via interactions at the cell surface. HA is held at the surface of cells by two mechanisms: interaction with cell surface HA receptors such as RHAMM and CD44 or transmembrane retention of newly synthesized HA by HA synthase (Fig. 1). Further interactions at the cell surface between HA and HA-binding proteoglycans can give rise to an extensive cell coat that creates a unique, highly hydrated, pericellular milieu (Fig. 2).

Purification

HA can be extracted and purified by a variety of methods, depending on the source. Common methods are extraction by protease digestion or with chaotropic agents, followed by ion exchange chromatography, molecular sieve chromatography,and/or CsCl density gradient centrifugation.[6]

Activities

HA has three major activities which have been measured in numerous ways. First, HA occupies an enormous hydrodynamic domain that greatly influences the hydration and physical properties of tissues.[1] Second, HA interacts

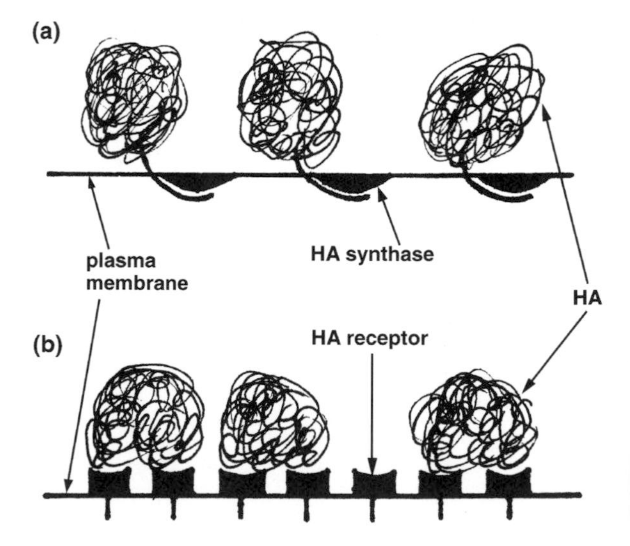

Figure 1. Two mechanisms whereby HA is retained at the cell surface. (a) HA is retained via interaction with the cytoplasmic active site of HA synthase while being extruded through the plasma membrane. (b) HA interacts with the extracellular binding site of HA receptors such as CD44 or RHAMM.

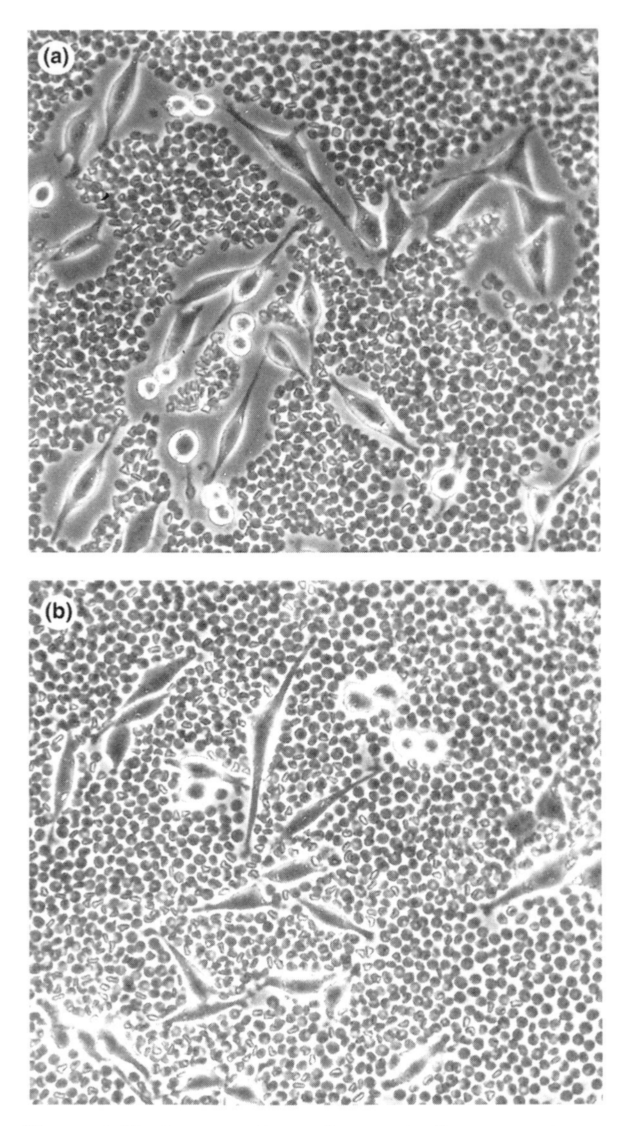

Figure 2. Hyaluronan-dependent pericellular matrices. (a) The pericellular matrices or 'coats' can be visualized by particle exclusion (the particles here are fixed red blood cells). These pericellular matrices can assemble by at least two mechanisms: (1) interactions between free HA, HA receptor, and HA-binding proteoglycan;[5] (2) interactions between 'nascent' HA, still bound to HA synthase (see Fig. 1(a)), and HA-binding proteoglycan.[29] (b) Streptomyces hyaluronidase treatment removes the pericellular matrices illustrating dependence of their structure on HA.

with other extracellular macromolecules (proteoglycans and link proteins) contributing to the structure of many tissues.[2,3] Third, HA interacts with cell surface receptors (RHAMM and CD44), influencing cell behaviour in several ways.[4,5,7–9]

Hyaluronan synthases

HA is synthesized at the cytoplasmic face of the plasma membrane by HA synthase and is concomitantly extruded through the plasma membrane[10] (Fig. 1(a)). HA synthases are a small family of multi-pass plasma membrane proteins related to chitin synthases and nodC from *Rhizobium*; *Xenopus* DG42 has also been shown to be a member of the HA synthase family. Each of the HA synthases has a large cytoplasmic loop that contains the active site for HA synthesis but the mechanism of transport of newly synthesized HA across the plasma membrane is not known.[10]

Purification

Streptococcal HA synthase has been purified by immunoaffinity chromatography.[11]

Activities

HA synthase activity is usually measured in membrane preparations by incorporation from isotopically labelled UDP-glucuronic acid or UDP-*N*-acetylglucosamine to form high molecular weight HA.[11]

Antibodies

Polyclonal antibodies to synthetic peptides of bacterial HA synthase have been made.[11]

Genes

Complete cDNA sequences for *Streptococcal*, *Xenopus*, murine, and human HA synthases have been published.[10] These studies indicate that there are at least three separate mammalian genes for HA synthase.[12]

Hyaluronan-binding proteins (hyaladherins)

HA-binding proteins mediate many of the structural and cellular functions of HA. The major HA-binding macromolecules of extracellular matrices are proteoglycans and link proteins. The HA-binding domains of link protein and the core proteins from HA-binding proteoglycans, such as aggrecan, versican, brevican, and neurocan, contain extensive homologies.[2,3] HA interacts with these proteoglycans and link proteins to form ternary complexes important to the structure of several tissues.[2,3]

Two classes of cell surface HABPs that act as HA receptors have been well characterized, namely RHAMM[9] and

Laminins

Overview of family

Laminins represent a protein family of α, β, γ chain heterotrimers primarily located in basement membranes but also in some mesenchymal compartments. So far 11 different chains have been identified giving rise to laminins-1–11. Major activities include the formation of networks and filaments, heterotypic binding in supramolecular assemblies, and the binding to cells through integrin and other receptors. The latter activities determine cell adhesion and migration, differentiation, gene expression, and cell fate. Several natural and experimental mutations underscore the importance of laminins for normal development.

Laminins are a growing family of related proteins (400–1000 kDa) characterized by a heterotrimeric chain assembly ($\alpha\beta\gamma$), a preferred localization in basement membranes, and a multitude of biological activities.[1] So far 11 different assembly forms, laminins-1 to -11, have been identified or suggested based on five different α chains, three β chains, and two γ chains (Table 1). This complexity has led to a novel nomenclature[2] which will be used throughout this review. This nomenclature has replaced several previous synonyms for laminins or their chains (s-laminin, k-laminin, merosin, kalinin/nicein, epiligrin).

Complete sequences are available for the ten laminin chains, mostly from human and mouse but also a few from *Drosophila*[3–33,119] and demonstrate sizes ranging from 1200 to 3600 residues (Table 1). The existence of two splice variants has been reported for the α3 chain (α3A, α3B)[8,9] and the γ2 chain.[30] All of the chains have a mosaic structure consisting of several types of modules

Table 1 List of laminin chains and several of their characteristics, including the number of amino acid residues (signal peptide in brackets), corresponding GenBank numbers; chromosomal localizations (CL); and their occurrence in distinct laminin (L) types with their chain compositions ($\alpha\beta\gamma$)

Chain	Species	Amino acid residues	Ref.	GenBank	CL	Laminin type
α1	Human	3075(17)	3, 4	S14458	18p11.3	L-1 ($\alpha1\beta1\gamma1$)
	Mouse	3084(24)	5	J04064	17	L-3 ($\alpha1\beta2\gamma1$)
α2	Human	3110(22)	6	Z26653	6q22–23	L-2 ($\alpha2\beta1\gamma1$)
	Mouse	3118(23)	7, 130	U12147	10	L-4 ($\alpha2\beta2\gamma1$)
α3A	Human	1713(20)	8	L34155	18q11.2	L-5 ($\alpha3\beta3v2$)
				L34156		L-6 ($\alpha3\beta1\gamma1$)
α3A	Mouse	1711(21)	9	X84013	18	L-7 ($\alpha3\beta2v1$)
α3B	Mouse	3284*	9, 33	X84014, U88353		
α4	Human	1816(24)	10, 11	X91171	6q21	L-8 ($\alpha4\beta1\gamma1$)
	Mouse	1815(24)**	12, 118	U69176, U59865	10	L-9 ($\alpha4\beta2\gamma1$)
α5	Mouse	~3630	13	U37501	2	L-10 ($\alpha5\beta1\gamma1$)
						L-11 ($\alpha5\beta2\gamma1$)
	Drosophila	3712(22)	14, 15	M96388	65A	L-10 ($\alpha5\beta1\gamma1$)
β1	Human	1786(21)	16	M61916	7q22	
	Mouse	1786(20)	17	M15525	12	
	Drosophila	1784(26)	18	M19525	28D	
β2	Human	1798(32)**	19, 20	X79683	3p21	
	Rat	1801(35)	21	X16563		
β3	Human	1170(17)**	22, 23	L25541	1q32	
				U17760		
	Mouse	1168(17)	24	U43298	1	
				S75986		
γ1	Human	1609(33)	25	J03202	1q25–31	
	Mouse	1607(33)**	26, 27	J02930	1	
				J03484		
	Drosophila	1639(33)	28, 29	X07806	67C	
γ2	Human	1193(21)	30, 31	Z15008	1q25–31	
				Z15009		
	mouse	1192(18)	32	U43327	1	

* Still lacks a signal peptide and a few N-terminal residues.[33]
** Slight differences between different reports.

(Fig. 1). Common to all of them is a coiled-coil domain with about 80 heptad sequence repeats at or close to the C-terminal end. This domain is crucial for heterotrimer assembly and thus a hallmark for classification as a laminin. The N-terminal regions typically consist of tandem arrays of LE modules (50–60 residues, eight Cys) with a remote similarity to EG modules. They are interrupted by globular modules of about 250 residues including LN and L4 modules and some unclassified domains. Individual modules or groups of modules have also been classified as domains I to VI and G (Fig. 1). Each class of laminin chains also has its unique characteristics, which, along with sequence comparisons, allows their unequivocal identification. The α chains have an invariable set of five LG modules (each about 180–200 residues) at the C terminus and the β chains contain a short cysteine-rich insert (α) in the middle of their coiled-coil region which separates domains I and II (Fig. 1).

Electron microscopical visualization of laminins usually shows a cross- or Ψ-shaped structure consisting of a long arm (80 nm) and two or three short arms (25–40 nm) (Fig. 2). Based on the analysis of laminin-1 fragments, the long arm rod consists of the coiled-coil domain of three chains and the C-terminal globule of the α chain LG modules. The short arms consist of single chains with the LE modules forming rods and other modules forming globular domains.[34] The assembly of intact laminin occurs in the coiled-coil domains of the constituent chains and is a fast process in which a three-stranded super α helix is formed through hydrophobic and polar interactions between all three chains.[34] This process occurs entirely

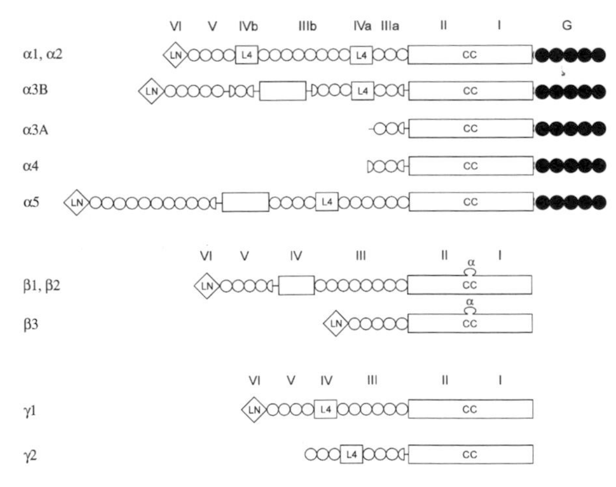

Figure 1. Modular structure of laminin α, β, and γ chains, and their assignment to different rod-like or globular domains (I to VI). Modules include the coiled-coil domain (CC), LE modules (open circles; half circles indicate incomplete modules), LG modules (filled circles), the LN (diamonds), and L4 (squares) modules, and unclassified modules (open rectangles). α3A and α3B denote two different splice variants, and α above a loop is an inserted segment unique to β chains. The α3B represents the composite of two partial sequences.[9,33] The N-terminal end is to the left.

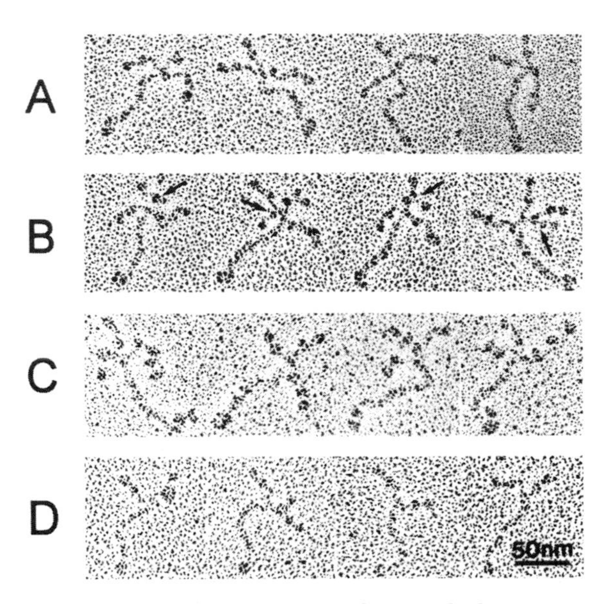

Figure 2. Electron micrographs of rotary shadowing images of individual laminin particles. (A) Mouse laminin-1, (B) laminin-1 complexed to nidogen, (C) *Drosophila* laminin-10, (D) Ψ-shaped laminin from rat Schwannoma.

intracellularly and is required for secretion, with the amounts of α chains usually being rate limiting.[35,36,119] The process also involves chain selection, since only a limited number of heterotrimeric assembly forms have so far been observed for the laminins (Table 1). So far, all five α chains have been shown or are supposed to associate with either $\beta1\gamma1$ chain pairs or with $\beta2\gamma1$, with laminin-5 ($\alpha3\beta3\gamma2$) being the only exception. Chain selection is considered to be due to distinct sequence requirements for interactions in the coiled-coil domain.[37]. Studies of the assembly of recombinant fragments from the coiled-coil domain indicate a minimum size of 50 residues for efficient assembly.[38,39] Laminins are eventually stabilized by interchain disulphide bridges, involving two cysteines at the N terminus of domain II in each chain at the centre of the cross and single cysteines at the C-terminal end of the β and γ chains.[34,120]

All laminins are glycoproteins as predicted by a moderate to large number of potential *N*-linked acceptor sites from sequence analyses. A precise carbohydrate analysis exists only for laminin-1 and indicates a content of 15–25 per cent. The structural characterization[40,41] demonstrated a large number of complex bi- to tetra-antennary oligosaccharides, some with additional lactosamine tails. The content of *O*-linked oligosaccharides seems to be low but may include glycosaminoglycan attachment to domain IV of the β1 chain (Sasaki and Timpl; unpublished observations), indicating that laminins can also act as proteoglycans.

The various laminin isoforms have been reported to have a large variety of functional and biological activities.[1,42] These include laminin interactions either in homotypic fashion or between different isoforms, heterotypic interactions with many other extracellular ligands, and cellular interactions through several integrin and non-integrin receptors. All of these interactions may have strong effects on cellular phenotypes such as cell shape, gene regulation, and other parameters. There is also increasing evidence of specific functions for particular laminin isoforms, providing some clues as to why a protein family has evolved. This is in part also reflected by distinct but frequently overlapping distributions of laminin isoforms [33,43–45,121,122]

Antibodies

Many studies of laminins have reported the production of antibodies and/or used them for analysis. Polyclonal and monoclonal antibodies have been produced against laminins, proteolytic and recombinant fragments, or individual chains. In most cases, these antibodies have been useful for immunohistology, immunoprecipitation, blotting, and radioimmunoassays with a few examples referred to in the following sections. Antibodies against individual chains are often highly specific[33,43,44,121,122] which reflects the relatively low sequence identity (usually below 50 per cent) between the various isoforms of the laminin chains. Because of the overlapping assembly patterns (Table 1), most of these antibodies cannot be used to identify a particular laminin isoform, particularly in immunohistological studies. This will require the generation of antibodies against assembly-specific epitopes of the coiled-coil domain which so far has not been accomplished.

Structure

Because of the large size of the laminins, information on their three-dimensional structures is usually restricted to electron microscopic resolution. Recombinant production of domains or modules recently opened a new approach for elucidation at atomic resolution. This was achieved for three LE modules of the γ1 chain by X-ray crystallography[46] and one LE module by NMR,[47] both spanning the nidogen-binding epitope. This established the disulphide connections within the module and demonstrated tight intermodular contacts, which explains why tandem arrays of LE modules form rod-like structures.

Laminin-1 and laminin-3

These laminins share the α1 chain either in the combination $\alpha1\beta1\gamma1$ (laminin-1) or $\alpha1\beta2\gamma1$ (laminin-3). Laminin-1 was the first laminin to be purified from a mouse tumour (EHS) basement membrane[48] and has a distinct cross-shaped structure as shown by electron microscopy after rotary shadowing (Fig. 2(A)) and negative staining.[34] Laminin-3 has not yet been obtained on a preparative scale and proof of its existence relies on biosynthetic data.[49] Interpretation of its domain structure indicates that it will very likely have the same assembly pattern

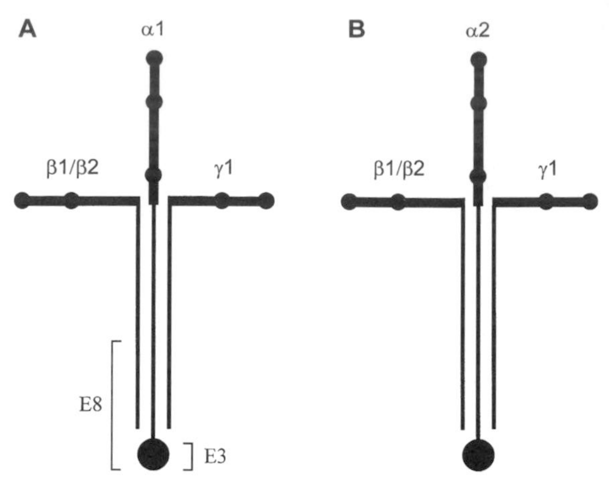

Figure 3. Schematic shape structure, and chain arrangements of laminin-1, and laminin-3 (A) and laminin-2, and laminin-4 (B). The long arm points to the bottom, and the β and γ chains terminate just before the distal globular domain. The positions of proteolytic fragments E3, and E8 are shown for laminin-1.

and cross-shaped topology (Fig. 3(A)). This topology has been confirmed for laminin-1 using its many proteolytic fragments, which have also been useful for mapping distinct activities.[34] Of particular interest are the heparin-binding fragment E3, which consists of the two C-terminal LG4 and LG5 modules of the α1 chain (Fig. 1) and fragment E8 containing the other LG modules and part of domain I.[50] The various fragments also show distinct differences in their *N*-glycosylation patterns[40] and hence lectin binding.[51]

Laminin-1 is the first laminin produced during mouse development, before the blastocyst stage,[52] and is subsequently found in many epithelial tissues during organogenesis, as demonstrated using α1-chain specific antibodies.[53] Such α1 chain expression can, however, be transient and locally restricted both during development and in adult tissues.[3,43–45,54] The α1 chain may be present either in laminin-1 or laminin-3, an assumption which is underscored by distinct differences in the expression patterns of β1 and β2 chains,[19–21,43–45] but normally cannot be detected in muscle and peripheral nerves.[121,122]

Purification

Extraction of the mouse EHS tumour with EDTA-containing neutral buffer, which yields laminin-1 complexed to nidogen, has become a standard procedure[50] and has replaced previous protocols.[48] Application to other tissues is often hampered by laminin-1 being contaminated by other laminin isoforms.[55] Purified laminin-3 has not yet been described. A recent approach is the preparative recombinant production of domains or chains including γ1 chain segments involved in nidogen binding,[56] the

N-terminal domains IV–VI[57,123] and domain G,[58,119] and an L4 module[59] of the α1 chain and several fragments of domain I.[38,39,120] It has also been possible to obtain β1 chains in secreted form,[60] to study extracellular deposition of assembled recombinant β2 chains[61] and to obtain secreted recombinant laminin-1 after triple transfection of mammalian cells.[119]

Activities

Laminin-1 and its complex with nidogen were shown to polymerize in a calcium-, temperature-, and concentration-dependent manner into networks with a quasi-hexagonal pattern.[62,63] Self assembly occurs through the interaction between the LN modules (domain VI) of all three chains (Fig. 4).[62] *In situ* this may also include cell receptors since β1 integrin-deficient teratomas show aberrant basement membrane formation and down-regulation of laminin-1.[124] Calcium binding to laminin-1 indicates 16 binding sites with k_d = 10–300 μM which have not yet been localized.[63] The most remarkable heterotypic binding activity of laminin-1 is with the C-terminal globular domain of nidogen, with k_d = 0.5 nM.[64] This high affinity allows the laminin–nidogen complex to be isolated from tissues[50] and has facilitated demonstration of a 1:1 stoichiometry by electron microscopy (Fig. 2(B)). The nidogen binding site has been mapped to a single LE module of the laminin γ1 chain domain III and its structure was elucidated by site-directed mutagenesis[65] and X-ray crystallography.[46] Nidogen was also shown to use a separate epitope to connect laminin-1 to collagen IV and is considered to provide the link between the networks formed by laminin and collagen IV in basement membranes.[64] Binding of laminin-1 to heparin and the heparan sulphate chains of perlecan occurs through the LG modules of fragment E3[58,66] and the LN module of the α1 chain.[57,123] Laminin-1 was also shown to bind fibulin-1 by its fragment E3[67] and fibulin-2 probably by an α1 chain domain.[68]

Laminin-1 is also a major cell-adhesive component and this and other cell-modulating activities are mediated primarily by binding to several integrin[69] and non-integrin receptors.[70] Most cells adhere to laminin-1 by $\alpha6\beta1$ integrin.[71] The binding site is localized on fragment E8, presumably on its LG modules which, however, need to be supported by domain I.[58,72] Further integrins ($\alpha7\beta1$, $\alpha9\beta1$, $\alpha6\beta4$) also bind laminin-1, possibly through similar E8 epitopes.[69] The dual collagen receptors, integrins $\alpha1\beta1$ and $\alpha2\beta1$, however, bind to the N-terminal region of the laminin α1 chain.[57,73,123] Mouse laminin α1 chain also contains a cryptic RGD-dependent integrin binding site[69] which is apparently masked by an adjacent L4 module.[59] Non-integrin laminin-1 receptors include a 67 kDa lectin, galactosyltransferases, 5′-nucleotidase, galactosidase, several microbial surface proteins,[70] and galectins.[51] The latter are considered to have a mediator role by binding lactosamine structures on both laminins and cell surfaces. The α-dystroglycan receptor is apparently specific for the α1 chain fragment E3[74] as is binding of sulfatides.[75] A large number of synthetic laminin peptides have also been claimed to bind to cell surfaces[34,76,125,126] but their receptors and/or folding state has remained unclear.

Several neuronal cells have also been shown to interact with laminin-1 which enhances survival and neurite outgrowth. This is considered to be important for nerve development and regeneration and agrees with expression of several laminin isoforms in the central and peripheral nervous system.[77] Neurite outgrowth is mediated through conformational epitopes of laminin fragment E8[72] but involves more complex processes than just integrin binding.[77] In contrast, an LRE sequence in domain I of the laminin β2 chain provides a selective stop signal for motor neurones,[21,78] in agreement with the restricted localization of β2 chain in muscle synapses.[43] Folding of this sequence into a coiled-coiled domain, however, abolishes this activity[79] but may determine the synaptic localization.[61] Laminin-1 has also been shown to cluster acetylcholine receptors into synapses in an agrin-independent manner.[127]

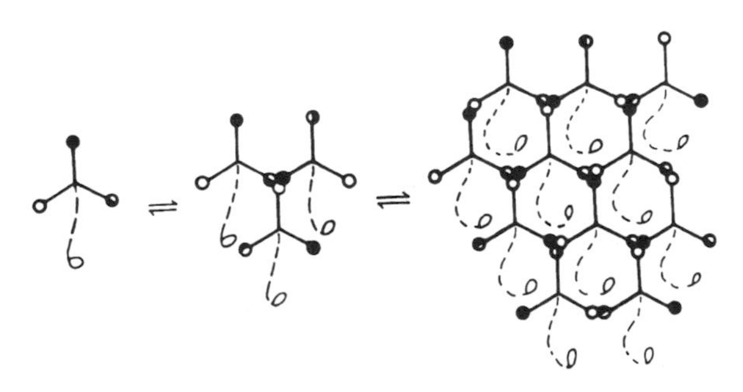

Figure 4. Schematic outline of the three-arm interaction model of the self-assembly of laminin-1 into a quasihexagonal network. Interactions occur through the LN modules of α1, β1, and γ1 chains (circles), and do not involve the long arm of laminin (broken line). (Reproduced with permission from ref. 1.)

Antibodies

Several monoclonal antibodies specific for α1, β1, β2 and γ1 chains have been described and used for tissue localization and immunoblotting of the chains and for immunoprecipitation and affinity purification of intact laminin.[43–45,80,81] Versatile polyclonal α1, β1, γ1 chain-specific antibodies could also be obtained with single chain fragments of laminin-1.[53,54,82] A γ1 chain-specific antiserum which blocks nidogen binding to laminin-1[56] was shown to inhibit basement membrane formation in several organ cultures.[83,84] An α1 chain specific antibody which blocks fragment E8 mediated cell adhesion[71,72] produced similar phenotypes,[80] as did monoclonal antibodies blocking α-dystroglycan binding to fragment E3.[85]

Genes

cDNA probes are available for α1, β1, β2 and γ1 chains (Table 1) and have also been used for chromosomal localization[86] (see Table 1). Exon structures of the genes encoding the human $\beta1$ (34 exons, 80 kb) and $\gamma1$ chains (28 exons, 60 kb) have been established.[86] A complete gene structure (33 exons, 12 kb) is also known for the human and mouse $\beta2$ chains.[87]

Mutants/phenotypes

Elimination of the laminin $\beta2$ chain gene in transgenic mice by homologous recombination caused abnormal neuromuscular junctions[88] and a massive proteinuria in homozygous animals.[89] The glomerular basement membrane remained rich in $\beta1$ chain, as occurs during fetal life, and although it was morphologically normal, it did not function properly as a filtration barrier. This caused an early postnatal death. Mice have also been generated which lack the $\gamma1$ chain and die during early embryonic development (N. Smyth, D. Edgar; personal communication). Several mutants deficient in nidogen-binding ability have been reported for a recombinant LE module of the $\gamma1$ chain.[65] Deletion of this nidogen-binding module from the $\gamma1$ chain does not impair assembly and secretation of laminins[128] but causes embryonic lethality in mice homozygous for this deletion.

Laminin-2 and laminin–4

These laminins are characterized by the $\alpha2$ chain and have the chain compositions $\alpha2\beta1\gamma1$ (laminin-2) or $\alpha2\beta2\gamma1$ (laminin-4). Both laminins show the same topological arrangement (Fig. 3(B)) and the same electron microscopical shapes[81,90] which are indistinguishable from that of laminin-1. The $\alpha2$ chain was originally discovered by a monoclonal antibody against its LG modules which bound to a 65 kDa fragment (merosin) of placental pepsin extracts.[91] A similar fragment (80 kDa) is found in purified laminin-2 and laminin-4,[90] indicating proteolytic processing not observed with laminin-1. Yet both laminins appear to be otherwise more resistant to proteolysis than laminin-1, particularly in their long arms.[92] Immunohistology of the $\alpha2$ chain demonstrated a distinct localization to basement membranes with only a partial colocalization with the laminin $\alpha1$ chain.[43,44,91,122] A broad tissue expression of the $\alpha2$ chain was also demonstrated by mRNA analyses.[6,7] Together the data indicate that laminin-2 and laminin-4 are prominent in basement membranes of striated and smooth muscle, vessels, nerves, and placenta but they also occur in some mesenchymal compartments.

Purification

Human laminin-2 and laminin-4 can readily be solubilized from placenta with EDTA-containing neutral buffer and are separated in the final step on a MonoQ anion exchanger.[90] Both laminins are complexed to nidogen to a variable degree, although the nidogen can be lost by endogenous proteolysis. Laminin-2 and laminin-4 were also purified from chick heart and separated from each other by affinity chromatography on lectins and a monoclonal antibody specific for laminin $\beta2$ chain.[81] Several other purification protocols exist which do not, however, separate laminin-2 and laminin-4.[92] Recombinant fragments of the $\alpha2$ chain are only available from its domain I and have been shown to assemble with $\beta1/\gamma1$ chain fragments into a coiled-coil structure.[39] Domains IV–VI[129] and G[130,131] have been also obtained in recombinant form and a single proteolytic cleavage has been demonstrated for the LG3 module.[130]

Activities

Because of their solubility in EDTA, laminin-2 and laminin-4 are likely to bind calcium and were shown to polymerize into networks, and to copolymerize with laminin-1.[132] The two $\alpha2$ chain laminins were also shown to bind to nidogen with high affinity but not to collagen IV and perlecan.[90] Binding to the latter two ligands can, however, be mediated by nidogen. There was only a moderate binding of laminin-2 and laminin-4 to heparin.[81,90] but a strong cell adhesiveness mediated through $\beta1$ integrins.[90] The receptors involved are presumably $\alpha3\beta1$, $\alpha6\beta1$, and $\alpha7\beta1$ integrins and α-dystroglycan[68] but also include $\alpha1\beta1$ and $\alpha2\beta1$ integrins.[72] which bind to the LN module of the $\alpha2$ chain.[129] Various LG modules of the $\alpha2$ chain, however, bind to other $\beta1$ integrins, α-dystroglycan, heparin and sulfatides (Talts and Timpl; unpublished). Laminins-2 and -4 also strongly stimulate neurite outgrowth[78,80] and are essential for the stabilization of myotube basement membranes.[93] Their G domain mediates the neural targeting of *Mycobacterium leprae*.[131]

Antibodies

Several monoclonal antibodies[44,81,91] and polyclonal antisera[92,130] specific for the $\alpha2$ chain have been described and have been useful for immunohistology, immunoprecipitation, blotting, and purification. Antibodies which specifically block $\alpha2$ chain functions have not yet been identified.

Genes

Human and mouse laminin $\alpha2$ chains have been cloned and their genes mapped to chromosomes[86] (Table 1). The exon structure (64 exons within more than 260 kb) has been determined for the human $\alpha2$ chain gene.[94]

Mutants/phenotypes

Two inbred dystrophic mouse strains, *dy/dy* and dy^{2J}/dy^{2J} were shown to either lack the laminin $\alpha2$ chain (*dy*)[95] or to have a deletion in its LN module (dy^{2J})[96]. The latter mutation does not prevent assembly and secretion but apparently interferes with laminin polymerization. Homozygous members of a family with congenital muscular dystrophy were shown to have a premature stop codon in the coiled-coil domain of the $\alpha2$ chain[97] which very probably prevents the assembly of laminin-2 and laminin-4. The experimental knock-out of the $\alpha2$ chain gene in mice leads to apoptosis and lack of intact basement membranes in muscle and causes early lethality.[133]

Laminins 5, 6, and 7

This is, in several respects, the most unique group of laminin isoforms which were previously referred to as kalinin, nicein, epiligrin, and k-laminin.[98] They are now known as laminin-5 ($\alpha3\beta2\gamma2$), laminin-6 ($\alpha3\beta1\gamma1$), and laminin-7 ($\alpha3\beta2\gamma1$) and are shown in their established topological arrangements in Fig. 5(A). Electron microscopy demonstrated a dumb-bell shape for laminin-5 (Fig. 6(a)) and a Ψ-shaped structure for laminin-6/-7 (Fig. 6(b)) when they contain the short α3A chain.[98–100] These shapes reflect the reduced size of the short arm of α3A (Fig. 1) and further proteolytic processing. A longer splice variant α3B has also been identified by cDNA sequencing[8,9] and after completion of the mouse[33] and part of the human[134] sequence shown to be similar to $\alpha5$ (Fig. 1). Laminins containing the α3B chain variant have not been isolated and their topological arrangements remain hypothetical (Fig. 5(B)). Northern (10–11 kb mRNA) and Western blots (280–300 kDa) demonstrated the expression of α3B in tissue extracts[9,134] and thus indicated assembly into a laminin. This also allowed the shorter α3A variant (5.5 kb mRNA, 190–200 kDa protein) to be distinguished. Laminin-5 assembles as a precursor within cells[36] but undergoes extensive proteolytic processing of the α3A and $\gamma2$ chains in the extracellular environment[98,101] (Fig. 5(A)). This processing includes conversion of the α3A chain to a 165 kDa protein by removal of the short arm and subsequent shortening in the G domain

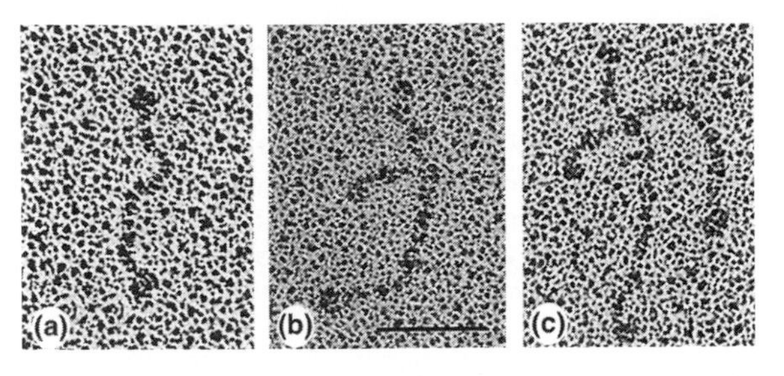

Figure 6. Electron micrographs of single molecules of laminin-5 (a), laminin-6 (b), and a covalent complex between these two laminins (c). The bar represents 50 nm. (Reproduced with permission, in modified form, from ref. 98.)

(145 kDa) as well as cleavage in domain III of the $\gamma2$ chain to yield a final 105 kDa product. Processing of the α3B chain has so far not been described. A further modification of processed laminin-5 involves disulphide-mediated connection of its N-terminal region to a central region in laminin-6 or laminin-7,[102] yielding a characteristic four-armed structure (Fig. 6(c)).

Laminin-5 was shown by immunogold staining to localize to anchoring filaments of the dermal–epidermal junction which connect hemidesmosomes with the basement membrane[98,99] and to anchoring plaques of the papillary dermis.[135] It also occurs in filaments of intestinal villi and presumably in many more anchoring structures,[98] as indicated by the broad expression of all three constituent chains of laminin-5 in the epithelium of amnion and of respiratory, digestive and urinary organs.[9,24,30,103,104] The α3B chain variant[9] and the truncated $\gamma2$ chain[104] show somewhat different and restricted expression patterns. Large tissue differences have also been observed in α3A and α3B mRNA expression.[134] Laminin-6 and laminin-7 have been identified by isolation from skin, lung, and some tumour cells.[98,100,102]

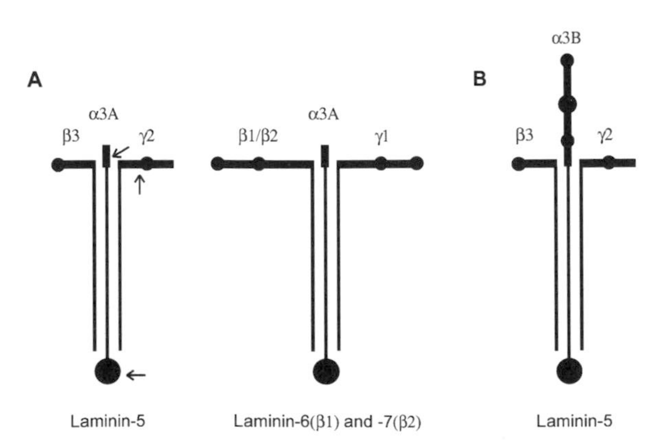

Figure 5. Schematic shape structure for laminins 5, 6, and 7 (A), and a hypothetical structure for laminin-5 containing the larger α3B chain (B). Laminin-5 is trimmed by proteolytic processing of its α3A, and $\gamma2$ chain (arrows).

Purification

Laminin-5 has been purified from keratinocyte medium or collagenase-digested amnion, mainly using affinity chromatography with an $\alpha3$ chain-specific monoclonal antibody.[99,101,102] Similar protocols also using $\gamma1$-chain-specific antibodies were used for the purification of laminin-6 and laminin-7 and their complexes with laminin-5.[100,102] Several modules of the $\gamma2$ chain have been produced by recombinant methods.[68,105] Laminins containing the α3B chain have so far not been purified.

Activities

A major function of the complex between laminin-5 and laminins-6/-7 seems to be the formation of anchoring filaments in the epidermis and some other organs which stabilizes the junction between cells and basement membranes.[98] These laminins do not seem to participate in the formation of networks,[132] probably because they contain less than three terminal LN modules (Fig. 1). This may, however, be different for the α3B variants. Laminin-7 but not laminin-5 binds nidogen with high affinity.[105] The LE module of γ2 chain domain III, which is homologous to the nidogen-binding LE module of the γ1 chain,[56] was shown to be inactive because of two substitutions in the binding epitope.[105] The insertion of anchoring filaments into basement membranes may therefore occur through nidogen binding to their laminin-6 or -7 constituents.[98] A second connection may occur through laminin-5 binding to the NC1 domain of collagen VII in the anchoring fibrils.[135] Fibulin-2 was shown to bind to the L4 module of the γ2 chain.[68] Laminin-5 is a strong adhesive substrate for keratinocytes and several other cells. The binding is apparently mediated by $\alpha3\beta1$, $\alpha6\beta4$ and $\alpha6\beta1$ integrins[69] which may generate different isoform-specific focal adhesions.[136] Induction of cell migration on laminin-5 requires further proteolytic processing of the γ2 chain.[137] Firm cell adhesion and hemidesmosome formation, however, depend on proteolytic processing of the α3 chain.[138]

Antibodies

Several monoclonal antibodies and polyclonal antisera against fusion proteins that are specific for α3, β3, or γ2 chains are available.[24,32,99,100,103] They have been used for immunohistology, immunoprecipitation, and blotting. A monoclonal antibody against the α3 chain was shown to inhibit cell adhesion.[99]

Genes

The chromosomal localization of α3, β3, and γ2 genes is known[86] and cDNA probes are available (Table 1). Gene structures have been determined for the β3 chain (23 exons, 29 kb)[23] and the γ2 chain (23 exons, 55 kb).[104] The shorter α3A chain splice variant is apparently transcribed from a different promoter to α3B.[139]

Mutants/phenotypes

A severe and lethal skin blistering disease, Herlitz's junctional epidermolysis bullosa, has been shown to be associated with mutations in the α3,[106,107] β3,[108] and γ2 chain[109,110] genes. All mutations cause a premature stop codon which should prevent chain assembly.

Laminin-8 to laminin-11

These laminins are predicted from recently published mammalian αchain cDNA sequences[10–13] and their identification as laminin-8 ($\alpha4\beta1\gamma1$), laminin-9 ($\alpha4\beta2\gamma1$), laminin-10 ($\alpha5\beta1\gamma1$), and laminin-11 ($\alpha5\beta2\gamma1$) is still incomplete.[33,112,113] Their topological arrangements (Fig. 7) indicated a Ψ-shaped structure for laminin-8 and laminin-9 because of their short α4 chain (Fig. 1) which was demonstrated by electron microscopy for laminin-8[112] and a Schwannoma laminin (Fig. 2(D)). A cross-shaped structure is likely for laminins 10 and 11, which possess the largest α chain, α5 (Fig. 1). An α5 chain homologue has previously been sequenced from *Drosophila*[14,15] and the corresponding laminin-10 was isolated[114] and showed a distinct cross-shape by electron microscopy (Fig. 2(C)). The α4[10–12,33,118] and α5 chain[13,33,140] mRNAs are strongly expressed in a large number of tissues (lung, heart, muscle, vessels, peripheral nerves) and expression of α1 and α5 chains appear in part overlapping or mutually exclusive during embryonic and postnatal development.[33,45,113,121,122] Laminin-8 and laminin-10 are particularly prominent in various endothelial cells.[112,113] Laminins containing α4 or α5 chains appear to have the broadest tissue expression in the mouse.[33]

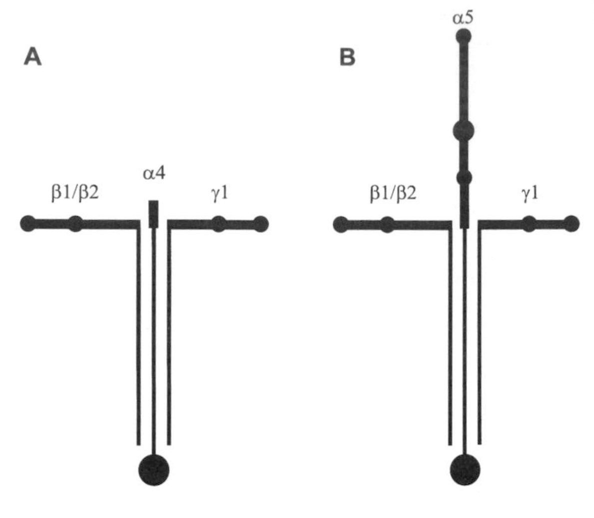

Figure 7. Topological arrangements of laminin-8 and laminin-9 (A), and of laminin-10 and -11 (B). An electron micrograph of *Drosophila* laminin-10 is shown in Fig. 2. The latter arrangement is also likely to exist for mouse laminins 10 and 11.

Purification

Bovine laminin-8 was isolated from aortic endothelial cell culture medium by immunoprecipitation with an antiserum to laminin-1 followed by anion exchange and heparin chromatography.[112] A similar procedure can be used to prepare laminin-10 from a lymph node endothelial cell line.[113] It seems also possible to obtain laminin-11 from Schwannoma cell medium.[122,141] *Drosophila* laminin-10 has been purified from the culture medium of K_c cells by sucrose gradient centrifugation and chromatography.[114] In addition to the 400 kDa α5 chain, it contained β1 and γ1 chains for which sequence analyses showed a distinct identity with the corresponding mammalian β1 and $\gamma 1$ chains[18,28,29] (Table 1). The existence of assembled laminins 8 and 9 (500–600 kDa) and of laminins 10 and 11 (700–800 kDa) was also indicated from immunoprecipitation and blotting of semipurified preparations obtained from lung EDTA extracts.[33]

Activities

Laminin-8 was shown to have affinity for heparin[112] and a similar Schwannoma laminin stimulates neurite outgrowth.[115] *Drosophila* laminin-10 was shown to bind heparin and to be cell-adhesive.[114] It also binds mammalian nidogen with the same affinity as mammalian laminins and interacts most strongly with heparin through LG modules of the $\alpha 5$ chain.[116]

Antibodies

A monoclonal antibody against mouse $\alpha 5$ chain[113] and a polyclonal antiserum against *Drosophila* laminin-10[114] are available and were used for immunohistology. Rabbit antisera against laminin $\alpha 4$ and $\alpha 5$ chains or recombinant fragments have been useful for immunohistology and immunoblotting.[33,121,122] A previously widely used monoclonal antibody against human placenta (4C7) was shown to be specific for the laminin $\alpha 5$ chain and not the $\alpha 1$ chain.[121]

Genes

cDNA clones are available for mammalian $\alpha 4$ and $\beta 5$ chains as well as the chromosomal localization of the human $\alpha 4$ gene[86] (Table 1). The localization of mouse laminin $\alpha 4$ and $\alpha 5$ to chromosomes 10 and 2, respectively,[33] and of human $\alpha 5$ to chromosome 20q13.2–3[140] has been established. The gene structure has been determined for human $\alpha 4$ chain (39 exons, 122 kb)[117] and the *Drosophila* $\alpha 5$ chain.[14]

Mutant/phenotypes

Deletion of the $\alpha 5$ chain gene in *Drosophila* caused late embryonic lethality or abnormal structures in many organs.[15]

References

1. Ekblom, P. and Timpl, R. (1996). *The laminins.* Harwood Academic, Reading, MA.
2. Burgeson, R. E., Chiquet, M., Deutzmann, R., Ekblom, P., Engel, J., Kleinmann, H., *et al.* (1994). *Matrix Biol.*, **14**, 209–11.
3. Nissinen, M., Vuolteenaho, R., Boot-Handford, R., Kallunki, P., and Tryggvason, K. (1991). *Biochem. J.*, **276**, 369–79.
4. Haaparanta, T., Uitto, J., Ruoslahti, E., and Engvall, E. (1991). *Matrix*, **11**, 151–60.
5. Sasaki, M., Kleinman, H. K., Huber, H., Deutzmann, R., and Yamada, Y. (1988). *J. Biol. Chem.*, **263**, 16536–44.
6. Vuolteenaho, R., Nissinen, M., Sainio, K., Byers, M., Eddy, R., Hirvonen, H., *et al.* (1994). *J. Cell Biol.*, **124**, 381–94.
7. Bernier, S. M., Utani, A., Sugiyama, S., Doi, T., Polistina, C., and Yamada, Y. (1994). *Matrix Biol.*, **14**, 447–55.
8. Ryan, M. C., Tizard, R., Van Devanter, D. R., and Carter, W. G. (1994). *J. Biol. Chem.*, **269**, 22779–87.
9. Galliano, M. -F., Aberdam, D., Aguzzi, A., Ortonne, J. P., and Meneguzzi, G. (1995). *J. Biol. Chem.*, **270**, 21820–26.
10. Iivanainen, A., Sainio, K., Sariola, H., and Tryggvason, K. (1995). *FEBS Lett.*, **365**, 183–8.
11. Richards, A., Al-Imara, L., and Pope, F. M. (1996). *Eur. J. Biochem.*, **238**, 813–21.
12. Liu, J. and Mayne, R. (1996). *Matrix Biol.*, **15**, 433–7.
13. Miner, H. H., Lewis, R. M., and Sanes, J. R. (1995). *J. Biol. Chem.*, **270**, 28523–6.
14. Kusche-Gullberg, M., Garrison, K., MacKrell, A. J., Fessler, L. I., and Fessler, J. H. (1992). *EMBO J.*, **11**, 4519–27.
15. Henchcliffe, C., Garcia-Alonso, L., Tang, J., and Goodman, C. S. (1993). *Development*, **118**, 325–37.
16. Pikkarainen, T., Eddy, R., Fukushima, Y., Byers, M., Shows, T., Pihlajaniemi, T., *et al.* (1987). *J. Biol. Chem.*, **262**, 10454–62.
17. Sasaki, M., Kato, S., Kohno, K., Martin, G. R., and Yamada, Y. (1987). *Proc. Natl Acad. Sci. USA*, **84**, 935–9.
18. Montell, D. J. and Goodman, C. S. (1988). *Cell*, **53**, 463–73.
19. Iivanainen, A., Vuolteenaho, R., Sainio, K., Eddy, R., Shows, T. B., Sariola, H., and Tryggvason, K. (1994). *Matrix Biol.*, **14**, 489–97.
20. Wewer, U. M., Gerecke, D. R., Durkin, M. E., Kurtz, K. S., Mattei, M. -G., Champliaud, M. F., *et al.* (1994). *Genomics*, **24**, 243–52.
21. Hunter, D. D., Shah, V., Merlie, J. P., and Sanes, J. R. (1989). *Nature*, **338**, 229–34.
22. Gerecke, D. R., Wagman, D. W., Champliaud, M. -F., and Burgeson, R. E. (1994). *J. Biol. Chem.*, **269**, 11073–80.
23. Pulkkinen, L., Gerecke, D. R., Christiano, A. M., Wagman, D. W., Burgeson, R. E., and Uitto, J. (1995). *Genomics*, **25**, 192–8.
24. Utani, A., Kopp, J. B., Kozak, C. A., Matsuki, Y., Amizuka, M., Sugiyama, S., and Yamada, Y. (1995). *Lab. Invest.*, **72**, 300–10.
25. Pikkarainen, T., Kallunki, T., and Tryggvason, K. (1988). *J. Biol. Chem.*, **263**, 6751–8.
26. Sasaki, M. and Yamada, Y. (1987). *J. Biol. Chem.*, **262**, 17111–17.
27. Durkin, M. E., Bartos, B. B., Liu, S. -H., Phillips, S. L., and Chung, A. E. (1988). *Biochemistry*, **27**, 5198–204.
28. Montell, D. J. and Goodman, C. S. (1989). *J. Cell Biol.*, **109**, 2441–53.
29. Chi, H.-C. and Hui, C.-F. (1989). *J. Biol. Chem.*, **264**, 1543–50.
30. Kallunki, P., Sainio, K., Eddy, R., Byers, M., Kallunki, T., Sariola, H., *et al.* (1992). *J. Cell Biol.*, **119**, 679–93.
31. Vailly, J., Verrando, P., Champliaud, M.-F., Gerecke, D., Wagman, D. W., Baudoin, C., *et al.* (1994). *Eur. J. Biochem.*, **219**, 209–18.
32. Sugiyama, S., Utani, A., Yamada, S., Kozak, C. A., and Yamada, Y. (1995). *Eur. J. Biochem.*, **228**, 120–8.
33. Miner, J. H., Patton, B. L., Lentz, S. I., Gilbert, D. J., Snider, W. D., Jenkins, N. A., Copeland, N,G., and Sanes, J. R. (1997). *J. Cell Biol.*, **137**, 685–701.
34. Beck, K., Hunter, I., and Engel, J. (1990). *FASEB J.*, **4**, 148–60.
35. Peters, B. P., Hartle, R. J., Krzesick, R. F., Kroll, T. G., Perini, F., Balun, J. E., *et al.* (1989). *J. Biol. Chem.*, **260**, 14732–42.
36. Matsui, C., Wang, C. K., Nelson, C. F., Bauer, E. A., and Hoeffler, W. K. (1995). *J. Biol. Chem.*, **270**, 23496–503.
37. Beck, K., Dixon, T. W., Engel, J., and Parry, D. A. D. (1993). *J. Mol. Biol.*, **231**, 311–23.

38. Kammerer, R., Antonsson, P., Schulthess, T., Fauser, C., and Engel, J. (1995). *J. Mol. Biol.*, **250**, 64–73.
39. Nomizu, M., Utani, A., Beck, K., Otaka, A., Roller, P. P., and Yamada, Y. (1996). *Biochemistry*, **35**, 2885–93.
40. Fujiwara, S., Shinkai, H., Deutzmann, R., Paulsson, M., and Timpl, R. (1988). *Biochem. J.*, **252**, 453–61.
41. Arumugham, R. G., Hsieh, T. C. -Y., Tanzer, M. L., and Laine, R. A. (1986). *Biochim. Biophys. Acta*, **883**, 112–26.
42. Ryan, M. C., Christiano, A. M., Engvall, E., Wewer, U. M., Miner, J. H., Sanes, J. R., and Burgeson, R. E. (1996). *Matrix Biol.*, **15**, 369–81.
43. Sanes, J. R., Engvall, E., Butkowski, R., and Hunter, D. D. (1990). *J. Cell Biol.*, **111**, 1685–99.
44. Engvall, E., Earwicker, D., Haaparanta, T., Ruoslahti, E., and Sanes, J. R. (1990). *Cell Regul.*, **1**, 731–40.
45. Durbeej, M., Fecker, L., Hjalt, T., Zhang, H. -Y., Salmivirta, K., Klein, G., *et al.* (1996). *Matrix Biol.*, **15**, 397–413.
46. Stetefeld, J., Mayer, U., Timpl, R., and Huber, R. (1996). *J. Mol. Biol.*, **257**, 644–57.
47. Baumgartner, R., Czisch, M., Mayer, U., Pöschl, E., Huber, R., Timpl, R., and Holak, T. A. (1996). *J. Mol. Biol.*, **257**, 658–68.
48. Timpl, R., Rohde, H., Gehron Robey, P., Rennard, S. I., Foidart, J. M., and Martin, G. R. (1979). *J. Biol. Chem.*, **254**, 9933–7.
49. Green, T. L., Hunter, D. D., Chan, W., Merlie, J. P., and Sanes, J. R. (1992). *J. Biol. Chem.*, **267**, 2014–22.
50. Paulsson, M., Aumailley, M., Deutzmann, R., Timpl, R., Beck, K., and Engel, J. (1987). *Eur. J. Biochem.*, **166**, 11–19.
51. Kishore, U., Eggleton, P., and Reid, K. B. M. (1997). *Matrix Biol.*, **15**, 583–92.
52. Dziadek, M. and Timpl, R. (1985). *Dev. Biol.*, **111**, 372–82.
53. Klein, G., Ekblom, M., Fecker, L., Timpl, R., and Ekblom, P. (1990). *Development*, **110**, 823–37.
54. Ekblom, M., Klein, G., Mugrauer, G., Fecker, L., Deutzmann, R., Timpl, R., and Ekblom, P. (1990). *Cell*, **60**, 337–46.
55. Paulsson, M. (1996). In *The laminins* (ed. P. Ekblom and R. Timpl), pp. 1–25. Harwood Academic, Reading, MA.
56. Mayer, U., Nischt, R., Pöschl, E., Mann, K., Fukuda, K., Gerl, M., *et al.* (1993). *EMBO J.*, **12**, 1879–85.
57. Sung, U., O'Rear, J. J., and Yurchenco, P. D. (1993). *J. Cell Biol.*, **123**, 1255–68.
58. Yurchenco, P. D., Sung, U., Ward, M. D., Yamada, Y., and O'Rear, J. J. (1993). *J. Biol. Chem.*, **268**, 8356–65.
59. Schulze, B., Mann, K., Pöschl, E., Yamada, Y., and Timpl, R. (1996). *Biochem. J.*, **314**, 847–51.
60. Pikkarainen, K., Schulthess, T., Engel, J., and Tryggvason, K. (1992). *Eur. J. Biochem.*, **209**, 571–82.
61. Martin, P. T., Ettinger, A. J., and Sanes, J. R. (1995). *Science*, **269**, 413–16.
62. Yurchenco, P. D. and Cheng, Y. -S. (1993). *J. Biol. Chem.*, **268**, 17286–99.
63. Paulsson, M. (1988). *J. Biol. Chem.*, **263**, 5425–30.
64. Fox, J. W., Mayer, U., Nischt, R., Aumailley, M., Reinhardt, D., Wiedemann, H., *et al.* (1991). *EMBO J.*, **10**, 3137–46.
65. Pöschl, E., Mayer, U., Stetefeld, J., Baumgartner, R., Holak, T. A., Huber, R., and Timpl, R. (1996). *EMBO J.*, **15**, 5154–59.
66. Battaglia, C., Mayer, U., Aumailley, M., and Timpl, R. (1992). *Eur. J. Biochem.*, **208**, 359–66.
67. Pan, T. -C., Kluge, M., Zhang, R. -Z., Mayer, U., Timpl, R., and Chu, M. -L. (1993). *Eur. J. Biochem.*, **215**, 733–40.
68. Utani, A., Nomizu, M., and Yamada, Y. (1997). *J. Biol. Chem.*, **272**, 2814–20.
69. Aumailley, M., Gimond, C., and Rouselle, P. (1996). In *The laminins* (ed. P. Ekblom and R. Timpl), pp. 127–158. Harwood Academic, Reading, MA.
70. Mecham, R. P. and Hinek, A. (1996). In *The laminins* (ed. P. Ekblom and R. Timpl), pp. 159–183. Harwood Academic, Reading, MA.
71. Sonnenberg, A., Linders, C. J. T., Modderman, P. W., Damsky, C. H., Aumailley, M., and Timpl, R. (1990). *J. Cell Biol.*, **110**, 2145–55.
72. Deutzmann, R., Aumailley, M., Wiedemann, H., Pysny, W., Timpl, R., and Edgar, D. (1990). *Eur. J. Biochem.*, **191**, 513–22.
73. Pfaff, M., Göhring, W., Brown, J. C., and Timpl, R. (1994). *Eur. J. Biochem.*, **225**, 975–84.
74. Brancaccio, A., Schulthess, T., Gesemann, M., and Engel, J. (1995). *FEBS Lett.*, **368**, 139–42.
75. Taraboletti, G., Rao, C. N., Krutzsch, H. C., Liotta, L. A., and Roberts, D. D. (1990). *J. Biol. Chem.*, **265**, 12253–8.
76. Nomizu, M., Kim, W. H., Yamamura, K., Utani, A., Song, S.-Y., Otaka, A., *et al.* (1995). *J. Biol. Chem.*, **270**, 20583–90.
77. Edgar, D. (1996). In *The laminins* (ed. P. Ekblom and R. Timpl), pp. 251–75. Harwood Academic, Reading, MA.
78. Porter, B. E., Weis, J., and Sanes, J. R. (1995). *Neuron*, **14**, 549–59.
79. Brandenberger, A., Kammerer, R. A., Engel, J., and Chiquet, M. (1996). *J. Cell Biol.*, **135**, 1583–92.
80. Sorokin, L., Conzelmann, S., Ekblom, P., Battaglia, C., Aumailley, M., and Timpl, R. (19920). *Exp. Cell Res.*, **201**, 137–144.
81. Brandenberger, R. and Chiquet, M. (1995). *J. Cell Sci.*, **108**, 3099–108.
82. Brown, J. C., Spragg, J. H., Wheeler, G. N., and Taylor, P. W. (1990). *Biochem. J.*, **270**, 463–8.
83. Ekblom, P., Ekblom, M., Fecker, L., Klein, G., Zhang, H.-Y., Kadoya, Y., *et al.* (1994). *Development*, **120**, 2003–14.
84. Kadoya, Y., Salmivirta, K., Talts, J. F., Kadoya, K., Mayer, U., Timpl, R., and Ekblom, P. (1997). *Development*, **124**, 683–91.
85. Durbeej, M., Larsson, E., Ibraghimov-Beskrovnaja, O., Roberds, S. L., Campbell, K. P., and Ekblom, P. (1995). *J. Cell Biol.*, **130**, 79–91.
86. Tryggvason, K., Haakana, H., Airenne, T., Iivanainen, A., and Kallunki, T. (1996). In *The laminins* (ed. P. Ekblom and R. Timpl), pp. 51–63, Harwood Academic, Reading, MA.
87. Durkin, M. E., Gantam, M., Loechel, F., Sanes, J. R., Merlie, J. P., Albrechtsen, R., and Wewer, U. M. (1996). *J. Biol. Chem.*, **271**, 13407–16.
88. Noakes, P. G., Gautam, M., Mudd, J., Sanes, J. R., and Merlie, J. P. (1995). *Nature*, **374**, 258–62.
89. Noakes, P. G., Miner, J. H., Gautam, M., Cunningham, J. M., Sanes, J. R., and Merlie, J. P. (1995). *Nature Genet.*, **10**, 400–6.
90. Brown, J. C., Wiedemann, H., and Timpl, R. (1994). *J. Cell Sci.*, **107**, 329–38.
91. Leivo, I. and Engvall, E. (1988). *Proc. Natl Acad. Sci. USA*, **85**, 1544–8.
92. Paulsson, M., Saladin, K., and Engvall, E. (1991). *J. Biol. Chem.*, **266**, 17545–51.
93. Vachon, P. H., Loechel, F., Xu, H., Wewer, U. M., and Engvall, E. (1996). *J. Cell Biol.*, **134**, 1483–97.
94. Zhang, X., Vuolteenaho, R., and Tryggvason, K. (1996). *J. Biol. Chem.*, **271**, 27664–9.
95. Xu, H., Christmas, P., Wu, X.-R., Wewer, U. M., and Engvall, E. (1994). *Proc. Natl Acad. Sci. USA*, **91**, 5572–6.
96. Xu, H., Wu., X-R., Wewer, U. M., and Engvall, E. (1994). *Nature Genet.*, **8**, 297–301.
97. Helbling-Leclerc, A., Zhang, X., Topaloglu, H., Cruaud, C., Tesson, F., Weissenbach, J., *et al.* (1995). *Nature Genet.*, **11**, 216–18.

98. Burgeson, R. E. (1996). In *The laminins* (ed. P. Ekblom and R. Timpl), pp. 65–96. Harwood Academic, Reading, MA.
99. Rouselle, P., Lunstrum, G. P., Keene, D. R., and Burgeson, R. E. (1991). *J. Cell Biol.*, **114**, 567–76.
100. Marinkovich, M. P., Lunstrum, G. P., Keene, D. R., and Burgeson, R. E. (1992). *J. Cell Biol.*, **119**, 695–703.
101. Marinkovich, M. P., Lunstrum, G. P., and Burgeson, R. E. (1992). *J. Biol. Chem.*, **267**, 17900–6.
102. Champliaud, M. -F., Lunstrum, G. P., Rouselle, P., Nishiyama, T., Keene, D. R., and Burgeson, R. E. (1996). *J. Cell Biol.*, **132**, 1189–98.
103. Aberdam, D., Aguzzi, A., Baudoin, C., Galliano, M. F., Ortonne, J. P., and Meneguzzi, G. (1994). *Cell Adhes. Commun.*, **2**, 115–29.
104. Airenne, T., Haakana, H., Sainio, K., Kallunki, P. Sariola, H., and Tryggvason, K. (1996). *Genomics*, **32**, 54–64.
105. Mayer, U., Pöschl, E., Gerecke, D. R., Wagman, D. W., Burgeson, R. E., and Timpl, R. (1995). *FEBS Lett.*, **365**, 129–32.
106. Kivirikko, S., McGrath, J., Baudoin, C., Aberdam, D., Ciatti, S., Dunnill, M. S. G., *et al.* (1995). *Hum. Mol. Genet.*, **4**, 959–62.
107. Vidal, F., Baudoin, C., Miquel, C., Galliano, M. -F., Christiano, A. M., Uitto, J., *et al.* (1995). *Genomics*, **30**, 273–80.
108. Pulkkinen, L., Christiano, A. M., Gerecke, D., Wagman, D. W., Burgeson, R. E., Pittelkow, M. R., and Uitto, J. (1994). *Genomics*, **24**, 357–60.
109. Pulkkinen, L., Christiano, A. M., Airenne, I., Haakana, H., Tryggvason, K., and Uitto, J. (1994). *Nature Genet.*, **6**, 293–7.
110. Aberdam, D., Galliano, M. -F., Vailly, J., Pulkkinen, L., Bonifas, J., Christiano, A. M., *et al.* (1994). *Nature Genet.*, **6**, 299–304.
112. Frieser, M., Nöckel, H., Pausch, F., Röder, C., Hahn, A., Deutzmann, R., and Sorokin, L. (1997). *Eur. J. Biochem.*, **246**, 727–35.
113. Sorokin, L. M., Pausch, F., Frieser, M., Kröger, S., Ohage, E., and Deutzmann, R. (1997). *Dev. Biol.*, **189**, 285–300.
114. Fessler, L. I., Campbell, A. G., Duncan, K. G., and Fessler, J. H. (1987). *J. Cell Biol.*, **105**, 2383–91.
115. Edgar,D., Timpl, R., and Thoenen, H. (1988). *J. Cell Biol.*, **106**, 1299–306.
116. Mayer, U., Mann, K., Fessler, L. I., Fessler, J. H., and Timpl, R. (1997). *Eur. J. Biochem.*, **245**, 745–50.
117. Richards, A., Luccarini, C., and Pope, F. M. (1997). *Eur. J. Biochem.*, **248**, 15–23.
118. Iivanainen, A., Kortesmaa, J., Sahlberg, C., Morita, T., Bergmann, U., Thesleff, I., and Tryggvason, K. (1997). *J. Biol. Chem.*, **272**, 27862–8.
119. Yurchenco, P. D., Quan, Y., Colognato, H., Mathus, T., Harrison, D., Yamada, Y., and O'Rear, J. J. (1997). *Proc. Natl Acad. Sci. USA*, **94**, 10189–94.
120. Niimi, T. and Kitagawa, Y. (1997). *FEBS Lett.*, **400**, 71–4.
121. Tiger, C.-F., Champliaud, M.-F., Pedrosa-Domellof, F., Thornell, L.-E., Ekblom, P. and Gullberg, D. (1997). *J. Biol. Chem.*, **272**, 28590–95.
122. Patton, B. L., Miner, J. H., Chiu, A. Y., and Sanes, J. R. (1997). *J. Cell Biol.*, **139**, 1507–21.
123. Ettner, N., Göhring, W., Sasaki, T., Mann, K., and Timpl, R. (1998). *FEBS Lett.*, (In press).
124. Sasaki, T., Forsberg, E., Bloch, W., Addicks, K., Fässler, R., and Timpl, R. (1998). *Exp. Cell Res.*, **238**, 70–81.
125. Takagi, Y., Nomizu, M., Gullberg, D., MacKrell, A. J., Keene, D. R., Yamada, Y., and Fessier, J. H. (1996). *J. Biol. Chem.*, **271**, 18074–81.
126. Nomizu, M., Kuratomi, Y., Song, S.-Y., Ponce, M.-L., Hoffman, M. P., Powell, S. K., Miyoshi, K., Otaka, A., Kleinman, H. K., and Yamada, Y. (1997). *J. Biol. Chem.*, **272**, 32198–205.
127. Sugiyama, J. E., Glass, D. J., Yancopolous, G. D., and Hall, Z. W. (1997). *J. Cell Biol.*, **139**, 181–91.
128. Mayer, U., Kohfeldt, E., and Timpl, R. (1998). *Ann. N.Y. Acad. Sci.*, (In press).
129. Colognato, H., MacCarrick, M., O'Rear, J. J., and Yurchenco, P. D. (1997). *J. Biol. Chem.*, **271**, 29330–36.
130. Talts, J. F., Mann, K., Yamada, Y., and Timpl, R. (1998). *FEBS Lett.*, **426**, 71–6.
131. Rambukkana, A., Salzer, J. L., Yurchenco, P. D., and Tuomanen, E. I. (1997). *Cell*, **88**, 811–21.
132. Cheng, Y.-S., Champliaud, M.-F., Burgeson, R. E., Marinkovich, M. P., and Yurchenco, P. D. (1997). *J. Biol. Chem.*, **272**, 31525–32.
133. Miyagoe, Y., Hanaoka, K., Nonaka, I., Hayasaka, M., Nabeshima, Y., Arahata, K., Nabeshima, Y., and Takeda, S. (1997). *FEBS Lett.*, **415**, 33–9.
134. Doliana, R., Bellina, I., Bucciotti, F., Mongiat, M., Perris, R., and Colombatti, A. (1997). *FEBS Lett.*, **417**, 65–70.
135. Rouselle, P., Keene, D. R., Ruggiero, F., Champliaud, M.-F., van der Rest, M., and Burgeson, R. E. (1997). *J. Cell Biol.*, **138**, 719–28.
136. Dogic, D., Rousselle, P., and Aumailley, M. (1998). *J. Cell Sci.*, **111**, 793–802.
137. Gianelli, G., Falk-Marzillier, J., Schiraldi, O., Stetler-Stevenson, W. G., and Quaranta, V. (1997). *Science*, **277**, 225–8.
138. Goldfinger, L. E., Stack, M. S., and Jones, J. C. R. (1998). *J. Cell Biol.*, **141**, 255–65.
139. Ferrigno, O., Virolle, T., Galliano, M.-F., Chauvin, N., Ortonne, J. P., Meneguzzi, G., and Aberdam, D. (1997). *J. Biol. Chem.*, **272**, 20502–7.
140. Durkin, M. E., Loechel, F., Mattai, M.-G., Gilpin, B. J., Albrechtsen, R., and Wewer, U. M. (1997). *FEBS Lett.*, **411**, 296–300.
141. Chiu, A. Y. Ugozolli, M., Meiri, K., and Ko, J. (1992). *J. Neurochem*, **59**, 10–17.

Takako Sasaki and Rupert Timpl
Max-Planck-Institut für Biochemie,
D-82152 Martinsried, Martinsried,
Germany

Link protein (LP)

Link protein[1] (LP) is an extracellular metalloprotein[2] found abundantly in cartilage where it forms a ternary complex with the cartilage proteoglycan, aggrecan, and hyaluronic acid. The function of LP is to stabilize the interaction between aggrecan and hyaluronic acid. LP is also present in numerous non-cartilaginous tissues where it stabilizes the interaction between proteoglycans other than aggrecan and HA.[3]

Synonymous names

None.

Homologous proteins

BEHAB[4] (EMBL Z28366). The expression of BEHAB is restricted to the central nervous system.

Protein properties

LP consists of an immunoglobulin-like domain and two tandemly repeated domains. The immunoglobulin-like domain places LP in the immunoglobulin superfamily.[1] Although BEHAB is the only protein that is homologous to LP, the structural motif of an entire LP is found in the amino terminal domains of the proteoglycans aggrecan, versican/PG-M, neurocan, and brevican.[5] All these proteoglycans interact with hyaluronic acid through an amino terminal domain (G1). The tandemly repeated domains of LP are also homologous to the second globular domain (G2) of aggrecan.[5] A structural motif homologous to one of the tandemly repeated domains of LP is present in CD44[6,7] (hyaluronic acid receptor, hermes antigen, pgp-1) and TSG-6.[8]

The complete sequence has been determined for LP of chicken,[9] rat,[10,11] horse,[12] and human,[13,14] either from protein or cDNA data. The primary structure derived from protein sequencing[10] has allowed the precise determination of the disulphide bonds. Although there are multiple forms of LP (i.e. 41, 46, and 51 kDa for rat), these result from post-translational modification involving differential glycosylation or proteolysis. LP has also been shown to be able to form oligomers in a variety of concentrated salt solutions.[15]

The gene for LP of chicken,[16] rat,[13] and human[17] is large and is present in a single copy in the genome. The relationship between chicken LP and the structure of the LP gene is given in Fig. 1. A 15 amino acid signal peptide together with approximately 19 amino acids of the N terminus of the mature protein is encoded by a single exon. Each subsequent domain is also encoded by a single exon. A similar situation exists in the organization of the rat[13] and human[17] genes. In addition, transcripts resulting from alternative splicing of RNA have been reported in the rat[3] and horse and pig. A complex pattern of alternative splicing in the leader sequence of the chicken LP mRNA has been reported.

The tandemly repeated domains of LP are involved in its interaction with hyaluronic acid (hyaluronan).[19–22] Evidence for this is derived from inhibitions of the interactions with the monoclonal antibody 8A4 and with synthetic peptides covering sequences in that region, and from expression of recombinant tandem repeat domains. In addition, the tandemly repeated domains are protected from proteolysis in mixtures of LP with hyaluronic acid[23] whereas the immunoglobulin-like domain of LP is protected when aggrecan and LP are mixed. The latter observation suggests that the immunoglobulin like domain of link protein is involved in its interaction with aggrecan.

Although the major source of LP is cartilage, the protein has also been found in numerous non-cartilaginous tissues.[3,24,25] The coding region of the LP isolated from these non-cartilaginous tissues is identical to that present in cartilage.[3] In non-cartilaginous tissues LP stabilizes the interaction of HA binding proteoglycans other than aggrecan with HA.

Purification

The purification of LP from cartilage is based on extractions in 4 M guanidine–HCl solutions in the presence of protease inhibitors.[15] Under these conditions the LP–aggrecan–hyaluronic acid complex is dissociated. Dialysis of the extracts to 0.4 M or less result in the re-establishment of the aggregate. Centrifugation of these extracts in CsCl gradients with an initial density of 1.6 g/ml results in the sedimentation of LP-stabilized proteoglycan aggregates to the lower densities of the gradient (A1 fraction). When the A1 fraction is returned to dissociative conditions (4 M guanidine–HCl) and subjected to CsCl gradient centrifugation the aggregate dissociates and aggrecan sediments to the bottom of the gradient (A1D1 fractions) and LP to the top (A1D6 fraction). Hyaluronic acid sediments in the middle region of the gradient. LP can be further purified by DEAE chromatography to remove remaining traces of low buoyant density proteoglycans. A method for isolating link protein using wheat germ agglutinin has also been reported.

Activities

The biological function of LP in cartilage is to stabilize the interaction of aggrecan with hyaluronic acid.[1] These

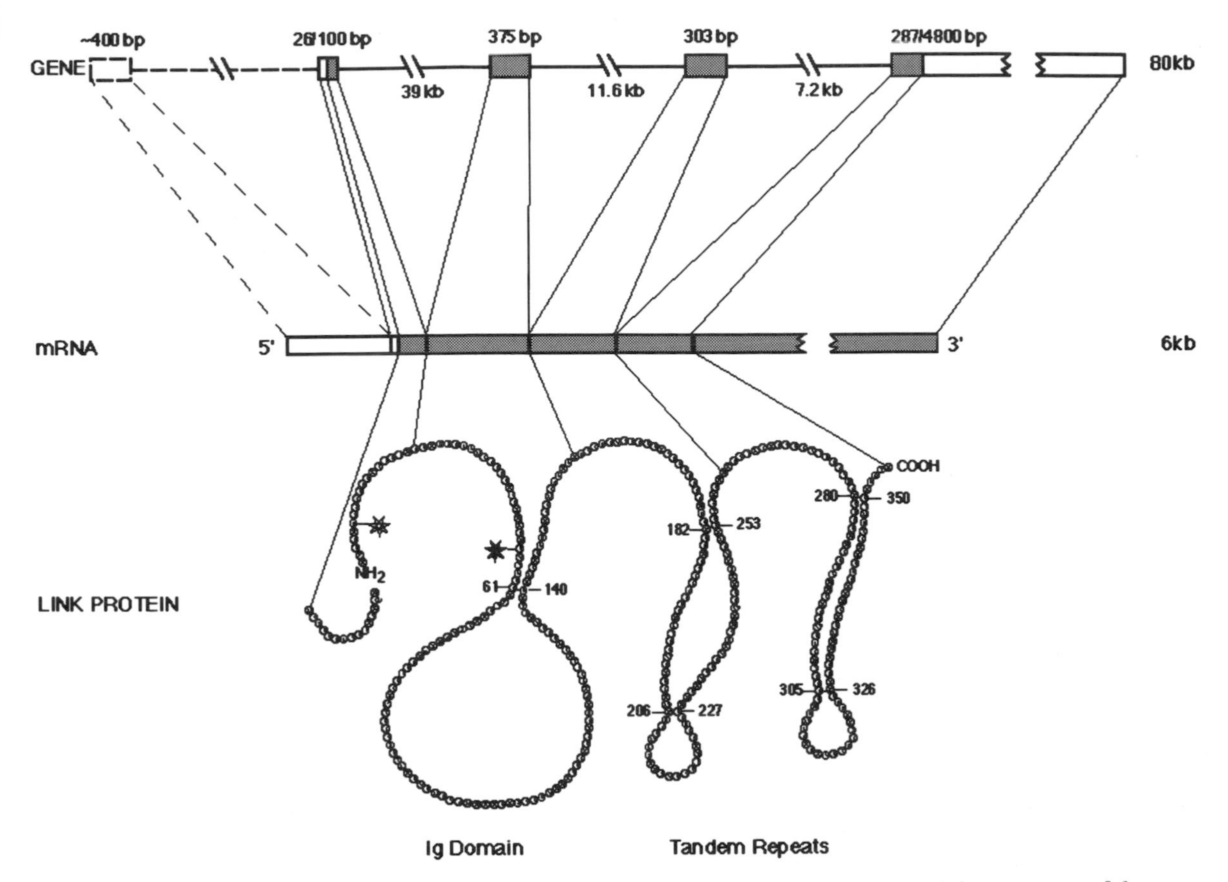

Figure 1. Diagrammatic representation showing the relationship between chicken LP and the structure of the chicken LP gene. LP consists of an immunoglobulin-like domain and two tandemly repeated domains. Each domain of LP is encoded by a separate exon. The disulfide bonds are based on the structure of LP determined from amino acid sequence. The stars on the LP diagram indicate glycosylation sites.

LP stabilized proteoglycan aggregates contribute to the growth of embryonic cartilaginous rudiments and to the compressibility of adult cartilage. In non-cartilaginous tissues LP performs the same function but with a different hyaluronic acid binding proteoglycan (i.e. versican/PG-M).[3]

Antibodies

Polyclonal antisera against bovine[26,27] and chicken[28] L.P have been described. Monoclonal antibody (mAb) 8A4[13] is well characterized and can be purchased from the Developmental Studies Hybridoma Bank (The Johns Hopkins University School of Medicine, Baltimore, MD). Two epitopes are recognized by 8A4. One epitope is in each of the tandem repeats of LP. A polyclonal antiserum against a synthetic peptide covering a sequence of the immunoglobulin domain of chicken LP has been reported.[19] mAbs 4B6/A5 and 3H8/C5 were generated against chicken LP.[3]

Genes

Complete cDNA sequences are available for chicken[9] (GenBank M13212), rat[11] (GenBank M13191), horse[12] (EMBL X78077) and human[13,14] (EMBL X17405) LP. The human LP gene has been mapped to chromosome 5q13–q14.1.[30]

Mutant phenotype/disease states

LP has been excluded as mutations in a number of heritable chondrodysplasias or as a cause of generalized osteoarthritis.[31,32]

Structure

X-ray and neutron solution scattering and electron microscopy indicate that the G1 domain of aggrecan, LP and the ternary complex involving HA have compact structures.[33–35] Consensus secondary structure predictions of 59 repeats indicate the presence of two α helices and seven β strands.[36] A similar structure was reported for the single repeat of TSG-6.[37] Preliminary X-ray diffraction studies of bovine LP have been reported.[38]

References

1. Neame, P. J. and Barry, F. P. (1994). *EXS*, **70**, 53–72.
2. Rosenberg, L.,Choi, H. U., Tang, L. H., Pal, S., Johnson, T., Lyons, D. A., and Lau, T. M. (1991). *J. Biol. Chem.*, **266**, 7016–24.
3. Binette, F., Cravens, J., Kahoussi, B., Haudenschild, D. R., and Goetinck, P. F. (1994). *J. Biol. Chem.*, **269**, 19116–22.
4. Jaworski, D. M., Kelley, G. M., and Hockfield, S. (1994). *J. Cell Biol.*, **125**, 495–509.
5. Iozzo, R. V. and Murdoch, A. D. (1996). *FASEB J.* **10**, 598–614.
6. Stamenkovic, I. M., Amiot, M., Pesando, J. M., and Seed, B. (1989). *Cell*, **56**, 1057–62.
7. Goldstein, L. A., Zhou, D. F. H., Picker, L. J., Minty, C. N., Bargatze, R. F., Ding, J. F. and Butcher, E. C. (1989). *Cell*, **56**, 1063–72.
8. Lee, T. H. Wisniewski, H. G., and Vilcek, J. (1992). *J. Cell Biol.*, **116**, 545–57.
9. Deák, F., Kiss, I., Sparks, K., Argraves, W. S., Hampikian, G., and Goetinck, P. F. (1986). *Proc. Natl Acad. Sci. USA*, **83**, 3766–70.
10. Neame, P. J., Christner, J. E., and Baker, J. R. (1986). *J. Biol. Chem.*, **261**, 3519–35.
11. Rhodes, C., Doege, K., Sasaki, M., and Yamada, Y. (1988). *J. Biol. Chem.*, **263**, 6063–7.
12. Dudhia, J. and Platt, D. (1995). *Am. J. Vet. Res.*, **56**, 959–65.
13. Doege, K., Rhodes, C., Sasaki, M., Hassell, J. R., and Yamada, Y. (1990). In: *Extracellular matrix genes*, p.137–55.
14. Dudhia, J. and Hardingham, T. E. (1990). *Nucl. Acids Res.*, **18**, 1292.
15. Baker, J. R. and Neame, P. J. (1987). *Meth. Enzymol.*, **144**, 401–13.
16. Kiss, I., Deák, F., Mestric, S., Delius, H., Soos, T., Dékány, K., *et al.* (1987). *Proc. Natl Acad. Sci. USA*, **84**, 6399–403.
17. Dudhia, J., Bayliss, M. T., and Hardingham, T. E. (1994). *Biochem. J.* **303**, 329–33.
18. Deak, F., Barta, E., Mestric, S., Biesold, M., and Kiss, I. (1991). *Nucl. Acids Res.*, **19**, 4983–90.
19. Goetinck, P. F., Stirpe, N. S., Tsonis, P. A., and Carlone, D. (1987). *J. Cell Biol.*, **105**, 2403–8.
20. Grover, J. and Roughley, P. J. (1994). *Biochem. J.*, **300**, 317–24.
21. Verelas, J. B., Kollar, J., Huynh, T. D., and Hering, T. M. (1995). *Arch. Biochem. Biophys.*, **321**, 21–30.
22. Hofer, U., Kahoussi, B. Trelstad, J., Zavaterelli, M., and Goetinck, P. F. (1996). *Ann. NY Acad. Sci.*, **785**, 271–3.
23. Périn, J. P., Bonnet, F., Thurleau, C., and Jollés, P. (1987). *J. Biol. Chem.*, **262**, 13269–72.
24. Gardell, S., Baker, J. R., Caterson, B., Heinegard, D., and Rodén, L. (1980). *Biochem. Biophys. Res. Commun.*, **95**, 1823–931.
25. Stirpe, N. S., Dickerson, K. T., and Goetinck, P. F. (1990). *Dev. Biol.*, **137**, 419–24.
26. Baker, J. R., Caterson, B. and Christner, J. E. (1982). *Meth. Enzymol.*, **83**, 216–35.
27. Poole, A. R. and Reiner, A. (1980). *J. Biol. Chem.*, **255**, 9295–305.
28. McKeown-Longo, P. J., Sparks, K. J., and Goetinck, P. F. (1982). *Collagen Rel. Res.*, **2**, 232–44.
29. Caterson, B., Calabro, T., and Hampton, A. (1987). In *Biology of proteoglycans.* (ed. T. Wight and R. Mecham), pp. 1–16.
30. Osborne-Lawrence, S. L., Sinclair, A. K., Hicks, R. C., Lacey, S. W., Eddy Jr, R. L., Byers, M. G., *et al.* (1990). *Genomics*, **8**, 562–7.
30. Loughlin, J., Irven, C., Fergusson, C., and Sykes, B. (1994). *Br. J. Rheumatol.*, **33**, 1103–6.
32. Loughlin, J., Irven, C., and Sykes, B. (1994). *Hum. Genet.*, **94**, 698–700.
33. Perkins, S. J., Nealis, A. S., Dunham, D. G., Hardingham, T. E., and Muir, I. H. (1991). *Biochemistry*, **30**, 10708–16.
34. Perkins, S. J., Nealis, A. S., Dunham, D. G., Hardingham, T. E., and Muir, I. H. (1992). *Biochem. J.*, **285**, 263–8.
35. Morgelin, M., Paulsson, M., Hardingham,T. E., Heinegard, D., and Engel, J. (1988). *Biochem. J.*, **253**, 175–85.
36. Brisett, N. C. and Perkins, S. J. (1996). *FEBS Lett.*, **388**, 211–16.
37. Khoda, D., Morton, C. J., Parkar, A. A., Hatanaka, H., Inagaki, F. M., Campbell, I. D., and Day, A. J. (1996). *Cell*, 767–75.
38. Jedrzejas, M. J., Baker, J. R., and Luo, M. (1995). *Proteins*, 76–8.

Paul F. Goetinck
Cutaneous Biology Research Center,
Massachusetts General Hospital,
Harvard Medical School, Building 149,
13th Street, Charlestown, MA 02129, USA

Mucins

The term mucin refers to a diverse group of glycoproteins (Table 1) ranging from the very large secreted gel-forming mucins of the gastrointestinal tract to the simpler integral membrane mucin expressed by many simple epithelial cells. Mucin-like membrane glycoproteins such as CD43 (leukosialin) and selectin ligands such as PSGL-1 are also found on leukocytes and/or endothelial cells. Mucin-type proteins form a family of glycoproteins that contain a large amount of carbohydrate in *O*-linkage.[1] The first sugar in mucin-type glycans in higher eukaryotes is *N*-acetylgactosamine which is attached to the hydroxyl group of serine or threonine. The peptide core of mucins therefore has a high content of serine and threonine and most mucin genes exhibit variable numbers of tandem repeats (VNTR) within their coding regions. This summary will be restricted to a description of the epithelial mucins.

Protein properties

The apparent molecular mass of native mucin can range from 39 kDa for the exceptionally small MUC7 to about 10 000 kDa for the gastric and tracheal mucins. However, due to the vast amounts of carbohydrate and the possibility of intra- and intermolecular aggregates the native molecular weight is difficult to interpret.[12] Moreover, the molecular weight of the core protein of each mucin may be different from individual to individual as many of the mucin genes show a genetic polymorphism due to varying numbers of tandem repeats. In addition, the same mucin may be differently glycosylated when it is expressed by different tissues.

A translated tandem repeat domain is common to all epithelial mucins (see Table 1). While all the repeats are rich in proline, serine, and/or threonine, the size and amino acid composition of the repeats vary widely among mucin types. However, this element appears to have a common function as serving as a scaffold for the *O*-linked carbohydrate, the serines and threonines forming the attachment sites and the prolines preventing α-helix formation and so allowing the necessary conformation for the close packing of the carbohydrates.

With the exception of MUC1, the epithelial mucins are secreted and are the major glycoproteins in mucus giving it its viscoelastic property. MUC2, MUC5, and MUC7 and the porcine and bovine submaxillary mucins all contain a cysteine-rich domain upstream of the carboxyl terminus in which the arrangement of cysteines is nearly identical to that in human von Willebrand factor.[4] It is thought that this region allows the oligomerization of mucin molecules to occur via disulphide bonds.

In constrast, MUC1 is an integral membrane protein which forms a rod-like structure that extends far above the glycocalyx of the cell.[13] The high content of sialic acid within the glycans of MUC1 give the molecule a strong negative charge, which can cause the repulsion of other cells. However, MUC1 may carry specific recognition signals which will attract other cells. In this context MUC1 has been shown to be a highly efficient ligand for sialoadhesin,[14] a lectin found on macrophages.

Purification

Most of the procedures used to purify mucins take advantage of their physical characteristics, that is, their large size and high buoyant densities. Gel filtration is normally used to select large molecules and caesium chloride density gradient ultracentrifugation is employed to select for highly glycosylated glycoproteins.[15] In the case of some of the very large mucins, for example those of the respiratory tract, it may be necessary initially to stir the mucin with guanidinium hydrochloride for solubilization.[16] In contrast, the membrane-bound MUC1 mucin can be relatively easily purified, particularly from human milk, by immunoaffinity chromatography.[17]

Activities

Secreted mucins are the glycoproteins found in mucus giving this material its elasticity and viscosity. Mucus represents one of the main interfaces between the organism and its environment and protects the tissues from physical damage, toxins, microorganisms, and dehydration.

Table 1 Human epithelial mucins

Mucin	Other names	Main tissue expression	Secreted or integral membrane	Size of tandem repeat	Ref.
MUC1	PEM, PUM, episialin, MAM6, DF3 antigen, EMA	Luminal surface of simple epithelial cells cells	Membrane	20 amino acids	2, 3
MUC2		Colon, small intestine, bronchus, trachea	Secreted	23 amino acids	4
MUC3		Colon, small intestine	Secreted	23 amino acids	5
MUC4		Trachea, colon	Secreted	16 amino acids	6
MUC5AC		Stomach, respiratory tract	Secreted	8 amino acids	7, 8
MUC5B		Stomach, respiratory tract	Secreted	169 amino acids	9
MUC6		Stomach, colon, gall bladder, endocervix	Secreted	169 amino acids	9
MUC7	MG2	Sublingual salivary gland	Secreted	23 amino acids	10
MUC8		Trachea	Secreted	?	11

The biological activity of MUC1 appears to be different from the secreted mucins (see above) although it too seems to have a protective role as it can be expressed by the skin in response to injury. However, MUC1 has a 69 amino acid cytoplasmic tail that is highly conserved among species, indicating functionality, and this domain has been implicated in actin binding[18] and signal transduction.[19]

Antibodies

There are a whole range of antibodies reacting with the different mucins, some that recognize carbohydrate epitopes, others that recognize peptide.[20] In the case of the large secreted mucins the peptide-reactive antibodies have often been raised to the mucin chemically stripped of its carbohydrate.[21] How the reactivity of some antibodies relates to the expression of the mucin core must be interpreted with care as epitopes may by masked by changes in the glycans attached to mucins when they are expressed by different tissues, or may be unmasked in malignant tissue. In the case of MUC1 there appears to be a highly immunodominant region between glycoslation sites within the tandem repeat.[22]

Genes

Table 2 provides information on the cDNA available and the chromosomal mapping of human MUC1 through to MUC8.

Disease states

In epithelial cancers many of the phenotypic markers for malignant cells have been found to be carried on mucins. In the change to malignancy there is often a change in the glycosylation of mucins which creates novel carbohydrate epitopes and exposes peptide epitopes that may be masked when the mucin is expressed by normal cells.[28] This has been most clearly demonstrated in the case of MUC1 which is aberrantly glycosylated in breast cancer due to the upregulation of α2,3-sialyltransferase (the glycosyltransferase responsible for the sialylation of core 1 carbohydrate chains), therefore preventing further chain elongation.[29] In addition, dysregulation of tissue and cell-specific regulation occurs in epithelial cancers.

Structure

No X-ray crystallographic or NMR structural information is available.

References

1. Brockhausen., I. (1996). In *New comprehensive biochemistry. Vol. 29a. Glycoproteins* (ed. J. Montreuil, F.G. Vligenthart, and H. Schachter), pp. 201–59. Elsevier.
2. Gendler, S., Lancaster, C., Taylor-Papadimitriou, J., Duhig, T., Peat, N., Burchell, J. *et al.* (1990). *J. Biol. Chem.*, **265**, 15286–93.
3. Zotter, S., Hageman, C., Lossnitzer, A., Mooi, W. J., and Hilgers, J. (1988). *Cancer Rev.*, **11–12**, 55–101.
4. Gum, J. R., Hicks, J. W., Toribara, N. W., Siddiki, B., and Kim, Y. S. (1994). *J. Biol. Chem.*, **269**, 2440–6.
5. Gum, J. R., Hicks, J. W., Swallow, D., Lagace, R. L., Byrd, J. C., Lamport, D. T. A., *et al.* (1990). *Biochem. Biophys. Res. Commun.*, **171**, 407–15.
6. Porchet, N., Cong, N. V., Dufosse, J., Audie, J. P., Guyonnet-Duperat, V., Gross, M. S. *et al.* (1991). *Biochem. Biophys. Res. Commun.*, **175**, 414–22.
7. Porchet, N., Pigny, P., Buisine, M.-P., Debailleul, V., Laine, A., and Aubert, J. P. (1995). *Biochem. Soc. Trans.*, **23**, 800–5.
8. Meerzaman, D., Charles, P., Daskal, E., Polymeropoulos, M. H., Martin, B. M., and Rose, M. C. (1994). *J. Biol. Chem.*, **269**, 12932–9.
9. Ho, S. B., Roberton, A. M., Shekels, L. L., Lyftogt, C. T., Niehans, G. A., and Toribara, N. W. (1995). *Gastroenterology*, **109**, 735–47.
10. Troxler, R. F., Offner, G. D., Zhang, F., Iontcheva, I., and Oppenheim, F. G. (1995). *Biochem. Biophys. Res. Commun.*, **217**, 1112–9.
11. D'Cruz, O. J., Dunn, T. S., Pichan, P., Hass, G. G., and Sachdev, G. P. (1996). *Fertil. Steril.*, **66**, 316–2.
12. Sheehan, J. K., Hanski. C., Corfield, A. P., Paraskeva, C., and Thornton, D. (1995). *Biochem. Soc. Trans.*, **23**, 819–22.
13. Hilkens, J., Marjolÿn, J., Ligtenberg, L., Vox, H. L., and Litvinov, S. V. (1992). *TIBS*, **17**, 359–63.
14. Crocker, P. R., Mucklow, S., Bouckson, V., McWilliams, A., Willis, A. C., Gordon, S., *et al.* (1994). *EMBO J.*, **13**, 4490–503.
15. Hovenberg, H. W., Davies, J. R., Herrmann, A., Linden, C. J., and Carlstedt, I. (1996). *Glycoconj. J.*, **13**, 839–47.
16. Thornton, D. J., Sheehan, J. K., Lindgren, H., and Carlstedt, I. (1991). *Biochem. J.*, **276**, 667–75.

Table 2 Epithelial mucins, clones available, and chromosomal location

Mucin	Partial or full length cDNA	Chromosome location	Ref.
MUC1	Full length	1q2	23
MUC2	Full length	11p15.5	24
MUC3	Partial	7q22	25
MUC4	Partial	3q29	26
MUC5AC	Partial	11p15.5	24
MUC5B	Partial	11p15.5	24
MUC6	Partial	11p15.5	24
MUC7	Full length	4q13–21	27
MUC8	Partial	?	11

17. Burchell, J., Gendler, S. J., Taylor-Papadimitriou, J., Girling, A., Lewis, A., Millis, R., and Lamport, D. (1987). *Cancer Res.*, **47**, 5476–82.
18. Parry, G., Beck, J. C., Moss, L., Bartley, J., and Ojakian, G. K. (1990). *Exp. Cell Res.*, **188**, 302–11.
19. Pandey, P., Kharbanda S., and Kufe, D. (1995). *Cancer Res.*, **55**, 4000–3.
20. Thornton, D. J., Carlstedt, I., Howard, M., Devine, P. L., Price, M. R., and Sheehan, J. K. (1996). *Biochem. J.*, **316**, 967–75.
21. Sotozono, M., Okada, Y., Sasagawa, T., Nakatou, T., Yoshida, A., Yokoi, T., *et al.* (1996). *J. Immunol. Meth.*, **192**, 87–96.
22. Burchell, J., Taylor-Papadimitriou, J., Boshell, M., Gendler, S. J., and Duhig, T. (1989). *Int. J. Cancer*, **44**, 691–6.
23. Swallow, D. M., Gendler, S., Griffiths, B., Kearney, A., Povey, S., Sheer, D., *et al.* (1987). *Ann. Hum. Genet.*, **51**, 289–95.
24. Pigny, P., Guyonnet-Duperat, V., Hill, A. S., Pratt, W.S., Galiegue-Zouitina, S., d'Hooge, M. C., *et al.* (1996). *Genomics*, **38**, 340–52.
25. Fox, M. F., Lahbib, F., Pratt, W., Attwood, J., Gum, J., Kim, Y., and Swallow, D. M. (1992). *Ann. Hum. Genet.*, **56**, 281–7.
26. Gross, M. S., Guyonnet-Duperat, V., Porchet, N., Bernheim, A., Aubert, J. P. and Nguyen, V. C. (1992). *Ann. Genet.*, **35**, 21–6.
27. Bobek, L. A., Liu, J., Sait, S. N., Shows, T. B., Bobek, Y. A., and Levine, Y. A. (1996). *Genomics*, **31**, 277–82.
28. Kim, Y. S., Gum, J. Jr, and Brockhausen, I. (1996). *Glycoconj. J.* **13**, 693–707.
29. Whitehouse, C., Burchell, J., Gschmeissner, S., Brockhausen, I., Lloyd, K., and Taylor-Papadimitriou, J. (1997). *J. Cell Biol.* **137**, 1229—41.

Joy Burchell and Joyce Taylor-Papadimitriou
Imperial Cancer Research Fund,
London, UK

Nectinepsin

Nectinepsin is a member of the 'pexin family', including haemopexin, vitronectin, and most matrix metalloproteases, which all contain a similar 'haemopexin repeat' domain.The deduced nectinepsin amino acid sequence contains the RGD cell binding motif of the integrin ligand and a substantial homology with vitronectin. However, the presence of a specific sequence and the lack of the heparin- and collagen-binding domains of vitronectin indicate that nectinepsin is an extracellular matrix protein of about 54 kDa.

Homologous proteins

Vitronectin and other members of the pexin family.

Protein properties

The alignment of the domain organization of nectinepsin and the five proteins of the pexin family (Fig. 1) shows an important homology with vitronectin. Almost all the mature protein sequence of haemopexin is constituted of two border units (referred to as 0) and eight repeated units organized into two domains (1 to 4; 'haemopexin repeat') joined by a short hinge region. The N-terminal half of metalloproteases consists of an enzymatic domain that is unique to each molecule; and the C-terminal half is the haemopexin repeat domain.

Vitronectin and nectinepsin have a haemopexin repeat domain without the fourth repeat and the C-terminal part of the third repeat. Nectinepsin has the N-terminal part of the second haemopexin repeat domain, but lacks the third and fourth repeats and the C-terminal part of the second repeat. The second haemopexin repeat of vitronectin contains an additional heparin-binding domain which replaces the C-terminal part of the second repeat and the N-terminal part of the third repeat. The domain specific to nectinepsin contains the consensus motif of the active site of eukaryotic aspartyl protease, and a potential *N*-glycosylation site at Asn-167; potential phosphorylation sites for protein kinase C at Ser-155, -185, -262, -328, -358 and Thr 85; for casein kinase II at Thr-62, -113, -118, -136, -287 and Ser-145, -146, -310; and for tyrosine kinase at Tyr-264, a potential *N*-myristoylation site at Gly-168, -247; and potential microbody C-terminal targeting signal at Cys-57, Ala-59, and Gly-349. Nectinepsin appears to have a similar, if not identical, structure to the biochemically characterized yolk vitronectin[1] (Fig. 2). Indeed, yolk vitronectin has been described as a distinct form of vitronectin, found preferentially in chick egg yolk plasma, with a molecular mass smaller than that of blood vitronectin (45 and 54 kDa for the yolk vitronectin, compares to 65 and 70 kDa for the 'standard' vitronectin). Yolk vitronectin contains the RGD cell-binding sequence but not the heparin- nor the collagen-binding domains of vitronectin. Moreover, the N-terminal microsequence of the 45 kDa yolk vitronectin protein, derived from the 54 kDa protein by cleavage of the somatomedin B peptide, is very similar to the beginning of the nectinepsin-specific domain[2] but not to vitronectin's corresponding sequences in human,[3,4] rabbit,[5] and mouse.[6] Nectinepsin is expressed in quail neuralretina early during development and in adult retina, but not in lung. Immunoreactivity is also seen in quail liver, brain, cerebellum, and in rabbit and mouse retina.

The presence of the 54 and 45 kDa proteins described by Nagano *et al.*[1] in the chick egg yolk indicates that nectinepsin is probably expressed very early, even before

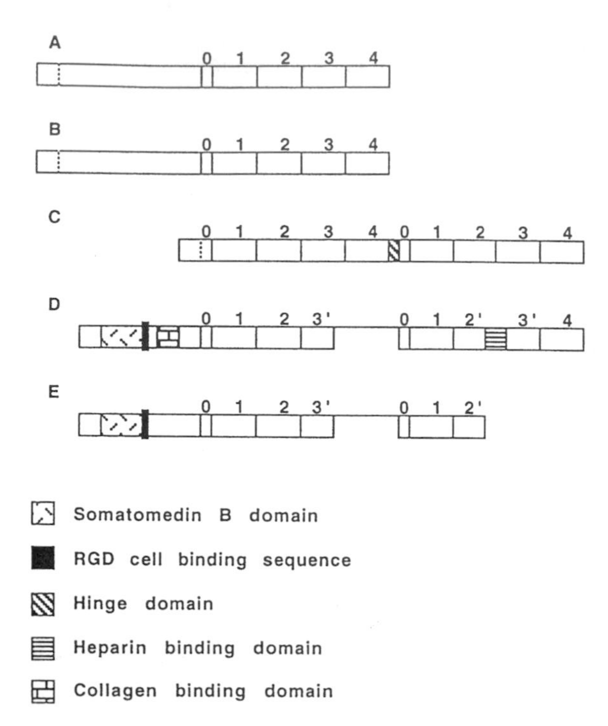

Figure 1. Alignment of the domain organization of the five proteins of the pexin family. (A) rat transin,[7] (B) human interstitial collagenase,[8] (C) human haemopexin,[9,10] (D) human vitronectin,[3,4] (E) quail nectinepsin.[2] The positions of the somatomedin B domain, RGD cell-binding sequence, hinge domain, heparin- and collagen-binding domains are indicated. Dimensions of the domains and other regions are approximately to scale.

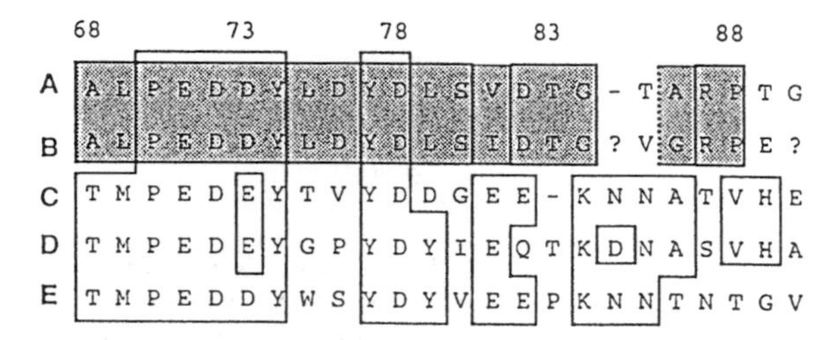

Figure 2. N-terminal amino acid sequences. (A) Nectinepsin,[2] (B) yolk vitronectin,[1] (C) human vitronectin,[3,4] (D) rabbit vitronectin,[5] and (E) mouse vitronectin.[6] Highly conserved amino acids are enclosed in boxes. The numbering identifies the amino acid position within nectinepsin.

development. It is noteworthy that Nagano *et al.* did not detect vitronectin and fibronectin in the chick egg yolk, and thus nectinepsin appears to be expressed first. In egg-laying species, the developing embryo depends completely on the egg's components for its physiological and nutritional requirements. Cell adhesion and migration are important during early development, and the abundance of nectinepsin, but not vitronectin and fibronectin, suggests that this new adhesion protein may serve as a main component and play an important role in the organization of distinct territories in the developing embryo.

Purification

Nectinepsin, like yolk vitronectin,[1] can be purified by three types of column chromatography: hydroxyapatite, DEAE–cellulose, anti-vitronectin–Sepharose, and heparin–Sepharose.

Activity

Not known.

Antibody

We have produced a recombinant protein, and are in the process of making a specific antibody. Anti-human vitronectin (polyclonal antiserum from Life Technologies, Inc.) can be used to detect nectinepsin, since this antiserum recognizes both 70 kDa vitronectin and 54 kDa nectinepsin.

Gene

It is very difficult to perform chromosomal localization in the quail since the nectinepsin-specific domain is too short to enable specific hybridization to mouse. We therefore performed hybridizations of digested genomic DNA with the entire cDNA nectinepsin probe, containing a region common to both nectinepsin and vitronectin, and with the domain specific to the nectinepsin probe. Our results suggest that nectinepsin and vitronectin are encoded by two different genes.

References

1.. Nagano, Y., Hamano, T., Nakashima, N., Ishikawa, M., Miyazaki, K., and Hayashi, M., (1992). *J. Biol. Chem.*, **267**, 24863–70.
2. Blancher, C., Omri, B., Bidou, L., Pessac, B., and Crisanti, P. (1996). *J. Biol. Chem.*, **271**, 26220–6.
3. Jenne, D. and Stanley, K. K. (1985). *EMBO J.*, **4**, 3153–7.
4. Suzuki, S., Oldberg, A., Hayman, E. G., Pierschbacher, M. D, and Ruoslahti, E. (1985). *EMBO J.*, **4**, 2519–24.
5. Sato, R., Komine,Y., Imanaka, T., and Takano, T. (1990). *J. Biol. Chem.*, **265**, 21232–6.
6. Seiffert, D., Keeton, M., Eguchi, Y., Sawdey, M., and Loskutoff, D. J. (1991). *Proc. Natl Acad. Sci. USA*, **88**, 9402–6.
7. Matrisian, L. M., Glaichenhaus, N., Gesnel, M. C., and Breathnach, R. (1985). *EMBO. J.* **4**, 1435–40.
8. Goldberg, G. I., Wilhelm, S. M., Kronberger, A., Bauer, E. A., Grant, G. A., and Eisen, A. Z., (1986). *J. Biol. Chem.*, **261**, 6600–5.

9. Altruda, F., Poli, V., Restagno, G., Argos, P., Cortese, R., and Silengo, L., (1985). *Nucleic. Acids. Res.* **13**, 3841–59.
10. Takahashi, N., Takahashi, Y., and Putman, F. W., (1985). *Proc. Natl Acad. Sci. USA*,. **82**, 73–7.

Patricia Crisanti
CNRS, Development and Immunity of the Central Nervous System, Université Paris VI, Faculté de Médecine Broussais, 15 Rue de L'Ecole de Médecine, 75270 Paris Cedex 06, France

Netrins

Netrins are phylogenetically conserved, laminin-related proteins that act as chemoattractants and chemorepellents for migrating cells and axonal growth cones in organisms as diverse as vertebrates, *C. elegans,* and *Drosophila.* The tripartite netrins consist structurally of two laminin-related N-terminal domains followed by a unique C-terminal domain.

Synonymous names

Netrin, unc-6.

Homologous proteins

Laminins

A portion of a netrin polypeptide is homologous to two contiguous domains in laminin polypeptides: domains homologous to domains VI and V of laminin polypeptides are present as roughly the first two-thirds of a netrin polypeptide.

Protein properties

Netrins were discovered independently in *C. elegans* and vertebrates: in the former, first as the predicted product of a gene, *unc-6,* which when mutated led to defects in cell and pioneer axon growth cone migrations along the dorsoventral axis of the epidermis of *C. elegans,*[1,2] and in the latter as two purified proteins that could elicit spinal commissural axon outgrowth and turning *in vitro.*[3,4] Mature netrin proteins are approximately 600 amino acids in length; the N-terminal two-thirds of a netrin polypeptide is homologous to two contiguous domains of laminin polypeptides, domains VI and V.[5] Chick netrin-1 and netrin-2 are approximately 50 per cent identical to UNC-6, although the identity rises to approximately 70 per cent within domain V. Netrins are more similar to one another in domains VI and than they are to any laminin polypeptide. Although these domains are more similar to laminin γ subunits than to other laminin subunits, netrins

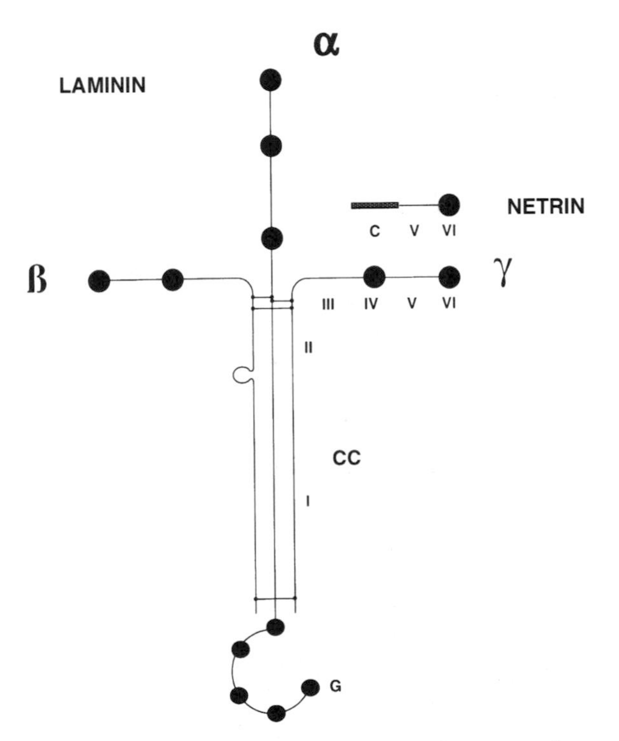

Figure 1. Netrins are diffusible axon guidance proteins related to laminins. The netrins define a family of phylogenetically conserved secreted factors related to extracellular matrix protein laminin, required for the proper migration of developing axons. The N-terminal two-thirds of the netrins are homologous to the N termini of the poolypeptide chains of laminin (domains VI and V). The C-terminal third of the netrins (domain C) is unique to the netrin family and is enriched in basic residues. This diagram illustrates the relative sizes and structural relationship between the netrins and laminin. α, β, and γ refer to the individual polypeptides that constitute the laminin heterotrimer; G, C and numerals I–VI refer to named domains of the laminin and netrin polypeptides. The figure is drawn only approximately to scale.

do share hallmarks of both laminin γ and β subunits (see ref. 2). Netrin domain V consists of three EGF-type repeats (V-1, V-2, V-3), with conservation of cysteine residues.[6,7] The C-terminal netrin domain (domain C) is not conserved between netrins and laminins, but the domain has six cysteine residues that are conserved among netrin proteins. The domain C is very basic (predicted p*I* of 10.6 and 10.5 for chick netrin-1 and netrin-2, respectively). *Drosophila* netrins (netrin-A and netrin-B) have some additional sequences relative to vertebrate netrins and unc-6.[8,9]

Chick netrins are presumably glycosylated, as evidenced by the use of a lectin-affinity chromatography step in their purification (see below) and by their higher than expected molecular weight on SDS–PAGE gels (79 kDa and 75 kDa for netrin-1 and netrin-2, respectively);[3] other netrins have potential *N*-glycosylation sites. Chick netrins and unc-6 are secreted when produced as recombinant proteins in COS cells[3] (C. Mirzayan, W. Wadsworth, and M. Tessier-Lavigne; unpublished results), and the proteins do not have predicted transmembrane domains or glycosyl phosphatidylinositol (GPI) lipid anchor addition sites. The recombinant chick netrins bind very tightly to cell surfaces and the ECM.[3,4] The basic domain C or N-terminal domain VI could mediate these interactions, especially as the latter has been shown to mediate multimerization of laminin heterotrimers into larger polymeric laminin matrices.[10] The multimeric state of netrin proteins has not yet been determined.

The tissue distribution of unc-6 has been examined by making transgenic animals that express a haemagglutinin-tagged version of the protein that is able to rescue all of the null mutant defects.[11] *Unc-6* is expressed dynamically in a variety of neurones and glia that are found in distinct regions of the animal. In each of these regions, unc-6 is spatially restricted to themost ventral cells. In the body wall, for example, unc-6 is expressed by ventral, but not lateral epidermoblasts. In the nerve ring, *unc-6* is expressed by ventral, but not dorsal, cephalic sheaths. In both the pharynx and the ventral nerve cord, *unc-6* is expressed by ventral midline neurones. In each region, the timing of *unc-6* expression is consistent with its known role in guiding cell and growth cone migrations. Thus unc-6 has the potential to provide dorsoventral polarity information in *C. elegans*, which fits the requirement for unc-6 to orient and guide migrating cells and growth cones on the dorsoventral axis.

Identification of the netrins in vertebrates derived from the analysis of the growth of spinal commissural axons from dorsally located cell bodies to the ventrally located floor plate of the spinal cord (reviewed in ref. 12). *In vitro* analysis showed that the floor plate was capable of reorienting the growth of commissural axons *in vitro*.[13] By *in situ* hybridization, *netrin-1* in both chick and mouse is

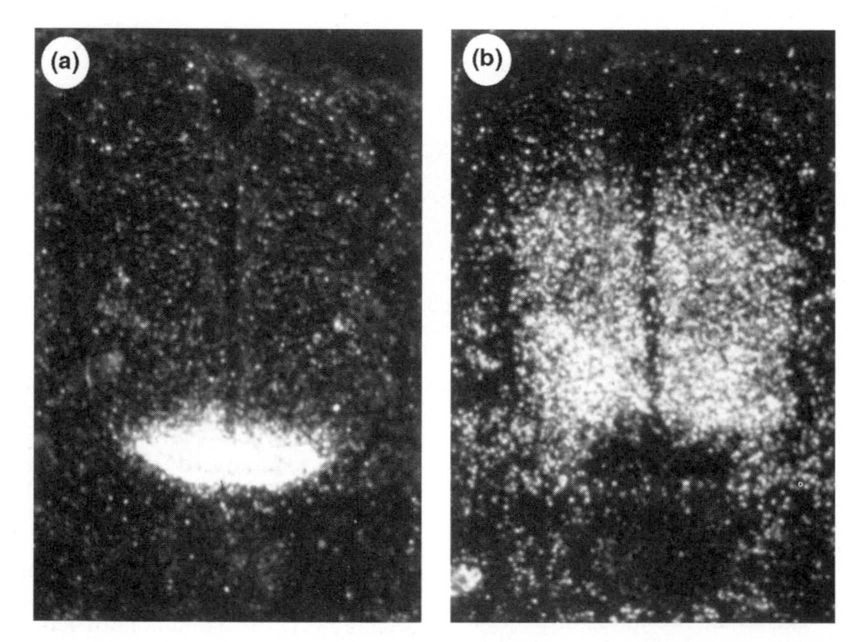

Figure 2. Expression of netrin-1 (a) and netrin-2 (b) in the developing chick spinal cord. Netrin-l and netrin-2 have been implicated in the ventral migration of spinal commissural axons in vertebrates. These neurones, born dorsally in the spinal cord, extend their axons ventrally towards the floor plate. Both netrin-1 and netrin-2 are expressed in the developing spinal cord during the period of commissural axon extension. Netrin-1 is expressed at the ventral midline in a region slightly broader than the morphological floor plate, whereas netrin-2 is expressed at lower levels in the ventral two-thirds of the spinal cord. This expression pattern suggests that these two proteins may cooperate to establish a long-range attractive gradient of netrin protein that directs commissural axon growth to the ventral midline of the spinal cord.

expressed at relatively high levels in the floor plate of the spinal cord during the period of commissural axon growth to the floor plate.[4,14] In addition, chick *netrin-2* and mouse *netrin-1* are also expressed in the ventral two thirds of the spinal cord during this period [mouse *netrin-2* is not expressed within the central nervous system at early stages (H. Wang and M. Tessier-Lavigne; unpublished results); it appears that netrin-1 in the mouse performs the functions of both netrin-1 and netrin-2 in the chick]. Additional studies show that within the vertebrate central nervous system, netrins are expressed along the midline into the caudal diencephalon and in various other regions (e.g. ganglionic eminences, basal ganglia, medial telencephalic walls, retina, and optic nerve) in a highly dynamic fashion.[4,14,15] Some of these sites of expression are consistent with various *in vitro* and *in vivo* analyses of netrin function (see below). Northern as well as *in situ* hybridization analysis indicates that outside of the nervous system, netrins are expressed in notochord, dennamyotome, developing limb, lung, heart, spleen, gut, ovary, and testis.[4]

Ventral midline expression within the nervous system is also conserved in *Drosophila,* where *netrin-A* and *netrin-B* are expressed in midline glia and neurones during the period of commissure formation.[8,9] Earlier, netrin expression is observed in presumptive mesoderm and visceral mesoderm, and in the somatic mesoderm layer in a more limited fashion. The two netrins are also expressed in a dynamic fashion in partially overlapping sets of CNS neurones, and they are also expressed in epidermis, several muscles, and imaginal discs in the periphery. The expression in the ventral midline and in muscles is consistent with the phenotypes observed in mutant flies lacking both netrins (see below).

Netrin effects are mediated through receptor mechanisms involving members of the DCC and unc-5 subfamilies of the immunoglobulin (Ig) superfamily and cAMP levels have been shown to be important in determining whether signalling through the DCC receptors causes an attractive or repulsive growth cone response[25] (see the relevant chapters, this volume).

Purification

The netrins were originally purified from embryonic day 10 (E10) chick brain, using commissural axon outgrowth from E13 rat dorsal spinal cord explants into collagen gels as an assay.[3] The netrins were purified to apparent homogeneity by subjecting E10 chick brain homogenates sequentially to differential centrifugation, differential salt extraction, heparin-affinity chromatography (with stepwise NaCl elution), lectin (WGA)-affinity chromatography, heparin-affinity chromatography (with gradient NaCl elution), and immobilized metal adsorption chromatography (with a Zn^{2+}-charged resin). In the last step, two proteins separately cofractionated with outgrowth activity: a protein of 75 kDa by SDS–PAGE (netrin-2) eluted isocratically at a pH of 6.5, and a protein of 78 kDa by SDS–PAGE (netrin-1) eluted a pH of 6.1.

Activities

The netrins were defined originally by their ability to elicit the outgrowth of commissural axons from E11 and E13 rat dorsal spinal cord explants into collagen gels (the assay used in their purification).[3,16] Recombinant netrins have detectable effects in this assay at a concentration of 20 ng/ml. Recombinant netrin-1 or netrin-2, presented from a point source, are also capable of reorienting the growth of axons towards the point source at a distance, thus defining the netrins as chemoattractants.[4,17,18,26] Netrins, however, are bifunctional proteins, in that they are also chemorepellents (notably, for trochlear motoneurones[19] and branchiomotor neurones[20]), repelling the growth of axons at a distance *in vitro.* Analysis of the phenotypes of animals mutant in genes encoding various netrin family members verify that these chemoattractant and chemorepellent activities of the netrins are important *in vivo* (see below).

Antibodies

There are at present no commercial sources of antibodies recognizing netrins. Several antipeptide antisera that can recognize vertebrate netrins have been raised in rabbits[21] (T. Kennedy, *et al.*; unpublished results).

Genes

The complete *C. elegans unc-6* gene has been cloned and sequenced (GenBank M80241),[2] cDNAs encoding the complete chicken netrin-1 protein (GenBank L34549)[3] and all but a few amino acids of the signal sequence of chicken netrin-2 (GenBank L34550)[3] have been cloned. Complete cDNAs encoding *Drosophila*, netrin-A (GenBank U60316 and U63736)[8,9] and netrin-B (GenBank U60317 and U63737)[8,9] have also been obtained; both genes lie close to one another (within 150 kb) on the X chromosome. A cDNA encoding the complete mouse netrin-1 protein (GenBank U65418)[14] has been cloned and sequenced. A human gene (GenBank U86758)[22] and its corresponding complete cDNA (GenBank U86759)[22] encoding a netrin-2-like protein have been sequenced; the gene maps to chromosome 16p13.3.

Mutant phenotypes/disease states

Unc-6 mutants have defects in a variety of cell and pioneer axon growth cone migrations, all of which are oriented on the dorsoventral axis of the epidermis toward or away from the ventral midline.[1] Affected cells include motor neurones (AS, DA, DB, DD, VD), the neurone-like excretory cell, and three mesodermal cell types (distal tip cell, head mesodermal cell, and male linker cell), plus a variety of sensory neurones with lateral cell bodies and pioneer axons that normally extend to the ventral midline. In unc-6 mutants, migrations still occur on the epidermis but are frequently misoriented.

In *Drosophila* mutants lacking both netrin-A and netrin-B, commissures in the ventral nerve cord are much thinner than normal and sometimes completely absent, and there are occasional breaks in the longitudinal tracts.[8,9] The posterior commissure is more severely affected than the anterior commissure. Midline expression of either netrin can rescue the commissural defects, while expression throughout the developing CNS leads to a loss-of-function phenotype, suggesting that the spatial pattern of netrin protein is vital to its guidance function. In addition, flies lacking both netrins have defects in motoneurone projections consistent with netrin acting as a targeting molecule in the periphery.[8,27]

Mice deficient in netrin-1 have been obtained through a novel gene-trap strategy.[14,23] Homozygous mutant mice are born but die perinatally; they do not suckle, and they display disorders of movement. In the spinal cord, these mice exhibit defects in commissural axon projections consistent with a chemoattractant role for netrin-1 in guiding these axons to the floor plate. These mice also have defects more rostrally which include agenesis of the corpus callosum, the hippocampal commissure, and the anterior commissure, indicating roles for netrin in forming other midline structures in the nervous system. In addition, these mice lack pontine nuclei (whose cells migrate to their final position during development). Furthermore, these mice exhibit optic nerve hypoplasia due to a failure of retinal ganglion cell axons to exit into the optic nerve, reflecting a guidance role for netrin-1 at the optic disc during eye development.[24]

Structure

Not available.

References

1. Hedgecock, E. M., Culotti, J. G. and Hall, D. H. (1990). *Neuron*, **4**, 61–85.
2. Ishii, N., Wadsworth, W. G., Stern, B. D., Culotti, J. G., and Hedgecock, E. M. (1992). *Neuron*, **9**, 873–81.
3. Serafini, T., Kennedy, T. E., Galko, M. J., Mirzayan, C., Jessell, T. M., and Tessier-Lavigne, M. (1994). *Cell*, **78**, 409–24.
4. Kennedy, T. E., Serafini, T., de la Torre, J. R., and Tessier-Lavigne, M. (1994). *Cell*, **78**, 425–35.
5. Sasaki, M., Kleinman, H. K., Huber, H., Deutzmann, R., and Yamada, Y. (1988). *J. Biol. Chem.*, **263**, 16536–44.
6. Blomquist, M., Hunt, L., and Barker, W. (1984). *Proc. Natl Acad. Sci. USA*, **81**, 7363–7.
7. Engel, J. (1989) *FEBS Lett.*, **251**, 1–7.
8. Mitchell, K. J., Doyle, J. L., Serafini, T., Kennedy, T. E., Tessier-Lavigne, M., Goodman, C. S., and Dickson, B. J. (1996). *Neuron*, **17**, 203–15.
9. Harris, R., Sabatelli, L. M., and Seeger, M. A. (1996). *Neuron*, **17**, 217–28.
10. Yurchenco, P. D. and Cheng, Y. S. (1993). *J. Biol. Chem.*, **268**, 17286–99.
11. Wadsworth, W. G., Bhatt, H., and Hedgecock, E. M. (1996). *Neuron*, **16**, 35–46.
12. Colamarino, S. A. and Tessier-Lavigne, M. (1995). *Ann. Rev. Neurosci.*, **18**, 497–529.
13. Tessier-Lavigne, M., Placzek, M., Lumsden, A. G., Dodd, J., and Jessell, T. M. (1988). *Nature*, 336, 775–8.
14. Serafini, T., Colamarino, S. A., Leonardo, E. D., Wang, H., Beddington, R., Skarnes, W.C., and Tessier-Lavigne, M. (1996). *Cell*, **87**, 1001–14.
15. Livesey, F. J. and Hunt, S. P. (1997). *Mol. Cell. Neurosci.*, **8**, 417–29.
16. Placzek, M., Tessier-Lavigne, M., Jessell, T., and Dodd, J. (1990), *Development*, **110**, 19–30.
17. Shirasaki, R., Mirzayan, C., Tessier-Lavigne,. M., and Murakami, F. (1996). *Neuron*, **17**, 1079–88.
18. Richards, L. J., Koester, S. E., Tuttle, R., and O'Leary, D.D.M. (1997). *J. Neurosci.*, **17**, 2445–58.
19. Colamarino, S. A. and Tessier-Lavigne, M. (1995). *Cell*, **81**, 621–9.
20. Varela-Echavarria, A., Tucker, A., Puschel, A. W., and Guthrie, S. (1997). *Neuron*, **18**, 193–207.
21. MacLennan, A. J., McLaurin, D. L., Marks, L., Vinson, E. N., Pfeifer, M., Szulc, S. V., *et al.* (1997). *J. Neurosci.*, **17**, 5466–79.
22. Van Raay, T. J., Foskett, S. M., Connors, T. D., Klinger, K. W., Landes, G. M., and Burn, T. C. (1997). *Genomics*, **41**, 279–82.
23. Skarnes, W. C., Moss, J. E., Hurtley, S. M., and Beddington, R. S. (1995). *Proc. Natl Acad. Sci. USA*, **92**, 6592–6.
24. Deiner, M. S., Kennedy, T. E., Fazeli, A., Serafini, T., Tessier-Lavigne, M., and Sretavan, D. W. (1997) *Neuron* **19**, 575–89.
25. Ming, G. L., Song, H. J., Berninger, B., Holt, C. E., Tessier-Lavigne, M., and Poo, M. M. (1997). *Neuron*, **19**, 1225–35.
26. Métin, C., Deléglise, D., Serafini, T., Kennedy, T. E., and Tessier-Lavigne, M. (1997). *Development*, **124**, 5063–74.
27. Winberg, M. L., Mitchell, K. J., and Goodman, C. S. (1998). *Cell*, **93**, 581–91.

Tito Serafini
Department of Molecular and Cell Biology,
University of California,
Berkeley, CA 94270–3200, USA

Marc Tessier-Lavigne
Howard Hughes Medical Institute,
Departments of Anatomy and of
Biochemistry and Molecular Biology,
University of California,
San Francisco, CA 94143–0452, USA

Joseph G. Culotti
Division of Molecular Immunology and
Neurobiology, Samuel Lunenfeld
Research Institute, Mount Sinai Hospital,
Torono M5G 1X5, Canada

Nidogen

The extracellular matrix protein nidogen is a typical basement membrane component and an early embryonic product. It binds to laminins, collagen IV, perlecan, fibulins, and calcium. Nidogen is considered as the essential component for connecting the networks of laminins and collagen IV in basement membranes.

Synonomous name

Entactin.

Protein properties

Nidogen-1, also known as entactin,[1] consists of single polypeptide chain of about 150 kDa[2] and is folded into three globular domains (G1–G3) connected by a 15 nm long rod or a flexible link (Fig. 1(a), Fig. 2(a)).[3] Sequences[4–6] are available for human and mouse nidogen (Table 1) indicating a mosaic structure based on at least three different modules identified so far (Fig. 1(b)). Analogues with some variations in module arrangement are known for an ascidian species[7] and in incomplete form for *C. elegans* (Table 1). A distinct isoform, osteonidogen (nidogen-2), was reported for human bone (Table 1). Post-translational modifications include in mouse nidogen-1 oligosaccharide attachment to two defined *N*-linked and seven O-linked acceptor sites[8] and *O*-sulphation of one or two tyrosines[4,5,9] mainly in the link region. Immunohistology shows the ubiquitous occurrence of nidogen-1 in basement membranes[1,10–13] and its appearance at the 8–16 cell stage of mouse development.[12] Yet during organogenesis and tissue repair, a substantial portion of nidogen-1 is derived from the mesenchyme but disappears from this compartment after basement membrane formation has been completed.[13–15] Nidogen was also shown to be highly sensitive to tissue proteases[16,17] which could modulate its biological activity.

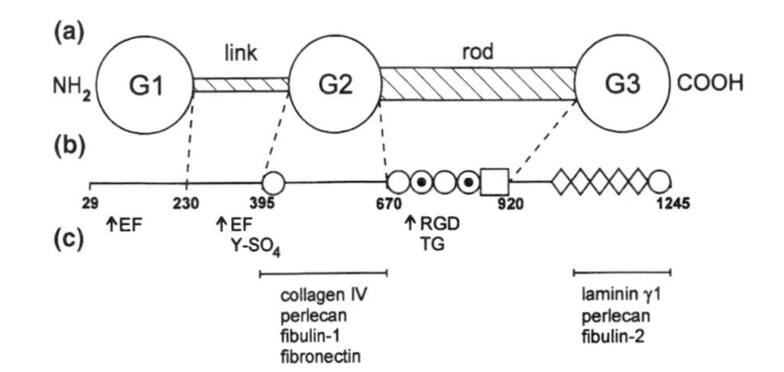

Figure 1. Correlation between shape (a), and modular structure (b) of mouse nidogen (position 29–1245), and localization of binding activities (c). (a) The shape was determined by electron microscopy, and includes three globular domains (G1–G3), and two connecting elements. (b) Identified modules include epidermal growth factor-like (EG) modules (circles), a larger thyroglobulin-like (TY) EG module (square), and LDL receptor-like LY modules (diamonds). (c) Putative calcium-binding sites include two predicted EF hands, and two EG modules (marked with a dot). Y-SO_4 marks a region for Tyr sulphation, RGD a potential cell adhesion site, and TG a transglutaminase-catalysed cross-linking site. Binding of several protein ligands occurs primarily through domains G2, and G3.

Purification

The most convenient way to obtain nidogens in native form is by recombinant production in mammalian cell clones and the isolation from culture medium (10–20 μg/ml) by binding to a cobalt-loaded chelating column.[3,17,31] Partially denatured forms are obtained by dissociating the laminin–nidogen complex from the mouse EHS tumour with 2–6 M guanidine–HCl followed by one to three chromatographic steps.[2,18]

Table 1 Gene and cDNA sequences of nidogens

Protein	Species	Residues*	GenBank no	Refs
Nidogen	Human	1247	M 30269	4
	Mouse	1245	X 14194	5, 6
	Ascidian	1251	D 14038	7
	*C. elegans***		Z 79696	
Osteonidogen	Human	1376	D 86425	
Nidogen-2	Human	1375	AJ223500	

*Residues of the polypeptide including signal peptide.
**Partial sequence within a cosmid clone.

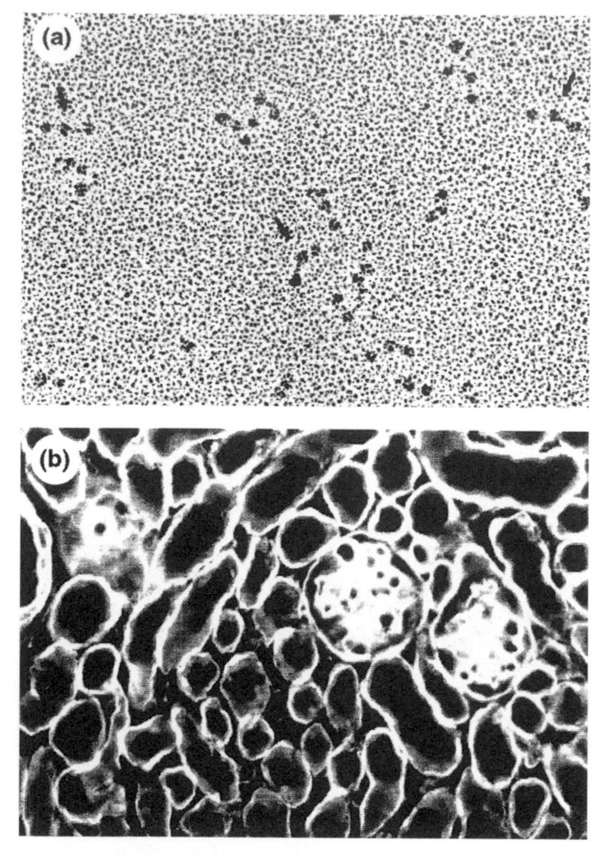

Figure 2. Rotary shadowing images of recombinant mouse nidogen-1 (a), and immunofluorescence detection of nidogen on a mouse kidney section (b). Arrows in (a) denote particles with a well-resolved three-globe structure.[3] Monoclonal antibody staining[27] in (b) includes glomerular, and tubular basement membranes.

Activities

A major function of nidogen-1 is apparently the connection of the networks of laminins and collagen IV in basement membranes[1,3,19] through domains G2 and G3 which are separated by 15nm (Fig. 1(c)). Because of its high affinity for laminin (k_d=0.7 nM),[3,20] nidogen-1 is frequently solubilized as a tight complex with laminin-1, -2, and -4 from tissues[18,21] but can also be easily lost through endogenous proteolysis. Based on the structure of the nidogen-binding epitope of laminin-1[22] it is obvious that all laminins possessing a γ1 chain but not those with a γ2 chain should bind nidogen-1 with high affinity. Further binding epitopes of nidogen for perlecan,[19] fibulins[20] and fibronectin[23] have been mapped to either domains G2 or G3 (Fig. 1(c)). A moderate cell adhesion activity of nidogen-1 through a single RGD sequence in the rod[3,24] includes, presumably, binding to $\alpha3\beta1$ integrin.[25] Transglutaminase-activated cross-linking involves Gln-726 in the rod of nidogen.[26] Nidogen-1 binds calcium[1] presumably through EF hands and EG modules (Fig. 1) and zinc[17,19] through several His-rich sequences. Recombinant human nidogen-2 has recently been obtained;[31] it showed a similar three-globe structure and a binding repertoire comparable but not completely identical to that of nidogen-1. Both nidogens showed a similar tissue distribution.

Antibodies

Several polyclonal rabbit antisera against mouse and human nidogen-1[2,3,10–12,17] and nidogen-2[31] have been characterized by immunofluorescence, immunoprecipitation, blotting and radioimmunoassay. A rat monoclonal antibody to mouse nidogen-1 has been described[27] (Fig. 2(b)).

Genes

cDNA probes covering the coding and some untranslated regions of mouse, human, and ascidian nidogen are available (Table1). The single copy gene was localized to chromosome 1q43.[28] The genomic structures of mouse[29] and human[30] nidogen include 20 exons spread over a distance of 65 to 100 kb.

Mutant/phenotype

Work on a nidogen knockout in transgenic mice is in progress. Antibodies which block nidogen binding to laminin perturb organogenesis in culture by interference with basement membrane formation.[14,15]

Structure

Not available.

References

1. Mayer, U. and Timpl, R. (1994). In *Extracellular matrix assembly and structure* (eds. P. D. Yurchenco, D. Birk and R. P. Mecham), pp. 389–416. Academic Press, Orlando, FL.
2. Paulsson, M., Deutzmann, R., Dziadek, M., Nowack, H., Timpl, R., Weber, S., and Engel, J. (1986). *Eur. J. Biochem.*, **156**, 467–78.
3. Fox, J. W., Mayer, U., Nischt, R. Aumailley, M., Reinhardt, D., Wiedemann, H., *et al.* (1991). *EMBO J.*, **10**, 3137–46.
4. Nagayoshi, T., Sanborn, D., Hickok, N. J., Olsen, D. R., Fazio, M. J., Chu, M.-L., *et al.* (1989). *DNA*, **8**, 581–94.
5. Durkin, M. E., Chakravarti, S., Bartos, B. B., Liu, S.-H., Friedman, R. L., and Chung, A. E. (1988). *J. Cell Biol.*, **107**, 2749–56.
6. Mann, K., Deutzmann, R., Aumailley, M., Timpl, R., Raimondi, L., Yamada, Y., *et al.* (1989). *EMBO J.*, **8**, 65–72.
7. Nakae, H., Sugano, M., Ishimori, Y., Endo, T., and Obinata, T. (1993). *Eur. J. Biochem.*, **213**, 11–19.
8. Fujiwara, S., Shinkai, H., Mann, K., and Timpl, R. (1993). *Matrix*, **13**, 215–22.

9. Paulsson, M., Dziadek, M., Suchanek, C., Huttner, W. B., and Timpl, R. (1985). *Biochem. J.*, **231**, 571–9.
10. Wu, T. C., Wan, Y. J., Chung, A. E., and Damjanov, I. (1983). *Dev. Biol.*, **100**, 496–505.
11. Schittny, J. C., Timpl, R., and Engel, J. (1988). *J. Cell Biol.*, **197**, 1599–610.
12. Dziadek, M. and Timpl, R. (1985). *Dev. Biol.*, **111**, 372–82.
13. Dziadek, M. (1995). *Experientia*, **51**, 901–13.
14. Ekblom, P., Ekblom, M., Fecker, L., Klein, G., Zhang, H.-Y., Kadoya, Y., *et al.* (1994). *Development*, **120**, 2003–14.
15. Kadoya, Y., Salmivirta, K., Talts, J. F., Kadoya, K., Mayer, U., Timpl, R., and Ekblom, P. (1997). *Development*, **124**, 683–91.
16. Mayer, U., Mann, K., Timpl, R., and Murphy, G. (1993). *Eur. J. Biochem.*, **217**, 877–84.
17. Mayer, U., Zimmermann, K., Mann, K., Reinhardt, D., Timpl, R., and Nischt, R. (1995). *Eur. J. Biochem.*, **227**, 681–6.
18. Paulsson, M., Aumailley, M., Deutzmann, R., Timpl, R., Beck, K., and Engel, J. (1987). *Eur. J. Biochem.*, **116**, 11–19.
19. Reinhardt, D., Mann, K., Nischt, R., Fox, J. W., Chu, M.-L., Krieg, T., and Timpl, R. (1993). *J. Biol. Chem.*, **268**, 10881–7.
20. Sasaki, T., Göhring, W., Pan, T. -C., Chu, M. -L., and Timpl, R. (1995). *J. Mol. Biol.*, **254**, 892–9.
21. Brown, J. C., Wiedemann, H., and Timpl, R. (1994). *J. Cell Sci.*, **107**, 329–38.
22. Pöschl, E., Mayer, U., Stetefeld, J., Baumgartner, R., Holak, T. A., Huber, R., and Timpl, R. (1996). *EMBO J.*, **15**, 5154–9.
23. Hsieh, J.-C., Wu, C., and Chung, A. E. (1994). *Biochem. Biophys. Res. Commun.*, **199**, 1509–17.
24. Dong, L.-J., Hsieh, J. -C., and Chung, A. E. (1995). *J. Biol. Chem.*, **270**, 15838–43.
25. Dedhar, S., Jewell, K., Rojiani, M., and Gray, V. (1992). *J. Biol. Chem.*, **267**, 18908–14.
26. Aeschlimann, D., Paulsson, M., and Mann, K. (1992). *J. Biol. Chem.*, **267**, 11316–21.
27. Dziadek, M., Clements, R., Mitrangas, K., Reiter, H., and Fowler, K. (1988). *Eur. J. Biochem.*, **172**, 219–25.
28. Olsen, D. R., Nagayoshi, T., Fazio, M., Mattei, M.-G., Passage, E., Weil, D., *et al.* (1989). *Am. J. Hum. Genet.*, **44**, 876–885.
29. Durkin, M. E., Wewer, U. M., and Chung, A. E. (1995). *Genomics*, **26**, 219–28.
30. Zimmermann, K., Hoischen, S., Hafner, M., and Nischt, R. (1995). *Genomics*, **27**, 245–50.
31. Kohfeldt, E., Sasaki, T., Göhring, W., and Timpl, R. (1998). *J. Mol. Biol.*, **282**, 99–109.

Rupert Timpl
Max-Planck-Institut für Biochemie, Martinsried, Germany

Osteopontin

Osteopontin[1–4] is a secreted, acidic glycoprotein normally found in mineralized matrices as well as in urine, kidney and most epithelia. In tissues undergoing inflammation, fibrosis, oncogenesis, or dystrophic calcification, osteopontin expression is increased due to *de novo* expression by macrophages, epithelial, tumour, and/or mesenchymal cell populations responding to injury. *In vitro*, osteopontin regulates cell adhesion, migration, and survival, NF-κB activity, NO synthesis, and calcium crystal formation. Osteopontin is encoded by a single gene and at least three different transcripts have been described which arise by differential splicing.

Synonymous names

Osteopontin is also called uropontin, bone sialoprotein I (BSPI), 44 kDa bone phosphoprotein, 2aR, 2B7, eta-1, and secreted phosphoprotein I (SPP).

Homologous proteins

No closely related proteins have been reported, but the cell adhesion domain (GRGDS) in osteopontin is similar to that found in a number of other adhesive proteins including fibronectin and vitronectin. Osteopontin also shares limited sequence homology in its poly-aspartate-containing region to aspartate-rich calcium-binding proteins such as calsequestrin.

Protein properties

Osteopontin is a protein of 264–301 amino acids (depending on the species) containing highly conserved regions which show homology to various previously described functional domains (Fig. 1). These include a signal sequence, calcium- and heparin binding regions, the RGD-containing cell adhesion motif, and sites for serine phosphorylation. Post-translational modification of osteopontin can include serine and threonine phosphorylation, *N*- and *O*-linked glycosylation, and sulphation. Sedimentation equilibrium analyses indicated a molecular weight of 44 kDa for rat bone osteopontin. Due to a high content of aspartate and glutamate, and phosphorylated residues, osteopontin is a highly electronegative protein (isoelectric point 4.9–5.1). This feature leads to anomolous migration of the protein on SDS–PAGE, such that apparent molecular weights of 44–88 kDa have been reported.[1–4]

Multiple osteopontin forms are often observed by Western blot, and most probably occur as a result of post-

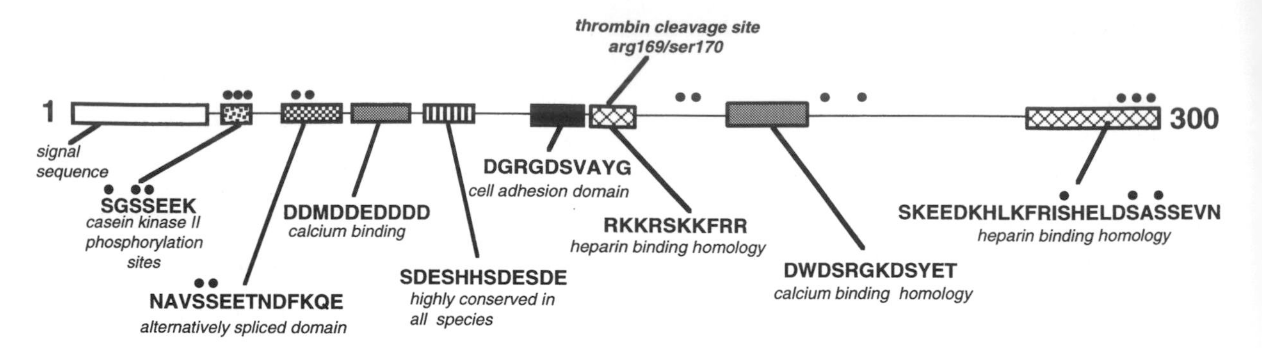

Figure 1. Primary structure of human osteopontin. The conserved domains of osteopontin found in most species are highlighted. Filled circles indicate sites proposed for serine phosphorylation.[19]

translational modifications including phosphorylation, sulphation, and/or glycosylation. High molecular weight forms may represent heat and SDS-stable osteopontin aggregates. How these modification affect function is not yet clearly established, although phosphorylation appears to be critical for hydroxyapatite inhibition. An alternatively spliced cDNA has been detected in human bone and decidual cells, but a corresponding protein has not yet been described[1–4]. Osteopontin can also be cleaved by thrombin generating two fragments of roughly equal size (Fig. 1). Thrombin cleavage has been reported to inhibit adhesive activity,[5] stimulate adhesive and migratory activity,[6] and alter osteopontin receptor specificity.[7] Allelic variants of osteopontin have been described in the mouse.[8]

Normally, osteopontin has a limited tissue distribution, and is found in bone, teeth, kidney, inner ear, and epithelial linings. In bone, osteopontin is synthesized by osteoblasts and osteoclasts, and localized in the osteoid, as well as in interfacial regions of calcified bone including various bone surfaces, lamina limitantes and cement lines. In teeth, osteopontin is expressed by odontoblasts during cementogenesis. In the kidney, osteopontin is synthesized primarily by renal epithelial cells, and secreted into the urine in soluble form. In gall bladder columnar epithelial cells, osteopontin was present in the external filamentous glycocalyx, as well as within cells in membrane bound vesicles and Golgi complex. In addition to these sites, early in development, osteopontin is synthesized by cells forming the notochord and head process.[1–4]

In tissues undergoing injury and repair, including atherosclerotic blood vessels, infarcted hearts, tumours, nephritic kidneys, and calcifying aortic valves, osteopontin levels are increased and correlated with fibrotic, inflammatory and calcified regions. Under these conditions, osteopontin expression is observed in macrophages, endothelial, epithelial, smooth muscle, and tumour cells. In macrophages responding to cardiac necrosis, osteopontin was localized in secretory vesicles.[1–4]

Osteopontin is regulated by hormonal and cytokine stimulation. In bone cells, osteopontin is regulated by the osteotropic hormones, parathyroid hormone, vitamin D, oestrogen, and dexamethasone. Cytokines have been shown to induce osteopontin expression in various cell types and include TGFβ, PDGF, angiotensin II, EGF, bFGF, LIF, TNF. Osteopontin expression is also induced by phorbol esters, retinoic acid, and lipopolysaccharide.[1–4]

While the precise biological roles for osteopontin have not yet been completely identified, the abnormal phenotypes of osteopontin mutant mice (*Ric*$^-$ and OPN$^{-/-}$ mice) indicates that osteopontin is involved in genetic resistance to bacteria in mice, as well as cutaneous wound repair.[8,20] *In vitro*, osteopontin has both adhesive and cytokine-like properties, in addition to calcium-binding activity. Osteopontin promotes adhesion, migration, and survival of a number of different cell types by interaction with integrins. Osteopontin inhibits cytokine-induced NO synthesis in several cell types. Osteopontin has been shown to inhibit calcium oxalate and hydroxyapatite crystal formation, suggesting that it may function as an inhibitor of calcium precipitation in urine and other body fluids. Based on *in vitro* and *in vivo* activities, osteopontin is proposed to play a biological role in biomineralization, inflammation, tumorigenesis, angiogenesis, and wound repair.[1–4]

Purification

Osteopontin is typically purified based on its charge and affinity for divalent cation salts. A typical method of purification from solution includes DEAE anion exchange chromatography followed by adsorption to a barium citrate precipitate, and elution into water. From bone, osteopontin is purified by guanidine–HCl/EDTA extraction, followed by DEAE anion exchange chromatography.[1–4]

Activities

Osteopontin promotes cell adhesion and migration by binding to av-containing integrins. Osteopontin promotes endothelial survival by inducing NF-κB activity in endothelial cells.[21] Osteopontin promotes macrophage mediated B-cell antibody production and inhibits NO synthesis. Osteopontin inhibits growth of hydroxyapatite and calcium oxalate crystals *in vitro*.[1–4]

Antibodies

Osteopontin is highly immunogenic and antibodies are easily raised. Both polyclonal and monoclonal antibodies tend to be species specific. A monoclonal antibody directed against rat bone osteopontin, MPIIIB10 (1), is currently available from the Developmental Studies Hybridoma Bank (Iowa city, IA).

Genes

Osteopontin is encoded by a single gene (*spp1*) located on chromosome 4 in humans, chromosome 5 in mouse, and chromosome 8 in pigs. The gene contains seven exons and six introns. Splice variants arise by variable usage of exon 5 and cell-type specific splicing of exon 1. Complete cDNA sequences have been published for mouse (GenBank X51834, S78177),[8,9] rat (GenBank RATOSP),[10] chicken (GenBank CHKBPP),[11] rabbit (GenBank RABOPN),[12] bovine (GenBank BOVCOST),[13] porcine (EMBL X16575),[14] and human (GenBank HUMU105MGA, HUMU105MGB)[15] osteopontin. Genomic clones for mouse (GenBank MMOESTEOP),[16] chicken (GenBank U01844),[17] pig (GenBank SSSPP1), and human (GenBank HSU20758)[18] have been described.

Mutant phenotype/disease state

The osteopontin gene has been mapped to the *Ric* locus, an autosomal dominant gene that controls natural resistance in mice. *Ric*$^-$ mice contain the eta-1b osteopontin allele, and die when challenged with an intraperitoneal injection of *Rickettsia tsutsugamushi*.[8] In addition, osteopontin mutant mice have been generated by targeted mutagenesis in embryonic stem cells. OPN$^{-/-}$ mice develop normally and are fertile, but show altered wound healing of skin incisional wounds.[20]

Structure

Not available.

References

1. Giachelli, C. M., Schwartz, S. M., and Liaw, L. (1995). *Trends Cardiovasc. Med.*, **5**, 88–95.
2. Butler, W. T. (1989). *Connect. Tiss. Res.*, **23**, 123–36.
3. Butler, W. T., Ridall, A. L., McKee, M. D. (1996). In *Principles of bone biology*, pp. 167–81.
4. Denhardt, D. T. and Guo, X. (1993). *FASEB J.*, **7**, 1475–82.
5. Xuan, J. W., Hota, C., and Chambers, A. F. (1994). *J. Cell Biochem.*, **54**, 247–55.
6. Senger, D. R. and Perruzzi, C. A. (1996). *Biochim. Biophys. Acta*, **1314**, 13–24.
7. Smith, L. L., Cheung, H. K., Ling, L. E., Chen, J., Sheppard, D., Pytela, R., and Giachelli, C. M. (1996). *J. Biol. Chem.*, **271**, 28485–91.
8. Patarca, R., Saavedra, R. A., and Cantor, H. (1989). *Crit. Rev. Immunol.*, **13**, 225–46.
9. Ono, M., Yamamoto, T., and Nose, M. (1995). *Mol. Immunol.*, **32**, 447–8.
10. Oldberg, A., Franzen, A., and Henegard, D. (1986). *Proc. Natl Acad. Sci. USA*. **83**, 8819–23.
11. Moore, M. A., Gotoh, Y., Rafidi, K., and Gerstenfeld, L. C. (1991). *Biochemistry*, **30**, 2501–8.
12. Nasu, K., Ishida, T., Setoguchi, M., Higuchi, Y., Akizuki, S., and Yamamoto, S. (1995). *Biochem. J.*, **307**, 257–65.
13. Kerr, J. M., Fisher, L. W., Termine, J. D., and Young, M. F. (1992). *Gene*, **108**, 237–43.
14. Wrana, J. L., Zhanag, Q., and Sodek, J. (1989). *Nucl. Acids Res.*, **17**, 10119.
15. Saitoh, Y., Kuratsu, J., Takeshima, H., Yamamoto, S., and Ushio, Y. (1995). *Lab. Invest.*, **72**, 55–63.
16. Miyazaki, Y., Setoguchi, M., Yoshida, S., Higuchi, Y., Akizuki, S., and Yamamoto, S. (1990). *J. Biol. Chem.*, **265**, 14432–8.
17. Rafidi, K., Simikina, I., Johnson, E., Moore, M. A., and Gerstenfeld, L. C. (1994). *Gene*, **140**, 163–9.
18. Crosby, A. H., Edwards, S. J., Murray, J. C., and Dixon, M. J. (1995). *Genomics*, **27**, 155–60.
19. Sorenson, E. S. and Petersen, T. E. (1994). *Biochem. Biophys. Res. Commun.*, **198**, 200–5.
20. Liaw, L., Birk, D. E., Ballas, C. B., Whitsitt, J. S., Davidson, J. M., and Hogan, B. L. M. (1998). *J. Clin. Invest.*, **101**, 1468–78.
21. Scatena, M., Almeida, M., Chaisson, M. L., Fausto, N., Nicosia, R. F., and Giachelli, C. M. (1998). *J. Cell Biol.*, **141**, 1083–93.

Cecilia M. Giachelli
Associate Professor, Pathology Department, University of Washington, Seattle, WA, USA.

Osteoprotegerin

Osteoprotegerin (OPG) is a novel, secreted member of the tumour necrosis factor receptor (TNFR) superfamily of proteins.[1] Overexpression of OPG in transgenic mice results in increased bone mass, due to the accumulation of newly synthesized bone matrix. Recombinant OPG blocks the terminal stages of osteoclast maturation *in vitro*, and *in vivo* blocks bone resorption, suggesting that it acts as a key regulator of bone metabolism.

Synonymous names

Osteoclastogenesis inhibitory factor (OCIF).[2]

Homologous proteins

All members of the TNFR superfamily contain repeats of a cysteine-rich motif[3] found in OPG, including, for example TNFR1, TNFR2, CD40, FAS, CD30, OX40, HVEM/ATAR, and DR3. The cysteine-rich motif is usually repeated 3–4 times in tandem, and together forming the ligand-binding domain of the respective receptor.

Protein properties

The cDNA sequence has been determined for mouse, rat, and human OPG. In each case a single predominant ~2.5–3.0 kb mRNA is expressed which encodes a 401 amino acid residue open reading frame (Fig. 1). The predicted products are highly conserved during evolution and share 85–90 per cent identity throughout the length of the protein. All cysteine residues within the TNFR repeats are conserved and align without gaps.

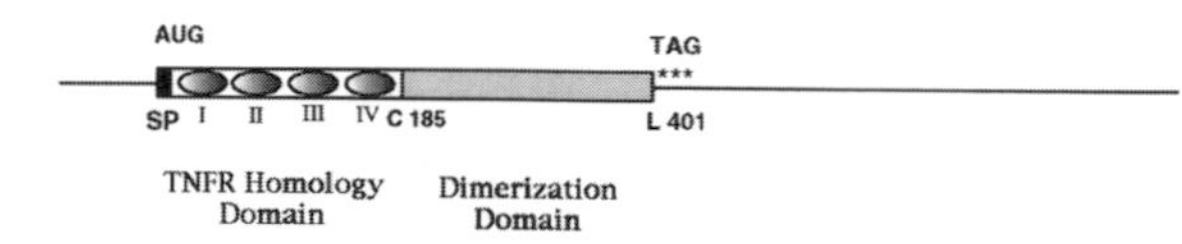

Figure 1. Structure of OPG protein. The full length rat cDNA isolate of OPG is 2.4 kb and is represented by the thin line. The 401 amino acid long open reading frame encoding the rat OPG product is represented by the coded box. The first 21 amino acids (black) indicate the position of the signal peptide sequence. The N-terminal half of OPG contains four tandem cysteine-rich repeat sequences (shaded ellipses), representing domains I–IV. The C-terminal dimerization domain is indicated by a grey box. The coding frame begins at a methionine codon (AUG) and terminates at a stop codon (TAG) following leucine 401. C 185 represents the last cysteine residue of the TNFR homology domain.

The protein can be structurally and functionally divided into two halves: an N-terminal domain of about 200 residues which is homologous to all TNFR family members and a C-terminal region that shares no homologies to other known proteins and has no recognizable motifs. A 21 a.a. signal peptide is predicted at the extreme N terminus, and sequence analysis of purified proteins indicates that this is an authentic cleavage site. The predicted mature product is a 40 kDa monomer, whereas a 55–60 kDa monomer is detected when expressed in mammalian cells. Treatment with *N*-glycanase reduces its relative molecular weight to ~40 kDa, suggesting that OPG is glycosylated.

A major interesting feature of these proteins is the noticeable lack of a hydrophobic transmembrane region, such as those found in other members of this family. This suggested that OPG was a secreted protein, and not a cell surface receptor. The OPG protein is found to be secreted from transfected mammalian cells and accumulates in conditioned media over time. OPG forms disulphide-linked dimers during secretion and both dimers (120 kDa) and monomers (55–60 kDa) are recovered during purification. The C-terminal domain of OPG was found to be involved in dimerization of the protein, which appears to involve several conserved cysteine residues within this region. This C-terminal domain can be replaced by the IgG1 Fc domain to produce dimers. Although the C-terminal dimerization domain is highly conserved during evolution, it is dispensable for activity. The N-terminal region spanning residues 22–185 represent the core active region of OPG (Fig. 1).

OPG is expressed in the cartilaginous bone primordia during mouse embryonic development and in adult tissues, including the kidney, lung, intestine, liver, and heart. No alternatively spliced mRNAs have been identified that encode altered forms of the protein, such as a transmembrane isoform.

Overexpression of OPG in transgenic mice results in osteopetrosis. Phenotype characterized by dense trabeculae consisting of newly synthesized bone matrix (Fig. 2). Bones are of normal length and shape and grow throughout development. The increase in bone density seen in transgenic mice appears to be dependent on the level of OPG present in the circulation. Mature osteoclasts are few in number or lacking completely, while osteoclast precursor populations are at normal levels. This suggests that osteoclast maturation and/or activation is impaired in OPG transgenic mice. Recombinant OPG protein blocks osteoclast maturation in an *in vitro* bone marrow and stromal cell co-culture system.[4] Normally, multinucleated osteoclasts are generated from monocyte/macrophage precursors that express tartrate-resistant acid phosphatase (TRAP), the calcitonin receptor, and the integrin

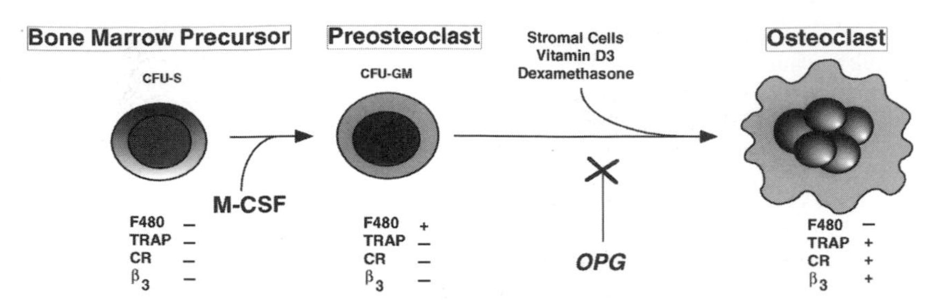

Figure 2. OPG effect on osteoclastogenesis *in vitro*. Diagram depicting the hypothetical action of OPG on osteoclast development *in vitro* using the bone marrow and stromal cell co-culture system. Bone marrow precursor cells treated with M-CSF (CSF-1) give rise to bone marrow macrophage cells capable of differentiating into mature osteoclasts. Co-culture of bone marrow macrophages with ST-2 stromal cells[1,4] in the presence of 1,25-dihydroxyvitamin D_3 and dexamethasone differentiate into mature osteoclasts. OPG blocks osteoclast development at the stage which requires stromal cells, vitamin D_3, and dexamethasone. F480, a monocyte/macrophage surface marker; TRAP, tartrate resistant acid phosphatase; CR, calcitonin receptor; β_3; integrin $\alpha V\beta 3$; CFU-S, colony forming unit–stem cells; CFU-GM, colony forming unit–granulocyte/macrophage.

$\alpha V\beta 3$. OPG blocks osteoclastogenesis at a point prior to the acquisition of these mature markers by an unknown mechanism. Recently a polypeptide ligand for OPG has been identified (OPG ligand or OPGL); it is related to TNF.[6] Soluble recombinant OPGL stimulates osteoclastogenesis from haematopoietic precursor cells, and activates mature osteoclasts to resorb bone, both *in vitro* and *in vivo*. This suggests that OPG acts as a secreted neutralizing receptor that sequesters a key osteoclast differentation and activation factor.

Administration of recombinant OPG blocks osteoclast function in growing mice and leads to the accumulation of bone similar to that observed in transgenic mice. OPG protein thus has the effect of an anti-resorptive compound capable of regulating bone mass. In an animal model where pathological increases in bone resorption were induced by ovariectomy, OPG blocked bone loss due to decreased oestrogen levels. This suggests a potential therapeutic role for OPG in treating osteopenic disorders characterized by increased osteoclast activity, such as osteoporosis.

Purification

OPG can be purified from conditioned media by heparin affinity chromatography, followed by ion exchange chromatography.[2] It exists as almost equimolar amounts of monomeric and dimeric protein. Monomers and dimers can be resolved by reverse-phase HPLC.

Activities

Inhibits osteoclast maturation in bone marrow co-culture assays in a dose-dependent manner.[1,2] Full length protein has a half-maximal effective dose of about 1–2 ng/ml. Using this assay, the minimal active region of OPG has been mapped so far to residues 22–185, which encodes the predicted ligand binding domain of this protein (I–IV).[1]

Antibodies

Rabbit polyclonal antibodies have been produced to the mouse and human OPG protein.[1] Anti-mouse OPG antibodies cross-react with the mouse, rat, and human proteins. Monoclonal antibodies to various regions of the protein are being developed.

Genes

The GenBank accession numbers for the mouse, rat, and human OPG sequences are U94331, U94330, and U94332, respectively.

Mutant phenotype/disease states

OPG transgenic mice have osteopetrosis.[1] Recombinant protein blocks ovariectomy-induced bone loss in rat (see above). No correlation in the levels or regulation of this protein in human disease has been made so far.

Structure

OPG structure for residues 22–190 can be predicted by superimposing its sequence on to the known structure of TNFR1 extracellular domain.[5] No crystal structure is available at this time.

References

1. Simmonet, W. S., Lacey, D. L., Dunstan, C. R., *et al.* (1997). *Cell*, **89**, 309–19.
2. Tsuda, E., Goto, M., Mochizuki, S. -I., Yano, K., Koboyashi, F., Morinaga, T., and Higashio, K. (1997). *Biochem. Biophys. Res. Commun.*, **234**, 137–42.
3. Smith, C. A., Farrah, T., and Goodwin, R. G. (1994). *Cell*, **76**, 959–62.
4. Lacey, D. L., Erdmann, J. M., Teitelbaum, S. L., Tan, H.-L., Ohara, J., and Shoi, A. (1995). *Endocrinology*, **136**, 2367–76.

5. Banner, D. W., D'Arcy, A., Janes, W., Gentz, R., Schoenfeld, H. -J., Loetscher, H., and Lessiauer, W. (1993). *Cell*, **73**, 431–45.
6. Lacey, D. L., Timms, E., Tan, H.-L. *et al.* (1998). *Cell*, **93**, 165–76.

■ *William J. Boyle*
Department of Cell Biology, Amgen, Inc., 1840 DeHavilland Drive, Thousand Oaks, CA 91320, USA

Perlecan

Perlecan is the largest and most common proteoglycan of basement membranes. The multidomain core protein contains two to three heparan sulphate/chondroitin sulphate side chains at its N-terminal end. The glycosaminoglycan side chains and the core protein manifest multiple functions: glomerular ionic filtration, possible control of serine protease activity, and binding of basic fibroblast growth factor (side chains); cell attachment and basement membrane assembly (core protein).

Synonymous names

Low density heparan sulphate proteoglycan, $HSPG_2$.

Homologous proteins

Murine and human perlecan are highly similar at the amino acid level. The *unc-52* gene product of *Caenorhabditis elegans* is a nematode homologue of perlecan.[1] Perlecan shares structural motifs with agrin and laminin.[2] Agrin, also a heparan sulphate proteoglycan, is produced by motor neurones, and participates in a signalling cascade involving protein tyrosine kinases.[3] Laminin is a basement membrane glycoprotein. At their C-terminal end all three proteins contain globular subdomains interspersed with one or two epidermal growth factor (EGF) repeat motifs. The genomic structures for perlecan and agrin show very different exon/intron structures and thus, from an evolutionary standpoint, these molecules are quite distant.

Properties

Perlecan (M_r = 396 000–467 000 Da) is a multidomain, heparan sulphate/chondroitin sulphate proteoglycan.[4,5] The primary structure of the core protein, derived from mouse and human cDNA clones, indicates five distinct domains (Fig. 1). Cysteine-free domain I, resembling no other proteins, is rich in acidic amino acids and contains three conserved Ser–Gly–Asp sequences at its N-terminal end for attachment of glycosaminoglycan (GAG) side chains; domain II contains four LDL receptor-like cysteine repeats; domain III, the second largest of the domains, with three globular subdomains separated by short cys-

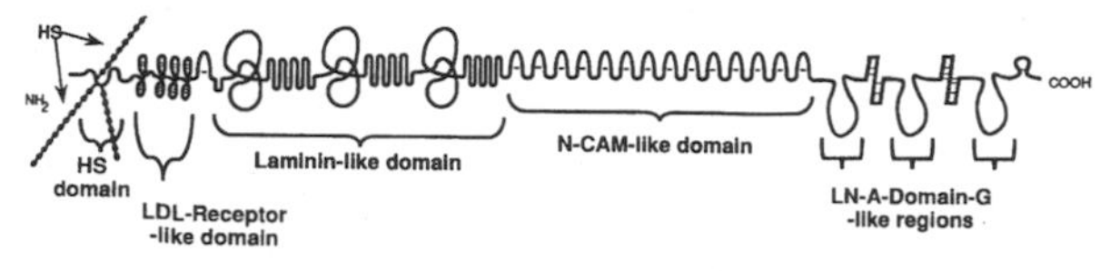

Figure 1. Diagrammatic representation of the domain structure of perlecan. The domains are numbered I–V as follows: domain I (HS domain); II (LDL-receptor-like domain); domain III (laminin-like domain): domain IV (NCAM-like domain); domain V (laminin domain-G-like).[4]

teine rich segments, is remarkably similar to domains III and IV of laminin; domain IV is made up of 14 (predominant form in the mouse) or 21 (human) NCAM-like IgG repeats; domain V is similar to the C-terminal domains of laminin and agrin and contains three globular subdomains, interspersed with EGF repeat motifs.[2,4] The *unc-52* gene product of nematode contains domains II, III, and IV of mammalian perlecan.[1]

The heparan sulphate/chondroitin sulphate chains are added to the three Ser–Gly–Asp sequences of domain I via *O*-glycosylation. Further downstream, this domain also has an SEA module, an 80 bp conserved sequence found in diverse extracellular matrix molecules that may enhance the glycosylation process.[5] Although perlecan is primarily a heparan sulphate proteoglycan, it may also be found as a chondroitin sulphate–heparan sulphate hybrid, as in an EHS tumour cell line.[2] Acidic amino acids, aspartic and glutamic acids at the N-terminal end of the conserved Ser–Gly–Asp seem to assure addition of heparan sulphate side chains, and their targeted mutagenesis to asparagines and glutamines, dictated addition of chondroitin sulphate.[5]

Perlecan is an integral component of most basement membranes, and like other basement membrane components, it is expressed very early during development.[2] Perlecan expression, however, is not limited to basement membranes. It is also present in hyaline cartilage, in the interterritorial matrix of articular cartilage, and in all the zones of the growth plate during endochondral bone formation (Fig. 2).[6] In human placenta, large amounts of perlecan transcript was detected by *in situ* hybridization in the syncytiotrophoblasts and in the endothelial cells of blood vessels.[7] Similarly in mouse embryos, perlecan tran-

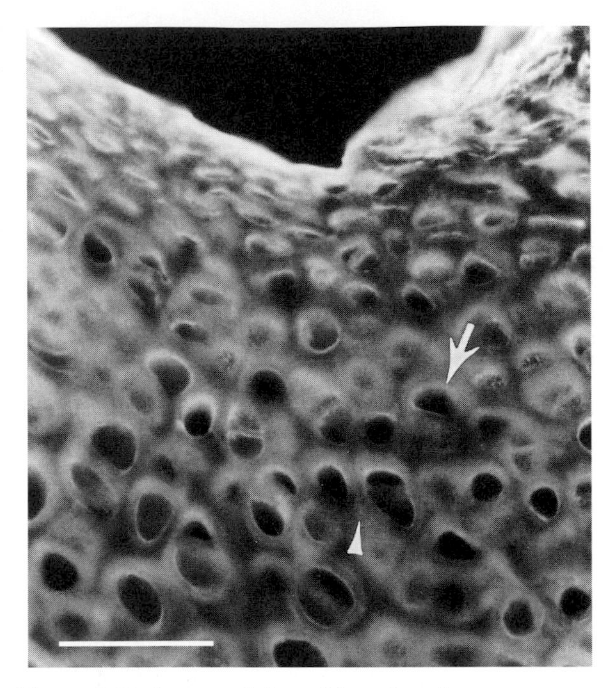

Figure 2. Perlecan staining with anti-perlecan antibodies in the articular surface of a bone section. Only the pericellular region (arrowhead) stained intensely with the antibodies.[6]

scripts were detected in endothelial cells lining the developing blood islands of the maternal deciduum (S. Chakravarti, unpublished observation). Perlecan is also present in the bone marrow extracellular matrix surrounding developing haematopoietic cells.[8] In colon carcinomas, perlecan is present not only in the basement membrane but in the fibrovascular tumour stroma as well.[2]

The multidomain core protein and its GAG side chains combine an array of biological properties that allow perlecan to play a diverse set of roles in the extracellular matrix. The polyanionic glycosaminoglycan side chains regulate selective filtration of macromolecules in glomerular basement membranes and basement membranes of other vascular beds.[4] Interactions of the side chains and the core protein with other extracellular matrix components (see Activities) are likely to stabilize the matrix and may constitute necessary intermediate steps in the matrix-assembly process. Interactions of the GAG side chains and the core protein also contribute significantly to the cell anchoring function of the basement membrane and embryo attachment to the uterine wall.[4,9] In *C. elegans*, perlecan's interactions with cell surface integrin receptors may play a role in myogenic differentiation (see Mutant phenotype, discussion of *C. elegans*). More recently, perlecan is considered to play a regulatory role in cell growth by binding to and modulating growth factor activities.[10] The heparan sulphate side chains serve as a low affinity accessory receptor for basic FGF.[11] In a rabbit ear model for angiogenesis, perlecan induced formation of interconnected blood vessels and elicited an even higher angiogenic response when complexed with bFGF.[11] Increased expression of perlecan in the vicinity of developing blood vessels, colon carcinomas, and liver neoplastic nodules, taken together with the findings of the rabbit angiogenesis model, suggest a close link between perlecan expression, cell proliferation, and vascularization.

Purification

Perlecan was originally isolated from the murine Engelbreth–Holm–Swarm (EHS) tumour.[12] It can be purified from these EHS tumours or from other sources such as epithelial-like cell cultures, or glomeruli, by using dissociative solvents and further purified by CsCl density centrifugation and chromatography on DEAE–sepharose and sepharose C1–4B.[13]

Activities

The purified core protein supports attachment of cultured epithelial-like cells, hepatocytes, and chondrocytes to tissue culture dishes coated with the core protein. Murine domain III, expressed as a recombinant protein, also supports cell attachment through an RGDS site that is absent in the human homologue.[14] The core perlecan self associates into higher order forms, binds fibronectin and, possibly, entactin in solid phase binding assays, while the HS side chains bind laminin and immobilize bFGF as discussed above.[11,15–17]

Antibodies

Several polyclonal and monoclonal antibodies are available through individual investigators.[6,18] There are some polyclonal antibodies against the EHS-tumor core protein that work well in immunoprecipititation and Western blotting.[12] Some of these, however, do cross-react with laminin. A monoclonal antibody (HS42) against human placental basement membrane perlecan has been used in immunohistochemical staining.[7] Also available are four additional epitope-specific antibodies against human perlecan, purified from lung fibroblast extracellular matrix. These are effective in Western blotting and immunohistochemical staining.[19]

Genes

To date cDNA sequences of human,[7] mouse,[20] and nematode[1] perlecan are available (human: GenBank number M85289, mouse: M77174, nematode: L13458). The gene has been localized to human chromosome 1p36.1[21] and to a syntenic region on distal mouse chromosome 4.[22] The human gene, consisting of 94 exons, spans over 120 kb of genomic DNA. Exons coding for domain IV are alternatively spliced to yield two different sized domain IVs.[23]

The promoter, contained in a CpG island, has an overall organization of housekeeping and growth factor gene promoters.[24]

Mutant phenotype/disease states

Mutations in the *unc-52* gene of *C. elegans* have been reported which lead to viable and lethal phenotypes, displaying disorganized body wall muscle. The lethal *st549* allele causes a complete loss of function for at least one of the alternatively spliced *unc-52* gene products, and disrupts myofilament assembly at the earliest stages of body wall muscle development. The viable alleles consist of mutations that affect the C-terminal domain IV and are likely to yield partially functional truncated proteins associated with a paralysed phenotype. The *C. elegans* study emphasizes that perlecan interactions with integrins are essential for body wall muscle cell differentiation. No disease-associated mutations have been reported so far in the mammalian genes. However, abnormal expression or deposits of perlecan are implicated in various renal diseases, diabetes-related alterations of the basement membrane, in the pathogenesis of islet amyloidosis[26] and in amyloid deposits in the brains of patients with Alzheimer's disease.[27]

Structure

Rotary shadowing of the intact proteoglycan revealed the core protein to be 83 nm long, consisting of five to seven continuous globules.[25]

References

1. Rogalski, T. M., Williams, B. D., Mullen, G. P., and Moerman, D. G. (1993). *Gene Dev.*, **7**, 1471–84.
2. Iozzo, R. V. and Murdoch, A. D. (1996). *FASEB J.*, **10**, 598–614.
3. Kleiman, R. J. and Reichardt, L. F. (1996). *Cell*, **85**, 461–4.
4. Hassell, J. R., Blochberger, T. C., Rada, J. A., Chakravarti, S., and Noonan, D. (1993). *Proteoglycan gene families* **1**, 69–113.
5. Dolan, M., Horchar, T., Rigatti, B., and Hassell, J. R. (1997). *J. Biol. Chem.*, **272**, 4316–22.
6. SundarRaj, N., Fite, D., Ledbetter, S., Chakravarti, S., and Hassell, J. R. (1995). *J. Cell Sci.*, **108**, 2663–72.
7. Murdoch, A. D., Dodge, G. R., Cohen, I., Tuan, R. S., and Iozzo, R. V. (1992). *J. Biol. Chem.*, **267**, 8544–57.
8. Klein, G., Conzelmann, S., Beck, S., Timpl, R., and Muller, C. A. (1994). *Matrix Biol.*, **14**, 457–65.
9. Carson, D. D., Tang, J.-P., and Julian, J. (1993). *Dev. Biol.*, **155**, 97–106.
10. Ruoslahti, E. and Yamaguchi, Y. (1991). *Cell*, **64**, 867–9.
11. Aviezer, D., *et al.* (1991). *Cell*, **79**, 1005–13.
12. Ledbetter, S. R., Tyree, B., Hassell, J. R., and Horigan, E. A. (1985). *J. Biol. Chem.*, **260**, 8106–13.
13. Haralson, M. A. and Hassell, J. R. (1995). *Extracellular matrix. A practical approach.* Oxford University Press, New York.
14. Chakravarti, S., Horchar, T., Jefferson, B., Laurie, G. W., and Hassell, J. R. (1995). *J. Biol. Chem.*, **270**, 404–9.
15. Yurchenco, P. D., Cheng, Y. S., and Ruben, G. C. (1987). *J. Biol. Chem.*, **262**, 17668–76.
16. Battaglia, C., Mayer, U., Aumailley, M., and Timpl, R. (1992). *Eur. J. Biochem.*, **208**, 359–66.
17. Heremans, A., Cock, B. D., Cassiman, J.-J., Berghe, H. V. D., and David, G. (1990). *J. Biol. Chem.*, **265**, 8716–24.
18. Couchman, J. R., Kapoor, R., Sthanam, M., and Wu, R.-R. (1996). *J. Biol. Chem.*, **271**, 9595–602.
19. Heremans, A. and Schueren, B. V. D. (1989). *J. Cell Biol.*, **109**, 3199–211.
20. Noonan, D., *et al.* (1991). *J. Biol. Chem.*, **266**, 22939–47.
21. Dodge, G. R., *et al.* (1991). *Genomics*, **10**, 673–80.
22. Chakravarti, S., Phillips, S. L., and Hassell, J. R. (1991). *Mamm. Genome*, **1**, 270–2.
23. Noonan, D. M. and Hassell, J. R. (1993). *Kidney Int.*, **43**, 53–60.
24. Cohen, I. R., Grassel, S., Murdoch, A. D., and Iozzo, R. V. (1993). *Proc. Natl Acad. Sci. USA*, **90**, 10404–8.
25. Laurie, G. W., Inoue, S., Bing, J. T., and Hassell, J. R. (1988). *Am. J. Anat.*, **181**, 320–6.
26. Castillo, G. M. *et al.* (1998). *Diabetes*, **47**, 612–20.
27. Castillo, G. M. *et al.* (1997). *J. Neurochem.*, **69**, 2452–65.

S. Chakravarti
Departments of Medicine and Genetics,
Case Western Reserve University,
10900 Euclid Avenue,
Cleveland, OH 44106–4952, USA

J. R. Hassell
Shriners Hospital for Children,
12502 North Pine Drive,
Tampa, Fl 33612–9466, USA

Reelin

Reelin[1] is a secreted glycoprotein expressed by specific neuronal populations during and after development of the vertebrate central nervous system. It is required for the formation of laminar structures in the brain such as the cerebral and cerebellar cortices and the hippocampus. Mutation of *reelin* results in neuroanatomical abnormalities and an ataxic phenotype in reeler mice.[1,7,8,30]

Protein properties

Reelin is a single polypeptide of 3461 amino acids (approximately 400 kDa on SDS–PAGE).[1] The N-terminal domain contains a cleavable signal peptide, a 190 amino acid region of modest similarity (25 per cent) to F-spondin,[2] and the epitope recognized by the neutralizing monoclonal antibody CR-50.[3,4] The majority of the protein comprises a series of eight consecutive repeats (reelin repeats) of 350–390 amino acids. Each repeat consists of two related subdomains flanking an EGF-like motif. The C terminus contains a short stretch of positively charged amino acids which is required for secretion.[4] Reelin is *N*-glycosylated and has potential sites of amidation and myristylation.

During mouse embryogenesis, *reelin* is expressed at high levels in the brain, spinal cord, kidney, and liver.[1,3,6,31,40] and transiently in other non-neuronal tissues.[31] In embryonic brain, *reelin* is present in the laminar structures disrupted in *reeler* mice, such as the neocortex, the cerebellum, and the hippocampus, and in other regions such as thalamic nuclei, striatum, hypothalamus, and the retina that do not appear disrupted in the mutant mice.[1,6,31,46] The earliest generated neurones of the cerebral cortex, the Cajal–Retzius cells, secrete Reelin into the marginal zone (layer I) where it provides a signal for proper positioning of migrating cortical neurones.[1,3,21,40] Cajal–Retzius-like cells provide a similar function in the developing hippocampus[17,40] and in the cerebellum Reelin, secreted by granule neurones, is required for Purkinje cell migration into the Purkinje cell layer.[19,32] In the adult, *reelin* is expressed at low levels in the brain, but also in non-neuronal tissues.[31,40]

The pattern of expression in laminar structures of the brain suggests that Reelin may provide a signal for terminating the migration of neurones moving radially that have reached the most superficial layers. An alternative view is that Reelin actively promotes cell migration. Although the exact mechanism of action is currently unknown, it is clear that Reelin, directly or indirectly, provides a signal that directs the formation of neuronal layers.[7,8,31] Furthermore, Reelin appears to be necessary for axonal branching in the developing hippocampus.[20,40] In adult organs, Reelin may be involved in stabilizing existing structures or in remodelling processes.[7,31,40]

Recently, mutations in the mouse disabled gene (*mDab1*) gene have been described which result in a phenotype very similar to reeler.[9–13] Spontaneous mutations resulting in aberrant mRNA expression and drastic protein reduction in scrambler[9,10] and yotari[9,14] mice, and targeted disruption of *mDab1*[15] cause anatomical alterations identical to those in reeler. mDab1 is a cytoplasmic protein that appears to function in signal transduction events involving phosphorylation.[16] Since mDab1 is expressed in the presumed target populations of Reelin, it has been proposed that Reelin activates a phosphorylation cascade involving mDab1 in the cells that go astray in the mutant mice.[7,9]

Purification

Native Reelin has not been purified, but it can be immunoprecipitated from supernatants of primary neuronal cultures and transfected COS cells.[4]

Activities

The activities of Reelin have been investigated *in vivo* and *in vitro* using the CR-50 neutralizing antibody.[3]

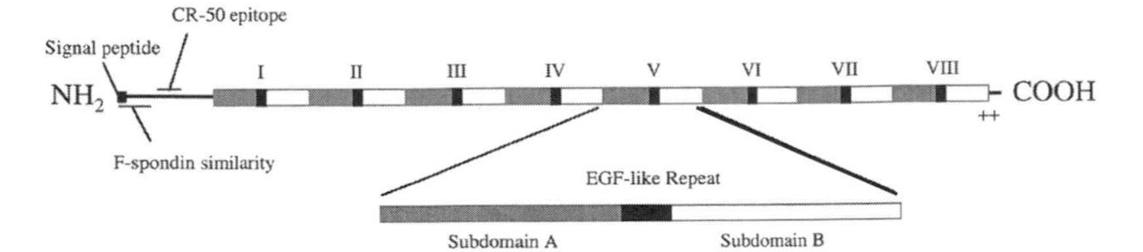

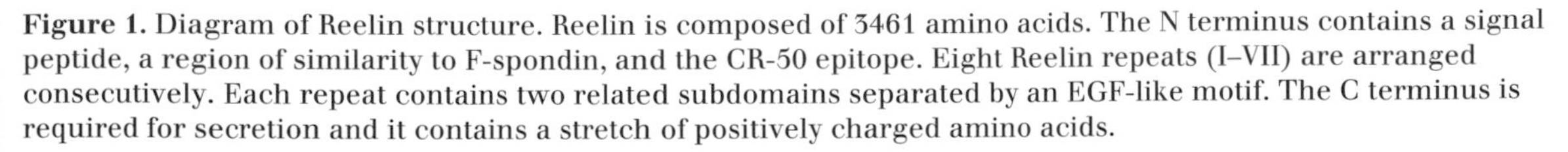
Figure 1. Diagram of Reelin structure. Reelin is composed of 3461 amino acids. The N terminus contains a signal peptide, a region of similarity to F-spondin, and the CR-50 epitope. Eight Reelin repeats (I–VII) are arranged consecutively. Each repeat contains two related subdomains separated by an EGF-like motif. The C terminus is required for secretion and it contains a stretch of positively charged amino acids.

Injection of CR-50 into the ventricles of embryonic mice results in the formation of a disorganized hippocampal pyramidal cell layer.[17] In rotation cultures of normal embryonic brain, cells aggregate according to an orderly pattern that resembles corticogenesis whereas cultures of neurones obtained from reeler mice form irregular aggregates.[18] Incubation of normal cultures with CR-50 causes a disruption similar to that seen with reeler cells.[3] Similarly, in embryonic cerebellar explants, CR-50 interferes with the formation of the Purkinje cell layer.[19] Furthermore, Reelin-producing granule cells rescue Purkinje cell layer formation in reeler explants.[19] This suggests that Reelin plays a direct role in the process of neuronal lamination.

In addition to abnormalities in cell position, *in vitro* culture studies with CR-50 indicate that Reelin affects branching and maturation of axonal terminals of entorhinohippocampal fibres.[20,35]

Antibodies

Several antibodies against Reelin have been generated in mice and rabbits. The first antibody was produced by the non-traditional route of immunizing reeler mice with normal embryonic brain extracts. Monoclonal antibodies were screened by immunohistochemistry for the ability to detect an antigen present in normal, but not in reeler mice. One such monoclonal, CR-50,[3] which strongly stained Cajal–Retzius cells in normal mice, was subsequently demonstrated to recognize an N-terminal epitope of Reelin.[4] CR-50 works well in immunohistochemistry[3] and immunoprecipitation[4] assays and it was demonstrated to be a neutralizing antibody in functional assays.[3,17,19,20] A panel of additional monoclonal antibodies was raised by immunizing reeler mice with regions of Reelin expressed in bacteria.[21,34] Two rabbit polyclonal anti-peptide antibodies, rp4 (amino acids 381–399) and rp5 (amino acids 3443–3461), also recognize native Reelin by immunohistochemistry[21] and in immunoprecipitation assays.[4]

Genes

The mouse *reelin* gene (*Reln*) maps to the proximal region of chromosome 5[27] and contains 65 exons spanning a region of over 450 kbp.[22] The mRNA is approximately 12 kb and it contains an open reading frame of 10 383 bp.[1] The human gene (*RELN*) is highly conserved (94.2 per cent amino acid identity) and it maps to chromosome 7q22.[23] Other vertebrate homologues have been detected[23,36,37] but the sequences have not been reported to date.

Mutant phenotype

Reeler is an autosomal recessive mutation in which mice exhibit ataxia, tremors, imbalance, and a typical reeling gait that becomes apparent two weeks after birth.[24] The anatomical defects include severe hypoplasia of the cerebellum and neuronal ectopia in laminated brain structures such as the cerebral and cerebellar cortices.[25,26] Five alleles of reeler, caused by deletions or insertions in the mouse *reelin* gene, result in a failure to produce or secrete protein.[1,5,21,27–29] Analysis of mutant mice suggests that Reelin is required for the correct positioning of neuronal cell bodies within laminar structures of the brain. For example, in contrast to the normal situation, cortical neurones in reeler do not form six distinct layers according to their time of origin. In the cerebellum, Purkinje cells fail to form a distinct layer underneath the external granular layer, instead they remain deep in an amorphic nuclear mass. In the hippocampus, pyramidal cells form an aberrant layer and granule cells in the dentate gyrus appear scattered. The absence of Reelin does not alter the time of origin, the differentiation, or the ability of specific neuronal populations to connect with the correct axonal projections. However, functional abnormalities related to altered synaptogenesis result from the widespread neuronal dysplasia.[38] Finally, Reelin affects the sprouting of entorhinohippocampal fibres in the outer molecular layer of the hippocampus.[20,33,35]

A mutant rat with reeler-like phenotype has recently been described.[39] In humans, many forms of neuronal ectopia have been described, but none closely resembles the reeler abnormalities.

Structure

Not available.

References

1. D'Arcangelo, G., Miao, G. G., Chen, S. C., Soares, H. D., Morgan, J. I., and Curran, T. (1995). *Nature*, **374**, 719–23.
2. Klar, A., Baldassare, M., and Jessel, T. M. (1992). *Cell*, **69**, 95–110.
3. Ogawa, M., Miyata, T., Nakajima, K., Yagyu, K., Seike, M., Ikenaka, K., *et al.* (1995). *Neuron*, **14**, 899–912.
4. D'Arcangelo, G., Nakajima, K., Miyata, T., Ogawa, M., Mikoshiba, K., and Curran, T. (1997). *J. Neurosci.*, **17**, 23–31.
5. Hirotsune, S., Takahara, T., Sasaki, N., Hirose, K., Yoshiki, A., Ohashi, T., *et al.* (1995). *Nature Genet.*, **10**, 77–83.
6. Schiffmann, S. N., Bernier, B., and Goffinet, A. M. (1997). *Eur. J. Neurosci.*, **9**, 1055–71.
7. D'Arcangelo, G. and Curran, T. (1998). *BioEssays*, **20**, 235–44.
8. Curran, T. and D'Arcangelo, G. (1997). *Brain Res. Rev.*, **26**, (In press.)
9. Sheldon, M., Rice, D. S., D'Arcangelo, G., Yoneshima, H., Nakajima, K., Mikoshiba, K., *et al.* (1997). *Nature*, **389**, 730–3.
10. Sweet, H. O., Bronson, R. T., Johnson, K. R., Cook, S. A., and Davisson, M. T. (1996). *Mamm. Genome*, **7**, 798–802.
11. Goldowitz, D., Cushing, R. C., Laywell, E., D'Arcangelo, G., Sheldon, M., Sweet, H. O., *et al.* (1997). *J. Neurosci.*, **17**, 8767–77.
12. Gonzalez, J. L., Russo, C. J., Goldowitz, D., Sweet, H. O., Davisson, M. T., and Walsh, C. A. (1997). *J. Neurosci.*, **17**, 9204–11.

13. Ware, M. L., Fox, J. W., Gonzales, J. L., Davis, N. M., Lambert de Rouvroit, C., Russo, C. J., *et al.* (1997). *Neuron*, **19**, 239–49.
14. Yoneshima, H., Nagata, E., Matsumoto, M., Yamada, M., Nakajima, K., Miyata, T., *et al.* (1997). *Neurosci. Res.* **29**, 217–23.
15. Howell, B. W., Hawkes, R., Soriano, P., and Cooper, J. A. (1997). *Nature*, **389**, 733–6.
16. Howell, B. W., Gertler, F. B., and Cooper, J. A. (1997). *EMBO J.*, **16**, 121–32.
17. Nakajima, K., Mikoshiba, K., Miyata, T., Kudo, C., and Ogawa, M. (1997). *Proc. Natl Acad. Sci. USA*, **94**, 8196–201.
18. DeLong, G. R. and Sidman, R. L. (1970). *Dev. Biol.*, **563**, 584–600.
19. Miyata, T., Nakajima, K., Mikoshiba, K., and Ogawa, M. (1997). *J. Neurosci.*, **17**, 3599–609.
20. Del Rio, J. A., Heimrich, B., Borrell, V., Forster, E., Drakew, A., Alcantara, S., *et al.* (1997). *Nature*, **385**, 70–4.
21. de Bergeyck, V., Nakajima, K., Lambert de Rouvroit, C., Naerhuyzen, B., Goffinet, A. M., Miyata, T., *et al.* (1997). *Mol. Brain Res.*, **50**, 85–90.
22. Royaux, I., Lambert de Rouvroit, C., D'Arcangelo, G., Demirov, D., and Goffinet, A. M. (1997). *Genomics*, **46**, 240–50.
23. DeSilva, U., D'Arcangelo, G., Braden, V. V., Chen, J., Miao, G., Curran, T., and Green, E. D. (1997). *Genome Res.*, **7**, 157–64.
24. Falconer, D. S. (1951). *J. Genet.*, **50**, 192–201.
25. Caviness, V. S., Jr. and Rakic, P. (1978). *Ann. Rev. Neurosci.*, **1**, 297–326.
26. Goffinet, A. M. (1984). *Brain Res.*, **319**, 261–96.
27. Bar, I., Lambert De Rouvroit, C., Royaux, I., Krizman, D. B., Dernoncourt, C., *et al.* (1995). *Genomics*, **26**, 543–9.
28. Flaherty, L., Messer, A., Russell, L. B., and Rinchik, E. M. (1992). *Proc. Natl Acad. Sci. USA*, **89**, 2859–63.
29. Royaux, I., Bernier, B., Montgomery, J. C., Flaherty, L., and Goffinet, A. M. (1997). *Genomics*, **42**, 479–82.
30. Rakic, P. and Caviness, V. S., Jr (1995). *Neuron*, **14**, 1101–4.
31. Ikeda, Y. and Terashima, T. (1998). *Dev. Dyn.*, **210**, 157–72.
32. Miyata, T., Nakajima, K., Aruga, J., Takahashi, S., Ikenaka, K., Mikoshiba, K., and Ogawa, M. (1996). *J. Comp. Neurol.*, **372**, 215–28.
33. Ghosh, A. (1997). *Nature*, **385**, 23–4.
34. de Bergeyck, V., Naerhuyzen, B., and Goffinet, A. M. (1998). *J. Neurosci. Methods*. (In press).
35. Frotscher, M., Heimrich, B., and Deller, T. (1997). *Trends Neurosci.*, **20**, 218–23.
36. Drakew, A., Frotscher, M., Deller, T., Ogawa, M., and Heimrich, B. (1998). *Neuroscience*, **82**, 1079–86.
37. Pesold, C., Impagnatiello, F., Pisu, M. G., Uzunov, D. P., Costa, E., Guidotti, A., and Caruncho, H. J. (1998). *Proc. Natl Acad. Sci. USA*, **95**, 3221–6.
38. Mariani, J., Crepel, F., Mikoshiba, K., Changeux, J. P., and Sotelo, C. (1977). *Phil. Trans. R. Soc. London. Biol.*, **281**, 1–28.
39. Ikeda, Y. and Terashima, T. (1998). *J. Comp. Neurol.*, **383**, 370–80.
40. Alcantara, S., Ruiz, M., D'Arcangelo, G., Ezan, F., de Lecea, L., Curran, T., Sotelo, C., and Soriano, E. (1998). *J. Neurosci.*, **18**, 7779–99.

Gabriella D'Arcangelo and Tom Curran
Department of Developmental Neurobiology,
St. Jude Children's Research Hospital,
Memphis, TN, USA

Sea urchin ECM molecules

Most of what is known about the biology of the extracellular matrix in the sea urchin is from the embryo. This embryo is easily obtained, cultured, and visualized, and because the developmental mechanisms of the embryo are dependent on cell interactions, investigators have a great interest in the identity and biology of the extracellular matrix (ECM). This article describes ECM proteins that have been described uniquely in sea urchins as well as sea urchin homologues of vertebrate ECM proteins.

Overview

The sea urchin embryo has two basic types of ECM environments.

1. The basal lamina and blastocoel matrix within the embryo: the blastocoel is formed early in development. By the 4–8 cell stage, a space between the blastomeres is apparent and it already contains proteins indicative of the blastocoel of larvae.
2. Outside of the embryo: the extraembryonic matrix consists of three distinct parts:

(a) the fertilization envelope, constructed during fertilization from cortical granule contents that functions in the block to polyspermy; this extracellular matrix is functionally analogous to the zona pellucida in mammals, and the chorion in fish and flies;

(b) the hyaline layer, also made from exocytosis of cortical granules during the fertilization reaction: the major protein of this layer is hyalin and it appears to be important for cell adhesion during early development; and

(c) the apical lamina, which is constructed during cleavage stages by polarized apical secretion of the blastomeres. This ECM is not part of the fertilization process, but is an important matrix for early development.

Procedures have been developed for the isolation of each of these extraembryonic matrix structures (see ref. 1 for procedures to isolate the fertilization envelope and

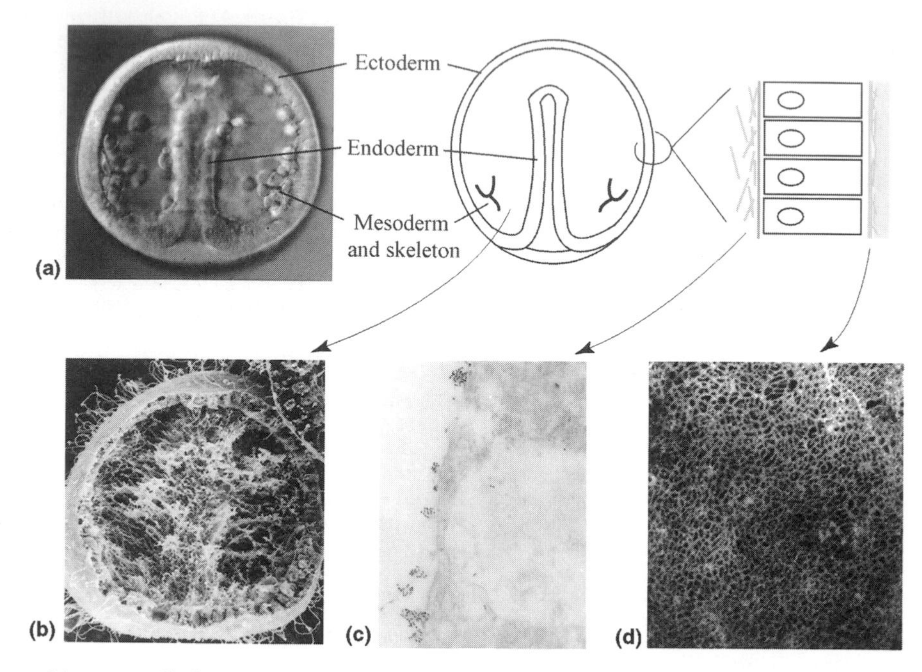

Figure 1. Sea urchin extracellular matrices. (a) DIC photo of embryo gastrulating showing the three primary germ layers; ectoderm, endoderm, and mesoderm. The embryo is approximately 125 μm in diameter. (b) Scanning electron micrograph (SEM) of the blastocoel matrix. The embryo was fixed and processed for SEM then cracked open to expose the blastocoel. The ectoderm is apparent on the outside, and is ciliated, whereas the endoderm and mesoderm is obscured by the extracellular matrix of the blastocoel. (Photo courtesy of Dr John B. Morrill.) (c) Immunolocalization of ECM 18 in the basal lamina using a colloidal gold conjugate. Each colloidal gold particle is 15 nm in diameter. (d) Immunofluorescence localization of the hyalin protein, the major protein of the hyaline layer. The entire field represents approximately 10 μm square.

its constituents; ref. 2 for isolation of the hyaline layer; ref. 3 for isolation of the apical lamina; and for the basal lamina/blastocoel matrix see ref. 4, 5). Because of its ease of isolation and conserved constituents, the blastocoel ECM from the sea urchin embryo has even been used in assays of invasion potential for mammalian metastatic tumour cells.[6]

Homologues: collagen, laminin, fibronectin

Members of the collagen, laminin and fibronectin families have each been shown to play important roles in defined morphogenetic events in the sea urchin embryo. Each of these proteins functions during gastrulation, an early morphogenetic event that results in the basic body plan of the larva. Several different collagen genes have been identified and these include both fibrillar and non-fibrillar types. The characterization of laminin has not been completed, but available evidence shows it to contain conserved A and B1 chains. Fibronectin appears to participate in the migration and adhesion of the primary mesenchyme cells during gastrulation but the identity of the fibronectin homologue in this animal has not been confirmed by cDNA cloning.

Protein properties

Collagens

The first collagen gene identified in the sea urchin embryo was a type IV-like collagen containing characteristic fibrillar and non-fibrillar regions.[7–9] It contains a triple helical domain of 1390 residues with 23 imperfections in the Gly–X–Y motif, and a 226 residue carboxyl-terminal non-collagenous region. It is most similar to the α1 (IV) and α5 (IV) chains of vertebrate collagen. This non-fibrillar collagen is expressed selectively by mesenchyme cells[10] and when the protein is blocked from functioning using antibodies, skeletogenesis is inhibited.[8]

The mesenchyme cells express another non-fibrillar collagen gene, most similar to vertebrate α2(IV)- collagen chains (Exposito *et al.*, 1994). In addition to a 31 residue, putative signal peptide, it contains a 14 amino acid N-terminal non-collagenous segment, a discontinuous 1477 amino acid triple helical domain, and a 225 residue carboxyl-terminal domain that is rich in cysteines. The amino and carboxyl-terminal non-collagenous regions of the echinoid molecule are very similar to the 7S and carboxyl-terminal non-collagenous (NC1) regions of the $\alpha1$ and $\alpha2$ chains of vertebrate type IV collagen. Thus, at least two non-fibrillar collagen genes are expressed by the mesenchyme cells, and each is distinct from the epithelial-generated basal laminar type IV collagen in vertebrates.

Two complete sequences have been reported that encode fibrillar collagen proteins. A pro-$\alpha2$(I) collagen homologue was shown to contain a 124 amino acid pre-propeptide, a 1064 amino acid α chain, and a 226 residue carboxyl-end propeptide.[11] The distribution of the conserved cysteines and the structure of the amino-end propeptide supports the homology of this collagen to the vertebrate $\alpha2$(I) gene and the cysteine motif also appears to be present in several other collagen genes of this embryo.[12] A second fibrillar collagen was reported[13] to contain a 786 amino acid collagenous domain with uninterrupted Gly–X–Y repeats and a 265 amino acid carboxyl-terminal globular extension. The C-terminal globular domain is very reminiscent of its vertebrate counterparts containing a putative peptidase cleavage site, a series of similarly arranged cysteines, and *N*-linked glycosylation sites. This collagen gene is also selectively expressed by both the primary and secondary mesenchyme cells.

These studies show that at least four different collagen genes are expressed by the embryo, and a fifth gene, a type III collagen molecule, was detected by immunological cross-reactivity with vertebrate type III antibodies,[5] though no sequence data are available to support this evidence. In addition, multiple fibrillar collagen α chains have been reported in several genuses of sea urchins, for example *Strongylocentrotus, Hemicentrotus, Asthenosoma*,[14] supporting a complexity of collagen throughout the echinoid family. Surprisingly, of the four genes whose mRNAs have been localized *in situ*, each of the genes are expressed selectively by the same cells, the primary and secondary mesenchyme cells, though with slightly different accumulation profiles. No collagen genes have yet been reported that are expressed by endoderm or ectoderm, but evidence exists for several other collagen genes in this embryo.[12]

In addition to the collagens, enzyme activities involved in the processing of collagen proteins have been reported in the embryo. Prolyl hydroxylase activity was found in the embryo by a conventional ^{3}H-water release method from ^{3}H protocollagen. This activity was detectable throughout development, but increased substantially during gastrulation.[15] Lysyl oxidase activity has also been detected in the embryo. This activity was very low but detectable early in development and, like prolyl hydroxylase, the activity increased several-fold during gastrulation. The activity was inhibited selectively with the lathrytic agent β-amino propionitrile.[16] Enzymes that target degradation of collagens have also been reported. These include a gelatinase[17] that is sensitive to 1,10-phenanthroline and EDTA. The hatching enzyme also has strong sequence conservation to collagenase (HE6)[18] although the target of this enzyme appears to be free of collagen.

Cell interactions with collagen have a major impact on the early development of the sea urchin embryo, especially during gastrulation when cell rearrangements and new gene expression are required for proper morphogenesis. Treating embryos with inhibitors of collagen biogenesis, especially BAPN, α, α'-dipyridyl, and proline analogues, results in defects in morphogenesis of the embryonic skeletal system,[19] and of the endoderm.[20–22] A select effect on gene expression is also seen.[22,23] It is not clear what the affected mechanism of collagen perturbation is, whether it is a direct effect mediated by collagen receptors, or whether a loss of collagen alters the ECM in general.

Laminin

Recent cloning of the A and B1 chains of a laminin homologue in two sea urchin species, *Strongylocentrotus purpuratus* (S. Benson; personal communication) and *Lytechinus variegatus*,[24] show strong conservation of molecular structure. A monoclonal antibody (BL1[25]), believed to bind specifically to sea urchin laminin, immunoprecipitates two proteins of 260 and 480 kDa from basal lamina preparations, and shows laminin accumulation specifically in the basal lamina. Rotary shadowing of this sea urchin laminin reveals the characteristic cruciform structure of vertebrate laminin. Injection of the BL1 monoclonal antibody into the blastocoel of embryos disrupted epithelial morphogenesis.[26] In treated embryos the epithelial cells thickened and lost intimate lateral cell contact. The effect was transient, presumably from clearing of the antibody, and the embryos recovered to develop normally.

Synthetic peptides of putative cell binding sites to laminin have also been used to test laminin function *in vivo*. Peptides containing the sequence YIGSR and IKVAV inhibited endoderm morphogenesis at gastrulation but had no effect on earlier or later development.[27] Using similar peptides, mesenchyme cells were stimulated to migrate out of isolated basal lamina 'bags'.[28] These studies are suggestive of a selective role for laminin in early morphogenesis but their exact interpretation is unclear, especially since the recently cloned cDNAs show imperfect conservation in the putative cell-binding sequences used.[29]

Fibronectin

Fibronectin-like molecules have been identified in the sea urchin embryo which cross-react with antibodies made against vertebrate fibronectins, and are of the expected size (220 kDa), and location in the embryo (blastocoel) to be part of the fibronectin family.[30,31] However, no cDNA or genomic DNA sequences have been reported to encode fibronectins in this animal. The location of proteins identified by different investigators using immunological cross-reactivity have some consistencies; presence in the blastocoel and in association with migratory cells, but significant immunostaining differences have also been reported causing uncertainties as to the true nature of the fibronectin-like molecule(s) (see ref. 32).

Primary mesenchyme cells (PMC) in the embryo respond strongly to exogenous treatments of vertebrate fibronectin. Fibronectin both binds to and promotes migration of the PMCs,[33,34] and these activities are blocked by competing peptides of the sequence RGDSPASSKP and PASS but not by RGDS.[35] Although the receptors for fibronectin in the embryo have not been biochemically characterized, several α and β integrins are known to be present in the embryo and on the PMCs, such as βG.[36] It is also noteworthy that PMCs acquire an ability to bind to fibronectin coincident with their epithelial–mesenchymal transition at the mesenchyme blastula stage[33] and with expression of the βG subunit.[36]

Purification

Collagens

Partial biochemical purification of collagens from sea urchins has been reported using conventional purification schemes.[37,38]

Laminin

Laminin has been affinity purified (using monoclonal antibody BL1) and the molecule visualized by rotary shadowing with the electron microscope.[25]

Fibronectin

A fibronectin-like molecule was isolated from sea urchin ovaries which bound to gelatin, fibrin, and fibrinogen, and mediated the spreading of baby-hamster kidney cells on plastic dishes.[39] Antiserum made to this fibronectin-like protein reacted strongly to the basal lamina of the ovarian capsule and to the basal lamina of embryos. Curiously though, this protein did not cross-react with antibodies to vertebrate fibronectin, which in other studies is used as the major evidence for the existence of fibronectin in this animal. The purification of fibronectin from embryos has not been reported.

Antibodies

Collagens

Significant conservation of collagens between sea urchins and vertebrates has enabled the use of anti-vertebrate collagen antibodies to visualize collagens in sea urchins.[5,22] Antibodies generated against sea urchin collagens[38] and to recombinant proteins[8] or synthetic peptides[10] have also been reported.

Laminin

Monoclonal antibody BL1 was reported to bind specifically to the sea urchin laminin homologue but not to vertebrate laminin.[25] Antibodies to vertebrate laminins have been reported to cross-react with sea urchin laminin.[5,40]

Fibronectin

Fibronectin is detected in the sea urchin embryo by cross-reactivity using antibodies generated against a variety of vertebrate fibronectins.[5,30,31,41] Antibodies have also been made to a fibronectin-like molecule isolated from ovaries which does not react with antibodies to vertebrate fibronectins.[39]

Genes

Collagens

The gene structure of several collagens has been reported[12,13,42] see also above for references to cDNA clones).

Laminin

The structure of the laminin gene has not been reported though cDNA sequences have been identified[24] (S. Benson; personal communication).

Fibronectin

Neither the gene nor cDNAs have been reported for fibronectin from the sea urchin.

Non-homologues

Several non-homologous proteins of the extracellular matrix have been identified in the sea urchin embryo which have important functions in morphogenesis. Some, like Endo 16 and the family of spicule matrix proteins are found only in one cell type during development, and have a major impact on the differentiation of that cell type. Others, like ECM 3 and ECM 18, also illustrate important morphogenetic regions or regulation features making them important research targets. Finally, the first extracellular matrix molecule shown to interact with cells, hyalin from the late 1890s, is discussed here. Although the mechanism of its effect remains elusive, much has recently been learned about this classically described protein. It is not yet clear how well these proteins are conserved in phylogeny, but it is clear that these molecules have critical roles in the biology of the sea urchin embryo.

Protein properties

Endo 16

Endo 16 is a 1560 amino acid protein with discrete domains. From N to C terminus it contains a consensus signal sequence, a cysteine rich region, two potential heparin-binding domains, an acidic region of five clustered repeats, an RGD cell binding motif, and a second acidic domain of 12 clustered repeats.[43] Based on gel mobility measurements, the protein (300 kDa vs. the predicted 175 kDa by primary sequence analysis) appears to be modified significantly post-translationally, though the putative modifications are not known. The cysteine-rich region contains 13 peptide repeats with strongly conserved cysteine positioning. The two acidic domains have sequence similarity to two types of calcium-binding proteins. The N-terminal acidic domain is similar to the low affinity, high capacity calcium-binding domains in other extracellular, calcium-binding proteins such as prothrombin and osteocalcin.[44] The C-terminal acidic domain has a different structure for calcium binding. Each of these repeats resembles the calcium-binding loop of the helix–loop–helix motif of the EF hand.[45] Using recombinant expressed proteins in bacteria it was also shown that Endo 16 binds ^{45}Ca, supporting the sequence analysis.[43]

Although its function is not known, Endo 16 has several intriguing properties for an extracellular matrix molecule. First, it is transcribed by cells only in the restricted region that will undergo major morphogenetic changes, the invaginating endoderm. Second, during this morphogenesis, the protein accumulates only in the basal lamina surrounding the differentiating endoderm, and later in development is confined to the stomach. From other experiments, this is crucial since endoderm–ECM interactions are known to be important for endoderm gene expression and morphogenesis. Finally, the modular design of the protein suggests that it performs an integrating function in the extracellular matrix.

ECM 3

A doublet of 240 kDa and 180 kDa are the major forms of the protein. During gastrulation, ECM 3 protein accumulates along the walls of the blastocoel, *except* where the invaginating endoderm will contact the overlying ectoderm to form the mouth. Thus, absence of ECM 3 predicts the position of endoderm/ectoderm tissue fusion, well in advance of targeting the invaginating endoderm during gastrulation. This pattern is reflected by the cells accumulating the ECM 3 mRNA, so it is believed that the cells at the site of mouth formation are already identified/specialized during gastrulation (see also ref. 46).

ECM 3 represents a population of ECM proteins in this animal, most identified only by antibodies, that are made and stored in the developing oocyte and then secreted during early development. Of interest, but currently not known is how the ECM molecules are stored (with other ECM or in mixed compartments?) and how the secretory pathway is reactivated and directed to the nascent blastocoel in early development. It is hypothesized that secretion of such ECM proteins could contribute to formation of the blastocoel. ECM 3 has no sequence similarities in the GenBank database and no function has been ascribed to it.[47]

ECM 18

ECM 18 was identified in an immuno-cDNA screen of ECM proteins from the late gastrula stage.[48] The protein is present in the basal lamina of embryos beginning at the late blastula stage and remains through development to larvae. Curiously though, the mRNA encoding the ECM 18 protein is present in eggs and all embryonic stages examined. The possibility exists that the mRNA is not translated in early development and is supported by the finding that the mRNA does not associate with polysomes until late blastula stage, the same stage as the protein is first detectable. Thus, entry into polysomes appears to be a mechanism of regulating ECM 18 protein expression. A single mRNA of approximately 6.5 kb is detected by RNA gel blots throughout development, and subsequently a 7 kb band, but the protein is present in a population of bands from as small as 50 kDa to as large as 200 kDa. It is thought that this population of bands is reflective of proteolytic processing important for the remodelling of ECM components during morphogenesis.

ECM 18 protein sequence derived from cDNA clones reveals a cysteine-rich repeat with perfectly conserved cysteine residues in approximately two-thirds of the protein. This sequence is most closely similar (42 per cent sequence identity) to domain D of the von Willebrand blood clotting factor, thought to be involved in intra- and interchain disulphide bonding.[49] The function of ECM 18 is not known but it is believed to interact with cells to permit morphogenetic changes at gastrulation. Blocking function of the ECM 18 protein by microinjecting antibody Fab fragments into the blastocoel blocks the ability of the embryos to gastrulate.[48]

Spicule matrix proteins

The primary mesenchyme cells (PMCs) of the embryo construct an elaborate set of crystalline rods (spicules) that serve as the larval skeleton. The spicules are composed of calcium carbonate and a small amount of magnesium carbonate deposited within a proteinaceous framework called the spicule matrix. This biomineralization occurs within a membrane-bound sheath of syncytial cables of the PMCs which serves as a direct template for the pattern of the species-characteristic larval skeleton. The growing spicule is not completely ensheathed within the cytoplasm of the PMCs[50] thus making this structure a true *extra*cellular matrix. Growth of the spicules occurs largely at the tips,[51,52] though cross-sections of the demineralized spicule reveals a lamellar pattern of an irregular fibrillar nature, suggesting a continuous, concentric increase in girth.[53] The organic structural matrix can now be visualized within an intact spicule which will allow characterization of the biomineralization process.[54] A useful property of the PMCs for these studies is that the skeletons are readily isolated, as well as the PMCs. When plated *in vitro*, the PMCs will replicate their *in vivo* ability for skeletal biomineralization, though the pattern is simplified relative to that of the intact embryo.

The proteinaceous spicule matrix contains at least 10 polypeptides. Although their function is not formally known, they are believed to play an important role in the nucleation of carbonate crystals and in the directed growth of the spicule. This matrix is composed of four major proteins of 47, 50, 57, and 64 kDa, and several minor components. Overall, the matrix is rich in Asx, Glx, Gly, Ser, and Ala, a composition similar to that of the matrix proteins of biomineralized tissues in molluscs, sponges, and arthropods.

Two of the major polypeptides of the spicule matrix have been studied extensively, as described below.

SM50

The original reported sequence[55,56] has been corrected.[57,58] The encoded protein is 46.3 kDa, contains a typical signal sequence of 15 amino acids, a repeating 13 amino acid motif at its 3′ end of QPG(F/M/W)G(N/G) QPG(V/M)GG(R/Q), a pI of 12 and no *N*-glycosylation consensus sites. SM50 transcripts are not detectable in eggs or in early embryos, and increase in abundance dramatically in PMCs during ingression.

SM30

The protein encoded by cDNAs suggests a 30.6 kDa protein, with a signal sequence, and a single possible *N*-glycosylation consensus site. The protein has a pI of approximately 5.9 with a very basic C terminus and no other similar sequences in databases. The middle of the protein has a short stretch of proline-rich sequence, as does SM50. SM30 accumulates several hours later in development than does SM50 and it accumulates approximately three-fold more transcripts than does SM50. Both SM50 and SM30 are expressed normally in cells that undergo biomineralization: the PMC lineage of the embryo and cells of the adult spine and tube feet.[59,60]

Homologous proteins in other sea urchins include LSM34,[61] the *L. pictus* homologue of SM30 from *S. purpuratus*; and HPSMC,[58] the *H. pulcherrimus* homologue of SM50 from *S. purpuratus*.

Fibropellins

Fibropellins Ia, Ib, and III are the three major proteins of the apical lamina. These proteins are derived from two different genes, *SpEGF I* and *SpEGF III*; the *SpEGF I* transcript is differentially spliced to give two different mRNA species and two different proteins, Ia and Ib, while the *SpEGF III* gene yields one transcript and one protein. The three fibropellin proteins differ in the number of EGF-like repeats; fibropellin 1a has the most (21) followed by fibropellin 1b (13) and fibropellin III (8). The two genes are otherwise 75 per cent identical. Other domains present in these proteins include a complement-like region (C1s) following the EGF-repeats, followed at the C terminus by an avidin-like domain.

Antibodies to the fibropellins reveal a meshwork of the proteins in the apical lamina of the embryo. Fibropellin Ia and Ib are present in eggs and are secreted early after fertilization to make up the apical lamina. Fibropellin III is not expressed until much later in development, just prior to gastrulation.

The fibropellins appear to function in the cellular movements associated with early gastrulation. Burke *et al.*[62] showed that cellular movements associated with primary invagination are blocked when the embryos are treated exogenously with either whole monoclonal antibodies or with Fabs against the fibropellins. This inhibition is manifest at primary invagination of the embryo, but not later steps of gastrulation. The three fibropellins also appear to interact functionally with each other in a ratio of 1:1:1, and may interact with hyalin, the major protein of the hyaline layer. These proteins are thought to serve as an extracellular substrate for cellular movement during early gastrulation.[62]

Synonymous names

EGF I and EGF II;[63] SpEGF I and SpEGF III;[64] AL-1, AL-2.[62]

Homologous proteins

The fibropellins contain domains related to epidermal growth factor.[65,66] Each of the fibropellins contains a different number of EGF repeats, with the sequence $X_4CX_4CX_5CX_8CXCX_8CX_2$ (where X is any amino acid). These proteins have the greatest similarity to the EGF-repeats of Notch, a neurogenic gene of *Drosophila*,[67] and its vertebrate homologues.

Hyalin

Hyalin is a large (in excess of 300 kDa), acidic glycoprotein. It contains approximately 25 per cent acidic residues, only 4 per cent basic residues, and 2–3 per cent carbohydrate by weight.[68] It has a sedimentation coefficient of 11.6 S and has a rod-like structure in solution[69] of 75 nm with a globular head of 12 nm in diameter.[70] These rods appear to aggregate head to head in association with a core particle to form large aggregates that polymerize in the presence of calcium.

Hyalin is synthesized in oocytes, stored in cortical granules, and exocytosed at fertilization to form the hyaline layer. This layer is visible in cleavage-stage embryos as an opaque covering immediately opposed to the embryonic cells, which remains throughout development. The major activities of hyalin are in structural support by a calcium sensitive gelation[71] and in cell binding. Its cell-binding activities have been observed in a centrifugal adhesion assay[33] and by treatment of embryos with function-blocking monoclonal antibodies which results in a separation of the hyaline layer from the epithelial cells.[72] In support of its role in morphogenesis is the observation that cells undergoing an epithelial–mesenchymal transition lose their ability to bind to hyalin.[33] The molecular mechanism of cell–hyalin interaction is not known.

Purification

Neither Endo 16, ECM 3 or ECM 18 has been purified, though each can be greatly enriched by making ECM 'bag' preparations.[4,5]

Spicule matrix proteins

The family of spicule matrix proteins are readily purified from spicules.[73] Spicules are isolated from an embryo by washing the embryos in ionic and non-ionic detergents, followed by brief exposure to sodium hypochlorite. The isolated spicules are then demineralized and the integral matrix collected.

Fibropellins

A highly enriched preparation of the apical lamina can be made according to Hall and Vacquier,[3] the mass of which is mostly the fibropellins.

Hyalin

The hyalin protein is easily purified, which has lent itself to extensive biochemical studies. The purification procedure relies on hyalin's solubility in buffers without calcium, and insolubility in the presence of calcium.[2,74]

Antibodies

Endo 16

A polyclonal antibody to an Endo 16 fusion protein has been reported to be useful in immunolocalizations and immunoblotting.[43,75]

ECM 3

Polyclonal antibodies to two different regions of recombinant ECM 3 have been reported.[47]

ECM 18

Polyclonal antibodies to recombinant ECM 18 have been made and used according to Berg *et al.*,[48] These antibodies are useful in immunoblotting, immunolocalization, and immuno-cDNA screening. Immunoprecipitations have not been successful probably due to the insolubility of the ECM 18 protein.

Spicule matrix proteins

Antibodies have been generated to the entire spicule matrix,[73] to an individual cloned component SM50,[59] and to specific epitopes of the matrix glycoproteins.[54]

Fibropellins

Both monoclonal and polyclonal antibodies to the fibropellins have been reported. Monoclonal antibodies AL-1 and AL-2[62] and polyclonal antiserum SpEGF 1[76] recognize both fibropellins 1a and 1b. A polyclonal antibody that specifically recognizes fibropellin III was also reported.[64]

Hyalin

Many polyclonal and monoclonal antibodies have been reported (see refs 33, 72, 77, 78). Most of these are useful in cross-species detection, probably indicative of a conservation of hyalin sequence.

Genes

Endo 16

The promoter of the Endo 16 gene has been cloned and characterized and is used extensively to study endoderm-specific transcriptional regulation.[43,79,80]

ECM 3

Over 4 kb of cDNA sequence is stored in GenBank under accession number U34202.

ECM 18

Several overlapping cDNA clones representing 5.4 kb of the approximately 6.5 kb mRNA have been reported.[48] Available sequence may be accessed with GenBank accession number U40065.

Spicule matrix proteins

The cDNAs and genes for both SM50 and SM30 have been identified and characterized.

The SM50 gene exists once in each haploid genome and contains a single intron located within the 35th codon. A unique transcription start site exits 110 nucleotides 5' of the start of translation. The mRNA is 1895 nucleotides in length and contains a single open reading frame of 445 amino acids.[56] The message accumulates to approximately 1 per cent of the mRNA in primary mesenchyme cells during gastrulation.

SM30 exists as a small gene family of two to four members. A cDNA encoding one of the possible SM30 transcripts[60] GenBank accession number of SM30, M63840) and two of the SM30 gene family members, SM30-α and SM30-β, have been identified.[81] The SM30-α

gene contains a single intron and is expressed at the time of spicule formation, significantly later than SM50.

Fibropellins

The structure of the two SpEGF genes has not been reported although the complete cDNA structure is known.[63,64]

Hyalin

Neither the cDNA sequence nor the gene structure of hyalin has been reported.

Others

Several proteins of the sea urchin extracellular matrix are mentioned here. They each have important ECM features but lack extensive characterization. If antibodies have been reported, they are identified by their types (mAb, monoclonal antibodies; pAb, polyclonal antibodies).

Basal lamina ECM

1B10

Present throughout development in all basal laminae, mAb available.[5]

1G9

Exclusive accumulation in basal laminae. mAb available.[5]

8D8

Present throughout basal lamina of early embryo, but in larvae it is restricted to the basal lamina associated with the mouth; mAb available.[5]

ECM 1

An *N*-linked carbohydrate present on several ECM proteins that is enriched in the basal lamina in the vegetal region of the embryo prior to gastrulation. A monoclonal antibody to the determinant, or glycopeptides bearing the ECM 1 carbohydrate moiety, inhibits endoderm morphogenesis selectively. A mAb is available.[82]

Pamlin

A large (255 kDa) ECM glycoprotein bound by primary mesenchyme cells, present in the basal lamina; A mAb available.[83]

Apical lamina/hyalin layer ECM

The apical lamina is an extraembryonic extracellular layer distinct from the hyaline layer both structurally[84] and biochemically.[3] The major proteins of the apical lamina, termed fibropellins, have been characterized and are described above, though other minor, uncharacterized proteins are present.[3]

8D11

Protein of the apical lamina, mAb available.[85]

HLC-32

A 32 kDa protein of the hyaline layer with sequence similarity to butanol-extract proteins, suggestive of a gene family. The gene is transcribed selectively in oocytes, the protein persists through development, and is modified prior to gastrulation by removal of four amino acids from the N terminus. pAb available.[86]

Ecto V

A glycoprotein of the hyaline layer that is concentrated in the embryo at the tips of the microvilli,[87] Ecto V is a large (350 kDa), detergent-insoluble (non-ionic detergents), filamentous (47 nm) glycoprotein that co-sediments with hyalin, the major protein of the hyaline layer. Monoclonal antibody 1C12 recognizes Ecto V specifically and is useful in immunolocalizations, immunoblots, and immunoprecipitations. The function of Ecto V is not known.

References

1. Weidman, P. J. and Kay, E. S. (1986). In *Methods in cell biology* Vol. 27 (ed. T. E. Schroeder), pp. 111–38. Academic Press, San Diego.
2. McClay, D. R. (1986). In *Methods in cell biology* (ed. T. E. Schroeder), pp. 309–23. Academic Press, San Diego.
3. Hall, G. and Vacquier, V. (1982). *Dev. Biol.*, **89**, 160–78.
4. Harkey, M. A. and Whitely, A. H. (1980). *Wilhelm Roux's Arch. Dev. Biol.*, **189**, 111–22.
5. Wessel, G. M., Marchase, R. B., and McClay, D. R. (1984). *Dev. Biol.*, **103**, 235–45.
6. Livant, D. L., Linn, S., Markwart, S., and Shuster, J. (1995). *Cancer Res.*, **55**, 5085–93.
7. Venkatesan, M., DePablo, F., Vogelli, G., and Simpson, R. T. (1986). *Proc. Natl Acad. Sci. USA*, **83**, 3351–5.
8. Wessel, G. M., Etkin, M., and Benson, S. (1991). *Dev. Biol.*, **148**, 261–72.
9. Exposito, J. Y., D'Alessio, M., Diliberto, M., and Ramirez, F. (1993). *J. Biol. Chem.*, **268**, 5249–54.
10. Angerer, L., Chambers, S., Yang, Q., Venkatesan, M., Angerer, R., and Simpson, R. (1988). *Genes Dev.*, **2**, 239–46.
11. Exposito, J. Y., D'Alessio, M., Solursh, M. Ramirez, F. (1992). *J. Biol. Chem.*, **267**, 15559–62.
12. Exposito, J. Y., Boute, N., Deleage, G., and Garrone, R. (1995). *Eur. J. Biochem.*, **234**, 59–65.
13. D'Allessio, M., Ramirez, F., Suzuki, H. R., and Gambino, R. (1992). *Proc. Natl Acad. Sci. USA*, **86**, 9303–7.
14. Tomita, M., Kinoshita, T., Izumi, S., Tomino, S., and Yoshizato, K. (1994). *Biochim. Biophys. Acta.*, **1217**, 131–40.
15. Benson, S. and Sessions, A. (1980). *Exp. Cell Res.*, **130**, 467–70.

16. Butler, E., Hardin, J., and Benson, S. (1987). *Exp. Cell Res.* **173**, 174–82.
17. Karakiulakis, G., Papakanstantinou, E., Maragoudakis, M. E., and Misevic, G. N. (1993). *J. Cell. Biochem.*, **52**, 92–106.
18. Lepage, T. and Gache, C. (1990). *EMBO J.*, **9**, 3003–12.
19. Benson, S., Smith, L., Wilt, F., and Shaw, R. (1990). *Exp. Cell Res.*, **188**, 141–6.
20. Mizoguichi, H., Fujiwara, A., and Yasumasu, I. (1983). *Differentiation*, **25**, 106–12.
21. Blankenship, J. and Benson, S. (1984). *Exp. Cell Res.*, **152**, 98–104.
22. Wessel, G. M. and McClay, D. R. (1987). *Dev. Biol.*, **121**, 149–65.
23. Wessel, G. M., Zhang, W., Tomlinson, C., Lennarz, W. J., and Klein, W. H. (1989). *Development*, **106**, 335–47.
24. Laxson, B. and Hardin, J. (1993) *Mol. Biol. Cell*, **5**, 6a.
25. McCarthy, R., Beck, K., and Burger, M. (1987). *EMBO J.*, **6**, 1587–93.
26. McCarthy, R. and Burger, M. (1987). *Development*, **101**, 659–71.
27. Hawkins, R., Fan, J., and Hille, M. (1995). *Cell Adhesion Commun.*, **3**, 163–77.
28. Crawford, B. and Burke, R. (1994). *Development*, **120**, 3227–34.
29. Hardin, J. (1996). In *Current topics in developmental biology* (ed. R. Pedersen and G. Schatten), Vol. 33, pp. 159–262.
30. Spiegel, E., Burger, M., and Spiegel, M. (1980). *J. Cell Biol.*, **87**, 309–13.
31. DeSimmone, D. W., Spiegel, E., and Spiegel., M. (1985). *Biochem. Biophys. Res. Commun.*, **133**, 183–8.
32. Ettensohn, C. A. and Ingersoll, E. P. (1992). In *Morphogenesis: an analysis of the development of biological form* (ed. E. F. Rossomando and S. Alexander), pp. 189–262. Marcel Decker, New York.
33. Fink, R. D. and McClay, D. R. (1985). *Dev. Biol.*, **107**, 66–74.
34. Katow, H. (1986). *Exp. Cell Res.*, **162**, 401–10.
35. Katow, H., Yazawa, S., and Sofuku, S. (1990). *Exp. Cell Res.*, **190**, 17–24.
36. Marsden, M. and Burke, R. D. (1997). *Dev. Biol.*, **181**, 234–45.
37. Pucci-Minafra, I., Casano, C., and LaRosa, C. (1972). *Cell Differ.*, **1**, 157–65.
38. Shimizu, K., Amemiya, S., and Yoshizato, K. (1990). *Biochim. Biophys. Acta*, **1038**, 39–46.
39. Iwata, M. and Nakano, E. (1981). *Roux's Arch. Dev. Biol.*, **90**, 83–6.
40. Spiegel, E., Burger, M., and Spiegel, M. (1983). *Exp. Cell Res.*, **144**, 47–55.
41. Katow, H., Yamada, K. M., and Solursh, M. (1982). *Differentiation*, **22**, 120–4.
42. Saitta, B., Buttice, G., and Gambino, R. (1989). *Biochem. Biophys. Res. Commun.*, **158**, 633–9.
43. Soltysik-Espanola, M., Klinzing, D. C., Pfarr, K., Burke, R. D., and Ernst, S. G. (1994). *Dev. Biol.*, **165**, 73–85.
44. Wylie, D. C. and Vanaman, T. C. (1988) In *Calmodulin* (ed. P. Cohen and C. Biklee), Elsevier, New York.
45. Moncrief, M., Kretsinger, R., and Goodman, M. (1990). *J. Mol. Evol.*, **30**, 522–62.
46. Hardin, J. and Armstrong, N. (1997). *Dev. Biol.*, **182**, 134–49.
47. Wessel, G. M. and Berg, L. (1995). *Dev. Growth Differ.*, **37**, 517–27.
48. Berg, L., Chen, S., and Wessel, G. M. (1995). *Development*, **122**, 703–13.
49. Mancuso, D. J., Tiley, E. A., Westfield, L., Worrall, N. K., Shelton-Inloes, B. B., Sorace, J. M., *et al.* (1989). *J. Biol. Chem.*, **264**, 19514–27.
50. Decker, G., Morrill, J., and Lennarz, W. (1987). *Development*, **101**, 297–308.
51. Decker, G. and Lennarz, W. (1985). *Dev. Biol.*, **126**, 433–46.
52. Ettensohn, C. A. and Malinda,K. M. (1993). *Development* **119**:155–67.
53. Benson, S. C., Jones, E., Crise-Benson, N., and Wilt, F. (1983). *Exp. Cell Res.*, **148**, 249–53.
54. Cho, J. W., Partin, J. S., and Lennarz, W. J. (1996). *Proc. Natl Acad. Sci. USA*, **93**, 1282–6.
55. Benson, S. C., Sucov, H. M., Stephens, L., Davidson, E. H., and Wilt, F. (1987). *Dev. Biol.*, **120**, 499–506.
56. Sucov, H., Benson, S., Robinson, J., Britten, R., Wilt, F., and Davidson, E. (1987). *Dev. Biol.*, **120**, 507–19.
57. Katoh-Fukui, *et al.* (1991). *Dev. Biol.*, **145**, 210–2.
58. Katoh-Fukui, Y., Noce, T., Ueda, T., Fujiwara, Y., Hashimoto, N., Tanaka, S., and Higashinakagawa, T. (1992). *Int. J. Dev. Biol.*, **36**, 353–61.
59. Richardson, W. R., Kitajima, T., Wilt, F., and Benson, S. (1989). *Dev. Biol.*, **132**, 266–9.
60. George, N., Killian, C., and Wilt, F. (1991). *Dev. Biol.*, **147**, 334–42.
61. Livingston, B., Shaw, R., Bailey, A., and Wilt, F. (1991). *Dev. Biol.*, **148**, 463–80.
62. Burke, R. D., Myers, R. L., Sexton, T. L., and Jackson, C. (1991). *Dev. Biol.*, **146**, 542–57.
63. Grimwade, J. E., Gagnon, M. L., Yang, Q., Angerer, R. C., and Angerer, L. M. (1991). *Dev. Biol.*, **143**, 44–57.
64. Bisgrove, B. W. and Raff, R. A. (1993). *Dev. Biol.*, **157**, 526–38.
65. Hursh, D. A., Andrews, M. A., and Raff, R. A. (1987). *Science*, **237**, 1487–90.
66. Yang, Q., Angerer, L., and Angerer, R. (1989). *Science*, **246**, 806–8.
67. Wharton, K. A., Johansen, K. M., Xu, T., and Artavanis-Tsakonas, S. (1985). *Cell*, **43**, 567–81.
68. Stephens, R. E. and Kane, R. E. (1970). *J. Cell. Biol.*, **44**, 611–17.
69. Gray, J., Justice, R., Nagel, G. M., and Carroll, E. J. (1986). *J. Biol. Chem.*, **261**, 9282.
70. Adelson, D. L., Alliegro, M. C., and McClay, D. R. (1992). *J. Cell Biol.*, **116**, 1283–9.
71. Robinson, J., Hall, D., Brennan, C., and Kean, P. (1992). *Arch. Biochem. Biophys.*, **298**, 129–34.
72. Adelson, D. L. and Humphreys, T. (1988). *Development*, **104**, 391–402.
73. Benson, S. C., Benson, N. C., and Wilt, F. (1986). *J. Cell Biol.*, **102**, 1878–86.
74. Kane, R. E. (1970). *J. Cell. Biol.*, **45**, 615–22.
75. Nocente-McGrath, C., Brenner, C. A., and Ernst, S. G. (1989). *Dev. Biol.*, **136**, 264–73.
76. Bisgrove, B., Andrews, M., and Raff, R. (1991). *Dev. Biol.*, **146**, 89–99.
77. Hylander and Summers (1982). *Dev. Biol.*, **93**, 368–80.
78. Vater and Jackson (1989). *Dev. Biol.*, **135**, 111–23.
79. Yuh, C-H., Ransick, A. Martinez, P., Britten, R. J., and Davidson, E. H. (1996). *Mech. Dev.*, **47**, 165–86.
80. Yuh, C-H., Moore, J. G., and Davidson, E. H. (1996). *Development*, **122**, 4045–56.
81. Akasaka, K., Frudakis, T., Killian, C., George, N., Yamasu, K., Khaner, O., and Wilt, F. (1994). *J. Biol. Chem.*, **269**, 20592–8.
82. Ingersoll, E. P. and Ettensohn, C. A. (1994). *Dev. Biol.*, **163**, 351–66.

83. Katow, H. (1995). *Exp. Cell Res.*, **218**, 469–78.
84. Spiegel, E., Howard, L., and Spiegel, M. (1989). *J. Morphol.*, **199**, 71–92.
85. Alliegro, M. C. and McClay, D. R. (1988). *Dev. Biol.*, **125**, 208–16.
86. Brennan, C. and Robinson, J. J. (1994). *Dev. Biol.*, **165**, 556–65.
87. Coffman, J. and McClay, D. R. (1990). *Dev. Biol.*, **140**, 93–104.

Note added in proof

During the preparation of the page proofs for this chapter, the following articles important to this chapter were published:

Ingersoll, E. P. and Wilt, F. H. (1998). Matrix metalloproteinase inhibitors disrupt spicule formation by primary mesenchyme cells in the sea urchin embryo. *Dev. Biol.*, **196**, 95–106.

Miller, J. R. and McClay, D. R. (1997). Changes in the pattern of adherens junction-associated beta-catenin accompany morphogenesis in the sea urchin embryo. *Dev. Biol.*, **192**, 310–22.

Miller, J. R. and McClay, D. R. (1997). Characterization of the role of cadherin in regulating cell adhesion during sea urchin development. *Dev. Biol.*, **192**, 323–39.

Mitsunaga-Nakatsubo, K., Akasaka, K., Akimoto, Y., Akiba, E., Kitajima, T., Tomita, M., Hirano, H. and Shimada, H. (1998). Arylsulfatase exists as a non-enzymatic cell surface protein in sea urchin embryos. *J. Exp. Zool.*, **280**, 220–30.

Wessel, G. M., Berg, L., Adelson, D. L., Cannon, G., and McClay, D. R. (1998). A molecular analysis of hyalin—a substrate for cell adhesion in the hyaline layer of the sea urchin embryo. *Dev. Biol.*, **193**, 115–26.

Zito, F., Tesoro, V., McClay, D. R., Nakano, E., and Matranga, V. (1998). Ectoderm cell–ECM interaction is essential for sea urchin embryo skeletogenesis. *Dev. Biol.*, **196**, 184–92.

Gary Wessel
Department of Molecular and Cell Biology and Biochemistry, Box 6, Brown University, Providence, RI 02912, USA

Semaphorins

The semaphorins constitute a family of secreted and transmembrane signalling proteins, found in invertebrates and vertebrates.[1] In the nervous system members of this family have been shown to play a role in axon pathfinding, branching, and targeting.[2–8] Some members of this family function as chemorepellents of specific growth cones.[4,6–8] This raised the possibility that axon guidance is mediated by both attractive and repulsive cues. Evidence also exists to suggest that semaphorins function outside the nervous system in the immune and cardiovascular systems.[5,9–12]

Protein properties

Semaphorins are ~750 amino acids in length.[1,4,13] Immediately following an amino-terminal signal sequence, each molecule contains ~500 amino acid extracellular semaphorin domain (sema domain) with 14–16 conserved cysteine residues.[1,13] The sema domain is the distinguishing feature of this protein family. Family members are classified according to their structural layout beyond the semaphorin domain.[14] Class I includes invertebrate transmembrane molecules. Class II includes invertebrate secreted molecules with a carboxyl terminal immunoglobulin domain. Class III includes vertebrate secreted molecules with an immunoglobulin domain and a carboxyl terminal positively charged basic region. Class IV includes vertebrate transmembrane molecules with an immunoglobulin domain followed by a transmembrane domain and cytoplasmic domain. Class V includes vertebrate transmembrane molecules with seven thrombospondin repeats in place of the immunoglobulin domain (see Table 1). The intracellular domain of the transmembrane family members (50–150 amino acids) shares little homology either within the family or with any other known intracellular domains.[1] Secreted semaphorins migrate between 90 and 100 kDa on reducing SDS-PAGE.[4]

Invertebrate semaphorins: G-semaphorin I (G-sema I)

The first invertebrate semaphorin family member was cloned during a monoclonal antibody screen. The screen identified surface glycoproteins in grasshopper which were expressed in developmentally interesting patterns in the developing central nervous system (CNS). Originally named fasciclin IV, G-sema I is a transmembrane protein dynamically expressed on a subset of axon pathways in the developing central nervous system and on a circumferential band of epithelial cells in the developing limb bud.[2] G-sema I cannot mediate cell aggregation *in vitro*, and thus is not a homophilic cell adhesion molecule. Antibody blocking experiments indicate that G-sema I functions *in vivo* during growth cone guidance. More specifically, G-sema I functions in the limb bud to steer specific pioneer axon growth cones as the axons encounter the epithelial stripe expressing the protein.[2]

Table 1 Invertebrate and vertebrate members of the semaphorin family

Invertebrate	Domains	Insect	*C. elegans*
Class 1 Transmembrane	Sema, TM	G, D-sema I[1, 2] D-sema III	Ce-sema I
Class II Secreted	Sema, Ig	G, D-sema II[3]	Ce-sema II

Vertebrate	Domains	Human	Mouse	Chick
Class III Secreted	Sema. Ig, +++	H-sema III[1]	Sema D[15]	Coll-1[4] Coll-2[13]
		H-sema V[16, 17] H-sema III/F[18]	Sema A[15]	
			Sema E[15]	Coll-3[13] Coll-5[13]
			Sema H	
Class IV Transmembrane	Sema, Ig, TM	CD100[9]	M-sema G[10] M-sema F[19] Sema C[15] Sema B[15]	Coll-4[13]
Class V Transmembrane	Sema, Thspdn, TM		Sema G[14] Sema F[14]	

Domains: Sema, semaphorin; Ig, immunoglobulin; +++, basic tail; Thspdn, thrombospondin; TM, transmembrane.
G, grasshopper; D, *Drosophila*; Ce, *C. elegans*; H, human; M, mouse.

D-Semaphorins I and II (D-sema I and II)

Shortly after G-sema I was characterized, two related *Drosophila* molecules were cloned: transmembrane D-sema I and secreted D-sema II.[1] Mutant embryo experiments show that D-sema I is required in neurones for correct guidance of their axons and appropriate formation of CNS pathways; sema I functions as a repulsive guidance cue during neurodevelopment and is required in neurones to make axon guidance decisions.[25] The secreted invertebrate semaphorin, D-sema II, is dynamically expressed during embryonic development by a subset of neurones in the CNS and by a single thoracic muscle in the periphery of *Drosophila*.[3] Transgenic flies were created in which D-sema II was ectopically expressed in normally non-expressing muscle. The phenotype of these transgenics showed that D-sema II functions *in vivo* to inhibit formation of specific synaptic terminal arbors.[3] D-sema II is thought to be a selective target-derived signal that regulates formation of synaptic connections in the developing fly.

Vertebrate semaphorins: collapsin-1 (coll-1)

The first vertebrate semaphorin was biochemically purified from adult chick brain using the *in vitro* growth cone collapse assay. The collapse assay was based on the observation that CNS axons had an inhibitory effect on the advance of PNS growth cones when the two types of explants were co-cultured.[20] Following biochemical enrichment of a growth cone collapsing factor from chick brain membranes that could induce the collapse of dorsal root ganglia (DRG) growth cones at about 10 pM, a molecule named collapsin-1 was cloned and sequenced.[4] The secreted protein collapsin-1 repels advancing axons,[6,15] paralyses their movements,[4] and collapses their growth cones *in vitro*.[4] Migrating growth cones whose filopodia encounter a collapsin laden bead turn to avoid the repulsive signal.[21] These *in vitro* results suggest that collapsin-1 might function *in vivo* to steer migrating axons.

Semaphorin signalling

Neuropilin is a neuronal, type 1 transmembrane protein implicated in neurodevelopment that has been shown to bind various secreted vertebrate semaphorins with high affinity (K_d = 100–300 pM).[26,27] Data suggest that neuropilin plays a significant role in transducing the semaphorin signal. Antibodies against an extracellular region of neuropilin block the ability of class III semaphorins from repelling sensory axons and inducing collapse of their growth cones.[26] Dorsal root ganglia from neuropilin-deficient mutant mice are not affected by class III semaphorins.[28] However, it seems unlikely that neuropilin is the sole component of a semaphorin receptor since multiple secreted semaphorins all bind to recombinant neuropilin with approximately equal affinities yet these different semaphorin family members have distinct *in situ* binding patterns and unique activity profiles.[29,30] Neuropilin is probably a necessary component of a semaphorin receptor complex. Interestingly, neuropilin is also expressed by endothelial cells and is an isoform-specific binding protein for vascular endothelial growth factor, a major regulator of angiogenesis.[31] In this system neuropilin modulates VEGF binding to receptor tyrosine kinases.

Although there is no link yet found between neuropilin and intracellular signalling, some studies have begun to determine how the semaphorin family members transduce their signal. One study found that the growth cone collapse caused by collapsin-1 is mediated through the depolymerization of F-actin.[22] Since growth cone collapse appears to be associated with actin depolymerization, another group showed that Rac-1, a small GTP-binding protein of Rho subfamily, may mediate the growth cone motility effects of collapsin-1. Constitutively active Rac-1 augments growth cone collapse, while dominant negative Rac-1 inhibits collapsin-1 induced growth cone collapse.[32] Another study identified a potential intracellular molecular component of collapsin-1 signalling, collapsin response mediator protein (CRMP-62).[23] In *Xenopus* oocytes, the presence of CRMP-62 is reported to be necessary for collapsin-1 induced current response.

Some studies have begun to determine how the semaphorin family members transduce their signal. One study found that the growth cone collapse caused by collapsin-1 is mediated through the depolymerization of F-actin.[22] Another study identified a potential intracellular molecular component of collapsin-1 signalling, collapsin response mediator protein (CRMP-62).[23] In *Xenopus* oocytes, the presence of CRMP-62 is necessary for collapsin-1-induced current response.

Sema D/coll-1 sensory patterning in spinal cord

Sensory neurones from the DRG project centrally into the spinal cord and peripherally to a variety of tissue types. This is heterogeneous population of axons that differ in their sensory modalities, dorsoventral termination locations in the spinal cord, and response profiles to neurotrophins. Large diameter proprioceptive fibres (1A) innervate muscle in the periphery, terminate in the ventral cord, and are NT-3 responsive. Small diameter pain and temperature fibres (C) innervate skin in the periphery, terminate in the dorsal cord, and are NGF responsive. Sema D/coll-1 are expressed in the ventral cord in a developmentally regulated interval in human/mouse/chick, respectively.[6–8,24] Ventral cord explants have a differential effect on explanted DRG axons, inhibiting the NGF-responsive population but having no effect on the NT-3-responsive population. Similarly, cells transfected with either sema D or coll-1 repel NGF-responsive axons. Thus, sema D/coll-1 expression in the ventral spinal cord during a developmentally defined interval may help to exclude NGF-responsive sensory axons but allow entry of NT-3-responsive sensory axons into the ventral cord.[6–8]

Human transmembrane semaphorin: CD100

CD100 is a transmembrane semaphorin in the immune system which activates leukocytes.[9] CD100 induces B cells to aggregate and improves their survival *in vitro*. In addition, activating antibodies to CD100 inhibit CD3-induced peripheral blood lymphocyte proliferation while increasing CD2-induced proliferation. However, when T cells are separated out, the activating antibody increases both CD2- and CD3-induced proliferation.[11]

Purification

Collapsin-1 can be extracted with cholate from adult chick brains. Its purification is based on its positive charge and glycosylation state using conventional column chromatography. The flow-through of Q-Sepharose anion exchange column is applied to an S-Sepharose cation exchange column. The high-salt eluate of the Sepharose cation exchange column is applied to a WGA column. The *N*-acetyl-D-glucosamine eluate of the WGA column is applied to an HPLC MonoS cation exchange column. A salt gradient is used to elute collapsin-1 enriched fractions from the MonoS cation exchange column.[13] Fractions with collapsin-1 are identified with *in vitro* collapse assay on DRG growth cones.

Recombinant collapsin-1 can be purified by ultracentrifugation. Conditioned media from baculovirus infected insect cells are centrifuged for 1 h at 28 000 r.p.m. The pellet is then resuspended in a high salt Tris buffer with CHAPS and centrifuged again for 1 h at 28 000 r.p.m. The supernatant from the second spin is diluted in low salt (1:10) and run on an S-Sepharose cation exchange column. Protein is eluted in high salt (50 mM Tris pH 7.4, 1 M NaCl, 0.1 per cent CHAPS).[23] Alternatively, the recombinant protein can be tagged with hexahistidine and purified on a nickel column.

Activities

There are two types of *in vitro* assay commonly used to determine the activity of semaphorins: collapse assay and collagen matrix co-culture assay. In the growth cone collapse assay neuronal tissue is explanted overnight on a glass coverslip in F12 media supplemented with various growth factors and additives. Recombinant protein or native tissue samples are applied to the tissue. Analysis involves comparing the number of collapsed to spread growth cones on the periphery of the explant. Approximately 30 pM of recombinant collapsin-1 is sufficient to cause 50 per cent collapse of a DRG explant cultured in the presence of NGF.[13]

In the collagen matrix co-culture assay a neuronal explant is embedded in collagen alongside an aggregate of cells transfected with a semaphorin. The assay in this case involves measuring neurite outgrowth toward the source of recombinant protein versus away from the source of recombinant protein.[6]

Antibodies

The monoclonal antibody 6F8 against G-sema I can be used for Western blots, immunohistochemistry, and functional blocking experiments.[2] A rabbit polyclonal antibody exists against collapsin-1. It can be used in Western blot analysis, but does cross-react with other family members like collapsin-2. It has been shown to block activity of a floor plate repellent presumed to be collapsin-1.[8] Two activating monoclonal antibodies exist to CD100: BD16 and BB18.[11]

Genes

See Table 1.

Mutant phenotypes

The mouse homologue of human semaphorin III and chick collapsin-1 is semaphorin D (sema D). Like coll-1, sema D both repels sensory and sympathetic growth cones and causes their collapse *in vitro*.[4,6,15] A mouse knockout of sema D, in which amino acids 1–38 have been replaced by a neo cassette, has moderate nervous system defects in addition to skeletal and cardiovascular abnormalities.[5] Seventy per cent of the homozygote knockout mice die within the first three days, while an additional 17 per cent die at weaning. Only 12 per cent are viable into adulthood. Sensory axons that ordinarily synapse in the dorsal spinal cord are reported to project into ventral cord where sema D is normally expressed. This is consistent with the above discussion on sensory patterning in spinal cord. Another nervous system defect is a paucity of neuropil in the cortex of homozygote knockouts, and the neuronal processes of the pyramidal neurones were not appropriately oriented. Skeletal defects included vertebral fusions and partial rib duplications. Mice that survived early postnatal development had severe right ventricular hypertrophy with right atrial dilatation.[5]

A second mouse, in which sema D expression was eradicated by targeted disruption of its gene, showed severe abnormalities in peripheral nerve projections. Aberrent projections were detected in trigeminal, facial, vagus, accessory, and glossopharyngeal nerves, but not in the oculomotor nerve.[33] It is interesting to note that sema D/coll-1 has been shown to bind and collapse growth cones of those specific axons affected in the knockout mouse but not to those axons unaffected.[34]

References

1. Kolodkin, A. L., Matthes, D. J., and Goodman, C. S. (1993). *Cell*, **75**, 1389–99.
2. Kolodkin, A. L., *et al.* (1992). *Neuron*, **9**, 831–45.
3. Matthes, D. J., Sink, H., Kolodkin, A. L., and Goodman, C. S. (1995). *Cell*, **81**, 631–9.
4. Luo, Y., Raible, D., and Raper, J. A. (1993). *Cell*, **75**, 217–27.
5. Behar, O., Golden, J. A., Mashimo, H., Schoen, F. J., and Fishman, M. C. (1996). *Nature*, **383**, 525–8.
6. Messersmith, E. K., *et al.* (1995). *Neuron*, **14**, 949–59.
7. Puschel, A. W., Adams, R. H., and Betz, H. (1996). *Mol. Cell. Neurosci.*, **7**, 419–31.
8. Shepherd, I. T., Luo, Y., Lefcort, F., Reichardt, L. F., and Raper, J. A. (1997). *Development*, **124**, 1377–85.
9. Hall, K.T., *et al.* (1996). *Proc. Natl Acad. Sci. USA*, **93**, 11780–5.
10. Furuyama, T., *et al.* (1996). *J. Biol. Chem.*, **271**, 33376–81.
11. Herold, C., Bismuth, G., Bensussan, A., and Boumsell, L. (1995) *Int. Immunol.*, **7**, 1–8.
12. Herold, C., Elhabazi, A., Bismuth, G., Bensussan, A., and Boumsell, L. (1996). *Int. Immunol.*, **157**, 5262–8.
13. Luo, Y., *et al.* (1995). *Neuron*, **14**, 1131–40.
14. Adams, R. H., Betz, H., and Puschel, A. W. (1996). *Mech. Dev.*, **57**, 33–45.
15. Puschel, A. W., Adams, R. H., and Betz, H. (1995). *Neuron*, **14**, 941–8.
16. Roche, J., *et al.* (1996). *Oncogene*, **12**, 1289–97.
17. Sekido, Y. *et al.* (1996). *Proc. Natl Acad. Sci. USA*, **93**, 4120–5.
18. Xiang, R., *et al.* (1996). *Genomics*, **32**, 39–48.
19. Inagaki, S., Furuyama, T., and Iwahashi, Y. (1995). *FEBS Lett.*, **370**, 269–72.
20. Kapfhammer, J. P. and Raper, J. A. (1987). *J. Neurosci.*, **7**, 201–12.
21. Fan, J. and Raper, J. A. (1995). *Neuron*, **14**, 263–74.
22. Fan, J., Mansfield, S. G., Redmond, T., Gordon-Weeks, P. R. (1993). *J. Cell Biol.*, **121**, 867–78.
23. Goshima, Y., Nakamura, F., Strittmatter, P., and Strittmatter, S. M. (1995). *Nature*, **376**, 509–14.
24. Shepherd, I., Luo, Y., Raper, J. A, and Chang, S. (1996). *Dev. Biol.*, **173**, 185–99.
25. Yu, H. H., Araj, H. H., Ralls, S. A., and Kolodkin, A. L. (1998). *Neuron*, **20**, 207–20.
26. He, Z. and Tessier-Lavigne, M. (1997). *Cell*, **90**, 739–51.
27. Kolodkin, A. L., Levengood, D. V., Rowe, E. G., Tai, Y. T., Giger, R. J. and Ginty, D. D. (1997). *Cell*, **90**, 753–62.
28. Kitsukawa, T., Shimizu, M., Sanbo, M. Hirata, T., Taniguchi, M., Bekku, Y., Yagi, T., and Fujisawa, H. (1997). *Neuron*, **19**, 995–1005.
29. Feiner, L., Koppel, A. M., Kobayashi, H., and Raper, J. A. (1997). *Neuron*, **19**, 539–45.
30. Koppel, A. M., Feiner, L., Kobayashi, H., and Raper, J. A. (1997). *Neuron*, **19**, 531–7.
31. Soker, S., Takashima, S., Miao, H. Q., Neufeld, G., and Klagsbrun, M. (1998). *Cell*, **92**, 735–45.
32. Jin, Z. and Strittmatter, S. M. (1997). *J. Neurosci.*, **17**, 6256–63.
33. Taniguchi, M., Yuasa, S., Fujisawa, H., Naruse, I., Saga, S., Mishina, M., and Yagi, T. (1997). *Neuron*, **19**, 519–30.
34. Kobayashi, H., Koppel, A. M., Luo, Y., and Raper, J. A. (1997). *J. Neurosci.*, **17**, 8339–52.

Adam M. Koppel and Jonathan A. Raper
Department of Neuroscience,
University of Pennsylvania School of Medicine,
Philadelphia, PA 19104, USA

SPARC (osteonectin, BM40)

SPARC (secreted protein acidic and rich in cysteine, also denoted osteonectin and BM40) is a matricellular glycoprotein, that is, a protein that binds both to extracellular matrix proteins and to cells and thereby regulates cell–matrix interactions. SPARC is expressed by most tissues of vertebrates during remodelling in development and disease. SPARC binds divalent cations, several matrix proteins, growth factors, and serum albumin. Addition of SPARC and SPARC-derived peptides to cultured cells modulates cell spreading, cell migration, cell-cycle progression, and gene expression. Thus, SPARC regulates cellular responses to developmental cues and to injury.

Synonyms

Osteonectin, BM40; formerly 43 kDa protein.

Homologues

Hevin is a SPARC homologue that was cloned from high endothelial venule endothelial cells isolated from human tonsils.[1] Its rodent homologue, SC1, was characterized initially as mRNA enriched in neuronal synapses.[2] In analogy to SPARC, hevin was shown to be counteradhesive for endothelial cells. The SC1 protein has not been isolated, but the distribution of SC1 mRNA in embryonic and adult mice was reported recently[3] (Fig. 1).

QR1 is an avian SPARC homologue that is expressed in quail embryonic retina and is believed to be involved in growth arrest and in the establishment of photoreceptor differentiation.[4]

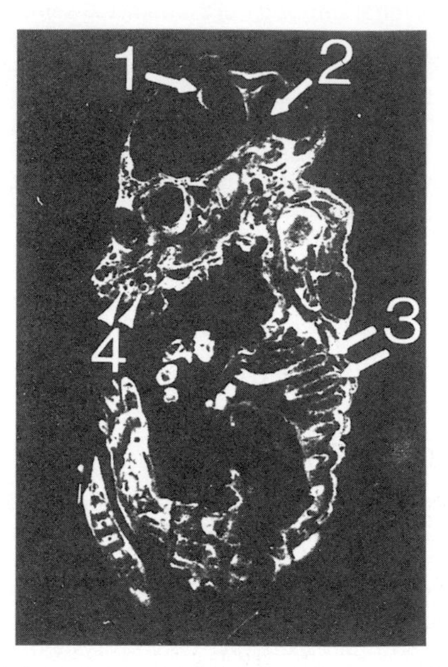

Figure 1. Distribution of SPARC in a day 17 embryo: 1, hippocampus; 2, semicircular canals and cochlea; 3, developing vertebrae; 4, vibrissae. (Reprinted from ref. 6, with permission.)

Protein properties

Vertebrate SPARC consists of 298–304 amino acids, including a signal sequence of 17 amino acids. The protein comprises four domains (denoted I–IV, see Fig. 2(a)): an acidic N-terminal domain (I), a cysteine-rich, follistatin-like domain (II), a protease-sensitive, α-helical domain (III), and a Ca^{2+}-binding domain containing an EF-hand (IV). The three-dimensional structure of a recombinant fragment consisting of domains III and IV was determined recently[5] (Fig. 2(b)). The mature SPARC protein has a molecular mass of 32 517 Da. Tissue-specific glycosylation gives rise to at least two isoforms: a 'secreted' form and a 'bone' form. The secreted form contains a complex glycan with sialic acid and fucose residues, whereas the form extracted from bone contains a high mannose glycan.[6] The SPARC homologue hevin, SC1, and QR1 differ from SPARC mainly in the length of their N-terminal domains (Fig. 2(a)).

In embryonic and adult tissues, the expression of SPARC is spatially and temporally regulated. Major sites of protein expression are remodelling and mineralizing tissues. A comparison of SPARC and SC1 transcription in kidney and adrenal gland is shown in Fig. 3 (the distribution of SPARC mRNA in a day 17 embryo is shown in Fig. 1; for a comprehensive review of the tissue distribution of SPARC, see ref. 6). In attachment-dependent cultured cells such as fibroblasts, endothelial cells, and glial cells, the expression of SPARC is usually increased as part of a 'culture shock' response upon tissue dissociation. *In vivo* and *in vitro,* SPARC is found both intracellularly and extracellularly. In addition, SPARC is detected in various sarcomas and carcinomas, in passive Heymann nephritis, and in atherosclerotic plaques.[6] The role of SPARC in these diseases is currently not understood. Suppression of SPARC expression with antisense oligonucleotides abrogates tumorigenicity of human melanoma cells in mice.[7] Thus, it is possible that inappropriate expression of SPARC is associated with hyperplasia in cancer, nephritis, and atherosclerosis. This notion is corroborated by experiments indicating that inappropriate expression of SPARC in *Xenopus* embryos[8] and in *C. elegans*[9] is associated with growth defects, including accumulation of disorganized cell masses and axial perturbation.

SPARC is induced by retinoic acid, steroids, and cytokines. Progesterone and dexamethasone activate the SPARC promoter. In addition, transforming growth factor

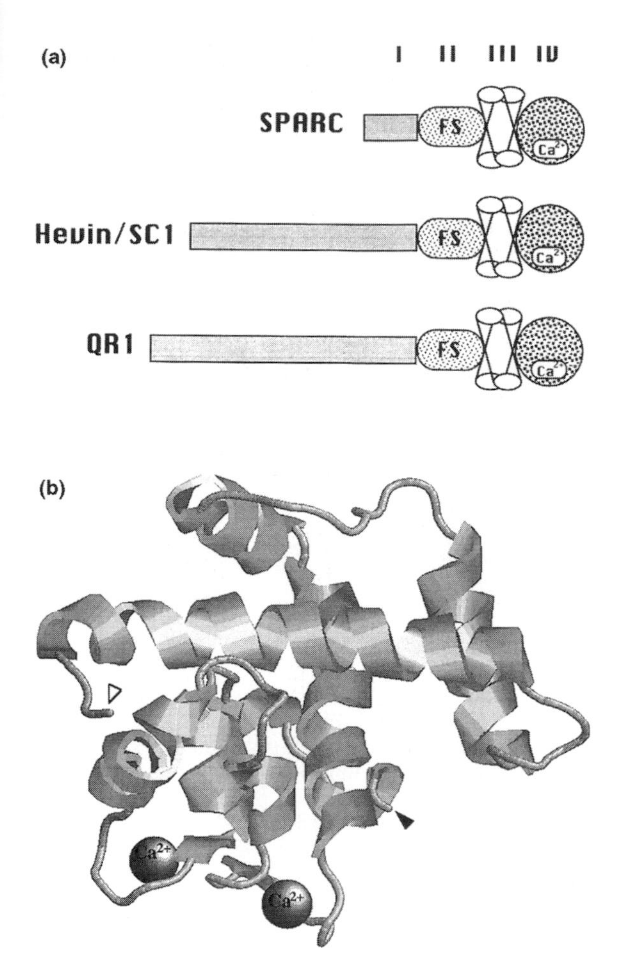

Figure 2. (a) Schematic domain structure of SPARC and its homologues, hevin/SC1 and QR1 (not to scale). The four domains are indicated by roman numerals. Domain I in all three proteins is acidic; FS, follistatin-like domain II; the location of Ca^{2+} bound within the EF-hand region of domain IV is indicated. (b) Ribbon model of domains III and IV of SPARC, which together comprise the extracellular, Ca^{2+}-binding (EC) domain.[5] Open arrowhead: N terminus; filled arrowhead: C terminus. (The image was prepared by Dr J. Bassuk with coordinates provided by Dr E. Hohenester.)

β1, interleukin 1, and colony-stimulating factor 1 increase steady-state levels of SPARC mRNA.[6]

Consistent with its expression in tissue remodelling in growth and disease, SPARC affects cell adhesion, cell-cycle progression, and gene expression *in vitro.* The cell types that have been studied in some detail are fibroblasts, vascular endothelial cells, and smooth muscle cells. In the presence of SPARC, cell spreading is inhibited and rounding of attached cells is induced through dissolution of focal contacts.[10,11] A secondary effect of the destabilization of focal adhesions in endothelial cells is a decrease in barrier function of confluent monolayers, an assay of vascular permeability *in vitro.*[12] SPARC delays G_1 to S transition of the cell cycle. For some growth-stimulatory cytokines [i.e. platelet-derived growth factors AB and BB,[13] and vascular endothelial growth factor (Kupprion and Sage; unpublished results)], this effect appears to result from their physical interaction with SPARC. Furthermore, SPARC binds to cells[14] and might therefore affect the cell cycle directly. Digestion of SPARC by extracellular proteases, such as plasmin, releases two peptides from the follistatin-like domain, lysyl-glycyl-histidyl-lysine (KGHK) and glycyl-histidyl-lysine (GHK)[15] which bind Cu^{2+}, stimulate proliferation of vascular endothelial cells, and increase angiogenesis *in vivo.* It is interesting in this respect that GHK has been described as a peptide in plasma which stimulates cell growth and enhances wound healing. Studies with vascular endothelial cells and fibroblasts indicate that SPARC induces the expression of proteinases (MMP-1, MMP-3, and MMP-9) and proteinase inhibitors (plasminogen activator inhibitor 1), and decreases the expression of fibronectin and thrombospondin-1.[6,16,17] These changes in gene expression occur late after the addition of SPARC and might involve a secondary mediator that is induced by SPARC. Bone SPARC (osteonectin) inhibits hydroxyapatite crystal growth *in vitro* and is believed to regulate bone mineralization.[18] More recently, a modest stimulation of plasminogen activation by SPARC was described.[19] Some of these functions can be mimicked by synthetic peptides of SPARC.[6]

SPARC also binds to serum albumin[20] and matrix proteins, including collagen I, III, IV, V, and VIII, thrombospondin-1, and vitronectin.[6] The significance of these interactions is not understood but might be related to matrix assembly and/or the modulation of cell adhesion to matrices containing these proteins.

SPARC contains several Ca^{2+}-binding sites. One of these sites, an EF-hand-like sequence, resides in domain IV and is stabilized by a disulphide bond. In addition, the N-terminal acidic domain I contains several low-affinity binding sites for Ca^{2+}.[21] Experiments in Ca^{2+}-free medium indicate that Ca^{2+} regulates the counteradhesive properties of SPARC, most likely because it promotes α-helix formation in the protein and thereby stabilizes the native conformation. However, the counteradhesive effect of SPARC on cultured cells cannot be attributed to the chelation of extracellular Ca^{2+} by this secreted protein.[22]

Purification

SPARC was purified initially from serum-free conditioned media of bovine aortic endothelial cells.[20] Certain tumour cell lines, such as murine parietal yolk sac teratocarcinoma cells (PYS-2), secrete high levels of SPARC and are thus used routinely as a source of this protein. PYS-SPARC is commercially available from Sigma (St Louis, MO) The protein is purified from conditioned medium by chromatography on DEAE–cellulose in the presence of urea, hypotonic precipitation at pH 5.5, and gel filtration on Sephacryl S-200. Recombinant SPARC and some of its domains have been obtained from 293 cells (a human embryonal kidney cell line), *S. cerevisiae,* and *E. coli.*[22] Bone SPARC (osteonectin) can be extracted from non-col-

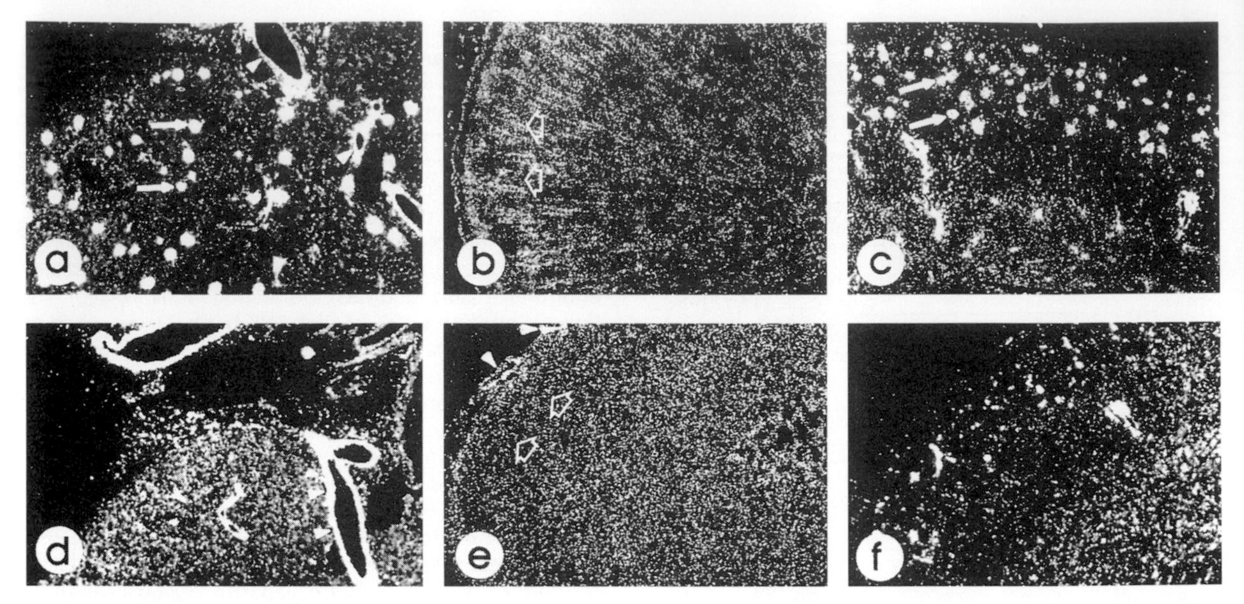

Figure 3. Distribution of SPARC and SC1 mRNA in murine tissues revealed by *in situ* hybridization. Comparison of SPARC and SC1 expression in kidney and adrenal gland of rats (a, b, d, and e) and in kidney of mice (c and f). SPARC mRNA is expressed in the glomeruli of the kidney (arrows in a and c). Both SPARC and SC1 are expressed in large vessels of the kidney (arrowheads in a and d). In the adrenal gland, SPARC transcripts are found predominantly in the medulla (arrows in b), whereas SC1 is expressed throughout the cortex and medulla (arrows in e) and in small- to medium-sized vessels of the adrenal capsule (arrowheads in e). (Reprinted from ref. 3, with permission.)

lagenous bone protein with Ca^{2+}-chelating agents, followed by chromatography on DEAE–Sephadex and gel filtration on Sephadex G-100.[23] However, the extraction of SPARC with chelating agents might lead to denaturation of the protein. Purity of the protein is assessed by denaturing polyacrylamide gel electrophoresis. SPARC migrates at an apparent molecular weight of 43 000 Da.

Activities

The counteradhesive effect of SPARC is assessed on bovine aortic endothelial cells, plated on plastic tissue culture dishes in the presence of SPARC, by a 'rounding index' for the proportion of rounded cells.[10,24] Focal adhesion disassembly is assessed by immunohistochemistry for a focal contact protein, such as vinculin, or by reflection interference microscopy.[11] The effect of SPARC on endothelial cell barrier function is demonstrated by measurement of the flux of ^{14}C-BSA across a confluent monolayer of bovine pulmonary artery endothelial cells.[12] Delay of G_1 to S transition of the cell cycle is assessed by thymidine incorporation by fibroblasts, endothelial cells, or vascular smooth muscle cells.[25] Cells synchronized by serum/growth factor deprivation are replated at subconfluent densities in the absence or presence of SPARC for 16–18 h, and are subsequently pulse-labelled for 2 h with ^{3}H-thymidine. For bone-derived SPARC, inhibition of hydroxyapatite crystal growth has been used to assess function in bone mineralization.[18]

Antibodies

Because the acidic domain I in SPARC is highly immunogenic, both polyclonal and monoclonal antibodies have been produced against this region of the protein. PYS-SPARC, bone SPARC, and SPARC-derived peptides have been used successfully to generate mono- and polyclonal antibodies. Frequently used antibodies against SPARC are listed, with references, in Table 1. Antibody $IIIA_3A_8$ (anti-osteonectin) is available through Hematologic Technologies Inc. (Essex Junction, VT); antibodies ON 1–1 and ON 4–2 are available through Pan Vera Corp. (Madison, WI).

Genes

SPARC is the product of a single gene in all species investigated. Chromosomal localization studies indicate that the SPARC locus is proximal to several cytokine, growth factor, and growth factor-receptor gene loci. Accession numbers for SPARC sequences in the SwissProt database are P09486 (human), 20112 (porcine), P13213 (bovine), P16975 (rabbit), P16975 (rat), P07214 (murine), S36377 (chicken), P36378 (*X. laevis*), and P34714 (*C. elegans*).

Mutant phenotypes/disease states

No mutant phenotypes or disease states have been identified to date. Mice rendered deficient in SPARC by

Table 1 Anti-SPARC antibodies

Antibody	Antigen	Applications[a]	Refs
Monoclonal			
ON 1–1, ON 4–2	Bone SPARC	ELISA, WB, IHC	26
MAb SSp2	Domain I peptide	ELISA, IHC	27
Mab 2	Bone SPARC	ELISA, IHC	28
AON 1,2,5,6	Bone SPARC	ELISA, WB, IHC, IP	29
$IIIA_3A_8$	Bone SPARC	ELISA, WB, IHC, IP	30
Polyclonal			
Anti-1.1	Domain I peptide	ELISA, WB	24
Anti-4.2	Domain IV peptide	WB	24

[a]ELISA, enzyme-linked immunosorbent assay; WB, Western blot; IHC, immunohistochemistry; IP, immunoprecipitation.

gene inactivation are viable and do not display gross physical abnormalities. However, preliminary data indicate that dermal wound healing is impaired in these animals, consistent with the proposed role of SPARC in tissue remodelling.

References

1. Girard, J. P. and Springer, T. A. (1995). *Immunity*, **2**, 113–23.
2. Johnston, I. G., Paladino, T., Gurd, J. W., and Brown, I. R. (1990). *Neuron*, **4**, 165–76.
3. Soderling, J. A., Reed, M. J., Corsa, A., and Sage, E. H. (1997). *J. Histochem. Cytochem.*, **45**, 1–13.
4. Casado, F. J., Pouponnot, C., Jeanny, J. C., Lecoq, O., Calothy, G., and Pierani, A. (1996). *Mech. Dev.*, **54**, 237–50.
5. Hohenester, E., Maurer, P., Hohenadl, C., Timpl, R., Jansonius, J. N., and Engel, J. (1996). *Nature Struct. Biol.*, **3**, 67–73.
6. Lane, T. F. and Sage, E. H. (1994). *FASEB J.*, **8**, 163–73.
7. Ledda, M. F., Adris, S., Bravo, A. I., Kairiyama, C., Bover, L., Chernajovsky, Y., *et al.* (1997). *Nature Med.*, **3**, 171–6.
8. Purcell, L., Gruia-Gray, J., Scanga, S., and Ringuette, M. (1993). *J. Exp. Zool.*, **265**, 153–64.
9. Schwarzbauer, J. E. and Spencer, C. S. (1993). *Mol. Cell. Biol.*, **4**, 941–52.
10. Sage, E. H., Vernon, R. B., Funk, S. E., Everitt, E. A., and Angello, J. (1989). *J. Cell Biol.*, **109**, 341–56.
11. Murphy-Ullrich, J. E., Lane, T. F., Pallers, M. A., and Sage, E. H. (1995). *J. Cell. Biochem.*, **57**, 341–50.
12. Goldblum, S. E., Ding, X., Funk, S. E.. and Sage, E. H. (1994). *Proc. Natl Acad. Sci. USA*, **91**, 3448–52.
13. Raines, E. W., Lane, T. F., Iruela-Arispe, M. L., Ross, R., and Sage, E. H. (1992). *Proc. Natl Acad. Sci. USA*, **89**, 1281–5.
14. Yost, J. C. and Sage, E. H. (1993). *J. Biol. Chem.*, **268**, 25790–6.
15. Lane, T. F., Iruela-Arispe, M. L., Johnson, R. S., and Sage, E. H. (1994). *J. Cell Biol.*, **125**, 929–43.
16. Hasselaar, P., Loskutoff, D. J., Sawdey, M., and Sage, E. H. (1991). *J. Biol. Chem.*, **266**, 13178–84.
17. Lane, T. F., Iruela-Arispe, M. L., and Sage, E. H. (1992). *J. Biol. Chem,*. **267**, 16736–45.
18. Romberg, R. W., Werness, P. G., Riggs, B. L., and Mann, K. G. (1986). *Biochemistry*, **25**, 1176–80.
19. Kelm, R. J., Swords, N. A., Orfeo, T., and Mann, K. G. (1994). *J. Biol. Chem.*, **269**, 30147–53.
20. Sage, H., Johnson, C., and Bernstein, P. (1984). *J. Biol. Chem.*, **259**, 3993–4007.
21. Engel, J., Taylor, J., Paulsson, M., Sage, H., and Hogan, B. (1987). *Biochemistry*, **26**, 6958–65.
22. Motamed, K., Bassuk, J., and Sage, E. H. (1996). In *Tenascin and counteradhesive molecules of the extracellular matrix* (ed. K.L. Crossin), pp. 127–144. Harwood Academic Publishers, Amsterdam.
23. Romberg, R. W., Werness, P. G., Lollar, P., Riggs, B. L., and Mann, K. G. (1985). *J. Biol. Chem.*, **260**, 2728–36.
24. Lane, T. F. and Sage, E. H. (1990). *J. Cell Biol.*, **111**, 3065–76.
25. Funk, S. E. and Sage, E. H. (1991). *Proc. Natl Acad. Sci. USA*, **88**, 2648–52.
26. Nakamura, S., Kamihagi, K., Satakeda, H., Katayama, M., Pan, H., Okamoto, H., *et al.* (1996). *Arthritis Rheum.*, **39**, 539–51.
27. Porter, P. L., Sage, E. H., Lane, T. F., Funk, S. E., and Gown, A. M. (1995). *J. Histochem. Cytochem.*, **43**, 791–800.
28. Malaval, L., Darbouret, B., Preaudat, C., Jolu, J. P., and Delmas, P. D. (1991). *J. Bone Miner. Res.*, **6**, 315–23.
29. Bolander, M. E., Robey, P. G., Fisher, L. W., Conn, K. M., Prabhakar, B. S., and Termine, J. D. (1989). *Calcif. Tissue Int.*, **45**, 74–80.
30. Stenner, D. D., Romberg, R. W., Tracy, R. P., Katzmann, J. A., Riggs, B. L., and Mann, K. G. (1984). *Proc. Natl Acad. Sci. USA*, **81**, 2868–72.

Alexander Riedlitz and E. Helene Sage
Department of Biological Structure,
University of Washington, Box 357420,
Seattle, WA 98195–7420, USA

Sponge ECM adhesion proteins

Two recently purified and characterized sponge extracellular matrix glycoproteins may represent the earliest extracellular matrix adhesion proteins to have arisen in metazoan evolution. These 68 and 210 kDa extracellular matrix proteins are ligands for the major proteoglycan of the sponge, which is named aggregation factor. The two extracellular matrix proteins have high affinities for aggregation factor and for unidentified receptors on the cell surface. Both participate in mediating the attachment of sponge cells to the extracellular matrix that surrounds them.

Homologous proteins

The 210 kDa extracellular matrix protein may be similar to a six-armed, tenascin-like sponge extracellular matrix adhesion protein that was previously isolated from the sponge species *Oscarella tuberculata*.[1]

Protein properties

Two extracellular matrix adhesion proteins from the sponge *Microciona prolifera* have been isolated from extracts of cell membranes and of the extracellular matrices of sponges.[2–4] A 68 kDa extracellular matrix protein is a monomeric ligand for the major sponge proteoglycan, the aggregation factor.[5–13] This 68 kDa adhesion protein is a cell-associated, extracellular matrix adhesion protein based on extensive studies which include intact cell proteolysis studies, Triton X-114 phase separation analyses, liposome incorporation studies and tissue distribution studies.[2] Peptide *N*-glycosidase digestion studies and lectin affinity chromatography analyses indicate that this protein is a glycoprotein. The protein has an affinity for peanut agglutinin but not concanavalin A, suggesting the presence of complex-type glycans[2–3] The presence of internal disulphide bonds is indicated by significantly altered migration in polyacrylamide gels under reducing conditions, suggesting a globular structure. The 68 kDa sponge adhesion protein binds to the sponge cell surface with a high affinity ($K_d = 6 \times 10^{-8}$ M) and to the extracellular matrix proteoglycan aggregation factor with a high affinity ($K_d = 2 \times 10^{-9}$ M). This protein participates in mediating the attachment of sponge cells to aggregation factor. It competitively inhibits the adhesion of cells to aggregation factor when present in excess quantities.[3]

A 210 kDa extracellular matrix protein has also been isolated from cell membrane and extracellular matrix extracts. This protein is more predominantly found in the extracellular matrix of sponges than on cell surfaces. This protein was also shown to be a cell-associated, extracellular matrix adhesion protein based on extensive studies which include intact cell proteolysis studies, Triton X-114 phase separation analyses, liposome incorporation studies, and tissue distribution studies.[2] Peptide *N*-glycosidase digestion studies and lectin affinity chromatography analyses indicate that this protein is a glycoprotein with a very small glycan content. The protein has an affinity for peanut agglutinin and lentil lectin, but not concanavalin A, suggesting the presence of complex-type glycans. The presence of intramolecular disulphide bonds is indicated by a loss in binding of the 210 kDa protein to the aggregation factor after reduction, but not by a shift in migration on SDS–polyacrylamide gels. The 210 kDa extracellular matrix protein is found primarily as a pentamer of 210 kDa monomers, but may also exist in other multimeric states. This protein has an isoelectric point of 4.3. The 210 kDa protein inhibits aggregation factor binding to the cells and binds directly to aggregation factor with a high affinity ($K_d = 7 \times 10^{-9}$ M). This protein mediates the attachment of sponge cells to aggregation factor. It competitively inhibits the adhesion of cells to aggregation factor when present in excess quantities.[4]

Purification

The 210 kDa sponge extracellular matrix protein can be purified by column chromatography of detergent extracts of sponge cell membranes.[2,4] A yield of 500 μg of purified 210 kDa protein was obtained from a starting material of membranes from 10^{12} sponge cells. A mild detergent extract of membranes is clarified by dialysis and chromatographed successively on MonoQ anion exchange, Sepharose 6B and Sepharose 3B columns. The purification of the 210 kDa protein can be followed by a non-reducing far-Western blotting assay in which iodinated, purified aggregation factor is used as a probe.[2,4]

The 68 kDa extracellular matrix protein can be purified to homogeneity by affinity chromatography on aggregation factor–Sepharose[3] or by traditional chromatographic methods from detergent extracts of sponge cell membrane preparations using a far-Western blotting strategy with iodinated aggregation factor as a probe.[2,4]

Activities

The 68 and 210 kDa extracellular matrix proteins have high affinities for aggregation factor and for the cell surface. Both participate in mediating the attachment of sponge cells to the extracellular matrix that surrounds them and is composed of proteoglycans, collagen and other structural proteins.[14]

Purified 68 kDa and the 210 kDa extracellular matrix proteins bind to aggregation factor with high affinities,

$K_d = 2 \times 10^{-9}$ M and $K_d = 7 \times 10^{-9}$ M, respectively.[3–4] This high affinity interaction is preserved in solution, as well as on non-reducing Western blots and dot blots of cell extracts and serves as the basis for a convenient means of protein detection.[2–4] The 68 kDa protein has been shown to bind directly to the cell surface with a high affinity, $K_d = 6 \times 10^{-8}$ M^2. This activity has not been determined for the 210 kDa protein but it, like the 68 kDa protein, can be isolated from extracts of sponge cell membranes.

Both the 210 kDa and the 68 kDa proteins inhibit cellular adhesion mediated by aggregation factor when preincubated with aggregation factor prior to addition of cells, presumably by blocking available cell binding sites on the aggregation factor molecule. The 210 kDa protein inhibits cellular adhesion with an IC_{50} of 8 nM and the 68 kDa protein inhibits cell adhesion with an IC_{50} of 5.5 nM.[3,4]

Although the interaction of these extracellular matrix proteins with the purified aggregation factor is well characterized,[2–4] little is known about the molecules on the cell surface that bind to these proteins. It is possible that integrin-like molecules may anchor these extracellular matrix molecules to the cell surface of sponges. Integrin subunits have been recently sequenced from sponges.[15] A potential model suggests that aggregation factor binds to the 210 and 68 kDa structural glycoproteins discussed in this review and these then bind to specific cell surface receptors. Together with the fibrillar and non-fibrillar collagens,[16–18] these extracellular matrix molecules are likely to play roles in cellular migration through the extracellular matrix and in sponge tissue homeostasis.

Antibodies

None available.

Genes

No information available.

Mutant phenotypes

No information available.

Structure

Not available.

References

1. Humbert-David, N. and Garrone, R. (1993). *Eur. J. Biochem.*, **216**, 225–60.
2. Varner, J. A., Burger, M. M., and Kaufman, J. F. (1988). *J. Biol. Chem.*, **263**, 8498–508.
3. Varner, J. A. (1995). *J. Cell Sci.*, **108**, 3119–126.
4. Varner, J. A. (1996). *J. Biol. Chem.*, **271**, 16119–125.
5. Humphreys, T. (1963). *Dev. Biol.*, **8**, 27–47.
6. Henkart, P., Humphreys, S., and Humphreys, T. (1973). *Biochemistry*, **12**, 3045–50.
7. Misevic, G., Jumblatt, J. and Burger, M. M. (1982). *J. Biol. Chem.*, **257**, 6931–36.
8. Misevic, G. and Burger, M. M. (1986). *J. Biol. Chem.*, **261**, 2853–9.
9. Misevic, G. and Burger, M. M. (1993). *J. Biol. Chem.*, **268**, 4922–9.
10. Misevic, G. and Burger, M. M. (1990). *J. Biol. Chem.*, **265**, 20577–84.
11. Jumblatt, J., Schlup, V., and Burger, M. M. (1980). *Biochemistry*, **12**, 3045–50.
12. Muller, W. E. G., Conrad, G., Pondeljack, V., Steffan, R., and Zahn, R. (1982). *Tiss. Cell*, **14**, 219–33.
13. Muller, W. E. G. (1982). *Int. Rev. Cytol.*, **77**, 129–81.
14. Garrone, R., Exposito, J. Y., Franc, J. M., Humberrt-David, N., Qin, L., and Tillet, E. (1993). *Comptes Rendus des Séances de la Sociéte de Biologie et de Sesfiliales*, **187**, 114–23.
15. Pancer, Z., Kruse, M., Muller, I., and Muller, W. E. G. (1997). *Mol. Biol. Evol.*, **14**, 391–8.
16. Aho, S., Turakainen, H. Onnela, M. -L. and Boedtker, H. (1993). *Proc. Natl Acad. Sci. USA*, **90**, 7288–92.
17. Exposito, J. -Y., Le Guellec, D., Lu, Q., and Garrone, R. (1991). *J. Biol. Chem.*, **266**, 21923–8.
18. Exposito, J. Y., and Garrone, R. (1990). *Proc. Natl Acad. Sci. USA*, **87**, 6669–73.

Judith A. Varner
University of California, San Diego,
Department of Medicine/Cancer Center,
9500 Gilman Drive, La Jolla, CA 92093–0684,
USA.

Tenascins

To date five members of the tenascin gene family have been described: tenascin-C, tenascin-R, tenascin-X, tenascin-Y and tenascin-W.[1–3,36] They represent large multimeric extracellular matrix proteins each of them consisting of identical subunits built from variable numbers of repeated domains. These include heptad repeats, EGF-like repeats, fibronectin type III domains and a C-terminal globular domain shared with the fibrinogens. Tenascin-R is an extracellular matrix component of the nervous system, whereas tenascin-X and tenascin-Y are prominently expressed in muscle connective tissues. Tenascin-C is present in a large number of developing tissues including the nervous system. It is abundant in adult ligaments and tendons but is absent from skeletal and heart muscle. Tenascin-C was the first tenascin to be discovered, partly because of its overexpression in tumours.

Synonymous names

Tenascin-C: GMEM (glial mesenchymal extracellular matrix antigen), myotendinous antigen, tenascin, cytotactin, hexabrachion, brachionectin, neuronectin. Tenascin-R: restrictin, J1, J1–160/180, janusin. Tenascin-X: human gene X, tenascin-like gene, flexilin. Tenascin-Y and tenascin-W: none.

Protein properties

The prototype of the tenascins is tenascin-C (for reviews see refs 1, 2, 4). It is a hexameric extracellular matrix protein with subunit molecular weights in the range of 190–300 kDa depending on the species analysed. Different subunit variants are generated by alternative splicing of one common primary transcript.[5] The model and the electron micrograph shown in Figs. 1 and 2(a), respectively, reveal the domain organization of tenascin-C. Each subunit is built from repeated domains arranged like beads on a string. These include heptad repeats that enable the tenascins to trimerize through generation of triple coiled-coil α helices. The adjacent tenascin-type EGF-like repeats are the shortest versions found in any protein (the only other proteins with the same types of EGF-like repeats are reelin,[6] and the *Drosophila* proteins Ten-a[7] and Ten-m[8]). Next are fibronectin type III domains, some of which can be alternatively spliced. Finally, a domain homologous to the fibrinogens (which is also found in other proteins, such as i.e. ficolins[9] or angiopoietin-1[10]) is present at the C terminus.

The other tenascins are constructed from the same types of domains which are arranged in the same order, but vary in number, as can be seen in Fig. 1. Tenascin-C and tenascin-R are equally related to each other as is tenascin-X to tenascin-Y. Tenascin-Y is peculiar in that it contains an additional domain rich in the tripeptide SPX which is linked to one half of a fibronectin type III repeat. The fibronectin type III repeats of tenascin-X and tenascin-Y are closely related, and in addition show an extremely high conservation within each molecule. In the extreme case of tenascin-Y their sequence identity is in the range of 70 to 100 per cent. The fibronectin type III repeats of tenascin-C and tenascin-R are more diverse and in contrast to the other two tenascins can be aligned collinearly according to highest similarity.[11,12] In contrast to tenascin-C which occurs as hexamers, tenascin-R exists as trimers as can be seen on the electron micrographs in Fig. 2(c). The oligomeric structure of tenascin-X and tenascin-Y has not been fully analysed yet.

The distinctive tissue distribution of tenascin-C has provoked much interest in this protein. During embryogene-

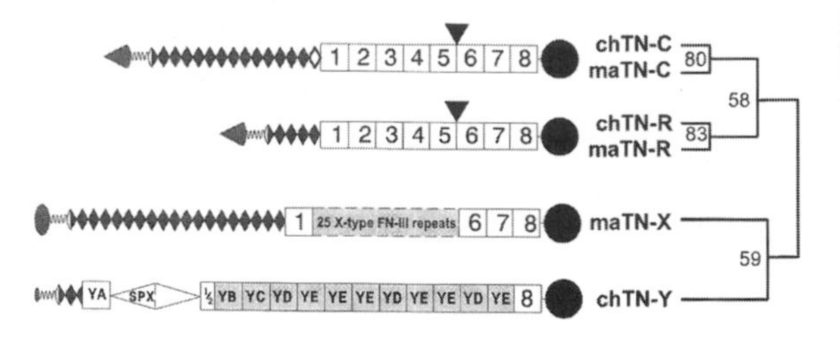

Figure 1. Modular structure of one subunit of each of the four types of tenascins: chicken, and mammalian tenascin-C (chTN-C, and maTN-C), chicken, and mammalian tenascin-R (chTN-R, and maTN-R), mammalian tenascin-X (maTN-X), and chicken tenascin-Y (chTN-Y). Their relationship is depicted in the dendrogram to the right, and the numbers refer to the percentage identity found within the common C-terminal region. The following domains are represented: N termini (pie segments, or ellipses), heptad repeats (wavy line), EGF-like repeats (diamonds), fibronectin type III repeats (boxes), fibrinogen homology (filled circle). The filled triangles between the fibronectin type III repeats 5, and 6 of TN-C, and TN-R mark the position where additional fibronectin type III repeats can be included by alternative splicing. For tenascin-C up to seven additional repeats have been found, whereas in tenascin-R only one extra repeat is known. The shaded fibronectin type III repeats in TN-X, and TN-Y show an exceptionally high similarity, in the range of 60–100 per cent. Tenascin-Y contains an additional domain rich in the tripeptide sequence SPX (rhomboid labelled SPX) followed by a half fibronectin type III repeat.

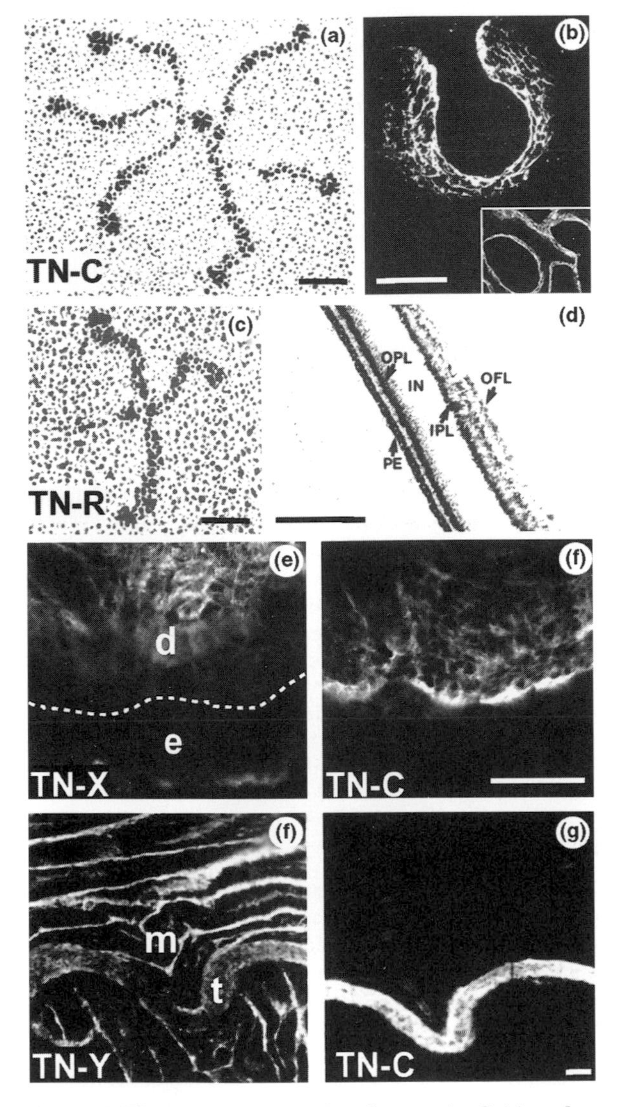

Figure 2. Electron micrographs of tenascin-C (a) and tenascin-R (c) after rotary shadowing. Bars represents 20 μm. Immunostaining of tenascin-C in the mesenchyme around an embryonic rat mammary gland (b), and in the stroma surrounding mammary tumour epithelium (insert), and of tenascin-R (d) in the outer plexiform layer (OPL), inner plexiform layer (IPL), and optic fibre layer (OFL) of E13 embryonic chicken retina. Bars represent 200 μm. Reciprocal distribution of tenascin-X (e) and tenascin-C (f) in E15 mouse skin. Whereas tenascin-X is present in the deeper part of the dermis ('d') tenascin-C is most concentrated below the epithelium ('e'). Bar represents 100 μm. Tenascin-Y (g) is present between muscle fibres ('m') and along the tendon ('t'). Tenascin-C (H) is absent from the muscle tissue, but is strongly stained within the tendon. Bar represents 100 μm.

sis tenascin is transiently expressed in the dense mesenchyme surrounding many developing organs such as the mammary gland (Fig. 2(b)), tooth, and kidney. It is present in embryonic cartilage and in the adult becomes confined to the perichondrium and periosteum, ligaments, tendons (Fig. 2(h)) and myotendinous junctions. Also smooth muscle expresses tenascin-C, whereas other adult tissues such as heart and skeletal muscle or epithelial organs contain very little tenascin-C. In contrast, epithelial tumours show very strong staining in their surrounding tumour stroma (Fig. 2(b), insert). In the nervous system, tenascin-C is made by glial cells and it is also highly expressed in gliomas.

Tenascin-R shows a very restricted expression pattern during development of the nervous system. In the spinal cord of chickens, for example, tenascin-R is only found around motor neurones and in lower concentration on motor axons. In the embryonic cerebellum it appears to be expressed primarily in the prospective white matter, and in the developing retina (Fig. 2(d)) it is localized to the outer and inner plexiform layers and, at later stages, also to the optic fibre layer.[13] In the mouse, highest tenascin-R expression has been observed in oligodendrocytes at periods of myelinization.[12]

Tenascin-X was originally discovered as a gene partially overlapping with the opposite strand of the human steroid 21-hydroxylase gene in the major histocompatibility complex (MHC) class III locus.[14] In human tissues the highest expression of tenascin-X transcripts was found in muscle and testis.[14] Tenascin-Y in the chicken shows characteristics highly similar to tenascin-X in mammals. Immunohistochemistry and *in situ* hybridization of chicken[3] and mouse[15] tissues have shown expression of tenascin-Y and tenascin-X, respectively, in the connective tissue between muscle fibres (Fig. 2(g)) as well as in the epicardium and around connective tissue cells interspersed in heart muscle. In the skin, tenascin-X is present in the deeper part of the dermis, while tenascin-C is most prominent just below the epithelium (Fig. 2(e,f)). No expression of tenascin-X nor tenascin-Y is found in the nervous system.

Purification

Tenascin-C is usually purified from conditioned media of cells that express high levels of tenascin-C, such as melanoma, glioma, or fibroblastic cells, either by immunoaffinity chromatography using monoclonal anti-tenascin-C antibodies,[16] or by a more conventional biochemical method involving affinity chromatography to hydroxyapatite.[17]

Tenascin-R can be purified from brain extracts by immunoaffinity chromatography using immobilized monoclonal anti-tenascin-R antibodies.[13]

Tenascin-X and tenascin-Y have been purified in very small quantities only from conditioned medium of a renal carcinoma cell line,[15] fetal bovine skin[37] or from heart extract,[3] respectively, using immunoaffinity chromatography.

Activities

A common feature of tenascin-C, -X, and -Y is their binding to heparin, which in the case of tenascin-C has been shown to be mediated by the fibrinogen globe. The physiological consequence of that could be the binding of these tenascins to heparan sulphate proteoglycans, as has been reported for the binding of tenascin-C to syndecan.

Tenascin-C has been shown to be anti-adhesive for many cells and to promote cell rounding. On the other hand it is able to promote neurite outgrowth. These activities were shown to depend on the concerted action of different types of domains of tenascin-C.[18] Growth promotion, haemagglutination, immunosuppression of T-cells, and the promotion of angiogenesis and chondrogenesis are other reported effects of tenascin-C. These activities of tenascin-C are assumed to be mediated by its binding to the following types of cellular receptors: integrins (α2b1, αvβ3, α8β1, α9β1), syndecan, contactin/F11, annexin II or the receptor type tyrosine phosphatase β.

Tenascin-R has been shown to bind to F11/F3 through its fibronectin type III domains 2 and 3, and to either enhance the F11-mediated neurite outgrowth[19] or to promote the repulsion of cerebellar neurones by this interaction.[20] Both tenascin-R and tenascin-C bind the differentiation factor CALEB, a neural transmembrane protein involved in neurite formation.[21]

Antibodies

Polyclonal antibodies raised against bacterial fusion proteins of parts of mouse tenascin-X[15] and chicken tenascin-Y[3] have been reported. Polyclonal and monoclonal antibodies to chicken tenascin-R[13] and monoclonal antibodies against mouse tenascin-R[22] have been described. A large number of polyclonal antisera as well as monoclonal antibodies against human, mouse, and chicken tenascin-C can be found in the literature. Monoclonal antibodies against human (DAKO, Denmark), mouse (Sigma, St Louis, Missouri), and chicken (antibody M1; Developmental Studies Hybridoma Bank, The University of Iowa) tenascin-C are commercially available. The monoclonal anti-chicken tenascin-C (antibody M1) cross-reacts with amphibian and fish tenascin-C.

Genes

Complete cDNA sequences (accession numbers in brackets) are available of the following tenascins: human (X78565), mouse (D90343), pig (X61599), and chicken (M23121) tenascin-C; human (Z67996), rat (Z18630), and chicken (X64649) tenascin-R; bovine tenascin-X (Y11915); chicken tenascin-Y (X99062) and zebrafish tenascin-W (AJ001423). The entire genomic region encoding mouse tenascin-X (AB010266) has been sequenced and a large part of genomic sequences containing the exons encoding human tenascin-X have been reported.[14] In addition partial sequences are available of rat (U15550), zebra fish (U14940), frog (X68620), and newt (M76615) tenascin-C as well as of mouse (X73959) and rat (U24489) tenascin-X.

Genomic clones containing the 5′ regions of human, mouse, and chick tenascin-C have been isolated and the promoter sequences have been compared.[23] The chromosomal location of tenascin-C is in humans 9q32–q34,[24] in pig 1q21.1–1q21.3,[25] and in mouse, chromosome 4.[26] The gene structure of human tenascin-R[27] has been elucidated and the chromosomal location has been assigned to 1q23–24.[28] The genes of human,[14] pig,[29] and mouse[15] tenascin-X are located within the MHC class III locus of either species.

Mutant phenotype/disease states

It is unclear what the consequences of mutations in tenascin genes could be. Transgenic mice lacking tenascin-C have no obvious phenotype,[30] but recently certain deficiencies in these mice have been reported.[31,38,39] Zebrafish with a chromosomal deletion including their tenascin-C gene show an embryonic lethal phenotype,[32] but involvement of other genes cannot be excluded yet. Indirect evidence through the analysis of patients with deletions within the MHC III locus has indicated that an intact tenascin-X gene is required for survival.[33] In contrast, a patient with a tenascin-X deficiency was found to suffer from Ehlers–Danlos syndrome.[40]

Structure

The structure of such large proteins as the tenascins cannot easily be solved, but it is assumed that each domain of the tenascins is an independently folding unit and therefore the structure of the entire protein can be inferred from the substructures of its domains. So far the structure of only one tenascin domain, namely the third fibronectin type III repeat of human tenascin-C, has been determined by MAD phasing at a 1.8 Å resolution.[34] This structure can be used as a basis on which to model all other fibronectin type III repeats. The C-terminal globular domain of all tenascins is homologous to the respective part in γ-fibrinogen and may therefore resemble in structure the one determined for the C-terminal fragment of this fibrinogen subunit.[35] The EGF-like repeats are assumed to adopt a similar structure as epidermal growth factor[4] and the short regions with heptad repeats are assumed to function in coiled-coil formation,[5] resulting in an oligomeric structures of tenascins as visualized by electron microscopy after rotary shadowing of the molecules (cf. Fig. 2(a,c)).

References

1. Erickson, H. P. (1993). *Curr. Opin. Cell Biol.*, **5**, 869–76.
2. Chiquet-Ehrismann, R. (1995). *Experientia*, **51**, 853–62.
3. Hagios, C., Koch, M., Spring, J., Chiquet, M., and Chiquet-Ehrismann, R. *J. Cell Biol.*, **134**, 1499–512.
4. Chiquet-Ehrismann, R. (1990). *FASEB J.*, **4**, 2598–604.

5. Spring, J., Beck, K., and Chiquet-Ehrismann, R. (1989). *Cell*, **59**, 325–34.
6. D'Arcangelo, G., Miao, G. G., Chen, S. C., Soares, H. D., Morgan, J. I., and Curran, T. (1995). *Nature*, **374**, 719–23.
7. Baumgartner, S. and Chiquet-Ehrismann, R. (1993). *Mech. Dev.*, **40**, 165–76.
8. Baumgartner, S., Martin, D., Hagios, C., and Chiquet-Ehrismann, R. (1994). *EMBO J.*, **13**, 3728–40.
9. Ichijo, H., Hellman, U., Wernstedt, C., Gonez, L. J., Claesson-Welsh,L., Heldin, C. H., and Miyazono, K. (1993). *J. Biol. Chem.*, **268**, 14505–13.
10. Davis, S., Aldrich, T. H., Jones, P. F., Acheson, A., Compton, D. L., Jain, V.,*et al.* (1996). *Cell*, **87**, 1161–9.
11. Fuss, B., Wintergerst, E. S., Bartsch, U., and Schachner, M. (1993). *J. Cell Biol.*, **120**, 1237–49.
12. Noerenberg, U., Wille, H., Wolff, J. M., Frank, R., and Rathjen, F. G. (1992). *Neuron*, **8**, 849–63.
13. Rathjen, F. G., Wolff, J. M., and Chiquet-Ehrismann, R. (1991). *Development*, **113**, 151–64.
14. Bristow, J.,Tee, M. K., Gitelmann, S. E., Mellon, S. H., and Miller, W. L. (1993). *J. Cell Biol.*, **122**, 265–78.
15. Matsumoto, K., Saga, Y., Sakakura, T., and Chiquet-Ehrismann, R. (1994). *J. Cell Biol.*, **125**, 483–93.
16. Schenk, S. and Chiquet-Ehrismann, R. (1994). *Meth. Enzymol.*, **245**, 52–61.
17. Saginati, M., Siri, A., Balza, E., Ponassi, M., and Zardi L. (1992). *Eur. J. Biochem.*, **205**, 545–9.
18. Fischer, D., Brown-Luedi, M., Schulthess, T., and Chiquet-Ehrismann, R. (1997). *J. Cell Sci.*, **110**, 1513–22.
19. N'renberg, U., Hubert, M., Brümmendorf, T., Tarnok, A., and Rathjen, F. G. (1995). *J. Cell Biol.*, **130**, 473–84.
20. Pesheva, P., Gennarini, G., Goridis, C., and Schachner, M. (1993). *Neuron*, **10**, 69–82.
21. Schumacher, S., Volkmer, H., Buck, F., Otto, A., Tarnok, A., Roth, S., and Rathjen, F. G. (1997). *J. Cell Biol.*, **136**, 805–906.
22. Pesheva, P., Spiess, E., and Schachner, M. (1989). *J. Cell Biol.*, **109**, 1765–78.
23. Chiquet-Ehrismann, R., Hagios, C., and Schenk, S. (1995). *BioEssays*, **17**, 873–8.
24. Gulcher, J. R., Alexakos, M. J., Le Beau, M. M., Lemons, R. S., and Stefansson, K. (1990). *Genomics*, **6**, 616–22.
25. Awata, T., Yamakuchi, H., and Yasue, H. (1995). *Cytogenet. Cell Genet.*, **69**, 33–4.
26. Pilz, A., Moseley, H., Peters, J., and Abbott, C. (1992). *Mamm. Genome*, **3**, 247–9.
27. Leprini,A., Gherzi, R., Querze, G., Viti, F., and Zardi, L. (1996). *J. Biol. Chem.*, **271**, 31251–4.
28. Carnemolla, B., Leprini, A., Borsi, L., Querze, G., Urbini, S., and Zardi, L. (1996). *J. Biol. Chem.*, **271**, 8157–60.
29. Burghelle-Mayeur, C., Geffrotin, C., and Vaiman, M. (1992). *Biochim. Biophys. Acta*, **1171**, 153–61.
30. Saga, Y., Yagi, T., Ikawa, Y., Sakakura, T., and Aizawa, S. (1992). *Genes Dev.*, **6**, 1821–31.
31. Fukamauchi, F., Mataga, N., Wang, Y. J., Sato, S., Yoshiki, A., and Kusakabe, M. (1997). *Biochem. Biophys. Res. Commun.*, **231**, 356–9.
32. Fritz, A., Rozowski, M., Walker, C., and Westerfield, M. (1996). *Genetics*, **144**, 1735–45.
33. Morel, Y., Bristow, J., Gitelman, S. E., and Miller, W. L. (1989). *Proc. Natl Acad. Sci. USA*, **86**, 6582–6.
34. Leahy, D. J., Hendrickson, W. A., Aukhil, I., and Erickson, H. P. (1992). *Science*, **258**, 987–91.
35. Yee, V. C., Pratt, K. P., Cote, H. C., Trong, I. L., Chung, D. W., Davie, E. W., *et al.* (1997). *Structure*, **5**, 125–38.
36. Weber, P., Montag, D., Schachner, M., and Bernhardt, R. R. (1998). *J. Neurobiol.*, **35**, 1–16.
37. Elefteriou, F., Exposito, J.-Y., Garrone, R., and Lethias, C. (1997). *J. Biol. Chem.*, **272**, 22866–74.
38. Nakao, N., Hiraiwa, N., Yoshika, A., Ike, F., and Kusakabe, M. (1998). *Am. J. Pathol.*, **152**, 1237–45.
39. Ohta, M., Sakai, Y., Saga, Y., Aizawa, S., and Saito, M. (1998). *Blood*, **91**, 4074–83.
40. Grant, H., Gong, Y., Liu, W., Dettman, R. W., Curry, C. J., Smith, L., Miller, W. L., and Bristow, J. (1997). *Nature Genet.*, **17**, 104–8.

Ruth Chiquet-Ehrismann
Friedrich Miescher Institute,
PO Box 2543, CH-4002 Basel,
Switzerland

Thrombospondins

The thrombospondins are a family of five extracellular calcium- and glycosaminoglycan-binding proteins that function during embryogenesis, angiogenesis, and wound healing.[1-4] Comparison of the sequences of the thrombospondins reveals that calcium binding sites and some sites for protein–protein interaction have been highly conserved during vertebrate evolution. These data imply that a common function of the thrombospondins is to direct the formation of molecular assemblies at the cell surface which modulate protease and growth factor activity to evoke specific cellular responses.

Protein properties

Molecular cloning of thrombospondin-1 revealed that each subunit of the trimer consists of multiple domains: amino- and carboxyl-terminal globular domains; a region of sequence similarity to procollagen; and three types of repeated sequence motifs, designated type 1, type 2, and type 3 repeats (Fig. 1). Molecular cloning of thrombospondin-2, -3, -4, and cartilage oligomeric matrix protein (COMP) indicates that the thrombospondins can be divided into two subgroups on the basis of their mole-

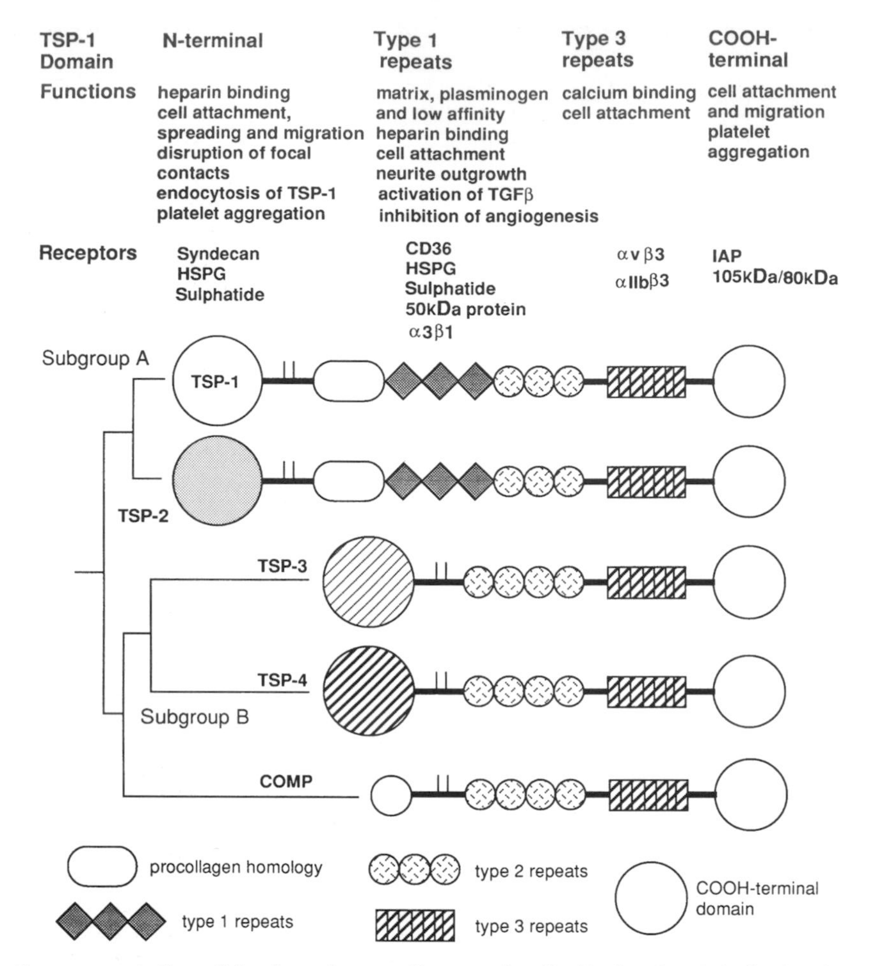

Figure 1. Schematic representation of the thrombospondin gene family. Each subunit is depicted as composed of multiple types of repeated amino acid sequence motifs. The phylogenic tree is shown at the left with the correct branching order, and arbitrary branch lengths. The functions of thrombospondin-1 are shown at the top of the figure along with the receptors that have been shown to interact with each domain.

cular architecture. Thrombospondin-2 has the same set of structural domains as thrombospondin-1 (subgroup A). By contrast, thrombospondin-3, -4, and COMP (subgroup B) lack the type 1 repeats and the region of homology with procollagen and contain an additional type 2 repeat. COMP also lacks the N-terminal domain. In addition, the subgroup A proteins are trimeric and the Subgroup B proteins are pentameric. The subgroup A proteins are members of the TSR (thrombospondin repeat) supergene family which is composed of a diverse group of proteins that contain the type 1 repeats. The most recent additions to this superfamily are proteins that are involved in axon guidance: the *unc*-5 gene product of *C. elegans*, F-spondin, and SCO-spondin.[5–7] In addition, the WSXWSXW sequence that is included in the type 1 repeats is found in the cytokine receptor family. The type 1 repeats are involved in cell binding and protein–protein interactions.

The type 3 repeats and the C-terminal domain display the highest level of conservation, both between a given type of thrombospondin from different species and between different family members. The type 3 repeats are a contiguous set of calcium-binding sites that resemble calmodulin in the positioning of oxygenated residues.[8] Thrombospondin-1 has been shown to bind 36 calcium ions.[9] A similar value would be expected for thrombospondin-2; the subgroup B thrombospondins, which are pentamers, may be able to bind 60 calcium ions.

The various functional activities of thrombospondin-1 have been mapped to specific structural regions (Fig. 1). The N-terminal domain of thrombospondin-1 contains a high-affinity binding site for heparin sulphate-containing proteoglycans, including syndecan (see chapter on Syndecan, this volume). Site-directed mutagenesis of thrombospondin-1 indicates that two areas of sequence that are rich in basic residues participate in heparin binding. Mutagenesis of either of these areas results in a decrease in affinity for heparin. Although the sequence identity in the heparin-binding motifs of the thrombospondins is relatively low, thrombospondin-2, -3, and -4 have all been shown to bind to heparin.

Thrombospondin-1 binds to CD36, the $\alpha_v\beta_3$ integrin, the integrin-associated protein (IAP), and one other partially characterized receptor, in addition to cell-surface proteoglycans.[1,10–12] The $\alpha_3\beta_1$ integrin functions as a thrombospondin-1 receptor on neurones.[13] The CSVTCG sequence within the type 1 repeats of thrombospondin-1 has been shown to interact with CD36 and a 50 000 Da polypeptide.[14] In addition, this sequence mediates the binding of thrombospondin-1 to gp120 of HIV.[26] Through this interaction, thrombospondin-1 suppresses HIV infectivity.[26] The CD36 binding sequence is conserved in the subgroup A thrombospondins but is not present in the subgroup B members.

The $\alpha_v\beta_3$ integrin on platelets and several types of cultured cells has been found to interact with the RGD sequence within the last type 3 repeat of thrombospondin-1 and -2.[1] The RGD sequence is conserved in the subgroup A thrombospondins. In the subgroup B thrombospondins, the RGD sequence is found in some species of thrombospondin-4 and COMP, but it has not been found in thrombospondin-3 in any species.

The ability of G361 melanoma cells, human intestinal smooth muscle cells, epidermal keratinocytes, and MG-63 osteosarcoma cells to attach to thrombospondin-1 is inhibited by antibodies to the C-terminal domain. In addition, one of these monoclonal antibodies inhibits platelet aggregation. Two peptide sequence motifs that are present in the C-terminal domain of thrombospondin-1, FYVVMWK and IRVVM, function as cell binding sites.[10,12] The FYVVMWK sequence is present in 16 of the 17 thrombospondin sequences that have been determined to date. Bovine COMP has the sequence FYVLMWK in this region. The IAP has recently been shown to interact with the VVM sequence. The conservation of the FYVVMWK sequence in most of the thrombospondins raises the possibility that a single receptor could bind to them all.

Purification

Thrombospondin-1 can be purified from the supernatant of thrombin or ionophore A23187 treated platelets.[15] Most isolation procedures utilize heparin–Sepharose affinity chromatography and a step that separates on the basis of size (gel filtration or sucrose gradient centrifugation). Since all of the thrombospondins are high molecular weight heparin-binding proteins, these approaches have also been used to purify various recombinant thrombospondins. MonoQ anion exchange chromatography is also very effective for purifying the released platelet thrombospondin-1 and for purification from the medium from cultured endothelial cells and fibroblasts. Affinity chromatography on fibrinogen–Sepharose has also been used to purify thrombospondin-1. COMP is purified from cartilage by EDTA extraction followed by affinity chromatography on wheat germ agglutinin– Sepharose and gel filtration chromatography.[16]

Activities

All of the thrombospondins bind calcium and heparin. The high level of sequence conservation of the type 3 repeats indicates that calcium sequestration at the cell surface and within the extracellular matrix is a common function of the thrombospondins. Thrombospondins, like other adhesive proteins, have been shown to modulate cell attachment, migration and proliferation, and neurite outgrowth.[1,17] Thrombospondin-1 is the endogenous platelet lectin which has been shown to agglutinate trypsinized, formalin-fixed erythrocytes. Platelet–monocyte interactions, as measured by rosetting assays, also involve thrombospondin-1. Thrombospondin-1 also supports attachment of normal and transformed cells, as well as malaria parasitized erythrocytes. Whereas most cells do not spread on thrombospondin-1, some myoblast, melanoma, and carcinoma cell lines do spread on thrombospondin-1-coated substrates. Thrombospondin-2 and -4 have also been shown to support cell attachment.[18,19] Thrombospondin-1 reportedly supports migration of human melanoma, carcinoma, and granule cells. It has also been reported to support smooth muscle cell proliferation. In contrast, thrombospondin-1 inhibits endothelial cell proliferation and thrombospondin-1 and –2 inhibit angiogenesis.[20,21] The anti-angiogenic activity of thrombospondin-1 is reportedly mediated by CD36.[27] The ability to activate transforming growth factor β seems to be specific to thrombospondin-1.[22] Thrombospondin-1 'knockout' mice have a mild curvature of the spine and abnormalities in the epithelium of the lung.[28] All of the mice with a 129 Sv background develop pneumonia. The lung abnormalities can be corrected by daily injections of a synthetic peptide that contains the transforming growth factor β (TGFβ)-activating sequence of thrombospondin-1.[29] These data indicate that thrombospondin-1 is an important activator of TGFβ *in vivo*.

Thrombospondin-2 'knockout' mice exhibit abnormalities in the structure of collagen fibrils in skin and tendon.[30] The skin of these mice has decreased tensile strength and skin fibroblast attachment to various adhesive proteins is diminished. Thrombospondin-2-deficient mice also have a significant increase in bleeding time.

Antibodies

Many laboratories have produced polyclonal and monoclonal antibodies to human platelet thrombospondin-1 and some of these reagents are commercially available (AMAC, ATCC, Sigma and GIBCO). Very little information about species cross-reactivity is available since most laboratories have focused on human tissue. One monoclonal fusion was performed with rat cells and at least one of the resulting antibodies, designated 5G11, reacts with mouse tissue.

Genes

All or most of the sequence for the coding regions of human (X04665), mouse (M87276), cow (X87618), and frog (L04278) thrombospondin-1; human (L12350), mouse (L07803), and chicken (M60853) thrombospondin-2; human (L38969) and mouse (L24434, L04302) thrombospondin-3; human (Z19585), rat (X89963), chicken (L27263), and frog (Z19091) thrombospondin-4; and human (L32137), rat (X72914), and cow (X74326) COMP have been determined. Two groups have reported sequences for the human thrombospondin-1 promoter (J04447 and J04835). The structure of the mouse and human thrombospondin-1 (J05605, J05606, M62449 to M62470), mouse thrombospondin-2 (L06421 and L06422), and mouse and human thrombospondin-3 (L38970) genes have been reported. The human thrombospondin-1 (chromosome 15q15), -2 (chromosome 6q27), -3 (chromosome 1q21–24), and COMP (chromosome 19p13.1) genes, and the mouse thrombospondin-1 (chromosome 2), -2 (chromosome 17), -3 (chromosome 3), and COMP (chromosome 8) genes have been mapped.

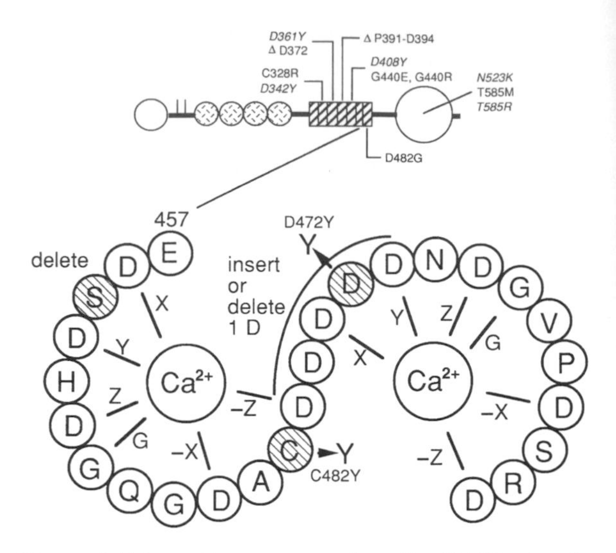

Figure 2. Schematic representation of a COMP subunit (top) with the position of mutations in PSACH, and MED (italics) indicated. The sequence of the calcium-binding sites in the sixth type 3 repeat is shown (bottom) with the position of the mutations indicated.

Mutant phenotype/disease states

Mutations in COMP have been shown to cause pseudoachondroplasia (PSACH) and multiple epiphyseal dysplasia (MED).[23,24] PSACH and MED are autosomal, dominant bone dysplasias. PSACH patients have shortened stature, marked joint laxity, joint erosion and pain, and early onset osteoarthritis.[25] The radiographic findings are distinctive, with characteristic vertebral epiphyseal and metaphyseal abnormalities. Most of the COMP mutations in PSACH and MED result in single amino acid deletions, insertions, or substitutions in the conformational-sensitive type 3 repeats (Fig. 2). The mutations probably disrupt the conformation of the type 3 repeats leading to defective calcium-binding, biosynthesis, and function. Since calcium binding is cooperative, the point mutations in a single calcium-binding site probably affect the activity of other sites. Since COMP is a pentamer, the abnormal subunits may have a dominant negative effect on the normal COMP subunits and other thrombospondins that form mixed multimers with COMP. As a result, mutations in single amino acids can give rise to a severe phenotype.

References

1. Adams, J. C., Tucker, R. P., and Lawler, J. (1995). *The thrombospondin gene family*. R. G. Lander Co., Austin.
2. Frazier, W. A. (1991). *Curr. Opin. Cell Biol.*, **3**, 792–9.
3. Bornstein, P. (1992). *FASEB J.*, **6**, 3290–9.
4. Roberts, D. D. (1996). *FASEB J.*, **10**, 1183–91.
5. Gobron, S., Monnerie, H., Meinel, R., Creveaux, I., Lehmann, W., Lamalle, D., *et al.* (1996). *J. Cell Sci.*, **109**, 1053–61.
6. Klar, A., Baldasse, M., and Jessell, T. M. (1992). *Cell*, **69**, 95–110.
7. Leung-Hagesteijn, C., Spense, A. M., Stern, B. D., Zhou, Y., Su, M. -W., Hedgecook, E. M., and Culotti, J. G. (1992). *Cell*, **71**, 289–99.
8. Lawler, J. and Hynes, R. O. (1986). *J. Cell Biol.*, **103**, 1635–48.
9. Misenheimer, T. M. and Mosher, D. F. (1995). *J. Biol. Chem.*, **270**, 1729–33.
10. Gao, A.-G., Lindberg, F. P., Finn, M. B., Blystone, S. D., Brown, E. J., and Frazier, W. A. (1996). *J. Biol. Chem.*, **271**, 21–4.
11. Tsao, P. W. and Mousa, S. A. (1995). *J. Biol. Chem.*, **270**, 23747–53.
12. Gao, A.-G., Lindberg, F. P., Dimitry, J. M., Brown, E. J., and Frazier, W. A. (1996). *J. Biol. Chem.*, **135**, 533–44.
13. DeFreitas, M. F., Yoshida, C. K., Frazier, W. A., Mendrick, D. L., Kypta, R. M., and Reichardt, L. F. (1995). *Neuron*, **15**, 333–45.
14. Tuszynski, G. P., Rothman, V. L., Papale, M., Hamilton, B. K., and Eyal, J. (1993). *J. Cell Biol.*, **120**, 513–21.
15. Lawler, J., Derick, L. H., Connelly, J. E., Chen, J., and Chao, F. C. (1985). *J. Biol. Chem.*, **260**, 3762–72.
16. DiCesare, P. E., Morgelin, M., Carlson, C. S., Pasumarti, S., and Paulsson, M. (1995). *J. Orthopaedic Res.*, **13**, 422–8.
17. Arber, S. and Caroni, P. (1995). *J. Cell. Biol.*, **131**, 1083–94.
18. Chem, H., Sottile, J., O'Rourke, K. M., Dixit, V. M., and Mosher, D. F. (1994). *J. Biol. Chem.*, **269**, 32226–32.
19. Adams, J. C. and Lawler, J. (1994). *Mol. Biol. Cell*, **5**, 423–37.
20. Tolsma, S. S., Volpert, O. V., Good, D. J., Frazier, W. A., Polverini, P. J., and Bouck, N. (1993). *J. Cell Biol.*, **122**, 497–511.
21. Volpert, O. V., Tolsma, S. S., Pellerin, S., Feige, J.-J., Chen, H., Mosher, D. F., and Bouck, N. (1995). *Biochem. Biophys. Res. Commun.*, **217**, 326–32.
22. Schultz-Cherry, S., Chen, H., Mosher, D. F., Misenheimer, T. M., Krutzsch, H. C., Roberts, D. D., and Murphy-Ullrich, J. E. (1995). *J. Biol. Chem.*, **270**, 7304–10.
23. Hecht, J. T., Nelson, L. D., Crowder, E., Wang, Y., Elder, F. F. B., Harrison, W. R., *et al.* (1995). *Nature Genet.*, **10**, 325–9.
24. Briggs, M. D., Hoffman, S. M. G., King, L. M., Olsen, A. S., Mohrenweiser, H., Leroy, J. G., *et al.* (1995). *Nature Genet.*, **10**, 330–6.
25. Wynne-Davis, R., Hall, C. M., and Young, I. D. (1986). *J. Med. Genet.*, **23**, 425–34.
26. Crombie, R., Silverstein, R. L., MacLow, C., Pearce, S. F. A., Nachman, R. L., and Laurence, J. (1998). *J. Exp. Med.*, **187**, 25–35.

27. Dawson, D. W., Pearce, S. F. A., Zhong, R., Silverstein, R. L., Frasier, W. A., and Bouck, N. P. (1997). *J. Cell Biol.*, **138**, 701–17.
28. Lawler, J., Sunday, M., Thibert, V., Duquette, M., George, E. L., Rayburn, H., and Hynes, R. O. (1998). *J. Clin. Invest.*, **101**, 982–92.
29. Crawford, S. E., Stellmach, V., Murphy-Ullrich, J. E., Ribeiro, S. F., Lawler, J., Hynes, R. O., Boivin, G. P., and Bouck, N. (1998). *Cell*, **93**, 1159–70.
30. Kyriakides, T. R., Zhu, Y.-H., Smith, L. T., Bain, S. D., Yang, Z., Lin, M. T., Danielson, K. G., Iozzo, R. V., LaMarca, M., McKinney, C. E., Ginns, E. I., and Bornstein, P. (1998). *J. Cell Biol.*, **140**, 419–30.

Jack Lawler
Department of Pathology, Beth Israel Deaconess Medical Center, Research North, Room 270C, 99 Brookline Avenue, Boston, MA 02215, USA

Versican

Versican[1,2] belongs to the family of large chondroitin sulphate proteoglycans. Its multidomain structure includes, at the N terminus, a hyaluronan binding domain and in the C-terminal region two EGF-like repeats, a lectin-like domain and a sushi-element. The variable content of glycosaminoglycan carrying domains in the middle of the core proteins gives rise to four splice-variants of human, mouse, and bovine versican[3,4,30] and to six isoforms of chick versican.[31] The expression of these splice-variants is partly tissue specific.[3] Although there are few functional data available, there is evidence that versican modulates cell adhesion and migration.[5,15]

Synonymous name

PG-M[6,7] Hyaluronectin[8] and GHAP[9] are probably proteolytic fragments of versican (identical N-terminal amino acid sequences).

Homologous proteins

Aggrecan, neurocan, brevican (family name: hyalectans[10] or lecticans[11]). Common features of these hyaluronan-binding proteoglycans are highly similar N- and C-terminal modular structures interrupted by chondroitin sulphate carrying middle portions that greatly differ in size and amino acid sequence.

Protein properties

In mammals, four different isoforms of versican have been identified (Fig. 1). All versican variants include, close to their N terminus, a globular link-protein-like structure composed of an immunoglobulin-like loop and two tandem repeats, which mediate the interaction with hyaluronan. A second common globular domain is localized at the C terminus. It contains two EGF-like repeats, a C-type lectin domain and a sushi module (also known as CRP domain). The differences between the versican isoforms result from the alternative splicing of two large exons, which encode the central glycosaminoglycan-carrying core protein portions named GAG-α and GAG-β.[3,12] Both of these domains are present in the largest V_0 splice-variant which may carry between 17 and 23 chondroitin sulphate side chains. The smaller V_1 and V_2 isoforms either contain only the GAG-β or GAG-α domain, reducing the number of putative attachment sites for glycosaminoglycan side chains to 12–15 in V_1 and 5–8 in V_2. The fourth versican isoform V_3 lacks both central domains and therefore seems to be devoid of glycosaminoglycan chains.[4] Six versican splice-variants exist in the chicken due to the presence (or absence) of a so-called PLUS domain, which is localized N-terminally of the GAG-α domain.[31] This species-specific PLUS domain displays some similarity to the KS domain in aggrecan.

The V_2, V_1 and V_0 splice-variants of versican are very large molecules, with molecular masses between 6×10^5 and 1.5×10^6 Da.[3,30] Like all proteoglycans they are rather heterogeneous in size due to the variability of their glycosaminoglycan side chains. After chondroitinase ABC digestion, versicans V_0, V_1, and V_2 migrate on SDS–polyacrylamide gels at about 550, 500, and 400 kDa, respectively. The discrepancy from the core protein sizes predicted from the amino acid sequences (370, 262, and 180 kDa, respectively) may partly result from the presence of *N*- and *O*-linked oligosaccharides. Currently, there are no protein chemical data available for versican V_3 (calculated molecular mass: 72 kDa). Its existence is inferred from Northern blot and RT-PCR experiments.[31]

Immunohistochemical studies with polyclonal antibodies against the GAG-β domain have revealed a relatively wide distribution of versican V_0 and/or V_1 in adult human tissues.[13] Versican V_0/V_1 is present in the loose connective tissues of various organs and is often associated with the elastic fibre network. Furthermore, it is localized in most smooth muscle tissues as well as in fibrous and elastic cartilage. Versican V_0/V_1 staining has been observed in the central and peripheral nervous systems, in the basal layer of the epidermis and on the luminal surfaces of some glandular epithelia. In blood vessels, versican V_0/V_1 is

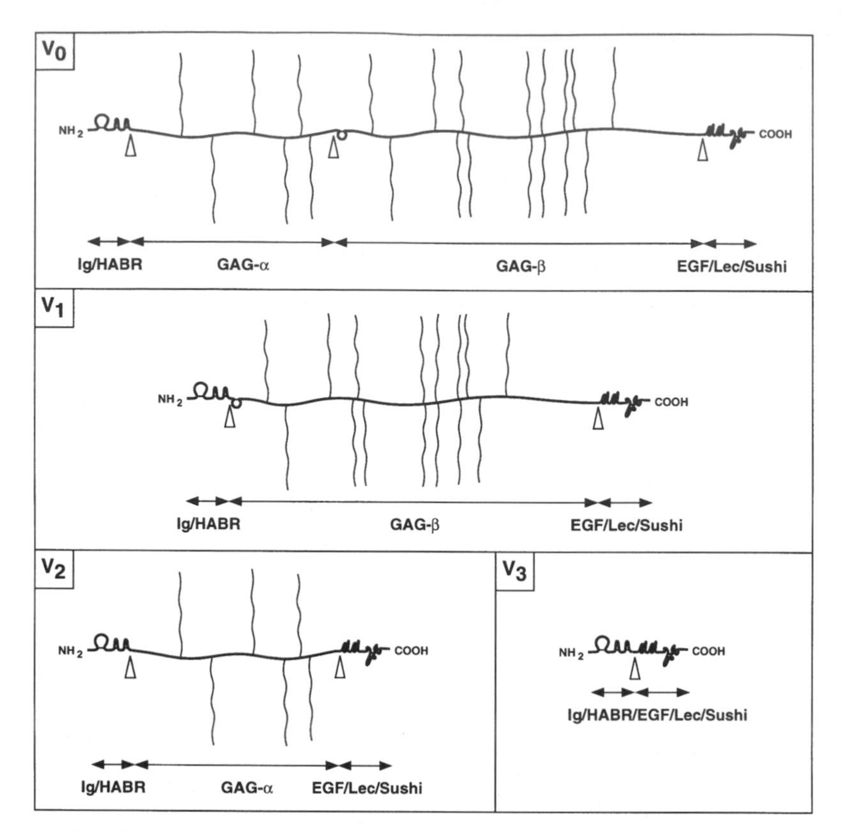

Figure 1. Schematic model of splice variants of human versican. The vertical wavy lines depict glycosaminoglycan side chains. Domain junctions that result from alternative splicing are indicated by open arrowheads. Ig: immunoglobulin-like domain; HABR: hyaluronan-binding region; GAG: glycosaminoglycan attachment domain; EGF: epidermal growth factor-like repeats; Lec: lectin-like domain; Sushi: sushi module (CRP domain).

present in all three wall layers of veins and elastic arteries. In muscular arteries the immunoreactivity is normally restricted to the tunica adventitia. RT-PCR experiments show a similarly wide expression pattern for the shortest versican splice variant V_3. In contrast, versican V_2 seems to be restricted to the central nervous system, where it represents the predominant versican splice-variant.[14,30]

In early embryonic development, versican V_0/V_1 is expressed in the subectodermal tissue neighbouring the neural tube and in the posterior halves of the sclerotomes.[15] At later stages it is found in the condensing mesenchyme of prechondrogenic tissues and in the developing gut.

In vitro experiments have demonstrated that the expression of versican is controlled by several growth factors and cytokines. Increased levels of versican V_0/V_1 have been measured in culture medium of aortic smooth muscle cells after addition of TGF-β1 or PDGF.[16] TGF-β1 also upregulates the expression of versican by cultured human fibroblasts,[17] whereas IL-1β reduces the versican steady-state mRNA levels.[18]

A number of ligands have been identified which interact with versican or recombinantly expressed fragments of its core protein. Best characterized is the interaction of the N-terminal globular domain of versican with hyaluronan.[19] It appears that the complex is stabilized by link protein.[20] Other potential ligands include tenascin-R (mediated by the C-type lectin-domain),[21] fibronectin, and collagen I.[6]

Purification

Versicans are usually purified by combinations of CsCl gradient centrifugation, anion exchange, and gel filtration and/or hyaluronan affinity chromatography. Its isolation from various tissue and cell culture sources has been described.[1,6,22,23,30]

Activities

Versican V_0/V_1 interferes *in vitro* with the attachment of cells to various extracellular matrix components such as collagen I, fibronectin, and laminin.[5] A close correlation between versican expression and the formation of tissues that act as barriers to migratory neural crest cells and outgrowing axons[15] during embryonic development has been observed. In addition to this putative role in the inhibition of cell adhesion and migration, versican may also participate in the control of keratinocyte and dermal fibroblast proliferation.[24]

Based on our current knowledge, it is conceivable that the selective expression of a particular versican splice variant regulates conjointly with hyaluronan, link protein, and CD44, the hydration properties of hyaluronan-rich pericellular matrices and thereby modulates cell–cell and cell–matrix interactions during migration and proliferation processes.

Antibodies

Various polyclonal antibodies have been prepared against core proteins of human,[3,19,24] chick,[15] and bovine[22,30] versicans. Monoclonal antibodies against human versican are commercially available from Seikagaku (Tokyo, Japan; mAb 2-B-1) and from the Developmental Studies Hybridoma Bank (Iowa City, USA; mAb 12C5).

Genes

The entire cDNA sequences of human,[2,3] chick,[7] mouse[25] and bovine[30] versicans and partial sequences of the monkey[32] and the axolotl homologue[26] have been determined. The genes (*Cspg2*) for human[12] and mouse[27] versicans have been characterized. *Cspg2* is localized on chromosome 5q12–q14 in the human and on chromosome 13 in the mouse genome.

Mutant phenotype/disease states

There is increasing evidence for a role of versican in atherosclerosis. The appearance of versican in intimal and medial layers of muscular arteries correlates with the thickening of the luminal vessel wall in early atherosclerotic lesions.[13,28] Versican is also present in neointimal tissues of rat carotids after balloon injury.[29] This emergence of versican seems to be partly controlled by TGF-β, since administration of neutralizing anti-TGF-β antibodies to rats after balloon injury significantly reduces the expression of versican and the size of the neointimal lesions.[29]

Structure

In rotary shadowing electron microscopy, aorta-derived versican (probably V_1) appears as a long extended filament flanked by a single globular domain at either end. The core protein has a length of approximately 240 nm. The glycosaminoglycan side chains are approximately 60 nm long.[20]

References

1. Krusius, T., Gehlsen, K. R., and Ruoslahti, E. (1987). *J. Biol. Chem.*, **262**, 13120–5.
2. Zimmermann, D. R. and Ruoslahti, E. (1989). *EMBO J.*, **8**, 2975–81.
3. Dours-Zimmermann, M. T. and Zimmermann, D. R. (1994). *J. Biol. Chem.*, **269**, 32992–8.
4. Zako, M., Shinomura, T., Ujita, M., Ito, K., and Kimata, K. (1995). *J. Biol. Chem.*, **270**, 3914–8.
5. Yamagata, M., Suzuki, S., Akiyama, S. K., Yamada, K. M., and Kimata, K. (1989). *J. Biol. Chem.*, **264**, 8012–8.
6. Kimata, K., Oike, Y., Tani, K., Shinomura, T., Yamagata, M., Uritani, M., and Suzuki, S. (1986). *J. Biol. Chem.*, **261**, 13517–25.
7. Shinomura, T., Nishida, Y., Ito, K., and Kimata, K. (1993). *J. Biol. Chem.*, **268**, 14461–9.
8. Delpech, B., Maingonnat, C., Delpech, A., Maes, P., Girard, N., and Bertrand, P. (1991). *Int. J. Biochem.*, **23**, 329–337.
9. Perides, G., Lane, W. S., Andrews, D., Dahl, D., and Bignami, A. (1989). *J. Biol. Chem.*, **264**, 5981–7.
10. Iozzo, R. V. and Murdoch, A. D. (1996). *FASEB J.*, 10, 598–614.
11. Ruoslahti, E. (1996). *Glycobiology*, **6**, 489–92.
12. Naso, M. F., Zimmermann, D. R., and Iozzo, R. V. (1994). *J. Biol. Chem.*, **269**, 32999–3008.
13. Bode-Lesniewska, B., Dours-Zimmermann, M. T., Odermatt, B. F., Briner, J., Heitz, P. U., and Zimmermann, D. R. (1996). *J. Histochem. Cytochem.*, **44**, 303–12.
14. Paulus, W., Baur, I., Dours-Zimmermann, M. T., and Zimmermann, D. R. (1996). *J. Neuropathol. Exp. Neurol.*, **55**, 528–33.
15. Landolt, R. M., Vaughan, L., Winterhalter, K. H., and Zimmermann, D. R. (1995) *Development*, **121**, 2303–12.
16. Schönherr, E., Järveläinen, H. T., Sandell, L. J., and Wight, T. N. (1991). *J. Biol. Chem.*, **266**, 17640–7.
17. Kähäri, V. M., Larjava, H., and Uitto, J. (1991) *J. Biol. Chem.*, **266**, 10608–15.
18. Qwarnström, E. E., Järveläinen, H. T., Kinsella, M. G., Ostberg, C. O., Sandell, L. J., Page, R. C., and Wight, T. N. (1993). *Biochem. J.*, **294**, 613–20.
19. LeBaron, R. G., Zimmermann, D. R., and Ruoslahti, E. (1992). *J. Biol. Chem.*, **267**, 10003–10.
20. Mörgelin, M., Paulsson, M., Malmström, A., and Heinegård, D. (1989). *J. Biol. Chem.*, **264**, 12080–90.
21. Aspberg, A., Binkert, C., and Ruoslahti, E. (1995). *Proc. Natl Acad. Sci. USA*, **92**, 10590–4.
22. Heinegård, D., Björne, P. A., Cöster, L., Franzén, A., Gardell, S., Malmström, A., *et al.* (1985). *Biochem. J.*, **230**, 181–94.
23. Perides, G., Rahemtulla, F., Lane, W. S., Asher, R. A., and Bignami, A. (1992). *J. Biol. Chem.*, **267**, 23883–7.
24. Zimmermann, D. R., Dours-Zimmermann, M. T., Schubert, M., and Bruckner-Tuderman, L. (1994). *J. Cell Biol.*, **124**, 817–25.
25. Ito, K., Shinomura, T., Zako, M., Ujita, M., and Kimata, K. (1995). *J. Biol. Chem.*, **270**, 958–65.
26. Stigson, M. and Kjellén, L. (1996) *In* Large chondroitin sulphate proteoglycans in axolotl embryo. Stigson, M. (Dissertation). Uppsala, Sweden. Section III, pp. 1–26.
27. Shinomura, T., Zako, M., Ito, K., Ujita, M., and Kimata, K. (1995). *J. Biol. Chem.*, **270**, 10328–33.
28. Lark, M. W., Yeo, T.-K., Mar, H., Lara, S., Hellström, I., Hellström, K.-E., and Wight, T. N. (1988). *J. Histochem. Cytochem.*, **36**, 1211–21.
29. Wolf, Y. G., Rasmussen, L. M., and Ruoslahti, E. (1994). *J. Clin. Invest.*, **93**, 1172–8.

30. Schmalfeldt, M., Dours-Zimmermann, M. T., Winterhalter, K. H., and Zimmermann, D. R. (1998). *J. Biol. Chem.*, **273**, 15758–64.
31. Zako, M., Shinomura, T., and Kimata, K. (1997). *J. Biol. Chem.*, **272**, 9325–31.
32. Yao, L. Y., Moody, C., Schönherr, E., Wight, T. N., and Sandell, L. J. (1994). *Matrix Biol.*, **14**, 213–25.

■ *Dieter R. Zimmermann*
Research Laboratory, Institute of Clinical Pathology, Department of Pathology, University of Zürich, Schmelzbergstrasse 12, 8091 Zürich, Switzerland

Vitronectin

Vitronectin is an abundant blood plasma glycoprotein which is also present in the extracellular matrix of many tissues. Vitronectin binds collagens and heparin-like glycosaminoglycans in the extracellular matrix and acts as a cell adhesion molecule. In the blood it regulates the complement, fibrinolytic and coagulation systems.[1]

■ Protein properties

Vitronectin, also known as S-protein and serum spreading factor, is a multifunctional protein that is abundant in blood plasma. Vitronectin is synthesized as a single chain which is cleaved by a signal peptidase, *N*-glycosylated, phosphorylated and sulphated prior to secretion. Human vitronectin circulates as one-chain and two-chain forms, the latter due to a proteolytic cleavage, probably by an intracellular processing protease. The one-chain form migrates as a 75 kDa band upon SDS–PAGE under reducing or non-reducing conditions whereas the two-chain form dissociates into 65 and 10 kDa bands upon reduction. The relative amount of one-chain and two-chain forms is determined by a polymorphism near the cleavage site.[2,3] Circulating vitronectin is synthesized in the liver and synthesis is upregulated during systemic inflammation.[4] Significant amounts of vitronectin mRNA are present in many other tissues, indicating *in situ* synthesis.[4]

Vitronectin has several different binding domains (Fig. 1). A somatomedin B domain (a.a. 1–44) at the amino terminus of the molecule contains a binding site for plasminogen activator inhibitor-1 (PAI-1).[5] There is some evidence to suggest that other domains also bind PAI-1. Immediately carboxyl-terminal from the somatomedin B domain is a region which contains the cell adhesion sequence (RGD, a.a. 45–47), an acidic region, a putative cross-linking site, a sulphation site, and one of two collagen-binding sites. The acidic region is hypothesized to interact with the positively charged glycosaminoglycan-binding domain to contribute to the tertiary structure of the molecule.[1] Two copies of a homology unit also found in haemopexin, interstitial collagenase, and stromolysin comprise the rest of the molecule. The glycosaminoglycan-binding sites are within this domain: two consensus heparin binding sequences have been localized between amino acids 347 and 352 and 354 and 362.[6] More recently, two novel heparin-binding domains have been identified outside this region using a phage display technique.[7] Vitronectin-binding sites for the thrombin–antithrombin III complex and β-endorphin have not been unambiguously identifed. Several types of evidence suggest that vitronectin exists in at least three different conformational states depending on specific ligand interactions.

Vitronectin is present in the supporting stroma of a variety of normal tissues but not in the basement membrane *per se*. Interaction with glycosaminoglycans and/or collagens in extracellular matrix presumably allows exposure of the cell attachment domain. Vitronectin is present in many different tissues during development, including brain and heart. Ultrastructural examination showed that in brain vitronectin is localized around the spines and postsynaptic densities of dendrites, suggesting a role in the formation, maintenance, or plasticity of synapses.[8]

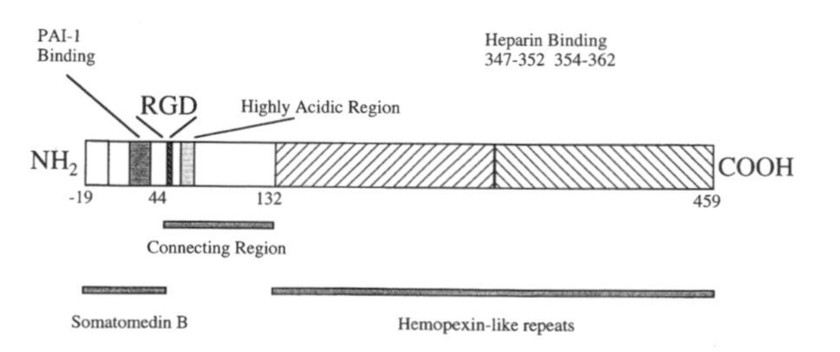

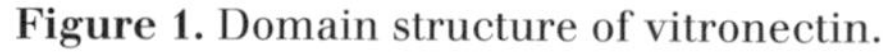

Figure 1. Domain structure of vitronectin.

■ Purification

Vitronectin can be purified in a native or denatured state from blood plasma where it has a concentration between 200 and 300 μg/ml. The native form is purified by differential precipitation with either polyethylene glycol[9] or ammonium sulphate[10] followed by sequential chromatography on DEAE–Sephacel, Blue Sepharose, and Sephacryl S-200. Contaminating albumin remaining after these steps can be removed by heparin–Sepharose chromatography. Reduced glutathione must be present in all buffers to prevent formation of disulphide-bonded multimers. Denatured vitronectin can be purified by a two-step procedure based on the fact that it binds heparin under physiological salt concentration only following denaturation with urea.[11]

Activities

Vitronectin acts both as an adhesive glycoprotein and as a pivotal regulator of the complement, fibrinolytic, and coagulation systems in the blood. In the extracellular matrix vitronectin binds heparin-like glycosaminoglycans and collagens. Cells attach to vitronectin via six distinct integrin receptors; αIIBβ3,[12] αvβ1,[13] αvβ3,[14] αvβ5,[15] αvβ8,[16] and α8β1.[17] Recent evidence[18] has implicated another integrin, αvβ6, in cell attachment to vitronectin though earlier work suggested that it acted as receptor for fibronectin but not vitronectin.[19] Vitronectin binds to the complement C5b–7 complex thus inhibiting complement-mediated cell lysis.[20] In the presence of glycosaminoglycans, vitronectin inhibits the rapid inactivation of thrombin and factor Xa by antithrombin III.[21,22] In addition, PAI-1, the regulator of both urokinase-type and tissue-type plasminogen activators, is stabilized in its active conformation by binding to vitronectin.[23,24] Vitronectin exists in different conformations and in monomeric and multimeric forms. The conformation and degree of multimerization appear to be critical to protein function. Plasma vitronectin circulates mainly as a non-heparin-binding form in which the cell binding domain is cryptic.[25] Interaction with either PAI-1[26] or heparin[27] induces multimer formation and exposes binding sites that are not available in the native unbound molecule. Interaction with thrombin–antithrombin III induces exposure of the heparin binding site. Denaturation irreversibly exposes a number of different binding sites, including the heparin binding site, and induces multimerization.[28] Vitronectin adsorbs strongly to glass and tissue culture plastic, even in the presence of high concentrations of other proteins, and mediates cell attachment to these substrates. Specific interactions of vitronectin with bacteria have been observed[29,30] and vitronectin mediates bacterial adherence to host cells.[31]

Antibodies

Human vitronectin is highly antigenic and high titred, broadly reactive antisera can be produced in rabbits. Commercial reagents are widely available. A number of monoclonal antibodies have been produced that recognize different vitronectin domains and these have been used for structural studies.[5,32]

Genes

cDNAs for human vitronectin have been isolated and sequenced and are available in GenBank (XO3168).[33,34] Sequences are also now available for rabbit,[35] pig,[36] rat,[37] and mouse[38] vitronectin. The gene for human vitronectin is small, 5.3 kb (GenBank XO5006)[39] and its chromosomal location has not been reported.

Mutant phenotype/disease state

A vitronectin knockout mouse, shown to be completely deficient in serum spreading factor and PAI-1 binding activities, had normal development and no major defects in haemostasis.[40] Vitronectin may be involved in tumour metastasis through its function as a regulator of the plasminogen activation cascade.[41] Arteriosclerotic lesions are rich in vitronectin[35] and kidney tissue from patients with glomerulonephritis had accumulated deposits of vitronectin colocalized with C5b–9 complex of complement.[42,43]

Web site

There is a very useful web site at http://www./bioinformatics.weizmann.ac.il/bioscience/knockout/cd401.htm

References

1. Preissner, K. T. (1991). *Ann. Rev. Cell. Biol.*, **7**, 275–310.
2. Tollefsen, D. M., Weigel, C. J., and Kabeer, M. H. (1990). *J. Biol. Chem.*, **265**, 9778–81.
3. Kubota, K., Hayashi, M., Oishi, N., and Sakaki, Y. (1990). *Biochem. Biophys. Res. Comm.*, **167**, 1355–460.
4. Seiffert, D., Crain, K., Wagner, N. V., and Loskutoff, D. J. (1994). *J. Biol. Chem.*, **269**, 19836–42.
5. Seiffert, D., Ciambrone, G., Wagner, N. V., Binder, B. R., and Loskutoff, D. J. (1994). *J. Biol. Chem.*, **269**, 2659–6.
6. Cardin, A. D. and Weintraub, H. J. R. (1989). *Arteriosclerosis*, **9**, 21–30.
7. Liang, O. D., Rosenblatt, S., Chhatwal, G. S., and Preissner, K. T. (1997). *FEBS Lett.*, **407**, 169–72.
8. Einhaber, S., Schnapp, L. M., Salzer, J. L., Cappiello, Z. B., and Milner, T. A. (1996). *J. Comp. Neurol.*, **370**, 105–34.
9. Dahlback, B. and Podack, E. R. (1985). *Biochemistry*, **24**, 2368–74.
10. Preissner, K. T., Wassmuth, R., and Mueller-Berhaus, G. (1985). *Biochem. J.*, **231**, 349–55.
11. Yatohgo, T., Izumi, M., Kashiwagi, H., and Hayashi, M. (1988). *Cell Struct. Funct.*, **13**, 281–92.
12. Thiagarajan, P. and Kelly, K. (1988). *Thromb. Haemost.*, **60**, 514–17.
13. Marshall, J. F., Rutherford, D. C., McCartney, A. C. E., Mitjans, F., Goodman, S. L., and Hart, I. R. (1995). *J. Cell Sci.*, **108**, 1227–38.
14. Brooks, P. C., Clark, R. A. F., and Cherish, D. A. (1994). *Science*, **264**, 569–71.
15. Smith, J. W., Vestal, D. J., Irwin, S. V., Burke, T. A., and Cherish, D. A. (1990). *J. Biol. Chem.*, **265**, 11008–13.
16. Nishimura, S. L., Sheppard, D., and Pytela, R. (1994). *J. Biol. Chem.*, **269**, 28708–15.
17. Schnapp, L. M., Hatch, N., Ramos, D. M., Klimanskaya, I. V., Sheppard, D., and Pytela, R. (1995). *J. Biol. Chem.*, **270**, 23196–202.
18. Dean Sheppard, personal communication.

19. Busk, M., Pytela, R., and Sheppard, D. (1992). *J. Biol. Chem.*, **267**, 5790–6.
20. Podack, E. R., Kolb, W. P., and Mueller-Eberhard, H. J. (1977). *J. Immunol.*, **119**, 2024–9.
21. Ill, C. R. and Ruoslahti, E. (1985). *J. Biol. Chem.*, **260**, 15610–15.
22. Preissner, K. T. and Mueller-Berghaus, G. (1987). *J. Biol. Chem.*, **262**, 12247–53.
23. Declerk, P. J., deMol, M., Alesi, M. -C., Baudner, S., Paques, E. -P., Preissner, K. T., *et al.* (1988). *J. Biol. Chem.*, **263**, 15454–61.
24. Seiffert, D. and Loskutoff, B. R. (1991). *J. Biol. Chem.*, **266**, 2824–30.
25. Seiffert, D. and Smith, J. W. (1997). *J. Biol. Chem.*, **272**, 13705–10.
26. Seiffert, D. and Loskutoff, D. J. (1996). *J. Biol. Chem.*, **271**, 29644–51.
27. Seiffert, D. (1997). *J. Biol. Chem.*, **272**, 9971–8.
28. Bittdorf, S. V., Williams, E. C., and Mosher, D. F. (1993). *J. Biol. Chem.*, **268**, 24838–46.
29. Fuquay, J. I., Loo, D. T., and Barnes, D. W. (1986). *Infect. Immunol.*, **52**, 714–7.
30. Chhatwal, G. S., Preissner, K. T., Mueller-Berghaus, G., and Blobel, H. (1987). *Infect. Immunol.*, **55**, 1878–83.
31. Duensing, T. D. and van Putten, J. P. (1997). *Infect. Immunol.*, **65**, 964–70.
32. Morris, C. A., Underwood, P. A., Bean, P. A., Sheehan, M., and Charlesworth, J. A. (1994). *J. Biol. Chem.*, **269**, 23845–52.
33. Jenne, D. and Stanley, K. K. (1985). *EMBO J.*, **4**, 3135–57.
34. Suzuki, S., Oldberg, A., Hayman, E. G., Peirschbacher, M. D., and Ruoslahti, E. (1985). *EMBO J.*, **4**, 2519–24.
35. Sato, R., Komine, Y., Imanaka, T., and Takano, T. (1990). *J. Biol. Chem.*, **265**, 21232–6.
36. Yoneda, A., Kojima, K., Matsumoto, I., Yamamoto, K., and Ogawa, H. (1996). *J. Biochem.*, **120**, 954–60.
37. Otter, M., Kuiper, J., Rijken, D., and von Zonneveld, A.-J. (1995). *Biochem. Mol. Biol. Int.*, **37**, 563–72.
38. Seiffert, D., Poenninger, J., and Binder, B. R. (1993). *Gene*, **134**, 303–4.
39. Jenne, D. and Stanley, K. K. (1987). *Biochemistry*, **26**, 6735–42.
40. Zheng, X., Saunders, T. L., Camper, S. A., Samuelson, L. C., and Ginsburg, D. (1995). *Proc. Natl Acad. Sci. USA*, **92**, 12426–30.
41. Stahl, A. and Mueller, B. M. (1997). *Int. J. Cancer*, **71**, 116–22.
42. Falk, R. J., Podack, E., Dalmasso, A. P., and Jennette, J. C. (1987). *Am. J. Pathol.*, **127**, 182–90.
43. Bariety, J., Hinglais, N., Bhakdi, S., Mandet, C., Rouchon, M., and Kazatchkine, M. D. (1989). *J. Clin. Exp. Immunol.*, **75**, 76–81.

Deborah E. Hall
Department of Neurology, Rm C215,
Box 0114, 505 Parnassus Avenue,
San Francisco, CA 94143–0114, USA

Von Willebrand factor

von Willebrand factor (vWF)[1] is a large multimeric glycoprotein with two distinct biological roles: (i) It mediates platelet adhesion and thrombus formation at sites of vascular injury; and (ii) it serves as the carrier for procoagulant factor VIII in blood, circulating as the factor VIII–vWF complex. Both these functions are essential for the normal arrest of bleeding (haemostasis) following tissue trauma. In pathological conditions, vWF is involved in the processes leading to thrombotic arterial occlusion. Congenital abnormalities of vWF result in the most common inherited human disorder of haemostasis, von Willebrand disease.[2]

Synonymous names

vWF was initially detected with immunochemical assays in the circulating factor VIII–vWF complex identified functionally by the procoagulant activity of factor VIII; thus, it became known as the factor VIII–related antigen. The term has been abandoned after it was definitively shown that factor VIII and vWF are two distinct gene products that do not share structural or immunological relationships.

Homologous proteins

The presence of vWF type A-like domains (called I domains in integrins) identifies a superfamily of proteins[3] with diverse activities in extracellular matrices, haemostasis, cellular adhesion, and defense mechanisms. Among the members of this group are integrin α subunits, notably αL, αM, and αX in the β2 integrins, α1 and α2 in the β1 integrins; collagens containing non-triple-helical domains, such as types VI (α1, α2, and α3 subunits), VII, and XIV; components C2 and factor B of the complement system; dihydropyridine-sensitive calcium channel;[4] inter-α-trypsin inhibitor;[4] thrombospondin-related anonymous protein (TRAP) of *Plasmodium falciparum*;[4] cartilage matrix protein, and matrilin-2.[5] Moreover, the presence of a vWF type A-like domain within the ligand-binding region of integrin β subunits has been predicted[6] on the basis of the known crystallographic structure of the I domains of αL[7] and αM.[8] vWF type A-like domains usually participate in establishing specific protein–protein interactions, in some cases dependent on, but in others independent of, divalent cations. As shown by three-dimensional structural analysis, metal coordination, when present, involves the contribution of a DXSXS motif as well as of a Ser and an Asp residue, all from non-contiguous regions. A sequence homologous to vWF type D domains, present in the propeptide sequence of pro-vWF and at the amino terminus of the mature vWF subunit (D1–D2–D′–D3), appears in regions of invertebrate vitellogenin[9] as well as in mucins.[10] In multimeric mucins, the common D domain structural motif, characterized by several conserved Cys residues, may mediate homodimerization as an early biosynthetic event, in analogy with the role played in vWF. Conserved vicinal Cys residues in the D domain sequence motif CGLCG are similar to those at the active site of disulphide isomerase, and correspond to the residues that have been proposed to mediate the multimerization of pro-vWF.[11] The spacing of these vicinal Cys residues is the key to function, as increasing the relative distance by inserting an additional Gly residue results in impaired vWF multimer assembly.

Protein properties

The vWF mRNA is ~9 kb in size and codes for a 2813 amino acid residue protein. This protein represents pre-pro-vWF.[12] The vWF precursor sequence consists of four distinct types of repeated domains (A–D) that account for > 90 per cent of the sequence[13] (Fig. 1). The mature vWF subunit is composed of 2050 amino acids and has an estimated molecular mass of 275 kDa.[14] Approximately 18.7 per cent (w/w) of the vWF molecular mass is carbohydrate. There are 22 probable glycosylation sites within the mature subunit, of which 12 are *N*-linked and 10 are *O*-linked.[14] Nucleotides in the vWF cDNA are numbered from the major transcription cap site, located 250 nucleotides before the first nucleotide in the ATG codon for the initiating methionine. Amino acid residues numbered from His[1] in mature vWF correspond to those in pre-pro-vWF after adding 763, which accounts for the 22 residues of the signal peptide and the 741 residues of the vWF propeptide (Fig. 1).

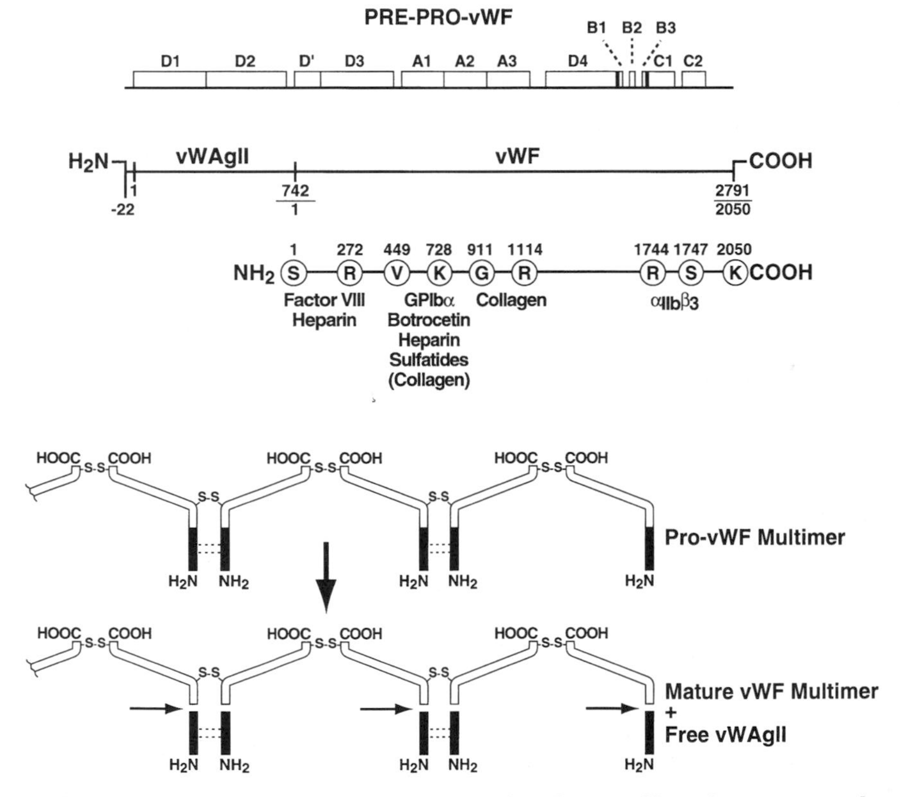

Figure 1. Schematic representation of the vWF molecule. Top: distribution of homologous repeat domains (A–D) in the pre-pro-vWF cDNA. Middle: linear representation of the pre-pro-vWF polypeptide, including the 22 residue signal peptide, 741 residue vWAgII (propeptide) region, and 2050 residue mature vWF subunit. The location of functional domains within the mature subunit is shown in detail, with the known first, and last amino acid residue of each domain identified by one-letter notation, and the corresponding position in the sequence. Bottom: schematic representation of the assembly of vWF subunits into dimers, and multimers. The process initiates with formation of pro-vWF dimers through disulphide bonding of two subunits at the respective carboxyl termini. This step takes place in the cellular endoplasmic reticulum. In the Golgi apparatus, dimers assemble into multimers first by non-covalent association of propeptide (vWAgII) domains, followed by covalent disulphide bonding between amino terminal domains of mature subunits. The propeptide regions are then proteolytically removed form the mature multimers, and maintain the non-covalent dimeric association characteristic of free vWAgII.

vWF is synthesized in endothelial cells and megakaryocytes as disulphide-linked oligomers of varying size,[15] ranging in estimated molecular mass from ~400 kDa to in excess of 10 000 kDa.[16] Large, fully extended multimers can reach lengths of > 1300 nm.[17] The initial vWF translation product, pre-pro-vWF, forms dimers in the endoplasmic reticulum by interchain disulphide bonding between C termini of two subunits, and undergoes post-translational processing including removal of the signal peptide and glycosylation. Pro-vWF dimers form multimers of increasing size by interchain disulphide bonds between N termini (Fig. 1). This process requires initial glycosylation and the acidic cellular environment of the Golgi apparatus, and is directed by the 741 residue propeptide sequence that is then cleaved after multimer assembly; thus, mature vWF multimers contain the 2050 residue subunit and no propeptide.

Unlike most other proteins, vWF can follow two pathways of secretion. The constitutive one is linked directly to synthesis, while the regulated one involves storage of mature molecules in appropriate granules, Weibel–Palade bodies in endothelial cells, or α-granules in platelets, and release after stimulation by secretagogues.[15] In megakaryocytes, and in platelets derived from fragmentation of their cytoplasm, only the regulated pathway is effectively operative *in vivo*, and platelet vWF is rapidly released from α-granules following activation. In contrast, the majority of the vWF synthesized by cultured endothelial cells is secreted through the constitutive pathway. It has been calculated that only about five per cent of the newly synthesized vWF is stored inside endothelial cells. Secretion by the vascular endothelium can be either into the circulation or abluminal, in which case vWF becomes part of the subendothelial matrix. Studies with *in vitro* culture have provided discordant results with regard to the amount of constitutively secreted vWF directed into circulating blood as opposed to that incorporated into the subendothelial matrix. The estimated values vary from 50:50[18] to three times higher for abluminal than luminal secretion.[19] The vWF propeptide, retaining the structure of a non-covalently associated homodimer, is removed from assembled multimers by proteolytic cleavage and is secreted from endothelial cells or released from the α-granules of platelets as a distinct protein also known as von Willebrand antigen II.[20] This molecule, which may become cross-linked to laminin by factor XIIIa,[21] interacts with collagen and may function as a negative regulator of platelet adhesion. The possible biological significance of such an activity remains unknown at present.

The mature subunit of circulating plasma vWF undergoes proteolytic cleavage under physiological conditions, while endothelial and platelet vWF is not proteolysed. As a result, non-identical subunits are present in plasma vWF multimers and cause their electrophoretic heterogeneity.[22] Because the degree of polymerization is directly correlated to the prothrombotic potential of the vWF molecule, proteolytic processing of the subunit may be an important regulatory mechanism controlling the size of multimers exposed to platelets in blood, thus limiting the possibility of vWF-mediated platelet aggregation. Preliminary information has become available on the enzyme responsible for cleaving circulating vWF,[23] opening a potentially important field of research on thrombotic disorders possibly linked to accumulation of large vWF multimers in blood.

The function of vWF is to promote thrombus formation by mediating adhesion of platelets to the injured vessel wall and to one another. This requires, in temporal sequence, binding to collagen – and possibly other matrix components – and to two platelet receptors, GP Ibα in the GP Ib–IX–V complex and αIIbβ3. Two vWF domains, A1 and A3, have been shown to interact with collagen in different experimental models.[24,38] The unique mechanism supporting platelet adhesion to immobilized vWF under high flow conditions has been elucidated.[25] The first step of this dynamic process is mediated by the binding of platelet GP Ibα to the vWF A1 domain, an interaction characterized by a fast association rate that can tether platelets to exposed thrombogenic surfaces even when the velocity of flowing blood relative to the vessel wall is elevated. This event is favoured by the multimeric nature of vWF, providing a high local density of active A1 domain sites and resulting in formation of multiple bonds. However, the vWF–GP Ibα interaction is also characterized by a fast dissociation rate and cannot provide bonds supporting irreversible adhesion, so that platelets tethered to the vessel wall in this manner move constantly in the direction of flow, albeit at a fraction of the free flow velocity (Fig. 2). During the slow translocation platelets become activated, and αIIbβ3 can eventually mediate irreversible adhesion by binding to the Arg–Gly–Asp (RGD) sequence corresponding to residues 1744–1747 in the C1 domain of the mature vWF subunit.

Purification

vWF is purified from plasma by cryoprecipitation, differential polyethylene glycol precipitation, and size exclusion chromatography. This method allows isolation of the larger molecular forms present in plasma, with a relative depletion of the smaller ones.[26]

Activities

The concentration of vWF in plasma and platelets can be measured with immunochemical assays, usually ELISA. The distribution of vWF multimers, crucial to assess function, can be analysed using SDS-agarose gel electrophoresis. The specific interactions of vWF with components of the vessel wall, particularly collagen, with GP Ibα and αIIbβ3 on the platelet surface, and with factor VIII, can all be measured with direct binding assays. The commonly used test of platelet agglutination in the presence of the antibiotic ristocetin reflects vWF binding to GP Ibα but is not, as usually stated, a global assay of vWF function. Evaluation of shear-induced platelet aggregation provides a reliable estimate of vWF interaction with

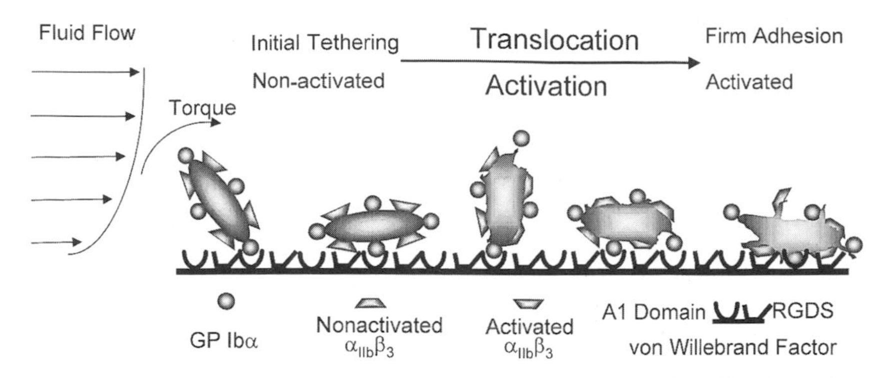

Figure 2. Schematic representation of the dual step mechanism supporting platelet adhesion to immobilized vWF. The first contact established between platelets, and immobilized vWF is mediated by GP Ibα. Non-activated αIIbβ3 cannot pair with the RGDS sequence in vWF. The bond between vWF A1 domain, and GP Ibα must form rapidly, since platelets can be tethered to a vWF-coated surface regardless of flow conditions, and must also have high resistance to tensile stress, since adherent platelets oppose the drag created by rapid flow. This bond, however, has an intrinsically high dissociation rate, thus a limited half-life, resulting in detachment where tension is greatest. The torque imposed by the flowing fluid causes a forward rotational movement (rolling), and new bonds are formed as different regions of the membrane of rolling platelets come in closer contact with the surface. Translocation at low velocity continues until αIIbβ3 becomes activated, and binds to the RGDS sequence in the vWF C1 carboxyl-terminal domain. The latter bond must have low dissociation rate, and can mediate irreversible adhesion, dependent also on cytoskeletal reorganization. An important aspect of this mechanism is that, for tethering to be apparent, the surface density of vWF must be sufficiently high to support the formation of multiple interactions at any given time.

platelets, involving both GP Ibα and αIIbβ3 as well as activation.

Antibodies

Polyclonal antibodies against human vWF (commercially available from several suppliers) show good cross-reactivity with vWF from other mammals, including porcine, bovine, canine, and feline vWF. Monoclonal antibodies usually exhibit more restricted species specificity. Numerous monoclonal antibodies have been prepared against each of the functional domains of the molecule and have provided relevant information on structure–function relationships. Anti-vWF antibodies are commonly used to identify vascular structures by immunofluorescent staining of endothelial cells.

Genes

The vWF gene is located on the short arm of chromosome 12, region 12p12–12pter (GenBank/EMBL accession number M25716).[27] It is approximately 178 kb in size and contains 52 exons. All exons and intron/exon boundaries have been sequenced.[28] A partial vWF pseudogene[29] has been located on chromosome 22. The complete sequence of human pre-pro-vWF cDNA is published[12] (accession number X04385). Portions of the vWF cDNA from numerous species have been sequenced and published. Complete updated information in this regard is available on the Internet at http://www.ncbi.nlm.nih.gov/. Targeted deletion of the mouse vWF gene is being accomplished, but has not been reported to date.

Mutant phenotypes/disease states

Abnormalities of vWF function result in the phenotype of congenital von Willebrand disease,[2] a bleeding disorder identified in several mammalian species and usually transmitted in autosomal dominant fashion. In the most common human cases (~70 per cent of patients), the phenotype is caused by decreased plasma concentration of what appears to be a structurally and functionally normal vWF molecule. The genetic lesions that may causes this form of the disease, designated type 1, have not been defined with certainty. The plasma levels of vWF in a mouse model of type I von Willebrand disease, the inbred strain RIIIS/J,[30] have been shown to be controlled by the product of a modifier gene located on mouse chromosome 11, thus distinct from the murine vWF locus on chromosome 6.[31] This finding suggests the possibility that the human homologue of this mouse gene may account, at least in part, for the variable expression and penetrance observed in human von Willebrand disease. Rare patients (type 3) have no demonstrable synthesis of the protein, and complete deletions of the corresponding gene have been reported in some cases. The remainder of the patients (type 2) present various distinct functional and structural abnormalities of the vWF molecule, resulting from the most part from single amino acid substitutions that alter the interaction with one of the receptors or substrates that mediate vWF participation in platelet thrombus formation. Several different variant forms of von Willebrand disease have been recognized and classified. A database containing polymorphisms in the vWF gene as well as mutations associated with various subtypes of von Willebrand disease is accessible on the world-wide web at http://mmg2.im.med.umich.edu/vWF/.

Structure

The three-dimensional structures of recombinant vWF A1[32,33] and A3[34,35] domains have been solved. The crystal structures of the homologous I (vWF type A-like) domains of integrin subunits α_L,[7] α_M,[8] α_2[36] have also been reported. All show an α/β 'Rossman' fold characterized by a central β-sheet core surrounded on both faces by α-helices. In the vWF A1 domain, helix $\alpha3$ forms a shallow groove with sides represented by helix $\alpha4$ and the N-terminal residues of strand $\beta3$, including Gly561 and the loop preceding it. This extended surface may represent the GP Ibα binding site, suggesting a functional role for residues in the helix $\alpha3$ groove whose side chains are oriented towards the surface of the molecule, notably Glu596 and Lys599. Of note, the substitution of Gly561 leads to loss of GP Ibα binding function.[37]

Web sites

http://www.ncbi.nlm.nih.gov/
http./mmg2.im.med.umich.edu/vWF/

References

1. Ruggeri, Z. M., Ware, J. L., and Ginsburg, D. (1994). In *Thrombosis and hemorrhage* (ed. J. Loscalzo and A. Schafer), pp. 305–29. Blackwell Scientific Publications, Boston.
2. Ware, J. L. and Ruggeri, Z. M. (1995). In *Molecular basis of thrombosis, and hemostasis* (ed. K. A. High and H. R. Roberts), pp. 197–214. Marcel Dekker, New York.
3. Colombatti, A. and Bonaldo, P. (1991). *Blood*, **77**, 2305–15.
4. Bork, P. and Rohde, K. (1991). *Biochem. J.*, **279**, 908–10.
5. Deak, F., Piecha, D., Bachrati, C., Paulsson, M., and Kiss, I. (1997). *J. Biol. Chem.*, **272**, 9268–74.
6. Tuckwell, D. S. and Humphries, M. J. (1997). *FEBS Lett.*, **400**, 297–303.
7. Qu, A. and Leahy, D. J. (1995). *Proc. Natl Acad. Sci. USA*, **92**, 10277–281.
8. Lee, J. -O., Rieu, P., Arnaout, M. A., and Liddington, R. (1995). *Cell*, **80**, 631–8.
9. Baker, M. E. (1988). *Biochem. J.*, **256**, 1059–61.
10. Joba, W. and Hoffmann, W. (1997). *J. Biol. Chem.*, **272**, 1805–10.
11. Mayada, T. N. and Wagner, D. D. (1992). *Proc. Natl Acad. Sci. USA*, **89**, 3531–5.
12. Bonthron, D., Orr, E. C., Mitsock, L. M., Ginsburg, D., Handin, R. I., and Orkin, S. H. (1986). *Nucleic Acids Res.*, **14**, 7125–7.
13. Shelton-Inloes, B. B., Titani, K., and Sadler, J. E. (1986). *Biochemistry*, **25**, 3164–71.
14. Titani, K., Kumar, S., Takio, K., Ericsson, L. H., Wade, R. D., Ashida, K., *et al*. (1986). *Biochemistry*, **25**, 3171–84.
15. Wagner, D. D. (1990). *Ann. Rev. Cell Biol.*, **6**, 217–46.
16. Ruggeri, Z. M. and Zimmerman, T. S. (1981). *Blood*, **57**, 1140–3.
17. Fowler, W. E. and Fretto, L. J. (1989). In *Coagulation and bleeding disorders. The role of factor VIII, and von Willebrand factor* (ed. T. S. Zimmerman and Z. M. Ruggeri), pp. 181–93. Marcel Dekker, New York.
18. Sporn, L. A., Marder, V. J., and Wagner, D. D. (1989). *J. Cell Biol.*, **108**, 1283–9.
19. van Buul-Wortelboer, M. F., Brinkman, H. M., Reinders, J. H., Van Aken, W. G., and van Mourik, J. A. (1989). *Biochim. Biophys. Acta*, **1011**, 129–33.
20. Fay, P. J., Kawai, Y., Wagner, D. D., Ginsburg, D., Bonthron, D., Ohlsson-Wilhelm, B. M., *et al*. (1986). *Science*, **232**, 995–8.
21. Usui, T., Takagi, J., and Saito, Y. (1993). *J. Biol. Chem.*, **268**, 12311–16.
22. Dent, J. A., Galbusera, M., and Ruggeri, Z. M. (1991). *J. Clin. Invest.*, **88**, 774–82.
23. Furlan, M., Robles, R., and Lammle, B. (1996). *Blood*, **87**, 4223–34.
24. Cruz, M. A., Yuan, H., Lee, J. R., Wise, R. J., and Handin, R. I. (1995). *J. Biol. Chem.*, **270**, 10822–7.
25. Savage, B., Saldivar, E., and Ruggeri, Z. M. (1996). *Cell*, **84**, 289–97.
26. De Marco, L. and Shapiro, S. S. (1981). *J. Clin. Invest.*, **68**, 321–8.
27. Ginsburg, D., Handin, R. I., Bonthron, D. T., Donlon, T. A., Bruns, G. A. P., Latt, S. A., and Orkin, S. H. (1985). *Science*, **228**, 1401–6.
28. Mancuso, D. J., Tuley, E. A., Westfield, L. A., Worrall, N. K., Shelton-Inloes, B. B., Sorace, J. M., *et al*. (1989). *J. Biol. Chem.*, **264**, 19514–27.
29. Mancuso, D. J., Tuley, E. A., Westfield, L. A., Lester-Mancuso, T. L., Le Beau, M. M., Sorace, J. M., and Sadler, J. E. (1991). *Biochemistry*, **30**, 253–69.
30. Sweeney, J. D., Novak, E. K., Reddington, M., Takeuchi, K. H., and Swank, R. T. (1990). *Blood*, **76**, 2258–65.
31. Mohlke, K. L., Nichols, W. C., Westrick, R. J., Novak, E. K., Cooney, K. A., Swank, R. T., and Ginsburg, D. (1996). *Proc. Natl Acad. Sci. USA*, **93**, 15352–7.
32. Celikel, R., Varughese, K. I., Madhusudan, A., Yoshioka, A., Ware, J., and Ruggeri, Z. M. (1998). *Nature Struct. Biol.*, **5**, 189–94.
33. Emsley, J., Cruz, M., Handin, R., and Liddington, R. (1998). *J. Biol. Chem.*, **273**, 10396–401.
34. Huizinga, E. G., van der Plas, R. M., Kroon, J., Sixma, J. J., and Gros, P. (1997). *Structure*, **5**, 1147–56.
35. Bienkowska, J., Cruz, M. A., Handin, R. I., and Liddington, R. C. (1997). *J. Biol. Chem.*, **272**, 25162–7.
36. Emsley, J., King, S. L., Bergelson, J. M., and Liddington, R. C. (1997). *J. Biol. Chem.*, **272**, 28512–17.
37. Rabinowitz, I., Tuley, E. A., Mancuso, D. J., Randi, A. M., Firkin, B. G., Howard, M. A., and Sadler, J. E. (1992). *Proc. Natl Acad. Sci. USA*, **89**, 9846–9.
38. Hoylaerts, M. F., Yamamoto, H., Nuyts, K., Vreys, I., Deckmyn, H., and Vermylen, J. (1997). *Biochem J.*, **324**, 185–91.

Zaverio M. Ruggeri and Jerry Ware
Roon Research Center for Arteriosclerosis and Thrombosis, Departments of Molecular and Experimental Medicine and of Vascular Biology, The Scripps Research Institute, La Jolla, California, USA

3

ECM proteinases

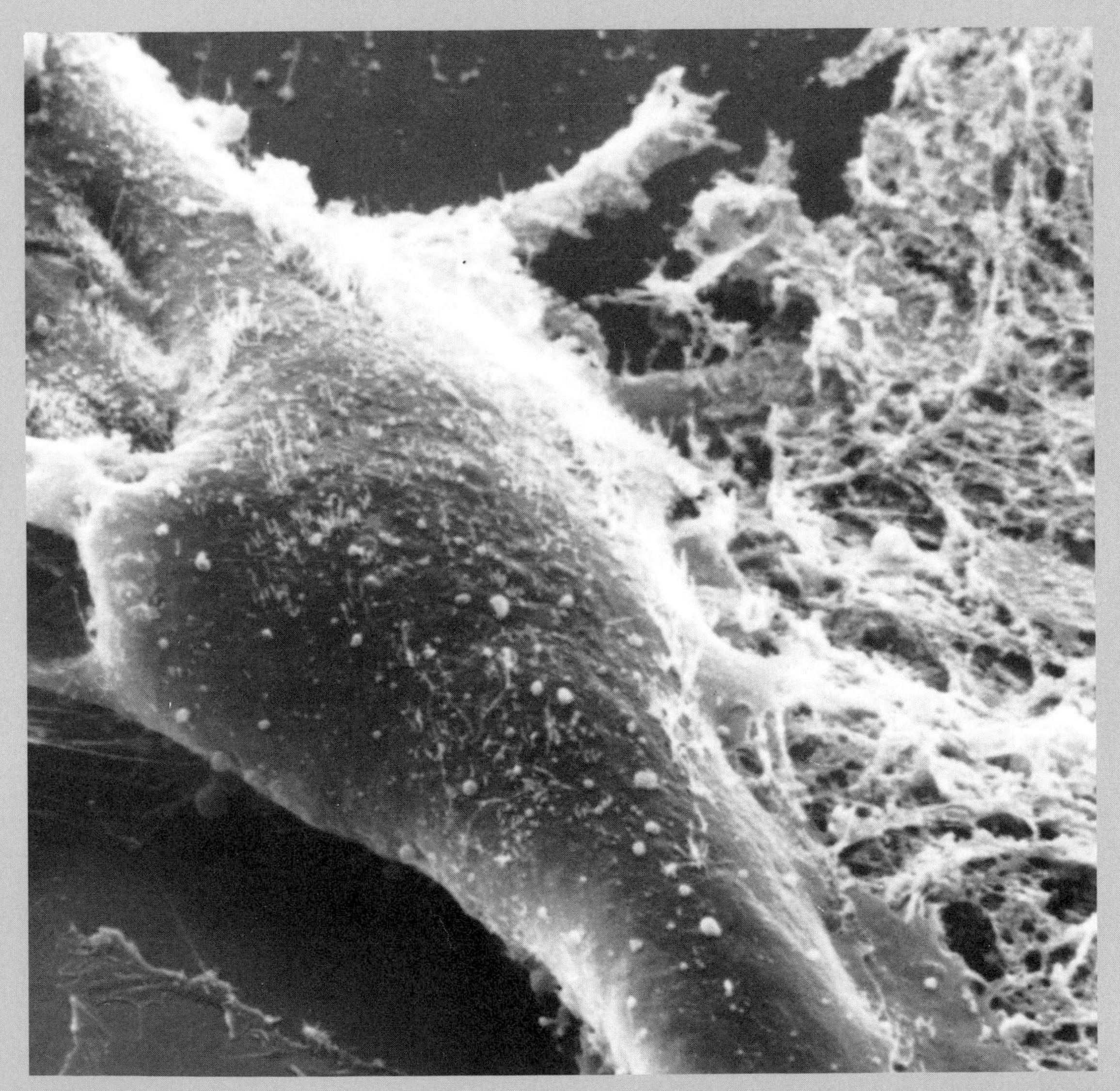

Proteolytic matrix degradation in the wake of a migrating trophoblast cell. (Courtesy of M. Sternlicht and Z. Werb, University of California, San Francisco.)

ECM proteinases

Several classes of proteolytic enzymes are involved in the hydrolysis of ECM and cell surface molecules. Thus, like ECM and cell surface molecules themselves, the ECM proteinases and their respective endogenous inhibitors play a critical role in diverse physiologic and pathologic processes.

Introduction

Proteinases that modify ECM and cell surface molecules are important in numerous normal and pathologic processes. For instance, considerable evidence indicates that ECM proteinases take part in normal fertilization, implantation, embryogenesis, morphogenesis, and differentiation, as well as in inflammation, wound repair, arthritis, liver fibrosis, emphysema, atherosclerosis, and cancer. Proteins are degraded by exopeptidases that hydrolyse peptide bonds at their amino or carboxyl termini, or by endopeptidases that cleave internal peptide bonds. The endopeptidases can, in turn, be mechanistically separated into four major classes, the 'serine', 'cysteine', 'aspartic' and 'metallo-' proteinases, by virtue of their essential catalytic group and specific inhibitor sensitivities. Thus serine proteinases (EC 3.4.21) are usually recognized by a characteristic catalytic triad composed of histidine, apartic acid, and serine residues; cysteine (thiol) proteinases (EC 3.4.22) possess an active site cysteine as well as a nearby histidine, and require activation by thiol reagents; aspartic (acid or carboxyl) proteinases (EC 3.4.23) possess two active site aspartic acids and have acidic pH optima; and metalloproteinases (EC 3.4.24) coordinate a transition metal (usually zinc) within their catalytic site. The serine and cysteine proteinases act by forming covalent enzyme–substrate intermediates, while the aspartic proteinases and metalloproteinases utilize other means of substrate degradation.[1] More specifically, serine and cysteine proteinases initially attack peptide bond carbonyl groups using the side-chain oxygen of the active site serine or the sulphur atom of the active site cysteine as their respective nucleophiles. Aspartic proteinases, on the other hand, are thought to form non-covalent tetrahedral intermediates with their substrates via the carboxyl side chains of the two active site aspartic acid residues. They then are thought to cleave their substrates by general acid–general base catalysis. It is likely that the coordinated metal ion of the metalloproteinases provides electrophilic pull so that an adjacent water molecule can attack substrate peptide bonds.

The classification of proteinases can also be based on their sensitivity to standard class-specific inhibitors such as phenylmethylsulphonyl fluoride (PMSF) for serine proteinases, E-64 for cysteine proteinases, pepstatin A for aspartic proteinases, and 1,10-phenanthroline for metalloproteinases.[1] Serine proteinases can also be inhibited by organophosphates such as diisopropylphosphofluoridate (DFP), sulfonyl fluorides such as 4-amidino-PMSF, coumarins such as 3,4-dichloroisocoumarin (3,4-DCI), boronic acid peptides, chloromethyl ketones, protein inhibitors such as aprotinin (Trasylol, BPTI) or soybean trypsin inhibitor, and endogenous inhibitors such as serpins. Other cysteine proteinase inhibitors include peptide diazomethanes, E-64 derivatives, and endogenous inhibitors of the cystatin superfamily; and other metalloproteinase inhibitors include chelating agents such as EDTA and EGTA, phosphoramidon and thiorphan which are limited in their specificity to certain vertebrate enzymes such as neprolysin and to small bacterial enzymes such as thermolysin, and TIMPs which are specific to the MMPs. Still other subclass-specific and naturally occuring inhibitors will be addressed in subsequent sections. Indeed, the design of specific proteinase inhibitors to combat various human diseases is an important undertaking that has recently received considerable research attention.

Each of the proteinase classes has members that can degrade specific ECM and cell surface molecules, although aspartic proteinases are of questionable physiologic relevance in this regard. It is also worth noting that in addition to the ECM-degrading proteinases, exo- and endoglycosidases contribute to ECM degradation by removing glycosaminoglycan and amino sugar moieties from proteoglycans; however, such enzymes lie outside the scope of this section.

Metalloproteinases

The metalloproteinases can be divided into five 'clans' or superfamilies and about 30 subfamilies.[2] One superfamily, 'the metzincins', is distinguished by a conserved structural topology, a consensus motif containing three histidines that bind zinc at the catalytic site, and a conserved 'Met-turn' motif that sits below the active site zinc[3] (Fig. 1). Their signature three-histidine zinc-binding motif is HEBXHXBGBXHZ, where H is histidine, E is glutamic acid, G is glycine, B is a bulky hydrophobic residue, X is a variable amino acid, and Z is a family-specific residue. Thus, the metzincins can be further subdivided into four distinct families, the ADAMs/adamalysins, the astacins, the serralysins, and the matrix metalloproteinases (MMPs, matrixins), based on the identity of the ultimate Z residue: that is, aspartic acid (D) in adamalysins, E in astacins, proline (P) in serralysins, and serine (S) or occasionally threonine or valine in the MMPs.[3] Likewise, the Met-turn consensus sequence UBMOJ (where M is methionine and U, O, and J are conserved family-specific residues) also distinguishes the various metzincin families from one another.[3] The U

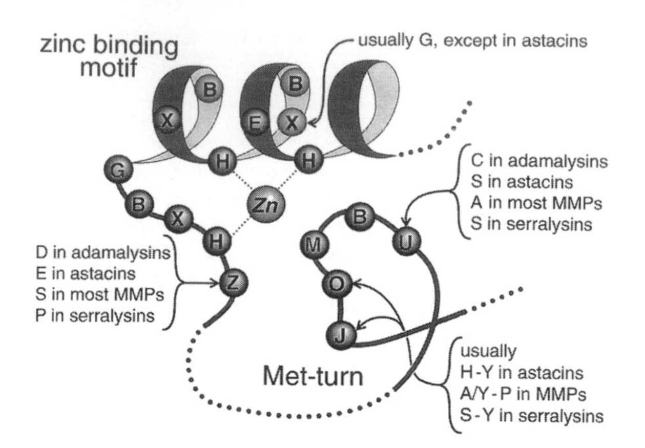

Figure 1. Structural model of the conserved zinc-binding region of the metzincins. Invariant and distinguishing residues of the three-histidine zinc-binding motif and Met-turn are indicated. Single-letter designations are explained in the text. (Modified from Stocker *et al.*[3])

residue is a disulphide-bonded cysteine (C) in adamalysins, S in astacins and serralysins, and alanine (A) in most but not all MMPs; whereas the O amino acid is usually H in astacins, S in serralysins, and A or tyrosine (Y) in the MMPs; and the J residue is a Y in astacins and serralysins and P in all MMPs other than human stromelysin-3. Finally, O and J are both variable residues in the ADAMs/adamalysins.[3,4] While all metzincins are inhibited by zinc ion chelators, natural endogenous inhibitors have only been identified for the MMPs. Another superfamily of metalloproteinases, the 'gluzincins,' has a five amino acid zinc-binding consensus motif HEXXH and a downstream glutamic acid as the third zinc-binding residue.[5] Members of this superfamily include microsomal endopeptidase, angiotensin-converting enzyme, neprolysin (enkephalinase), aminopeptidase A, and the *Bacillus* metalloproteinase thermolysin. In addition, a novel family member has recently been cloned which is immunologically related to type III procollagen N-proteinase,[6] thus raising the possibility that this latter enzyme may prove to be a gluzincin once it, too, is cloned.

The serralysins are large metzincins that are secreted by various bacteria of the genera *Serratia*, *Pseudomonas* and *Erwinia*.[3,7] No members of this family have yet been identified in higher eukaryotes. Serralysins and other metalloproteinases of bacterial origin can degrade extracellular substrates such as collagens, elastin, laminin, fibronectin, fibrin, complement components, and immunoglobulins.[8] Thus, several metalloproteinases have either been shown or suggested to play an important part in the virulence and pathogenicity of bacterial organisms. For example, the haemagglutinin metalloproteinase of *V. cholerae* can activate the A subunit of cholera toxin; botulinum B and tetanus neurotoxins are both metalloproteinases that block neurotransmitter release by cleaving the integral membrane protein synaptobrevin; and the anthrax toxin lethal factor displays metalloproteinase homology and activity.[8] In addition, some bacterial metalloproteinases, such as *Clostridium histolyticum* collagenases, have garnered widespread use as research tools. Owing to their prokaryotic origin, however, these metalloproteinases will not be considered further in this text.

The matrix metalloproteinase (MMP) family currently has 18 known members in vertebrates, one in the sea urchin, and another in soybean leaves. The vertebrate MMPs each have distinct but often overlapping substrate specificities, and together they possess enzymatic activity against virtually all ECM components. In addition to their dependence on zinc and calcium, the MMPs share several other common features. They show extensive sequence homology; they are all synthesized as inactive zymogens that are activated by limited proteolysis or exposure to organic mercurial compounds; they are either secreted or expressed as transmembrane proteins; they are universally inhibited by 1,10-phenanthroline, chelating agents, and naturally occuring tissue inhibitors of metalloproteinases (TIMPs); they are transcriptionally responsive to cytokines, growth factors, hormones, and ECM-derived signals; and, as mentioned, they can hydrolyse ECM proteins.[9] Individual MMPs have been variously named, grouped, and subdivided based on their substrate specificities and the presence or absence of specific functional protein domains. In the present text, MMPs will be classified as either 'collagenases', 'gelatinases', 'stromelysins', 'membrane-type MMPs', or 'other MMPs', although other schema could also have been adopted.[9]

The adamalysins (reprolysins) represent a family of soluble snake-venom enzymes that aggressively degrade structural ECM components. Together with snake-venom disintegrins that contain the integrin-binding RGD sequence and prevent platelet aggregation, the adamalysins can cause rapid haemorrhage, tissue damage, and necrosis. The family name, adamalysins, derives not from any domain considerations, but from the prototypic enzyme adamalysin II of the diamondback rattlesnake *Crotalus adamanteus*. The ADAMs, on the other hand, are cell surface rather than secreted proteins that share a disintegrin and metalloproteinase domain and are thought to play important roles in fertilization, spermatogenesis, myogenesis, neurogenesis, and somatogenesis.[4] There are presently at least 17 ADAMs that have been cloned from various species and tissue sources, and each contains a signal sequence followed in order by a prodomain, a metalloproteinase or metalloproteinase-like domain, a disintegin-like domain, a cysteine-rich domain, EGF-like repeats, a transmembrane domain, and a cytoplasmic tail. Accordingly, ADAMs could carry out four distinct but complementary functions: proteolysis via the metalloproteinase domain, adhesion via the disintegrin domain, cell–cell fusion via a candidate hydrophobic fusion peptide in the cysteine-rich domain, and cell signalling via the intracellular domain which often contains sequences that could bind the SH3 domain of src. The metalloproteinase domain of the ADAMs shows close homology to that of the adamalysins, with the highly conserved D at the end of the zinc-binding motif remaining invariant. However, not all ADAMs possess the active-site consensus sequence described above for the metzincins, and thus not all are likely to have proteolytic activity. Those ADAMs that do contain the signature three-histidine zinc-binding motif are fertilin α (which has been implicated in early spermatogenesis and sperm–egg binding), MADM (a bovine ADAM that can cleave type IV collagen and which has been implicated in myelin degradation), KUZ (a *Drosophila* homologue of MADM implicated in neurogenesis), Ms2 (which is specifically expressed on monocytic cells), MDC9 (which is widely expressed), meltrin α (which is implicated in myoblast fusion), metargidin (the only known ADAM with an RGD motif in its disintegrin domain), and human tumour necrosis factor α (TNF-α) converting enzyme (TACE) which cleaves membrane-bound TNF-α to generate the active secreted form of TNF-α.[4,10,11,37] Identification of the latter ADAM raises the likelihood that other members of this growing multigene family may process other cell surface proteins. Indeed, KUZ and its mouse orthologue cleave notch (D.J. Pan: personal communication), and metalloproteinase inhibitors can block the shedding of a number of cell surface molecules. For instance, they can inhibit the shedding of the L-selectin adhesion molecule, soluble fas ligand, β-amyloid precursor protein A4, and the receptors for interleukin 6, vasopressin, transforming growth factor-α, and TNF.[12–14] Likewise, metalloproteinase inhibitors can block the conversion of inactive paracrine factors, such as endothelin-1, to their mature active forms.[15] In most cases, however, the responsible 'sheddase' or 'convertase' remains unidentified. Also unidentified is the 'aggrecanase' that is responsible for the cleavage of one of two sites within the interglobular domain of aggrecan that occurs during arthritic disease (the other site being susceptible to several MMPS). The recent findings that a snake venom reprolysin (atrolysin C) can cleave this site and that three ADAMs are expressed in human articular chondrocytes suggests that an ADAM may play a significant role in arthritic chondrodegeneration.[38]

The astacin family of over 20 metzincin metalloproteinases derives its name from the small prototype digestive enzyme astacin of the crayfish *Astacus astacus* L.[3,16] Mammalian family members that fall within the purview of this text are bone morphogenetic protein-1 (BMP-1) and meprins A and B. Human BMP-1 was originally isolated from bone in a complex with BMPs 2 and 3 based on their combined ability to induce bone formation *in vivo*.[17] The latter BMPs are members of the TGF-β superfamily of growth factors, and BMP-1 was originally thought to activate latent TGF-β-like growth factors.[16] Indeed, the *Drosophila* BMP-1 counterparts, tolloid and tolkin, activate the TGF-β superfamily member decapentaplegic (dpp), and each is essential during fly development to specify the dorsal–ventral axis in the embryo.[18] More recently, however, BMP-1 was shown to be the procollagen C-proteinase that cleaves the C-terminal propeptides of fibrillar procollagens.[19] Thus BMP-1 is essential for the proper formation of supramolecular collagen fibrils within the ECM. Although BMP-1 may potentially activate TGF-β-like proteins, mammalian tolloid, which is encoded by an alternatively spliced transcript from the BMP-1 gene, is another likely candidate in this regard, because its domain structure is identical to that of *Drosophila* tolloid. All astacins have an N-terminal secretory signal domain, a prodomain that keeps the enzyme inactive until it is removed, and a conserved catalytic domain.[16] In addition, BMP-1 and mammalian tolloid have complement C1r/C1s-like and EGF-like domains that follow the catalytic domain. Meprins, on the other hand, belong to another branch of the astacin family tree.[16] Their catalytic domain is followed by a putative adhesion domain, a unique 'extender' domain, an inserted domain in the meprin α subunit, an EGF-like domain, a transmembrane domain, and a cytoplasmic domain. The meprin α and β subunits form dimers of disulphide-linked dimers resulting in either membrane-bound meprins B or A, or secreted meprin A.[16] These are abundant in the brush border membranes of kidney and intestine, and are found in lesser amounts in various other tissues. There they are thought to process various peptide hormones, and possibly ECM proteins such as collagen IV, laminin, and fibronectin.[20]

While the type I procollagen C-proteinase (BMP-1) has been cloned, procollagen N-proteinases have not. A human type III procollagen N-proteinase has been isolated from placenta, and determined to be a 70 kDa metalloproteinase.[21] Utilizing antibodies raised against this enzyme, Scott *et al.*[6] cloned a novel 33 kDa protein, termed PRSM1 (for protease, metallo 1), from a human placental cDNA expression library. Evidence indicates, however, that the 70 kDa enzyme is not a dimer or post-translationally modified product of PRSM1, and thus PRSM1 is unlikely to be the type III procollagen N-proteinase. Nevertheless, the two may share sequence homology owing to their apparent immunologic similarity. Interestingly, PRSM1 has a two-histidine zinc-binding sequence (HELGH) and the third zinc-coordinating amino acid is a glutamic acid that sits outside the consensus HEXXH motif. Thus PRSM-1 belongs to the gluzincin rather than metzincin superfamily of metalloproteinases. In addition, PRSM1 has a putative 13 amino acid signal sequence and contains a collagen-like domain with six Gly–X–Y repeats. Finally, the human PRSM1 gene is located on chromosome 16q24.3 and encodes a 2.5 kb message.

Serine proteinases

Well over 20 serine peptidase families are now recognized.[22] Serine proteinases are synthesized as inactive zymogens which are activated by removal of an N-terminal pro sequence of variable length. For those serine proteinases with known genomic sequences, each has a His–Asp–Ser catalytic triad where each residue is encoded by a different exon; each is predicted to have signal and pro-N peptides; and their catalytic domains are spread over several exons.[23] Serine proteinases that can degrade

ECM molecules include neutrophil elastase, cathepsin G, proteinase 3, mast cell chymases, granzymes, tissue kallikreins, trypsin, chymotrypsin, fibroblast activation protein-α, dipeptidyl peptidase IV, seprase (a recently discovered dimeric membrane-bound gelatinase),[24] and the enzymes of the fibrinolytic (plasminogen activator/plasmin) system. In this latter system, tissue-type and urokinase-type plasminogen activators (tPA and uPA) cleave the proenzyme plasminogen to form active plasmin. Plasmin, in turn, can activate latent MMPs and can itself degrade fibrin and ECM proteins such as laminin and fibronectin. The plasminogen activators are primarily controlled by plasminogen activator inhibitors (PAI-1 and PAI-2) and protease nexin-1, and plasmin can be inhibited by α_2-antiplasmin, α_2-macroglobulin, α_1-proteinase inhibitor (α_1-antitrypsin) and protease nexin-2.[25–27] In addition, a specific GPI-linked uPA receptor (CD87) has been identified which localizes uPA to the cell surface and modulates its activity.[28]

In addition to degrading ECM proteins directly, certain serine proteinases can activate still other ECM proteinases. Some serine proteinases such as plasmin, uPA, and elastase activate a number of metalloproteinases and do so outside the cell or in association with the cell surface. Other transmembrane serine proteinases, termed furin/PACE/kex 2-like proprotein convertases, apparently activate another set of metalloproteinases, namely meprin A, stromelysin-3, the membrane-type MMPs, and probably BMP-1 as well.[29] These furin-like enzymes belong to the large clan of subtilisin serine peptidases; they act at multiple cellular sites including the *trans*-Golgi network, cell surface and early endosomes; and they cleave a wide array of precursor proteins at the consensus sequence RXK/RR.[29] Since furin-like enzymes can activate membrane-type MMPs, which, in turn, activate secreted MMPs, the furin-like enzymes appear to be critical upstream initiators of further activation events.

Cysteine proteinases

Cysteine proteinases that can hydrolyse ECM proteins include the lysosomal cathepsins B, K, L and S. Although these enzymes have acidic pH optima, they remain partially active at neutral pH and when secreted they may also act in locally acidic microenvironments.[30–32] Cathepsin B can activate latent MMPs and uPA, and can itself degrade proteoglycans, fibronectin, and non-helical regions of collagens I, II, III, and IV.[33] Cathepsin K, which is highly expressed in osteoclasts, can degrade type I collagen, elastin, osteocalcin, and osteonectin/SPARC, and mutations in its gene cause pycnodysostosis, an autosomal recessive bone resorption disorder.[34,35] Cathepsins L and S are both secreted by human alveolar macrophages and both can cleave elastin and collagen.[30,31] In addition, cathepsins B and L are expressed by thyrocytes and are involved in the extracellular solubilization of cross-linked thyroglobulin.[32] The lysosomal cysteine peptidase cathepsin C (dipeptidyl-peptidase I) is the likely activator of lysosomal serine proteinases that have a 2-amino acid propeptide domain, such as neutrophil elastase, cathepsin G, mast cell chymases, and lymphocytic granzymes.[35] These enzymes, together with cathepsin D (an aspartic proteinase implicated in ECM degradation and the activation of cysteine-type cathepsins), are further described in Lysosomal proteinases.

Acknowledgement

This work was supported by an Institutional National Research Service Award from the National Institute of Environmental Health Sciences (T32 ES07106) and grants from the National Institutes of health (CA57621, DE10306, and HD26732).

References

1. Beynon, R. J. and Bond, J. S. (1989). *Proteolytic enzymes. A practical approach*. Oxford University Press, New York.
2. Rawlings, N. D. and Barrett, A. J. (1995). *Meth. Enzymol.*, **248**, 183–228.
3. Stocker, W., Grams, F., Baumann, U., Reinemer, P., Gomis-Ruth, F. X., McKay, D. B., and Bode, W. (1995). *Protein Sci.*, **4**, 823–40.
4. Wolfsberg, T. G. and White, J. M. (1996). *Dev. Biol.*, **180**, 389–401.
5. Hooper, N. M. (1994). *FEBS Lett.*, **354**, 1–6.
6. Scott, I. C., Halila, R., Jenkins, J. M., Mehan, S., Apostolou, S., Winqvist, R., *et al.* (1996). *Gene*, **174**, 135–43.
7. Maeda, H. and Morihara, K. (1995). *Meth. Enzymol.*, **248**, 395–413.
8. Hase, C. C. and Finkelstein, R. A. (1994). *Microbiol. Rev.* **57**, 823–37.
9. Coussens, L. M. and Werb, Z. (1996). *Chem. Biol.*, **3**, 895–904.
10. Black, R. A., Rauch, C. T., Kozlosky, C. J., Peschon, J. J., Slack, J. L., Wolfson, M. F., *et al.* (1997). *Nature*, **385**, 729–33.
11. Moss, M. L., Jin, S. L. C., Milla, M. E., Burkhart, W., Carter, H. L., Chen, W. J., *et al.* (1997). *Nature*, **385**, 733–6.
12. Crowe, P. D., Walter, B. N., Mohler, K. M., Otten-Evans, C., Black, R. A., and Ware, C. F. (1995). *J. Exp. Med.*, **181**, 1205–10.
13. Kojro, E. and Fahrenholz, F. (1995). *J. Biol. Chem.*, **270**, 6476–81.
14. Arribas, J., Coodly, L., Vollmer, P., Kishimoto, T. K., Rose-John, S., and Massague, J. (1996). *J. Biol. Chem.*, **271**, 11376–82.
15. Patel, K. V. and Schrey, M. P. (1995). *Br. J. Cancer*, **71**, 442–7.
16. Bond, J. S. and Beynon R. J. (1995). *Protein Sci.*, **4**, 1247–61.
17. Wozney, J. M., Rosen, V., Celeste, A. J., Mitsock, L. M., Whitters, M. J., Kriz, R. W., *et al.* (1988). *Science*, **242**, 1528–31.
18. Finelli, A. L., Xie, T., Bossie, C. A., Blackman, R. K., and Padgett, R. W. (1995). *Genetics*, **141**, 271–81.
19. Kessler, E., Takahara, K., Biniaminov, L., Brusel, M., and Greenspan, D. S. (1996). *Science*, **271**, 360–2.
20. Kaushal, G. P., Walker, P. D., and Shah, S. V. (1994). *J. Cell Biol.*, **126**, 1319–27.
21. Halila, R. and Peltonen, L. (1986). *Biochem. J.*, **239**, 47–52.
22. Rawlings, N. D. and Barrett, A. J. (1994). *Meth. Enzymol.*, **244**, 19–61.
23. Takahashi, H., Nukiwa, T., Yoshimura, K., Quick, C. D., States, D. J., Holmes, M. D., *et al.* (1988). *J. Biol. Chem.*, **263**, 14739–47.

24. Pineiro-Sanchez, M. L., Goldstein, L. A., Dodt, J., Howard, L., Yeh, Y., and Chen, W. T. (1997). *J. Biol. Chem.*, **272**, 7595–601.
25. Kwaan, H. C. (1992). *Cancer Metastasis Rev.* **11**, 291–311.
26. Baker, J. B., Low, D. A., Simmer, R. L., and Cunningham, D. D. (1980). *Cell*, **21**, 37–45.
27. Van Nostrand, W. E., Wagner, S. L., Farrow, J. S., and Cunningham, D. D. (1990). *J. Biol. Chem.*, **265**, 9591–4.
28. Stahl, A. and Mueller, B. M. (1995). *J. Cell Biol.*, **129**, 335–44.
29. Basbaum, C. B. and Werb, Z. (1996). *Curr. Opin. Cell Biol.*, **8**, 731–8.
30. Shi, G. P., Munger, J. S., Meara, J. P., Rich, D. H., and Chapman, H. A. (1992). *J. Biol. Chem.*, **267**, 7258–62.
31. Reilly, J. J., Mason, R. W., Chen, P., Joseph, L. J., Sukhatme, V. P., Yee, R., and Chapman, H. A. (1989). *Biochem. J.*, **257**, 493–8.
32. Brix, K., Lemansky, P., and Herzog, V. (1996). *Endocrinology*, **137**, 1963–74.
33. Sloane, B. F. and Honn, K. V. (1990). *Cancer Metastasis Rev.* **3**, 249–63.
34. Bossard, M. J., Tomaszek, T. A., Thompson, S. K., Amegadzie, B. Y., Hanning, C. R., Jones, C., *et al.* (1996). *J. Biol. Chem.*, **271**, 12517–24.
35. Gelb, B. D., Shi, G. P., Chapman, H. A., and Desnick, R. J. (1996). *Science*, **273**, 1236–8.
36. Rao, N. V., Rao, G. V., Marshall, B. C., and Hoidal, J. R. (1996). *J. Biol. Chem.*, **271**, 2972–8.
37. Millichip, M. I., Dallas, D. J., Wu, E., Dale, S., and McKie, N. (1998). *Biochem. Biophys. Res. Commun.*, **245**, 594–8.
38. Tortorella, M. D., Pratta, M. A., Fox, J. W., and Arner, E. C. (1998). *J. Biol. Chem.*, **273**, 5846–50.

Mark D. Sternlicht and Zena Werb
University of California, San Francisco,
Department of Anatomy, LR-208,
3rd and Parnassus Avenues, San Francisco,
CA 94143–0452, USA

Astacin/tolloid proteinases

The astacins constitute a family of metzincin metalloproteinases found in species from hydra to humans. Relevant mammalian members are bone morphogenetic protein-1 (BMP-1) which is the type I procollagen C-proteinase (PCP), mammalian tolloid (mTld) which is an alternative splice product of the same gene and which may activate latent TGF-β-like growth factors, and meprins A and B which are abundant at intestinal and renal brush borders and are thought to process peptide hormones and possibly ECM molecules as well.

Protein properties

The astacin family of metzincin metalloproteinases derives its name from the prototype digestive enzyme astacin (EC 3.4.24.21) of the crayfish *Astacus astacus* L.[1,2] The astacins are distinguished from other metzincins by the final glutamic acid in their signature three-histidine zinc-binding sequence HEXXHXXGFXHE.[1,2] Astacins are ususally secreted, but in the case of the meprins, they can be either membrane-associated or secreted. All fully characterized family members contain a transient N-terminal signal sequence that directs their synthesis to the endoplasmic reticulum. The prodomain that follows shows considerable heterogeneity among the astacins and is thought to regulate enzyme activation and possibly enzyme expression as well. A highly conserved catalytic domain follows the prodomain and other subsequent domains may or may not be present, depending upon the particular enzyme (Fig. 1). Astacin family members of particular interest include bone morphogenetic protein-1 (BMP-1), mammalian tolloid (mTld), and the meprin α and β subunits which form hetero- and homooligomeric meprins A and B.

Human BMP-1 was first isolated from demineralized bone in a complex with two added proteins (BMPs 2 and 3) that turned out to be members of the TGF-β superfamily of growth factors.[3] Full-length cDNAs for each BMP were then obtained using amino acid sequences from tryptic peptides of the isolated BMP complex. Because BMP-1 complexes with TGF-β-like proteins and can induce ectopic bone formation *in vivo*, and because its *Drosophila* counterparts, tolloid and tolkin, are requisite activators of the fruitfly TGF-β-like protein decapentaplegic,[4] it has been suggested that BMP-1 may function to activate latent TGF-β-like proteins. Direct evidence for such a role, however, has not been forthcoming. Recently, it was determined that BMP-1 is identical to the procollagen C-proteinase (PCP) that removes the C-terminal propeptides of fibrillar procollagens, thus enabling their self-assembly into supramolecular collagen aggregates.[5] Recent evidence also indicates that BMP-1 can cleave laminin-5 as well (R.E. Burgeson, personal communication). Although BMP-1 could still potentially activate TGF-β-like proteins, an alternatively spliced transcript from the same BMP-1 gene encodes mTld, which is perhaps a more likely candidate in this regard since its domain structure is identical to that of *Drosophila* tolloid.[6] Following their pre-, pro-, and catalytic domains, BMP-1 and mTld contain three or five so-called 'CUB' domains (for complement subcomponent C1r/C1s, embryonic sea urchin uegf, BMP-1) and one or two intervening EGF-like domains, respectively (Fig. 1). In mice and

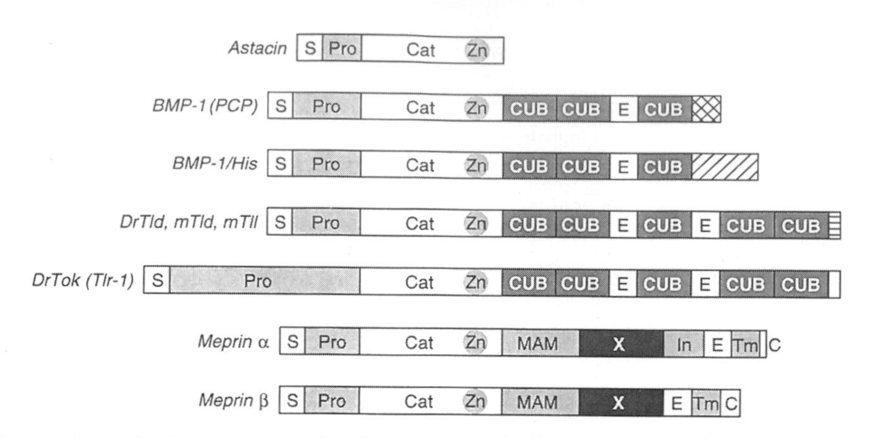

Figure 1. Domain structure of relevant astacin family members. S, signal sequence; Pro, propeptide; Cat, catalytic domain; Zn, zinc-binding site; CUB, complement C1r/C1s-like domains; E, EGF-like domains; MAM, putative adhesion domain; X, extender domain; In, inserted domain; Tm, transmembrane domain; C, cytoplasmic tail; Dr, *Drosophila*. Modified from refs 2, 6, 7.)

humans, a third alternative splice product of the same gene encodes BMP-1/His, a protein with a novel C-terminal domain rich in Ser, Thr, Leu, and His residues.[6] A mTld-specific probe recognizes 5.5 and 4.1 kb transcripts, a BMP-1/mTld probe recognizes the same two transcripts plus a 3.0 kb band, and a BMP-1/His-specific probe detects 4.9 and 3.4 kb bands, where underlined sizes are those expected from corresponding cDNA clones. BMP-1 and mTld are both highly expressed in placenta and developing bone. Low mRNA levels of each have also been detected in adult human heart, lung, liver, kidney, pancreas, and muscle; and low mTld, but not BMP-1, levels have been detected in adult brain. Likewise, *in situ* hybridization has revealed high mRNA levels for only mTld in the neural tube floor plate of the developing mouse nervous sytem. The third alternative splice product has not been detected in adult human tissues but has been found, together with the other two isoforms, in placental trophoblast giant cells, with strongest expression being near the maternal interface. A related mammalian tolloid-like (mTll) gene product has also been identified.[7] It has an identical domain structure to mTld and overlapping but distinct developmental expression patterns compared to the BMP-1 gene products. Only BMP-1 has been characterized at the protein level, and only because it has turned out to be identical to the already purified PCP enzyme. A 125 kDa PCP has been isolated from mouse fibroblast cultures, and 110 and 95 kDa forms of PCP have been purified from chick organ cultures.[8]

Meprins A (EC 3.4.24.18) and B (EC 3.4.24.63) are tetramers of α and β subunits.[2] Each subunit has pre-, pro-, and catalytic domains that are followed by a so-called MAM domain (for meprin, a-5 protein, receptor protein tyrosine phosphatase mu) which is thought to play a role in adhesion. The MAM domain is then followed by a unique 'extender' domain, a 56 amino acid inserted domain in meprin-α alone, an EGF-like domain, a transmembrane domain, and a short cytoplasmic tail domain (Fig. 1). The α and β subunits form disulphide-linked homo- or heterodimers which then dimerize to form either membrane-bound meprins A or B, or secreted meprin A (Fig. 2). Secreted meprin A is a dimer of disulphide-linked α–α homodimers. Membrane-bound meprin A, on the other hand, is a hetero-oligomer with at least one disulphide-linked α–β dimer, and meprin B is a membrane-bound dimer of disulphide-linked β–β dimers. For the membrane-bound meprins, the β subunits retain their propeptide, EGF-like, transmembrane, and cytoplasmic domains, whereas the α subunits do not. In addition, the unique 56 amino acid insert in the α subunit is necessary and sufficient for the C-terminal processing to take place in the endoplasmic reticulum, as indicated using mutant meprin subunits.[2] Meprins are abundant in the brush border membranes of kidney and intestine, and are found in lesser amounts in various other tissues. There they are thought to process various peptide hormones as well as ECM proteins.[2,9] Solubilized mouse meprin A has a molecular mass of 270–320 kDa and individual subunits have masses of 72–100 kDa, depending on the species of origin.[10] Each subunit has several potential *N*-glycosylation sites and deglycosylation studies indicate that they contain 20–30 per cent *N*-linked carbohydrate.[10]

Purification

In the original description of BMP-1, only 40 μg of a 30 kDa fragment containing polypeptides from three BMPs could be obtained from 40 kg of bovine bone powder.[3] Recombinant BMP-1 has, however, been produced using a baculovirus expression system and thus determined to be identical to PCP.[5] Thus, although BMP-1 (PCP) is apparently quite difficult to extract from bone, it can readily be purified from chick embryo tendon and mouse fibroblast conditioned medium.[8] The first method utilizes organ culture-derived medium and green A Matrex, concanavalin A–Sepharose, heparin–Sepharose, and Sephacryl S-300 and S-200 chromatographies; and

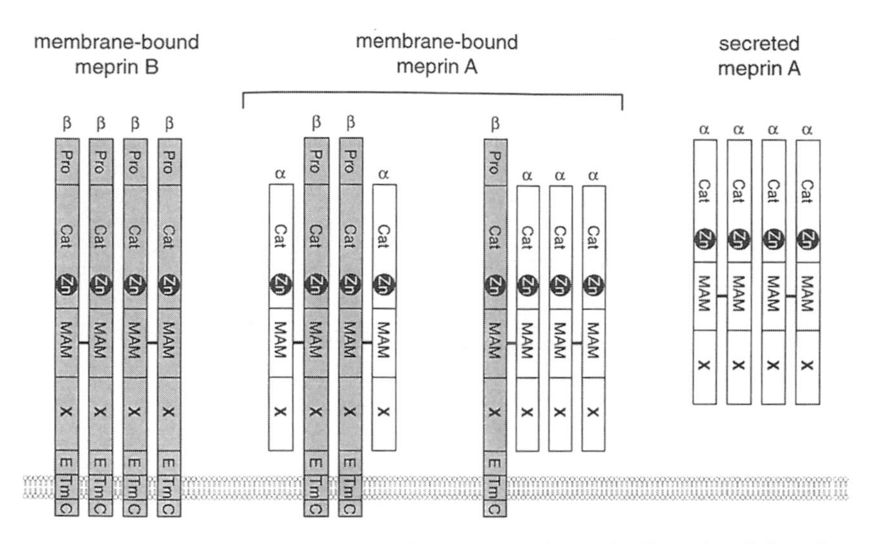

Figure 2. Oligomeric structure of membrane-associated and secreted meprin. Functional domains are shown for the unshaded α and the shaded β subunits. Horizontal lines between adjacent MAM domains represent disulphide bridges. Membrane anchoring is via the β subunits since mature α subunits are truncated at the end of their X domain. Modified from ref. 2.)

the latter utilizes ammonium sulphate precipitation and fractionation followed by Sephadex G-150, lysyl–Sepharose, and C-propeptide–Sepharose 4B chromatographies.

Meprin A is most often purified from mouse kidney homogenates (strain ICR) by sequential ultracentrifugation to sediment brush border membranes, limited papain or trypsin solubilization, ammonium sulphate precipitation, and MonoQ anion exchange and Superose gel-filtration chromatographies.[10] Meprin B is usually purified from mouse kidney using strains (e.g. C3H/He) that only express meprin B.[10] In this case, however, active fractions obtained by the same method as outlined above are further purified by MonoQ and angiotensin I-converting enzyme affinity chomatographies.

Activities

PCP (BMP-1) catalyses the removal of C-propeptides from fibrillar procollagens I, II, and III and probably procollagens V and XI as well.[8] Its activation occurs via prodomain removal at the sequence RSRR, suggesting that intracellular furin-like enzymes are responsible.[5] Once activated, PCP cleaves specifically at the x–Asp bond between the C-telopeptide and C-propeptide of the respective fibrillar procollagen α chains, and does so at neutral pH.[8] In addition, a 55 kDa fibroblast-derived glycoprotein, termed PCP enhancer protein, binds to the type I procollagen C-propeptide via CUB domains and enhances PCP activity 4–7-fold.[5,8]

Meprin A can cleave a number of bioactive peptides, including parathyroid hormone, α melanocyte-stimulating hormone, bradykinin, neurotensin, gonadotropin (LHRH), substance P, TGF-α, α-atrial natriuretic peptide, angiotensins, and endothelins.[2,10] In addition, meprins have been implicated in the processing of ECM components such as collagen IV, laminin, fibronectin, and gelatin.[9] Substrates used to monitor meprin activity include azocasein, ^{125}I-insulin B chain, *N*-benzoyltyrosyl-*p*-aminobenzoate, bradykinin derivatives, and chromogenic and fluorogenic peptide substrates.[10] Meprins A and B differ in terms of their relative activities and specificities.[10] For instance, meprin A cleaves the insulin B chain at 10 sites, whereas meprin B cleaves it at one unique site and at three of the same sites as does meprin A. Before trypsin treatment, meprin A has 5–10-fold greater activity against azocasein and the insulin B chain than does meprin B. Trypsin activation increases the activity of meprin A against these substrates less than two-fold, but increases the activity of meprin B 5–20-fold. Also, meprin B fails to cleave nitrobradykinin, whereas meprin A cleaves it readily with or without trypsin activation. Meprin A prefers substrates with a proline in the P2′ or P3′ position. Both enzymes are inhibited by standard metalloproteinase inhibitors, but not by phosphoramidon or captopril. Amino acid hydroxamates with aromatic side chains and the naturally occuring peptide hydroxamate actinonin are apparently potent inhibitors of meprin A.

Antibodies

Guinea pig antiserum to a human BMP-1-polyhistidine fusion protein recognizes both recombinant human BMP-1 and purified mouse PCP by immunoblotting.[5] Such blots have shown that the human and murine proteins have identical electrophoretic mobilities once they are deglycosylated. Rabbit antisera have been raised against purified rat kidney meprin and against an 82 kDa rat meprin polypeptide.[11] Polyclonal antibodies to the mouse meprin α subunit have been raised using deglycosylated meprin A and using a peptide from the α subunit EGF-like domain.[12] Polyclonal antibodies against the C-terminal cytoplasmic domain of the the mouse β subunit and against its catalytic domain have also been described.[12]

Genes

The human BMP-1/mTld gene (GenBank L40486) has been mapped to chromosome 8p21 between markers D8S298 and D8S5; it spans 46 kbp, it has 22 exons, its 5' region contains the polyadenylation site of another nearby gene and nine potential GC box/SP-1-binding sites, but lacks canonical CAAT and TATA boxes; it has two potential transcription start sites; and its major start site sits within a 69 bp GAGA box.[13] Full length cDNAs and alternative splice variants are published for BMP-1/mTld in humans (M22488, Y08723–25, L35278–81) and mice (L24755). The complete mouse cDNA for mammalian tolloid-like protein (mTll; U34042) has been published and its gene mapped to the middle of mouse chromosome 8, whereas the mouse BMP-1/mTld gene maps to the proximal half of mouse chromosome 14.[2,7] Complete cDNAs are also published for *Drosophila* tolloid (M76976) and tolloid-related-1 (tolkin) (U12634, U34777), which are the products of two separate genes. Two distinct genes have also been identified for *Xenopus* BMP-1 and tolloid (xolloid).[2]

Complete cDNAs have been obtained for the mouse meprin α subunit (M74897) and for the β subunit in mice (L15193) and rats (M88601). Structural genes for the meprin α and β subunits map to mouse chromosomes 17 and 18, and to human chromosomes 6p11–p12 and 18q12.2–q12.3, respectively.[2] The α subunit gene is linked to the major histocompatability complex in both species. Gene expression varies depending on the particular tissue, species, and strain.

Mutant phenotype/disease states

Mouse embryos with homozygous deletions in the BMP-1/mTld active site developed until late gestation with few morphologic defects, but then failed to close their ventral body wall and died at birth.[14] Abnormal collagen fibrils and collagen processing were observed in the amnions and cultured fibroblasts of homozygous mutant embryos. The related *Drosophila* genes tolloid and tolkin are required for proper dorsal–ventral patterning during larval and pupal development.[4,15,16]

Structure

The X-ray crystal structure of crayfish astacin has been solved to a resolution of 1.8 Å[17–19] and used to model the structure of other astacin family members.[1,20] Astacin itself has a kidney-like shape with a long, deep active-site cleft separating the N- and C-terminal lobes. The required zinc ion is present at the bottom of the cleft and is penta-coordinated in a trigonal bipyramidal manner by the three histidines of the conserved zinc-binding motif, a tyrosine within the so-called 'Met turn', and a water molecule. In addition, the N-terminal alanine is buried and forms a water-linked salt bridge with the glutamic acid that resides at the end of the three-histidine zinc-binding motif and which distinguishes the astacins from other metzincins. Protein Data Bank (PDB) identifiers are 1AST for astacin, 1IAA–1IAE for astacin with zinc removed or replaced by other metals, and 1IAF for a theoretical model of the catalytic domain of meprin α.

References

1. Stocker, W. and Zwilling, R. (1995). *Meth. Enzymol.*, **248**, 305–25.
2. Bond, J. S. and Beynon, R. J. (1995). *Protein Sci.*, **4**, 1247–61.
3. Wozney, J. M., Rosen, V., Celeste, A. J., Mitsock, L. M., Whitters, M. J., Kriz, R. W., *et al.* (1988). *Science*, **242**, 1528–31.
4. Finelli, A. L., Xie, T., Bossie, C. A., Blackman, R. K., and Padgett, R. W. (1995). *Genetics*, **141**, 271–81.
5. Kessler, E., Takahara, K., Biniaminov, L., Brusel, M., and Greenspan, D. S. (1996). *Science*, **271**, 360–2.
6. Takahara, K., Lyons, G. E., and Greenspan, D. S. (1994). *J. Biol. Chem.*, **269**, 32572–8.
7. Takahara, K., Brevard, R., Hoffman, G. G., Suzuki, N., and Greenspan, D. (1996). *Genomics*, **34**, 157–65.
8. Kadler, K. E. and Watson, R. B. (1995). *Meth. Enzymol.*, **248**, 771–81.
9. Kaushal, G. P., Walker, P. D., and Shah, S. V. (1994). *J. Cell Biol.*, **126**, 1319–27.
10. Wolz, R. L. and Bond, J. S. (1995). *Meth. Enzymol.*, **248**, 325–45.
11. Johnson, G. D. and Hersh, L. B. (1992). *J. Biol. Chem.*, **267**, 13505–12.
12. Marchand, P., Tang, J., and Bond, J. S. (1994). *J. Biol. Chem.*, **269**, 15388–93.
13. Takahara, K., Lee, S., Wood, S., and Greenspan, D. S. (1995). *Genomics*, **29**, 9–15.
14. Suzuki, N., Labosky, P. A., Furuta, Y., Harfett, L., Dunn, R., Fogo, A. B *et al.* (1996). *Development*, **122**, 3587–95.
15. Shimell, M. J., Ferguson, E. L., Childs, S. R., and O'Connor, M. B. (1991). *Cell*, **67**, 469–81.
16. Nguyen, T., Jamal, J., Shimell, M. J., Arora, K., and O'Connor, M. B. (1994). *Dev. Biol.*, **166**, 569–86.
17. Bode, W., Gomis-Ruth, F. X., Huber, R., Zwilling, R., and Stocker, W. (1992). *Nature*, **358**, 164–7.
18. Gomis-Ruth, F. X., Stocker, W., Huber, R., Zwilling, R., and Bode, W. (1993). ??? **229**, 945–68.
19. Gomis-Ruth, F. X., Grams, F., Yiallouros, I., Nar, H., Kusthardt, Zwilling, R., *et al.* (1994). *J. Biol. Chem.*, **269**, 17111–17.
20. Stocker, W., Gomis-Ruth, F. X., Bode, W., and Zwilling, R. (1993). *Eur. J. Biochem.*, **214**, 215–31.

Mark D. Sternlicht and Zena Werb
University of California, San Francisco, Department of Anatomy, LR-208, 3rd and Parnassus Avenues, San Francisco, CA 94143–0452, USA

Lysosomal proteinases

Several lysosomal enzymes of the cysteine, serine and aspartic classes can be secreted and can degrade ECM proteins directly. For example, the cysteine-type cathepsins B, H, K, L and S and the aspartic proteinase cathepsin D can each degrade ECM components, and often at neutral pH. In addition, cathepsin B may activate other ECM proteinases such as stromelysin-1 and uPA, cathepsin D may activate serine-type cathepsins, and the cysteine exopeptidase cathepsin C (dipeptidyl-peptidase I) may activate serine proteinases such as cathepsin G and neutrophil elastase. Accordingly, the above enzymes have each been implicated in a number of physiologic and pathologic processes.

Protein properties

Several lysosomal enzymes are likely to act as ECM-degrading enzymes and/or as activators of other ECM proteinases. These include cysteine proteinases such as cathepsins B, C, K, L, and S, the aspartic proteinase cathepsin D, and the serine proteinases of specialized leukocyte azurophilic granules such as neutrophil elastase, proteinase 3, cathepsin G, and possibly mast cell chymases and lymphocytic granzymes as well. Several are called 'cathepsins' after the Greek word χαθεψειν, which means 'to digest.' This term was first used to describe an acid proteinase of the gastric mucosa that was distinct from pepsin, and later used to describe a splenic enzyme that is now known as cathepsin D.[1] Since then, enzymes of each class have been called cathepsins based usually on their acidic pH optima and lysosomal origin (Table 1). Most, however, belong to the papain superfamily of cysteine proteinases. These particular cathepsins have signal peptides of 15–28 amino acids that direct their synthesis to the endoplasmic reticulum, propeptides of usually 100 amino acids or so that until removed maintain the enzymes in a latent inactive state, and mature active forms containing the invariant catalytic triad of Cys, His and Asn residues (Fig. 1). Like other lysosomal proteins, the cathepsins receive carbohydrate moieties that target them to the lysosomes via the mannose 6-phosphate receptor pathway. Despite their lysosomal origin and acidic pH optima, however, they can still be secreted, and many retain a large portion of their proteolytic activity above pH 7.0 and/or act within acidic extracellular microenvironments. Under certain conditions, macrophages secrete cathepsins B, H, L, and S (and probably K as well), resulting in an elastinolytic capacity that is unsurpassed by other cell types, and is only inhibited by cathepsin-specific inhibitors.[2] Accordingly, the cathepsins have been implicated in a number of physiologic and pathologic processes, including bone resorption,[3,4] macrophage-mediated tissue destruction,[2,5] periodontal disease,[6] Alzheimer's disease,[7,8] and cancer.[9–11]

Cathepsin B

Cathepsin B has a latent 45 kDa pro-form and active single- and double-chain forms of 31 and 25 kDa, respectively.[2] The two-chain form results from limited proteolysis and has an N-terminal light chain (amino acids 1–47) and C-terminal heavy chain (amino acids 50–254) that remain linked by a disulphide bond and the inter-foldings of the enzyme itself.[12] Multiple isoforms can be resolved by ion-exchange chromatography and isoelectric focusing (IEF), with the human enzyme having pI values of 4.5–5.5 and a major form of pI 5.0–5.2.[1] Cathepsin B is constitutively expressed in a wide array of tissues and is especially abundant in liver and spleen. It can be selectively inhibited by a 56 amino acid peptide from its own propeptide domain.[12] Multiple mRNA splice variants that yield enzymes lacking signal peptides and/or propeptide sequences are associated with certain tumours; these may help explain the selective overexpression, aberrant localization, and increased stability and activity of cathepsin B in such cancers.[11,12]

Cathepsin L

Cathepsin L has a 43 kDa pro-form and mature one- and two-chain forms of 34 and 25 kDa, respectively.[2] It is a widely expressed enzyme that is particularly abundant in liver and alveolar macrophages and it has been found to be identical to the major excreted protein (MEP) of transformed mouse fibroblasts.[13] Its synthesis and secretion are also induced in untransformed cells by phorbol esters and growth factors such as PDGF and EGF. The selective diversion of the enzyme from the lysosomal to secretory pathway is thought to arise from its lower affinity for the mannose 6-phosphate receptor, which prefers two phospho-mannose moieties over one, coupled with a redistribution of the receptor to the cell surface so that the receptor concentration in the *trans*-Golgi network becomes limiting.[13]

Cathepsin S

Cathepsin S has a 37 kDa pro-form and a 24–28 kDa mature form.[14,15] Its only two potential glycosylation sites are within the propeptide region.[12] Cathepsin S is only detected in human spleen, heart, and lung and, more specifically, in the alveolar macrophages of the latter tissue.[16] It is also strongly expressed in rat and cow brain,

Table 1 Lysosomal proteinases

Cathepsin	EC no	Other names	Activity
Serine proteinases			
Cathepsin A	3.4.16.1	Lysosomal carboxypeptidase A	Removes single C-terminal residues with broad specificity
Cathepsin G	3.4.21.20		Chymotrypsin-like endopeptidase
Cysteine proteinases			
Cathepsin B	3.4.22.1	Cathepsin B1	Broad specificity endopeptidase that also removes C-terminal dipeptides; cleaves small peptides after Phe–Arg, Arg–Arg or Leu–Arg
Cathepsin B2	3.4.18.1	Lysosomal carboxypeptidase B, cathepsin IV	removes single C-terminal residues other than Pro
Cathepsin C	3.4.14.1	Dipeptidyl-peptidase I (DPP-I), Dipeptidyl- aminopeptidase I, dipeptidyl transferase cathepsin J	cleaves N-terminal dipeptides between P2–P1 and P1′, unless P2 is Arg or Lys, or P1 or P1′are Pro; activates lysosomal serine proteases such as neutrophil elastase
Cathepsin H	3.4.22.16	Cathepsin B3	Weak endopeptidase; removes N-terminal residues (especially Arg)
Cathepsin K	3.4.22.38	Cathepsin O, O2, OC-2, X	Potent endopeptidase; favours small peptides with hydrophilic P1 and small, hydrophobic P2 residues
Cathepsin L	3.4.22.15	Major excreted protein (MEP)	Endopeptidase; prefers small substrates with Phe–Arg or Leu–Arg
Cathepsin O	3.4.22.–		Endopeptidase with poor activity against substrates of cathepsins B and L
Cathepsin S	3.4.22.27		Endopeptidase that readily cleaves elastin; can act at neutral pH
Cathepsin T	3.4.22.24		Acts on tyrosine aminotransferase
Aspartic proteinases			
Cathepsin D	3.4.23.5		Endopeptidase with a narrowed pepsin-like specificity
Cathepsin E	3.4.23.34	Erythrocyte membrane aspartic proteinase	Non-lysosomal endopeptidase with pepsin-like activity
Metalloproteinases			
Carboxycathepsin	3.4.15.1	Peptidyl-dipeptidase A, angiotensin I-converting enzyme, dipeptidyl-carboxypeptidase I	Cleaves C-terminal dipeptides between P1 and P1′-P2′, unless P1 or P1′ are Pro

Two enzymes have been called cathepsin O; one with a novel sequence,[27] the other with a sequence identical to that of cathepsin K.[44] Data are from refs 27, 31, 44, 45.

but not in normal human brain, although it is found in the neurones of patients with Alzheimer's disease.[16] Indeed, cathepsins B, D, E and S have each been implicated in the processing of Alzheimer's β-amyloid precursor protein.[7,8] Cathepsin S mRNA is also upregulated in thyrocytes by thyroid stimulating hormone and may play a role, together with cathepsins B and L, in the processing of thyroglobulin.[16] Unlike many other cathepsins, cathepsin S retains much of its proteolytic activity at neutral pH.

Cathepsin K

Considerable evidence indicates that the cysteine proteinase cathepsin K plays an important role in osteoclast-mediated bone resorption. First, it is abundantly and selectively expressed in osteoclasts, and although cysteine proteinase inhibitors can inhibit bone resorption by osteoclasts *in vitro*, no other cathepsins are readily detected in these cells.[17] Second, a cathepsin K antisense

```
               18  ↓  *      156 *          170    *
     Papain ...NQGSCGSCWAFS...KVDHAVAAVGYG...YILIKNSWGTGWG...
Cathepsin B ...DQGSCGSCWAFG...MGGHAIRILGWG...YWLVANSWNTDWG...
Cathepsin C ...DQESCGSCYSFA...LTNHAVLLVGYG...YWIVKNSWGSQWG...
Cathepsin H ...NQGACGSCWTFS...KVNHAVLAVGYG...YWLVKNSWGPEWG...
Cathepsin K ...NQGQCGSCWAFS...NLNHAVLAVGYG...HWIIKNSWGENWG...
Cathepsin L ...NQGQCGSCWAFS...DMDHGVLVVGYG...YWLVKNSWGEEWG...
Cathepsin O ...NQQMCGGCWAFS...EANHAVLITGFD...YWIVRNSWGSSWG...
Cathepsin S ...YQGSCGACWAFS...NVNHGVLVVGYG...YWLVKNSWGHNFG...
Cathepsin W ...DQKNCNCCWAMA...LVDHSVLLVGFG...YWILKNSWGAQWG...
```

Figure 1. Conserved sequences surrounding the catalytic Cys, His, and Asn residues of the human papain-like cathepsins. Aligned residues are numbered according to papain and identical amino acids are shaded. The three catalytic residues are indicated by asterisks, a conserved Cys residue that forms one of six intramolecular disulphide bonds is indicated by an arrow.

construct inhibits osteoclastic bone resorption,[4] and the autosomal recessive bone resorption disorder pycnodysostosis, which results in short stature, is caused by cathepsin K deficiency.[3] Cathepsin K is synthesized as a preproenzyme of approximately 44 kDa and is targeted to lysosomes as a 38–43 kDa proenzyme.[17,18] Treatment with pepsin yields a mature active enzyme of 27–29 kDa, and although two potential glycosylation sites exist, the enzyme is probably not glycosylated.[17,18] Although cathepsin K is predominantly expressed in osteoclasts and chondroclasts, its transcripts have also been seen at low levels in lung, heart, and skeletal muscle.[19,20] It has also been detected in primary breast carcinomas and may contribute to invasion and bone metastasis.[53]

Cathepsin C

Cathepsin C is better known as dipeptidyl-peptidase I, a widely distributed aminopeptidase that has been implicated in intracellular protein degradation, cell growth, neuraminidase activation, platelet factor XIII activation, and the activation of serine proteinases with dipeptide prodomains such as cathepsin G.[21,22] It is unique among the papain superfamily members in that it has an exceptionally long propeptide (206 residues in humans) and exists as an oligomeric enzyme of about 200 kDa.[21] The 55 kDa pro-enzyme is rapidly processed by unknown enzymes to a two-chain form with 25 and 10 kDa heavy and light chains and assembled into the larger oligomeric complex.[21,23] Evidence suggests that the complex itself is made up of four identical subunits which, in turn, are composed of disulphide-linked heavy and light chains plus a significant portion of the prodomain.[23] Three potential *N*-glycosylation sites (two in the prodomain and one in the mature peptide) are completely conserved in mice, rats, and humans, suggesting that they are indeed glycosylated.[21]

Cathepsin D

Cathepsin D is a ubiquitous aspartic proteinase with a mature form that shares 49 per cent amino acid identity with pepsin. The human enzyme has a 20 amino acid signal sequence, a 44 amino acid propeptide, and an active single-chain enzyme with 348 residues and two glycosylated *N*-glycosylation sites.[24] Intra-chain cleavages also yield an active two-chain form, with one active site Asp residue being found in the N-terminal light chain and the other in the C-terminal heavy chain. Thus the human enzyme has a 52 kDa pro-form, a 48 kDa active one-chain form, and an active two-chain form with 34 and 14 kDa heavy and light chains.[10] Cathepsin D is likely to be an important activator of papain-like cathepsins such as cathepsin K in osteoclasts, since such activation is inhibited by the aspartic proteinase inhibitor pepstatin,[25] and because cathepsin D is the only known aspartic proteinase of lysosomal origin.[24] It can also degrade extracellular proteins such as fibronectin and cartilage proteoglycan, but it requires a more acidic environment than other cathepsins, and is completely inactive at neutral pH.[1] In addition, cathepsin D exhibits autocrine mitogenic activity for cells.[10] Such activity may be an indirect result of its enzymatic activity or may result from its interaction with cell surface mannose-6-phosphate/IGF-II receptors.[10] Cathepsin D is also a mediator of apoptosis; its overexpression is a poor prognostic indicator in human cancers; and recent evidence suggests that it may play a role in p53-dependent tumor suppression and chemosensitivity.[54]

Other cathepsins

Cathepsin H is a widely distributed lysosomal glycoprotein with greater aminopeptidase than endopeptidase activity.[5] It has a 41 kDa pro-form, a 28 kDa single-chain form, and a two-chain form containing 22 and 6 kDa heavy and light chains.[25,26] Cathepsin O was recently cloned from a human breast carcinoma cDNA library,[27] cathepsin L2 was cloned from a human brain cDNA library,[55] and another cathepsin (W) has been identified from an EST databank.[28] Cathepsin O has a major 2.5 kb mRNA that is ubiquitously expressed and the recombinant cathepsin O from *E. coli* exhibits E-64-inhibitable activity against two cysteine proteinase substrates, albeit at levels well below those of cathepsins B and L. Cathespin L2 has a single 1.8 kb mRNA that is predominantly expressed in thymus and testis. It is also widely expressed in breast and colorectal carcinomas, but not in normal peritumoral tissues. Its mRNA encodes a protein with 334 residues, a predicted molecular weight of 37 kDa, and 78 per cent identity with cathepsin L. Recombinant cathepsin L2 can cleave a cysteine proteinase substrate and is inhibited by E-64. Cathepsin W has a 1.45 kb mRNA that is predominantly expressed in CD8+ T-cells, a relatively low sequence homology to other cathepsins (21–31 per cent), and pro- and mature forms with predicted weights of 39 and 27 kDa, respectively. Whether or not these latter cathepsins can act as ECM proteinases remains to be determined. Finally, the serine proteinase cathepsin G is dealt with in another section.

Purification

Cathepsins B and D are readily purified from pH 6.5 liver extracts using CM–cellulose ion exchange and organomercurial–Sepharose affinity chromatography.[1] The latter step separates the two cathepsins since only cathepsin B is adsorbed via its free sulphydryl group. The cathepsin B is then eluted using a thiol reagent such as cysteine, the reagent is removed by Ultrogel AcA-54 or Sephadex G-75 gel filtration, and final purification is achieved by CM– or DEAE–cellulose chromatography. Unadsorbed cathepsin D is further purified by DEAE–cellulose chromatography and isoelectric focusing. Methods have also been described for the purification of cathepsins B, H, and L from human and rat liver,[29] and human kidney.[30] These take advantage of the continued binding of cathepsin L, but not B or H, to CM–Sephadex in 0.2 M KCl, and the binding of cathepsins B and L, but not H, to DEAE–cellulose. Purification of cathepsin C from human kidney takes advantage of its large size relative to other cathepsins.[23] Cathepsin S has been purified from human, bovine and rat spleen and from yeast expressing the human enzyme.[12] Its purification from spleen involves acid extraction, ammonium sulphate fractionation, CM–Sephadex C-50 chromatography, Sephacryl S-200 gel filtration, and chromatofocusing. Recombinant human cathepsin K has been purified from insect cells.[18,31]

Activities

The catalytic mechanism of papain-like cysteine proteinases involves non-covalent enzyme–substrate binding, nucleophilic attack by the active site thiol group to form a covalent intermediate (the acyl-enzyme), deacylation involving the formation of transient tetrahedral intermediates, and enzyme release.[32] Among the cathepsins, only B and H exhibit both endo- and exopeptidase activities. Together, the cathepsins can degrade a wide array of ECM components. For instance, cathepsin B can degrade aggrecan, collagens, and gelatins, but it has poor activity against elastin.[33] Cathepsin K, on the other hand, is an extremely potent elastase, even at neutral pH; cathepsin S is somewhat less potent, but also remains active at neutral pH; and cathepsin L is less potent still, only acting at acid pH.[18] Specifically, the elastinolytic activity of cathepsin K is 1.7–3.5-fold greater than that of cathepsin S between pH 4.5 and 7, almost 9-fold greater than that of cathepsin L at pH 5.5, and almost 2.4-fold greater than that of pancreatic elastase at pH 7. The elastinolytic activity of cathepsin S at pH 4.5 is about 1.7-fold greater than that of pancreatic elastase at pH 7, and at pH 7 its activity is about 2/3 that of pancreatic elastase. In turn, the activity of pancreatic elastase is about 3- and 10-fold greater than that of neutrophil elastase and macrophage metalloelastase, respectively.[5] Cathepsin K also readily degrades type I collagen and gelatin, whereas cathepsin L degrades these less readily, and cathepsin S has only weak collagenolytic and gelatinolytic activity.[18] Cathepsin K can also cleave within the helical domain of type II collagen.[56] In addition, cathepsin K has potent activity against the bone matrix proteins osteocalcin, α-2HS-glycoprotein, and osteonectin/SPARC, whereas cathepsins B and L only cleave osteonectin.[31,34] Cathepsins B and L can also degrade cartilage components such as aggrecan, link proteins, and collagens II, IX and XI, albeit at different sites, with cathepsin L being considerably more active than cathepsin B.[35,36] In addition, cathepsin L can degrade fibronectin and laminin.[37] Furthermore, cathepsins can activate one another and other ECM proteinases as well. For example, cathepsin B can activate MMPs such as stromelysin-1[38] and serine proteinases such as urokinase-type plasminogen activator;[9] cathepsin C (DPP-I) can activate lysosomal serine proteinases that have 2-amino acid propeptides (e.g. neutrophil elastase, cathepsin G, azurocidin, chymases, and granzymes, but not proteinase 3),[22] and cathepsin D is probably the physiologic activator of the papain-like cathepsins.[18,25]

The S2 subsite pocket of the cysteine proteinases is an important determinant of their respective substrate specificities. Using fluorogenic dipeptides with an R at the P1 position, it has been determined that only cathepsin B will accept substrates with another R in the P2 position and that the preferred P2 residues are F >> L = R > V for cathepsin B, F >> L >> V for cathepsin L, L >> F > V for cathepsin K, and L >> F >> V for cathepsin S.[18] Indeed, site-directed mutagenesis involving specific amino acids of the S2 subsite can transform the substrate specificity of cathepsin S to either that of cathepsin B or L, depending on the amino acid relacements made. Substrates containing RR or KK sequences are the only truly specific substates, because they are quite sensitive to cathepsin B, but are resistant to other cathepsins.[12] True specificity can be approached for cathepsin S, however, if substrates containing FVR or VVR are used at pH 7.5.[12] A number of peptide substrates have also been developed for the aspartic proteinase cathepsin D. Chromogenic and/or fluorogenic substrates for the various cathepsins are available from Bachem, Calbiochem-Novabiochem, Enzyme Systems Products, Molecular Probes, and Sigma.

The lysosomal cysteine proteinases are inhibited by standard cysteine proteinase inhibitors such as leupeptin and E-64, but not by inhibitors that are specific to other proteinase classes. In addition, several classes of synthetic inhibitors irreversibly inhibit the various papain-like cathepsins with efficiencies that essentially follow the substrate specificities of the respective enzymes.[12,18,27,31] The cysteine proteinases are also inhibited by natural endogenous inhibitors, namely the cystatins (stefins), plasma kininogens, and α_2-macroglobulin.[12] Interestingly, cathepsins K, L, and S are also inhibited by squamous cell carcinoma antigen 1 (SCCA1) with a potency that rivals that of the prototype cysteine proteinase inhibitor cystatin C even though SCCA1 is a member of the serpin superfamily of *serine* proteinase inhibitors.[57] The aspartic proteinase cathepsin D can be irreversibly inhibited by diazoketones that esterify carboxyl groups and by other synthetic inhibitors and it can be reversibly inhibited by pepstatin.[1] It is also inhibited by the universal endogenous inhibitor α_2-macroglobulin. Chloromethyl ketone based inhibitors are used for both aspartic and serine proteinases, whereas fluoromethyl ketone derivatives are selective for cysteine proteinases only.

Antibodies

Monoclonal antibodies against human procathepsins B and L and mature cathepsins B and D are available from Calbiochem. Monoclonal anti-cathepsin D antibodies are also available from Accurate Chemical and Scientific, Becton Dickinson, Biogenesis, Biogenex, Biomeda, Caltag Labs, Chemicon, Scripps Labs and Vector Labs/Novocastra. Polyclonal antibodies against cathepsins B and D are available from several commercial suppliers and polyclonal antibodies against human cathepsin L are available from Calbiochem and Athens Research and Technology. Rabbit antibodies have also been raised against recombinant human procathepsin K and a cathepsin K-specific peptide[17,18] and affinity-purified rabbit antibodies to bovine cathepsin S have been described which also recognize human cathepsin S.[14]

Genes

Characteristics of the various human cathepsin genes are given in Table 2. Although the structure of the human cathepsin H gene has not yet been determined, the rat gene spans over 21.5 kbp and has 12 exons.[39] The human cathepsin C gene structure has also not been determined, but the mouse gene spans 20 kbp, has seven exons, resides on mouse chromosome 7, and is constitutively expressed.[21] Like their human counterparts, the mouse genes for cathepsins B, H, L, and S also map to different chromosomes. The 5'-flanking region of the human cathepsin B gene resembles that of a housekeeping gene in that it is > 80 per cent GC-rich, has no TATA box, and contains 15 SP1-binding sites.[16] The human cathepsin S promoter also lacks classical TATA and CAAT boxes, but has only two SP1 sites, a GC content near 40 per cent, and at least 18 AP1 binding sites, thus mirroring its restricted expression pattern.[16] Likewise, the human cathepsin K gene lacks TATA and CAAT boxes, also has a GC content near 40 per cent and numerous potential regulatory sequences, including SP1, AP1, AP3, H-APF-1, Pu.1, Ets-1, and PEA-3 sites.[40] The cathepsin D promoter has features similar to those of both a housekeeping gene (high GC content and potential SP1 sites) and a hormonally regulated gene (oestrogen-responsive TATA-dependent transcription).[41] This mixed promoter may allow for both constitutive expression under some conditions and gene induction under others.

Complete cDNA sequences are available through GenBank for cathepsins B in humans (L16510, M14221), mice (M14222), rats (X82396), cows (L06075) and chickens (U18083); C in humans (X87212), mice (U89269) and rats (D90404); D in humans (X05344, M11233), mice (X52886, X53337), and rats (X54467); E in humans (J05036, M82847), rats (D38104), and guinea pigs (M88653); H in humans (X07549, X16832), mice (U06119), and rats (Y00708); K in humans (X82153, U13665, U20280), mice (X94444), and rabbits (D14036); L in humans (M20496, X12451), mice (M20495), rats (Y00697), cows (X91755), and pigs (D37917); L2 in humans (Y14734); O in humans (X77383); and S in humans (M90696, M86553), rats (L03201), and cows (M95211). Genomic sequences are available for cathepsins B in mice (M65263–67) and cows (U16337–43); D in humans (M63134–38) and mice (X68379–83); E in humans (M84413, M84417–24); L in mice (L06427) and rats (X51648); and S in humans (U07369–73). Probes for human cathepsins D and G are available from ATCC and a mouse cathepsin L plasmid is available from University Technologies Intenational (Calgary).

Mutant phenotype/disease states

Nonsense, mis-sense and stop codon mutations in the cathepsin K gene cause pycnodysostosis, a rare autosomal recessive bone disorder characterized by short stature, osteosclerosis, wide cranial sutures, and bone fragility.[3] The potential involvement of the cathepsins in other pathologies, including emphysema, periodontal disease, Alzheimer's disease, and cancer, has also been described.[5–11]

Structure

The X-ray crystal structure of cathepsin B reveals a disc-shaped molecule composed of two domains with a V-shaped active site cleft between the two domains.[42] Some of the primed subsites are occluded by a loop that is absent from other cathepsins, thus explaining the dipep-

Table 2 Human cathepsin genes

Enzyme	Gene	Location	Size (kbp)	mRNA (kb)	Refs
Cathepsin B	*CTSB*	8p22	27 (13 exons)	4.0, 2.2	11
Cathepsin D	*CTSD*	11p15.5	11 (9 exons)	2.2	24, 46
Cathepsin E	*CTSE*	1q31	17.5 (9 exons)	3.6, 2.6, 2.1	47
Cathepsin G	*CTSG*	14q11.2	2.7 (5 exons)	0.9	48
Cathepsin H	*CTSH*	15q24–25		1.7	49
Cathepsin K	*CTSK*	1q21	9 (8 exons)	1.8	19, 40, 50
Cathepsin L	*CTSL*	9q22.1–22.2	5.1 (8 exons)	4.0, 1.5	51, 52
Cathepsin S	*CTSS*	1q21	> 20 (> 6 exons)	4.0, 1.9	16

tidyl carboxypeptidase activity of cathepsin B.[42] Crystal structures have also been determined for pepstatin-inhibited human and bovine cathepsin D.[43] These reveal carbohydrate moieties at the expected lysosomal targeting region and two domains with a deep active site cleft between them. Each domain contains several β sheets, two disulphide bonds, and a single carbohydrate group, and each contributes an Asp residue to the centre of the active-site cleft. Brookhaven PDB identifiers are 1THE, 1CSB, 1CPJ, 1CTE, 1HUC for cathepsin B and 1LYB and 1LYA for cathepsin D.

References

1. Barrett, A. J. (1977). In *Proteinases in mammalian cells and tissues* (ed. A. J. Barrett), pp. 1–55 and 181–248. North-Holland Publishing, New York.
2. Reddy, V. V., Zhang, Q. Y., and Weiss, S. J. (1995). *Proc. Natl Acad. Sci. USA*, **92**, 3849–53.
3. Gelb, B. D., Shi, G. P., Chapman, H. A., and Desnick, R. J. (1996). *Science*, **273**, 1236–8.
4. Inui, T., Ishibashi, O., Inaoka, T., Origane, Y., Kumegawa, M., Kokubo, T., and Yamamura, T. (1997). *J. Biol. Chem.*, **272**, 8109–112.
5. Chapman, H. A., Munger, J. S., and Shi, G. P. (1994). *Am. J. Respir. Crit. Care Med.* **150**, S155–9.
6. Trabandt, A., Muller-Ladner, U., Kriegsmann, J., Gay, R. E., and Gay, S. (1995). *Lab. Invest.*, **73**, 205–12.
7. Munger, J. S., Haass, C., Lemere, C. A., Shi, G. P., Wong, W. S., Teplow, D. B., *et al.* (1995). *Biochem. J.*, **311**, 299–305.
8. Mackay, E. A., Ehrhard, A., Moniatte, M., Guenet, C., Tardif, C., Tarnus, C., *et al.* (1997). *Eur. J. Biochem.*, **244**, 414–25.
9. Duffy, M. J. (1992). *Clin. Exp. Metastasis*, **10**, 145–55.
10. Rochefort, H., Capony, F., and Garcia, M. (1990). *Cancer Metastasis Rev.*, **9**, 321–31.
11. Keppler, D. and Sloane, B. F. (1996). *Enzyme Protein*, **49**, 94–105.
12. Kirschke, H. and Wiederanders, B. (1994). *Meth. Enzymol.*, **244**, 500–11.
13. Prence, E. M., Dong, J., and Sahagian, G. G. (1990). *J. Cell Biol.*, **110**, 319–26.
14. Wiederanders, B., Bromme, D., Kirschke, H., Figura, K., Schmidt, B., and Peters, C. (1992). *J. Biol. Chem.*, **267**, 13708–13.
15. Shi, G. P., Munger, J. S., Meara, J. P., Rich, D. H., and Chapman, H. A. (1992). *J. Biol. Chem.*, **267**, 7258–62.
16. Shi, G. P., Webb, A. C., Foster, K. E., Knoll, J. H. M., Lemere, C. A., Munger, J. S., and Chapman, H. A. (1994). *J. Biol. Chem.*, **269**, 11530–6.
17. Drake, F. H., Dodds, R. A., James, I. E., Connor, J. R., Debouck, C., Richardson, S., *et al.* (1996). *J. Biol. Chem.*, **271**, 12511–6.
18. Bromme, D., Okamoto, K., Wang, B. B., and Biroc, S. (1996). *J. Biol. Chem.*, **271**, 2126–32.
19. Bromme, D. and Okamoto, K. (1995). *Biol. Chem. Hoppe-Seyler*, **376**, 379–84.
20. Inaoka, T., Bilbe, G., Ishibashi, O., Tezuka, K. I., Kumegawa, M., and Kokubo, T. (1995). *Biochem. Biophys. Res. Commun.*, **206**, 89–96.
21. Pham, C. T. N., Armstrong, R. J., Zimonjic, D. B., Popescu, N. C., Payan, D. G., and Ley, T. J. (1997). *J. Biol. Chem.*, **272**, 10695–703.
22. Rao, N. V., Rao, G. V., Marshall, B. C., and Hoidal, J. R. (1996). *J. Biol. Chem.*, **271**, 2972–8.
23. Dolenc, I., Turk, B., Pungercic, G., Ritonja, A., and Turk, V. (1995). *J. Biol. Chem.*, **270**, 21626–31.
24. Faust, P. L., Kornfeld, S., and Chirgwin, J. M. (1985). *Proc. Natl Acad. Sci. USA*, **82**, 4910–4.
25. Nishimura, Y. and Kato, K. (1988). *Arch. Biochem. Biophys.*, **260**, 712–8.
26. Fuchs, R., Machleidt, W., and Gassen, H. G. (1988). *Biol. Chem. Hoppe-Seyler*, **369**, 469–75.
27. Valesco, G., Ferrando, A. A., Puente, X. S., Sanchez, L. M., and Lopez-Otin, C. (1994). *J. Biol. Chem.*, **269**, 27136–42.
28. Linnevers, C., Smeekens, S. P., and Bromme, D. (1997). *FEBS Lett.*, **405**, 253–9.
29. Barrett, A. J. and Kirschke, H. (1981). *Meth. Enzymol.*, **80**, 535–61.
30. Popovic, T., Puizdar, V., Ritonja, A., and Brzin, J. (1996). *J. Chromatogr. Biomed. Appl.*, **681**, 251–262.
31. Bossard, M. J., Tomaszek, T. A., Thompson, S. K., Amegadzie, B. Y., Hanning, C. R., Jones, C., *et al.* (1996). *J. Biol. Chem.*, **271**, 12517–24.
32. Storer, A. C. and Menard, R. (1994). *Meth. Enzymol.*, **244**, 486–500.
33. Mason, R. W., Johnson, D. A., Barrett, A. J., and Chapman, H. A. (1986). *Biochem. J.*, **233**, 925–7.
34. Page, A. E., Hayman, A. R., Andersson, L. M., Chambers, T. J., and Warburton, M. J. (1993). *Int. J. Biochem.*, **25**, 545–50.
35. Nguyen, Q., Mort, J. S., and Roughley, P. J. (1990). *Biochem. J.*, **266**, 569–73.
36. Maciewicz, R. A., Wotton, S. F., Etherington, D. J., and Duance, V. C. (1990). *FEBS Lett.*, **269**, 189–93.
37. Gal, S. and Gottesman, M. M. (1986). *J. Biol. Chem.*, **261**, 1760–5.
38. Murphy, G., Ward, R., Gavrilovic, J., and Atkinson, S. (1992). *Matrix*, **S1**, 224–30.
39. Ishidoh, K., Kominami, E., Katunuma, N., and Suzuki, K. (1989). *FEBS Lett.*, **253**, 103–7.
40. Rood, J. A., van Horn, S., Drake, F. H., Gowen, M., and Debouck, C. (1997). *Genomics*, **41**, 169–76.
41. Cavailles, V., Augereau, P., and Rochefort, H. (1993). *Proc. Natl Acad. Sci. USA*, **90**, 203–7.
42. Musil, D., Zucic, D., Turk, D., Engh, R. A., Mayr, I., Huber, R., *et al.* (1991). *EMBO J.*, **10**, 2321–30.
43. Metcalf, P. and Fusek, M. (1993). *EMBO J.*, **12**, 1293–302.
44. Shi, G. P., Chapman, H. A., Bhairi, S. M., DeLeeuw, C., Reddy, V. Y., and Weiss, S. J. (1995). *FEBS Lett.*, **357**, 129–34.
45. *Enzyme nomenclature* (1992). pp. 379–408. Academic Press, New York.
46. Redecker, B., Heckendorf, B., Grosch, H. W., Mersmann, G., and Hasilik, A. (1991). *DNA Cell Biol.*, **10**, 423–31.
47. Azuma, T., Liu, W. G., Vander Laan, D. J., Bowcock, A. M., and Taggart, R. T. (1992). *J. Biol. Chem.*, **267**, 1609–14.
48. Hohn, P. A., Popescu, N. C., Hanson, R. D., Salvesen, G., and Ley, T. J. (1989). *J. Biol. Chem.*, **264**, 13412–9.
49. Fong, D., Chan, M. M. Y., and Hsieh, W. T. (1991). *Biomed. Biochim. Acta*, **50**, 595–8.
50. Gelb, B. D., Shi, G. P., Heller, M., Weremowicz, S., Morton, C., Desnick, R. J., and Chapman, H. A. (1997). *Genomics*, **41**, 258–62.
51. Joseph, L. J., Chang, L. C., Stamenkovich, D., and Sukhatme, V. P. (1988). *J. Clin. Invest.*, **81**, 1621–9.
52. Chauhan, S. S., Popescu, N. C., Ray, D., Fleischmann, R., Gottesman, M. M., and Troen, B. R. (1993). *J. Biol. Chem.*, **268**, 1039–45.
53. Littlewood-Evans, A. J., Bilbe, G., Bowler, W. B., Farley, D., Wlodarski, B., Kokubo, T., Inaoka, T., Sloane, J., Evans, D. B., and Gallagher, J. A. (1997). *Cancer Res.*, **57**, 5386–90.
54. Wu, G. S., Saftig, P., Peters, C., and El-Deiry, W. S. (1998). *Oncogene*, **16**, 2177–83.

55. Santamaria, I., Valeso, G., Cazorla, M., Fueyo, A., Campo, E., and Lopez-Otin, C. (1998). *Cancer Res.*, **58**, 1624–30.
56. Kafienah, W., Bromme, D., Buttle, D. J., Croucher, L. J., and Hollander, A. P. (1998). *Biochem. J.*, **331**, 727–32.
57. Schick, C., Pemberton, P. A., Shi, G. P., Kamachi, Y., Cataltepe, S., Bartuski, A. J., Gorstein, E. R., Bromme, D., Chapman, H. A., and Silverman, G. A. (1998). *Biochemistry*, **37**, 5258–66.

Mark D. Sternlicht and Zena Werb
University of California, San Francisco, Department of Anatomy, LR-208, 3rd and Parnassus Avenues, San Francisco, CA 94143–0452, USA.

Matrix metalloproteinases (MMPs)

The matrix metalloproteinases (MMPs) constitute a multigene family of zinc- and calcium-dependant endopeptidases with extensive sequence homology. To date, 18 vertebrate MMPs and 16 human homologues have been identified. Those MMPs that have been well characterized display distinct, but overlapping, substrate specificities and collectively they can cleave virtually all structural ECM molecules. The MMPs are regulated by positive and negative transcriptional controls, by their activation from a latent state, and through the influence of endogenous tissue inhibitors of metalloproteinases (TIMPs).

Protein properties

Matrix metalloproteinases (MMPs, matrixins) are key mediators of the normal ECM remodelling that takes place, for instance, during development, tissue morphogenesis, and repair. In addition, MMPs have been implicated in the untoward ECM degradation that occurs in diseases such as arthritis and cancer.[1] Indeed, nine MMPs were first cloned from tumour cell lines. All human MMPs have been detected in one tumour cell line or another, but in solid tumours they are most often expressed by nearby stromal cells rather than by the tumour cells themselves.[1] Their expression in cancers often correlates with poor prognosis and tumour cells can be made more aggressive both *in vitro* and *in vivo* by over-expression of MMPs or down-regulation of their endogenous inhibitors, or they can be made less aggressive by MMP down-regulation, exogenous MMP inhibitors, or the over-expression of endogenous inhibitors.

As their name implies, the MMPs form a multigene family within the metalloproteinase class of endopeptidases and they can hydrolyse essentially all ECM molecules (Table 1). Other features that distinguish the MMPs include

(1) a requirement that zinc be bound at their catalytic site;

(2) a family-specific zinc-binding motif, usually **HE**F/L**GH**S/AL**G**LX**HS**, where bold residues are invariant (Fig. 1);

Collagenases			
Hu MMP-1	YNLHRVAA	HELGHSLGLSHS	TDIGALMYPSY
Por MMP-1	YNLYRVAA	HELGHSLGLSHS	TDIGALMYPNY
Bov MMP-1	YNLYRVAA	HEFGHSLGLAHS	TDIGALMYPSY
Rab MMP-1	YNLYRVAA	HELGHSLGLSHS	TDIGALMYPNY
Bfr MMP-1	YNLYRVAA	HELGHSLGLSHS	TDIGALMYPTY
Hu MMP-8	YNLFLVAA	HEFGHSLGLAHS	SDPGALMYPNY
Hu MMP-13	YNLFLVAA	HEFGHSLGLDHS	KDPGALMFPIY
Mus MMP-13	YNLFIVAA	HELGHSLGLDHS	KDPGALMFPIY
Rat MMP-13	YNLFIVAA	HELGHSLGLDHS	KDPGALMFPIY
Xe MMP-13	YNLFVVAA	HEFGHALGLDHS	RDPGSLMFPVY
Xe MMP-18	YNLFLVAA	HEFGHSLGLSHS	TDQGALMYPTY
Gelatinases			
Hu MMP-2	YSLFLVAA	HEFGHAMGLEHS	QDPGALMAPIY
Mus MMP-2	YSLFLVAA	HEFGHAMGLEHS	QDPGALMAPIY
Rat MMP-2	YSLFLVAA	HEFGHAMGLEHS	QDPGALMAPIY
Rab MMP-2	YSLFLVAA	HEFGHAMGLEHS	QDPGALMAPIY
Ch MMP-2	YSLFLVAA	HEFGHAMGLEHS	EDPGALMAPIY
Hu MMP-9	YSLFLVAA	HEFGHALGLDHS	SVPEALMAPMY
Rab MMP-9	YSLFLVAA	HEFGHALGLDHS	SVPERLMAPMY
Bov MMP-9	YSLFLVAA	HEFGHALGLDHT	SVPEALMAPMY
Mus MMP-9	YSLFLVAA	HEFGHALGLDHS	SVPEALMAPLY
Stromelysins and Matrilysin			
Hu MMP-3	TNLFLVAA	HEIGHSLGLFHS	ANTEALMYPLY
Rab MMP-3	TNLFLVAA	HELGHSLGLFHS	ANPEALMYPVY
Rat MMP-3	TNLFLVAA	HELGHSLGLFHS	ANAEALMYPVY
Mus MMP-3	TNLFLVAA	HELGHSLGLYHS	AKAEALMYPVY
Hu MMP-7	INFLYAAT	HELGHSLGMGHS	SDPNAVMYPTY
Mus MMP-7	VNFLFAAT	HEFGHSLGLQHS	NNPKSVMYPTY
Hu MMP-10	TNLFLVAA	HELGHSLGLFHS	ANTEALMYPLY
Rat MMP-10	TNLFLVAA	HELGHSLGLFHS	NNKESLMYPVY
Hu MMP-11	TDLLQVAA	HEFGHVLGLQHT	TAAKALMSAFY
Mus MMP-11	TDLLQVAA	HEFGHVLGLQHT	TAAKALMSPFY
Xe MMP-11	TDLLQVAA	HEFGHMLGLQHS	SISKSLMSPFY
Membrane-type MMPs			
Hu MMP-14	NDIFLVAV	HELGHALGLEHS	SDPSAIMAPFY
Mus MMP-14	NDIFLVAV	HELGHALGLEHS	NDPSDIMSPFY
Rat MMP-14	NDIFLVAV	HELGHALGLEHS	NDPSAIMAPFY
Rab MMP-14	NDIFLVAV	HELGHALGLEHS	NDPSAIMAPFY
Hu MMP-15	NNLFLVAV	HELGHALGLEHS	SNPNAIMAPFY
Hu MMP-16	NDLFLVAV	HELGHALGLEHS	NDPTAIMAPFY
Ch MMP-16	NDLFLVAV	HELGHALGLEHS	NDPTAIMAPFY
Hu MMP-17	MDLFLVAV	HEFGHAIGLSHV	AAAHSIMRPYY
Other MMPs			
Hu MMP-12	TNLFLTAV	HEIGHSLGLGHS	SDPKAVMFPTY
Mus MMP-12	TNLFLVAV	HELGHSLGLQHS	NNPKSIMYPTY
Hu MMP-19	VNLRIIAA	HEVGHALGLGHS	RYSQALMAPVY
Por MMP-20	FNLFTVAA	HEFGHALGLAHS	TDPSALMYPTY
Xe MMP-x	ISLLKVAA	HEIGHVLGLSHI	HRVGSIMQPNY
Ur MMP-y	TNLFQVAA	HEFGHSLGLYHS	TVRSALMYP-Y
Sb MMP-z	FDLESVAV	HEIGHLLGLGHS	SDLRAIMYPSI

Figure 1. Amino acid sequences surrounding the zinc-binding region of various MMPs. Conserved residues are shaded and the zinc-binding and Met-turn motifs are underlined. Abbreviations: Hu, human; Por, porcine; Bov, bovine; Mus, mouse; Rab, Rabbit; Ch, Chicken; Xe, frog (*Xenopus*); Bfr, bullfrog (*Rana*); Ur, sea urchin; Sb, soybean. (Data are from refs 2–6 and corresponding GenBank entries.)

(3) a highly conserved 'Met turn', usually AL**MYP**, which sits below the active site zinc;

(4) a dependence on calcium for structural integrity;

Table 1 The MMP multigene family

MMP (EC no)	Common names	Substrates
Collagenases		
MMP-1 (3.4.24.7)	Collagenase-1 (fibroblast, interstitial, or tissue Collagenase)	Collagens I, II, III, VII, X, gelatins, entactin, link protein, aggrecan, tenascin, L-selectin, IGF-binding proteins, proMMPs 2 and 9
MMP-8 (3.4.24.34)	Collagenase-2 (neutrophil or PMN collagenase)	Collagens I, II, III, gelatins, aggrecan
MMP-13	Collagenase-3	Collagens I, II, III, gelatins, aggrecan
MMP-18	Collagenase-4	Collagen I, gelatins
Gelatinases		
MMP-2 (3.4.24.24)	Gelatinase A (72 kDa gelatinase, 72 kDa type IV collagenase)	Gelatins, collagens I, IV, V, VII, X, elastin, fibronectin, laminin, link protein, aggrecan, galectin-3, IGF-binding proteins, vitronectin, fibulin-2, FGF receptor-1, proMMPs 9 and 13
MMP-9 (3.4.24.35)	Gelatinase B (92 kDa gelatinase, 92 kDa type IV collagenase)	Gelatin, collagens IV, V, XI, elastin, aggrecan, link protein, vitronectin, galectin-3, proMMP-2
Stromelysins 1 and 2		
MMP-3 (3.4.24.17)	Stromelysin-1 (transin-1, proteoglycanase, procollagenase activating protein, CAP)	Proteoglycans, laminin, fibronectin, gelatins, collagens III, IV, V, IX, X, XI, link protein, fibrin, entactin, SPARC, tenascin, vitronectin, proMMPs 1, 8, 9, and 13, antithrombin III, PAI-2, α_1-proteinase inhibitor, α_1-antichymotrypsin, α2-macroglobulin, L-selectin, E-cadherin, HB-EGF
MMP-10 (3.4.24.22)	Stromelysin-2 (transin-2)	Proteoglycans, laminin, gelatins, elastin, collagens III, IV, V, IX, fibronectin, link protein, proMMP-1
Membrane-type MMPs		
MMP-14	MT1-MMP (MT-MMP-1, membrane-type MMP-1)	Collagens I, II, III, gelatins, fibronectin, laminin, vitronectin, proteoglycans, proMMPs 2 and 13, α_1-proteinase inhibitor, α_2-macroglobulin
MMP-15	MT2-MMP (MT-MMP-2, membrane-type MMP-2)	Unknown
MMP-16	MT3-MMP (MT-MMP-3, membrane-type MMP-3)	Pro-MMP-2
MMP-17	MT4-MMP (MT-MMP-4, membrane-type MMP-4)	Unknown
Matrilysin and other MMPs		
MMP-7(3.4.24.23)	Matrilysin (matrin, PUMP-1, small uterine metalloproteinase)	Proteoglycans, laminin, gelatins, collagen IV, entactin, fibronectin, link protein, vitronectin, elastin, tenascin, fibulins, proMMPs 1, 2, and 9
MMP-11	Stromelysin-3	Laminin, fibronectin, aggrecan, α_1-proteinase inhibitor, α_2-macroglobulin
MMP-12 (3.4.24.65)	Metalloelastase (macrophage elastase)	Elastin, fibrinogen, fibronectin, laminin, entactin, collagen IV, proteoglycans, myelin basic protein, IgGs, plasminogen, α_1-proteinase inhibitor
MMP-19	None	Unknown, fluorescent MMP substrates
MMP-20	Enamelysin	Amelogenin

EC, Enzyme Commission.
Data are from refs 1, 2, 5, 25, 31–41, 56, 57.

(5) inhibition by chelating agents, 1,10-phenanthroline, α2-macroglobulin, and endogenous tissue inhibitors of metalloproteinases (TIMPs);

(6) an N-terminal signal sequence that directs their synthesis to the endoplasmic reticulum such that most MMPs are secreted and others expressed as transmembrane proteins;

(7) a propeptide domain that sits N-terminal to the catalytic domain and maintains the enzyme as an inactive zymogen until it is removed by limited proteolysis; and

(8) their *in vitro* activation by diverse reagents including organomercurials (e.g. 4-aminophenylmercuric acetate, APMA), chaotropes, oxidants, heavy metals, disulphide compounds, sulphhydryl-alkylating agents, and detergent denaturation and renaturation.[2]

To date, 18 distinct MMPs and numerous homologues have been identified in the vertebrates (Table 1), another (envelysin, hatching enzyme, EC 3.4.24.12) has been cloned from embryonic sea urchin,[3] and yet another has been cloned in plants (soybean leaf metalloendopeptidase).[4] To keep pace with the frequent discovery of new family members, a sequential MMP numbering system has been adopted, with MMP numbers 4–6 having been abandoned due to initial redundancy and inappropriate classification. Recent comparison of the primary sequences of 30 cloned MMPs from 8 species revealed 30–99 (46 ± 12) per cent identity and 49–99 (63 ± 10) per cent similarity among all analysed family members.[2] The more recently cloned MMP-19, however, shows only 28–35 per cent identity when compared with other family members,[5] and a transiently expressed *Xenopus* MMP (XMMP) is most closely related to stromelysin-3 at only 20 per cent identity.[6] Although this type of homology analysis may offer a more appropriate means of grouping MMP family members, common MMP names and groupings have been based primarily on substrate specificity and protein domain considerations. Thus MMPs are conventionally classified as 'collagenases', 'gelatinases', 'stromelysins', 'membrane-type MMPs', and 'other MMPs'.

Figure 2 offers an alternative classification scheme based on domain structure. As already indicated, all MMPs possess an N-terminal signal sequence or 'pre' domain, a propeptide 'pro' domain that is removed during activation, and a conserved catalytic domain. With the exception of matrilysin, all MMPs also contain a haemopexin/vitronectin-like domain and a hinge (or linker) domain that connects the catalytic and haemopexin-like domains. For gelatinases, the haemopexin-like domain is thought to mediate enzyme–TIMP interactions and association with cell surface 'receptors', whereas for collagenases this domain has been associated with inhibitor and substrate binding. The hinge region varies in length and composition among different family members and may also be important in determining substrate specificity. A cysteine-rich domain with homology to the collagen-binding region of fibronectin splits the catalytic domain of gelatinases A and B, and is required for collagen binding and cleavage.[7] Finally, the membrane-type MMPs (MT-MMPs) are unique in that they possess a membrane spanning domain and a short C-terminal cytoplasmic tail.

All MMPs are synthesized as inactive zymogens (or pro-MMPs). Crystallographic studies have confirmed that pro-MMP latency is maintained by coordination of the active site zinc by an unpaired cysteine thiol group within a highly conserved sequence motif (usually P**RC**GVP**D**) at the carboxyl end of the prodomain.[8] Thus, whether initiated by normal proteolytic removal of the propeptide, by ectopic perturbants, or by site-directed mutagenesis of

Minimal Domain MMPs *(Matrilysin)*

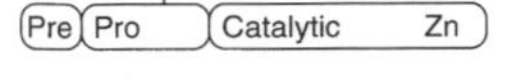

Hemopexin/Vitronectin Domain MMPs

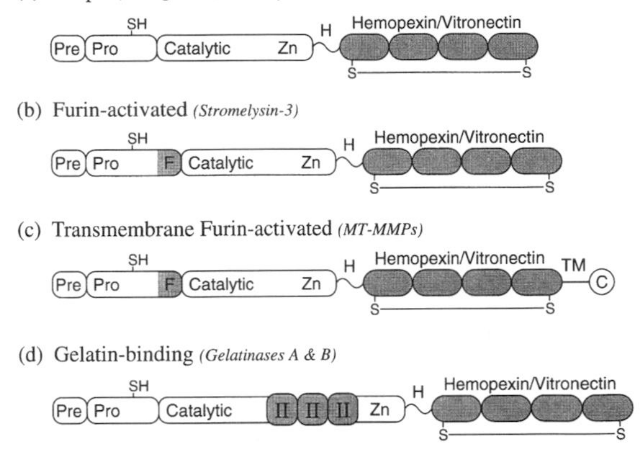

Figure 2. Domain structure of the MMPs. Pre, signal sequence; Pro, propeptide with a free zinc-ligating thiol group; F, furin-like enzyme-recognition motif; Zn, zinc-binding site; II, collagen-binding fibronectin type II inserts; H, hinge region; TM, transmembrane domain; C, cytoplasmic tail. The haemopexin/vitronectin-like C-terminal domain contains four repeats, the first and fourth being connected by a disulphide bridge. (Modified from ref. 1.)

residues within this motif, MMPs are ultimately activated by interruption of this cysteine–zinc interaction, allowing for replacement of the blocking cysteine by water.[9] In addition, once this so-called 'cysteine switch' is opened, further autocatalytic processing of the propeptide results in final cleavage at a highly conserved site.

Although most MMPs are secreted as latent zymogens, stromelysin-3 and all MT-MMPs contain an added furin-like enzyme recognition motif, **R**XK/R**R**, in their propeptide domain, and can be activated intracellularly by calcium-dependent transmembrane serine proteinases of the subtilisin group called furin/PACE/kex-2-like proprotein convertases.[10] Indeed, insertion of this stromelysin-3 sequence into procollagnease-1 results in intracellular activation of the latter proenzyme. All other MMPs lack this recognition motif and are thus activated outside the cell. For collagenase-1, stromelysin-1, gelatinase B, and MT1-MMP, extracellular serine proteinases such as plasmin, urokinase-type plasminogen activator, elastase, and trypsin can cleave their prodomains and incite further autocatalytic activation. By contrast, no serine proteinase tested thus far can activate gelatinase A. It can, however, be activated by a multi-step pericellular pathway involving MT1-MMP. First, MT1-MMP is activated intracellularly by a furin-like enzyme in the *trans*-Golgi network, or it is activated at the cell surface by plasmin. Active cell surface MT1-MMP, in turn, acts as a receptor for TIMP-2, and the activated binary complex acts as a further receptor for progelatinase A, which binds to TIMP-2 via its C-terminal hemopexin domain.[11] Finally, MT1-MMP cleaves and activates the tethered progelatinase A. In addition, other activated MMPs can activate still other proMMPs, or they can inactivate endogenous serine proteinase inhibitors (Table 1). Thus intracellular and extracellular serine proteinases can potentially initiate an array of complex activation cascades involving the MMPs. Other potential means of localizing MMPs to the cell surface include the binding of gelatinase A by $\alpha v \beta 3$ integrin,[12] and the presence of cell surface receptors for MMP-activating enzymes such as uPA, plasmin(ogen), and elastase. Such cell surface-localized proteolysis can potentially confer enhanced proenzyme activation, protection against inhibitors, concentration of required molecules, and spatial limits on the extent of proteolysis.[1]

Another obvious means of regulating MMPs is at the level of transcription. For example, the collagenase-1, stromelysin-1, and gelatinase B genes can be induced by 12-*O*-tetradecanoylphorbol-13-acetate (TPA), EGF, basic FGF, TNF-α, leukotriene D4, integrin-derived signals, alterations in cell shape, and ECM proteins such as SPARC. Collagenase-1 expression is also induced by extracellular matrix metalloproteinase inducer (EMMPRIN, tumour cell-derived collagenase stimulatory factor) and MT1-MMP expression is induced by concanavalin A.[13–18] Many of these first induce the expression of c-*fos* and c-*jun* proto-oncogene products, which in turn recognize an activator protein-1 (AP-1) site (or TPA-responsive element) within the MMP gene promoter. Other motifs found in MMP promoter regions include PEA-3 sequences (which bind *ets* gene family products), AP-2 sites, GC box/SP-1-binding sites, CA-rich sequences, and TGF-β inhibitory elements.[19] Thus MMPs can be strictly controlled at the level of transciption, as well as at the protein level via enzyme activators and inhibitors.

Assays of functional activity

Enzymatic activity can be defined and monitored by several means.[20–22] Substrate gel electrophoresis (zymography) methods are highly sensitive, provide molecular mass and proteinase class information, provide for dissociation of non-covalent enzyme-inhibitor complexes, and detect both latent and active forms of an enzyme since activation occurs by denaturation–renaturation. Though numerous variations of this technique exist,[20] a protein mixture is usually separated on a substrate-containing SDS–PAGE gel under non-reducing conditions. Following SDS removal and incubation in a buffer that allows (or disallows) enzymatic activity to take place, the gel can be stained for protein, thus revealing proteolytic activity as clear bands against a background of uniformly staining substrate. Net enzymatic activity in a sample can also be assayed using radiolabelled, chromogenic or fluorogenic substrates.[20,21,23,24] Radiolabelled collagens are available from DuPont-NEN; chromogenic azo-conjugated (gelatin and casein) substrates and more sensitive and specific thiopeptilide substrates are available from Bachem, Amersham, and Calbiochem. ELISA-based kits are also available from the latter two sources for MMPs 1, 2, 3 and 9. Methods for determining active-site molarity using TIMP-1 and TIMP-2 as titrants have been described for MMPs 1, 2, 3, 7 and 9.[22] In addition to standard metalloproteinase inhibitors, a number of peptide- and hydroxamic acid-based inhibitors optimized to specific MMPs have been developed for therapeutic purposes. Producers of the latter inhibitors include Agouron, British Biotech, Celltech Therapeutics, Ciba-Geigy, Glycomed, and Merck. Finally, once an enzyme has been purified, its activity against a given substrate can be determined and its specific cleavage sites mapped by electrophoresis and amino acid microsequence analysis.

Genes

Full length cDNAs and genomic sequences are published for the MMPs of multiple species. Unless otherwise indicated, human genes, their chromosome locations, and their gene and transcript sizes are listed in Table 2. Mammalian homologues have not yet been found for *Xenopus* collagenase-4 (MMP-18)[25] or XMMP.[6] Eight human MMP genes are clustered on chromosome 11q22.3, and fine mapping indicates that their order is centromere/MMP-8/MMP-10/MMP-1/MMP3/MMP12/MMP-7/MMP-13/telomere.[26] The relative location of the MMP-20 gene, which is also found within this cluster,[55] has not yet been described. Gene intron–exon structure is well conserved among the MMPs, in particular among those within the MMP gene cluster. For those MMP genes that have been characterized so far, the signal sequence is defined in exon 1, the prodomain is defined by the remaining portion of exon 1 and most of exon 2, the catalytic domain regions that lie N-terminal to the zinc-binding domain are defined by the remainder of exon 2 plus exons 3 and 4, three additional exons in gelatinases A and B encode their fibronectin-like domains, the zinc-binding domain is defined by exon 5 (or exon 8 in the gelatinases), and any C-terminal domains or, in the case of matrilysin, any extra amino acids are defined by subsequent exons.[27] The multiple transcripts of MMPs 13, 20, and probably 17 as well, are due to alternative polyadenylation.[28–30] Additional details concerning genes in other species, promoter elements and intron–exon structure are considered in subsequent chapters.

References

1. Coussens, L. M. and Werb, Z. (1996). *Chem. Biol.*, **3**, 895–904.
2. Sang, Q. A. and Douglas, D. A. (1996). *J. Protein Chem.*, **15**, 137–60.
3. Lepage, T. and Gache, C. (1990). *EMBO J.*, **9**, 3003–12.
4. McGeehan, G., Burkhart, W., Anderegg, R., Becherer, J. D., Gillikin, J. W., and Graham, J. S. (1992). *Plant Physiol.*, **99**, 1179–83.
5. Pendas, A. M., Knauper, V., Puente, X. A., Llano, E., Mattei, M. G., Apte, S., *et al.* (1997). *J. Biol. Chem.*, **272**, 4281–6.

Table 2 MMP genes

Enzyme	GenBank no	Location	Size (kbp)	mRNA (kb)	Refs
MMP-1 (collagenase-1)	M16567, X05231	11q22.3	8.2 (10 exons)	2.5	13, 42
MMP-2 (gelatinase A)	J03210, J05471	16q12.2	27 (13 exons)	3.2	43, 44
MMP-3 (stromelysin-1)	X05232	11q22.3	14 (10 exons)	2.2	42
MMP-7 (matrilysin)	X07819	11q21–22	9.65 (6 exons)	1.2	45
MMP-8 (collagenase-2)	J05556	11q22.3		3.3	46
MMP-9 (gelatinase B)	J05070, D10051	20q12–13	7.7 (13 exons)	2.8	44, 47
MMP-10 (stromelysin-2)	X07820, M30461	11q22.3	14 (10 exons)	1.8	45
MMP-11 (stromelysin-3)	X57766	22q11.2	11.5 (8 exons)	2.4	48
MMP-12 metalloelastase)	U78045	11q22.3	13 (10 exons)	1.8	49
MMP-13 (collagenase-3)	X75308	11q22.3	12.5 (10 exons)	3.0, 2.5, 2.0	29
MMP-14 (MT1-MMP)	D26512, X83535	14q11–12		4.5	50, 51
MMP-15 (MT2-MMP)	Z48482	16q12.2–21		4.4	52
MMP-16 (MT3-MMP)	D50477	8q13–21		12	53
MMP-17 (MT4-MMP)	X89576			7.5, 2.7	30
MMP-18 (collagenase-4)	NA (*Xenopus laevis*)			1.9	25
MMP-19	Y08622, X92521	12q14		2.4	5, 54
MMP-20 (enamelysin)	Y12779	11q22.3		4.3, 2.5	26, 55
MMP-x (XMMP)	U82541 (*Xenopus*)			3.0	6

6. Yang, M., Murray, M. T., and Kurkinen, M. (1997). *J. Biol. Chem.*, **272**, 13527–33.
7. Allan, J. A., Docherty, A. J. P., Barker, P. J., Huskisson, N. S., Reynolds, J. J., and Murphy, G. (1995). *Biochem. J.*, **309**, 299–306.
8. Birkedal-Hansen, H. (1995). *Curr. Opin. Cell Biol.*, **7**, 728–35.
9. Van Wart, H. E. and Birkedal-Hansen, H. (1990). *Proc. Natl Acad. Sci. USA*, **87**, 5578–82.
10. Pei, D. and Weiss, S. (1995). *Nature*, **375**, 244–7.
11. Strongin, A. Y., Collier, I., Bannikov, G., Marmer, B. L., Grant, G. A., and Goldberg, G. I. (1995). *J. Biol. Chem.*, **270**, 5331–8.
12. Brooks, P., Stromblad, S., Sanders, L. von Scalscha, T., Aimes, R., Stetler-Stevenson, W., *et al.* (1996). *Cell*, **85**, 683–93.
13. Angel, P., Baumann, I., Stein, B., Delius, H., Rahmsdorf, H. J., and Herrlich, P. (1987). *Mol. Cell. Biol.*, **7**, 2256–66.
14. Edwards, D. R., Murphy, G., Reynolds, J. J., Whitham, S. E., Docherty, A. J. P., Angel, P., and Heath, J. K. (1987). *EMBO J.*, **6**, 1899–904.
15. Brenner, D. A., O'Hara, M., Angel, P. Chojkier, M., and Karin, M. (1989). *Nature*, **337**, 661–3.
16. Roskelley, C., Srebrow, A., and Bissell, M. (1995). *Curr. Opin. Cell Biol.*, **7**, 736–47.
17. Biswas, C., Zhang, Y., DeCastro, R., Guo, H., Nakamura, T., Kataoka, H., and Nabeshima, K. (1995). *Cancer Res.*, **55**, 434–9.
18. Tremble, P., Damsky, C. H., and Werb, Z. (1995). *J. Cell Biol.*, **129**, 1707–20.
19. Gaire, M., Magbanua, Z., McDonnell, S., McNiel, L., Lovett, D. H., and Matrisian, L. M. (1994). *J. Biol. Chem.*, **269**, 2032–40.
20. Fisher, S. J. and Werb, Z. (1995). In *Extracellular matrix. A practical approach*. (ed. M. A. Haralson and J. R. Hassell), pp. 261–87. IRL Press, New York.
21. Woessner, J. F. (1995). *Meth. Enzymol.*, **248**, 510–28.
22. Murphy, G. and Willenbrock, F. (1995). *Meth. Enzymol.*, **248**, 496–510.
23. Netzel-Arnett, S., Mallya, S. K., Nagase, H., Birkedal-Hansen, H., and Van Wart, H. E. (1991). *Anal. Biochem.*, **195**, 86–92.
24. Knight, C. G., Willenbrock, F., and Murphy, G. (1992). *FEBS Lett.*, **296**, 263–6.
25. Stolow, M. A., Bauzon, D. D., Li, J., Sedgwick, T., Liang, V. C. T., Sang, Q. A., and Shi, Y. B. (1996). *Mol. Biol. Cell*, **7**, 1471–83.
26. Pendas, A. M., Santamaria, I., Alvarez, M. V., Pritchard, M., and Lopez-Otin, C. (1996). *Genomics*, **37**, 266–9.
27. Pendas, A. M., Balbin, M., Llano, E., Jimenez, M. G., and Lopez-Otin, C. (1997). *Genomics*, **40**, 222–33.
28. Bartlett, J. D., Simmer, J. P., Xue, J., Margolis, H. C., and Moreno, E. C. (1996). *Gene*, **183**, 123–8.
29. Freije, J. M. P., Diez-Itza, I., Balbin, M., Sanchez, L. M., Blasco, R., Tolivia, J., and Lopez-Otin, C. (1994). *J. Biol. Chem.*, **269**, 16766–73.
30. Puente, X. S., Pendas, A. M., Llano, E., Valesco, G., and Lopez-Otin, C. (1996). *Cancer Res.*, **56**, 944–9.
31. Knauper, V., Will, H., Lopez-Otin, C., Smith, B., Atkinson, S. J., Stanton, H., *et al.* (1996). *J. Biol. Chem.*, **271**, 17124–131.
32. Fosang, A. J., Last, K., and Maciewicz, R. A. (1996). *J. Clin. Invest.*, **98**, 2292–9.
33. Imai, K., Shikata, H., and Okada, Y. (1995). *FEBS Lett.*, **369**, 249–51.
34. Rajah, R., Nunn, S. E., Herrick, D. J., Grunstein, M. M., and Cohen, P. (1996). *Am. J. Physiol.*, **271**, L1014–22.
35. Ohuchi, E., Imai, K., Fujii, Y., Sato, H., Seiki, M., and Okada, K. (1997). *J. Biol. Chem.*, **272**, 2446–51.
36. Levi, E., Fridman, R., Miao, H. Q., Ma, Y. S., Yayon, A., and Vlodavsky, I. (1996). *Proc. Natl Acad. Sci. USA*, **93**, 7069–74.
37. Sasaki, T., Mann, K., Murphy, G., Chu, M. L., and Timpl, R. (1996). *Eur. J. Biochem.*, **240**, 427–34.
38. Mayer, U., Mann, K., Timpl, R., and Murphy, G. (1993). *Eur. J. Biochem.*, **217**, 877–84.
39. Ochieng, J., Fridman, R., Nangia-Makker, P., Kleiner, D. E., Liotta, L. A. Stetler-Stevenson, W. G., and Raz, A. (1994). *Biochemistry*, **33**, 14109–114.
40. Preece, G., Murphy, G., and Ager, A. (1996). *J. Biol. Chem.*, **271**, 11634–640.
41. Banda, M. J., Clark, E. J., and Werb, Z. (1985). In *Mononuclear phagocytes: characteristics, physiology, and function*. (ed. R. van Furth), pp. 295–301. Martinus Nijhoff, The Hague.
42. Whitham, S. E., Murphy, G., Angel, P., Rahmsdorf, H.-J., Smith, B. J., Lyons, A. *et al.* (1986). *Biochem. J.*, **240**, 913–6.

43. Collier, I. E., Wilhelm, S. M., Eisen, A. Z., Marmer, B. L., Grant, G. A., Seltzer, J. L., *et al.* (1988). *J. Biol. Chem.*, **263**, 6579–87.
44. Huhtala, P., Tuuttila, A., Chow, L. T., Lohi, J., Keski-Oja, J., and Tryggvason, K. (1991). *J. Biol. Chem.*, **266**, 16485–90.
45. Muller, D., Quantin, B., Gesnel, M. C., Millon-Collard, R., Abecassis, J., and Breathnach, R. (1988). *Biochem. J.*, **253**, 187–92.
46. Hasty, K. A., Pourmotabbed, T. F., Goldberg, G. I., Thompson, J. P., Spinella, D. G., Stevens, R. M., and Mainardi, C. M. (1990). *J. Biol. Chem.*, **265**, 11421–4.
47. Wilhelm, S. M., Collier, I. E., Marmer, B. L., Eisen, A. Z., Grant, G. A., and Goldberg, G. I. (1989). *J. Biol. Chem.*, **264**, 17213–21. [Erratum in *J. Biol. Chem.*, (1990), **265**, 22570].
48. Basset, P., Bellocq, J. P., Wolf, C., Stoll, I., Hutin, P., Limacher, J. M., *et al.* (1990). *Nature*, **348**, 699–704.
49. Shapiro, S. D., Kobayashi, D. K., and Ley, T. J. (1993). *J. Biol. Chem.*, **268**, 23824–9.
50. Sato, H., Takino, T., Okada, Y., Cao, J., Shinagawa, A., Yamamoto, E., and Seiki, M. (1994). *Nature*, **370**, 61–65.
51. Okada, A., Bellocq, J. P., Rouyer, N., Chenard, M. P., Rio, M. C., Chambon, P., and Basset, P. (1995). *Proc. Natl Acad. Sci. USA*, **92**, 2730–4.
52. Will, H. and Hinzmann, B. (1995). *Eur. J. Biochem.*, **231**, 602–8.
53. Takino, T., Sato, H., Shinagawa, A., and Seiki M. (1995). *J. Biol. Chem.*, **270**, 23013–20.
54. Cossins, J., Dudgeon, T. J., Catlin, G., Gearing, A. J. H., and Clements, J. M. (1996). *Biochem. Biophys. Res. Commun.*, **228**, 494–8.
55. Llano, E., Pendas, A. M., Knauper, V., Sorsa, T., Salo, T., Salido, E., Murphy, G., Simmer, J. P., Bartlett, J. D., and Lopez-Otin, C. (1997). *Biochemistry*, **36**, 15101–8.
56. Giannelli, G., Falk-Marzillier, J., Schiraldi, O., Stetler-Stevenson, W. G., and Quaranta, V. (1997). *Science*, **277**, 225–8.
57. Suzuki, M., Raab, G., Moses, M. A., Fernandez, C. A., and Klagsbrun, M. (1997). *J. Biol. Chem.*, **272**, 31730–7.

■ *Mark D. Sternlicht and Zena Werb*
University of California, San Francisco, Department of Anatomy, LR-208, 3rd and Parnassus Avenues, San Francisco, CA 94143–0452, USA.

Collagenases (MMPs 1, 8, 13, and 18)

MMP collagenases are the only endogenous enzymes that can readily cleave the triple-helical domain of native fibrillar collagens I, II, and III. Cleavage of each collagen α-chain occurs at a specific site three-quarters of the distance from the N terminus and is followed by spontaneous collagen denaturation. At present, four distinct collagenases have been identified, including an enzyme that has only been identified in *Xenopus laevis*.

■ Protein properties

Four distinct MMP collagenases have been identified: collagenase-1 (MMP-1, fibroblast, interstitial, or tissue collagenase), collagenase-2 (MMP-8, neutrophil, leukocyte, or PMN collagenase), collagenase-3 (MMP-13), and *Xenopus* collagenase-4 (MMP-18). Each cleaves fibrillar collagens at a single Gly–Ile/Leu bond three-quarters of the distance from the N to C terminus and generates characteristic three-quarters and quarter-length collagen fragments.[1] Once this occurs, the helical collagens spontaneously denature and become susceptible to further degradation by diverse gelatinolytic enzymes. Each collagenase has a typical domain structure consisting of conserved pre, pro, catalytic, hinge, and C-terminal haemopexin-like domains. Insight into the function of various MMP domains has been gained by comparing the activities and affinities of various deletion mutants and chimeric constructs with those of their full length counterparts. For instance, C-terminally truncated collagenase-1 can still cleave gelatin, casein, and peptide substrates, and it remains sensitive to TIMP inhibition, but it can no longer cleave native triple-helical collagens, indicating the importance of the hinge and haemopexin-like domains in this regard.[2,3] Chimeras containing the catalytic domain of collagenase-1 and the C-terminal domains of stromelysin-1 or -2, and those containing the stromelysin catalytic domain and the collagenase C-terminal domain, show no triple-helicase activity, but retain their caseinase and gelatinase activity, suggesting that the triple-helicase activity of collagenase-1 derives from both ends of the enzyme.[2,3] Furthermore, as indicated by gelatin cleavage product fingerprints from various chimeras, the cleavage specificity of a given MMP is confered by the zinc-binding domain that sits within its larger catalytic domain.[3] This specificity may be conferred by three conserved residues that flank the three-histidine zinc-binding motif and distinguish all four collagenases from other MMPs: a Tyr that sits eight residues towards the N terminus from the first His residue and Asp and Gly residues that sit three and five residues towards the C terminus from the third His residue, respectively.[3]

With the exception of bullfrog collagenase-1, which is a somewhat shorter but highly homologous enzyme, known collagenases from other species have 48–97 per cent amino acid identity, 67–98 per cent sequence similarity, latent forms with 447–452 amino acids and deduced molecular weights of 51–52 kDa, and active forms with 368–369 residues and predicted weights of about 42 kDa.[1,4,5] Interesting differences also exist. Depending on the presence or absence of Asn-linked glycosylation, collagenase-1 has two latent forms of 52 and 57 kDa and a 42 kDa active form; collagenase-2 has a 42 kDa active form and a 75 kDa latent form that is readily proteolysed on extraction to a 58 kDa form that retains its latency; and collagenase-3 has 60 and 48 kDa latent and active forms, respectively.[6,7] While collagenases can degrade all fibrillar collagens, the preferred substrates for collagenases 1, 2, and 3 are collagen types III, I, and II, respectively,[6,8,9] while collagenase-4 specificity has only been characterized for type I collagen.[1] Collagenase-1 is produced and immediately secreted from a variety of cell types primarily of mesenchymal origin in response to specific inducers; collagenase-2 is predominantly produced by neutrophils during their maturation in bone marrow and is stored until the cells are stimulated to degranulate; collagenase-3 is found in bone, normal and pathologic cartilage, and various epithelial cancers, and is the predominant collagenase in mice and rats (collagenase-1 being undetectable in these species); and *Xenopus* collagenase-4 is expressed only transiently during tadpole tail resorption, hindlimb morphogenesis, and intestinal remodelling.[1,7,10–12] Indeed, the literature on collagenases is somewhat confused in that mouse and rat collagenase were not initially recognized to be what is now called collagenase-3. Thus the enzyme referred to as mouse or rat MMP-1 in the literature is actually MMP-13.

Purification

Collagenases 1, 2 and 3 have been purified from cells and tissues and each collagenase has been purified as a recombinant enzyme. Collagenase-1 can be purified from fibroblast-conditioned medium using cation-exchange (phosphocellulose or carboxymethylcellulose) and gel filtration chromatography.[13] Methods for purifying latent and active collagenase-2 from human neutrophils have also been described.[14] Leukocyte extracts are passed over concanavalin A–Sepharose and active collagenase-2 is purified from the unbound eluant by collagen–Sepharose adsorption, elution with salt, and sequential Sephacryl S-300 and DEAE–Sephacel chromatography. Latent collagenase-2 is purified by applying fractions not adsorbed onto the collagen–Sepharose to the latter two columns directly. More recently modified methods are described in the *Methods in Enzymology* series.[7,15] Recombinant collagenase-3 has been produced using prokaryotic, vaccinia virus, baculovirus and mammalian cDNA expression systems, and purified by methods involving heparin–agarose and SP–Sepharose, or S-Sepharose and Sephacryl S-200 chromatography.[6,10,12] His-tagged collagenase-4 has been produced using an *E. coli* overexpression system and purified from insoluble inclusion bodies using a metal chelation column.[1]

Activities

Collagenases cleave triple-helical collagens I, II, and III at a single Gly–Ile/Leu bond. Collagen I fibrils composed of wild-type α2(I) chains and collagenase-resistant α1(I) chains generated by site-directed mutagenesis are themselves collagenase resistant, indicating a requirement for normal α1(I) sequences in the native triple helix.[16] Active-site specificity mapping with a variety of synthetic substrates has indicated that collagenase-1 is an esterase that prefers lipophilic sequences, and has revealed a particularly sensitive substrate, Ac–Pro–Leu–Gly–[2-mercapto-4-methyl-pentanoyl]-Leu–Gly–OEt, which is available from Bachem.[17] In addition to cleaving native fibrillar collagens at a single site, collagenases can degrade a number of other matrix molecules and collagenase-1 can activate progelatinases A and B (Table 1, p. 520). Their own activation is thought to occur through proteolytic removal of the prodomain by other MMPs (Table 1, p. 520) and by serine proteinases such as plasmin; although the latter may also cleave the C-terminal domain and thereby abolish collagenolytic activity.[18]

Antibodies

Mouse monoclonal antibodies against human collagenases 1 (41-IE5) and 2 (115–1302) are available from Amersham, Calbiochem, Fuji Chemicals, and ICN Immunologics. Function blocking and non-neutralizing monoclonal anti-human collagenase-1 antibodies are also available from Chemicon and a monoclonal antibody against rabbit collagenase-1 is available from the Developmental Studies Hybridoma Bank (University of Iowa). Polyclonal antibodies against collagenase-1 are available from Biogenesis, Chemicon, Cortex Biologics, The Binding Site, and Triple Point Biologics. Mouse monoclonal anti-human collagenase-3 antibodies are available from Fuji Chemicals and Oncogene Research Products. An affinity purified rabbit anti-human collagenase-3 antibody is available from Triple Point Biologics.

Genes

In addition to the human sequences listed in Table 2 (p. 523), cDNA and/or genomic sequences have been obtained for MMP-1 in rabbit (M17820–3, M19240), pig (X54724), cow (X58256), and bullfrog (S75623); and for MMP-13 in mouse (X66473), rat (M60616, M36452), rabbit (AF059201), newt (D82055), and frog (L49412, U41824). A mammalian homologue of frog MMP-18[1] has not yet been discovered, nor has a non-human MMP-8 homologue been found so far. All three known human collagenase genes map to the MMP gene cluster on chromosome 11q22.3.

Analysis of 5′-flanking regions has revealed the presence of a TATA box, an AP-1 site, a PEA-3 sequence, and a TGF-β inhibitory element upstream of both the human collagenase-1 and collagenase-3 genes, as well as an osteoblast-specific element (OSE-2) within the collagenase-3 gene promoter.[19–21] Functional analysis indicates that the AP-1 site of both genes confers phorbol ester inducibility, but that the AP-1 and PEA-3 elements only act synergistically in the collagenase-1 promoter, presumably because they are separated by nine nucleotides rather than 20, as is the case for the collagenase-3 gene promoter.[19,21] Human breast cancer cells can induce fibroblasts to express collagenase-3, with IL-1α and β being potential inducers.[25] Interestingly, collagenase-3 expression is induced by TGF-β, whereas the expression of collagenase-1 and other MMPs is repressed (J. Uria; personal communication). On the other hand, collagenases 1 and 3 are both upregulated in rabbit chondrocytes and fibroblasts by IL-1, TNF-α and PMA, but the induction is more rapid and transient for collagenase-3 than for collagenase-1.[26]

Mutant phenotype/disease states

Although mutant phenotypes have not been described for any of the collagenase genes, mice carrying a collagenase-resistant *COL1A1* transgene show late embryonic lethality, and introduction of the same mutation into the endogenous *COL1A1* gene leads to marked dermal fibrosis resembling human scleroderma and impaired postpartum uterine involution.[22] These studies suggest that collagen resorption can also be accomplished through collagenase cleavage at a novel site in the non-helical N-telopeptide domain. In another animal model, expression of a collagenase-1 transgene directed to squamous epithelium led to hyperproliferative skin lesions and increased sensitivity to chemical carcinogenesis.[23] Collagenase-3 expression is induced in chronically inflamed oral mucosal epithelium and may thus contribute to the migratory behaviour of these cells and the matrix degradation seen during chronic periodontitis.[27]

Structure

X-ray crystal structures have been solved for truncated human collagenase-1, full-length porcine collagenase-1, substrate analogue inhibited catalytic domains of human collagenases 1 and 2, and the C-terminal haemopexin-like domain of human collagenase-3 (see ref. 24 and references therein). Such analyses indicate that a spherical catalytic domain with a zinc-containing active-site cleft is attached via the hinge domain to an ellipsoid disc-shaped C-terminal domain (Fig. 1). The latter domain is composed of four β-sheets arranged sequentially about a central funnel-like tunnel containing two calcium and two chloride ions, thus giving it a 'four-bladed β-propeller-like' appearance. The first and last blades are covalently linked by a disulphide bridge between cysteines at the leading portion of blade I and the C-terminal end of blade IV. Beyond this, several subtle structural differences probably help define the binding and cleavage characteristics of the collagenases when compared with one another and with other MMPs. Protein Data Bank (PDB) identifiers are 1CGE, 1CGF, 1CGL, 1HFC, 1FBL and 2TCL for collagenase-1; 1MNC, 1MMB and 1JAN to 1JAQ for collagenase-2.

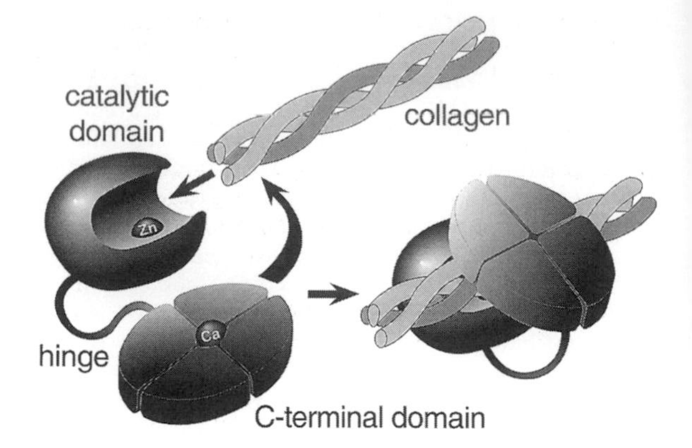

Figure 1. Putative structure–function model for collagenase MMPs. Once triple-helical collagen enters the catalytic active-site cleft, the flexible hinge probably bends so that the haemopexin-like C-terminal portion entraps the collagen substrate. (Modified from ref. 24.)

References

1. Stolow, M. A., Bauzon, D. D., Li, J., Sedgwick, T., Liang, V. C. T., Sang, Q. A., and Shi, Y. B. (1996). *Mol. Biol. Cell*, **7**, 1471–83.
2. Murphy, G., Allan, J. A., Willenbrock, F., Cockett, M. I., O'Connel, J. P., and Docherty, A. J. P. (1992). *J. Biol. Chem.*, **267**, 9612–8.
3. Sanchez-Lopez, R., Alexander, C. M., Behrendtsen, O., Breathnach, R., and Werb, Z. (1993). *J. Biol. Chem.*, **268**, 7238–47.
4. Sang, Q. A. and Douglas, D. A. (1996). *J. Protein Chem.*, **15**, 137–60.
5. Oofusa, K., Yomori, S., and Yoshizato, K. (1994). *Int. J. Dev. Biol.*, **38**, 345–50.
6. Knauper, V., Lopez-Otin, C., Smith, B., Knight, G., and Murphy, G. (1996). *J. Biol. Chem.*, **271**, 1544–50.
7. Dioszegi, M., Cannon, P., and Van Wart, H. E. (1995). *Meth. Enzymol.*, **248**, 413–31.
8. Hasty, K. A., Jeffrey, J. J., Hibbs, M. S., and Welgus, H. G. (1987). *J. Biol. Chem.*, **265**, 11421–4.
9. Welgus, H. G., Jeffrey, J. J., and Eisen, A. Z. (1981). *J. Biol. Chem.*, **265**, 9511–15.
10. Freije, J. M. P., Diez-Itza, I., Balbin, M., Sanchez, L. M., Blasco, R., Tolivia, J., and Lopez-Otin, C. (1994). *J. Biol. Chem.*, **269**, 16766–73.
11. Reboul, P., Pelletier, J. P., Tardif, G., Cloutier, J. M., and Martel-Pelletier, J. (1996). *J. Clin. Invest.*, **97**, 2011–19.
12. Mitchell, P. G., Magna, H. A., Reeves, L. M., Lopresti-Morrow, L. L., Yocum, S. A., Rosner, P. J., *et al.* (1996). *J. Clin. Invest.*, **97**, 761–8.
13. Stricklin, G. P., Bauer, E. A., Jeffrey, J. J., and Eisen, A. Z. (1977). *Biochemistry*, **16**, 1607–15.
14. Macartney, H. W. and Tschesche, H. (1983). *Eur. J. Biochem.*, **130**, 71–8.
15. Tschesche, H. (1995). *Meth. Enzymol.*, **248**, 431–49.
16. Wu, H., Byrne, M. H., Stacey, A., Goldring, M. B., Birkhead, J. R., Jaenisch, R., and Krane, S. M. (1990). *Proc. Natl Acad. Sci. USA*, **87**, 5888–92.

17. Weingarten, H., Martin, R., and Feder, J. (1985). *Biochemistry*, **24**, 6730–4.
18. Knauper, V., Will, H., Lopez-Otin, C., Smith, B., Atkinson, S. J., Stanton, H., *et al*. (1996). *J. Biol. Chem.*, **271**, 17124–31.
19. Pendas, A. M., Balbin, M., Llano, E., Jimenez, M. G., and Lopez-Otin, C. (1997). *Genomics*, **40**, 222–33.
20. Angel, P., Baumann, I., Stein, B., Delius, H., Rahmsdorf, H. J., and Herrlich, P. (1987). *Mol. Cell. Biol.*, **7**, 2256–66.
21. Gutman, A. and Wasylyk, B. (1990). *EMBO J.*, **9**, 2241–6.
22. Liu, X., Wu, H., Byrne, M., Jeffrey, J., Krane, S., and Jaenisch, R. (1995). *J. Cell Biol.*, **130**, 227–37.
23. D'Armiento, J., DiColandrea, T., Dalal, S., Okada, Y., Huang, M. T., Conney, A., and Chada, K. (1995). *Mol. Cell. Biol.*, **15**, 5732–9.
24. Gomis-Ruth, F. X., Gohlke, U., Betz, M., Knauper, V., Murphey, G., Lopez-Otin, C., and Bode, W. (1996). *J. Mol. Biol.*, **264**, 556–66.
25. Uria, J. A., Stahle-Backdahl, M., Seiki, M., Fueyo, A., and Lopez-Otin, C. (1997). *Cancer Res.*, **57**, 4882–8.
26. Vincenti, M. P., Coon, C. I., Mengshol, J. A., Yocum, S., Mitchell, P., and Brinckerhoff, C. E. (1998). *Biochem. J.*, **331**, 341–6.
27. Uitto, V. J., Airola, K., Vaalamo, M., Johansson, N., Putnins, E. E., Firth, J. D., Salonen, J., Lopez-Otin, C., Saarialho-Kere, U., and Kahari, V. M. (1998). *Am. J. Pathol.*, **152**, 1489–99.

Mark D. Sternlicht and Zena Werb
University of California, San Francisco,
Department of Anatomy, LR-208,
3rd and Parnassus Avenues, San Francisco,
CA 94143–0452, USA.

Stromelysins 1 and 2 (MMPs 3 and 10)

Stromelysins 1 and 2 can degrade several structural substrates, activate other MMPs, and inactivate several serine proteinase inhibitors. They are distinguished from other MMPs by their close sequence similarity to one another, their extended hinge domain, and their similar substrate specificities.

Protein properties

Stromelysin-1 (SL-1, MMP-3, transin, proteoglycanase, procollagenase activating protein, acid metalloproteinase of cartilage) was initially isolated by biochemical means in several independent laboratories.[1,2] It is secreted as a 56 kDa inactive zymogen that is processed to yield 45–48 and 28 kDa active forms with identical N termini and indistinguishable activities and specificities (although only the larger form binds to collagen I via its C-terminal domain).[1] Single Asn glycosylation sites are present in its catalytic and C-terminal domains such that about 20 per cent of fibroblast stromelysin-1 is secreted as a 59 kDa glycoprotein, the remainder being unglycosylated.[1] Stromelysin-2 (SL-2, MMP-10, transin-2) was later discovered by molecular cloning,[3] and has likewise been found to have a 57 kDa latent form, with active forms of 45–47 and 28 kDa, the latter also being C-terminally truncated.[1] Both stromelysins share considerable sequence homology with one another and they are further distinguished by a nine-residue insert in their hinge domain that is absent from most other MMPs.

Purification

SL-1 can be purified from media conditioned by TPA-stimulated fibroblasts using DEAE–Sephacel and heparin–Sepharose chromatography.[2] Another method uses anti-SL-1 immunoaffinity chromatography, yet another uses sequential DEAE–cellulose, Green A Matrex, Gelatin–Sepharose, and Sephacryl S-200 chromatographies plus anti-MMP-1 immunoadsorption or CM-cellulose chromatography to remove contaminating MMP-1.[1] Glycosylated and unglycosylated forms can be separated on a ConA–Sepharose column, active and latent forms can be separated on a green A Matrex gel column. SL-2 from stimulated keratinocyte cultures can be purified in similar fashion.[1] Purification of recombinant SL-2 from mouse myeloma cells grown in the presence of high amounts of BSA requires that the green A Matrex step be replaced by Zn-chelation chromatography.

Activities

Like other secreted MMPs, SL-1 and SL-2 are activated extracellularly. Their activation can be accomplished using perturbants such as APMA or serine proteinases such as plasmin, plasma kallikrein, neutrophil elastase, chymases, tryptase, chymotrypsin, or trypsin.[1] SL-1 can hydrolyse numerous structural proteins, activate procollagenases 1–3 and progelatinase B, and inactivate a number of serpins (Table 1, p. 520). In addition, SL-1 can digest the telopeptides of collagens I and II.[1] Activity against collagen III and elastin is weak, however, and activation of procollagenase-1 requires prior removal of a more proximal portion of its prodomain by another enzyme.[1] Still, the activation of procollagenases by SL-1 leaves the N-terminal Phe that is required for expression of their full enzymatic activity. SL-1 can directly or indirectly cleave cell surface E-cadherin;[16] it can release soluble L-selectin from leukocytes;[17] and it releases active heparin-binding EGF-like growth factor (HB-EGF) from the cell surface by cleaving the Glu151–Asn152 bond within the HB-EGF juxtamembrane domain.[18] Cleavage-site and synthetic substrate analyses indicate that SL-1 readily cleaves between P1 and P1′ residues when a hydrophobic residue is in the P1′ position, while there are no strict requirements for the P1 residue.[1] SL-1 is most active against aggrecan, gelatin and Azocoll at pH 5.3–5.5, but remains 30–50 per cent active at pH 7.5–8.0.[1] The enzymatic activity of SL-2 against various substrates is generally weaker than that of SL-1. The activities of both enzymes can be measured by zymography or by methods involving Azocoll, fluorogenic peptide substrates, polyacrylamide-entrapped aggrecan, or radiolabelled transferrin or casein.[1] SL activity can also be assessed based on the appearance of other substrate breakdown products.[2]

Antibodies

A mouse monoclonal anti-human SL-1 antibody (55–2A4) is available from Amersham, Calbiochem, Fuji Chemicals and ICN. A neutralizing monoclonal antibody (11N13) that recognizes the active site epitope of SL-1 has also been described.[4] Polyclonal antibodies against SL-1 are available from Biogenesis, Biodesign International, Cambio, Chemicon, The Binding Site, and Triple Point Biologics, and a rabbit anti-human SL-2 antibody is available from Triple Point Biologics.

Genes

In addition to the human sequences listed in Table 2 (p. 523), cDNA and/or genomic sequences have been obtained for SL-1 in mouse (X63162, X66402), rat (M13012), rabbit (M25664), and horse (U62529); and for SL-2 in mouse (X76537) and rat (X64020–1, X05085). Both human genes are found within the MMP gene cluster on chromosome 11. Depending on the particular SL and cell type involved, SL gene expression can be enhanced by TPA, IL-1, TNF-α, and growth factors such as EGF, KGF, TGF-α, NGF, and basic FGF.[1,5,6] For example, TPA induces expression of the SL-2 gene, but not the SL-1 gene, in keratinocytes, whereas the opposite occurs in fibroblasts.[6] The SL-1 gene promoter contains a functional AP1-binding site (TPA responsive element), two oppositely oriented PEA3 sites, and an NGF-IA protein-binding site of questionable significance.[5,7] In addition, the TGF-β inhibitory element and a novel NGF-responsive element were first identified in the SL-1 gene promoter.[5,8] The SL-2 gene promoter also has a single consensus AP1-binding site.[6] These and other sites in the murine SL-1 gene promoter indicate that the promoter should show activity in cells with a mesenchymal phenotype.[14]

Mutant phenotype/disease states

SL-1 knockout mice exhibit hypomorphic virgin and pregnant mammary glands and delayed closure of excisional wounds (unpublished observations). On the other hand, mice that express an autoactivating SL-1 transgene directed to mammary epithelium by the whey acidic protein gene promoter display precocious virgin glandular development, unscheduled apoptosis during mid-pregnancy, formation of a reactive stroma, and development of progressive mammary gland lesions.[9] Chemical carcinogenesis in the mammary gland is impeded, however, in SL-1 transgenic mice where the transgene is under the control of the MMTV LTR.[10]

Several MMPs have also been implicated in coronary atherosclerosis. Interestingly, a common polymorphism (6A) in the SL-1 gene promoter leads to reduced gene expression, and atherosclerosis patients who are homozygous for 6A exhibit more rapid disease progression.[11] SL-1 is also considered an important mediator of cartilage loss in rheumatoid arthritis and osteoarthritis, yet SL-1 knockout and wild-type mice are similarly susceptible to collagen-induced arthritis and both show cleavage of the aggrecan Asn341–Phe342 peptide bond.[15] Thus other redundant and compensatory 'aggrecanases' are also likely to play a role in arthritic cartilage destruction.

Structure

X-ray structures for the inhibited catalytic domain of SL-1 and C-terminally truncated proSL-1 have been published and are available through the Brookhaven Protein Data Bank (PDB accession numbers 1UMS, 1UMT, and 2SRT).[12,13] Crystallographic studies also indicate that the Cys-75 residue in the prodomain of SL-1 ligates the active site zinc and thus maintains enzyme latency.[1]

References

1. Nagase, H. (1995). *Meth. Enzymol.*, **248**, 449–70.
2. Chin, J. R., Murphy, G., and Werb Z. (1985). *J. Biol. Chem.*, **260**, 12367–76.
3. Muller, D., Quantin, B., Gesnel, M. C., Millon-Collard, R., Abecassis, J., and Breathnach, R. (1988). *Biochem. J.*, **253**, 187–92.
4. Su, J. L., Becherer, J. D., Edwards, C., Bukhart, W., McGeehan, G. M., and Champion, B. R. (1995). *Hybridoma*, **14**, 383–90.
5. deSouza, S., Lochner, J., Machida, C. M., Matrisian, L. M., and Ciment, G. (1995). *J. Biol. Chem.*, **270**, 9106–114.
6. Windsor, L. J., Grenett, H., Birkedal-Hansen, B., Bodden, M. K., Engler, J. A., and Birkedal-Hansen, H. (1993). *J. Biol. Chem.*, **268**, 17341–7.
7. Gaire, M., Magbanua, Z., McDonnell, S., McNiel, L., Lovett, D. H., and Matrisian, L. M. (1994). *J. Biol. Chem.*, **269**, 2032–40.
8. Kerr, L. D., Miller, D. B., and Matrisian, L. M. (1990). *Cell*, **62**, 1189–1204.
9. Sympson, C. J., Talhouk, R. S., Bissell, M. J., and Werb, Z. (1995). *Persp. Drug Discovery Design*, **2**, 401–11.
10. Witty, J. P., Lempka, T., Coffey, R. J., and Matrisian, L. M. (1995). *Cancer Res.*, **55**, 1401–6.
11. Ye, S., Eriksson, P., Hamsten, A., Kurkinen, M., Humphries, S. E., and Henney, A. M. (1996). *J. Biol. Chem.*, **271**, 13055–60.

12. Becker, J. W., Marcy, A. I., Rokosz, L. L., Axel, M. G., Burbaum, J. J., Fitzgerald, P. M. D., *et al.* (1995). *Protein Sci.* **4**, 1966–76.
13. Dhanaraj, V., Ye, Q. Z., Johnson, L. L., Hupe, D. J., Ortwine, D. F., Dubar, J. B., *et al.* (1996). *Structure*, **4**, 375–86.
14. Yee, J., Kuncio, G. S., Bhandari, B., Shinhab, F. S., and Nielson, E. G. (1997). *Kidney Int.*, **52**, 120–9.
15. Mudgett, J. S., Hutchinson, N. I., Chartrain, N. A., Forsyth, A. J., McDonnell, J., Singer, I. I., Bayne, E. K., Flanagan, J., Kawka, D., Shen, C. F., Stevens, K., Chen, H., Trumbauer, M., and Visco, D. M. (1998). *Arthritis Rheum.*, **41**, 110–21.
16. Lochter, A., Galosy, S., Muschler, J., Freedman, N., Werb, Z., and Bissell, M. J. (1977). *J. Cell Biol.*, **139**, 1861–72.
17. Preece, G., Murphy, G., and Ager, A. (1996). *J. Biol. Chem.*, **271**, 11634–40.
18. Suzuki, M., Raab, G., Moses, M. A., Fernandez, C. A., and Klagsbrun, M. (1997). *J. Biol. Chem.*, **272**, 31730–7.

Mark D. Sternlicht and Zena Werb
University of California, San Francisco,
Department of Anatomy, LR-208,
3rd and Parnassus Avenues, San Francisco,
CA 94143–0452, USA.

Gelatinases (MMPs 2 and 9)

Gelatinases A and B (GelA and GelB) are distinguished from other MMPs by a fibronectin type II repeat insert that mediates their particular ability to bind to and degrade denatured collagens. Although they share fairly broad overlapping substrate specificities, they differ considerably in terms of their transcriptional regulation, their glycosylation, their modes of activation, and their tendency to form GelA–TIMP-2 and GelB–TIMP-1 enzyme–inhibitor complexes preferentially.

Protein properties

Like other MMPs, gelatinase A (GelA, MMP-2, 72 kDa type IV collagenase) and gelatinase B (GelB, MMP-9, 92 kDa type IV collagenase) each have pre, pro, catalytic and haemopexin-like domains. They are distinguished, however, by three head-to-tail repeats inserted within their catalytic domain which resemble the type II repeats of the gelatin-binding region of fibronectin.[1–3] These repeats are encoded by three extra exons,[4,5] and they play an important role in the marked ability of gelatinases to degrade denatured collagens. Indeed, deletion of these repeats from GelA abolishes its ability to bind collagen and lowers its gelatinolytic activity by over 90 per cent, but has little or no effect on its membrane-mediated activation, its TIMP-1 or TIMP-2-binding capacity, or its ability to cleave small peptide substrates.[6] Furthermore, GelA, GelB, and fibronectin compete with one another for binding to collagen, whereas collagenase-1 and stromelysin-1, which bind collagen via their C-terminal domains, do not.[7] In addition to this collagen-binding domain, GelB has an extended proline-rich hinge region of unknown significance that shows homology to the $\alpha 2$ chain of collagen V.[3] ProGelA is 72 kDa, which is comparable to its predicted size of 71 kDa, whereas human and mouse ProGelB have predicted molecular weights of 76 and 79 kDa, but apparent molecular weights of 92 and 105 kDa due to *N*- and *O*-linked glycosylation as well as 16 additional residues in the murine collagen-like hinge domain.[1,3] Activated GelA is 62 kDa, and slow autolytic cleavages eventually yield an active C-terminally truncated form of 42 kDa and several inactive breakdown products.[1] For GelB, active forms of 82 to 65 kDa have been described, where proteolytic and autolytic processing of the N and C termini ultimately generates the smaller form.[1,8,9]

Gelatinases A and B have similar substrate specificities (Table 1, p. 520), but differ in terms of their transcriptional regulation, extracellular activation, and inhibition.[1,5] For example, macrophages, neutrophils, and keratinocytes have been noted to express GelB but not GelA, whereas melanoma cells and fibroblasts express primarily GelA.[3] In addition, constitutive GelA expression in HT-1080 fibrosarcoma cells is unaffected by TPA, whereas GelB expression is strongly induced.[5] Although tumour cell lines of diverse origin can produce GelA and GelB, their mRNA in epithelial cancers is generally limited to stromal fibroblasts for GelA, and to monocytes, neutrophils, and endothelial cells for Gel B.[10] GelB is also expressed by cytotrophoblasts during implantation and by osteoclasts during bone remodeling.

Both the latent and active forms of GelA and GelB are bound and inhibited by TIMPs 1, 2, and 3, but TIMP-1 binds more strongly to GelB, TIMP-2 binds GelA more strongly, and TIMP-3 inhibits either enzyme equally well.[10] This binding results from strong interactions between the C-terminal regions of both the enzyme and inhibitor, and inhibition results from interactions between the N-terminal domain of the inhibitor and the catalytic domain of the activated enzyme.[1] Because proenzyme-inhibitor complexes that are formed by combining purified TIMP and progelatinase dissociate more readily than their cell-derived counterparts, the latter may fold together within the cell prior to their co-secretion.[1]

Exposure of MMPs to organomercurial agents and other perturbants initiates self-processing of their prodomains, but unlike GelA, GelB can not remove the final 13 residues of its own prodomain including the cysteine switch residue.[8] The peptide bond separating the pro and catalytic domains of GelB can, however, be cleaved by collagenase-1, stromelysin-1, matrilysin, and trypsin.[8] Other enzymes that activate GelB *in vitro* and thus might do so *in vivo* include plasmin, cathepsin G, tissue kallikrein, and GelA.[1,9] GelA itself, however, cannot be activated directly by any serine proteinase tested thus far, although neutrophil elastase cleaves within the haemopexin-like domain of APMA-treated GelA, yielding a 40 kDa product with four-fold greater gelatinolytic activity.[11] In the absence of gelatin, however, at least two sites within the fibronectin-like gelatin-binding domain of GelA are cleaved by neutrophil elastase, thus inactivating the enzyme. Another serine proteinase, thrombin, can induce the activation of GelA in endothelial cells.[12] This thrombin-mediated induction requires the presence of the cells themselves, and is inhibited by thrombin inhibitors, TIMP-2 and a C-terminal fragment of GelA. Studies showing that pro-GelA binds to cell membranes via its C-terminal domain, and that it can be activated by cell membrane fractions,[6,13,14] presaged the discovery of the MT-MMPs and the demonstration that MT1- and MT3-MMP could indeed bind GelA to the cell surface via TIMP-2 and then induce its activation there.[15,16] In addition, GelA can be bound at the cell surface by integrin $\alpha v\beta 3$,[17] and a 190 kDa cell surface binding protein for GelB has recently been identified as the $\alpha 2$ (IV) chain of type IV collagen.[18]

Purification

ProGelA from fibroblast- or tumour cell-conditioned media and ProGelB from monocytic leukaemia U937 cells, degranulated neutrophils, or other cell sources can be purified by DEAE–Sepharose or green A agarose chromatography followed by gelatin–Sepharose affinity chromatography to isolate both gelatinases, and ConA–Sepharose chromatography to bind and separate away glycosylated GelB from unglycosylated GelA.[1] Pro-GelA–TIMP-2 complexes can be separated from free pro-GelA using heparin–Sepharose, and pro-GelB-TIMP-1 complexes can be separated from free pro-GelB by anti-TIMP-1 immunoaffinity or green A agarose chromatography.[1] Commercial sources of purified human GelA and GelB include Amersham, Biogenesis, Calbiochem, Chemicon, and Triple Point Biologics.

Activities

Gelatinases A and B can be assayed using radiolabelled gelatin, fluorescent peptide substrates, active-site titration, ELISA-based methods, and gelatin zymography.[1] Both gelatinases are active at neutral pH and each prefers to cleave peptide bonds between small residues (Gly or Ala) and aliphatic or hydrophobic residues.[1] GelA and GelB share a number of common substrates, which they both degrade in an apparently similar manner. These are gelatins, elastin, fibrillar collagen V, and basement membrane collagen IV, the latter being cleaved at a single unknown site to yield quarter and three-quarter length fragments.[1] Both enzymes also cleave cartilage link protein and aggrecan, but with some differences in their respective activities.[19,20] GelA, but not GelB, also cleaves triple helical collagen I to yield three-quarter and quarter length fragments at about 1/10 the k_{cat}/K_M of collagenase-1.[21] Subtle differences in the substrate specificities of the two enzymes are most likely to be due to the minor differences in their active sites.

Antibodies

Monoclonal antibodies against human GelA (75–7F7 and 42–5D11) and GelB (7–11C, IM37L, 56–2A4, and neutralizing clone 6–6B) are available from Amersham, Calbiochem, Fuji Chemicals, and ICN Immunologics. Polyclonal antibodies raised against various domains of GelA and GelB are also available from Biogenesis, Biodesign International, Cambio, Chemicon, The Binding Site, and Triple Point Biologics.

Genes

In addition to the human sequences listed in Table 2 (p. 523), cDNA and/or genomic sequences have been obtained for GelA in mouse (M84324), rat (U30822, U65656, X71466), rabbit (D63579), and chicken (U07775); and for GelB in mouse (D15060), rat (U36476, U24441), rabbit (D26514, L36050), dog (U68533), and cow (X78324). The human GelA gene lacks a TATA box and AP-1 binding site, but contains two upstream GC boxes and a potential AP-2 binding site in its first exon.[4] No other MMP gene lacks a TATA box (whereas certain housekeeping, viral, and basement membrane protein genes do), and the stromelysin-3 gene is the only other MMP gene without an AP-1 site. In addition, GelA expression appears to be upregulated by the tumour suppressor p53 owing to an active p53 binding site in the GelA gene promoter.[25] By contrast, the human GelB promoter has a TATA-like motif, two AP-1 sites, a consensus TGF-β inhibitory element, a 42 bp alternating CACA sequence, and a single GC box.[5] In addition to their unique exons 5–7 which encode the three fibronectin-like type II inserts, exon 9 of GelB and the final exon 13 of GelA are unusually large due to coding sequences for the collagen V-like insert in GelB and the presence of a long 3'-untranslated region for GelA.[4,5]

Mutant phenotype/disease states

GelB knockout mice exhibit delayed growth plate vascularization and endochondral ossification, subtle shortening of the long bones, and abnormal implantation sites[22] (unpublished data).

Structure

The crystal structure of the isolated C-terminal haemopexin-like domain of human GelA has been solved (PDB identifiers 1GEN and 1RTG) and, like that of porcine collagenase-1 and human collagenase-3, it reveals a 'four-bladed β-propeller' configuration with an ion-containing central channel.[23,24]

References

1. Murphy, G., and Crabbe, T. (1995). *Meth. Enzymol.*, **248**, 470–84.
2. Collier, I. E., Wilhelm, S. M., Eisen, A. Z., Marmer, B. L., Grant, G. A., Seltzer, J. L., *et al.* (1988). *J. Biol. Chem.*, **263**, 6579–87.
3. Wilhelm, S. M., Collier, I. E., Marmer, B. L., Eisen, A. Z., Grant, G. A., and Goldberg, G. I. (1989). *J. Biol. Chem.*, **264**, 17213–21. [Erratum in *J. Biol. Chem.*, (1990). **265**, 22570.]
4. Huhtala, P., Chow, L. T., and Tryggvason, K. (1990). *J. Biol. Chem.*, **265**, 11077–82.
5. Huhtala, P., Tuuttila, A., Chow, L. T., Lohi, J., Keski-Oja, J., and Tryggvason, K. (1991). *J. Biol. Chem.*, **266**, 16485–90.
6. Murphy, G., Nguyen, Q., Cockett, M. I., Atkinson, S. J., Allan, J. A., Knight, C. G., *et al.* (1994). *J. Biol. Chem.*, **269**, 6632–6.
7. Allan, J. A., Docherty, A. J. P., Barker, P. J., Huskinsson, N. S., Reynolds, J. J., and Murphy, G. (1995). *Biochem. J.*, **309**, 299–306.
8. Sang, Q. X., Birkedal-Hansen, H., and Van Wart, H. E. (1995). *Biochim. Biophys. Acta.*, **1251**, 99–108.
9. Fridman, R., Toth, M., Pena, D., and Mobashery. (1995). *Cancer Res.*, **55**, 2548–55.
10. Coussens, L. M. and Werb, Z. (1996). *Chem. Biol.*, **3**, 895–904.
11. Rice, A. and Banda, M. J. (1995). *Biochemistry*, **34**, 9249–56.
12. Zucker, S., Conner, C., DiMassmo, B. I., Ende, H., Drews, M., Seiki, M., and Bahou, W. F. (1995). *J. Biol. Chem.*, **270**, 23730–8.
13. Murphy, G., Willenbrock, F., Ward, R. V., Cockett, M. I., Eaton, D., and Docherty, A. J. P. (1992). *Biochem. J.*, **283**, 637–41.
14. Strongin, A. Y., Marmer, B. L., Grant, G. A., and Goldberg, G. I. (1993). *J. Biol. Chem.*, **268**, 14033–9.
15. Strongin, A. Y., Collier, I., Bannikov, G., Marmer, B. I., Grant, G. A., and Goldberg, G. I. (1995). *J. Biol. Chem.*, **270**, 5331–8.
16. Takino, T., Sato, H., Shinagawa, A., and Seiki M. (1995). *J. Biol. Chem.*, **270**, 23013–20.
17. Brooks, P., Stromblad, S., Sanders, L. von Scalscha, T., Aimes, R., Stetler-Stevenson, W., *et al.* (1996). *Cell*, **85**, 683–93.
18. Olson, M. W., Toth, M., Gervasi, D., Sado, Y., Ninomiya Y., and Fridman, R. (1998). *J. Biol. Chem.*, **273**, 10672–81.
19. Nguyen, Q., Murphy, G., Hughes, C. E., Mort, J. S., and Roughley, P. J. (1993). **295**, 595–8.
20. Lark, M. W., Williams, H., Hoernner, L. A., Weidner, J., Ayala, J. M., Harper, C. F., *et al.* (1995). *Biochem. J.*, **307**, 245–52.
21. Aimes, R. T. and Quigley, J. P. (1995). *J. Biol. Chem.*, **270**, 5872–6.
22. Vu, T. H., Shipley, J. M., Bergers, G., Berger, J. M., Helms, J. A., Hanahan, D., Shapiro, S. D., Senior, R. M., and Werb, Z. (1998). *Cell*, **93**, 411–22.
23. Gohlke, U., Gomis-Ruth, F. X., Crabbe, T., Murphy, G., Docherty, A. J. P., and Bode, W. (1996). *FEBS Lett.*, **378**, 126–30.
24. Libson, A. M., Gittis, A. G., Collier, I. E., Marmer, B. L., Goldberg, G. I., and Lattman, E. E. (1995). *Nature Struct. Biol.*, **2**, 938–42.
25. Bian, J. and Sun, Y. (1997). *Mol. Cell. Biol.*, **17**, 6330–8.

Mark D. Sternlicht and Zena Werb
University of California, San Francisco,
Department of Anatomy, LR-208,
3rd and Parnassus Avenues, San Francisco,
CA 94143–0452, USA.

Matrilysin (MMP-7)

Matrilysin differs from other MMPs in several regards. It is the smallest MMP, owing to its unique lack of a haemopexin-like C-terminal domain. It readily cleaves several substrates and, unlike other MMPs, its expression is limited to glandular epithelium and macrophages. Finally, its gene promoter shows both similarities and differences when compared with other MMP promoters.

Protein properties

Matrilysin (also known as MMP-7, matrin, small uterine metalloproteinase (ump), and pump-1 for 'putative' and later 'punctuated' metalloproteinase-1) was first identified in involuting rat uterus,[1] and later purified from the same source[2] and cloned from a human tumour cDNA library.[3,4] Matrilysin is the only MMP synthesized without a haemopexin-like domain, thus it is comparably small in size; having 28 and 19 kDa latent and active forms, respectively [5]. Nevertheless, it can cleave a wide array of ECM substrates and can activate procollagenase-1 and both progelatinases (Table 1, p. 520). Indeed, it cleaves aggrecan, versican, link protein, and entactin more readily than do other MMPs, including stromelysin-1.[6,7] The tissue-specific pattern of matrilysin expression differs from that of most other MMPs in that it is restricted to glandular epithelium, whereas other MMPs are primarily expressed by stromal cells. Its expression is highest in Paneth cells, where its function in processing the microbial peptides cryptidins (defensins) has been suggested. Matrilysin has also been implicated in early tumour formation[8,9,17] and late-stage invasion[10] and, unlike other MMPs which are only expressed by the stromal cells that surround epithelial tumours, only matrilysin and possibly collagenase-3 are expressed by the tumour cells themselves.[11] Reports of matrilysin expression in osteosarcomas and in fibroblasts surrounding breast carcinomas, however, indicate that their expression may not be limited to epithelial cells.[11] Surprisingly, matrilysin immunolocalizes to the apical rather than basal surface of premalignant cells, suggesting that it may promote tumour development by degrading molecules other than those of the ECM.[9] It is also apically located in normal epithelium, suggesting the existence of normal apical substrates.[17,18] Matrilysin has also been implicated in the rupture of atherosclerotic plaques, because it is expressed by lipid-laden macrophages at potential rupture sites where it colocalizes with versican, one of matrilysin's proteoglycan substrates.[12]

Purification

Matrilysin can be extracted from involuting rat uterus and purified by Ultrogel AcA 54, blue Sepharose, and Zn-chelating Sepharose chromatographies followed by a second round of Ultrogel AcA 54 chromatography using a different buffer.[5] Two methods for purifying human matrilysin from cell culture medium and two methods for purifying recombinant matrilysin from mammalian cell cultures have been described.[5] In addition, large quantities of recombinant matrilysin bearing a C-terminal poly-His tag can be purified from *E. coli* inclusion bodies using urea solubilization and a Ni–NTA resin.[13]

Activities

Once activated by APMA or propeptide cleavage with trypsin or other enzymes, matrilysin can cleave a large number of substrates. Rat and human enzymes have similar enzymatic activities, although the latter digests elastin reasonably well, while rat matrilysin does not.[5] Cleavage-site analyses indicate only that the presence of a large hydrophobic side chain is prefered at the P1′ position, where cleavage occurs between P1 and P1′ residues. Matrilysin can be assayed using chromogenic Azocoll, radiolabelled transferrin, a fluorescent peptide, or zymography.[5]

Antibodies

Affinity purified polyclonal antibodies that specifically recognize latent and active matrilysin have been raised against the final 40 amino acids of mouse matrilysin and against a 16 amino acid peptide from the propeptide-to-catalytic domain junction of human matrilysin.[9,12] A mouse monoclonal anti-human matrilysin antibody that is specific for the active enzyme is available from Oncogene Research Products.

Genes

In addition to the human matrilysin cDNA listed in Table 2 (p. 523) (GenBank X07189),[3] human genomic sequences have also been obtained (L22519-L22525),[14] as have sequences for mouse (L36238-L36243), rat (L24374), and cat (U04444) matrilysin. The human gene maps to the MMP gene cluster on chromosome 11 and the first five exons conform to the conserved organization of all other MMP genes, whereas the sixth and final exon does not.[14] This exon encodes the final nine residues of the protein and the 3′-untranslated sequences, whereas other MMP genes have exons encoding further C-terminal domains. The matrilysin promoter contains three upstream TGF-β inhibitory elements, two PEA3 sites, an AP-1 site, and a TATA box.[14] As with other MMP promoters, the AP-1 and PAE3 elements cooperate to enhance TPA and EGF responsiveness, yet while upstream stromelysin-1 sequences further enhance such responsiveness, those of the matrilysin promoter do not.[14] Evidence also indicates that TGF-β produced by stromal cells in response to progesterone suppresses the expression of epithelial matrilysin during endometrial cycling.[15] Other promoter elements and tissue-specific transcription factors are also likely to play a role in the unique pattern of matrilysin gene expression.

Mutant phenotype/disease states

Matrilysin-null mice have no apparent phenotype; however, when they are mated with *Min* (multiple intestinal neoplasia) mice, the resulting matrilysin-deficient *Min* mice display diminished intestinal tumorigenesis compared to matrilysin-expressing *Min* mice.[9] Mice that over-express a matrilysin transgene under the control of the MMTV LTR exhibit precocious mammary gland differentiation and male infertility.[17,19] In addition, approximately half of the multiparous transgenic mice developed preneoplastic mammary lesions, and when these mice were crossed with MMTV-*neu* transgenic mice, mammary tumorigenesis was accelerated.[17]

Structure

The X-ray crystal structures of matrilysin complexed with carboxylate, hydroxymate, and sulphodiimine inibitors have been solved (PDB identifiers 1MMP-1MMR).[16] When compared, these complexes indicate the importance of the zinc-coordinating group in determining inhibitor efficacy.

References

1. Sellers, A. and Woessner, J. F. Jr. (1980). *Biochem. J.*, **189**, 521–31.
2. Woessner, J. F. Jr. and Taplin, C. J. (1988). *J. Biol. Chem.*, **263**, 16918–25.
3. Muller, D., Quantin, B., Gesnel, M. C., Millon-Collard, R., Abecassis, J., and Breathnach, R. (1988). *Biochem. J.*, **253**, 187–92.

4. Quantin, B., Murphey, G., and Breathnach, R. (1989). *Biochemistry*, **28**, 5327–33.
5. Woessner, J. F. Jr. (1995). *Meth. Enzymol.*, **248**, 485–95.
6. Nguyen, Q., Murphy, G., Hughes, C. E., Mort, J. S., and Roughley, P. J. (1993). *Biochem. J.*, **295**, 595–8.
7. Sires, U. I., Griffin, G. L., Broekelmann, T. J., Mecham, R. P., Murphy, G., Chung, A. E., *et al.* (1993). *J. Biol. Chem.*, **268**, 2069–74.
8. Witty, J. P., McDonnell, S., Newell, K., Cannon, P., Navre, M., Tressler, R., and Matrisian, L. M. (1994). *Cancer Res.*, **54**, 4805–12.
9. Wilson, C. L., Heppner, K. J., Labosky, P. A., Hogan, B. L. M., and Matrisian, L. M. (1997). *Proc. Natl Acad. Sci. USA*, **94**, 1402–7.
10. Powell, W. C., Knox, J. D., Navre, M., Grogan, T. M., Kittelson, J., Nagle, R. B., and Bowden, G. T. (1993). *Cancer Res.*, **53**, 417–22.
11. Coussens, L. M. and Werb, Z. (1996). *Chem. Biol.*, **3**, 895–904.
12. Halpert, I., Sires, U. I., Roby, J. D., Potter-Perigo, S., Wight, T. N., Shapiro, S. D., *et al.* (1996). *Proc. Natl Acad. Sci. USA*, **93**, 9748–53.
13. Itoh, M., Masuda, K., Ito, Y., Akizawa, T., Yoshioka, M., Imai, K., *et al.* (1996). *J. Biochem.*, **119**, 667–73.
14. Gaire, M., Magbanua, Z., McDonnell, S., McNiel, L., Lovett, D. H., and Matrisian, L. M. (1994). *J. Biol. Chem.*, **269**, 2032–40.
15. Bruner, K. L., Rodgers, W. H., Gold, L. I., Korc, M., Hargrove, J. T., Matrisian, L. M., and Osteen, K. G. (1995). *Proc. Natl Acad. Sci. USA*, **92**, 7362–6.
16. Browner, M. F., Smith, W. W., and Castelhano, A. L. (1995). *Biochemistry*, **34**, 6602–10.
17. Rudolph-Owen, L. A. and Matrisian, L. M. (1998). *J. Mammary Gland Biol. Neoplasia*, **3**, 177–89.
18. Saarialho-Kere, U. K., Crouch, E. C., and Parks, W. C. (1995). *J. Invest. Dermatol.*, **105**, 190–6.
19. Rudolph-Owen, L. A., Cannon, P., and Matrisian, L. M. (1998). *Mol. Biol. Cell*, **9**, 421–35.

Mark D. Sternlicht and Zena Werb
University of California, San Francisco, Department of Anatomy, LR-208, 3rd and Parnassus Avenues, San Francisco, CA 94143–0452, USA.

Membrane-type MMPs (MMPs 14, 15, 16, and 17)

Membrane-type MMPs (MT-MMPs) are unique among the MMPs in that they possess a plasma membrane-spanning domain and short cytoplasmic tail at their C-terminal end, and can be activated intracellularly by furin-like enzymes in the Golgi apparatus or extracellularly by plasmin. Once activated, they act as a receptor for TIMP-2 or TIMP-3, which in turn bind secreted pro-MMPs 2 and 13. The pro-MMPs are then activated at the cell surface, perhaps by the same complexed MT-MMPs or perhaps by monomeric MT-MMPs lacking bound inhibitor.

Protein properties

A significant step towards understanding the mechanisms that underlie the pericellular activation of some MMPs came with the discovery of the membrane-type MMPs (MT-MMPs). Using degenerate PCR primers, Sato *et al.* cloned the first of these integral plasma membrane enzymes (MT1-MMP; MMP-14).[1] Since then, three more MT-MMPs have been cloned (MT2-, MT3- and MT4-MMP; MMPs 15–17) [2–4]. In addition to their signal, pro, catalytic, hinge, and haemopexin-like domains, the MT-MMPs have a 10–12 residue insert with a potential furin-like enzyme recognition motif (RRK/RR) situated between their pro and catalytic domains similar only to that seen in SL-3, an eight-residue insert of unknown significance in the mid–proximal portion of their catalytic domain (not seen in MT4-MMP), and a final 75–105 residue insert containing a transmembrane domain of about 24 residues and a short cytoplasmic C-terminal tail. MT1-MMP is synthesized as a 63 kDa latent enzyme that is processed to a 60 kDa active form. Such processing may be catalysed by furin-like proprotein convertases in the Golgi apparatus, although unprocessed 63 kDa MT1-MMP has also been seen at the cell surface.[1] Using recombinant and transmembrane domain-deleted MT1-MMP, it was shown that furin could indeed activate the enzyme.[5,6] In another study, however, MT1-MMP remained activatable despite the use of a furin inhibitor or the substitution of its RRKR residues with ARAA by site-directed mutagenesis.[7] More recently, it was shown that pro-MT1-MMP can be activated at the cell surface by extracellular plasmin, but not by any other extracellular trypsin-like proteinase.[8] Thus MT-MMP activation may occur both within the cell and at the cell surface. Once activated, MT1-MMP binds TIMP-2, and this complex can then bind and activate pro-MMP-2 or -13 (Fig. 1).[8–11] Because a hydroxamic acid-derived MMP inhibitor blocks TIMP-2 binding, it is likely that the MT1-MMP active site is involved in TIMP-2 binding.[12] It is also possible, therefore, that another MT1-MMP without bound TIMP-2 activates the pro-MMP that is bound via TIMP-2 to an otherwise inhibited MT1-MMP.[12] MT1-MMP and TIMP-2 can also be secreted as a bimolecular complex that can then interact with pro-MMP-2.[9] By deleting the MT1-MMP transmembrane domain or replacing it with the IL-2 receptor α chain transmembrane domain, it has been shown that the transmembrane domain is required for recruitment of MT1-MMP to sites of active ECM degradation (invadopodia).[13] Apparent molecular masses of 72, 64, and 70 kDa have been obtained for latent MT2-, MT3-, and MT4-MMPs, respectively.[2–4] Each MT-MMP shows a unique pattern of expression by Northern analysis. MT1-MMP is expressed in numerous normal tissues, but is undetectable in brain and leukocytes.[2,3] MT2-MMP expression is also undetectable in brain and leukocytes, but is stronger than MT1-MMP expression in liver, heart, and skeletal muscle, and is weak or absent in ovary, prostate, thymus, and spleen.[2] On the other hand, MT3-MMP mRNA is detectable in brain, as well as placenta, lung and heart,[3] and MT4-MMP expression is strongest in brain, leukocytes, colon, ovary, and testis.[4] MT1-MMP expression is upregulated in invasive tumours as compared to adjacent normal tissues[1] and although its transcripts are most often seen in adjacent stromal cells by *in situ* hybridization,[14] carcinoma cells themselves may also overexpress this enzyme.[15]

Purification

Secreted MT1-MMP/TIMP-2 complexes have been isolated from the conditioned media of concanavalin A-stimulated MDA-MB-231 human breast carcinoma cells by DEAE–cellulose and green A Dye Matrex chromatographies followed by an anti-TIMP-2–IgG immunoaffinity step using a monoclonal antibody (67–4H11) against the C-terminal tail of TIMP-2, and then anti-TIMP-2-IgG chromatography using another clone (68–6H4) against the first loop of the N-terminal domain of TIMP-2 in order to remove free TIMP-2.[9] The complexes can also be reapplied to the first anti-TIMP-2 column and MT1-MMP can then be separated from the antibody-bound TIMP-2 using EGTA in a $CaCl_2$-free elution buffer.[16] A recombinant glutathione S-transferase–MT1–MMP fusion protein has been purified using glutathione–Sepharose beads.[5] Otherwise, most analyses have been performed using transfected cells and plasma membrane fractions.[1,2,7]

Activities

Membrane-associated MT1- and MT3-MMP can both activate pro-MMP-2.[1,3] MT1-MMP can also activate pro-MMP-13[10] and it has been implicated as a key player in a proteinase cascade involving MMPs 2, 9 and 13.[22] Secreted MT1-MMP and an MT1-MMP deletion mutant lacking the transmembrane domain can also cleave gelatin, cartilage proteoglycan, fibronectin, vitronectin, laminin-1, α1-proteinase inhibitor, and α2-macroglobulin.[6,9,16] Furthermore, like collagenases 1–4, they can cleave collagens I, II, and III at specific Gly-Ile/Leu bonds to yield characteristic three-quarter and quarter length fragments.[16] Despite this activity, they lack the active site-flanking Tyr, Asp, and Gly residues that distinguish the collagenases and are thought to confer their cleavage specificities. Since MT1-MMP is sensitive to inhibition by TIMP-2 and TIMP-3, but not TIMP-1, it can still function as a proenzyme activator and broad-spectrum proteinase even in the presence of high levels of TIMP-1.[17] A soluble

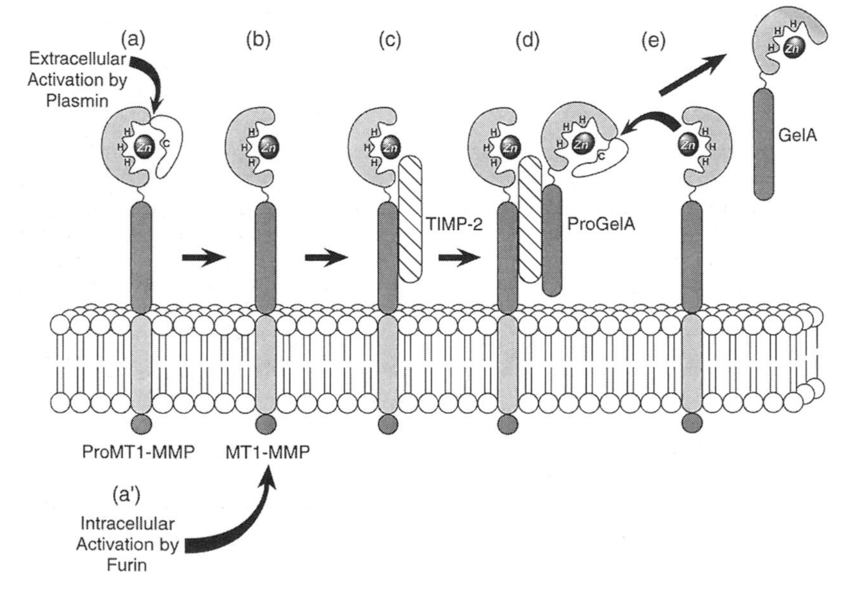

Figure 1. Cell-surface activation of pro-GelA by MT1-MMP. Pro-MT1-MMP is activated at the cell surface by plasmin (a), or it is activated in the *trans*-Golgi network by a furin-like enzyme during its passage to the cell surface (a′). Active MT1-MMP (b) interacts with TIMP-2 (c) and then pro-GelA to form a trimolecular complex (d). Pro-GelA may then be activated by the same MT1-MMP, or since TIMP-2 may block the MT1-MMP active site, another uninhibited MT1-MMP may remove the pro-GelA propeptide (e). (Modified from ref. 21.)

MT3-MMP that is formed by alternative mRNA splicing and is able to cleave type III collagen and fibronectin has also been described.[23] The activities of MT2- and MT4-MMP remain to be characterized, but it is noteworthy that MT4-MMP has a Val rather than Ser residue following the final His of the three-histidine zinc-binding motif, a feature that sets it apart from all other metzincins.

Antibodies

A mouse monoclonal anti-human MT1-MMP antibody (114-IF2) is available from Amersham, Calbiochem and Fuji Chemicals, and an affinity purified rabbit anti-human antibody is available from Triple Point Biologics. Other monoclonal antibodies have been raised against unique regions of human MT1-MMP (clones 113-5B7 and 114-1F2) and MT3-MMP (117-4E1 and 117-13B6).[1,3] Polyclonal antibodies against the MT-MMPs have also been generated,[2,4] and anti-human MT2-, MT3- and MT4-MMP antibodies are available from Calbiochem.

Genes

In addition to the human sequences listed in Table 2 (p. 523), cDNA and genomic sequences have been obtained for MT1-MMP in mouse (X83536), rat (X83537, X91785), and rabbit (U73940, U83918); and for MT3-MMP in chicken (U66463). The human MT1-, MT2-, and MT3-MMP genes have been mapped to chromosomes 14q11–12, 16q12.2–21 and 8q21.3–22.1, respectively.[18] The mouse MT1-MMP gene has 10 exons extending over 10 kbp and maps to mouse chromosome 14.[24] Exons encoding the pro and catalytic domains have a unique structure as compared with other known MMP genes, and the novel C-terminal domains and 3'-untranslated region are encoded by a single large exon. There appears to be strict co-expression of the MT1-MMP and TIMP-2 genes during development, suggesting common regulatory pathways.[24] Concanavalin A and TPA can induce MT1-MMP mRNA expression, this induction is blocked by cycloheximide, cAMP-inducing agents, and dexamethasone; and IL-1α, EGF, basic FGF, TGF-α and calcium ionophore have negligible effects on mRNA levels.[19,20]

Structure

Studies using hydroxamate inhibitors suggest that the S1' subsite of MT1-MMP is larger than that of MMP-1 and similar to those of MMPs 2, 3, and 9.[25] These studies also indicate that the S1 and S2' subsites of MT1-MMP are narrower than those of other MMPs.

References

1. Sato, H., Takino, T., Okada, Y., Cao, J. Shinagawa, Yamamoto, E., and Seiki, M. (1994). *Nature*, **370**, 61–5.
2. Will, H. and Hinzmann, B. (1995). *Eur. J. Biochem.*, **231**, 602–8.
3. Takino, T., Sato, H., Shinagawa, A., and Seiki M. (1995). *J. Biol. Chem.*, **270**, 23013–20.
4. Puente, X. S., Pendas, A. M., Llano, E., Valesco, G., and Lopez-Otin, C. (1996). *Cancer Res.*, **56**, 944–9.
5. Sato, H., Kinoshita, T., Takino, T. Nakamura, K., and Seiki, M. (1996). *FEBS Lett.*, **393**, 101–4.
6. Pei, D. and Weiss, S. J. (1996). *J. Biol. Chem.*, **271**, 9135–40.
7. Cao, J., Rehemtulla, A., Bahou, W., and Zucker, S. (1996). *J. Biol. Chem.*, **271**, 30174–80.
8. Okumura, Y., Sato, H., Seiki, M., and Kido, H. (1997). *FEBS Lett.*, **402**, 181–4.
9. Imai, K., Ohuchi, E., Aoki, T., Nomura, H., Fujii, Y., Sato, M., and Okada, Y. (1996). *Cancer Res.*, **56**, 2707–10.
10. Knauper, V., Will, H., Lopez-Otin, C., Smith, B., Atkinson, S. J., Stanton, H., *et al.* (1996). *J. Biol. Chem.*, **271**, 17124–31.
11. Strongin, A. Y., Collier, I., Bannikov, G., Marmer, B. I., Grant, G. A., and Goldberg, G. I. (1995). *J. Biol. Chem.*, **270**, 5331–8.
12. Zucker, S., Drews, M., Conner, C., Foda, H. D., DeClerck, Y. A., Langley, K. E., Bahou, W. F., Docherty, A. J. P., and Cao, J. (1998). *J. Biol. Chem.*, **273**, 1216–22.
13. Nakahara, H, Howard, L., Thompson, E. W., Sato, H., Seiki, M., Yeh, Y., and Chen, W. T. (1997). *Proc. Natl Acad. Sci. USA*, **94**, 7959–64.
14. Okada, A., Bellocq, J. P., Rouyer, N., Chenard, M. P., Rio, M. C., Chambon, P., and Besset, P. (1995). *Proc. Natl Acad. Sci. USA*, **92**, 2730–4.
15. Ohtani, H., Motohashi, H., Sato, H., Seiki, M., and Nagura, H. (1996). *Int. J. Cancer*, **68**, 565–70.
16. Ohuchi, E., Imai, K., Fujii, Y., Sato, H., Seiki, M., and Okada, Y. (1997). *J. Biol. Chem.*, **272**, 2446–51.
17. Atkinson, S. J. Crabbe, T., Cowell, S., Ward, R. V., Butler, M. J., Sato, H., *et al.* (1995). *J. Biol. Chem.*, **270**, 5331–8.
18. Mattei, M. G., Roeckel, N., Olsen, B. R., and Apte, S. S. (1997). *Genomics*, **40**, 168–9.
19. Yu, M., Sato, H., Seiki, M., Spiegel, S., and Thompson, E. W. (1998). *Clin. Exp. Metastasis*, **16**, 185–91.
20. Lohi, J., Lehti, K., Westermarck, J., Kahari, V. M., and Keski-Oja, J. (1996). *Eur. J. Biochem.*, **239**, 239–47.
21. Coussens, L. M. and Werb, Z. (1996). *Chem. Biol.*, **3**, 895–904.
22. Cowell, S., Knauper, V., Stewart, M. L., d'Ortho, M. P., Stanton, H., Hembry, R. M., Lopez-Otin, C., Reynolds, J. J., and Murphy, G. (1998). *Biochem. J.*, **331**, 453–8.
23. Matsumoto, S. I., Katoh, M., Saito, S., Watanabe, T., and Masuho, Y. (1997). *Biochim. Biophys. Acta*, **1354**, 159–70.
24. Apte, S. S., Fukai, N., Beier, D. R., and Olsen, B. R. (1997). *J. Biol. Chem.*, **272**, 25511–7.
25. Yamamoto, M., Tsujishita, H., Hori, N., Ohishi, Y., Inove, S., Ikeda, S., and Okada, Y. (1998). *J. Med. Chem.*, **41**, 1209–17.

Mark D. Sternlicht and Zena Werb
University of California, San Francisco,
Department of Anatomy, LR-208,
3rd and Parnassus Avenues, San Francisco,
CA 94143–0452, USA.

Other MMPs (MMPs 11, 12, 19, and 20)

MMPs that fall outside of the conventional subgroups are macrophage metalloelastase (MMP-12), an MMP with a unique acidic hinge region (MMP-19), the enamel organ-specific enzyme enamelysin (MMP-20), and a recently discovered *Xenopus* MMP (XMMP). In addition, stromelysin-3 (MMP-11) exhibits early evolutionary divergence from other MMPs in terms of sequence homology and gene structure, and is the only secreted MMP that is activated intracellularly by the Golgi-associated enzyme furin.

Protein properties

Stromelysin-3 (SL-3, MMP-11) was initially identified from a subtracted breast cancer cDNA library and found to be expressed specifically by stromal cells surrounding invasive tumours.[1] Like stromelysins 1 and 2, it too has a broad substrate specificity, a nine-residue insert in its hinge region, a latent form of about 60 kDa, and active forms of 45–47 and 28 kDa, the latter being C-terminally truncated.[2,3] This, however, is where the similarities end. On average, SL-1 and –2 from various species show 76 per cent sequence identity and 86 per cent similarity to one another, whereas SL-3 shares only 39 per cent identity and 56 per cent similarity with the other SLs and only 36 per cent identity and 55 per cent similarity with all other MMPs.[4] Furthermore, whereas all other secreted MMPs are activated extracellularly, processed SL-3 can be found within cells, and both latent and active SL-3 are rapidly detected in cell-conditioned medium, yet their relative amounts remain stable over time, suggesting intracellular rather than extracellular activation.[3] As it turns out, SL-3 is the only MMP other than the MT-MMPs with a 10 residue insert at the end of its prodomain that harbours an RXK/RR furin-like enzyme recognition motif, and the Golgi-associated enzyme furin indeed activates SL-3 prior to its secretion by cleaving the bond between the final Arg of the insert and the subsequent Phe residue.[3] Although most furin-processed proteins contain only an RXK/RR site, the SL-3 sequence (RNRQKR) contains an added Arg residue that potentiates its furin-catalysed activation.[3] Indeed, substitution of the upstream Arg decreases SL-3 processing by some 90 per cent, and mutation of the Lys or other Arg residues abolishes processing altogether. Processing is also blocked by a furin-specific inhibitor and is only seen in the LoVo cell line that lacks functional furin if such cells are cotransfected with SL-3 plus furin, thus indicating the obligate nature of this activation mechanism.

It has long been recognized that stimulated mouse macrophages secrete a metalloenzyme with elastinolytic activity.[5] Since then, a 22 kDa murine macrophage metalloelastase (MMP-12) has been purified[6] and its murine and human genes cloned.[7,8] The molecular mass of the proenzyme is 53–54 kDa and the mature 22 kDa active form apparently derives from both standard propeptide removal (which yields a 45 kDa active intermediate) and atypical C-terminal processing.[7] In addition to elastin, MMP-12 can cleave fibrinogen, fibronectin, laminin, proteoglycans, myelin basic protein, immunoglobulins, plasminogen, and α_1-proteinase inhibitor. In addition, MMP-12 is required for tissue invasion by macrophages[9] and it can generate the angiogenesis inhibitor angiostatin by partially hydrolysing plasminogen.[10] MMP-12 may also play a role in pulmonary emphysema and vascular aneurysm formation.[23]

Human MMP-19 cDNAs have been cloned from a liver cDNA library,[11] from a rheumatoid arthritic, inflamed synovium cDNA library,[21] and by performing 5′-RACE on an expressed sequence tag with MMP sequence homology.[12] The predicted protein has 508 amino acids, a molecular weight of 57.4 kDa, two potential *N*-glycosylation sites, and characteristic pre, pro, catalytic, and haemopexin-like MMP domains. MMP-19, however, lacks specific features that would otherwise distinguish it as a collagenase, gelatinase, or membrane-type MMP. Furthermore, the hinge region of MMP-19 contains a 16 residue acidic insertion rather than the nine-residue hydrophobic insertion typical of the SLs. MMP-19 also has a unique threonine-rich C-terminal region and the highly conserved MMP propeptide sequence PRCGVPD reads PRCGLED in MMP-19. Its mRNA is expressed in a variety of human tissues, particularly placenta, lung, pancreas, ovary, spleen, and intestine. MMP-19 has also been detected on the surface of activated peripheral blood mononuclear cells.[21]

The porcine enamelysin (MMP-20) cDNA was isolated from a porcine enamel organ cDNA library.[13] The human cDNA was subsequently cloned[20] and an orthologous bovine cDNA has also recently been obtained (P.K. DenBesten; personal communication). The predicted protein has 483 amino acids, a molecular weight of 54.1 kDa, and a characteristic MMP domain structure, but lacks distinctive features of the collagenase, gelatinase, stromelysin, and MT-MMP subgroups. It also diverges from all other MMPs in terms of three otherwise invariant amino acids found in the haemopexin-like domain. Human enamelysin has no potential N-glycosylation sites.[20] Enamelysin mRNA transcripts of 2.5 and 4.3 kb have only been detected in the enamel organ. During tooth enamel formation, ameloblast cells of this organ secrete an organic matrix made up mostly of amelogenin that is continuously turned over as biomineralization proceeds. This suggests that enamelysin may be a key player in enamel matrix processing.

Recently, a novel *Xenopus* MMP cDNA (XMMP) was cloned, and found to be expressed transiently in gastrula and neurula stage embryos.[14] XMMP is distantly related to other MMPs, with stromelysin-3 being most related at only 20 per cent sequence identity. It has a signal peptide suggesting that it is secreted and a predicted molecular weight of 70 kDa excluding this domain and any post-translational modifications. Unlike other MMPs, it has a 37 amino acid vitronectin-like insert in its propeptide domain and it lacks a proline-rich hinge region between its catalytic and C-terminal domains. Like stromelysin-3, it has an RRKR furin-like enzyme recognition motif at the C-terminal end of its pro domain, suggesting that it too may be secreted as an active enzyme.

Purification

MMP-11

Recombinant SL-3 has been expressed in *E. coli* and recovered from inclusion bodies, it has been purified from the conditioned medium of transfected myeloma cells by S, Q, and Zn-chelate Sepharose chromatographies, and it has been immunoprecipitated from the conditioned medium of other transfected mammalian cells.[2,3]

MMP-12

Active 22 kDa mouse metalloelastase has been purified from dialysed and lyophilized medium conditioned by thioglycollate-elicited peritoneal macrophages using anion exchange (DEAE–Sephadex A-25) and gel filtration (Utrogel AcA54) chromatographies.[6] Metalloelastase has also been purified from human alveolar macrophage-conditioned medium by gelatin– and heparin–agarose chromatographies in the presence of EDTA to prevent proenzyme activation.[8] Recombinant murine and human metalloelastases have been purified using heparin–agarose chromatography.[7,8,23]

MMP-19

Recombinant pro-MMP-19 has been expressed in *E. coli* and purified from inclusion bodies by gel filtration (Sephacryl S-200) chromatography.[11] Recombinant MMP-19 has also been obtained from Sf9 insect cells.[21] Native MMP-19 has not yet been isolated.

MMP-20

Active 21–25 kDa forms of enamelysin have been extracted from bovine secretory enamel matrix under neutral conditions and purified by ammonium sulphate precipitation and sequential ion exchange (DEAE–cellulose), affinity (CA–Sepharose), and reverse phase (C4) HPLC chromatographies.[15] Recombinant human enamelysin has also been purified from *E. coli* inclusion bodies.[20]

Activities

MMP-11

SL-3 is a very weak proteinase that can hydrolyse α1-proteinase inhibitor and several structural proteins with a specificity similar to those of stromelysins 1 and 2 (Table 1, p. 00).

MMP-12

The substrate specificities and inhibitor sensitivities of human and mouse metalloelastase have been described.[16,23] Assays of elastinolytic activity have been described using elastin-containing agarose gels and using ^{3}H- and rhodamine-labelled elastin.[5–8] MMP-12 is ~30 per cent as active as neutrophil elastase at degrading elastin, but it can also inactivate the major neutrophil elastase inhibitor α_1-proteinase inhibitor.[23] Studies using synthetic substrates reveal a preference for leucine at the MMP-12 P_1' site, although both small and large residues are accepted.[23]

MMP-19

Recombinant MMP-19 can hydrolyse fluorogenic MMP substrates only after trypsin activation and this activity is abolished by TIMP-2 and EDTA.[11] Substrate mapping further indicates that its enzymatic activity is most like that of the stromelysins and not at all similar to that of the collagenases. MMP-19 was recently shown to have weak but distinct gelatinolytic activity; however, other substrates remain to be determined.[21]

MMP-20

Gelatin zymograms done on acid extracts of developing porcine enamel matrix show two major metalloproteinases at 50–65 kDa that may or may not represent enamelysin.[13] Multiple 21–25 kDa enzymes that probably do represent truncated forms of enamelysin have been purified from neutral extracts of bovine enamel matrix.[15] These have poor enzymatic activity against gelatin, but readily cleave casein and isoforms of amelogenin, the ameloblast-derived protein that forms the organic matrix of tooth enamel (P.K. DenBesten; personal communication). Recombinant enamelysin can also hydrolyse amelogenin and is fully inhibited by TIMP-2.[20]

Antibodies

Polyclonal antibodies (Ab 349) raised against the 25 C-terminal amino acids of human SL-3 and monoclonal antibodies specific to the SL-3 catalytic (5ST-4C10) and C-terminal (5ST-4A9) domains have also been described.[2,17] A rabbit anti-human SL-3 antibody is also available from Triple Point Biologics. A human metalloelastase-specific rabbit polyclonal antibody raised against the first 12 amino acids of the catalytic domain detects latent (54 kDa), active (45 kDa) and trypsin-activated (22 kDa) recombinant enzyme but no other MMPs by Western hybridization.[8] Rabbit and chicken polyclonal antibodies have been raised against a peptide from the unique hinge region of human MMP-19.[21] Antibodies that recognize MMP-20 have not yet been described.

Genes

Complete cDNAs and/or genomic sequences have been obtained for SL-3 in humans (X57766), rats (X07821–4), and frogs (Z27093); for metalloelastase in humans (U78045), mice (M82831), and rats (X98517); for MMP-19 in humans (Y08622, X92521, U37791, U38321); for enamelysin in humans (Y12779) and pigs (U54825); and for XMMP in frogs (U82541). Because the GenBank sequences submitted as human MMPs 18 (Y08622)[12] and 19 (X92521)[11] are identical to one another, but diverge considerably from the *Xenopus* MMP-18 sequence, they have been designated as MMP-19. The human metalloelastase and enamelysin genes map to the MMP gene cluster on chromosome 11q22.3, whereas the SL-3 and MMP-19 genes map to chromosomes 22q11.2 and 12q14, respectively.[11,18] SL-3 gene expression is induced by PDGF, EGF, basic FGFa, and TPA.[1] However, whereas most MMP gene promoters have at least one consensus AP1-binding site, the SL-3 promoter lacks such a site.[19] The SL-3 gene promoter also differs from those of other MMPs in that it contains a functional retinoic acid receptor element, two GC and GT boxes, a non-functional PEA-3 site, a nuclear factor-1 binding motif, and a putative silencer binding site.[19] In addition, the intron–exon structure is similar for all defined MMP genes except the SL-3 gene which shows major differences in the exons defining its haemopexin domain, thus again suggesting early evolutionary divergence.[19]

Mutant phenotype/disease states

In addition to contributing to late cancer progression, MMPs 1, 3 and 7 have also been implicated in early tumorigenesis (see relevant sections). Likewise, SL-3 knockout mice form fewer and smaller chemically induced tumours than do wild-type mice, and whereas wild-type fibroblasts foster the tumorigenicity of human breast cancer cells in nude mice, SL-3-deficient fibroblasts do not.[22] Thus MMPs may act early in tumorigenesis by altering cell–cell and cell–matrix interactions and microenvironmental signals.

As already noted, an MMP-19 cDNA has been cloned from an arthritic synovium cDNA library and MMP-19 is present on the surface of activated inflammatory cells.[21] In addition, MMP-19 is recognized by autoantibodies in the sera of about a quarter of rheumatoid arthritis patients, further suggesting that it may play a role in arthritic joint destruction.[21]

References

1. Basset, P., Bellocq, J. P., Wolf, C., Stoll, I., Hutin, P., Limacher, J. M., *et al.* (1990). *Nature*, **348**, 699–704.
2. Murphy, G., Segain, J. P., O'Shea, M., Cockett, M., Ioannou, C., Lefebvre, O., *et al.* (1993). *J. Biol. Chem.*, **268**, 15435–41.
3. Pei, D. and Weis, S. J. (1995). *Nature*, **375**, 244–7.
4. Sang, Q. A., and Douglas, D. A. (1996). *J. Protein Chem.*, **15**, 137–60.
5. Werb, Z. and Gordon, S. (1975). *J. Exp. Med.*, **142**, 361–77.
6. Banda, M. J. and Werb, Z. (1981). *Biochem. J.*, **193**, 589–605.
7. Shapiro, S. D., Griffen, G. L., Gilbert, D. J., Jenkins, N. A., Copeland, N. G., Welgus, H. G., *et al.* (1992). *J. Biol. Chem.*, **267**, 4664–71.
8. Shapiro, S. D., Kobayashi, D. K., and Ley, T. J. (1993). *J. Biol. Chem.*, **268**, 23824–9.
9. Shipley, J. M., Wesselschmidt, R. L., Kobayashi, D. K., Ley, T. J., and Shapiro, S. D. (1996). *Proc. Natl Acad. Sci. USA*, **93**, 3942–6.
10. Dong, Z., Kumar, R., Yang, X., and Fidler, I. J. (1997). *Cell*, **88**, 801–10.
11. Pendas, A. M., Knauper, V., Puente, X. A., Llano, E., Mattei, M. G., Apte, S., *et al.* (1997). *J. Biol. Chem.*, **272**, 4281–6.
12. Cossins, J., Dudgeon, T. J., Catlin, G., Gearing, A. J. H., and Clements, J. M. (1996). *Biochem. Biophys. Res. Commun.*, **228**, 494–8.
13. Bartlett, J. D., Simmer, J. P., Xue, J., Margolis, H. C., and Moreno, E. C. (1996). *Gene*, **183**, 123–128.
14. Yang, M., Murray, M. T., and Kurkinen, M. (1997). *J. Biol. Chem.*, **272**, 13527–33.
15. Punzi, J. S. and DenBesten, P. K. (1995). *J. Dent. Res.*, **74**, 95.
16. Banda, M. J., Clark, E. J., and Werb, Z. (1985). In *Mononuclear phagocytes: characteristics, physiology and function*. (ed. R. van Furth), pp. 295–301. Martinus Nijhoff, The Hague.
17. Santavicca, M., Noel, A., Chenard, M. P., Lutz, Y., Stoll, I., Segain, J. P., *et al.* (1995). *Int. J. Cancer*, **64**, 336–41.
18. Belaaouaj, A., Shipley, J. M., Kobayashi, D. K., Zimonjic, D. B., Popescu, N., Silverman, G. A., and Shapiro, S. D. (1995). *J. Biol. Chem.*, **270**, 14568–75.
19. Anglard, P., Melot, T., Guerin, E., Thomas, G., and Basset, P. (1995). *J. Biol. Chem.*, **270**, 20337–44.
20. Llano, E., Pendas, A. M., Knauper, V., Sorsa, T., Salo, T., Salido, E., Murphy, G., Simmer, J. P., Bartlett, J. D., and Lopez-Otin, C. (1997). *Biochemistry*, **36**, 15101–8.
21. Sedlacek, R., Mauch, S., Kolb, B., Schatzlein, C., Eibel, H., Peter, H. H., Schmitt, J., and Krawinkel, U. (1997). *Immunobiology*, **198**, 408–23.
22. Masson, R., Lefebvre, O., Noel, A., El Fahime, M., Chenard, M. P., Wendling, C., Kebers, F., LeMeur, M., Dierich, A., Foidart, J. M., Basset, P., and Rio, M. C. (1998). *J. Cell Biol.*, **140**, 1535–41.
23. Gronski, T. J., Martin, R. L., Kobayashi, D. K., Walsh, B. C., Holman, M. C., Huber, M., Van Wart, H. E., and Shapiro, S. D. (1997). *J. Biol. Chem.*, **272**, 12189–94.

Mark D. Sternlicht and Zena Werb
University of California, San Francisco,
Department of Anatomy, LR-208,
3rd and Parnassus Avenues, San Francisco, CA
94143–0452, USA.

Tissue inhibitors of metalloproteinases (TIMPs)

The tissue inhibitors of metalloproteinases (TIMPs) constitute a family of at least four 22–29 kDa proteins that specifically and reversibly inhibit the matrix metalloproteinases (MMPs). Twelve completely conserved Cys residues form six intrachain disulphide bonds to yield a six-loop two-domain overall structure. The more conserved N-terminal domains are thought to interact with MMP active sites and both TIMP domains are thought to influence MMP–TIMP complex formation, which occurs with 1:1 stoichiometry. Their gene structure is highly conserved, but they exhibit differences in their preferred targets, their gene regulation, and their patterns of expression.

Protein properties

Just as the matrix metalloproteinases (MMPs) are instrumental in normal and pathologic ECM remodelling, so too are their natural endogenous inhibitors, the tissue inhibitors of metalloproteinases (or TIMPs). At present, four members of the TIMP multigene family (TIMPs 1–4) have been cloned in humans and other species.[1–3] The respective TIMPs show 37–51 per cent overall peptide sequence identity, with TIMPs 2, 3, and 4 being more similar to one another than to TIMP-1. Features that distinguish the TIMPs are

(1) their confirmed extracellular secretion which is mediated by a transient 23–29 amino acid N-terminal signal sequence;

(2) their ability to bind to latent pro-MMPs or active MMPs with 1:1 stoichiometry;

(3) their ability thereby to inhibit the autocatalytic activation of latent enzymes and the proteolytic capacity of active ones;

(4) their retained ability to bind and inhibit other MMPs following dissociation;

(5) a conserved gene structure in terms of number of coding exons, correspondence between protein domains and exons, and conserved splice-site locations, but not in terms of overall gene size or the size of their untranslated regions;

(6) the required presence of 12 similarly spaced Cys residues that form six disulphide bonds and a conserved six-loop structure (Fig. 1);

(7) their inactivation by disulphide bond reduction; and

(8) a highly conserved N-terminal domain (consisting of loops 1–3) that is both necessary and sufficient for MMP inhibition.[2,4]

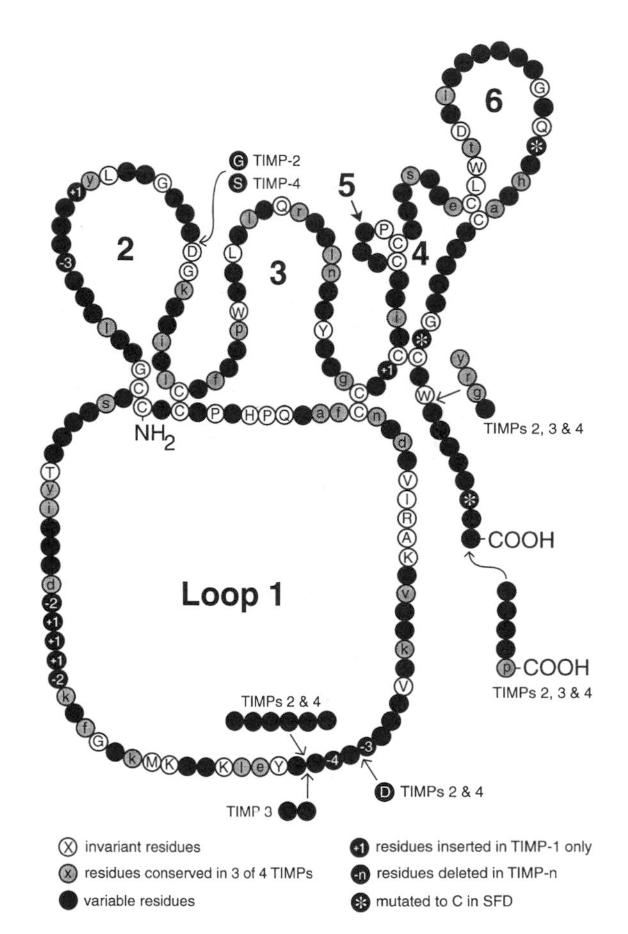

Figure 1. Structural model of human TIMPs 1–4. The composite model is based on a structural model of human TIMP-1[14] and the aligned amino acids of the other human TIMPs. Unshaded upper-case residues are common to all human TIMPs, and shaded lower-case residues are common to any three of the four cloned TIMPs. Gaps and insertions in the aligned primary sequences are also indicated, as are the three mutated residues in TIMP-3 that have been linked to Sorsby's fundus dystrophy (SFD).

Differences between individual family members exist in terms of

(1) their C-terminal domains (loops 4–6 and a free tail);

(2) their affinities for and specific activities against particular MMPs;

(3) their tissue-specific expression;
(4) their transcriptional regulation; and
(5) their gene promoters.[5]

Human TIMP-1 is a 184 residue, 28.5 kDa glycoprotein with heterogeneous *N*-linked glycosylation at two sites which is apparently not required for its inhibitory activity.[5] Upon cloning, TIMP-1 was found to be identical to erythroid-potentiating activity (EPA), indicating a growth factor role that may be independent of its MMP inhibitory function.[6] TIMP-1 expression is limited largely to adult bone and ovary and to tissues undergoing remodelling or inflammation.[7] In cultured cells, its expression is regulated by serum, viruses, phorbol esters, cytokines, and growth and differentiation factors.[7] TIMP-2, on the other hand, is an unglycosylated, 194 residue protein of 21 kDa with an extended and negatively charged C terminus.[5] The C-terminal domains of TIMPs 1 and 2 interact strongly with the C-terminal domains of gelatinases B and A, respectively, resulting in tight, preferential binding of TIMP-1 to gelatinase B and TIMP-2 to gelatinase A.[5] Like TIMP-1, TIMP-2 has been reported to have growth factor-like activity.[5] On the other hand, TIMP-2 mRNA levels are similar in tumours and adjacent normal tissues, whereas TIMP-1 mRNA levels are greater in tumours than matched normal tissues.[8,9] TIMP-2 expression in cultured cells remains unchanged by serum and TPA and is decreased by TGF-α1, whereas TIMP-1 expression is increased by each of these treatments.[8,9] TIMP-3 is a 188 residue, 27 kDa glycoprotein with a single *N*-glycosylation site near its C terminus;[4] and TIMP-4 is a 195 amino acid protein with both predicted and apparent molecular masses of 22 kDa.[2] TIMP-3 is expressed in a variety of developing tissues, most notably uterine decidua during implantation, and cartilage, muscle, various epithelia, and placental trophoblasts during late gestation.[4] It is also expressed in a variety of adult normal tissues and in certain cancers and diseases of the eye.[10–12] In the eye, TIMP-3 is a normal component of Bruch's membrane and is present in drusen which contain a variety of ECM molecules.[30] Like TIMP-1, TIMP-3 expression is induced by mitogenic stimuli (indeed, TIMP-3 was cloned as mitogen-inducible gene 5 (*mig*-5), a serum-inducible gene of the delayed–early type), but unlike other TIMPs, TIMP-3 is subject to cell cycle-specific regulation, being upregulated during G_1 progression.[13] Finally, TIMP-4 is distinguished by its abundant expression in adult heart tissue and its relatively low expression in certain other adult tissues.[2,3] It is also expressed in stromal fibroblasts of normal and benign breast tissue, but not in malignant breast cancer.

A portion of the N-terminal domain of the TIMPs clearly binds to the MMP active site, because recombinant C-terminally truncated 'tiny TIMPs', containing only the first three loops of TIMPs 1 or 2 can still inhibit MMPs.[4,14] Indeed, NMR chemical shift analyses indicate that much of one face of the TIMP-2 N-terminal domain binds to the catalytic domain of stromelysin-1.[31] X-ray crystallography has further revealed that the disulphide-linked segments Cys-1-Thr-2-Cys-3-Val-4 and Ser-68-Val-69 of TIMP-1 bind opposite sides of the active site zinc, that Cys-1 coordinates the zinc ion, and that the side chain of Thr-2 occupies the large S1′ specificity pocket of stromelysin-1.[32] Single-site mutations at several conserved sites within the N-terminal domain of TIMP-1 have had little effect on its ability to bind and inhibit the minimal domain MMP matrilysin, although single-residue replacements in the anchored region between Cys-3 and Cys-13 do diminish its affinity for matrilysin up to six-fold.[15] However, most of this region does not bind the active site itself, but instead helps maintain overall TIMP conformation.[5,32] TIMP-1 peptides that can compete with TIMP-1 for binding to collagenase-1 and can themselves inhibit collagenase-1 directly, tend to surround the second so-called 'disulphide knot' between loops 3 and 4 (Fig. 1).[14] Furthermore, two neutralizing TIMP-1 antibodies recognize epitopes in this same region of loop 3, whereas non-blocking antibodies recognize portions of loops 1, 4, and 6.[14] Analysis of single residue mutations in and around the collagenase-1 active site suggests that TIMP-1 binding to this particular MMP requires a properly folded, but not necessarily functional, active site.[16] Taken together, these findings indicate that both the N- and C-terminal domains of the TIMPs contribute to their binding properties, whereas TIMP inhibitory activity derives from the N-terminal domain alone and in particular, from the region surrounding the first disulphide knot.

Purification

TIMP-1 is best purified from the conditioned medium of TPA- or IL-1-stimulated fibroblasts.[5] Purification involves chromatography over zinc acetate-saturated iminodiacetic acid–Sepharose to remove MMPs, heparin–Sepharose to isolate the TIMP-1, and ConA–Sepharose to separate its differentially glycosylated forms.[5] Monoclonal immunoaffinity matrices have also been used to purify TIMP-1.[15] TIMP-2 is generally complexed with progelatinase A in cell-conditioned media. Thus it can be purified using gelatin–agarose to bind such complexes, ConA–Sepharose to remove contaminating progelatinase B–TIMP-1 complexes, and reverse-phase HPLC or gel filtration to recover the pure inhibitor after complex dissociation or partial progelatinase A denaturation, respectively.[5] Unlike TIMPs 1 and 2, TIMP-3 has been found in the subcellular matrix of several human cell lines, rather than in their conditioned medium.[17] Methods for purifying TIMP 4 have not yet been described. Owing to their disulphide bonds, TIMPs are highly thermostable and even remain active after boiling.

Activities

The TIMPs specifically and reversibly inhibit MMPs, including the plant MMP soybean leaf metalloendopeptidase, but they are not known to inhibit any other metalloproteinases.[5] Methods for assaying TIMP activity include measuring the inhibition of radiolabelled collagen degradation by collagenase-1, measuring inhibition using fluorogenic

MMP substrates, and reverse zymography.[5] In the latter method, samples are separated on non-reducing SDS–PAGE gels containing copolymerized gelatin, the SDS is removed using a Triton X-100-containing buffer, and the gels are impregnated with activated MMPs, incubated in a buffer that allows the MMPs to digest the gelatin, and then stained for protein. In this way, stained bands indicate retained gelatin at sites of MMP inhibition.

Antibodies

Monoclonal antibodies against human TIMPs 1 (7–6C1, 50–3D2) and 2 (T2–101, 67–4H11) are available from Amersham, Calbiochem, Fuji Chemicals, and ICN Immunologics. Neutralizing and non-neutralizing monoclonal anti-human TIMP-1 antibodies are also available from Cambio and Chemicon. A monoclonal antibody prepared against a synthetic peptide from the C-terminal end of human TIMP-3 has also been described.[30] Polyclonal antibodies against TIMPs 1–4 are available from Biogenesis, Cambio, Chemicon, and Triple Point Biologics. Rabbit anti-rat TIMP-2 antibodies are available from BioSource International. Recombinant bovine and human TIMP-1 and human TIMP-2 proteins are available from Amersham, Calbiochem, Fuji Chemicals, and ICN Immunologics.

Genes

Full length cDNA sequences are available in GenBank for TIMP-1 in humans (X02598, X03124, A10416), mice (M28312), rats (L31883, U06179), cows (S70841), horses (U95039), baboons (L37295), sheep (S67450), and rabbits (J04712); for TIMP-2 in humans (M32304, J05593, S48568), mice (X62622, M93954), rats (L31884, S82718, U14526), and cows (M32303); for TIMP-3 in humans (U02571, S78453, X76227, U14394, U67195, Z30183), mice (L27424, Z30970, L19622, M82858), rats (U27201), and chickens (M94531); and for TIMP-4 in humans (U76456) and mice. Genomic sequences (including promoter sequences which are underlined) are also available for TIMP-1 in humans (Y09720, D26513), mice (X69413, M21162, M28308-M28311), and rats (X90486); for TIMP-2 in humans (S68860, U44381-U44385); and for TIMP-3 in humans (S79779, U38952-U38955, L15078) and mice (U19462, U26433-U26437, L15078).

The genes for TIMPs 1, 2 and 3 have been localized to human chromosomes Xp11.23–11.3, 17q25 and 22q12.1–13.2, respectively.[10,18,19]. The TIMP-1 gene spans 4.3 kbp and has six exons, the first of which encodes only 5′ untranslated sequence.[4,20] The TIMP-2 and TIMP-3 genes are considerably larger at 83 and 55 kbp, respectively, and each has five exons and an extended 3′ untranslated region in the final exon.[4,10,21,22] Despite their divergence in size and the extra non-coding exon in the TIMP-1 gene, exon–intron splice sites are preserved for all three genes. The TIMP-1 gene has a TATA-less promoter with multiple Ets binding sites, a high-affinity AP-1 site that sits adjacent to an Ets binding site, and a second AP-1 site whose activity is only revealed after deletion of the first.[7] Indeed, functional interaction of AP-1 and Ets-1 transcription factors has been shown to induce synergistic transcriptional activation of the promoter.[7] In addition, virus-responsive elements have been demonstrated within the first intron of the mouse TIMP-1 gene,[20], and the 5′ region of the rat gene has been shown to contain potential binding sites for Sp1, C/EBP, STAT3, AP-1, and Ets transcription factors, the latter three being contained within an IL-6-responsive element.[23] The 5′ flanking end of the TIMP-2 gene has several consensus sequences, including a TATA-like motif, five Sp1, two AP-2, one AP-1, one NF1, and three PEA-3 binding sites plus a CpG island.[21] The AP-1 site, however, is non-responsive to TPA due to its position, and gene expression is not altered by CpG methylation. The TIMP-3 gene has a cell cycle-regulated promoter with four Sp1 sites and potential NF1 and C/EBP sites.[10,22] Thus, unlike the TIMP-1 and TIMP-3 promoters, the TIMP-2 promoter resembles that of a housekeeping gene.[21] The TIMP-3 gene, on the other hand, is the only TIMP that is subject to cell cycle regulation, with its peak expression being near mid-G_1.[13] Major mRNA transcript sizes are approximately 0.9 kb for TIMP-1;[6] 1.2 and 3.8 kb for TIMP-2;[22] 2.2, 2.5, and 4.5 kb for TIMP-3;[10] and 1.4 kb for TIMP-4.[2]

Antisense oligonucleotides and *in situ* hybridization probes for TIMPs 1–3 are available from Chemicon. Partial and complete cDNAs for human and mouse TIMPs 1–3 in pBluescript and expression plasmids suitable for transient expression of human TIMP-1, mouse TIMP-2, or mouse TIMP-3 in COS cells are also available from University Technologies International (University of Calgary).

Mutant phenotype/disease states

The importance of TIMP-3 in the eye has been demonstrated by the observation of increased TIMP-3 mRNA levels in simplex retinitis pigmentosa,[12] by its immunolocalization to Bruch's membrane and drusen,[30] and by the discovery of TIMP-3 mutations in patients with Sorsby's fundus dystrophy (SFD), an autosomal dominant disorder characterized by progressive degeneration of the central retina with relatively early onset.[24,25] Specifically, three point mutations that introduce an additional Cys residue into the C-terminal domain of TIMP-3 have been found in the affected members of three SFD families (Fig. 1). These could result in either intra- or inter-molecular disulphide bonds that disrupt normal TIMP-3 function, thus further indicating the importance of the conserved disulphide links in bringing about proper folding and inhibitory activity. TIMP-1 knockout mice have exhibited altered susceptibility to *Pseudomonas* infections, but no other obvious phenotypes have yet been observed (unpublished data). Targeted mutagenesis of TIMP-1 has also indicated that tumour cell lung invasion depends on the TIMP-1 genotype of the tumour cells themselves rather than that of the host.[26] TIMP-1 overexpression in transgenic mice has been shown to inhibit SV40 T antigen-induced hepatocellular carcinoma initiation and progression, whereas its antisense-mediated reduction accelerates these processes.[27] TIMP-1 overexpression also alters implantation and mammary epithelial cell apoptosis.[28] TIMP-2 knockout mice are viable, but have not yet been well characterized (P.D. Soloway; personal communication).

Structure

The twelve completely conserved Cys residues of the TIMPs form six intrachain disulphide bonds that fold the TIMPs into six-loop, two-domain molecules.[5] NMR analysis of the active, N-terminal domain of TIMP-2 (loops 1–3) has provided secondary and low resolution tertiary structure information on this domain.[5,29] These determinations indicate the presence of a closed α barrel that is formed by a rolled up five-stranded α sheet, plus two short, closely packed α helices that lie on the barrel's outer surface. The α-barrel topology is homologous to that seen in the oligosaccharide/oligonucleotide binding (OB) fold protein family, and is likely to hold for the other TIMP N-terminal domains as well.[29] The X-ray crystal structure of human TIMP-1 complexed with the catalytic domain of stromelysin-1 (PDB ID code IUAC) indicates that TIMP-1 has an elongated wedge shape with a long edge that occupies the entire MMP active site cleft.[32] The N-terminal domain of TIMP-1 indeed forms an OB fold, but other aspects of this model differ considerably from those of the less refined NMR model. In particular, the first four N-terminal residues of TIMP-1 bind to the S1 to S3′ subsites of the active site cleft and at Val-4 the chain turns inward so that the region around His-7 forms part of the intra- rather than extra-molecular subdomain. Cys-1 sits directly above and coordinates the active site zinc, and Ser-68 and Val-69, which are fixed nearby by the Cys-1 to Cys-70 disulphide bridge, bind to the opposite side of the catalytic zinc (subsites S2 and S3). Altogether, four separate segments from the TIMP-1 N-terminal domain and two short segments from the C-terminal domain contact the MMP catalytic domain.

References

1. Coussens, L. M. and Werb, Z. (1996). *Chem. Biol.*, **3**, 895–904.
2. Greene, J., Wang, M., Liu, Y. E., Raymond, L. A., Rosen, C., and Shi, Y. E. (1996). *J. Biol. Chem.*, **271**, 30375–80.
3. Leco, K. J., Apte, S. S., Taniguchi, G. T., Hawkes, S. P., Khokha, R., Schultz, G. A., and Edwards, D. R. (1997). *FEBS Lett.*, **401**, 213–7.
4. Apte, S. S., Olsen, B. R., and Murphy, G. (1995). *J. Biol. Chem.*, **270**, 14313–8.
5. Murphy, G. and Willenbrock, F. (1995). *Meth. Enzymol.*, **248**, 496–510.
6. Docherty, A. J. P., Lyons, A., Smith, B. J., Wright, E. W., Stephens, P. E., Harris, T. J. R., *et al.* (1985). *Nature*, **318**, 66–9.
7. Logan, S. K., Garabedian, M. J., Campbell, C. E., and Werb, Z. (1996). *J. Biol. Chem.*, **271**, 774–82.
8. Stetler-Stevenson, W. G., Brown, P. D., Onisto, M., Levy, A. T., and Liotta, L. A. (1990). *J. Biol. Chem.*, **265**, 13933–8.
9. Leco, K. J., Hayden, L. J., Sharma, R. R., Rocheleau, H., Greenberg, A. H., and Edwards, D. R. (1992). *Gene*, **117**, 209–17.
10. Wick, M., Haronen, R., Mumberg, D., Burger, C., Olsen, B. R., Budarf, M. L., *et al.* (1995). *Biochem. J.*, **311**, 549–54.
11. Uria, J. A., Ferrando, A. A., Velasco, G., Freije, J. M. P., and Lopez-Otin, C. (1994). *Cancer Res.*, **54**, 2091–4.
12. Jones, S. E., Jomary, C., and Neal, M. J. (1994). *FEBS Lett.*, **352**, 171–4.
13. Wick, M., Burger, C., Brusselbach, S., Lucibello, F. C., and Muller, R. (1994). *J. Biol. Chem.*, **269**, 18953–60.
14. Bodden, M. K., Harber, G. J., Birkedal-Hansen, B., Windsor, L. J., Caterina, N. C. M., Engler, J. A., and Birkedal-Hansen, H. (1994). *J. Biol. Chem.*, **269**, 18943–52.
15. O'Shea, M., Willenbrock, F., Williamson, R. A., Cockett, M. I., Freedman, R. B., Reynolds, J. J., *et al.* (1992). *Biochemistry*, **31**, 10146–52.
16. Windsor, L. J., Bodden, M. K, Birkedal-Hansen, B., Engler, J. A., and Birkedal-Hansen, H. (1994). *J. Biol. Chem.*, **269**, 26201–7.
17. Kishnani, N., Staskus, P., Yang, T., Masiarz, F., and Hawkes, S. (1995). *Matrix Biol.*, **14**, 479–88.
18. Willard, H. F., Durfy, S. J., Mahtani, M. M., Dorkins, H., Davies, K. E., and Williams, B. R. (1989). *Hum. Genet.*, **81**, 234–8.
19. De Clerck, Y., Szpirer, C., Aly, M. S., Cassiman, J. J., Eckhout, Y., and Rousseau, G. (1992). *Genomics*, **14**, 782–4.
20. Coulombe, B., Ponton, A., Daigneault, L., Williams, B. R. G., and Skup, D. (1988). *Mol. Cell Biol.*, **8**, 3227–34.
21. Hammani, K., Blakis, A., Morsette, D., Bowcock, A. M., Schmutte, C., Henriet, P., and DeClerck, Y. A. (1996). *J. Biol. Chem.*, **271**, 25498–505.
22. Stohr, H., Roomp, K., Felbor, U., and Weber, B. H. F. (1995). *Genome Res.*, **5**, 483–487.
23. Bugno, M., Graeve, L., Gatsios, P., Koj, A., Heinrich, P. C., Travis, J., and Kordula, T. (1995). *Nucl. Acids Res.*, **23**, 5041–7.
24. Weber, B. H. F., Vogt, G., Pruett, R. C., Stohr, H., and Felbor, U. (1994). *Nature Genet.*, **8**, 352–6.
25. Felbor, U., Stohr, H., Amann, T., Schonherr, U., and Weber, B. H. F. (1995). *Hum. Mol. Genet.*, **4**, 2415–6.
26. Soloway, P. D., Alexander, C. M., Werb, Z., and Jaenisch, R. (1996). *Oncogene*, **13**, 2307–14.
27. Martin, D. C., Ruther, U., Sanchez-Sweatman, O. H., Orr, F. W., and Khokha, R. (1996). *Oncogene*, **13**, 569–76.
28. Alexander, C. M., Howard, E. W., Bissell, M. J., and Werb, Z. (1996). *J. Cell Biol.*, **135**, 1669–77.
29. Williamson, R. A., Martorell, G., Carr, M. D., Murphy, G., Docherty, A. J. P., Freedman, R. B., and Feeney, J. (1994). *Biochemistry*, **33**, 11745–59.
30. Fariss, R. N., Apte, S. S., Olsen, B. R., Iwata, K., and Milam, A. H. (1997). *Am. J. Pathol.*, **150**, 323–8.
31. Williamson, R. A., Carr, M. D., Frenkiel, T. A., Feeney, J., and Freedman, R. B. (1997). *Biochemistry*, **36**, 13882–9.
32. Gomis-Ruth, F. X., Maskos, K., Betz, M., Bergner, A., Huber, R., Suzuki, K., Yoshida, N., Nagase, H., Brew, K., Bourenkov, G. P., Bartunik, H., and Bode, W. (1997). *Nature*, **389**, 77–81.

Mark D. Sternlicht and Zena Werb
University of California, San Francisco,
Department of Anatomy, LR-208,
3rd and Parnassus Avenues, San Francisco, CA
94143–0452, USA.

Neutrophil elastase and cathepsin G

Neutrophil elastase and cathepsin G are serine proteinases stored in the azurophilic granules of neutrophils. They can readily degrade a wide array of ECM and humoral proteins and can potentially activate still other ECM proteinases. Both enzymes share a number of features with other granule-associated serine proteinases found in circulating and fixed cells of bone marrow origin. Neutrophil elastase in particular has been implicated in a number of inflammatory disorders.

Protein properties

Neutrophil elastase (NE, leukocyte elastase, EC 3.4.21.37) and cathepsin G (EC 3.4.21.20) are serine proteinases found in the dense azurophilic (or 'primary') granules of mature polymorphonuclear leukocytes. Here, they are likely to assist in the intracellular killing of engulfed pathogens since they can hydrolyse components of bacterial cell walls. They are also released upon cell stimulation or lysis and can degrade numerous humoral and ECM molecules, although cathepsin G is apparently less potent in this regard.[1–4] Thus, in addition to being critical antimicrobial enzymes, they are also thought to act as mediators of coagulation, immune responses, and wound debridement.[1–4] In addition, NE has been strongly implicated in a number of inflammatory diseases, including idiopathic pulmonary fibrosis, adult respiratory distress syndrome, cystic fibrosis, glomerulonephritis, emphysema, and rheumatoid arthritis.[1–4]

Human NE is a 218 amino acid, 28 kDa glycoprotein with two *N*-linked carbohydrate side chains and four disulphide bridges.[3–5] An additional 20 amino acid C-terminal extension (or 'pro-C' peptide), identified by 3' cDNA and genomic sequence analysis, is probably removed during post-translational lysosomal packaging of the enzyme.[4,5] Analysis of coding exons also predicts an added 29 residue N-terminal precursor peptide with a 27 residue 'pre' signal peptide followed by a 'pro-N' dipeptide which can be removed by cathepsin C (dipeptidyl peptidase I).[5] The pre signal peptide probably targets the nascent protein to the endoplasmic reticulum, while the pro-N and/or pro-C peptides may play a role in targeting NE to the azurophilic granules or in its latency and activation. Human NE is 36–40 per cent homologous to pancreatic elastases of various species and shows lesser homology to other serine proteinases.[6] Small amounts of human NE can also be found in blood monocytes and alveolar macrophages and it is expressed in leukemic cell lines of myelomonocytic origin.[4] Indeed, NE was cloned from one such cell line (U937) by two groups[4,7] and found to be identical to the already cloned bone marrow serine proteinase medullasin.[8] Although mature neutrophils, monocytes, and macrophages carry the enzyme, its mRNA is undetectable in them. Its 1.3 kb transcripts are present, however, in bone marrow myelomonocytic progenitor cells, indicating that the gene is turned off before neutrophils and monocytes exit the marrow.[4] In addition to its enzymatic activity, recent evidence indicates that NE can bind to integrin CR3 (CD11b/CD18, Mac-1, $\alpha M\beta 2$) at the cell surface of neutrophils and thereby regulate their integrin-mediated adhesion.[9] The principal endogenous inhibitors of NE are α_1-proteinase inhibitor (α_1-PI, α_1-antitrypsin), secretory leukocyte protease inhibitor (SLPI, antileukoprotease, mucous proteinase inhibitor) and α_2-macroglobulin.

Cathepsin G is a 26 kDa enzyme with a single potential *N*-glycosylation site.[2,10] Like NE, it is found in the azurophilic granules of neutrophils and monocytes and is abundantly expressed in U937 cells. In addition, the expression of both enzymes drops after promyelocytic or promonocytic cell lines are treated with TPA, thus inducing their differentiation to a more mature myelocytic or monocytic phenotype.[2,4] Also like NE, cloning of human cathepsin G from the same U937 cell line predicts the presence of an 18 residue N-terminal signal peptide and a 2-amino acid pro-N peptide.[2] It is also noteworthy that both of these enzymes, as well as human mast cell chymase, other known mast cell chymases of dog, rat, and mouse, and most lymphocyte granzymes (i.e. other granule-stored serine proteinases of bone marrow-derived cells) have a 2-amino acid pro-N peptide that ends in Glu, whereas serine proteinases from other cell types have considerably longer pro-N peptides that end in Arg or Lys.[5,11] In addition, cathepsin G has a predicted pro-C peptide that is probably removed during lysosomal packaging.[5]

Purification

Pure NE and cathepsin G can be obtained from leukocytes or spleen by diverese methods.[1] They are generally extracted from granule preparations using high salt, urea, or detergent and are then purified using affinity adsorbents such as Sepharose-linked bovine pancreatic trypsin inhibitor (BPTI, Aprotinin, Trasylol), DEAE–cellulose, or other matrices. After their separtion on DEAE–cellulose, elastase can be further purified using CM–cellulose and Sephadex G-75 gels, and cathepsin G can be purified using hydroxyapatite and Sephadex G-75. Purified human NE is available from several commercial sources.

Activities

The active site of NE contains a His-41/Asp-88/Ser-173 'catalytic triad' typical of the serine proteinases. The His and Asp residues are thought to bind transiently a proton from the Ser residue, transforming the latter into a potent nucleophile that attacks substrate peptide bonds. NE, in particular, cleaves P1–P1′ peptide bonds where the P1 amino acid residue contains a small alkyl side chain.[6] NE is one of the few enzymes that can cleave insoluble elastin. Areas of elastin susceptibility are alanine rich and a large number of alanine derivatives, such as Boc-Ala-nitrophenyl ester, *Z*-Ala-2-napthyl ester, and Ac-$(Ala)_3$-nitroanilide have been developed as substrates for NE.[1] Fluorogenic alanine-rich substrates for the assay of NE are also available from Enzyme Systems Products and Molecular Probes. Other natural substrates of NE include collagens I–IV, fibronectin, proteoglycans, entactin/nidogen, fibulins, tenascin, immunoglobulins, haemoglobin, coagulation factors, complement components, the insulin B chain, and components of bacterial cell walls.[1,4,12–14] Cleavage of fibrillar collagens is at regions of extra-helical cross-linking, thus resulting in the release of individual α chains from otherwise insoluble collagen fibrils.[1] In addition, NE can activate a number of matrix metalloproteinases and can cleave a number of sites within gelatinase A, a matrix metalloproteinase not normally activated by serine proteinases.[15] Here, sequential treatment of gelatinase A with 4-aminophenylmercuric acetate and NE yields a truncated 40 kDa enzyme with four-fold increased gelatinolytic activity. In the absence of gelatin, however, at least two sites within the fibronectin-like gelatin-binding domain of gelatinase A are cleaved, thus inactivating the enzyme. Furthermore, NE is a preferred inactivator of TIMP-1 and is probably able to inactivate other TIMPs as well.[16] A variety of natural and synthetic NE inhibitors are available, including α_1-PI, soybean trypsin inhibitor, BPTI, elastinal, ovomucoid inhibitors, peptide chloromethyl ketones and fluoromethyl ketones, heterocyclics, isocoumarin inhibitors, β-lactam inhibitors, and others.[6]

Cathepsin G exhibits chymotrypsin-like enzymatic activity in that it cleaves a number of synthetic substrates of chymotrypsin.[1] Still, its activity is low compared to that of chymotrypsin itself, and little else is known concerning its natural substrates. It readily cleaves tenascin and is fairly active against proteoglycans, fibrinogen, casein, azocasein, haemoglobin, and collagens I and II, which it hydrolyses from the ends.[1,13,17] In addition, it may activate certain matrix metalloproteinases such as gelatinase B.[18] Cathepsin G can be inhibited by α_1-PI, α_1-antichymotrypsin and the recently cloned serpin serine proteinase inhibitor squamous cell carcinoma antigen 2.[21]

Antibodies

Monoclonal antibodies against human NE are available from Accurate Chemical and Scientific, Biodesign International, Biogenesis, Biomeda, Chemicon, DAKO, Dimension Labs, ICN Immunologics, and Pharmingen. Monoclonals against human cathepsin G are available from Biogenesis, Chemicon, and Pharmingen. Polyclonal antibodies against both antigens are available from numerous suppliers including some of those listed above.

Genes

Gene and cDNA sequences have been obtained for NE in humans (J040432, J03545, M27783, Y00477, X05875) and mice (U04962, U06076), and for cathepsin G in humans (M16117, J04990) and mice (X78544, M96801). The human NE gene is a single copy gene with five exons located on chromosome 11q14 [5]. Its 5′-flanking region contains typical TATA, CAAT, and GC motifs, a number of repetitive sequences of potential significance, and a 19 bp motif with 90 per cent homology to a segment in the promoter region of the human myeloperoxidase (MPO) gene, a gene also expressed in immature myeloid cells and whose product is also stored in neutrophil azurophilic granules.[5] More recent analysis of the murine MPO and NE genes indicates that polyomavirus enhancer-binding factor 2/core-binding factor (PEBP2/CBF), which is the murine homologue of the human AML1 and PEBP2β/CBF β proto-oncogene products, can bind and regulate both the murine MPO and NE genes.[19] In addition, the murine NE enhancer contains possible binding sites for an Ets family member, C/EBP, and Myb.[19] The human cathepsin G gene spans 2.7 kb, has five exons, and is organized like the NE, mast cell chymase, and cytotoxic T-cell granzyme genes.[10] The gene is located on chromosome 14q11.2 near the α/δ T-cell receptor complex and is clustered with the genes for human granzymes B and H.[10] Its 5′ region contains TATA and CAAT boxes, but no cAMP or serum-response elements, nor any consensus binding sites for SP-1, AP-1, AP-2, or octamer proteins. Likewise, the mouse cathepsin G gene is approximately 2.5 kb in length, has five exons, and shows high homology to its human counterpart.[20] The murine gene is tightly linked to the granzyme B (CTLA-1, CCP1) locus on mouse chromosome 14 where the genes for granzymes C, E, and F are also clustered.[20]

Mutant phenotype/disease states

Aberrant NE expression has been implicated in the high rate of infections seen in inherited Chediak–Higashi syndrome and in the beige (*bg/bg*) mouse, and hereditary α_1-PI deficiency results in emphysema due to inadequate inhibition of NE and chronic degradation of alveolar wall components.[5] Thus the proteolytic or oxidative inactivation of α_1-PI and the recruitment and degranulation of neutrophils are thought to play a major part in the pathogenesis of emphysema in general.

Structure

Crystal-based structures have been solved for human NE complexed with three different inhibitors and for numerous porcine pancreatic elastase derivatives.[6] Like other serine proteinases, both elastases have two similar and interacting antiparallel β-barrel cylindrical domains that form an active-site crevice containing the catalytic Ser, His, and Asp residues. Topological differences between the elastases and other serine proteinases are found mostly in their surface loops. Differences between the two elastases are also apparent and no doubt influence their mechanisms of action, substrate specificities, and inhibitor susceptibilities.[6] Brookhaven Protein Data Bank identifiers for human NE are 1HNE, 1PPF, and 1PPG.

References

1. Starkey, P. M. (1977). In *Proteinases in mammalian cells and tissues*. (ed. A. J. Barrett), pp. 57–89. North-Holland, New York.
2. Salvesen, G., Farley, D., Shuman, J., Przybyla, A., Reilly, C., and Travis, J. (1987). *Biochemistry*, **26**, 2289–93.
3. Sinha, S., Watorek, W., Karr, S., Giles, J., Bode, W., and Travis, J. (1987). *Proc. Natl Acad. Sci. USA*, **84**, 2228–32.
4. Takahashi, H., Nukiwa, T., Basset, P., and Crystal, R. G. (1988). *J. Biol. Chem.*, **263**, 2543–7.
5. Takahashi, H., Nukiwa, T., Yoshimura, K., Quick, C. D., States, D. J., Holmes, M. D., *et al*. (1988). *J. Biol. Chem.*, **263**, 14739–47.
6. Bode, W., Meyer, J. E., and Powers, J. C. (1989). *Biochemistry*, **28**, 1951–63.
7. Farley, D., Salvesen, G., and Travis, J. (1988). *Biol. Chem. Hoppe-Seyler*, **369** (Suppl.), 3–7.
8. Okano, K., Aoki, Y., Sakurai, T., Kajitani, M., Kanai, S., Shimazu, T., *et al*. (1987). *J. Biochem.*, **102**, 13–16.
9. Cai, T. Q. and Wright, S. D. (1996). *J. Exp. Med.*, **184**, 1213–23.
10. Hohn, P. A., Popescu, N. C., Hanson, R. D., Salvesen, G., and Ley, T. J. (1989). *J. Biol. Chem.*, **264**, 13412–19.
11. Caughey, G. H., Zerweck, E. H., and Vanderslice, P. (1991). *J. Biol. Chem.*, **266**, 12956–63.
12. Mayer, U., Mann, K., Timpl, R., and Murphy, G. (1993). *Eur. J. Biochem.*, **217**, 877–84.
13. Imai, K., Kusakabe, M., Sakakura, T., Nakanishi, I., and Okada, Y. (1994). *FEBS Lett.*, **352**, 216–18.
14. Sasaki, T., Mann, K., Murphy, G., Chu, M. L., and Timpl, R. (1996). *Eur. J. Biochem.*, **240**, 427–34.
15. Rice, A. and Banda, M. J. (1995). *Biochemistry*, **34**, 9249–56.
16. Itoh, Y. and Nagase, H. (1995). *J. Biol. Chem.*, **270**, 16518–21.
17. Roughly, P. J. and Barrett, A. J. (1977). *Biochem. J.*, **167**, 629–37.
18. Murphy, G. and Crabbe, T. (1995). *Meth. Enzymol.*, **248**, 470–84.
19. Nuchprayoon, I., Meyers, S., Scott, L. M., Suzow, J., Hiebert, S., and Friedman, A. D. (1994). *Mol. Cell. Biol.*, **14**, 5558–68.
20. Heusel, J. W., Scarpati, E. M., Jenkins, N. A., Gilbert, D. J., Copeland, N. G., Shapiro, S. D., and Ley, T. J. (1993). *Blood*, **81**, 1614–23.
21. Schick, C., Kamachi, Y., Bartuski, A. J., Cataltepe, S., Schechter, N. M., Pemberton, P. A., and Silverman, G. A. (1997). *J. Biol. Chem.*, **272**, 1849–55.

Mark D. Sternlicht and Zena Werb
University of California, San Francisco,
Department of Anatomy, LR-208,
3rd and Parnassus Avenues, San Francisco,
CA 94143–0452, USA.

Plasminogen

Plasminogen is the zymogen for the fibrinolytic serine proteinase plasmin. Conversion of the single-chain zymogen to its active disulphide-linked two-chain form is catalysed by other serine proteinases such as urokinase- and tissue-type plasminogen activators, or by non-enzymatic activators such as streptokinase. Plasminogen binds to certain ECM proteins as well as unidentified cell surface molecules. Plasmin, in turn, plays an important role in fibrinolysis and has been implicated in the direct and indirect degradation of other ECM molecules. In addition, the angiogenesis inhibitor angiostatin is a breakdown product of plasminogen.

Protein properties

Native plasminogen (or 'Glu-plasminogen' after its N-terminal glutamic acid) is a single-chain glycoprotein of about 92 kDa that is produced mainly by hepatocytes and found in plasma and interstitial fluids at a concentration of around 2 μM. It is the inactive zymogen for the serine proteinase plasmin (fibrinolysin, fibrinase, EC 3.4.21.7) and contains 790 amino acids, 24 intrachain disulphide bonds, and five so-called 'kringles' (K1–K5) after the Danish pastry of similar two dimensional representational appearance (Fig. 1).[1] These five triple-loop structures each have three disulphide bridges and are homologous to the kringle domains of prothrombin, tissue-type, and urokinase-type plasminogen activators (tPA and uPA), apolipoprotein A (which has 37 such structures), hepatocyte growth factor, defensins, and other proteins.[2] Plasminogen contains a single high-affinity (k_d = 9 μM) lysine binding site in K1 that mediates its interaction with fibrin and the fast-acting plasma inhibitor α_2-antiplasmin.[1] In addition, four low-affinity (k_d = 5 μM) lysine binding sites may also play a role in ligand binding. Two major plasminogen glycoforms (type 1 and type 2) can be distinguished by ion exchange chromatography and elec-

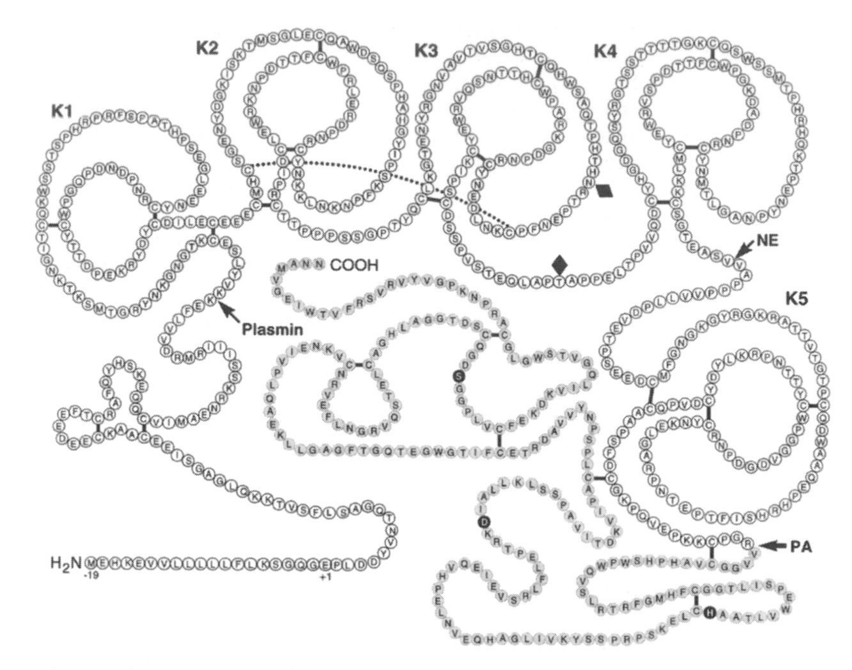

Figure 1. Primary and secondary structure of human plasminogen. The signal peptide is shown and the first residue of Glu-plasminogen is labelled +1. Plasmin and neutrophil elastase (NE) can cleave plasminogen at the sites indicated to yield Lys- and Val-plasminogen; plasminogen activators (PA) cleave the Arg-560–Val-561 peptide bond. A and B chain residues and catalytic residues are indicated by opened, shaded, and closed circles, respectively. Also indicated are kringles 1–5 (K1–K5), potential glycosylation sites (black diamonds), the disulphide bond linking kringles 2 and 3 (dotted line), and all other disulphide bonds (solid lines). (Modified from Petersen *et al.*[28])

trophoresis.[3] Each is *O*-glycosylated at Thr-345, but only type 1 plasminogen is *N*-glycosylated at Asn-288 on K3.[4] *N*-glycosylated type 1 plasmingen has less affinity for cell surface plasminogen receptors, a lower rate of activation by tPA and uPA, and a lower lysine binding affinity than does type 2 plasminogen.[4] Nevertheless, both glycoforms have similar fibrin binding constants. Removal of the common Thr-linked trisaccharide has also been noted to promote activation, albeit slightly. Native 92 kDa Glu-plasminogen is readily converted to 84 kDa 'Lys-plasminogen' by limited plasmin hydrolysis of the Arg-67–Met-68, Lys-76–Lys-77, or Lys-77–Val-78 peptide bonds just N-terminal to K1.[5] The altered conformation of Lys-plasminogen gives it a higher affinity for fibrin than Glu-plasminogen and enables its more rapid conversion to plasmin.[1] In addition, neutrophil elastase cleavage of the Val-441–Val-442 peptide bond between K4 and K5 yields a truncated yet functional 38 kDa pro-enzyme termed 'Val-plasminogen' or 'mini-plasminogen'.[3] Like Lys-plasminogen, Val-plasminogen is more readily activated than Glu-plasminogen. Also, active 'Val-' or 'mini-plasmin' is relatively resistant to α_2-antiplasmin inhibition due to its lack of K1–K4, but it still binds fibrin and remains readily inhibited by α_2-macroglobulin.[6] Its physiologic role, however, is unclear.

The conversion of plasminogen to plasmin can be catalysed by several physiologic and pathologic activators. These include the serine proteinases uPA, tPA, kallikrein, and factors IXa and XIIa, and non-enzymatic activators such as streptokinase and staphylokinase. The non-enzymatic activators form an equimolar complex with plasminogen which can activate other plasminogen molecules, whereas the enzymatic activators cleave plasminogen directly. Either way, plasminogen is converted to plasmin by specific hydrolysis of its Arg-560–Val-561 peptide bond. This yields a two-chain molecule (plasmin) composed of an N-terminal heavy (A) chain and a C-terminal light (B) chain; both chains remaining linked by two disulphide bonds (Fig. 1). The A chain contains kringles K1–K5 and has a molecular mass of 12–65 kDa depending on the type of plasminogen from which it was derived and the B chain contains the catalytic triad (His-602/Asp-645/Ser-740) and has a molecular mass of 25 kDa. Conversion of Glu-plasminogen to plasmin is much slower than the conversion of Lys-plasminogen, but both reactions ultimately yield 83 kDa Lys-plasmin. Glu-plasmin can, however, be obtained from Glu-plasminogen if it is converted in the presence of α2-antiplasmin. Thus both Glu- and Lys-plasmin can arise *in vivo*.

Plasminogen binds to various substrates via its kringle domains and such binding can be disrupted by lysine analogues such as ε-aminocaproic acid (EACA). In addition, the 4 kDa neutrophil defensins I and II exhibit structural similarity to the plasminogen kringles and can inhibit tPA-mediated plasminogen-dependent fibrinolysis by competing with plasminogen for fibrin binding.[7] Proteins other than fibrin that bind plasminogen include tetranectin,[8] fibronectin,[9] vitronectin,[10] thrombospondin,[11] laminin,[12] histidine-rich glycoprotein,[13] platelet glycoproteins IIb/IIIa (αIIb-β3 integrin), proteoglycans, IgG, and very low density lipoprotein (VLDL).[1] In addition, virtually all cells bind plasmin(ogen) via abundant, low-affinity cell surface receptors that recognize the plasminogen kringle domains.[14] Such receptors include non-protein gangliosides and proteins with C-terminal Lys residues (e.g. enolase and annexin II).[14] Cell-bound plasminogen is more readily activated than free plasminogen and once activated it is protected from plasma proteinase inhibitors [15]. Likewise, ECM-bound plasminogen is much more readily activated by uPA and tPA and less readily inhibited by α_2-antiplasmin than is unbound plasminogen.[16]

Amino acid analysis of angiostatin, a recently discovered 38 kDa angiogenesis inhibitor, has shown it to be the K1–K4 fragment of plasminogen.[17,18] Angiostatin inhibits the neovascularization and growth of lung metastases *in vivo* and the proliferation of endothelial cells but not tumour cells *in vitro*. Intact plasminogen, however, has no such activity. The anti-proliferative activity of angiostatin derives from K1–K3 and may be tempered somewhat by K4.[19] In addition, the lysine binding capacity of the kringles is probably not required for antiproliferative activity. Neutrophil elastase can cleave plasminogen to yield one form of angiostatin *in vitro*, and macrophage metalloelastase (MMP-12) may be responsible for its production *in vivo*.[20]

Purification

Plasminogen is readily purified from human plasma by gel filtration and lysine–Sepharose affinity chromatography in the presence of proteinase inhibitors.[3] The two plasminogen glycoforms are separated by gradient elution from the affinity column using a lysine analogue such as EACA.

Activities

Plasmin is a trypsin-like serine proteinase with broad substrate specificity. It hydrolyses lysyl and arginyl peptide bonds and has esterolytic activity against basic amino acid esters and amidolytic activity against basic amino acid amides.[3] In addition to its clear role in fibrinolysis and fibringenolysis, plasmin can also cleave a large number of other molecules. It can activate several matrix metalloproteinases (e.g. collagenase-1, gelatinase B, stromelysin-1, and MT1-MMP) and complement zymogens (eg., C1, C3 and C5); it can inactivate C1 inhibitor and factor XIIa; it can directly degrade several ECM proteins (e.g. fibrin, fibronectin, thrombospondin, and laminin); and it can degrade several non-ECM molecules (eg., proaccelerin, ACTH, glucagon, and IgG).[3,21,22]

Plasmin can be inhibited by standard serine proteinase inhibitors and by the endogenous inhibitors α_2-plasmin inhibitor (α_2-antiplasmin), α_2-macroglobulin, α_1-proteinase inhibitor (α_1-antitrypsin), and protease nexin-2 (β-amyloid precursor protein A4).[23,24] The most important physiologic inhibitor of plasmin is the 70 kDa Arg-serpin α_2-antiplasmin which has a plasma concentration of around 1 μM and a second-order inhibitory rate constant of 2–4 ◊ 10^7 M^{-1} s^{-1} for plasmin.[25]

A variety of chromogenic and fluorogenic peptide substrates have been described for the analysis of plasmin and plasminogen activators.[26] Chromogenic substrates can be obtained from Pharmacia, Hepar-Chromogenix (formerly Kabi Diagnostica), American Diagnostica, and Calbiochem-Novabiochem. Fluorogenic substrates can be obtained from Bachem, Calbiochem-Novabiochem, Enzyme Systems Products, Molecular Probes, and Sigma. Casein, azocasein, and fibrin plate assays have also been described.[27]

Antibodies

Monoclonal antibodies against the N-terminal portion of Glu-plasminogen, plasminogen kringles K1–3 and K4, and other plasminogen epitopes are available from Advanced Immunochemical, American Research Products, American Diagnostica, Biodesign International, Calbiochem, Cortex Biochem, Enzyme Research Labs, Research Diagnostics, and Technoclone GmbH. An inhibitory antibody is also available from Affinity Biologics and polyclonal antibodies are available from numerous suppliers. Monoclonal and polyclonal antibodies against human α_2-antiplasmin are also commercially available.

Genes

The human plasminogen gene (*PLG)* is located on chromosome 6q26–27, spans 52.5 kbp, and has 19 exons.[28] The human α_2-antiplasmin gene *(PLI)* is located on chromosome 17p13, spans 16 kbp, and has 10 exons.[29] Plasminogen cDNA sequences are available for humans (X05199, M74220, K02922), monkeys (J04697), mice (J04766), rats (M62832), and cows (X79402). Genomic sequences are also available for human plasminogen (J05286, M62890, K02921, U07744) and genomic and cDNA sequences are available for human α_2-antiplasmin (J03830, D00174).

Mutant phenotype/disease states

Bleeding tendencies have been observed in patients undergoing thrombolytic therapy, in patients with acquired conditions such as liver cirrhosis, and in individuals with congenital conditions such as α_2-antiplasmin deficiency, PAI-1 deficiency or excessive tPA production.[30,31] On the other hand, congenital plasminogen defects yielding either a non-functional reactive centre or altered activation kinetics tend to result in thromboembolic disease.[30] Gene-targeted mice that lack plasminogen are viable and capable of reproduction, but develop spontaneous thrombotic lesions leading to severe organ damage.[32] These mice also undergo wasting after 2 months of age and exhibit gastrointestinal ulcerations, rectal prolapse, and high mortality. In addition, they exhibit severely impaired wound healing due to diminished keratinocyte migration from the wound edges[33] and have delayed mammary gland involution (L.R. Lund; personal communication). Interestingly, in mice that are deficient in both plasminogen and fibrinogen, the removal of fibrinogen alleviates the diverse untoward effects of plasminogen deficiency and yields a phenotype like that of mice deficient in fibrinogen alone, suggesting that the primary role of plasminogen is in fibrinolysis.[34]

The plasminogen activator/plasmin system has also been implicated in atherosclerosis, vascular remodelling and restenosis, aneurysm formation, brain function, and malignant disease.[31,38] Its components may foster tumour invasion and metastasis by direct or indirect degradation of ECM proteins including fibrin, or they may hinder tumour progression through the anti-angiogenic effects of angiostatin. Interestingly, metastatic mammary carcinomas develop rapidly in mice with an MMTV promoter-driven polyoma virus middle T transgene, but when such mice are crossed with plasminogen-null mice, their plasminogen-deficient offspring develop significantly fewer and smaller lung metastases, but show no difference in primary tumour growth.[39]

Plasminogen activators have also been exploited for the treatment of myocardial infarction and stroke. These include streptokinase, anisoylated plasminogen–streptokinase complex (Eminase), and recombinant tPA (Alteplase) and uPA (Saruplase). In addition, a number of uPA and tPA variants have been genetically engineered in efforts to produce more efficient and less inhibitable enzymes.

Structure

Crystal structures have been determined for ligand-complexed human plasminogen kringles 1 (PDB identifiers 1CEA, 1CEB, and 1PKR) and 4 (1PMK, 1PK4 and 2PK4).[28] Both kringles fold in a similar manner, but the lysine/fibrin binding site of K1 differs from that of K4 at its cationic centre. Kringles 2 and 3 are unique in that they are linked by a disulphide bridge, thus forming a 'supermodule' that has been characterized by NMR spectroscopy.[36] The structure of kringles derives from a number of conserved residues, including three disulphide bridges that usually connect cysteines 1–6, 2–4, and 3–5.[37]

References

1. Collen, D. and Lijnen, H. R. (1985). In *Thrombolysis, biological and therapeutic properties of new thrombolytic agents.* (ed. D. Collen, H. R. Lijnen, and M. Verstraete), pp. 1–14. Churchhill Livingstone, New York.
2. Bork, P., Downing, A. K., Kieffer, B., and Campbell, I. D. (1996). *Quart. Rev. Biophys.*, **29**, 119–67.
3. Castellino, F. J. and Powell, J. R. (1981). *Meth. Enzymol.*, **80**, 365–78.
4. Mori, K., Dwek, R. A., Downing, A. K., Opdenakker, G., and Rudd, P. M. (1995). *J. Biol. Chem.*, **270**, 3261–7.
5. Wallen, P. and Wiman, B. (1972). *Biochim. Biophys. Acta*, **257**, 122–34.
6. Moroz, L. A. (1981). *Blood*, **58**, 97–104.
7. Higazi, A. A., Barghouti, I. I., and Abu-Much, R. (1995). *J. Biol. Chem.*, **270**, 9472–7.
8. Clemmensen, I., Petersen, L. C., and Kluft, C. (1986). *Eur. J. Biochem.*, **156**, 327–33.
9. Salonen, E. M., Saksela, O., Vartio, T., Vaheri, A., Nielsen, L. S., and Zeuthen, J. (1985). *J. Biol. Chem.*, **260**, 12302–7.
10. Preissner, K. (1990). *Biochem. Biophys. Res. Commun.*, **168**, 966–71.
11. Silverstein, R. L., Leung, L. L. K., Harpel, P. C., and Nachman, R. L. (1984). *J. Clin. Invest.*, **74**, 1625–33.
12. Salonen, E. M., Zitting, A., and Vaheri, A. (1984). *FEBS Lett.*, **172**, 29–32.
13. Lijnen, H. R., Hoylaerts, M., and Collen, D. (1980). *J. Biol. Chem.*, **255**, 10214–22.
14. Redlitz, A. and Plow, E. F. (1995). *Baillière's Clin. Haematol.*, **8**, 313–27.
15. Stephens, R. W., Pollanen, J., Tapiovaara, H., Leung, K. C., Sim, P. S., Salonen, E. M., *et al.* (1989). *J. Cell Biol.*, **108**, 1987–95.
16. Knudsen, B. S., Silverstein, R. L., Leung, L. L., Harpel, P. C., and Nachman, R. L. (1986). *J. Biol. Chem.*, **261**, 10765–71.
17. O'Reilly, M. S., Holmgren, L., Shing, Y., Chen, C., Rosenthal, R. A., Moses, M., *et al.* (1994). *Cell*, **79**, 315–28.
18. Folkman, J. (1995). *Nature Med.*, **1**, 27–31.
19. Cao, Y., Ji, R. W., Davidson, D., Schaller, J., Marti, D., Sohndel, S., *et al.* (1996). *J. Biol. Chem.*, **271**, 29461–7.
20. Dong, Z., Kumar, R., Yang, X., and Fidler, I. J. (1997). *Cell*, **88**, 801–10.
21. Dano, K., Andreasen, P. A., Grondahl-Hansen, J., Kristensen, P., Nielsen, L. S., and Skriver, L. (1985). *Adv. Cancer Res.*, **44**, 139–266.
22. Liotta, L. A., Goldfarb, R. H., Brundage, R., Siegal, G. P., Terranova, V., and Garbisa, S. (1981). *Cancer Res.*, **41**, 4629–36.
23. Kwaan, H. C. (1992). *Cancer Metastasis Rev.* **11**, 291–311.
24. Van Nostrand, W. E., Wagner, S. L., Farrow, J. S., and Cunningham, D. D. (1990). *J. Biol. Chem.*, **265**, 9591–4.
25. Lijnen, H. R. and Collen, D. (1990). In *Serine proteases and their serpin inhibitors in the nervous system.* (ed. B. W. Festoff), pp. 9–20. Plenum Press, NY.
26. Hemker, H. C. (1983). *Handbook of synthetic substrates for the coagulation and fibrinolytic system.* Martinus Nijhoff, Boston.
27. Fisher, S. J. and Werb, Z. (1995). In *Extracellular matrix. A practical approach.* (ed. M. A. Haralson and J. R. Hassell), pp. 261–287. IRL Press, New York.
28. Petersen, T. E., Martzen, M. R., Ichinose, A., and Davie, E. W. (1990). *J. Biol. Chem.*, **265**, 6104–11.
29. Hirosawa, S., Nakamura, Y., Miura, O., Sumi, Y., and Aoki, N. (1988). *Proc. Natl Acad. Sci. USA*, **85**, 6836–40.
30. Juhan-Vague, I., Alessi, M. C., and Declerck, P. J. (1995). *Baillière's Clin. Haematol.*, **8**, 329–43.
31. Carmeliet, P. F. M. (1995). *Baillière's Clin. Haematol.*, **8**, 391–401.
32. Bugge, T. H., Flick, M. J., Daugherty, C. C., and Degen, J. L. (1995). *Genes Dev.*, **9**, 794–807.
33. Romer, J., Bugge, T. H., Pyke, C., Lund, L. R., Flick, M. J., Degen, J. L., and Dano, K. (1996). *Nature Med.*, **2**, 287–92.
34. Bugge, T. H., Kombrinck, K. W., Flick, M. J., Daugherty, C. C., Danton, M. J. S., and Degen, J. L. (1996). *Cell*, **87**, 709–19.
35. Mathews, I. I., Vanderhoff-Hanaver, P., Castellino, F. J., and Tulinsky, A. (1996). *Biochemistry*, **35**, 2567–76.

36. Sohndel, S., Hu, C. K., Marti, D., Affolter, M., Schaller, J., Llinas, M., and Rickli, E. E. (1996). *Biochemistry*, **35**, 2357–64.
37. Bork, P., Downing, A. K., Kieffer, B., and Campbell, I. D. (1996). *Quart. Rev. Biophys.*, **29**, 119–67.
38. Carmeliet, P., Moons, L., Lijnen, R., Baes, M., Lemaitre, V., Tipping, P., Drew, A., Eeckhout, Y., Shapiro, S., Lupo, F., and Collen, D. (1997). *Nature Genet.*, **17**, 439–44.
39. Bugge, T. H., Lund, L. R., Kombrinck, K. K., Nielsen, B. S., Holmback, K., Drew, A. F., Flick, M. J., Witte, D. P., Dano, K., and Degan, J. L. (1998). *Oncogene*, **16**, 3097–104.

Mark D. Sternlicht and Zena Werb
University of California, San Francisco,
Department of Anatomy, LR-208,
3rd and Parnassus Avenues, San Francisco,
CA 94143–0452, USA.

Plasminogen activator inhibitors

Type 1 and type 2 plasminogen activator inhibitors (PAI-1 and PAI-2) are the principle physiologic inhibitors of the serine proteinase plasminogen activators tPA and uPA. Protein C inhibitor (PAI-3) is an important inhibitor of uPA in urine. The tumour suppressor maspin specifically interacts with tPA in a unique bifunctional manner. Other broad-spectrum inhibitors such as protease nexin I, α_2-antiplasmin, α_1-proteinase inhibitor, C1-inhibitor and α_2-macroglobulin can inhibit tPA and uPA, albeit with efficiencies well below those of PAI-1 and PAI-2. Except for α_2-macroglobulin, each of these is a member of the serpin superfamily of serine proteinase inhibitors.

Plasminogen activator inhibitor type 1 (PAI-1)

The serine proteinase inhibitor PAI-1 was first purified from bovine endothelial cells and thus initially called 'endothelial-type' plasminogen activator inhibitor.[1] It is produced by a wide variety of other cell types and has a plasma concentration of around 20 ng/ml.[2] PAI-1 contains 379 amino acids, not including a 23 residue signal peptide. It contains three potential *N*-glycosylation sites and no cysteines for disulphide bonding and it has predicted and apparent molecular weights of 42 and 52 kDa, respectively.[2] It is a member of the serpin superfamily of both enzyme inhibitory and non-inhibitory proteins, and given the presence of Arg-346 at its reactive site P1 position, it can be further classified as an Arg-serpin. Because the target specificity of serpins is primarily dictated by their P1 residue, this is consistent with the specificity of PAI-1 for tPA and uPA, which cleave a single Arg–Val bond in plasminogen.

In general, inhibitory serpins exhibit a rapid initial association with their targets, followed by a slow second step that culminates in the formation of a stable enzyme–inhibitor complex.[3,4] This complex is refractory to boiling in SDS, but sensitive to nucleophiles. The exact mechanism of complex formation, however, remains controversial. PAI-1 apparently forms an acyl-enzyme linkage with its target enzymes, whereas α_2-antiplasmin and other serpins form a stable tetrahedral intermediate with theirs.[3,4] Either way, the reactive site loop is irreversibly cleaved at the P1–P1′ peptide bond during complex formation, rendering the cleaved inhibitor inactive against additional enzyme molecules. For this reason, serpins are considered 'suicide substrate inhibitors.' Because the reactive site loop is exposed at the protein surface, it is also susceptible to cleavage (and thus inactivation) by various non-target enzymes.[4] Individual serpins can exist as either intact active inhibitors, intact latent inhibitors, intact non-inhibitory substrates, cleaved inactive proteins, or inactive dimers, each with a distinct spatial conformation.[2–4]

Plasminogen activator inhibitor type 2 (PAI-2)

Like PAI-1, PAI-2 (or 'placental-type' plasminogen activator inhibitor) is an Arg-serpin in that it possesses Arg-380 as its P1 residue. Unlike PAI-1 however, PAI-2 has both an intracellular 47 kDa form and a secreted 60 kDa (glycosylated) form.[5] Both forms contain 415 amino acids and lack an N-terminal signal peptide. The three potential *N*-linked glycosylation sites are apparently each occupied in the extracellular form. Two common PAI-2 variants (A and B) also exist, with type A having Asn, Asn, and Ser and type B having Asp, Lys, and Cys at residues 120, 404, and 413, respectively.[5] The extra (sixth) Cys residue in type B PAI-2 may mediate its dimerization or the disulphide linkage of PAI-2 to ECM proteins such as vitronectin. PAI-2 shares considerable sequence homology with chicken ovalbumin (a non-inhibitory serpin) and other members of the 'ov-serpin' subfamily.[5] PAI-2 and ovalbumin also share an identical gene structure; and both lack an N-terminal signal sequence, but have an internal signal sequence that may regulate the intra- and extracellular localization of PAI-2.[5–7] Indeed, the internal signal sequence is relatively inefficient and can be made more efficient by increasing its hydrophobicity.[5] In addition, PAI-2 spontaneously forms polymers, possibly by the insertion of the reactive site loop of one molecule into the so-called A sheet of another.[8] PAI-2 is produced by monocyte/macrophages and placenta, as well as by fibroblast-like cells of various tissues, endothelial cells exposed to inflammatory stimuli, mesothelial cells, and microglia.[5] It has been detected in gingival crevicular fluid and saliva and is not normally detectable in plasma except during pregnancy.[5] Indeed, decreased PAI-2 levels during pregnancy are associated with intrauterine growth retardation, and elevated PAI-1 levels or PAI-1/PAI-2 ratios are predictive markers for preeclampsia.[5] PAI-2 has also been detected in some solid tumours, and plasma levels of PAI-2 are profoundly increased in certain myelomonocytic leukaemias. PAIs are also thought to play a role in acute and chronic inflammation, diseases of the skin, ophthalmic disease, arthritis and Alzheimer's disease.[5]

Other inhibitors

Protease nexin I (PN-I, glia-derived neurite promoting factor, glia-derived nexin, protease inhibitor 7) is a 51 kDa Arg-serpin that was first identified in fibroblasts and later found in astrocytes, skeletal and cardiac muscle cells, endothelial cells, and other cell types.[9,10] Its molecular weight decreases to 46 kDa, however, following its dissociation from a target enzyme. The name 'protease nexin' derives from its ability to link with thrombin and plasminogen activators, and thereby mediate their binding, uptake, and degradation by cells.[9] PN-I is a potent inhibitor of thrombin, trypsin, uPA, and plasmin. It also has a heparin-binding site that has been localized to its so-called D and A helices, and heparin binding has been shown to accelerate its rate of thrombin inhibition 500-fold.[11] In addition, binding of PN-I to the surface of cells and to ECM potentiates its inhibitory activity against thrombin, but blocks its inhibition of uPA and plasmin.[12,13] Specifically, ECM and cell-surface heparan sulphate proteoglycan are most responsible for the accelerated inhibition of thrombin by PN-I, yet they have no effect on the rate of uPA and plasmin inhibition.[12] On the other hand, type IV collagen slows the inhibition of uPA and plasmin by PN-I, but has no effect on thrombin inhibition.[12] Other ECM proteins (type I collagen, vitronectin, fibronectin, and denatured type IV collagen) have no effect on PN-I inhibitory activity. Thus because much of the available PN-I is cell- and ECM-associated, its principal role *in vivo* may be to regulate thrombin rather than uPA or plasmin. In addition, the tight association of PN-I with ECM molecules that regulate its activity further implicates it as a mediator of ECM degradation. Furthermore, by inhibiting serine proteinases, PN-I can modulate the mitogenic and chemotactic effects of thrombin on certain cell types, promote neurite outgrowth in neuronal cells, and apparently protect motoneurones against programmed cell death.[10,13]

Protein C inhibitor (PAI-3, urinary urokinase inhibitor) is a 54 kDa Arg-serpin that primarily inhibits the potent anticoagulant enzyme protein C.[14] It can also inhibit uPA, tPA, thrombin, tissue and plasma kallikrein, and factors Xa and XIa.[14,15] Furthermore, its inhibitory activity is enhanced by heparin and other glycosaminoglycans.[16] Human PAI-3 has a 19 residue signal peptide and the mature inhibitor has 387 amino acids, with Arg-354–Ser-355 as its P1–P1′ residues. It also contains a single Cys residue, three potential *N*- and two potential *O*-glycosylation sites, and a heparin-binding domain.[14] It is found in plasma, urine, and seminal plasma at around 5.3, 0.75, and 200 μg/ml, respectively.[14] Thus despite its relatively slow activity against uPA and tPA, its concentrations in these fluids are orders of magnitude greater than those of PAI-1 and PAI-2. Furthermore, its abundance in seminal plasma and its absence in patients with seminal vesicle dysfunction are suggestive of its importance in reproduction.

Maspin (mammary serpin) is a 42 kDa Arg-serpin with tumour suppressor activity; i.e., its expression is repressed during cancer progression, it inhibits tumour cell motility and invasion *in vitro*, and it inhibits tumour growth and metastasis *in vivo*.[41–45] Maspin is present both in the cell cytoplasm, where it partitions into secretory vesicles, and at the cell surface.[46] Exogenous recombinant maspin also associates with the cell membrane.[43] Maspin transcripts are detected in human mammary epithelium, prostate, testis, and small intestine,[46] as well as in rat mammary gland, vagina, bladder, thymus, small intestine, skin, prostate, seminal vesicles, and thyroid gland.[47] Maspin is also particularly abundant in mammary and salivary gland myoepithelial cells and their tumours where it contributes to the anti-invasive nature of these cells and to the low-grade behaviour of their tumours.[41,45,48] It was originally thought that maspin acted as a ligand-binding rather than proteinase-inhibitory serpin based on structural features of its reactive site loop, its failure to undergo a stressed to relaxed transition, and initial failures at identifying a target proteinase.[49–51] More recently, however, maspin was found to interact specifically with single-chain (sc) tPA in a unique bifunctional manner.[52] These data suggest that two separate maspin domains interact with the catalytic and activating domains of sc-tPA so that maspin may help regulate plasminogen activation at the cell surface.

Purification

PAI-1 has been purified from the culture media of bovine endothelial cells and human melanoma cells using various affinity and gel filtration methods[1,17] and from human fibrosarcoma cells using a one-step immunoaffinity method.[18] Human PAI-2 has been purified from human placenta by an eight-step procedure[19] and from phorbol ester-treated U937 monocytic leukaemia cells by preparative IEF plus Affi-Gel blue–Sepharose affinity chromatography.[20] PAI-2 has also been purified by immunoaffinity chromatography,[21,22] and recombinant PAI-2 has been obtained from bacteria, yeast, and CHO cells.[5,8] Human PAI-3 can be purified from plasma and urine by various multi-step methods.[14,15,23] PN-I can be purified from fibroblast-conditioned media by heparin–Sepharose and octyl–agarose[24] or dextran sulphate– and DEAE–Sepharose chromatographies.[25] Recombinant human maspin has been purified from *E. coli*, Sf9 insect cells, and yeast.[42] Recombinant mouse maspin has been obtained from *E. coli*.[44]

Activities

Active inhibitory serpins act as 'pseudo-substrates' of their target enzymes. Accordingly, they form equimolar complexes with their targets via rapid association followed by slow formation of the final stable complex. The overall efficiency of such inhibition can be expressed as a second-order rate constant (k). Thus PAI-1 is a more efficient inhibitor of uPA and tPA than is PAI-2, with k values for inhibition of uPA, tc-tPA, and sc-tPA that are 8.5-, 85-, and 1200-fold greater than those of PAI-2 (Table 1).[26] Still, PAI-2 rapidly inhibits both high and low molecular weight uPA with a k of 1–2 $\times 10^6$ M^{-1} s^{-1}.[5,26] Its k value for inhibition of uPA is 17- and 435-fold higher than for tc- and sc-tPA, respectively (Table 1). Thus PAI-2 is only a fair inhibitor of tc-tPA and a poor inhibitor of sc-tPA. Intracellular non-glycosylated PAI-2 and extracellular glycosylated PAI-2 each exhibit similar inhibitory activity towards uPA and tPA.[8] Although PAI-2 can inhibit other serine proteinases such as plasmin and acrosin, it is unlikely to play an important role in their regulation.[5] For example, PAI-2 and PAI-3 both inhibit acrosin with equal efficiency ($k \approx 3.5 \times 10^4$ M^{-1} s^{-1}), but the concentration of PAI-3 in seminal plasma (5.3 μM) is almost 27 000 times greater than that of PAI-2.[5] The lesser k values for PAI-3 and PN-I are also listed in Table 1. Heparin accelerates the inhibitory activity of PN-I against thrombin (from k = 6.0 $\times 10^5$ to 1.2 $\times 10^8$ M^{-1} s^{-1}), but has no effect on the inhibition of other enzymes.[24] On the other hand, heparin increases the k value of PAI-3 from 8 $\times 10^3$ to 9 $\times 10^4$ M^{-1} s^{-1} for the inhibition of uPA.[23]

PAIs are conventionally assayed by measuring their ability to inhibit plasminogen activators using standard chromogenic or fluorogenic substrates.[14,24,27] A reverse zymographic method for detecting PAI-1 has also been developed which utilizes fibrin–agar indicator gels containing plasminogen and uPA.[28] PAI-2 is not detected by this method because, unlike PAI-1, it is not reactivated following SDS denaturation.[22,27] Thus a modified non-denaturing method of reverse zymography has been employed to detect PAI-2.[22] Interestingly, PA–PAI-1 complexes, but not PA–PAI-2 complexes, are detectable as proteolytic bands by direct fibrin and casein zymography.[5] Electrophoretic detection of enzyme–inhibitor complexes can also be obtained using radiolabelled target enzyme.[9]

PAIs also have activities that are independent of their ability to inhibit uPA and tPA. For example, they play a role in the cellular uptake of uPA and its cell surface receptor (uPAR). Indeed, uninhibited, uPAR-bound uPA is not internalized, whereas uPAR-bound uPA/serpin complexes are.[29] PAI-1 also binds to the N-terminal somatomedin B domain of both closed and extended vitronectin.[29] In so doing, PAI-1 competes with uPAR for binding to the same site on vitronectin and with integrins for binding to the nearby RGD sequence.[29] Such binding also stabilizes the active conformation of PAI-1 against conversion to latent PAI-1. Because PAI-1 binds to ECM molecules such as vitronectin, it may form a link between the ECM and cell surface uPA/uPAR complexes, which in turn can be linked to the cytoskeleton via interactions with integrins. Indeed, immobilized plasminogen promotes myogenic cell adhesion and spreading in a dose-dependent manner that is inhibited by removal of GPI-anchored uPAR from the cell surface, and by antibodies against PAI-1 or $\alpha v\beta 3$ integrin.[30] Other studies have shown that PAI-1 can inhibit the migration of certain cell types by blocking $\alpha v\beta 3$ integrin binding to vitronectin.[29,31] Mutant PAI-1 that is unable to inhibit plasminogen activators can still inhibit migration, indicating that this inhibition is independent of its antiproteolytic activity.[31] In addition, PAI-1 inhibits the migration of bovine aortic endothelial cells, but stimulates the migration of bovine aortic smooth muscle cells.[29] Thus PAI-1 can either promote migration (eg., by protecting vitronectin and stimulating the internalization and recycling of uPAR) or inhibit it (e.g. by blocking adhesion to vitronectin and by inhibiting plasmin production).[29] Unlike PAI-1, PAI-2 does not bind to vitronectin in purified systems, although this does not necessarily preclude such binding *in vivo*.[8] However, intracellular PAI-2 may modulate apoptosis.[29]

As already noted, maspin exerts tumour-suppressive activity, yet its molecular target(s) remained elusive for some time. Recent biochemical studies, however, indicate that its reactive site loop binds specifically to sc-tPA; that it forms a stable detergent-resistant complex with sc-tPA; that it activates free sc-tPA but inhibits poly-D-lysine-bound (preactivated) sc-tPA; that it competitively inhibits fibrinogen/gelatin-preactivated sc-tPA at low concentra-

Table 1 Second-order rate constants (k; M^{-1} s^{-1}) for inhibition of plasminogen activators and plasmin by four plasminogen activator inhibitors[5, 14, 23, 24, 26]

Enzyme	k (M^{-1} s^{-1})			
	PAI-1	PAI-2	PAI-3	PN-I
uPA	1.7×10^7	2.0×10^6	8×10^3	1.5×10^5
tc-tPA	1.7×10^7	1.2×10^5	0.8×10^3	3.0×10^4
sc-tPA	5.5×10^6	4.6×10^3		1.5×10^3
Plasmin	1×10^2	9×10^4		1.3×10^5

tions but acts as a stimulator at higher concentrations; that its 38 kDa C-terminally truncated form, a form which may accumulate during plasminogen activation, further stimulates sc-tPA; and that it only inhibits fibrinogen/gelatin-associated sc-tPA and not uPA, plasmin, trypsin, chymotrypsin or elastase.[52] These results suggest that maspin is a bifunctional serpin with an N-terminal domain that interacts with the activation site of sc-tPA and a reactive site loop region that interacts with the catalytic site of sc-tPA when the sc-tPA activation domain is already bound. Maspin inhibits plasminogen activation by sc-tPA with an efficacy comparable to that of PAI-2 and less than that of PAI-1; however, its inhibitory activity would be expected to increase if the sc-tPA activation domain or the N-terminal maspin domain were blocked. Although other intracellular and extracellular targets may still remain undiscovered, maspin may exert much of its tumour-suppressive activity by inhibiting sc-tPA, which can otherwise stimulate tumour cell motility[52] and perhaps other aspects of tumour progression.

Antibodies

Monoclonal antibodies against PAI-1 that neutralize, augment, or have no effect on its inhibitory activity are available from American Diagnostica. Monoclonal anti-PAI-1 antibodies are also available from Alexis Corp., Biogenesis, Celsus Laboratories, Cortex Biochem, Technoclone, and York Biologicals. Rabbit and goat polyclonal anti-PAI-1 antibodies are also readily available. Goat anti-human PAI-2 and a mouse monoclonal antibody against human PAI-2 are available from American Diagnostica, as is a murine monoclonal antibody against human PAI-3. Other polyclonal and monoclonal antibodies against PAI-3 and uPA–PAI-3 complexes have also been developed.[23] Mouse monoclonal and rabbit polyclonal anti-human maspin antibodies are available from Pharmingen and Transduction Laboratories.

Genes

The human PAI-1 gene (*PLANH1)* is located on chromosome 7q22.1–3, spans 12.3 kbp, and has nine exons.[33] Two mRNA transcripts of 2.4 and 3.2 kb probably arise from alternative polyadenylation.[2] PAI-1 cDNA sequences are available for humans (M16006, M18082, X04429, X12701, X04744), mice (M33960, X16490), rats (M24067, J05206), pigs (Y11347), cows (X16383), and mink (X58541). Human genomic sequences are also available (J03764, M17121, X06692, J03836, M33136, X13323, X13338–45). An important feature of the PAI-1 gene promoter is its exquisite sensitivity to upregulation by TGF-β, thus forming the basis for various TGF-β bioassays.[34]

The human PAI-2 gene (*PLANH2)* is located on chromosome 18q21.3 within a cluster containing five other ov-serpins.[53] It spans 16.5 kbp and has eight exons and a number of promoter regulatory elements, including two essential AP-1 sites and a CRE-like element.[5–7] PAI-2 cDNA sequences are available for humans (J03603, J02685, Y00630), mice (X16490), and rats (X64563). Human genomic sequences are also available (J04752, M24651–7, L19065–6, M22469, M23092, J04606).

The human PAI-3 gene (*PCI)* is located on chromosome 14q32.1, spans 11.5 kbp, and has five exons.[34] Furthermore, its gene structure is similar to that of the α_1-antitrypsin and α_1-antichymotrypsin genes which are also found on chromosome 14. Genomic and cDNA sequences are available for PAI-3 in humans (M68516, M64880–4, U35464, J02639, S69366) and mice (U67877–8).

The human PN-I gene (*PI7)* is located on chromosome 2q33–q35 and on syntenic regions of mouse chromosome 1 and sheep chromosome 2.[35] Genomic and cDNA sequences are available for PN-I in humans (U33453, A03911, E01330), mice (X70296, X70946), and rats (X71791, X71010, A03913, E01331).

The human maspin gene (PI5; protease inhibitor 5) is the serpin gene closest to the centromere in the serpin gene cluster at 18q21.3.[53] Major (3.0 kb) and minor (1.6 kb) transcripts have been described.[41,45,48] An Ets element and an AP-1 site in the maspin gene promoter cooperate to upregulate expression in normal but not malignant mammary and prostrate epithelial cells, and a hormone responsive element recognized by androgen receptor represses expression in both normal and malignant prostate cells.[54,55] Maspin expression is also upregulated by gamma linolenic acid.[56] Complete maspin cDNAs are available for humans (U04313), mice (U54705), and rats (U58857).

Mutant phenotype/disease states

PAI-1 null mice exhibit normal fertility and produce normal offspring.[36] Whereas PAI-1 deficiency in humans causes delayed rebleeding, this is not observed in PAI-1 deficient mice. As compared to wild-type controls, however, PAI-1 null mice lyse pulmonary clots at a significantly faster rate and are more resistant to endotoxin-induced thrombosis. Double-mutant mice lacking PAI-1 and either uPA or tPA exhibit phenotypes like those of mice lacking either tPA or uPA alone and lack apparent gross, microscopic or hemostatic abnormalities.[37] On the other hand, combined uPA/tPA/PAI-1 null mice show significantly fewer, smaller, and less calcified hepatic fibrin deposits than do uPA/tPA null mice, which exhibit profound fibrin deposition and calcification. The added inactivation of PAI-1 in the double and triple mutants enhances the lysis of exogenous labelled fibrin in uPA null mice, but fails to restore the fibrinolytic capacity of tPA- or uPA/tPA-null mice.

Thrombotic disorders have been noted in individuals with elevated PAI-1 levels. Likewise, transgenic PAI-1-overexpressing mice develop venous but not arterial occlusions.[38] In addition, transgenic neonates that overexpress PAI-1 in their skin exhibit hyperkeratosis and delayed hair growth, indicating that plasminogen activators play a role in desquamation and follicular neogenesis.[39]

Recent evidence indicates that PAI-1 is required for tumour angiogenesis in a mouse model in which malignant keratinocytes are placed on a collagen gel and covered by silicone chambers on the skin (J. M. Foidart; personal communication). Whereas angiogenesis normally proceeds into the gel allowing for tumour growth, the angiogenic vessels are unable to penetrate the gel in PAI-1 null mice and tumours fail to grow unless angiogenesis is rescued by intravenous injection of a PAI-1 retrovirus.

Structure

As a family, serpins exhibit a highly conserved tertiary structure composed of three β sheets (A–C) and nine α helices (A–I).[32] Unlike smaller proteinase inhibitors with a fixed reactive centre that tightly fits within the active site of their target enzymes, the serpins have a reactive centre loop that is mobile and able to adopt various conformations. PAI-1 has at least four distinct structural forms: an active form, a latent form, an inactivated form, and a non-inhibitory substrate form.[2] Models of these have been deduced from the solved spatial structures of uncleaved serpins (e.g. ovalbumin, α_1-antichymotrypsin, and antithrombin); several inactive serpins with cleaved reactive site peptide bonds; and latent, non-functional PAI-1 itself.[3,4,40] Thus the active (inhibitory) form of PAI-1 is thought to have a portion of its reactive site loop inserted within the so-called central β-sheet A, with the reactive site peptide bond (P1–P1′) remaining exposed.[4] Active PAI-1 spontaneously converts to a latent (non-inhibitory) form with a $t_{1/2}$ of 4 h in plasma.[2] In this case, the reactive site remains intact, but it is entirely inserted into β-sheet A, causing the displacement of a strand from β-sheet C.[3,4] Thus the entire reactive site loop is buried and inaccessible to the target enzymes as well as other inactivating enzymes. The inhibitory activity of latent PAI-1 can be restored *in vitro* by protein denaturation and renaturation, and limited reactivation may also occur *in vivo* on negatively charged membrane phospholipids.[2] The release of PAI-1 from inhibited PA/PAI-1 complexes yields the inactive (cleaved) PAI-1 conformation, wherein the entire cleaved reactive site strand (P1 to P15) is thought to be inserted into the centre of β-sheet A.[3,4] In this conformation the free P1 and P1′ termini are separated by 70 Å on opposite sides of the molecule and are thus unable to rejoin. A non-inhibitory substrate form of PAI-1 has also been described which can be cleaved at the P1–P1′ peptide bond, but does not form a stable complex with tPA.[2] Similar properties exist for engineered and pathologic mutant serpins as well as for natural non-inhibitory substrate serpins such as ovalbumin. The intact form of the latter serpin has an entirely exposed α-helical reactive site segment, and upon cleavage to form plakalbumin, the reactive site strand is not inserted into β-sheet A.[4,40] The structure of an inactive antithrombin dimer has also been obtained, and it has been noted that PAI-2 spontaneously undergoes similar 'loop–sheet' polymerization.[8,40] Theoretical models have also been obtained for PAI-3 (PDB identifiers 1PAI and 2PAI) and maspin.[51]

References

1. van Mourik, J. A., Lawrence, D. A., and Loskutoff, D. J. (1984). *J. Biol. Chem.*, **259**, 14914–21.
2. Rijken, D. C. (1995). *Baillière's Clin. Haematol.*, **8**, 291–312.
3. Wright, H. T. (1996). *Bioessays*, **18**, 453–64.
4. Potempa, J., Korzus, E., and Travis, J. (1994). *J. Biol. Chem.*, **269**, 15957–60.
5. Kruithof, E. K. O., Baker, M. S., and Bunn, C. L. (1995). *Blood*, **86**, 4007–24.
6. Ye, R. D., Ahern, S. M., Le Beau, M. M., Lebo, R. V., and Sadler, J. E. (1989). *J. Biol. Chem.*, **264**, 5495–502.
7. Samia, J. A., Alexander, S. J., Horton, K. W., Auron, P. E., Byers, M. G., Shows, T. B., and Webb, A. C. (1990). *Genomics*, **6**, 159–67.
8. Mikus, P., Urano, T., Liljestrom, P., and Ny, T. (1993). *Eur. J. Biochem.*, **218**, 1071–82.
9. Baker, J. B., Low, D. A., Simmer, R. L., and Cunningham, D. D. (1980). *Cell*, **21**, 37–45.
10. Houenou, L. J., Turner, P. L., Li, L., Oppenheim, R. W., and Festoff, B. W. (1995). *Proc. Natl Acad. Sci. USA*, **92**, 895–9.
11. Evans, D. L. L., Christy, P. B., and Carrell, R. W. (1990). In *Serine proteinases and their serpin inhibitors in the nervous system*. (ed. B. W. Festoff), pp. 69–78. Plenum Press, New York.
12. Donovan, F. M., Vaughan, P. J., and Cunningham, D. D. (1994). *J. Biol. Chem.*, **269**, 17199–205.
13. Guttridge, D. C., Lau, A. L., and Cunningham, D. D. (1993). *J. Biol. Chem.*, **268**, 18966–74.
14. Suzuki, K. (1993). *Meth. Enzymol.*, **222**, 385–99.
15. Ecke, S., Geiger, M., Resch, I., Jerabek, I., Sting, L., Maier, M., and Binder, B. R. (1992). *J. Biol. Chem.*, **267**, 7048–52.
16. Geiger, M., Priglinger, U., Griffin, J. H., and Binder, B. R. (1991). *J. Biol. Chem.*, **266**, 11851–7.
17. Wagner, O. F. and Binder, B. R. (1986). *J. Biol. Chem.*, **261**, 14474–81.
18. Nielsen, L. S., Andreasen, P. A., Grondahl-Hansen, J., Huang, J. Y., Kristensen, P., and Dano, K. (1986). *Thromb. Haemost.*, **55**, 206–12.
19. Wun, T. C. and Reich, E. (1987). *J. Biol. Chem.*, **262**, 3646–53.
20. Kruithof, E. K. O., Vassalli, J. D., Schleuning, W. D., Mattaliano, R. J., and Bachmann, F. (1986). *J. Biol. Chem.*, **261**, 11207–13.
21. Astedt, B., Lecander, I., Brodin, T., Lundblad, A., and Low, K. (1982). *Thromb. Haemost.*, **53**, 122–5.
22. Cajot, J. F., Kruithof, E. K. O., Schleuning, W. D., Sordat, B., and Bachmann, F. (1986). *Int. J. Cancer*, **38**, 719–27.
23. Stump, D. C., Thienpont, M., and Collen, D. (1986). *J. Biol. Chem.*, **261**, 12759–66.
24. Scott, R. W., Bergman, B. L., Bajpai, A., Hersh, R. T., Rodriguez, H., Jones, B. N., *et al.* (1985). *J. Biol. Chem.*, **260**, 7029–34.
25. Farrell, D. H., van Nostrand, W. E., and Cunningham, D. D. (1986). *Biochem. J.*, **237**, 907–12.
26. Thorsen, S., Philips, M., Selmer, J., Lecander, I., and Astedt, B. (1988). *Eur. J. Biochem.*, **175**, 33–9.
27. Andreasen, P. A., Georg, B., Lund, L. R., Riccio, A., and Stacey, S. N. (1990). *Mol. Cell. Endocrinology*, **68**, 1–19.
28. Erickson, L. A., Lawrence, D. A., and Loskutoff, D. J. (1984). *Anal. Biochem.*, **137**, 454–63.
29. Andreasen, P. A., Kjoller, L., Christensen, L., and Duffy, M. J. (1997). *Int. J. Cancer*, **72**, 1–22.

30. Planus, E., Barlovatz-Meimon, G., Rogers, R. A., Bonavaud, S., Ingber, D. E., and Wang, N. (1997). *J. Cell Sci.*, **110**, 1091–8.
31. Kjoller, L., Kanse, S. M., Kirkegard, T., Rodenburg, K. W., Ronne, E., Goodman, S. L., *et al.* (1997). *Exp. Cell Res.*, **232**, 420–9.
32. Loskutoff, D. J., Linders, M., Keijer, J., Veerman, H., van Heerikhuizen, H., and Pannekoek, H. (1987). *Biochemistry*, **26**, 3763–8.
33. van Waarde, M. A., van Assen, A. J., Kampinga, H. H., Konings, A. W., and Vujaskovic, Z. (1997). *Anal. Biochem.*, **247**, 45–51.
34. Meijers, J. C. and Chung, D. W. (1991). *J. Biol. Chem.*, **266**, 15028–34.
35. Carter, R. E., Cerosaletti, K. M., Burkin, D. J., Fournier, R. E., Jones, C., Greenberg, B. D., *et al.* (1995). *Genomics*, **27**, 196–9.
36. Carmeliet, P., Bouche, A., De Clercq, C., Janssen, S., Pollefeyt, S., Wyns, S., *et al.* (1995). *Ann. NY Acad. Sci.*, **748**, 367–82.
37. Lijnen, H. R., Moons, L., Beelen, V., Carmeliet, P., and Collen, D. (1995). *Thromb. Haemost.*, **74**, 1126–31.
38. Erickson, L. A., Fici, G. J., Lund, J. E., Boyle, T. P., Polites, H. G., and Marotti, K. R. (1990). **346**, 74–6.
39. Lyons-Giordano, B. and Lazarus, G. S. (1995). *Dev. Biol.*, **170**, 289–98.
40. Carrell, R. W. and Stein, P. E. (1996). *Biol. Chem. Hoppe-Seyler*, **377**, 1–17.
41. Zou, Z., Anisowicz, A., Hendrix, M. J., Thor, A., Neveu, M., Sheng, S., Rafidi, K., Seftor, E., and Sager, R. (1994). *Science*, **263**, 526–9.
42. Sheng, S., Pemberton, P. A., and Sager, R. (1994). *J. Biol. Chem.*, **269**, 30988–93.
43. Sheng, S., Carey. J., Seftor, E. A., Dias, L., Hendrix, M. J., and Sager, R. (1996). *Proc. Natl Acad. Sci. USA*, **93**, 11669–74.
44. Zhang, M., Sheng, S., Maass, N., and Sager, R. (1997). *Mol. Med.*, **3**, 49–59.
45. Sternlicht, M. D., Kedeshian, P., Shao, Z. M., Safarians, S., and Barsky, S. H. (1997). *Clin. Cancer Res.*, **3**, 1949–58.
46. Pemberton, P. A., Tipton, A. R., Pavloff, N., Smith, J., Erickson, J. R., Mouchabeck, Z. M., and Kiefer, M. C. (1997). *J. Histochem. Cytochem.*, **45**, 1697–706.
47. Umekita, Y., Hiipakka, R. A., and Liao, S. (1997). *Cancer Lett.*, **113**, 87–93.
48. Sternlicht, M. D., Safarians, S., Rivera, S. P., and Barsky, S. H. (1996). *Lab. Invest.*, **74**, 781–96.
49. Hopkins, P. C. R. and Whisstock, J. (1994). *Science*, **265**, 1893–4.
50. Pemberton, P. A., Wong, D. T., Gibson, H. L., Kiefer, M. C., Fitzpatrick, P. A., Sager, R., and Barr, P. J. (1995). *J. Biol. Chem.*, **270**, 15832–7.
51. Fitzpatrick, P. A., Wong, D. T., Barr, P. T., and Pemberton, P. A. (1996). *Protein Eng.*, **9**, 585–9.
52. Sheng, S., Truong, B., Fredrickson, D., Wu, R., Pardee, A. B., and Sager, R. (1998). *Proc. Natl Acad. Sci. USA*, **95**, 499–504.
53. Bartuski, A. J., Kamachi, Y., Schick, C., Overhauser, J., and Silverman, G. A. (1997). *Genomics*, **43**, 321–8.
54. Zhang, M., Magit, D., and Sager, R. (1997). *Proc. Natl Acad. Sci. USA*, **94**, 5673–8.
55. Zhang, M., Maass, N., Magit, D., and Sager, R. (1997). *Cell Growth Differ.*, **8**, 179–86.
56. Jiang, W. G., Hiscox, S., Horrobin, D. F., Bryce, R. P., and Mansel, R. E. (1997). *Biochem. Biophys. Res. Commun.*, **237**, 639–44.

■ *Mark D. Sternlicht and Zena Werb*
University of California, San Francisco,
Department of Anatomy, LR-208,
3rd and Parnassus Avenues, San Francisco,
CA 94143–0452, USA.

Tissue-type plasminogen activator (tPA)

Tissue-type plasminogen activator (tPA) is a serine proteinase with high specificity for plasminogen, which it converts to plasmin by cleaving a single Arg–Val peptide bond. Like plasminogen, tPA binds to fibrin and fibrin fragments, thus enabling the colocalization of the enzyme activator, enzyme, and substrate. In addition, fibrin promotes plasminogen activation by increasing both the affinity of tPA for plasminogen and its catalytic rate. tPA is an important mediator of intravascular thrombolysis and of apoptosis in the CNS.

Protein properties

Tissue-type plasminogen activator (tPA, EC 3.4.21.68) is a highly specific serine proteinase which cleaves plasminogen at a single site to form plasmin. It is secreted into the circulation by vascular endothelial cells and rapidly cleared by hepatocytes ($t_{1/2} \approx 5$ min), thus yielding normal plasma levels of 5–10 ng/ml.[1] It is also produced by other cell types, including melanoma and neuroblastoma cells in culture. In addition, tPA can be released rapidly into the bloodstream in response to various stimuli that are known to activate the sympathoadrenal system.[2] Accordingly, it has been shown that the catecholamine storage vesicles of chromaffin cells act as a depot for the rapid, regulated release of tPA.[2] Nucleic acid sequences reveal leader, signal peptide, and propeptide sequences N-terminal to the mature tPA protein and although Ser1 is generally refered to as the N-terminal residue, mature tPA probably begins at Gly-3 (Fig. 1).[1] Mature tPA has 527 amino acids, a predicted (carbohydrate-free) molecular mass of 59 kDa, and an apparent molecular mass of 70 kDa (with carbohydrates). It has a fibronectin finger-like domain (Cys-6–Cys-43), an EGF-like domain (His-44–Thr-91), two kringle domains (Cys-92–Cys-173 and Cys-180–Cys-261), and a serine proteinase domain (Ile-

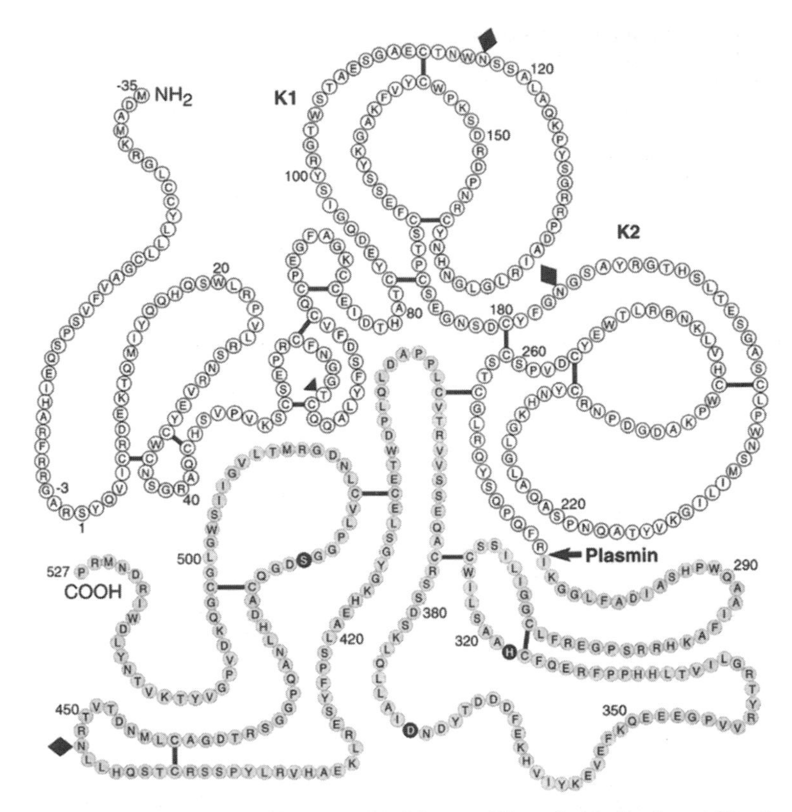

Figure 1. Primary and secondary structure of human tPA. Mature tPA probably begins at Gly-3, although conventional amino acid numbering starts at Ser-1. Plasmin can cleave sc-tPA at the Arg-275–Ile-276 peptide bond to yield tc-tPA, but this is not required for tPA activation. Also indicated are the two kringles (K1 and K2), the three *N*-glycosylation sites (diamonds), the fucosylation site at Thr-61 (triangle), and the catalytic His, Asp, and Ser residues (filled circles). (Modified from Rijken.[1])

276–Pro-527). Of the two kringles, only the latter kringle (2) binds lysine. tPA is secreted as a single-chain molecule (sc-tPA) and is readily cleaved by plasmin or other enzymes at its Arg-275–Ile-276 peptide bond to yield two-chain (tc) tPA, with the 40 kDa N-terminal (A) and 30 kDa C-terminal (B) chains remaining linked by a single disulphide bridge. Nevertheless, both sc- and tc-uPA can activate plasminogen, and if fibrin is present, they can do so with equal ability. Three potential *N*-glycosylation sites are present at Asn-117 on kringle 1, Asn-184 on kringle 2, and Asn-448 in the proteinase domain. Two tPA glycoforms are usually present in nearly equal amounts, with type I tPA being glycosylated at all three sites and type II tPA lacking carbohydrate at Asn-84.[1,3] Both variants contain an *O*-linked fucose at Thr-61 of the growth factor domain.[1,3] In addition, a minor 120 kDa variant (type D tPA) has been identified, and determined to be a tPA dimer with particularly high affinity for lysine and increased fibrinolytic activity.[3] In addition to binding fibrin, tPA can also bind to heparin,[4] fibronectin, laminin,[5] and Thy-1.[6] The latter molecule is a phosphatidylinositol-linked cell surface antigen that is abundantly expressed on thymocytes and mature brain cells, and its binding of tPA may leave the tPA active site exposed.[6] Maspin, a serpin that can both stimulate and inhibit sc-tPA, is located at the cell surface where it may thus regulate sc-tPA activity.[22] A number of other receptors have been implicated in tPA uptake and clearance, including a mannose receptor on liver endothelial cells, a calcium-dependent receptor on hepatocytes, and LDL receptor-related protein/α_2-macroglobulin receptor on hepatocytes.[1] The latter receptor is able to recognize both free tPA and tPA/PAI-1 complexes. In addition, tPA binding can occur via cell surface receptors that also recognize plasminogen and other kringle-containing proteins.[7] Such receptors include annexin II on endothelial cells and possibly amphoterin on neuronal cells.[7,8]

Purification

tPA can be purified from tissues using various multi-step methods, but these yield low amounts of pure enzyme.[9] Excellent yields, however, are obtained from the serum-free conditioned medium of the Bowes human melanoma cell line using similar but simplified methods.[9] Purification of tPA from this same cell line can also be accomplished by immunoaffinity chromatography.[10] The type I, II, and D variants of tPA can be separated by lysine–Sepharose affinity chromatography.[3] Recombinant tPA (Alteplase) is also available for therapeutic use.

Activities

tPA is a trypsin-like serine proteinase. It cleaves peptide bonds with Arg or (to a lesser extent) Lys in the P1 position and, as such, it specifically cleaves the Arg-560–Val-561 peptide bond of plasminogen to yield plasmin. The direct amidolytic activity of tPA can be assayed using chromogenic substrates such as S-2288 and indirect amidolytic activity (either stimulated by fibrin or not) can be assayed in the presence of plasminogen using a plasmin substrate such as D-Val–Leu–Lys–pNA (S-2251).[3] Direct plasminogen–casein zymography or antibody-based methods can also be used to assay tPA. Ser-478 of the catalytic triad is irreversibly blocked by PMSF and DFP and the catalytic His-322 residue is methylated upon irreversible inhibition by D-Phe–Pro–Arg chloromethyl ketone.[1] Both insoluble fibrin and soluble fibrin breakdown products promote the activation of plasminogen by tPA by dramatically increasing the affinity ($1/k_M$) of tPA for plasminogen and mildly increasing the catalytic rate constant (k_{cat}) such that the overall efficiency of activation (k_{cat}/k_M) is increased 1000-fold.[11] This stimulation of tPA by fibrin relies on the lysine binding kringle 2 and the finger-like domain of chain A, the latter being unaffected by lysine or lysine analogues.[1] Unlike other serine proteinases which have zymogenic single-chain forms, sc-tPA is not inactive, but has the same plasminogen activating activity as tc-tPA when on fibrin, and up to six times less activity than tc-tPA when fibrin is absent.[1] A zymogenic tPA mutant has, however, been engineered with a mechanism for maintaining latency that is similar to that of chymotrypsinogen.[12] All three tPA variants (I, II and D) have similar direct amidolytic activities, but the indirect (plasminogen-mediated) amidolytic activity of type II tPA is two-fold greater than that of type I tPA when in the presence of fibrin, and type D tPA is several-fold more active than tPA types I and II whether or not fibrin is present.[3]

Antibodies

Murine monoclonal antibodies against several epitopes of tPA are available from American Diagnostica and Technoclone GmbH, including antibodies that inhibit plasminogen activation or fibrin stimulation, antibodies that stimulate plaminogen activation, those that recognize the kringles or the finger and EGF-like domains, and others that can distinguish sc- and tc-tPA. Still other monoclonal anti-tPA antibodies are available from Affinity Biologicals, Alexis, Biodesign International, Biogenesis, Celsus Laboratories, Coulter Cytometry, Enzyme Research Labs, Fitzgerald Industries, Immunotech, Kamiya Biomedical, Paesel + Lorei, and York Biologicals. Several polyclonal antibodies are also available.

Genes

The human tPA gene (*PLAT*) is located on chromosome 8p12, spans 33 kbp, and has 14 exons.[13,14] Complete cDNA sequences for tPA are available for humans (M15518, M18182, K03021, X07393, U63828), mice (J03520), rats (M23697), and cows (X85800). Genomic sequences are available for humans (J00278, S77144, M11888–90, Z48484, S78429, S83170), mice (M26065), and rats (J05226, S73569).

Mutant phenotype/disease states

Genetic deficiencies have not been described for either tPA or uPA in humans. Gene-targeted mice that lack tPA are viable, but such mice lyse exogenous clots at a significantly slower rate than do wild-type mice and they exhibit an increased incidence and extent of venous thrombosis after endotoxin challenge.[15] Combined tPA/uPA-null mice are also viable, but have a significantly shorter lifespan than do wild-type mice or mice deficient in either enzyme alone.[15] Double-mutant mice also exhibit growth retardation, facial ulcerations, reduced fertility, and fibrin deposits in multiple tissues.[15] In addition, ovulatory efficiency in double-mutant mice is 26 per cent lower than in single-mutant and wild-type mice.[16] These mice also lyse radiolabelled pulmonary clots at a significantly slower rate than do tPA-null mice (which, in turn, lyse clots at a slower rate than wild-type and uPA null mice). However, the fact that their thrombotic phenotype is not worse than it is, suggests that plasminogen activators other than uPA and tPA can carry out overlapping, compensatory functions.

tPA may play an important role in neuroneal cell death and seizure. Evidence for such a role includes

(1) the immediate–early induction of tPA mRNA in microglia of the hippocampus by inducers of neuronal activity;

(2) the resistance of tPA-null mice to neuroneal destruction after administration of glutamate analogues (excitotoxins);

(3) the isolated occurrence of substantial neuroneal death in tPA-null mice in hippocampal regions receiving both exogenous tPA and an excitotoxin, but not in those regions receiving either agent alone; and

(4) the resistance of PAI-1 infused wild-type mice to excitotoxin-induced neurodegeneration.[17]

Plasminogen-null mice also resist excitotoxin-induced neurodegeneration, indicating that the excitotoxin effect is probably mediated via tPA-activated plasmin.[18] However, whereas the impaired wound healing and other effects of plasminogen deficiency are alleviated by the added deletion of fibrinogen, double-null mice still resist excitotoxin-induced neurodegeneration, suggesting that the nerve cell death derives from plasmin cleavage of an ECM substrate other than fibrin (T.H. Bugge; personal communication). Thus the requirement for tPA in excitotoxin-mediated neuronal cell death, together with the involvement of excitotoxic mechanisms in stroke, raises the possibility that tPA may be an inappropriate agent for the treatment of thrombotic stroke, even though it is approved for this purpose.[17]

Structure

The tPA molecule has a relatively compact ellipsoid shape with an overall globular appearance.[19] NMR studies indicate that the isolated finger domain of tPA is most similar in structure to the seventh type I finger domain of fibronectin, but with hydrophobic residues covering the exposed surface of its primary β sheet.[20] The subsequent EGF-like domain has an NMR structure similar to that of other isolated EGF-like modules, and the finger and EGF-like domains are apparently fixed together by hydrophobic interactions.[20] This is consistent with earlier melting studies that showed these two domains to be tightly linked.[21] These studies also revealed that the proteinase domain of tPA is composed of two independent domains, and that these interact strongly with the finger and/or EGF domains.[23] Because these domains are from opposite termini, their interactions provide for a stable compact overall structure. NMR and X-ray crystal structures have also been obtained for tPA kringle 2 (PDB identifiers 1PML, 1TPK, and 1PK2) and confirm its similarity to the kringles of other proteins.

References

1. Rijken, D. C. (1995). *Baillière's Clin. Haematol.* **8**, 291–312.
2. Parmer, R. J., Mahata, M., Mahata, S., Sebald, M. T., O'Connor, D. T., and Miles, L. A. (1997). *J. Biol. Chem.*, **272**, 1976–82.
3. Mori, K., Dwek, R. A., Downing, A. K., Opdenakker, G., and Rudd, P. M. (1995). *J. Biol. Chem.*, **270**, 3261–7.
4. Andrade-Gordon, P., and Strickland, S. (1986). *Biochemistry*, **25**, 4033–40.
5. Salonen, E. M., Saksela, O., Vartio, T., Vaheri, A., Nielsen, L. S., and Zeuthen, J. (1985). *J. Biol. Chem.*, **260**, 12302–7.
6. Liesi, P., Salonen, E. M., Dahl, D., Vaheri, A., and Richards, S. J. (1990). *Exp. Brain Res.*, **79**, 642–50.
7. Redlitz, A. and Plow, E. F. (1995). *Baillière's Clin. Haematol.*, **8**, 313–27.
8. Hajjar, K. A., Jacovina, A. T., and Chacko, J. (1994). *J. Biol. Chem.*, **269**, 21191–7.
9. Rijken, D. C., Wijngaards, G., and Collen, D. (1985). In *Thrombolysis, biological and therapeutic properties of new thrombolytic agents*. (ed. D. Collen, H. R. Lijnen, and M. Verstraete), pp. 15–30. Churchhill Livingstone, New York.
10. Andreasen, P. A., Nielsen, L. S., Grondahl-Hansen, J., Skriver, L., Zeuthen, J., Stephens, R. W., and Dano, K. (1984). *EMBO J.*, **3**, 51–6.
11. Zamarron, C., Lijnen, H. R., and Collen, D. (1984). *J. Biol. Chem.*, **259**, 2080–3.
12. Madison, E. L., Kobe, A., Gething, M. J., Sambrook, J. F., and Goldsmith, E. J. (1993). *Science*, **262**, 419–21.
13. Degen, S. J. F., Rajput, B., and Reich, E. (1986). *J. Biol. Chem.*, **261**, 6972–85.
14. Ny, T., Elgh, F., and Lund, B. (1984). *Biochemistry*, **81**, 5355–9.
15. Carmeliet, P., Bouche, A., De Clercq, C., Janssen, S., Pollefeyt, S., Wyns, S., *et al.* (1995). *Ann. NY Acad. Sci.*, **748**, 367–82.
16. Leonardsson, G., Peng, X. R., Liu, K., Nordstrom, L., Carmeliet, P., Mulligan, R., *et al.* (1995). *Proc. Natl Acad. Sci. USA*, **92**, 12446–50.
17. Tsirka, S. E., Rogove, A. D., and Strickland, S. (1996). *Nature*, **384**, 123–4.
18. Tsirka, S. E., Rogove, A. D., Bugge, T. H., Degen, J. L., and Strickland, S. (1997). *J. Neurosci.*, **17**, 543–52.
19. Margossian, S. S., Slayter, H. S., Kaczmarek, E., and McDonagh, J. (1993). *Biochim. Biophys. Acta*, **1163**, 250–6.
20. Smith, B. O., Downing, A. K., Dudgeon, T. J., Cunningham, M., Driscoll, P. C., and Campbell, I. D. (1994). *Biochemistry*, **33**, 2422–2429.
21. Novokhatny, V. V., Ingham, K. C., and Medved, L. V. (1991). *J. Biol. Chem.*, **266**, 12994–3002.
22. Sheng, S., Truong, B., Fredrickson, D., Wu, R., Pardee, A. B., and Sager, R. (1998). *Proc. Natl Acad. Sci. USA*, **95**, 499–504.
23. Bode, W. and Renatus, M. (1997). *Curr. Opin. Struct. Biol.*, **7**, 865–702.

Mark D. Sternlicht and Zena Werb
University of California, San Francisco,
Department of Anatomy, LR-208,
3rd and Parnassus Avenues, San Francisco,
CA 94143–0452, USA.

Urokinase-type plasminogen activator (uPA) and its receptor (uPAR)

Urokinase-type plasminogen activator (uPA) is a serine proteinase with high specificity for plasminogen, which it cleaves at a single site to form plasmin. It is secreted as a single-chain molecule with poor activity and is activated by plasmin or by binding to its high affinity, glycolipid-anchored, cell surface receptor (uPAR). Binding of uPA to uPAR enhances its activity, protects it against plasminogen activator inhibitors and against plasmin cleavage, and localizes it to focal contacts. In this way, it is thought to indirectly catalyse pericellular ECM degradation, disrupt cell–matrix interactions, and promote cell migration and invasion. In addition, uPA/uPAR complexes interact with integrins; and by so doing they can modulate cellular signalling, adhesion, and migration by means that are independent of proteolytic activity.

Protein properties

Urokinase-type plasminogen activator (uPA, urokinase, EC 3.4.21.73) is a serine proteinase with high specificity towards plasminogen, which it cleaves at a single peptide bond to form the active serine proteinase plasmin. It is found in plasma at about 3.5 ng/ml and in urine at over 200 ng/ml.[1] It is secreted as a minimally active single-chain glycoprotein (sc-uPA, pro-uPA) and is activated either by plasmin proteolysis, which yields a disulphide-linked two-chain form (tc-uPA), or by binding as sc-uPA to its cell surface receptor, uPAR.[2] Studies indicate that sc-uPA/uPAR complexes have greater affinity for plasminogen and are less sensitive to inhibition by plasminogen activator inhibitor type 1 (PAI-1) than tc-uPA.[2] Also, receptor-bound sc-uPA is less readily cleaved by plasmin than is free sc-uPA.[2] Thus, binding of sc-uPA to uPAR enhances its activity and shields it against both inhibition and conversion to tc-uPA (a step that may precede its ultimate inactivation and clearance).

Human sc-uPA (Fig. 1) has a molecular weight of 54 kDa and contains 411 amino acids (not including its 20 residue signal peptide).[1] Cleavage of sc-uPA by plasmin or plasma kallikrein occurs at its Lys-158–Ile-159 peptide bond and yields 54 kDa high molecular weight (HMW) tc-uPA.[3] The resulting N-terminal A chain (Ser-1–Lys-158) contains a growth factor-like domain (His-5–Thr-49) with homology to the receptor-binding regions of EGF and TGF-α. This domain is responsible for the binding of uPA to uPAR and is followed by a single kringle domain (Cys-50–Cys-131) without a lysine/fibrin-binding site. The C-terminal B chain (Ile-159–Lys-411) contains the catalytic triad (His-204/Asp-255/Ser-356) and remains linked to the A chain by a single disulphide bridge between Cys-148 and Cys-279. Nevertheless, no portion of the A chain is required for full catalytic activity. Indeed, HMW-tc-uPA can be further cleaved by plasmin at its Lys-134–Lys-135 peptide bond, thus releasing the kringle and growth factor domains to form 33 kDa low molecular weight (LMW) tc-uPA. This LMW form has a truncated A chain, but retains its capacity to convert plasminogen to plasmin. In addition, matrilysin (MMP-7) can cleave the nearby Glu-143–Leu-144 peptide bond of uPA, causing the release of a 143 residue amino-terminal fragment (ATF) that is mitogenic for some cells.[4,5] This mitogenic activity relies on the fucosylation of Thr-18 in the uPA growth factor domain.[5] A single occupied *N*-glycosylation site is also found in human uPA at Asn-302, whereas no glycosylation sites exist in mouse uPA. Non-glycosylated sc-uPA, however, is more readily activated by plasmin than glycosylated sc-uPA and non-glycosylated tc-uPA is more active against plasminogen than glycosylated tc-uPA.[1] Plasmin, plasma kallikrein, mast cell tryptase, and cathepsin G can all cleave and activate sc-uPA, whereas cleavage of the nearby Arg-156–Phe-157 peptide bond by thrombin yields an inactive tc-uPA that can be activated by amino dipeptidyl peptidase I, cathepsin C, or high amounts of plasmin.[1] Other enzymes that can activate uPA include factor XIIa, T-cell associated serine proteinase, cathepsins B and L, nerve growth factor-γ, and prostate specific antigen.[6]

The cell surface uPA receptor (uPAR, CD87) is a 55–60 kDa three-domain glycoprotein that is anchored to the cell membrane by a glycosyl-phosphatidylinositol (GPI) moiety rather than by any transmembrane domain (Fig. 2).[7,8] It is found on several cell types and binds sc- and HMW- tc-uPA (but not LMW -uPA or tPA) with high affinity (k_d = 0.1–1 nM).[6,7] Human uPAR has a transient 22 residue signal peptide and 30 C-terminal residues are post-translationally removed within the endoplasmic reticulum, then replaced by the GPI anchor, thus yielding the mature 283 amino acid protein. Furthermore, uPAR has five potential *N*-glycosylation sites and is highly and heterogeneously glycosylated. Indeed, its molecular weight decreases to 35 kDa after deglycosylation.[3] uPAR has three repeated domains with homology to a module found in proteins of the Ly-6 superfamily (e.g. the murine leukocyte Ly-6 antigens, membrane inhibitor of reactive lysis (MIRL, CD59), and a squid brain glycoprotein (Sgp-2)).[7,8] Each Ly-6 family member is a GPI-anchored membrane protein and each has a domain-specific gene structure like that of the three uPAR domains.[8] However, other members of this gene family have only one such domain, whereas uPAR has three.[7,8] The region linking the N-terminal domain of uPAR (domain I) and

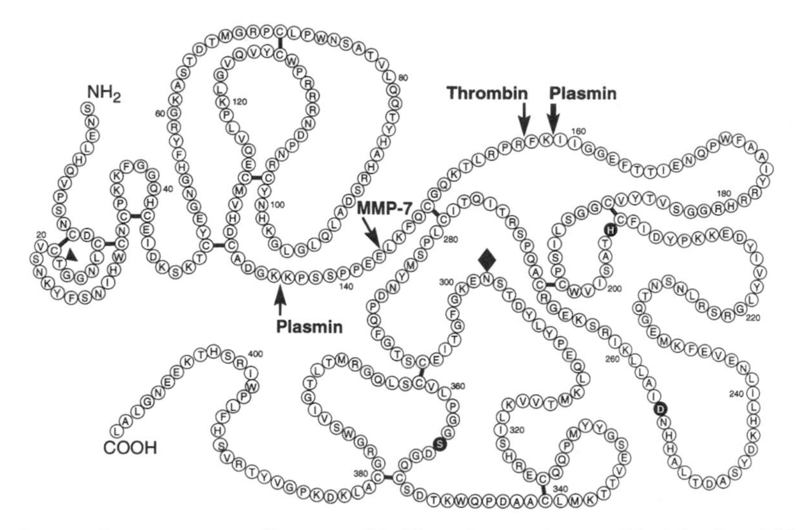

Figure 1. Primary and secondary structure of human uPA. Plasmin can cleave uPA at the Lys-158–Ile-159 peptide bond (to form active tc-uPA) and at the Lys-134–Lys-135 peptide bond (to form LMW uPA lacking the growth factor and kringle domains). Also indicated are the matrilysin (MMP-7) and thrombin cleavage sites, the Asn-302 glycosylation site (diamond), the Thr-18 fucosylation site (triangle), and the catalytic His, Asp, and Ser residues (filled circles). (Modified from Rijken.[1])

the middle domain (II) can be cleaved by chymotrypsin, trypsin, neutrophil elastase, cathepsin G, plasmin, and uPA itself, whereas the region linking domains II and III is relatively resistant to proteolytic cleavage.[7,8] Structural similarities at both the gene and protein level also exist between members of the Ly-6/uPAR family and those of the large family of secreted snake venom α-neurotoxins such as the acetylcholine receptor antagonist α-bungarotoxin.[8] Various methods for detecting uPAR that rely on its high affinity binding to the uPA ATF or to DFP-inactivated uPA have been described.[9]

Because of their different binding properties, uPA and tPA localize to different extracellular sites: uPA is directed to the surface of uPAR-expressing cells by its growth factor-like domain and tPA is directed to fibrin and other ECM components by its finger and kringle domains. Thus uPA is thought to mediate pericellular matrix degradation and tPA is thought to primarily influence intravascular fibrinolysis. The N-terminal growth factor-like domain of uPA has been shown to bind uPA to uPAR. Indeed, the liberated uPA growth factor domain itself and chimeric proteins containing this domain are able to bind uPAR and compete with uPA for cell surface receptor.[7] Recently, the binding of human uPA to uPAR was shown to specifically involve the Asn-22, Asn-27, His-29, and Trp-30 residues of uPA.[33] The binding of uPA also involves uPAR domain I, as indicated by uPA–uPAR cross-linking studies. However, uPA shows over 1000-fold lower affinity toward both isolated domain I alone and the fragment containing only domains II and III than toward intact uPAR, indicating that the intact three-domain receptor is required for efficient binding.[7,8] Binding of uPA to uPAR can be inhibited by uPAR-blocking antibodies, by the non-competitive inhibitor suramin, and by a potent 15 residue peptide that was isolated by bacteriophage display, but which has no sequence similarity to any region of uPA.[7] In addition, uPAR binds vitronectin with high affinity (k_d = 0.2–2 nM) and such binding is stimulated by uPA and uPA/PAI-1 complexes, but inhibited by PAI-1, which competes for binding to the same somatomedin B domain on vitronectin.[6] Recent evidence also indicates that soluble uPAR exists and that vitronectin concentrates complexes between it and uPA to the cell surface and ECM.[34]

The internalization and clearance of uPAR-bound uPA/PAI-1 complexes is mediated by LDL receptor-related protein/α_2-macroglobulin receptor (LRP) in some cell types, and by the related epithelial glycoprotein gp330 or the VLDL receptor in other cell types.[1,6,7,10] These members of the LDL receptor family have extracellular clusters of complement-like repeats that bind ligand and short cytoplasmic tails that mediate clathrin-coated pit formation and endocytosis. That uPAR plays a role in this process is indicated by studies showing that transfection of uPAR-deficient cells with uPAR enables them to internalize uPA/PAI-1 and that internalization is partially prevented if uPAR is blocked.[7] Internalization of uPA/PAI-2, uPA/PAI-3, and uPA/protease nexin I complexes is also mediated by uPAR, although the exact mechanisms may be different. Uninhibited uPA, however, is not internalized. Thus it is thought that the inhibition of uPAR-bound uPA by serpins promotes the binding of these complexes to LRP and their internalization. During the internalization of uPA/serpin complexes, cell surface uPAR levels decrease and then return to normal due to uPAR recycling rather than to recruitment of intracellular receptors or *de novo* uPAR synthesis.[11]

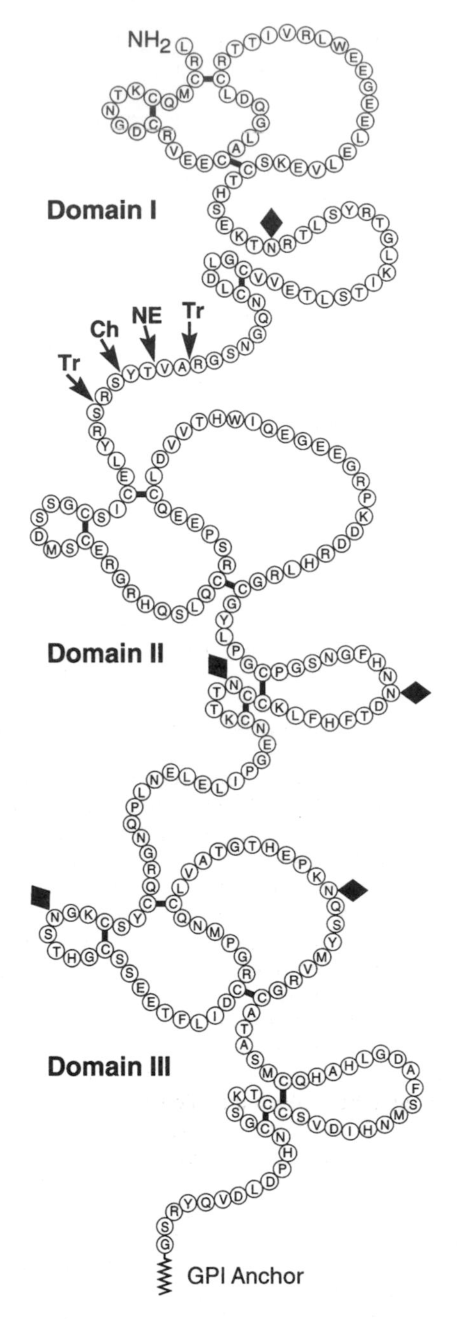

Figure 2. Primary and secondary structure of human uPAR. uPAR can be cleaved between domains 1 and 2 by trypsin (Tr), chymotrypsin (Ch), and neutrophil elastase (NE) at the sites shown. It can also be cleaved by uPA and plasmin. uPA is predicted to cleave at the most N-terminal trypsin site and plasmin can potentially cleave after each of the three Arg residues within the region linking domains 1 and 2. Potential *N*-glycosylation sites are indicated by black diamonds. (Modified from Plough and Ellis.[8])

Purification

Urokinase has been purified from urine, cultured kidney cells, and various tumour cell lines, and recombinant uPA has been isolated from bacteria and eukaryotic cells. Purification of uPAR can be accomplished by Triton X-114 phase separation of cell extracts and affinity chromatography using immobilized DFP-inactivated uPA.[9]

Activities

Urokinase is a highly specific serine proteinase with activity against the Arg-560–Val-561 peptide bond of plasminogen. In addition, uPA can indirectly activate various latent MMPs through plasminogen.[3] On its own, however, uPA is unable to cleave laminin, fibronectin, or native collagens.[12] It can, however, activate hepatocyte growth factor/scatter factor (HGF) by cleaving a single peptide bond within inactive single-chain HGF to yield active two-chain HGF.[6] Like HGF, other growth factors such as macrophage stimulating protein also have high sequence similarity to plasminogen; these too may be substrates of uPA. uPA can also cleave uPAR between domains I and II, although the significance of such cleavage remains unknown.[13] Plasminogen activators can be detected using chromogenic and fluorogenic substrates, or by adding plasminogen to casein-containing zymographic gels or fibrin–agar indicator gels.[14,15] Interestingly, 110 kDa uPA/PAI-1 complexes can also be detected by such zymographic methods.

Because tc-uPA converts plasminogen to active plasmin, which in return converts sc-uPA to tc-uPA, these enzymes can participate in a positive feedback activation cascade. This activation loop is further potentiated by the presence of cell surface (but not soluble) uPAR, and it is not enhanced if cells from uPAR null mice are used.[6,7] This acceleration is also abolished by neutralizing anti-uPAR antibodies or by lysine analogues that inhibit the binding of plasminogen to cells.[7] Thus receptors for both uPA and plasminogen are required and these may induce conformational changes in their ligands and/or provide orientation and concentration effects that promote plasminogen activation. Other studies indicate that the amount of bound plasminogen is rate-limiting for uPA-mediated activation of plasminogen at the cell surface.[16] In addition, when compared to their free counterparts, bound uPA is moderately protected against inhibition by PAI-1 and PAI-2 and bound plasmin is substantially protected against inhibition by α_2-antiplasmin.

Several studies have suggested that uPAR takes part in cellular signal transduction, chemotaxis and adhesion by pathways that are independent of uPA proteolytic activity.[6,7] Both active and inactivated uPA can induce c-fos expression in certain cells, and this effect can be blocked by soluble uPAR.[17] In addition, both uPA and its non-catalytic ATF can induce epidermal cell migration and diacylglycerol formation, and this effect is augmented in transfectants that overexpress uPAR.[18] The pro-migratory effect of uPA can also be competitively inhibited by recombinant uPA containing a mutated growth factor-like domain.[35] In addition, the binding of uPA or its ATF to uPAR causes the rapid phosphorylation and activation of extracellular signal-regulated kinases 1 and 2 (ERK1 and ERK2) and this signalling is required for uPA to stimulate breast cancer cell migration.[36] However, because uPAR lacks a cytoplasmic domain, it is unable to directly transmit signals to the cell interior, thus it must interact with other cellular proteins in order to do so. Accordingly, it has been shown that uPAR/uPA (but not unoccupied uPAR) colocalizes with β1 integrins at focal contacts, that uPAR co-caps with the myelomonocytic cell β2 integrin Mac-1 (CD11b/CD18, complement receptor type 3) and promotes Mac-1-mediated cell adhesion, and that chemical cross-linking of sc-uPA to an unknown high molecular weight protein is hindered by blocking antibodies against uPAR.[3,6,7,19] In addition, uPAR and integrins form stable complexes that interfere with integrin-mediated adhesion (e.g. to fibronectin) and promote adhesion to vitronectin via a site on uPAR other than the uPA binding site.[20] Myogenic cell adhesion and spreading is also mediated by the binding of uPAR-bound uPA to PAI-1, which is often associated with the ECM.[21] In these experiments, the promotion of cell adhesion and spreading by immobilized PAI-1 was dose-dependent and could be inhibited by removal of GPI-linked proteins using phosphatidylinositol-phospholipase C and by antibodies against PAI-1 or $\alpha v\beta 3$ integrin. Thus various integrins can link uPAR to the cytoskeleton and the intracellular signal transduction machinery.

Antibodies

Several monoclonal antibodies to uPA, including some that block plasminogen activation or uPAR binding are available from Accurate Chemical and Scientific, American Diagnostica, Biodesign International, Biogenesis, Cedarlane Laboratories, Cortex Biochem, Coulter Cytometry, Immunotech, Sanbio, Technoclone, and York Biologicals. Polyclonal anti-uPA antibodies are also available from numerous suppliers. Blocking and non-blocking monoclonal antibodies against uPAR and its individual domains, and rabbit and goat anti-human uPAR antibodies are available from American Diagnostica.

Genes

The human uPA gene (*PLAU)* is located on chromosome 10q24, spans 6.4 kbp, and has 11 exons.[22] The human uPAR gene (*PLAUR)* is located on chromosome 19q13.2, spans 23 kbp, and has seven exons.[23,24] Complementary DNA sequences are available for uPA in humans (M18182, D00244, M15476, K03226), mice (X62700–01, M17922, X02389), rats (X65651, X63434, X66907), baboons (X51935), pigs (X74381, X02724), cows (L03546, X85801), and chickens (J05187–8). Genomic sequences are available for humans (X02419, X12641, K03027, K02286), pigs (X92447, X01648, L27481), and mice (X63356, X62702, X52971). Complementary DNA sequences are also available for uPAR in humans (U08839, U09346–7, X74039, X51675), mice (U12235), rats (X71898–9) and cows (L03545). Genomic sequences are available for human uPAR (U07842, S78532).

Mutant phenotype/disease states

Deficiencies in uPA or uPAR have not been described in humans. However, uPAR is found in plasma rather than bound to cell membranes in patients with paroxysmal nocturnal haemoglobinuria due to their inability to form glycolipid anchors. Soluble uPAR can also be found in the ascitic fluid of ovarian cancer patients and in normal plasma, although the reasons for its presence are unclear. Gene-targeted mice lacking uPA alone, uPAR, tPA, tPA plus uPAR, and both uPA and tPA are all viable.[25,26] Hepatic fibrin deposits are absent in uPAR null mice, rare in uPA-null and tPA-null mice, and common in both uPAR/tPA and uPA/tPA double-null mice. However, the combined uPAR/tPA-null mice lack the multi-organ fibrin deposits, severe tissue damage, low fertility, impaired wound healing, and shorter survival seen in uPA/tPA double-null mice.[26] Thus uPAR plays an important role in the clearance of hepatic fibrin deposits by uPA, but uPA can still effectively lyse fibrin elsewhere despite the absence of both uPAR and tPA. uPA-deficient mice are also susceptible to staphylococcal infections, pleuritis, and lymphoid follicular effacement.[27] After vascular insult, neointimal formation and accumulation is delayed in uPA- but not tPA-deficient mice and accelerated in PAI-1-deficient mice, thus implicating uPA in vascular healing and restenosis.[28] Plasminogen activators have also been implicated in atherosclerosis, but their role has not yet been explored adequately using gene knockout mice. Plasmin activation by uPA but not tPA appears also to play a role in atherosclerotic aneurysm formation, because only when uPA-null mice were crossed with apolipoprotein E-null mice was there protection against medial damage and aneurysm formation.[37] In addition, uPA can fully promote the pericellular activation of plasminogen that is required for vascular wound healing to take place, regardless of the presence or absence of its receptor, as indicated by comparing healing after electric injury in uPAR-null and wild-type mice.[38] Embryos with homozygous inactivation of the LRP gene exhibit intraperitoneal bleeding and are only viable to mid-gestation.[29] Because LRP is an important mediator of uPA and tPA clearance, their build-up in the absence of LRP may be responsible for the observed embryonic lethality.

The plasminogen activator/plasmin system has also been strongly implicated in angiogenesis[30] and cancer progression.[6] For example, increased expression of uPA, uPAR, and PAI-1 is associated with poor prognosis in several forms of cancer, whereas high PAI-2 levels tend to predict a more favourable outcome. In addition, tumour cell invasion and metastasis are increased by uPA or uPAR transfection and decreased by commercial serine proteinase inhibitors, anti-uPA antibodies, exogenous PAI-1, PAI-2, or PN-I, and the expression of recombinant PAI-2 or mutant uPA (which competitively displaces active uPA from its receptor). Furthermore, diminished local invasion of chemically induced nevi and their reduced progression to malignancy has been observed in uPA-null mice and lung metastases in a transgenic mouse mammary carcinoma model are reduced when these mice are rendered plasminogen-deficient.[6]

Structure

Whole uPA has proven difficult to crystalize, but NMR structures have been obtained for the isolated uPA kringle domain (PDB identifiers 1KDU and 1URK). NMR data have also revealed considerable motion between the different uPA domains,[31] and neutron scattering and circular dichroism studies further indicate that sc- and tc-uPA have similar asymmetric overall structures.[32] Adequate crystallization of uPAR has also not yet been achieved, due in large part to its considerable glycosylation. However, the NMR structure for CD59, another member of the Ly-6 protein family, has been obtained (PDB identifiers 1ERG, 1ERH, and 1CDQ).[8] These data also reveal structural similarity between members of the Ly-6 family and the single-domain snake venom α-neurotoxins for which there are over 15 solved X-ray and NMR structures.[8] Taken together, the data suggest a model wherein the three uPAR domains each have a flattened spherical shape dominated by three long and two short antiparallel β strands that form three adjacent loops.[8] These loops emerge from a small globular core that is stabilized by four conserved disulphide bonds joining cysteines 1 to 5, 4 to 6, 7 to 8, and 9 to 10. Conserved cysteines 7 and 8, however, are uniquely absent in the N-terminal domain I of uPAR. An additional disulphide bond in the middle of the first loop joins cysteines 2 and 3 and is present in all Ly-6 family members but only a few 'weak' snake venom toxins.

References

1. Rijken, D. C. (1995). *Baillière's Clin. Haematol.*, **8**, 291–312.
2. Higazi, A. A., Mazar, A., Wang, J., Reilly, R., Henkin, J., Kniss, D., and Cines, D. (1996). *Blood*, **87**, 3545–9.
3. Kwaan, H. C. (1992). *Cancer Metastasis Rev.* **11**, 291–311.
4. Marcotte, P. A., Dudlak, D., Leski, M. L., and Henkin, R. J. (1992). *Fibrinolysis*, **6**, 57–62.
5. Rabbani, S. A., Mazar, A. P., Bernier, S. M., Haq, M., Bolivar, I., Henkin, J., and Goltzman, D. (1992). *J. Biol. Chem.*, **267**, 14151–6.
6. Andreasen, P. A., Kjoller, L., Christensen, L., and Duffy, M. J. (1997). *Int. J. Cancer*, **72**, 1–22.
7. Behrendt, N., Ronne, E., and Dano, K. (1995). *Biol. Chem. Hoppe-Seyler*, **376**, 269–79.
8. Plough, M. and Ellis, V. (1994). *FEBS Lett.*, **349**, 163–8.
9. Behrendt, N., Plough, M., Ronne, E., Hoyer-Hansen, G., and Dano, K. (1993). *Meth. Enzymol.*, **223**, 207–22.
10. Kounnas, M. Z., Henkin, J., Argraves, W. S., and Strickland, D. K. (1993). *J. Biol. Chem.*, **268**, 21862–7.
11. Nykjaer, A., Conese, M., Christensen, E. I., Olson, D., Cremona, O., Gliemann, J., and Blasi, F. (1997). *EMBO J.*, **16**, 2610–20.
12. Liotta, L. A., Goldfarb, R. H., Brundage, R., Siegal, G. P., Terranova, V., and Garbisa, S. (1981). *Cancer Res.*, **41**, 4629–36.
13. Hoyer-Hansen, G., Ronne, E., Solberg, H., Behrendt, N., Plough, M., Lund, L. R., *et al.* (1992). *J. Biol. Chem.*, **267**, 18224–9.
14. Heussen, C. and Dowdle, E. B. (1980). *Anal. Biochem.*, **102**, 196–202.
15. Erickson, L. A., Lawrence, D. A., and Loskutoff, D. J. (1984). *Anal. Biochem.*, **137**, 454–63.
16. Namiranian, S., Naito, Y., Kakkar, V. V., and Scully, M. F. (1995). *Biochem. J.*, **309**, 977–82.
17. Dumler, I., Petri, T., and Schleuning, W. D. (1994). *FEBS Lett.*, **343**, 103–6.
18. Del Rosso, M., Anichini, E., Pedersen, E., Blasi, F., Fibbi, G., Pucci, M., and Ruggiero, M. (1993). *Biochem. Biophys. Res. Commun.*, **190**, 347–52.
19. Sitrin, R. G., Todd, R. F., Petty, H. R., Brock, T. G., Shollenberger, S. B., Albrecht, E., and Gyetko, M. R. (1996). *J. Clin. Invest.*, **97**, 1942–51.
20. Wei, Y., Lukashev, M., Simon, D. I., Bodary, S. C., Rosenberg, S., Doyle, M. V., and Chapman, H. A. (1996). *Science*, **273**, 1551–5.
21. Planus, E., Barlovatz-Meimon, G., Rogers, R. A., Bonavaud, S., Ingber, D. E., and Wang, N. (1997). *J. Cell Sci.*, **110**, 1091–8.
22. Riccio, A., Grimaldi, G., Verde, P., Sebastio, G., Boast, S., and Blasi, F. (1985). *Nucleic Acids Res.*, **13**, 2759–71.
23. Casey, J. R., Petranka, J. G., Kottra, J., Fleenor, D. E., and Rosse, W. F. (1994). Blood **84**, 1151–6.
24. Roldan, A. L., Cubellis, M. V., Masucci, M. T., Behrendt, N., Lund, L. R., Dano, K., *et al.* (1990). *EMBO J.*, **9**, 467–74.
25. Carmeliet, P., Bouche, A., De Clercq, C., Janssen, S., Pollefeyt, S., Wyns, S., *et al.* (1995). *Ann. NY Acad. Sci.*, **748**, 367–82.
26. Bugge, T. H., Flick, M. J., Danton, M. J., Daugherty, C. C., Romer, J., Dano, K., *et al.* (1996). *Proc. Natl Acad. Sci. USA*, **93**, 5899–904.
27. Shapiro, R. L., Duquette, J. G., Nunes, I., Roses, D. F., Harris, M. N., Wilson, E. L., and Rifkin, D. B. (1997). *Am. J. Pathol.*, **150**, 359–69.
28. Carmeliet, P. F. M. (1995). *Baillière's Clin. Haematol.*, **8**, 391–401.
29. Herz, J., Clouthier, D. E., and Hammer, R. E. (1992). *Cell*, **71**, 411–21.
30. Mignatti, P. and Rifkin, D. B. (1996). *Enzyme Protein*, **49**, 117–37.
31. Nowak, U. K., Li, X., Teuten, A. J., Smith, R. A., and Dobson, C. M. (1993). *Biochemistry*, **32**, 298–309.
32. Mangel, W. F., Lin, B., and Romakrishna, V. (1991). *J. Biol. Chem.*, **266**, 9408–12.
33. Quax, P. H., Grimbergen, J. M., Lansink, M., Bakker, A. H., Blatter, M. C., Belin, D., van Hinsbergh, V. W., and Verheijen, J. H. (1998). *Atheroscler. Thromb. Vasc. Biol.*, **18**, 693–701.
34. Chavakis, T., Kanse, S. M., Yutzy, B., Lijnen, H. R., and Preissner, K. T. (1998). *Blood*, **91**, 2305–12.
35. Stepanova, V., Bobik, A., Bibilashvily, R., Belogurov, A., Rybalkin, I., Domogatsky, S., Little, P. J., Goncharova, E., and Tkachuk, V. (1997). *FEBS Lett.*, **414**, 471–4.
36. Nguyen, D. H., Hussaini, I. M., and Gonias, S. L. (1998). *J. Biol. Chem.*, **273**, 8502–7.

37. Carmeliet, P., Moons, L., Lijnen, R., Baes, M., Lemaitre, V., Tipping, P., Drew, A., Eeckhout, Y., Shapiro, S., Lupu, F., and Collen, D. (1997). *Nature Genet.*, **17**, 439–44.
38. Carmeliet, P., Moons, L., Dewerchin, M., Rosenberg, S., Herbert, J. M., Lupu, F., and Collen, D. (1998). *J. Cell Biol.*, **140**, 233–45.

Mark D. Sternlicht and Zena Werb
University of California, San Francisco,
Department of Anatomy, LR-208,
3rd and Parnassus Avenues, San Francisco,
CA 94143–0452, USA.

Index

Note: Proteins are listed under their current names, with page references to the section(s) in which they are discussed, and to the page where their synonyms, if any, are given. Synonyms are followed by the corresponding current name in brackets.